SEMICONDUCTING DEVICES

A Bibliography of
Fabrication Technology,
Properties, and Applications

SEMICONDUCTING DEVICES

A Bibliography of Fabrication Technology, Properties, and Applications

A. H. Agajanian
IBM System Products Division
Hopewell Junction, New York

IFI/PLENUM • NEW YORK–WASHINGTON–LONDON

Library of Congress Cataloging in Publication Data

Agajanian, A H
Semiconducting devices.

Includes index.
1. Semiconductors – Bibliography. I. Title.
Z5838.S4A4 [TK7871.85] 016.6213815'2 76-42313

ISBN 978-1-4684-6092-6 ISBN 978-1-4684-6090-2 (eBook)
DOI 10.1007/978-1-4684-6090-2

Softcover reprint of the hardcover 1st edition 1976
A Division of Plenum Publishing Corporation
227 West 17th Street, New York, N.Y. 10011

To my children,
Linda, Anita, and Aram

PREFACE

Since the advent of planar technology the literature on semiconductor devices has been growing exponentially. This book of over 14000 references is intended to make the world literature available to workers in the field (beginners as well as experienced) to assist them in finding out what has been done by others in their fields of endeavor.

The literature on theory, preparation, properties, characterization, packaging, instrumentation and applications of semiconductor devices is thoroughly covered; the only aspect of preparation not covered is diffusion. However, several books and comprehensive review articles on this subject are given in Chapter A. Radiation damage due to ion implantation is given in Chapter F. Radiation damage due to all other types of radiations is given in Chapter J. The chapters on processing technology, radiation damage, structural defects and analysis, modeling and packaging are quite general and apply to most types of semiconductor devices. Special attention is given to electrical properties and applications of FET, Schottky and charge-coupled devices in Chapters M, N, O and P.

The literature from 1967 to 1975 is covered - fairly exhaustively for the 1970-1975 period. Selected references prior to 1967 are also included. The main sources searched for references are: Physics Abstracts (PA), Electrical and Electronics Abstracts (EEA), Chemical Abstracts (CA), Nuclear Science Abstracts (NSA), Metals Abstracts (MA), Ceramics Abstracts and Engineering Index. In addition to these sources, current journals, Books in Print, Cumulative Book Index and the IBM Technical Information Retrieval Center were searched for material.

To facilitate access to the needed references the book is divided into eighteen chapters arranged alphabetically A to R. Each chapter is subdivided into sections and subsections (see Contents). This division is based on the abstract of each reference and not just the title. In addition, a comprehensive subject index is also given. Following each subheading, the references are listed in alphabetical order of the names of the principal authors in a conventional format. The language in which the article is published, if not English, is given. The volumes and numbers of the abstracts are also given for easy access to the abstracts.

This book is the integrated result of some 40 bibliographies on various aspects of semiconductor devices which were published as IBM reports; some were also published as journal articles. I would like to thank the IBM Corporation for permission to incorporate these bibliographies in my book. I also would like to thank Mr. R. B. Murphy, the IBM East Fishkill library manager, for initiating the compilation of the bibliographies - without his continued support and encouragement this book would not have been possible; my wife Else for her assistance and patience during the course of the preparation of this book; my colleagues Dr. R. A. Gdula, Dr. L. V. Gregor, Dr. S. M. Hu, Dr. C. G. Jambotkar, Dr. D. R. Kerr, Dr. B. J. Masters, Mrs. K. A. Murley, Dr. G. Neirotti, Dr. H. B. Pogge, Dr. V. L. Rideout and Mr. R. C. Turnbull for many helpful discussions; Mr. L. A. Plaushin for editorial assistance and the IBM East Fishkill library staff for their helpful assistance. Special thanks are due to Mrs. F. Sisson of Graphic Techniques for typing the manuscript.

Availability of Documents

The references in this book consist of technical articles presented in journals, books, symposia proceedings, technical meetings, reports, patents and theses. If a reader does not have access to a library to obtain copies of the references we suggest the following sources:

1. Journal articles, books, symposia proceedings:
   Library of Congress
   Photoduplication Service
   Washington, D. C. 20540

   Engineering Societies Library
   345E 47th Street
   New York, New York 10017

2. U. S. government contract reports:
   National Technical Information Service
   Springfield, Virginia 22151

3. IAEC reports
   International Atomic Energy Agency
   Kaerntnerring A 1010
   Vienna, Austria

   National Lending Library
   Boston Spa, England

4. Theses listed as Dissertation Abstracts + number (U. S.)
   University Microfilms
   Dissertation Copies
   P.O. Box 1764
   Ann Arbor, Michigan 48106

   Others
   University Microfilms, Ltd.
   St. John's Road
   Penn, Buckinghamshire
   England

5. Patents: U. S. and foreign
   Patent Office
   Box 9
   Washington, D. C. 20231

CONTENTS

A. BOOKS AND REVIEW ARTICLES (GENERAL)

I. Books

1. Boltaks, B. I., "Diffusion in semiconductors", Academic Press, New York, 1963.
2. Burger, R. M. and Donovan, R. P., (Eds.), "Fundamentals of silicon integrated device technology, Vol. 1, oxidation, diffusion and epitaxy", Prentice Hall, Inc., Englewood Cliffs, N. J., 1967.
3. Coutts, T. J., "Electrical conduction in thin metal films", Elsevier Scientific Book Co., New York, 1974, PA78-1354.
4. Fistul, V. I., "Heavily doped semiconductors", Plenum Press, New York, 1969.
5. Fogiel, M., "Modern microelectronics; basic principles, circuit design, fabrication technology", Research and Education Association, New York, 1972.
6. Goldsmith, H. J., (Ed.), "Diffusion in semiconductors", Academic Press, New York, 1963.
7. Grove, A. S., "Physics and technology of semiconductor devices", Wiley, New York, 1967.
8. Holonyak, N., Jr., "Integrated electronic systems", Prentice-Hall International, New York, 1970, EEA73-30423.
9. Hunter, L. P., "Handbook of semiconductor electronics", McGraw-Hill Book Co., New York, 1970.
10. Lindmayer, J. and Wrigley, C. Y., "Fundamentals of semiconductor devices", Van Nostrand, Princeton, N. J., 1965.
11. Luecke, G., Mize, J. P. and Carr, W. N., "Semiconductor memory design and application", McGraw-Hill Book Co., New York, 1973.
12. Maissel, L. I. and Glang, R., (Eds.), "Handbook of thin film technology", McGraw-Hill Book Co., New York, 1970.
13. Meaden, G. T., "Electrical resistance of metals", Plenum Press, New York, 1965.
14. Myamlin, V. A. and Pleskov, Y. V., "Electrochemistry of semiconductors", Plenum Publishing Co., New York, 1967.
15. Nowick, A. S. and Burton, J. J., "Diffusion in solids: Recent developments", Academic Press, New York, 1975.
16. Phillips, A. B., "Transistor engineering and introduction to integrated semiconductor circuits", McGraw-Hill Book Co., New York, 1962.
17. Runyan, W. R., "Silicon semiconductor technology", McGraw-Hill Book Co., New York, 1965.
18. Seeger, K., "Semiconductor physics", Springer-Verlag, New York, 1973.
19. Sharma, B. L., "Diffusion in semiconductors", Trans Tech Publications, Clausthal-Zellerfeld, Germany, 1970.

20. Shaw, D., (Ed.), "Atomic diffusion in semiconductors", Plenum Press, New York, 1973.
21. Stern, L., "Fundamentals of integrated circuits", Hayden Book Co. Inc., New York, 1968.
22. Sze, S. M., "Physics of semiconductor devices", Wiley-Interscience, New York, 1969.
23. Tuck, B., "Introduction to diffusion in semiconductors", Peter Peregrinus Ltd., Southgate House, Stevenage, Herts, England, 1974.
24. Warner, R. M., Jr. and Fordemwalt, J. N., (Eds.), "Integrated circuits: design principles and fabrication", McGraw-Hill Book Co., Inc., New York, 1965.

II. Review Articles

25. Baccarani, G., Calzolari, P. U., Graffi, S. and Soncini, G., "Recent developments in silicon-device technology at 'LAMEL'", Alta Freq., 43, 853-64 (1974), EEA78-5800.
26. Bean, K. E. and Gleim, P. S., "The influence of crystal orientation on silicon semiconductor processing", Proc. IEEE, 57, 1469-76 (1969), EEA73-1086.
27. Bergh, A. A. and Dean, P. J., "Light-emitting diodes", IEEE Proc., 60, 156-224 (1972).
28. Blech, I. A., "Electromigration and crevice formation in thin metallic films", Thin Solid Films, 13, 117-29 (1972).
29. Brothers, J. S., "A high performance bipolar LSI technology", International Conference on Large Scale Integrated Circuits, 1974, p. 131-8, EEA78-11749.
30. Carlson, J. W., "Industry standardization of silicon substrates", Solid State Technol., 16 (9), 49-53 (1973), PA77-1449.
31. Davies, L. W., "Semiconductors", In: Information, Computers, Machine and Man, Huey, R. M., (Ed.), Wiley, New York, p. 152-60 (1971), EEA75-7430.
32. Daw, A. N. and Mitra, R. N., "Electroless plating on semiconductors - a survey", J. Sci. & Ind. Res., 33, 23-6 (1974), PA78-7469.
33. Doyle, N., "State of the art in semiconductor technology", Aust. Electron. Eng., 6 (6), 12-5 (1973), EEA77-14239.
34. Elliott, D. J., "Impact of technological change in the semiconductor industry", Microelectronics, 6 (2), 29-36 (1974), EEA78-24327.
35. Goklaney, S. M., "Capital equipment and primary material requirements for the semiconductor industry", Electron. Inf. & Plann., 1, 876-86 (1974), EEA78-16.
36. Gregor, L. V., "Thin-film processes for microelectronic applications", Proc. IEEE, 59, 1390-1403 (1971).
37. Grunberg, G., "Advances in bipolar technology", International Conference on Large Scale Integrated Circuits, 1974, p. 117-30, French, EEA78-11748.

38.Gunshor, R. L., "The interaction between semiconductors and acoustic surface waves - a review", Solid-State Electron., 18 (12), 1089-93 (1975).
39.Hodges, D. A., "Review and projection of semiconductor components for digital storage", Proc. IEEE, 63, 1136-47 (1975).
40.Hoffman, D. M., "The structure and properties of thin metal films", Proceedings of the 28th Annual Frequency Control Symposium, 1974, p. 85-8, EEA78-11718.
41.Horiuchi, S. and Blanchard, R., "Boron diffusion in polycrystalline silicon layers", Solid-State Electron., 18, 529-32 (1975).
42.Hu, S. M., "Diffusion in silicon and germanium", In: Atomic Diffusion in Semiconductors, Shaw, D., (Ed.), Plenum Press, New York, p. 217-350 (1973).
43.Jakits, O., "Problems in highly integrated semiconductor circuits", Elektrotech Z. (ETZ) A, 89, 569-75 (1968), German, EEA73-448.
44.John, H. F., "Silicon power device material problems", Proc. Instn. Elect. Engrs., 55, 1249-71 (1967), EEA70-22468.
45.Kakihana, S., "Current status and trends in high frequency transistors", Microelectronics, 5 (4), 7-12 (1974), EEA78-37520.
46.Kamins, T. I., "Chemically vapor deposited polycrystalline-silicon films", IEEE Trans. Parts, Hybrids & Packag., PHP-10 (4), 221-9 (1974), PA78-26298.
47.Kendall, D. L. and DeVries, D. B., "Diffusion in silicon", In: Semicond. Silicon, Haberecht, R. R. and Kern, E. L., (Eds.), The Electrochem. Soc., Inc., New York, p. 358-421 (1969).
48.Klaassen, F. M., "Device physics of integrated injection logic", IEEE Trans. Electron Devices, ED-22, 145-52 (1975).
49.Klitzke, J., "Survey of the work of standardisation of semiconductor devices within the framework of the RGW", Nachrichtentech. Elektron., 23, 185-6 (1973), German, EEA76-29054.
50.Lapidus, G., "Amorphous devices: at the threshold", IEEE Spectrum, 10 (1), 44-52 (1973), EEA76-16193.
51.Lepselter, M. P., "Silicon integrated circuit technology", Proceedings of the 5th International Conference on Solid State Devices, 1973, p. 13-17.
52.Lloyd, P., "A review of thin film techniques for microelectronics", Microelectron. & Reliab., 6, 177-87 (1967).
53.Matick, R. E., "Review of current proposed technologies for mass storage systems", IEEE Proc., 60, 266-89 (1972).
54.Mattox, D. M., "Thin film metallization of oxides in microelectronics", Thin Solid Films, 18, 173-86 (1973).
55.Moeller, A., "Micro-electronics, present state of the art", Fernmelde-Praxis, 50, 845-54 (1973), German, EEA77-2579.
56.Newman, P. C., "High performance bipolar technology for LSI", AGARD Lecture Series No. 75 Custom Design for LSI, 1975, p. 2/1-22, EEA78-19928.
57.Riley, W. B., "Semiconductor memories are taking over data-storage applications", Electronics, 46 (16), 75-90 (1973), EEA76-32286.

58. Rosinski, W., "Semiconductor technology", Elektronika, no. 1-2, 15-20 (1973), Polish, EEA76-23256.
59. Schmid, H., "Making LSI circuits a comparison of processing techniques", IEEE Trans. Manuf. Technol., MFT-1, 19-31 (1972), EEA76-8762.
60. Schnable, G. L. and Keen, R. S., "On failure mechanisms in large-scale integrated circuits", In: Advances in Electronics and Electron Physics, Vol. 30, Academic Press, New York, 79-138 (1971).
61. Seeger, A. and Chik, K. P., "Diffusion mechanisms and point defects in silicon and germanium", Phys. Status Solidi, 29, 455-542 (1968).
62. Shaw, D., "Self and impurity diffusion in Ge and Si", Phys. Status Solidi B, 72 (1), 11-39 (1975).
63. Smollett, M., "The technology of semiconductor manufacture", Radio & Electron. Eng., 43, 29-38 (1973), EEA76-12246.
64. Spandorfer, L. M., "Large scale integration - an appraisal", In: Advances in Computers, Alt, F. L. and Rubinoff, M., (Eds.), Academic Press, New York, p. 179-238 (1968), EEA73-20104.
65. Spenke, E., "History and future needs in silicon technology", In: Semiconductor silicon, Electrochem. Soc., Inc., New York, p. 1-35 (1969), EEA73-10407.
66. Steslow, J. J., Rapp, J. E. and White, P. L., "Advances in solid planar dopant sources for silicon (semiconductors device manufacture)", Solid State Technol., 18 (1), 31-4 (1975), PA78-33972.
67. Terry, L. E. and Wilson, R. W., "Metallization systems for silicon integrated circuits", Proc. IEEE, 57, 1580-6 (1969).
68. Vossen, J. L., Stephens, A. W. and Schnable, G. L., "Bibliography on metallization materials and techniques for silicon devices", Published by: Thin Film Division of American Vacuum Society, November 1974.
69. Wallmark, J. T., "Fundamental physical limiations in integrated electronic circuits", Proceedings of the International Conference on Solid State Devices, 1974, p. 133-67, EEA78-24530.
70. Wilcock, J. D., "Semiconductor memory review", New Electron., 8 (10), 39-40, 43, 45 (1975), EEA78-28866.
71. Witt, G. R., "The electrochemical properties of thin films and thin film strain gauge", Thin Solid Films, 22, 133-56 (1974).

B. CRYSTAL GROWTH AND WAFER PREPARATION

I. Bibliographies

1. Agajanian, A. H., "A bibliography on semiconductor wafer preparation", Solid State Technol., 16 (12), 73-4, 104-5 (1973), EEA77-10124.
2. Connolly, T. F., (Ed.), "Solid state physics literature guides, Vol. 2. Semiconductor preparation, crystal growth and selected properties", Plenum Publishing Co., New York, 1972, EEA75-36174.

II. Books

3. Bond, W. I., "Crystal technology", Wiley-Interscience, New York, 1975.
4. Brice, J. C., "The growth of crystals from liquids", North-Holland/American Elsevier, New York, 1973.
5. Goodman, C. H., (Ed.), "Crystal growth: Theory and techniques", Plenum Publishing Co., New York, 1974.
6. Hartman, P., "Crystal growth: an introduction", American Elsevier, New York, 1973.
7. Pamplin, B. R., "Crystal growth", Pergamon Press, New York, 1974.
8. Sittig, M., "Semiconductor crystal manufacture", Noyes Development Corporation, New Jersey, 1969.

III. Review Articles

9. Barrett, J. and Heil, R. H., Jr., "Trends in crystal growing", Solid State Technol., 17 (2), 37-42, 48 (1974), EEA77-18723.
10. Benson, K. E., "Silicon crystals for microelectronic applications", Chem. Engng. Progr., 64 (4), 93-101 (1968), EEA71-20528.
11. Bialek, R. T., "Silicon crystal growth - a production system review", Presented at Second National Conference on Crystal Growth (abstracts only), Princeton, N. J., July-August 1972, EEA76-8567.
12. Carlson, J. W., "Industry standardization of silicon substrates", Solid State Technol., 16 (9), 49-53 (1973), EEA77-2484.
13. Harrison, J. W., Jr., "A survey of single crystal growth methods for gallium arsenide", Solid State Technol., 16 (1), 48-53 (1973), EEA76-19847.
14. Herrmann, H., Herzer, H. and Sirtl, E., "Modern silicon technology", In: Festkorper Probleme XV; Advances in Solid State Physics, Queisser, H. J., (Ed.), Pergamon Press, New York, p. 279-316 (1975).
15. Hoselitz, K., "The crystals we need", J. Cryst. Growth, 3-4, Spe., 5-12 (1968), EEA72-20171.

16. Law, J. T., "Silicon wafer technology state of the art", Solid State Technol., 14 (1), 25-9 (1971), EEA74-13195.
17. Lenzing, J., "Survey of semiconductor crystal-growing processes and equipment", Solid State Technol., 18 (2), 34-9 (1975), EEA78-15239.
18. Lorenzini, R. E., Neff, F. S. and Blair, D. J., "An overview of silicon crystal growing processes", Solid State Technol., 17 (2), 33-6, 55 (1974), EEA77-18722.
19. Mendel, E., "Measurements in silicon planar technology: mechanical properties of semiconductor surfaces", J. Test. & Eval., 1, 312-16 (1973), EEA76-35235.
20. Moore, G. E., Sr. and Law, T., "Worldwide wafer standardization - fact or fancy?", Solid State Technol., 18 (2), 40-3 (1975), EEA78-15240.
21. Pamplin, B. R., "Critical review of the growth of II-IV-V_2 compounds", J. Cryst. Growth, 26, 239-42 (1974), PA78-26310.
22. Podolsky, J. L., "Optimization of silicon wafer yield toward electrical parameter ranges", Semiconductor Prod. Solid State Technol., 10 (4), 51-2 (1967), EEA70-14848.
23. Soper, R. B., "Mechanical damage - its role in silicon surface preparation", Presented at Symposium on Silicon Device Processing, Gaithersburg, Md., June 1970, EEA73-37934.
24. Suzuki, M., "A survey of ultra-precision on silicon wafer", J. Inst. Electron. & Commun. Eng. Jap., 54, 1090-5 (1971), Japanese, EEA75-998.
25. Townley, D. O., "Optimum crystallographic orientation for silicon device fabrication", Solid State Technol., 16 (1), 43-7 (1973), EEA76-19846.

IV. Automation

26. Bardsley, W., Green, G. W., Holliday, C. H., Hurle, D. T. J., Joyce, G. C., MacEwan, W. R. and Tufton, P. J., "Automated Czochralski growth of III-V compounds", Gallium Arsenide and Related Compounds, 1974, p. 355-61, 1975, EEA78-28675.
27. Blumberg, H. and Wilke, K. T., "The automisation of Czochralski equipment", Krist. & Tech., 9, 447-55 (1974), German, PA77-67717.
28. Bochman, R. A., Cormier, Y. J., Dessauer, R. G., Guggenheim, R. C., Jen, D. P., Seksinsky, J. J. and Slocum, R. A., "Computerized slicing saw and sorter", IBM Tech. Disclosure Bull., 13, 2317-18 (1971), EEA74-18544.
29. Corburn, D. S., Seksinsky, J. J. and Tucker, R. B., "Monitoring lateral dimensions of semiconductor-crystals during automated growth", IBM Tech. Disclosure Bull., 15, 1334-5 (1972), EEA76-8517.
30. Digges, T. G., Jr., Hopkins, R. H. and Seidensticker, R. G., "The basis of automatic diameter control utilizing 'bright ring' meniscus reflections", J. Cryst. Growth, 29, 326-8 (1975), PA78-86375.

31.Eisenstadt, B. M., "Semiconductor positioning system", IBM Tech. Disclosure Bull., 15, 1363-5 (1972).
32.Eisenstadt, B. M., "Automatic semiconductor positioning system using an image dissector", IBM Tech. Disclosure Bull., 14, 3367-9 (1972), EEA75-27745.
33.Eisenstadt, B. M. and Petvai, S. I., "Automatic semiconductor positioning system", IBM Tech. Disclosure Bull., 14, 3370-1 (1972), EEA75-27746.
34.Forslund, D. C., "Site selectable pneumatic wafer chuck", IBM Tech. Disclosure Bull., 15, 1105-6 (1972).
35.Fredriksen, T. R., "Microprocessor controls fully automatic wafer prober", Solid State Technol., 18 (9), 49-55 (1975).
36.Glenn, D. A., "Automatically indexed probe assembly for testing semiconductor wafers and the like", Patent USA 3437929, Publ. April 1969, EEA73-4357.
37.Hannan, J., Musits, B., Pavone, P. and Stanley, R., "Automatic wafer contamination measurement system", IBM Tech. Disclosure Bull., 14, 3411-2 (1972), EEA75-27748.
38.Isoda, K., "Automation in semiconductor device production", J. Soc. Instrum. Control Engrs., 8, 80-6 (1969), Japanese, EEA73-15432.
39.James, B. D., "Mechanisation of the wafer fabrication process using air bearing handling techniques", Presented at Semiconductor Integrated Circuit Processing and Production Conference, New York, June 1971, EEA74-24524.
40.Jen, D. P., Slocum, R. A. and Valentino, C. R., "Programmed digital-computer-controlled system for automatic growth of semiconductor crystals", Patent USA 3621213, Publ. November 1971, EEA75-10149.
41.Leibovich, V. S., Sukharev, V. A., Shushkov, V. M. and Fedorov, V. A., "Automatic control of the process of growing single crystals", Prib. & Sist. Upr., no. 5, 7-9 (1975), Russian, PA78-86380.
42.Markstein, H. W., "Trends in automated wafer processing", Electron. Packag. & Prod., 15 (4), 23-32 (1975).
43.Patzner, E. J., Dessauer, R. G. and Poponiak, M. R., "Automatic diameter control of Czochralski crystals", Semiconductor Prod. Solid State Technol., 10 (10), 25-30 (1967), EEA71-16899.
44.Pedrotti, D. G., "Equipment for automatic processing", 1969 Western Electronic Show and Convention (Wescon technical papers), 6 pp., EEA73-31277.
45.Romanenko, V. N., Bessonova, N. V., Vasilevskii, S. A., Lavinzon, D. I., Malorossiyanov, V. S., Ostanina, K. V. and Smirnov, Y. M., "Programmed growth of single crystals of Ge", Inorg. Mater., 10, 457-9 (1974), PA78-81978.
46.Wiesler, M., "Automation systems in IC die handling", Presented at Semiconductor Integrated Circuit Processing and Production Conference, New York, June 1971, EEA74-25625.

47.Zinnes, A. E., Nevis, B. E. and Brandle, C. D., "Automatic diameter control of Czochralski grown crystals", J. Cryst. Growth, 19, 187-92 (1973).

V. Crystal Growth
1. General

48.Allison, D. F. and Schweizer, J. E., Jr., "Method for processing semiconductor wafers", Patent USA 3829335, Publ. August 1974.
49.Biedermann, E., "Producing Si wafers without crystal defects", IBM Tech. Disclosure Bull., 18, 156 (1975), EEA78-37400.
50.Biedermann, E. and Briska, M., "Producing Si wafers without crystal defects", IBM Tech. Disclosure Bull., 18, 444 (1975), EEA78-37640.
51.Billings, A. R., "Microzone recrystallization of semiconductor compound films", J. Vacuum Sci. Technol., 6, 757-65 (1969), EEA73-10372.
52.Clark, K. G., "Application of 'ISO' chemicals in wafer processing", Solid State Technol., 16 (8), 10, 12 (1973), EEA76-38312.
53.Davies, R., "Liquid encapsulation important to crystal growth", Electronic Engng., 42 (503), 58 (1970), EEA73-10334.
54.Dziuba, Z., "Analysis of single crystal pulling methods with respect to application of crystal preparing from semiconductor compounds and alloys", Przeglad Elektron., 10, 581-4 (1969), Polish, EEA73-21064.
55.Givargizov, E. I., "Periodic instability in whisker growth", J. Cryst. Growth, 20, 217-26 (1973), PA77-3669.
56.Heil, R. H., Jr., "Advances in zone refining", Solid State Technol., 11 (1), 21-8 (1968), EEA71-13170.
57.Hoffman, H., "Quality requirements for high purity materials used in semiconductor technology", Przeglad Elektron., 7, 417-23 (1966), Polish, EEA70-8474.
58.Horvath, P., Jedlovszkyn, P. and Gadanyi, P., "Technological researches for producing silicon of great diameter. I. Microcrystals", Elektrotechnika, 65, 23-7 (1972), Hungarian, EEA76-24190.
59.Kaganovskii, I. P., Filippov, M. A., Smirnov, V. A., Freidlina, V. L., Mikhailovna, E. E., Ratner, A. I. and Lepikhova, E. E., "Use of the theory of stochastic processes in investigating the thermal conditions of growing single crystals", Ind. Lab., 39, 84-8 (1973), EEA76-38187.
60.Keller, W., "Method of growing a rod-shaped monocrystal of semiconductor material by crucible free floating zone melting", Patent USA 3498846, Publ. March 1970, EEA73-21105.
61.Kim, K. M., Witt, A. F. and Gatos, H. C., "Crystal growth from the melt under destabilizing thermal gradients", J. Electrochem. Soc., 119, 1218-26 (1972), PA75-74708.

62.Knittel, A. Z., "Vapour growth of crystals with a steady state source (semiconductors)", J. Cryst. Growth, 26, 33-6 (1974), EEA78-8461.
63.Kostiner, E., "Crystal growth", In: Solid state chemistry and physics, vol. 2, P. F. Weller (Ed.), Marcel Dekker, New York, p. 773-95 (1973), PA78-3995.
64.Lafeuille, D., Roche, D. and Juleff, E. M., "Purity of chemicals for semiconductor processing", Solid State Technol., 18 (1), 43-8 (1975), EEA78-15237.
65.Lozovskii, V. N. and Popov, V. P., "Zone melting with a temperature gradient as a technique in semiconductor technology", Izv. VUZ Radioelektron., 15, 3-12 (1972), Russian.
66.Maeda, K. and Sato, J., "Method of growing semiconductor crystals", Patent USA 3484302, Publ. December 1969, EEA73-17514.
67.Muhlbauer, A., "The effect of axial currents when growing rod-shaped semiconductor crystals", Elektrotech. Z. (ETZ) A, 87, 718-23 (1966), German, EEA70-4889.
68.Mullin, J. B., MacEwan, W. R., Holliday, C. H. and Webb, A. E. V., "Pressure balancing: a technique for suppressing dissociation during the melt-growth of compounds", J. Cryst. Growth, 13-14, 629-34 (1972), EEA75-24222.
69.Pamplin, B. R., "Crystal growth of II-IV-V_2 semiconducting compounds", Presented at ACCG Conference on Crystal Growth, Gaithersburg, Md., August 1969, EEA73-28087.
70.Pruett, H. D. and Lien, S. Y., "X-ray imaging technique for observing liquid encapsulation Czochralski crystal growth", J. Electrochem. Soc., 121, 822-6 (1974), EEA77-30182.
71.Romanenko, V. N. and Kheifets, V. S., "Phenomena arising on the arrest of crystal growth", Sov. Phys.-Crystallogr., 18, 663-6 (1974), PA77-53602.
72.Sandman, H., "Comparison of three single crystal techniques", In: Semiconductor silicon, Electrochem. Soc., Inc., New York, 124-31 (1969), EEA73-10412.
73.Scheel, H. J., "Production of high-quality crystals from high-temperature solutions", IBM Tech. Disclosure Bull., 18, 1177-8 (1975), PA78-81960.
74.Spear, K. E., "Growth of single crystals", Earth Miner. Sci., 43 (3), 17-20, 22 (1974).
75.Tiller, W. A., "Theoretical analysis of requirements for crystal growth from solution", J. Cryst. Growth, 2, 69-79 (1968), EEA71-18082.
76.Trompa, K., "Semiconductor crystals", Patent East German 98833, Publ. July 1973, CA80-53780.
77.Volpyan, A. E., Gnilov, S. V. and Nashelskii, A. Y., "Calculation of a program for the growth of single-crystals by the zone-levelling method with a uniform distribution of impurities along the length", Izv. Akad. Nauk. SSR. Neorg. Mater., 7, 1301-4 (1971), Russian, PA75-1211.

78.Wang, E. Y., "Analysis of zone melting processes", J. Electrochem. Soc., 121, 1671-2 (1974), EEA78-8456.
79.Witt, A. F., "Thermo-hydrodynamic effects during crystal growth", Presented at Second National Conference on Crystal Growth, Princeton, N. J., July-August 1972, EEA76-8556.
80.Witt, A. F. and Gatos, H. C., "Homogeneous impurity incorporation during crystal growth from the melt", J. Electrochem. Soc., 116, 511-13 (1969), EEA72-15454.
81.Zarabi, M. J., "A normalised demand profile for electronic materials. I. Semiconductor devices", Electron. Inf. & Plann., 1, 393-6 (1974), EEA77-26210.

2. Group IV Semiconductors

82.Abe, T., "The growth of Si single crystals from the melt and impurity incorporation mechanisms", J. Cryst. Growth, 24-25, 463-7 (1974), PA78-7479.
83.Anon., "Silicon single crystals of extreme perfection", Microelectron. & Reliab., 13, 149 (1974), PA77-57711.
84.Antonov, P. I., Grigorev, N. S. and Stepanov, A. V., "The growth of ribbon crystals of germanium by the Stepanov method", Izv. Akad. Nauk SSSR Ser. Fiz., 36, 494-500 (1972), Russian, PA75-47269.
85.Artyshevskii, P. P., Arendarchuk, A. V., Zatulovskii, L. M., Kravetskii, D. Y., Maergoiz, I. D., Medvedovskaya, L. A. and Chaikin, P. M., "Investigation of electromagnetic field and temperature distribution in the working zone for the growth of shaped silicon crystals", Izv. Akad. Nauk SSSR Ser. Fiz., 37, 2271-4 (1973), Russian, EEA77-5873.
86.Artyshevskii, P. P., Kravetskii, D. Y., Zatulovskii, L. M., Chaikin, P. M., Smelyanskii, M. Y., Falkevich, F. S. and Malorossiyanov, V. S., "Investigation of various adaptations of the process of silicon crystal growth from a pedestal, using combined heating", Izv. Akad. Nauk SSSR Ser. Fiz., 36, 519-21 (1972), Russian, PA75-53601.
87.Artyshevskii, P. P., Kravetskii, D. Y., Zatulovskii, L. M., Chaikin, P. M., Zhvirblyanskii, V. Y. and Freiman, A. K., "Development of new methods for growing shaped silicon crystals from a melt in a crucible and from a supercooled melt on a mounting", Izv. Akad. Nauk SSSR Ser. Fiz., 37, 2275-6 (1973), Russian, EEA77-5874.
88.Barrett, D. L., Myers, E. H., Hamilton, D. R. and Bennett, A. I., "Growth of wide, flat crystals of silicon web", J. Electrochem. Soc., 118, 952-7 (1971), EEA74-26580.
89.Benson, K. E., "Growth of silicon crystals", In: Semiconductor silicon, Electrochem. Soc., Inc., New York, p. 97-123 (1969), EEA73-10411.

90. Bessonova, N. V., Levinzon, D. I., Sachkov, G. V. and Smirnov, Y. M., "Growth of germanium single crystals from a molten layer", Izv. Akad. Nauk SSSR Ser. Fiz., 36, 534-6 (1972), Russian, PA75-47270.
91. Borle, W. N., "Growth features of silicon filamentary crystals grown from solution", J. Appl. Phys., 41, 3184-6 (1970), EEA73-31237.
92. Borle, W. N., "A new growth form of silicon micro crystals grown from solution", Indian J. Pure & Appl. Phys., 10, 317-18 (1972), PA75-74684.
93. Borle, W. N., Bal, M., Sharda, G. D., Seth, G. L., Nangia, O. P. and Bagai, R. K., "Growth of large diameter float zone silicon crystals", Indian J. Pure & Appl. Pbys., 13, 406-7 (1975), EEA78-37395.
94. Borle, W. N., Tata, S. and Prakash, V., "Report on the work on growth of semiconductor grade silicon single crystals", J. Instn. Telecomm. Engrs., 14, 370-3 (1968), EEA72-4912.
95. Borle, W. N., Tata, S., Varma, S. K., Chatterjee, S. N. and Agarwal, S. K., "On doping of silicon crystals", Indian J. Pure & Appl. Phys., 9, 136-7 (1971), EEA74-26572.
96. Chartier, C. P. and Sibley, C. B., "Czochralski silicon crystal growth at reduced pressures", Solid State Technol., 18 (2), 31-3 (1975), PA78-33973.
97. Ciszek, T. F., "Solid-liquid interface morphology of float zoned silicon crystals", In: Semiconductor silicon, Electrochem. Soc., Inc., New York, p. 156-68 (1969), EEA73-10415.
98. Ciszek, T. F., "Growth of 40 mm diameter silicon crystals by a pedestal technique using electron beam heating", J. Cryst. Growth, 12, 281-7 (1972), EEA75-24204.
99. Ciszek, T. F., "Characteristics of (115) dislocation-free float-zoned silicon crystals", J. Electrochem. Soc., 120, 799-802 (1973), EEA76-29167.
100. Ciszek, T. F., "Edge-defined, film-fed growth (EFG) of silicon ribbons", Presented at Second National Conference on Crystal Growth (abstracts only), Princeton, N. J., July-August 1972, EEA76-8564.
101. Ciszek, T. F., "Continuous growth of semiconductor crystal ribbons without a crucible", IBM Tech. Disclosure Bull., 17, 2338-9 (1975), EEA78-19636.
102. Corley, R. A., "An improved technique for the growth of ultra-pure germanium", IEEE Trans. Nucl. Sci., NS-21, 342-3 (1974), PA77-49582.
103. Daey Ouwens, C. and Heijligers, H., "Recrystallization processes in polycrystalline silicon (for solar cells)", Appl. Phys. Lett., 26, 569-71 (1975), EEA78-32992.
104. Digges, T. G., Jr., "A solution for melt-level control when growing Czochralski silicon crystals in hemispherical quartz crucibles", Solid State Technol., 18 (2), 44-6 (1975), PA78-33974.

105.Digges, T. G., Jr. and Yaws, C. L., "A vacuum float-zone process for the production of high-resistivity silicon crystals", Presented at Electrochemical Society Meeting, Chicago, Ill., May 1973, EEA78-23292.

106.Egorov, L. P., Okun, L. S., Zatulovskii, L. M., Chaikin, P. M., Gulyaev, V. V., Zhvirblyanskii, V. Y., Levinzon, D. I., Smirnov, Y. M. and Sachkov, G. V., "Investigation of the group growing process for monocrystalline germanium rods by Stepanov's method in a rectilinear thermal zone", Izv. Akad. Nauk SSSR Ser. Fiz., 37, 2268-70 (1973), Russian, PA77-10542.

107.Egorov, L. P., Zatulovskii, L. M. and Chaikin, P. M., "Special features of growing thin germanium strips at elevated melt pressures in a former by Stepanov's method", Izv. Akad. Nauk SSSR Ser. Fiz., 37, 2277-9 (1973), Russian, PA77-13570.

108.Gillessen, K. and von Muench, W., "Growth of silicon carbide from liquid silicon by a traveling heater method", J. Cryst. Growth, 19, 263-68 (1973).

109.Gotz, G. and Klinge, K. D., "Production and analysis of single crystal Si films", Exp. Tech. Phys., 21, 473-83 (1973), German, PA77-17177.

110.Haas, E., Brandt, W. and Martin, J., "The behaviour of carbon in a zone-refining process of silicon", Solid-State Electron., 12, 915-21 (19-9), German, EEA73-10392.

111.Hearn, E. W., Schwuttke, G. H. and Tekaat, E. H., "Reducing crystallographic defects in semiconductor structures", Patent USA 3737282, Publ. June 1973, CA79-59188.

112.Law, J. T., "Recent advances in silicon materials", Western Electronic Show and Convention (Vol. 16), 1972, 19/1 5 pp., EEA76-8579.

113.Murgai, A., Witt, A. F. and Gatos, H. C., "Elimination of random compositional inhomogeneities in Czochralski grown silicon", J. Electrochem. Soc., 122, 1276-7 (1975).

114.Noack, J. and Berg, G., "Semiconductor single crystal of high crystallographic perfection", Patent German (East), 97554, Publ. May 1973, CA79-150804.

115.Pogodin, A. I., Tumin, I. M. and Eidenzon, A. M., "Forming silicon monocrystals during growth by Czochralski's method", Izv. Akad. Nauk SSSR Ser. Fiz., 37, 2292-4 (1973), Russian, PA77-13571.

116.Ramakrishna, V., "Manufacture of semi-conductor grade silicon", Defence Sci. J., 18 (2a), 13-20 (1968), EEA72-688.

117.Reithard, E. M., "Growing two-inch silicon crystals", Semiconductor Prod. Solid State Technol., 10 (5), 46-7 (1967), EEA70-16480.

118.Reuschel, K., "Method of crucible-free zone melting of semiconductor material, particularly silicon", Patent USA 3454367, Publ. July 1969, EEA73-10460.

119. Robertson, D. S. and Young, I. M., "Observations on the unrestrained growth of germanium crystals", J. Phys. D, 8 (5), L59-61 (1975), PA78-30241.
120. Rubenstein, M., "Preparation and analysis of homogeneous solid solution crystals of the system Si-Ge", J. Cryst. Growth, 17, 149-54 (1972), EEA76-4523.
121. Sandulova, A. V., Bogoyavlenskaya, I. P. and Vasilina, V. S., "Doping of Si crystals during growth by the method of gas transport reaction", Izv. Akad. Nauk SSSR Neorg. Mater., 8, 630-3 (1972), Russian, PA75-47430.
122. Sharma, V., "Semiconductor silicon substrate for high yield", IBM Tech. Disclosure Bull., 18, 336 (1975), EEA78-37401.
123. Shashkov, Y. M., Silkin, V. B. and Bletskan, N. I., "Calculation of the temperature distribution in square and rectangular shaped silicon ingots grown by Stepanov's method", Izv. Akad. Nauk SSSR Ser. Fiz., 37, 2458-61 (1973), Russian, PA77-13048.
124. Siemens A. G., "Semiconductor monocrystals", Patent UK 1321204, Publ. June 1973, CA79-84837.
125. St. John, C., "The brittle-to-ductile transition in pre-cleaved silicon single crystals", Philos. Mag., 32, 1193-212 (1975).
126. Vlasenko, N. A. and Voshchinkin, V. V., "Production of silicon carbide films and their properties", Poluprovodn. Tekh. & Mikroelektron., no. 8, 98-109 (1972), Russian, EEA76-16301.
127. Vojdani, S., Dabiri, A. E. and Ashoori, H., "Diameter control of pulled germanium crystals by means of Peltier cooling", J. Cryst. Growth, 24-25, 374-5 (1974), PA78-10537.
128. Voltmer, F. W. and Digges, T. G., Jr., "Anomalous resistivity profiles in long silicon crystals grown by the Czochralski method", J. Cryst. Growth, 19, 215-17 (1973), EEA76-35120.
129. Wolfe, G., "Multilayer silicon wafer production methods", Patent USA 3821045, Publ. June 1974.

3. Group III-V Semiconductors

130. Bachmann, K. J. and Buehler, E., "The growth of InP crystals from the melt", J. Electron. Mater., 3, 279-302 (1974), PA77-29018.
131. Bachmann, K. J., Buehler, E., Shay, J. L. and Strnad, A. R., "Liquid encapsulated Czochralski pulling of InP crystals", J. Electron. Mater., 4, 389-406 (1975), EEA78-19671.
132. Blom, G. M., "Growth of constant diameter LEC GaP", Presented at Second National Conference on Crystal Growth, Princeton, N. J., July-August 1972, EEA76-8559.
133. Blom, G. M. and Zwicker, W. K., "The growth of GaP single crystals by liquid encapsulated Czochralski pulling", Acta Electron., 16, 315-18 (1973), EEA76-38179.
134. Blum, S. E. and Chicotka, R. J., "Liquid encapsulation-Czochralski and vertical gradient freeze", Presented at Electrochemical Society Meeting, Chicago, Ill., May 1973, EEA76-23268.

135.Cerrina, F., Margadonna, D. and Perfetti, P., "Growth of $GaAs_{1-x}P_x$ crystals by pulling from gallium-rich solutions", J. Cryst. Growth, 18, 202-4 (1973), EEA76-16294.
136.Choe, B. D., Kim, K. S. and Lee, K. S., "Crystal growth of GaAs and its properties", New Phys. (Korean Phys. Soc.), 13 (1), 19-23 (1973), Korean, EEA77-18716.
137.Fahrig, R. H., Webb, G. N. and Hart, R. L., "Melt growth of GaP", Presented at Second National Conference on Crystal Growth, Princeton, N. J., July-August 1972, EEA76-8558.
138.Fischer, P., Kuhn, G., Bindemann, R., Rheinlander, B. and Horig, W., "Growth and physical characteristics of $Al_{1-x}Ga_xAs$ crystals", Krist. & Tech., 8, 167-76 (1973), German, EEA76-29172.
139.Grabmaier, B. C. and Grabmaier, J. G., "Dislocation-free GaAs by the liquid encapsulation technique", J. Cryst. Growth, 13-14, 635-9 (1972), EEA75-24223.
140.Johnson, E. S., "Liquid encapsulated floating zone melting of GaAs", J. Cryst. Growth, 30, 249-56 (1975).
141.Johnson, E. S. and Allred, W. P., "Liquid encapsulated zone refining of gallium arsenide", Presented at Second National Conference on Crystal Growth, Princeton, N. J., July-August 1972, EEA76-8563.
142.Klausutis, N., Hunt, M., Adamski, J., Weiner, J. and Buckmelter, J., "Preparation and growth of indium phosphide", Presented at Second National Conference on Crystal Growth, Princeton, N. J., July-August 1972, EEA76-8561.
143.Larkin, J., Hilton, R., Bruce, J., Posen, H., Armington, A. and Lipson, H., "Some studies on the growth of gallium phosphide", Presented at Second National Conference on Crystal Growth, Princeton, N. J., July-August 1972, EEA76-8560.
144.Leung, P. C. and Allred, W. P., "Liquid-seal Czochralski growth of gallium arsenide", J. Cryst. Growth, 19, 356-58 (1973).
145.Miller, E. A., Weisberg, L. R. and Liebman, W. K., "On the Czochralski growth of GaP", J. Electrochem. Soc., 118, 825-9 (1971), EEA74-22500.
146.Moulin, M., Faure, M. and Bichon, G., "Growth of GaAs single crystals by a rotating liquid seal method", J. Cryst. Growth, 24-25, 376-9 (1974), PA78-10538.
147.Mullin, J. B., Royle, A., Straughan, B. W., Tufton, P. J. and Williams, E. W., "Crystal growth and properties of Group IV doped indium phosphide", J. Cryst. Growth, 13-14, 640-6 (1972), PA75-47272.
148.Nichols, D. R. and Thompson, A. G., "The growth of large diameter gallium arsenide", Presented at Second National Conference on Crystal Growth, Princeton, N. J., July-August 1972, EEA76-8562.
149.Nishizawa, J., Suto, K. and Okuno, Y., "Method for producing a single crystal of $In_xGa_{1-x}P$", Patent USA 3880677, Publ. April 1975.

150.Nygren, S. F., "Liquid encapsulated Czochralski growth of 35 mm diameter single crystals of GaP", J. Cryst. Growth, 19, 21-32 (1973), EEA76-25973.
151.Osamura, K. and Murakami, Y., "Preparation of $GaAs_{1-x}P_x$ compounds from the melt under controlled vapour pressures", Mem. Fac. Eng. Kyoto Univ., 34, 84-102 (1972), EEA75-28409.
152.Ovchinnikov, S. Y., Sorokin, V. S. and Yaskov, D. A., "Phase equilibria in the system GaAs-GaP. (Crystal growth)", Inorg. Mater., 10, 200-4 (1974), EEA78-8444.
153.Pekarek, L., "Preparation of GaAs monocrystals and InP and GaP polycrystals", Elektrotech. Cas., 25, 319-27 (1974), Czech, PA77-67719.
154.Poiblaud, G. and Jacob, G., "Synthesis of GaP by gallium solution growth in a horizontal sealed ampoule", Gallium Arsenide and Related Compounds, 1974, p. 134-44, 1975, PA78-54982.
155.Russ, M. J., "Liquid encapsulated Czochralski growth of GaAs and GaP", Solid State Technol., 15 (8), 29-32 (1972), EEA75-36168.
156.Segawa, K. and Miki, H., "Liquid-phase growth of gallium antimonide crystals", Patent Japan 74-114349, Publ. October 1974, CA82-105473.
157.Thyagarajan, R., Narula, R. C. and Mohan, M., "Synthesis and growth of gallium arsenide single crystals", Indian J. Technol., 10 (1), 25-8 (1972), PA75-60724.
158.Tranchart, J. C., "Preparation of indium arsenide single crystals", Acta Electronica, 13 (1), 13-20 (1970), French, EEA73-31163.
159.van Dijk, H. J. A., Joehem, C. M. G., Scholl, G. J. and van der Werf, P., "Diameter control of LEC grown GaP crystals", J. Cryst. Growth, 21, 310-12 (1974), EEA77-18706.
160.von Neida, A. R., Oster, L. J. and Nielsen, J. W., "Liquid encapsulated growth of GaP from non-stoichiometric melts (luminescence)", J. Cryst. Growth, 13-14, 647-50 (1972), PA75-47268.
161.Vyborny, Z., "A new method of GaP crystal growth for electroluminescent diodes", Cesk. Cas. Fis. A, 25, 186-7 (1975), Czech, PA78-69594.
162.Wagner, J. W., "Preparation and properties of bulk $In_{1-x}Ga_xAs$ alloys", J. Electrochem. Soc., 117, 1193-6 (1970), EEA73-37876.
163.Weiner, M. E. and Lassota, D. T., "High resistivity crystals of GaAs prepared by encapsulate Czochralski growth", Presented at ACCG Conference on Crystal Growth, Gaithersberg, Md., August 1969, EEA73-28076.

4. Group II-VI Semiconductors

164.Bartlett, B. E., Deans, J. and Ellen, P. C., "Growth and properties of $Cd_xHg_{1-x}Te$ crystals", J. Materials Sci., 4, 266-70 (1969), EEA73-1077.
165.Baubinas, R. V., Vishchakas, Y. K. and Smilga, A. A., "Production of CdSe crystals by the static sublimation method", Izv. Akad. Nauk SSSR Neorg. Mater., 8, 648-52 (1972), Russian, PA75-47217.

166. Baughman, R. J. and Lefever, R. A., "Czochralski encapsulation growth of GeTe, SnTe and PbTe single crystals", Materials Res. Bull., 4, 721-6 (1969), EEA73-7357.
167. Bell, R. O. and Wald, F., "Growth of CdTe from solution-status, problems and potential (Doping considerations)", Proceedings of the International Symposium on Cadmium Telluride a Material for Gamma-Ray Detectors, 1971, 6/15 pp., EEA75-20098.
168. Bell, R. O. and Wald, F. V., "Recent advances in the preparation of CdTe for nuclear detectors", IEEE Trans. Nucl. Sci., NS-19, 334-8 (1972), PA75-49668.
169. Bourassa, F. J., "Production of optical grade cadmium telluride by directional solidification", Am. Ceram. Soc. Bull., 51, 338 (1972), PA75-50607.
170. Fujisaki, H., "Crystal growth from the melt of sulfide with induction furnace", Presented at International Union of Crystallography, Kyoto, Japan, August-September 1972, EEA76-12430.
171. Gutknecht, K. H., Preier, H. and Hesse, J., "Growth of lead sulfide single crystals by the Bridgman method", Mater. Res. Bull., 10, 277-80 (1975), EEA78-24215.
172. Hussain, S. B., "Dependence of preferred orientation on thickness: the 'stifling process' method", Thin Solid Films, 23, S21-3 (1974), PA77-72662.
173. Jones, K. A., "Growth of CdS platelets", J. Cryst. Growth, 19, 33-44 (1973).
174. Klausutis, N., Adamski, J. A., Collins, C. V., Hunt, M., Lipson, H. and Weiner, J. R., "Growth of $CdTe_{1-x}Se_x$ by the LEC and Bridgman techniques", J. Electron. Mater., 4, 625-34 (1975), PA78-58823.
175. Kozielski, M. J., "Polytype single crystals of $Zn_{1-x}Cd_xS$ and $ZnS_{1-x}Se_x$ solid solutions grown from the melt under high argon pressure by Bridgman's method", J. Cryst. Growth, 30, 86-92 (1975).
176. Kuznetsov, V. A., Lobachev, A. N., Kotelnikov, A. R. and Efremova, E. P., "Crystallization of CdS under hydrothermal conditions", Sov. Phys.-Crystallogr., 19, 531-4 (1975), EEA78-24228.
177. Kyle, N. R., "Growth of semi-insulating cadmium telluride", Proceedings of the International Symposium on Cadmium Telluride a Material for Gamma Ray Detectors, 1971, 4/16 pp., EEA75-20096.
178. Kyle, N. R., "Phase equilibria considerations in the preparation and properties of single crystal CdTe", Am. Ceram. Soc. Bull., 51, 338 (1972), PA75-50593.
179. Lupin, V. M., Petrov, M. A. and Ramazanov, P. E., "Production and properties of single-crystal CdS films on GaAs crystals", Izv. VUZ Fiz., no. 2, 93-8 (1974), Russian, EEA77-18696.
180. Medvedev, S. A., Klevkov, Y. V., Kiseleva, K. V. and Sentyurina, N. N., "Growth of CdTe crystals from molten solutions and a study of their properties", Izv. Akad. Nauk SSSR Neorg. Mater., 8, 1210-13 (1972).

181. Polivka, P., "Preparation and properties of doped and undoped monocrystals of cadmium telluride", Elektrotech. Cas., 24, 36-41 (1973), Czech, EEA76-12434.
182. Strauss, A. J., "Metallurgical and electronic properties of CdTe related to preparation of single crystals", Proceedings of the International Symposium on Cadmium Telluride a Material for Gamma Ray Detectors, 1971, 1/31 pp., EEA75-20094.
183. Triboulet, R. and Marfaing, Y., "Growth of high purity CdTe single crystals by vertical zone melting", J. Electrochem. Soc., 120, 1260-5 (1973), EEA77-2455.
184. Vandekerkhof, J., "Some problems in the preparation of single crystals of cadmium telluride by the Czochralski liquid encapsulation technique", Proceedings of the International Symposium on Cadmium Telluride a Material for Gamma Ray Detectors, 1971, 8/7 pp., EEA75-20100.
185. Wald, F. V. and Bell, R. O., "Natural and forced convection during solution growth of CdTe by the traveling heater method (THM)", J. Cryst. Growth, 30, 29-36 (1975).

5. Impurity Distribution

186. Akai, S., "Doping III-V semiconductor compound with oxygen", Patent Japan 73-30701, Publ. September 1973, CA80-75731.
187. Antonov, P. I., "Properties of profiled semiconductor single crystals grown by Stepanov's method", J. Cryst. Growth, 23, 318-22 (1974), PA78-16091.
188. Boltaks, B. I. and Kulikov, C. S., "Solubility of impurities in silicon", In: Kompensirovannyi Kremnii, B. I. Boltaks, (Ed.), "Nauka," Leningrad, Otd.: Leningrad, USSR, p. 20-8 (1972), Russian, CA79-10937.
189. Fabre, E., Bhargava, R. N. and Zwicker, W. K., "Thermally-stimulated-current measurements in n-type LEC (liquid-encapsulated-Czochralski) gallium phosphide", J. Electron. Mater., 3, 409-30 (1974), CA81-18271.
190. Hicks, H. G. B. and Greene, P. D., "Control of silicon contamination in solution growth of gallium arsenide in silica", Gallium Arsenide Related Compounds, 1970, p. 92-9, Publ. 1971, CA80-64889.
191. Katz, L. E., "Gas doping of float zone silicon in vacuum from a solid source", Presented at Second National Conference on Crystal Growth, Princeton, N. J., July-August 1972, EEA76-8554.
192. Kuznetsov, V. A., Sachkov, G. V., Bessonova, N. V. and Levinzon, D. I., "Features of the distribution of impurities in cylindrical germanium monocrystals of different diameters grown by Stepanov's method", Izv. Akad. Nauk SSSR Ser. Fiz., 37, 2299-2301 (1973), Russian, PA77-10548.
193. Mentzel, F., Reinke, H. and Eckardt, D., "Semiconductor single crystals with adjustable impurity concentration", Patent German 1444541, Publ. June 1973, CA79-109627.

194.Okun, L. S., Egorov, L. P., Zatulovskii, L. M., Levinzon, L. I., Sachkov, G. V. and Chaikin, P. M., "Distribution of impurities in monocrystalline Ge rods growth in a rectangular working zone by a group method", Izv. Akad. Nauk SSSR Ser. Fiz., 37, 2297-8 (1973), Russian, EEA77-5880.

195.Shadeev, N. I., Ilinykh, V. E. and Grekhova, T. I., "Copper doping of germanium single crystals grown by the Czochralski method", Tekhnol. Mater. Elektron. Tekh., no. 2, 175-80 (1972), Russian, CA80-8292.

196.Stepanov, A. V. and Antonov, P. I., "Growth of controlled profile germanium crystals", Presented at 9th International Congress of Crystallography of the International Union of Crystallography, Kyoto, Japan, August-September 1972, EEA76-12428.

197.Tochikubo, H. and Sugawara, K., "Growing of semiconductor single crystals with uniform distribution of resistance", Patent Japan 73-00374, Publ. January 1973, CA79-11275.

198.Witt, A. F. and Gatos, H. C., "Impurity heterogeneities in semiconductor single crystals", In: Semiconductor silicon, Electrochem. Soc., Inc., New York, p. 146-55 (1969), EEA73-10414.

6. Equipment

199.Bletskan, N. I., Berezenko, L. E., Stepanova, G. M. and Shashkov, Y. M., "Apparatus for studying the growth of dislocation free silicon single crystals from a melt", Prib. Tekh. Eksp., no. 5, 234-5 (1973), Russian, CA80-31159.

200.Clement, D. R., Helber, C. A. and Mason, L. D., "Apparatus for pulling a crystal from a melt", Patent German 2449275, Publ. April 1975, CA83-124558.

201.Dessauer, R. G., Patzner, E. J. and Poponiak, M. R., "Radiation sensitive control system for crystal growing apparatus", Patent USA 3493770, Publ. February 1970, EEA73-31279.

202.Forner, H. G. and Schmugge, K., "Apparatus for thermal treatment of semiconductor crystals grown from the melt by crucible-free zone melting", Patent German (East), 100637, Publ. October 1973, CA80-53761.

203.Gautherie, M., "Experimental apparatus for the study of crystal growth", Rev. Phys. Appl., 3, 131-42 (1968), French, EEA71-28746.

204.Gehring, J., "Apparatus for pulling crystals from the melts", Patent German 2349718, Publ. April 1975, CA83-124557.

205.Hall, R. N., "Non-clamping seed holder for Czochralski crystal growth", J. Cryst. Growth, 22, 65-6 (1974), PA77-45429.

206.Henkel, H. and Matthes, H., "Seed holder for Czochralski growth of high melting compounds", J. Cryst. Growth, 29, 215-16 (1975), PA78-69584.

207.Keller, W., "Device for the crystal growth of semiconductors by floating zone melting", Patent German 2306755, Publ. August 1974, CA82-50680.

208.Kokusai Denki KK, "Crystal pulling device", Patent UK 1287789, Publ. September 1972.
209.Lenard, J. A. and Mendel, E., "Holder for crystal growth seed", IBM Tech. Disclosure Bull., 27, 1352 (1974), PA78-16063.
210.Marshall, R. C., "High pressure, high temperature crystal growth system", Air Force Cambridge Res. Lab. Rep., no. 135, 1-25 (1967), EEA71-18079.
211.Mendel, E., Lenard, J. A. and Seeley, G., "Seed holder for crystal growth", IBM Tech. Disclosure Bull., 27, 1353 (1974), PA78-16064.
212.Muraoka, H., Katsuki, A., Kato, I. and Kato, T., "Toshiba semiconductor crystal growing furnaces", Toshiba Rev. (Internat. Ed.), no. 40, 44-8 (1969), EEA73-13371.
213.Nishizawa, J., "Semiconductor single crystal manufacturing equipment", Oyo Buturi, 43, 220-9 (1974), Japanese, EEA78-2739.
214.Plovnick, R. H. and Schmidt, G. E., "A dopant injector for crystal pulling furnaces", Rev. Sci. Instrum., 41, 1248-9 (1970), EEA74-1414.
215.Satulowsky, L. M. and Tschaikin, P. M., "Process and equipment for growing single-crystal germanium ribbon", Elektrowarme Int. B, 30 (3), B123-9 (1972), German, EEA75-32510.
216.Schnoeller, M., Mladenovic, T., Regus, E. and Schilling, E., "Device for floating zone melting of semiconductors", Patent German 2228779, Publ. January 1974, CA80-75737.
217.Suzuki, T., Hoshi, K. and Tamate, Y., "Apparatus for pulling up semiconductor crystals", Patent USA 3822111, Publ. July 1974.
218.Tairov, Y. M. and Tsvetkov, V. F., "Monitoring a gas medium during the growth of silicon carbide crystals", Istrum. & Exp. Tech., 14, 1533-5 (1971), PA75-60704.
219.Weinberg, E., "Viewing port window for crystal pullers", Rev. Sci. Instrum., 46, 494-5 (1975), PA78-66557.
220.Wenckus, J. F., Castonguay, R. A., Hanley, B. C. and Mallahan, F. J., "Pressure and temperature-controlled apparatus for large-scale production of crystals by the Czochralski technique", Patent USA 3865554, Publ. February 1975.

VI. Wafer Processing

1. Slicing

221.Forman, S. E. and Rhines, W. J., "Vibration characteristics of crystal slicing ID saw blades", J. Electrochem. Soc., 119, 686-90 (1972), EEA75-27847.
222.Hakes, G. L., "Abrasive dicing of semiconductor wafers", Patent USA 3693302, Publ. September 1972.
223.Hargraves, D., "High yield method for breaking wafer into dice", Patent USA 3870196, Publ. March 1975, EEA78-37664.
224.Hoshi, K. and Sugita, K., "Machine for grinding an edge contour on a semiconductor wafer", Patent USA 3834083, Publ. September 1974.

225. Ikawa, T., "New approach to wafer dicing", Ind. Diamond Rev., 136-38 (1972).
226. Kachajian, G. S., "A systems approach to semiconductor slicing to improve wafer quality and productivity", Solid State Technol., 15 (9), 59-64 (1972), EEA76-83.
227. Klatskin, J. and Rosen, A., "Simple method for separating completed diodes from semiconductor wafers", J. Electrochem. Soc., 122, 1565-6 (1975).
228. Kobayashi, A., Komine, N. and Kuniyoshi, S., "Improving cutoff characteristics of semiconductor materials using new diamond wheel", Toshiba Rev. (Int. Ed.), no. 73, 29-34 (1972), EEA76-653.
229. Kobayashi, A., Nagano, Y. and Sumiya, M., "Scribing of silicon wafers with YAG laser", Proceedings of the International Conference on Product Engineering, Pt. 1, 1974, p. 22-25.
230. Lane, R. L., "Billet holder for cutting and slicing apparatus (crystals)", Patent USA 3802412, Publ. April 1974.
231. Mansbridge, D. S. and Miller, J., "Electroformed multiple bladed diamond saw (for semiconductor wafer dicing)", IBM Tech. Disclosure Bull., 17, 2448-9 (1975), EEA78-15411.
232. Marshall, J. F., "Scribing and breaking of semiconductor wafers", Solid State Technol., 17 (9), 70-5 (1974), EEA78-2891.
233. Norton Company, "Semiconductor dicing machine", Patent UK 1140874, Publ. January 1969, EEA72-15479.
234. Powell, W. J. A. and Fynn, G. W., "The slow speed multi-sawing of brittle semiconducting compounds", J. Sci. Instrum. (J. Phys. E.), 2, 1122-3 (1969), EEA73-10323.
235. Richter, H., Siebert, R., Kieback, D. and Mohr, U., "Orientation procedure for semi-conducting crystals", Krist. & Tech., 8, 1333-6 (1973), German, PA77-29019.
236. Seyfarth, H., Richter, F., Borchardt, W., Artopee, G. and Frahnow, D., "A new method of slicing, semiconductor single crystals", Krist. & Tech., 10, F89-94 (1975), German, PA78-86410.
237. Shah, A. S. and St. Pierre, E. J., "Two-pass laser cutting", IBM Tech. Disclosure Bull., 16, 3237 (1974), EEA77-30384.

2. Polishing

238. Bakin, N. N., Dedkov, V. D. and Porokhovnichenko, L. P., "Polishing gaseous etching of GaAs", Izv. Akad. Nauk SSSR Neorg. Mater., 8, 1364-7 (1972), Russian, PA75-83853.
239. Basi, J. S., "Polishing semiconductor gallium arsenide planar surfaces", Patent USA 3738882, Publ. June 1973, CA79-59189.
240. Basi, J. S., "Polishing gallium phosphide surfaces", Patent German 2252045, Publ. May 1973, CA79-36427.
241. Basi, J. S., "Method of polishing zinc selenide", Patent USA 3869323, Publ. March 1975.
242. Basi, J. S., "Polishing solution for silicon", Patent German 2456244, Publ. July 1975, CA83-187374.

243.Basi, J. S. and Mendel, E., "Method of polishing cadmium telluride", Patent USA 3869324, Publ. March 1975.
244.Blake, L. H. and Mendel, E., "Chemical-mechanical polishing of silicon", Solid State Technol., 13 (1), 42-6 (1970), EEA73-17501.
245.Colacino, J. J. and Bartush, T. A., "Analysis of velocity as a cause of thickness variation in a chemical-mechanical polishing process", Solid State Technol., 16 (8), 30 (1973), EEA76-38313.
246.Cronkhite, P. W., Voege, R. D. and Poletti, J. W., "Apparatus and process for polishing semiconductor or similar materials", Patent USA 3708921, Publ. January 1973.
247.Davidson, J. L. and Rouse, G. V., "Precision polishing of semiconductor crystal wafers", Patent USA 3699722, Publ. October 1972.
248.Deckert, H., Heinke, W., Jacob, H. and Kirschner, H., "Polishing suspensions for silicon semiconductor surfaces", Patent German 2305188, Publ. August 1974, CA81-178917.
249.Deckert, H. and Jacob, H., "Method of polishing semiconductor surfaces", Patent USA 3877183, Publ. April 1975.
250.Dyment, J. C. and Rozgonyi, G. A., "Evaluation of a new polish for gallium arsenide using a peroxide-alkaline solution", J. Electrochem. Soc., 118, 1346-50 (1971), EEA74-33961.
251.Goetz, F. E. and Hause, J. R., "Wafer polishing machine", Patent USA 3691694, Publ. September 1972.
252.Hajkova, E. and Fremunt, R., "Effective chemical polishing of n-type GaP surfaces", Phys. Status Solidi A, 10, K35-7 (1972), EEA75-16648.
253.Hallas, C. E., "Electropolishing silicon", Solid State Technol., 14 (1), 30-2 (1971), EEA74-13196.
254.Iiyama, S., Ida, I. and Furumoto, S., "Electropolishing GaAs substrates with a rotating disc", Rev. Elec. Commun. Lab., 18, 670-80 (1970), EEA74-13193.
255.Jensen, E. W., "Polishing silicon wafers", Semiconductor Prod. Solid State Technol., 10 (2), 19-24 (1967), EEA70-11100.
256.Jensen, E. W., "Polishing compound semiconductors", Solid State Technol., 16 (8), 49-52 (1973), EEA76-38208.
257.Jentzsch, J., "Polishing composition for semiconductor surfaces", Patent German 2336735, Publ. February 1975, CA83-20808.
258.Keen, J. M., "An investigation of a chemical polishing process for silicon", R. R. E. Newsletter Res. Rev., no. 7, 4 pp.,(1968), EEA72-13367.
259.Kiszkurno, J., "Polishing of a GaAs sample with perhydrolalkaline mixture", Elektronika, 14, 326-30 (1973), Polish, EEA76-38183.
260.Lazarovici, D., "Single crystal silicon targets for blocking and channeling experiments", Inst. Fiz. At. (Rep.), NR-54, 12 pp. (1974), CA82-24530.
261.Liaw, H. M. and Faust, J. W., Jr., "Studies on the surface preparation of optically flat and damage-free silicon surfaces", Presented at Electrochemical Society Meeting, Chicago, Ill., May 1973, EEA76-23286.

262. Lim, M. J., "Method of vapor-phase polishing a surface of a semiconductor", Patent USA 3855024, Publ. December 1974.
263. Mayburg, S., "Optimum abrasive sizes for minimizing the polishing times of semiconductors", J. Electrochem. Soc., 116, 509-10 (1969), EEA72-15453.
264. Mendel, E. and Basi, J. S., "Chem-mech polishing of cadmium telluride", Presented at Electrochemical Society Fall Meeting, Dallas, Texas, October 1975.
265. Mendel, E. and Yang, K. H., "Polishing of silicon by the cupric ion process", Proc. IEEE, 57, 1476-80 (1969), EEA73-1087.
266. Novozhilov, N. O., Bykov, V. A. and Tsepalov, V. S., "Composition for polishing semiconductor materials", Patent USSR 457712, Publ. January 1975, CA83-20826.
267. Orlov, P. N., Skvortsov, K. F. and Yakovlev, A. T., "Influences of treatment and polishing method on the surface layer characteristics of a silicon substrate", Izv. VUZ Priborostr., 17 (2), 92-6 (1974), Russian, PA77-41504.
268. Osipyan, Y. A. and Strukova, G. K., "Composition for polishing semiconductor materials", Patent USSR 428485, Publ. May 1974, CA81-178923.
269. Paola, C. R., "Fixture for lapping and polishing semiconductor wafers", Patent USA 3842544, Publ. October 1974.
270. Pickhardt, V. Y. and Smith, D. L., "Chemomechanical polishing of CdS", J. Electrochem. Soc., 121, 1064-6 (1974), EEA77-38754.
271. Regh, J. and Silvey, G. A., "Silicon planar chemical polishing", Electrochem. Technol., 6 (3-4), 155-8 (1968), EEA71-28943.
272. Richter, E. and Nittner, A., "Planar etching of semiconductor wafers", Patent German (East) 99614, Publ. August 1973, CA80-138479.
273. Seltzer, R., "Guide to wafer scrubbing", Circuit Manufac., 15 (11), 56-63 (1975).
274. Tanimoto, S., "Surface treatment of semiconductors", Jitsumu Hyomen Gijutsu, 253, 60-6 (1975), Japanese, CA82-179012.
275. Tuck, B., "The chemical polishing of semiconductors", J. Mater. Sci., 10, 321-39 (1975), EEA78-11493.
276. Unvala, B. A., San, A. and Holt, D. B., "The production of clean, flat, strain-free germanium surfaces", J. Sci. Instrum (J. Phys. E), 2, 119-20 (1969), EEA72-6490.
277. Walsh, R. J., "Apparatus and method for the chemical treatment of semiconductor wafers", Patent German 2459892, Publ. September 1975, CA83-211926.
278. Walsh, R. J., "Apparatus for waxless polishing of thin wafers", Patent USA 3857123, Publ. December 1974.
279. Williamson, W. J., "Preparation of gallium arsenide surfaces for Gunn effect devices", Inst. Eng. Aust. Electr. Eng. Trans., EE9 (1-2), 26-7 (1973), EEA77-2530.
280. Wolkenberg, A., "Laboratory technological method of silicon wafer polishing by electrochemical processing", Elektronika, no. 7-8, 294-6 (1972), Polish.

3. Cleaning

281.Agnew, J., "Cleaning guide for hybrid (integrated) circuit manufacturing", Insul./Circuits, 20 (9), 39-42 (1974), EEA77-38852.
282.Anon., "Cleaning wafers after scribing", Electron. Prod. Methods & Equip., 4 (4), 10 (1975), EEA78-28848.
283.Bochkans, P. Y., Kalnach, Y. V., Klotynsh, E. E., Nesaule, V. P., Ratnietse, R. V. and Feltyn, I. A., "Influence of chemico-mechanical treatment and cleaning on silicon surface recombination rate", Latv. PSR Zinat. Akad. Vestis Fiz. Teh. Ser., no. 3, 35-9 (1974), Russian, EEA77-38769.
284.Bochkans, P. Y., Orupe, A. V. and Feltyn, I. A., "A unit for cleaning semiconductor surfaces and devices", Latv. PSR Zinat. Akad. Vestis Fiz. Teh. Ser., no. 5, 23-7 (1973), Russian, EEA77-10119.
285.Boonstra, A. H., "Some investigations on germanium and silicon surfaces", Philips Res. Rep. Suppl., no. 3, 106 pp. (1968), EEA71-25939.
286.Carbonel, M. B., "The problems of water. Ultra pure de-ionised water production and wafer rinsing. (Semiconductor device manufacture)", International Symposium on Materials for Electronic Components, 1975, p. 209, French, EEA78-32922.
287.Evans, A., "Machine for cleaning semiconductive wafers", Patent USA 3760822, Publ. September 1973.
288.Guitard, B. J., "Cleaning mixture", IBM Tech. Disclosure Bull., 17, 3305 (1975), EEA78-37396.
289.Hart, R. K., Hicklin, W. H. and Phillips, L. A., "Methods of cleaning contaminants from quartz surfaces during resonator fabrication", Proceedings of the 28th Annual Frequency Control Symposium, 1974, p. 89-95, EEA78-7870.
290.Henderson, R. C., "Silicon cleaning with hydrogen peroxide solutions: a high energy electron diffraction and Auger electron spectroscopy study", J. Electrochem. Soc., 119, 772-5 (1972), PA75-53555.
291.Humpfner, K. H., "Cleaning of electronic components", Elektroniker, 12 (9), EL13-15 (1973), German, EEA77-9287.
292.Ii, L. B. and Shang, D. C. T., "Environmental filter", IBM Tech. Disclosure Bull., 17, 2626-7 (1975), EEA78-24156.
293.Ii, L. B. and Shang, D. C., "Electrostatic cleaner for integrated circuit devices", IBM Tech. Disclosure Bull., 18, 348-50 (1975), EEA78-37574.
294.Jacquemet, J. C., "The use of ultrafiltration for the final purification of high resistivity wafer (semiconductor device manufacture)", International Symposium on Materials for Electronic Components, 1975, p. 211-16, French, EEA78-32923.
295.Kiewit, D. A., D'Haenens, I. J. and Roth, J. A., "Germanium surface cleaning - an Auger analysis", J. Electrochem. Soc., 121, 310-11 (1974), PA77-28304.

296.Kohout, R., "Ultrapure water and reverse osmosis technology", Solid State Technol., 17 (6), 60-5 (1974), EEA77-34104.
297.Kohout, R. and Goodwin, W. D., "Ultrapure water in the microelectronics industry contemporary quality characteristics and the method of preparation", Presented at 4th International Conference on Atomic Spectroscopy and the 20th Canadian Spectroscopy Symposium, Toronto, Canada, October-November 1973, EEA77-6060.
298.Lal, K., "A new approach to cleaning and its relative merit", Electron. Today, 4 (12), 18, 20 (1971), EEA75-31897.
299.Mottram, J. D., Thanailakis, A. and Northrop, D. C., "P-type surface layers on n-type silicon heat-cleaned in UHV", J. Phys. D, 8, 1316-20 (1975), PA78-66006.
300.Savola, W. G. and Wallace, J. T., "Deionized water for integrated circuit fabrication", Solid State Technol., 16 (11), 47-9 (1973), EEA77-10302.
301.Schmeid, H. and Ziffermayer, G., "Investigation into methods of cleaning semiconductor materials", Des. Electron., 8 (1), 48-51 (1970), EEA74-1304.
302.Schneider, A., "Nonflammable azeotropic cleaning composition for semiconductor devices", Patent German 2158075, Publ. June 1973.
303.Smith, F. J., Jr., "Semiconductor chip cleaning fixture", IBM Tech. Disclosure Bull., 16, 3159 (1974), EEA77-30380.
304.Sowell, R. R., Cuthrell, R. E., Mattox, D. M. and Bland, R. D., "Surface cleaning by ultraviolet radiation", J. Vac. Sci. & Technol., 11, 474-5 (1974), PA77-67112.
305.Spencer, N. E., "Reverse osmosis for semiconductor processing facilities", Microelectronics, 4 (4), 36-8 (1973), EEA76-29177.
306.Sugata, E. and Kim, H., "Cleaning of silicon surfaces by high field and heating", Technol. Rep. Osaka Univ., 17, 237-47 (1967), EEA71-18132.
307.Suzuki, C. K., "Semiconductor wafer cleaning", Patent USA 3728151, Publ. April 1973.
308.Tolliver, D., "LSI wafer cleaning techniques", Solid State Technol., 18 (11), 33-3 (1975).
309.Unvala, B. A., San, A. and Holt, D. B., "The production of clean, flat, strain-free germanium surfaces", J. Sci. Instrum (J. Phys, E), 2, 119-20 (1969), EEA72-6490.
310.Von Wrangell, M. C. P. E., "The cleaning of semiconductor wafers", International Symposium on Materials for Electronic Components, 1975, p. 193, EEA78-33241.

4. Contamination

311.Agnew, J., "Problem-solving production techniques for handling semiconductor surfaces", Insul./Circuits, 19 (12), 32-6 (1973), PA77-6111.

312.Biedermann, E., "Travelling solvent defects on silicon wafers", J. Electrochem. Soc., 114, 107-8 (1967), EEA70-10430.
313.Faust, J. W., Jr., "Surface contamination", Presented at Symposium on Silicon Device Processing, Gaithersburg, Md., June 1970, EEA73-35137.
314.Franz, I. and Langheinrich, W., "Influence of quartz impurities on the cleanliness of annealed semiconductor surfaces", Telefunken Ztg, 39, 365-9 (1966), German, EEA71-8184.
315.Galenzer, R. H. and Jordan, A. G., "The effects of contamination on the electrical properties of edge dislocations in silicon", Phil. Mag., 18, 717-23 (1968), EEA72-2704.
316.Greene, P. D., "The kinetics of silicon contamination of liquid gallium in silica boats", J. Phys. D, 6, 1550-4 (1973), EEA76-35231.
317.Hope, D. A., "Contamination problems in the semiconductor industry", Environmental Engng., no. 35, 9-12 (1968), EEA72-2687.
318.Htoo, M. S. and Schmitt, H. F., "Facedown horizontal position wafer processing", IBM Tech. Disclosure Bull., 17, 2896 (1975), EEA78-19646.
319.Kern, W., "Semiconductor surface contamination investigated by radioactive tracer techniques. II", Solid State Technol., 15 (2), 39-45 (1972), EEA75-28416.
320.Luhrs, F. H. and Straub, R. J., "Wafer, clean air drying apparatus", IBM Tech. Disclosure Bull., 17, 2863-4 (1975), EEA78-19644.
321.Meek, R. L., "High sensitivity detection of solution contaminants by MeV ion scattering", J. Electrochem. Soc., 121, 172-3 (1974), EEA77-14380.
322.Meek, R. L., Buck, T. M. and Gibbon, C. F., "Silicon surface contamination: polishing and cleaning", J. Electrochem. Soc., 120, 1241-6 (1973), EEA77-2501.
323.North, J. C. and Lightowlers, E. C., "Determination of boron and oxygen surface contamination due to common solvents by nuclear microanalysis. (Silicon wafer cleaning)", J. Electrochem. Soc., 121, 593-5 (1974), EEA77-22891.
324.Sato, Y. and Uchida, M., "Surface contamination of GaAs during photo-etching", Jap. J. Appl. Phys., 11, 1805-9 (1972), EEA76-4540.
325.Thompson, D. A. and Barber, H. D., "The determination of surface contamination on silicon by large angle ion scattering", Appl. Phys. Lett., 14, 102-3 (1969), EEA72-15440.
326.Usami, A. and Kato, Y., "Damage coefficient and defect level of copper-contaminated silicon $N^{+}P$ diode", IEEE Trans. Nucl. Sci., NS-22, 820-4 (1975), EEA78-33111.
327.Weiner, M. E., "Si contamination of open flow quartz systems for the growth of GaAs and GaP", J. Electrochem. Soc., 119, 496-504 (1972), EEA75-24201.
328.Wey, S. J. and Lewis, G., "Mass spectrometric studies of gold-silicon interface contamination", J. Vac. Sci. & Technol., 10, 413-15 (1973), EEA76-26007.

329. White, M. L., "Clean surface technology", Proceedings of the 27th Annual Frequency Control Symposium, 1973, p. 79-88, EEA77-10281.
330. Yang, M. G., Koliwad, K. M. and McGuire, G. E., "Auger electron spectroscopy of cleanup-related contamination on silicon surfaces", J. Electrochem. Soc., 122, 675-8 (1975), EEA78-24197.

5. Holding/Positioning

331. Baker, M. R. and Thomson, W. G., "Semiconductor wafer to frame aligner", IBM Tech. Disclosure Bull., 18, 811-12 (1975).
332. Bridgegam, D. D., "Vacuum control system for wafer aligning machine", Tech. Dig., no. 20, 15-6 (1970), CCA74-8744.
333. Cachon, R. P., "Semiconductor wafer chuck", Patent USA 3711081, Publ. January 1973.
334. Chitayat, A. K., "High speed positioning", Solid State Technol., 16 (6), 42-4 (1973), EEA76-35342.
335. Chough, E., Deverse, F. T., Konrad, F., Kurtz, J. A. and et al., "Mounting block for semiconductor wafers", Patent USA 3809050, Publ. May 1974.
336. Christiansen, A. D., "Electronic component mounting wafers for repeated connection in a variety of circuit designs", Patent USA 3833838, Publ. September 1971.
337. Fischl, J. J. and Jorgensen, R. R., "Precision minimum contacting wafer orientation system", IBM Tech. Disclosure Bull., 16, 3235-6 (1974), EEA77-30383.
338. Forslund, D. C., "Site selectable pneumatic wafer chuck", IBM Tech. Disclosure Bull., 15, 1105-6 (1972), EEA76-8758.
339. Gagne, M. and Ringer, F. C., Jr., "No-edge-contact wafer orientor", IBM Tech. Disclosure Bull., 17, 2220- (1975), EEA78-15410.
340. Kato, N. and Momose, K., "Semiconductor wafer positioning device", Patent USA 3830592, Publ. August 1974.
341. Keenan, W. A. and Seirmarco, J. A., "Variable-angle wafer holder", IBM Tech. Disclosure Bull., 17, 1969-70 (1974), EEA78-19632.
342. Khoury, H. A., "Analog automatic wafer alignment for fine positioning", IBM Tech. Disclosure Bull., 17, 2895 (1975), EEA78-19949.
343. Khoury, H. A., "Linear photodiode feedback automatic alignment system", IBM Tech. Disclosure Bull., 17, 2887-9 (1975), EEA78-19947.
344. Khoury, H. A., "2-D area array solid-state feedback automatic wafer alignment system", IBM Tech. Disclosure Bull., 17, 2890-2 (1975), EEA78-19948.
345. Khoury, H. A., "Scattered light feedback detector for automatic wafer alignment", IBM Tech. Disclosure Bull., 18, 414-5 (1975), EEA78-37637.
346. Krogmann, G. and Malin, K., "Wafer holder to simultaneously treat horizontally positioned wafers", IBM Tech. Disclosure Bull., 27, 1400-1 (1974), EEA78-8614.

347.Leverett, T. W., Marks, R. E. and Nurczynski, A. W., "Semiconductor wafer aligner", IBM Tech. Disclosure Bull., 13, 2983-4 (1971), EEA74-18545.
348.Monsanto Co., "Semiconductor wafer mounting system", Patent UK 1195879, Publ. June 1970, EEA74-1546.
349.Pritchard, E. J., "Modified polypropylene mounting medium", Tech. Dig., no. 37, 27 (1975), EEA78-24430.
350.Valdik, L., "Holding device for semiconductor wafers", Patent USA 3708851, Publ. January 1973.
351.Wardley, G. A., "Electrostatic wafer, chuck for electron beam microfabrication", Rev. Sci. Instrum., 44, 1506-9 (1973), EEA77-18571.
352.Waugh, J. H., Jr. and Fenn, D. S., "Slice prealigner (for Si or Ge)", Patent USA 3820647, Publ. June 1974.
353.White, M. L., "Mounting media for semiconductor fabrication", Patent USA 3752717, Publ. August 1973, CA79-98495.

6. Transport

354.Cochran, T. J., Gasparrii, A. S. and Rance, W. G., "Delivery mechanism for small parts", IBM Tech. Disclosure Bull., 17, 2272-3 (1975), EEA78-24404.
355.Gagne, M. and Gosselin, R. P., "Vacuum wafer stop-release apparatus", IBM Tech. Disclosure Bull., 16, 2799 (1974), EEA77-25645.
356.Galli, R. J. and Kelemen, D. G., "Semiconductor chip carriers and strips thereof", Patent USA 3781596, Publ. December 1973.
357.Htoo, M. S., Schneider, R. V. and Wirtz, L. H., "Rotating hot plate (semiconductor wafer processing)", IBM Tech. Disclosure Bull., 16, 2437-8 (1974), EEA77-18568.
358.Hurlbrink, R. W., III, "Semiconductor wafer transfer fixture", Patent USA 3841689, Publ. October 1974.
359.Jorgensen, R. R. and Klein, H. A., "Precision wafer orientation and transfer system", IBM Tech. Disclosure Bull., 16, 2910-11 (1974), EEA77-23139.
360.Kehagioglou, T. and Lonser, D. E., "Air conveyor wafer transport", IBM Tech. Disclosure Bull., 17, 2319-20 (1975), EEA78-19634.
361.Lucas, J., "Transporting and testing device for use with semiconductor wafers", Patent UK 1284742, Publ. August 1972.
362.Mack, A., "Wafer carrier", IBM Tech. Disclosure Bull., 16, 2904-5 (1974), EEA77-23136.
363.Mack, A. and Rio, R., "Wafer stop", IBM Tech. Disclosure Bull., 16, 2906 (1974), EEA77-23137.
364.Mack, A. and Rio, R., "Plastic film side guides for semiconductor wafer conveyor", IBM Tech. Disclosure Bull., 16, 2907 (1974), EEA77-23138.
365.Rajac, T. J., "Variable-pitch transfer mechanism", IBM Tech. Disclosure Bull., 17, 1680-1 (1974), EEA78-11731.

366.Rajac, T. J., "Article transfer/cleaning/inspection station", IBM Tech. Disclosure Bull., 17, 2703-4 (1975), EEA78-19914.
367.Scacciaferro, F. and Yakubowski, C., "Wafer drive assembly", IBM Tech. Disclosure Bull., 16, 2878-80 (1974), EEA77-23134.
368.Smith, F. J., Jr., "Chip transfer tool", IBM Tech. Disclosure Bull., 17, 1565-6 (1974), EEA78-11727.
369.Stoner, C. L., "Protective carrier for semiconductor chips", Patent USA 3823350, Publ. July 1974.

7. Handling

370.Bachmann, A. H. and Kostenko, A., Jr., "Wafer loader/unloader", IBM Tech. Disclosure Bull., 17, 2908-9 (1975), EEA78-24159.
371.Barnett, J. A. and Bengner, L. H. O., "Multispeed carriage stepper motor control and addressing system for wafer handling", IBM Tech. Disclosure Bull., 17, 3395-6 (1975), EEA78-32916.
372.Baumann, D. P., "Touchless wafer handling", Solid State Technol., 16 (3), 45-9 (1973), EEA76-23357.
373.Boyer, L. F. and Johnson, A. F., "Manually indexed wafer magazine loader and unloader", Tech. Dig., no. 34, 1-2 (1974), EEA77-35865.
374.Gagne, M. and White, W. F., "Semiconductor air track handling apparatus", IBM Tech. Disclosure Bull., 16, 2800 (1974), EEA77-25646.
375.Gardiner, G. G., Ponte, F. J., Jr. and Stenabaugh, D. D., "Mechanical wafer pickup", IBM Tech. Disclosure Bull., 16, 3253-4 (1974), EEA77-30388.
376.Mitsubishi Denki KK, "Manufacture of single-crystal semiconductor substrates", Patent UK 1383165, Publ. February 1975, PA78-66563.
377.Musits, B. and Wernikowski, R., "Wafer handling system", IBM Tech. Disclosure Bull., 17, 2904 (1975), EEA78-24158.
378.Vokali, A. G., "Wafer handler assembly (air driven)", IBM Tech. Disclosure Bull., 16, 3317-18 (1974), EEA77-29369.
379.Vokali, A. G., "Wafer handler assembly (with bidirectional motor drive)", IBM Tech. Disclosure Bull., 16, 3319-20 (1974), EEA77-29370.

8. Inspection

380.Abbe, R. C., "Semiconductor wafer measurements", Solid State Technol., 17 (3), 47-50 (1974), EEA77-26378.
381.Adley, J. M. and Gorey, E. F., "The measurement of specular defects on semiconductor surfaces", J. Electrochem. Soc., 117, 971-5 (1970), EEA73-35165.
382.Affleck, J. H., "Auger spectroscopy and silicon surfaces", Presented at Symposium on Silicon Device Processing, Gaithersburg, Md., June 1970, EEA73-35165.
383.Baxter, G. A., Otto, W. G. and Wanesky, W. R., "Mass alignment and inspection", Tech. Dig., no. 38, 5-6 (1975), EEA78-28550.

384.Beckmann, K. H., "Optical investigations of semiconductor surfaces", Philips Tech. Rev., 29, 129-42 (1968), EEA71-20583.
385.Cook, J. S., Miller, J. C. and Palmer, M. D., "Dark-field inspection tool for semiconductor devices", IBM Tech. Disclosure Bull., 17, 1147-8 (1974), EEA78-5357.
386.Corl, E. A., Silverman, S. L. and Kim, Y. S., "Non-destructive detection of phosphorus oxide layers on semiconductor wafers", Solid-State Electron., 9, 1009-17 (1966), EEA70-2197.
387.Gruber, L. W. and Musits, B., "Wafer notch locator pin and wafer sensor ports", IBM Tech. Disclosure Bull., 17, 2332-3 (1975), EEA78-19905.
388.Kharlanov, N. A. and Kharlanov, E. A., "Method of determining the depth of a defective layer in semiconductor wafers", Ind. Lab., 40 (6), 875-6 (1974), EEA78-28618.
389.Matare, H. F., "Wafer testing", Solid State Technol., 18 (8), 58-62 (1975), EEA78-41694.
390.Mayer, A., "The quality of starting silicon", Solid State Technol., 15 (4), 38-45 (1972), EEA75-28418.
391.Munro, D. F., "A silicon wafer scanner", 1973 IEEE/OSA Conference on Laser Engineering and Applications Digest of Technical Papers, p. 56, EEA77-10306.
392.Pennington, K. S. and Will, P. M., "Spatial frequency filtering for chips and masks", IBM Tech. Disclosure Bull., 13, 3269 (1971), EEA74-29752.
393.Seebe, N. B., "Double-pass grazing incidence scanner (for semiconductor wafers)", IBM Tech. Disclosure Bull., 17, 2628-9 (1975), EEA78-24157.
394.Theis, W., "Lattice defects in semiconductor wafers", Telefunken Ztg, 39, 315-39 (1966), German, EEA71-8185.
395.Wolkenberg, A. and Szymanska, L., "Classification and chemical evaluation methods of surface defects in GaAs and Si wafers", Elektronika, no. 4, 149-5 (1972), Polish, EEA75-24272.
396.Wood, D. R. and Morgan, D. V., "Characterization of gallium arsenide surfaces", J. Electrochem. Soc., 122, 773-7 (1975), CA83-89691.
397.Woodard, O. C., "Method and apparatus for detecting a registration mark on a target such as a semiconductor wafer", Patent USA 3875415, Publ. April 1975.

9. Others

398.Basi, J. S. and Hull, E. M., "Method for protecting GaAs wafer surfaces", Patent USA 3776789, Publ. December 1973.
399.Bickford, H. G. and Peredy, S. K., "Semiconductor wafer wet processing rotating apparatus", IBM Tech. Disclosure Bull., 16, 2788 (1974), EEA77-23130.
400.Chesnokov, Y. A., "The transient thermal resistance of thyristors with silicon wafers of various thicknesses", Elektrotekhnika, no. 9, 25-7 (1973), Russian, EEA77-8855.

401.Cocca, F. J., "Method of dividing wafers", Patent USA 3699644, Publ. October 1972.
402.Dipressi, T. J. and Spinicci, D. A., "Minienvironmental enclosure", IBM Tech. Disclosure Bull., 16, 3257-8 (1974), EEA77-30011.
403.Emeis, R., "Method and apparatus for subdividing a crystal wafer", Patent USA 3694972, Publ. October 1972.
404.Flowers, D. L. and Rice, L., "Semiconductor wafer corrosion-protection process and solution", Patent USA 3904784, Publ. September 1975, CA83-171688.
405.Gates, G. A. and Ryan, W. J., "Particle collector", IBM Tech. Disclosure Bull., 17, 2589 (1975), EEA78-19638.
406.Guggenheim, R. C., Sr. and Liucci, F. V., "Crystal indexing fixture", Patent USA 3855738, Publ. December 1974.
407.Harrison, H. D., Seeley, G. and Wuestenhoefer, W. C., "Matrix manifolds (for semiconductor processing)", IBM Tech. Disclosure Bull., 17, 2317-18 (1975), EEA78-19633.
408.Hebard, R. R. and Patzner, E. J., "Protecting semiconductor wafer surfaces", IBM Tech. Disclosure Bull., 11, 303 (1968), EEA72-706.
409.Hirata, T., Ohga, S. and Kaya, T., "Device and method for accommodating semiconductor wafers", Patent USA 3850296, Publ. November 1974.
410.Holzinger, G., Kosanke, K. and Menz, W., "Printing of part numbers using a high power laser beam", Opt. & Laser Technol., 5, 256-65 (1973), EEA77-19426.
411.Hoshide, M., "Optimization of lot size for quality assurance of silicon wafers", Rep. Stat. Appl. Res. UJSE, 19 (2), 8-23 (1972), EEA76-12464.
412.Jacksen, N. F., "Method of making semiconductor wafer having concave rim", Patent USA 3723210, Publ. March 1973.
413.Keller, J. H., Ko, W. C., Uman, M. F. and Winnard, J. R., "Light system for integrated circuit chip-to-chip or wafer-to-wafer communication", IBM Tech. Disclosure Bull., 16, 3251-2 (1974), EEA77-30376.
414.Makhijani, M. P. and Scacciaferro, F., "Semiconductor chip fixture", Patent USA 3851758, Publ. December 1974.
415.Maslov, A., "On the problem of the mechanical strength of semiconductor slices", Feingeraete Tech., 23, 357-60 (1974), German, EEA78-2664.
416.Mayer, A. and Puotinen, D. A., "The preparation of practical, stabilized surfaces for silicon device fabrication", Presented at Symposium of Silicon Device Processing, Gaithersburg, Md., June 1970, EEA73-35136.
417.Moore, A. H., "Wafer spreader", Patent USA 3714704, Publ. February 1973.
418.Moore, A. H., "Wafer spreader", Patent USA 3766638, Publ. October 1973.

419.Muhlbauer, A., "Resistivity topography and growth facets of float-zoned silicon single crystals", J. Electrochem. Soc., 120, 94C (1973), EEA76-23298.
420.Nestler, K. R. and Pavlovic, G. F., "Wafer mapping for chip picking", IBM Tech. Disclosure Bull., 17, 1053-4 (1974), EEA78-5356.
421.Orr, R. F., "Semiconductor wafer processing apparatus", Patent USA 3772980, Publ. November 1973.
422.Peterson, K. E., "Apparatus for counting chips in compliant tape", Tech. Dig., no. 38, 29-30 (1975), EEA78-28882.
423.Ward, W. C., "Wafer lead frame technique", IBM Tech. Disclosure Bull., 17, 2593 (1975), EEA78-19943.
424.Wersant, P. D., "Wafer spray etching, washing and drying apparatus", Tech. Dig., no. 21, 41-2 (1971), EEA74-17653.
425.Wu, L. L., "Wafer structure permitting distribution from the back side of the wafer", IBM Tech. Disclosure Bull., 16, 2898-9 (1974), EEA77-23135.
426.Yeich, C. R., Jr., "Automatic article counter", Tech. Dig., no. 37, 39-40 (1975), EEA78-24091.
427.Yoshinaka, A. and Sugita, Y., "Relaxation process of thermal stress in silicon wafers during heating or cooling", Presented at Electrochemical Society Meeting, San Francisco, Calif., May 1974, EEA77-26340.

C. EPITAXY

I. Books and Symposia

1. Matthews, J. W., (Ed.), "Epitaxial growth, Part A", Materials Science Series, Academic Press, New York, 1975.
2. Matthews, J. W., (Ed.), "Epitaxial growth, Part B", Materials Science Series, Academic Press, New York, 1975.
3. Schneider, H. G. and Ruth, V., (Eds.), "Advances in epitaxy and endotaxy", VEB Deutscher Verlag fur Grundstoff industrie, Leipzig, Germany, 1971.
4. Proceedings of the First National Conference on Crystal Growth and Epitaxy from the Vapor Phase, Held at: Zurich, Switzerland, September 1970, Published in: J. Crystal Growth, Vol. 9, May 1971.
5. Proceedings of the Second International Conference on Vapor Growth and Epitaxy, Held at: Jerusalem, Israel, May 1972, Published in: J. Crystal Growth, Vol. 17, December 1972, EEA76-4518.
6. Proceedings of the Third International Conference on Vapor Growth and Epitaxy, Held at: Amsterdam, the Netherlands, August 1975, Published in: J. Crystal Growth, Vol. 31, December 1975.

II. Review Articles

7. Abe, T., Nishi, O. Y. and Sato, K., "Recent development in epitaxy", Toshiba Rev. (Internat. Ed.), no. 28, 36-41 (1966), EEA70-8473.
8. Aleksandrov, L. N., "Epitaxial semiconductor layers for microelectronics", Elektrotech. Casopis, 21, 558-66 (1970), Slovak, EEA74-7549.
9. Burmeister, R. A., "Recent developments in liquid phase epitaxial growth", Presented at Second National Conference on Crystal Growth, Princeton, N. J., July-August 1972, EEA76-8555.
10. Doo, V. Y. and Ernst, E. O., "A survey of epitaxial growth processes and equipment", Semiconductor Prod. Solid State Technol., 10 (10), 31-9 (1967), EEA71-10516.
11. Foxon, C. T., "Molecular beam epitaxy", Acta Electron., 16, 323-8 (1973), EEA76-38180.
12. Gupta, D. C. and Wang, P., "Advances in silicon epitaxial technology", Solid State Technol., 11 (10), 48-52 (1968), EEA72-2720.
13. Hartnagel, H. L., "GaAs liquid epitaxy - a useful aid in teaching electronic materials technology", Int. J. Electr. Eng. Educ., 11, 252-7 (1974), EEA78-2126.

14.Holonyak, N., Jr., "Epitaxial growth and related processes", In: Integrated electronic systems, Prentice-Hall International, New York, p. 262-71 (1970), EEA73-31271.
15.Joyce, B. A., "The growth and structure of semiconducting thin films", Rep. Prog. Phys., 37, 363-420 (1974), EEA77-18720.
16.Manasevit, H. M., "A survey of the heteroepitaxial growth of semiconductor films on insulating substrates", J. Cryst. Growth, 22, 125-48 (1974), EEA77-26363.
17.Minden, H. T., "A comparison of liquid phase epitaxy and chemical vapor epitaxy of III-V compound semiconductors", Solid State Technol., 16 (1), 31-8, 47 (1973), EEA76-19844.
18.Nelson, H., "Liquid-phase epitaxy - its role in crystal growth technology", J. Cryst. Growth, 27, 1-5 (1974), EEA78-11479.
19.Runyan, W. R., "The status of silicon epitaxy", In: Semiconductor silicon, Electrochem. Soc., Inc., New York, p. 169-88 (1969), EEA73-10416.
20.Takahashi, K., "Molecular beam crystal growth", Oyo Buturi, 43 (6), 547-61 (1974), Japanese, EEA78-5261.
21.Van Ruyven, L. J., "Phenomena at heterojunctions:, In: Annual review of materials science, Vol. 2, R. A. Huggins (Ed.), Annual Review Inc., Palo Alto, Calif., p. 501-28 (1972), EEA76-16344.
22.Von Munch, W., "Crystallization and epitaxy of GaAs from the vapour phase", J. Cryst. Growth, 9, 144-8 (1971), EEA74-18569.
23.Wang, C. C., "Vapor phase epitaxy by reactions involving organometallics", Presented at Second National Conference on Crystal Growth, Princeton, N. J., July-August 1972, EEA76-8566.
24.Watts, B. E., "The preparation and properties of elemental semiconductor thin films", Thin Solid Films, 18, 1-23 (1973), EEA77-331.
25.Zemel, J. N., "Recent developments in epitaxial IV-VI films", J. Lumin., 7, 525-41 (1973), EEA76-32185.

III. Equipment

26.Aengenheister, J., "Apparatus for producing epitaxial layers", Patent USA 3889635, Publ. June 1975.
27.Akai, S., Mori, H., Takahashi, N. and Iguchi, S., "Method for continuously growing epitaxial layers of semiconductors from liquid phese", Patent USA 3809584, Publ. May 1974.
28.Andrews, R. W., Rynne, D. M. and Wright, E. G., "Effect of reactor geometry on growth rate of epitaxial silicon", Solid State Technol., 12 (10), 61-5 (1969), EEA73-14052.
29.Balain, K. S., Vyas, P. D. and Dixit, B. B., "System and growth studies for depositing uniform epitaxial silicon films on silicon in horizontal reactor", Radiat. Eff., 6, 23-31 (1970), EEA73-37904.
30.Balain, K. S., Vyas, P. D. and Dixit, B. B., "Design and development of a lab-model silicon epitaxial reactor", IEE-IERE Proc. India, 13 (2), 47-51 (1975), PA78-51107.

31. Barybin, A. A., Zakharov, A. A. and Nedev, N. K., "An apparatus for growing multilayer epitaxial gallium arsenide structures from the liquid state", Instrum. & Exp. Tech., 17, 1514-16 (1974), EEA78-28615.
32. Bauser, E., Frik, M., Loechner, K. S. and Schmidt, L., "Liquid phase epitaxy apparatus for thin layers and multiple layers", Gallium Arsenide and Related Compounds, 1974, p. 10-15, Publ. 1975, EEA78-28660.
33. Chistyakov, Y. D., Palienko, A. N., Danilkov, N. K., Rainov, Y. A., Yulidov, D. N., Rainova, Y. P. and Seliverstov, V. I., "Apparatus for obtaining epitaxial semiconductor layers", Patent French 2243522, Publ. April 1975, CA83-156869.
34. Flaks, S., Saintesprit, R. and Constantin, C., "Isoepitaxy of silicon by chemical means", Cables Transm., 24, 62-72 (1970), French, EEA73-20978.
35. Graper, E. B., "Selecting thin film deposition equipment", Res./Dev., 26 (2), 49-50, 52, 55 (1975), EEA78-41495.
36. Hammond, M. L. and Cox, W. P., "Equipment considerations for silicon epitaxy reactors", Symposium on Silicon Device Processing, 1970, p. 12, EEA73-35107.
37. Kakati, D., "A focused molecular source for beam epitaxy", Presented at International Electron Devices Meeting Supplement, Washington, D. C., December 1973, PA77-33628.
38. Kappelmeyer, R., Scheuter, K. and Steggewentz, H., "Method and apparatus for epitaxial deposition of semiconductor material", Patent USA 3471326, Publ. October 1969, EEA73-10456.
39. Kirkman, E. L., "Apparatus for chemically depositing epitaxial layers on semiconductor substrates", Patent USA 3865072, Publ. February 1975.
40. Kobayashi, H., "Apparatus for deposition of semiconductor thin layers", Patent USA 3854447, Publ. October 1974, EEA78-19696.
41. Kobayasi, H., "Apparatus for epitaxial growth from the liquid state", Patent USA 3858551, Publ. January 1975.
42. Kobayasi, H. and Akasaki, I., "Apparatus for the liquid-phase epitaxial growth of multilayer wafers", Patent USA 3783825, Publ. January 1974.
43. Kovda, A. V., "A technique of growing single-layer epitaxial systems", Kristallografiya, 13, 915-17 (1968), Russian, EEA73-4317.
44. Legros, B. and Marie, C., "Epitaxial growth of semiconductors from the liquid phase", Patent German 2452197, Publ. May 1975, CA83-124559.
45. Minden, H. T., "A quartz sliding boat for GaAs liquid phase epitaxy", J. Cryst. Growth, 23, 79-80 (1974), PA77-67736.
46. Ravi, H. N., Guha, S. and Sarpangal, S., "New rotating-boat technique for liquid phase epitaxial growth of gallium arsenide", J. Cryst. Growth, 18, 212-16 (1973).

47.Rosinski, A., Stiennicki, A. and Wisniewski, R., "Device for nondestructive measurements of the resistivity of epitaxial layers by a three-point method", Pomiary Autom. Kontr., 18, 246-7 (1972), Polish, EEA75-32532.
48.Sakai, R., "Vapor-phase reaction apparatus", Patent UK 1334796, Publ. October 1973, CA80-53775.
49.Scheel, H. J., "Apparatus for the epitaxial growth of semiconducting material by liquid phase epitaxy from at least two source solutions", Patent USA 3858553, Publ. January 1975.
50.Siebert, W., "Technological equipment for the production of epitaxial layers", Technik, 24, 375-7 (1969), German, EEA73-4251.
51.Siemens AG, "Apparatus for the epitaxial deposition of semiconductor material", Patent UK 1298006, Publ. November 1972.
52.Sol, N. and Merenda, P., "Improvements to apparatus for epitaxy in liquid phase, of the vertical axis type", Patent French 2213802, Publ. August 1974, CA82-67314.
53.Spring-thorpe, A. J., "Apparatus for growing of epitaxial layers", Patent USA 3809010, Publ. May 1974.
54.Vago, G., Glaser, P. and Herman, A., "Experiences with electron beam evaporation source", Presented at Proceedings of the 5th Czechoslovakia Conference on Electronics and Vacuum Physics, Brno, Czechoslovakia, October 1972, EEA76-8572.
55.Van Oirschot, T. G. J. and Nijman, W., "Improved boat for multiple-bin liquid phase epitaxy", J. Cryst. Growth, 20, 301-5 (1973), EEA77-5893.
56.Winstel, G. and Jochen, P., "Apparatus for the production of homogeneous and plane parallel epitactic growth layers of semiconducting compounds by melt epitaxy", Patent USA 3839991, Publ. October 1974.

IV. Automation

57.Benzing, W. C. and Fisk, R., "Automation in CVD wafer processing", Solid State Technol., 18 (1), 39-42 (1975), EEA78-15236.
58.Califano, F. P. and Luciano, A., "An automatic test set for measuring the doping profile of semiconductor epitaxial layers", Rev. Sci. Instrum., 41, 865-9 (1970), EEA74-1413.
59.Esaki, L., "Computer-controlled molecular beam epitaxy", Jap. J. Appl. Phys., Suppl. 2 (pt. 1), 821-8 (1974), PA78-37608.
60.Germano, C. A., Ellett, M. R. and Russ, M. J., "Minicomputer controls evaluation of epitaxial deposition for ICs", Electronics, 46 (14), 102-6 (1973), EEA76-29309.
61.Hammond, M. L., Cox, W. P., Bowers, G. M. and Roberts, R. K., "Automation techniques for silicon epitaxy reactors", Presented at Semiconductor Integrated and Production Conference and Exhibition, Anaheim, Calif., February 1971, EEA74-22560.

62.Melka, D. P., "Computer controlled epitaxial silicon production facility", Instrumentation in the Chemical and Petroleum Industries, vol. 11 and Instrument Maintenance Management, vol. 10, 1975, p. 67-73, EEA78-33057.
63.Ogawa, H., Nishinaga, T. and Arizumi, T., "A continuous stational reactor", Jap. J. Appl. Phys., 13, 594-8 (1974), PA77-45480.
64.Reutlinger, G. W., Regas, S. J., Sidor, D. J. and Schwartz, B., "An automatic test set for measuring dopant concentration profiles in epitaxial films", Solid State Electron., 12, 31-9 (1969), EEA72-9566.
65.Roe, C. C., "The future of automatic processing of semiconductors", 1969 Wescon technical papers presented at the Western Electronic Show and Convention, 5 pp., EEA73-31278.
66.Stinson, L. and Howard, J. A., "Minicomputer control of a silicon epitaxial reactor", 1973 IEEE Region 6 Conference on Minicomputers and their Applications, p. 77-81, PA77-4051.

V. Growth Kinetics

67.Akita, K., Nakai, S., Kotani, T., Dazai, K. and Ryuzan, O., "Nitrogen doping of GaP during LPE growth - a kinetic study", J. Cryst. Growth, 28, 263-6 (1975), EEA78-24203.
68.Aleksandrov, L. V. and Sidorov, Y. G., "Crystallization kinetics and structure of epitaxial films of two-component semiconductors", In: Growth of crystals, Consultants Bureau, London, England, p. 241-9 (1969), Russian, EEA73-7613.
69.Astles, M. G. and Rowland, M. C., "Nucleation and growth of $Ga_{1-x}Al_xAs$ on (111) GaP", J. Cryst. Growth, 27, 142-7 (1974), EEA78-15220.
70.Batijewski, R. and Herman, M. A., "Kinetics of crystallization of vapor-phase epitaxial GaN deposited on sapphire substrates", Acta-Phys. Pol. A, 47, 563-67 (1975).
71.Bradshaw, S. E., "The kinetics of epitaxial silicon deposition by the hydrogen reduction of chlorosilanes", Internat. J. Electron., 21, 205-27 (1966), EEA70-4881.
72.Bradshaw, S. E., "The effects of gas pressure and velocity on epitaxial silicon deposition by the hydrogen reduction of chlorosilanes", Internat. J. Electron., 23, 381-91 (1967), EEA71-13169.
73.Bryskiewicz, T. and Herman, M. A., "A method of determining the liquid phase GaAs growth mechanism based on growth measurements", Krist. & Tech., 9, 771-8 (1974), PA77-75725.
74.Chistyakov, Y. D. and Rainova, Y. P., "Role of the oxidizer in the chloride process of preparation of autoepitaxial layers of Si", Inorg. Mater., 10, 336-9 (1974), EEA78-41471.
75.Dilorenzo, J. V., "Vapor growth of epitaxial GaAs: a summary of parameters which influence the purity and morphology of epitaxial layers", J. Cryst. Growth, 17, 189-206 (1972), EEA76-4527.

76. Dismukes, J. P. and Curtis, B. J., "A survey of transient effects in silicon CVD systems", Presented at Electrochemical Society Meeting, Chicago, Ill., May 1973, EEA76-23310.
77. Foxon, C. T., Joyce, B. A., Farrow, F. C. and Griffiths, R. M., "Identification of species evolved in the evaporation of III-V compounds", J. Phys. D, 7, 2422-35 (1974).
78. Fujii, E., Nakamura, H., Haruna, K. and Koga, Y., "A quantitive calculation of the growth rate of epitaxial silicon from $SiCl_4$ in a barrel reactor", J. Electrochem. Soc., 119, 1106-13 (1972), EEA75-36148.
79. Gonda, S., Matsushima, Y. and Nishikawa, T., "Spatial distribution of the molecular beam effused from a cell (molecular beam epitaxy)", Bull. Electrotech. Lab., 37, 914-22 (1973), Japanese, EEA77-30143.
80. Ikonnikova, G. M. and Ivleva, O. M., "Effect of a constant electric field on epitaxial growth of GaAs", Inorg. Mater., 10, 343-6 (1974), EEA78-41473.
81. Ivanyutin, L. A., Nishanov, D. N., Dyachkova, N. N., Sabinin, A. G. and Andreev, V. M., "Migration of silicon during growth of epitaxial layers of gallium arsenide from the gas phase", Inorg. Mater., 9, 1869-71 (1973), PA78-3661.
82. Jona, F., "Study of the early stages of the epitaxy of silicon on silicon", Appl. Phys. Lett., 9, 235-7 (1966), EEA70-531.
83. Joyce, B. A., Bradley, R. R., Watts, B. E. and Booker, G. R., "A study of nucleation in chemically grown epitaxial silicon films using molecular beam techniques. V. Nucleation kinetic measurements on (100) surfaces", Phil. Mag., 19, 403-13 (1969), EEA72-9541.
84. Kay, P. M. R., Lilley, P. and Litting, C. N. W., "The influence of growth conditions on the deposition of thick epitaxial (100) ZnS layers in $HCl:H_2$ gas flow", J. Phys. D, 7, 1206-13 (1974), EEA77-26369.
85. Khitova, L. N., Lazarova, V. B. and Ryabtsev, N. G., "Epitaxy kinetics of GaAs in a flow-type system $GaAs-I_2-He$", Inorg. Mater., 9, 699-702 (1973), PA77-44795.
86. Konis, U., "GaAs-Sn liquid-phase epitaxy: growth mechanism and production of holohedral and selective n^{++}-layers", Presented at 2nd National Conference on Crystal Growth, Freiburg im Breisgau, Germany, September 1972, German, EEA76-4554.
87. Kordos, P., "Epitaxial efficiency coefficient for growth from the liquid phase", J. Cryst. Growth, 20, 307-9 (1973), PA77-13579.
88. Lavrenteva, L. G., Ivonin, I. V., Krasnilnikova, L. M., Rumyantsev, Y. M. and Yakubenya, M. P., "Influence of crystallization temperature on growth and doping of autoepitaxial layers of gallium arsenide. I", Izv. VUZ Fiz., no. 6, 68-71 (1973), Russian, EEA76-32167.
89. Lavrenteva, L. G., Kataev, Y. G., Rumyantsev, Y. M. and Shumkov, A. D., "Influence of crystallization temperature on growth and doping of autoepitaxial layers of gallium arsenide. II", Izv. VUZ Fiz., no. 6, 71-5 (1973), Russian, EEA76-32168.

90.Lavrenteva, L. G., Porokhovnichenko, L. P., Ivonin, I. V., Krasilnikova, L. M. and Ivanova, N. N., "Influence of the method of preparing the surface of a substrate on the growth kinetics of autoepitaxial layers of gallium arsenide and their properties", Izv. VUZ Fiz., no. 1, 20-4 (1974), Russian, PA77-33613.
91.Lavrenteva, L. G., Zakharov, I. S., Ivonin, I. V., Kolodeshnikova, N. P. and Opryshko, T. I., "Anisotropy of growth and alloying for epitaxial germanium layers in the Ge-I_2 system", Sov. Phys.-Crystallogr., 18, 228-31 (1973), EEA76-38211.
92.Loginova, R. G., Rubtsova, R. A. and Ovsyannikov, M. I., "Transport of boron from a source to an epitaxial layer during sublimation of silicon in vacuum", Sov. Phys.-Crystallogr., 18, 232-3 (1973), EEA76-38212.
93.Lovyagain, P. N., Fedina, L. I., Aleksandrov, L. N. and Krivorotov, E. A., "Electron microscopic investigation of the initial stages in the growth of epitaxial silicon films on the real surface of silicon and germanium during ionic sputtering", Izv. Akad. Nauk SSSR Ser. Fiz., 38, 1528-32 (1974), Russian, PA77-73259.
94.Lozovskii, V. N. and Marev, V. B., "Kinetics of the growth of epitaxial layers of gallium phosphide during zone recrystallization with a temperature gradient", Izv. VUZ Fiz., no. 7, 115-18 (1974), Russian, PA77-73260.
95.Lyutovich, A. S., Pashkudenko, V. P. and Kharchenko, V. V., "An investigation of the kinetics of epitaxial growth of silicon in the chloride process", Izv. Akad. Nauk SSSR Neorg. Mater., 8, 1713-6 (1972), Russian, EEA76-8520.
96.Manasevit, H. M. and Simpson, W. I., "The use of metalorganics in the preparation of semiconductor materials. V. The formation of In-group V compounds and alloys", J. Electrochem. Soc., 120, 135-7 (1973), EEA76-12456.
97.Mattes, B. L. and Route, R. K., "LPE growth of GaAs: formation of nuclei and surface terraces", J. Cryst. Growth, 27, 133-41 (1974), PA78-29521.
98.Metonidze, Z. A. and Sokolov, E. V., "Kinetic characteristics of gallium arsenide epitaxy in the hydride process", Sb. Nauch. Tr. Probl. Mikroelektron., Mosk. Inst. Elektron. Tekh., no. 13, 144-51 (1972), Russian, CA80-31161.
99.Mitsuhata, T., "Phenomenological theory on liquid phase epitaxy", Japan. J. Appl. Phys., 9, 90-4 (1970), EEA73-14026.
100.Mizuno, O. and Watanabe, H., "Vapor growth kinetics of III-V compounds in a hydrogen-inert gas mixed carrier system", J. Cryst. Growth, 30, 240-8 (1975).
101.Mogab, C. J. and Leamy, H. J., "Epitaxial growth of β-SiC on Si: kinetics and growth mechanism", Presented at International Conference on Silicon Carbide, Miami Beach, Fla., September 1973, PA77-10529.

102.Moon, R. L., "The influence of growth solution thickness on the LPE layer thickness and constitutional supercooling requirement for diffusion-limited growth", J. Cryst. Growth, 27, 62-9 (1974), EEA78-15216.
103.Muszynski, Z., "On the kinetics and conditions of optimum growth of epitaxial layers of GaAs", Elektronika, no. 8, 324-9 (1971), Polish, EEA75-1006.
104.Muszynski, Z., "Optimum growth kinetics and conditions of gallium arsenide obtained by the liquid phase epitaxy method", Electron. Technol., 5 (1), 23-33 (1972), PA75-71384.
105.Nakamura, H., Haruna, K. and Fujii, F., "A macroscopic model approach to silicon epitaxial process analysis", Tokyo 1971 AICA Symposium on Simulation of Complex Systems, F-10, 5 pp., EEA75-7437.
106.Nicolaus, W., Seidowski, E. and Voronin, V. A., "Kinetics of the transport processes and the crystallization of InAs from the gas phase", Kristall & Tech., 7, 589-99 (1972), German, PA75-74673.
107.Nishinaga, T. and Mizutani, T., "Effect of growth parameters on the epitaxial growth of BP on Si substrate", Jap. J. Appl. Phys., 14, 753-60 (1975), EEA78-28633.
108.Nishizawa, J. and Shimbo, M., "Mechanism of crystal growth in the $SiCl_4$-H_2 system", J. Cryst. Growth, 24-25, 215-19 (1974), PA78-6665.
109.Nishizawa, J., Terasaki, T. and Shimbo, M., "Silicon epitaxial growth", J, Cryst. Growth, 17, 241-8 (1972), EEA76-4533.
110.Nishizawa, J., Yagi, K., Miyamoto, N. and Terasaki, T., "Perfect crystal growth of silicon by vapour deposition", J. Electrochem. Soc., 122, 664-9 (1975), EEA78-24195.
111.Petzke, W. H., Gottschalch, V. and Butter, E., "Epitaxial deposition of GaAs in the $Ga(CH_3)_3$-AsH_3-H_2 system. IV. Thermodynamic and kinetic considerations", Krist. & Tech., 9, 763-70 (1974), PA77-75724.
112.Pogge, H. B. and Kemlage, B. M., "Doping behavior of silicon in vapor-grown III-V epitaxial films", J. Cryst. Growth, 31, 183-9 (1975).
113.Popov, V. P., Raynov, Y. A., Skoropanov, Y. S., Merkulova, N. A., Gleb, L. K. and Gazaryan, V. S., "A study of heat and mass transfer during growing of silicon films in horizontal reactors", Heat Transfer - Sov. Res., 5 (5), 29-36 (1973), PA77-32826.
114.Price, J. B. and Goldman, J., "Junction migration during silicon epitaxial growth", Presented at Electrochemical Society Meeting, San Francisco, Calif., May 1974, PA77-49600.
115.Racheva-Stambolieva, T. M., Krasulin, G. A., Chistyakov, Y. D., Dragieva, I. D. and et al., "Thermodynamic model of epitaxial growth of CdSe layers", Phys. Status Solidi A, 22, 593-97 (1974).
116.Rawlins, T. G. R. and Brosseland, L. E., "Electron-microscope replica study of epitaxial silicon nucleation on silicon", Trans. Metall. Soc. AIME, 236, 280-4 (1966), EEA70-4717.

117. Riedl, W., "Conditions for the elimination of crystal nuclei deteriorating structure of epitaxial silicon films", Prace Inst. Tele-i Radiotech., 10, 71-7 (1966), Polish, EEA70-4085.
118. Semiletov, S. A., Vlasov, V. A. and Magomedov, Z. A., "Formation of epitaxial films by vacuum deposition", In: Growth of crystals, Consultants Bureau, London, England, p. 184-7 (1969), Russian, EEA74-7610.
119. Shaw, D. W., "Kinetics and thermodynamics of vapor phase epitaxy", Electrochemical Society Fall Meeting (Extended abstracts), 1974, p. 508-9, EEA78-37446.
120. Shintani, A. and Minagawa, S., "Kinetics of the epitaxial growth of GaN using A, HCl and NH_3", J. Cryst. Growth, 22, 1-5 (1974), PA77-45477.
121. Small, M. B., Blakeslee, A. E., Shih, K. K. and Potemski, R. M., "A phenomenological study of meniscus lines on the surfaces of GaAs layers grown by LPE", J. Cryst, Growth, 30, 257-66 (1975).
122. Valiev, K. A., Volkova, O. V., Volkov, A. F., Lukasevich, M. I. and Mashkova, V. S., "Microanisotropy of the growth of auto-epitaxial silicon layers on substrates with photolithographic relief", Sb. Nauch. Tr. Po Probl. Mikroelektron. Mosk. In-T Elektron. Tekhn., no. 20, 189-99 (1975), Russian, CA83-187300.
123. van der Merwe, J. H. and Auret, F. D., "The significance of certain crystal parameters in epitaxial growth", Jap. J. Appl. Phys., suppl. 2 (pt. 1), 583-7 (1974), PA78-37598.
124. Weinstein, M. and Wolff, G. A., "Mechanisms of epitaxial growth of compound semiconductors", J. Phys. Chem. Solids, 28, 537-41 (1967), EEA70-16425.
125. Wu, T. Y., "Some morphological studies for optimum growth conditions of vapor epitaxial $GaAs_{1-x}P_x$", J. Cryst. Growth, 21, 85-92 (1974), PA77-21202.
126. Wu, T. Y., "Vapor phase epitaxy and thermodynamic calculations of $GaAs_{1-x}P_x$ (0.3 x 0.5)", J. Electrochem. Soc., 121, 1357-61 (1974), EEA78-550.
127. Zeveke, T. A., Buzynin, Y. N., Kuznetsov, O. A. and Zntsheva, L. N., "Morphology and growth kinetics of autoepitaxial germanium in the hydride process as a function of substrate orientation", Sov. Phys.-Crystallogr., 19, 528-30 (1975), EEA78-19689.

VI. Homoepitaxy

1. General

128. Alexander, J. H. and Sterling, H. F., "Semiconductor epitaxy-gas phase doping by electric discharge", Solid-State Electron. 10, 485-90 (1967), EEA70-13925.
129. Bailey, L. G. and Rogers, G. G., "Method for producing graded composition epitaxial layers by vapor deposition", J. Electrochem. Soc., 118, 834-6 (1971), EEA74-22501.

130.Baliga, B. J., "Improved materials for high frequency field-effect transistors", Thesis, Rensselaer Polytech. Inst., Troy, N. Y., 1974, 156 pp., Order No. 74-26063, EEA78-15371.
131.Bergh, A. A., Paola, C. R. and Saul, R. H., "Semiconductor epitaxial growth from solution", Patent USA Reissued #28140, Publ. August 1974.
132.Berkenblit, M. and Reisman, A., "Method for producing improved epitaxial films", Patent USA 3476592, Publ. November 1969, EEA73-14117.
133.Bloem, J., "Trends in the chemical vapor deposition of silicon", Presented at Electrochemical Society Meeting (abstracts only), Chicago, Ill., May 1973, EEA76-23305.
134.Choudhury, P. R. and Schroder, D. K., "Selective growth of epitaxial silicon and gallium arsenide", J. Electrochem. Soc., 118, 107-10 (1971), EEA74-15450.
135.Curtis, B. J., Emmenegger, F. P. and Nitsche, R., "The preparation of ternary and quaternary compounds by vapor-phase growth", RCA Rev., 31, 647-61 (1970), EEA74-13188.
136.Deitch, R., "Liquid-phase epitaxial growth under transient thermal conditions", Patent USA 3791887, Publ. February 1974, CA80-88787.
137.Distler, G. I. and Vlasov, V. P., "Problems of epitaxy", Izv. Akad. Nauk SSSR, Ser. Fiz., 38, 1438-42 (1974), Russian.
138.Ettenberg, M. and Cannuli, V. M., "Method for epitaxially growing layers of a semiconductor material from the liquid phase", Patent USA 3767841, Publ. October 1973.
139.Eversteyn, F. C., "Chemical-reaction engineering in the semiconductor industry", Philips Res. Rep., 29, 45-66 (1974), PA77-49619.
140.Findeisen, K. E., Petermann, K., Haase, K. H., Speck, D., Kumert, S., Hiller, R. and Johne, J., "Epitaxial growth of semiconductors", Patent German 2300021, Publ. November 1973, CA80-42205.
141.Fukugawa, Y., Mori, T. and Fujii, S., "Epitaxial growth from liquid phase", Patent Japan 73-14548, Publ. May 1973, CA80-53783.
142.Gittler, F. L., "Materials for diffusion and epitaxy", J. Test. & Eval., 1, 310-11 (1973), EEA76-35234.
143.Golubev, L. V., Khachaturyan, O. A., Shik, A. Y. and Shmartsev, Y. V., "Growth of films of semiconducting alloys by liquid-phase epitaxy with a temperature gradient (LPETG)", Phys. Status Solidi A, 22, K203-K204 (1974).
144.Hiscocks, S. E. R., "The vapour growth of IV-VI compounds", J. Cryst. Growth, 17, 222-9 (1972), EEA76-4531.
145.Hollan, L., "Liquid solution method of epitaxially depositing a semiconductor compound", Patent USA 3755013, Publ. August 1973.
146.Honma, K., Ogirima, M., Hirao, M. and Kurata, K., "Growth of epitaxial semiconductor films from the liquid phase", Patent Japan 73-102779, Publ. December 1973, CA80-150260.
147.Ichiki, E., Iida, K., Ogura, M., Seki, Y. and et al., "Chemical vapor deposition for epitaxial growth", Patent USA 3867202, Publ. February 1975.

148. Joyce, B. D., "Epitaxial growth and growth defects", Electronic Compon., 9, 1389-95 (1968), EEA73-10354.
149. Knight, S., "Heat flow in n^{++}-n-n^{+} epitaxial GaAs bulk effect devices", Proc. IEEE, 55, 112-13 (1967), EEA70-17001.
150. Lebwohl, P. A. and Carlson, D. G., "Transient heat flow from an epitaxial layer into the substrate", Proc. IEEE, 59, 1030-2 (1971), EEA74-18601.
151. LeDuc, J. M. and Leroy, D., "Liquid-phase epitaxy", Patent French 2145830, Publ. March 1973, CA79-98481.
152. Lockwood, H. F. and Marinelli, D. P., "Method of depositing epitaxial semiconductor layers from the liquid phase", Patent USA 3753801, Publ. August 1973.
153. Meinders, H., "An alternative method for liquid phase epitaxy", J. Cryst. Growth, 26, 180-2 (1974), PA78-26309.
154. Melngailis, I. and Calawa, A. R., "Solution growth of epitaxial layers of semiconductor material", Patent USA 3463680, Publ. August 1969, EEA73-10452.
155. Migitaka, M., Doi, A. and Miyazaki, M., "Liquid phase epitaxial growth of semiconductor compounds", Patent Japan 74-15687, Publ. February 1974, CA81-43030.
156. Miyazaki, M. and Migitaka, M., "Liquid-phase growing of compound semiconductor crystals", Patent Japan 73-14588, Publ. February 1973, CA79-47068.
157. Pelevin, O. V. and Milividskii, M. G., "Some basic features of liquid-phase epitaxy of solid solutions of semiconductors", Inorg. Mater., 10, 784-5 (1974), PA78-72717.
158. Petrosyan, S. G., Shik, A. Y. and Shmartsev, Y. V., "Liquid epitaxy with a temperature gradient", Sov. Phys.-Solid State, 16, 256-8 (1974), PA78-3663.
159. Popov, V. P. and Pamplin, B. R., "Epitaxial growth of solid solutions of $ZnSiP_2$ in Si", J. Cryst. Growth, 15, 129-32 (1972), EEA75-32515.
160. Potemski, R. M. and Woodall, J. M., "A new technique for terminating liquid phase epitaxial growth", J. Electrochem. Soc., 119, 277-9 (1972), EEA75-13556.
161. Silvestri, V. J., "Conditions for obtaining smooth epitaxial surfaces in the $GeCl_4$-H_2 system", Presented at Electrochemical Society Fall Meeting, Chicago, Ill., October 1967, EEA72-20267.
162. Sutton, R. P., "The choice of chlorosilane source for epitaxy reactors", In: Semiconductor silicon, Electrochem. Soc., Inc., New York, p. 214-21 (1969), EEA73-10420.
163. Unno, Y., Yamamoto, M. and Kurihara, H., "Method for the liquid-phase epitaxial growth of semiconductor crystals", Patent USA 3765959, Publ. October 1973.
164. Weissmantel, C., Fiedler, O., Hecht, G. and Reisse, A. G., "Ion beam sputtering and its application for the deposition of semiconducting films", Thin Solid Films, 13, 359-66 (1972), EEA76-8542.

165.Yasufuku, M., Maeda, K. and Yamagami, E., "Epitaxial growth by the mass transfer method", Fujitsu Sci. Tech. J., 1 (2), 51-71 (1965), EEA70-2245.
166.Yaws, C. L., Hopper, J. R. and Swinderman, E. M., "Physical and thermodynamic properties of semiconductor industry chemical materials. P-type gas phase dopants (diborane and pentaborane)", Solid State Technol., 17 (11), 31-5 (1974), EEA78-8389.

2. Group IV Semiconductors

a. Silicon

167.Abdurakhmanov, B. M., Aladina, Z. N., Pashkudenko, V. P. and Kharchenko, V. V., "Effect of crystallization conditions on parameter uniformity in epitaxial silicon", Appl. Sol. Energy, 11, 82-5 (1975), PA78-66546.
168.Aleksandrov, L. N. and Lovyagin, R. N., "Step motion of the growth surface in the initial stage of semiconductor film epitaxy with ion sputtering", Thin Solid Films, 20, 1-10 (1974), PA77-20498.
169.Arizumi, T., Nishingaga, T. and Kasuga, M., "A continuous procedure for growing epitaxial silicon films", Proceedings of the Second International Congress on Microelectronics, 1966, p. 259-68, Publ. 1967, EEA71-10513.
170.Arizumi, T., Nishingaga, T., Kasuga, M. and Gawa, H., "A new method for the epitaxial growth of silicon and its thermodynamical analysis", Jap. J. Appl. Phys., 8, 32-40 (1969), EEA72-6492.
171.Avigal, Y. and Schieber, M., "Chemical transport epitaxy of silicon by organic compounds", J. Electrochem. Soc., 117, 1585-6 (1970), EEA74-10486.
172.Avigal, Y. and Schieber, M., "Silicon carbide contamination of epitaxial silicon grown by pyrolysis of tetramethyl silane", J. Cryst. Growth, 9, 127-31 (1971), EEA74-18567.
173.Benedict, T. S. and Norman, H. W., "Single crystal silicon rods", Patent USA 3447902, Publ. June 1969, EEA73-4342.
174.Bloem, J., "Silicon epitaxy from mixture of SiH_4 and HCl", J. Electrochem. Soc., 117, 1397-401 (1970), EEA74-7575.
175.Bloem, J., "High chemical vapour deposition rates of epitaxial silicon layers", J. Cryst. Growth, 18, 70-6 (1973), EEA76-16292.
176.Canali, C., Campisano, S. U., Lau, S. S., Liau, Z. L. and Mayer, J. W., "Solid-phase epitaxial growth of Si through palladium silicide layers", J. Appl. Phys., 46, 2831-6 (1975), EEA78-37410.
177.D'Asaro, L. A. and Furnange, R. A., "Low defect silicon by liquid phase epitaxy", In: Semiconductor silicon, Electrochem. Soc., Inc., New York, p. 233-42 (1969), EEA73-10422.
178.Delong, D. J., "Advances in dichlorosilane epitaxial technology", Solid State Technol., 15 (10), 29-34, 41 (1972), EEA76-4548.

179.Drum, C. M. and Clark, C. A., "Anisotropy of macrostep motion and pattern edge-displacements during growth of epitaxial silicon on silicon near (100)", J. Electrochem. Soc., 117, 1401-5 (1970), EEA74-7576.
180.Dyer, L. D., "The influence of methane on the growth of silicon crystals from the vapor", Presented at ACCG Conference on Crystal Growth, Gaithersburg, Md., August 1969, EEA73-28081.
181.Eversteyn, F. C., Severin, P. J. W., v.d. Brekel, C. H. J. and Peek, H. L., "A stagnant layer model for the pitaxial growth of silicon from silane in a horizontal reactor", J. Electrochem. Soc., 117, 925-31 (1970), EEA73-35065.
182.Frieser, R. G., "Low-temperature silicon epitaxy", J. Electrochem. Soc., 115, 401-5 (1968), EEA71-20527.
183.Frieser, R. G. and Casey, J. J., "Method of producing silicon body", Patent USA 3490961, Publ. January 1970, EEA73-21103.
184.Gittler, F. L., "Epitaxial deposition of silicon in nitrogen", J. Cryst. Growth, 17, 271-5 (1972), EEA76-4537.
185.Goldsmith, N. and Robinson, P. H., "Epitaxial growth of silicon using dichlorosilane: growth on single-crystal hemispheres", RCA Rev., 34, 358-68 (1973), EEA77-327.
186.Goral, H., "Epitaxial growth of silicon layers hydrogen reduction of $SiCl_4$ on silicon substrates", Przeglad Elektron., 9, 216-25 (1960), Polish, EEA71-20521.
187.Gupta, D. C., "Improved methods of depositing vapour-phase homoepitaxial silicon", Solid State Technol., 14 (10), 33-40 (1971), EEA75-28414.
188.Gupta, D. C. and Porter, J. L., "Techniques for depositing highly uniform and defect-free epitaxial silicon", Symposium on Silicon Device Processing, 1970, p. 15, EEA73-37932.
189.Gupta, D. C. and Yee, R., "Low-temperature deposition of silicon epitaxial layers using pyrolysis of silane", Presented at 133rd Meeting of the Electrochemical Society, Boston, Mass., May 1968, EEA73-14086.
190.Hasegawa, S., Nomura, K., Komiya, Y., Teshima, H. and Tarui, Y., "Epitaxial growth of high purity silicon using SiH_4", Bull. Electrotech. Lab., 36, 77-84 (1972), Japanese, EEA75-36279.
191.Hovel, H. J., "Oxide prevention on Si surfaces using iodine saturated immiscible solvents", IBM Tech. Disclosure Bull., 13, 3284 (1971), EEA74-30737.
192.Ishii, T., Takahashi, K., Kondo, A. and Shirahata, K., "Silicon epitaxial wafer with abrupt interface by two-step epitaxial growth technique", J. Electrochem. Soc., 122, 1523-31 (1975).
193.Itoh, T. and Nakamura, T., "Epitaxial growth of silicon assisted by ion implantation", Radiat. Eff., 9 (1-2), 1-4 (1971), EEA74-26614.
194.Joyce, B. A., "Epitaxial growth of silicon", Symposium on Silicon Device Processing, 1970, p. 6-7, EEA73-35104.
195.Kim, H. J., "Liquid phase epitaxial growth of silicon in selected areas", J. Electrochem. Soc., 119, 1394-8 (1972), EEA76-727.

196.Kimura, A., "Effect of thermal etching on silicon epitaxial growth by vacuum sublimation", Solid-State Electron., 18, 901-5 (1975), EEA78-37422.
197.Kimura, A., "Silicon epitaxial growth by sublimation in high vacuum system", Thesis, Cornell Univ., Ithaca, N. Y., 1974, 167 pp., Order No. 74-27546, PA78-62575.
198.Kohler, W. A., "A comparison of a resistance heated reactor for silicon epitaxial growth with other epitaxial systems", Symposium on Silicon Device Processing (programme and abstracts), 1970, p. 13-14, EEA73-35108.
199.Kramer, J., "Vacuum deposition of epitaxial silicon layers", Elektrotech. Obzor., 57, 476-80 (1968), EEA72-2686.
200.Kumagawa, M., Sunami, H., Terasaki, T. and Nishizawa, J., "Epitaxial growth with light irradiation", Japan J. Appl. Phys., 7, 1332-41 (1968), EEA72-4914.
201.Lekholm, A., "Epitaxial growth of silicon from dichlorosilane", J. Electrochem. Soc., 119, 1122-3 (1972), EEA75-36152.
202.Lucarini, V. J., Bratter, R. L. and Basso, J. E., "Epitaxial growth from SiH_4 in a reduced pressure system", Presented at Electrochemical Society Meeting, San Francisco, Calif., May 1974, PA77-49599.
203.Mikhailov, Y. A. and Marina, G. F., "Calculation of supersaturation and rate of growth of Si from the mixture of $SiCl_4$-HCl-H_2", Inorg. Mater., 9, 1449-50 (1973), PA77-67722.
204.Murthy, B. S., Kesavan, R. and Ratnarati, K. V., "Silicon vapour epitaxy", J. Instn. Telecomm. Engrs., 14, 486-95 (1968), EEA72-6491.
205.Ogawa, H., Nishinaga, T., Kasuga, M. and Arizumi, T., "The selective epitaxial growth of silicon by using silicon nitride film as a mask", Jap. J. Appl. Phys., 10 (12), 1675-9 (1971), EEA75-7413.
206.Pchelyakov, O. P., Lovyagin, R. N., Krivorotov, E. A., Toropov, A. I., Aleksandrov, L. N. and Stenin, S. I., "Silicon homoepitaxy with ion sputtering. I. Mechanism of growth", Phys. Status Solidi A, 17, 339-51 (1973), EEA76-25983.
207.Rai-Choudhury, P., "Substrate surface preparation and its effect on epitaxial silicon", J. Electrochem. Soc., 118, 1183-9 (1971), EEA74-30660.
208.Rai-Choudhury, P. and Schroder, D. K., "Selective silicon epitaxy and orientation dependence of growth", J. Electrochem. Soc., 120, 664-8 (1973), EEA76-25970.
209.Richman, D. and Arlett, R. H., "Low-temperature epitaxial growth of single crystalline silicon from silane", J. Electrochem. Soc., 116, 872-3 (1969), EEA73-4309.
210.Richman, D., Chang, Y. S. and Robinson, P. H., "Low temperature vapor growth of homoepitaxial silicon", RCA Rev., 31, 613-19 (1970), EEA74-13185.
211.Robinson, P. H. and Goldsmith, N., "Silicon epitaxial growth using dichlorosilane", J. Electron. Mater., 4, 313-28 (1975), PA78-47465.

212.Rundle, P. C., "The epitaxial growth of silicon in horizontal reactors", Internat. J. Electron., 24, 405-13 (1968), EEA72-2698.
213.Sankur, H., McCaldin, J. O. and Devaney, J., "Solid-phase epitaxial growth of Si mesas from Al metallization", Appl. Phys. Lett., 22, 64-6 (1973), EEA76-12431.
214.Seki, Y., Tanno, K., Matsui, J. and Kawamura, T., "Dislocation free silicon epitaxy", In: Semiconductor silicon, Electrochem. Soc., Inc., New York, p. 653-61 (1969), EEA73-10442.
215.Shimbo, M., Nishizawa, J. and Terasaki, T., "Defect-free nucleation of silicon on (111) silicon surfaces", J. Cryst. Growth, 23, 267-74 (1974), PA78-15607.
216.Sirtl, E. and Seiter, H., "Selective epitaxy of silicon under quasi-equilibrium conditions", In: Semiconductor silicon, Electrochem. Soc., Inc., New York, p. 189-99 (1969), EEA73-10417.
217.Smeltzer, R. K., "Epitaxial deposition of silicon in deep grooves", J. Electrochem. Soc., 122, 1666-71 (1975).
218.Spiro, A. and Deines, J. L., "Low pressure silicon epitaxy", Presented at Electrochemical Society Meeting, San Francisco, Calif., May 1974, PA77-49598.
219.Sugawara, K., "Silicon epitaxial growth by rotating disk method", J. Electrochem. Soc., 119, 1749-60 (1972), EEA76-8534.
220.Tuft, E. A., "Epitaxial growth of doped silicon using an iodine cycle", J. Electrochem. Soc., 118, 1535-8 (1971), EEA74-37144.
221.Tikhonov, Y. N. and Chistyakov, Y. D., "Silicon auto-epitaxial layers for high-powered semiconductor devices", Elektrotekhnika, no. 9, 14-17 (1966), Russian, EEA70-11089.
222.Townsend, W. G. and Uddin, M. E., "Epitaxial growth of silicon from SiH_4 in the temperature range 800-1150°C", Solid-State Electron., 16, 39-42 (1973), EEA76-8540.
223.Williams, J. H. and Jackson, D. M., Jr., "The growth of epitaxial silicon films in a radiantly heated reactor", In: Semiconductor silicon, Electrochem. Soc., Inc., New York, p. 208-13 (1969), EEA74-10419.
224.Yim, E. W., "Growth and characterization of ultra-thin epitaxial silicon films", Electrochemical Society Fall Meeting (extended abstracts), 1970, p. 439-40, EEA74-7625.

b. Germanium

225.Avigal, Y., Itzhak, D. and Schieber, M., "Low carbon contamination of epitaxial germanium films produced by pyrolysis of alkyl germanium compounds", J. Electrochem. Soc., 122, 1226-9 (1975).
226.Grigorev, V. E., Malkin, G. M. and Tumanov, A. A., "Transport of Ge by iodine in a flow stream", Inorg. Mater., 9, 1168-70 (1973), PA77-61300.

227. Inoue, M., Asahi, K., Takahashi, J. and Teramoto, I., "Epitaxial growth of Ge layers from $GeCl_4$ and GeH_4", Nat. Tech. Rep., 18, 311-19 (1972), Japanese, EEA75-32530.
228. Kamadjiev, P. R. and Simeonev, S. S., "Certain dependence during growth of germanium epitaxial layers by close-spaced system", C. R. Acad. Bulg. Sci., 21, 419-22 (1968), EEA71-28744.
229. Khan, I. H., "Low-temperature epitaxy of Ge films by sputter deposition", J. Appl. Phys., 44, 14-19 (1973), EEA76-12448.
230. Laugier, A., Garand, M. and Mesnard, G., "Germanium epitaxial layers from solution in tin or mixtures of lead and tin", Solid-State Electron., 10, 77-9 (1967), EEA71-5824.
231. Marrello, V., Caywood, J. M., Mayer, J. W. and Nicolet, M. A., "Solid-phase epitaxial growth of Ge layers", Phys. Status Solidi A, 13, 531-6 (1972), EEA76-4546.
232. Tramposch, R. F., "Method for improving perfection of epitaxially grown germanium films", Patent USA 3783009, Publ. January 1974.
233. Vervyka, A. V., Zhemchuzhina, E. A., Kuznetsov, A. S. and Levinzon, D. I., "Deposition of autoepitaxial germanium coatings on rectangular substrates", Izv. Akad. Nauk SSSR Ser. Fiz., 37 (11), 2310-12 (1973), Russian, EEA77-5883.

c. Silicon and Germanium

234. Aharoni, H. and Bar-Lev, A., "Some practical remarks on the design of experimental systems for epitaxial growth of Si, Ge and Si-Ge", Vacuum, 24 (2), 89-93 (1974), PA77-57721.
235. Canali, C., Ottaviani, G., Mayer, J. W., Sigurd, D. and Van der Weg, W., "Solid-phase transport and epitaxial growth of Ge and Si", Appl. Phys. Lett., 25, 3-5 (1974), PA77-70414.
236. Cullis, A. G. and Booker, G. R., "The epitaxial growth of silicon and germanium films on (111) silicon surfaces using UHV sublimation and evaporation techniques", J. Cryst. Growth, 9, 132-8 (1971), EEA74-18568.
237. Lau, S. S., Mayer, J. W. and Nicolet, M. A., "Solid-phase epitaxial growth of Ge and Si through metal films", Electrochemical Society Spring Meeting (Extended abstracts), 1975, p. 261-2, PA78-86418.

d. Silicon Carbide

238. Berman, I., Marshall, R. C. and Ryan, C. E., "Annealing of sputtered SiC", Presented at International Conference on Silicon Carbide, Miami Beach, Fla., September 1973, PA77-10528.
239. Brander, R. W., "Epitaxial growth of SiC layers", Presented at International Conference on Silicon Carbide, Miami Beach, Fla., September 1973, PA77-10525.
240. Callaghan, M. P. and Brander, R. W., "A new system for the chemical vapour deposition of SiC", J. Cryst. Growth, 13-14, 394-401 (1972), EEA75-24221.

241.Haq, K. E., "Single-crystal β-SiC films by reactive sputtering", Appl. Phys. Lett., 26, 255-6 (1975), EEA78-19622.

242.Learn, A. J. and Hag, K. E., "Low-temperature epitaxy of β-SiC by reactive deposition", Appl. Phys. Lett., 17, 26-9 (1970), EEA73-31167.

243.Minagwa, S. and Gatos, H. C., "Epitaxial growth of a β-SiC from the vapor phase (light emitting p-n junctions)", Jap. J. Appl. Phys., 10, 1680-90 (1971), EEA75-7414.

244.Mogab, C. J. and Leamy, H. J., "Conversion of Si to epitaxial SiC by reaction with C_2H_2", J. Appl. Phys., 45, 1075-84 (1974), PA77-57005.

245.Muench, W. V. and Gillessen, K., "Liquid-phase epitaxy of silicon carbide by the traveling heater method", Presented at International Conference on Silicon Carbide, Miami Beach, Fla., September 1973, PA77-10537.

246.Pellegrini, P. W. and Feldman, J. M., "Liquid phase epitaxial growth of SiC from transition-metal silicide solvents", J. Cryst. Growth, 27, 320-4 (1974), EEA78-24202.

247.Powell, J. A. and Will, H. A., "Epitaxial growth of 6H SiC in the temperature range 1320-1390°C", J. Appl. Phys., 44, 5177-8 (1973), EEA77-14375.

248.Ryan, C. E., Berman, I. and Marshall, R. C., "Thin films of alpha and beta silicon carbide prepared by liquid epitaxy and by sputtering", Report AFCRL-TR-74-0484, Air Force Cambridge Res. Lab., Bedford, Mass., 1974, 10 pp., EEA78-24233.

249.Suzuki, A., Ikeda, M., Matsunami, H. and Tanaka, T., "Liquid-phase epitaxial growth of 6H-SiC by vertical dipping technique", J. Electrochem. Soc., 122, 1741-2 (1975).

250.Swiderski, I., Szczutowski, W. and Niemyski, T., "Preparation of pure and boron doped epitaxial α-SiC layers on silicon carbide crystals", J. Cryst. Growth, 21, 125-34 (1974), PA77-21204.

251.Tairov, Y. M. and Tsvetkov, V. F., "Studies of silicon carbide crystal growth from the vapor phase", Presented at International Conference on Silicon Carbide, Miami Beach, Fla., September 1973, PA77-10535.

252.Wessels, B., Gatos, H. C. and Witt, A. F., "Epitaxial growth of silicon carbide by chemical vapor deposition (Electroluminescent junction preparation)", Presented at International Conference on Silicon Carbide, Miami Beach, Fla., September 1973, PA77-10526.

253.Will, H. A. and Powell, J. A., "Improved epitaxial process for fabricating silicon carbide semiconductor devices", Solid State Technol., 18 (4), 70 (1975), PA78-51132.

254.Yamada, S. and Kumagawa, M., "Epitaxial growth of SiC using Al as an accelerator", J. Cryst. Growth, 9, 309-13 (1971), EEA73-18577.

3. Group III-V Semiconductors
a. General

255.Akai, S., Hayashi, M., Iguchi, S. and Shimoda, T., "Method and apparatus for continuous preparation of epitaxial layers of semiconducting III-V compounds from vapor phase", Patent USA 3893876, Publ. July 1975.
256.Bryskiewicz, T., "On the problems of epitaxial growth of $A^{III}B^{V}$ components from liquid phase", Elektronika, 14, 277-81 (1973), EEA76-35199.
257.Cho, A. Y. and Panish, M. B., "Molecular beam epitaxy method for fabricating magnesium-doped thin films of group III(A)-V(A) compounds", Patent USA 3839084, Publ. October 1974, CA81-178946.
258.Chu, T. L. and Smeltzer, R. K., "Epitaxial growth of III-V compounds for electroluminescent sources", IEEE Trans. Parts, Hybrids & Packag., PHP-9, 208-15 (1973), PA77-25266.
259.Dawson, L. R., Lorimor, O. G. and Saul, R. H., "Epitaxial crystal growth of Group III-V compound semiconductors from solution", Patent USA 3859148, Publ. January 1975, CA82-118489.
260.Dorfman, V. F. and Pypkin, B. N., "Epitaxial growth in systems GaAs-I and GaP-I under a wide range of crystallization conditions", Izv. Akad. Nauk SSSR Neorg. Mater., 8, 1539-45 (1972).
261.Enstrom, R. E., Miller, E. A. and Sigai, A. G., "Vapor phase growth system and its use in the preparation of several III-V compound semiconductors", Report NASA CR-1886, RCA Labs., N. J., 1971, 112 pp., EEA75-24254.
262.Gisdakis, S., "Sources for gas phase epitaxy of Group III-V compounds", Patent German 2247976, Publ. April 1974, CA81-7614.
263.Hallasi, J., "Vapour phase epitaxy of the III-V pseudo-binary compounds", Acta Electron., 17 (1), 13 pp. (1974), French, PA77-25263.
264.Kishino, S., Ogirima, M. and Kurata, K., "Epitaxial growth of IIIb-Vb group compound semiconductor crystals", Patent Japan 73-44197, Publ. June 1973, CA79-98489.
265.Kulish, U. M., "Effects of the substrate orientation and of impurities on the perfection of the epitaxial layers of $A^{III}B^{V}$ (gallium arsenide and gallium antimonide) semiconductors and their solid solutions", Fiz. Poluprovodn. Soedin. Slozhn. Sostava, 3-18 (1974), Russian, CA83-88854.
266.Kurihara, Y. and Ura, M., "Group III-V compound semiconductor crystals", Patent Japan 75-71272, Publ. June 1975, CA83-186773.
267.Naganuma, M. and Takahashi, K., "GaAs, GaP and $GaAs_{1-x}P_x$ films deposited by molecular beam epitaxy", Phys. Status Solidi A, 31, 187-200 (1975).
268.Sugawara, Y., Kurihara, Y., Ura, M., Yagyu, S. and Ogawa, T., "Group III-V-type compound semiconductor crystals", Patent Japan 75-74970, Publ. June 1975, CA83-186774.

269.Tietjen, J. J., "Vapor-phase growth of several III-IV compound semiconductors", J. Vac. Sci. & Technol., 8 (1), 242 (1971), EEA74-13181.

270.Tietjen, J. J., Ban, V. S., Enstrom, R. E. and Richman, D., "Vapor-phase growth of several III-V compound semiconductors", J. Vac. Sci. & Technol., 8 (5), S5-11 (1971), EEA75-4577.

271.Tietjen, J. J., Clough, R., Dreeben, A. B., Enstrom, R. and Richman, D., "Vapour phase growth technique and system for several III-V compound semiconductors", Report NASA-CR-86192, RCA, Princeton, N. J., 1969, 29 pp., EEA72-31275.

272.Tietjen, J. J., Enstrom, R. E. and Richman, D., "Vapor-phase growth of several III-V compound semiconductors", RCA Rev., 31, 635-46 (1970), EEA74-13187.

273.Tietjen, J. J., Enstrom, R. E., Ban, V. S. and Richman, D., "Vapor-phase growth of several III-V compound semiconductors", Solid-State Technol., 15 (10), 42-9 (1972), EEA76-4549.

274.Tyagi, R. C., Debnam, W. J., Jr., McNear, M. F., Crouch, R. K. and Breckenridge, R. A., "Vapor phase growth of groups III-V compounds by hydrogen chloride transport of the elements", Patent USA 3888705, Publ. June 1975, PA78-82001.

275.Verleur, H. W., "Epitaxial growth of Group III-V semiconductors from solution", Patent USA 3853643, Publ. December 1974.

276.Watanabe, H., Nishinaga, T. and Arizumi, T., "Vapour transport equations for III-V compound semiconductors", J. Cryst. Growth, 17, 183-8 (1972), EEA76-4526.

b. Gallium Arsenide

277.Andre, E., Deyris, E. and LeDuc, J. M., "Growth of pure GaAs from stoichiometric and gallium rich melts", Presented at 133rd Meeting of the Electrochemical Society, Boston, Mass., May 1968, EEA73-14080.

278.Aoki, T., "High purity epitaxial GaAs", Jap. J. Appl. Phys., 14, 1267-72 (1975), PA78-86401.

279.Aoki, T. and Yamaguchi, M., "High purity GaAs epitaxial growth", Elec. Commun. Lab. Tech. J., 20, 1787-96 (1971), Japanese, EEA75-20050.

280.Araki, H., Iwane, G. and Aoki, T., "Vapor phase transport in the epitaxial growth of GaAs", Rev. Elec. Commun. Lab., 18, 608-17 (1970), EEA74-13191.

281.Bapat, D. R., Guha, S., Ravi, H. N. and Roy, S. K., "Liquid phase epitaxial growth of gallium arsenide", Presented at Symposium on Electronics (Electron Devices) (abstracts only), Hyderabad, India, February 1972, EEA75-13592.

282.Barybin, A. A., Zakharov, A. A. and Nedev, N. K., "Effect of growth temperature on the physical properties of gallium arsenide epitaxial films doped with oxygen", Izv. Akad. Nauk SSSR, Neorg. Mater., 9, 884-6 (1973), Russian, CA79-98407.

283. Berkova, A. V., Gran, Y. M. and Sabanova, L. D., "Structural features of epitaxial layers of gallium arsenide grown from liquid phase", Nauch. Tr. Nauch.-Issled. Inst. Redkometal. Prom., no. 46, 49-53 (1973), Russian, CA80-31163.
284. Bobb, L. C., Holloway, H., Maxwell, K. H. and Zimmerman, E., "Oriented growth of semiconductors. II. Homoepitaxy of gallium arsenide", J. Phys. Chem. Solids, 27, 1679-85 (1966), EEA70-503.
285. Bolkhovityanov, Y. B., Bolkhovityanova, R. I. and Melnikov, P. L., "Selective growth of epitaxial layers of gallium arsenide from solution in a melt", Inorg. Mater., 9, 1485-6 (1973), PA77-67726.
286. Boucher, A. and Easton, B. C., "Epitaxial growth of gallium arsenide", Philips Tech. Rev., 32, 380-4 (1971), EEA76-16299.
287. Bradshaw, A. and Knappet, J. E., "Epitaxial growth of device quality gallium arsenide layers", Solid State Technol., 13 (10), 45-50 (1970), EEA74-4829.
288. Cho, A. Y. and Ballamy, W. C., "GaAs planar technology by molecular beam epitaxy (MBE)", J. Appl. Phys., 46, 783-5 (1975), PA78-47432.
289. Davey, J. E. and Pankey, T., "Epitaxial GaAs films deposited by vacuum evaporation", J. Appl. Phys., 39, 1941-8 (1968), EEA71-13175.
290. Deitch, R. H., "Liquid-phase epitaxial growth of gallium arsenide under transient thermal conditions", Presented at ACCG Conference on Crystal Growth, Gaithersburg, Md., August 1969, EEA73-28086.
291. Fabri, G. and Schiarini, G. M., "Epitaxial growth of gallium arsenide", Alta Frequenza, 39 (6), 34-6 (1970), Italian, EEA73-34366.
292. Faktor, M. M., Garrett, I. and Moss, R. H., "Growth of crystals from the gas phase. IV. Growth of gallium arsenide by chemical vapour transport, and the influence of compositional convection", J. Chem. Soc. Faraday Trans. I, 69 (pt. 11), 1915-25 (1973), EEA77-18711.
293. Giannini, F., Ottavi, C. M. and Salsano, A., "Epitaxial growth at high efficiency on the (111) Ga face of gallium arsenide", Alta Frequenza, 39, 816-17 (1970), Italian, EEA74-4775.
294. Gonda, S., Matsushima, Y., Makita, Y. and Mukai, S., "Molecular beam epitaxial growth of GaAs", Bull. Electrotech. Lab., 39, 361-9 (1975), Japanese, EEA78-41461.
295. Goodwin, A. R., Gordon, J. and Dobson, C. D., "High-mobility gallium arsenide grown by liquid-phase epitaxy", Brit. J. Appl. Phys. (J. Phys. D), 1, 115-16 (1968), EEA71-15682.
296. Hammerling, H., "Gas phase composition in the Ga-$AsCl_3H_2$ system for epitaxial deposition of GaAs", J. Cryst. Growth, 9, 171-4 (1971), EEA74-18574.
297. Harada, H., "GaAs vapour phase epitaxial growth", Rev. Electr. Commun. Lab., 20, 1077-86 (1972), EEA76-19843.

298.Hicks, H. G. B. and Manley, D. F., "High purity GaAs by liquid phase epitaxy", Solid State Commun., 7, 1463-5 (1969), EEA73-7382.
299.Hollan, L., "Vapour phase epitaxial growth of GaAs for multiple applications", Gallium Arsenide and Related Compounds, 1974, p. 22-30, Publ. 1975, EEA78-28662.
300.Hollan, L. and Schiller, C., "Difference between the (001) facet and the vicinal planes in vapour phase epitaxial growth of GaAs", J. Cryst. Growth, 22, 175-80 (1974), EEA77-30184.
301.Hsieh, J. J., "LPE growth of GaAs from supercooled solutions", Presented at Electrochemical Society Meeting, San Francisco, Calif., May 1974, PA77-49604.
302.Ihara, M., Dazai, K. and Ryuzan, O., "Vapour-phase epitaxial growth of GaAs in a nitrogen atmosphere", J. Appl. Phys., 45, 528-31 (1974), PA77-53611.
303.Ishibashi, Y. and Yamaguchi, M., "Anisotropy in etching and deposition of selective epitaxial growth of GaAs", Jap. J. Appl. Phys., 9, 1007-8 (1970), EEA73-37889.
304.Ishibashi, Y. and Yamaguchi, M., "Anisotropy in etching and deposition of selective epitaxial growth of GaAs II", Jap. J. Appl. Phys., 10, 525-6 (1971), EEA74-18587.
305.Ishibashi, Y. and Yamaguchi, M., "Selective epitaxial GaAs vapour growth", Elec. Commun. Lab. Tech. J, 20, 1769-86 (1971), Japanese, EEA75-16620.
306.Ishibashi, Y., Yamaguchi, M. and Uchida, M., "Selective epitaxial GaAs vapor growth", Rev. Electr. Commun. Lab., 22, 1035-42 (1974), EEA78-24218.
307.Joyce, B. D. and Mullin, J. B., "The facet effect in epitaxial GaAs", Solid State Commun., 5, 727-9 (1967), EEA71-1142.
308.Kaneko, Y. and Isozaki, H., "Vapor-phase epitaxial formation of gallium arsenide", Patent Japan 73-16436, Publ. May 1973, CA79-150813.
309.Kordos, P. and Benc, V., "Preparation and properties of epitaxial GaAs from the liquid phase", Sloboproudy Obz., 35, 355-9 (1974), Slovak, PA77-82823.
310.Kordos, P. and Benc, V., "Epitaxial growth of p-GaAs from the liquid phase", Elektrotech. Cas., 25, 254-61 (1974), Czech, PA77-61293.
311.Lavrenteva, L. G., Nesteryuk, L. G. and Gaidareva, S. P., "Influence of excess gallium and arsenic in the vapour phase on the growth of gallium arsenide in the (111) direction", Izv. VUZ Fiz., no. 10, 43-8 (1972), Russian, EEA76-4507.
312.Lavrenteva, L. G., Nesteryuk, L. G. and Sennikora, V. M., "Orientation effects during the growing of gallium arsenide layers in the gas transport system GaAs-$ZnCl_2$", Izv. VUZ Fiz., no. 6, 111-16 (1972), Russian, EEA75-36125.
313.LeDuc, J. M. and Leroy, D., "Epitaxial deposition of semiconductor compounds from solution", Patent French 2145829, Publ. March 1973, CA79-98482.

314. Long, S. I., Ballantyne, J. M. and Eastman, L. F., "Steady-state LPE growth of GaAs", J. Cryst. Growth, 26, 13-20 (1974), PA78-16093.
315. Magomedov, K. A. and Magomedov, N. N., "Influence of the preliminary treatment of the surface of a gallium arsenide substrate on the perfection of autoepitaxial layers growing on it", Kristallografiya, 12, 343-7 (1967), Russian, EEA70-13938.
316. Magomedov, N. N., Rodionov, A. I. and Sheftal, N. N., "Epitaxial gallium arsenide obtained by a transport reaction with hydrochloric acid", Sov. Phys.-Crystallogr., 17, 604-7 (1972), EEA76-8541.
317. Maruyama, S., "GaAs epitaxial growth liquid phase", Elec. Commun. Lab. Tech. J., 20, 1807-20 (1971), Japanese, EEA75-16621.
318. Maruyama, S., "Liquid phase epitaxy of gallium arsenide", Jap. J. Appl. Phys., 11, 424-5 (1972), EEA75-13566.
319. Maruyama, M., Kikuchi, S. and Hasegawa, F., "Preparation and properties of epitaxial gallium arsenide", Presented at 133rd Meeting of the Electrochemical Society, (abstracts), Boston, Mass., May 1968, EEA73-14081.
320. Merenda, P., "A new method of vapour phase epitaxy of high purity gallium arsenide", J. Cryst. Growth, 13-14, 331-5 (1972), French.
321. Migitaka, M. and Miyazaki, A. D. M., "Continuous growth of multilayer GaAs crystal by liquid phase epitaxy (mm wave transferred electron diode)", Proceedings of the 4th International Symposium on Gallium Arsenide and Related Compounds, 1972, p. 249-55, Publ. 1973, EEA76-22933.
322. Mihara, M., Toyoda, N. and Hara, T., "Liquid-phase epitaxial growth of thin GaAs layers from supercooled solutions", Appl. Phys. Lett., 27, 131-3 (1975), EEA78-41458.
323. Miki, H., Ito, M. and Oda, T., "Silicon contamination in vapor-grown gallium arsenide", Jap. J. Appl. Phys., 11, 623-7 (1972), EEA75-20068.
324. Munoz, E., Snyder, W. I. and Moll, J. L., "Effect of arsenic pressure on heat treatment of liquid epitaxial GaAs", Appl. Phys. Lett., 16, 262-5 (1970), EEA73-24731.
325. Nagunuma, M. and Takashashi, K., "Epitaxial GaAs films deposited by molecular beam method", Electr. Eng. Jap., 93 (5), 19-24 (1973), PA78-33571.
326. Otsubo, M. and Miki, H., "Liquid phase epitaxial growth of GaAs crystals under a mixed gas atmosphere", Jap. J. Appl. Phys., 14, 621-8 (1975), EEA78-24210.
327. Otsubo, M., Segawa, K. and Miki, H., "The influence of oxygen on the properties of GaAs grown by liquid phase epitaxy", Jap. J. Appl. Phys., 12, 797-803 (1973), EEA76-29171.
328. Piskorski, M., "GaAs layers grown by liquid phase epitaxy", Electron Technol., 5 (1), 103-31 (1972), PA75-71385.
329. Piskorski, M., "Thin GaAs layers grown by liquid phase epitaxy", Electron. Technol., 6 (1-2), 115-24 (1973), PA77-53606.

330.Piskorski, M. M. and Stareev, G. D., "Selective epitaxial growth of GaAs from the liquid phase", Solid-State Electron., 18, 859-61 (1975), EEA78-37420.
331.Pretrescu-Prhaova, I. B. and Mihailovici, P., "The influence of a temperature gradient on the LPE growth process of GaAs", J. Cryst. Growth, 21, 214-18 (1974), PA77-33616.
332.Rai-Choudhury, P., "Epitaxial gallium arsenide from trimethyl gallium and arsine", J. Electrochem. Soc., 116, 1745-6 (1969), EEA73-17482.
333.Ravi, H. N. and Guha, S., "A new rotating-boat technique for liquid phase epitaxial growth of gallium arsenide", J. Cryst. Growth, 18, 212-16 (1973), EEA76-19836.
334.Ryuzan, O., Dazai, K. and Ihara, M., "Epitaxial GaAs crystals. I. Vapor-phase growth of GaAs", Fujitsu Sci. Tech. J., 8 (2), 115-29 (1972), PA75-71386.
335.Schlachetzki, A. and Salow, H., "High resistivity layers of GaAs grown by liquid phase epitaxy", Appl. Phys., 7, 195-201 (1975), EEA78-32998.
336.Shaw, D. W., "Selective epitaxial deposition of gallium arsenide in holes", J. Electrochem. Soc., 113, 904-8 (1966), EEA70-509.
337.Shaw, D. W., "Influence of substrate temperature on GaAs epitaxial deposition rates", J. Electrochem. Soc., 115, 405-8 (1968), EEA71-20523.
338.Stangel, J. E., "Growth of epitaxial gallium arsenide by the vapor phase process", Report AD-742428, Air Force Inst. Tech., Wright-Patterson AFB, Ohio, 1972, 113 pp., EEA76-735.
339.Steinberg, R. F. and Scruggs, D. M., "Preparation of epitaxial GaAs films by vacuum evaporation of the elements", J. Appl. Phys., 37, 4586-7 (1966), EEA70-4888.
340.Taylor, R. C., "Epitaxial deposition of GaAs in an argon atmosphere", J. Electrochem. Soc., 114, 410-12 (1967), EEA70-14843.
341.Teshima, H., Tarui, Y. and Takeda, O., "Epitaxial growth of GaAs(I)", Bull. Electrotech. Lab., 33, 692-9 (1969), Japanese, EEA73-17414.
342.Toyoda, N., Mihara, M. and Hara, T., "LPE growth of GaAs by supercooling technique: Elimination of surface terraces", Appl. Phys. Lett., 27 (11), 627-28 (1975).
343.Watanabe, H. and Seki, Y., "A new method of vapor growth of GaAs", J. Cryst. Growth, 24-25, 220-4 (1974), PA78-10536.
344.Wilson, O. W. and Cronin, G. R., "Epitaxial gallium arsenide", Patent USA 3471324, Publ. October 1969, EEA73-10455.
345.Yamaguchi, M., "Multi-layer GaAs epitaxial vapour growth", Elec. Commun. Lab. Tech. J, 20, 1797-805 (1971), Japanese, EEA75-20051.
346.Yamaguchi, M., "Multi-layer GaAs epitaxial vapor growth. II", Rev. Electr. Commun. Lab., 21, 482-7 (1973), PA77-33623.

c. Gallium Phosphide

347. Andrle, C., "The influence of temperature distribution in a reaction zone on the growth of GaP epitaxial layers prepared by the chloride transport method", Elektrotech. Cas., 25, 357-63 (1974), Czech, EEA77-33924.

348. Asao, I., Ohki, Y., Akasaki, I. and Hashimoto, M., "Defect-free epitaxial semiconductor", Patent German 2449305, Publ. April 1975, CA83-70303.

349. Bergh, A. A. and Saul, R. H., "Growth of GaP layers from thin aliquot melts: Liquid phase epitaxy as a commerical process", J. Electrochem. Soc., 120, 1558-63 (1973), EEA77-2460.

350. Blom, G. M. and Bhargava, R. M., "Radiative complex formation in vapor phase epitaxial GaP", J. Cryst. Growth, 17, 38-45 (1972), EEA76-4520.

351. Galginaitis, S. V., Barnett, A. M. and Heumann, F. K., "Selected liquid epitaxy of GaP", Proceedings of the Third International Symposium on Gallium Arsenide and Related Compounds, 1970, p. 80-5, Publ. 1971, EEA76-4563.

352. Kamath, G. S. and Bowman, O., "Preparation and properties of epitaxial gallium phosphide", J. Electrochem. Soc., 114, 192-5 (1967), EEA70-10431.

353. Lien, S. Y. and Bestel, J. L., "Liquid-phase epitaxy growth of gallium phosphide by a centrifugal tipping technique", J. Electrochem. Soc., 120, 1571-3 (1973), EEA77-2463.

354. Lim, M. J., "Vapor phase growth of epitaxial gallium phosphide", Electrochemical Society Fall Meeting (Extended abstracts), 1970, p. 432, EEA74-7622.

355. Lorimor, O. G., Haszko, S. E. and Dapkus, P. D., "Observations on Si contamination of GaP LPE", J. Electrochem. Soc., 122, 1230-3 (1975).

356. Nishizawa, J. and Okuno, Y., "Liquid phase epitaxy of GaP by a temperature difference method under controlled vapor pressure", IEEE Trans. Electron Devices, ED-22, 716-21 (1975), EEA78-41466.

357. Nohavica, D., "Epitaxial growth of GaP from the fluid phase", Elektrotech. Cas., 25, 287-93 (1974), Czech, PA77-61297.

358. Nohavica, D., "Large-area epitaxial growth of GaP from the gaseous phase", Elektrotech. Cas., 25, 262-71 (1974), Czech, PA77-61294.

359. Roy, A. S., "Closed tube chemical vapor transport of gallium phosphide", Presented at Israel Physical Society 1972 Annual Meeting, Jerusalem, Israel, May 1972, PA75-71372.

360. Seki, H. and Takizawa, Y., "A new vapor growth for GaP using a single flat temperature zone", Jap. J. Appl. Phys., 12, 1112-13 (1973), EEA76-35227.

361. Smith, B. L. and Abbott, M., "Minority carrier diffusion length in liquid epitaxial GaP", Solid-State Electron., 15, 361-70 (1972), EEA75-16650.

362. Sudlow, P. D. and Saunders, I. J., "Molecular beam growth of gallium phosphide", Phys. Status Solidi A, 16, K147-51 (1973), EEA76-23356.

363. Tomek, K., "Use of GaN for doping epitaxial GaP layers with nitrogen", Elektrotech. Cas., 25, 281-6 (1974), Czech, EEA77-30148.

d. Indium Phosphide

364. Astles, M. G., Smith, F. G. H. and Williams, E. W., "Indium phosphide. II. Liquid epitaxial growth", J. Electrochem. Soc., 120, 1750-7 (1973), PA77-17104.

365. Brown, K. E., "Criteria for tipping during the liquid phase epitaxial growth of indium phosphide", J. Cryst. Growth, 20, 161-4 (1973), EEA77-315.

366. Hess, K., Stath, N. and Benz, K. W., "Liquid phase epitaxy of InP", J. Electrochem. Soc., 121, 1208-12 (1974), PA78-1694.

367. Mizuno, O., "Vapor growth of InP", Jap. J. Appl. Phys., 14, 451-7 (1975), EEA78-24207.

368. Sol, N., Clariou, J. P., Linh, N. T., Bichon, G. and Moulin, M., "New technique for epitaxial growth of InP", Gallium Arsenide and Related Compounds, 1974, p. 369-72, Publ. 1975, French, EEA78-28677.

369. Wrick, V. L., III, "Liquid phase epitaxy in indium phosphide", Thesis, Cornell Univ., Ithaca, N. Y., 1974, 130 pp., Order No. 75-4272, PA78-66561.

e. Ternaries

370. Andre, E., LeDuc, J. M. and Mahieu, M., "Liquid phase epitaxy of $Ga_{1-x}Al_xAs$ (for p-n junction formation)", J. Cryst. Growth, 13-14, 663-7 (1972).

371. Antypas, G. A., "Liquid phase epitaxy of $In_xGa_{1-x}As$", J. Electrochem. Soc., 117, 1393-7 (1970), EEA74-7574.

372. Aulombard, R. L., Jouillie, A. and Bougnot, G., "Influence of preparation temperature on epitaxial deposition of $Ga_xIn_{1-x}Sb$ solutions. Conditions for homogeneous growth", C. R. Hebd. Seances Acad. Sci. B, 278, 893-5 (1974), French, PA77-53605.

373. Bedair, S. M., "Growth and characterization of $Al_xGa_{1-x}Sb$", J. Electrochem. Soc., 122, 1150-2 (1975), EEA78-41482.

374. Bleicher, M., "Thermodynamic calculations for $GaAs_{1-x}P_x$ vapour growth", J. Electrochem. Soc., 119, 613-17 (1972), PA75-47220.

375. Blom, G. M., "The In-Ga-P ternary phase diagram and its application to liquid phase epitaxial growth", J. Electrochem. Soc., 118, 1834-6 (1971), EEA75-7404.

376. Blum, J. M. and Shih, K. K., "Growth of smooth uniform epitaxial layers by liquid-phase-epitaxial method (LED's), J. Appl. Phys., 43, 1394-5 (1972), EEA75-16631.

377. Chang, L. L., Esaki, L., Howard, W. E., Ludeke, R. and Schul, G., "Structures grown by molecular beam epitaxy (GaAs and GaAs-$Ga_{1-x}Al_xAs$)", J. Vac. Sci. & Technol., 10, 655-62 (1973), PA77-25272.
378. Conrad, R. W., Hoyt, P. L. and Martin, D. D., "Preparation of epitaxial $Ga_xIn_{1-x}As$", J. Electrochem. Soc., 114, 164-6 (1967), EEA70-11094.
379. Daniele, J. J. and Michel, C., "Peltier-induced growth of GaAlAs", Gallium Arsenide and Related Compounds, 1974, p. 155-64, Publ. 1975, PA78-54984.
380. Golubev, L. V., Khachaturyan, O. A., Shik, A. Y. and Shmartsev, Y. V., "Growth of films of semiconducting alloys by liquid-phase epitaxy with a temperature gradient (LPETG)", Phys. Status Solidi A, 22, K203-4 (1974), PA77-45481.
381. Goncharova, T. S., Konnikov, S. G., Tretyakov, D. N. and Aleksandrova, T. P., "Epitaxial build-up of Al_x-$Ga_{1-x}Sb$ from a metallic melt", Izv. VUZ Fiz., no. 9, 146-8 (1973), Russian, PA77-1440.
382. Gonda, S. and Ijuin, H., "Liquid-phase-epitaxial growth of $Al_xGa_{1-x}As$. I. Calculation of phase diagram and crystal growth process", Bull. Electrotech. Lab., 39, 490-504 (1975), Japanese, EEA78-41462.
383. Hakki, B. W., "Growth of $In_{(1-x)}Ga_xP$ p-n junctions by liquid phase epitaxy", J. Electrochem. Soc., 118, 1469-73 (1971), EEA74-37142.
384. Hallais, J., Schemali, C. and Fabre, E., "Vapour epitaxial growth and characterization of $InAs_{1-x}P_x$", J. Cryst. Growth, 17, 173-82 (1972), EEA76-4525.
385. Hojo, A. and Kuru, I., "Liquid phase epitaxy of high purity n-GaSb and $In_xGa_{1-x}Sb$ and their electron transfer effects", Oyo Buturi, 43 (Suppl.), 226-31 (1974), EEA78-527.
386. Huber, H. and Winstel, G. H., "Preparation of epitaxial $GaAs_{1-x}P_x$ layers using a solid GaAs source and the chlorides of As and P", Siemens Forsch.- & Entwicklungsber., 3, 55-60 (1974), EEA77-18724.
387. Inoue, M. and Asahi, K., "Epitaxial growth of $GaAs_{1-x}P_x$ by thermal decomposition method. I. Growth experiments", Nat. Tech. Rep., 18, 303-10 (1972), Japanese, EEA75-32529.
388. Inoue, M. and Asahi, K., "Vapour growth of $GaAs_{1-x}P_x$ by the pyrolysis of $Ga(CH_3)_3$, AsH_3 and PH_3", Jap. J. Appl. Phys., 11, 919-20 (1972), EEA75-28407.
389. Joullie, A., Aulombard, R. and Bougnot, G., "Growth of $Ga_xIn_{1-x}Sb$ liquid phase epitaxial layers from a knowledge of the ternary phase diagram", J. Cryst. Growth, 24-25, 276-81 (1974), EEA78-2731.
390. Joyce, B. D., Fairhurst, K. M., Clarke, R. C. and Born, P. J., "The vapour phase growth of $Ga_xIn_{1-x}P$ epitaxial layers", J. Cryst. Growth, 11, 243-8 (1971), EEA75-10128.

391.Kawakami, T. and Sugiyama, K., "Selective liquid phase epitaxy of AlGaAs", Jap. J. Appl. Phys., 12, 1808-9 (1973), EEA77-2473.

392.Kulish, U. M., "The effect of substrate orientation in the range (111)-(100) on growth rate and parameters of GaAs and $In_xGa_{1-x}As$ films (liquid epitaxy)", Izv. VUZ Fiz., no. 6, 121-3 (1973), Russian, EEA76-35200.

393.Kusumoto, H., Kurata, K., Ogirina, M., Ono, Y., Honma, K., Gejo, T., Minagawa, S., Ataka, S., Manabe, T., Miya, N. and Seki, H., "Epitaxial growth of $GaAs_{1-x}P_x$", Hitachi Hyoron, 53, 891-5 (1971), Japanese, EEA75-24194.

394.Manasevit, H. M., "The use of metal-organics in the preparation of semiconductor materials. III. Studies of Epitaxial III-V aluminum compound formation using trimethylaluminum", J. Electrochem. Soc., 118, 647-50 (1971), EEA74-18564.

395.Marev, V. B. and Popov, V. P., "Liquid epitaxy of $GaAs_{1-x}P_x$ at constant temperature", Sov. Phys.-Crystallogr., 19, 118 (1974), PA77-82830.

396.Markov, E. I., "Crystal growth and properties of indium gallium antimonide ($In_{1-x}Ga_xSb$) epitaxial films", Itogi Nauch. Issled. Kishinev, Politekhn. In-Ta Za 1973 G., 101-2 (1974), Russian, CA83-51933.

397.Mau, N. V., Gouskov, A., Averous, M. and Bougnot, G., "Liquid phase epitaxy of $Ga_{1-x}Al_xSb$", C. R. Hebd. Seances Acad. Sci. B, 278, 15-17 (1974), French, PA77-33611.

398.Miki, H., Segawa, K., Otsubo, M., Shirahata, K. and Fubjibayashi, K., "Growth in $In_{1-x}Ga_xSb$ by liquid phase epitaxy (current oscillations)", Gallium Arsenide and Related Compounds, 1974, p. 16-21, Publ. 1975, EEA78-28661.

399.Minagawa, S., Seki, H. and Eguchi, H., "Thermodynamic calculation for the vapor growth of $In_xGa_{1-x}As$: The In-Ga-As-Cl-H system", Jap. J. Appl. Phys., 11, 855-9 (1972), PA75-47222.

400.Mizuno, O. and Arai, K., "Vapor growth of $InAs_xP_{1-x}$", Jap. J. Appl. Phys., 13, 1955-8 (1974), EEA78-11494.

401.Moon, R. L. and Antypas, G. A., "Surface irregularities due to spiral growth in LPE (liquid phase epitaxy) layers of AlGaAs and InGaAsP", J. Cryst. Growth, 19, 109-12 (1973).

402.Mullin, J. B. and Hurle, D. T. J., "Epitaxy of mixed III-V compounds", J. Lumin., 7, 176-91 (1973), EEA76-32184.

403.Naganuma, M. and Takahashi, K., "GaAs, GaP, and $GaAs_{1-x}P_x$ films deposited by molecular beam epitaxy", Phys. Status Solidi A, 31, 187-200 (1975), EEA78-41494.

404.Nahory, R. E., Pollack, M. A. and DeWinter, J. C., "Growth and characterization of liquid-phase epitaxial $In_xGa_{1-x}As$", J. Appl. Phys., 46, 775-82 (1975), PA78-47431.

405.Rado, W. G. and Crawley, R. L., "The variation of the solid composition during the LPE growth of $Ga_{1-x}Al_xAs$", J. Electrochem. Soc., 119, 1779-80 (1972), EEA76-8536.

406. Saitoh, T. and Minagawa, S., "Epitaxial growth and characterization of pyrolytic-grown $GaAs_{1-x}P_x$ for electroluminescent diodes", J. Electrochem. Soc., 120, 656-9 (1973), EEA76-26545.
407. Shih, K. K., "Epitaxial growth of $GaAs_xP_{1-x}$, from the liquid phase", J. Electrochem. Soc., 117, 387-9 (1970), EEA73-20760.
408. Shotgrin, I. I., Maslov, V. N., Sakharov, B. A. and Firsanova, L. A., "Epitaxial growth of solid solutions $In_xGa_{1-x}P$ from the gaseous phase", Izv. Akad. Nauk SSSR Neorg. Mater., 8, 1755-7 (1972), Russian, EEA76-8524.
409. Sigai, A. G., Nuese, C. J., Enstrom, R. E. and Zamerowski, T., "Vapor growth of $In_{1-x}Ga_xP$ for p-n junction electroluminescence. I. Material preparation", J. Electrochem. Soc., 120, 947-55, EEA76-32174.
410. Silhavy, J., "On the growth of $GaAs_{1-x}P_x$ epitaxial layer using the sandwich method", Elektrotech. Cas., 25, 346-56 (1974), Czech, PA77-67720.
411. Stringfellow, G. B. and Greene, P. E., "Liquid phase epitaxial growth of $InAs_{1-x}Sb_x$", J. Electrochem. Soc., 118, 805-10 (1971), EEA74-22499.
412. Sugano, T., Katoda, T. and Osaka, F., "Vapor growth and properties of $In_{1-x}Ga_xAs$ epitaxial layer", Annu. Rep. Eng. Res. Inst. Fac. Eng. Univ. Tokyo, 33, 139-45 (1974), PA78-38372.
413. Takahashi, R., Ono, Y. and Kasano, H., "Vapor phase epitaxial growth of $GaAs_{1-x}P_x$ crystals", Patent Japan 74-132000, Publ. December 1974, CA83-89831.
414. Teramoto, I. and Inoue, M., "Epitaxial growth of $GaAs_{1-x}P_x$ by thermal decomposition method. II. Analysis of growth rate", Natl. Tech. Rep., 19, 191-5 (1973), Japanese, EEA76-29178.
415. Vigdorovich, E. N., Plavich, L. F. and Andreev, V. M., "Gallium arsenide phosphide epitaxial layer obtained by a gas transport method", Protsessy Rosta Sint. Poluprovodn. Krist. Plenok, Mater. Vses. Simp., 3rd, 1972, p. 1, 238-40, Russian, Aleksandrov, L. N. and Kuznetsov, F. A. (Eds.), "Nauka", Sib. Otd.: Novosibirsk, USSR, 1975, CA83-211392.
416. Wu, T. Y., "On the optimum conditions for vapor epitaxy of $GaAs_{1-x}P_x$ with $0.3 < x < 0.5$", J. Cryst. Growth, 23, 348-50 (1974), PA78-16092.
417. Wu, T. Y., "Thermochemical analyses and optimum conditions for vapor epitaxy of $GaAs_{1-x}P_x (0.7 < x < 0.9)$", J. Electrochem. Soc., 122, 778-80 (1975), PA78-62549.

f. Others

418. Arizumi, T., Kakchi, M. and Shimokawa, R., "Epitaxial vapor growth of gallium antimonide", J. Cryst. Growth, 9, 151-7 (1971), EEA74-18571.
419. Cardetta, V. L., Mancini, A. M., Manfredotti, C. and Rizzo, A., "Growth and habit of GaSe crystals obtained from vapour by various methods", J. Cryst. Growth, 17, 155-61 (1972), EEA76-4524.

420. Chu, T. L., Ing. D. W. and Noreika, A. J., "Epitaxial growth of aluminum nitride", Solid-State Electron., 10, 1023-7 (1967), EEA71-1138.
421. Godinho, N. and Brunnschweiler, A., "Epitaxial indium arsenide by vacuum evaporation", Solid-State Electron., 13, 47-52 (1970), EEA73-14054.
422. Holloway, H., Richards, J. L., Bobb, L. C., Perry, J., Jr. and Zimmerman, E., "Oriented growth of semiconductors. IV. Vacuum deposition of epitaxial indium antimonide", J. Appl. Phys., 37, 4694-8 (1966), EEA70-12536.
423. Kakehi, M., Shimokawa, R. and Arizumi, T., "Epitaxial growth of gallium antimonide", Japan J. Appl. Phys., 9, 1039-44 (1970), EEA73-37890.
424. McCarthy, J. P., "Preparation and properties of epitaxial InAs", Solid State Electron., 10, 649-55 (1967), EEA70-18827.
425. Miki, H., Segawa, K. and Fujibayashi, K., "Undoped n-type GaSb grown by liquid phase epitaxy", Jap. J. Appl. Phys., 13, 203-4 (1974), PA77-29012.
426. Mizuno, O., Watanabe, H. and Shinoda, D., "Vapor growth of InAs", Jap. J. Appl. Phys., 14, 184-91 (1975), EEA78-19655.
427. Nicolaus, W., Voronin, V. A. and Seidowski, E., "Determination of supersaturation in the crystallization of InAs from the gas phase", Kristall & Tech., 7, 395-402 (1972), German, PA75-74672.
428. van Egmond, G. E. and Lieth, R. M. A., "Transport and growth of GaSe from the vapour phase", Mater. Res. Bull., 9, 763-74 (1974), EEA77-33961.
429. Woelk, C. and Benz, K. W., "Gallium antimonide LPE growth from Ga-Rich and Sb-rich solutions", J. Cryst. Growth, 27, 177-82 (1974), PA78-26314.

4. Group II-VI

a. General

430. Blanconnier, P. and Henoc, P., "Epitaxial growth of II-VI compounds from vapour phase", J. Cryst. Growth, 17, 218-21 (1972), French, EEA76-4530.
431. Nassau, K. and Shiever, J. W., "Vapor growth of II-VI compounds and the identification of donors and acceptors (Luminescence)", J. Cryst. Growth, 13-14, 375-9 (1972), EEA75-24217.
432. Smith, D. L. and Pickhardt, V. Y., "Molecular beam epitaxy of II-VI compounds (for integrated optics application)", J. Appl. Phys., 46, 2366-74 (1975), EEA78-37409.

b. Cadmium Telluride

433. Ezhovskii, Y. K. and Kalinkin, I. P., "Effect of the vapor composition on epitaxial growth of films in a quasicloud volume", Inorg. Mater., 9, 1487-91 (1973), PA77-67727.

434. Paorici, C., "Vapour-phase transport properties of the Cd:Te:I:H system", Proceedings of the International Symposium on Cadmium Telluride a Material for Gamma-Ray Detectors, 1971, 7/3 pp., EEA75-20099.

c. Cadmium Sulphide

435. Ezhovskii, Y. K., Kalinkin, I. P. and Bogomolov, N. S., "Epitaxy of CdS and CdSe films in a quasienclosed space", Ingor. Mater., 10, 1196-200 (1974), PA78-85676.
436. Gabor, T. and Dierssen, G. H., "Epitaxial growth of solid-state diffusion", J. Electrochem. Soc., 122, 430-3 (1975), EEA78-24183.

d. Ternaries

437. Cadoz, J., Faure, M. and Moulin, M., "Study of the liquid phase growth and characterization of epitaxial layers of $Pb_{1-x}Sn_xTe$", J. Cryst. Growth, 29, 241-7 (1975), French, PA78-86400.
438. Harris, J. S., Longo, J. T., Gertner, E. R. and Clarke, J. E., "The Pb-Sn-Te phase diagram and its application to the liquid phase epitaxial growth of $Pb_{1-x}Sn_xTe$", J. Cryst. Growth, 28, 334-42 (1975), EEA78-28627.
439. Kanamori, A., Ota, T. and Takahashi, K., "Liquid-phase epitaxial growth of ZnTe and $Zn_{1-x}Cd_xTe$ (LED fabrication)", J. Electrochem. Soc., 122, 1117-22 (1975), EEA78-41477.
440. Kasai, I. and Bassett, D. W., "Liquid phase epitaxial growth of $Pb_{1-y}Sn_ySe$", J. Cryst. Growth, 27, 215-20 (1974), EEA78-11487.
441. Laugier, A., Cadoz, J., Faure, M. and Moulin, M., "Ternary phase diagram and liquid phase epitaxy of Pb-Sn-Se and Pb-Sn-Te", J. Cryst. Growth, 21, 235-42 (1974), PA77-33617.
442. Rufer, H. and Bille, J., "Epitaxial growth and luminescence of CdS_xSe_{1-x} graded gap single crystal layers", Phys. Status Solidi A, 14, 147-51 (1972), EEA76-4547.
443. Saraie, J., Furukawa, S., Sawa, B. and Tanaka, T., "Epitaxial growth of $Cd_xHg_{1-x}Te$", Jap. J. Appl. Phys., 12, 1259-60 (1973), EEA76-35228.
444. Thomson, A. G. and Wagner, J. W., "Growth and characterization of lead-tin telluride epitaxial layers", Phys. Status Solidi A, 5, 439-48 (1971), EEA74-26610.
445. Tomasetta, L. R. and Fonstad, C. G., "Thallium and indium doped $Pb_{1-x}Sn_xTe$ by liquid phase epitaxy", Mater. Res. Bull., 9, 799-802 (1974), EEA77-33962.
446. Ueda, R. and Ohtsuki, O., "Crystal growth of narrow gap semiconductors", Oyo Buturi, 43, 562-87 (1974), Japanese, EEA78-5262.

e. Others

447. Calow, J. T., Kirk, D. L. and Owen, S. J. T., "The growth of epitaxial ZnSe upon germanium substrates", Thin Solid Films, 9, 409-29 (1972), EEA75-10143.
448. Jones, P. L., "Substrate effects in autoepitaxial thick films of single-crystal zinc sulphide", J. Phys. D, 7, 147-8 (1974), PA77-70427.
449. Lilley, P., Jones, P. L., Litting, C. N. W. and Kay, P. M. R., "A technique for the growth of thick films of zinc sulphide", J. Phys. E, 6, 1238-8 (1973), PA77-4047.
450. Lopez, Otero, A., "The use of a phase diagram as a guide for the growth of PbTe films", Appl. Phys. Lett., 26, 470-2 (1975), EEA78-28604.
451. Paic, M., Paic, V., Dukh, K. and Zemel, J. N., "Quasi-static growth for PbS epitaxic films", Thin Solid Films, 12, 419-25 (1972), EEA76-4552.
452. Parker, S. G., "Preparation of high purity epitaxial ZnSe", Presented at Electrochemical Society Fall Meeting, Chicago, Ill., October 1967, EEA72-20270.
453. Unvala, B. A., Woodcock, J. M. and Holt, D. B., "Epitaxial growth of cubic ZnS by evaporation in ultra-high vacuum", Brit. J. Appl. Phys. (J. Phys. D), 1, 11-14 (1968), EEA71-13177.
454. Wagner, J. W. and Thompson, A. G., "Growth and characterization of lead telluride epitaxial layers", J. Electrochem. Soc., 117, 936-40 (1970), EEA73-35067.
455. Yu, H. J. and Weng, T. H., "Preparation of thin film CdSe-CdS systems", Solid-State Electron., 13, 93-4 (1970), EEA73-14057.

VII. Autodoping

456. Bolzer, C. O., "Reduction of autodoping", J. Electrochem. Soc., 122, 1705-9 (1975).
457. Boss, D. W., Kemlage, B. M., Lyons, V. J. and Pogge, H. B., "Minimizing autodoping in epitaxial deposition", Patent USA 3765960, Publ. October 1973, CA80-20705.
458. Dyer, L. D. and Padovani, F., "Autodoping and other problems encountered in establishing a low-temperature silicon epitaxial manufacturing process", In: Semiconductor silicon 1973, The Electrochemical Society, Inc., Princeton, N. J., 1973, p. 201-12.
459. Grove, A. S., Roder, A. and Sah, C. T., "Impurity distribution in epitaxial growth", J. Appl. Phys., 36, 802-10 (1965).
460. Joyce, B. A., Weaver, J. C. and Maule, D. J., "Impurity redistribution process in epitaxial silicon layers", J. Electrochem. Soc., 112, 1100-106 (1965).
461. Kasano, H., "Autodoping effects of Ge in vapor-grown $GaAs_{1-x}P_x$ layers on the Ge substrates (electroluminescence)", Solid-State Electron., 16, 913-20 (1973), EEA76-29181.

462.Krause, H., "Autodoping effect in silicon epitaxial layers on Mg-Al spinel", Phys. Status Solidi A, 1, K53-6 (1970), EEA74-1408.
463.Langer, P. H. and Goldstein, J. I., "Impurity redistribution during silicon epitaxial growth and semiconductor device processing. (Computer model)", J. Electrochem. Soc., 121, 563-71 (1974), EEA77-22887.
464.Pogge, H. B., Boss, D. W. and Ebert, E., "Autodoping during silicon epitaxial growth", In: Chemical Vapor Deposition, Second International Conference, J. M. Blocher, Jr. and J. C. Withers (Eds.), The Electrochemical Society, Inc., New York, 1970, p. 767-93.
465.Rose, J. W., "Masking to prevent autodoping of epitaxial deposits", Patent USA 3663319, Publ. May 1972, CA77-40610.
466.Shachev, V. I., "Autodoping of epitaxial silicon layers", Inorg. Mater., 10, 176-9 (1974), EEA78-8439.
467.Shepherd, W. H., "Doping of epitaxial silicon films", J. Electrochem. Soc., 115, 541-5 (1968), EEA71-20530.
468.Shepherd, W. H., "Autodoping of epitaxial silicon", J. Electrochem. Soc., 115, 652-6 (1968), EEA71-25878.
469.Skelly, G. and Adams, A. C., "Impurity atom transfer during epitaxial deposition of silicon", J. Electrochem. Soc., 120, 116-25 (1973), EEA76-12453.
470.Sladkov, I. B., Tuchkevich, V. V., Schmidt, N. M. and Danchik, M. I., "Self-doping of silicon with boron in the chloride epitaxial growth method", Fiz. & Tekh. Poluprov., 4, 2385-7 (1970), Russian, EEA74-10477.
471.Suzuki, T. and Ura, M., "Properties of high purity Si epitaxial layers", In: Semiconductor silicon 1973, The Electrochemical Society, Inc., Princeton, N. J., 1973, p. 191-200.
472.Thomas, C. O., Kahng, D. and Manz, R. C., "Impurity distribution in epitaxial silicon films", J. Electrochem. Soc., 109, 1055-61 (1962).
473.Yoshikawa, A. and Sakai, Y., "Autodoping effects near the interface of cadmium sulfide-gallium arsenide junction", Jap. J. Appl. Phys., 13, 547-8 (1974), CA80-138315.

VIII. Impurities

1. Profiles

474.Arizumi, T., Nishinaga, T., Kakehi, M. and Shimokawa, R., "The distribution of the imperfection centres in an epitaxially-grown germanium layer", J. Cryst. Growth, 3-4, (Spe), 226 (1968), EEA72-20216.
475.Bennett, R. J. and Parish, C., "Control of impurity density in homoepitaxial semiconductor layers grown by sublimation at UHV", Solid-State Electron., 16, 497-501 (1973), EEA76-16307.

476.Bolkhovityanov, Y. B., Bolkhovityanova, R. I., Zembatov, N. B. and Marchenko, N. E., "Investigation of doping profiles and incorporation of Sn and Te in GaAs and $In_xGa_{1-x}As$ films grown from thin solution layer", Phys. Status Solidi A, 22, 349-55 (1974), PA77-33622.
477.Brown, K. E., "Effective distribution coefficients of some group VI elements in indium phosphide grown by liquid phase epitaxy", Solid-State Electron., 17, 505-7 (1974), PA77-45482.
478.Cho, A. Y., "Impurity profiles of GaAs epitaxial layers doped with Sn, Si, and Ge grown with molecular beam epitaxy", J. Appl. Phys., 46, 1733-5 (1975), EEA78-24179.
479.Cho, A. Y. and Reinhart, F. K., "Interface and doping profile characteristics with molecular-beam epitaxy of GaAs:GaAs voltage varactor", J. Appl. Phys., 45, 1812-17 (1974), PA77-61311.
480.Decker, D. R., "Measurement of epitaxial doping density vs. depth", J. Electrochem. Soc., 115, 1085-9 (1968), EEA72-11198.
481.Frank, H., "Some problems of measurements of epitaxial and diffused layers of semiconductors", Cesk. Cas. Fis. A, 23, 253-71 (1973), Czech, EEA76-38182.
482.Greisukh, M. R., Tuchkevich, V. V. and Kharchenko, V. V., "Influence of initial conditions on the distribution of impurities at an interface between two semiconductor films", Sov. Phys.-Semicond., 7, 1491-2 (1974), PA77-67124.
483.Ikoma, H. and Nakagawa, M., "Relationship between carrier and impurity concentrations in vapor-epitaxial n-GaAs crystals", Jap. J. Appl. Phys., 11, 338-42 (1972), PA75-29685.
484.Jacobs, K., Jacobs, B., Streubel, P. and Butter, E., "Information about the growth process from aluminum concentration profiles in $Ga_{1-x}Al_xAs$ liquid phase epitaxial layers", Krist. & Tech., 9, 1243-48 (1974), PA78-26328.
485.Johnson, W. J., Rado, W. G., Crawley, R. L. and Amey, J. E., "Carrier concentration profiles in Si-doped liquid-phase epitaxial $Ga_{1-x}Al_xAs$", J. Appl. Phys., 44, 1311-15 (1973), EEA76-16289.
486.Langer, P. H. and Pearce, C. W., "Epitaxial resistivity", J. Test. Eval., 1, 305-9 (1973).
487.Lorenz, G. and Scheffer, G., "Impurity distribution in epitaxial films after oxidation", Solid-State Electron., 10, 73-5 (1967), EEA71-5843.
488.Maeda, K. and Takayama, O., "The effect of vapour etching on 'diffuse-up' of buried impurities into epitaxial layer", Jap. J. Appl. Phys., 13, 1903-4 (1974), EEA78-5257.
489.Mayer, J. W., Ziegler, J. F., Chang, L. L., Tsu, R. and Esaki, L., "Profiling of periodic structures (GaAs-GaAlAs) by nuclear backscattering", J. Appl. Phys., 44, 2322-5 (1973), EEA76-23258.
490.Miki, H., Ito, M., Otsubo, M. and Fujibayashi, K., "Impurity profile in a vapor-grown GaAs layer", Electron. & Commun. Jap., 5 (12), 104-10 (1972), PA77-6978.

491. Mitsuhashi, G., "Impurity profile in silicon epitaxial wafer with buried layers", NEC Res. & Dev., no. 36, 68-74 (1975), EEA78-28645.

492. Moest, R. R. and Lassota, D. T., "Carrier-concentration profiles of n-type Sn and Te doped epitaxial GaAs films", J. Electrochem. Soc., 114, 110-12 (1967), EEA70-8477.

493. Muszynski, Z., "The distribution of Al in epitaxial layers $Ga_xAl_{1-x}As$ obtained by liquid phase epitaxy", Elektronika, no. 6, 230-5 (1971), Polish, EEA74-33951.

494. Nanishi, Y., Aoki, T. and Sato, Y., "Liquid phase epitaxial growth of n-GaAs on i-GaAs", Jap. J. Appl. Phys., 13, 387-8 (1974), PA77-29014.

495. Nihira, H., Shirasu, T., Terasaki, T. and Nishizawa, J., "Abnormal impurity distributions in high-purity epitaxial silicon layers", J. Electrochem. Soc., 122, 781-6 (1975), EEA78-33014.

496. Okamoto, H., Sakata, S. and Sakai, K., "Correlation between impurities and Ga vacancies in the vapor phase epitaxial GaAs", Acta Crystallogr. A, A28 (pt. 4 suppl.), S135 (1972), EEA76-12427.

497. Okamoto, H., Sakata, S. and Sakai, K., "Depth profile of concentration of deep-level impurities in vapor-phase epitaxial gallium-arsenide grown under various arsenic vapor pressures", J. Appl. Phys., 44, 1316-26 (1973), EEA76-16290.

498. Peters, R. C. and Vink, A. T., "On the oxygen donor concentration in GaP grown from a solution of GaP in Ga", Gallium Arsenide and Related Compounds, 1974, p. 254-65, Publ. 1975, PA78-54988.

499. Plajner, O., "The distribution of impurities in silicon epitaxial films", TESLA Electron., 2 (4), 117-23 (1969), EEA73-21075.

500. Price, J. B. and Goldman, J., "Junction migration during silicon epitaxial growth", J. Electrochem. Soc., 121, 91C (1974), EEA77-26337.

501. Rausch, H. and Vertesy, M., "Investigation of diffusion boron profiles in epitaxial silicon films", Proceedings of the 5th Colloquium on Microwave Communication, Vol. V. System Measurements, Semiconductor and Magnetic Materials for Microwave Applications and Plasma Phenomena, 1974, p. SM-22/291-300, EEA77-33968.

502. Sato, H. and Yida, S., "A thin GaAs N on N^+ epitaxial film with abrupt interface in carrier concentration profile", Japan. J. Appl. Phys., 9, 156-7 (1970), EEA73-14027.

503. Saul, R. H. and Hackett, W. H., Jr., "Distribution of impurites in Zn, O-doped GaP liquid phase epitaxy layers", J. Electrochem. Soc., 117, 921-4 (1970), EEA73-35064.

504. Stareev, G., Andrelowicz, M. and Piskorski, M., "Impurity profile in tin doped GaAs epitaxial layers", Electron Technol., 6, 125-34 (1973), PA77-49579.

505. Suzuki, T., Ura, M. and Ogawa, T., "Properties of high-purity silicon epitaxial layers", Hitatchi Hyoron, 56 (4), 25-30 (1974), EEA78-537.
506. Tsunoda, Y., "Estimation of impurity profile near the interface in N on N^+ silicon epitaxial layers", Rev. Elec. Commun. Lab., 19, 1038-43 (1971), EEA75-10142.

2. Incorporation
   a. Group IV Semiconductors

507. Bloem, J., "The effect of trace amounts of water vapor on boron doping in epitaxially grown silicon", J. Electrochem. Soc., 118, 1837-41 (1971), EEA75-7405.
508. Bloem, J., "Doping in chemically vapour deposited epitaxial silicon", J. Cryst. Growth, 13-14, 302-5 (1972).
509. Bloem, J., Giling, L. J. and Graef, M. W. M., "The incorporation of phosphorus in silicon epitaxial layer growth", J. Electrochem. Soc., 121, 1354-7 (1974), PA78-1700.
510. Boss, D. W. and Spiro, A., "Epitaxial deposition on a semiconductor substrate", Patent UK 1314149, Publ. April 1973, CA79-36428.
511. Crooke, M., "Some metallurgical aspects of double diffused semiconductor manufacture", Trans. S. Afr. Inst. Eng., 62, 237-47 (1971), EEA75-7425.
512. Dobson, P. S. and Filby, J. D., "Precipitation of boron in silicon", J. Cryst. Growth, 3-4, Spe, 209-13 (1968), EEA72-20214.
513. Francombe, M. H. and Thomas, R. N., "Low-temperature vacuum deposition of homoepitaxial doped silicon", Presented at ACCG Conference on Crystal Growth, Gaitherburg, Md., August 1969, EEA73-28079.
514. Hall, L. H., Koliwad, K. M. and Swink, L. N., "Preparation and properties of boron-doped silicon films grown at low temperature by chemical vapor deposition", Thin Solid Films, 18, 145-55 (1973), EEA77-333.
515. Kulikov, G. S. and Givargizov, E. I., "Diffusion of antimony in epitaxial germanium layers", Sov. Phys. Solid State, 8 (11), 2670-4 (1967), EEA70-13917.
516. Manasevit, H. M., Simpson, W. I. and Erdmann, F. M., "Trimethylstibine as a source of Sb for doping epitaxial Si layers", J. Electrochem. Soc., 121, 967-8 (1974), PA77-70423.
517. Rai-Choudhury, P., "Gallium doped epitaxial silicon", J. Cryst. Growth, 8, 165-71 (1971), EEA74-10488.
518. Rai-Choudhury, P. and Salkovitz, E. I., "Doping of epitaxial silicon. Effect of dopant partial pressure", J. Cryst. Growth, 7, 361-7 (1970), EEA74-4795.
519. Rai-Choudhury, P. and Salkovitz, E. I., "Doping of epitaxial silicon. Equilibrium gas phase and doping mechanism", J. Cryst. Growth, 7, 353-60 (1970), EEA74-4794.

520. Shachev, V. I., "Preparation of doped autoepitaxial layers of silicon", Inorg. Mater., 9, 768-70 (1973), PA77-45445.
521. Shewchun, J. and King, F. D., "Gas doping of vacuum evaporated epitaxial silicon films", Solid-State Electron., 15, 775-81 (1972), EEA75-24250.
522. Sladkov, I. B. and Tuchkevich, V. V., "Epitaxial growth of silicon which was heavily doped with boron", Fiz. & Tekh. Poluprov., 4, 2383-4 (1970), Russian, EEA74-10476.
523. Starodubstev, S. V., Lyutovich, A. S., Kharchenko, V. V. and Prutkin, V. P., "Entry of impurities into epitaxial silicon films during growth", In: Growth of crystals, Consultants Bureau, London, England, p. 264-70 (1969), Russian, EEA74-7614.
524. Swiderski, I., Szezutowski, W. and Niemyski, T., "Preparation of nitrogen-doped and phosphorus-doped epitaxial α-SiC layers on silicon carbide crystals", J. Cryst. Growth, 23, 185-9 (1974), EEA78-554.

b. Group III-V Semiconductors

525. Akita, K., Nakai, S., Kinugasa, T., Kotani, T., Dazai, K. and Ryuzan, O., "GaP liquid phase epitaxial growth using a vertical furnace system (light emitting diode fabrication)", Jap. J. Appl. Phys., 12, 631-5 (1973), EEA76-26549.
526. Andrews, A. M. and Holonyak, N., Jr., "Properties of N-type Ge-doped epitaxial GaAs layers grown from Au-rich melts", Solid State Electron., 15, 601-4 (1972), EEA75-24244.
527. Ashen, D. J., Dean, P. J., Hurle, D. T. J., Mullin, J. B., White, A. M. and Greene, P. D., "Incorporation and characterization of acceptors in epitaxial gallium arsenide", J. Phys. Chem. Solids, 36, 1041-53 (1975), CA83-156715.
528. Cho, A. Y. and Panish, M. B., "Magnesium-doped GaAs and $Al_xGa_{1-x}As$ by molecular beam epitaxy", J. Appl. Phys., 43, 5118-23 (1972), EEA76-12447.
529. Conrad, R. W., Haisty, R. W. and Mehal, E. W., "Method of doping a semiconductor material", Patent USA 3492175, Publ. January 1970, EEA73-21104.
530. DiLorenzo, J. V. and Moore, G. E., Jr., "Effects of the $AsCl_3$ mole fraction on the incorporation of germanium, silicon, selenium, and sulfur into vapor grown epitaxial layers of GaAs", J. Electrochem. Soc., 118, 1823-30 (1971), EEA75-7402.
531. Dyachkova, N. N., Ivanyutin, L. A., Malinin, A. Y., Nichanov, D. and Sabinin, A. G., "Transfer of Se from $AsCl_3$ to an epitaxial layer of GaAs", Inorg. Mater., 9, 696-8 (1973), EEA77-22847.
532. Esaki, L. and McGroddy, J. C., "Incorporation of dopant species in molecular beam epitaxial materials by low energy ion implantation", IBM Tech. Disclosure Bull., 17, 3108-9 (1975), PA78-51106.

533. Harris, J. S. and Snyder, W. I., "Homogeneous solution grown epitaxial GaAs by tin doping", Solid-State Electron., 12, 337-40 (1969), EEA73-7377.
534. Ilegems, M. and Dingle, R., "Acceptor incorporation in GaAs grown by beam epitaxy", Gallium Arsenide and Related Compounds, 1974, p. 1-9, Publ. 1975, EEA78-28659.
535. Jordan, A. S., Trumbore, F. A., Wolfstirn, K. B., Kowalchik, M. and Roccasecca, D. D., "The incorporation of tellurium in liquid phase epitaxial (LPE) GaP: implications for oxygen incorporation", J. Electrochem. Soc., 120, 791-9 (1973), EEA76-29166.
536. Keller, K. and Munch, W. V., "Orientation dependence of the zinc distribution coefficient in the liquid-phase epitaxy of gallium arsenide", Solid-State Electron., 14, 526-8 (1971), EEA74-26618.
537. Kirovskaya, I. A. and Ustalova, T. N., "Adsorption of GaI_3, H_2, and their mixtures on GaAs during epitaxial growth", Inorg. Mater., 9, 1456-8 (1973), PA77-67724.
538. Kordos, P., Jansak, L. and Benc, V., "Preparation and properties of Mn-doped epitaxial gallium arsenide", Solid-State Electron., 18, 223-6 (1975), PA78-30246.
539. Kowalchik, M., Jordan, A. S. and Read, M. H., "Coprecipitation of Ga_2O_3 in the liquid-phase epitaxial growth of GaP", J. Electrochem. Soc., 119, 756-61 (1972), EEA75-28400.
540. Kumagawa, M., Witt, A. F., Lichtensteiger, M. and Gato, H. C., "Current-controlled growth and dopant modulation in liquid phase epitaxy", J. Electrochem. Soc., 120, 583-4 (1973), EEA76-23346.
541. Lavrenteva, L. G., Vilisova, M. D., Kataev, Y. G., Rumyantsev, Y. M. and Shumkov, A. D., "Transition layers in epitaxial gallium arsenide. Effect of the substrate treatment on the distribution of electrons and impurities", Izv. Vyssh. Ucheb. Zaved., Fiz., 16 (5), 138-9 (1973), Russian, CA79-98380.
542. Lawrence, D. J. and Eastman, L. F., "Electric current controlled growth and doping modulation in GaAs liquid phase epitaxy", J. Cryst. Growth, 30, 267-75 (1975).
543. Luther, L. C. and Roccasecca, D. D., "Epitaxial growth of zinc- and cadmium-doped gallium phosphide by gallium chloride vapor transport", J. Electrochem. Soc., 115, 850-6 (1968), EEA71-25875.
544. Manchon, D. D., Jr., Gross, B. J. and Reese, B. A., "Nitrogen incorporation in LPE GaP - a systematic study of temperature dependence above and below saturation", Presented at Electrochemical Society Meeting, San Francisco, Calif., May 1974, PA77-53617.
545. Maruyama, S. and Amano, T., "Liquid phase epitaxy of Si-doped GaAs", Electr. Commun. Lab. Tech. J., 24, 355-61 (1975), Japanese, EEA78-32999.

546.Minden, H. T., Premo, R. and Collins, C. V., "The effect of germanium on the distribution coefficient of aluminium in the system solid $Al_xGa_{1-x}As$ liquid Ga-Al-As-Ge", J. Cryst. Growth, 27, 316-19 (1974), PA78-51120.
547.Naganuma, M. and Takahashi, K., "Ionized Zn doping of GaAs molecular beam epitaxial films", Appl. Phys. Lett., 27, 342-3 (1975).
548.Nakanisi, T. and Kasiwagi, M., "Effects of the growth temperature and substrate orientation on the incorporation of Si, Ge and Sn into vapour epitaxial GaAs", Jap. J. Appl. Phys., 13, 484-94 (1974), EEA77-18709.
549.Otsubo, M. and Miki, H., "Chromium-doped semi-insulating gallium arsenide crystals grown by liquid phase epitaxy", Jap. J. Appl. Phys., 13, 1655-6 (1974), PA78-1708.
550.Panish, M. B., "Phase equilibria in the system Al-Ga-As-Sn and electrical properties of Sn-doped liquid phase epitaxial $Al_xGa_{1-x}As$", J. Appl. Phys., 44, 2667-75 (1973), EEA76-25968.
551.Rosztoczy, F. E., Antypas, G. A. and Casau, C. J., "Distribution coefficients of Ge, Sn, and Te InP grown by liquid phase epitaxy", Proceedings of the Third International Symposium on Gallium Arsenide and Related Compounds, 1970, p. 86-91, Publ. 1971, EEA76-4564.
552.Saul, R. H. and Hackett, W. H., Jr., "On the incorporation of oxygen in GaP liquid phase epitaxy layers", J. Electrochem. Soc., 119, 542-5 (1972), EEA75-19854.
553.Stillman, G. E., Wolfe, C. M. and Dimmock, J. O., "Donor magnetospectroscopy in high purity epitaxial GaAs", Solid State Commun., 7, Pap. No. 5-6 (1969), EEA72-19950.
554.Sudlow, P. D., Mottram, A. and Peaker, A., "Doping gradients in layers of gallium phosphide grown by liquid epitaxy", J. Mater. Sci., 7, 168-75 (1972), EEA75-13559.
555.Volkov, A. S., Galavanov, V. V. and Saimkulov, Z. A., "Solubility and distribution coefficient of Te and Se in epitaxial layers of InSb", Izv. Akad. Nauk SSSR Neorg. Mater., 8, 2033-4 (1972), Russian, EEA76-8530.

IX. Heteroepitaxy

1. Semiconductor on Semiconductor

556.Abdalla, M. I. and Holt, D. B., "The epitaxial growth and structure of films of CdTe evaporated onto Ge", Phys. Status Solidi A, 17, 267-79 (1973), EEA76-25982.
557.Aleksandrov, L. N., Lovyagin, R. N., Pchelyakov, O. P. and Stenin, S. I., "Heteroepitaxy of germanium thin films on silicon by ion sputtering", J. Cryst. Growth, 24-25, 298-301 (1974), PA78-9934.

558. Alekseenko, L. I., Bilenkii, B. F., Grechukh, Z. G. and Savitskii, V. G., "Properties of epitaxial layers of HgTe deposited onto CdTe from the gas phase", Inorg. Mater., 10, 1041-3 (1974), EEA78-37408.
559. Allen, H. A., "The orientation dependence of epitaxial $InAs_xP_{(1-x)}$ on GaAs", J. Electrochem. Soc., 117, 1417-19 (1970), EEA74-7577.
560. Antypas, G. A. and Moon, R. L., "Growth and characterization of InP-InGaAsP lattice-matched heterojunctions", J. Electrochem. Soc., 120, 1574-7 (1973), EEA77-2464.
561. Antypas, G. A., Rosztoczy, F. E., Stein, W. W. and von Szeremy, S., "Preparation of GaP-Si heterojunction by liquid phase epitaxy", Patent USA 3810794, Publ. May 1974.
562. Aoki, M. and Kasano, H., "Studies on vapour growth of GaP. II. Epitaxial growth on the Ge substrate", Annu. Rep. Eng. Res. Inst. Fac. Eng. Univ. Tokyo, 28, 97-103 (1969), Japanese, EEA74-4779.
563. Avigal, Y., Schieber, M. and Levin, R., "The growth of hetero-epitaxial SiC films by pyrolysis of various alkyl-silicon compounds", J. Cryst. Growth, 24-25, 188-92 (1974), PA78-10534.
564. Bachmann, K. J., Buehler, E., Kammlott, G. W. and Shay, J. L., "The preparation of $CdSnP_2$/InP heterojunctions by liquid phase epitaxy from Sn-solution", J. Electron. Mater., 3, 451-73 (1974), PA77-57708.
565. Balch, J. W., "Epitaxial growth of gallium arsenide on zinc selenide (diode characteristics)", Thesis, Ohio State Univ., Columbus, Ohio, 1971, 127 pp., Order No. 75-4415, EEA72-28420.
566. Baliga, B. J. and Ghandhi, S. K., "Heteroepitaxial InAs grown on GaAs from triethylindium and arsine. I. Growth characterization", J. Electrochem. Soc., 121, 1642-6 (1974), PA78-19087.
567. Baliga, B. J. and Ghandhi, S. K., "Growth and properties of heteroepitaxial GaInAs alloys on GaAs substrates using trimethylgallium, triethylindium and arsine", J. Electrochem. Soc., 122, 683-7 (1975), EEA78-24199.
568. Berkenblit, M., Reisman, A. and Light, T. B., "Epitaxial growth of mirror smooth Ge on GaAs and Ge by the low temperature GeI_2 disproportionation reaction", J. Electrochem. Soc., 115, 966-9 (1968), EEA71-28747.
569. Bielis, I. Y. and Patmalnieks, A. A., "On possibilities of liquid-phase epitaxy by zone-melting", Latv. PSR Zinat. Akad. Vestis Fiz. Teh. Ser., no 2, 12-17 (1973), Russian, EEA76-32187.
570. Bolkhovityanov, Y. B. and Zembatov, K. B., "Preparation of solid solutions $Al_xGa_{1-x}As$ on gallium phosphide substrates by the liquid epitaxy method", Izv. VUZ Fiz., no. 3, 146-7 (1974), Russian, PA77-45439.
571. Bougnot, G., Ettienne, D., Chevrier, J. and Bohe, C., "Epitaxial growth of zinc selenide on gallium arsenide", Mater. Res. Bull., 6 (3), 145-51 (1971), French, EEA74-18595.

572.Buehler, E., Bachmann, K. J., Kammlott, G. W. and Shay, J. L., "Phase relations, crystal growth and heteroepitaxy in the quaternary system Cd, Sn, In and P", J. Cryst. Growth, 24-25, 260-5 (1975), PA78-7476.
573.Burnham, R. D., Holonyak, N., Jr., Korb, H. W., Macksey, H. M., Scifres, D. R., Woodhouse, J. B. and Alferov, Z. I., "AlGaAsP solid solutions and their use in two-heterojunction injection lasers", Sov. Phys.-Semicond., 6, 77-81 (1972), EEA75-32396.
574.Butter, E., Jacobs, B. and Stary, J., "LPE-growth of InAs on GaAs substrates", Phys. Status Solidi A, 26, K105-7 (1974), PA78-29524.
575.Chan, W. S., "Vacuum epitaxial growth of $Pb_{1-x}Sn_xTe$ films on cleaved BaF_2 substrates", Infrared Phys., 14 (3), 177-81 (1974), EEA77-36219.
576.Chang, L. L., Esaki, L., Howard, W. E. and Ludeke, R., "The growth of a GaAs-GaAlAs superlattice", J. Vac. Sci. & Technol., 10, 11-16 (1973), EEA76-23355.
577.Chevrier, J., Etienne, D., Camassel, J., Auvergne, D., Pons, J. C., Mathieu, H. and Bougnot, G., "Hetero-epitaxy of ZnSe on GaAs by hydrogen transport (and luminescence centres)", Mater. Res. Bull., 7, 1485-92 (1972), EEA76-12459.
578.Chevrier, J., Galibert, G., Etienne, D. and Bougnot, G., "Heteroepitaxy of ZnSe on GaAs by open tube transport", J. Cryst. Growth, 28, 109-16 (1975), French, PA78-26320.
579.Cho, A. Y., Panish, M. B. and Hayashi, I., "Molecular beam epitaxy of GaAs, $Al_xGa_{1-x}As$ and GaP", Proceedings of the Third International Symposium on Gallium Arsenide and Related Compounds, 1970, p. 18-29, Publ. 1971, EEA76-4560.
580.Christmann, M. H., Jones, K. A. and Olsen, K. H., "Formation of hexagonal pyramids and hexagonal flat tops on the surface of heteroepitaxial (0001) CdS films", J. Appl. Phys., 45, 4295-302 (1974), EEA78-8446.
581.Cucmo, J. J. and Gambino, R. J., "The synthesis and epitaxial growth of GaP by fused salt electrolysis", J. Electrochem. Soc., 115, 755-9 (1968), EEA71-28745.
582.Dawson, L. R., "Near-equilibrium LPE growth of GaAs-$Ga_{1-x}Al_xAs$ double heterostructures", J. Cryst. Growth, 27, 86-96 (1974), PA78-30238.
583.Dingle, R., Gossard, A. C. and Wiegmann, W., "Direct observation of superlattice formation in a semiconductor heterostructure", Phys. Rev. Lett., 34, 1327-30 (1975), PA78-72720.
584.Donnelly, J. P. and Milnes, A. G., "The epitaxial growth of Ge on Si by solution growth techniques", J. Electrochem. Soc., 113, 297-8 (1966), EEA70-516.
585.Duh, K. and Preier, H., "Properties of $PbS_{1-x}Se_x$ epilayers deposited onto PbS substrates by hot-wall epitaxy", J. Mater. Sci., 10, 1360-6 (1975), EEA78-41484.
586.Duh, K. and Preier, H., "PbSe heteroepitaxy by the hot wall technique", Thin Solid Films, 27, 247-50 (1975), EEA78-28657.

587.Espinosa, G. P., Lim, T. C., Cory, E. S., Zucca, R. and Hengstenberg, D. H., "Mutual doping effects between epitaxial ZnS films and GaAs substrates", Appl. Phys. Lett., 26, 514-16 (1975), PA78-61937.
588.Ettenberg, M., "Epitaxially depositing a semiconductor material on a substrate", Patent USA 3821039, Publ. June 1974, CA81-70136.
589.Ettenberg, M. and McFarlane, S. H., III, "Effect of substrate preparation on the smoothness of liquid phase epitaxial (AlGa)As on GaP", J. Cryst. Growth, 23, 233-6 (1974), PA78-1707.
590.Fujita, S., Itoh, K., Arai, S. and Sakaguchi, T., "Epitaxial solution growth of ZnTe on ZnSe", Jap. J. Appl. Phys., 10, 516-17 (1971), EEA74-18585.
591.Fujita, S., Moriai, F., Arai, S. and Sakaguchi, T., "Preparation of ZnSe-ZnTe heterojunctions by liquid-phase epitaxial growth", Jap. J. Appl. Phys., 12, 1841-9 (1973), PA77-17182.
592.Furukawa, Y., "Epitaxial growth of gallium arsenide by using silicon tetrachloride", Japan. J. Appl. Phys., 6 (11), 1344 (1967), EEA71-10511.
593.Gavand, M., Mayet, L., Montegu, B. and Castet, L., "Metallurgical aspects of liquid phase epitaxy of InSb on CdTe", J. Cryst. Growth, 19, 310-16 (1973), French, EEA76-35222.
594.Gejji, F. H. and Holt, D. B., "Epitaxial growth and structure of CdSe evaporated in vacuum onto Ge", J. Electrochem. Soc., 122, 535-41 (1975), EEA78-24189.
595.Hill, J. S. and Simpson, G. N., "A technique for the epitaxial growth of zinc sulphide on silicon", J. Phys. E, 6, 2-9-302 (1973), EEA76-16295.
596.Hohnke, D. K. and Kaiser, S. W., "Epitaxial PbSe and $Pb_{1-x}Sn_xSe$: growth and electrical properties", J. Appl. Phys., 45, 892-7 (1974), EEA77-26326.
597.Holloway, H., Kohnke, D. K. and Logothetis, E. M., "Epitaxial growth of small bandgap semiconductors", J. Vac. Sci. & Technol., 8, 146 (1971), EEA74-13179.
598.Holt, D. B., "Comment on 'The epitaxial growth of sphalerite and wurtzite ZnSe on a (111) Ge surface'", J. Appl. Phys., 45, 966-7 (1974), PA77-53615.
599.Holt, D. B. and Abdalla, M. I., "Epitaxial growth and structure of films of CdTe evaporated in vacuum on to silicon", Phys. Status Solidi A, 26, 507-12 (1974), PA78-15610.
600.Huber, H. and Winstel, G. H., "Growth of heteroepitaxial GaP on Si substrates by a chloride transport process", Siemens Forsch.- & Entwicklungsber., 2, 171-4 (1973), EEA76-32196.
601.Igarashi, O., "Heteroepitaxial growth of GaP on Si substrates by evaporation method", J. Appl. Phys., 41, 3190-2 (1970), EEA73-31238.
602.Igarashi, O., "Selective growth of heteroepitaxial GaP on Si substrates", J. Electrochem. Soc., 119, 1430-1 (1972), PA75-83864.
603.Ito, K. and Iida, S., "Epitaxial growth of semiconductor single crystals", Patent Japan 73-00281, Publ. January 1973, CA79-36429.

604.Ito, K. and Takahashi, K., "Epitaxial growth of Ge layers on Si substrates by vacuum evaporation", Jap. J. Appl. Phys., 7, 821-6 (1968), EEA72-722.
605.Jackson, D. M., Jr. and Howard, R. W., "Epitaxial growth of germanium", Patent USA 3473978, Publ. October 1969, EEA73-24870.
606.Jain, F. C., "Low-temperature epitaxially-grown Ge-GaAs and Ge-ZnSe heterojunctions", Thesis, Univ. Connecticut, Storrs, 1973, 121 pp., Order No. 74-21817, EEA78-19695.
607.Jones, P. L., Litting, C. N. W., Mason, D. E. and Williams, V. A., "The epitaxial growth of the zinc sulphide on silicon by vacuum evaporation", Brit. J. Appl. Phys. (J. Phys. D), 1, 283-9 (1968), EEA71-15684.
608.Kamadjiev, P. R., Mladjov, L. K., Gospodinov, M. M. and Vassilev, I. S., "Preparation, morphology and structure of cadmium telluride epitaxial layers on germanium substrates", Krist. & Tech., 9, 1249-54 (1974), PA78-26329.
609.Kennedy, J. K. and Potter, W. D., "Epitaxial deposition of degenerate n-type layers of Ge on GaAs", J. Cryst. Growth, 13-14, 315-18 (1972).
610.Kocsis, Z. and Gorog, T., "Growth and electrical properties of heteroepitaxial GaP layers", Kozl. Magy. Tud. Akad. Musz. Fiz., Kut. Intez., no. 0-14, 47-64 (1974), Hungarian, PA78-10544.
611.Kopf, L. and Sumski, S., "Reduction of Al contamination in the GaAs layer of LPE-grown $Al_xGa_{1-x}As$-GaAs heterostructures", J. Cryst. Growth, 28, 365-6 (1975), PA78-54962.
612.Kotelyanskiy, Y. M., Mityagin, A. Y. and Orlov, V. P., "Growth and crystal structures in heteroepitaxial CdS deposited on GaP", Proceedings of the International Conference on the Physics and Chemistry of Semiconductor Heterojunctions and Layer Structures, 1970, p. 231-8, Publ. 1971, EEA74-22562.
613.Kuroiwa, K. and Sugano, T., "Vapor-phase deposition of beta-silicon carbide on silicon substrates", J. Electrochem. Soc., 120, 138-40 (1973), EEA76-12457.
614.Lever, R. F. and Huminski, E. J., "Epitaxy of germanium films on gallium arsenide by vacuum evaporation", J. Appl. Phys., 37, 3638-9 (1966), EEA70-523.
615.Lilley, P., Jones, P. L. and Litting, C. N. W., "The epitaxial growth of zinc sulphide on silicon by forced vapour transport in argon flow", J. Cryst. Growth, 13-14, 371-4 (1972).
616.Lilley, P., Kay, P. M. R. and Litting, C. N. W., "The vapour phase deposition of thick epitaxial (100) ZnS layers on elemental and compound substrates in H_2 gas flow", J. Mater. Sci., 10, 1317-22 (1975), EEA78-41483.
617.Lockwood, H. F. and Ettenberg, M., "Thin solution multiple layer epitaxy, (Semiconductor heterojunction preparation)", J. Cryst. Growth, 15, 81-3 (1972).

618.Ludeke, R. and Paul, W., "Growth and optical properties of epitaxial thin films of some II-VI compounds", II-VI Semiconducting Compounds 1967 International Conference, p. 123-36, EEA73-4343.
619.Matare, H. F., "Heteroepitaxy", Scientia Electica, 15 (3), 95-109 (1969), EEA73-21068.
620.Matare, H. F., "Heteroepitaxy", Scientia Electrica, 15 (4), 129-45 (1969), EEA73-35077.
621.Moriizumi, T. and Takahashi, K., "Lattice match in the heteroepitaxy of 3-5 compound alloys", J. Appl. Phys., 43, 4606-8 (1972), EEA76-4509.
622.Morimoto, Y., Uchiho, K. and Ushio, S., "Vapor phase epitaxial growth of GaN on GaAs, GaP, Si, and sapphire substrates from $GaBr_3$ and NH_3", J. Electrochem. Soc., 120, 1783-5 (1973), PA77-17181.
623.Moulin, M., Huber, A. and Dugue, M., "Contamination of CdS heteroepitaxial layers during vapour growth on GaAs substrates", J. Cryst. Growth, 17, 212-17 (1972), EEA76-4529.
624.Nagai, H. and Noguchi, Y., "A new grading layer for liquid epitaxial growth of $Ga_xIn_{1-x}As$ on a GaAs substrate", Appl. Phys. Lett., 26, 108-10 (1975), PA78-33963.
625.Nakamura, M., Aiki, K., Umeda, J., Yariv, A., Yen, H. W. and Morikawa, T., "Liquid phase epitaxy of GaAlAs on GaAs substrates with fine surface corrugations", Appl. Phys. Lett., 24, 466-8 (1974), PA77-61290.
626.Ohki, Y., Asao, I. and Akasaki, I., "Vapor phase epitaxial growth of nitrogen-doped $In_{1-x}Ga_xP$ alloys", J. Cryst. Growth, 24-25, 244-8 (1974), PA78-7474.
627.Oleszek, G. M. and Anderson, R. L., "Heteroepitaxial growth of Si_xGe_{1-x} alloys by thermal decomposition of SiH_4 and GeH_4", J. Electrochem. Soc., 120, 554-9 (1973), EEA76-23343.
628.Olshanetskii, B. Z. and Repindkii, S. M., "Study of the heteroepitaxial growth of silicon on an atomically pure germanium surface (111)", Sov. Phys.-Solid State, 16, 1104-7 (1974), EEA78-19688.
629.Ota, T., Kanamori, A. and Takahashi, K., "Heterodiodes of ZnTe thin films on GaAs, InAs, and GaP single crystals prepared by vacuum-deposition epitaxy", Phys. Status Solidi A, 17, K5-7 (1973), EEA76-26047.
630.Papazian, S. A. and Reisman, A., "Epitaxial deposition of germanium onto semi-insulating GaAs", J. Electrochem. Soc., 115, 961-5 (1968), EEA71-28749.
631.Parker, S. G., "Single crystals and epitaxial films of ZnSe by chemical transport", J. Cryst. Growth, 9, 177-82 (1971), EEA74-18575.
632.Popov, V. P. and Pamplin, B. R., "Epitaxial growth of solid solutions of $ZnSiP_2$ in Si", J. Cryst. Growth, 15, 129-32 (1972), EEA75-32515.

633. Richter, K., "On the early deposition of GaP onto GaAs", J. Cryst. Growth, 17, 207-11 (1972), EEA76-4528.
634. Rode, D. L. and Sobers, R. G., "Growth of AlGaAs-GaAs heterostructures from step-cooled solutions (for laser)", J. Cryst. Growth, 29, 61-4 (1975), EEA78-33023.
635. Rosztoczy, F. E. and Stein, W. W., "The growth of Ge-GaAs and GaP-Si heterojunctions by liquid phase epitaxy", J. Electrochem. Soc., 119, 1119-21 (1972), EEA75-36151.
636. Rozgonyi, G. A. and Panish, M. B., "Stress compensation in $Ga_{1-x}Al_xAs_{1-y}P_y$ LPE layers on GaAs substrates", Appl. Phys. Lett., 23, 533-5 (1973), PA77-5869.
637. Sharma, B. L. and Mukerjee, S. N., "(p)GaP-(n)GaP-(p)GaAs heterostructure", Phys. Status Solidi A, 29, K141-3 (1975), EEA78-33085.
638. Shay, J. L., Bachman, K. J. and Buehler, E., "Preparation and properties of $CdSnP_2$/InP heterojunctions grown by LPE from Sn solution", J. Appl. Phys., 45, 1302-10 (1974), PA77-61304.
639. Shumskii, V. N., Aseev, A. L., Pogorlev, Y. N. and Stenin, S. I., "Structural characteristics of Ge-GaAs heterojunctions obtained by various different methods", Izv. Akad. Nauk SSSR Ser. Fiz., 38, 1522-7 (1974), Russian, PA77-73258.
640. Sleger, K. J. and Milnes, A. G., "The growth and electrical characteristics of epitaxial layers of zinc sulphide and of zinc selenide on p-type gallium phosphide", Int. J. Electron., 33, 565-81 (1972), EEA75-39808.
641. Sugiyama, K. and Saito, H., "GaAsSb-AlGaAsSb double heterojunction lasers, (Preparation by liquid phase epitaxy)", Jap. J. Appl. Phys., 11, 1057-8 (1972), EEA75-32393.
642. Takakura, H., Kitamura, H. and Hamakawa, Y., "Epitaxial vapor growth of $GaAs_{1-x}P_x$ on germanium substrates and their optoelectronic properties", J. Cryst. Growth, 24-25, 249-52 (1974), PA78-7475.
643. Takenaka, T. and Shohno, K., "Double-layer epitaxial growth of Si and $B_{13}P_2$n Si substrates and some electrical properties of Si layers", Jap. J. Appl. Phys., 13, 1211-15 (1974), PA77-70426.
644. Tamura, T., Moriizumi, T. and Takahashi, K., "ZnTe-InAs Heterojunctions prepared by liquid-phase epitaxy", Jap. J. Appl. Phys., 11, 1024-31 (1972), EEA75-28461.
645. Tao, T. F. and Wang, C. C., "Epitaxial growth of $Pb_{0.918}Sn_{0.082}$ Se films on CaF_2 and BaF_2 substrates", J. Appl. Phys., 43, 1313-16 (1972), EEA75-13547.
646. Ugai, Y. A., Kavetskii, V. S., Dzampaev, K. S. and Yatsenko, N. G., "Production of epitaxial indium phosphide films on gallium arsenide substrates", Protseesy Rosta Sint. Poluprovodn. Krist. Plenok, Mater. Vses. Simp., 34d, 1972, p. 1, 247-50, Russian, Aleksandrov, L. N. and Kuznetsov, F. A. (Eds.), "Nauka", Sib. Otd.: Novosibirsk, USSR, 1975, CA83-211394.

647. van Oirschot, T. G. J., "Liquid-phase epitaxial growth of (AlGa)As on polished and roughened GaP substrates for transmission photocathodes", Appl. Phys. Lett., 24, 211-13 (1974), PA77-41498.
648. van Oirschot, T. G. J., Acket, G. A. and Bartels, W. J., "LPE growth and properties of GaP-(AlGa)As-GaAs heterostructures as a function of the GaP substrate roughness", J. Appl. Phys., 46, 1893-1900 (1975), PA78-58817.
649. Vanyukov, A. V., Krotov, I. I. and Mikolyuk, E. A., "Epitaxial growth of CdTe from the elements in a current of hydrogen", Inorg. Mater., 9, 1334-8 (1973), PA77-70418.
650. Vohl, P., Buchan, W. R. and Genthe, J. E., "Thick epitaxial films of cubic ZnSe and ZnS by vapor phase transport", J. Electrochem. Soc., 118, 1842-7 (1971), EEA75-7406.
651. Wickenden, D. K., Faulkner, K. R., Brander, R. W. and Isherwood, B. J., "Growth of epitaxial layers of gallium nitride on silicon carbide and corundum substrates", J. Cryst. Growth, 9, 158-64 (1971), EEA74-18572.
652. Yabe, M., "Heteroepitaxial growth of trigonal selenium crystal on the (1010) cleavage surface of tellurium crystal", Jap. J. Appl. Phys., 12, 1317-22 (1973), EEA77-317.
653. Yamamoto, A. and Yamaguchi, M., "Epitaxial growth of ZnSe on Ge by fused salt electrolysis", Jap. J. Appl. Phys., 14, 561-2 (1975), EEA78-24209.
654. Yoshikawa, A. and Sakai, Y., "Growth and properties of CdS epitaxial layers by the close-spaced technique", J. Appl. Phys., 45, 3521-9 (1974), EEA78-541.

2. Semiconductor on Insulator

655. Aksenova, L. L., Distler, G. I., Kovda, A. V. and Kortukova, E. I., "Oriented crystallization of CdS", Sov. Phys.-Solid State, 15, 1100 (1973), PA77-28313.
656. Aoki, M., Matsumoto, T. and Sano, M., "Crystal growth and optical properties of GaN", Annu. Rep. Eng. Res. Inst. Fac. Eng. Univ. Tokyo, 32, 115-20 (1973), Japanese, PA77-45438.
657. Batijewski, R. and Herman, M. A., "Some problems concerning the kinetics of crystallization of vapor phase epitaxial GaN deposited on sapphire substrates", Acta Phys. Pol. A, A47, 563-7 (1975), PA78-46503.
658. Bozler, C. O., "Indium arsenide films grown on aluminum oxide ceramic", Solid-State Electron., 17, 251-5 (1974), PA77-41509.
659. Bunton, G. V. and Day, S. C. M., "Epitaxial thin films of ZnS and GaAs prepared by r.f. sputtering on NaCl substrates", Thin Solid Films, 10, 11-20 (1972), EEA75-16655.
660. Chiang, Y. S. and Looney, G. W., "The epitaxial growth and properties of silicon on stoichiometric spinel and sapphire substrates using silane-helium mixtures vs. silane-hydrogen mixtures", J. Electrochem. Soc., 120, 550-3 (1973), EEA76-23342.

661.Corsi, C., Alfieri, I. and Petrocco, G., "$Pb_xSn_{1-x}Te$ epitaxial layers by r.f. multicathode sputtering", Appl. Phys. Lett., 24, 484-5 (1974), PA77-61291.

662.Cullen, G. W., "The preparation and properties of chemically vapor deposited silicon on sapphire and spinel", J. Cryst. Growth, 9, 107-25 (1971), EEA74-18566.

663.Cullen, G. W., Gottlieb, G. E. and Wang, C. C., "The epitaxial growth of silicon on sapphire and spinel substrates: suppression of changes in the film properties during device processing", RCA Rev., 31, 355-71 (1970), EEA74-4825.

664.Dexler, R. J., Watelski, S. B. and Picraux, S. T., "Epitaxial silicon layers grown on ion-implanted silicon nitride layers", Appl. Phys. Lett., 23, 455-7 (1973), EEA77-5868.

665.Druminski, M. and Schlotterer, H., "The combination of two growth methods of the epitaxial deposition of silicon films on insulating substrates", J. Cryst. Growth, 17, 249-53 (1972), EEA76-4534.

666.Duffy, M. T., Wang, C. C., O'Clock, G. D., Jr., McFarlane, S. H., III and Zanzucchi, P. J., "Epitaxial growth and piezoelectric properties of AlN, GaN, and GaAs on sapphire or spinel", J. Electron. Mater., 2, 359-72 (1973), EEA77-324.

667.Dumin, D. J., "Hydride sources for diffusion of dopants into silicon-on-sapphire films", J. Electrochem. Soc., 116, 133-7 (1969), EEA73-1074.

668.Dunning, W. J., Fox, P. G. and Parker, D. W., "Observations on the epitaxial growth of ammonium iodide on mica substrates", J. Phys. Chem. Solids, 28, 509-13 (1967), EEA70-17283.

669.Dumin, D. J., Robinson, P. H., Cullen, G. W. and Gottlieb, G. E., "Heteroepitaxial growth of germanium and silicon on insulating substrates", RCA Rev., 31, 620-34 (1970), EEA74-13186.

670.Egerton, R. F. and Juhasz, C., "Epitaxial films of PbTe, PbSe and PbS grown on mica substrates", Brit. J. Appl. Phys., 18, 1009 et sec. (1967), EEA71-1139.

671.Farinre, T. O., "Preparation and properties of $Pb_{1-x}Sn_xTe$ epitaxial films", J. Vacuum Sci. Technol., 7, 121-6 (1970), EEA73-31247.

672.Filby, J. D. and Nielsen, S., "Single-crystal films of silicon on insulators", Brit. J. Appl. Phys., 18, 1357-82 (1967), EEA71-8594.

673.Freik, D. M., Voitkiv, V. V., Gaiduchok, G. M., Maksimyak, L. B. and Klyuchevskii, P. P., "Epitaxial growth of films of $(SnTe)_{1-x}(PbSe)_x$ on mica", Inorg. Mater., 10, 22-4 (1974), EEA78-8437.

674.Galli, G. and Coker, J. E., "Epitaxial ZnO on sapphire", Appl. Phys. Lett., 16, 439-41 (1970), EEA73-28012.

675.Gassmann, F., Della Casa, A. and Heschlimann, R., "Epitaxial silicon films on spinel by evaporation in ultra-high vacuum", Mater. Res. Bull., 6, 817-22 (1971), German, EEA74-37150.

676.Goswami, A. and Singh, P., "Epitaxial growth of CoTe films", Indian J. Pure & Appl. Phys., 11, 373-4 (1973), PA77-20493.

677. Gutierrez, W. A., Pommerrenig, H. D., Jasper, M. A. and Mantzouranis, A. P., "Epitaxial growth of GaAs on insulating substrates using HCl-H_2 vapor transport", Solid-State Electron., 13, 1199-200 (1970), EEA73-35083.

678. Haldinger, W., Zega, B., Steiner, H. and Pongratz, M., "Selective growth of Si single crystals on to quartz and sapphire", Proceedings of the International Conference on the Physics and Chemistry of Semiconductor Heterojunctions and Layer Structures, 1970, p. 103-14, Publ. 1971, EEA74-26626.

679. Harman, R. and Urmin, J., "Evaporation of GaAs thin films on sapphire substrates", Presented at Proceedings of the 5th Czechoslovakia Conference on Electronics and Vacuum Physics, Brno, Czechoslovakia, October 1972, EEA76-8569.

680. Hawrylo, F. Z. and Pankove, J. I., "Method of epitaxially depositing gallium nitride from the liquid phase", Patent USA 3811963, Publ. May 1974.

681. Holt, D. B., "Epitaxial films of ZnTe on ionic substrates", Brit. J. Appl. Phys., 17, 1395-9 (1966), EEA70-4093.

682. Hutchinson, T. E. and Olsen, K. H., "Substrate-condensate chemical interaction and the vapor deposition of epitaxial niobium films", J. Appl. Phys., 38, 4933-7 (1967), EEA71-5821.

683. Ilegems, M., "Vapour epitaxy of gallium nitride", J. Cryst. Growth, 13-14, 360-4 (1972).

684. Jackson, D. M., Jr. and Noiling, J. A., "Semiconductor growth on dielectric substrates", Patent USA 3496037, Publ. February 1970, EEA73-24872.

685. Kasai, I., Hornung, J. and Baars, J., "$Pb_{1-x}Sn_xTe$ epitaxial layers prepared by the hot-wall technique", J. Electron. Mater., 4, 299-311 (1975), EEA78-24213.

686. Kenty, J. L., "Physical vapor deposition of GaAs on single crystal sapphire", J. Electron. Mater., 2, 239-54 (1973), EEA77-320.

687. Khan, I. H. and Learn, A. J., "Formation of epitaxial β-SiC films on sapphire", Appl. Phys. Lett., 15, 410-14 (1969), EEA73-13966.

688. Kovda, A. V., Aksyonova, L. L., Distler, G. I., Gerasimov, Y. M. and Kortukova, E. I., "Growth of monocrystalline CdS films on mica surfaces through amorphous interfacial layers of silicon monoxide and carbon by chemical transport reactions", Thin Solid Films, 20, 217-22 (1974), PA77-29023.

689. Kroko, L. J. and Shaw, G. L., "The mechanism of silicon vapor deposition on sapphire", In: Semiconductor silicon, Electrochem. Soc., Inc., New York, p. 316-28 (1969), EEA73-10426.

690. Kuhn, G. L. and Rhee, C. J., "Thin silicon film on insulating substrate", J. Electrochem. Soc., 120, 1563-6 (1973), EEA77-2461.

691. Logan, R. A. and Thurmond, C. D., "Heteroepitaxial thermal gradient solution growth of GaN", J. Electrochem. Soc., 119, 1727-35 (1972), EEA76-8532.

692.Lowney, J. R. and Cammack, D. A., "A new technique for growing epitaxial films of $Pb_{1-x}Sn_xTe$", Mater. Res. Bull., 9, 1639-46 (1974), PA78-22499.
693.Manasevit, H. M., Erdmann, F. M. and Thorsen, A. C., "The preparation and properties of (111) Si films on sapphire", Electrochemical Society Fall Meeting (Extended abstracts), 1974, p. 332-3, EEA78-37433.
694.Manasevit, H. M., Forbes, D. H. and Cadoff, I. B., "Growth of single-crystal silicon on beryllium oxide", Trans. Metall. Soc. AIME, 236, 275-80 (1966), EEA70-4054.
695.Manasevit, H. M. and Simpson, W. I., "The use of metal-organics in the preparation of semiconductor materials on insulating substrates. I. Epitaxial III-V gallium compounds", Presented at 133rd Meeting of the Electrochemical Society (abstracts), Boston, Mass., May 1968, EEA73-14082.
696.Manasevit, H. M. and Thorsen, A. C., "Heteroepitaxial GaAs on aluminum oxide. The formation and electrical properties of Zn- and Cd-doped films", J. Electrochem. Soc., 119, 99-103 (1972), EEA75-13551.
697.Matare, H. F., "Heteroepitaxy of silicon on insulator crystal substrates", In: Semiconductor silicon, Electrochem. Soc., Inc., New York, p. 249-90 (1969), EEA73-10423.
698.Mercier, J., "Origin and nature of involuntary contamination of epitaxial silicon layers on a substrate of corindon (Al_2O_3-α)", Rev. Phys. Appl., 3, 127-30 (1968), French, EEA71-28753.
699.Kalinkin, I. P., Aleskovsky, V. B. and Bogomotov, N. S., "Growth and electrophysical properties of monocrystalline films of cadmium and zinc chalcogenides", Thin Solid Films, 5, 7-14 (1970), EEA73-28055.
700.Namba, S., Kawazu, A. and Maruyama, T., "Evaporated silicon single crystal films on vacuum sapphire", Finommechanika, 7 (2), 45-9 (1968), Hungarian, EEA71-28748.
701.Nickl, J. J., Just, W. and Bertinger, R., "Preparation of epitaxial gallium nitride", Mater. Res. Bull., 9, 1413-20 (1974), EEA78-2738.
702.O'Neal, J. E. and Leonhard, F. W., "Epitaxy of GaAs on vacuum-cleaved rocksalt", J. Cryst. Growth, 2, 80-4 (1968), EEA71-18081.
703.Price, J. B. and Davidson, U. S., "Manufacture of thin single-crystal silicon layers on insulating substrates", Patent German 2401380, Publ. July 1974, CA81-161230.
704.Rakova, E. V. and Semiletov, S. A., "Some special features of the annealing of epitaxial layers of lead telluride on mica in tellurium vapour", Izv. Akad. Nauk SSSR Ser. Fiz., 38, 2371-6 (1974), Russian, PA78-22485.
705.Ratcheva, T. M. and Dragieva, I. D., "The epitaxial growth of ZnSe on sapphire substrates", Phys. Status Solidi A, 29, 579-85 (1975), PA78-62565.

706. Ratcheva-Stambolieva, T. M., Djoglev, D. H., Tchistyakov, Y. D., Krasulin, G. A. and Vanyukov, A. V., "Growth of epitaxial CdSe upon sapphire", Phys. Status Solidi A, 16, 315-22 (1973), EEA76-19842.

707. Ratcheva-Stambolieva, T. M., Dragieva, I. D., Djoglev, D. H., Tchistyakov, Y. D. and Krasulin, G. A., "Preparation of CdSe epitaxial layers by the vapour transport method", Phys. Status Solidi A, 21, 703-10 (1974), PA77-29020.

708. Robinson, P. H. and Mueller, C. W., "The deposition of silicon upon sapphire substrates", Trans Metall. Soc. AIME, 236, 268-74 (1966), EEA70-4055.

709. Scott, B. A., Nichols, K. H., Potemski, R. M. and Woodall, J. M., "Magnesium gallate spinel: a substrate for the direct liquid-phase epitaxial growth of (Ga,Al)As", Appl. Phys. Lett., 21, 121-2 (1972), EEA75-36111.

710. Seifert, W. and Tempel, A., "The epitaxial growth of gallium nitride on sapphire from $GaCl/HN_3/Ar$ system", Krist. & Tech., 9, 1213-21 (1974), German, EEA78-11502.

711. Seifert, W. and Tempel, A., "Cubic phase gallium nitride by chemical vapour deposition", Phys. Status Solidi A, 23, K39-40 (1974), PA77-53633.

712. Stepchenkova, O. V., Chistyakov, Y. D., Andreev, V. M., Galikin, B. D., Markova, T. I. and Pechkina, L. D., "Initial growth stage of heteroepitaxial layers, of gallium arsenide on sapphire", Sov. Phys.-Crystallogr., 19, 418-19 (1974), PA78-46507.

713. Sumner, G. G. and Reynolds, L. L., "Preparation and properties of single-crystal lead telluride films", J. Vac. Sci. Technol., 6, 493-7 (1969), EEA73-14028.

714. Szomody, K., "Preparation of samples of epitaxial silicon films for measurements by the Van der Pauw method", Instrum. & Exp. Tech., 16, 1859-60 (1972), EEA76-38186.

715. Tramposch, R. F., "Epitaxial films of germanium deposited on sapphire via chemical vapor transport", J. Electrochem. Soc., 116, 654-8 (1969), EEA73-4306.

716. Unvala, B. A., Woodcock, J. M. and Holt, D. B., "Epitaxial growth of cubic ZnS by evaporation in ultra-high vacuum", Brit. J. App. Phys. (J. Phys. D), 1, 11-14 (1968), EEA71-13177.

717. Wang, C. C., "Growth and properties of GaAs on spinel", Presented at Electrochemical Society Meeting, Chicago, Ill., May 1973, EEA76-23289.

718. Wang, C. C., Dougherty, F. C., Zanzucchi, P. J. and McFarlane, S. H., III, "Epitaxial growth and properties of GaAs on magnesium aluminate spinel", J. Electrochem. Soc., 121, 571-82 (1974), PA77-45472.

719. Wang, C. C., Ladany, I., McFarlane, S. H., III and Dougherty, F. C., "Two-stage epitaxial growth of GaP on spinel", J. Cryst. Growth, 24-25, 239-43 (1974), EEA78-2727.

720.Wang, C. C. and McFarlane, S. H., III, "Epitaxial growth and characterization of GaP on insulating substrates", J. Cryst. Growth, 13-14, 262-7 (1972), PA75-47111.
721.Yasuda, Y., "Epitaxial growth of silicon films on sapphire and spinel by vacuum evaporation", Jap. J. Appl. Phys., 10, 45-51 (1971), EEA74-13177.
722.Yasuda, Y. and Ohmura, Y., "Epitaxial growth of silicon films evaporated on sapphire", Japan J. Appl. Phys., 8, 1098-106 (1969), EEA73-7356.
723.Yezhovsky, Y. K. and Kalinkin, I. P., "Epitaxy of cadmium telluride in a quasi-closed cell", Thin Solid Films, 18, 127-36 (1973), EEA77-332.
724.Yim, W. M. and Stofko, E. J., "Vapor-phase epitaxial growth of CdSe on sapphire and CdS", J. Electrochem. Soc., 121, 965-7 (1974), PA77-70422.
725.Zytecki, J. and Lappa, R., "Growth of epitaxial layers of GaN from $GaBr_3.4NH_3$ complex", Krist. & Tech., 9, 1089-1094 (1974), PA78-19089.

X. Film Properties

1. Electrical

726.Aboltin, E. E., Grigulis, Y. K. and Porin, V. M., "Study of epitaxial and diffusion semiconductor layers using microwave e.m. fields", In: Semiconductor and their applications in electronics IV, Zinatne Publications, Riga, USSR, p. 109-36 (1970), Russian, EEA74-4841.
727.Aleksandrova, G. A., Vilkotskii, V. A., Kornilov, B. V., Marchukov, L. V. and Skvotsov, I. M., "Anomalous magnitude and temperature dependence of the mobility of electrons in epitaxial gallium arsenide films", Sov. Phys.-Semicond., 6, 857-9 (1972), EEA76-713.
728.Amrinov, N. M., Guro, G. M., Kovtonyuk, N. F. and Levedev, E. N., "Measurement of the bulk lifetime in epitaxial films by an insulated crystal method", Fiz. Tekh. Poluprov., 4, 1617-21 (1970), Russian, EEA74-1325.
729.Aoki, T. and Yamaguchi, M., "Dependence of the properties of epitaxial gallium arsenide on the temperature of $AsCl_3$", Jap. J. Appl. Phys., 11, 1775-82 (1972), EEA76-4539.
730.Arizumi, T., Nishinaga, T. and Kakehi, M., "Sheet resistivity of the epitaxially grown germanium layer", Japan J. Appl. Phys., 7, 468-75 (1968), EEA71-23238.
731.Atakova, M. M., Ramazanov, P. E. and Salman, E. G., "Local levels in ZnS films", Izv. VUZ Fiz., no. 10, 95-8 (1973), Russian, PA77-6591.
732.Baliga, B. J. and Ghandhi, S. K., "Heteroepitaxial InAs grown on GaAs from triethylindium and arsine. II. Electrical properties", J. Electrochem. Soc., 121, 1646-50 (1974), PA78-16088.

733. Batavin, V. V., Rusakov, V. V., Zukkov, N. M., Prilpko, V. I. and Shevelkov, M. A., "Measurement of leakage currents through insulating pn-junction in epitaxial structures", Ind. Lab., 38, 1046-7 (1972), EEA76-26045.

734. Bindemann, R., Buchheiser, G. K. and Zehe, A., "Donor-acceptor pair and bound exciton recombination in epitaxially grown $Al_xGa_{1-x}P$", Phys. Status Solidi A, 24, K189-92 (1974).

735. Blinnikova-Vyazemskaya, E. B. and Konisevoi, Y. A., "Note on the procedure for investigating the electrophysical parameters of epitaxial silicon", Ind. Lab., 38, 1044-5 (1972), EEA76-25875.

736. Blood, P. and Orton, J. W., "The electrical properties of n-type epitaxial InP in the temperature range 5K to 700K", J. Phys. C, 7, 893-904 (1974), EEA77-18631.

737. Bochkaryova, L. V. and Simashkevich, A. V., "Electrical properties of ZnTe-ZnSe heterostructures", Thin Solid Films, 20, 329-37 (1974), PA77-28486.

738. Conrad, R. W. and Wacker, R. W., "Low frequency current oscillations in epitaxial GaAs", Solid-state Electron., 13, 91-2 (1970), EEA73-14056.

739. Corsi, C. and Miller, M., "Electrical properties and spectral response of PbTe-Ge heterojunctions", Thin Solid Films, 20, S41-3 (1974), PA77-28487.

740. Cullen, G. W., Gottlieb, G. E., Wang, C. C. and Zaininger, K. H., "Epitaxial growth and properties of silicon on alumina-rich single-crystal spinel", J. Electrochem. Soc., 116, 1444-9 (1969), EEA73-4304.

741. Cullen, G. W., Gottlieb, G. E., Wang, C. C. and Zaininger, K. H., "Epitaxial growth and electrical properties of silicon films on spinel", In: Semiconductor silicon, Electrochem. Soc., Inc., New York, p. 291-8 (1969), EEA73-10424.

742. Cullen, G. W. and Wang, C. C., "A comparison of the hole mobility and early growth of epitaxial growth on flame fusion, flux, and Czochralski spinel", J. Electrochem. Soc., 118, 640-4 (1971), EEA74-18562.

743. Dalal, V. L., Hicinbothem, W. A., Jr. and Kressel, H., "Carrier lifetimes in epitaxial InAs", Appl. Phys. Lett., 24, 184-5 (1974), EEA77-22765.

744. Davey, J. E., "Electrical characteristics of epitaxial germanium films vacuum deposited on semi-insulating GaAs up to thickness of 10^6 A", Appl. Phys. Lett., 8, 164-6 (1966), EEA70-2199.

745. Deasley, P. J., Webb, P. W. and Owen, S. J. T., "The growth and electrical characteristics of epitaxial layers of zinc sulphide on p-type germanium", 21st Annual Southwestern IEEE Conference and Exhibition, 1969, 8 pp., EEA73-4356.

746. DiLorenzo, J. V. and Machala, A. E., "Orientation effects on the electrical properties of high purity epitaxial GaAs", J. Electrochem. Soc., 118, 1516-17 (1971), EEA74-37143.

747. Dmitruk, N. L., Lyashenko, V. I. and Maeva, O. I., "Some peculiarities of the field effect in gallium arsenide epitaxial films", Ukr. Fiz. Zh., 19, 1375-80 (1974), Russian, EEA77-36197.
748. Dmitruk, N. L., Lyashenko, V. I., Tereshenko, A. K. and Spektor, S. A., "Investigation of surface recombination on epitaxial GaAs films", Phys. Status Solidi A, 20, 53-62 (1973), EEA77-2425.
749. Dubey, G. C., "Electrical properties of epitaxial germanium films on germanium", Thin Solid Films, 24, 71-8 (1974), PA78-6926.
750. Dumin, D. J., "Deep levels within the forbidden gap of silicon-on-sapphire films", Solid-State Electron., 13, 415-24 (1970), EEA73-21190.
751. Dumin, D. J., "Interdependence of the electrical properties and the defect structure in heteroepitaxial germanium and silicon", J. Vac. Sci. & Technol., 8, 235-41 (1971), EEA74-13180.
752. Dumin, D. J. and Robinson, P. H., "Electrically and optically active defects in silicon-to-sapphire films", J. Cryst. Growth, 3-4, 214-18 (1968), EEA72-20215.
753. Dumin, D. J. and Ross, E. C., "Temperature dependence of the Hall mobility and carrier concentration in silicon-on-sapphire films", J. Appl. Phys., 41, 3139-43 (1970), EEA73-28026.
754. Eernisse, E. P. and Norris, C. B., "Electrical effects of clustered defects in heteroepitaxial Si films", Solid-State Electron., 16, 315-25 (1973), EEA76-16255.
755. Emelyanenko, O. V., Lagunova, T. S., Radu, R. K., Talalakin, G. N. and Telegin, A. A., "Some features of the electrical properties of epitaxial tin-doped GaAs", Sov. Phys.-Semicond., 7, 665-6 (1973), EEA77-10059.
756. Epstein, A. S., "Electron scattering mechanisms in n-type epitaxial GaP", J. Phys. Chem. Solids, 27, 1611-21 (1966), EEA70-504.
757. Favennec, P. N. and Diguet, D., "Electrical properties of proton-bombarded $Ga_{1-x}Al_xAs$. (semiconductor technology applications)", Appl. Phys. Lett., 23, 546-7 (1973), EEA77-5870.
758. Gashin, P. A. and Simashkevich, A. V., "ZnTe-CdSe heterojunctions. I. Electrical properties", Phys. Status Solidi A, 19, 379-86 (1973), EEA76-38249.
759. Glover, G. H., "Microwave measurement of the velocity-electric field characteristic of inhomogeneous epitaxial GaAs with conducting substrate", J. Appl. Phys., 42, 4025-34 (1971), EEA74-37140.
760. Glover, G. H., "Addendum to 'Microwave measurement of the velocity-electric field characteristic of inhomogeneous epitaxial GaAs with conducting substrate", J. Appl. Phys., 43, 2905 (1972), EEA75-24199.
761. Gorodnichii, O. P., Seitov, E. P. and Shavrin, A. G., "Transverse magnetoresistance of epitaxial films of n-type gallium arsenide", Sov. Phys.-Semicond., 7, 694-5 (1973), EEA77-10061.

762.Grant, W. N., "Electron and hole ionization rates in epitaxial silicon at high electric fields", Solid-State Electron., 16, 1189-203 (1973), EEA76-35290.
763.Hart, P. B., Etter, P. J., Jervis, B. W. and Flanders, J. M., "Electrical properties of epitaxial silicon films on α-alumina", Brit. J. Appl. Phys., 18, 1389-98 (1967), EEA71-8288.
764.Hasegawa, F., "Origin of the high resistance region at the epitaxial GaAs layer-substrate interface", J. Electrochem. Soc., 119, 930-5 (1972), EEA75-36140.
765.Hasegawa, S., Kaminaka, N. and Nakamura, T., "Analysis of carrier transport in vacuum-evaporated epitaxial films of silicon on spinel", J. Appl. Phys., 40, 4620-3 (1969), EEA73-10363.
766.Hasegawa, F. and Majerfield, A., "Majority-carrier traps in n- and p-type epitaxial GaAs", Electron. Lett., 11, 286-8 (1975), EEA78-28578.
767.Hooge, F. N., van Dijk, H. J. A. and Hoppenbrowers, A. M. H., "I/f noise in epitaxial silicon", Philips Res. Rep., 25, 81-6 (1970), EEA73-28043.
768.Hsueh, Y. W. and Preece, J. B., "The effect of buried layers on epitaxial resistivity profiles determined by spreading resistance probe techniques", Presented at 133rd Meeting of the Electrochemical Society (abstracts), Boston, Mass., May 1968, EEA73-14089.
769.Ipri, A. C., "Variations in electrical properties of silicon films on sapphire using the MOS Hall technique", Appl. Phys. Lett., 20, 1-2 (1972), EEA75-10172.
770.Ito, S., Shinohara, T. and Seki, Y., "Properties of epitaxial gallium arsenide from trimethylgallium and arsine", J. Electrochem. Soc., 120, 1419-23 (1973), EEA77-271.
771.Itoh, T., Hasegawa, S. and Kaminaka, N., "Electrical properties of n-type epitaxial films of silicon on sapphire formed by vacuum evaporation", J. Appl. Phys., 39, 5310-13 (1968), EEA72-23740.
772.Itoh, T., Hasegawa, S. and Watanabe, H., "Epitaxial films of silicon on sapphire formed by vacuum evaporation and their electrical properties", J. Appl. Phys., 39, 2969-71 (1969), EEA71-23380.
773.Jain, F. C. and Melchy, M. A., "Abrupt avalanche breakdown in GaAs/Ge heterojunctions", Int. J. Electron., 38, 807-8 (1975), EEA78-33063.
774.Jain, F. C. and Melchy, M. A., "High-forward-voltage phenomenon in injection GaAs/Ge heterojunctions", Appl. Phys. Lett., 27, 36-8 (1975), EEA78-33060.
775.Kalevich, I. N., Kompan, M. E., Masagutov, K. G. and Timchenko, I. N., "Influence of the thermal size effect on the current-voltage characteristics of epitaxial films on n-type GaAs in the impurity breakdown region", Sov. Phys.-Semicond., 8, 1047-8 (1975), EEA78-15196.

776.Katoda, T., Osaka, F. and Sugano, T., "Electron mobility in $In_{1-x}Ga_xAs$ epitaxial layer", Jap. J. Appl. Phys., 13, 561-2 (1974), EEA77-22779.
777.Kemarskii, V. A., Kmita, A. M. and Telegin, A. A., "Acousto-electric domains in epitaxial films of GaAs", Sov. Phys.-Semicond., 6, 1287-91 (1973), EEA76-16266.
778.Kennedy, J. K., Potter, W. D. and Davies, D. E., "The effect of the hydrogen carrier gas flow rate on the electrical properties of epitaxial GaAs prepared in a hydride system", J. Cryst. Growth, 24-25, 233-8 (1974), PA78-7472.
779.Khokhlov, V. I., Sidorov, Y. G. and Dvoretskii, S. A., "Electrophysical properties of non-doped epitaxial GaAs in the range from 10 to 1100K", Phys. Status Solidi A, 25, 311-21 (1974), EEA77-38728.
780.Klimenko, A. P., Kurilo, P. M. and Tkhorik, Y. A., "Investigation of germanium heteroepitaxial films. III. Hall effect and magnetoresistance", Ukr. Fiz. Zh., 17, 1797-802 (1972), Russian, EEA76-8489.
781.Klimenko, A. P., Matveeva, L. A. and Tkhorik, Y. A., "Electrophysical properties and energetical spectrum of heteroepitaxial germanium films", Czech. J. Phys. B, B24, 1139-48 (1974), PA77-79419.
782.Kordos, P., "Using GaAs epitaxial layers for preparing Hall probes", Slaboproudy Obzor., 33, 354-8 (1972), Slovak, EEA75-36187.
783.Kuroiwa, K., Aoki, T. and Fujimoto, M., "Electrical properties of GaAs vapor phase epitaxial layers", J. Cryst. Growth, 24-25, 229-32 (1974), EEA78-2726.
784.Langer, P.H. and Pearce, C. W., "Epitaxial resistivity", J. Test. & Eval., 1, 305-9 (1973), EEA76-35233.
785.Langmann, U., "Minority carrier diffusion lengths in GaAs epitaxial layers - comparison of different measuring methods", Solid-State Electron., 16, 1011-15 (1973), EEA76-35160.
786.Lavrenteva, L. G., Kataev, Y. G. and Moskovkin, V. A., "Anisotropy of the electrophysical properties of epitaxial n-type gallium arsenide", Izv. VUZ Fiz., no. 11, 159-60 (1969), Russian, EEA73-21022.
787.Lavrenteva, L. G., Vilisova, M. D., Ivonin, I. V., Krasilnikova, L. M., Kuznetsov, F. A., Rumyantsev, Y. M. and Yakubenya, M. P., "Investigation of transition layers in epitaxial gallium arsenide. Micro morphology and electron distribution in layers as affected by growth time in an iodide system", Izv. VUZ Fiz., no. 2, 63-9 (1973), Russian, EEA76-19829.
788.Li, S. S., Anderson, J. R. and Kennedy, J. K., "Electrical properties of epitaxially grown $InAs_{0.61}P_{0.39}$ films", J. Appl. Phys., 46, 1223-8 (1975), EEA78-24123.
789.Lupin, V. M. and Ramazanov, P. E., "Electrical characteristics of epitaxial layers of CdS grown on GaAs crystals", Izv. VUZ. Fiz., no. 2, 127-9 (1974), Russian, EEA77-18628.

790. Lysenko, V. S., Litovchenko, V. G., Kosenko, V. E. and Ignatkov, V. D., "Electrical characteristics of surface of single-crystal films on sapphire", Ukrayin Fiz. Zh., 15, 410-5 (1970), Russian, EEA74-21079.

791. Majerfeld, A., "Transport properties of $GaAs_{1-x}P_x$ with applications to high-speed current limiters", Thesis, Stanford Univ., Calif., 1971, 96 pp., Order No. 72-1161, EEA75-36176.

792. Majerfeld, A., Potter, K. E. and Robson, P. N., "Temperature dependence of the subthreshold velocity/field characteristic for epitaxial InP", Electron. Lett., 11, 81-2 (1975), EEA78-15174.

793. Matulenis, A. Y., Pozhela, Y. K. and Yutsene, V. Y., "Negative resistance of $Al_xGa_{1-x}As$", Litov. Fiz. Sb., 13, 387-91 (1973), Russian, PA77-1240.

794. Mercier, J., "Origin and nature of involuntary contamination of epitaxial silicon layers on a substrate of corindon (Al_2O_3-α)", Rev. Phys. Appl., 3, 127-30 (1968), French, EEA71-28753.

795. Metz, L. S. and Gandhi, O. P., "DC electric-field profile control in long epitaxial layers of GaAs", Proc. IEEE, 61, 1048-50 (1973), EEA76-29222.

796. Mircea, A. and Mitonneau, A., "A study of electron traps in vapour-phase epitaxial GaAs", Appl. Phys., 8, 15-21 (1975), EEA78-41459.

797. Morvic, M., Kordos, P., Simko, L. and Benc, V., "The influence of the preparation technology on the I-V characteristics of epitaxial PSN-structures from GaAs", Elektrotech. Cas., 26, 318-22 (1975), Slovak, PA78-47423.

798. Nishiyama, A., "The magnetoresistance in epitaxial SnTe", J. Phys. Soc. Jap., 38, 1216 (1975), EEA78-28558.

799. Norris, C. B., "Carrier recombination and trapping in heteroepitaxial Si/spinel", Appl. Phys. Lett., 20, 187-90 (1972), EEA75-13536.

800. Ota, Y. and Zemel, J. N., "Piezoresistance effect of SnTe epitaxic films", J. Nonmet., 2, 25-44 (1973), PA78-1358.

801. Parker, E. H. C. and Williams, D., "The properties of thin epitaxial PbTe films prepared under very clean and closely monitored conditions", Jap. J. Appl. Phys., suppl. 2 (pt. 1), A-865 (1974), EEA78-24205.

802. Paxman, D. H., Tree, R. J. and Wood, C. F. C., "Unstable domains in solution-grown epitaxial InP", Electron. Lett., 8, 241-3 (1972), EEA75-20117.

803. Podor, B., Nador, N. and Bartoti, I., "Investigation of the electrical properties of epitaxial GaAs layers for microwave devices", Proceedings of the 5th Colloquium on Microwave Communication, Vol. V, System Measurements, Semiconductor and Magnetic Materials for Microwave Applications and Plasma Phenomena, 1974, p. SM-21/279-90, EEA77-33911.

804.Podor, B., Nador, N., Bertoti, I. and Hagyo, J., "Current carrier scattering and electrical properties of liquid phase epitaxial GaAs", Kozl. Magy. Tud. Akad. Musz. Fiz. Kut. Intez., no. 0-9, 61-74 (1973), PA77-20661.

805.Pokorny, J., "Measuring electron mobility in thin epitaxial layers of GaAs", Slaboproudy Obzor, 33, 316-18 (1972), Czech, EEA75-32536.

806.Reichl, H. and Bernt, H., "Lifetime measurements in silicon epitaxial materials", Solid-State Electron., 18, 453-8 (1975), PA78-42245.

807.Reynolds, R. A., "Impact ionization breakdown of n-type epitaxial GaAs at liquid helium temperatures", Solid-State Electron., 11, 385-90 (1968), EEA71-15755.

808.Rheinlander, B., Neumann, H. and Gropp, M., "Electron effective masses in direct-gap $Al_xGa_{1-x}As$ epitaxial layers from Faraday rotation measurements", Exp. Tech. Phys., 23, 33-9 (1975), PA78-42837.

809.Rosinski, A., Siennicki, A. and Wisinewski, R., "Device for non-destructive measurements of the resistivity of epitaxial layers by a three-point method", Pomiary Autom. Kontr., 18, 246-7 (1972), Polish.

810.Ross, E. C. et al., "Superlinear current density vs electric field in p-type silicon-on-sapphire", J. Appl. Phys., 41, 2657-64 (1970), EEA73-28023.

811.Rybka, V., Seveik, Z., Krejci, P. and Dudrova, E., "Some electrical properties of epitaxial germanium films deposited on semi-insulating gallium arsenide substrate", Thin Solid Films, 9 (1), 83-96 (1972), EEA75-7427.

812.Ryuzan, O., Kotani, T. and Akita, K., "Epitaxial GaAs crystals. II. Electrical properties of GaAs epitaxial layer", Fujitsu Sci. Tech. J., 8, 131-47 (1972), EEA75-36124.

813.Rzhanov, A. V., Lisenker, B. S., Maronchuk, Y. E. and Sherstyakov, A. P., "Epitaxial gallium arsenide films with a high electron mobility", Fiz. Tekh. Poluprovodnikov, 2, 714-19 (1968), Russian, EEA71-23212.

814.Schlotterer, H., "Mechanical and electrical properties of epitaxial silicon films on spinel", Solid-State Electron., 11, 947-56 (1968), EEA72-2721.

815.Seki, T., Tanno, K., Iida, K. and Ichiki, E., "Properties of epitaxial GaAs layers from a triethyl gallium and arsine system", J. Electrochem. Soc., 122, 1108-12 (1975), EEA78-41437.

816.Severin, P. J., "On the interpretation of some measurement methods for epitaxially grown layers", Symposium on Silicon Device Processing (programme and abstracts), 1970, p. 41, EEA73-35124.

817.Severin, P. J., "Four point-probe resistivity measurements on silicon heterotype epitaxial layers with altered probe order", Philips Res. Rep., 26, 279-97 (1971), EEA74-37152.

818. Severin, P. J., Bulle, H., Poodt, G. and Wasscher, J. D., "On the technique and evaluation of angle-beveling silicon epitaxial layers", J. Electrochem. Soc., 122, 440-3 (1975), EEA78-24185.
819. Sidorov, Y. G., Dvoretskii, S. A., Aleksandrov, L. N., Kraychenko, A. F., Ryndina, L. N. and Leutina, A. P., "Influence of the composition of the gas phase on the electrical properties of epitaxial layers of GaAs", Izv. Akad. Nauk SSSR Neorg. Mater., 8, 1373-8 (1972), Russian, EEA76-722.
820. Sigai, A. G., Abrahams, M. S. and Blanc, J., "Properties of vapor-deposited aluminum arsenide", J. Electrochem. Soc., 119, 952-6 (1972), EEA75-36142.
821. Simonic, M. and Jevic, S., "Resistance profiles of epitaxial layers", Elektroteh. Vestn., 28, 219-23 (1971), Slovene, EEA75-10113.
822. Skouson, G. W. and White, M. H., "Semiconductor fabrication technique permitting examination of epitaxially grown layers", Patent USA 3473977, Publ. October 1969, EEA73-24869.
823. Sloope, B. W. and Tiller, C. O., "Electrical properties of epitaxial Ge films deposited on (111) CaF_2 substrates", J. Appl. Phys., 38, 140-8 (1967), EEA70-8433.
824. Stefanakos, E. K., Abul-Fadl, A. and Workman, M. D., "Measurements of Peltier cooling at a Ga-GaAs interface using a liquid-phase epitaxy system", J. Appl. Phys., 46, 3002-5 (1975), EEA78-77763.
825. Sytenko, T. N. and Tyagulskii, I. P., "Residual conductance of epitaxial films of gallium arsenide", Sov. Phys.-Semicond., 8, 109-10 (1974), EEA77-36194.
826. Thomas, H. and Townsend, W. G., "Low resistivity epitaxial layers of silicon", Solid-State Electron., 9, 1137-9 (1966), EEA70-2217.
827. Thorsen, A. C. and Manasevit, A. C., "Heteroepitaxial GaAs on aluminum oxide: electrical properties of undoped films", J. Appl. Phys., 42, 2519-27 (1971), EEA74-22490.
828. Tufte, O. N. and Stelzer, E. L., "Growth and properties of $Hg_{1-x}Cd_xTe$ epitaxial layers", J. Appl. Phys., 40, 4559-68 (1969), EEA73-10360.
829. van der Does-de-Bye, J. A. W. and Peters, R. C., "Preparation and properties of epitaxial gallium phosphide grown by HCl-gas transport", Philips Res. Rep., 24, 210-30 (1969), EEA73-4319.
830. Viktora, B., "Measuring electrical parameters of GaP epitaxial films at low temperatures", TESLA Electron., 7 (4), 107-15 (1974), PA78-18699.
831. Viktora, B. and Cukr, M., "Investigation of properties of GaP epitaxial layers by electrical and mass-spectral methods", Elektrotech. Cas., 25, 390-0 (1974), Czech, EEA77-33928.

832.Voronina, T. I., Emelyanenko, O. V., Nasledov, D. N. and Nedeoglo, D. D., "Magnetoresistance and scattering of electrons by polar optical phonons in pure GaAs and InP crystals", Sov. Phys.-Semicond., 7, 925-8 (1974), EEA77-18642.

833.Wang, C. C. and McFarlane, S. H., "Epitaxial growth and characterization of GaP on insulating substrates", J. Cryst. Growth, 13-14, 262-7 (1972).

834.Wen, C. P., Chiang, Y. S., Young, A. F. and Presser, A., "High-frequency electron conductivity mobility in vapor-phase homoepitaxial silicon", Appl. Phys. Lett., 23, 390-1 (1973), EEA77-247.

835.Wrigley, C. Y. and Kroko, L. J., "Properties of the silicon-sapphire interface in heteroepitaxy", In: Semiconductor silicon, Electrochem. Soc., Inc., New York, p. 329-38 (1969), EEA73-10427.

836.Wu, K. and Zemel, J. N., "Properties of $Cd_xPB_{1-x}S$ epitaxial films", J. Vacuum Sci. Technol., 6, 497 (1969), EEA73-14030.

837.Yim, E. W., "Silicon epitaxial films evaluation", In: Semiconductor silicon, Electrochem. Soc., Inc., New York, p. 222-32 (1969), EEA73-10421.

838.Yurova, E. S., Kistova, E. M., Kevorkov, M. N., Konstantinova, M. A., Shtogrin, I. I. and Maslov, V. N., "Influence of inhomogeneity on the electrical properties of $GaAs_{1-x}P_x$. (Epitaxial films)", Inorg. Mater., 10, 14-18 (1974), PA78-21982.

839.Zuleeg, R. and Knoll, P., "Space-charge-limited currents in heteroepitaxial films of silicon grown on sapphire", Appl. Phys. Lett., 11, 183-5 (1967), EEA71-3352.

2. Optical

840.Alferov, Z. I., Zimogorova, N. S., Korol'kov, V. I. and Trukan, M. K., "Electrical and photoelectric properties of GaAs epitaxial films", Fiz. Tekh. Poluprovodnikov, 1, 283-5 (1967), Russian, EEA70-12526.

841.Beppu, T., "Luminescent properties variation toward growth direction in nitrogen doped GaP n-LPE layer", Jap. J. Appl. Phys., 14, 761-8 (1975), EEA78-28634.

842.Campbell, J. C., Hitchens, W. R., Holonyak, N., Jr., Lee, M. H., Ludowise, M. J. and Coleman, J. J., "Luminescence, laser, and carrier-lifetime behaviour of constant-temperature LPE $In_{1-x}Ga_xP$(x=0.52) grown on (100) GaAs", Appl. Phys. Lett., 24, 327-30 (1974), EEA77-26888.

843.Duh, K. Y. and Zemel, J. N., "Properties of PbS epitaxial films grown by a hot wall method", Thin Solid Films, 26, 165-79 (1975), EEA78-19555.

844.Fujita, S., Arai, S., Itoh, F. and Sakaguchi, T., "Electrical properties and injection luminescence in ZnSe-ZnTe heterojunctions prepared by liquid-phase epitaxy", J. Appl. Phys., 46, 3070-5 (1975), EEA78-37464.

845. Ghosh, A. K. and Addiss, R. R., Jr., "Luminescence characteristics of epitaxially grown cubic ZnS", J. Appl. Phys., 44, 4431-6 (1973), PA77-21086.

846. Gorelenok, A. T., Nasledov, D. N., Negreskul, V. and Tsarenkov, B. V., "Photoluminescence of epitaxial n-type GaAs prepared from p-type GaAs", Fiz. Tekh. Poluprovodnikov, 2, 659-65 (1968), Russian, EEA71-23287.

847. Hegems, M., Dingle, R. and Rupp, L. W., Jr., "Optical and electrical properties of Mn-doped GaAs grown by molecular-beam epitaxy", J. Appl. Phys., 46, 3059-65 (1975), EEA78-37371.

848. Herzog, A. H., Groves, W. O. and Craford, M. G., "Electroluminescence of diffused $GaAs_{1-x}P_x$ diodes with low donor concentrations", J. Appl. Phys., 40, 1830-8 (1969), EEA72-15305.

849. Joyce, B. D. and Williams, E. W., "The preparation and photoluminescent properties of high purity vapour grown indium phosphide layers (donor concentration obs.)", Proceedings of the Third International Symposium on Gallium Arsenide and Related Compounds, 1970, p. 57-63, Publ. 1971, EEA76-4562.

850. Kasano, H. and Hosoki, S., "Optical characteristics of GaAs layers grown on Ge substrates", J. Electrochem. Soc., 122, 112-8 (1975).

851. Kressel, H. and Nelson, H., "Electrical and optical properties of n-type Si-compensated GaAs prepared by liquid phase epitaxy", J. Appl. Phys., 40, 3720-5 (1969), EEA73-4300.

852. Kressel, H., Nicoll, F. H., Ettenberg, M., Yim, W. M. and Sigai, A. G., "Photo and cathodoluminescence in vapor-grown AlAs", Solid State Commun., 8, 1407-9 (1970), EEA73-37605.

853. Kressel, H. and von Philipsborn, H., "Electrical and optical properties of vapor-grown GaAs:Si", J. Appl. Phys., 41, 2244-6 (1970), EEA73-24803.

854. Kulshreshta, A. P. and Saunders, I. J., "Hall effect and photoluminescence in vapor-phase epitaxial gallium arsenide heavily doped with sulfur", Solid State Commun., 16, 1031-3 (1975), EEA78-19590.

855. Lisenker, B. S., Lisinker, S. S., Maronchuk, Y. E. and Sherstyakova, V. N., "Impurity photoconductivity in epitaxial layers of gallium arsenide", Izv. VUZ Fiz., no. 11, 19-23 (1974), Russian, EEA78-15177.

856. Lozykowski, H. J., Oczlowski, H. L. and Firszt, F., "Electroluminescence in ZnSe-ZnTe heterojunctions", Bull. Acad. Pol. Sci. Ser. Sci. Math. Astron. & Phys., 21, 485-9 (1973), PA77-4012.

857. Macksey, H. M., Lee, M. H., Holonyak, N., Jr., Hitchens, W. R. and Dupuis, R. D., "Crystal and luminescence properties of constant-temperature liquid-phase-epitaxial $In_{1-x}Ga_xP$ ($x \sim 0.7$) grown on (100) $GaAs_{1-x}P_x$ ($x \sim 0.4$)", J. Appl. Phys., 44, 5035-40 (1973), PA77-28888.

858.Miki, H., Ito, M. and Otsubo, M., "Effects of impurities on electrical and optical properties of epitaxial GaAs layers", Spring Meeting of the Electrochemical Society (Extended abstracts), 1971, p. 201-2, EEA75-1034.

859.Nishino, S., Matsunami, H. and Tanaka, T., "Optical properties of β-SiC crystals prepared by chemical vapor deposition", Jap. J. Appl. Phys., 14, 1833-34 (1975).

860.Nuese, C. J., Sigai, A. G., Abrahams, M. S. and Gannon, J. J., "Vapor growth in $In_{1-x}Ga_xP$ for p-n junction electroluminescence. II. Luminescence characteristics", J. Electrochem. Soc., 120, 956-65, EEA76-32175.

861.Pasechnik, Y. A., Ignatkov, V. D. and Snitko, O. V., "Impurity photoconductivity of epitaxial silicon films on sapphire", Ukr. Fiz. Zh., 18, 726-31 (1973), EEA76-29107.

862.Philips, W. E., "Interpretation of steady-state surface photovoltage measurements in epitaxial semiconductor layers", Solid-State Electron., 15, 1097-102 (1972).

863.Ryuzan, O., Dazai, K. and Ozeki, M., "Epitaxial GaAs crystals. III. Optical properties of GaAs epitaxial layer", Fujitsu Sci. Tech. J., 8, 149-66 (1972), EEA75-35921.

864.Shay, J. L., Bachmann, K. J. and Buehler, E., "Efficient electroluminescence from InP diodes grown by LPE from Sn solutions", Appl. Phys. Lett., 24, 192-4 (1974), PA77-41496.

865.Swiderski, J., "Photoelectric measurements of electrical parameters of silicon epitaxial layers", Przeglad Elektron., 9, 168-72 (1968), Polish, EEA71-20536.

866.Takakura, H., Kitamura, H. and Hamakawa, Y., "Epitaxial vapor growth of $GaAs_{1-x}P_x$ on germanium substrates and their optoelectronic properties", J. Cryst. Growth, 24-25, 249-52 (1974), EEA78-2729.

867.Vekua, V. L., Dzhioev, R. I., Zakharchenya, B. P., Ivchenko, E. L. and Fleisher, V. G., "Determination of the sign of the g-factor and observation of epitaxial film deformation in the transverse optical orientation effect in semiconductors", Zh. Eksp. & Teor. Fiz., 66, 1790-7 (1974), Russian, PA77-70381.

868.Vishchakas, Y. K., Vaitkus, Y. Y., Mishkinis, Y. Y., Senulis, F. D., Sergeeva, L. A., Kharlamov, O. A., Kalinkin, I. P. and Aleskovskii, V. B., "Structure, composition and photoelectric properties of heat treated epitaxial CdS and CdSe films", Litov. Fiz. Sb., 12, 1007-19 (1972), Russian, EEA76-19749.

869.Williams, E. W., "Photoluminescence of epitaxial n-type GaAs at 20°K", Solid State Commun., 4, 585-7 (1966), EEA70-2229.

3. Structural

870.Abrahams, M. S., Blanc, J. and Buiocchi, C. J., "Interdependence of strain, precipitation, and dislocation formation in epitaxial Se-doped GaAs", J. Appl. Phys., 45, 3277-87 (1974), PA77-82329.

871. Abrahams, M. S. and Buiocchi, C. J., "Twins and stacking faults in vapor grown GaAs", J. Phys. Chem. Solids, 28, 927-30 (1967), EEA70-18815.

872. Afromowitz, M. A. and Rode, D. L., "Limitations on stress compensation in $Al_xGa_{1-x}As_{1-y}P_y$-GaAs LPE layers", J. Appl. Phys., 45, 4738-40 (1974), EEA78-11455.

873. Aharoni, H. and Bar-Lev, A., "Crystal defects in epitaxial layers", Proceedings of the 7th Convention of Electrical and Electronics Engineers in Israel, 1971, p. 96-109, EEA75-1040.

874. Antypas, G. A. and Moon, R. L., "Growth and characterization of GaAsSb-GaAlAsSb lattice-matched heterojunctions", J. Electrochem. Soc., 121, 416-18 (1974), PA77-45457.

875. Baliga, B. J. and Ghandhi, S. K., "Hillocks on epitaxial GaAs grown from trimethylgallium and arsine", J. Cryst. Growth, 26, 314-16 (1974), EEA78-11476.

876. Brown, A. S. and Springthorpe, A. J., "Defects in epitaxial gallium phosphide layers (p-n junctions)", Phys. Status Solidi A, 7, 495-502 (1971), EEA75-4582.

877. Brown, R. L. and Sobers, R. G., "Stress compensation in GaAs-$Al_{0.24}Ga_{0.76}As_{1-y}P_y$ LPE binary layers", J. Appl. Phys., 45, 4735-7 (1974), EEA78-11454.

878. Butter, E., Jacobs, B., Kramer, H., Schmidt, W., Stary, J. and Wolff, W., "Real structure of $Ga_{1-x}In_xAs$ LPE-layers", Krist. & Tech., 10, 553-60 (1975), German, PA78-76994.

879. Chopra, K. L. and Khan, I. H., "Polymorphic transformation in epitaxial CdS films", Surface Sci., 6, 33-41 (1967), EEA70-6189.

880. D'Asaro, L. A., Landorf, R. W. and Furnanage, R. A., "Low defect silicon by liquid phase epitaxy", In: Semiconductor silicon, Electrochem. Soc., Inc., New York, p. 233-42 (1969), EEA73-10422.

881. Dorfman, V. F., Pypkin, B. N. and Ocheretyanskii, A. L., "Effect of growth rate and crystallization conditions on defect formation in GaAs and GaP epitaxial layers", Sov. Phys.-Crystallogr., 17, 1070-4 (1973), EEA77-329.

882. Golikov, V. I., Kaphan, V. I., Onipko, A. F., Svechnikov, S. V. and Skripoko, G. F., "Examination of monocrystalline substrates for epitaxy after mechanical working", Poluprovodn. Tekh. & Mikroelektron., no. 13, 3-16 (1973), Russian, PA77-20497.

883. Gonda, S., Matsushima, Y., Makita, Y. and Mukai, G., "Characterization and substrate-temperature dependence of crystalline state of GaAs grown by moleculer beam epitaxy", Jap. J. Appl. Phys., 14, 935-42 (1975), PA78-66553.

884. Gorelik, S. S., Mironenko, V. A. and Ukrainskii, Y. M., "Effect of heterogeneities in the substrates of semiconductive gallium arsenide on the structural perfection of epitaxial gallium arsenide layers", Sb. Nauch. Tr. Po Probl. Mikroelektron. Mosk. In-T Elektron. Tekhn., no. 20, 256-60 (1975), Russian, CA83-200406.

885. Grenning, D. A. and Herzog, A. H., "Dislocations and their relation to irregularities in zinc-diffused GaAsP p-n junctions", J. Appl. Phys., 39, 2783-90 (1968), EEA71-23273.

886.Hollan, L., Hallais, J. and Schiller, C., "Interface study of epitaxial vapour grown GaAs", J. Cryst. Growth, 9, 165-70 (1971), EEA74-18573.
887.Holloway, H., "Oriented growth of semiconductors-VI. Stacking disorder in epitaxial indium phosphide", J. Phys. Chem. Solids, 29, 1977-86 (1968), EEA72-4913.
888.Holt, D. B. and Abdalla, M. I., "Some effects of annealing and non-stoichiometry on the structure of epitaxial films of CdTe on Ge", Phys. Status Solidi A, 29, 129-40 (1975), PA78-58283.
889.Holt, D. B. and Wilcox, D. M., "Crystallographic defects in epitaxial layers of cadmium sulphide", J. Cryst. Growth, 9, 193-208 (1971), EEA74-18576.
890.Ishida, K., Matsui, J., Kamejima, T. and Sakuma, I., "X-ray study of $Al_xGa_{1-x}As$ epitaxial layers", Phys. Status Solidi A, 31 (1), 255-62 (1975).
891.Jesser, W. A. and van der Merwe, J. H., "The size dependence of equilibrium elastic strain in finite epitaxial islands", Surface Sci., 31, 229-40 (1972), PA75-56696.
892.Joyce, B. D., "Epitaxial growth and growth defects", Electronic Compon., 9, 1389-95 (1968), EEA73-10354.
893.Kachalkina, M. N., Mironenko, V. A. and Ukrainskii, Y. M., "Effects on autoepitaxial film perfection from deposition of a second phase on chromium-doped GaAs substrates", Sov. Phys.-Crystallogr., 19, 265 (1974), PA78-12907.
894.Kawakami, T. and Saito, H., "Defect observation of GaAs-$Al_xGa_{1-x}As$ heterostructures by transmission infrared microscopy", Jap. J. Appl. Phys., 14, 2073-74 (1975).
895.Kennedy, J. K. and Potter, W. D., "The effect of various growth parameters on the formation of pits and hillocks on the surface of epitaxial GaAs layers", J. Cryst. Growth, 19, 85-9 (1973), EEA76-32180.
896.Kenty, J. L., "Limitations of the lattice matching theory in predicting GaAs/BeO epitaxy", Thin Solid Films, 26, 181-95 (1975), PA78-37616.
897.Khokhlov, V. I., Dvoretskii, S. A., Sidorov, Y. G. and Kravchenko, A. F., "Investigation of electrophysical characteristics of defects arising in epitaxial n-type gallium arsenide under uncontrolled doping", Izv. VUZ Fiz., no. 10, 92-7 (1974), Russian, PA78-18696.
898.Kishino, S. and Ida, S., "Plane defect at the influence and dislocations in epitaxially grown GaAs", J. Electrochem. Soc., 119, 1113-18 (1972), EEA75-36149.
899.Kishino, S., Ogirima, M., Kajimura, T. and Kurata, K., "Lattice mismatch at the interface in GaP-GaP and GaAlAs-GaAs epitaxial growth", J. Cryst. Growth, 24-25, 266-71 (1974), PA78-6666.
900.Kressel, H., Nelson, H., McFarlane, S. H., Abrahams, M. S. and LeFur, P., "Effect of substrate imperfections on GaAs injection lasers prepared by liquid-phase epitaxy", J. Appl. Phys., 40, 3587-96 (1969), EEA73-4144.

901. Kumagai, H. Y., Thompson, J. M. and Krauss, G., "Properties and structure of thin silicon films sputtered on fused quartz substrates", Trans Metall. Soc. AIME, 236, 295-9 (1966), EEA70-4053.

902. Kuznetsov, Y. N., Markova, T. I., Olkhovikova, T. I. and Khashimov, F. R., "Influence of the degree of imperfection of substrates on the structure of epitaxial films of gallium arsenide", Sov. Phys.-Solid State, 15, 1067-8 (1973), PA77-28312.

903. Laugier, A., Durupt, P., Vicario, E. and Pitaval, M., "Pseudo-Kikuchi patterns: A new method for orientation, structure, and quality control of epitaxial layers", Proceedings of the International Conference on the Physics and Chemistry of Semiconductor Heterojunctions and Layer Structures, 1970, p. 239-51, Publ. 1971, EEA74-22563.

904. Lavrenteva, L. G., Nesteryuk, L. G. and Sennikova, V. M., "Orientation effects during the growing of gallium arsenide layers in the gas transport system GaAs-$ZnCl_2$", Izv. VUZ Fiz., no. 6, 111-16 (1972), Russian, PA75-71387.

905. Leamy, H. J. and Mogab, C. J., "Epitaxial growth of SiC on Si: Growth defects", Proceedings of the 31st Annual Meeting of the Electron Microscopy Society of America, 1973, p. 36-7, PA77-1452.

906. Ledneva, F. I., Aoramov, B. G. and Ramazanov, P. E., "Structural perfection of epitaxial films $In_xGa_{1-x}As$", Izv. VUZ Fiz., no. 2, 62-5 (1974), Russian, PA77-37122.

907. Lee, Y. T., Miyamoto, N. and Nishizawa, J., "The lattice misfit and its compensation in the Si-epitaxial layer by doping with germanium and carbon", J. Electrochem. Soc., 122, 530-5 (1975), EEA78-24188.

908. Levin, E. R. and Coutts, M. D., "Growth defects in epitaxial semiconductor layers", Proceedings of the 7th International Congress on Electron Microscopy, 1970, p. 407-8, EEA74-33991.

909. Maev, K. L., Stadnik, A. V., Kosenko, V. Y. and Grigorev, O. N., "Mechanical stresses in silicon films on sapphire", Poluprovodn. Tekh. & Mikroelektron., no. 17, 60-5 (1974), Russian, PA78-4012.

910. Magden, I. N., Krasotova, L. N. and Turchaninov, Y. N., "A new type of fault in epitaxial layers", Poluprovodn. Tekh. & Mikroelektron., no. 19, 79-82 (1975), Russian, EEA78-28706.

911. Malyujiov, B. A., Korolev, V. E. and Vigdorovich, E. N., "Misorientation of GaP_xAs_{1-x} epitaxial films with respect to a GaAs substrate", Sov. Phys.-Crystallogr., 19, 775-6 (1975), PA78-85683.

912. Mariano, A. N. and Chopra, K. L., "Polymorphism in some II-VI compounds induced by high pressure and thin-film epitaxial", Appl. Phys. Lett., 10, 282-4 (1967), EEA70-17288.

913. Michel, C., "Effect of substrate microdefects due to thermal etching on the near-junction properties of GaP:Zn,O grown by liquid phase epitaxy", J. Electrochem. Soc., 122, 678-82 (1975), PA78-51116.

914.Miloslvaskii, V. K., Naboikina, E. N. and Kiyan, T. S., "Effect of structure defects on the optical properties of thin ZnS films in the fundamental absorption edge region", Fiz. Tekh. Poluprovodnikov, 1, 629-37 (1967), Russian, EEA79-20407.
915.Ogirima, M., "Vapor phase epitaxial growth and defects of gallium phosphide", Jap. J. Appl. Phys., 11, 281-7 (1972), EEA75-13562.
916.Pomerantz, D., "A cause and cure of stacking faults in silicon epitaxial layers", J. Appl. Phys., 38, 5020-6 (1967), EEA71-5841.
917.Pomerantz, D. I., "Effects of grown-in and process-induced defects in single crystal silicon", J. Electrochem. Soc., 119, 255-65 (1972), PA75-25248.
918.Rode, D. L., Brown, R. L. and Afromowitz, M. A., "Measurement of lattice dilation in LPE GaAs due to Ge acceptor doping", J. Cryst. Growth, 30, 299-303 (1975).
919.Rozgonyi, G. A. and Saul, R. H., "Effect of substrate preparation on the perfection of GaP liquid-phase epitaxy layers", J. Appl. Phys., 43, 1186-90 (1972).
920.Saito, H. and Kawakami, T., "Etch hills at the heterointerface in GaAs-$Al_xGa_{1-x}As$ epitaxial wafers prepared by LPE", Jap. J. Appl. Phys., 14, 2071-72 (1975).
921.Saito, H. and Kawakami, T., "Observation of etched features of GaAs-$Al_xGa_{1-x}As$ DH wafer", Jap. J. Appl. Phys., 14, 2075-76 (1975).
922.Schultz, M. and Thiel, L., "Effects of substrate surface on the perfection of epitaxial silicon films", In: Growth of crystals, Consultants Bureau, London, England, p. 204-10 (1969), Russian, EEA74-7612.
923.Shaw, D. W., "Faceted defects in GaAs epitaxial layers", J. Cryst. Growth, 12 , 249-53 (1972), PA75-25200.
924.Shinoda, Y. and Kawakami, T., "X-ray topographic observations of defects on the interface of GaAs-$Al_xGa_{1-x}As$ epitaxial wafers", Jap. J. Appl. Phys., 14, 1855-6 (1975).
925.Shumskii, M. G., Zhukova, L. A., Kistova, E. M., Maslov, V. N. and Fomin, V. G., "Structural perfection, homogeneity, and composition of single-crystal gallium arsenide-gallium phosphide films obtained by the sandwich method", Protsessy Rost. Sint. Poluprovodn. Krist. Plenok, Mater. Vses. Simp., 3rd, 1972, p. 240-3, Russian, Aleksandrov, L. N. and Kuznetsov, F. A. (Eds.), "Nauka", Sib. Otd.: Novosibirsk, USSR, 1975, CA83-211393.
926.Shumskii, V. N., Aseev, A. L., Pogorelov, Y. N. and Stenin, S. I., "Structural characteristics of Ge-GaAs heterojunctions obtained by various different methods", Izv. Akad. Nauk SSSR Ser. Fiz., 38, 1522-7 (1974), Russian, EEA77-36216.
927.Simpson, J., Adams, A. C. and Hanes, M. H., "The growth of submicron single and multilayer silicon epitaxy", Symposium on Silicon Device Processing, 1970, p. 19, EEA73-35110.

928.Sverdlina, O. A., Kuzovlev, I. A., Solomatin, V. S., Yurkova, V. Y., Kovykova, N. V., Nishanov, D. and Shatalina, L. G., "An atomic-absorption method of layer analysis for epitaxial GaAs structures with chromatographic separation of the substrate", Ind. Lab., 41, 211-13 (1975), EEA78-41470.
929.Thomas, R. N. and Francombe, M. H., "A LEED study of the homoepitaxial growth of thick silicon films", Appl. Phys. Lett., 11, 108-10 (1967), EEA71-1144.
930.Tolomasov, V. A., Abrosimova, L. N., Sergievskaya, T. N. and Zotova, T. M., "Alloying of epitaxial silicon layers with phosphorus and arsenic in a vacuum", Sov. Phys.-Crystallogr., 18, 558-9 (1974), EEA77-26380.
931.Tucker, R. N., "Limitations of current epitaxial evaluations", Symposium on Silicon Device Processing, 1970, p. 40, EEA73-35123.
932.Vasilevskaya, V. N., Datsenko, L. I., Osadchaya, N. V., Prokopenko, I. V., Soldatenko, N. N. and Tkhorik, Y. A., "Structural perfection of the Ge-Si and Si-Ge heteroepitaxial systems", Thin Solid Films, 22, 221-9 (1974), PA77-67125.
933.Wang, P., Pink, F. X. and Gupta, D. C., "Structural faults in epitaxial and diffused silicon material in device fabrication", Symposium on Silicon Device Processing, 1970, p. 48, EEA73-35128.
934.Wolley, E. D., Stickler, R. and Chu, T. L., "Effects of phosphorus diffusions in epitaxial silicon layers", J. Electrochem. Soc., 115, 409-15 (1968), EEA71-20532.

4. Morphology

935.Barabtarlo, P. O., Maksimov, Y. I. and Radautsan, S. I., "Growth conditions and morphological peculiarities of gallium phosphide whiskers obtained by thermal evaporation in vacuum", Phys. Status Solidi A, 8, 521-9 (1971), PA75-9294.
936.Bauser, E., Frik, M., Loechner, K. S., Schmidt, L. and Ulrich, R., "Substrate orientation and surface morphology of gallium arsenide liquid phase epitaxial layers", J. Cryst. Growth, 27, 148-53 (1974), CA82-78877.
937.Blom, G. M., Daniele, J. J., Kyros, T. and Witt, A. F., "Interface demarkation during LPE growth of GaAs", J. Electrochem. Soc., 122, 1541-4 (1975).
938.Bolkhovityanov, Y, B., Bolkhovityanova, R. I. and Melnikov, P. L., "Effect of liquid epitaxy conditions on the morphology of gallium arsenide films", Izv. Akad. Nauk SSSR, Neorg. Mater., 9, 1320-4 (1973), Russian, CA79-140365.
939.Bolkhovityanov, Y. B., Bolkhovityanova, R. I. and Melnikov, P. L., "Dependence of the morphology of GaAs films on liquid epitaxy conditions", Inorg. Mater., 9, 1174-7 (1973), EEA77-30157.
940.Crossley, I. and Small, M. B., "Surface morphologies of GaAs layers grown by liquid phase epitaxy", J. Cryst. Growth, 19, 160-68 (1973).

941.Fukuda, H. and Sakai, Y., "Crystalline morphology and electrical properties of selenium epitaxially grown on tellurium", Jap. J. Appl. Phys., 9, 429-40 (1970), EEA73-35068.
942.Gudz, E. S., Lisenker, B. S., Maronchuk, I. E. and Maronchuk, Y. E., "Heterogeneities in GaAs epitaxial layers", Izv. Akad. Nauk SSSR Neorg. Mater., 8, 1529-34 (1972).
943.Guha, S., Majerfeld, A., Moyes, N. and Robson, P. N., "Surface morphology of liquid-phase-epitaxial InP", Electron. Lett., 11, 303-4 (1975), EEA78-28607.
944.Hsieh, J. J., "Thickness and surface morphology of GaAs LPE layers grown by supercooling, step-cooling, equilibrium-cooling, and two-phase solution techniques", J. Cryst. Growth, 27, 49-61 (1974), EEA78-15215.
945.Lavrenteva, L. G., Ivonin, I. V., Krasilnikova, L. M., Moskovkin, V. A. and Yakubenya, M. P., "Investigation of the micromorphology of autoepitaxial gallium arsenide. Dependence on substrate orientation", Izv. VUZ Fiz., no. 10, 148-50 (1973), Russian, PA77-6564.
946.Lavrenteva, L. G., Yakubenya, M. P., Ivleva, O. M. and Moskovkin, V. A., "Morphology of autoepitaxial gallium arsenide in the crystallographic range of (111)A-(100)-(111)B", Rost Krist., 9, 218-24 (1972), Russian, CA79-58500.
947.Liu, J. K., Lakin, K. M. and Wang, K. L., "Growth morphology and surface-acoustic-wave measurements of AlN films on sapphire", J. Appl. Phys., 46, 3703-6 (1975).
948.Longo, J. T., Harris, J. S., Jr., Gertner, E. R. and Chu, J. C., "Improved surface quality of solution grown GaAs and $Pb_{1-x}Sn_xTe$ epitaxial layers: A new technique", J. Cryst. Growth, 15, 107-16 (1972), EEA75-32514.
949.Peters, R. C., "Growth characteristics and morphology of GaP liquid phase epitaxy", Proceedings of the 4th International Symposium on Gallium Arsenide and Related Compounds, 1972, p. 55-62, Publ. 1973, EEA76-23370.
950.Peters, R. C. and Andre, E., "Morphology and thicknesses of GaP layers and composition of $Ga_{1-x}Al_xAs$ layers grown by liquid phase epitaxy", Acta Electron., 17 (1), 9 pp. (1974), PA77-25262.
951.Runyan, W. R., Alexander, E. G. and Craig, S. E., Jr., "Behavior of large-scale surface perturbations during silicon epitaxial growth", J. Electrochem. Soc., 114, 1154-8 (1967), EEA71-8150.
952.Smorodina, T. A., Georgiev, A. I. and Rodevich, V. A., "Effects of details of the gas flow on the morphology of epitaxial germanium films", In: Growth of crystals, Consultants Bureau, London, England, p. 198-203 (1969), Russian, EEA74-7611.
953.Tietze, H. J. and Butter, E., "The growth of $A^{III}B^{V}$-compounds from nonstoichiometric melts. III. Investigation of surface morphology of homoepitaxial GaAs layers", Krist. & Tech., 8, 1303-11 (1973), German, PA77-10063.

XI. Applications

1. Bipolar Transistors

954. Belanovskii, E. A. and others, "Planar-epitaxial p-n-p germanium transistors and a method for manufacture of such transistors", Patent UK 1296638, Publ. November 1972.

955. Bhattacharyya, A. B., Srivastava, A. and Kumar, R., "Switching properties of epitaxial planar transistors operating in saturation", Solid-State Electron., 18, 277-86 (1975), EEA78-15331.

956. Dimitrov, G. I., "Diffusion processes in making silicon planar-epitaxial p-n-p transistors", Elektro Prom.-st. & Priborostr., 8 (5), 169-71 (1973), Bulgarian, EEA77-5993.

957. Gonauser, E., "Signal delays of integrated silicon NPN planar epitaxial transistors", Siemens Forsch.- & Entwicklungsber., 3, 149-54 (1974), German, EEA77-30315.

958. Grutchfield, H. B. and Moutoux, T. J., "Current mode second breakdown in epitaxial planar transistors", IEEE Trans. Electron Devices, ED-13, 743-8 (1966), EEA70-9090.

959. Heiman, F. P. and Robinson, P. H., "Silicon-on-sapphire epitaxial bipolar transistors", Solid-State Electron., 11, 411-18 (1968), EEA71-18239.

960. Konagi, M. and Takahashi, K., "Formation of GaAs-(GaAl)As heterojunction transistors by liquid phase epitaxy", Electr. Eng. Jap., 94 (4), 97-103 (1974), EEA78-37391.

961. Luisi, J. A., "SOS brings new life to read-only units", Electronics, 42 (4), 114-19 (1969), EEA73-9605.

962. Miller, A., "Silicon-on-sapphire approach affords freedom and flexibility", Electronics, 40 (4), 171-6 (1967), EEA70-13103.

963. Shideler, E., Konnan, Y. A., Spicer, L. and Longo, T. A., "NPN and PNP silicon epitaxial power transistors", IEEE Northeast Electronics Research and Engineering Meeting, 1965, p. 198-9, EEA70-9113.

964. Zuleeg, R. and Knoll, P., "Thin-film lateral bipolar transistor in silicon-on-sapphire structure", Electron. Lett., 3, 137-9 (1967), EEA79-14809.

2. P-N Junctions

965. Antonov, V. B., Kurbanova, E. I. and Saldev, E. Y., "Some characteristics of epitaxial p-n junctions in $Pb_{1-x}Sn_xSe$", Sov. Phys.-Semicond., 8, 146 (1974), PA77-75956.

966. Asano, T., Doi, A. and Migitaka, M., "Delineation of GaAs junctions made by liquid phase epitaxy", Jap. J. Appl. Phys., 14, 575-6 (1975), PA78-47460.

967. Batukova, L. M., Karpovich, I. A. and Yankova, T. N., "Epitaxial periodic p-n structures on PbS", Izv. VUZ Fiz., no. 1. 53-6 (1974), Russian, PA77-33027.

968.Brown, A. S. and Springthorpe, A. J., "Defects in epitaxial gallium phosphide layers (p-n junctions)", Phys. Status Solidi A, 7, 495-502 (1971), EEA75-4582.
969.Chu, T. L. and Gruber, G. A., "Large area silicon junctions by epitaxial growth technique", J. Electrochem. Soc., 114, 522-5 (1967), EEA70-14857.
970.Deml, F. and Talpova, J., "On the preparation of GaAs epitaxial layer and GaAsP and GaP p-n junctions by the growth from the gaseous phase", Elektrotech. Cas., 25, 338-45 (1974), Czech, EEA77-33922.
971.Enstrom, R. E., Nuese, C. J., Appert, J. R. and Gannon, J. J., "Influence of gas-phase stoichiometry on the defect morphology, impurity doping, and electroluminescence efficiency of vapor-grown GaAs p-n junctions", J. Electrochem. Soc., 121, 1516-23 (1974), EEA78-5243.
972.Klamka, J., "Silicon epitaxial power varactors with diffused p-n junctions", Przeglad Elektron., 8, 520-5 (1967), Polish, EEA71-10620.
973.Kressel, H., Robinson, P., McFarlane, S. H. and D'Aiello, R. V., "Epitaxial silicon p-n junctions on polycrystalline 'ribbon' substrates", Appl. Phys. Lett., 25, 197-9 (1974), PA77-76392.
974.Kurihara, M., Moriizumi, T. and Takahashi, K., "Ge-doped $In_xGa_{1-x}As$ p-n junctions", Solid-State Electron., 16, 763-71 (1973), EEA76-29244.
975.Lee, T. P., "Calculations of cutoff frequency, breakdown voltage, and capacitance for diffused junctions in thin epitaxial silicon layers", IEEE Trans. Electron Devices, ED-13, 881-96 (1966), EEA70-10852.
976.Peaker, A. R., Sudlow, P. D. and Mottram, A., "Solution growth of gallium phosphide p-n junctions by liquid phase-epitaxy", J. Cryst. Growth, 13-14, 651-6 (1972).
977.Saget, P., "Formation of mosaics of germanium p-n junctions by epitaxy", Acta Electronics, 10, 181-94 (1966), French, EEA71-3255.
978.Taubkin, I. I., "On the possibility of determining the electro-physical parameters of regions of epitaxial p-n junctions with the aid of galvanomagnetic and photomagnetic measurements", Radio Engng Electronic Phys., 11 (2), 267-76 (1966), EEA70-4651.

3. LED

979.Craford, M. G. and Groves, W. O., "Vapor phase epitaxial materials for LED applications", Proc. IEEE, 61, 862-80 (1973), EEA76-26551.
980.Dazai, K., Akita, K. and Nakai, S., "GaP liquid phase epitaxial growth and light emitting diodes. II. GaP liquid phase epitaxial growth for green LED's", Fujitsu Sci. & Tech. J., 9, 149-72 (1973), PA77-6980.

981.Hackett, W. H., Jr., Dapkus, P. D., Lorimor, O. G., Kammlott, G. W. and Haszko, S. E., "Characterization of nitrogen-doped GaP for efficient green light-emitting diodes", Presented at Electrochemical Society Meeting, Chicago, Ill., May 1973, EEA76-23755.

982.Henkel, H. J. and Ziegler, G., "Epitaxial hemispherical electroluminescent diode", Solid-State Electron., 10, 158 (1967), EEA70-6774.

983.Herzog, A. H. and McLain, H. F., "Green emitting Zn-diffused LPE GaP diodes", J. Electron. Mater., 3, 265-77 (1974), PA77-29017.

984.Keil, J. G., "Diffusion process for heteroepitaxial germanium device fabrication utilizing a polycrystalline silicon mask", Patent USA 3723201, Publ. March 1973, CA78-141661.

985.Kim, C. K., Min, S., Kim, O. K., Chea, Y. H. and Jo, H. S., "The fabrication of gallium phosphide red light emitting diode by liquid phase epitaxy", J. Korean Inst. Electron. Eng., 10 (3), 89-97 (1973), Korean, EEA77-2892.

986.Kotani, T., Segi, K. and Ryuzan, O., "GaP liquid phase epitaxial growth and light emitting diodes. I. GaP liquid phase epitaxial growth for red LED", Fujitsu Sci. & Tech. J., 9, 133-48 (1973), PA77-6979.

987.Kravitz, L. C., Womac, J. F., Heumann, F. K. and Woodbury, H. H., "Monolithic LED arrays by selective liquid-phase epitaxy in GaP", Proc. IEEE, 61, 891-4 (1973), EEA76-26572.

988.Ladany, I. and Wang, C. C., "Properties of GaP light-emitting diodes grown on spinel substrates", Solid-State Electron., 17, 573-6 (1974), PA77-49624.

989.Lorimor, O. G., Saul, R. H., Dawson, L. R. and Paola, C. R., "High capacity liquid phase epitaxy apparatus utilizing thin metals (light emitting diodes)", Solid-State Electron., 16, 1289-95 (1973), EEA77-5898.

990.Mottram, A. and Peaker, A. R., "The growth of gallium phosphide layers of high surface quality by liquid phase epitaxy: A commercial process for green light emitting diodes", J. Cryst. Growth, 27, 193-204 (1974), EEA78-11485.

991.Nabory, R. E., Pollack, M. A. and DeWinter, J. C., "Efficient LPE-grown $In_xGa_{1-x}As$ LEDs at 1-1.1μm wavelengths", Appl. Phys. Lett., 25, 146-8 (1974), EEA77-39073.

992.Saitoh, T. and Minagawa, S., "Epitaxial growth and characterization of pryolytic-grown $GaAs_{1-x}P_x$ for electroluminescent diodes", J. Electrochem. Soc., 120, 656-69 (1973).

993.Saul, R. H. and Lorimore, O. G., "Liquid phase epitaxy processes for GaP LED's", J. Cryst. Growth, 27, 183-92 (1974), EEA78-11484.

994.Saul, R. H. and Roccasecca, D. D., "Vapor-doped multislice LPE for efficient GaP green LED's", J. Electrochem. Soc., 120, 1128-31 (1973), EEA76-35581.

995. Spring-Thorpe, A. J., "A simple multiple-slice liquid-phase epitaxy technique for the production of efficient GaP LEDs", Electrochemical Society Fall Meeting (Extended abstracts), 1974, p. 451-2, PA78-73310.
996. Wagner, S., "Preparation and properties of green-light-emitting $CdS-CuGaS_2$ heterodiodes", J. Appl. Phys., 45, 246-51 (1974), PA77-45452.
997. Weyrich, C., Winstel, G. H., Mettler, K. and Plihal, M., "A simple LPE process for efficient red and green GaP diodes", Gallium Arsenide and Related Compounds, 1974, p. 145-54, Publ. 1975, EEA78-28667.

4. Lasers

998. Burnham, R. D. and Scifres, D. R., "Etched buried heterostructure GaAs/GaAlAs injection lasers", Appl. Phys. Lett., 27 (9), 510-11 (1975).
999. Coleman, J. J., Hitchens, W. R., Holonyak, N., Jr., Ludowise, M. J., Groves, W. O. and Keune, D. L., "Liquid phase epitaxial $In_{1-x}Ga_xP_{1-z}As_z/GaAs_{1-y}P_y$ quaternary (LPE)-ternary (VPE) heterojunction laser ($x \sim 0.70$, $z \sim 0.01$, $y \sim 0.40$; $\lambda < 6300$ A, 77K)", Appl. Phys. Lett., 25, 725-7 (1974), EEA78-12170.
1000. Eliseev, P. G., Ismailov, I., Krasil'nikov, A. I., Man'ko, M. A. and Strakhov, V. P., "Spontaneous and coherent emission from epitaxial p-n junctions in gallium arsenide", Fiz. Tekh. Poluprovodnikov, 1, 1315-19 (1967), Russian, EEA71-3298.
1001. Ettenberg, M., Nuese, C. J., Appert, J. R., Gannon, J. J. and Enstron, R. E., "Metallurgical and electroluminescence characteristics of vapor-phase and liquid-phase epitaxial junction structures of $In_xGe_{1-x}As$ (and laser-diode parameters)", J. Electron. Mater., 4, 37-66 (1975), EEA78-15225.
1002. Hitchens, W. R., Holonyak, N., Jr., Lee, M. H. and Campbell, J. C., "Liquid phase epitaxial growth and photoluminescence characterization of laser-quality (100)$In_{1-x}Ga_xP$", J. Cryst. Growth, 27, 154-65 (1974), PA78-26313.
1003. Iga, K., Arai, R. and Fukuyo, H., "Some characteristics in the preparation of GaAlAs epitaxial layers for laser materials", Bull. Tokyo Inst. Technol., no. 125, 1-7 (1974), PA78-33964.
1004. McLane, G. F. and Sleger, K. J., "Vacuum deposited epitaxial layers of $PbS_{1-x}Se_x$ for laser devices", J. Electron. Mater., 4, 465-79 (1975), EEA78-28641.
1005. Miller, B. I. and Johnston, W. D., Jr., "Successful liquid phase epitaxial growth and optically pumped laser operation of $In_{0.5}Ga_{0.05}P-Ga_{0.4}Al_{0.6}As$ double-heterostructure material", Appl. Phys. Lett., 25, 216-18 (1974), EEA77-39140.
1006. Miller, B. I., Pinkas, E., Hayashi, I. and Capik, R. J., "Reproducible liquid-phase-epitaxial growth of double heterostructure $GaAs-Al_xGa_{1-x}As$ laser diodes", J. Appl. Phys., 43, 2817-26 (1972).

1007.Nahory, R. E. and Pollack, M. A., "Low-threshold room-temperature double-heterostructure $GaAs_{1-x}Sb_x/Al_yGa_{1-y}As_{1-x}Sb_x$ injection lasers at 1-μm wavelengths", Appl. Phys. Lett., 27 (10), 562-3 (1975).

1008.Nannichi, Y. and Hayashi, I., "Degradation of (Ga.Al)As double heterostructure diode lasers. (Improved liquid phase epitaxy technique)", J. Cryst. Growth, 27, 126-32 (1974), EEA78-15807.

1009.Shiozawa, L. and Kuhl, G. E., "Physical vapour deposition of CdTe for high power laser windows at 10.6 μm", Am. Ceram. Soc. Bull., 51, 648 (1972), EEA76-9262.

1010.Thompson, G. H. B. and Kirkby, P. A., "Liquid phase epitaxial growth of six-layer GaAs/(GaAl)As structures for injection lasers with 0.04 μm thick centre layer", J. Cryst. Growth, 27, 70-85 (1974), EA78-26312.

1011.Tomasetta, L. R. and Fonstad, C. G., "Liquid phase epitaxial growth of laser heterostructures in $Pb_{1-x}Sn_xTe$", Appl. Phys. Lett., 24, 567-70 (1974), EEA77-30141.

1012.Yang, L. and Ballantyne, J. M., "Epitaxial growth over optical gratings on GaAs", Appl. Phys. Lett., 25, 67-8 (1974), EEA77-36208.

5. IMPATT

1013.Bar-Lev, A. and Aharoni, H., "Measurements on light emission and current distribution in avalanching epitaxial diodes", IEEE Trans. Electron Devices, ED-21, 537-9 (1974), EEA77-34024.

1014.Bar-Lev, A. and Aharoni, H., "Pre-secondary breakdown behaviour of silicon epitaxial diodes", Int. J. Electron., 38, 353-65 (1975), EEA78-15306.

1015.Bouvet, J. V., "Silicon avalanche diode with an epitaxial double drift structure for millimetric applications", 1973 European Microwave Conference, Vol. I, A8.1/4 pp., EEA77-5977.

1016.Cho, A. Y., Dunn, C. N., Kuvas, R. L. and Schroder, W. E., "GaAs IMPATT diodes prepared by molecular beam epitaxy", Appl. Phys. Lett., 25, 224-6 (1974), EEA77-38817.

1017.Doi, A., Toyabe, T. and Migitaka, M., "GaAs DDR (double-drift-region) IMPATT diodes made by a successive liquid phase epitaxy", Oyo Buturi, 43 (suppl.), 217-21 (1974), EEA78-2829.

1018.Gall. A., "Determination of doping concentration in epitaxial IMPATT diodes with diffusion profile", Nachrichtentech. Z. (NTZ), 27, 93-7 (1974), German, EEA77-18780.

1019.Goldwasser, R. E. and Rosztoczy, F. E., "High efficiency GaAs lo-hi-lo IMPATT devices by liquid phase epitaxy for X-band", Appl. Phys. Lett., 25, 92-4 (1974), EEA77-36286.

1020.Howard, A. M., Smith, D. J. and Purcell, J. J., "Epitaxially grown double-drift silicon IMPATT diodes at 60 to 90 GHz", Electron. Lett., 10, 443-5 (1974), EEA78-2417.

1021. Luther, L. C. and DiLorenzo, J. V., "Growth of epitaxial GaAs structures for high efficiency IMPATT's", J. Electrochem. Soc., 122, 760-9 (1975), PA78-62547.

1022. Miller, B. I. and Casey, J. C., Jr., "Preparation of GaAs p-n junctions by multiple-layer liquid phase epitaxy (IMPATT structures)", Proceedings of the 4th International Symposium on Gallium Arsenide and Related Compounds, 1972, p. 231-9, Publ. 1973, EEA76-23428.

1023. Niizuma, H. and Suzuki, K., "Si epitaxial crystals and IMPATT diode characteristics", Electr. Commun. Lab. Tech. J., 23, 191-202 (1974), Japanese, EEA78-8530.

1024. Tantraporn, W. and Yu, S. P., "GaAs double-drift avalanche diode from vapour-phase epitaxy", Electron. Lett., 11, 196-8 (1975), EEA78-24308.

1025. Wen, C. P., Weller, K. P. and Young, A. F., "Multilayer vapor-phase epitaxial silicon millimeter-wave IMPATT diodes", IEEE Trans. Electron Devices, ED-19, 891-3 (1972), EEA75-24312.

6. Waveguides

1026. Kamath, G. S., "GaAs-(GaAl)As system for integrated optics", Electrochemical Society Fall Meeting (Extended abstracts), 1974, p. 356-7, PA78-69618.

1027. Kleinknecht, H. P. and Widmer, A. E., "(GaAl)P optical-waveguide modulators fabricated by liquid-phase epitaxy", J. Appl. Phys., 45, 3453-9 (1974), PA78-7467.

1028. Merz, J. L., Logan, R. A., Wiegmann, W. and Gossard, A. C., "Taper couplers for GaAs-$Al_xGa_{1-x}As$ waveguide layers produced by liquid phase and molecular beam epitaxy", Appl. Phys. Lett., 26, 337-40 (1975), PA78-47419.

1029. Ralston, R. W., Walpole, J. N., Harman, T. C. and Melngailis, I., "Double-heterostructure $Pb_{1-x}Sn_xTe$ waveguides at 10.6 μm", Appl. Phys. Lett., 26, 64-6 (1975), PA78-29519.

1030. Widmer, A. E. and Kleinnecht, H. P., "Liquid-phase epitaxy for the preparation of $Ga_{1-x}Al_xP$-heterostructures for optical waveguide modulators", Z. Angew. Math. & Phys., 24, 452 (1973), German, EEA77-10581.

1031. Wolfe, C. M., "GaAs-GaInAs epitaxy for integrated optics applications", Electrochemical Society Fall Meeting (Extended abstracts), 1974, p. 362-3, PA78-69619.

1032. Wolfe, C. M., Stillman, G. E. and Melngailis, I., "Epitaxial growth of $In_xGa_{1-x}As$ waveguide detectors for integrated optics", J. Electrochem. Soc., 121, 1506-9 (1974), EEA78-5857.

7. Microwave Devices

1033.Anderson, S. J. and Robinson, G. Y., "The effect of dielectric surface loading on the stability of epitaxial coplanar Gunn diodes", IEEE Trans. Electron Devices, ED-21, 377-8 (1974), EEA77-26432.

1034.Bass, J. C., Edridgg, A. L. and Knight, J. R., "Gunn-effect oscillators with vapour-grown contact layers", Electron. Lett., 3, 24 (1967), EEA70-9184.

1035.Berson, B. E. and Narayan, S. Y., "L-band epitaxial Gunn oscillators", Proc. Inst. Elect. Electron. Engrs., 55, 1078 (1967), EEA70-20956.

1036.Brady, D. P., Knight, S., Lawley, L. and Uenchara, M., "Recent results with epitaxial GaAs Gunn effect oscillators", Proc. Inst. Elect. Electron. Engrs., 54, 1497-8 (1966), EEA70-2747.

1037.Henry, R., Gibeau, P. and Duchemin, J. P., "Epitaxial techniques improvements for silicon microwave diodes", Proceedings of the 5th Colloquium on Microwave Communication, Vol. V, System Measurements, Semiconductor and Magnetic Materials for Microwave Applications and Plasma Phenomena, 1974, p. SM-9/175-81, EEA77-34034.

1038.Hollan, L., Easton, B. C. and le Duc, J. M., "Epitaxial layers of GaAs for Gunn diodes", Acta Electron., 15, 37 pp. (1972), French, EEA75-16614.

1039.King, G. and Wasse, M. P., "High-power Gunn-effect oscillators using epitaxial GaAs", Electron. Lett., 2, 314-15 (1966), EEA70-9182.

1040.Long, S. I., Jeppsson, B. and Johnson, E., "Solution regrowth epitaxial gallium arsenide for Gunn hybrid and LSA devices", Nachrichtentech. Fachber. (NTF), 35, 447-53 (1968), EEA72-15457.

1041.Mottram, A. and Peaker, A. R., "Liquid phase epitaxial growth", Patent UK 1369023, Publ. October 1974, CA82-50674.

1042.Rosztoczy, F. E., Caldwell, J. F., Kinoshita, J. and Omori, M., "Ge-doped p-type epitaxial GaAs for microwave device application", Appl. Phys. Lett., 22, 525-6 (1973), EEA76-25961.

1043.Rosztoczy, F. E., Goldwasser, R. E. and Kinoshita, J., "New structures by liquid phase epitaxy for microwave devices", Gallium Arsenide and Related Compounds, 1974, p. 37-45, Publ. 1975, EEA78-28664.

1044.Rosztoczy, F. E. and Kinoshita, J., "Growth and evaluation of epitaxial GaAs for microwave devices", J. Electrochem. Soc., 121, 439-44 (1974), EEA77-22869.

1045.Rosztoczy, F. E., Long, S. I. and Kinoshita, J., "LPE GaAs for microwave applications", J. Cryst. Growth, 27, 205-14 (1974), EEA78-11486.

1046.Teszner, J. L. and Boccon-Gibod, D., "Influence of electron trapping on domain dynamics in epitaxial Gunn diodes", J. Appl. Phys., 44, 2765-74 (1973), EEA76-26034.

1047.Wrick, V. L., III and Eastman, L. F., "Diffusion-limited LPE growth of InP for microwave devices", Gallium Arsenide and Related Compounds, 1974, p. 31-6, Publ. 1975, EEA78-28663.
1048.Wolfe, C. M., Fogt, A. G. and Lindley, W. T., "Epitaxial gallium arsenide for high-efficiency Gunn oscillators", Electrochem. Technol., 6, 208-14 (1968), EEA71-25877.

8. Others

1049.Azimov, S. A., Mirzabaev, M. M. and Rasulov, K., "Investigation of epitaxial silicon p-i-n structures exhibiting negative resistance regions", Sov. Phys.-Semicond., 8, 650-1 (1974), EEA78-11585.
1050.Cho, A. Y., "Optical devices utilizing single-crystal GaP on GaAs films epitaxially grown on CaF_2 substrates and method of fabricating", Patent USA 3830654, Publ. August 1974.
1051.Gutierrez, W. A., Wilson, H. L. and Yee, E. M., "GaAs transmission photocathode grown by hybrid epitaxy", Appl. Phys. Lett., 25, 482-3 (1974), PA78-26295.
1052.Hoss, P. A., Murray, L. A. and Rivera, J. J., "The close-spaced growth of degenerate p-type GaAs, GaP, and $Ga(As_x, P_{1-x})$ by $ZnCl_2$ transport for tunnel diodes", J. Electrochem. Soc., 115, 553-6 (1968), EEA71-20529.
1053.Kressel, H., "Gallium arsenide and (AlGa)As devices prepared by liquid-phase epitaxy", J. Electron. Mater., 3, 747-90 (1974), PA78-26325.
1054.Kressel, H., Robinson, P., D'Aiello, R. V. and McFarlane, S. H., "Properties of high-voltage silicon epitaxial diodes", J. Appl. Phys., 45, 3930-3 (1974), EEA78-2819.
1055.Liu, Y. Z., Hollish, C. D., Stein, W. W., Bolger, D. E. and Greene, P. D., "LPE GaAs/(Ga,Al)As/GaAs transmission photocathodes and a simplified formula for transmission quantum yield", J. Appl. Phys., 44, 5619-21 (1973), PA77-37734.
1056.Migitaka, M., "Transferred electron oscillator diodes made by liquid phase epitaxy", Proceedings of the 1971 European Microwave Conference, 4 pp., EEA75-1060.
1057.Naganuma, M., Kamimura, K., Takahashi, K. and Sakai, Y., "GaAs FET prepared with molecular beam epitaxial films", Jap. J. Appl. Phys., 14, 581-2 (1975), PA78-47461.
1058.Nakanuma, S. and Mitsuhashi, G., "Silicon variable capacitance diodes by low temperature epitaxial growth", J. Inst. Elect. Commun. Engrs., 50, 819-20 (1967), EEA71-1146.
1059.Portele, F. and Schliep, K. P., "Fusion-epitaxial production of Si-doped GaAs layers for solid state lamps", Tech.-Wiss. Abh. Osram-Ges., 11, 235-9 (1973), German, PA77-49629.
1060.Rai-Choudhury, P., "Chemical-vapor deposited silicon and its device applications", Presented at Electrochemical Society Meeting, Chicago, Ill., May 1973, EEA76-23309.

1061.Roy, K., "Silicon epitaxy for power devices", Presented at Electrochemical Society Meeting, Chicago, Ill., May 1973, EEA76-23409.

1062.Thorsen, A. C., Manasevit, H. M. and Harada, R. H., "Material and device properties of heteroepitaxial GaAs on BeO", Solid-State Electron., 17, 855-62 (1974), PA77-60889.

1063.Todd, C. D., "Silicon epitaxial F.E.T.'s: IV. F.E.T. as a voltage variable resistor", Electronic Compon., 7, 347-50 (1966), EEA70-17044.

1064.Todd, C. D., "Silicon epitaxial F.E.T.'s: IX. Follower circuits with bipolar transistors", Electronic Compon., 7, 943-8 (1966), EEA70-6812.

1065.Todd, C. D., "Silicon epitaxial F.E.T.'s: X. Negative resistance circuits", Electronic Compon., 7, 1041-5 (1966), EEA70-6813.

1066.Uddin, M. E., "A study of the design and construction of an improved epitaxial silicon solar cell", Thesis, Univ. Coll., Swansea, Wales, 1973, PA77-57722.

D. DIELECTRIC THIN FILMS

I. Bibliographies

1. Agarwal, V. K., "Breakdown conduction in thin dielectric films: A bibliographical survey", Thin Solid Films, 24 (1), 55-70 (1974), EEA78-2942.
2. Hampshire, M. J., "Insulating films", In: Digest of literature on dielectrics Vol. 35, 1971, Budenstein, P. P., (Ed.), Nat. Acad. Sci., Washington, D. C., p. 372-411 (1973), EEA77-2489.

II. Books

3. Diggle, J. W., (Ed.), "Oxides and oxide films, Volume 1", Marcel Dekker, Inc., New York, 1972.
4. Diggle, J. W., (Ed.), "Oxides and oxide films, Volume 2", Marcel Dekker, Inc., New York, 1973.
5. Heavens, O. S., "Thin film physics", Barnes and Noble Inc., New York, 1970.
6. Kooi, E., "Surface properties of oxidized silicon", Springer-Verlag, New York, 1967.
7. Leaver, K. D. and Chapman, B. N., "Thin films", Springer-Verlag, New York, 1971.
8. Milek, J. T., "Silicon nitride for microelectronic applications, Part 1, Preparation and properties", IFI/Plenum, New York, 1971.
9. Milek, J. T., "Silicon nitride for microelectronic applications, Part 2, Applications and devices", IFI/Plenum, New York, 1972.
10. Murt, E. M. and Guldner, W. G., (Eds.), "Physical measurement and analysis of thin films", Plenum Press, New York, 1969.
11. Vratny, F., (Ed.), "Thin film dielectrics", The Electrochemical Society, Inc., New York, 1969.

III. Review Articles

12. Balk, P., "Layered dielectrics in the MOS technology", 3rd European Solid State Device Research Conference, 1973, p. 51-82, Publ. 1974, PA77-45011.
13. Deal, B. E., "The current understanding of charges in the thermally oxidized silicon structure", J. Electrochem. Soc., 121, 198C-205 (1974), EEA77-30227.
14. Deal, B. E., Snow, E. H. and Grove, A. S., "Properties of the silicon dioxide-silicon system", Semiconductor Prod. Solid State Technol., 9 (11), 25-33 (1966), EEA70-2637.
15. Donovan, R. P., "Formation and properties of oxides on silicon", Proc. 4th Ann. Microelectronics Symp., 1965, p. 4A-1-4A-23.

16. Grove, A. S., Deal, B. E., Snow, E. H. and Sah, C. T., "Investigation of thermally oxidized silicon surfaces using metal-oxide-semiconductor structures", Solid-State Electron., 8, 145-163 (1965).
17. Hseuh, Y. W., "Titanium dioxide thin film in microelectronics", Thesis, Univ. Maryland, College Park, 1974, 138 pp., Order No. 74-22675, EEA78-5269.
18. Kerr, D. R., "A review of instability mechanisms in passivation films", The Annual Reliability Physics Symposium, 1970, p. 2-3, EEA73-28224.
19. McMillan, R. E. and Misra, R. P., "Insulating materials for semiconductor surfaces", IEEE Trans. Elect. Insulation, EI-5, 10-18 (1970), EEA73-21185.
20. Nicollian, E. H., "Surface passivation of semiconductors", J. Vac. Sci. & Technol., 8, S39-49 (1971), EEA75-4633.
21. Pliskin, W. A., "The evaluation of thin film insulators", Thin Solid Films, 2, 1 (1968).
22. Pliskin, W. A., "Use of infrared spectroscopy for the characterization of dielectric films on silicon", In: Semiconductor silicon, Huff, H. R. and Burgess, R. R., (Eds.), Electrochemical Society, New York, p. 506-29 (1973).
23. Pliskin, W. A., Kerr, D. R. and Perri, J. A., "Thin glass films", Phys. Thin Films, 4, 257 (1967).
24. Reimer, H., "Survey of the properties of insulating layers used in solid state electronics and their applications", Z. Elektr. Inf. - & Energietech., 2, 180-2 (1972), German, EEA76-8401.
25. Revesz, A. G., "Noncrystalline silicon dioxide films on silicon - a review", J. Non-Cryst. Solids, 11, 309-30 (1973), EEA76-12515.
26. Schnable, G. L., Kern, W. and Comizzoli, R. B., "Passivation coatings on silicon devices", J. Electrochem. Soc., 122, 1092-103 (1975), EEA78-41396.
27. Schwartz, B., "The chemistry of oxides on semiconductors", Electrochemical Society Fall Meeting (Extended abstracts), 1974, p. 494, EEA78-37687.
28. Szedon, J. R., "Charge effects in dielectric films on silicon", In: Thin film dielectrics, Vratny, F., (Ed.), Electrochem. Soc., New York, p. 84-113 (1969), EEA73-10539.
29. Tarui, Y., "The importance and research problems of silicon oxide", Bull. Electrotech. Lab., 30 (1), 1-7 (1966), Japanese, EEA70-2253.
30. Zaininger, K. H. and Wang, C. C., "Thin film dielectric materials for microelectronics", Proc. IEEE, 57, 1564-70 (1969), EEA73-456.

IV. Film Preparation

1. Equipment

31. Benzing, W. C., Rosler, R. S. and East, R. W., "A production reactor for continuous deposition of silicon dioxide", Solid State Technol., 16 (11), 37-42 (1973), EEA77-10123.
32. Budo, Y., Kulak, J. E. and Law, J., "Automated RF sputtering apparatus", IBM Tech. Disclosure Bull., 17, 2324-5 (1975), EEA78-19858.
33. Davidse, P. D., "R.F. cathodic sputtering apparatus having an electrically conductive housing", Patent USA 3471396, Publ. October 1969, EEA73-9669.
34. Fabian, L., "Processes and equipment for chemical vapour deposition of thin films", Finommech. & Mikrotech., 12, 201-5 (1973), Hungarian, PA77-10524.
35. Kaminski, L., "An arrangement for programmed oxidation of silicon materials", Elektronika, 15 (1), 14-16 (1974), Polish, EEA79-30150.
36. Leibenguth, R. E., "Self-regulating evaporator for growth of silicon dioxide films on semiconductor materials", Tech. Dig., no. 23, 35-6 (1971), EEA74-33981.
37. Olmstead, J. A. and Ronen, R. S., "Simple, efficient, clean-steam generator for oxidation of semiconductor wafers", Rev. Sci. Instrum., 46, 110-12 (1975), EEA78-19681.
38. Wohlheiter, V. D. and Whitner, R. A., "High production system for the deposition of silicon dioxide", Presented at the Electrochemical Society Spring Meeting, Toronto, Canada, May 1975.

2. Oxidation Kinetics

39. Burkhardt, P. J. and Gregor, L. V., "Kinetics of the thermal oxidation of silicon in dry oxygen", Trans. Metal. Soc. AIME, 236, 299-305 (1966).
40. Deal, B. E. and Grove, A. S., "General relationship for the thermal oxidation of silicon", J. Appl. Phys., 36, 3770-3778 (1965).
41. Ghez, R. and van der Meulen, Y. J., "Kinetics of thermal growth of ultra-thin layers of SiO_2 on silicon. II. Theory", J. Electrochem. Soc., 119, 1100-6 (1972), PA75-71337.
42. Grigonis, A. I. and Pranyavichyus, L. I., "The formation and removal of SiO_2 films on silicon in gas discharges", Litov. Fiz. Sb., 13 (3), 433-7 (1973), Russian, EEA77-2480.
43. Hess, D. W. and Deal, B. E., "Kinetics of the thermal oxidation of silicon in O_2/N_2 mixtures at 1200°C", J. Electrochem. Soc., 122, 579 (1975).
44. Junghans, R., "The heterojunction Si-SiO_2 in thermodynamical equilibrium", Z. Elektr. Inf. & Energietch., 2, 221-3 (1972), German, EEA76-8655.

45.Kang, K. D., "The mechanisms for oxidation of silicon and diffusion of donor and acceptor impurities into the oxide for a silicon-silicon dioxide system", Thesis, Ohio State University, Columbus, Ohio, 1962, Order No. 63-4672.
46.Laverty, S. J. and Ryan, W. D., "The kinetics of oxide growth during thermal oxidation of silicon", Internat. J. Electronics, 26, 519-28 (1969), EEA73-10357.
47.Ligenza, J. R. and Spitzer, W. G., "The mechanisms for silicon oxidation in steam and oxygen", J. Phys. Chem. Solids, 14, 131-6 (1960).
48.Nakayama, T. and Collins, F. C., "Kinetics of thermal growth of SiO_2 films in water vapor-oxygen-Argon mixtures", J. Electrochem. Soc., 113 (7), 706-13 (1966).
49.Oakley, R. E. and Godber, G. A., "Growth of very thin oxide films on silicon for use in MNOS charge storage devices", Thin Solid Films, 9, 287-91 (1972), EEA75-7554.
50.Pranevychius, L. and Grigonis, A., "The formation of dielectric films in gas discharges", 2nd International Conference on Gas Discharges, 1972, p. 388-90, EEA75-35822.
51.Revesz, A. G. and Evans, R. J., "Kinetics and mechanism of thermal oxidation of silicon with special emphasis on impurity effects", J. Phys. Chem. Solids, 30, 551-64 (1969).
52.van der Meulen, Y. J., "The kinetics of thermal growth of ultrathin layers of SiO_2 on silicon", J. Electrochem. Soc., 119, 530-4 (1972), EEA75-698.
53.van der Meulen, Y. J. and Cahill, J. G., "Effects of HCl and Cl_2 additions on silicon oxidation kinetics", J. Electron. Mater., 3, 371-89 (1974), PA77-57707.
54.Virtmanis, A. S., Zhagata, L. A., Kalnynya, R. P., Feltyn, I. A. and Freiberga, L. A., "Growth rate influence on SiO_2 film properties", Latv. PSR Zinat. Akad. Vestis Fiz. Teh. Ser., no. 4, 8-13 (1973), Russian, EEA77-18750.

3. Techniques

a. Thermal Oxidation

55.Aboaf, J. A., "Formation of 20-25A thermal oxide films on silicon at 950°-1140°C", J. Electrochem. Soc., 118, 1370-1 (1971), EEA74-33646.
56.Arnold, H., "The electrochemistry of the thermal oxidation of silicon", Wiss. Z. Tech. Hochsch. Ilmenau, 20, 171-80 (1974), German, EEA78-19692.
57.Bruchez, J. A., "Silicon semiconductor devices utilizing enhanced thermal oxidation", Patent USA 3886004, Publ. May 1975, CA83-107220.
58.Cahill, J. G., van der Meulen, Y. J. and Zirinsky, S., "Control over the growth of thin SiO_2", IBM Tech. Disclosure Bull., 16, 3058-9 (1974), EEA77-22842.

59. Cohen, R. A. and Wheeler, R. K., "Method and apparatus for stable silicon dioxide layers on silicon grown in silicon nitride ambient", Patent USA 3796592, Publ. March 1974.
60. Devereux, O. F., Wang, R. Y. and Chien, K. H., "The effect of sodium oxide on the thermal oxidation of silicon", J. Electrochem. Soc., 118, 1147-52 (1971).
61. Esch, R. P. and Huang, P. C. S., "Controlling insulating film thickness across a semiconductor wafer", Patent USA 3899372, Publ. August 1975, CA83-171676.
62. Evitts, H. C., Cooper, H. W. and Flaschen, S. S., "Rate of formation of thermal oxides of silicon", J. Electrochem. Soc., 111, 688-90 (1964).
63. Fehlner, F. P., "Formation of ultrathin oxide films on silicon", J. Electrochem. Soc., 119, 1723-7 (1972), EEA76-8531.
64. Frantsuzov, A. A. and Makrushin, N. I., "Temperature dependence of oxidation rate in clean Ge(111)", Surf. Sci., 40, 320-42 (1973), PA77-6563.
65. Fritzsche, C. R. and Rothermund, W., "Thermal oxidation of silicon after ion implantation", J. Electrochem. Soc., 120, 1603-5 (1973), EEA77-2467.
66. Goodman, A. M. and Breece, J. M., "Thin tunnelable layers of silicon dioxide formed by oxidation of silicon", J. Electrochem. Soc., 117, 982-4 (1970), EEA73-35199.
67. Hass, G. A. and Gray, H. F., "Oxidation of Si surfaces", J. Appl. Phys., 46, 3885-7 (1975).
68. Hile, J. W. and Chen, M. C., "Thermal oxidation of silicon", Patent USA 3837905, Publ. September 1974, CA81-143306.
69. Hirabayashi, K. and Iwamura, I., "Kinetics of thermal growth of $HCl-O_2$ oxides on silicon", J. Electrochem. Soc., 120, 1595-1601 (1973).
70. Hopper, M. A., Clarke, R. A. and Young, L., "Thermal oxidation of silicon: in situ measurement of the growth rate using all ellipsometry", J. Electrochem. Soc., 122 (9), 1216-22 (1975).
71. Irene, E. A., "The effects of trace amounts of water on the thermal oxidation of silicon in oxygen", J. Electrochem. Soc., 121, 1613-16 (1974), EEA78-8450.
72. Kassabov, J. D. and Georgiev, V. K., "SiO_2 for MOST and MOS IC obtained by oxidation in water vapour", C. R. Acad. Bulg. Sci., 25, 893-6 (1972), EEA75-852.
73. Kemnitz, K. H. and Malin, K., "Water injection system for thermal oxidation of semiconductors", IBM Tech. Disclosure Bull., 17, 2064 (1975), EEA78-19640.
74. Korwin-Pawlowski, M. L. and Heasell, E. L., "Thermal oxide layers on indium antimonide", Phys. Status Solidi A, 27, 339-46 (1975), EEA78-11518.
75. Kriegler, R. J., Cheng, Y. C. and Colton, D. R., "The effect of HCl and Cl_2 on the thermal oxidation of Si", J. Electrochem. Soc., 119, 388-92 (1972).

76.Law, J. T., "The interaction of oxygen with clean silicon surfaces", J. Phys. Chem. Solids, 4, 91-100 (1958).
77.Lukes, F., "Oxidation of Si and GaAs in air at room temperature", Surface Sci., 30, 91-100 (1972).
78.Mohr, T. O., "Silicon and silicon-dioxide processing for high-frequency M.E.S.F.E.T. preparation", IBM J. Res. Develop., 14, 142-7 (1970), EEA73-28215.
79.Margalit, S., Bar-Lev, A., Kuper, A. B., Aharoni, H. and Neugroschel, A., "Oxidation of silicon-germanium alloys", Presented at Israel Physical Society 1972 Annual Meeting Program and Abstracts, Jerusalem, Israel, May 1972, PA75-75132.
80.Murarka, S. P., "Thermal oxidation of GaAs", Appl. Phys. Lett., 26, 180-1 (1975), EEA78-19619.
81.Nagasima, N., "Structure analysis of silicon dioxide films formed by oxidation of silane", J. Appl. Phys., 43, 3378-86 (1972).
82.Ono, K. and Sunohara, Y., "Characteristics of thermal grown silicon oxide film", Rev. Elec. Commun. Lab., 20, 105-15 (1972), EEA75-26208.
83.Ono, K., Sunohara, Y. and Inoue, T., "Characteristics of thermal-grown silicon dioxide films", Electron. & Commun. Jap., 53 (10), 158-65 (1970), EEA75-10176.
84.Oren, R. and Ghandhi, S. K., "Ultraviolet-enhanced oxidation of silicon", J. Appl. Phys., 42, 752-6 (1971).
85.Ota, Y. and Butler, S. R., "Reexamination of some aspects of thermal oxidation of silicon", J. Electrochem. Soc., 121, 1107-11 (1974), EEA77-38758.
86.Pammer, E., "Superficial oxidation of germanium crystals", Patent German 2208806, Publ. September 1973, CA80-8333.
87.Phillips, G. J., "Fabrication of GaAsP MIS capacitors using a thermal-oxidation dielectric-growth process", J. Electrochem. Soc., 120, 1089-91 (1973), EEA76-35211.
88.Pliskin, W. A., "Separation of the linear and parabolic terms in the steam oxidation of silicon", IBM J. Res. & Develop., 10, 198-206 (1966).
89.Pliskin, W. A. and Gnall, R. P., "Evidence for oxidation growth at the oxide-silicon interface from controlled etch studies", J. Electrochem. Soc., 111, 872 (1964).
90.Powell, R. J., Ligenza, J. R. and Schneider, M. S., "Selective oxidation of silicon in low-temperature high-pressure steam", IEEE Trans. Electron Devices, ED-21, 636-40 (1974), EEA77-38751.
91.Sharma, K. K., Floerke, O. W., Gugel, E. and Sahl, K., "Behavior of silicon carbide under high-temperature oxidizing and hydrothermal conditions", Sprechsaal Keram., Glas. Baustoffe, 107, 418-20, 422, 424 (1974).
92.Singh, B. R., Tyagi, B. D. and Marathe, B. R., "On the role of protons in thermally grown silicon dioxide films", Phys. Status Solidi A, 19, K143-7 (1973), EEA77-325.

93. Smith, T. and Carlan, A. J., "Oxidation of n-type Si in the 10-1400A oxide thickness range", J. Appl. Phys., 43 (5), 2455-9 (1972).
94. Smithies, S. A., "A practical low temperature MOS gate oxidation technique", Microelectronics, 6 (4), 49-50 (1975).
95. Spitzer, W. G. and Legenza, J. R., "Oxygen exchange between silica and high pressure steam", J. Phys. Chem. Solids, 17, 196-202 (1961).
96. van der Meulen, Y. J., Osburn, C. M. and Ziegler, J. F., "Properties of SiO_2 grown in the presence of HCl or Cl_2", J. Electrochem. Soc., 122, 284-90 (1975), EEA78-19710.
97. von Munch, W. and Pfaffeneder, I., "Thermal oxidation and eletrolytic etching of silicon carbide", J. Electrochem. Soc., 122, 642-3 (1975), PA78-51112.
98. Zeto, R. J., Hryckowian, E., Bosco, C. D., Iafrate, G. J., Brower, R. W. and Thornton, C. G., "High-pressure dry thermal oxidation of silicon", Electrochemical Society Fall Meeting (Extended abstracts), 1974, p. 504-5, EEA78-37688.

b. Sputtering

99. Davidse, P. D. and Maissel, L. I., "Dielectric thin films through rf sputtering", J. Appl. Phys., 37, 574-9 (1966).
100. Erdesz, E., "Silicon-nitride thin films produced by reactive sputtering", Finommech. & Mikrotech., 12, 263-6 (1973), Hungarian, EEA77-18693.
101. Erler, H. J. and Weissmantel, C., "Production of nitride layers by reactive ion beam sputtering", Finommech. & Mikrotech., 13, 1-4 (1974), Hungarian, EEA77-18694.
102. Jorgenson, G. V. and Wehner, G. K., "Sputtering studies of insulators by means of langmuir probes", J. Appl. Phys., 36, 2672-3 (1965).
103. Kalnina, R., Penezis, I., Feltin, I., Freiberga, L., Eglitis, I. and Elimanis, I., "Properties of boron-doped silicon dioxide film during plasma deposition", Latv. PSR Zinat. Akad. Vestis, Fiz. Teh. Zinat Ser., no. 6, 31-4 (1974), Russian, CA82-164088.
104. Kennedy, T. N., "Preparation and properties of rf (radio frequency) sputtered aluminum oxide", Proc. Tech. Program, Semicond. Process. Conf., 1973, p. 63-70, CA81-83476.
105. Kominiak, G. J., "Silicon nitride films by direct RF sputter deposition", J. Electrochem. Soc., 122, 1271-3 (1975).
106. Kushmir, A. J., "Electrical properties of RF sputtered silicon dioxide films", Presented at Electrochemical Society Fall meeting, Atlantic City, N. J., October 1970, EEA74-7193.
107. Lechaton, J. S., "Depositing a sputtered SiO_2 film having maximum planarization", IBM Tech. Disclosure Bull., 17, 2270-1 (1975), EEA78-24463.

108.Mogab, C. J., Petroff, P. M. and Sheng, T. T., "Effect of reactant nitrogen pressure on the microstructure and properties of reactively sputtered silicon nitride films", J. Electrochem. Soc., 122, 815-22 (1975), PA78-62554.
109.Ohwada, K., Sakuma, K., Kiuchi, K. and Tomimuro, H., "SiO_2 thin-film capacitor prepared by RF sputtering", Electr. Commun. Lab. Tech. J., 24, 555-74 (1975), Japanese, EEA78-37598.
110.Orlinov, V. I., Goranchev, B. G. and Choubriev, Z. R., "Determination of minimum oxygen content in gas flow for the preparation of SiO_2 films by D.C. reactive sputtering", C. R. Acad. Bulg. Sci., 27, 325-8 (1974), PA77-67709.
111.Orris, E. and Shiota, P., "Low-temperature deposition of silicon oxynitride and silicon oxide by decomposition of organo-silicates at an RF electrode", Presented at 133rd meeting of the Electrochemical Society, Boston, Mass., May 1968, EEA73-13548.
112.Pliskin, W. A., Davidse, P. D., Lehman, H. S. and Maissel, L. I., "Properties of insulating thin films deposited by RF sputtering", IBM J. Res. & Develop., 11, 461-7 (1967).
113.Reindl, K. and Widmann, D., "Study of sputtered SiO_2 layers on Si single crystal", Solid State Electron., 13, 1445-50 (1970), German, EEA74-1511.
114.Sugano, T. and Mori, Y., "Oxidation of gallium arsenide phosphide ($GaAs_{1-x}P_x$) surface by oxygen plasma and properties of oxide film", J. Electrochem. Soc., 121, 113-18 (1974), CA80-53614.
115.Terry, L., "Sputtering and its effects on semiconductor devices", Presented at Electrochemical Society Meeting, San Francisco, Calif., May 1974, EEA77-26481.
116.Valletta, R. M., Perri, J. A. and Riseman, J., "Reactively sputtered silicon dioxide films", Electrochem. Technol., 4, 402-6 (1966).
117.Wehner, G. K., "Sputter deposition of dielectric films", In: Thin film dielectrics, Vratny, F., (Ed.), Electrochem. Soc., New York, p. 117-29 (1969), EEA73-9897.

c. Pyrolytic (CVD)

118.Airey, A. C., Clarke, S. and Popper, P., "Pyrolytic silicon nitride coatings", Proc. Brit. Ceram. Soc., no. 22, 305-20 (1973).
119.Avigal, Y., Beinglass, I. and Schieber, M., "A new method for chemical vapor deposition of silicon dioxide", J. Electrochem. Soc., 121, 1103-7 (1974), EEA77-38757.
120.Bratter, R. L. and Gaind, A. K., "Method of chemical vapor deposition to provide silicon dioxide films with reduced surface state charge on semiconductor substrates", Patent USA 3887726, Publ. June 1975.
121.Chu, T. L. and Kelm, R. W., Jr., "Preparation and properties of aluminum nitride films", J. Electrochem. Soc., 122, 995-1000 (1975), CA83-88843.

122. Feltl-Pohl, U., Herbst, J. and Splittgerber, H., "The influence of deposition parameters on growth rates and pinholes of pryolytic SiO_2 layers and on interface states", Electrochemical Society Fall Meeting (Extended abstracts), 1974, p. 498-500, PA78-73315.
123. Fremunt, R., "Preparation of thin silicon dioxide layers on silicon substrates by thermal decomposition of tetraethoxisilane", Elektrotech. Cas., 24, 77-86, Czech, EEA76-12435.
124. Gaind, A. K., Ackermann, G. K., Nagarajan, A. and Bratter, R. L., "Preparation and properties of CVD oxides with low charge levels from SiH_4-CO_2-HCl-H_2", Presented at Electrochemical Society Spring Meeting, Toronto, Canada, May 1975, EEA78-37451.
125. Georgiev, G. I., Dzhoglev, D. and Stamenov, Z. I., "Preparation of pyrolytic silicon dioxide. I. Preparation under vacuum", Khim.-Tekhnol. Inst., Sofia, 1972, 20 (2), 77-85 (1974), Bulgarian, CA83-171571.
126. Grabowska, K. and Kosmowska, B., "Methods used to obtain the SiO_2 layers by silicon compound decomposition", Elektronika, no. 7-8, 283-90 (1972), Polish, EEA75-39480.
127. Gurovic, J. and Vrana, M., "Al_2O_3 layers and their use in semiconductor engineering", Elektrotech. Cas., 25, 53-65 (1972), PA77-41500.
128. Gurskii, L. I., Koleshko, V. M., Reznikov, B. S. and Kovalevskii, A. A., "Influence of the substrate on the rate of growth of silicon nitride films", Fiz. & Khim. Obrab. Mater., no. 4, 45-7 (1974), Russian, PA78-10494.
129. Ishii, E., Tabata, H. and Okuda, H., "Growth of γ-Al_2O_3 single crystal by chemical vapor reaction method. I. Formation of γ-Al_2O_3 by vapor phase reaction of $AlCl_3$ with H_2O", Rep. Gov. Ind. Res. Inst., 22, 401-8 (1973), Japanese, PA77-79868.
130. Kanbara, G., Koike, S., Matsuda, T. and Inoue, M., "Deposition and properties of silicon nitride films", Natl. Tech. Rep., 20, 209-16 (1974), Japanese, PA77-82821.
131. Kern, W. and Fisher, A. W., "Deposition and properties of silicon dioxide and silicate films prepared by low-temperature oxidation of hydrides", RCA Rev., 31, 715-27 (1970), EEA74-13190.
132. Kijima, K., Setaka, N. and Tanaka, H., "Preparation of silicon nitride single crystals by chemical vapor deposition", J. Cryst. Growth, 24-25, 183-7 (1974), EEA78-5247.
133. Kroll, W. J., Jr., Titus, R. L. and Wagner, J. B., Jr., "Formation of silica films on silicon using silane and carbon dioxide", J. Electrochem. Soc., 122, 573 (1975), PA78-47452.
134. Leuenberger, F., "Vapour-deposited silicon dioxide for device applications", Thin Solid Films, 22, 245-53 (1974), PA77-67742.
135. Lisenker, B. S., Lisovenko, V. D., Maronchuk, I. Y., Maronchuk, Y. Y. and Senoshenko, O. V., "Growth and electrical characteristics of silicon nitride on gallium arsenide", Poluprovodn. Tekh. & Mikroelektron., no. 16, 73-7 (1974), Russian, PA77-76405.

136.Mellottee, H. and Delbourgo, R., "Deposition of silicon nitride films by pyrolysis of silane-ammonia mixtures", Bull. Soc. Fr. Ceram., no. 102, 65-73 (1974), French, PA77-79860.
137.Micheletti, F. B., Norris, P. E., Schlesier, K. M. and Shaw, J. M., "Pyrohydrolytic Al_2O_3 for MOS applications", Solid State Technol., 14 (12), 37-42 (1971), EEA75-27762.
138.Mollimard, D., Mellottee, H. and Delbourgo, R., "Effect of deposit conditions on the microhardness of pyrolytic silicon nitride", Bull. Soc. Fr. Ceram., no. 105, 11-15 (1974).
139.Mukherjee, S. P. and Evans, P. E., "The deposition of thin films by the decomposition of tetra-ethoxy silane in a radio frequency glow discharge", Thin Solid Films, 14, 105-18 (1972), EEA76-12470.
140.Nakai, Y., "Low-temperature deposition of silica films", Presented at 133rd meeting of the Electrochemical Society, Boston, Maws., May 1968, EEA73-13547.
141.Rand, M. J. and Roberts, J. F., "Silicon oxynitride films from the NO-NH_3-SiH_4 reaction", J. Electrochem. Soc., 120, 446-53 (1973), EEA76-23265.
142.Sato, J. and Maeda, K., "Very low temperature CVD of SiO_2 films using Ozone and Organosilane", Presented at spring meeting of the Electrochemical Society, Washington, D. C., May 1971, EEA74-36766.
143.Swann, R. C. G. and Pyne, A. E., "The preparation and properties of silica films deposited from silane and carbon dioxide", J. Electrochem. Soc., 116, 1014-17 (1969), EEA73-4417.
144.Teshima, H. and Takeda, O., "SiO_2 films grown from the tetraethoxysilane system", Bull. Electrotech. Lab., 33, 748-54 (1969), Japanese, EEA73-14719.
145.Teshima, H., Tarui, Y., Yoshida, Y. and Katayama, H., "Vapor growth of SiO_2 and TiO_2 films", Bull. Electrotech. Lab., 30 (1), 45-58 (1966), Japanese, PA70-1345.
146.Tietjen, J. J., "Chemical vapor deposition of electronic materials", In: 1973 Annual review of materials science, vol. 3, Huggins, R. A., (Ed.), Annual Reviews, Palo Alto, California, p. 317-26 (1973), EEA77-10134.
147.Tsokov, I. S., Tsalev, D. K., Faitondzhiev, L. P., Kotseva, I. K. and Pesheva, D. P., "Accelerated method for obtaining lamellar silicon nitride and silicon dioxide from the vapor and gas phase", Khim. Ind., 43, 302-3 (1971), Bulgarian, CA76-105321.
148.Yanazawa, H., Utsugi, H., Hashimoto, N. and Ashikawa, M., "Hydrophobic conversion of the chemical-vapor-deposited SiO_2 surface", Jap. J. Appl. Phys., Suppl. 2 (Pt. 1), 753-6 (1974), EEA78-19541.

d. Anodic

149.Berlicki, T. and Osadnik, S., "Selective oxidation of titanium films by anodization", Pr. Nauk. Inst. Technol. Elektron. Politech. Wrodaw, no. 5, 47-9 (1971), Polish, CA76-91550.

150. Card, F. E., "Anodic aluminum oxide on silicon", Presented at Electrochemical Society Meeting, Chicago, Ill., May 1973, EEA76-23267.
151. Choo, Y. H. and Devereux, O. F., "Barrier-type aluminum oxide films formed under prolonged anodizing. I. Influence of anodizing parameters on film morphology", J. Electrochem. Soc., 122 (12), 1645-53 (1975).
152. Dell'Oca, C. J., "Properties of anodic oxide films formed in the anodization of silicon nitride", J. Electrochem. Soc., 120, 1225-30 (1973), EEA77-2454.
153. Ermanis, F. and Schwartz, B., "Anodic oxidation of gallium phosphide in aqueous hydrogen peroxide", J. Electrochem. Soc., 121, 1665-7 (1974), PA78-16089.
154. Haden, C. R., Barrett, J. L. and Stone, J. L., "High-quality dielectric suitable for use with amorphous semiconductors", IEEE J. Solid-State Circuits, SC-9, 118-24 (1974), PA77-49581.
155. Hasegawa, H., Forward, K. E. and Hartnagel, H., "Improved method of anodic oxidation of GaAs", Electron. Lett., 11, 53-4 (1975), PA78-30233.
156. Hasegawa, H., Forward, K. E. and Hartnagel, H., "New anodic native oxide of GaAs with improved dielectric and interface properties", Appl. Phys. Lett., 26, 567-9 (1975), PA78-62534.
157. Huez, M., Restelli, G. and Manara, A., "On the anodic oxide grown on silicon carbide", Thin Solid Films, 23, S33-6 (1974), PA77-72663.
158. Kuhn, M. and Ligenza, J. R., "Plasma grown Al_2O_3-SiO_2 MOS structures", Presented at Electrochemical Society Fall Meeting, Atlantic City, N. J., October 1970, EEA74-7672.
159. Libsch, T. A. and Devereux, O. F., "Barrier-type aluminum oxide films formed under prolonged anodizing. II. Dielectric characteristics", J. Electrochem. Soc., 122 (12), 1654-60 (1975).
160. Ligenza, J. R., "Anodic plasma oxidation of silicon", Presented at Spring meeting of the Electrochemical Society, Washington, D. C., May 1971, EEA74-33652.
161. Ligenza, J. R., Kuhn, M. and Lopez, A. D., "Surface preparation of silicon for plasma anodization", Presented at Electrochemical Society Fall meeting, Atlantic City, N. J., October 1970, EEA74-7679.
162. Muller, H., Eisen, F. H. and Mayer, J. W., "Anodic oxidation of GaAs as a technique to evaluate electrical carrier concentration profiles", J. Electrochem. Soc., 122, 651-5 (1975), EEA78-24124.
163. Perkins, K. D. and Sanders, I. R., "Anodic process for semiconductor fabrication", Patent UK 1346496, Publ. February 1974, CA81-57610.
164. Pulfrey, D. L. and Reche, J. J. H., "Preparation and properties of plasma-anodized silicon dioxide films", Solid-State Electron., 17 (6), 627-32 (1974), EEA77-22971.

165.Restelli, G., Ostidich, A. and Manara, A., "Anodic oxidation of silicon carbide (oxide film growth)", Thin Solid Films, 23, 23-9 (1974), EEA77-36237.
166.Rhines, W. C., Reinberg, A. R., Hewes, C. R. and Collins, C. A., "Growth and characterization of insulator films on Ge and Si by plasma annodization", Presented at Electrochemical Society Meeting, San Francisco, Calif., May 1974, PA77-45464.
167.Schwartz, B., "Surface oxidation of gallium phosphide and gallium arsenide semiconductors", Patent German 2259829, Publ. July 1973, CA79-98471.
168.Schwartz, B., "Electrolytic oxidation of silicon on semiconductor surfaces", Patent USA 3764491, Publ. October 1973, CA79-150795.
169.Spitzer, S. M., Schwartz, B. and Weigle, G. D., "Preparation and stabilization of anodic oxides on GaAs", J. Electrochem. Soc., 122, 397-402 (1975), EEA78-19650.

e. Evaporation

170.Delzer, D. R., "Vacuum evaporated dielectrics in MOS structures", Thesis, Oregon State Univ., Corvallis, 163 pp., Order No. 69-5826, EEA73-21194.
171.Grannemann, W. W., Kim, T. W., Deokar, V. D. and Cheng, C. C., "A comparative study of silicon and $GaAs_{0.5}P_{0.5}$ MIS capacitors fabricated by using electron-beam-evaporated Al_2O_3", 23rd Electronic Components Conference, 1973, p. 273-6, EEA76-29215.
172.Gosney, W. M. and Muller, R. S., "Aluminum oxide films made from evaporated sapphire", Thin Solid Films, 14, 255-66 (1972), EEA76-16343.
173.Kassabov, J., Popova, L., Toncheva, L. and Stoeva, R., "Fast-forming of thick SiO_2 layers on silicon wafers", C. R. Acad. Bulg. Sci., 22, 1003-6 (1969), EEA73-17422.
174.Kusaka, M., Kanakura, M. and Okazaki, S., "Aluminium oxide films prepared by electron beam evaporation of sapphire", Jap. J. Appl. Phys., Suppl. 2 (pt. 1), A-864 (1974), EEA78-24249.
175.Kusaka, M., Tamura, S., Kanakura, M. and Okazaki, S., "Al_2O_3 films prepared by electron-beam evaporation of sapphire", Rep. Res. Lab. Surf. Sci. Okayama Univ., 3, 327-31 (1972), EEA76-23391.
176.Munro, P. C. and Thompson, H. W., Jr., "Electron-beam evaporated Al_2O_3 on Si", J. Electrochem. Soc., 122, 127-32 (1975), PA78-26306.
177.Pliskin, W. A. and Castrucci, P. P., "Reactivity and bond strain of films formed by electron-gun-evaporation of silicon dioxide", Electrochem. Technol., 6, 85-8 (1968).

f. Others

178. Alessandrini, E. I. and Campbell, D. R., "Catalyzed crystallization in SiO_2 thin films", J. Electrochem. Soc., 121, 1115-18 (1974), PA77-75730.
179. Bautsch, H. J. and Ickert, L., "Structure and preparation of alumina films", Z. Elektr. Inf.- & Energietech., 2, 196-9 (1972), German, EEA76-8544.
180. Bykov, V. I., Volkov, V. L., Kovarskii, V. A., Lepilina, R. G. and Fratkin, Z. G., "Crystallization of synthetic silicon dioxide", Tr. Leningrad. Nauch.-Issled. Prockt. Inst. Osn. Khim. Prom., no. 6, 122-6 (1972), Russian, CA80-7808.
181. Goodman, A. M. and Breece, J. M., "Provision of reproducible thin layers of silicon dioxide", Patent USA 3892891. Publ. July 1975.
182. Irene, E. A., "Preparing thin ($<$100Å) SiO_2 films", IBM Tech. Disclosure Bull., 17, 3094-5 (1975), EEA78-24162.
183. MacRae, A. U. and Moline, R. A., "Production of SiO_2 tapered films", Patent USA 3769109, Publ. October 1973.
184. Ramsey, T. H. and Rice, S. B., "Effects of heat treatment on electron-gun-deposited SiO_2 films", J. Amer. Ceram. Soc., 52 (4), 225-6 (1969), EEA73-789.
185. Ruzicka, M., "Microwave glow discharge deposition of thin silicon nitride films", Czech. J. Phys. B, B24, 465-8 (1974), PA77-61292.
186. Sterling, H. F., Alexander, J. H. and Hughes, S. Y., "Deposition of a cohesive, solid layer of silicon nitride on a substrate", Patent German 1621234, Publ. November 1974, CA83-52065.
187. Sulimin, A. D., Ostashkin, L. P., Neustroev, S. A., Yakovenko, V. G. and Lukanov, N. M., "Investigation into certain properties of a Si-SiO_2 system produced by the plasma-chemical deposition of silicon dioxide", Fiz. & Khim. Obrab. Mater., no. 2, 57-60 (1975), Russian, EEA78-37461.

V. Physical Measurements

1. Structural Defects

188. Abrahams, M. S., Buiocchi, C. J., Rayl, M. and Wojtowicz, P. J., "Transmission electron microscopy of granular nickel/SiO_2 cermet films", J. Appl. Phys., 43, 2537-41 (1972).
189. Amendola, A., Anolick, E. S. and Levitz, P. J., "Predicting SiO_2 failure rates from oxide defect densities", IBM Tech. Disclosure Bull., 14, 3345-7 (1972), EEA75-27744.
190. Anolick, E. S., "The use of self-healing breakdown to determine the distribution of defect densities in SiO_2 films", Presented at Electrochemical Society Meeting, San Francisco, Calif., May 1974, PA77-44529.

191. Beezhold, W., "Proton-induced characteristic X-ray analysis of Na and Cl impurity atoms in SiO_2 thin films", Appl. Phys. Lett., 24, 540-2 (1974), EEA77-30140.
192. Benes, O. and Cermy, A., "The influence of the SiO_2 structure on the properties of a MOS system", Wiss. Z. Elektrotech., 11, 213-32 (1968), EEA72-2833.
193. Blumenthal, R. N., Baukus, J. and Hirthe, W. M., "Studies of the defect structure of nonstoichiometric rutile, TiO_{2-x}", J. Electrochem. Soc., 114, 172-6 (1967), EEA70-10434.
194. Chou, N. J. and Eldridge, J. M., "On the statistics governing the defect density determination in SiO_2 films (integrated circuit yield)", J. Electrochem. Soc., 120, 1139-40 (1973), EEA76-35363.
195. Edel'man, F. L., Mel'nik, I. G., Latuta, V. Z., Ayupov, B. M. and et al., "Nature of flowerlike defects in SiO_2 on planar structures", Izv. Akad. Nauk SSSR Neorg. Mater., 8, 719-23 (1972).
196. Galstyan, V. G., Nosikov, S. V., Press, F. P. and Pastushkov, V. V., "The use of a scanning electron microscope for studying defects in insulator films and semiconductor structures", Radio Eng. & Electron. Phys., 17, 1522-8 (1972), EEA76-38227.
197. Gorban, H. P., Litovchenko, V. G., Romanyuk, B. N., Lerintsi, A., Nemet-Shallai, M. and Pataki, G., "Heterogeneity of surface characteristics of semiconductor/dielectric structures", Poluprovodn. Tekh. & Mikroelektron., no. 20, 35-45 (1975), Russian, PA78-69246.
198. Gorbatov, A. G. and Kamyshov, V. M., "Nature of the defects of silicon and aluminum nitrides", Sb. Nauch. Tr. Sverdlovsk. Inst. Nar. Khoz., no. 18, 53-74 (1970), Russian, CA77-67122.
199. Hickmott, T. W., "Thermoluminescence and color centers in rf-sputtered SiO_2 films", J. Appl. Phys., 43 (5), 2339-51 (1972).
200. Hickmott, T. W., "Defect centers in oxygen-deficient r.f.-sputtered SiO_2 films. I. Electron spin resonance", J. Appl. Phys., 45, 1050-9 (1974), PA77-56738.
201. Hickmott, T. W., "Defect centers in oxygen-deficient RF-sputtered SiO_2 films. II. Thermoluminescence", J. Appl. Phys., 45, 1060-70 (1974), PA77-56739.
202. Hufault, A. D. and Stabell, D. E., "Location of electrically conducting regions in SiO_2 films", IBM Tech. Disclosure Bull., 27, 1358 (1974), EEA78-8386.
203. Kasprzak, L., Laibowitz, R., Herd, S. and Ohring, M., "Nucleation of small metal particles on ultrathin SiO_2 films on Si", Thin Solid Films, 22, 189-96 (1974), EEA77-30201.
204. Kato, K., Inoue, Z., Kijima, K., Kawada, I. and et al., "Structural approach to the problem of oxygen content in alpha silicon nitride", J. Amer. Ceram. Soc., 58 (3-4), 90-91 (1975).
205. Kern, W., "Detection and characterization of localized defects in dielectric films", RCA Rev., 34, 655-90 (1973), EEA77-18718.

206.Kern, W., "Characterization of localized defects in dielectric films for electron devices. I", Solid State Technol., 17 (3), 35-42 (1974), EEA77-26377.
207.Kern, W., "Characterization of localized defects in dielectric films for electron devices. II", Solid State Technol., 17 (4), 78-84 (1974), EEA77-30014.
208.Litovchenko, V. G., Marchenko, R. I. and Romanova, G. F., "Application of secondary ion emission technique for the analysis of Si-SiO_2 structures", Poluprovodn. Tekh. & Mikroelektron., no. 11, 20-7 (1973), Russian, EEA76-29199.
209.Litovchenko, V. G., Zuev, V. A., Ilyushin, B. N., Mazurok, I. G. and Torchun, N. M., "Study of a transient destructive silicon layer in Si-SiO_2", Poluprovodn. Tekh. & Mikroelektron., no. 15, 24-9 (1974), Russian, PA78-3757.
210.Mayer, A., "Detection of damage on silicon surfaces: origin and propagation of defects", RCA Rev., 31, 414-30 (1970), EEA74-4866.
211.Meek, R. L. and Braub, R. H., "Devitrification of steam-grown silicon dioxide films", J. Electrochem. Soc., 119, 1538-44 (1972), EEA76-4512.
212.Morgan, D. V. and Gittens, R. P., "An investigation of the stoichiometry and impurity content of thin SiO_2 films using Rutherford scattering of MeV γ-particles", Phys. Status Solidi A, 13, 517-26 (1972).
213.Pliskin, W. A. and Lehman, H. S., "Structural evaluation of silicon oxide films", J. Electrochem. Soc., 112, 1013-9 (1965).
214.Prokhorov, V. I. and Sorokin, L. M., "Inheritance of defects in silicon by a growing oxide film", Inorg. Mater., 9, 28-30 (1973), EEA76-38188.
215.Ritter, E., "Knowledge of the SiO and Si_2O_3 phase in thin films", Optica Acta, 9, 197-202 (1962), German.
216.Schmidt, M. and Glaefeke, H., "Correlation between short range order defects and electron trap level spectra in amorphous SiO_2 layers obtained by TSEE technique", 4th International Symposium on Exoelectron Emission and Dosimetry, 1973, p. 122-6, Publ. 1974, PA77-49548.
217.Sigle, G. H., Jr., Friebele, E. J., Ginther, R. J. and Griscom, D. L., "Effects of stoichiometry on the radiation response of SiO_2", IEEE Trans. Nucl. Sci., NS-21, 56-61 (1974), PA78-41754.
218.Sigman, W. T., Chu, W. K. and Lugujjo, E., "Stoichiometry of thin silicon oxide layers on silicon", Appl. Phys. Lett., 24, 105-7 (1974), PA77-40991.
219.Swaroop, B., "Characterization of silicon dioxide films by the electron probe", J. Electrochem. Soc., 118, 913-18 (1971), EEA74-26057.
220.Thiemt, K. and Kaltofen, R., "Determination of pinhole density in dielectric layers by chemical etching. I. Silicon-silicon dioxide system", Krist. Tech., 10, 305-12 (1975), German, CA82-179227.

221. Williams, R. and Goodman, A. M., "Wetting of thin layers of SiO_2 by water", Appl. Phys. Lett., 25, 531-2 (1974), EEA78-11445.
222. Yamazaki, S. and Sugimura, Y., "Some properties of Si_3N_4 produced by SiH_4-NH_3 catalysis", Extended abstracts of the Spring meeting of the Electrochemical Society, 1971, p. 218-19, PA75-9257.
223. Yang, K. H., "Growth kinetics of stacking faults in thermally oxidized silicon", Presented at Electrochemical Society Spring Meeting, Toronto, Canada, May 1975.
224. Zak, J., Lis, D. and Maczynski, S., "Investigation of defects in SiO_2 layer by determining the activation energy of charge carrier diffusion", Electron Technol., 4 (3), 29-35 (1971), EEA75-13642.

2. Thickness

225. Adams, J. R. and Bashara, N. M., "SiO_2 thicknesses determination by reflection ellipsometry: substrate effects", Surf. Sci., 47, 655-60 (1975), PA78-33397.
226. Anand, K. V. and Momodu, S. K., "Applications of ellipsometry in semiconductor technology", Electron. Engineering, 47 (563), 51-3 (1975), EEA78-8383.
227. Archer, R. J., "Determination of the properties of films on silicon by the method of ellipsometry", J. Opt. Soc. Amer., 52, 970-7 (1962).
228. Archer, R. J. and Gobeli, G. W., "Measurement of oxygen adsorption on silicon by ellipsometry", J. Phys. Chem. Solids, 26, 343-51 (1965).
229. Bohg, A. and Briska, M., "Measuring the layer thickness of extremely thin SiO_2 films", IBM Tech. Disclosure Bull., 17, 2662-3 (1975), EEA78-19913.
230. Corl, E. A. and Wimpfheimer, A., "Thickness measurement of silicon dioxide layers by ultraviolet-visible interference method", Solid-State Electron., 7, 755-61 (1964).
231. Daneu, V., "Optical thickness monitor for thin-film deposition", Appl. Opt., 14, 962-69 (1975).
232. Ferguson, I. F., "An X-ray diffraction method for determining the thickness of coatings", UKAEA Diffraction Analysis Conference, 1972, p. 29-36, Publ. 1973, EEA76-39580.
233. Frenzel, M., "A contribution to the determination of penetration depths of dopants in silicon dioxide films on silicon", Krist. & Tech., 8, 965-73 (1973), German, EEA77-2476.
234. Golanski, A., "The utilization of neutron activation to determine the thickness of oxide layers on silicon", Elektronika, no. 11, 448-53 (1972), Polish, EEA76-8396.
235. Hauge, P. S. and Dill, F. H., "Design and operation of ETA, an automated ellipsometer (f.e.t. gate insulator thickness measurement)", IBM J. Res. & Dev., 17, 472-89 (1973), EEA77-18809.

236. Konnerth, K. L. and Dill, F. H., "IOTA, a new computer controlled thin film thickness measurement tool (applied to silicon semiconductor technology)", Solid-State Electron., 15, 371-80 (1972), EEA75-16651.
237. Konnerth, K. L. and Dill, F. H., "In-situ measurement of dielectric thickness during etching or developing processes", IEEE Trans. Electron Devices, ED-22, 452-6 (1975), EEA78-24395.
238. Navratil, K. and Ohlidal, I., "Control of the deposition of silicon nitride layers by 2537Å radiation. Comments", J. Electrochem. Soc., 120, 847-8 (1973), CA79-59084.
239. Patrin, N. A., "Optoelectronic determination of insulator thickness", IBM Tech. Disclosure Bull., 16, 2674 (1974), EEA77-18569.
240. Pliskin, W. A. and Aboaf, J. A., "Thickness and porosity measurement of porous silicon and porous silicon oxide", Presented at Electrochemical Society Spring Meeting, Toronto, Canada, May 1975.
241. Pliskin, W. A. and Conrad, E. E., "Nondestructive determination of thickness and refractive index of transparent films", IBM J Res. & Develop., 8, 43-51 (1964).
242. Pompei, J. and Vacon, Y. L., "Method of determining the thickness of a layer of dielectric material during its growth", Patent USA 3775277, Publ. November 1973.
243. Raider, S. I., Flitsch, R. and Palmer, M. J., "Oxide growth on etched silicon in air at room temperature", J. Electrochem. Soc., 122, 413 (1975).
244. Rand, M. J., "Spectrophotometric thickness measurement for very thin SiO_2 films on Si", J. Appl. Phys., 41, 787-90 (1970), EEA73-16983.
245. Sandera, M., "Dielectric film thickness measurement from the reflection relations", Solid-State Electron., 18, 110-12 (1975), EEA78-13225.
246. Sugawara, K., Yoshimi, T., Okuyama, H. and Shirasu, T., "In situ monitoring of film deposition using He-Ne laser system. I. Measurements of CVD insulating film at 6328 Å", J. Electrochem. Soc., 121, 1233-5 (1974), PA78-1698.
247. Van den Brekel, C. H. J. and Severin, P. J., "Control of the deposition of silicon nitride layers by 2537 Å radiation", J. Electrochem. Soc., 120, 848 (1973), CA79-59085.
248. Wesson, R. A., Phillips, R. P. and Pliskin, W. A., "Phase-shift-corrected thickness determination of silicon dioxide on silicon by ultraviolet interference", J. Appl. Phys., 38, 2455-60 (1967).
249. Wesson, R. A., Young, H. W. and Pliskin, W. A., "Reflectivity thickness corrections for silicon dioxide films on silicon", Appl. Phys. Lett., 11, 105-6 (1967).

3. Contamination and Impurities

250. Bogenschutz, A. F. and George, U., "The analysis investigation of doped silicon dioxide coats and doping gas mixture", Feinwerk-tech. & Messtech., 83, 114-16 (1975), German, EEA78-28609.
251. Burkhardt, P. J., "Tracer evaluation of hydrogen in steam-grown SiO_2 films", J. Electrochem. Soc., 114, 196-201 (1967).
252. Chang, C. C., Adams, A. C., Quintana, G. and Sheng, T. T., "Phosphorus concentration profiles in p-doped silicon dioxide measured using Auger spectroscopy", J. Appl. Phys., 45, 252-6 (1974), PA77-45453.
253. Chou, N. J., Osburn, C. M., van der Meulen, Y. J. and Hammer, R., "Auger analysis of chlorine in 'HCl- or Cl_2-grown' SiO_2 films", Appl. Phys. Lett., 22, 380-1 (1973), EEA76-23254.
254. Colton, D. R., Cheng, Y. C. and Kriegler, R. J., "Reducing the mobile ion contamination in thermally grown silicon dioxide", Patent Canada 936791, Publ. November 1973, CA80-53782.
255. Deal, B. E., Grove, A. S., Snow, E. H. and Sah, C. T., "Observation of impurity redistribution during thermal oxidation of silicon using the MOS structure", J. Electrochem. Soc., 112, 308-314 (1965).
256. Declerck, G. J., Hattori, T., May, J. A., Beaudouin, J. and Meindl, J. D., "Some effects of 'trichloroethylene oxidation' on the characteristics of MOS devices", J. Electrochem. Soc., 122, 436 (1975).
257. DiStefano, T. H. and Lewis, J. E., "The influence of sodium on the Si-SiO_2 interface", J. Vac. Sci. & Technol., 11, 1020-4 (1974), EEA78-24261.
258. Franz, I. and Langheinrich, W., "Tracer investigations on the diffusion of phosphorus from doped oxide sources", Solid-State Electron., 18, 209 (1975).
259. Kern, W., "Semiconductor surface contamination investigated by radioactive tracer techniques. I", Solid State Technol., 15 (1), 34-8 (1972), EEA75-28449.
260. Kerr, D. R., Logan, J. S., Burkhardt, P. J. and Pliskin, W. A., "Stabilization of SiO_2 passivation layers with P_2O_5", IBM J. Res. & Develop., 8, 376 (1964).
261. Komiya, Y. and Tarui, Y., "Comparison between the experimental and theoretical sodium distribution in silicon oxide", Japan. J. Appl. Phys., 8, 272 (1969), EEA72-11327.
262. Kotas, P. and Simkova, M., "Determination of the concentration profile of sodium in SiO_2 and Si_3N_4 layers from the MNOS technology by the method of neutron activation analysis", Elektrotech. Cas., 25, 46-52 (1972), EEA77-22833.
263. Kriegler, R. J., "Neutralization of Na^+ ions in 'HCl-grown' SiO_2", Appl. Phys. Lett., 20, 449-51 (1972), EEA75-24270.
264. Kriegler, R. J., "Dielectric loss and current-voltage measurements in sodium-contaminated Si-SiO_2-Cr structures", IEEE Trans. Electron Devices, ED-20, 722-31 (1973), EEA76-29191.

265. Kriegler, R. J., Cheng, Y. C. and Colton, D. R., "Silicon dioxide structure having resistance to mobile ion contamination", Patent Canada 936790, Publ. November 1973, CA80-53779.
266. Kucirek, J. and Melmed, A. J., "Influence of an initial (contamination) film on the determination of film properties by ellipsometry", J. Opt. Soc. Am., 65, 611-2 (1975), PA78-58776.
267. Kumar, S. and Gregor, L. V., "Distribution of sodium in sputtered SiO_2 as determined by neutron activation and u.v. spectrographic analyses", J. Electrochem. Soc., 120, 1285-7 (1973), PA77-6939.
268. Kushner, R. A., McCaughan, D. V., Murphy, V. T. and Heilig, J. A., "Mobilization of sodium in SiO_2 films by ion bombardment", Phys. Rev. B, 10, 2632-41 (1974), PA78-18466.
269. MacKenna, E. and Kodama, P., "The suppression of ionic contamination during silicon nitride deposition", J. Electrochem. Soc., 119, 1094-9 (1972), PA75-71336.
270. Marciniak, W. and Przewlocki, H. M., "Equilibrium distribution of the uncompensated mobile charge in the dielectric layer of a MOS structure", Phys. Status Solidi A, 24, 359-66 (1974), EEA77-33990.
271. Meek, R. L., "Residual chlorine in O_2:HCl grown SiO_2", J. Electrochem. Soc., 120, 308-10 (1973), EEA76-19834.
272. Nicollian, E. H., "Instabilities in thermally grown SiO_2", Presented at the Electrochemical Society Spring Meeting, Toronto, Canada, May 1975.
273. Paddock, A. D. and Harper, J. G., "Tolerance of packaged MOS devices to sodium ion contamination", Extended abstracts of the Spring Meeting of the Electrochemical Society, 1971, p. 93-4, EEA74-37314.
274. Raider, S. I., "Alkali ions in SiO_2 and their effect on electronic conduction and time-dependent capacitor breakdowns", Presented at Electrochemical Society Meeting, San Francisco, California, May 1974, BEA77-22939.
275. Raider, S. I. and Flitsch, R., "Mobile ion transfer to SiO_2 films from ethanol (MOS structures)", J. Electrochem. Soc., 118, 1011-12 (1971), EEA74-26708.
276. Sixt, G., Schulz, M. and Goetzberger, A., "Determination of cesium distributions in oxides of MOS structures by photo-injection studies", Appl. Phys., 4, 217-23 (1974), PA77-76393.
277. Splittgerber, H. and Takacs, D., "Influence of phosphorus content in pyrolytic oxide on n-silicon-gate device parameters", Electrochemical Society Fall Meeting (Extended abstracts), 1974, p. 464-6, EEA78-37684.
278. Swaroop, B., "The effects of various ambients on dielectric films (SiO_2-Si_3N_4) deposited on silicon. II. Ionic charge", 1972 Annual Conference on Electrical Insulation and Dielectric Phenomena, p. 75-83, Publ. 1973, EEA76-26026.

279. Tsang, P. J., Doo, V. Y. and Kastl, R. H., "Sodium adsorption on silicon and silicon dioxide", In: Semiconductor silicon, Haberecht, R. R. and Kern, E. L., (Eds.), Electrochem. Soc., New York, p. 758-66 (1969), EEA73-10558.
280. van der Meulen, Y. J., Osburn, C. M. and Ziegler, J. F., "Properties of SiO_2 grown in the presence of HCl or Cl_2", J. Electrochem. Soc., 122, 284 (1975).
281. Vigdorovich, V. N., Budkevich, G. B. and Zavodyan, A. V., "Use of IR spectroscopy to investigate phosphorus-, boron-, and zinc-containing silicon dioxide films", Inorg. Mater., 8, 1881-4 (1972), EEA76-35204.
282. White, J. J., "Sodium contamination of Si and SiO_2 from alcohol and water", Presented at Spring Meeting of the Electrochemical Society, Washington, D. C., May 1971, EEA75-4605.
283. Yudelevich, I. G., Shelpakova, I. P., Buyanova, L. M., Chuchalina, L. A., Shcherbakova, O. I., Serednyakova, T. P. and Zelentsova, L. V., "Determination of impurities in silicon oxide and nitride films on a silicon substrate", Zh. Anal. Khim., 29, 518-21 (1974), Russian, CA81-32912.
284. Yurash, B. and Deal, B. E., "A method for determining sodium content of semiconductor processing materials", J. Electrochem. Soc., 115, 1191-6 (1968), PA72-31223.

VI. Film Properties

1. Optical

285. Bauer, J. and Riemann, M., "Optical constants and band spacing of amorphous Si_3N_4 layers in u.v. and nearly u.u.v. region", Wiss. Z. Tech. Hochsch. Ilmenau, 20, 181-9 (1974), German, PA78-42840.
286. Bess, M., Oswald, R. and Ohring, M., "Contact potential measurements on thin SiO_2 films", Solid-State Electron., 17, 813-17 (1974), EEA77-30246.
287. Dell'oca, C. J., "Properties and anodization of evaporated aluminum films studied by ellipsometry", Thin Solid Films, 26, 371-80 (1975), EEA78-24500.
288. Frieser, R. G., "Characterization of thermally grown SiO_2 surface by contact angle measurements", J. Electrochem. Soc., 121, 669-72 (1974), EEA77-30174.
289. Goodman, A. M., Hang, K. W. and Breece, J. M., "The relative dielectric constant of two semiconductor passivation glasses", J. Electrochem. Soc., 122, 822-4 (1975), EEA78-33019.
290. Hoffmann, G., Nemeth-Sallay, M. and Schanda, J., "Optical investigation of the Si-SiO_2 system", Acta Phys. Acad. Sci. Hung., 36, 351-64 (1974), PA78-22414.
291. Kalmynya, R. P., Feltyn, I. A., Eglitis, I. E. and Eimanis, I., "Internal photoemission of r.f. plasma growth SiO_2 films (Al-SiO_2-Si structures)", Latv. PSR Zinat. Akad. Vestis Fiz. Teh. Ser., no. 5, 113-16 (1974), Russian, PA78-33558.

292. Kobayashi, H., Tanaka, S. and Saskura, H., "Comparison of electroluminescent Ta-Ta_2O_5-ZnS:Tb^{3+}-SiO_2-Au films with Ta-Ta_2O_5-ZnS:Tb^{3+}-Au films", Jap. J. Appl. Phys., 12, 759-60 (1973), EEA76-29192.
293. Kubota, T., "Effects of substrate temperature during RF-sputtering of silica on optical and chemical properties and re-emission coefficient", Jap. J. Appl. Phys., 11, 1413-21 (1972), EEA76-651.
294. Mashita, M. and Matsushima, K., "Optical properties of reactively-evaporated silicon oxynitride films", Jap. J. Appl. Phys., Suppl. 2 (pt. 1), 761-4 (1974), PA78-38274.
295. Meissner, G. H. and Wilmsen, C. W., "Conduction mechanisms in thermally grown SiO_2", 21st Annual Southwestern IEEE Conference and Exhibition, 1971, p. 289-94, EEA75-7432.
296. Mirsch, S. and Bauer, J., "Properties of silicon nitride and silicon oxynitride films prepared by reactive sputtering", Phys. Status Solidi A, 26, 579-84 (1974), EEA78-15274.
297. Philipp, H. R., "Optical properties of non-crystalline Si, SiO, SiO_x and SiO_2", J. Phys. Chem. Solids, 32, 1935-45 (1971).
298. Philipp, H. R., "Optical properties of silicon nitride", J. Electrochem. Soc., 120, 295-300 (1973).
299. Pliskin, W. A., "Simplified refractive index dispersion relationships for dielectric films used in semiconductor technology", J. Electrochem. Soc., 121, 88C (1974), PA77-49588.
300. Pliskin, W. A. and Esch, R. P., "Refractive index of SiO_2 films grown on silicon", J. Appl. Phys., 36, 2011-3 (1965).
301. Protschka, H. A. and Framkovsky, F., "Determination of internal electric fields in silicon dioxide layers by optical probing", Appl. Phys. Lett., 14, 304-5 (1969), EEA73-4552.
302. Rabinovitch, K. and Pagis, A., "Polarization effects in multilayer dielectric thin films", Opt. Acta, 21, 963-80 (1974), PA78-1642.
303. Rand, M. J. and Standley, R. D., "Silicon oxynitride films on fused silica for optical waveguides", Appl. Opt., 11, 2482-8 (1972), EEA76-1591.
304. Sorokin, O. M., Blank, V. A. and Lebedeva, G. A., "Optical properties of SiO_2 in the spectral region 5-40 eV", Opt. & Spectrosc., 35, 291-4 (1973), PA77-25196.
305. Vertesy, A., "The effect of MOS-IC technology on the optical characteristics of silicon-nitride layers", Hiradstech. Ipari Kut. Intez. Kozl., 15, 7-15 (1975), Hungarian, EEA78-28854.
306. Weissman, I., Antypas, G., Gerlach, R., Hoffman, V. and Taylor, N., "Deposition and Auger analysis of deposited SiO_2 on $Al_xGa_{(1-x)}As$", Vacuum, 24 (2), 81-8 (1974), EEA77-26385.

2. Ionic Diffusion and Conductivity

307. Abbott, R. A. and Kamins, T. I., "Sodium migration through electron-gun evaporated Al_2O_3 and double layer Al_2O_3-SiO_2 structures", Solid-State Electron., 13, 565-76 (1970), EEA73-24933.
308. Beezhold, W., "Effects on ion bombardment on Na and Cl motion in SiO_2 thin films", IEEE Trans. Nucl. Sci., NS-21, 62-6 (1974), PA78-41995.
309. Campbell, D. R., Alessandrini, E. I., Tu, K. N. and Lewis, J. E., "Phosphorus diffusion in partially crystallized films of SiO_2", J. Electrochem. Soc., 121, 1081-5 (1974), EEA77-38756.
310. Chaudhari, P. K., Frey, W. J. and Quinn, R. M., "Arsenosilicate glass source arsenic diffusion in silicon through SiO_2 layer", Presented at Electrochemical Society Meeting, Chicago, Ill., May 1973, EEA76-23283.
311. Clemens, J. T. and Labuda, E. F., "Impurity diffusion in SiO_2 layers and related effects on the MOS properties of the Si-gate technology", Presented at Electrochemical Society Meeting, San Francisco, California, May 1974, EEA77-26478.
312. Conti, M. and Vanin, A., "Oxide and nitride conduction mechanisms in MNOS structures", Thin Solid Films, 14, 211-20 (1972), EEA76-16342.
313. De Mey, G. and De Wilde, W., "Influence of mobile ions on the electronic conduction in silicon oxide films", Thin Solid Films, 23, 301-304 (1974).
314. Fair, R. B., "High concentration arsenic diffusion in silicon from a doped oxide source", J. Electrochem. Soc., 119, 1389-94 (1972), EEA76-726.
315. Ghezzo, M. and Brown, D. M., "Diffusivity summary of B, Ga, P, As, and Sb in SiO_2", J. Electrochem. Soc., 120, 146-8 (1973), EEA76-12458.
316. Ghezzo, M. and Brown, D. M., "Arsenic glass source diffusion in Si and SiO_2", J. Electrochem. Soc., 120, 110-16 (1973).
317. Ghoshtagore, R. N., "Phosphorus diffusion processes in SiO_2 films", Thin Solid Films, 26, 501 (1975).
318. Ghoshtagore, R. N., "Silicon dioxide masking of phosphorus diffusion in silicon", Solid-State Electron., 18, 399 (1975).
319. Grimberg, G. A., "Diffusion of a foreign substance through a SiO_2-Si type system or through other stratified media having an interfacial segregation coefficient not equal to unity", Pis'ma V Zh. Tekh. Fiz., 1, 481-5 (1975), PA78-72654.
320. Gyulai, J., Mayer, J. W., Mitchell, I. V. and Rodriguez, V., "Outdiffusion through silicon oxide and silicon nitride layers on gallium arsenide", Appl. Phys. Lett., 17, 332-4 (1970), EEA74-4436.
321. Hickmott, T. W., "Thermally stimulated ionic conductivity of sodium in thermal SiO_2", J. Appl. Phys., 46, 2583-98 (1975), PA78-66034.

322. Huntley, F. A. and Willoughby, A. F. W., "Surface diffusion of gold on silicon", Solid-State Electron., 14, 641-3 (1971), EEA74-30748.
323. Kirillova, A. A. and Pashintsev, Y. I., "Investigation of the drift of ions in silicon dioxide", Fiz. Tverdogo Tela, 11, 3570-3 (1969), Russian, EEA73-16979.
324. Liblich, S. and Nassibian, A. G., "Effects of diffused nickel on the silicon-silicon dioxide interface", Solid-State Electron., 16, 1495-9 (1973), EEA77-10181.
325. Madams, C. J., Morgan, D. V. and Howes, M. J., "Migration of gold atoms through thin silicon oxide films", J. Appl. Phys., 45, 5088-90 (1974), EEA78-11547.
326. McCaughan, D. V. and Kushner, R. A., "Ion migration effects in r.f. 'sputter cleaning' of dielectric films", Thin Solid Films, 22, 359-63 (1974), EEA77-33997.
327. Morris, J. D. and White, J. J., "Sodium redistribution in SiO_2 films on silicon during immersion in hot electrolytes", Presented at Spring Meeting of the Electrochemical Society, Washington, D. C., May 1971, EEA75-4607.
328. Nakajima, Y. and Okhawa, S., "The effect of the oxide layer on the diffusion of Ga in Si", Jap. J. Appl. Phys., 11, 1742 (1972), EEA76-12585.
329. Navik, G., "The ionic conductivity in silicon oxide films", Thin Solid Films, 6, 145-59 (1970), EEA74-7192.
330. Pankove, J. I., "Method for diffusing impurities into nitride semiconductor crystals", Patent USA 3865655, Publ. February 1975.
331. Pepper, M., "Electrical studies of ion transport in SiO_2 films on Si", Symposium on Mass Transport in Amorphous Solids, 1973, 4 pp., EEA76-38233.
332. Pepper, M., "The influence of oxygen annealing on ion drift in SiO_2", Phys. Status Solidi A, 18, K19-22 (1973), EEA76-35257.
333. Pepper, M. and Eccleston, W., "Cation transport in SiO_2", Phys. Status Solidi A, 12, 199-207 (1972), EEA75-32595.
334. Raider, S. I., Gregor, L. V. and Flitsch, R., "Transfer of mobile ions from aqueous solutions to the silicon dioxide surface", J. Electrochem. Soc., 120, 425-31 (1973), EEA76-23261.
335. Schmidt, P. F. and Adda, L. P., "Interface-state generation by gold diffusion through SiO_2 films on silicon: MOS and neutron-activation results", J. Appl. Phys., 45, 1826-33 (1974), EEA77-30219.
336. Schrager, M. and Collins, F. C., "Ion migration in titanium dioxide films", J. Appl. Phys., 46, 1934-42 (1975), EEA78-28690.
337. Shimakura, K., Suzuki, T. and Yadoiwa, Y., "Boron and phosphorous diffusion through an SiO_2 layer from a doped polycrystalline Si source under various drive in ambients", Solid-State Electron., 18, 991-8 (1975), PA78-81999.

338.Silversmith, D. J., "Nonuniform lateral ionic impurity distributions at Si-SiO_2 interfaces", J. Electrochem. Soc., 119, 1585-8 (1972), EEA76-4576.
339.Singh, B. R., Rai, S. S. and Srivastava, R. S., "Transport of sodium ions in silicon dioxide films using MOS structure", Phys. Status Solidi A, 13, 51-9 (1972), EEA75-39869.
340.Snow, E. H., Grove, A. S., Deal, B. E. and Sah, C. T., "Ion transport phenomena in insulating films", J. Appl. Phys., 36, 1664-73 (1965).
341.Swaroop, B., "Ambient effect on ionic charges in dielectric films", IEEE Trans. Parts, Hybrids & Packag., PHP-9, 234-6 (1973), PA77-24831.
342.Vigdorovich, V. N. and Zavodyan, A. V., "Properties of diffusion structures in silicon: diffusion from doped SiO_2", Inorg. Mater., 9, 1164-7 (1973), EEA77-30155.
343.Wagner, S. and Povilonis, E. I., "Diffusion of gallium through silicon dioxide films into silicon", J. Electrochem. Soc., 121, 1487-96 (1974), EEA78-5274.
344.Williams, R. and Woods, M. H., "Image forces and the behavior of mobile positive ions in silicon dioxide", Appl. Phys. Lett., 22, 458-9 (1973), EEA76-23374.
345.Wilson, P. R., "The diffusion of boron in the Si-SiO_2 system", Solid-State Electron., 15, 961-70 (1972), EEA75-36170.
346.Wojtowicz, P. J., "Two dimensional phase transitions of mobile ions at SiO_2-Si interface", RCA Rev., 36, 132-48 (1975).
347.Woods, M. H. and Williams, R., "Injection and removal of ionic charge at room temperature through the interface of air with SiO_2", J. Appl. Phys., 44, 5506-10 (1973), EEA77-18699.
348.Zuther, G., Prandtke, H. and Schmidt, M., "Thermally stimulated currents of SiO_2 thin films doped with nitrogen and phosphorus", Thin Solid Films, 21, S7-9 (1974), EEA77-18665.

3. Charge Injection, Trapping and Storage

349.Aboaf, J. A., Kerr, D. R. and Bassous, E., "Charge in SiO_2-Al_2O_3 double layers on silicon", J. Electrochem. Soc., 120, 1103-6 (1973), EEA76-35253.
350.Adams, P. M., Jones, P. L., Taylor, E. F. and Litting, C. N. W., "Electron-beam-induced conductivity in thin sputtered films of Al_2O_3", J. Phys. D, 7, L165-6 (1974), PA77-82505.
351.Arnett, P. C. and DiMaria, D. J., "Hole injection into silicon nitride: Dark current dependence on electrode materials and insulator thickness", Appl. Phys. Lett., 27, 34-6 (1975), PA78-66032.
352.Arnett, P. C. and Yun, B. H., "Silicon nitride trap properties as revealed by charge-centroid measurements on MNOS devices", Appl. Phys. Lett., 26, 94-6 (1975), CA82-148763.
353.Balk, P. and Stephany, F., "Charge injection in MAOS system", J. Electrochem. Soc., 118, 1634-8 (1971), EEA75-1078.

354.Bosselaar, C. A., "Charge injection into dielectric coatings covering reverse biased junctions", Presented at 1972 European Solid State Devices Research Conference, Lancaster, England, September 1972, EEA76-32250.
355.Bosselaar, C. A., "Charge injection into SiO_2 from reverse-biased junctions", Solid-State Electron, 16, 648-51 (1973), EEA76-19919.
356.Breed, D. J. and Kramer, R. P., "Charge trapping in MOS systems", Thin Solid Films, 13, 1-4 (1972), EEA76-4669.
357.Buff, W. and Greiner, H., "Charge storage at the boundaries interfaces of insulator layers", Wiss. Z. Tech. Hochsch. Ilmenau, 17 (4), 87-96 (1971), German, EEA75-16689.
358.Bulucea, C., Postolache, C. and Rusu, A., "Avalanche-injected electron currents in thermally grown SiO_2", Rev. Roum. Phys., 19, 1015-27 (1974), EEA78-11573.
359.Burkhardt, P. J., "Effects of intermediate temperature annealing on mobile charge generation in thermally grown SiO_2 films", Presented at Electrochemical Society Spring Meeting, Toronto, Canada, May 1975.
360.Card, H. C., "Factors affecting avalanche injection into insulating layers from a semiconductor surface", Solid-State Electron., 17, 501-2 (1974), PA77-45009.
361.Chou, N. J., Aboaf, J. A., Hammer, R. and Crowder, H. P., "Effect of insulator thickness fluctuations on MNOS charge storage characteristics", IEEE Trans. Electron Devices, ED-19, 198-204 (1972), EEA75-7536.
362.Chowdry, A., "The role of bulk traps and surface states in the contact charging of insulators", Thesis, Johns Hopkins Univ., Baltimore, Md., 1974, 146 pp., Order No. 74-29000, EEA78-19747.
363.Csabay, O. and Frank, H., "The influence of silicon surface treatment and the oxide layer on the effective defect charge density of MOS structure", Elektrotech. Cas., 24 (4), 228-43 (1973), Slovak, EEA76-25994.
364.Curtis, O. L., Jr., Srour, J. R. and Chiu, K. Y., "Hole and electron transport in SiO_2 films", J. Appl. Phys., 45, 4506-13 (1974), EEA78-11543.
365.Dearnaley, G., "Stored charge in oxide layer structres", Int. J. Electron., 29, 299-300 (1970), EEA73-37857.
366.Declerck, G., Van Overstraeten, R. and Broux, G., "Discussion of the surface-potential fluctuations caused by oxide-charge fluctuations", J. Appl. Phys., 45, 2593-5 (1974), PA77-67280.
367.DiMaria, D. J. and Arnett, P. C., "Hole injection into silicon nitride: Interface barrier energies by internal photoemission", Appl. Phys. Lett., 26, 711-13 (1975), PA78-72908.
368.DiMaria, D. J., Feigl, F. J. and Butler, S. R., "Capture and emission of electrons at 2.4 eV deep trap levels in SiO_2 films", Phys. Rev. B, 11, 5023-30 (1975), EEA78-41531.

369.Eisele, K. M., "Charge storage and stoichiometry in electron beam evaporated alumina", J. Electrochem. Soc., 122, 148-52 (1975), EEA78-11552.

370.Fanet, J. M. and Poirier, R., "Charge storage in SiO_2 under low-energy electron bombardment", Appl. Phys. Lett., 25, 183-5 (1974), PA77-79424.

371.Fowler, A. B., "Oxide-charge-induced impurity level in silicon inversion layers", Phys. Rev. Lett., 34, 1435-7 (1975), EEA78-28699.

372.Franz, J. M., Fitzgerald, D. and DiStefano, T., "Charge trapping in MNOS capacitors", Presented at Electrochemical Society Meeting, San Francisco, California, May 1974, EEA77-22938.

373.Gangatirkar, P. D., "Recovery of FET characteristics after electron injection into gate dielectrics", IBM Tech. Disclosure Bull., 17, 3528 (1975), EEA78-28802.

374.Gould, R. D., Hogarth, C. A. and Collins, R. A., "Hot electron attenuation in silicon oxide", J. Non-Cryst. Solids, 12, 131-34 (1973).

375.Hillery, R. V., "Change in the surface charge density of thermally oxidized silicon due to silicon nitride deposition", In: Semiconductor silicon, Haberecht, R. R. and Kern, E. L., (Eds.), Electrochem. Soc., New York, p. 339-49 (1969), EEA73-10517.

376.Hornung, A. and Kasprzak, L. A., "Removal of charge in the gate region of a FET", IBM Tech. Disclosure Bull., 17, 1042-3 (1974), EEA78-5332.

377.Howson, R. P. and Taylor, A., "Charge storage in evaporated silicon oxide films", Thin Solid Films, 6, 31-50 (1970), EEA74-4444.

378.Hughes, R. C., "Charge-carrier transport phenomena in amorphous SiO_2: direct-measurement of the drift mobility and lifetime", Phys. Rev. Lett., 30, 1333-6 (1973), EEA76-29138.

379.Johnson, N. M., "Electron trapping in ion-implanted silicon dioxide films on silicon", Thesis, Princeton Univ., N. J., 1974, 83 pp., Order No. 74-25956.

380.Kahng, D., Sundburg, W. J., Boulin, D. M. and Ligenza, J. R., "Interfacial dopants for dual-dielectric, charge-storage cells", Bell Syst. Tech. J., 53, 1723-39 (1974), EEA78-8564.

381.Knowles, T. J., "Flat band shifts in m.i.s. structures under controlled charge", Presented at Spring Meeting of the Electrochemical Society, Washington, D. C., May 1971, EEA75-4642.

382.Kovtonyuk, N. F. and Morozov, V. A., "Effect of discharge of a stored charge in MDPDM structures", Radio Eng. & Electron. Phys., 18, 315-17 (1973), EEA77-10176.

383.Maes, H. and Van Overstraeten, R. J., "Simple technique for determination of centroid of nitride charge in MNOS structures", Appl. Phys. Lett., 27 (5), 282-4 (1975).

384.McLean, F. B., Boesch, H. E., Jr., Winokur, P. S., McGarrity, J. M. and Oswald, R. B., Jr., "Rapid annealing and charge injection in Al_2O_3 MIS capacitors", IEEE Trans. Nucl. Sci., NS-21, 47-55 (1974), EEA78-19701.
385.Mehta, D. A., Butler, S. R. and Feigl, F. J., "Effects of post-deposition annealing treatments on charge trapping in CVD Al_2O_3 films on Si", J. Electrochem. Soc., 120, 1707 (1973), EEA77-10152.
386.Nagai, K., Hayashi, Y. and Tarui, Y., "Injection, trapping and release from SiO_2 of photo-generated hole charge for an erasable non-volatile optical memory", Jap. J. Appl. Phys., 14, 159-60 (1975), PA78-21965.
387.Nagai, K., Hayashi, Y. and Tarui, Y., "Carrier injection into silicon dioxide from silicon surface driven to avalanche breakdown by a linear ramp pulse, and trapping, distribution, and thermal annealing of injected holes in silicon dioxide", Jap. J. Appl. Phys., 14, 1539-45 (1975), CA83-187295.
388.Neugebauer, C. A., Bargess, J. F., Joynson, R. E. and Mundy, J. L., "Electron trapping in thin SiO_2 films due to avalanche currents", Thin Solid Films, 13, 5-9 (1972), EEA76-4587.
389.Nicollian, E. H. and Berglund, C. N., "Avalanche injection of electrons into insulating SiO_2 using MOS structures", J. Appl. Phys., 41, 3052-7 (1970), EEA74-31320.
390.Nicollian, E. H., Bergland, C. N., Schmidt, P. F. and Andrews, J. M., "Electrochemical charging of thermal SiO_2 films by injected electron currents", J. Appl. Phys., 42, 5654-64 (1971), EEA75-7157.
391.Nicollian, E. H., Goetzberger, A. and Berglund, C. N., "Avalanche injection currents and charging phenomena in thermal SiO_2", Appl. Phys. Lett., 15, 174-7 (1969), EEA73-7340.
392.Ning, T. H., Osburn, C. M. and Yu, H. N., "Electron trapping at positively charge centres in SiO_2", Appl. Phys. Lett., 26, 248-50 (1975), EEA78-19814.
393.Ohta, K. and Hamano, K., "Carrier transport and space charge in MAOS structure", Jap. J. Appl. Phys., 11, 546.54 (1972), EEA75-20140.
394.Pepper, M., "Electron injection into SiO_2 from an avalanching p-n junction", J. Phys. D, 6, 2124-30 (1973), EEA77-348.
395.Pepper, M., "Electron injection into SiO_2 in the N channel stacked gate MOS tetrode", IEEE Trans. Electron Devices, ED-21, 174-5 (1974), EEA77-6035.
396.Powell, R. J. and Hughes, G. W., "Charge injection and trapping in Al_2O_3 gate insulators", IEEE Trans. Nucl. Sci., NS-21, 179-85 (1974), PA78-42306.
397.Raider, S. I., Gdula, R. A. and Petrak, J. R., "Nitrogen reaction at a silicon-silicon dioxide interface", Appl. Phys. Lett., 27, 150-2 (1975), PA78-86386.

398. Salama, C. A. T., "Charge storage in Si-SiO_2-AlO_3 memory structures", Presented at Electrochemical Society Fall Meeting, Atlantic City, N. J., October 1970, EEA74-7678.
399. Saminadayar, K., Pautrat, J. L. and Pfister, J. C., "Injection and migration of charges in the oxide of a metal-oxide-semiconductor structure", J. Phys. Chem. Solids, 29 (10), 1709-13 (1968), French, PA72-5741.
400. Shuskus, A. J., Quinn, D. J. and Cullen, D. E., "Charge storage characteristics of MIS structures employing dual-insulator composites of HfO_2-SiO_2 and $SrTiO_3$-SiO_2", Appl. Phys. Lett., 23, 184-5 (1973), EEA76-35317.
401. Svensson, C. and Lundstrom, I., "Trap-assisted charge injection in MNOS structures", J. Appl. Phys., 44, 4657-63 (1973), PA77-20691.
402. Thomas, J. H., III and Feigl, F. J., "Electron trapping levels in silicon dioxide thermally grown on silicon", J. Phys. & Chem. Solids, 33, 2197-216 (1972), EEA76-4578.
403. Tsujide, T. and Iida, K., "Influence of Al_2O_3 deposition temperature on charge-storage and retention in MAOS structures", Jap. J. Appl. Phys., 11, 1599-604 (1972), EEA76-746.
404. Verwey, J. F., "Avalanche-injected current in MNOS structures (breakdown in npn planar transistors)", Philips Res. Rep., 26, 382-90 (1971), EEA75-1084.
405. Vinetskii, V. L., Cahika, G. E. and Shevchenko, E. S., "Charge and potential distribution in dielectric layers of MOS structures under ionization", Phys. Status Solidi A, 26, 743-52 (1974), EEA78-15276.
406. Walker, L. G. and Pratt, G. W., Jr., "Long-term charge storage in interface states of ZnS MOS capacitors", J. Appl. Phys., 46, 2992-7 (1975), PA78-77190.
407. Weinberg, Z. A., "Hole injection and transport in SiO_2 films on Si", Appl. Phys. Lett., 27 (8), 437-9 (1975).
408. Wintle, H. J., "Transient charging currents in insulators", Solid-State Electron., 18, 1039-42 (1975).

4. Pressure Sensitivity

409. Ashcroft, W., "Mechanical properties of silicon nitride at elevated temperatures", Proc. Brit. Ceram. Soc., no. 22, 169-79 (1973).
410. Brotherton, S. D., Read, T. G., Lamb, D. R. and Willoughby, A. F. W., "Surface charge and stress in the Si/SiO_2 system", Solid-State Electron., 16, 1367-75 (1973), EEA77-5930.
411. Friedrich, H., "Surface charges induced by mechanical stresses in the silicon-silicon oxide interface", Solid State Electron., 14, 639-41 (1971), EEA74-30747.
412. Fonash, S. J., "Effects of stress on metal-oxide-semiconductor structures", J. Appl. Phys., 44, 4607-15 (1973), PA77-20690.

413. Gadzhiev, N. D., Kasimov, F. D. and Mamikonova, V. M., "Influence of local pressure on the properties of the Si-SiO_2 interface in metal-oxide-semiconductor structures", Fiz. & Tekh. Poluprov., 5, 950-1 (1971), Russian, EEA74-26704.
414. Lin, S. C. H. and Pugacz-Muraszkiewicz, I., "Local stress measurement in thin thermal SiO_2 films on Si substrates", J. Appl. Phys., 43, 119-25 (1972).
415. Serebrinsky, J. H., "Stress concentration in silicon-insulator interfaces", Solid-State Electron., 13, 1435-44 (1970), EEA74-1510.
416. Verkovskii, E. I., Azhazha, E. G., Epifanov, G. I. and Sanzharovskii, A. T., "Internal stress in SiO_2 films obtained by the method of high-temperature oxidation of silicon", Sov. Phys.-Dokl., 18, 73-4 (1973), EEA76-32192.
417. Wilmsen, C. W., Thompson, E. G. and Meissner, G. H., "Buckling of thermally-grown SiO_2 thin films", IEEE Trans. Electron Devices, ED-19, 122 (1972), EEA75-7156.
418. Wlodarski, W., "The effect of hydrostatic pressure on MOS field-effect transistors", Bull. Acad. Pol. Sci. Ser. Sci. Tech., 22, 529-36 (1974), EEA78-2868.
419. Wonsiewicz, B. C. and McCaughan, D. V., "Electrical properties of metal-SiO_2-silicon structures under mechanical stress", J. Appl. Phys., 44, 5476-9 (1973), EEA77-18739.
420. Zhagata, L. A., Kalnach, Y. B., Neimane, I. P. and Feltyn, I. A., "Stresses in semiconductor-SiO_2 (or SiC) systems with vapour-phase deposited films", Latv. PSR Zinat. Akad. Vestis Fiz. Teh. Ser., no. 4, 33-41 (1974), Russian, EEA78-5260.

VII. Double Dielectrics

421. Brown, D. M., Garfinkel, M., Ghezzo, M., Taft, E. A., Tenney, A. and Wong, J., "Characteristics of doped oxides and their use in silicon device fabrication", J. Cryst. Growth, 17, 276-87 (1972), EEA76-4538.
422. Clemens, J. T., Labuda, E. F. and Berglund, C. N., "Aluminum oxide/silicon dioxide, double-insulator, MOS structure", Bell Syst. Tech. J., 54, 687-719 (1975), EEA78-28679.
423. Litovchenko, V. G., "Three layer model of semiconductor-dielectric structures", Poluprovodn. Tekh. & Mikroelektron., no. 12, 3-12 (1973), Russian, EEA76-38223.
424. Lundkvist, L., "Ellipsometry of multilayered dielectrics on silicon, applied to MNOS structures", J. Electrochem. Soc., 120, 1140-2 (1973), EEA76-35220.
425. Ono, M. and Momoi, T., "Silicon nitride on silicon oxide coatings for semiconductor devices", Patent USA 3788913, Publ. January 1974, CA80-75751.

426.Pileur, P. and Viret, J. C., "Production and characterization of the silicon dioxide-silicon nitride sandwich formed by reactive sputtering", Colloq. Int. Pulverisation Cathodique Ses Appl., C. R., 1st, 1973, p. 123-35, Publ. 1974, French, CA81-83503.
427.Sato, S., Sato, A. and Okamoto, E., "An SiO_2-Ta_2O_5 thin film capacitor", 23rd Electronic Components Conference, 1973, p. 302-8, EEA76-29373.
428.Schipanski, D., "SiO_2-Si_3N_4 combination layers on germanium", Nachrichtentech. Elektron., 23, 23-7 (1973), German, EEA76-19840.

VIII. Dielectric-Metal Interface

429.Agranovich, V. M., Lozovik, Y. E., Malshukov, A. G. and Mekntiyev, M. A., "Electron and exciton states of a dielectric near the dielectric-metal interface", Presented at 11th European Congress on Molecular Spectroscopy, Tallinn, USSR, May-June 1973, EEA76-38234.
430.Armigliato, A. and Morettini, L., "Preparation of Au-SiO_2 RF-cosputtered films and combined electron microscopy-microanalysis observation", Proceedings of the 5th European Congress on Electron Microscopy, 1972, p. 668-9, EEA76-849.
431.Barruel, F. and Pfister, J. C., "Study of an Al-Si_3N_4 interface by photoinjection measurements in MNS and MNOS sandwiches", Thin Solid Films, 28, 323-35 (1975), EEA78-41542.
432.Burykina, A. L. and Kosteruk, V. P., "Influence of ambient atmosphere on the reaction of silicon nitride with titanium", Sov. Powder Metall. Met. Ceram., 12, 983-87 (1973).
433.Garton, C. G., "Charge transfer from metal to dielectric by contact potential", J. Phys. D, 7, 814-23 (1974).
434.Gershinskii, A. E., Khoromenko, A. A. and Edelman, F. L., "Reduction of thermally grown SiO_2 by Al films", Phys. Status Solidi A, 25, 645-51 (1974).
435.Hiraki, A., Lugujjo, E. and Mayer, J. W., "Formation of silicon oxide over gold layers on silicon substrates", J. Appl. Phys., 43, 3643-9 (1972), EEA75-36135.
436.Kurdov, L. E., Pospelov, V. V. and Tsilibin, B. I., "Conductivity of silicon nitride (Si_3N_4) films in metal-silicon nitride silicon dioxide and metal-silicon nitride-silicon", Mikroelektronika (Akad. Nauk SSSR), 3, 363-5 (1974), Russian, CA81-112710.
437.Morettini, L., Armigliato, A. and Ruffini, G., "Preparation of Au-SiO_2 rf cosputtered films and combined electron microscopic microanalysis", Met. Ital., 65, 349-52 (1973).
438.Powell, R. J. and Beairsto, R. C., "Properties of the tungsten-silicon dioxide interface", Solid-State Electron., 16, 265-7 (1973), EEA76-12529.
439.Swaroop, B., "Effect of metallization on dielectric-semiconductor structure", 1973 Annual Conference on Electrical Insulation and Dielectric Phenomena, p. 79-85, Publ. 1974, EEA78-602.

440.Ugai, Y. A., Kolesnikov, V. G., Anokhin, V. Z. and Fetisova, S. V., "Characteristics of heterogeneous reactions in systems involving SiO_2 and Al, V, Cr, or Mo", Inorg. Mater., 9, 156-8 (1973), EEA76-38220.

441.Ugai, Y. A., Tverdokhlebova, L. Y., Anokhin, V. Z., Loskutov, N. I. and Averbakh, E. M., "Solid-phase reaction of aluminum films with silicon dioxide", Inorg. Mater., 10, 888-90 (1974), PA78-76979.

IX. Passivation and PSG

442.Baliga, B. and Ghandhi, S. K., "Growth of silica and phosphosilicate films", J. Appl. Phys., 44, 990-94 (1973).

443.Berkner, R., "Forming stabilizing and/or insulating coatings on semiconductor surfaces", Patent German 2155849, Publ. May 1973, CA79-47059.

444.Borkun, L. P., Fuks, G. M. and Khludkov, S. S., "Protective properties of an SiO_2 film during diffusion of manganese in GaAs", Inorg. Mater., 10 (4), 623-4 (1974), EEA78-41475.

445.Bracken, R. C., Harper, J. G. and Paddock, A. D., "Anodic passivating of integrated circuits", Patent USA 3798135, Publ. January 1974, CA80-126044.

446.Burn, R. A., Hovel, H. J. and Woodall, J. M., "Surface passivation for decomposable substrates", IBM Tech. Disclosure Bull., 17, 2774 (1975), EEA78-19641.

447.Cherry, A. and Galli, G., "Boron passivation of silicon", IBM Tech. Disclosure Bull., 17, 3728 (1975), EEA78-33007.

448.Eldridge, J. M. and Kerr, D. R., "Sodium ion drift through phosphosilicate glass-SiO_2 films (MOS structures)", J. Electrochem. Soc., 118, 986-91 (1971), EEA74-26707.

449.Elgan, D. L. and Flahie, M. F., "Passivating mesa p-n junction structures", Patent British 1300186, Publ. December 1972, CA78-64156.

450.Fischer, H., Rau, S. and Trappe, P., "Passivating layers for silicon planar semiconductor devices", Patent German (East) 90185, Publ. May 1972, CA77-132422.

451.Goetzberger, A. and Eisele, K., "Surface passivation of p-n junction devices", Patent German 2205020, Publ. August 1973, CA79-150792.

452.Goodman, A. M., Hang, K. W. and Breece, J. M., "The relative dielectric constant of two semiconductor passivation glasses", J. Electrochem. Soc., 122, 822-4 (1975), PA78-62555.

453.Greenwood, J. C., "Passivation of semiconductor devices", Patent UK 1352959, Publ. May 1974, CA81-55739.

454.Gregor, L. V., "Passivation of semiconductor surfaces", Solid State Technol., 14 (4), 37-43 (1971), EEA74-25570.

455.Gregor, L. V., Grosewald, P. and Powlus, R. A., "Reducing radiation damage in in-channel IGFET's", IBM Tech. Disclosure Bull., 13, 3281 (1971), EEA74-30828.

456. Gregor, L. V. and Maissel, L. I., "Method for passivating semiconductor material and field-effect transistor formed thereby", Patent USA 3783119, Publ. January 1974.
457. Gregor, L. V. and Raniseski, J. W., "The role of dielectric films in device fabrication", Insul./Circuits, 17 (9), 27-30 (1971), EEA74-33734.
458. Grosewald, P., Gregor, L. V. and Powlus, R., "Radiation effects in f.e.t.s from sputtering passivation", Presented at 17th International Electron Devices Meeting, Washington, D. C., October 1971, EEA75-13777.
459. Guldan, A., Luby, S. and Cervenak, J., "Properties of cathode-sputtered silica layers used in the passivation of semiconductor devices", Slaboproudy Obz., 35, 405-9 (1974), Slovak, CA82-67154.
460. Hitachi, Ltd., "Semiconductor device having a passivation film and insulating films on a semiconductor substrate and method of making", Patent UK 1283769, Publ. August 1972.
461. Inoue, M., "Semiconductor device having a TiO_2-SiO_2 composite oxide layer", Patent USA 3614548, Publ. October 1971, EEA75-13658.
462. Ito, S., Ota, M., Yamada, E. and Sugawara, K., "Effect of oxide charge density in passivation films on breakdown voltage of planar diodes", Jap. J. Appl. Phys., 14, 733-4 (1975), PA78-54563.
463. Izidinov, S. O., "Photopassivation and surface recombination rate on n-silicon in the active state region in 10 N potassium hydroxide", Elektrokhimiya, 11, 1142-5 (1975), Russian, CA83-185422.
464. Kasprzak, L. and Hornung, A., "Polarization and depolarization in PSG films (MIS structures)", IBM J. Res. & Dev., 19, 127-32 (1975), CA78-62112.
465. Kern, W., "Chemical vapor deposition systems for glass passivation of integrated circuits", Solid State Technol., 18 (12), 25-33 (1975).
466. Konantz, M. L. and Lesure, R. K., "Applying silicone passivants to etch moats in mesa device wafers", Patent USA 3860448, Publ. January 1975, CA82-118485.
467. Kraus, H. and Breazeale, B. H., "Surface passivation of semiconductors", Patent USA 3799803, Publ. March 1974, CA80-138486.
468. Kriegler, R. J., "The role of HCl in the passivation of MOS structures", Thin Solid Films, 13, 11-14 (1972), EEA76-4588.
469. Kriegler, R. J., "The uses of HCl and Cl_2 for the preparation of electrically stable SiO_2", Presented at Electrochemical Society Meeting, Chicago, Ill., May 1973, EEA76-23332.
470. Kriegler, R. J., Aitken, A. and Morris, J. D., "The use of HCl passivation in the processing technology of MOSFET devices", Oyo Buturi, 43 (Suppl.), 341-7 (1974), EEA78-2879.
471. Langdon, J. L., "Process for fabricating passivated transistors", Patent USA 3771218, Publ. November 1973.
472. Lloyd, P., "The effects of r.f. sputter coating on MOS transistors", Thin Solid Films, 10, 159-60 (1972), EEA75-16756.

473. Lue, P. C., Jackson, J. A. and Han, Y. P., "Conduction at the silicon/organic interface (passivation polymers)", Electrochemical Society Fall Meeting (Extended abstracts), 1974, p. 490-1, EEA78-37341.
474. McCaughan, D. V., Kushner, R. A. and Wagner, S., "Complete removal of sodium from silicon dioxide films by formation of phosphosilicate glass", J. Electrochem. Soc., 121, 724-5 (1974), EEA77-30179.
475. Morzeck, P., "Preparation and investigation of silicon nitride passivating films", Z. Elektr. Inf. - & Energietech., 2 (4), 199 (1972), German, EEA76-8545.
476. Nagai, H., "Passivation of germanium devices. I. Preparation of passivation films by chemical reaction", Rev. Electr. Commun. Lab., 22, 1043-56 (1974), PA78-51128.
477. Pammer, E., "Coatings on silicon by surface reactions", Patent German 2025779, Publ. December 1971, CA76-39047.
478. Paulson, W. M. and Kirk, R. W., "The effects of phosphorus-doped passivation glass on the corrosion of aluminum", Proceedings of the 12th Annual Reliability Physics Conference, 1974, p. 172-9.
479. Pikor, A., "Glass passivation of high voltage power transistors", Presented at Electrochemical Society Meeting, San Francisco, Calif., May 1974, PA77-49589.
480. Pikor, A. J., "Glass passivation of semiconductor devices", Report TN-997, RCA, Princeton, N. J., 1975, 1 pp., EEA78-23276.
481. Saia, F. J. and Quetsch, J. G., "Surface-protected semiconductor devices and methods of manufacturing", Patent USA 3440496, Publ. April 1969, EEA73-7443.
482. Sanchez, D., Carchano, M. and Bui, A., "Electrical properties of metal-polymer (polystyrene) silicon devices", J. Appl. Phys., 45, 1233-8 (1974), PA77-57171.
483. Sato, K., Nishi, Y. and Abe, T., "Stabilization of MOS structures by phosphorus glass treatments of SiO_2 films", In: Thin film dielectrics, Vratny, F., (Ed.), Electrochem. Soc., New York, p. 378-406 (1969), EEA73-10542.
484. Sato, Y., "Formation of the conductive layer near the surface of semi-insulating GaAs covered with oxide film", Jap. J. Appl. Phys., 12, 242-51 (1973), EEA76-19863.
485. Seki, H., Ohosaka, S., Kanda, M., Kawasaki, Y., Yamazaki, H. and Fujimoto, M., "Passivation of gallium arsenide with silicon nitride", Rev. Electr. Commun. Lab., 20, 810-16 (1972), EEA76-12462.
486. Sharma, B. L., Gopal, V., Mukerjee, S. N. and Modi, J. K., "Phosphosilicate glass films on GaP", Thin Solid Films, 20, S21-S23 (1974), EEA77-10132.
487. Shibata, M. and Sugawara, K., "Deposition rate and phosphorus concentration of phosphosilicate glass films in relation to O_2/SiH_4+PH_3 mole fraction", J. Electrochem. Soc., 122, 155-6 (1975), PA78-26261.

488. Shibata, M., Yoshimi, T. and Sugawara, K., "Deposition rate and phosphorus concentration of phosphosilicate glass films in relation to PH_3/SiH_4+PH_3 mole fraction", J. Electrochem. Soc., 122, 157-8 (1975), PA78-26262.
489. Shibata, T., Igarashi, Y. and Yano, K., "Passivation of germanium devices. III. Fabrication and performance of germanium planar photodiodes", Rev. Electr. Commun. Lab., 22, 1069-77 (1974), EEA78-24221.
490. Snow, E. H. and Deal, B. E., "Polarization phenomena and other properties of phosphosilicate glass films on silicon", J. Electrochem. Soc., 113, 263-269 (1966).
491. Tenney, A. S. and Ghezzo, M., "Composition of phosphosilicate glass in infrared absorption", J. Electrochem. Soc., 120, 1276-9 (1973), EEA77-2457.
492. Vigdorovich, V. N., Sulimin, A. D., Pavlenko, L. S., Neustroev, S. A. and Yakovenko, V. G., "Determination of the protective characteristics of oxide films by micro-hardness measurement", Ind. Lab., 39, 402-4 (1973), EEA77-2283.
493. White, M. H., Blaha, F. C. and Herman, D. S., "A study relating m.o.s. processes to a model of the $Al-SiO_2-Si$ system", Presented at Symposium on Silicon Device Processing, Gaithersburg, Md., June 1970, EEA73-35177.
494. Wong, J. and Ghezzo, M., "High P_2O_5-content phosphosilicate glass films: Preparation and characterization by infrared spectroscopy", Presented at Electrochemical Society Meeting, San Francisco, Calif., May 1974, PA77-49597.
495. Yashiro, T., Nagai, H. and Yano, K., "Passivation of germanium devices. II. The electrical properties of $Ge-SiO_2$ and $Ge-Si_3N_4$ interfaces", Rev. Electr. Commun. Lab., 22, 1057-68 (1974), EEA78-24220.
496. Young, D. R., "Silicon semiconductor device passivation", J. Vac. Sci. & Technol., 7, 141 (1970), EEA74-31325.

E. LITHOGRAPHY

I. Books and Symposia

1. DeForest, W. S., "Photoresist: materials and processes", McGraw-Hill Book Co., New York, 1975.
2. DeRenzo, D., "Polymers in lithography", Noyes Data Corp., Park Ridge, N. J., 1971.
3. Dinaburg, M. S., "Photosensitive diazo compounds", Focal Press, New York, 1965.
4. Kosar, J., "Light-sensitive systems", John Wiley & Sons, New York, 1965.
5. Bakish, R., (Ed.), International Conference on Electron and Ion Beam Science and Technology (Volumes 1-6), The Electrochemical Soc., Inc., Princeton, N. J., (1964, 1966, 1968, 1970, 1972, 1974).
6. Kodak Seminars on Microminiaturization and Photoresists, Presented by Professional, Commercial, Industrial and Marketing Divisions, Eastman Kodak Co., Rochester, N. Y., 1965-1974.
7. SPE Symposium on Photopolymers: Principles, Processes and Materials, Society of Plastics Engineers, Mid-Hudson Section, (1967, 1970, 1973).
8. 13th Symposium on Electron, Ion, and Photo Beam Technology, Proceedings were published in: J. Vac. Sci. Technol., 12 (6), December 1975.

II. Review Articles

9. Bindell, J. B., "Scanning electron microscopy and related electron beam techniques in IC technology", Solid State Technol., 18 (4), 45-50 (1975), EEA78-24428.
10. Bobenrieth, A., Dubee, A. and Arnodo, C., "Electronic masking", Rev. Tech. Thomson-CSF, 5, 791-840 (1973), French, EEA77-23109.
11. Brewer, G. R., "Electron and ion beams in microelectronic fabrication processes", Solid State Technol., 15 (7), 36-9 (1972), EEA75-35405.
12. Brewer, G. R., "Electron and ion beams in microelectronic fabrication processes. II", Solid State Technol., 15 (8), 43-7 (1972), EEA75-35407.
13. Broers, A. N. and Dennard, R. H., "Impact of electron beam technology on silicon device fabrication", Presented at Electrochemical Society Meeting, Chicago, Ill., May 1973, EEA76-23150.
14. Brunner, F., Poetsch, H. and Schwartzbach, F., "Light-sensitive coating in microelectronics", Chem. Ztg., 96, 552-60 (1972), German, CA78-5456.

15. Capocaccia, F., "State of the art of integrated circuit mask fabrication", Elettrotecnica, 60 (2), 131-40 (1973), Italian, EEA76-26113.
16. Charman, P. A., "Electron-beam microfabrication. The prospects for a practical system", J. Vac. Sci. & Technol., 10, 1132 (1973), EEA77-23108.
17. Clark, K. G., "Properties and processes for photoresists in semiconductor manufacture", Solid State Technol., 14 (6), 52-6 (1971), EEA74-29766.
18. Clark, K. G., "Properties and processes for photoresists in semiconductor manufacture. II", Solid State Technol., 14 (9), 48-53 (1971).
19. Clark, K. G., "The semiconductor in manufacture", Microelectronics, 4 (3), 31-6 (1972), EEA73-26131.
20. Clark, K. G., "Negative and positive photoresist systems", Galvanotechnik, 65 (2), 94-104 (1974), German, CA80-149847.
21. Critchlow, D. I., "Microstructures in semiconductor technology", 6th International Congress on Microelectronics, 1974, 7 pp., EEA78-24511.
22. Crooke, M. and Wittstock, S., "Photolithography in semiconductor manufacture", Trans. S. Afr. Inst. Electr. Eng., 64 (pt. 10), 194-211 (1973), EEA77-2368.
23. Duncalf, B. and Dunn, A. S., "Photosensitized reactions of poly(vinyl alcohol) used in printing technology and other applications", In: Polyvinyl Alcohol, Finch, C. A., (Ed.), Wiley, New York, p. 461-91 (1973), CA81-162001.
24. Dunham, K. R., "Solid-state photoresist technology", Solid State Technol., 14 (6), 41-6 (1971), CA74-69524.
25. Gordon, E. I. and Herriott, D. R., "Pathways in devices lithography", IEEE Trans. Electron Devices, ED-22, 371-5 (1975), EEA78-24384.
26. Hatzakis, M., "Electron sensitive polymers as high resolution resists", Appl. Polym. Symp., 23, 73-86 (1974), CA81-129740.
27. Hatzakis, M., "Electron beam processing systems. State of the art review", Polym. Eng. Sci., 14, 516-17 (1974), CA81-129759.
28. Hatzakis, M., Ting, C. H. and Viswanathan, N., "Fundamental aspects of electron beam exposure of polymeric resist system", Electron Ion Beam Sci. Technol., Int. Conf., 6th, 1974, p. 542-79, CA83-35581.
29. Hook, H. O., "The capabilities and limitations of photolithography", J. Vac. Sci. & Technol., 11, 94 (1974), EEA77-38865.
30. Howland, F. L. and Poole, K. M., "An overview of the new maskmaking system", Bell Syst. Tech. J., 49, 1997-2015 (1970), EEA74-9793.
31. Moreau, W. and Viswanathan, N., "Applications of radiation-sensitive polymer systems", Presented at International Symposium on UV Induced Reactions, Philadelphia, Pa., April 1975.

32. Murrmann, H., "Development of masking technology for semiconductor devices", Nachrichtentech. Z. (NTZ), 21, 616-18 (1968), German, EEA72-714.
33. Naraoka, K., "Photoresists", Bussei, 14, 412-21 (1973), Japanese, CA79-59931.
34. Nonogaki, S., "Photoresist", Kobunshi, 24, 328-30 (1975), Japanese, CA83-35584.
35. O'Malley, A. J., "An overview of photomasking technology in the past decade", Solid State Technol., 14 (6), 57-62 (1971), EEA74-25573.
36. O'Malley, A. J., "Technological implications in the photo-masking process", Solid State Technol., 18 (6), 40-5 (1975), EEA78-33231.
37. Pease, R. F. W., "Electron lithography for integrated circuits", Electrochemical Society Fall Meeting (Extended abstracts), 1974, p. 460-4, EEA78-37683.
38. Rottmann, H. R., "Photolithography in integrated circuit mask metrology", Solid State Technol., 18 (6), 29-34 (1975), EEA78-33230.
39. Ryan, R. J., Davidson, E. B. and Hook, H. O., "Photofabrication", In: Handbook of Materials and Processes for Electronics, Harper, C. A., (Ed.), McGraw-Hill Book Co., New York, 133 pp. (1970), EEA74-29770.
40. Sato, K., "Application of photolacquers to the electronic industry", Kobunshi, 19 (215), 136-46 (1970), Japanese, CA70-72027.
41. Smith, H. I., "Electron beam, ion beam, X-ray optical techniques for fabricating surface-acoustic-wave and thin-film optical devices", Vide, 29 (173), 364-70 (1974), EEA78-23274.
42. Sobanski, J., "Influence of electron beam on organic polymers", Pr. Nauk. Inst. Technol. Elektron. Politech. Wroclaw., no. 10, 119-31 (1973), Polish, CA79-146867.
43. Stevens, G. W. W., "Photographic materials and processes in microelectronics", Galvanotechnik, 65, 375-88 (1974), German, CA81-162002.
44. Wakimoto, Z. and Kawakami, T., "IC and microphotography", Oyo Buturi, 40, 1014-18 (1971), Japanese.
45. Westphal, P., "Aligning and exposure devices - a survey of technical requirements and test results", Jena Rev., 18 (1), 30-2 (1973), EEA76-26128.

III. Equipment

46. Anon., "Artwork-making machines for integrated circuits", Int. Electron., 17 (6), 12-16 (1970), EEA74-17576.
47. Anon., "Photo-mask pattern forming device", JEE, no. 56, 68 (1971), EEA74-36459.
48. Asahi Kogaku Kogyo, "Projection exposure of printing integrated circuit", Patent UK 1382148, Publ. January 1975, EEA78-33234.
49. Bersin, R. L., "Automatic plasma machines for stripping photoresist", Solid State Technol., 13 (6), 39-45 (1970), CA73-82499.

50. Bok, H. F. and St. Onge, E., "Solution agitation", Patent USA 3824137, Publ. July 1974, CA81-107956.
51. Bondarenko, V. A., Voitovich, I. D. and Zholob, V. P., "A device for semiautomatic production of photo-masks", Mekh & Avtom. Upr., no. 1, 38-40 (1972), Russian.
52. Buechele, J. W., "Semiconductor leveling system", IBM Tech. Disclosure Bull., 16, 3255-6 (1974), EEA77-30351.
53. Cahen, O. and Trotel, J., "High performance step and repeat machine using an electron beam and laser interferometers", Electron and Ion Beam Science and Technology 4th International Conference, 1970, p. 581-7, EEA74-30643.
54. Canon KK, "Objective lens assembly for photomask printing on IC wafer", Patent UK 1377937, Publ. December 1974, EEA78-24505.
55. Chang, T. H. P., "An electron detection system for microfabrication registration", J. Vac. Sci. & Technol., 10, 1129 (1973), EEA77-23104.
56. Chang, T. H. P., "Instrumentation for electron beam lithography", 1974 Digests of the Intermag. Conference, 30.5/1 pp., EEA78-8599.
57. Cohen, L. G. and Schneider, M. V., "Microlenses for coupling junction lasers to optical fibers", Appl. Opt., 13, 89-94 (1974), PA77-30499.
58. Ekner, J., "Technology of master drawings for production of two-sided printed circuit boards", Pomiary Autom. Kontr., 16, 439-43 (1970), Polish, EEA74-7043.
59. Engelter, A. and Bierbaum, P., "A high precision alignment device", Solid State Technol., 11 (11), 43-5 (1968), EEA72-10633.
60. Frosch, A., "Optical gap meter", IBM Tech. Disclosure Bull., 16, 3284 (1974), EEA97-30391.
61. Gaylord, T. K., "A laboratory photomask production facility", Rev. Sci. Instrum., 43, 1268-71 (1972), EEA75-31816.
62. Gerber Sci. Instrument Co., "Integrated circuit making apparatus", Patent UK 1202617, Publ. August 1970, EEA74-9813.
63. Glendinning, W. B., "Micro-electronic form masking system", Patent USA 3422442, Publ. January 1969, EEA72-14925.
64. Goldsmith, S. H., "Techniques of phototooling: Selecting the right film type", Circuits Manuf., 14 (9), 96-101 (1974), EEA78-8585.
65. Handy, R. M., Malmberg, P. R. and O'Keeffe, T. W., "Apparatus for producing a precisely located, altered surface zone on a substrate", Patent German 2050763, Publ. May 1971, CA75-92497.
66. Harrell, S., "Practical automation for photolithographic wafer processing", Solid State Technol., 17 (8), 47-9 (1974), EEA77-38892.
67. Herriott, D. R., "Lenses for the photolithographic system", Bell Syst. Tech. J., 49, 2105-16 (1970), EEA74-9801.
68. Heyroth, A., "The E instrument system of VEB Carl Zeiss JENA", Jena Rev., 18 (1), 3-19 (1973), EEA76-26124.

69.Heyroth, A., "Equipments for the photo-lithographic production of micro-structures for the semiconductor industry", Radio Fernsehen Elektron., 24 (4), 109-13 (1975), German, EEA78-24426.
70.Hewlett Packard Co., "Micro circuit mask positioning system interferometrically controlled", Patent UK 1196281, Publ. June 1970, EEA73-9812.
71.Hock, F., "Optically scanned 2-dimensional measuring units as position backfeeders for step repeaters and mask measuring equipment", 6th International Congress on Microelectronics, 1974, 10 pp., German, EEA78-24435.
72.Ichinose, W., "Pattern generator for manufacturing higher capacity, higher density ICs", JEE, no. 92, 30-5 (1974), EEA78-2909.
73.Kerwin, R. F., "Thin photosensitive materials", Bell Syst. Tech. J., 49, 2179-92 (1970), EEA74-9803.
74.Klostermann, F. T., "A step and repeat camera for making photomasks for integrated circuits", Philips Tech. Rev., 30 (3), 57-70 (1969), EEA73-454.
75.Koops, H., "On electron projection systems (microcircuit production)", J. Vac. Sci. & Technol., 10, 909-12 (1973), EEA77-23088.
76.Kornfeld, W. H., "A practical photoresist projection printer system", Electrochemical Society Fall Meeting (Extended abstracts), 1970, p. 453-4, EEA74-7044.
77.Kornilov, A. P., Morgunov, A. S., Nedzvetskiy, Y. A. and Sysun, V. V., "Apparatus of exposing photoresists in the manufacture of integrated circuits", Sov. J. Opt. Technol., 39, 553-4 (1972), EEA76-32294.
78.Kornilov, A. P., Morgunov, A. S. and Nedzvetskiy, Y. A., "An optical apparatus for projection photolithography", Sov. J. Opt. Technol., 40, 396-7 (1973), EEA77-14574.
79.Kosanke, K., Kulcke, W. and Max, E., "Optical measuring method", IBM Tech. Disclosure Bull., 11, 902-3 (1969), EEA72-23732.
80.Lemmond, C. Q., Buschmann, E. C., Klotz, T. H. and White, G. M., Jr., "Electron fly's eye lens artwork camera", IEEE Trans. Electron Devices, ED-21, 589-603 (1974), PA77-69101.
81.Liebergeld, H. and Gallo, W., "Device for mounting the substrate and the appropriate masks in vacuum vapor-depositing equipment", Patent German (East) 75927, Publ. September 1970, CA75-65561.
82.Lin, L. H. and Beauchamp, H. L., "High speed beam deflection and blanking for electron lithography", J. Vac. Sci. & Technol., 10, 987-90 (1973), PA77-42866.
83.Livesay, W. R., "Electron-beam microfabrication equipment for microcircuit research and production", Solid State Technol., 17 (6), 37-42 (1974), EEA77-34101.
84.Livesay, W. R. and Fritz, R. B., "Electron image projection systems for microcircuit lithography", IEEE Trans. Electron Devices, ED-19, 647-51 (1972), EEA75-15980.

85. Mackawa, A., Munakata, C., Hirano, C., Morita, O., Kozawa, T., Isoda, K. and Monogaki, S., "Electron beam drafting machine for IC masks", Electron and Ion Beam Science and Technology 4th International Conference, 1970, p. 503-14, EEA74-29775.
86. Makosch, G., "Dark field illumination device", IBM Tech. Disclosure Bull., 17, 3264 (1975), EEA78-24408.
87. Markle, D. A., "A new projection printer", Solid State Technol., 17 (6), 50-3 (1974), EEA77-34102.
88. McCarthy, S. L. and Stanko, J., "Pattern generators and technique for electron-beam fabrication of submicron thin-film bridges for superconductivity studies", Rev. Sci. Instrum., 45, 335-40 (1974), PA77-57225.
89. Meingailis, J., Smith, H. I. and Efremow, N., "Instrumentation for conformable photomask lithography", IEEE Trans. Electron Devices, ED-22, 496-8 (1975), EEA78-24403.
90. Michl, U. and Nestler, K., "Automatic nine-barrel repeater ANR 4 of VEB Carl Zeiss JENA. A photolithographic mask-making system for the semiconductor industry", Jena Rev., 19, 56-62 (1974), EEA77-38859.
91. Morrow, C. T. and Brouns, A. J., "Acoustic impedance calibrator for mask and microphone measurements", J. Audio Eng. Soc., 18, 518-23 (1970), EEA74-2641.
92. Nippon Kogaku KK, "Apparatus for projection printing of photomasks on integrated circuits", Patent UK 1380074, Publ. January 1975, EEA78-24507.
93. O'Keeffe, T. W., Vine, J. and Handy, R. M., "An electron imaging system for the fabrication of integrated circuits", Solid-State Electron., 12, 841-8 (1969), EEA73-9563.
94. Ozdemir, F. S., Perkins, W. E., Yim, R. and Wolf, E. D., "Precision electron beam microfabrication", J. Vac. Sci. & Technol., 10, 1008-11 (1973), EEA77-24906.
95. Penkunas, J. R. and Spohn, D. Q., "Simultaneous exposure of photoresist on both sides of a wafer", Tech. Dig., no. 35, 47-8 (1974), EEA77-38868.
96. Pircher, T., Lacombat, F. and Dubee, D., "High precision microlithography: tools and technology", International Conference on Large Scale Integrated Circuits, 1974, p. 89-116, EEA78-11710.
97. Poole, K. M., "A rotating mirror pattern generator", 1970 IEEE International Convention Digest, p. 84-5, EEA73-37068.
98. Rawson, E. G., "Reduction cameras: optical design and adjustment", Bell Syst. Tech. J., 49, 2117-28 (1970), EEA74-9802.
99. Rosenau, M. D., Jr., Jones, R. A. and Contente, L., "Optical projection system design for photoresist exposure", Appl. Optics, 8, 75-7 (1969), EEA72-12800.
100. Rublack, W. D., "In-line plural beam cathode ray tube with an aspherical aperture mask", Patent USA 3435268, Publ. March 1969, EEA72-19891.

101. Schedewic, F., "Electro-optic laser microscope for high speed, high precision edge detection", Israel J. Technol., 9, 263-6 (1971), EEA74-38281.
102. Schilling, J., "Special lenses made by VEB Carl Zeiss JENA for imaging micropatterns in the automatic nine-barrel step and repeat camera ANR", Jena Rev., 18 (1), 24-7 (1973), EEA76-26126.
103. Schwartz, R. J., Ozdemir, F. S. and Ward, E. R., "An integrated circuit mask fabrication facility for universities", IEEE Trans. Educ., E-14, 140-1 (1971), EEA74-25279.
104. Scott, J. P., "An electron image projector with automatic alignment", IEEE Trans. Electron Devices, ED-22, 409-13 (1975), EEA78-24389.
105. Stevens, R. C., "Photolithographic pattern size monitor", IBM Tech. Disclosure Bull., 17, 2285-6 (1975), EEA78-24406.
106. Tarui, Y., Denda, S., Bab, H., Miyauchi, S. and Tanaka, K., "Electron beam exposure system for integrated circuits", Microelectron. and Reliab., 8, 101-11 (1969), EEA73-6735.
107. Tibbetts, R. E. and Wilczynski, J. S., "High performance reduction lenses for microelectronic circuit fabrication", IBM J. Res. Developm., 13, 192-6 (1969), EEA72-12791.
108. Trotel, J., "Automated electronic masking machine", Rev. Tech. Thomson-CSF, 5, 873-88 (1973), French, EEA77-23111.
109. Trotel, J., "The use of an electron beam pattern generator", Presented at Electrochemical Society Meeting, San Francisco, Calif., May 1974, EEA77-29156.
110. Wege, G. and Meier, B., "A few examples of using the MP 320 measuring projector of VEB Carl Zeiss JENA to best advantage in microelectronics", Jena Rev., 18 (1), 33-5 (1973), EEA76-26129.
111. Weibrecht, O., "The precision coordinatograph Prakordi E - its operation and efficiency in the electronic industry", Jena Rev., 18 (1), 20-3 (1973), EEA76-26125.
112. Western Electric Co., Inc., "Pattern generating apparatus", Patent UK 1259588, Publ. January 1972, EEA75-19270.
113. Yew, N. C., "A modular system for electron beam microfabrication", Presented at Electrochemical Society Meeting, San Francisco, Calif., May 1974, EEA77-29157.
114. Yew, N. C., "A new EB microfabrication system", Solid State Technol., 18 (8), 63-4 (1975), EEA78-41695.

IV. Photoresist

1. General

115. Baumann, G. A., "Non-liquid photo polymer resists", Electron. Mfr., 15 (4), 31-6 (1971), EEA74-25818.
116. Bendz, D. J., Bendz, G. A. and Wildman, A. M., "Photomechanical fabrication with photoresist", Patent German 2420589, Publ. November 1974, CA83-35724.

117.Bosch, R., "Photolithographic process for engraving, by chemical attack, semiconductor plates", Patent French 2010439, Publ. February 1970, CA74-93467.
118.Celeste, J. R., "Process for making photoresists", Patent USA 3469982, Publ. September 1969, CA72-49680.
119.Celeste, J. R. and Heiart, R. B., "Unique properties and uses of photo-polymer film resists", In: Appl. Photopolymers, Povinelli, R. J., (Ed.), Soc. Photogr. Sci. Eng., Inc., Washinton, D. C., p. 42-52 (1970), CA78-166901.
120.Clark, K. G., "Photoresists. II. Negative resist systems", Electron. Components, 14, 621-3, 625-39 (1973), EEA76-32287.
121.Clark, K. G., "Photoresists. IV. A. Application of photoresists in semiconductor manufacture", Electron. Components, 14, 711-15 (1973), EEA76-38074.
122.Clark, K. G., "Photoresists. IV. B. Application of photoresists in semiconductor manufacture", Electron. Components, 14, 751, 754-8 (1973), EEA76-38075.
123.Clark, K. G. and Turner, R. G., "Micro-image photoresist evaluation", Solid State Technol., 12 (7), 25-33 (1969), EEA73-467.
124.Dartsch, H., "The importance of photolithographic process in semiconductor technology", Radio Fernsehen Elektron., 20, 371-2 (1971), German, EEA74-29764.
125.Dathe, J., Miller, W. and Ghosh-Dastidar, S., "Method of forming semiconductor junctions, by etching, masking, and diffusion", Patent USA 3498853, Publ. March 1970, EEA73-24975.
126.Dhaka, V. A., Clark, E. R. and Pinck, A., "Photochemical fabrication process for preparing printed circuits and components", Patent USA 3573975, Publ. April 1971, CA75-13571.
127.Dill, F. H., "Optical lithography", IEEE Trans. Electron Devices, ED-22, 440-4 (1975), EEA78-24393.
128.Dolby, R. F., "The technology of negative working photosensitive resists", Electron. Mfr., 16 (6), 23-6 (1971), EEA74-33337.
129.Eltsov, A. V., Yurre, T. A. and Guk, E. G., "On the multifunctional use of photoresists", Pis'ma V Zh. Tekh. Fiz., 1, 571-4 (1975), Russian, EEA78-41645.
130.Gyombolai, G., "Diazo photocopy lacquers", Kolor. Ert., 16, 227-32 (1974), Hungarian, CA82-162897.
131.Lekas, T. C., "High purity photoresists for microelectronics production", Solid State Technol., 16 (12), 70 (1973), EEA77-10280.
132.Long, J. D. and Mann, M., "Dry film resists", Electroplating & Met. Finish, 24 (2), 20-5 (1971), EEA74-21770.
133.Lussow, R. O., "Photoresist materials and applications", J. Vac. Sci. Technol., 6, 18-24 (1969), CA70-119942.
134.Malikowski, K. and Piotrowska, M., "Photolithography application in obtaining deposited tantalum capacitors and resistors", Pr. Osrodka Badaw.-Rozwojowego Elektron. Prozniowej, 2 (2), 47-55 (1974), Polish, EEA78-2898.

135.Nakane, H., Yokota, A., Asaumi, S. and Hatazawa, T., "Product control and the thermostability of positive photoresist", Tech. Pap., Ref. Tech. Conf., Soc. Plast. Eng., Mid-Hudson Sect., 1973, p. 92-101, CA80-9035.
136.Orlovic, M. and Pearson, I. M., "Method for producing metallic, dielectric, or semiconductive layers on substrates in selected configurations", Patent S. Africa 69-00353, Publ. July 1969, CA72-72416.
137.Powell, H., "Dry film photoresist for additive plating", Insul./Circuits, 21 (13), 25-9 (1975).
138.Ritchie, K. and Hughes, H. G., "Comparative studies of photoresist processing of microelectronics regarding chemical costs, pollution and shortage", International Symposium on Materials for Electronic Components, 1975, p. 184-91, EEA78-33240.
139.Roman, W. and Smolski, R., "Technological production line for thin film hybrid photolithography processing", Elektronika, 16 (3), 97-101 (1975), Polish, EEA78-24531.
140.Rosshaupter, E. and Hundt, O., "Photoresists", Chem. Unscrer Zeit, 5 (5), 147-53 (1971), German, CA72-87229.
141.Stebbing, P. J., "Printed circuit resists", Electroplating & Met. Finish, 25 (2), 29-33 (1972), EEA75-16189.
142.Stevens, G. W. W., "Photographic materials and processes in microelectronics", Galvanotechnik, 65 (5), 375-88 (1974), German, EEA77-36316.
143.Stockley, F., "Positive working photo resists: application in printed circuits and microelectronics", Proceedings of the EIPC Conference on Printed Circuit Manufacturing, 1970, p. 93-107, EEA74-4316.
144.Troschle, E., "Photo-screening methods in the semiconductor technology", Elektro-Anzeiger, 23, 517-21 (1970), German, EEA74-10457.
145.Wakimoto, Z. and Kawakami, T., "IC and microphotography", Oyo Buturi, 40, 1014-18 (1971), Japanese, EEA75-19258.
146.Yurre, T. A., Shaburov, V. V. and Eltsov, A. V., "Silver-free photographic materials and photoresists", Zh. Vsev. Khim. O-va., 19, 412-23 (1974), CA81-129749.

2. Compositions

a. Positive

147.Abolafia, O. R., "Producing positive images from epoxy compositions", Patent USA 3752669, Publ. August 1973, CA79-151682.
148.Agnihotri, R. K., "Polydiacrylyl photoresists", Patent USA 3644118, Publ. February 1972, CA77-12344.
149.Amariti, L., "Diazo photoprinting materials", Patent USA 3582334, Publ. June 1971, CA75-103709.
150.Anderson, R. M. and Heimsch, R. A., "Photoresist compositions for making printed circuit boards", Patent USA 3656951, Publ. April 1972, CA77-54906.

151.Asami, T., "Diazo light-sensitive material", Patent USA 3793030, Publ. February 1974, CA80-139544.

152.Asano, T. and Takeuchi, T., "Photoresist composition", Patent Japan 75-08934, Publ. April 1975, CA83-200239.

153.Bonham, J. A. and Petrellis, P. C., "Dry positive-acting photoresist", Patent USA 3782939, Publ. January 1974, CA80-139561.

154.Bowden, M. J. S. and Chandross, E. A., "Positive photoresist comprising polysulfones formed by reacting vinyl aromatic hydrocarbons with sulfur dioxide", Patent USA 3884696, Publ. May 1975, CA83-139876.

155.Canestaro, M. J. and Orinik, M. T., "Epoxy resin-based positive photoresist and dielectric composition", Patent German 2231297, Publ. January 1973, CA78-112415.

156.Colom, L. A. and Levine, H. A., "Light-sensitive, positive photoresist compositions", Patent German 2149527, Publ. April 1972, CA77-35893.

157.Christensen, C. W. and Isaacson, C. M., "Composite photoresist materials", Patent German 2106574, Publ. September 1971, CA75-157068.

158.Christensen, C. W. and Isaacson, C. M., "Diazo-sensitized photoresists", Patent German 2329208, Publ. January 1974, CA81-19299.

159.Deutsch, A. S. and Loprest, F. J., "Positive-working photoresist composition", Patent S. Africa 70-01019, Publ. October 1970, CA74-28316.

160.Dietrich, R., "Light-sensitive polymeric layers", Patent German 1298414, Publ. June 1969, CA71-66049.

161.Dinaburg, V. A., Dinaburg, M. S. and Strelkova, N. M., "Hydroxy-containing resins", Patent USSR 300491, Publ. April 1971, CA75-89124.

162.Dunham, K. R., Jorden, D. G., Fields, D. L. and Miller, J. B., "6-Diazo-5-oxo-4a-azoniaanthracene photoresist compositions and lithographic printing plates", Patent S. Africa 68-01224, Publ. July 1968, CA72-30683.

163.Fukutani, H. and Miura, K., "Photosensitive polymeric material and its use", Patent German 2146414, Publ. March 1972, CA77-95379.

164.Jones, V. V., "Light-sensitive etchable copying material", Patent German 2023083, Publ. November 1970, CA74-48098.

165.Kalle, A. G., "Naphthoquinone diazide sulfonic esters in light-sensitive materials", Patent UK 1127996, Publ. September 1968, CA70-8120.

166.Komiya, H., Awada, Y. and Tsuchiya, M., "Photosensitive compositions based on phenolic resins and oxide compounds", Patent Japan 74-67701, Publ. July 1974, CA82-37341.

167.Levine, H. A., "Positive photoresist compositions", Patent French 2075039, Publ. November 1971, CA77-107738.

168.Makarov, N. V., Nicol'skii, V. G., Timerova, N. D. and Mozzhukhin, D. D., "Photoresist compositions", Patent USSR 357547, Publ. October 1972, CA78-91065.

169.Maksimov, B. Y., Verner, Y. G., Kabinetskii, O. E. and others, "Aerosol-type photoresists", Patent USSR 269313, Publ. April 1970, CA74-70228.
170.Martynenko, A. P., Makarov, N. V. and Nikol'skii, V. G., "Positive photoresist", Patent USSR 332413, Publ. March 1972, CA77-95380.
171.Morimoto, K., Minobe, A. and Kuroda, M., "Polymer compositions able to be insolubilized by light", Patent Japan 74-106585, Publ. October 1974, CA82-163043.
172.National Cash Register Co., "Photochromic imaging", Patent German 1571088, Publ. June 1972, CA78-78160.
173.Nomura, M., Tanaka, H. and Nagasawa, K., "Diazo photosensitive composition for lithography", Patent Japan 74-22930, Publ. February 1974, CA81-84442.
174.Ohmura, K., Toyomoto, K. and Komoto, H., "Epoxy resin positive photoresist compositions", Patent Japan 75-68320, Publ. June 1975, CA83-200251.
175.Paramonov, A. I. and Prokhotskii, Y. M., "Positive photoresist", Patent USSR 451978, Publ. November 1974, CA82-163064.
176.Rauner, F. J., Arcesi, J. A. and Guild, J. R., "Polymeric, light-sensitive o-quinonediazides", Patent German 2044869, Publ. April 1971, CA75-21506.
177.Roos, L., "Photoresists from diazosulfonate polymer colloids", Patent German 2028903, Publ. November 1971, CA76-29558.
178.Ruckert, H., "Positive-working photoresist systems. Fundamentals, methods, and special uses", Galvanotechnik, 65, 591-6 (1974), German, CA81-162005.
179.Ruckert, H. and Wildenhain, B., "Positively working photoresists", Patent German 2336941, Publ. February 1974, CA80-126806.
180.Sayuki, S. and Hata, Y., "Photosensitive resin compositions", Patent Japan 74-16504, Publ. February 1974, CA81-19292.
181.Slimowicz, C. and Welch, W. J., "Diazo materials for black line images", Patent USA 3785826, Publ. January 1974, CA80-139545.
182.Smith, G. H. and Bonham, J. A., "Material soluble upon irradiation and its use in light-sensitive elements", Patent German 2306248, Publ. September 1973, CA80-114851.
183.Ulrich, H., "Tris (m- or p-(axidosulfonvl)phenyl) isocyanurates and their uses as crosslinking and blowing agents", Patent USA 3562269, Publ. February 1971, CA75-6880.
184.Valiev, K. A., Lubashevskaya, A. V., Korsakov, V. S. and et al., "Positive photoresist compositions", Patent USSR 421548, Publ. March 1974, CA81-113725.
185.Van Loewis, G., "Positive photocopying lacquer", Patent German (East) 104633, Publ. March 1974, CA81-129901.
186.Voitovich, I. D., Kostyshin, M. T., Mikhailovskaya, E. V., Petrov, V. V., Romanenko, P. F. and Skuridin, V. P., "Photosensitive imageable layers", Patent French 2135736, Publ. January 1973, CA79-25729.

187. Yamada, H., Shingu, T. and Kitanishi, Y., "Photosensitive resin compositions", Patent Japan 74-44054, Publ. April 1974, CA81-113708.
188. Yamaguchi, H., Iwaki, A. and Tokura, H., "Photoresist materials", Patent German 2053287, Publ. May 1971, CA75-82431.
189. Yamaji, K. and Kurahashi, A., "Light-sensitive plate for electrophotography", Patent German 2011357, Publ. November 1970, CA74-125045.
190. Yonezawa, T., Kobayashi, K., Kita, N. and Matsumoto, H., "Diazo-type lithographic photoresist", Patent Japan 74-110403, Publ. October 1974, CA82-148530.

b. Negative

191. Abolafia, O. R., "Light-sensitive epoxy resins for photographic lacquers", Patent German 2258880, Publ. June 1973, CA79-60000.
192. Abolafia, O. R. and Canestaro, M. J., "Light-sensitive epoxy formulation", Patent USA 3726679, Publ. April 1973, CA79-12013.
193. Adams, W. R., "Photosensitizers for vinyl polymerization", Patent German 2426449, Publ. January 1975, CA82-140972.
194. Agnihotri, R. K., "Photolacquer compositions", Patent German 2151068, Publ. April 1972, CA77-41376.
195. Agnihotri, R. K., Falcon, D. L., Hood, F. P., Lesoine, L. G., Needham, C. D. and Offenbach, J. A., "Structure and behavior of cyclized rubber photoresist", Photogr. Sci. Eng., 16, 443-8 (1972), CA78-36223.
196. Akama, T. and Iwamoto, M., "Polyamide-type photoresists", Patent Japan 74-110401, Publ. October 1974, CA82-163054.
197. Arcesi, J. A., Petropoulos, C. and Rauner, F. J., "Photosensitive polymers containing o-quinone diazide", Patent French 2173074, Publ. November 1973, CA80-76688.
198. Arcesi, J. A. and Rauner, F. J., "Light-sensitive film-forming polysulfonate reaction product of an aromatic disulfonyl chloride and bisphenol containing the styryl ketone group", Patent USA 3640722, Publ. February 1972, CA77-63521.
199. Arcesi, J. A., Rauner, F. J., McConkey, R. G. and Noonan, J. M., "Photopolymerizable compositions", Patent USA 3748132, Publ. July 1973, CA80-83887.
200. Azami, K., Ohtami, H. and Fukutomi, H., "Photosensitive resin", Patent Japan 73-17000, Publ. May 1973, CA80-21444.
201. Bakos, P. and Kozak, G. S., "Photosensitive polychloroprene used as a photoresist for metal etching, electric insulation, relief images, and lithographic printing", Patent USA 3551160, Publ. December 1970, CA74-54913.
202. Berg, C. J., "Dry-film negative photoresist having amidized styrene-maleic anhydride binder materials", Patent USA 3873319, Publ. March 1975, CA83-35739.
203. Borden, D. G., "Photosensitive polymers", Patent USA 3737319, Publ. June 1973, CA79-110316.

204. Borden, D. G., "Photocrosslinkable polyphosphonates having unusual resistance to acids", Polym. Eng. Sci., 14, 487-90 (1974), CA81-113625.

205. Celeste, J. R. and Cohen, A. B., "Photoresists", Patent German 2162207, Publ. June 1972, CA78-91067.

206. Chang, C. T. L., Sheeto, J. J., Jr. and Beutel, J., "Photopolymerizable emulsions", Patent German 2327513, Publ. December 1973, CA80-65528.

207. Clecak, N. J. and Cox, R. J., "Arsenic-containing bisazide compounds as negative photoresists", Patent USA 3679670, Publ. July 1972, CA77-119852.

208. Clecak, N. J., Cox, R. J. and Moreau, W. M., "Photoresist compositions", Patent German 2148216, Publ. March 1972, CA77-35852.

209. Clecak, N. J., Cox, R. J. and Moreau, W. M., "Azide photoresists for projection photolithography", Polym. Eng. Sci., 14, 491-3 (1974), CA81-144203.

210. Crampton, K. J. A., "Photoresists", Patent UK 1325789, Publ. August 1973, CA79-151683.

211. Crowe, W. J. and Ritchie, K., "Photosensitive organic vitreous film", Patent German 2211017, Publ. October 1972, CA78-142452.

212. Davidson, E. B., "Coating material for the photoresist process", Patent German 2050859, Publ. April 1971, CA75-38042.

213. Davidson, E. B., "Photoresist composition for use in making etched patterns", Patent USA 3615952, Publ. October 1971, CA76-61085.

214. Davidson, E. B., "Stabilized photoresist composition", Patent USA 3776736, Publ. December 1973, CA80-102353.

215. Deviny, E. J., "Photocrosslinkable polymers for dielectric barriers", Patent German 2306531, Publ. August 1973, CA80-15623.

216. Dietrich, R., "Photopolymerizable compositions for copying materials", Patent German 2060576, Publ. June 1972, CA77-68568.

217. Dunham, K. R., Newcomb, R. A. and Yost, M. E., "Light-sensitive polymeric compositions for photoresist materials", Patent German 2135482, Publ. January 1972, CA76-154828.

218. Dunham, K. R. and Yost, M. E., "Photoresist compositions", Patent USA 3711287, Publ. January 1973, CA78-142458.

219. Eastman Kodak Co., "Photosensitive product for preparing chemically engraved resists", Patent French 2037739, Publ. December 1970, CA74-82432.

220. Eastman Kodak Co., "Cyclized rubber as a base for photosensitive compositions used in photography", Patent French 2084996, Publ. January 1972, CA77-107737.

221. Efros, L. S., Yurre, T. A., Korobitsina, I. K., Dinaburg, V. A. and Orlova, D. N., "Negative photoresist composition", Patent USSR 466480, Publ. April 1975, CA83-88744.

222. Feinberg, J. H., "4-Methoxybenzenediazonium hexafluorophosphate catalyst for photopolymers in epoxy systems", Patent USA 3829369, Publ. August 1974, CA82-10025.

223. Foster, J. S., Brandon, R. and Wagner, H. M., "Photosensitive material", Res. Discl., 137, 17-18 (1975), CA83-206945.
224. Foster, J. S., Holstead, C., Osborne, P. A. and Wagner, H. M., "Diazodihydrooxonaphthalene sensitizers for photoresists", Patent UK 1331441, Publ. September 1973, CA80-32526.
225. Frank, V. S. and Kehr, C. L., "Photoresist compositions", Patent German 2107286, Publ. September 1971, CA75-146262.
226. Freyermuth, H. B. and Field, N. D., "Light-sensitive, film-forming poly (vinyl alcohol esters)", Patent German 2062424, Publ. July 1971, CA75-130337.
227. Fukutani, H., Miura, K., Eguchi, C. and Takahashi, Y., "Light-sensitive epichlorohydrin polymers", Patent German 2410066, Publ. September 1974, CA82-58852.
228. Fukutani, H., Miura, K. and Takahashi, Y., "Photoresist composition with improved sensitivity", Patent Japan 74-12901, Publ. February 1974, CA81-56685.
229. Fukutani, H., Miura, K., Takahashi, Y. and Torige, K., "Light-sensitive materials for etchant-resistant relief images", Patent Japan 73-57619, Publ. August 1973, CA80-102350.
230. Fukutani, H., Miura, K., Takahashi, Y. and Torige, K., "Photo-polymerizable compositions", Patent German 2345095, Publ. March 1974, CA80-139548.
231. Fulkerson, B. and Mench, J. W., "Photosensitive and thermo-sensitive compositions based on cellulosic esters, for preparing photomechanical resists", Patent French 2027643, Publ. November 1970, CA74-13572.
232. Gant. G. A. L., "Preparations hardenable by uv light", Patent German 2417944, Publ. April 1975, CA82-172778.
233. Gates, A. P., "Light-sensitive polymeric esters", Patent German 2251828, Publ. April 1973, CA79-19676.
234. Giangualano, M. N., Martenson, I. W. and Ott, L. H., "Light-sensitive compositions for use on printing plates", Patent German 1547849, Publ. April 1971, CA75-13546.
235. Gilson, R. E. and Rauner, F. J., "Photosensitive composition for preparing colored photoresists", Patent French 2037523, Publ. December 1970, CA74-69560.
236. Goto, K., Muto, N. and Zokuno, M., "Photosensitive compositions based on N-vinyl compounds, organic polyhalides, and poly (glutamic acid) derivatives", Patent Japan 74-05022, Publ. January 1974, CA80-139546.
237. Grace, W. R., and Co., "Photoresist resist for solid surfaces", Patent Neth. 73-05548, Publ. October 1974, CA83-69131.
238. Grace, W. R., and Co., "Photoresist composition containing polyene and polythiol", Patent French 2227558, Publ. November 1974, CA83-35728.
239. Harada, K., Harita, Y., Ito, H. and Komatsu, K., "Wet processable photosensitive materials", Patent Japan 74-19819, Publ. February 1974, CA81-84448.

240.Harada, K., Inomata, S., Tsuchiya, T., Ono, H., Tsuchida, T., Harita, Y., Komatsu, K., Toyoda, H. and Minoura, T., "Photoresist compositions from cyclized polymers or cyclized rubber", Patent Japan 75-33826, Publ. April 1975, CA83-139874.
241.Hasegawa, E. and Tsuji, N., "Photosensitive compositions containing xylene resins", Patent German 2226008, Publ. December 1972, CA78-142461.
242.Hasegawa, E. and Tsuji, N., "Light-sensitive polymers", Patent German 2232185, Publ. January 1973, CA78-117633.
243.Hay, A. S., "Photosensitive polymeric films", Patent German 1949208, Publ. September 1970, CA73-125722.
244.Hay, A. S., "Photosensitive acetylenic polymers", Patent UK 1149697, Publ. April 1969, CA71-39738.
245.Heimsch, R. A., Bartosczewicz, J. S. and Slocombe, R. J., "Polymeric photosensitive compositions", Patent USA 3787212, Publ. January 1974, CA81-113723.
246.Hepher, M., "Photographic transfer materials", Patent USA 3778272, Publ. December 1973, CA80-126799.
247.Hepher, M. and Sperry, J. A., "Pigmented photoresists", Patent German 2109200, Publ. September 1971, CA76-20052.
248.Hirohashi, M., Tsunoda, T., Yamaoka, A. and Nagamatsu, G., "Photoresist compositions", Patent Japan 73-47826, Publ. July 1973, CA79-127037.
249.Hudgin, D. E., "Photoresists of sensitized protein", Patent German 2318286, Publ. October 1974, CA82-163063.
250.Hurley, R. B. and Lazarus, S., "Photopolymerizable compositions for producing photoresists", Patent German 2063571, Publ. July 1971, CA76-52997.
251.Ichikawa, M., Takeuchi, Y., Miura, T., Harita, Y., Tashiro, M., Tsunoda, T., Toyoda, H. and Minoura, T., "Solvent-developed-type light-sensitive materials", Patent German 2230969, Publ. December 1972, CA79-60001.
252.Ichimura, S. and Maruhashi, M., "Water soluble photoresist", Patent Japan 75-30602, Publ. March 1975, CA83-155780.
253.Ichimura, S. and Sumitomo, Y., "Photosensitive compositions for print-out images", Patent Japan 74-08213, Publ. January 1974, CA80-139563.
254.Ikeda, K., Takakura, K. and Imai, K., "Photosensitive resins", Patent Japan 74-40391, Publ. April 1974, CA81-136975.
255.Isobe, A., Tsukada, K., Abo, M., Ishimaru, T. and Ogawa, K., "Photohardenable resin compositions for photoresists", Patent Japan 75-48092, Publ. April 1975, CA83-155785.
256.Isobe, A., Tsukada, K., Hayashi, N. and Abo, M., "Photocurable resin compositions", Patent Japan 74-88929, Publ. August 1974, CA82-112758.
257.Iwama, H. and Iwaki, A., "Photoresist composition", Patent Japan 73-76521, Publ. October 1973, CA80-42942.
258.Iwama, H. and Iwaki, A., "Photoresist compositions", Patent Japan 73-73207, Publ. October 1973, CA80-114854.

259. Jahnke, R. W., "Photosensitive diacetone acrylamide resins", Patent USA 3725231, Publ. April 1973, CA79-6125.
260. Jones, V. V., "Light-sensitive copying materials for preparing photoresist layers", Patent German 2037345, Publ. February 1971, CA75-13574.
261. Jones, V. V., "Photoimaging material comprising a polyamide coating", Patent USA 3695887, Publ. October 1972, CA77-171260.
262. Kaetsu, I., Ito, A. and Hayashi, K., "Photopolymerization", Patent Japan 73-65287, Publ. September 1973, CA80-121508.
263. Kakuta, T., Ogura, I., Ichikawa, M. and Takeuchi, Y., "Light-sensitive material", Patent Japan 74-19162, Publ. May 1974, CA82-148533.
264. Kalle, A. G., "Glycidol ether polymer photoresists", Patent French 1573456, Publ. July 1969, CA72-95324.
265. Kaplan, L. H., Lounsbury, J. B. and Zimmerman, S. M., "Improved photoresist lacquers", Patent German 2410880, Publ. September 1974, CA82-49903.
266. Kashiwagi, T., "Photopolymerizable composition", Patent Japan 74-10843, Publ. March 1974, CA81-56667.
267. Kawada, H., Iwama, H., Yumiki, K., Kurita, Y. and Tokura, H., "Photosensitive resin composition", Patent Japan 75-55406, Publ. May 1975, CA83-186375.
268. Kerwin, R. E. and Goldrick, M. R., "Thermally stable photoresist polymer", Polym. Sci., <u>11</u>, 426-30 (1971), CA74-135803.
269. Killy, R., "Photosensitive polymeric coating material", Patent German 2142242, Publ. March 1973, CA80-9110.
270. Kita, N. and Yonezawa, T., "Photopolymerizable urethanes", Patent German 2404714, Publ. August 1974, CA81-179978.
271. Klein, D. L., MacIntyre, M. W. and Rotham, L. J., "Polyisoprene-based photoresists", Patent German 2153781, Publ. May 1972, CA77-41377.
272. Klein, D. L., MacIntyre, M. W. and Rotham, L. J., "Cyclized polyisoprene photoresist", Patent French 2109786, Publ. June 1972, CA78-91066.
273. Kokawa, T. and Iwase, S., "Photohardening resin compositions for imaging", Patent Japan 75-17827, Publ. February 1975, CA83-50810.
274. Koltsov, Y. I., Mozzhukhin, D. D., Nikolskii, V. G. and Petyushin, A. A., "Negative photoresistive material", Patent USSR 454526, Publ. December 1974, CA83-88745.
275. Konishiroku Photo Industry, Co., Ltd., "Photoresist-forming compositions", Patent UK 1223632, Publ. March 1971, CA74-133072.
276. Kornfeld, W. H. and Sinder, R. M., "4,4-Diazidobenzophenone-p-carboxyphenylhydrazone for use as a sensitizer in preparation of photoresist compositions", Patent USA 3538125, Publ. November 1970, CA74-59363.
277. Kurka, K. A., "Light-sensitive composition", Patent USA 3825430, Publ. July 1974, CA81-97815.

278.Lakiza, V. V. and Lobyntsev, S. A., "Epoxy-novolak composition", Patent USSR 394401, Publ. August 1973, CA81-50547.
279.Lazarus, S. and Turner, E. J., "Photoresist compositions containing diazoquinone siloxanes", Patent German 2312499, Publ. December 1973, CA80-126807.
280.Levy, M. F. and Schmitt, G. P., "Photosensitive dielectric composition", Patent USA 3776729, Publ. December 1973, CA80-102352.
281.Lewis, J. M., "Photoresist material", Patent German 2306353, Publ. August 1973, CA79-131394.
282.Lundsager, C. B., "Photoresist process and compositions for etching metals", Patent USA 3660088, Publ. May 1972, CA77-82225.
283.Magnotta, F., Ketley, A. D. and Kehr, C. L., "Prepolymer mixture hardenable to a solid", Patent German 2311124, Publ. September 1973, CA80-4282.
284.Martynenko, A. P., Glukhova, Y. V. and Koltsov, Y. I., "Light-sensitive 1,2-naphthoquinone diazide-(2)-5-sulfonyl chloride ester", Zh. Fiz. Khim., 47, 2820-2 (1973), Russian, CA80-151032.
285.Mayuzumi, T., Fujino, T., Ooto, M. and Noshiro, A., "Photopolymerizable organic silicon compounds", Patent Japan 73-47997, Publ. July 1973, CA80-27865.
286.Mayuzumi, T., Fujino, T., Ooto, M. and Noshiro, A., "Photopolymerizable silicone compounds", Patent Japan 73-48000, Publ. July 1973, CA80-15669.
287.Mench, J. W., Fulkerson, B. and Dulmage, W. J., "Light-sensitive compositions for photographic materials", Patent USA 3615628, Publ. October 1971, CA76-29552.
288.Michihiro, O., Mayuzumi, T., Fujino, T. and Noshiro, A., "Photopolymerizable silicone material for forming printing plates and etch-resistant resists", Patent German 2233514, Publ. January 1973, CA78-137311.
289.Milgrom, J., "Photoresist material for forming printing plates", Patent USA 3674486, Publ. July 1972, CA78-10197.
290.Miura, K. and Tomita, Y., "Light-sensitive preparation", Patent German 2439449, Publ. May 1975, CA83-124098.
291.Miura, K., Watanabe, N. and Tomita, Y., "Photolacquer preparation", Patent German 2439433, Publ. May 1975, CA83-106242.
292.Moreau, W. M., "Negative working photoresist compositions", Patent German 2132063, Publ. January 1972, CA76-134191.
293.Morgan, C. R., "Light-sensitive coatings", Patent German 2319848, Publ. November 1974, CA82-100333.
294.Morgan, C. R., "Polyene-polythiol photohardenable compositions", Patent German 2322051, Publ. November 1973, CA80-139559.
295.Morgan, C. R., "Solid photoresist comprised of a polyene and a polythiol", Patent USA 3767398, Publ. October 1973, CA81-31893.
296.Morita, J., Ijichi, I., Maeda, S. and Kuki, H., "Photoresist compositions based on acrylonitrile-acrylate copolymers", Patent Japan 74-27302, Publ. March 1974, CA81-113727.

297.Morita, J., Shimomura, T., Maeda, S. and Kubo, S., "Photo sensitive compositions for photoresists", Patent Japan 75-38524, Publ. April 1975, CA83-170925.
298.N. V. Philips' Gloeilampenfabrieken, "Photoresist compositions", Patent UK 1325974, Publ. August 1973, CA79-151673.
299.Nagashima, S. and Hasegawa, T., "Photoresist compositions for alkali-resistant images", Patent Japan 73-73148, Publ. October 1973, CA80-54539.
300.Nakane, H., Yokota, N. and Kotake, K., "Rubber photoresist composition", Patent Japan 74-06202, Publ. February 1974, CA82-49902.
301.Nakayama, T., Yokota, A., Nuino, K., Nakane, H. and Asaumi, S., "Photoresist materials for projection printing", Patent Japan 74-60525, Publ. June 1974, CA82-10026.
302.Nishikubo, T., Imaura, M. and Takaoka, T., "Photopolymer photographic materials", Patent Japan 75-21803, Publ. March 1975, CA83-106224.
303.Nishikubo, T., Ugai, S., Ichi, J. and Kishida, M., "Photo-hardenable photosensitive materials", Patent Japan 74-02601, Publ. January 1974, CA81-44138.
304.Noguchi, M., "Photoresist composition", Patent Japan 73-98904, Publ. December 1973, CA80-126805.
305.Oba, Y. and Tsunoda, T., "Photoresist compositions", Patent Japan 74-01225, Publ. January 1974, CA80-139560.
306.Ohba, Y., "Photoresist composition", Patent Japan 73-86601, Publ. November 1973, CA80-114858.
307.Ohba, Y., "Photoresist compositions based on organic diazide compounds and poly (sodium L-glutamate)", Patent Japan 73-90727, Publ. November 1973, CA80-114857.
308.Ohba, Y., "Photoresist composition based on organic diazide compounds and maleic acid-vinyl alcohol copolymers", Patent Japan 73-97602, Publ. December 1973, CA80-126804.
309.Ohba, Y., "Photoresist compositions", Patent Japan 74-68801, Publ. July 1974, CA82-37342.
310.Ohba, Y., Kimura, T. and Morishita, H., "Photoresist composition containing polymer and bisazide-type photocrosslinking agent", Patent Japan 73-79617, Publ. October 1973, CA80-54542.
311.Ohba, Y. and Nonogaki, S., "Photoresist compounds based on organic diazide compound and acrylamide-vinyl alcohol copolymers", Patent Japan 73-97603, Publ. December 1973, CA80-126802.
312.Ohba, Y. and Nonogaki, S., "Photoresist compositions based on organic diazide compounds and poly (vinyl butyral)", Patent Japan 73-98905, Publ. December 1973, CA80-126803.
313.Oleinik, A. V. and Karyakina, L. N., "Photocrosslinking of cyclorubber resists by bisazides", Zh. Nauchn. Prikl. Fotogr. Kinematogr., 19, 426-8 (1974), Russian, CA82-78756.
314.Oleinik, A. V., Karyakina, L. N., Smirnova, G. A., Belevich, G. M., Shapkin, G. A. and Kolmakov, O. A., "Negative photoresist", Patent USSR 475595, Publ. June 1975, CA83-155787.

315.Osada, C., Satamura, M. and Ono, H., "Light-sensitive β-(cinnamoyloxy) ethyl methacrylate polymer", Patent German 2164625, Publ. July 1972, CA77-107730.
316.Osada, C., Satomura, M. and Ono, H., "Photosensitive compositions", Patent Japan 73-08384, Publ. February 1973, CA79-80010.
317.Phlipot, G. A., Michiels, M. J. S. and De Villenuve, Y. P. M., "Photosensitive composition comprising a light-sensitive copolyester", Patent USA 3726685, Publ. April 1973, CA79-25726.
318.Powers Chemco Inc., "Photoresist coating material", Patent French 2151542, Publ. May 1973, CA80-49240.
319.Prasad, J. and Jain, P. K., "Phenol-formaldehyde photoresist", Res. Ind., 17, 4-6 (1972), CA78-50527.
320.Rao, D. V., Ulrich, H., Stuber, F. A. and Sayigh, A. A. R., "Photosensitive polymers derived from acryloyl-substituted alkoxy stilbenes", Patent USA 3583940, Publ. June 1971, CA75-77727.
321.Rao, D. V., Ulrich, H., Stuber, F. A. and Sayigh, A. A. R., "N-Stilbenylacrylamides", Patent USA 3718692, Publ. February 1973, CA78-125178.
322.Roos, L., "Light-sensitive bisdiazonium salts as crosslinking agents", Patent German 2059431, Publ. June 1972, CA77-102761.
323.Rosenkranz, H. J., Rudolph, H., Wolff, E. and Von Rintelen, H., "Crosslinkable vinyl group-containing polyurethanes", Patent German 2121253, Publ. November 1972, CA78-112782.
324.Rosenkranz, H. J., Rudolph, H., Wolff, E. and Von Rintelen, H., "Photocrosslinkable (meth)acrylamide-modified acrylic copolymers", Patent German 2125910, Publ. December 1972, CA78-73061.
325.Roteman, J., "Epoxy resin photoresist with iodoform and bismuth triphenyl", Patent USA 3895954, Publ. July 1975, CA83-155786.
326.Rowe, W. and Golda, E., "Photosensitive composition for photolithography", Patent German 2361931, Publ. June 1974, CA81-129907.
327.Rudneva, S. P., Trindyak, Z. V., Orlov, F. I., Shepelskaya, I. M., Spirin, Y. L., Magdinets, V. V. and Maslyuk, A. F., "Photopolymerizing composition", Patent USSR 412590, Publ. January 1974, CA81-71095.
328.Satomura, M., "Photohardenable polymers", Patent Japan 74-60390, Publ. June 1974, CA81-129920.
329.Sayigh, A. A. R., Stuber, F. A. and Ulrich, H., "Light-sensitive polyurethanes prepared from a hydroxyl-containing polymer and an isocyanatostilbene compound", Patent USA 3644289, Publ. February 1972, CA77-76001.
330.Sayigh, A. A. R., Stuber, F. A. and Ulrich, H., "Photoresist compositions comprising light-sensitive ester-amide polymers", Patent USA 3657199, Publ. April 1972, CA77-35286.
331.Schlesinger, S. I., "Photopolymerization of epoxy monomers", Patent USA 3708296, Publ. January 1973, CA78-85077.

332.Schlesinger, S. I., "Epoxy photopolymers in photoimaging and photofabrication", Tech. Pap., Reg. Tech. Conf., Soc. Plast. Eng., Mid-Hudson Sect., 85-91 (1973), CA80-9026.
333.Schlesinger, S. I. and Cochran, V., "Copolymers and their use as photoresists", Patent German 2329407, Publ. April 1974, CA81-71100.
334.Schreckendgust, J. G., "Powder developed transparencies for projection", Patent German 2146830, Publ. April 1972, CA77-107733.
335.Schwartz, B., "Ceramic photoresists", Patent USA 3502520, Publ. March 1970, CA73-9474.
336.Skoultchi, M., "Novel photosensitive coating systems", Patent USA 3574617, Publ. April 1971, CA75-28317.
337.Smets, G. J. and DeSchrijver, F. C., "Bismaleimides for photoresists", Patent German 1963731, Publ. September 1970, CA73-104343.
338.Srinivasan, R., "Photoresist compositions", Patent UK 1128119, Publ. September 1968, CA70-5039.
339.Steele, F. W., "Photopolymerizable polymers containing anhydride groups", Patent German 2422365, Publ. January 1975, CA82-140829.
340.Stiehler, H. and Loeckler, I., "Solvent combinations for photocopying lacquers", Patent German 2319159, Publ. March 1975, CA83-35741.
341.Stutz, H., "Light-sensitive copying materials", Patent German (East) 102823, Publ. December 1973, CA82-24373.
342.Takenouchi, O. and Ohtani, H., "Alkali soluble photoresists", Patent Japan 74-120703, Publ. November 1974, CA82-163053.
343.Takita, S., Hashimoto, E. and Ohkawa, K., "Photoresist composition", Patent Japan 75-19683, Publ. March 1975, CA83-170923.
344.Takita, S. and Ohkawa, K., "Photopolymerizable compositions", Patent Japan 74-04738, Publ. January 1974, CA81-56669.
345.Tanabe, K., Saito, S. and Takimoto, Y., "Photosensitive resin compositions", Patent Japan 74-39414, Publ. April 1974, CA81-97796.
346.Tanaka, H. and Tsuda, M., "Photosensitive composition comprising light-sensitive polymer", Patent USA 3767415, Publ. October 1973, CA80-21435.
347.Thomas, D. C., "Polycarbonate-cinnamate photopolymers for making photoresist plates and presensitized, negative-working printing plates", Patent USA 3622331, Publ. November 1971, CA76-100818.
348.Togura, H., Kawata, H. and Yamaguchi, H., "Light-sensitive material", Patent Japan 73-08876, Publ. March 1973, CA80-42941.
349.Tomita, K., Takagi, K. and Oonishi, S., "Photosensitive poly (vinyl alcohol) derivatives", Patent Japan 73-65292, Publ. September 1973, CA80-76684.
350.Tsukada, K., Isobe, A., Abo, M. and Ogawa, K., "Photosensitive resin compositions for photoresists", Patent Japan 74-107333, Publ. October 1974, CA83-19031.

351.Tsukada, K., Isobe, A., Hayashi, N. and Abo, M., "Photoresist resinous compositions", Patent Japan 73-57620, Publ. August 1973, CA79-131393.
352.Tsukada, K., Isobe, A., Hayashi, N. and Abo, M., "Photosensitive resin compositions", Patent Japan 73-65924, Publ. September 1973, CA80-15793.
353.Tsunoda, T., Maeda, S. and Hata, Y., "Photosensitive resin compositions containing 2-azidophenylbenzothiazole derivatives", Patent Japan 73-94419, Publ. December 1973, CA80-102348.
354.Tsunoda, T., Maeda, S. and Hata, Y., "Photosensitive resin composition containing 2-(4-azidophenyl)azothiazole derivatives", Patent Japan 73-9-420, Publ. December 1973, CA80-102338.
355.Ueno, K. and Tsunoda, T., "Epoxycinnamylidene acetate-type photoresist composition", Patent Japan 74-135616, Publ. December 1974, CA83-106227.
356.Ulrich, H., Stuber, F. A. and Broggi, A. J., "Radiation-sensitive polymers", Patent USA 3794626, Publ. February 1974, CA81-106575.
357.Umiki, K., Yamazaki, A. and Kawada, H., "Azide-type photoresists", Patent Japan 75-32922, Publ. March 1975, CA83-139873.
358.Von Bonin, W. and Wolff, E., "Photocrosslinkable photographic layers", Patent German 2100222, Publ. July 1972, CA77-153286.
359.von Unger, F., "Du Pont photopolymer resist film Riston", Proceedings of the EIPC Conference on Printed Circuit Manufacturing, 1970, p. 77-92, German, EEA74-4321.
360.Wade, R. C., "Photosensitive titanium-containing redox system", Patent German 2011304, Publ. February 1971, CA75-13518.
361.Wagner, H. M., Lowman-Riggs, C. and Foster, J. S., "Polymers and photoresist compositions containing them", Patent UK 1398088, Publ. June 1975, CA83-194323.
362.Wagner, H. M. and Przezdziecki, W. M., "Light-sensitive photographic element", Patent USA 3850646, Publ. November 1974, CA83-180212.
363.Walters, R. M. and Weisz, E. M., "Improved negative photoresist", Electrochemical Society Fall Meeting, Extended Abstracts, 1970, p. 458, EEA74-9937.
364.Watanabe, S. and Ichimura, K., "Photocrosslinkable resin compositions", Patent Japan 74-129736, Publ. December 1974, CA83-44206.
365.Watanabe, S. and Ichimura, K., "Novel photoresist composition", Patent Japan 74-130218, Publ. December 1974, CA82-178260.
366.Watt, W. R., "Controlled polymerization process and stabilized polymerizable compositions", Patent UK 1398226, Publ. June 1975, CA83-181248.
367.Watt, W. R., "Controlled polymerization process and stabilized polymerizable compositions", Patent UK 1398354, Publ. June 1975, CA83-207683.
368.Wiegand, D. A., "Inorganic azides as photosensitive materials at low temperatures", Patent USA 3778269, Publ. December 1973, CA80-126811.

369. Wolff, E., Laessig, W., Seelig, E. and Kolf, G., "Light-sensitive layers containing crosslinkable, film-forming polymers", Patent German 2030506, Publ. January 1972, CA77-12337.
370. Wuehrmann, H., "Photopolymerizable photoresists", Patent German 2307292, Publ. August 1973, CA80-21446.
371. Yamada, K., Ikeda, K., Imai, K. and Izumi, M., "Cyclized rubber photoresist compositions", Patent Japan 75-72702, Publ. June 1975, CA83-200252.
372. Yamamoto, Y., "Organic photosensitive materials base on resins containing styrene sulfonyl groups", Patent Japan 74-67705, Publ. July 1974, CA82-37340.
373. Yanagisawa, S., Kakiuchi, H. and Koyama, T., "Epoxy resin compositions", Patent Japan 74-16800, Publ. February 1974, CA81-154809.
374. Yonezawa, T., Kobayashi, K. and Kita, N., "Photopolymerizable compositions", Patent Japan 74-107083, Publ. October 1974, CA82-178246.
375. Yoshino, T. and Ohbayashi, G., "Photosensitive resin compositions", Patent Japan 75-85329, Publ. July 1975, CA83-211337.
376. Yumiki, K., Yamazaki, A. and Yamaguchi, H., "Photoresists containing diazides and leuco dyes", Patent Japan 74-69322, Publ. July 1974, CA82-66350.
377. Zahir, S. A. C., Rembold, H. and Losert, E., "Photoresist compositions for printed multilayer circuits", Patent German 2342407, Publ. March 1974, CA81-113724.
378. Zavarzina, Z. Y., Prokopova, T. V., Peikarova, A. A., Shibina, V. I. and Tsimmerman, L. D., "Photoresist composition for printed circuits", Patent USSR 398914, Publ. September 1973, CA81-44137.
379. Zollei, M., "Silver-sensitization of sintered photoresists prepared from colloidal cadmium sulfide solution", Acta Phys. Chem., 17, 29-32 (1971), German, CA76-20001.

3. Processing

a. Applying

380. Auerbach, I. A., Hecker, E. P. and Manthei, L. W., "Photoresist apply and dry machine", IBM Tech. Disclosure Bull., 17, 2275-9 (1975), EEA78-24405.
381. Celeste, J. R., "Photoresists which can be laminated to surfaces", Patent UK 1128850, Publ. October 1968, CA70-42851.
382. Christensen, C. W. and Isaacson, C. M., "Materials for forming photoresist layers on a support", Patent German 2046115, Publ. May 1971, CA75-93035.
383. Christensen, C. W. and Isaacson, C. M., "Applying an unexposed photoresist layer to a substrate", Patent USA 3649283, Publ. March 1972, CA76-160817.
384. Fredericks, E. C. and Moreau, W. M., "Resist thickness regulation", IBM Tech. Disclosure Bull., 17, 1048 (1974), EEA78-8587.

385.Goldmacher, J. E. and Dow, O. E., "Applying an N,N-diallylmelamine resist to a semiconductor surface", Patent USA 3627599, Publ. December 1971, CA76-65369.
386.Holihan, J. E., Sr. and Landstein, L., "Controlled gap photoresist spinning process", IBM Tech. Disclosure Bull., 17, 3281 (1975), EEA78-24410.
387.Iscoff, R. L. Z., "Photoresist application techniques in semiconductor processing", Electron. Packag. & Prod., 15 (11), 27-36 (1975).
388.Karp, J., "Curing photo resist using microwave energy", 1970 International Hybrid Microelectronic Symposium, 6 pp., EEA74-29784.
389.Kunz, C. O., Long, P. C. and Wright, A. N., "Positive and negative photoresist applications of thin films surface-photopolymerized from hexachlorobutadiene", Polym. Eng. Sci., 12, 209-12 (1972), CA77-12266.
390.LaRue, A. J., "Use of a venturi to accomplish resist splatter dispersal", IBM Tech. Disclosure Bull., 17, 3211-12 (1975), EEA78-28828.
391.Maurischat, K., "Application of nonliquid photopolymer resists", Galvanotechnik, 65, 659-72 (1974), German, CA81-162003.
392.Mondou, E. R. and Schmidt, P. R., "Striation-free resist coating process", IBM Tech. Disclosure Bull., 18, 391-2 (1975), EEA78-37576.
393.Simeonov, V. and Karadzhov, G., "Conditions for the application and processing of photoresists", Khim. Ind., 47, 212-16 (1974), Bulgarian, CA83-170777.
394.Stahl, K., Merk, H. and Mueller, H., "Application of a light-sensitive lead sulfide resist layer", Patent German (Amended) 1564871, Publ. December 1974, CA83-50808.
395.Stiglitz, M. R. and Kearns, W. J., "A photoresist film applicator for long and fragile crystal substrates", Rev. Sci. Instrum., 44, 1671-2 (1973), EEA77-18719.
396.Stockley, F., "Applications of positive-working photoresists", Galvanotechnik, 65, 183-91 (1974), German, CA81-162067.

b. Exposure

397.Akamatsu, K., "Control of exposure of photosensitive resin plates by using flash exposure", Patent Japan 75-43922, Publ. April 1975, CA83-155782.
398.Akiyama, K., "Photographic treatment for semiconductor devices or the like", Patent USA 3767397, Publ. October 1973, CA80-53776.
399.Bates, T. R., "Planographic mandrel", Patent USA 3432299, Publ. March 1969, CA70-102543.
400.Blome, E. R. and Fok, S. S. M., "Hard-surface transparent mask for use in exposing photopolymer areas", Patent USA 3625728, Publ. December 1971, CA76-39066.

401.Bolon, D. A. and Kunz, C. O., "Ultraviolet depolymerization of photoresist polymers", Polym. Eng. Sci., 12, 109-11 (1972), CA76-154542.

402.Bottomley, S. C. and Nightingale, E. G., "Illumination system for photoprocessing by the proximity printing method", Opt. & Laser Technol., 7 (2), 65-72 (1975), EEA78-33174.

403.Chapman, J. C., "Proximity printing", Electron. Prod. Methods & Equip., 2 (5), 17-20 (1973), EEA76-35333.

404.Clark, K. G. and Turner, R. G., "Microimage photoresist evaluation", J. Photogr. Sci., 17 (4), 126-36 (1969), CA71-107597.

405.Dill, F. H., Hornberger, W. P., Hauge, P. S. and Shaw, J. M., "Characterization of positive photoresist", IEEE Trans. Electron Devices, ED-22, 445-52 (1975), CA83-124016.

406.Dill, F. H., Nuereuther, A. R., Tuttle, J. A. and Walker, E. J., "Modeling projection printing of positive photoresists", IEEE Trans. Electron Devices, ED-22, 456-641 (1975), EEA78-24396.

407.Doering, A. P., "Exposure of strongly light-absorbing photographic coatings", Patent German 2249060, Publ. April 1973, CA79-47854.

408.Fried, L. J., Flachbart, R., Utlen, D. F., Raniseski, J. W., Anderson, F. W. and Patel, K. V., "Standing wave interference patterns in photoresists observed with a scanning electron microscope", J. Electrochem. Soc., 117, 1079-81 (1970), CA73-115010.

409.Frosch, A. and Schmackpheffer, A., "Exposure process for producing a fine-structure pattern on a photoresist coating", Patent German 2116713, Publ. March 1974, CA81-56684.

410.Goldrick, M. R. and Plauger, L. R., "Evaluation of exposure parameters for photoresist materials", Photogr. Sci. & Eng., 17, 386-9 (1973), EEA76-35335.

411.Hallman, R. W. and Kurtz, G. W., "Radiation sensitive materials, and methods for exposing and processing them for the manufacture of useful articles", Patent S. Africa 68-03224, Publ. November 1969, CA73-52989.

412.Heynick, L. N., Westerberg, E. R., Hartelius, C. C., Jr. and Lee, R. E., "Projection electron lithography using aperture lenses", IEEE Trans. Electron Devices, ED-22, 399-409 (1975), EEA78-24388.

413.Ilten, D. F. and Patel, K. V., "Standing wave effects in photoresist exposure", Image Technol., 13 (2), 9-14 (1971), CA74-149164.

414.Kakihama, S. and Stoneham, E. B., "Achieving uniform exposure in a photosensitive material on a semiconductor wafer", Patent USA 3884698, Publ. May 1975, CA83-107212.

415.Kiba, W. M., "Transparent masks for exposing photoresists on a semiconductor body", Patent USA 3561963, Publ. February 1971, CA74-92684.

416.Korka, J. E., "Standing waves in photoresists", Appl. Optics, 9, 969-70 (1970), EEA73-23850.

417.Krasovitskaya, T. I., Murav'ev, V. S. and Moskvitin, N. I., "Photoresist composition for imaging silicon semiconductors", Sb. Rab., Mosk. Lesotckh, Inst., no. 22, 96-8 (1969), Russian, CA74-118348.
418.Lawson, L. E. and Smith, P. J., "Exposure of light-sensitive materials", Patent USA 3890148, Publ. June 1975, CA83-200244.
419.Lundsager, C. B., "Photoresist process", Patent German 2146198, Publ. August 1973, CA79-131392.
420.Mathews, J. R. and Raetschi, P. A., "Experience in the use of photolithographic projection exposure", 1971 IEEE International Convention Digest, p. 292, EEA74-26658.
421.Morita, J., Shimomura, T. and Maeda, S., "Photoresist relief imaging", Patent Japan 75-43920, Publ. April 1975, CA83-155783.
422.Moritz, H. and Wuestenhagen, J. F., "Method of switching fields during wafer exposure", IBM Tech. Disclosure Bull., 16, 3282 (1974), EEA77-30390.
423.Narasimham, M. A., "Automatic exposure process for incoherent projection printing in positive resists", IBM Tech. Disclosure Bull., 18, 113 (1975), EEA78-37629.
424.Narasimham, M. A., "Projection printed photolithographic images in positive photoresists", IEEE Trans. Electron Devices, ED-22, 478-82 (1975), EEA78-24400.
425.Parker, A. L., "Backside exposure of positive photoresists", IBM Tech. Disclosure Bull., 13, 285 (1970), EEA73-37615.
426.Penkunas, J. R. and Spohn, D. Q., "Simultaneous exposure of photoresist on both sides of a wafer", Tech. Dig., no. 35, 47-8 (1974), EEA77-38868.
427.Roman, W., "Thin film hybrid microcircuits masker", Elektronika, 14, 479-82 (1973), Polish, EEA77-14624.
428.Ross, D. and Kiemle, H., "Production of semiconductor modules by a photoresist technique with holographic projection of etching patterns", Patent USA 3712813, Publ. January 1973, CA79-72323.
429.Sunners, B., "Monolithic circuit manufacture and photoresist exposure technique utilized therein", Patent USA 3524394, Publ. August 1970, EEA74-9833.
430.Veidenbakh, L. V. and Lyalikov, K. S., "Effect of additional exposure on positive-working photopolymers", Zh. Nauch. Prikl. Fotogr. Kinematogr., 17, 459-60 (1972), Russian, CA79-25646.
431.Westphal, P., "Relief images", Patent German (East) 109275, Publ. October 1974, CA83-186388.
432.Wilczynski, J. S., "Optical projection in microfabrication", J. Vac. Sci. & Technol., 10 (6), 1127 (1973), EEA77-23103.

c. Developing

433.Bartolini, R. A., "Improved development for holograms recorded in photoresist", Appl. Opt., 11, 1275-8 (1972), EEA75-24157.

434.Casler, N. L. C. and Vincentsen, B. E., "Photoresist developer extender baths containing poly (oxyalkylene ethers and esters)", Patent USA 3682641, Publ. August 1972, CA77-120782.
435.Chiu, G. T., "Developers for polysulfone terpolymer resists", IBM Tech. Disclosure Bull., 17, 3283 (1975), EEA78-24412.
436.Coombs, C. F., Jr., "Photoresist film solves production problems", Solid State Technol., 11 (9), 47-8 (1968), EEA72-2393.
437.Gallois, D. and Ducrocq, J., "Photolithographic developer", Patent French 2171864, Publ. November 1973, CA80-76693.
438.Gunina, N. M., Safonova, N. V., Tkacheva, R. I. and Tonkikh, N. N., "Organic developers for positive photoresists", Patent USSR 396664, Publ. August 1973, CA81-19236.
439.Haining, F. W., "Ultrasonics used for developing photoresist", IEEE Trans. Sonics Ultrasonics, SU-17, 57 (1970), EEA73-23857.
440.Hamatani, T., "Development of printing plates and other relief patterns obtained with a liquid photoresist", Patent Japan 75-64004, Publ. May 1975, CA83-211317.
441.Hasegawa, E. and Hirata, K., "Dry-development type photopolymerization imaging materials", Patent Japan 75-46315, Publ. January 1975, CA83-186389.
442.Ito, F., "Alkaline developers for photoresists", Patent Japan 74-34323, Publ. March 1974, CA81-97816.
443.Iwama, H., Iwarki, A. and Yamaguchi, H., "Water-developable photoresist materials", Patent German 2234511, Publ. January 1973, CA78-117639.
444.Iwamatsu, S., "Oxidation development of photoresists", Patent Japan 74-36333, Publ. April 1974, CA81-179985.
445.Kimoto, A. and Kameyama, K., "Development of photoresists", Patent Japan 73-101128, Publ. December 1973, CA80-139562.
446.Macy, E. H. and Sienkowski, R. P., "Nonthinning polymer for positive photoresists", IBM Tech. Disclosure Bull., 17 (11), 3284 (1975), EEA78-24413.
447.Mizuno, K., Takagi, K., Ohnishi, N. and Nagai, K., "Water-developable photoresists", Patent Japan 75-54405, Publ. May 1975, CA83-170916.
448.Osabe, Y., Watanabe, T. and Ito, F., "Development of photoresists", Patent Japan 75-65304, Publ. June 1975, CA83-200237.
449.Press, F. P., "Organic developers for positive photoresists based on naphthoquinone diazides", Elektron. Tekh. Nauch.-Tekh. Sb. Poluprov. Prib., no. 1, 175-81 (1970), Russian, CA76-160786.
450.Shimamura, I., Iijima, Y., Okutsu, E., Iwano, H., Shishido, T. and Ohi, R., "Development of lithographic films", Patent German 2128498, Publ. December 1971, CA76-119909.
451.Stevens, G. W. W., "Photographic materials and processes in microelectronics", Galvanotechnik, 65, 375-88 (1974), German, PA77-68714.
452.Xerox Corp., "Electrolytically reducing tin oxide to tin", Patent UK 1270845, Publ. April 1972, CA77-69480.

453.Zech, R. G., Dwyer, J. C., Fichter, H. and Lewis, M., "Heat-processed photoresist for holographic data storage applications", Appl. Opt., 12, 2822-7 (1973), PA77-14894.

d. Etching

454.Aramaki, I., Kawabata, T. and Arimoto, K., "Automation of etching pattern layout", Commun. ACM, 14, 720-30 (1971), EEA75-4134.
455.Brandes, R. G. and Dudley, R. H., "Wall profiles produced during photoresist-masked isotropic etching", J. Electrochem. Soc., 120, 140-2 (1973), CA78-90107.
456.Goell, J. E., "Selective multilayered etching", Patent USA 3751248, Publ. August 1973, CA79-151679.
457.Hashimoto, N., Yanazawa, H., Ashikawa, M., Douta, K. and Utsugi, H., "Application of surface conversion technique to photolithography", J. Electrochem. Soc., 121, 90C (1974), EEA77-26497.
458.Hatzakis, M., "Variable-depth etching of film layers using variable exposures of photoresists", Patent USA 3649383, Publ. March 1972, CA76-119939.
459.Hillis. B. F., Klint, R. D. and Cross, W. D., "Etching of polyimide layers", Patent German 2336511, Publ. March 1974, CA80-139566.
460.Horn, H., Blank, J., Hoepfner, E. and Delle, E., "After hardening of gelatin etch resists", Patent German (East) 91981, Publ. August 1972, CA78-91064.
461.Horst, R. S., Kaplan, L. H. and Merritt, D. P., "Photoresist of decreased undercutting", Patent USA 3873313, Publ. March 1975, CA83-35740.
462.Kluge, H. C., "Photoresist material for plasma etching of via holes into sputtered silicon dioxide", IBM Tech. Disclosure Bull., 17, 3270 (1975), EEA78-24470.
463.Kolf, G. and Laessig, W., "Vibration etching, a method of evaluation of photoresist characteristics", Polym. Eng. Sci., 11, 492-501 (1971), CA75-157019.
464.Koyano, T., Ikeda, K., Nakamoto, H. and Imai, K., "Photoresist compositions with reduced undercutting", Patent Japan 75-06321, Publ. January 1975, CA83-50809.
465.Kunz, C. O. and Long, P. C., "Selective etching of coatings by using photoresists", Patent German 2109108, Publ. September 1971, CA75-143470.
466.LaCombe, D. J., "Passivation of photoresist materials used in selective plasma etching", Patent USA 3816196, Publ. June 1974, CA81-43008.
467.LaCombe, D. J. and Babcock, G. L., "Selective plasma etching of organic materials employing photolithographic techniques", Patent USA 3816198, Publ. June 1974, CA81-43009.

468. Oto, M., Mayuzumi, T., Fujino, T. and Noshiro, A., "Resist film for ion etching", Patent Japan 73-38429, Publ. June 1973, CA79-85681.
469. Sugimoto, E., "Etch-resistant photoresists", Patent USA 3506440, Publ. April 1970, CA73-30702.
470. Verelst, J. L. and Willems, J. F., "Improving the etch resistance of polymeric resists used in photoengraving", Patent French 1572683, Publ. June 1969, CA72-116834.
471. Vossen, J. L. and Davidson, E. B., "The interaction of photoresists with metals and oxides during R.F. sputter-etching", J. Electrochem. Soc., 119, 1708-14 (1972), MA6-04-340326.
472. Yamashiro, Y. and Ojuma, K., "Photoresist for deep etch plate", Patent Japan 74-62205, Publ. October 1974, CA82-10027.

e. Removal

473. Abe, H., "Application of gas plasma to the fabrication of MOS LSI (metal-oxide-semiconductor large scale integrated arrays)", Proc. Conf. Solid State Devices, 1974, p. 287-95, Publ. 1975, CA83-140840.
474. Bakos, P. and Knight, D. J., "Compositions for photoresist-removing baths", Patent French 2072172, Publ. October 1971, CA78-77327.
475. Bakos, P. and Knight, D. J., "Method for stripping resists from substrates", Patent USA 3813309, Publ. May 1974.
476. Baltakmens, G. and Tourish, J. P., "Compositions for the removal of finish coatings", Patent USA 3650969, Publ. March 1972, CA77-50296.
477. Barile, C. A., Elie, G. T. and Kluge, H. C., "Resist removal process", IBM Tech. Disclosure Bull., 17, 3282 (1975), EEA78-24411.
478. Barton, O. A. and Murphy, K. P., "Ternary azeotropic compositions", Patent USA 3686130, Publ. August 1972, CA77-154324.
479. Bauer, S. and Sikora, H., "Stripping materials for light-sensitive resists", Patent German 2123702, Publ. November 1972, CA78-91068.
480. Beck, W., Brunner, F. C., Frasch, P. U., Ivancic, B., Schwerdt, F. W. and Vogtmann, T., "Stripping layers of organic material", Patent USA 3900337, Publ. August 1975, CA83-171681.
481. Bolon, D. A., "Removal of photoresist masks", Patent German 2341216, Publ. February 1974, CA80-138488.
482. Christensen, G. L. and Gittelman, D. H., "Cleaning photoresist patterns", Patent German 2145647, Publ. April 1972, CA77-12342.
483. Coates, A. E. and Vegh, B. J., "Photoresist stripping solution", Patent French 2028658, Publ. November 1970, CA75-13573.
484. Denis, C., "Removing an etching mask from the intermediate product of a semiconductor active component", Patent Swiss 499881, Publ. January 1971, CA74-92665.

485.Hawkes, R. P., Sykes, J. A. and Tucker, E. A., "Cleaning organic material from the surface of components made of an inorganic material", Patent UK 1241152, Publ. July 1971, CA75-114127.
486.Hori, B. and Kato, B., "Solvent composition for removing cinnamate photoresist film", Patent Japan 74-14681, Publ. April 1974, CA82-49904.
487.Hughes, H. G., Hunter, W. L. and Ritchie, K., "Residues remaining after RF plasma photoresist removal", J. Electrochem. Soc., 120, 99-101 (1973), EEA76-12664.
488.Irving, S. M., "A plasma oxidation process for removing photoresist films", Solid State Technol., 14 (6), 47-51 (1971), EEA74-29765.
489.Irving, S. M. and Lemons, K. E., "Removing photoresist in manufacture of semiconductor devices", Patent USA 3837856, Publ. September 1974, CA81-143307.
490.Jacob, A., "Semiconductor devices", Patent USA 3806365, Publ. April 1974, CA81-7627.
491.Jacob, A., "Material for manufacturing semiconductor devices", Patent USA 3867216, Publ. February 1975, CA83-52026.
492.Jomura, K., "Photoresist remover", Patent Japan 75-55405, Publ. May 1975, CA83-156853.
493.Kirk, R. W., "Applications of plasma technology to the fabrication of semiconductor devices", In: Techniques and applications of plasma chemistry, Hollohan, J. R., (Ed.), Wiley-Interscience, New York, p. 347-77 (1974), EEA78-24093.
494.Kodak-Pathe, "Composition for stripping photosensitive surfaces for preparation of photographic resists", Patent French 2221050, Publ. October 1974, CA83-19030.
495.Laming, F. P. and Staile, R. E., "Photoresist cleaning", J. Electrochem. Soc., 120, 292-5 (1973), CA78-90966.
496.Lenart, M. and Mihalik, B., "Plasma oxidation process for removing photoresist films", Hiradastech. Ipari Kutatointez. Kozlem., 12, 41-50 (1972), Hungarian, CA78-78084.
497.Lure, M. S., "Composition for removal of photoresists", Patent USSR 439035, Publ. August 1974, CA82-49905.
498.Motusevich, E. I., Rets, V. K. and Dmitrakov, D. G., "Removal of layers of photoresist materials", Patent USSR 333731, Publ. March 1972, CA77-68569.
499.Otani, H. and Ishitsu, T., "Removal of photohardened poly(vinyl alcohol)-type photoresists", Patent Japan 74-52001, Publ. May 1974, CA81-113706.
500.Robinson, P., "Methanol/methylene chloride mixtures for removing masks from printed circuits", Patent German 2138200, Publ. March 1972, CA77-50549.
501.Rodionova, A. A. and Evstifeeva, A. F., "Solution for removing the protective relief formed from a positive photoresist containing naphthoquinonediazides and phenolformaldehyde resins from the silicon plate surfaces", Patent USSR 455312, Publ. December 1974, CA83-50811.

502. Schevey, W. R. and Kremers, F. J., "Polymeric etch resist strippers", Patent USA 3871929, Publ. March 1975, CA83-20810.
503. Schroeder, K. H. and Kelly, G. J., Jr., "Chemical strippers and method of using", Patent USA 3787239, Publ. January 1974, CA80-75748.
504. Sowell, R. R., Cuthrell, R. E., Mattox, D. M. and Bland, R. D., "Surface cleaning by ultraviolet radiation", J. Vac. Sci. & Technol., 11, 474-5 (1974), EEA77-34073.
505. Stelter, M. K., "Removing photoresist mask residuals from a semiconductor substrate", Patent USA 3592691, Publ. July 1971, CA75-92495.
506. Takenouchi, O., Fukutomi, H., Norota, S. and Ohtani, H., "Removal of photoresist coatings from metal plates", Patent Japan 73-27938, Publ. April 1973, CA79-6907.
507. Wanesky, W. R., "Removal of resist material from a substrate in the manufacture of electric circuits and (or) devices", Patent UK 1217050, Publ. December 1970, CA74-92687.
508. Winter, R. E., "Metal deposition with polyimide lift-off technique", IBM Tech. Disclosure Bull., 27, 1309 (1974), EEA78-8588.
509. Wright, A. N. and Merrill, R. C., "Removing organic polymer resist films from laminated surfaces", Patent German 2063721, Publ. October 1971, CA79-141546.

f. Others

510. Arcesi, J. A., Rauner, F. J. and Williams, J. L. R., "Synthesis and characterization of photocrosslinkable polysulfonates", Tech. Pap. Reg. Tech. Conf. Soc. Plast. Eng. Mid-Hudson Sect., 1970, p. 19-23, CA74-4154.
511. Berardinelli, F. and Bervasi, J., "High-temperature photoresist containing cross-linked poly(2,6-dimethyl-1,4-phenylene oxide)", Patent USA 3597216, Publ. August 1971, CA75-125088.
512. Beuckman, F. O., "Photoresist technology for etching and electroplating", Proceedings of the Printed Circuit Plating Symposium, 1969, p. 38-51, EEA73-24066.
513. Blum, J. M. and Hoekstra, J. P., "Alloying semiconductor metals in the presence of reactive substances", Patent French 2014594, Publ. April 1970, CA74-58568.
514. Boyer, L. F. and Johnson, A. F., Jr., "Selective coating of semiconductor workpieces", Patent USA 3695928, Publ. October 1972, CA77-157907.
515. Brandes, R. G. and Curran, R. K., "Modulation transfer function of AZIII photoresist", Appl. Opt., 10, 2101-3 (1971), EEA75-3932.
516. Brill, F. W., "Printed circuits by use of metal resinate photoresists", Patent French 2017782, Publ. May 1970, CA74-70280.

517.Broers, A. N., "Forming highly soluble patterns on a photoresist-coated support", Patent German 2227342, Publ. January 1973, CA78-142463.
518.Buckbee-Mears Co., "Hardening of photoresists", Patent German 2259180, Publ. August 1973, CA79-120496.
519.Celeste, J. R., "Improvements relating to the preparation and use of photoresists", Patent UK (Amended) 1128850, Publ. October 19-8, CA77-133223.
520.Celeste, J. R., "Photocrosslinkable processes useful for imagewise modification of glass, ceramic, and metal surfaces, and making printed circuits", Patent USA 3526504, Publ. September 1970, CA73-104324.
521.Chumbres, L. G. and Rudy, G. J., "Landless plated through hole photoresist-making process", Patent USA 3772101, Publ. November 1973.
522.Cohen, A. B. and Tobias, A., "Photoresists by transfer", Patent German 2256448, Publ. June 1973, CA79-110323.
523.Gee, A., "Use of organosilicon subbing layers in the photoresist method in integrated circuit fabrication", Patent USA 3405017, Publ. October 1968, CA70-32753.
524.Gottfried, W. A., "Double photoresist processing", Patent USA 3506441, Publ. April 1970, CA73-9475.
525.Haas, H. C., "Photographic products and processes", Patent USA 3452182, Publ. June 1969, CA71-107622.
526.Hata, Y. and Toyofuku, K., "Photosensitive resins for photoresists", Patent Japan 74-03702, Publ. January 1974, CA81-8431.
527.Hayday, J. R., "Photofabrication", Patent UK 1376316, Publ. December 1974, CA82-148527.
528.Herzog, J. F., "Photo-resists and metallisation", Proceedings of the EIPC Conference on Printed Circuit Manufacturing, 1970, p. 108-15, French, EEA74-4322.
529.Hinz, W., Solow, G. and Mueller, D., "Light-sensitive glass sheets with smooth resist coatings of uniform thickness", Patent German (East) 70354, Publ. December 1969, CA73-69418.
530.Htoo, M. S., "Photoresist for silicon disks", Patent French 1577139, Publ. August 1969, CA72-138356.
531.Ichijo, T., "Undercutting-free photoresist compositions containing imidazoles", Patent Japan 74-96805, Publ. September 1974, CA82-178256.
532.Inoue, N., Kashiwagi, T. and Tanaka, S., "Thermal transfer of photoresist images", Patent Japan 73-93337, Publ. December 1973, CA80-114856.
533.Iwashita, M., Ueyama, M. and Hada, A., "Holing plastic films", Patent Japan 74-113872, Publ. October 1974, CA82-99520.
534.Kodak-Pathe, "Photographic resists by transfer", Patent French 2167375, Publ. September 1973, CA80-21447.
535.Kolb, G., Schmidt, A., Schmitz-Josten, R. and Wolff, E., "Lightcrosslinkable layers containing cyclopentadiene polymers", Patent German 2019598, Publ. November 1971, CA76-73294.

536. Laridon, U. L., Delzenne, G. A. and Peeters, H. K., "Positive working diazo printing plates", Patent Belgian 740597, Publ. April 1970, CA74-81781.
537. Lenart, M., "Questions in photolithographic technology using positive photoresist lacquers", Hiradastech. Ipari Kutatointez. Kozlem., 12, 40-9 (1972), Hungarian, CA79-37094.
538. Minami, S., Yamaguchi, Y., Yamagishi, H., Itoga, M. and Shimokawa, Y., "Photoresist for relief image formation in chemically decomposable polyester layer", Patent Japan 74-62625, Publ. May 1974, CA81-113736.
539. Morita, J., Shimomura, T., Maeda, S. and Kubo, S., "Photosensitive compositions for relief imaging and photoresists", Patent Japan 75-38522, Publ. April 1975, CA83-200250.
540. Parmer, W. F., "Field effect transistors", Patent USA 3436281, Publ. April 1969, EEA73-1219.
541. Patrick, E. L., "Coloring of anodized aluminum", Patent USA 3833374, Publ. September 1974, CA81-154817.
542. Prakash, O., "Micro-miniaturized solid state devices through photo-resist", Indian J. Phys., 48, 458-63 (1974), EEA78-2816.
543. Raine, J. S., "Optimising positive photo resist processing", Electron. Prod. Methods & Equip., 2 (10), 42, 44, 46 (1973-1974), EEA77-14571.
544. Skarvinko, E. R., "Photoresist composition useful for making chromium masks", Patent USA 3802889, Publ. April 1974, CA81-31894.
545. Smith, H. I., Bachner, F. J. and Efremow, N., "A high-yield photolithographic technique for surface wave devices", J. Electrochem. Soc., 118, 821-5 (1971), EEA74-28541.
546. Steinhart, F., "Photolithographic semiconductor production", Patent German 1521780, Publ. January 1970, CA72-94575.
547. Steppan, H., Moschel, A. and Lueders, W., "Light-sensitive layers from crosslinkable esters of β-substituted α-ajanoacrylic acids with polyfunctional hydroxy compounds", Patent S. Africa 69-01980, Publ. October 1969, CA72-116795.
548. TeVelde, T. S., "Electrical monograin layers", Patent USA 3764325, Publ. October 1973, CA79-140840.
549. Thielen, L. E., "Composition for copperizing aluminum offset plates coated with an exposed and developed water-soluble photoresist", Patent USA 3617340, Publ. November 1971, CA76-52988.
550. Timofeev, V. Y., Savotin, Y. I. and Levitman, Y. K., "Statistical analysis of the effect of different factors on the quality of photolithographic operations in the manufacture of integrated circuits", Izv. VUZ Radioelektron., 16 (4), 30-4 (1973), Russian, EEA76-26120.
551. Toyoda, H., Abe, H. and Minoura, T., "Baking a photoresist using microwave power", Nippon Shashin Gakkaishi, 38 (1), 14-19 (1975), Japanese, CA83-187303.
552. Verstraete, J. A., Noonan, J. M. and Neubert, R. W., "Photoresist article", Patent USA 3832176, Publ. August 1974, CA82-24379.

553.Wright, A. N. and Merrill, R. C., "Selective depolymerization of polymer films for electrical circuits", Patent German 1925496, Publ. November 1969, CA73-19723.

4. Properties
a. Adhesion

554.Clecak, N. J. and Cox, R. J., "Photoresists", Patent German 2314868, Publ. October 1973, CA80-76697.
555.Cortellino, C. A., "Resist adhesion promotors", IBM Tech. Disclosure Bull., 17, 3285 (1975), EEA78-24414.
556.Dafter, R. V. and Henrickson, J. F., "A method of improving adhesive properties of a cured epoxy-polyester surface", Tech. Dig., no. 35, 15-16 (1974), EEA78-76.
557.Davidson, E. B., "Photoresist adhesion", Tech. Pap., Teg. Tech. Conf., Soc. Plast. Eng., Mid-Hudson Sect., 1970, p. 141-6, CA74-4331.
558.Davidson, E. B., "Photoresist adhesion the characterization of photoresists by of the critical surface tension for spreading", Extended Abstracts of the Spring Meeting of the Electrochemical Society, 1971, p. 179-80, EEA75-3957.
559.Eltsov, A. V., Yurre, T. A. and Guk, E. G., "Epoxidized cyclo-rubber photoresists", Pis'ma Zh. Tekh. Fiz., 1, 571-4 (1975), Russian, CA83-186276.
560.Fredericks, E. C. and Moreau, W. M., "Positive resists with increased adhesion", IBM Tech. Disclosure Bull., 27, 1354 (1974), EEA78-8590.
561.Garbarini, V. C., "Increasing photoresist adherence", Patent USA 3716390, Publ. February 1973, CA78-104484.
562.Gittelman, D. H., "Adhering polymeric coatings to semiconductor substrates in particular", Patent German 2108336, Publ. September 1971, CA76-8963.
563.Hayashi, N., Isobe, A., Tsukada, K., Ogawa, K. and Abo, M., "Photothermographic process for polymer resist images", Patent German 2309062, Publ. September 1973, CA80-102335.
564.Johnson, C., Jr. and Palmer, M. D., "Improving photoresist adherence", Patent USA 3827908, Publ. August 1974, CA82-59308.
565.Jones, J. L. W., "Heterocyclics for improving bonding of a photoresist to copper", Patent USA 3645772, Publ. February 1972, CA76-106471.
566.Kerwin, R. E., "Photoresist method for forming patterns on sub-strates", Patent USA 3502683, Publ. July 1970, CA73-61243.
567.Kholod, G. I., Sitnik, A. B., Maksimchuk, E. N. and Martynyuk, M. P., "Adhesion of photopolymer duplicating layers to metals", Poligrafiya, no. 3, 23-4 (1973), Russian, CA79-47797.
568.Kletchenkov, I. I. and Terekhova, G. V., "Effect of chemical composition of the positive photoresist on contraction and adhesion to the substrate", Poluprovodn. Tekh. & Mikroelektron., no. 20, 95-7 (1975), Russian, EEA78-37584.

569.Kookootsedes, G. J., "Photosensitive masks for semiconductors", Patent French 2007530, Publ. January 1970, CA73-40043.
570.Lesaicherre, A. and Linguenheld, L., "Siloxane adhesives for photoresists on oxide surfaces", Patent German 2338839, Publ. February 1974, CA81-31892.
571.Morrison, R. J. and Ross, S. D., "Preparation of a silica surface of an electrical device for receipt of an adherent photoresist coating", Patent USA 3687718, Publ. August 1972, CA77-119854.
572.Oba, Y., Kimura, Y. and Morishita, H., "Photoresist layers", Patent German 2303671, Publ. August 1973, CA79-120497.
573.Oba, Y. and Tsunoda, T., "Diazide-sensitized photoresists", Patent German 2318855, Publ. October 1973, CA80-102349.
574.Roos, L., "Photopolymerizable compositions and elements containing organosilanes", Patent USA 3758306, Publ. September 1973, CA80-21445.
575.Rosenberg, R. M., "Use of photopolymerizable materials for noble metal patterns on dielectric substrates", Patent German 2115945, Publ. October 1971, CA77-12343.
576.Sadamori, M., "Surface treatment of substrates prior to formation of photosensitive resin coatings", Patent Japan 75-40126, Publ. April 1975, CA83-155781.
577.Scherber, W., "Use of photoresists in semiconductor component manufacture", Patent German 2153441, Publ. May 1973, CA79-47071.
578.Wolff, E., "Light-sensitive polymers", Patent German 2217744, Publ. October 1973, CA80-121507.

b. Resolution

579.Angerstein, J. and Voges, E., "Photoresist techniques for acoustic surface wave transducers", Arch. Elektron. & Ubertragungstech., 26, 387-90 (1972), EEA75-41297.
580.Anon., "Special microelectronics characteristics key to high resolution plate images", Solid State Technol., 16 (4), 69-70 (1973), EEA76-35338.
581.Bassous, E., "Fabricating submicrometer silicon devices", IBM Tech. Disclosure Bull., 15, 1823-5 (1972), EEA76-12676.
582.Binder, H., Rosshaupter, E. and Widmann, D., "Microstructure pattern generation in photoresist layers", Siemens Forsch.- & Entwicklungsber., 3 (6), 390-3 (1974), EEA78-4799.
583.Ito, N., Yoshida, H. and Eguchi, K., "Effect of the molecular weight of photoresists on resolution", Nagoya-Shi Kogyo Kenkyusho Kenkyu Hokoku, no. 46, 44-7 (1971), Japanese, CA76-141397.
584.Ito, N., Yoshida, H. and Eguchi, K., "Resolution of poly(vinyl cinnamate) photoresists", Nagoya-Shi Kogyo Kenkyusho Kenkyu Hokoku, no. 48, 11-15 (1972), Japanese, CA78-117568.
585.Jain, P. K. and Prasad, J., "Suitability of polyvinyl cinnamate photo-resist for graticule production", J. Opt., 2 (1), 21 (1973), PA77-18509.

586. Laming, F. P. and Staile, R. E., "Light scattering from photoresist films", J. Electrochem. Soc., 119, 1745-9 (1972), EEA76-8733.
587. Laming, F. P. and Straile, R. E., "Photoresist cleaning", J. Electrochem. Soc., 120, 292-5 (1973), EEA76-19946.
588. Licari, J. J. and DeForest, W. S., "Photographic formation of high-resolution, thick-layered electrical circuits", Patent French 2046153, Publ. April 1971, CA76-91969.
589. McGillis, D. A. and Fehrs, D. L., "Photolithographic line-width control", IEEE Trans. Electron Devices, ED-22, 471-7 (1975), EEA78-24399.
590. Mlynski, D. A., "Optical requirements for structure production in integrated circuit masking techniques", Solid-State Electron., 11, 933-46 (1968), German, EEA72-4453.
591. Moreau, W. M. and Schmidt, P. R., "Photoresist for high-resolution proximity printing", Electrochemical Society Fall Meeting, Extended Abstracts, 1970, p. 459-60, EEA74-9938.
592. Nakatsui, H., Hajimoto, Y. and Hanada, H., "High resolution photosensitive material", Patent Japan 73-79616, Publ. October 1973, CA80-42948.
593. National Cash Register Co., "Process for making high-resolution metal, dielectric, or semiconductive film patterns on a substrate", Patent UK 1197398, Publ. July 1970, CA73-82524.
594. Nicholas, K. H., Brockman, H. E. and Stemp, I. J., "Fabrication of submicron polysilicon lines by conventional techniques", Appl. Phys. Lett., 26, 398-9 (1974), EEA78-24377.
595. O'keeffe, T. W. and Vine, J., "High-resolution image tube for integrated-circuit fabrication", Advan. Electron. Electron Phys., 28A, 47-59 (1969), CA72-94527.
596. Sauter, G. F., Paul, M. C. and Oberg, P. E., "Photoresist exposure technique for high resolution etching of straight lines", Rev. Sci. Instrum., 41, 1514-15 (1970), EEA74-4182.
597. Schinke, D. P., Smith, R. G., Spencer, E. G. and Galvin, M. F., "Thin-film distributed-feedback laser fabricated by ion milling", Appl. Phys. Lett., 21, 494-6 (1972), CA78-22315.
598. Taylor, C. J., "High-resolution development of photosensitive resists", Patent USA 3458313, Publ. July 1969, CA71-86621.
599. Vavilov, A. G., Bycheva, V. A. and Mityaeva, L. A., "Resolving power of layers of photoresists based on cyclorubber with bisazides", Zh. Nauch. Prikl. Fotogr. Kinematogr., 17, 180-5 (1972), Russian, CA77-146181.
600. Wada, Y. and Uehara, K., "Optical limitation in fine pattern photolithography", Jap. J. Appl. Phys., 13, 2014-18 (1974), EEA78-11735.
601. Walker, E. J., "Optical interference effects on linewidth in positive photoresist films", Presented at Electrochemical Society Meeting, San Francisco, Calif., May 1974, EEA77-26496.
602. Walters, R. M. and Brecher, R., "High-resolution negative photoresists", J. Electrochem. Soc., 119, 1703-8 (1972), EEA76-8731.

603.Widmann, D. W. and Binder, H., "Linewidth variations in photoresist patterns on profiled surfaces", IEEE Trans. Electron Devices, ED-22, 467-71 (1975), EEA78-24398.

c. Spectral Sensitivity

604.Arcesi, J. A., Rauner, F. J. and Williams, J. L. R., "Photoreactive polymers. IV. Synthesis and characterization of photocrosslinkable polysulfonates", J. Appl. Polym. Sci., 15, 513-18 (1971), CA75-36791.
605.Beesley, M. J., Castledine, J. G. and Cooper, D. P., "Sensitivity of resist-coated silicon slices to argonlaser wavelengths", Electron Lett., 5, 257-8 (1969), CA70-55513.
606.Branin, P. B. and Fonger, W. H., "Wavelength dependence of PVA (poly(vinyl alcohol)-phosphor-dot photohardening", J. Electrochem. Soc., 122, 94-7 (1975), CA82-92013.
607.Broyde, B., "Positive photosensitive lacquer additives for increasing their sensitivity", Patent German 2028214, Publ. January 1971, CA74-118423.
608.Chibisov, K. V. and Trubnikora, A. A., "Photochemical properties of silver chloride, bromide, and iodide", Dokl. Akad. Nauk SSSR, 198, 898-900 (1971), Russian, CA75-56763.
609.Clecak, N. J., Cox, R. J. and Moreau, W. M., "Azide photoresists for projection photolithography", Tech. Pap., Reg. Tech. Conf., Soc. Plast. Eng., Mid-Hudson Sect., 1974, p. 43-9, CA80-9029.
610.Freund, T., "Less-than-bandgap photosensitivity of zinc oxide and dye sensitization by ferrocyanide ion", Surface Sci., 33, 295-305 (1972), CA78-10133.
611.Froeschle, E. and Backhus, R., "Effect of oxygen on the light sensitivity and image quality of the photoresist KTFR (Kodak Thin Film Resist)", Solid-State Electron., 14, 95-105 (1971), German, CA74-132994.
612.Fukutani, H., Miura, K., Eguchi, C. and Takahashi, Y., "Sensitizers for light-sensitive polymers", Patent German 2426645, Publ. December 1974, CA83-200249.
613.Gaeva, G. L., "Determination of the spectral sensitivity of positive photoresists", Tr. Leningrad. Inst. Kinoinzh., no. 19, 141-6 (1972), Russian, CA80-32483.
614.Gerischer, H., "Electrochemical techniques for the study of photosensitization", Photochem. Photobiol., 16, 243-60 (1972), CA78-10093.
615.Huebner, J. S. and Bruner, L. J., "Complex formation and photoreduction processes in sulfur-doped silver bromide", Z. Naturforsch. A, 26, 1860-4 (1971), CA76-66296.
616.Ilten, D. F., "Spectrophotometric determination of photoresist photosensitivity. I. Theoretical discussion of those photoresist reactions which entail the depletion of a photoactive species", J. Electrochem. Soc., 119, 537-9 (1972), CA77-12265.

617.Ilten, D. F. and Sutton, R. J., "Spectrophotometric determination of photoresist photosensitivity. II. Comparison of the theoretical model with experimental studies on AZ-1350", J. Electrochem. Soc., 119, 539-41 (1972), CA77-41330.
618.Loprest, F. J. and Fitzgerald, E. A., "Photochemical speed of positive photoresists", Photogr. Sci. Eng., 15, 260-1 (1971), CA75-43099.
619.Lyalikov, K. S., Kovaleva, K. A. and Gaeva, G. L., "Sensitometry of photosensitive polymers", Photogr. Korresp., 105 (4), 61-3 (1969), CA71-34997.
620.Martynenko, A. P., Prokhotskii, Y. M., Krasnora, G. P. and Soloveva, V. N., "Determining the spectral sensitivity of photoresists", Zh. Nauchn. & Prikl. Fotogr. & Kinematogr., 19, 225-7 (1974), Russian, PA77-68727.
621.Nishikubo, T., Ichijyo, T. and Takaoka, T., "Photosensitivities and rates of photocrosslinking of poly(vinyl α-cyanocinnamate) and poly(vinyl α-cyanocinnamoxy acetate)", J. Appl. Polym. Sci., 18, 2009-13 (1974), CA81-153250.
622.Norman, S. L. and Singh, M. P., "Spectral sensitivity and linearity of Shipley AZ-1350J photoresist", Appl. Opt., 14, 818-20 (1975), EEA78-28824.
623.Shankoff, T. A. and Trozzolo, A. M., "Spectral sensitization of negative and positive photoresists", Photogr. Sci. Eng., 19, 173-5 (1975), CA83-50680.
624.Skarvinko, E. R., "Preventing photographic speed loss in oxygen-sensitive photoresist layers", Patent USA 3669667, Publ. June 1972, CA77-133224.
625.Thomas, D. C. and Sorkin, J. L., "Lithographic plate and photoresist having two photosensitive layers", Patent USA 3808004, Publ. April 1974, CA81-56683.
626.Tsunoda, T. and Yamaoka, T., "Spectral sensitization of azido polymers", Chiba Daigaku Kogakubu Kenkyu Hokoku, 23 (43), 73-9 (1971), Japanese, CA77-12264.
627.Tsunoda, T., Yamaoka, T. and Nagamatsu, G., "Spectral sensitization of bisazide compounds", Photogr. Sci. & Eng., 17, 390-3 (1973), EEA76-35336.
628.Zollei, M., "Sensitization of cadmium sulfide photoresists prepared from alcohol suspensions with cobalt and copper", Acta Phys. Chem., 16, 147-53 (1970), German, CA75-43100.

d. Defects and Inspection

629.Alishoev, V. R., Berezkin, V. G., Mirzabaev, G. A. and Strizhkov, B. V., "Determination of nitrogen in positive photoresists by gas chromatography", Izv. Akad. Nauk SSSR, Ser. Khim., 2181-3 (1968), Russian, CA70-43904.
630.Engbert, R. and Conze, P., "Semiconductor device", Patent German 1764243, Publ. July 1971, CA75-134955.

631.Harada, K., Harita, Y., Ito, H., Komatsu, K. and Ichikawa, M., "Dyeing of hydrocarbon polymers used as photoresists", Patent Japan 74-63762, Publ. June 1974, CA82-66353.
632.Hashimoto, T., Yamagishi, S., Morioka, K. and Goto, T., "Generation of pinholes in rubber photoresists", Denki Kagaku, 41, 533-8 (1973), Japanese, CA80-42910.
633.Hecker, K. W., "The examination of microstructures in the semiconductor industry with the aid of PIMC systems", Elektron. Ind., 5 (10), 203-5 (1974), German, EEA78-8586.
634.Mimura, Y., "Decrease in pinhole density by improving photoresist coating conditions", Elec. Commun. Lab. Tech. J., 20, 2113-28 (1971), Japanese, EEA75-19245.
635.Moskalevskii, A. I., Terkhova, G. V., Kudina, A. V. and Yarandin, V. A., "Defect reduction in photoresist film", Dielektr. Poluprovodn., 5, 76-9 (1974), Russian, CA81-144207.
636.Sbar, N. L., "The application of laser scanning to detect detection in unexposed high-resolution photographic plates", 1973 IEEE/OSA Conference on Laser Engineering and Applications Digest of Technical Papers, p. 58, PA77-19159.
637.Wickham, R. S. and Russell, G. C., "Photoresist images", Patent UK 1386586, Publ. March 1975, CA83-50783.

e. Others

638.Blok, V. R., Prokhotskii, Y. M. and Alatyrtsev, L. A., "Determination of the quantum yield and threshold concentration of light-sensitive substances for negative photoresists", Zh. Nauch. Prikl. Fotogr. Kinematogr., 18 (1), 53-5 (1973), Russian, CA79-120450.
639.Brandes, R. G. and Curran, R. K., "Modulation transfer function of AZIII photoresist", Appl. Opt., 10, 2101-3 (1971).
640.Clark, K. G., "Photoresists. I. Photopolymerization", Electron. Components, 14, 553-4 (1973), EEA76-29286.
641.Dill, F. H., Hornberger, W. P., Hauge, P. S. and Shaw, J. M., "Characterization of positive photoresist", IEEE Trans. Electron Devices, ED-22, 445-52 (1975), EEA78-24394.
642.Gray, H. F. and Haas, G. A., "Emission properties of oxide cathodes using photoresist binder", Rev. Sci. Instrum., 44, 1616-17 (1973), PA77-37730.
643.Heinen, K. G. and Larrabee, G. B., "The determination of residual photoresist on silicon using radiotracer iodine-131", Solid State Technol., 12 (4), 44-57 (1969), EEA73-13152.
644.Heyroth, A., "On the inherent errors in the photolithographic process for the production of photostencils", Feingeraete Tech., 22, 507-13 (1973), German, PA77-26545.
645.Hunter, W. L. and Crabtree, P. N., "Characterization of Kodak Metal Etch Resist Resin", Photogr. Sci. Eng., 13, 271-5 (1969), CA71-92132.

646.Jones, W. N., Dennison, J. E. and Broyde, B., "Effects of ozone on the crosslinking of photoresists", Solid State Technol., 17 (1), 56-8 (1974), PA77-25282.
647.Kroll, H., "Comparison of negative and positive photoresists for semiconductor application", Mikroelektron., Manuskriptdruck Vortr. Mikroelektron.-Kongr., 5th, 1972, p. 136-52, Publ. 1973, German, CA81-144204.
648.Kuchta, A. A., "Photopolymerizable elements of low optical density containing thickeners with discrete, orderly orientation", Patent USA 3754920, Publ. August 1973, CA80-21434.
649.Mondou, E. R. and Schmidt, P. R., "Registration process with noise suppressor", IBM Tech. Disclosure Bull., 27, 1372 (1974), EEA78-8591.
650.Nakai, G., Yoshizawa, M., Ohtani, H. and Ijitsu, T., "Sensitization of water-soluble photoresist compositions", Patent Japan 74-53639, Publ. May 1974, CA81-144284.
651.Nakane, H., "Chemical characteristics of photosensitive resists", Denki Kagaku Oyobi Kogyo Butsuri Kagaku, 43 (3), 112-18 (1975), Japanese, CA83-106124.
652.Neustroev, S. A., "Effect of electron bombardment on a positive photoresist", Sb. Nauch. Tr. Probl. Mikroelektron., Mosk. Inst. Elektron. Tekh., no. 13, 33-7 (1972), Russian, CA80-139458.
653.Neustroev, S. A., "Polymerization of poly(vinyl cinnamate) photoresist", Khim. Vys. Energ., 7, 452-7 (1973), Russian, CA80-9038.
654.Norland Products, Inc., "Acid-resistant photoresist layers", Patent Neth. 73-05836, Publ. October 1974, CA83-88743.
655.Oleinik, A. V., Treushnikov, V. M., Frolova, N. V., Kalashnikov, B. P., Kiseleva, L. L. and Yatsenko, T. V., "Mechanism of relief formation on cyclized rubber films. II", Zh. Nauchn. Prikl. Fotogr. Kinematogr., 19, 350-3 (1974), Russian, CA82-105150.
656.Paramonov, A. I., Prokhotskii, Y. M. and Erlikh, R. D., "Photolysis and thermolysis of o-naphthoquinone diazides", Zh. Nauch. Prikl. Fotogr. Kinematogr., 19, 122-7 (1974), Russian, CA81-97720.
657.Petrova, T. V., Savvin, S. B. and Dzherayan, T. G., "Increase in the selectivity of the photometric determination of alkaline earth elements in aqueous-organic media", Zh. Anal. Khim., 30, 918-24 (1975), Russian, CA83-172099.
658.Romankiw, L. T., "Microcircuit device using a very thin insulation coating on the conducting bodies", Patent USA 3602635, Publ. August 1971, CA75-124338.
659.Schantz, L. E., "A photoresist-characterization approach to hybrid-microcircuit dimensional control", Microelectronics, 5 (4), 25-31 (1974), EEA78-37693.

660. Timofeev, V. Y., Savotin, Y. I. and Levitman, Y. K., "Statistical analysis of the effect of different factors on the quality of photolithographic operations in the manufacture of integrated circuits", Izv. VUZ Radioelektron., 16 (4), 30-4 (1973), Russian, EEA76-26120.
661. Tsunoda, T., "Mechanism of the photochemical reaction between azides and phenol resins", Tech. Pap., Reg. Tech. Conf., Soc. Plast. Eng., Mid-Hudson Sect., 1973, p. 21-6, CA80-32480.
662. Van Beck, H. C. A., Heertjes, P. M. and Schaafsma, K., "Photoreduction of azo dyes by reducing agents formed upon photoexcitation of some simple aliphatic and aromatic compounds", J. Soc. Dyers Colour, 87, 342-8 (1971), CA76-60903.
663. Vavilov, A. G., "Increasing the effectiveness of the photochemical vulcanization of cyclorubber molecules by the introducing aluminum chloride", Tr. Leningrad. Inst. Kinoinzh., no. 19, 127-33 (1972), Russian, CA80-32481.
664. Vavilov, A. G. and Bycheva, V. A., "Dielectric properties of films from a cyclorubber containing bisazides", Tr. Leningrad Inst. Kinoinzh., no. 19, 134-40 (1972), Russian, CA79-6481.
665. Vinikman, V., Samuel, A. and Schneider, D., "Surface desorption of gases from photoresist", Vacuum, 24 (2), 77-9 (1974), EEA77-26505.
666. Walker, E. J., "Reduction of photoresist standing-wave effects by post-exposure bake", IEEE Trans. Electron Devices, ED-22, 464-6 (1975), EEA78-24397.
667. Weber, H. W., Ulrich, R., Chandross, E. A. and Tomlinson, W. J., "Light-guiding structures of photoresist films", Appl. Phys. Lett., 20, 143-5 (1972), CA76-106421.
668. Widmann, D. W., "Quantitative evaluation of photoresist patterns in the 1-μm range", Appl. Opt., 14, 931-4 (1975), EEA78-28825.

V. Electron Resist

1. General

669. Blotekjaer, K. and Overoye, K., "The use of electron beams in the manufacture of micro circuits", Elektrotek. Tidsskr., 84 (8), 53 (1971), Norwegian, EEA74-29742.
670. Brewer, G. R., "Electron and ion beams in microelectronic fabrication processes", Solid State Technol., 15 (7), 36-9 (1972), EEA75-35405.
671. Broers, A. N. and Dennard, R. H., "Impact of electron beam technology on silicon device fabrication", J. Electrochem. Soc., 120, 101C (1973), EEA76-23150.
672. Broers, A. N. and Hatzakis, M., "Microcircuits by electron beam", Sci. Am., 227 (5), 34-44 (1972), EEA76-8746.
673. Chang, T. H. P., Wilson, A. D., Speth, A. J. and Kern, A., "Electron beam lithography for complex high density devices", Presented at Electrochemical Society Meeting, San Francisco, Calif., May 1974, EEA77-26521.

674.Dubois, J. C. and Gazard, M., "Electron sensitive resins", International Symposium on Materials for Electronic Components, 1975, p. 173-83, French, EEA78-33239.
675.Fang, F., Hatzakis, M. and Ting, C. H., "Electron beam fabrication of ion implanted high performance FET circuits", J. Vac. Sci. & Technol., 10, 1082-5 (1973), EEA77-23147.
676.Hatzakis, M., "Electron-sensitive resists for microfabrication", J. Vac. Sci. & Technol., 11, 95 (1974), EEA77-38888.
677.Henderson, R. C., Pease, R. F. W., Voshchenkov, A. M., Mallery, P. and Wadsack, R. L., "A high-speed p-channel random access 1024-bit memory made with electron lithography", 1973 International Electron Devices Meeting Technical Digest, p. 138-40, EEA77-14613.
678.Jones, G. A. C., Ahmed, H. and Nixon, W. C., "Electron optical system for microfabrication with 10000 lines per field", Electron. Lett., 11, 214-16 (1975), EEA78-33166.
679.Kanaya, K., Yamazaki, H., Okazaki, I. and Tanoka, K., "Fundamental characteristics of electron beam exposure of photoresist and application to manufacturing diodes", Bull. Electrotech. Lab., 31, 38-58 (1967), EEA71-6329.
680.Livesay, W. R., "Integrated circuit production with electron beams", J. Vac. Sci. & Technol., 10, 1028-32 (1973), EEA77-23097.
681.Pankratz, J. M., Yuan, H. T. and Creagh, L. T., "A high-gain, low-noise transistor fabricated with electron beam lithography", 1973 International Electron Devices Meeting Technical Digest, p. 44-6, EEA77-14501.
682.Pickar, K. A. and Thibault, L. R., "Fabrication of IGFETS (insulated-gate field-effect transistors)", J. Vac. Sci. & Technol., 10, 1074-7 (1973), CA80-88660.
683.Roberts, E. D., "Electron resists for the manufacture of integrated circuits", Philips Tech. Rev., 35 (2-3), 41-52 (1975), EEA78-33227.
684.Shackle, P. W. and Payne, R. S., "The fabrication of bipolar transistors using electron lithography, ion implantation, and nickel masked gold metallization", J. Vac. Sci. & Technol., 10, 1090-3 (1973), EEA77-23018.
685.Sopira, M. M., Malmberg, P. R. and Meiksin, Z. H., "Fabrication of integrated CMOS transistors using electron lithography and ion implantation", J. Vac. Sci. & Technol., 10, 1086-9 (1973), EEA77-23148.
686.Thompson, L. F., "Design of polymer resists for electron lithography. I", Solid State Technol., 17 (7), 27-30, 40 (1974), EEA77-38891.
687.Thompson, L. F., "Design of polymer resists for electron lithography. II", Solid State Technol., 17 (8), 41-6 (1974), EEA77-38867.
688.Wardly, G. A., "Potential of field emission cathodes for microfabrication", J. Vac. Sci. & Technol., 10, 975-8 (1973), EEA77-23089.

2. Compositions

689.Bowden, M. J. and Thompson, L. F., "Poly(styrene sulfone) - a sensitive ion-millable positive electron beam resist", J. Electrochem. Soc., 121, 1620-3 (1974), EEA78-8594.
690.Bowden, M. J. and Thompson, L. F., "Effect of olefin structure on the electron beam vaporization of poly(olefin sulfones)", Electron Ion Beam Sci. Technol., Int. Conf., 8th, 1974, p. 81-6, CA83-35632.
691.Brewer, T. L., "High speed electron resists", Electron Ion Beam Sci. Technol., Int. Conf., 6th, 1974, p. 71-80, CA83-35631.
692.Cole, H. S., Skelly, D. W. and Wagner, B. C., "Electron sensitive resists derived from vinyl ether-maleic anhydride copolymers", IEEE Trans. Electron Devices, ED-22, 417-20 (1975), EEA78-24391.
693.Davidson, E. B., "Unsaturated polyester resin electron beam resist composition", Patent USA 3825428, Publ. July 1974, CA81-97813.
694.Fredericks, E. C., Moreau, W. M. and Viswanathan, N. S., "Photo-electron-beam resist", IBM Tech. Disclosure Bull., 17, 3286 (1975), EEA78-24415.
695.Gipstein, E. and Hewett, W. A., "Terpolymers for electron beam positive resists", Patent USA 3898350, Publ. August 1975, CA83-195464.
696.Gipstein, E., Hewett, W. A. and Levine, H. A., "Electron beam-sensitive polymeric material", Patent German 2314124, Publ. October 1973, CA80-109975.
697.Gipstein, E. and Moreau, W. M., "Resists for microcircuits", Patent German 2355701, Publ. May 1974, CA81-97812.
698.Gipstein, E., Moreau, W. and Need, O., III, "Sensitive positive electron-beam resists", IBM Tech. Disclosure Bull., 16, 3028 (1974), EEA77-26495.
699.Harris, R. A., "Polymethyl methacrylate as an electron sensitive resist", J. Electrochem. Soc., 120, 270-4 (1973), EEA76-19945.
700.Harris, R. A., "Molecular weight effects in the use of poly (methyl methacrylate) as an electron sensitive resist", In: Ind. Polym.: Charact. Mol. Weight, Proc. Meet., Green, J. H. S. and Dietz, R., (Eds.), Transcripta Books, London, England, p. 12-15 (1973), CA83-180551.
701.Hirai, T., Hatano, Y. and Nonogaki, S., "Epoxide-containing polymers as highly sensitive electron-beam resists", J. Electrochem. Soc., 118, 699-72 (1971), EEA74-21553.
702.Johnson, D. E. and Miller, R. D., "Radiation sensitive polyesters", IBM Tech. Disclosure Bull., 17, 1476 (1974), EEA77-8593.
703.Ku, H. Y. and Scala, L. C., "Polymeric electron-beam resists", J. Electrochem. Soc., 116, 980-5 (1969), CA71-34269.
704.Lewis, J. M. and Wainer, E., "Light sensitive reproduction and electron beam-sensitive material", Patent USA 3820993, Publ. June 1974, CA81-97798.

705.Morishita, H., Hirai, T. and Nonagaki, S., "Cathode ray beam resist materials", Patent German 2232202, Publ. January 1973, CA78-117642.
706.Morishita, H. and Nonogaki, S., "Cathode ray-sensitive methyl methacrylate copolymer coatings", Patent German 2227008, Publ. December 1972, CA78-104483.
707.Nongaki, S., Morishita, H. and Saitou, N., "Epoxidized cis-1,4-polybutadiene, a highly sensitive electron beam resist", Appl. Polym. Symp., 23, 117-23 (1974), CA82-32569.
708.Paal, G., Strahle, U. D. and Kielhorn, G., "New high sensitive electron resist materials", J. Electrochem. Soc., 120, 1714-16 (1973), EEA77-10277.
709.Sekigawa, K. and Aoyama, S., "Positive-type resist for use with electron beams", Patent Japan 75-50041, Publ. May 1975, CA83-156854.
710.Shimokawa, Y., Takeda, A. and Miyama, H., "Electron- and ultra-violet-sensistive image recording materials contg. alkenyl maleate polymers", Patent Japan 73-92025, Publ. November 1973, CA80-114864.
711.Shimokawa, Y., Takeda, A. and Miyama, H., "Electron beam and light sensitive materials", Patent Japan 74-24437, Publ. March 1974, CA81-129902.
712.Thompson, L. F., "A new family of positive electron beam resists - poly(olefin sulfones)", J. Electrochem. Soc., 120, 1722 (1973), EEA77-10279.

3. Processing

713.Bartelt, J. L. and Ballantyne, J. P., "A phenomenological optimization of negative-electron resist exposure conditions", J. Vac. Sci. & Technol., 10, 1131 (1973), EEA77-23105.
714.Bartelt, J. L. and Feit, E. D., "The preparation and characterization of a new, highly sensitive, crosslinking electron resist", J. Electrochem. Soc., 122, 541-4 (1975), EEA78-24418.
715.Bowden, M. J. and Thompson, L. F., "Electron irradiation of poly(olefin sulfones). Application to electron beam resists", J. Appl. Polym. Sci., 17, 3209-19 (1973), CA80-32482.
716.Bowden, M. J. and Thompson, L. F., "Vapor development of poly(olefin sulfone) resists", Tech. Pap., Reg. Tech. Conf., Soc. Plast. Eng., Mid-Hudson Sect., 1973, p. 119-26, CA80-9034.
717.Broers, A. N., "Electron-beam fabrication", J. Vac. Sci. & Technol., 8 (5), S50-1 (1971), EEA75-3942.
718.Broyde, B., "Exposure of photoresists: electron beam exposure of negative photoresists", J. Electrochem. Soc., 116, 1241-5 (1969), EEA73-3579.
719.Broyde, B., "Exposure of photoresists. II. Electron and light exposure of a positive photoresist", 4th International Conference Electron Ion Beam Science and Technology, 1970, p. 575-80, CA74-118349.

720. Cecil, O. B., "Electron beam techniques in integrated circuits", Patent USA 3453723, Publ. July 1969, EEA73-9571.

721. Chang, T. H. P., "Device fabrication using a scanning electron beam system", Proceedings of the 4th Annual Scanning Electron Microscope Symposium and Workshop on Forensic-Applications of the Scanning Electron Microscope, 1971, p. 417-24, EEA74-25627.

722. Chiu, G. T. and Fredericks, E. C., "Electron-beam resist developers", IBM Tech. Disclosure Bull., 18, 395 (1975), EEA78-37578.

723. Dix, C. J., "Paper-tape-controlled electron probe resist exposure and direct metallic deposition", IEEE Trans. Electron Devices, ED-19, 641-6 (1972).

724. Ersoy, O., "A study of electron beam exposure of positive resists", Optik, 41, 479-87 (1975), EEA78-11707.

725. Goell, J. E., "Electron-resist fabrication of integrated optical circuits", Topical Meeting on Integrated Optics Guided Waves, Materials, and Devices, 1972, 4 pp., EEA75-19262.

726. Greeneich, J. S., "Developer characteristics of poly-(methyl methacrylate) electron resist", J. Electrochem. Soc., 122, 970-6 (1975), EEA78-37583.

727. Greeneich, J. S., "Time evolution of developed contours in poly-(methyl methacrylate) electron resist", J. Appl. Phys., 45, 5264-8 (1974), EEA78-11703.

728. Greeneich, J. S. and van Duzer, T., "An exposure model for electron-sensitive resists", IEEE Trans. Electron Devices, ED-21, 286-99 (1974), EEA77-18824.

729. Hawryluk, R. J. and Smith, H. I., "Electron beam exposure profiles in thin polymer films", Electron Ion Beam Sci. Technol., Int. Conf., 1972, p. 51-67, CA83-98268.

730. Heidenreich, R. D. and Thompson, L. F., "Electron beam generated patterns of metal-containing polymers", Patent USA 3885076, Publ. May 1975, CA83-156851.

731. Henderson, R. C. and Pease, R. F. W., "Variable development response of resists using electron beam lithography. Methods and applications", Tech. Pap., Reg. Tech. Conf., Soc. Plast. Eng., Mid-Hudson Sect., 1973, p. 146-52, CA80-9037.

732. Herriott, D. R., Collier, R. J., Alles, D. S. and Stafford, J. W., "EBES: A practical electron lithographic system", IEEE Trans. Electron Devices, ED-22, 385-92 (1975), EEA78-24386.

733. Herzog, R. F., Greeneich, J. S., Everhart, T. E. and Van Duzer, T., "Computer-controlled resist exposure in the scanning electron microscope", IEEE Trans. Electron Devices, ED-19, 635-41 (1972), CA77-41329.

734. Hirai, T., Hatano, Y., Nonogaki, S. and Kobayashi, T., "Cathode-ray sensitive polymeric resist", Patent USA 3885060, Publ. May 1975, CA83-155784.

735. Hlista, J. F., "Providing holes in a glass layer disposed on a substrate", Patent USA 3551176, Publ. December 1970, CA74-154786.

736.Hyde, R. J., "Producing semiconductor devices by photolithographic techniques", Patent UK 1201861, Publ. August 1970, CA73-92756.
737.Jewett, R. E., Van Duzer, T. and Greeneich, J. S., "Electron-resist exposure contours for various patterns", Presented at Electrochemical Society Meeting, San Francisco, Calif., May 1974, EEA77-23085.
738.Kanaya, K., Miyauchi, S. and Tanaka, K., "Micro pattern fabrication by electron beam exposure of photoresist polymer films", Bull. Electrotech. Lab., 33 (11), 83-104 (1969), EEA73-30311.
739.Kaplan, M. and Davidson, E. B., "Electron beam recording media", Patent USA 3893127, Publ. July 1975, CA83-139877.
740.Kielhorn, G., Straehle, U. D., Paal, G. and Wilhelm, A., "Electron-beam recording materials", Patent German 2336517, Publ. February 1975, CA83-35749.
741.Kilcoyne, M. K. and Fritz, R. B., "Electron image projection for LSI circuits", Electron. Packag. & Prod. (Int. Ed.), 7 (2), 5-10 (1971), EEA74-29744.
742.Koops, H., "On electron projection systems (microcircuit production)", J. Vac. Sci. & Technol., 10, 909-12 (1973), PA77-39778.
743.Levine, A. W., Kaplan, M. and Poliniak, E. S., "Interaction of 5 keV electrons with polymers of methyl isopropenyl ketone", Tech. Pap., Reg. Tech. Conf., Soc. Plast. Eng., Mid-Hudson Sect., 1973, p. 106-18, CA80-9025.
744.Marvzo, A. A., "Production of an image on a substrate", Patent UK 1154212, Publ. June 1969, CA71-66054.
745.Matsushita Electric Industrial Co., Ltd., "Direct-positive photographic recording process", Patent French 2102876, Publ. May 1972, CA78-142451.
746.Miyauchi, S. and Tanaka, K., "IC pattern exposure by scanning electron beam apparatus", Solid State Technol., 12 (7), 43-8 (1969), EEA73-469.
747.Miyauchi, S. and Tanaka, K., "IC pattern exposure by scanning electron beam apparatus", Solid State Technol., 13 (6), 56-9 (1970), EEA74-353.
748.Nakamura, K. and Sakata, T., "Electron beam exposure of silicone resins", Bull. Soc. Photogr. Sci. Technol. Jpn., 1973, p. 26-9, Publ. 1974, CA81-144205.
749.Newberry, S. P., "Electron beam fabrication of microelectronic circuit patterns", Patent USA 3491236, Publ. January 1970, EEA73-30360.
750.Odam, L., "Integrated circuits by electronbeams", Elteknik, no. 7-8, 36-7 (1970), Swedish, EEA74-339.
751.O'Keeffe, T. W., "Fabrication of integrated circuits using the electron image projection system (ELIPS)", IEEE Trans. Electron Devices, ED-17, 465-9 (1970), EEA73-23856.

752.Overoie, K., "Application of the electron beam in manufacture of micro circuits", Elektrotek. Tidsskr., 85 (1), 11-12 (1972), Norwegian.
753.Pease, R. F. W., Ballantyne, J. P., Henderson, R. C., Voshchenkov, A. M. and Yau, L. D., "Applications of the electron beam exposure system (EBES)", IEEE Trans. Electron Devices, ED-22, 393-9 (1975), EEA78-24387.
754.Rass, D. L. and Barton, L. A., "Photoresists for electron beam recording", Patent German 2405831, Publ. October 1974, CA81-179986.
755.Scala, L. C. and Ciliberti, D. F., "Electron-beam-sensitive resists for printed circuits", Patent French 2168593, Publ. October 1973, CA80-126810.
756.Schmidt, P. R., "Developer for electron-beam resist process", IBM Tech. Disclosure Bull., 18, 393 (1975), EEA78-37577.
757.Shimokawa, Y. and Miyama, H., "Electron beam image recording materials", Patent Japan 73-60636, Publ. August 1973, CA79-131401.
758.Shimokawa, Y. and Miyama, H., "Electron beam recording material", Patent Japan 73-61147, Publ. August 1973, CA79-131402.
759.Sobanski, J., Chabicowsky, R. and Schuler, A., "Electron beam exposure of polymeric resists", J. Vac. Sci. & Technol., 9, 358-60 (1972), EEA75-19253.
760.Thompson, L. F., Feit, E. D., Melliar-Smith, C. M. and Heidenreich, R. D., "Fundamental aspects of electron beam lithography. II. Low-voltage exposure of negative resists", J. Appl. Phys., 44, 4048-51 (1973), CA80-9030.
761.Varnell, G. L., Spicer, D. F., Rodger, A. C. and Holland, R. D., "High-speed electron-beam pattern generation", Presented at Electrochemical Society Meeting, San Francisco, Calif., May 1974, EEA77-29158.
762.Yatsui, Y., Nakata, T. and Umehara, K., "Electron-beam exposure of silicones", J. Electrochem. Soc., 116, 94-7 (1969), CA70-42839.

4. Properties

763.Bartelt, J. L., "Electron beam-hardenable resist lacquer", Patent German 2416543, Publ. October 1974, CA82-92057.
764.Brewer, T. L., "Factors affecting the sensitivity of e-beam resists", Polym. Eng. Sci., 14, 534-7 (1974), CA81-153433.
765.Brown, M. J., "High resolution electron beam polymerization patterns", Patent USA 3681103, Publ. August 1972, CA77-120781.
766.Broyde, B., "Electron-sensitive materials", Bell Syst. Tech. J., 49, 2095-104 (1970), CA74-118346.
767.Chorzewska, M. and Nowicki, S., "Sensitivity and contrast of a poly(methyl methacrylate) film activated by an electron beam", Elektronika, 16, 296-8 (1975), Polish, CA83-194169.

768.Dubois, J. C. and Gazard, M., "Poly(vinylsiloxane) (PVS), a highly sensitive electron-beam resist", Electron Ion Beam Sci. Technol., Int. Conf., 1972, 112-22, CA83-115553.
769.Feit, E. D., Heindenreich, R. D. and Thompson, L. F., "Radiation chemistry of epoxidized polybutadienes. Negative electron beam resists", Appl. Polym. Symp., 23, 125-38 (1974), CA81-144202.
770.Gazard, M., Dubois, J. C. and Duchesne, C., "Properties of some electron sensitive siloxane resists", Appl. Polym. Symp., 23, 107-16 (1974), CA81-56576.
771.Greeneich, J. S., "Solubility rate of poly-(methyl methacrylate), PMMA, electron-resist", J. Electrochem. Soc., 121, 1669-71 (1974), EEA78-8595.
772.Hartemann, P., Arnodo, C. and Gaudry, R., "Excitation of surface waves of frequency higher than 1000 MHz with transducers obtained by electron beam masking", J. Phys., 33, 266-70 (1972), French, EEA76-28079.
773.Heidenreich, R. D., Thompson, L. F., Feit, E. D. and Melliar-Smith, C. M., "Fundamental aspects of electron beam lithography. I. Depth-dose response of polymeric electron beam resists", J. Appl. Phys., 44, 4039-47 (1973), CA80-9028.
774.Ichikawa, M., Harita, Y., Harada, K., Komatsu, K., Inomata, S., Tsuchiya, T., Ono, H. and Tsunoda, T., "Cyclized polybutadiene-based electron beam", Patent Japan 74-84656, Publ. August 1974, CA82-132166.
775.Munakata, C., Kohno, H., MacKawa, A. and Honda, Y., "Distortion of electron-beam-recorded patterns on a photographic plate due to charge-up", Japan J. Appl. Phys., 9, 1187-8 (1970), EEA73-37055.
776.Nonogaki, S., Morishita, H. and Saitou, N., "Epoxidized cis-1, 4-polybutadiene, a highly sensitive electron-beam resist", Am. Chem. Soc., Div. Org. Coat. Plast. Chem., Pap., 33, 378-82 (1973), CA82-44495.
777.Roberts, E. D., "The preparation and properties of a polysiloxane electron resist", J. Electrochem. Soc., 120, 1716-21 (1973), EEA77-10278.
778.Roberts, E. D., "Modified methacrylate positive electron beam resist", Appl. Polym. Symp., 23, 87-98 (1974), CA81-113623.
779.Roberts, E. D., "Positive-working electron resists", Patent German 2363092, Publ. July 1974, CA82-132158.
780.Romankiw, L. T., Hatzakis, M. and Castellani, E. E., "Additive electroplating limits for gold conductors using PMMA electron beam resist", Presented at Electrochemical Society Meeting, San Francisco, Calif., May 1974, EEA77-23087.
781.Saitou, N., "Monte-Carlo simulation for the energy dissipation profiles of 5-20 keV electrons in layered structures", Jap. J. Appl. Phys., 12, 941-2 (1973), CA79-120451.

782.Schaudy, R., Hubeny, H., Schmitz, E., Proksch, E. and Schrott, S., "Radiation crosslinking of stabilized polyethylene in the production of extremely age-resistant profiles", Ber. Oesterr. Studienges. Atomenerg., no. 2272, 51 pp. (1974), CA81-153425.
783.Sedgwick, T. O., Broers, A. N. and Agule, B. J., "A novel method for fabrication of ultrafine metal lines by electron beams", J. Electrochem. Soc., 119, 1769-71 (1972), EEA76-8734.
784.Thompson, L. F. and Bowden, M. J., "Poly(olefin sulfones) of positive electron beam resists", J. Electrochem. Soc., 120, 1722-6 (1973), CA80-76645.
785.Thompson, L. F., Feit, E. D. and Heidenreich, R. D., "Lithography and radiation chemistry of epoxy containing negative electron resists", Polym. Eng. Sci., 14, 529-33 (1974), CA81-153432.
786.Wardly, G. A., "Electrode flatness requirements for cathode projection microfabrication systems", IEEE Trans. Electron Devices, ED-22, 414-17 (1975), EEA78-24390.
787.Wolf, E. D., Ozdemir, F. S., Perkins, W. E. and Coane, P. J., "Response of the positive electron resist Elvacite 2041 to kilovolt electron beam exposure", Rec. Symp. Electron, Ion, Laser Beam Technol., 11th, 1971, p. 331-6, CA79-99173.

VI. Other Resists

1. X-ray

788.Bernacki, S. E. and Smith, H. I., "X-ray lithography applied to silicon device fabrication", Presented at Electrochemical Society Meeting, San Francisco, Calif., May 1974, EEA77-34096.
789.Brault, R. G., "Properties of metal acrylate compositions as x-ray resists", Electron Ion Beam Sci. Technol., Int. Conf., 6th, 1974, p. 63-70, CA83-35630.
790.Feder, R., Hatzakis, M., Heritage, M. and Spiller, E., "Single exposure system for multilevel metallurgy", IBM Tech. Disclosure Bull., 17, 2460-1 (1975), EEA78-19859.
791.Feder, R. and Spiller, E., "Fabrication of overlapping patterns from a single exposure in X-ray lithography", IBM Tech. Disclosure Bull., 17, 2168-9 (1974), EEA78-15388.
792.Greeneich, J. S., "X-ray lithography: I. Design criteria for optimizing resist energy absorption; II. Pattern replication with polymer masks", IEEE Trans. Electron Devices, ED-22, 434-9 (1975), EEA78-24392.
793.Greeneich, J. S., "Efficient continuum sources for x-ray lithography", Appl. Phys. Lett., 27, 579-80 (1975).
794.Hesse, E., "Microstructure - generation for planar microwave devices", Nachrichtentech. Z. (NTZ), 28, 145-55 (1975), German, EEA78-24423.
795.Lenzo, P. V. and Spencer, E. G., "High-speed low-power X-ray lithography", Appl. Phys. Lett., 24, 289-91 (1974), EEA77-23073.

796. McCoy, J. H. and Sullivan, P. A., "Progress in X-ray lithography", Presented at Electrochemical Society Meeting, San Francisco, Calif., May 1974, EEA77-26502.
797. Neureuther, A. R. and Hagouel, P., "X-ray lithographic fabrication of blazed diffraction gratings", Presented at Electrochemical Society Meeting, San Francisco, Calif., May 1974, PA77-46637.
798. Schmidt, C. J., Lenzo, P. V. and Spencer, E. G., "Preparation of thin windows in silicon mask for x-ray lithography", J. Appl. Phys., 46, 4080-2 (1975).
799. Smith, H. I., "X-ray and optical replication of electron beam generated masks", 1974 Digests of the Intermag. Conference, 30-7/1 pp., EEA78-8600.
800. Smith, H. I., Spears, D. L. and Bernacki, S. E., "X-ray lithography. A complementary technique to electron beam lithography (surface wave transducer of production)", J. Vac. Sci. & Technol., 10, 913-17 (1973), EEA77-24905.
801. Spears, D. L. and Smith, H. I., "X-ray lithography - a new high resolution replication process", Solid State Technol., 15 (7), 21-6 (1972), EEA75-35403.
802. Thompson, L. F., Feit, E. D., Bowden, M. J., Lenzo, P. V. and Spencer, E. G., "Polymeric resists of X-ray lithography", J. Electrochem. Soc., 121, 1500-3 (1974), EEA78-5344.
803. Trzyna, C. J. and Trzyna, T. S., "Producing aligned configurations on both sides of an opaque sheet", Patent USA 3447924, Publ. June 1969, CA71-44453.

2. Ion-beam

804. Abe, H., Matsui, H., Sonobe, Y., Enomoto, T. and Demizu, S., "Application of gas plasma to the fabrication of semiconductor integrated circuits", Denki Kagaku, 41, 544-8 (1973), Japanese, CA80-53610.
805. Barr, R. W. and Bazzarre, D. F., "Micro-ion milling of electronic devices with 1μ periodicity", Presented at Electrochemical Society Meeting, San Francisco, Calif., May 1974, PA77-53626.
806. Cahen, O., "Electronic masking and cutting of thin films by an ion beam. Application to electronic and optical circuits", Vide, 28, 117-23 (1973), French, EEA77-2593.
807. Garvin, H. L., "High resolution fabrication by ion beam sputtering", Solid State Technol., 16 (11), 31-6 (1973), EEA77-10300.
808. Goell, J. E., Standley, R. D., Gibson, W. M. and Rodgers, J. W., "Ion bombardment fabrication of optical waveguides using electron resist masks", Appl. Phys. Lett., 21, 72-3 (1972), CA77-95260.
809. Grosewald, P. S. and Laming, F. P., "Fine-line pattern definition with sputter etch mask", IBM Tech. Disclosure Bull., 17, 3274 (1975), EEA78-24409.

810. Holland, L., "Scattering effects in ion beam exposure of photoresist polymer films", Brit. J. Appl. Phys., 2, 767-8 (1969), CA71-44430.
811. Makris, J. S. and Masters, B. J., "Forming fine-line geometries in integrated circuits", IBM Tech. Disclosure Bull., 16, 3240-1 (1974), EEA77-30385.
812. Ohto, M., Noshiro, A., Mayuzumi, T. and Fujino, T., "Photopolymerized organosiloxane resists for the ion etch process", Patent German 2246020, Publ. March 1973, CA79-25727.
813. Seliger, R. L. and Fleming, W. P., "Focused ion beams in fabrication", J. Appl. Phys., 45, 1416-22 (1974), PA77-60530.

3. Laser and Holography

814. Boivin, L. P., "Standing wave effect in grating production by laser beam interference on resist coated highly reflecting substrates", Opt. Commun., 9, 206-11 (1973), PA77-35.
815. Iwata, F. and Tsujiuchi, J., "Replication of a surface relief hologram", Nippon Shashin Gakkaishi, 37, 320-8 (1974), Japanese, CA83-88680.
816. Iwata, F. and Tsujiuchi, J., "Characteristics of a photoresist hologram and its replica", Appl. Opt., 13, 1327-36 (1974), PA77-58830.
817. Kalestynski, A., "The application of holography in electronics", Elektronika, 15, 196-201 (1974), Polish, EEA77-36315.
818. Kalvina, I. N., Moskalenko, V. F., Ostapchenko, E. P., Pavlovskii, L. L., Protsenko, T. V. and Rychkov, V. I., "Resonant action of laser radiation on polymerisation processes", Sov. J. Quantum Electron., 4, 1285-6 (1975), EEA78-28832.
819. Kiemle, H., "Projection of etching masks for semiconductor devices with the aid of holography", Internat. Elektronische Rundschau, 24 (7), 176-8 (1970), German, EEA73-37047.
820. Kirk, J. P., "Hologram photoresist noise", Appl. Opt., 11, 1429 (1972), EEA75-24161.
821. Lehmann, M., "Fabrication of metallic gratings on photoresist by holographic recording", Proceedings of the Engineering Applications of Holography Symposium, 1972, p. 327-30, EEA76-1381.
822. McDowell, M. W., Chinery, D. N. W. and Dierkse, A., "A fibre optical technique for production of reticle patterns on spherical surfaces", J. Phys. E, 8, 452-3 (1975), EEA78-24419.
823. Nakajima, J., Inagaki, T. and Nishimura, Y., "Copied phase hologram of photoresists", Fujitsu Sci. Tech. J., 6 (3), 69-90 (1970), CA74-36912.
824. Puech, C., "Copying holograms on photoresist", Opt. Commun., 4, 279-82 (1971), CA77-106419.
825. Raamot, J. and Zaleckas, V. J., "Laser pattern generation using X-Y beam deflection", 1973 IEEE/OSA Conference on Laser Engineering and Applications Digest of Technical Papers, p. 30, PA77-6584.

826. Shank, C. V. and Schmidt, R. V., "Optical technique for producing 0.1-μ periodic surface structures", Appl. Phys. Lett., 23, 154-5 (1973), EEA76-35854.
827. Tsang, W. T. and Wang, S., "Microfabrication of two-dimensional periodic arrays by laser beam interferometric technique", Appl. Phys. Lett., 27, 79-82 (1975), EEA78-41640.
828. Veiko, V. P., Kotov, G. A. and Libenson, M. N., "Thermal action of laser radiation on some polymeric materials", Fiz. Khim. Obrab. Mater., no. 2, 16-18 (1973), Russian, CA79-54122.
829. Vlasov, V. V., Slavinskaya, V. N. and Teleshov, B. V., "Holographic method for generating sets of high-quality identical images by the axial multiplication method", Sov. J. Quantum Electron., 4, 1310-14 (1975), EEA78-28833.

VII. Masks

1. General

830. Altman, J. H., "Photo-optical aspects of mask technology", Solid State Technol., 12 (7), 34-42 (1969), EEA73-468.
831. Anon., "Guidelines for selecting metals for photofabrication of electrical/electronic parts", Insul./Circuits, 20 (2), 41-4 (1974), EEA77-17840.
832. DeRamaix, M., "Major trends in masking for color improvement", Photogr. Sci. Eng., 15, 262-7 (1971), CA74-43028.
833. Dubee, A., "Reducing the dimensions of integrated circuits by the combination of electronic masking and ion etching", Colloq. Int. Pulverisation Cathodique Ses Appl., C. R., 1st, 1974, p. 273-81, Publ. 1974, French, CA81-70078.
834. Heyroth, A., "On the inherent errors in the photolithographic process for the production of photostencils", Feingeraete Tech., 22, 507-13 (1973), German, EEA77-14569.
835. Iscoff, R. L. Z., "Photomasking services face demand for still higher quality, lower prices", Electron. Packag. & Prod., 15 (8), 109-14 (1975).
836. Lemke, G. A., "Mask-making technology", 1970 IEEE International Solid State Circuits Conference, 1970, p. 148-9, EEA73-23954.
837. Murphy, J. P. and Henriksen, G. M., "Good masks, good yields", Electron. Packag. & Prod. (Int. Ed.), 7 (5), 17-25 (1971), EEA75-444.
838. Murphy, J. P., Henriksen, G. and Woods, L., "Photocomposition and LSI", Solid State Technol., 16 (6), 29-32 (1973), EEA76-35339.
839. Pircher, G., "Masking and duplication in the realm of one-micron-wide lines. State of the art and prospects in optics and electronics", Rev. Tech. Thomson-CSF, 7, 5-46 (1975), French, EEA78-24427.

840. Pyatnitskii, I. V. and Glushchenko, L. M., "Masking of iron, aluminum, titanium, and zirconium by tartaric and citric acids. I. Masking of precipitation reactions", Zh. Anal. Khim., 24, 1619-28 (1969), French, CA73-50593.
841. Rottmann, H. R., "Photolithography in integrated circuit mask metrology", IBM J. Res. & Dev., 18, 225-31 (1974), PA77-77068.
842. Sprague, J. L., "Production of planar semiconductor devices by masking and diffusion", Patent USA 3490963, Publ. January 1970, EEA73-28111.
843. Stebbing, P., "A brief survey into the use of solder masks (resists) on PCBs", Electron. Mfr., 17 (6), 22-5 (1972), EEA75-31993.
844. Stupp, W., "Reprographic methods for circuit manufacture", Elektron. Anz., 5 (5), 114-15 (1973), German, EEA76-31971.
845. Vancleemput, W. M. and Linders, J. G., "An improved graph-theoretic model for the circuit layout problem", 11th Design Automation Workshop, 1974, p. 82-90, EEA77-30362.
846. Zivkovic, D., "Microphotomasks for integrated circuits", Tehnika, 22, 270-6 (1967), Slovene, EEA73-6768.

2. Compositions

847. Ahn, K. Y. and Cuomo, J. J., "Photomask", Patent UK 1354809, Publ. May 1974, CA81-97800.
848. Bakos, P., "Photographic emulsions for forming etch-resistant masks on supports", Patent German 2053312, Publ. July 1971, CA75-125087.
849. Baliga, B. J. and Ghandhi, S. K., "PSG masks for diffusions in gallium arsenide", IEEE Trans. Electron Devices, ED-19, 761-4 (1972), EEA75-20055.
850. Bertens, T. C. J. M., "Etching masks", Patent German 2348779, Publ. April 1974, CA81-7642.
851. Bideau, M. A., Cauveton, C. E., Fruchard, A. and Laurens, M., "Printing from color transparencies", Patent USA 3692524, Publ. September 1972, CA78-22519.
852. Blome, E. R. and Fok, S. S. M., "Hard-surface transparent mask", Patent USA 3712816, Publ. January 1973, CA78-153708.
853. Bratter, R. L., "Silicon oxynitride masks for solid-state diffusion", IBM Tech. Disclosure Bull., 17, 2937-8 (1975), EEA78-19647.
854. Clark, K. G., "Processing of emulsion photomasks for semiconductor applications", Proceedings of the Symposium on Photographic Processing, 1971, p. 265-83, Publ. 1973, PA77-8408.
855. Costa, A. C. S. and Tarares, T. M., "Masking agents in metal titration with EDTA. IV. Thorium titration in the presence of some metals", Anal. Lett., 3, 549-61 (1970), CA74-94062.
856. Desai, K. S. and DiGregorio, G. J., "Photolithographic process using a disposable nonmetallic mask", IBM Tech. Disclosure Bull., 16, 2644-5 (1974), EEA77-18836.

857. Doo, V. Y., Regh, J. and Seto, D. K., "Photolithographic masks of semiconductor material", Patent USA 3701659, Publ. October 1972, CA78-35474.
858. Dunham, K. R. and Yost, M. E., "Ultraviolet absorptive masks", Def. Publ., U.S. Patent Off. 900009, July 1972, CA77-82226.
859. du Pont de Nemours, E. I., and Co., "Composite photohardenable recording element for mask formation", Patent French 2122578, Publ. October 1972, CA78-142462.
860. El-Hoshy, A., Brown, D. M. and Engeler, W. E., "Molybdenum films as partial diffusion masks in MOS processing", Appl. Phys. Lett., 17, 261-3 (1970), PA73-79027.
861. Frieser, H., "Photographic masks", Optik, 39, 268-83 (1974), German, PA77-18510.
862. Fujinami, T. and Hattori, I., "Chromium oxide photomasks", Patent German 2225105, Publ. December 1972, CA78-91063.
863. Ghoshtagore, R. N., "Silicon dioxide masking of phosphorus diffusion in silicon", Solid-State Electron., 18, 399-406 (1975), EEA78-19686.
864. Godfrey, J., "Semiconductor device fabrication using a combination of energy beams for masking and impurity doping", Patent USA 3737346, Publ. June 1973, CA79-59195.
865. Grebe, K. R., "Masking of deposited thin films by use of a masking layer photoresist composite", Patent USA 3849136, Publ. November 1974.
866. Grebe, K., Ames, I., and Ginzberg, A., "Masking of deposited thin films by means of an aluminum-photoresist composite", J. Vac. Sci. & Technol., 11, 458-60 (1974), CA81-8351.
867. Gribov, B. G., Koshchienko, A. V., Rumyantseva, V. P. and Koltsov, Y. I., "Iron oxide layers", Patent USSR 466045, Publ. April 1975, CA83-88767.
868. Griest, E. M., "Transparent iron oxide microcircuit mask", Patent USA 3824100, Publ. July 1974.
869. Henning, S. M., "Mesa diode with channel guard", Patent USA 3808058, Publ. April 1974, CA81-18391.
870. Hill, B. H., "Etch retarding metallic oxide films as masks for etching", Patent USA 3615953, Publ. October 1971, CA76-78424.
871. Horwath, R. S., Horvat, J. G. and Sadagopan, V., "Si_3N_4: an ultrahard protective layer for glass masks", Presented at Electrochemical Society Meeting, San Francisco, Calif., May 1974, EEA77-23083.
872. Ikari, K. and Kato, H., "Masking agent for masks used in photoresist imaging", Patent Japan 74-120630, Publ. November 1974, CA82-163056.
873. Janus, A. R., "Properties of iron oxide films for photomask applications", Solid State Technol., 16 (6), 33-6 (1973), EEA76-35340.
874. Janus, A. R., "Rapid crystallization of an iron oxide film used as a mask in semiconductor processing", Patent USA 3873341, Publ. March 1975, CA83-36595.

875.Kammlott, G. W. and Sinclair, W. R., "Fe_2O_3 - an inorganic electron resist material", J. Electrochem. Soc., 121, 929-32 (1974), EEA77-34099.
876.Kato, H., Tamai, M. and Kudo, S., "Photosensitive compositions for photomasks", Patent Japan 75-75421, Publ. June 1975, CA83-211335.
877.Kerwin, R. E., Sinclair, W. R. and Sullivan, M. V., "Photomasks used in photoresist and related photolithographic processes", Patent British 1277037, Publ. June 1972, CA77-68570.
878.Komatsu, E. and Yamazaki, S., "Etching treatment", Patent Japan 74-40777, Publ. November 1974, CA82-164225.
879.MacChesney, J. B., O'Connor, P. B. and Sullivan, M. V., "Iron oxide photomask", Patent S. African 71-02438, Publ. January 1972, CA78-153707.
880.MacRae, A. U. and Moline, R. A., "Semiconductor masking", Patent USA 3767492, Publ. October 1973, CA80-20721.
881.Marinace, J. C., "Durable seethrough photoresist mask", Patent USA 3815978, Publ. June 1974.
882.Muska, W. M., "Manganese sputter-etch masks for integrated optical circuits", J. Vac. Sci. & Technol., 11, 883-4 (1974), CA81-143209.
883.Orlinov, V. I. and Sarov, G. A., "Effect of oxygen concentration in a gas stream on deposition conditions and properties of iron (III) oxide films prepared by reactive d.c. sputtering", Bulg. J. Phys., 2, 156-66 (1975), Russian, CA83-211756.
884.Prakash, O. M., "Dielectric mask for gallium arsenide devices", Indian J. Pure Appl. Phys., 12, 160-2 (1974), CA81-31115.
885.RCA Corp., "Semiconductor device including a resist film", Patent UK 1230421, Publ. May 1971, CA75-42205.
886.Rysanek, V., "Thin films from organic compounds", Elektrotech. Cas., 22, 694-705 (1971), Czech, CA76-51381.
887.Sadagopan, V., "Anti-interference and antireflection coatings for chromium masks", IBM Tech. Disclosure Bull., 14, 795-6 (1971), EEA75-3941.
888.Sato, M., "Photosensitive plates for photomasks", Patent Japan 75-86336, Publ. July 1975, CA83-211336.
889.Sato, T. and Tsuda, K., "Colored transparent photomasks", Patent German 2016056, Publ. June 1973, CA79-131398.
890.Simenov, V. D., Mednikarov, B. D. and Malinovski, J. P., "Photographic material for production of photomasks", Patent French 2193212, Publ. February 1974, CA82-178265.
891.Sinclair, W. R., Rousseau, D. L. and Stancavish, J. J., "Iron oxide - an inorganic photoresist and mask material", J. Electrochem. Soc., 121, 925-8 (1974), EEA77-34098.
892.Sinclair, W. R., Sullivan, M. V. and Fastnacht, R. A., "Materials for use in a durable selectively semitransparent photomask", J. Electrochem. Soc., 118, 341-4 (1971), CA74-81744.

893. Soehibrand, H., "Aluminum oxide and silicon oxide masks for silicon doping", Patent German 2012080, Publ. September 1971, CA75-145348.
894. Spitzer, S. M., Schwartz, B. and Weigle, G. D., "Native oxide mask for zinc diffusion in gallium arsenide", J. Electrochem. Soc., 121, 820-2 (1974), CA81-55594.
895. Szupillo, R. E., "Chromium masks. I.", Solid State Technol., 12 (7), 49-52 (1969), EEA73-470.
896. Szupillo, R. E., "α-Iron oxide thin films for microcircuit masks", Patent USA 3695908, Publ. October 1972, CA77-157947.
897. Taylor, R. C. and Sadagopan, V., "Preparation and properties of hexaferrite films (photomask)", J. Electrochem. Soc., 119, 788-90 (1972), EEA75-27752.
898. Terry, L., "Use of platinum thin films as masks in semiconductor processing", Patent USA 3629022, Publ. December 1971, CA76-65368.
899. Thompson, L. F., "Poly(vinylferrocene) - Conversion to an oxidized iron system suitable for use as a semitransparent hard photomask", J. Electrochem. Soc., 122, 108-11 (1975), EEA78-11704.
900. Toyoda, H., Yada, T., Itasaka, T. and Ito, K., "High precision chromium masking", Denki Kagaku, 40, 531-6 (1972), Japanese, CA77-170246.
901. Tsutsumi, T., Ide, K. and Ichikawa, O., "Etching semiconductors with rare earth oxide masks", Patent Japan 73-41815, Publ. December 1973, CA81-7638.
902. Vicente-Perez, S., Hernandez, L. and Hernandez-Mendez, J., "PDTA (propylenediaminetetraacetic acid) as a masking agent in general reactions of cations", Inform. Quim. Anal., 26 (2), 77-83 (1972), Spanish, CA77-96521.
903. Voskoiboinikov, V. V., Rabinovich, L. N. and Repinskii, S. M., "Solution of silicon nitride films in phosphoric acid", Inorg. Mater., 9, 1629-30 (1973), EEA77-33942.
904. Whiteley, T. E. and Krueger, S. M., "Filter for negative diazo imaging", Res. Discl., 127, 15-17 (1974), CA83-186261.
905. Yurist, I. M., "Masking substances in complexometry", Zh. Anal. Khim., 26, 975-91 (1971), Russian, CA75-29507.

3. Computer-Aided-Design (CAD)

906. Astier, B. and Carry, G., "Simultaneous design of circuitry and masks for LSI", International Conference on Large Scale Integrated Circuits, 1974, p. 207-15, EEA78-11756.
907. Atiyah, J., "The graphic display as a drawing aid for circle layout", In: Computer aided design techniques, Wolfendale, E., (Ed.), Butterworth, London, England, p. 222-39 (1970), EEA74-6696.
908. Baldoni, R. and Capocaccia, F., "Automatic generation of masks for read only memories", Proceedings of the 18th International Congress on Electronics, 1971, p. 269-79, Italian, EEA74-36465.

909.Bastin, D., Girard, E., Rault, J. C., Rosenstiehl, P. and Souesme, P., "Automatic design of integrated circuit masks: a new locating algorithm", Rev. Tech. Thomson-CSF, 6 (1), 229-46 (1974), French, EEA78-8598.

910.Bazhenov, S. N., Evteev, G. V., Kanakov, A. I., Lapin, V. I., Lyubomudrov, A. S. and Ridiger, V. K., "A programmed scanning laser system for photomask synthesis", Izv. VUZ Priborostr., 16, 121-5 (1973), Russian, EEA76-35334.

911.Beardsley, C. W., "Computer aids for IC design, artwork, and mask generation", IEEE Spectrum, 8 (9), 63-79 (1971), EEA74-36457.

912.Beasley, J. P., "A computer controlled electron mask maker", Electron and Ion Beam Science and Technology 4th International Conference, 1970, p. 515-23, EEA74-29776.

913.Beasley, J. P. and Squire, D. G., "An electron beam mask-maker", IEEE Trans. Electron Devices, ED-22, 376-84 (1975), EEA78-24385.

914.Bertram, U., Hoffmann, H. and Neuhaus, W., "Highly integrated bipolar digital circuits", Internat. Elektronische Rundschau, 23, 300-4 (1969), German, EEA73-13013.

915.Bracchi, G. and Ferrari, D., "A graphic language for describing and manipulating two-dimensional patterns", International Symposium and Exhibition on Computer Graphics, 1970, 12 pp., EEA73-19800.

916.Cast, R., "Computer controlled artwork generation at 10X", Solid State Technol., 14 (2), 29-47 (1971), EEA74-17585.

917.Chang, T. H. P., "A computer-controlled electron-beam machine for microcircuit fabrication", IEEE Trans. Electron Devices, ED-19, 629-35 (1972), EEA75-15774.

918.Cospodnetic, R. and Ruzic, V., "Computer programme for drawing masks of integrated circuits", Proceedings of the 5th Yugoslavian International Symposium of Information Processing, 1969, p. 179-80, Croatian, EEA73-36846.

919.Crawford, D. O., Hale, R. J. and Nichols, K. G., "A digital generator for pattern drawing applications", Radio & Electron. Eng., 45, 63-73 (1975), EEA78-11708.

920.Dierick, P., Larie, R. and Martin, J. L., "Automatic plotting of masks for hybrid integrated circuits", Onde Elect., 49, 93-103 (1969), French, EEA72-10648.

921.Dowd, P. G., Cowan, M. J., Rosenfeld, P. E. and Zacharias, A., "The primary pattern generator. III. The control system", Bell Syst. Tech. J., 49, 2061-8 (1970), EEA74-9798.

922.Eberl, H. and Koller, K., "DIAMASK, an interactive programming system for the generation and modification of mask artworks for integrated circuits", Siemens Forsch. & Entwicklungsber., 1, 279-83 (1972), German, EEA72-39193.

923.Engel, A., Steffen, J. and Herziger, G., "Laser material processing with modulated zone plates", 1973 IEEE/OSA Conference on Laser Engineering and Applications Digest of Technical Papers, p. 29-30, PA77-8864.

924.Ferrari, D. and Mezzalira, L., "A computer-aided approach to integrated circuit layout design", Proceedings of the 2nd Biennial Cornell Electrical Engineering Conference, 1969, p. 358-68, EEA73-20116.
925.Fowler, B. R., "XYMASK", Bell Lab Record, 47, 204-9 (1969), EEA73-446.
926.Fowler, A. G., Gonzal, B. C. and Balyuzi, H. H. M., "Use of microfilm computer plotting facilities for generating optical-diffraction masks", J. Appl. Crystallogr., 6, 487-8 (1973), CA80-89527.
927.Friedmann, E. B., Livesay, W. R. and Rubiales, A. L., "An automated electron beam mask generator", J. Vac. Sci. & Technol., 10, 1020-4 (1973), EEA77-23095.
928.Gatehouse, J. S. and McGulfin, R., "Interactive graphics for mask design", Comput. Aided Des., 7, 196-7 (1975), EEA78-37567.
929.Goutte, G., Mutel, J. and Rochet, A., "Computer aided fabrication of integrated circuits - example: mask generation", Onde Elec., 50, 773-8 (1970), French, EEA74-4178.
930.Green, N. P., "Computer aided draughting", In: Mullard Research Laboratories annual review, Mullard Res. Labs., London, England, 44-50 (1969), EEA73-30335.
931.Gross, A. G., Raamot, J. and Watkins, S. B., "Computer systems for pattern generator control", Bell Syst. Tech. J., 49, 2011-30 (1970), EEA74-9794.
932.Hansen, W., "A graphic arts approach to printed circuit masks", Solid State Technol., 12 (11), 53-6 (1969), EEA73-16803.
933.Hariu, T. and Hartnagel, H. L., "New Gunn-effect three-terminal pulse regenerator by planar photolithography", Int. J. Electron., 37, 451-55 (1974), EEA77-38811.
934.Harrell, S., "Practical automation for photolithographic wafer processing", Solid State Technol., 17 (8), 47-9 (1974), EEA77-38892.
935.Herzog, R. F., Lewis, B. L. and Everhart, T. E., "Computer control and the scanning electron microscope. (Electron beam microfabrication and data acquistion)", Scanning Electron Microscopy 1974, p. 175-82, EEA78-693.
936.Hiebsch, W., "Experiences and findings of the manufacturer of photomasks in the industrial testing of the automatic nine-barrel step and repeat camera ANR 22 of VEB Carl Zeiss JENA", Jena Rev., 18 (1), 28-9 (1973), EEA76-26127.
937.Hubscher, A. J., "The automated production of printed circuit masks", Automation, 6 (3-4), 17-19 (1971), EEA74-17740.
938.Jarvis, J. F., "The design of interactive graphics aid to mask layout", Proc. IEEE, 60, 35-9 (1972), EEA75-6967.
939.Kawashima, K., "Computer aided design for mask layout", J. Inst. Electron. & Commun. Eng. Jap., 55, 526-30 (1972), Japanese, EEA76-35399.

940.Kim, J. J., Sampson, H. G. and Everhart, T. E., "Microfabrication using a computer controlled scanning transmission (electron microscope)", J. Vac. Sci. & Technol., 10, 1005-7 (1973), EEA77-23092.

941.Kojima, T., Yanagi, M. and Uyehara, T., "Computer-aided design of hybrid IC mask", Fujitsu Sci. & Tech. J., 10, 75-93 (1974), EEA77-30419.

942.Koller, K., "AUTOMASK-a programming system for the automatic plotting of mask patterns", Siemens Forsch. & Entwicklungsber., 1 (1), 45-9 (1971), German, EEA75-3951.

943.Kosanke, K., Schedewie, F. and Wagner, D., "A fast computer-controlled co-ordinate measuring system for semiconductor mask", 6th International Congress on Microelectronics, 1974, 7 pp., German, EEA78-24434.

944.Kozawa, T., Horino, H., Ishiga, T., Sakemi, J. and Sato, S., "Advanced LILAC - an automated layout generation system for MOS/LSIs", 11th Design Automation Workshop, 1974, p. 26-46, EEA77-30413.

945.Lee, S. C., Wyndrum, R. W. and Rosler, L., "Computer-aided integrated circuit design using the Graphic-2-GE 635 interactive system", Proceedings of the 2nd Biennial Cornell Electrical Engineering Conference, 1969, p. 217-26, EEA73-20114.

946.Luaria, J. and Polizzotto, J. F., "Computer controls LSI mask-making", Control Engng, 16, 100-2 (1969), EEA72-12801.

947.Malmberg, P. R., O'Keeffe, T. W., Sopira, M. M. and Levi, M. W., "LSI pattern generation and replication of electron beams", J. Vac. Sci. & Technol., 10, 1025-7 (1973), EEA77-23096.

948.Mattison, R. L., "Design automation of MOS artwork", Computer, 7 (1), 21-7 (1974), EEA77-18805.

949.Mays, C. H., "A brief survey of computer-aided integrated circuit layout", IEEE Trans. Circuit Theory, CT-18, 10-13 (1971), EEA74-12763.

950.Narud, J. A., Dertouzos, M. J. and Jessel, G. P., "Computer aided analysis for integrated circuits", Proceedings IEEE Electrical and Electronics Engineering Resources Roundup, 1969, p. 243-50, EEA73-3530.

951.Newberry, S. P. and Klotz, T. H., Jr., "Integrated circuit mask making by electron optical means", 1968 NEREM Record, p. 172-3, EEA72-17163.

952.Niessen, C. and Wulms, H. E. J., "Automatic drawing of masks for integrated circuits", Philips Tech. Rev., 30 (2), 29-35 (1969), EEA72-14921.

953.Nishikawa, S. and Okada, K., "Design automation system for IC mask patterns", Rev. Electr. Commun. Lab., 23, 271-8 (1975), EEA78-37585.

954.Nishikawa, S., Okada, K., Noshita, K., Hashimoto, A. and Naemura, K., "Musashino experimental design automation system for IC mask patterns MEDIC", Electr. Commun. Lab. Tech. J., 23, 287-300 (1974), Japanese, EEA78-8584.

955.Northwood, M. A., "Interactive graphics and artwork preparation", AGARD Lecture Series No. 75 Custom Design for Large Scale Integration (LSI), 1975, p. 5/1-6, EEA78-19930.
956.Owen, G. and Nixon, W. C., "Aberration correction for increased lines per field in scanning electron beam technology (IC fabrication)", J. Vac. Sci. & Technol., 10, 983-6 (1973), EEA77-23090.
957.Pascher, A. K. and Koller, K., "Generating IC artwork automatically saves time, prevents costly errors", Electronics, 42 (24), 117-8 (1969), EEA70-12982.
958.Pasiecznik, J., Jr. and Frey, J., "A high performance, low cost digitally driven SEM system for materials studies and microfabrication", J. Vac. Sci. & Technol., 10, 1012-15 (1973), EEA77-23093.
959.Piwczyk, B. P. and McQuhae, K. G., "A fast turn around electron beam pattern generator system", J. Vac. Sci. & Technol., 10, 1016-19 (1973), EEA77-23094.
960.Poole, K. M., "The primary pattern generator: introduction", Bell Syst. Tech. J., 49, 2031-2 (1970), EEA74-9795.
961.Potenza, G., "Automatic mask making for integrated circuits", Alta Frequenza, 39, 314-16 (1970), Italian, EEA73-27254.
962.Raamot, J. and Zaleckas, V. J., "Laser pattern generation using X-Y beam deflection", Appl. Opt., 13, 1179-83 (1974), EEA77-26490.
963.Ressler, D. G., "A simple computer-aided artwork system that works", 11th Design Automation Workshop, 1974, p. 92-7, EEA77-30363.
964.Richardson, F. K., "Computer aided design for photomask production", Solid State Technol., 13 (6), 60-4 (1970), EEA73-37060.
965.Rizek, S., "Computer aided mask generation for integrated circuits", Proceedings of the 5th Colloquium on Microwave Communication, Vol. II. Circuit Theory and Computer Aided Design, 1974, p. 213-17, EEA77-30359.
966.Rosenberg, L. M. and Benbassat, C., "CRITIC: an integrated circuit design rule checking program", 11th Design Automation Workshop, 1974, p. 14-18, EEA77-30361.
967.Rosenfeld, R. L., "Computer assisted mask production", Proc. IEEE, 57, 1629-34 (1969), EEA73-464.
968.Rowe, T. J. and Lawn, M., "Computer aided preparation of SCIC fabrication masks", International Conference on Microelectronics, 1969, p. 19-20, EEA72-22902.
969.Saillard, J. C. and Sarret, M. J., "Automatic design of masks: The desmag programme", Onde Elect., 49, 92-7 (1969), French, EEA72-9210.
970.Saitour, N., Munakata, C., Miura, Y. and Honda, Y., "Computer-controlled electron beam microfabrication machine with a new registration system", J. Phys. E, 7, 441-4 (1974), EEA77-26212.
971.Samaroo, W., Raamot, J., Parry, P. and Robertson, G., "The electron beam pattern generator", Bell Syst. Tech. J., 49, 2077-94 (1970), EEA74-9800.

972.Schwarz, H. J., "Automated artwork design realization", Proceedings of the 1st EIPC Technical Conference on Circuit Design and Artwork Preparation, 1969, p. 36-40, EEA74-4303.
973.Seliutin, V. A. and Gurevich, D. Z., "Computer-aided photomask design for LSI", International Symposium on Discrete Systems, Pt. V, 1974, p. 254-62, EEA78-15421.
974.Silverberg, M., "CLIC-Computer layout of integrated circuits", Proceedings of the Fifth Annual SHARE ACM-IEEE 1968 Design Automation Workshop, 5 pp., EEA72-10469.
975.Spicer, D. F., Rodger, A. C. and Varnell, G. L., "Computer controlled pattern generating system for use with electron beam writing instruments", J. Vac. Sci. & Technol., 10, 1052-5 (1973), EEA77-23099.
976.Sugiyama, N. and Hirano, M., "An automated integrated circuit layout design program of a systematic chip using building blocks", International Conference on Computer-Aided Design, 1972, p. 293-8.
977.Szabo, L., "Computer accelerates design and production of IC's artworks", Hiki, 8, 59-65 (1968), Hungarian, EEA72-251.
978.Tamaru, K., "Integrated circuits. VII. Computer aided design: The present status and its future", J. Inst. Electronics Commun. Engrs., 53, 345-51 (1970), Japanese, EEA73-30326.
979.Tsuboi, Y., Tsuchiya, R. and Asakawa, T., "Computer-aided design of MOS LSI", Proc. Inst. Radio Electron. Eng. Aust., 32 (16), 196-203 (1971), EEA74-29475.
980.Uyehara, T., Takahashi, O. and Shiraishi, H., "Program of automatic circuit drawing for mask making system", Fujitsu Sci. & Tech. J., 8 (3), 35-58 (1972), EEA76-4065.
981.Varnell, G. L., Spicer, D. F. and Rodger, A. C., "E-beam writing techniques for semiconductor device fabrication", J. Vac. Sci. & Technol., 10, 1048-51 (1973), EEA77-23098.
982.Vasseur, J. P., "Computer's aid for integrated circuit mask production", Onde Elect., 49, 90-1 (1969), French, EEA72-10629.
983.Wagner, R., "Shift the mask-planning burden to software", Solid State Technol., 18 (6), 14 (1975), EEA78-33233.
984.Wolf, E. D., "Computer-controlled electron beam microfabrication (Development of SEM concepts)", Seventh National Conference on Electron Probe Analysis, 1972, 24A/7 pp., EEA76-22828.
985.Wood, J., Newton, R. S., Snell, D. J. and Walmsley, M. B., "Computer aided production of masks for silicon integrated circuits", International Conference on Computer Aided Design, 1969, p. 122-9, EEA72-23160.
986.Zucker, J. S. and Crawford, B. J., "Graphical layout system for IC mask design", IEEE Trans. Circuit Theory, CT-18, 163-73 (1971), EEA74-12691.

4. Fabrication

987. Abe, H., Nagasawa, K., Kuramitsu, Y., Nishioka, K., Iwama, I. and Yamazaki, T., "Dried process with gas plasma for production of chromium and chromium oxide film masks", 6th International Congress on Microelectronics, 1974, 3 pp., EEA78-24432.
988. Aigner, E., "Masks for semiconductor technology", Elektron. Anz., 7 (4), 41-4 (1975), German, EEA78-24381.
989. Andreae, H. R., "Preparation of coinciding films", Proceedings of the 1st EIPC Technical Conference on Circuit Design and Artwork Preparation, 1969, p. 41-7, German, EEA74-4306.
990. Bajorek, C. H., Pfeiffer, A. T. and Romankiw, L. T., "Making iron oxide masks", IBM Tech. Disclosure Bull., 15, 1595-6 (1972), EEA76-12660.
991. Bondarenko, V. A., Voitovich, I. D. and Zholob, V. P., "A device for semiautomatic production of photo-masks", Mekh. & Avtom. Upr., no. 1, 38-40 (1972), Russian, EEA75-23376.
992. Bouwer, A. G., Bruel, R. H., van Heek, H. F., Klostermann, F. T. and Mannetje, J. J. T., "The Opthycograph (IC photomask production)", Philips Tech. Rev., 34, 257-69 (1974), EEA78-15393.
993. Burvenich, X., "Etching techniques for the manufacture of metal masks for micro-electronics", Galvano, 40, 851-9 (1971), French, MA5-04-34-0166.
994. Camenzind, H. R., "Making integrated circuit masks", Patent USA 3663223, Publ. May 1972, CA77-94593.
995. Catier, E., "More precise and more viable masks", Electron. & Microelectron. Ind., no. 202, 53-8 (1975), French, EEA78-24380.
996. Chand, N., "Production of photographic masks on substrates for semiconductor devices", Patent French 2028069, Publ. November 1970, CA74-148209.
997. Chorzewska, M., "Chrome masks and their fabrication technology", Przeglad Elektron, 10, 3-8 (1969), Polish, EEA72-12863.
998. Cohen, R. A., Mountain, R. W., Smith, H. I., Lemma, M. A. and Spears, D. L., "Fabrication procedure for silicon membrane x-ray lithography masks", Report 769857/4GA, Lincoln Lab., Massachusetts Inst. Technol., Lexington, Mass., 1973, 14 pp., CA81-8352.
999. Conner, M. E., "Mask Maker I", Report HDL-TR-1582, Harry Diamond Labs., Washington, D. C., 1972, 27 pp., EEA75-35414.
1000. Crimi, J. S., Franklin, R. H. and Marrone, G. M., "Fabrication of stainless steel screening masks", IBM Tech. Disclosure Bull., 18, 491-2 (1975), EEA78-37641.
1001. Crimi, J. S. and Johnson, A. H., "Composite mask fabrication", IBM Tech. Disclosure Bull., 15, 1073-4 (1972), EEA76-8728.
1002. Croset, M., "Masks for semiconductor preparation", Patent French 2135428, Publ. January 1973, CA79-11288.
1003. Depierre, J., "Making metallic masks of high definition", Patent French 2074740, Publ. November 1971, CA77-55729.

1004.Dinklage, J. B., "Glass mask fabrication", IBM Tech. Disclosure Bull., 13, 15 (1970), EEA73-37104.
1005.Fay, B., Ostrowsky, D. B., Roy, A. M. and Trotel, J., "Electron beam duplication of integrated optical circuits and masks", Opt. Commun., 9, 424-6 (1973), EEA77-11119.
1006.Feldstein, N., "Opaque covering matrix for luminescent screens in color television tubes", Patent German 2122258, Publ. November 1971, CA76-39037.
1007.Feldstein, N. and Weiner, J. A., "Fabrication of semitransparent masks", J. Electrochem. Soc., 120, 1654-7 (1973), EEA77-10276.
1008.Gioia, J. C., "How to prepare low cost artwork and masks for thin film hybrid integrated circuits", 1970 IEEE International Convention Digest, p. 418-19, EEA73-37101.
1009.Glendinning, W. B. and Pharo, W. B., "Forming a diffusion mask barrier on a silicon substrate", Patent USA 3718503, Publ. February 1973, CA78-116896.
1010.Heiman, F. P., "Making high-area-density-array photomasks having matching registry", Patent USA 3647438, Publ. March 1972, CA76-119907.
1011.Heinz, A., "Thin-film mask generation by light beam synthesis", Tech. Dig., no. 22, 21-2 (1971), EEA74-25720.
1012.Hinsenkamp-Feher, I. M., "Preparation of master masks for printed circuits by the help of a drafting machine. I", Finommechanika, 11, 119-22 (1972), Hungarian, EEA75-27968.
1013.Horne, D. F., "Recent advances in graticule and mask making", Opt. Acta, 20, 939-57 (1973), PA77-2310.
1014.Iizaka, Y., Sotodate, M. and Fukushima, Y., "Photomask for photoetching", Patent Japan 74-94401, Publ. September 1974, CA82-148535.
1015.Jackson, T. M., Bush, E. L., Rowe, T. J. and Moule, D. J., "Photolithographic masks", Patent USA 3650796, Publ. March 1972, CA77-12327.
1016.Jackson, T. M. and Rowe, T. J., "Laser cuts masks to size, eliminating most errors", Electronics, 42 (2), 81-4 (1969), EEA73-9575.
1017.Kane, J. and Schweizer, H., "Forming transition metal oxide films on a substrate for photomasks", Patent USA 3914515, Publ. October 1974, CA83-211943.
1018.Kutzer, E. and Martin, R., "Manufacturing of precise mask-prototypes for electronic components", Werkstattstechnik, 62 (2), 71-5 (1972), German, EEA75-19335.
1019.Land, L., "Photomask fabrication using transparent hard photomask blanks", Solid State Technol., 14 (6), 8, 15, 71 (1971), EEA74-25574.
1020.Lasky, D., "Emulsion mask processing techniques", Solid State Technol., 18 (6), 46-50 (1975), EEA78-33232.
1021.Lehrer, W. I., "Photomasks (Ge) and method of fabrication", Patent USA 3830686, Publ. August 1974.

1022.MacChesney, J. B., O'Connor, P. B. and Sullivan, M. V., "Chemical vapor deposition of iron oxide films for use as semitransparent masks", J. Electrochem. Soc., 118, 776-81 (1971), CA75-41336.
1023.Malone, D., "Fine position patterning techniques", Trans. Inst. Metal Finishing, 47 (Pt. 3), 122-4 (1969), EEA73-3622.
1024.Mimura, Y. and Ozawa, A., "High-resolution and high-precision metallic photomasks fabricated by projection exposure", Denki Kagaku, 41, 538-43 (1973), Japanese, CA80-42143.
1025.Minzenberger, T. C., "Back etching a perforated metal mask for color television tubes", Patent USA 3679499, Publ. July 1972, CA77-119897.
1026.Moreau, W. M. and Warnecke, A. J., "Dielectric photomasks", IBM Tech. Disclosure Bull., 13, 158 (1970), EEA73-37865.
1027.Murphy, J. P. and Henriksen, G. M., "Good masks, good yields", Electron Packag. & Prod., 7 (5), 17-25 (1971), EEA75-444.
1028.Nightingale, E. G. and Culli, R., "Photolithography", Patent UK 1391270, Publ. April 1975, CA83-88728.
1029.Osvath, I., "Preparation operations of master masks for printed circuits", Finommechanika, 11 (4), 115-18 (1972), Hungarian, EEA75-27967.
1030.Parsonage, H. N. and Kruger, M. I., "Color photographic process for producing visually transparent but photographically opaque photomasks", Patent USA 3592649, Publ. July 1971, CA75-157069.
1031.Peters, F. G., Sinclair, W. R. and Sullivan, M. V., "Iron oxide films by cathode sputtering", Patent German 2129000, Publ. December 1971, CA76-79352.
1032.Peters, F. G., Sinclair, W. R. and Sullivan, M. V., "Sputtered Fe_2O_3 films for use in 'see through' masks", J. Electrochem. Soc., 119, 305-9 (1972), EEA75-19495.
1033.Pritchard, J. P., Jr., "Method for producing sets of photomask having accurate registration", Patent USA 3477848, Publ. November 1969, EEA73-24074.
1034.Reinberg, A. R. and Fish, J. G., "Fabricating silicon photomask", Patent USA 3814641, Publ. June 1974, CA81-55743.
1035.Rice, E. J., "Etching patterns into solid-state devices", Patent USA 3542551, Publ. November 1970, CA74-36220.
1036.Saitou, N., Morishita, H., Nonogaki, S., Itoh, H. and Maekawa, A., "Mask fabrication with submicron linewidth by electron beam", Jap. J. Appl. Phys., 10, 1486-7 (1971), CA75-156174.
1037.Saitou, N. and Nanogaki, S., "Submicron-mask making by electron beam", Electron and Ion Beam Science and Technology Fifth International Conference, 1972, p. 68-79, EEA76-8718.
1038.Sato, M., "Formation of masks for photoresists", Patent Japan 75-02539, Publ. January 1975, CA83-35738.
1039.Sato, M., "Fabrication of transparent photomask from silver halide photographic film", Patent Japan 75-71335, Publ. June 1975, CA83-211334.

1040. Shima, K. and Ishihara, T., "Masking films", Patent Japan 75-51579, Publ. May 1975, CA83-116374.
1041. Strable, U. D., "Electron-optical methods for the production of exposure masks and masking structures in semiconductor technology", Wiss. Ber. AEG-Telefunken, 48 (1), 30-6 (1975), EEA78-24501.
1042. Taylor, R. C. and Sadagopan, V., "Preparation and properties of hexaferrite films", J. Electrochem. Soc., 119, 788-90 (1972), CA77-68521.
1043. Tetzner, W., "Photomask for the production of surface patterns, especially for printed circuits", Patent German 1797023, Publ. July 1974, CA81-143344.
1044. Wade, T. and Pabich, R., "On-line interactive processing of interconnect masks", Electron. Prog., 15 (4), 21-5 (1973), EEA77-30377.
1045. Western Electric Co., Inc., "Mask pattern manufacture in silicon dioxide dielectric layer of silicon solid state devices", Patent UK 1208598, Publ. October 1970, EEA74-9824.

5. Alignment

1046. Bachman, A. L. and Felker, D. E., "Method of aligning a separation-etch mask with beam-lead semiconductor devices formed in a wafer", Tech. Dig., no. 33, 5-6 (1974), EEA77-23169.
1047. Borner, M., Hennings, K. E. and Strahle, U. D., "Magnetostrictive stage for the line adjustment of a mask over a semiconductor disc", Mess Technik, 77, 244-7 (1969), German, EEA73-9561.
1048. Clark, K. C. and Juleff, E. M., "Projection mask alignment - problems and solutions", Microelectronics, 6 (2), 37-42 (1974), EEA78-24422.
1049. Clark, K. G., "Automatic mask alignment in MOS/LSI processing", Microelectronics, 3 (9), 47-55 (1970), EEA74-12766.
1050. Clark, K. G., "Automatic out-of-contact mask alignment", Solid State Technol., 18 (8), 47-53, 57 (1975), EEA78-41692.
1051. Clark, K. G. and Juleff, E. M., "Projection mask alignment - A production technique", Solid State Technol., 16 (6), 37-41 (1973), EEA76-35341.
1052. Clark, K. G. and Okutsu, K., "Improved techniques for proximity mask alignment", Microelectronics, 6 (4), 51-9 (1975).
1053. Finnila, R. M., Su, S. C. and Braunstein, A. I., "Projection printing", Proc. Soc. Photo-Opt. Instrum. Eng., 1974, p. 55, 68-75, Publ. 1975, CA83-155699.
1054. Gegenwarth, R. E. and Laming, F. P., "Process for photolithographic alignment", IBM Tech. Disclosure Bull., 17, 3271 (1975), EEA78-24471.
1055. Hudson, J. L. and Jordan, L. L., "Masking semiconductor wafers using a self-aligning mask", Patent USA 3887421, Publ. June 1975, CA83-107229.

1056.Jacobs, B. A. J. and Kramer, P., "Apparatus for aligning a mask with respect to a semiconductor substrate", Patent USA 3811779, Publ. May 1974.

1057.Johnson, A. M. and Zacharias, A., "The primary pattern generator. IV. Alignment and performance evaluation", Bell Syst. Tech. J., 49, 2069-76 (1970), EEA74-9799.

1058.Jorgensen, R. R. and Wojcik, W. J., "Mask-to-wafer alignment using exposure wavelength", IBM Tech. Disclosure Bull., 18, 385-6 (1975), EEA78-37635.

1059.Kerprt, E., "The use of opto-electronics in the manufacture of transistors and integrated circuits", Jemna Mech. & Opt., 17, 237-9 (1972), Czech, EEA76-12666.

1060.King, M. C., "Moire pattern aligning of photolithographic masks", Patent USA 3690881, Publ. September 1972, CA78-21671.

1061.King, M. C. and Berry, D. H., "Photolithographic mask alignment using Moire techniques", Appl. Opt., 11, 2455-9 (1972), EEA73-830.

1062.Klingenfeld, D., "Projection-type aligning and exposure device for micron patterns JUBPM 50", Jena Rev., 19 (spec. issue), 63-8 (1974), EEA77-38860.

1063.Kolb, G. A. and Su, L. S., "Autocorrelation codes for mask and wafer alignment design", IBM Tech. Disclosure Bull., 17, 2309-10 (1975), EEA78-24466.

1064.Levy, K., "Second generation automated wafer to mask aligners", Presented at Semiconductor Integrated Circuit Processing and Production Conference, June 1971, EEA74-25623.

1065.Livesay, W. R., "Computer controlled electron-beam projection mask aligner", Solid State Technol., 17 (7), 21-6 (1974), EEA77-38890.

1066.Livesay, W. R. and Wing, M. E., "Mask alignment system for electron beam pattern generator", Patent USA 3874916, Publ. April 1975, EEA78-37670.

1067.Oshima, Y., "Microservo (system for automatic mask alignment)", J. Soc. Instrum. & Control Eng., 14, 196-208 (1975), Japanese, EEA78-19862.

1068.Sotta, H. and Froschle, E., "Automatic adjustment technique for the illumination of microstructures", 6th International Congress on Microelectronics, 1974, 8 pp., German, EEA78-24437.

1069.Stein, K. U., "Apparatus for aligning a semiconductor wafer with respect to a mask for preparation of a structure in a photolacquer with x-irradiation", Patent German 2346720, Publ. April 1975, CA83-107193.

1070.Stein, K. U., "Alignment of a photomask for use in producing fine structure in semiconductor chips", Patent German 2422982, Publ. August 1975, CA83-211925.

1071.Tasch, A. F., Jr. and Porter, V. R., "Photomask array placement on slices in LSI processing", Microelectron. & Reliab., 12, 539-45 (1973), EEA77-23151.

1072. Walls, D. C., "Mask alignment methods in making semiconductible devices", Patent USA 3759767, Publ. September 1973, CA79-140864.
1073. Wilson, J. and Bottomley, S. C., "Non-contact printing and automatic alignment of chromium mask patterns on silicon slices", Solid State Technol., 15 (6), 37-41 (1972), EEA75-35402.
1074. Yamada, E., Kosa, Y. and Sato, E., "Alignment of diffusion masks for semiconductors", Patent USA 3801383, Publ. April 1974, CA80-150261.

6. Special Techniques

1075. Abita, J. L., "Improved conventional photolithography by relief mask processing", Solid State Technol., 17 (6), 48-9 (1974), EEA77-34074.
1076. Baker, T. H. and Ghafghaichi, M., "Masking method in semiconductor technology", Patent German 2453528, Publ. July 1975, CA83-140946.
1077. Barson, F., "Eliminating Si_3N_4 for emitter masking", IBM Tech. Disclosure Bull., 17, 3590-1 (1975), EEA78-33006.
1078. Bergh, A. A. and van Gelder, W., "Etch masks on semiconductor surfaces", Patent USA 3479237, Publ. November 1969, EEA73-14186.
1079. Beynon, J. D. E., Hall, B. P. and Smith, J. G., "Controlled diffraction as the basis of an out-of-contact masking system", Solid State Technol., 15 (11), 29-34 (1972), EEA76-8737.
1080. Brechling, G. E., Chadwick, J. E., Colton, W. F., Gogos, B., Marshall, R. R. and Sharp, F., "Truncating internal power columns on ordered grid arrays", IBM Tech. Disclosure Bull., 27, 1418-23 (1974), EEA78-8615.
1081. Brodsky, M. D. and Triacca, J. P., "An innovation in photomasks", Solid State Technol., 12 (7), 55-9 (1969), EEA73-472.
1082. Broom, R. F. J., "Masking semiconductor surfaces", Patent German 2234189, Publ. March 1973, CA78-166031.
1083. Cadek, V., "Indirect masks for vacuum sputtering of thin film circuits", Sdelovaci Tech., 16, 376-8 (1969), Czech, EEA70-3543.
1084. Calhoun, D. F. and McNamee, L. P., "A means of reducing custom LSI interconnection requirements", IEEE J. Solid-State Circuits, SC-7, 395-404 (1972), EEA75-35388.
1085. Couture, R. A., Lajza, J. J., Jr. and Wright, W. E., "Chromium complex primers for polymer coatings on metal", Patent German 2001548, Publ. July 1970, CA73-82027.
1086. DelMonte, L. A. and Grangroth, R. H., "Material deposition masking for microcircuit structures", Patent USA 3897324, Publ. Muly 1975, CA83-123938.
1087. Deneau, A. L., "Color correction by masking for a four-color printing process", Patent USA 3615441, Publ. October 1971, CA76-20046.
1088. Dickman, M. K., "Masking semiconductor systems", Patent German 2134385, Publ. February 1972, CA76-146750.

1089.Elgan, D. L. and Hays, R. G., "Masks for semiconductor bodies", Patent German 2131143, Publ. February 1972, CA76-133346.
1090.Etter, R. W., "Color kinescope production with a temporary mask", Patent USA 3653901, Publ. April 1972, CA77-40628.
1091.Fassett, J. R., "Photomasking", Patent USA 3823015, Publ. July 1974, CA81-113705.
1092.Gibbon, C. F., Povilonis, E. I. and Ketchow, D. R., "The effect of mask edges on dopant diffusion into semiconductors", J. Electrochem. Soc., 119, 767-72 (1972), EEA75-28401.
1093.Glinski, V. J., "A three mask bipolar integrated circuit structure", IEEE Trans. Electron Devices, ED-19, 182-90 (1972), EEA75-6956.
1094.Godfrey, J., "Semiconductor device fabrication using a combination of energy beams for masking and impurity doping", Patent USA 3737346, Publ. June 1973, CA79-59195.
1095.Heinecke, R. A. H., "Contact mask for semiconductor manufacture", Patent UK 1359543, Publ. July 1974, CA81-128757.
1096.Henning, W., Von Hoerschelman, K., Krueger, I. and Weidlich, H., "Maximum-frequency silicon planar transistors", Patent German 2020531, Publ. November 1971, CA76-91964.
1097.Hennings, K. E. and Meyer, H., "Projection mask printing - a new method for generating semiconductor devices", Nachrichtentech. Z. (NTZ), 21, 611-16 (1968), German, EEA72-713.
1098.Hofer, P. H., "Masking process with cold sink", Patent USA 3892892, Publ. July 1975, CA83-165982.
1099.Jarvis, J. F., "The design of interactive graphics aids to mask layout", Proc. IEEE, 60, 35-9 (1972).
1100.Johnson, W. S., "Multiple masking technique in ion implantation", IBM Tech. Disclosure Bull., 15, 660-1 (1972), EEA72-39812.
1101.Kalestynski, A., "The application of holography in electronics", Elektronika, 15, 196-201 (1974), Polish, PA77-68704.
1102.Keene, C. E. and Bradley, R. D., "Masking techniques for use in fabricating microelectronic components", Patent USA 3700510, Publ. October 1972, CA78-9414.
1103.Kiba, W. M., "Copper-stained transparent glass masks for making patterns on photoresists", Patent USA 3620795, Publ. November 1971, CA75-36924.
1104.Kiewit, D. A., "Lead sulfide ion-implantation mask for semiconductor materials", Patent USA 3607450, Publ. September 1971, CA75-134946.
1105.Koger, T. B. and Terry, L. E., "Electrolytic plating using a self-orienting mask", Patent German 2059333, Publ. June 1971, CA75-83579.
1106.Kooi, E., "Semiconductor device comprising an insulated-gate field-effect transistor", Patent USA 3676921, Publ. July 1972, CA78-35483.
1107.Kurez, E., "New masking method for microelectronics", Elektronika, no. 7, 273-9 (1971), Polish, EEA74-36469.

1108.Lawrence, H. and Schaefer, P. C., "Three-resist-masking-step process for fabricating insulated-gate field-effect transistors", Patent USA 3670403, Publ. June 1972, CA77-67755.
1109.Lehrer, W. I., "Double masking technique for integrated circuit", Patent USA 3498833, Publ. March 1970, EEA73-20129.
1110.Malek, L., "The metal-masking technique for thin-film evaporation", Elektronika, 14 (11), 477-9 (1973), Polish, EEA77-14582.
1111.Middelhock, S., "Projection masking, thin photoresist layers and interference effects", IBM J. Res. Developm., 14, 117-24 (1970), EEA73-27301.
1112.Penberg, M., "Trimetallic masks for use in deposition processes for construction of thin film devices", Patent USA 3574012, Publ. April 1971, CA74-145698.
1113.Pennington, K. S. and Will, P. M., "Spatial frequency filtering for chips and masks", IBM Tech. Disclosure Bull., 13, 3269 (1971), EEA74-29752.
1114.Pestie, J. P., "Manufacturing semiconductor components", Patent USA 3635772, Publ. January 1972, CA76-133320.
1115.Philips Electronic and Associated Industries Ltd., "Silk-screen-printing masks", Patent UK 1374468, Publ. November 1974, CA83-119912.
1116.Poole, K. M., "New masks - new method", Bell. Lab. Record, 48, 131-6 (1970), EEA73-37037.
1117.Rideout, V. L., "Masking for one-device cell memories using self-registering metal-to-polysilicon contacts", IBM Tech. Disclosure Bull., 17, 2802-4 (1975), EEA78-19915.
1118.Roumiguieres, J. L., Maystre, D. and Petit, R., "Diffraction by an infinitely conducting lamellar grating in the presence of dioptres parallel to the lamellae", Opt. Commun., 7, 402-5 (1973), French, EEA76-26132.
1119.Schnepf, H., "Forming microstructures on semiconductor substrates", Patent German 2115823, Publ. October 1972, CA78-49779.
1120.Shaw, P. W., "Aperturing thin members by photochemical masking, exposing, and etching techniques", Patent USA 3653898, Publ. April 1972, CA76-146762.
1121.Sinclair, W. R. and Sullivan, M. V., "See-through masks aid semiconductor device manufacturing", Bell Lab. Rec., 52, 338-44 (1974), EEA78-15378.
1122.STC Ltd., "Producing semiconductor components", Patent UK 1201861, Publ. August 1970, EEA74-9815.
1123.Terry, L. E., "Mask for etching a semiconductor arrangement", Patent German 2055982, Publ. May 1971, CA75-81744.
1124.Texas Instruments Inc., "Photomask for use in the manufacture of semiconductor devices", Patent UK 1146297, Publ. March 1969, CA71-25957.
1125.Thornley, R. F. M., "Electron-optical masking of semiconductor structures", IEEE Trans. Electron Devices, ED-17, 961-4 (1970), EEA74-1349.

1126.Tsang, W. T. and Wang, S., "Grating masks suitable for ion-beam machining and chemical etching", Appl. Phys. Lett., 25, 415-18 (1974), EEA78-7861.
1127.Tselinskii, Y. K., Kuselman, I. I., Somova, I. I. and Goitseva, V. S., "Masking of molybdenum", Patent USSR 460475, Publ. February 1975, CA83-71121.
1128.Vance, T. R., "Emulsion-coated chrome mask", IBM Tech. Disclosure Bull., 13, 34 (1970), EEA73-37548.
1129.Widmann, D., "Radiation mask for pattern production in photo-lacquer using x-irradiation", Patent German 2346719, Publ. May 1975, CA83-124968.
1130.Widman, D., Kappelmeyer, R., Schulter, K. and Steggeweht, Z., "Contact exposure mask for the selective exposure of photo-varnish coatings for semiconductor-making purposes", Patent USA 3644134, Publ. February 1972, CA76-133322.
1131.Williams, B. D. and Maleham, J., "Masking a surface of a semiconductive substrate", Patent UK 1327306, Publ. August 1973, CA79-150811.
1132.Wolter, H. G., "Masks for screen-printed circuits", Patent German 2049146, Publ. April 1972, CA77-68576.
1133.Zwicker, W. K., "Producing high-resolution patterns in single crystals", Patent USA 3770533, Publ. November 1973, CA80-42204.

7. Properties

1134.Anand, K. V., McKell, H. D. and Northrop, D. C., "The properties of silicon dioxide films on silicon as diffusion masks for boron", J. Phys. D, 4, 1722-30 (1971), EEA75-1020.
1135.Baker, A. G., "Adherent masks on oxide-coated semiconductor bodies", Patent USA 3482977, Publ. December 1969, CA72-49037.
1136.Blome, E. R., "Multilayer antireflective absorption film", Patent USA 3698928, Publ. October 1972, CA78-78103.
1137.Brewer, T. L., "Negative etch mask by polymerization using energetic radiation", Patent German 2446930, Publ. May 1975, CA83-98495.
1138.Burock, R., Switsky, D. M. and Whitner, R. A., "Plating or etching treatment to improve the adhesion of masking patterns on metallic surfaces", Patent German 2108327, Publ. October 1971, CA76-29097.
1139.Cahen, O., Sigelle, R. and Trotel, J., "Automatic control of an electron beam pattern generator", Electron and Ion Beam Science and Technology Fifth International Conference, 1972, p. 92-101, EEA76-8719.
1140.Cantagrel, M., "Considerations on high resolution patterns engraved by ion etching", IEEE Trans. Electron Devices, ED-22, 483-6 (1975), EEA78-24401.
1141.Cantagrel, M., "Comparison of properties of different materials used as masks for ion beam etching", J. Vac. Sci. Technol., 12, 1340-3 (1975).

1142.Chandra, R. and Mueller, J. L., "Selection of mask sets by partial exhaustive search", IBM Tech. Disclosure Bull., 12, 1350 (1970), EEA73-27260.
1143.DiFazio, J. J., Giuffre, G. J. and Zingerman, J. R., "Defect-free irradiation mask", Patent German 2419947, Publ. November 1974, CA82-178253.
1144.Frankson, G. D., "Ultraviolet-sensitive template", Patent USA 3510371, Publ. May 1970, CA73-30082.
1145.Franz, I. and Langhrich, W., "Silicon nitride as a mask in indium-diffusion", Solid-State Electron., 12, 63-4 (1969), EEA72-6211.
1146.Gibbon, C. F., Povilonis, E. I. and Ketchow, D. R., "Effect of mask edges on dopant diffusion into semiconductors", J. Electrochem. Soc., 119, 767-72 (1972), CA77-26266.
1147.Glashauser, W., "Congruent mask patterns on both sides of a foil", Patent German 2333284, Publ. January 1975, CA82-148541.
1148.Goldrick, M. R., Kerwin, R. E. and Skinner, J. G., "Laminar photomask", Patent USA 3674492, Publ. July 1972, CA78-22572.
1149.Gutteridge, C., Jr., "Method of computing the mask characteristics required for accurate color reproduction in photomechanical processes", Photogr. Sci. Eng., 16, 214-20 (1972), CA76-160774.
1150.Haens, J. D., "Dimensional stability of film masters for printed circuit production", Proceedings of the EIPC Conference on Printed Circuit Manufacturing, 1970, p. 1-6, EEA74-4315.
1151.Heimer, R. J., "Influence of process parameters on the final quality of photomasks", Topical Meeting on the Use of Optics in Microelectronics, 1971, 3 pp., EEA74-25605.
1152.Kinamari, K., "Relation between photomask and photoresist pattern", Bull. Jap. Soc. Precis. Eng., 7 (3), 72-6 (1973), EEA77-14570.
1153.Kloetzer, S., "Camera-sensitive photomasks", Patent German 2366278, Publ. August 1974, CA82-9986.
1154.Lichauco, R. J., Tuck, G. and Weed, D. R., "Cleaning process for semiconductor masks", IBM Tech. Disclosure Bull., 27, 1310 (1974), EEA78-8589.
1155.Kossyk, G. J. W., Laico, J. P., Rongred, L. and Stafford, J. W., "The primary pattern generator. II. Mechanical Design", Bell Syst. Tech. J., 49, 2043-60 (1970), EEA74-9797.
1156.Lepselter, M. P. and MacRae, A. U., "Self-supporting masks of high resolving power", Patent German 2162232, Publ. July 1972, CA77-119363.
1157.Loprest, F. J. and Barr, D. E., "Transparent photographic masks", Patent French 2096224, Publ. March 1972, CA78-117641.
1158.Makosch, G. and Schedewie, F., "New measurement and adjustment methods in mask technology", Presented at 4th International Microelectronics Conference, Munich, Germany, November 1970, German, EEA74-12771.

1159. McCreight, C. R. and Tien, C. L., "Interpretation problems in the correlation-mask sensing technique", Joint Conf. Sensing Environ. Pollutants, Collect. Tech. Pap., 1971, 5 pp., Publ. 1972, CA77-24310.

1160. Panteleev, V. A., Kamaeva, G. V. and Urman, P. N., "Diffusion of zinc over the gallium arsenide protecting mask grain boundary during selective doping", Izv. Vyssh. Ucheb. Zaved., Fiz., 14 (5), 132-4 (1971), Russian, CA75-92435.

1161. Skvortsov, I. M. and Nikolaeva, V. V., "Interaction of Si and SiO_2 during heat treatment of masked silicon plates in hydrogen", Izv. Akad. Nauk SSSR Neorg. Mater., 8, 1587-90 (1972), Russian, EEA76-8519.

1162. Tolstykh, S. A. and Synorov, V. F., "Masking properties of silicon nitride films", In: Radiofiz. Mikroelektron., Synorov, V. F., (Ed.), Voronezh. Gos. Univ., Voronezh, USSR, p. 86-8 (1970), Russian, CA77-170305.

1163. Tolstykh, S. A. and Vyglovskii, V. M., "Properties of silicon nitride", In: Fiz. Poluprov. Mikroelektron., Synorov, V. F., (Ed.), Izd. Voronezh Univ., Voronezh, USSR, p. 42-5 (1972), Russian, CA78-90078.

8. Defects and Inspection

1164. Axelrod, N. N., "Intensity spatial filtering applied to defect detection in integrated circuit photomasks", Proc. IEEE, 60, 447-8 (1972).

1165. Bruning, J. H., Feldman, M., Kinsel, T. S., Sittig, E. K. and Townsend, R. L., "An automated mask inspection system - AMIS", IEEE Trans. Electron Devices, ED-22, 487-95 (1975), EEA78-24402.

1166. Denk, H. H. and Thomas, R. A., "Chrome mask defect detection", IBM Tech. Disclosure Bull., 17, 3330-1 (1975), EEA78-37572.

1167. Denny, J. M. and Thomas, F., "Removing defects and modifying photomasks (laser application)", Electron. Prod. Methods & Equip., 4 (4), 12, 14 (1975), EEA78-28849.

1168. DiFazio, J. J., Giuffre, G. J. and Zingerman, J. R., "Photomask repair", Patent USA 3823016, Publ. July 1974, CA81-113726.

1169. Druschel, W. O., "Mask overlay-checking means", Patent USA 3674487, Publ. July 1972, CA78-9402.

1170. Fehrs, D. L. and Cuthbert, J. D., "A summation-gated optical comparator for automated pattern inspection", 1973 IEEE/OSA Conference on Laser Engineering and Applications Digest of Technical Papers, p. 57, EEA77-10307.

1171. Frenzel, M., "Localization of masking and passivation flaws in masking layers on silicon", Krist. Tech., 7, 265-70 (1972), German, CA76-159901.

1172. Fromer, M., "Mask defect burst counter", IBM Tech. Disclosure Bull., 14, 160 (1971), EEA74-33260.

1173. Gaffney, D. T., "Rapid photomask measurement using video scanning time intervals", Solid State Technol., 15 (3), 35-40 (1972).

1174.Giedd, G. R. and Perkins, M. H., "Apparatus for forming circuit masks on photosensitized screens and inspecting the same and/ or inspecting substrates having conductive circuit patterns", Patent USA 3523495, Publ. August 1970, EEA74-9950.
1175.Kobayashi, H. and Will, P. M., "Detection of defects in chips or wafers by use of incoherent imaging system", IBM Tech. Disclosure Bull., 13, 3270-1 (1971), EEA74-29753.
1176.Koller, K. and Lauther, U., "AUTOPRUEF: a program for checking the geometry of masks designed for ICs", Siemens Forsch.- & Entwicklungsber., 2 (2), 98-100 (1973), German, EEA76-26134.
1177.Laming, F. P. and Wojcik, W., "Mask defect inspection by holograms", IBM Tech. Disclosure Bull., 14, 28-9 (1971), EEA74-33340.
1178.Mitchell, C. L. and Gould, J. M., "MAP: a user-controlled automated mask analysis program", 11th Design Automation Workshop, 1974, p. 107-18, EEA77-30364.
1179.Pugacz-Muraszkiewicz, I. J., "Detection of discontinuities in passivating layers on silicon by sodium hydroxide anisotropic etch", IBM J. Res. Develop., 16, 523-9 (1972), CA78-165954.
1180.Shutoh, M., Mita, Y. and Sugiyama, N., "Application of the graphic COM recorder to creation of LSI mask pattern drawings for checking", NEC Res. & Dev., no. 32, 79-85 (1974), EEA77-30397.
1181.Stevens, G. W. W., "Trouble-shooting in microphotographic mask fabrication", Trans. Inst. Metal Finishing, 47 (Pt. 3), 111-13 (1969), EEA73-3620.
1182.Stratford, R. I., "Computer aided checking of integrated circuit layout constraints", International Conference on Computer Aided Design, 1972, p. 45-50, EEA75-19266.
1183.Treble, D. P., "Dimensional checking of M.O.S.L.S.I. layouts", International Conference on Computer-Aided Design, 1972, p. 7-11.
1184.Varnell, G., "A high resolution photomask inspection system using intensity spatial filters", Presented at Topical Meeting on the Use of Optics in Microelectronics, Las Vegas, Nev., January 1971, EEA74-25607.
1185.Watkins, L. S., "Inspection of periodic patterns with intensity spatial filters", Solid State Technol., 12 (2), 35-8 (1969), EEA72-12795.
1186.Watkins, L. S., "Integrated circuit photomask inspection by spatial filtering", Solid State Technol., 12 (9), 29-32 (1969), EEA73-6738.
1187.Watkins, L. S., "Inspection of integrated circuit photomasks with intensity spatial filters", Proc. IEEE, 57, 1634-9 (1969), EEA73-465.
1188.Watkins, L. S., "Application of spatial filtering subtraction to thin film and integrated circuit mask inspection", Appl. Opt., 12, 1880-4 (1973), EEA76-35345.

1189. Winzer, G., Douklias, N. and Kraft, H. P., "Problems of fault diagnosis in masks for integrated semiconductor circuits by coherent-optical filtering", Frenquenz, 25, 220-6 (1971), German, EEA74-33338.

F. ION IMPLANTATION

I. Bibliographies

1. Agajanian, A. H., "Ion implantation - an annotated bibliography", Radiat. Eff., 23, 73-126 (1974), EEA78-2755.
2. Morgan, R. and Greenbalgh, K. R., "Ion implantation - a bibliography", BIB.176, AERE, Harwell, England, February 1972.
3. Plunkett, J. C. and Stone, J. L., "A selected bibliography on ion implantation in solid state technology", Solid State Technol., 18 (12), 49-55 (1975).
4. Tsurushima, T. and Tanoue, H., "Ion implantation and its applications - review of current research and development", Circ. Electrotech. Lab., no. 171, 1-122 (1970).

II. Books and Symposia

5. Brice, D. K., "Ion implantation range and energy deposition distributions: High incident ion energies, Vol. I", IFI Plenum, New York, 1975.
6. Crowder, B. L., (Ed.), "Ion implantation in semiconductors and other materials", Plenum Press, New York, 1973. (Proceedings of the Third International Conference).
7. Dearnaley, G., Freeman, J. H., Nelson, R. S. and Stephen, J., "Ion implantation", Series: Defects in Crystalline Solids, Elsevier, New York, 1973.
8. Eisen, F. H. and Chadderton, L. T., (Eds.), "Ion implantation", Gordon and Breach Science Publishers, New York, 1971. (Proceedings of the First International Conference).
9. Gibbons, J. F., Johnson, W. S. and Mylroie, S. W., "Projected range statistics, semiconductor and related materials", 2nd Edition, Dowden, Hutchinson & Ross, Stroudsburg, Pa., 1975.
10. Mayer, J. W., Eriksson, L. and Davies, J. A., "Ion implantation in semiconductors: silicon and germanium", Academic Press, New York, 1970.
11. Morgan, D. V., (Ed.), "Channeling theory, observation and applications", Wiley-Interscience, New York, 1973.
12. Namba, S., (Ed.), "Ion implantation in semiconductors: science and technology", Plenum Press, New York, 1975, (Proceedings of the Fourth International Conference).
13. Picraux, S. T., EerNisse, E. P. and Vook, F. L., (Eds.), "Applications of ion beams to metals", Plenum Press, New York, 1974.
14. Ruge, I. and Graul, J., (Eds.), "Ion implantation in semiconductors", Springer-Verlag, New York, 1971. (Proceedings of the Second International Conference).

15.Wilson, R. G. and Brewer, G. R., "Ion beams with applications to ion implantation", Wiley-Interscience, New York, 1973.
16.European Conference on Ion Implantation, held at Reading, Berks, England, September 1970; Proceedings were published by: Peter Perregrinus, Stevenage, Herts, England, 1970.
17.Symposium on Ion Sources and Formation of Ion Beams, Conference held at Upton, New York, October 1971; Proceedings were published by: AIP and BNL, New York, 1971.
18.U.S. - Japan Seminar on Ion Implantation in Semiconductors, Conference held at Kyoto, Japan, August 1971; Proceedings were published by: Japan Society for Promotion of Science, Tokyo, Japan, 1972.
19.1975 Particle Accelerator Conference, held at Washington, D. C., March 1975; Proceedings were published in: IEEE Trans. Nucl. Sci., NS-22, June 1975.

III. Review Articles

20.Bertolini, G., Cappellani, F. and Restelli, G., "Ion implantation in semiconductors", Euro-Spectra, 12 (3), 58-72 (1973), EEA77-10100.
21.Brown, W. L. and MacRae, A. U., "Ion implantation", Bell Lab. Record, 53, 389-94 (1975).
22.Cerny, F., "The technology of ion implantation", Sdelovaci Tech., 20 (8), 291-2 (1972), Czech, PA75-74789.
23.Dearnaley, G., "Ion implantation", In: 1974 Annual review of materials science vol. 4, Huggins, R. A., (Ed.), Annual Reviews, Palo Alto, Calif., p. 93-123 (1974), PA78-15488.
24.Dearnaley, G., "Ion implantation in semiconductors", Conference on Solid State Devices, 1974, p. 75-96, Publ. 1975, PA78-51136.
25.Dearnaley, G., "Ion implantation", Vide, 29, 401-6 (1974), PA78-66559.
26.Dearnaley, G., "Ion implantation", Nature, 256, 701-5 (1975), PA78-76705.
27.Delecroix, P., "Ion implantation: principles and applications", Electron. and Microelectron. Ind., 141, 31-7 (1971), PA74-44718.
28.Eckstein, D., "Ion implantation as a doping process in semiconductor technology", Funk-Tech., no. 15, 525-8 (1974), German, EEA78-536.
29.Fiorini, F. and Gislon, R., "Ion implantation: advantages, possibilities, limits, problems, realizations", Ind. Ital. Electrotec and Electron., 24, 479-83 (1971), Italian, PA74-61724.
30.Gemmell, D. S., "Channeling and related effects in the motion of charged particles through crystals", Rev. Mod. Phys., 46, 129-227 (1974), PA77-56786.
31.Gibson, W. M., "Particle channeling principles and applications: A brief personal review", Report BNL-50336, Bell Telephone Labs., Inc., Murray Hill, New Jersey, 1973, p. 331-403, NSA28-6365.

32.Hamilton, G. W., "Ion source technology", Symposium on Ion Sources and Formation of Ion Beams, 1971, p. 171-6, EEA75-8247.
33.Hlavka, J., "Ion implantation: possibilities and exploitation", Cesk. Cas. Fis. A, 25, 128-38 (1975), Czech, EEA78-37386.
34.Jensen, S. E., "Application of ion implantation for the doping of semiconductors", Phys. Norv., 6, 203 (1972), PA76-41675.
35.Kostka, A. and Strack, H., "Ion implantation - a new semiconductor technology", Int. Elektron. Rundsch., 25, 246-8 (1971), German, PA74-79729.
36.Krimmel, E. F., "Microelectronics enters new dimensions", Int. Elektron. Rundsch., 27 (3), 53-7 (1973), German, PA76-34487.
37.Mayer, J. W., "Ion implantation in semiconductors", 1973 International Electron Devices Meeting Technical Digest, p. 3-5, PA77-29024.
38.Mayer, J. W. and March, O. J., "Ion implantation in semiconductors", In: Applied solid state science, vol. 1, Wolfe, R., (Ed.), Academic Press, New York, p. 239-342 (1969).
39.McDermott, J., "Enter, the ion-implantation age, where 'impossibles' are possible", Electron. Des., 19 (4), 25-6, 28 (1971), PA74-36540.
40.McElhinney, J. and Wolick, E. A., "Nuclear sciences and technology at NRL", Rep. NRL Prog., 1-14 (1973), PA77-5701.
41.Mezei, F., "Ion implanted impurities in solid state", Magyar Fiz. Folyoirat, 19, 419-25 (1971), Hungarian, PA75-12782.
42.Morehead, F. F., Jr. and Crowder, B. L., "Ion implantation", Sci. Am., 228 (4), 64-71 (1973), EEA76-25984.
43.Nagy, J., "Ion implantation procedure", Hiradastechnika, 22 (3), 80-4 (1971), Hungarian, PA74-33322.
44.Nishizawa, J., "Ion implantation into semiconductors", J. Inst. Electron and Commun. Eng. Jap., 54, 378-92 (1971), Japanese, PA74-48470.
45.Prehn, P., "Ion implantation and silicon technology", Elektronik, 10 (10), 14 (1972), Danish, EEA76-4505.
46.Richardt, A., "Ion implantation. A method for doping semiconductors", Vide, 24, 272-4 (1969), French, PA73-34544.
47.Rohde, H. J. and Schwabe, S., "The implantation of ions and its use in the technique of semiconductors. I", Nachrichtentechnik, 21, 151 (1971), German, PA74-33323.
48.Rohde, H. J. and Schwabe, S., "The implantation of ions and its use in the technique of semiconductors. II", Nachrichtentechnik, 21, 302-3 (1971), German, PA74-69144.
49.Runge, H., "Ion implantation - a doping method with a future", Elektron. Anz., 5, 200-4 (1973), German, EEA77-14369.
50.Rzewuski, H., "Ion implantation in semiconductors", Postepy Fiz., 26, 49-68 (1975), EEA78-24217.
51.Schulz, M., "Ion implantation: a useful tool for semiconductor research", Appl. Phys., 4, 91-104 (1974), EEA77-33917.
52.Shimizu, K., "Fundamental study for ion beam applications", Res. Electrotech. Lab., no. 748, 1-78 (1974), Japanese, EEA78-43626.

53.Thompson, M. W., "Applications of ion implantation outside the semiconductor area", European Conference on Ion Implantation, 1970, p. 109-15, PA74-18766.
54.Toi, A., Yoshida, M. and Tachibana, K., "Ion implantation and its application to semiconductor technology", Natl. Tech. Rep., 21, 93-108 (1975), Japanese, EEA78-28646.
55.Urli, N., "Ion implantation technique in electronic component technology", Elektrotenika, 14 (1), 5-15 (1971), Croatian, PA74-40433.
56.Wilson, R. G., "The challenges for ion implantation technology", Electron and Ion Beam Science and Technology Fifth International Conference, 1972, p. 217-24, EEA76-8577.
57.Yonas, G., "Review of ion acceleration research at Physics International Company", IEEE Trans. Nucl. Sci., NS-19, 297-300 (1972), EEA75-21395.

IV. Theory

58.Akkerman, A. F. and Akkerman, S. A., "Calculation of the spatial distribution of defects and of the profile of implanted atoms in silicon irradiated with boron ions up to 100 keV energy", Sov. Phys.-Semicond., 8, 629-30 (1974), PA78-22508.
59.Ali, S. P. and Gallaher, D. F., "Electronic stopping power of channelled ions in a model incorporating the Pauli principle", J. Phys. C, 7, 2434-46 (1974), PA77-69836.
60.Beloshitski, V. V. and Kumakhov, M. A., "The dechanneling of ions", Fiz. & Khim. Obrab. Mater., no. 3, 39-46 (1973), Russian, PA76-72107.
61.Beloshitski, V. V., Kumakhov, M. A. and Muralev, V. A., "Multiple scattering of channeling ions in crystals. II. Planar channeling", Radiat. Eff., 20, 95-109 (1973), NSA29-27401.
62.Belyakov, V. A. and Orlov, V. P., "Cherenkov structure radiation of particle chanelled in crystal", Acta Crystallogr. A, A28 (Pt. 4), S238 (1972), PA76-26916.
63.Biersack, J. P., "Depth distributions of implanted atoms under the influence of surface erosion and diffusion", Radiat. Eff., 19, 249-56 (1973), PA77-9925.
64.Bourgoin, J., Peak, D. and Corbett, J. W., "Ionization-enhanced diffusion: ion implantation in semiconductors", J. Appl. Phys., 44, 3022-7 (1973), EEA76-32171.
65.Brice, D. K., "Recoil contribution to ion-implantation energy-deposition distribution", J. Appl. Phys., 46, 3385-94 (1975), PA78-85384.
66.Briggs, J. S. and Pathak, A. P., "Momentum transfer cross sections and the Z_1 oscillations in stopping power", J. Phys. C, 6, L153-7 (1973), PA76-36338.
67.Bulgakov, Y. V. and Shulga, V. I., "Application of the impact parameter method to be channeled ion energy distribution calculation", Phys. Lett. A, 48A, 429-30 (1974), PA77-75550.

68. Bullough, R. and Perrin, R. C., "Theory of void formation and growth in irradiated materials", 1971 International Conference on Radiation Induced Voids in Metals, p. 769-97, PA75-53701.
69. Campisano, S. U., Foti, G., Grasso, F. and Rimini, E., "Investigation of channel-to-random transition by means of single crystals overlaid with amorphous layers", 6th Yugoslav Symposium on Physics of Ionized Gases, 1971, p. 93-6, PA75-61127.
70. Campisano, S. U., Foti, G., Grasso, F., Savio, M. L. and Rimini, E., "Lindhard's multiple scattering description justifies axial and planar dechanneling data", Radiat. Eff., 13, 157-66 (1972), PA75-47788.
71. Chu, W. K., Moruzzi, V. L. and Ziegler, J. F., "Calculations of energy loss of ^{4}He ions in solid elements", J. Appl. Phys., 46, 2817-20 (1975), PA78-72502.
72. Corciovei, A. and Apostol, M., "Transfer matrix approach to axial hyperchanneling", Phys. Status Solidi B, 67, 609-17 (1975), PA78-29307.
73. Corciovei, A., Apostol, M. and Mihalache, D., "Planar channeling and transfer matrix technique", Phys. Lett. A, 44A, 259-60 (1973), PA76-52653.
74. Dearnaley, G., Wilkins, M. A. and Goode, P. D., "Non-Gaussian implantation profiles", 2nd International Conference on Ion Implantation in Semiconductors, 1971, p. 439-54, PA75-25437.
75. Desalvo, A. and Rosa, R., "Computer analysis of the axial-to-planar channeling transition: linkage with continuum theory", Phys. Rev. B, 9, 4605-12 (1974), PA77-72505.
76. Dettmann, K. and Robinson, M. T., "Stopping power of fast protons under channelling conditions", Phys. Rev. B, 10, 1-9 (1974), PA78-6476.
77. Edge, R. D., Varelas, C. and Sizmann, R., "Channeling and blocking along polar lattice directions", Radiat. Eff., 16, 95-9 (1972), NSA27-6795.
78. Etingof, E. I., "Theory of the channeling effect in fully ordered alloys", Metallofizika. Kiev, no. 38, 12-17 (1971), Russian, PA75-44003.
79. Etingof, E. I., "Channeling in partly ordered alloys and in metals with impurities", Metallofizika. Kiev, no. 28, 17-19 (1971), Russian, PA75-44004.
80. Filippov, G. M., Vodyanov, N. G. and Korneev, V. I., "Theory of the shadow effect", Sov. Phys.-JETP, 36, 404-6 (1973), PA76-75553.
81. Firsov, O. B., "The effect of crystal atomic chain discontinuity upon channeling", Radiat. Eff., 21, 265-7 (1974), PA77-72509.
82. Furukawa, S. and Ishiwara, H., "Range distribution theory based on energy distribution of implanted ions", J. Appl. Phys., 43, 1268-74 (1972), PA75-25707.
83. Furukawa, S., Matsumara, H. and Ishiwara, H., "Theoretical considerations on lateral spread of implanted ions", Jap. J. Appl. Phys., 11, 134-42 (1972), PA75-19082.

84.Gartner, K., Grosser, H., Hehl, K., Kolditz, M. and Scheliga, B., "Calculation of projected ranges of implanted ions", Phys. Status Solidi B, 64, K127-30 (1974), PA77-69842.
85.Gibbons, J. F., "On the effectiveness of dislocation loops as sites for the adsorption of implanted ions", Radiat. Eff., 6, 313-8 (1970), PA74-21448.
86.Gibbons, J. F., "Range-energy relations for protons in heavy substrates", Jap. J. Appl. Phys., 12, 804-6 (1973), PA76-52651.
87.Gibbons, J. F. and Mylroie, S., "Estimation of impurity profiles in ion-implanted amorphous targets using joined half-gaussian distributions", Appl. Phys. Lett., 22, 568-9 (1973), EEA76-25962.
88.Golanski, A., "Implantation of ions. I. Elements of theory of ranges", Nukleonika, 18, 335-50 (1973), Polish, PA77-56788.
89.Haymann, P., "Theoretical study of electronic stopping power in the planar channeling of ions", Surf. Sci., 49, 606-26 (1975), PA78-54232.
90.Hiraki, A., "Backscattering of high energy He^+ ions. I. Back-scattering method and channeling effects", Solid State Phys., 7, 303-13 (1972), Japanese, PA75-84224.
91.Ichikawa, M. and Ohtsuki, Y. H., "Inelastic-scattering theory for dechanneling", Phys. Rev. B, 10, 1129-34 (1974), PA78-3546.
92.Ishiwara, H. and Furukawa, S., "Theoretical considerations on range distribution of ions implanted into multilayer substrates", Electron. & Commun. Jap., 56, 114-21 (1973), PA77-44579.
93.Jackson, J. D. and McCarthy, R. L., "Z^3 corrections to energy loss and range", Phys. Rev. B, 6, 4131-41 (1972), PA75-1659.
94.Kelly, R. and Nghi, L. Q., "Diffusion in ion-implanted solids with non-uniform diffusion parameters", Radiat. Eff., 6, 247-56 (1970), PA74-21460.
95.Kesselman, V. S., "Calculation of the range of ions implanted in semiconductors", Fiz. Tekh. Poluprov., 4, 228 (1970), Russian, PA73-34543.
96.Kitagawa, M. and Ohtsuki, Y. H., "Modified dechanneling theory and diffusion coefficients", Phys. Rev. B, 8, 3117-23 (1973), PA77-20343.
97.Kitagawa, M. and Ohtsuki, Y. H., "Inelastic scattering of slow ions in channeling", Phys. Rev. B, 9, 4719-23 (1974), PA77-72507.
98.Krimmel, E. F., "Maximum range of axial channeled ions", Phys. Lett., 53A, 231-2 (1975), PA78-58089.
99.Krimmel, E. F. and Pfleiderer, H., "Implantation profiles modified by sputtering", Radiat. Eff., 19, 83-5 (1973), PA76-79400.
100.Kucherov, V. I., "Dynamic theory of the acceleration of ions in powerful electron beams", Pis'ma V Zh. Tekh. Fiz., 1, 235-8 (1975), Russian, EEA78-30654.
101.Kumakhov, M. A., "On some adiabatic invariant in channeling experiments", Phys. Lett. A, 45A, 121-2 (1973), PA76-72109.

102.Kumakhov, M. A., "The theory of channeling at small depths", Radiat. Eff., 26, 43-8 (1975).
103.Kumakhov, M. A. and Muralev, V. A., "Theoretical problems of ion beam interaction with crystals", Izv. Akad. Nauk SSSR Ser. Fiz., 37, 2585-9 (1973), Russian, PA77-20337.
104.Kutcher, G. J. and Mittleman, M. H., "Atomic physics of channeled particles", Phys. Rev. A, 11, 125-34 (1975), PA78-41826.
105.Lindhard, J. and Scharff, M., "Energy dissipation by ions in the keV region", Phys. Rev., 124, 128-30 (1961), PA64-16441.
106.Lindhard, J., Scharff, M. and Schiott, H., "Range concepts and heavy ion ranges", Mat. Fys. Medd. Dan. Vid. Selsk., 33 (14), 1-39 (1963).
107.Mannami, M., Kudo, H. and Oshiyama, T., "Dechanneling of fast ions at dislocation", J. Phys. Soc. Jap., 37, 287 (1974), PA77-72504.
108.Martynenko, Y. V., "Distribution of the penetration depths of enhanced particles", Fiz. Tver. Tela, 13, 2580-6 (1971), Russian, PA74-80072.
109.Matsumara, H. and Furukawa, S., "Theoretical considerations on lateral damage distribution formed by ion-implantation", Jap. J. Appl. Phys., 14, 1783-90 (1975).
110.Mezey, G., "Channelling", Magy. Fiz. Foly., 20, 297-321 (1972), Hungarian, PA76-7802.
111.Mikhlin, E. Y., "Nature of changes in the concentration of implanted ions near the surface of a target caused by changes in the direction of bombardment", Fiz. Tver. Tela, 12, 3667-8 (1970), Russian, PA74-21447.
112.Moak, C. D., Datz, S., Appleton, B. R., Biggerstaff, J. A., Brown, M. D., Krause, H. F. and Noggle, T. S., "Influence of ionic charge state on the stopping power of 27.8- and 40-MeV oxygen ions in the (011) channel of silver", Phys. Rev. B, 10, 2681-6 (1974), PA78-18339.
113.Morita, K. and Sizmann, R., "Dechanneling by interstitial atoms", Radiat. Eff., 24, 281-5 (1975), PA78-41830.
114.Nelson, R. S., "The theory of recoil implantation", Radiat. Eff., 2, 47-50 (1969), PA73-11685.
115.Ohtsuki, Y. H., "Complex continuum potential for channeling", J. Phys. Soc. Jap., 34, 473-5 (1973), PA76-20772.
116.Ohtsuki, Y. H. and Kitagawa, M., "'Abnormal' stopping power for low velocity ion channeling", Phys. Lett. A, 40A, 313-14 (1972), PA75-68268.
117.Ohtsuki, Y. H., Kitagawa, M. and Mizuno, M., "Abnormal stopping power of proton and ion channeling", Phys. Status Solidi A, 51, K19-21 (1972), PA75-44014.
118.Pathak, A. P., "The Z_1 oscillations in stopping power of metals for low velocity channelled heavy ions", J. Phys. F, 4, 1883-8 (1974), PA78-12783.
119.Pathak, A. P., "Energy loss of low-velocity heavy ions in planar channeling", J. Phys. C, 8, L341-5 (1975), PA78-76726.

120. Pathak, A. P., "Stopping power of energetic light ions in planar channeling", Phys. Status Solidi B, 71, K35-8 (1975), PA78-85388.
121. Pathak, A. P. and Yussouff, M., "Channeling of charged particles in perfect crystals", Radiat. Eff., 16, 1-6 (1972), PA76-11665.
122. Perloff, D. S., "Impurity atom distribution resulting from thermal redistribution of an implanted impurity source", J. Electrochem. Soc., 120, 1135-8 (1973), PA76-64793.
123. Pokhil, G. P. and Tulinov, A. F., "Shape of the planar shadow for scattering of particles by a thick single crystal", Sov. Phys.-Solid State, 14, 1669-72 (1973), PA76-20776.
124. Quere, Y., "Dechanneling of fast particles by lattice defects", J. Nucl. Mater., 53, 262-7 (1974), PA77-79080.
125. Quere, Y., "Temperature-dependent planar potential for channeling", Phys. Rev. B, 11, 1818-22 (1975), PA78-33213.
126. Roosendaal, H. E., Kool, W. H., Van Der Weg, W. F. and Sanders, J. B., "Critical angles and minimum yields for planar channeling", Radiat. Eff., 22, 89-99 (1974), PA77-72513.
127. Ryabov, V. A., "Quantum theory of inelastic scattering of channeled particles", Zh. Eksp. & Teor. Fiz., 67, 150-60 (1974), Russian, PA77-79087.
128. Sanders, J. B. and Winterbon, K. B., "Time dependence of radiation damage cascades", Radiat. Eff., 22, 109-15 (1974), PA77-72500.
129. Schimko, R., Schwarz, G. and Rogge, K., "Diffusion from implanted layers with consideration of the out-diffusion on the surface", Phys. Status Solidi A, 28, K163-6 (1975), PA78-50433.
130. Stoneham, A. M., "The theory of defects in irradiated semiconductors", International Conference on Radiation Effects in Semiconductors, 1970, p. 7-13, PA75-47364.
131. Sugiyama, H., "Stopping power and range tables for heavy ions. Fundamental data on radiation physics. II", Circ. Electrotech. Lab., no. 181, 1-72 (1974), Japanese, PA78-54228.
132. Tsurushima, T. and Tanoue, H., "Spatial distribution of energy deposited by energetic heavy ions in semiconductors", J. Phys. Soc. Jap., 31, 1695-1711 (1971), PA75-15773.
133. Vasilev, A. V., "Kinetics of the formation and annealing of radiation defects in crystals", Sov. Phys.-Semicond., 6, 522-7 (1972), NSA27-16410.
134. Vlasov, A. A. and Kuraev, V. N., "Temperature distribution in a beam in the presence of excited structures: Theory of channeling and the shadow effect in the statistical model of a crystal", Fiz. Astron., 13, 431-7 (1972), Russian, NSA27-13848.
135. Vlasov, A. A. and Kuraev, V. N., "Channeling theory and shadow effects in excited states of a crystal", Fiz. Astron., 13, 471-4 (1972), Russian, NSA27-13849.
136. Waho, T. and Ohtsuki, Y. H., "Diffusion coefficients in channeling", Radiat. Eff., 21, 217-19 (1974), PA77-72508.

137. Wilson, W. D. and Bisson, C. L., "Particle re-emission during irradiation", J. Nucl. Mater., 53, 154-61 (1974), PA77-79074.
138. Winterbon, K. B., "Heavy-ion range profiles and associated damage distributions", Radiat. Eff., 13, 215-26 (1972), PA75-47776.
139. Wu, C. P., Douglas, E. C. and Mueller, C. W., "Limitations of the CV technique for ion-implanted profiles", IEEE Trans. Electron Devices, ED-22, 319-29 (1975), PA78-47424.
140. Youssouff, M., "Formal theory of electronic stopping of channeled ions", Solid State Commun., 17, 119-25 (1975), PA78-68972.
141. Yudin, V. V., "Theory of atomic displacements in ion implantation", Sov. Phys.-Dokl., 18, 707-8 (1974), PA77-56791.
142. Yudin, V. V., "Analytical calculation of the vacancy distribution and the semiconductor amorphization dose during ion implantation", Oyo Buturi, 43 (Suppl.), 484-7 (1974), PA78-3549.

V. Materials

1. General

143. Abel, F., Amsel, G., Bruneaux, M. and Cohen, C., "A stochastic interpretation of the oscillations observed in the backscattering spectra obtained in planar channeling", C. R. Hebd. Seances Acad. Sci. B, 276, 267-70 (1973), French, PA76-27589.
144. Akilov, Y. Z. and Lenchenko, V. M., "Atomic displacement cascades in Ge and Si (machine modeling)", Sov. Phys.-Semicond., 8, 18-22 (1974), PA77-75548.
145. Alexander, R. B., Dearnaley, G., Morgan, D. V. and Poate, J. M., "The interpretation of channelled-ion beam measurements for foreign atom location in crystals", Phys. Lett. A, 32A, 365-6 (1970), PA73-62231.
146. Alexander, R. B., Dearnaley, G., Morgan, D. V., Poate, J. and van Vliet, D., "Lattice site location measurements and their interpretation", European Conference on Ion Implantation, 1970, p. 181-6, PA74-18711.
147. Allen, R. M., "Implantation of compound semiconductors", European Conference on Ion Implantation, 1970, p. 127-34, PA74-18768.
148. Andersen, T. and Ebbesen, A., "A sectioning technique for sodium chloride single crystals and its application for ion implantation studies", Radiat. Eff., 11, 113-18 (1971), PA75-9594.
149. Andersen, J. U. and Laegsgaard, E., "Use of the channeling technique to locate interstitial impurities", Radiat. Eff., 12, 219-23 (1972), PA75-36051.
150. Andersen, N. and Sigmund, P., "Energy dissipation by heavy ions in compound targets", K. Dan. Vidensk. Selsk. Mat-Fys. Medd., 39 (3), 1-45 (1974), PA78-12784.
151. Anon., "Semiconductors with ion implants", Elettrificazione, no. 3, 132-6 (1973), Italian, EEA76-32165.

152.Anon., "Implantation and analysis by means of ionic beams", Ned. Tijdschr. Natuurk., 38 (19), 30-2 (1973), Dutch, EEA76-19684.
153.Anon., "Implantation of environmental atoms into solids by neutron collisions in reactors", Radiat. Eff., 18, 79-85 (1973), PA76-60272.
154.Baccarani, G. and Pickar, K. A., "Range and straggle of boron in photoresist (ion implantation)", Solid-State Electron., 15, 239-43 (1972), EEA75-13570.
155.Barrett, J. H., "Breakdown of the statistical-equilibrium hypothesis in channeling", Phys. Rev. Lett., 31, 1542-5 (1973), PA77-9935.
156.Beloshitsky, V. V. and Kumakhov, M. A., "On the rule of reversibility in channeling and blocking experiments", Phys. Lett. A, 47A, 217-8 (1974), PA77-32592.
157.Bernas, H. and Obert, J., "Isotope separator implantation of radioactive ions for hyperfine interaction studies", Nucl. Instrum. & Methods, 107, 423-8 (1973), PA76-32217.
158.Bjorklund, F. L. M. and Matzner, E., "Using ion implantation in combination with diffusion techniques", Patent USA 3892596, Publ. July 1975, CA83-171673.
159.Bjorkqvist, K. and Domeij, B., "Stopping powers of C, N and O ions in Cr, Fe, Co, Ni, Cu and Zn in the 1 MeV region", Radiat. Eff., 13, 191-6 (1972), PA75-47782.
160.Blank, P. and Wittmaack, K., "Range parameter distortion in heavy ion implantation", Phys. Lett. A, 54A, 33-4 (1975), PA78-81137.
161.Blewer, R. S., "Proton backscattering as a technique for light ion surface interaction studies in CTR materials investigations", J. Nucl. Mater., 53, 268-75 (1974), PA77-78250.
162.Boettinger, J., Davies, J. A., Morgan, D. V., Whitton, J. L. and Winterbon, K. B., "Damage profiles in ion implanted semiconductors at low (25°K) temperature", Proceedings of the 3rd International Conference on Ion Implanation in Semiconductors and Other Materials, 1972, p. 599-609, Publ. 1973, CA81-178840.
163.Boroffka, H., Krimmel, E. F. and Runge, H., "Method of fast change of dopants for ion implantation", Nucl. Instrum. & Methods, 107, 251-2 (1973), PA76-25941.
164.Bottiger, J. and Eisen, F. H., "On conversion from an energy scale to a depth scale in channelling experiments", Thin Solid Films, 19, 239-46 (1973), PA77-16536.
165.Brice, D. K., "Recoil contribution to ion-implantation energy-deposition distribution", J. Appl. Phys., 46, 3385-95 (1975).
166.Brown, W. L., "Analysis of crystalline defects by ion beam techniques", Proceedings of the International Conference on Radiation Damage and Defects in Semiconductors, 1972, p. 416-28, Publ. 1973, PA76-33577.

167. Bulgakov, Y. V., Nikolaev, V. S. and Shulga, V. I., "The experimental determination of the impact parameter dependence of inelastic energy loss of channeled ions", Phys. Lett. A, 45A, 477-8 (1974), PA77-28132.
168. Cahen, O. and Netange, B., "A new method of fast measurement of ion implantation profiles", European Conference on Ion Implantation, 1970, p. 192, PA74-18772.
169. Carter, G., Baruah, J. N. and Grant, W. A., "The collection of ions implanted in semiconductors. II. Range distributions derived from collection and sputter-etch curves", Radiat. Eff., 16, 107-14 (1972), PA76-11647.
170. Carter, G. and Grant, W. A., "Ion implantation of semiconductors, metals, and alloys", Report AFOSR-TR-74-104, Salford Univ., UK, 1973, 23 pp., NSA30-30088.
171. Cely, J., "Ion implantation: stopping and the range of ions in solids", Cesk. Cas. Fis. A, 24, 21-40 (1974), Czech, PA77-52797.
172. Chadderton, L. T., "Directional effects in radiation-induced damage: Ions", In: Physics of ionized gases, Kurepa, M. V., (Ed.), Belgrade; Inst. of Physics, p. 305-47 (1972), NSA28-8764.
173. Chemin, J. F., Mitchell, I. V. and Saris, F. W., "Lattice location of low-Z impurities in medium-Z targets using ion-induced X-rays. I. Analytical technique (with reference to P and S implants in Ge)", J. Appl. Phys., 45, 532-6 (1974), PA77-53612.
174. Chemin, J. F., Mitchell, I. V. and Saris, F. W., "Lattice location of low-Z impurities in medium-Z targets using ion-induced X-rays. II. Phosphorus and sulfur implants in germanium single crystals", J. Appl. Phys., 45, 537-41 (1974), PA77-53613.
175. Chu, W. K., Mayer, J. W., Nicolet, M. A., Buck, T. M., Amsel, G. and Eisen, F., "Principles and applications of ion beam techniques for the analysis of solids and thin films", Thin Solid Films, 17, 1-4 (1973), EEA76-32197.
176. Collins, L. E., O'Connell, P. A., Perkins, J. G., Pontet, F. R. and Stroud, P. T., "Effects produced by ion bombardment and implantation into thin films and surfaces", Nucl. Instrum. and Methods, 92, 445-9 (1971), PA74-40142.
177. Darwish, M. Y. and Luginbuhl, H. W., "Carrier mobility profiles for low-dose boron-implanted layers", Appl. Phys., Lett., 25, 390-1 (1974), PA78-16074.
178. Davies, J. A., "Location of implanted atoms within the unit cell of a crystal", European Conference on Ion Implantation, 1970, p. 172-80, PA74-18770.
179. Dearnaley, G., "Ion bombardment and implantation", Rep. On Prog. in Phys., 32, 409-92 (1969).
180. Dearnaley, G., "Ion penetration", European Conference on Ion Implantation, 1970, p. 162-71, PA74-18769.
181. Dearnaley, G., "Atomic-displacement mechanism in ion-bombarded semiconductors and insulators", Appl. Phys. Lett., 26, 499-501 (1975), PA78-58085.

182. Degen, P. L., "Ion implantation doping of compound semiconductors", Phys. Status Solidi A, 16, 9-42 (1973), PA76-30166.
183. Della Mea, G., Drigo, A. V., Mazzoldi, P. and Nardelli, G., "Disorder in implanted semiconductors: energy dependence and penetration depth", Phys. Status Solidi A, 4, 797-804 (1971), EEA74-22524.
184. Demichelis, F., Liscia, R. and Tartoolia, A., "Range of fission fragments in light solid materials", Nuovo Cimento B, 10B, 523-36 (1972), PA75-61125.
185. Desalvo, A., Rosa, R. and Zianini, F., "Computer studies of dechanneling effects on ion implantation profiles", Lett. Nuovo Climento, Ser. 2, 2, 390-4 (1971), PA71-69145.
186. Ducommun, J. P., Cantagrel, M. and Moulin, M., "Evolution of well-defined surface contour submitted to ion bombardment: computer simulation and experimental investigation", J. Mater. Sci., 10, 52-62 (1975), PA78-29292.
187. Dunning, K. L., Hubler, G. K., Comas, J., Luck, W. H. and Hughes, H. L., "Depth profiles of aluminium and sodium near surfaces nuclear resonance method", Thin Solid Films, 19, 145-56 (1973), EEA77-2487.
188. Edge, R. D., Varelas, C. and Sizman, R., "Channeling and blocking along polar lattice directions", Radiat. Eff., 16, 95-9 (1972), PA76-11666.
189. Eisen, F. H., "New aspects of atom location: Flux peaking", 2nd International Conference on Ion Implantation in Semiconductors, 1971, p. 287-96, PA75-25435.
190. Furukawa, S. and Ishiwara, H., "Mass dependence of critical amorphizing dose in ion implantation", Jap. J. Appl. Phys., 11, 1062-3 (1972), PA75-56677.
191. Gaillard, M. J., Poizat, J. C. and Remillieux, J., "Oscillating trajectories in planar channeling studies in a double alignment configuration", Phys. Lett. A, 45A, 325-6 (1973), PA76-79398.
192. Gibbons, J. F., "Ion implantation in semiconductors. I. Range distribution theory and experiments", Proc. IEEE, 56, 295-319 (1968).
193. Gibbons, J. F., "Ion implantation in semiconductors. II. Damage production and annealing", Proc. IEEE, 60, 1062-96 (1972), EEA75-36160.
194. Gusev, V. M., "Streltsov, L. N. and Khaibullin, I. B., "Interference method for measuring the effective thickness of ion implanted layers", Fiz. and Tekh. Poluprov., 5, 832-35 (1971), Russian, PA74-52302.
195. Hafemeister, D. W. and de Waard, H., "Mossbauer studies of implanted I^{129} ions in semiconductors and alkali halides", Phys. Rev. B, 7, 3014-7 (1973), PA76-34261.
196. Hasegawa, S., "Electronic stopping power for energetic implanted ions in amorphous substrates", Mem. Fac. Technol. Kanazawa Univ., 7, 177-84 (1973), Japanese, PA77-32595.

197.Hendrickson, T. E., "Ion implanted profiles from two point spreading resistance measurements", J. Electrochem. Soc., 122, 1539-41 (1975).
198.Hernandez, R., Vidal, G., Lanusse, P. and Slodzian, G., "Surface analysis of ion implantations by secondary-emission mass spectrography", Mem. Sci. Rev. Metall., 70, 47-51 (1973), French, PA76-40946.
199.Hlavka, J., "Ion implantation: Measurement of ion pressure", Phys. Lett. A, 45A, 407-8 (1973), PA77-2369.
200.Hudson, J. A., Mazey, D. J. and Nelson, R. S., "Void formation during high dose ion implantation at elevated temperatures", European Conference on Ion Implantation, 1970, p. 231-4, PA74-18721.
201.Hunsperger, R. G., Wolf, E. D., Shifrin, G. A., Marsh, O. J. and Lamb, D. M., "Measurement of lattice damage caused by ion-implantation doping of semiconductors", Radiat. Eff., 9, 133-8 (1971), PA74-52232.
202.Hutson, A. R., "Moire patterns of ion implantation by channeling", J. Appl. Phys., 42, 1239-40 (1971), PA74-30578.
203.Jech, C., "Channeling patterns recorded using radioactive recoil atoms", Phys. Lett. A, 39A, 417-18 (1972), PA75-54088.
204.Kola-zhik, V., "Investigation of ion channeling by means of emission electron microscopy", Izv. Akad. Nauk SSSR Ser. Fiz., 38, 350-3 (1974), Russian, PA77-32589.
205.Kumakhov, M. A., "Double peak in channeling experiments", Phys. Lett. A, 39A, 191-2 (1972), PA75-44005.
206.Kumakhov, M. A., "Exact determination of the location of an impurity atom by means of the channeling effect", Sov. Phys. Dokl., 17, 348-51 (1972), PA76-1660.
207.Kurata, M., Morikawa, Y., Nagami, K. and Kuroda, H., "Remarks on the vacancy mechanisms in ion implantation", Jap. J. Appl. Phys., 12, 472-3 (1973), PA76-33537.
208.Lee, H. Y., "Low temperature diffusion by ion collisions", J. Korean Inst. Electron. Eng., 11, 33-9 (1974), Korean, EEA77-36277.
209.Legrand, C., Bahezre, C., LeDuigou, J. and Trillat, J. J., "On the use of electron microprobe for the study of ion implantation", C. R. Acad. Sci. Ser. B, 271, 88-92 (1970), French, PA73-62223.
210.Lewis, R. K., Morabito, J. M. and Tsai, J. C. C., "Primary oxygen ion implantation effects on depth profiles by secondary ion emission mass spectrometry", Appl. Phys. Lett., 23, 260-2 (1973), PA76-66439.
211.Litvinov, V. L., Sadikov, V. P., Smilga, V. P. and Ukhin, N. A., "Asymmetry of the damage regions generated by high-energy particles in semiconductors", Sov. Phys.-Semicond., 6, 460-1 (1972), NSA27-16409.

212.Maddock, A. G., "Chemical studies of ion implantation", Report CONF-730618, Chemical Labs., Cambridge, Eng., 1973, p. 57-74, NSA30-32876.
213.Mannami, M., "Channeling and blocking of charged particles in single crystals", Butsuri, 27, 571-83 (1972), Japanese, PA75-14900.
214.Manning, I. and Mueller, G. P., "Depth distribution of energy deposition by ion bombardment", Comput. Phys. Commun., 7, 85-94 (1974), PA77-36921.
215.Masuda, K., Gamo, K., Imada, A. and Namba, S., "Atom location in the case of enhanced diffusion measured by backscattering method", 2nd International Conference on Ion Implantation, 1971, p. 455-8, PA75-25438.
216.Matzke, H. J., "Ion bombardment induced phase transformations and inert gas mobility in the semiconductors Ge, Si and GaAs", Radiat. Eff., 3, 93-105 (1970).
217.Mayer, J. W., "Ion implantation in semiconductors: lattice disorder and electrical effects", IEEE Trans. Nuclear Sci., NS-15 (7), 10-21 (1968), EEA72-11189.
218.Mayer, J. W., "Ion implantation lattice disorder", Radiat. Eff., 8, 269-77 (1971), PA74-52301.
219.Mayer, J. W., "Ion implantation and analysis of thin film structures by backscattering of MeV particles", J. Vac. Sci. & Technol., 9, 39 (1972), PA75-39708.
220.Mayer, J. W., "Semiconductor physics and channeling", Radiat. Eff., 12, 183-91 (1972), PA75-36044.
221.Mayer, J. W., Mitchell, I. V. and Nicolet, M. A., "Analysis of contact formation and surface layers on semiconductors", 2nd International Conference on Ion Implantation in Semiconductors, 1971, p. 274-86, PA75-25877.
222.Meek, R. L., Gibson, W. M. and Sellschop, J. P. F., "Dimpling - a new manifestation of ion produced lattice damage", 2nd International Conference on Ion Implantation in Semiconductors, 1971, p. 297-8, EEA75-25368.
223.Mory, J., "Heavy particle channeling", Report CEA-CONF-2110, Centre d'Etudes Nucleaires, 1972, 26 pp., French, NSA27-13825.
224.Namba, S., Masuda, K., Gamo, K., Ohba, S. and Murakami, K., "ESR and electrical properties of ion implanted layer in semiconductors", Sci. Pap. Inst. Phys. Chem. Res., 65 (1), 22-5 (1971), PA74-62139.
225.Navinsek, B., "Investigation of surface changes induced by ion bombardment", In: Physics of ionized gases, Kurepa, M. V., (Ed.), Inst. of Physics, Belgrade, p. 221-52 (1972), NSA28-8762.
226.Nelson, R. S., "The clustering of defects in irradiated semiconductors", Proceedings of the International Conference on Radiation Damage and Defects in Semiconductors, 1972, p. 140-58, Publ. 1973, PA76-33576.
227.Nelson, R. S., "Metallurgical applications of ion implantation and ion bombardment", Vacuum, 23 (3), 79-84 (1973), PA76-64759.

228.Ordette, G. R., Schwartz, D. M. and Ardell, A. J., "Particle range and energy deposition in materials containing voids", Radiat. Eff., 22, 217-23 (1974), PA77-82210.

229.Pabst, H. J., "Ionization effects on damage yield in ion irradiation of semiconductors", Radiat. Eff., 25, 279-82 (1975), PA78-41828.

230.Phillips, D. and Pringle, J. P. S., "Surface effects in the measurement of range profiles by oxide dissolution", J. Electrochem. Soc., 120, 1067-8 (1973), PA76-68300.

231.Picraux, S. T., "Channeling in semiconductors and its application to the study of ion implantation", Report AD-689187, Calif. Inst. Technol., Pasadena, 1969, 123 pp., PA73-43933.

232.Prince, J. L. and Schwettmann, F. N., "Diffusion of boron from implanted sources under oxidizing conditions", J. Electrochem. Soc., 121, 705-10 (1974), EEA77-30177.

233.Roosild, S., Dolan, R. and Buchanan, B., "Semiconductor doping by high energy 1-2.5 MeV ion implantation", J. Electrochem. Soc., 115, 307-11 (1968), PA71-49263.

234.Schmid, K., Fischer, G., Muller, H. and Ryssel, H., "Experimental data about dechanneling and channel stopping power", Radiat. Eff., 23, 145-9 (1974), PA78-9757.

235.Schmid, K., Muller, H., Ryssel, H. and Ruge, I., "The investigation of lateral damage effects of ion-implanted layers by back-scattering techniques", Thin Solid Films, 19, 313-8 (1973), PA77-16525.

236.Schmidt, H. and Ewald, H., "Computer simulation of the penetration of high energy heavy ions through thin gold single crystals", Z. Naturforsch. A, 29A, 1442-8 (1974), German, PA78-1174.

237.Schulson, E. M. and Marsden, D. A., "An assessment of ion implantation damage using the scanning electron microscope", Radiat. Eff., 24, 195-8 (1975), PA78-37277.

238.Shockley, W., "Forming semiconductor devices by ionic bombardment", Patent USA 2787564, Publ. April 1957.

239.Skakun, N. A., Dikii, N. P., Matyash, P. P. and Svetashev, P. A., "Investigation of amorphous films on single crystals by elastic scattering of channeled accelerated ions", Sov. Phys.-Solid State, 15, 1970-2 (1974), PA77-57014.

240.Stroud, P. T., Collins, L. E., Perkin, G. J. and Stephens, K. G., "The doping of thin films by recoil atom implantation", European Conference on Ion Implantation, 1970, p. 116-9, PA74-18767.

241.Torrens, I. M., "Computer simulation in radiation damage studies", Comput. Phys. Commun., 5, 32-43 (1973), NSA28-3304.

242.Tsurushima, T. and Tanoue, H., "Spatial distribution of deposited energy and lattice defects in semiconductors bombarded with energetic heavy ions", Denki Shikensho Iho, 36, 495-515 (1972), Japanese, NSA27-13852.

243.Vakselj, M. and Navinsek, B., "Measurement of penetration profiles of energetic heavy ions in solids using (p,γ) reactions", 6th Yugoslav Symposium on Physics of Ionized Gases, 1972, p. 105-8, PA75-6112.
244.Wada, Y., Usui, H. and Ashikawa, M., "Substrate temperature measurement during ion implantation", Jap. J. Appl. Phys., 14, 1351-6 (1975), PA78-86402.
245.Whitton, J. L., "Measurement of range profiles", In: Physics of ionized gases, Kupita, M. V., (Ed.), Inst. of Physics, Belgrade, p. 283-304 (1972), NSA28-8763.
246.Whitton, J. L., Carter, G., Baruah, J. N. and Grant, W. A., "The collection of ions implanted in semiconductors. I. Saturation effects", Radiat. Eff., 16, 101-5 (1972), PA76-11646.
247.Williams, J. S., "Range and stopping power effects obtained from high resolution Rutherford backscattering analysis of implanted targets", Radiat. Eff., 22, 211-13 (1974), PA78-1173.
248.Wittmaack, K., Schulz, F. and Maul, J., "Non-Gaussian range profiles in amorphous solids", Phys. Lett. A, 43A, 477-8 (1973), PA76-40934.
249.Wu, C. P., Douglas, E. C. and Mueller, C. W., "Limitations of the CV technique for ion-implanted profiles", IEEE Trans. Electron Devices, ED-22, 319-28 (1975), EEA78-19629.
250.Zander, K., "Interaction of radiation with condensed matter with special regard to semiconductors", Nachrichtentech. Z. (NTZ), 26, 305-11 (1973), German, EEA76-32191.
251.Ziegler, J. F. and Chu, W. K., "The stopping of ^{4}He ions in elemental matter", Thin Solid Films, 19, 281-7 (1973), PA77-16539.
252.Zielinska-Rohozinska, E., Lefeld-Sosnowska, M., Gronkowski, J. and Krylow, J., "Investigation of X-ray diffraction contrast asymmetries of the implanted region-nonimplanted region boundary", Phys. Status Solidi A, 20, 93-9 (1973), PA77-6416.

2. Group IV Semiconductors
a. Silicon

253.Ablyaev, S. A., Chirva, V. P. and Arsenin, V. V., "Influencc of ion bombardment on physical and physicochemical properties of silicon", In: Secondary emission and structural properties of solids, Arifov, U. A., (Ed.), Plenum Publishing Co., New York, p. 65-7 (1971), NSA27-13853.
254.Adams, J. R. and Bashara, N. M., "Determination of the complex refractive index profiles in P^+_{31} ion implanted silicon by ellipsometry", Surf. Sci., 49, 441-58 (1975), EEA78-28653.
255.Akasaka, Y. and Horie, K., "Precipitation of boron atoms implanted in silicon as detected by channeling analysis", J. Appl. Phys., 44, 3372-4 (1973), EEA76-32173.
256.Akasaka, Y., Horie, K. and Kawazu, S., "Lateral spread of boron ions implanted in silicon", Appl. Phys. Lett., 21, 128-9 (1972), EEA75-36112.

257. Akasaka, Y., Horie, K., Yoneda, K., Sakurai, T., Nishi, H., Kawabe, S. and Tohi, A., "Depth distributions of defects and impurities in 100-keV B^+ ion implanted silicon", J. Appl. Phys., 44, 220-4 (1973), PA76-17890.
258. Allen, C., Bicknell, R. W., Newman, R. C., Peart, R. F. and Stirland, D. J., "Beam-induced impurity displacement during channeling measurements on phosphorus diffused silicon", Phys. Status Solidi A, 22, K171-5 (1974), PA77-49620.
259. Allen, C. R. and Thomas, C. R., "Detection of phosphorus in heavily diffused silicon by He^+ back-scattering", Electron. Lett., 9, 475-7 (1973), EEA77-2449.
260. Altman, M. R., Feldman, L. C. and Gibson, W. M., "Dechanneling of 5 MeV protons from planar channels in silicon and its temperature dependence", Radiat. Eff., 18, 171-80 (1973), PA76-60753.
261. Andersson, A. and Swenson, G., "Annealing experiments on silicon layers implanted with 50 keV phosphorous ions", European Conference on Ion Implantation, 1970, p. 65-9, PA74-19175.
262. Andersson, A. and Swenson, G., "Additional ion bombardment of well annealed silicon crystals implanted with phosphorous and arsenic", 2nd International Conference on Ion Implantation in Semiconductors, 1971, p. 110-18, PA75-25434.
263. Andersson, A. and Swenson, G., "Electrical measurements on silicon layers implanted with 50 keV phosphorus ions", Radiat. Eff., 15, 231-41 (1972), EEA76-692.
264. Astrup, G., Johansen, A., Olsen, J. S. and Sarhoft-Kristensen, L., "Mobility, resistivity and carrier concentration measured on silicon implanted with channeled and nonchanneled indium ions", 2nd International Conference on Ion Implantation in Semiconductors, 1971, p. 473-84, EEA75-13606.
265. Auleytner, J., Drwiega, M., Furmanik, Z., Godwod, K., Krylow, J., Lazarski, S., Maciaszek, M. and Maydell-Ondrusz, E., "Microscope and X-ray investigations of the implanted silicon monocrystals", Elektronika, 15, 173-7 (1974), Polish, PA77-70416.
266. Auleytner, J., Furmanik, Z., Godwod, K. and Krylow, J., "X-ray diffraction effects in the case of implanted silicon single crystals", Acta Phys. Pol. A, A43, 507-12 (1973), PA76-38522.
267. Bader, R. and Kalbitzer, S. K., "Low energy boron and phosphorous implants in silicon: II. Doping profiles", Radiat. Eff., 6, 211-6 (1970), PA74-21455.
268. Baeri, P., Campisano, S. U., Ciavola, G., Foti, G. and Rimini, E., "Influence of implanted dose on the recrystallization of Si amorphous layers", Appl. Phys. Lett., 26, 154-5 (1975), PA78-42896.
269. Baker, J. F. C. and Ogden, R., "Crystal imperfections in silicon epitaxial layers grown on ion-implanted substrates", J. Mater. Sci., 10, 1259-61 (1975), EEA78-33027.

270.Baldo, E., Cappellani, F. and Restelli, G., "Annealing behaviour of arsenic implants in silicon", Radiat. Eff., 19, 271-5 (1973), PA77-12963.

271.Baranova, E. K., "Infrared absorption investigation of radiation defects in ion-doped silicon", Sov. Phys.-Semicond., 6, 322-3 (1972), PA75-71457.

272.Baranova, E. K., Gusev, V. M., Martynenko, Y. V., Starinin, C. V. and Haibullin, I. B., "On silicon amorphization during different mass ion implantation", Radiat. Eff., 18, 21-6 (1973), PA76-60754.

273.Baranova, E. K., Gusev, V. M. and Streltsov, L. N., "Distribution of defects with depth in silicon irradiated with 80 keV ions", Sov. Phys.-Semicond., 7, 1239-40 (1974), PA77-57719.

274.Barez, A., Rosinski, W., Turos, A., Wielunski, L. and Wojitowicz-Natanson, B., "Influence of preamorphization on concentration profiles of boron implanted into silicon", Phys. Status Solidi A, 27, K65-8 (1975), PA78-38392.

275.Barez, A., Turos, A., Wielunski, L., Rosinski, W. and Wojtowicz-Natanson, B., "Depth distribution of silver ions implanted in silicon and silica", Radiat. Eff., 25, 91-6 (1975), CA83-156707.

276.Baruch, P., Bouvatier, M., Picard, C., Epelboin, Y., Monnier, J. and Netange, B., "Diffusion accelerated by irradiation in silicon", Report CEA-CONF-2535, CEA Centre d'Etudes Nucleaires de Grenoble, France, 1973, 18 pp., French, NSA29-24800.

277.Batavin, V. V., Kuznetsov, Y. N., Perezhogin, G. A., Pantuev, V. S. and Popova, G. V., "Application of radiation doping in fabrication of uniform standard silicon specimens", Ind. Lab., 38, 1041-2 (1972), PA76-47353.

278.Bauer, L. O., "Enhanced annealing effects in boron-implanted layers in silicon by postimplantation of silicon ions", Appl. Phys. Lett., 20, 107-9 (1972), EEA75-10111.

279.Bauer, L. O., MacPherson, M. R., Robinson, A. T., and Dill, H. G., "Properties of silicon implanted with boron ions through thermal silicon dioxide", Solid-State Electron., 16, 289-300 (1973), EEA76-16330.

280.Bayadilov, E. M., Vavilov, V. S., Katsaurov, L. N., Krasnopevtsev, V. V., Milyutin, Y. V. and Nusupov, K. K., "Heat-treatment-induced redistribution of the concentration of implanted Li ions in silicon", Sov. Phys.-Semicond., 7, 1225-6 (1974), PA77-57718.

281.Beezhold, W. and Brower, K. L., "Electron paramagnetic resonance of the lattice damage in boron-implanted intrinsic silicon", IEEE Trans. Nucl. Sci., NS-20, 209-13 (1973), PA77-41501.

282.Beloshitsky, V. V., Bulgakov, Y. V. and Kumakhov, M. A., "Dechanneling of fast light ions", Phys. Lett. A, 40A, 181-2 (1972), PA75-61120.

283. Beloshitsky, V. V. and Kumakhov, M. A., "Use of the channeling technique and the theory of the flux peaking effect to determine the location of B in Si", Radiat. Eff., 25, 167-73 (1975), PA78-85390.
284. Bergamini, P., Fabri, G. and Pandarese, F., "Hall-effect measurements on indium-implanted silicon", Appl. Phys. Lett., 17, 18-20 (1970), PA73-59229.
285. Bertolotti, M., Sette, D., Stagni, L. and Vitali, G., "Electron microscope observation of lattice disorder in ion-implanted silicon", Appl. Phys. Lett., 18, 257-9 (1971), PA74-36489.
286. Bertolotti, M., Sette, D., Stagni, L. and Vitali, G., "The damage produced around ion tracks in implanted silicon", Proceedings of the International Conference on Radiation Damage and Defects in Semiconductors, 1972, p. 159-64, Publ. 1973, PA76-33588.
287. Bicknell, R. W., "Electrical and electron microscope studies of annealed ion-implanted silicon", European Conference on Ion Implantation, 1970, p. 57-60, PA74-18740.
288. Bicknell, R. W., "The annealing characteristics of phosphorous implanted silicon. I", Phil. Mag., 26, 273-86 (1972), PA75-64678.
289. Bicknell, R. W. and Allen, R. M., "Correlation of electron microscope studies with the electrical properties of boron implanted silicon", Radiat. Eff., 6, 45-9 (1970), PA74-21392.
290. Bilger, H. R., Nicolet, M. A. and Mayer, O., "Electrical properties of alkali implanted silicon", European Conference on Ion Implantation, 1970, p. 70-3, PA74-19176.
291. Bisaro, R. and Blet, G., "Influence of the thermal contact between a silicon (111) specimen and its substrate on the amorphisation phenomenon during medium energy phosphorus ion implantation", C. R. Hebd. Seances Acad. Sci. B, 279, 129-30 (1974), French, PA-7-75556.
292. Blamires, N. G., "The influence of ion bombardment damage on the annealing behaviour of boron implanted silicon", European Conference on Ion Implantation, 1970, p. 52-4, EEA71-10518.
293. Blamires, N. G., "The influence of ion beam current densities on the electrical properties of boron implanted silicon", 2nd International Conference on Ion Implantation in Semiconductors, 1971, p. 119-23, PA75-25868.
294. Blank, P. and Wittmaack, K., "The influence of implanted xenon on the sputtering yield of silicon", Radiat. Eff., 27, 29-34 (1975).
295. Blet, G., "Dosing the phosphorus in silicon and doped oxides by X-fluorescence", Rev. Tech. Thomson-CSF, 7, 47-63 (1975), French, EEA78-24222.
296. Blood, P., "Planar channelling in ion-implanted silicon", Phys. Status Solidi A, 25, K151-4 (1974), PA78-10549.

297.Blood, P., Dearnaley, G. and Wilkins, M. A., "The depth distribution of phosphorus ions implanted into silicon crystals", Radiat. Eff., 21, 245-51 (1974), PA77-73267.
298.Blood, P. and Wilkins, M. A., "The origin of non-Gaussian profiles in phosphorus-implanted silicon", J. Appl. Phys., 45, 5123-8 (1974), PA78-22492.
299.Borders, J. A. and Brower, K. L., "EPR of substitutional group V implants in silicon", Radiat. Eff., 6, 135-40 (1970), PA74-21445.
300.Botvin, V. A., Gorelkinskii, Y. V., Kudryashev, V. A. and Sigle, V. O., "Interaction of hydrogen atoms and radiation defects in the case of H^+-ion implantation in silicon", Sov. Phys.-Semicond., 8, 1049-50 (1975), PA78-38014.
301.Bourgoin, J. C., Morhange, J. F. and Beserman, R., "Amorphous layer formation in silicon by ion implantation", Radiat. Eff., 22, 205-8 (1974), CA82-10122.
302.Brack, K., Gorey, E. and Schwuttke, G. H., "Damage profiles in high energy carbon (+) ion bombarded silicon", Lattice Defects, 4, 109-21 (1973), CA79-98405.
303.Brack, K. and Schwuttke, G. H., "Damage profiles in high energy C^+ bombarded silicon (amorphous layer formation)", Acta Crystallogr., A, A28 (Pt. 4), S157 (1972), PA76-17887.
304.Brack, K. and Schwuttke, G. H., "Superlattice formation in annealed silicon after high-energy O^+-implantation", Lattice Defects in Semiconductors, 1974, p. 440-5, Publ. 1975, PA78-81130.
305.Brelot, A., Lagorsse, J. M., Dumetz, M. and Assemat, J. L., "Influence of defects on the diffusion of arsenic implanted in silicon", Vide, 29, 371-3 (1974), French, PA78-58223.
306.Brice, D. K., "Special distribution of energy deposited into atomic processes in ion-implanted silicon", Radiat. Eff., 6, 77-87 (1970), PA74-21439.
307.Brice, D. K., "Application of a stopping power formula to channeled ions", Radiat. Eff., 18, 13-6 (1973), PA76-60752.
308.Brodsky, M. H., Hartstein, A. M. and Morehead, F. F., "Fabrication of amorphous silicon/crystalline silicon heterojunctions by ion implantation", IBM Tech. Disclosure Bull., 18, 1211-12 (1975), PA78-81972.
309.Brower, K. L. and Beezhold, W., "Electron paramagnetic resonance of the lattice damage in oxygen-implanted silicon", J. Appl. Phys., 43, 3499-506 (1972), PA75-60780.
310.Brower, K. L. and Borders, J. A., "EPR of ion-implanted donors in silicon", Appl. Phys. Lett., 16, 169-72 (1970), PA73-31723.
311.Brower, K. L., Vook, F. L. and Borders, J. A., "Depth distribution of EPR centers in 400 keV O^+ ion-implanted silicon", Appl. Phys. Lett., 16, 108-10 (1970), PA73-30905.

312.Buck, T. M., Pickar, K. A., Poate, J. M. and Hsieh, C. M., "Gettering rates of various fast-diffusing metal impurities at ion-damaged layers on silicon", Appl. Phys. Lett., 21, 485-7 (1972), EEA76-4504.

313.Buck, T. M., Poate, J. M., Pickar, K. A. and Hsieh, C. M., "A Rutherford scattering study of the diffusion of heavy metal impurities in silicon to ion-damaged surface layers", Surf. Sci., 35, 362-79 (1973), EEA76-19851.

314.Bulgakov, Y. V., Kolomenskaya, T. I. and Kuznetsov, N. V., "Rate of removal of the majority carriers from silicon irradiated with heavy charged particles", Sov. Phys.-Semicond., 7, 1489-90 (1974), PA77-66950.

315.Bulgakov, Y. V., Kumakhov, M. A., Muralev, V. A. and Yatsenko, L. A., "Penetration of multicharged ions with energies in the MeV range into silicon. II. Distribution of implanted ions with depth", Mikroelektronika, 1, 270-3 (1972), Russian, CA79-98336.

316.Bulgakov, Y. V., Nikolaev, V. S. and Rogachev, Y. V., "Anomalous energy loss during passage of fast boron ions through plane channels in single-crystal silicon", Fiz. & Khim. Obrab. Mater., no. 5, 54-8 (1973), Russian, PA77-20335.

317.Bulgakov, Y. V., Nikolaev, V. S. and Shulga, V. I., "Impact parameter dependence of inelastic energy losses for He and N ions channeled in Si", Phys. Status Solidi A, 31, 341-50 (1975), PA78-85389.

318.Burkhardt, F., Mertens, A. and Wagner, C., "Concentration profiles of implanted phosphorus in silicon", Phys. Status Solidi A, 22, K45-7 (1974), PA74-32575.

319.Burr, P. and Whitehouse, J. E., "Electrical measurements on silicon following phosphorous and boron implantation and annealing", European Conference on Ion Implantation, 1970, p. 61-4, PA74-19174.

320.Cairns, J. A., Holloway, D. F. and Nelson, R. S., "Measurement of implanted boron concentration profiles in silicon by the use of heavy ion x-ray excitation", European Conference on Ion Implantation, 1970, p. 203-6, PA71-18785.

321.Cairns, J. A., Holloway, D. F. and Nelson, R. S., "Selective X-ray generation by heavy ions. II. Measurement of the concentration distribution of ion-implanted antimony in silicon by the use of selective heavy ions X-ray excitation", Radiat. Eff., 7, 167-72 (1971), PA74-40444.

322.Cairns, J. A., Nelson, R. S. and Briggs, J. S., "The use of ion-induced X-rays to investigate the concentration distribution and atom location of boron-implanted silicon", 2nd International Conference on Ion Implantation in Semiconductors, 1971, p. 299-306, PA75-25436.

323.Cappellani, F., Restelli, G. and Spinoni, L., "Isothermal annealing of arsenic-implanted silicon", J. Phys. C, 7, L50-53 (1974), NSA29-27408.

324. Cass, T. R. and Reddi, V. G. K., "Anomalous residual damage in silicon after annealing of through-oxide arsenic implantations", Appl. Phys. Lett., 23, 268-70 (1973), CA80-53617.
325. Cembali, F., Galloni, R., Mousty, F., Rosa, R. and Zignani, F., "Doping and radiation damage profiles of P^+ ions implanted in silicon along the (110) axis", Radiat. Eff., 21, 255-64 (1974), PA77-73268.
326. Cembali, F., Galloni, R. and Zignani, F., "Electrical activation processes of P^+ ions channeled along the (110) axis of silicon: effect of annealing on carriers profiles shape", Radiat. Eff., 26, 161-72 (1975).
327. Chadderton, L. T., "Nucleation of damage centres during ion implantation of silicon", Radiat. Eff., 8, 77-86 (1971), EEA74-22530.
328. Chadderton, L. T. and Eisen, F. H., "On the annealing of damage produced by boron ion implantation of silicon single crystal", Radiat. Eff., 7, 129-38 (1971), PA74-36493.
329. Chadderton, L. T. and Whitton, J. L., "On the annealing of damage produced by copper ion implantation of silicon single crystals", Radiat. Eff., 23, 63-6 (1974), EEA78-2754.
330. Cherki, M. and Kalma, A. H., "Interstitial defects in p-type silicon", IEEE Trans. Nucl. Sci., NS-16, 24-7 (1969), EEA78-14012.
331. Chou, S., "Investigation of diffused impurities in silicon using the ion backscattering technique", Thesis, Stanford Univ., Calif., 1971, 72 pp., Order No. 72-5899, EEA75-32550.
332. Chu, W. K., Muller, H., Mayer, J. W. and Sigmon, T. W., "Residual disorder in Si from oxygen recoils in annealed 'through-oxide' arsenic implants", Appl. Phys. Lett., 25, 297-9 (1974), PA78-12789.
333. Comer, J. J. and Rossild, S. A., "Electron microscope observations of precipitation in boron implanted silicon", Radiat. Eff., 25, 275-8 (1975).
334. Crowder, B. L., "The role of damage in the annealing characteristics of ion implanted Si", J. Electrochem. Soc., 117, 671-4 (1970), PA73-59233.
335. Crowder, B. L., "The influence of the amorphous phase on ion distributions and annealing behavior of group III and group V ions implanted into silicon", J. Electrochem. Soc., 118, 943-52 (1971), PA74-52306.
336. Crowder, B. L. and Fairfield, J. M., "High-dose implantation of P, As, and Sb in silicon: A comparison of room temperature implantations followed by a 550°C anneal and implantation conducted at 600°C", J. Electrochem. Soc., 117, 363-7 (1970), PA73-94256.
337. Crowder, B. L. and Title, R. S., "The distribution of damage produced by ion implantation of silicon at room temperature", Radiat. Eff., 6, 63-75 (1970), PA74-21393.

338. Davidson, S. M., "Electrical profiles of ion implanted silicon and their comparison with defect structures", 2nd International Conference on Ion Implantation in Semiconductors, 1971, p. 79-84, EEA75-13596.
339. Davidson, S. M., "Rapid profile measurements in ion implanted silicon", J. Phys. E., 5, 23-6 (1972), EEA75-10131.
340. Davidson, S. M. and Booker, G. R., "Damage produced by ion implantation in silicon", Radiat. Eff., 6, 33-43 (1970), PA74-18776.
341. Davies, A. L. M. and Carter, S., "Gas release studies of silicon implanted with low energy He^+ and Kr^+ ions", Radiat. Eff., 10, 227-33 (1971), PA74-82946.
342. Davies, D. E., "The implanted profiles of boron, phosphorous and arsenic in silicon from junction depth measurements", Solid-State Electron., 13, 229-37 (1970), PA73-25862.
343. Davies, D. E. and Roosild, S., "Irradiation defects and the electrical quality of ion implanted silicon", Solid-state Electron., 14, 975-83 (1971), EEA75-1052.
344. Davies, D. E. and Roosild, S., "Energy levels of defects in ion implanted silicon", 2nd International Conference on Ion Implantation in Semiconductors, 1971, p. 23-32, EEA75-13594.
345. Davies, J. A., Denhartog, J., Eriksson, L. and Mayer, J. W., "Ion implantation of silicon. I. Atom location and lattice disorder by means of 1.0-MeV helium ion scattering", Canad. J. Phys., 45, 4053-71 (1967), PA71-17414.
346. Davies, J. A., Eriksson, L. and Mayer, J. W., "Experimental evidence for interstitial In and Tl in ion-implanted silicon", Appl. Phys. Lett., 12, 255-6 (1968), EEA71-23240.
347. Davies, J. A., Foti, G., Howe, L. M., Mitchell, J. B. and Winterbon, K. B., "Polyatomic-ion implantation damage in silicon", Phys. Rev. Lett., 34, 1441-4 (1975), PA78-58088.
348. Dearnaley, G., Ward, G. A., Temple, W. and Wilkins, M. A., "Depth distribution of gallium ions implanted into silicon crystals", Appl. Phys. Lett., 27, 17-18 (1975), EEA78-32994.
349. Dearnaley, G., Wilkins, M. A., Goode, P. D., Freeman, J. H. and Guard, G. A., "The range distribution of radioactive ions implanted into silicon crystals", Conference on Atomic Collision Phenomena in Solids, 1969, p. 633-56, PA73-68858.
350. Della Mea, G., Drigo, A. V., Mazzoldi, P. and Nardelli, G., "Interaction between B atoms and defects produced by Hg bombardment in silicon", Appl. Phys. Lett., 16, 383-2 (1970), EEA73-28009.
351. Della Mea, G., Drigo, A. V., Mazzoldi, P., Nardelli, G. and Zannoni, R., "Hg ion implantation in silicon: lattice disorder created and Hg atom lattice location", Radiat. Eff., 3, 259-62 (1970), PA73-61945.

352. Della Mea, G., Drigo, A. V., Russo, S. L. and Mazzoldi, P., "Indirect determination of the energy loss of protons channeled in silicon single crystals", Atti Accad. Naz. Lincei, Rend. Cl. Sci. Fis. Mat. & Nat., 52, 727-33 (1972), PA76-64077.
353. Della Mea, G., Drigo, A. V., Russo, S. L., Mazzoldi, P., Bentini, G. G., Desalvo, A. and Rosa, R., "Axial- to planar-channeling transition", Phys. Rev. B, 7, 4029-41 (1973), PA76-38602.
354. Demichelis, F., Liscia, R. and Tartaglia, A., "Channeling of fission fragments in silicon", Lett. Nuovo Cimento, 5, 918-20 (1972), PA76-7801.
355. Dennis, J. R. and Hale, E. B., "Model correction for the formation of amorphous silicon by ion implantation (ESR data)", Radiat. Eff., 19, 67-8 (1973), PA76-68302.
356. Dennis, J. R., Woodward, G. K. and Hale, E. B., "Vacancy motion in ion-implanted silicon", Conference on Lattice Defects in Semiconductors, 1974, p. 467-73, Publ. 1975, PA78-81131.
357. Desalvo, A., Rosa, R. and Zignani, F., "Computer evaluation of primary deposited energy profiles in ion-implanted silicon under channeling conditions", J. Appl. Phys., 43, 3755-60 (1972), PA75-68017.
358. Desalvo, A., Rosa, R. and Zignani, F., "Dose dependence of penetration and damage profiles of P^+-channeled ions in silicon simulated by computer", Radiat. Eff., 27, 89-96 (1975).
359. Domeij, B., Fladda, G. and Johansson, N. G. E., "Anomalously high yields of elastically scattered ^{12}C-ions from Zr, Hf, Tl and Hg atoms implanted into silicon", Radiat. Eff., 6, 155-69 (1970), PA74-21446.
360. Donovan, R. P., Littlejohn, M. A. and Roosild, S. A., "Piezoresistive properties of ion implanted layers in silicon", 2nd International Conference on Ion Implantation in Semiconductors, 1971, p. 362-72, PA75-25880.
361. Drum, C. M., "Diffusion of ion implanted antimony and arsenic in silicon", Presented at Electrochemical Society Meeting (abstracts only), San Francisco, Calif., May 1974, PA77-49609.
362. EerNisse, E. P., "Investigation of ion implantation damage with stress measurements", 2nd International Conference on Ion Implantation in Semiconductors, 1971, p. 17-22, PA75-25379.
363. EerNisse, E. P., "Sputtering and strain of silicon by ion implantation", J. Appl. Phys., 42, 480-2 (1971), PA74-18775.
364. Eichhorn, F., "Neutron and X-ray diffraction investigations of silicon implanted by phosphorus ions", Phys. Status Solidi A, 30, 155-62 (1975), PA78-66555.
365. Eisen, F. H. and Bottiger, J., "Interpretation of channeling effect measurements of disorder in ion-implanted silicon", Appl. Phys. Lett., 24, 3-5 (1974), PA77-36923.
366. Eisen, F. H. and Welch, B., "Disorder produced during ^{11}B and ^{12}C implantation in silicon", European Conference on Ion Implantation, 1970, p. 227-30, PA74-18786.

367.Eisen, F. H. and Welch, B., "Flux and fluence dependence of disorder produced during implantation of ^{11}B in silicon", Radiat. Eff., 7, 143-8 (1971), PA74-36494.
368.Eisen, F. H., Welch, B., Westmorland, J. E. and Mayer, J. W., "Lattice disorder produced in silicon by boron ion implantation", Conference on Atomic Collision Phenomena in Solids, 1970, p. 111-27, PA73-68339.
369.Eklund, K. H. and Andersson, A., "Amorphization of silicon crystals bombarded by 30 keV phosphorus ions at different temperatures", 2nd International Conference on Ion Implantation in Semiconductors, 1971, p. 103-9, PA75-25381.
370.Eltekov, V. A., Karpuzov, D. S., Martynenko, Y. V., Rubakha, E. A., Simonov, V. A. and Yurasova, V. E., "Computer studies of boron ion channeling in silicon single crystals", Radiat. Eff., 13, 237-42 (1972), PA75-47439.
371.Eriksson, L., Davies, J. A., Johansson, N. G. E. and Mayer, J. W., "Implantation and annealing behavior of group III and V dopants in silicon as studied by the channeling technique", J. Appl. Phys., 40, 842-54 (1969), PA73-12142.
372.Eriksson, L., Davies, J. A. and Mayer, J. W., "Ion implantation studies in silicon", Science, 163, 627-33 (1969), PA73-4715.
373.Fair, R. B., "Cooperative effects between arsenic and boron in silicon during simultaneous diffusions from ion implanted and chemical source predepositions", Solid-State Electron., 17, 17-24 (1974), PA77-17185.
374.Fair, R. B. and Pappas, P. N., "Diffusion of ion-implanted B in high concentration P- and As-doped silicon", J. Electrochem. Soc., 122, 1241-4 (1975).
375.Fair, R. B. and Tsai, J. C. C., "The diffusion of ion-implanted arsenic in silicon", J. Electrochem. Soc., 122, 1689-96 (1975).
376.Feldman, L. C. and Rodgers, J. W., "Depth profiles of the lattice disorder resulting from ion bombardment of silicon single crystals", J. Appl. Phys., 41, 3776-82 (1970), PA73-68349.
377.Feuerstein, A., Kalbitzer, S. and Oetzmann, H., "Range parameters of heavy ions at 10 and 35 keV in silicon", Phys. Lett. A, 51, 165-6 (1975), PA78-41824.
378.Fiderkiewicz, A. and Golanski, A., "Directional effect for 1-MeV protons in a silicon single crystal", Nukleonika, 17, 369-70 (1972), Polish, NSA27-21505.
379.Fite, C. B. and Grannemann, W. W., "Experimental study of ultra-high fluence oxygen ion implantation in silicon", Report AD-760350, Bur. Eng. Res., Univ. New Mexico, Albuquerque, N. Mex., 1973, 77 pp., CA79-98411.
380.Fladda, G., Bjorkquist, K., Eriksson, L. and Sigurd, D., "The lattice location of boron ion implanted into silicon", Appl. Phys. Lett., 16, 313-5 (1970), PA73-43943.

381.Fladda, G., Mazzoldi, P., Rimini, E., Sigurd, D. and Eriksson, L., "Evidence of a replacement reaction between ion implanted substitutional Ti dopants and interstitial Si atoms", Radiat. Eff., 1, 249-56 (1969), PA73-7874.

382.Foster, C., Kool, W. H., Van der Weg, W. F. and Roosendaal, H. E., "Random stopping power for protons in silicon", Radiat. Eff., 16, 139-40 (1972), PA75-11667.

383.Frank, W., "Interstitial properties deduced from internal friction measurements on boron-implanted silicon", Radiat. Eff., 21, 119-33 (1974), PA77-66909.

384.Frank, W. F. J. and Berry, B. S., "Lattice location and atomic mobility of implanted boron in silicon", Radiat. Eff., 21, 105-111 (1974), PA77-67740.

385.Frey, H. and Mittenbacher, J., "Damage stimulated annealing behaviour of conductivity within boron implanted silicon layers", Exp. Tech. Phys., 23, 275-82 (1975), German, PA78-72879.

386.Fujimoto, F. and Komaki, K., "Displacement of arsenic atoms in silicon crystal during irradiation", Phys. Status Solidi A, 12, K7-8 (1972), PA75-60820.

387.Furukawa, S. and Matsumura, H., "Backscattering study on lateral spread of implanted ions", Appl. Phys. Lett., 22, 97-8 (1973), EEA76-12432.

388.Furukawa, S., Matsumura, H. and Ishiwara, H., "Backscattering study of heavy-ion distribution in semiconductors", Thin Solid Films, 19, 399-406 (1973), EEA77-10129.

389.Gamo, K., Iwaki, M., Kawasaki, K., Masuda, K., Namba, S., Ishihara, S. and Kimura, I., "Enhanced diffusion of ion implanted antimony in silicon", Annu. Rep. Res. Reactor Inst. Kyoto Univ., no. 4, 9-16 (1971), PA75-29192.

390.Gamo, K., Iwaki, M., Masuda, K., Ishihara, S., Kimura, I. and Namba, S., "Concentration profiles and enhanced diffusion of ion implanted impurities into silicon", Sci. Pap. Inst. Phys. Chem. Res. (Japan), 65 (1), 19-21 (1971), PA74-61729.

391.Gamo, K., Iwaki, M., Masuda, K., Namba, S., Ishihara, S. and Kimura, I., "Concentration properties of room temperature ion implanted indium in silicon", Jap. J. Appl. Phys., 10, 523-5 (1971), PA75-36557.

392.Gamo, K., Iwaki, M., Masuda, K., Namba, S., Ishihara, S. and Kimura, I., "Enhanced diffusion and electrical properties of ion implanted silicon", 2nd International Conference on Ion Implantation in Semiconductors, 1971, p. 459-65, PA75-25439.

393.Gamo, K., Iwaki, M., Masuda, K., Namba, S., Ishihara, S., Kimura, I., Mitchell, I. V., Ilic, G., Whitton, J. L. and Davies, J. A., "Enhanced diffusion and lattice location of indium and gallium implanted in silicon", Jap. J. Appl. Phys., 12, 735-41 (1973), PA76-46727.

394.Gamo, K., Masuda, K., Namba, S., Ishihara, S. and Kumara, I., "Enhanced diffusion of high-temperature ion-implanted antimony into silicon", Appl. Phys. Lett., 17, 391-3 (1970), PA74-12741.

395. Gerasimenko, N. N., Dvurechenskii, A. V., Romanov, S. I. and Smirnov, L. S., "Interaction of defects and impurities in silicon containing implanted ions", Sov. Phys.-Semicond., 6, 1969-4 (1973), PA76-40910.
396. Gerasimenko, N. N., Dvurechenskii, A. V., Romanov, S. I. and Smirnov, L. S., "Effects of large doses of implanted ions on a semiconductor", Sov. Phys.-Semicond., 7, 1461-4 (1974), PA77-66949.
397. Gerasimenko, N. N., Dvurechenskii, A. V. and Smirnov, L. S., "Paramagnetic centers generated by ion bombardment of silicon", Sov. Phys.-Semicond., 6, 965-8 (1972), NSA27-16406.
398. Gerasimenko, N. N., Dvurechenskii, A. V., Smirnov, L. S. and Shirokov, E. G., "On amorphisation of ion implanted crystals", Cryst. Lattice Defects, 2, 125-6 (1971), PA74-66180.
399. Gerasimov, A. I., Zorin, E. I., Pavlov, P. V. and Tetelbaum, D. I., "On the peculiarities of silicon amorphization at ion bombardment", Phys. Status Solidi A, 12, 679-85 (1972), PA75-71456.
400. Gerward, L., "X-ray study of lateral strains in ion-implanted silicon", Z. Phys., 259, 313-22 (1973), PA76-38603.
401. Gibbons, J. F., "Acceptor profiles obtained by diffusive redistribution of implanted impurities during annealing", Appl. Phys. Lett., 23, 49-51 (1973), EEA76-2-157.
402. Gittins, R. P., Morgan, D. V. and Dearnaley, G. J., "Application of the ion microprobe analyser to the measurement of the distribution of boron ions implanted into silicon crystals", J. Phys. D, 5, 165-63 (1972), NSA27-8899.
403. Golanski, A. and Dybczynski, R., "Implantation of ions. II. Study on distribution of effective ranges of phosphorus atoms implanted into silicon monocrystal", Nukleonika, 18, 351-60 (1973), Polish, PA77-56789.
404. Golanski, A. and Fiderkiewicz, A., "Annealing behaviour of silicon bombarded with 140 keV Si^{++} ions", Radiat. Eff., 25, 213-14 (1975), PA78-85381.
405. Goode, P. D., Wilkins, M. A. and Dearnaley, G., "The electrical activity of phosphorous channeled into silicon", Radiat. Eff., 6, 237-45 (1970), PA74-21458.
406. Gorelkinskii, Y. V., Sigie, V. O. and Takibaev, Z. S., "EPR of conduction electrons produced in silicon by hydrogen ion implantation", Phys. Status Solidi A, 22, K55-7 (1974), PA77-33266.
407. Gotz, G. and Endter, R., "Switching effects in amorphized silicon layers", J. Non-Cryst. Solids, 13, 286-94 (1974), EEA77-14357.
408. Gotz, G., Klinge, K. D. and Schwabe, F., "Measurement of the energy loss in thin silicon single crystals", Exp. Tech. Phys., 23, 167-9 (1975), PA78-54229.

409.Gverdtsiteli, I. G., Guldamashvili, A. I., Diasamidze, E. M., Zaslavsky, S. A., Kalinin, A. N., Kumakhov, M. A. and Muralev, V. A., "Implantation angle influence on penetration of boron channelled ions into silicon", Radiat. Eff., 19, 171-4 (1973), PA76-79365.
410.Gyulai, J., Csepregi, L., Nagy, T., Mayer, J. W. and Muller, H., "Drive-in diffusion studies on antimony and gallium implanted into silicon", Vide, 29, 416-18 (1974), PA78-66560.
411.Gyulai, J., Meyer, D., Pashley, R. D. and Mayer, J. W., "Lattice location and dopant behavior of group II and VI elements implanted in silicon", Radiat. Eff., 7, 17-27 (1971), PA74-36549.
412.Gyulai, J., Revesz, P., Zsoldos, L., Vertesi, G. and Gyimesi, J., "Defects and amorphization in ion-implanted silicon", Acta Phys. & Chem. Szeged, 20, 259-66 (1974), PA78-37253.
413.Hart, R. R., Dunlap, H. L., Mohr, A. J. and Marsh, O. J., "The detection sensitivity of heavy impurities in Si using 280 keV He^{2+} and C^{2+} back-scattering", Thin Solid Films, 19, 137-44 (1973), PA77-6951.
414.Hasegawa, S. and Shimizu, T., "Isothermal annealing in P^+ ion-implanted silicon", J. Phys. Soc. Jap., 38, 766-73 (1975), PA78-33714.
415.Haskell, J., Rimini, E. and Mayer, J. W., "Channeling measurements in As-doped Si", J. Appl. Phys., 43, 3425-31 (1972), PA75-60818.
416.Hauser, J. J. and Kimerling, L. G., "Electrical conduction in Si-implanted amorphous Si", Phys. Rev. B, 11, 4043-4 (1975), EEA78-32951.
417.Hirvonen, J. K., "Investigations of ion implantation in silicon using particle channeling", Thesis, Rutgers, State Univ., N. J., 1971, 162 pp., Order No. 72-9628, PA75-71511.
418.Hirvonen, J. K., Brown, W. L. and Glotin, P. M., "Structural differences in light and heavy ion disorder in Si studied by single and double alignment channeling techniques", 2nd International Conference on Ion Implantation in Semiconductors, 1971, p. 8-16, PA75-25378.
419.Hofker, W. K., "Determination of moments of experimental range distributions of boron implanted in silicon", Radiat. Eff., 25, 205-6 (1975), PA78-86414.
420.Hofker, W. K., Oosthoek, D. P., Koeman, N. J. and De Grefte, H. A. M., "Concentration profiles of boron implantations in amorphous and polycrystalline silicon", Radiat. Eff., 24, 223-31 (1975), EEA78-19682.
421.Hofker, W. K., Werner, H. W., Oosthoek, D. P. and De Grefte, H. A. M., "Influence of annealing on the concentration profiles of boron implantations in silicon", Appl. Phys., 2, 265-78 (1973), EEA77-2448.

422.Hofker, W. K., Werner, H. W., Oosthoek, D. P. and De Grefte, H. A. M., "Experimental analysis of concentration profiles of boron implanted in silicon", Proceedings of the 3rd International Conference on Ion Implantation in Semiconductors and Other Materials, 1972, p. 133-45, Publ. 1973, CA81-178837.
423.Hofker, W. K., Werner, H. W., Oosthoek, D. P. and Koeman, N. J., "Boron implantations in silicon: a comparison of charge carrier and boron concentration profiles", Appl. Phys., 4, 125-33 (1974), EEA77-33919.
424.Hurrle, A. and Schulz, M., "Anomalous outdiffusion of beryllium implanted into silicon", Conference on Lattice Defects in Semiconductors, 1974, p. 474-80, Publ. 1975, PA78-81255.
425.Igras, E., "Effect of gallium acceptor displacement in ion-bombarded silicon", Biul. Wojsk. Akad. Tech., 23, 81-8 (1974), Polish, CA82-132179.
426.Igras, E., "Effect of removing bismuth donors in ion bombarded silicon", Biul. Wojsk. Akad., 24, 47-52 (1975), Polish, CA83-186552.
427.Igras, E. and Krylov, Y., "Electron mirror microscopic investigation of silicon surface layers with implanted impurities", 3rd Symposium on the Reliability of Electronics, 1973, p. 135-43, EEA77-5941.
428.Igras, E., Krylov, Y. and Rosinski, W., "Investigation of surface layers of ion-implantation-doped silicon by mirror electron microscopy", Sov. Phys.-Semicond., 8, 1131-4 (1975), PA78-47472.
429.Igras, E., Krylov, Y. and Rosinski, W., "The effect of removal of gallium acceptors in ion bombarded silicon", Phys. Status Solidi A, 28, 321-5 (1975), EEA78-19678.
430.Ikeda, T., Tamura, M., Yoshihiro, N. and Tokuyama, T., "Defects in heavily phosphorus-implanted silicon observed after drive-in process", Oyo Buturi, 44, 311-16 (1975), EEA78-41491.
431.Ikeda, T., Tsuchimoto, T. and Tokuyama, T., "Defects in silicon during ion implantation", Record of the 10th Symposium on Electron, Ion, and Laser Beam Technology, 1969, p. 183-8, NSA27-8891.
432.Inada, T., Ohnuki, Y. and Nishimura, H., "Annealing study of ion-implanted silicon by photoelectromagnetic method", Appl. Phys. Lett., 21, 137-9 (1972), EEA75-36113.
433.Inoue, T., Ohmura, Y., Koike, K. and Yoshioka, T., "Observation of impurity profile in ion implanted silicon by Auger electron spectroscopy", Jap. J. Appl. Phys., 13 (Suppl.), 819-22 (1974), EEA78-24206.
434.Itoh, T., Inada, T. and Kanekawa, K., "Doping of silicon by ion implantation", Appl. Phys. Lett., 12, 244-6 (1968), EEA71-25876.
435.Itoh, T. and Nakamura, T., "Epitaxial growth of silicon assisted by ion implantation", Radiat. Eff., 9 (1-2), 1-4 (1971), PA74-52019.
436.Itoh, T. and Shinada, K., "Annealing behavior of boron and oxygen implanted silicon", Jap. J. Appl. Phys., 14, 1627-28 (1975).

437. Iwaki, M., Gamo, K., Masuda, K., Namba, S., Ishihara, S., Kimura, I. and Yokota, K., "Experimental method for measuring both atom and carrier concentration profiles in the same sample of ion-implanted silicon layers by radioactive-ion implantation", Nucl. Instrum. & Methods, 127, 93-8 (1975), PA78-72506.
438. Iwaki, M., Gamo, K., Masuda, K., Namba, S., Ishihara, S., Kimura, I. and Yokota, K., "Comparison between concentration profiles of arsenic implanted in silicon measured by means of neutron activation analysis and radioactive ion implantation", Jap. J. Appl. Phys., 14, 167-8 (1975), PA78-22497.
439. Johansson, N. G. E. and Mayer, J. W., "Influence of ultrasonic agitation on the electrical behavior of ion implanted silicon", Radiat. Eff., 2, 61-2 (1969), PA73-12144.
440. Johansson, N. G. E. and Mayer, J. W., "Hall effect measurements on Sb and Ga implanted silicon; anneal behavior and comparison with other species", Solid-State Electron., 13, 123-30 (1970), PA73-25861.
441. Karatsyuba, A. P., Kurinny, V. I., Rytchkova, S. V., Timashova, T. P. and Yudin, V. V., "Defect centre studies in boron-implanted silicon", Proceedings of the International Conference on Radiation Damage and Defects in Semiconductors, 1972, p. 81-6, Publ. 1973, EEA76-19900.
442. Kiselevich, M., Lyatushinski, A., Zhuk, V. and Osipenko, B. P., "Implantation of heavy ions into silicon single crystals", Sov. Phys. Solid State, 17, 687-9 (1975).
443. Kostka, A. and Kalbitzer, S., "Electrical properties of silicon implanted with boron ions of MeV energy", Radiat. Eff., 19, 77-82 (1973), EEA77-259.
444. Krasivina, L. E., Zverev, B. P., Simakhin, Y. F., Usmanova, M. M. and Zhumaev, N., "A neutron-activation-analysis investigation of the distributions of antimony and phosphorus introduced into silicon by ionic implantation", Zavod, Ind. Lab., 40, 1444-5 (1974), PA78-58815.
445. Krautle, H. and Kalbitzer, S., "Conductive properties of the ion implanted binary system $Si_{1-x}Al_x$", 2nd International Conference on Ion Implantation in Semiconductors, 1971, p. 499-506., PA75-25873.
446. Krylov, Y., Rosinski, W. and Saczuk, K., "Study of the concentration profiles of phosphorus and boron implanted into silicon", Electron Technol., 5, 109-17 (1972), PA76-53191.
447. Kumakhov, M. A. and Muralev, V. A., "Spatial distribution of radiation defects in silicon irradiated with boron and phosphorus ions", Sov. Phys. Semicond., 6, 1350-4 (1973), PA76-24166.
448. Large, L. N., Hill, H. and Ball, M. P., "Profiles of high conductivity shallow layers in silicon produced by boron ion implantation", Int. J. Electron., 22, 153-64 (1967), EEA70-18818.
449. Lazarovici, D., "Single crystal silicon targets for blocking and channeling experiments", Report IFA-NR-54-1974, Institutul de Fizica Atomica, Bucharest, Romania, 1974, 13 pp., NSA30-32877.

450.Lebedev, G. P., Gerasimenko, N. N. and Dvurechenskii, A. V., "Interaction of defects introduced by ion bombardment (ESR)", Sov. Phys.-Semicond., 7, 1530-1 (1974), PA77-82207.
451.Lebedev, S. Y. and Omelyanovskaya, N. M., "Annealing of defects at high temperature irradiation of silicon by accelerated ions", Radiat. Eff., 22, 135-7 (1974), PA77-72501.
452.Lecrosnier, D. P. and Pelous, G. P., "High energy ion implantation into silicon", European Conference on Ion Implantation, 1970, p. 102-6, PA74-18777.
453.Lecrosnier, D. P. and Pelous, G. P., "Distribution of secondary defects in boron-implanted silicon", Conference on Lattice Defects in Semiconductors, 1974, p. 487-93, Publ. 1975, PA78-81132.
454.Lee, T. F., Pashely, R. D., McGill, T. C. and Mayer, J. W., "Investigation of tellurium-implanted silicon", J. Appl. Phys., 46, 381-8 (1975), EEA78-15162.
455.Leneva, A. E. and Streltsov, L. N., "Amorphization of single crystals during irradiation by ions of average energies", Fiz. Tekh. Poluprovodn, 8, 2425-7 (1974), Russian, CA82-118302.
456.Linder, J. S., Stone, J. L. and Haden, C. R., "Electrical and physical measurements of boron ion-implanted layers in silicon", 21st Annual Southwestern IEEE Conference and Exhibition, 1971, P. 307-11, EEA75-7434.
457.Lowe, L. F., Kennedy, J. K., Davies, D. E., Deane, M. L. and Eyges, L. J., "Surface temperature measurements for ion-bombarded Si and GaAs at 1.0 to 2.0 MeV", Rev. Sci. Instrum., 46, 633-4 (1975), PA78-61736.
458.Lucke, W., Comas, J., Hubler, G. and Dunning, K., "Effects of annealing on profiles of aluminum implanted in silicon carbide", J. Appl. Phys., 46, 994-7 (1975), EEA78-24172.
459.Lugujjo, E. and Mayer, J. W., "Energy dependence of He^+ and H^+ channeling in Si overlaid with Au films", Phys. Rev. B, 7, 1782-91 (1973), PA76-33581.
460.MacDonald, C. and Galster, G., "Measurement of Hall-effect and sheet resistivity as a function of temperature on hot, phosphorous implants in silicon", Radiat. Eff., 6, 223-36 (1970), PA74-21457.
461.MacDonald, C. and Galster, G., "The evaluation of electrically active damage in hot, phosphorus implantation in silicon by means of Hall-effect measurements", 2nd International Conference on Ion Implantation in Semiconductors, 1971, p. 124-33, PA75-25869.
462.Maciaszek, M., "X-ray section topographical images of implanted silicon crystals", Phys. Status Solidi A, 30, K1-2 (1975), PA78-72496.
463.Maekawa, H., Kobayashi, T. and Matsumori, T., "Annealing characteristics of boron-implanted layers in silicon", Tokai Daigaku Kyyo Kogakubu, no. 1, 9-15 (1972), Japanese, NSA27-16438.

464.Marsden, D. A., Bellavance, G. R., Davies, J. A., Martini, M. and Sigmund, P., "The energy dependence and depth distribution of lattice disorder in ion-implanted silicon", Radiat. Eff., 6, 181-5 (1970), PA74-7882.
465.Masters, B. J., Fairfield, J. M. and Crowder, B. L., "Radiochemical determination of damage profiles in silicon (on ion implantation)", Radiat. Eff., 6, 57-61 (1970), PA74-18726.
466.Matsumori, T., Takeda, K., Kobayashi, T., Izumi, T. and Maekawa, H., "Electron spin resonance of nitrogen-ion-implanted silicon", Tokai Daigaku Kiyo Kogakubu, no. 1, 1-7 (1972), Japanese, NSA27-13837.
467.Matthews, M. D., "Electrical and electron microscope observation on antimony-implanted silicon", J. Materials Sci., 4, 997-1002 (1969), PA73-25858.
468.Mayer, J. W., "Ion implantation-lattice disorder", Proceedings of the International Conference on Radiation Effects in Semiconductors, 1972, p. 367-75, PA75-47367.
469.Mayer, J. W., Marsh, O. J., Shifrin, G. A. and Baron, R., "Ion implantation of silicon. II. Electrical evaluation using Hall-effect measurements", Canad. J. Phys., 45, 4073-89 (1967), PA71-18072.
470.McCaldin, J. O. and Widmer, A. E., "Silicon heavily doped by energetic cesium ions", J. Phys. Chem. Solids, 24, 1073-80 (1963), PA66-25349.
471.Melvin, J. D., "Energy loss of light ions channeling in silicon", Thesis, California Inst. of Tech., Pasadena, Calif., 1974, 125 pp., Order No. 74-24450, NSA31-10151.
472.Melvin, J. D. and Tombrello, T. A., "Energy loss of low energy protons channeling in silicon crystals", Radiat. Eff., 26, 113-26 (1975).
473.Meyer, O., Johansson, N. G. E., Picraux, S. T. and Mayer, J. W., "Lattice location and dopant behavior of group II and VI elements implanted in silicon", Solid State Commun., 8, 529-31 (1970), PA73-37512.
474.Mezey, G. and Szokefalvi-Nagy, Z., "Measurement of the boron distribution in ^{10}B-implanted silicon by the $(n\,\alpha)$ nuclear reaction", Thin Solid Films, 19, 173-5 (1973), EEA77-2488.
475.Mitchell, I. V., "Ion bombardment induced damage in silicon single crystals", Conference on Atomic Collision Phenomena in Solids, 1970, p. 130, PA73-68340.
476.Mitchell, J. B., Pronko, P. P., Shewchun, J., Thompson, D. A. and Davies, J. A., "Nitrogen-implanted silicon. I. Damage annealing and lattice location", J. Appl. Phys., 46, 332-4 (1975), PA78-29291.
477.Mitchell, J. B., Shewchun, J., Thompson, D. A. and Davies, J. A., "Nitrogen-implanted silicon. II. Electrical properties", J. Appl. Phys., 46, 335-43 (1975), EEA78-15161.
478.Moline, R. A., "Ion-implanted phosphorous in silicon: profiles using C-V analysis", J. Appl. Phys., 42, 3553-8 (1971), PA74-65150.

479. Moline, R. A. and Cullis, A. G., "Residual defects in Si produced by recoil implantation of oxygen", Appl. Phys. Lett., 26, 551-3 (1975), EEA78-28605.
480. Moline, R. A. and Reutlinger, G. W., "Phosphorus channeled in silicon: Profiles and electrical activity", 2nd International Conference on Ion Implantation in Semiconductors, 1971, p. 58-69, PA75-25866.
481. Morabito, J. M. and Tsai, J. C., "In-depth profiles of phosphorus-ion-implanted silicon by Auger spectroscopy and secondary ion emission", Surface Sci., 33 (2), 422-6 (1972), NSA27-23784.
482. Morehead, F. F. and Crowder, B. L., "A model for the formation of amorphous Si by ion bombardment", Radiat. Eff., 6, 27-33 (1970), PA74-21438.
483. Morehead, F. F., Crowder, B. L. and Title, R. S., "Formation of amorphous silicon by ion bombardment as a function of ion, temperature, and dose (carrier concentration)", J. Appl. Phys., 43, 1112-18 (1972), PA75-25425.
484. Morhange, J. F., Beserman, R. and Balkanski, M., "Raman study of the vibrational properties of implanted silicon", Phys. Status Solidi A, 23, 383-91 (1974), PA77-57541.
485. Morhange, J. F., Beserman, R., Sebenne, C. and Balkanski, M., "Vibrational properties of ion-implantation doped silicon", J. Phys., Colloq., no. 5, 99-102 (1973), French, NSA30-4032.
486. Morokhovskii, V. L., Koralenko, G. D., Grishaev, I. A., Kasilov, V. I., Fisun, A. N., Shramenko, P. I. and Krinitsyn, A. N., "Channeling of positions of 1 GeV energy", JETP Lett., 16, 112-14 (1972), PA76-1647.
487. Moss, S. C. and Flynn, P., "Structural disorder in ion-implanted silicon (amorphous layer formations)", Acta Crystallogr. A, A28 (Pt. 4), S157 (1972), PA76-17888.
488. Muller, H., Chu, W. K., Gyulai, J., Mayer, J. W., Sigmon, T. W. and Cass, T. R., "Crystal orientation dependence of residual disorder in As-implanted Si", Appl. Phys. Lett., 26, 292-4 (1975), PA78-46222.
489. Muller, H., Gyulai, J., Chu, W. K., Mayer, J. W. and Sigmon, J. W., "Influence of As, Sb, and Ga implanted into silicon", J. Electrochem. Soc., 122, 1234-8 (1975).
490. Muller, H., Kranz, H., Ryssel, H. and Schmid, K., "Electrical and backscattering measurements of arsenic implanted silicon", Appl. Phys., 4, 115-23 (1974), EEA77-33918.
491. Muller, H., Ryssel, H. and Ruge, I., "A new method for boron doping of silicon by implantation of BF_2-molecules", 2nd International Conference on Ion Implantation in Semiconductors, 1971, p. 85-95, EEA75-13597.
492. Muller, H., Ryssel, H. and Schmid, K., "Electrical properties of silicon layers implanted with BF_2 molecules", J. Appl. Phys., 43, 2006-8 (1972), PA75-33016.

493.Muller, H., Schmid, K., Ryssel, H. and Ruge, I., "Determination of the critical dose for different mass ions implanted into silicon", Proceedings of the 3rd International Conference on Ion Implantation in Semiconductors and Other Materials, 1972, p. 203-14, Publ. 1973, CA81-178846.

494.Muller, J. C., Stuck, R., Berger, R. and Siffert, P., "Thermally stimulated current measurements on silicon junctions produced by implantation of low energy boron ions", Solid-State Electron., 17, 1293-7 (1974), EEA78-597.

495.Namba, S., Masuda, K., Gamo, K., Doi, A., Ishihara, S. and Kimura, I., "Enhanced diffusion in ion implanted silicon", Radiat. Eff., 6, 115-20 (1970), PA74-21442.

496.Namba, S., Masuda, K., Gamo, K., Ishihara, S. and Kimura, I., "Concentration profiles of ion-implanted ions in silicon", 5th International Conference on Electron and Ion Beam Science and Technology, 1972, p. 254-64, PA75-15652.

497.Nagasawa, K. and Schulz, M., "Fast transient capacitance measurements for implanted deep levels in silicon", Appl. Phys., 8, 35-42 (1975), EEA78-41515.

498.Nelson, R. S., "Radiation damage in silicon during ion implantation", European Conference on Ion Implantation, 1970, p. 212-8, PA74-18778.

499.Nelson, R. S. and Cairns, J. A., "Antimony implanted silicon: a comparison between the total implanted concentration profile and the donor concentration profile", Radiat. Eff., 6, 131-4 (1970), PA74-21444.

500.Netange, B., Cherki, M. and Baruch, P., "Implanted interstitial boron atoms in silicon", Appl. Phys. Lett., 20, 349-51 (1972), PA75-44126.

501.Nishi, H., Sakurai, T. and Furuya, T., "Investigation of boron implanted silicon by backscattering method. Defect and impurity distributions", Fujitsu Sci. & Tech. J., 8, 123-35 (1972), EEA76-4506.

502.Niu, H., Matsuda, T., Yamauchi, K. and Takai, M., "Lateral photovoltaic effect in nitrogen-implanted p-type silicon", Appl. Phys. Lett., 21, 423-4 (1972), PA76-2124.

503.Noonan, J. R., Kirkpatrick, C. G. and Streetman, B. G., "Low-temperature photoluminescence from boron ion implanted Si", Radiat. Eff., 21, 225-8 (1974), PA77-75547.

504.Norris, C. B., Brower, K. L. and Vook, F. L., "Ionization effects in self interstitial migration and implant damage annealing in silicon", Radiat. Eff., 18, 1 (1973).

505.North, J. C. and Gibson, W. M., "Channelling study of boron-implanted silicon", Appl. Phys. Lett., 16, 126-9 (1970), PA73-30915.

506.North, J. C. and Gibson, W. M., "Channelling study of boron implanted silicon: liquid nitrogen temperature implantation", Radiat. Eff., 6, 199-203 (1970), PA74-21453.

507.Ohdomari, I., Ikeda, M., Ohno, K. and Itoh, T., "Structure of surface layers of silicon heavily implanted with Ar", J. Appl. Phys., 46, 5065-8 (1975).
508.Ohlidal, M. and Orel, V., "Preparation of ion-implanted silicon for transmission electron microscopy", Czech. J. Phys. B, B24, 349-54 (1974), PA77-40725.
509.Ohmura, Y. and Koike, K., "Evidence for electronic stopping in ion implantation: shallower profile of lighter isotope ^{10}B in Si", Appl. Phys. Lett., 26, 221-2 (1975), PA78-41823.
510.Okabayashi, H. and Shinoda, D., "Lateral spread of ^{31}P and ^{11}B ions implanted in silicon", J. Appl. Phys., 44, 4220-1 (1973), PA76-80179.
511.Okabayashi, H. and Shinoda, D., "Range and standard deviation of ion-implanted phosphorus in silicon", Jap. J. Appl. Phys., 13, 1187-8 (1974), EEA77-33957.
512.Okabayashi, H. and Shinoda, D., "Range and standard deviation of ion-implanted phosphorous and boron in silicon", NEC Res. & Dev., no. 35, 10-14 (1974), EEA78-15230.
513.Pabst, H. J. and Palmer, D. W., "Irradiation-induced defects studied by the channelling method in p- and n-type silicon", Proceedings of the International Conference on Radiation Damage and Defects in Semiconductors, 1972, p. 438-46, Publ. 1973, PA76-33579.
514.Pabst, H. J. and Palmer, D. W., "Light ion irradiation experiments suggest an ionization damage mechanism for silicon", Radiat. Eff., 21, 135-6 (1974), PA77-66948.
515.Pashley, R. D., "Ionization energy determination in indium implanted silicon", 2nd International Conference on Ion Implantation in Semiconductors, 1971, p. 485-8, PA75-25872.
516.Pavlov, P. V., Pashkov, V. I., Genkin, V. M., Kamaeva, G. V., Nikishin, V. I., Ogarkov, Y. N. and Uspenskaya, G. I., "Changes in the dislocation structure of silicon resulting from irradiation with ions of moderate energies", Sov. Phys.-Solid State, 15, 1914-15 (1974), PA77-56753.
517.Pavlov, P. V. and Popov, Y. S., "Surface properties of silicon after irradiation with ions of moderate energies", Sov. Phys. Solid State, 9, 2431-4 (1968), EEA71-8217.
518.Pavlov, P. V., Popov, Y. S. and Belich, T. V., "Distribution of recombination centers produced by bombardment of silicon with moderate-energy ions", Sov. Phys.-Semicond., 8, 602-4 (1974), PA78-21533.
519.Pavlov, P. V., Tetelbaum, D. I., Zorin, E. I. and Alekseev, V. I., "Distribution of implanted atoms and radiation defects in the ion bombardment of silicon (Monte Carlo calculation method)", Sov. Phys. Solid State, 8, 2141-6 (1967), EEA70-12817.
520.Pavlov, P. V., Zorin, E. I., Tetelbaum, D. I., Baranova, A. S. and Vasilev, V. K., "The influence of channeling and anomalous diffusion on the profiles of implanted phosphorus and boron atoms in silicon", Radiat. Eff., 13, 153-5 (1972), PA75-47432.

521.Peterstrom, S. and Holmen, G., "Isothermal annealing of boron implanted silicon", Phys. Scr., 10, 142-4 (1974), PA77-76404.

522.Petty, R. J. and Dearnaley, G., "Charge state ratios for helium ions channelled through silicon", Phys. Lett. A, 50A, 273-4 (1974), PA78-21536.

523.Picraux, S. T., "Ion channeling studies of the crystalline perfection of epitaxial layers", J. Appl. Phys., 44, 587-93 (1973), PA73-18039.

524.Picraux, S. T., Brown, W. L. and Gibson, W. M., "Lattice location by channeling angular distributions: Bi implanted in Si", Phys. Rev. B, 6, 1382-94 (1972), PA75-60819.

525.Picraux, S. T. and Stein, H. J., "Ion-channelling and optical absorption studies of implantation disorder in germanium", J. Appl. Phys., 45, 3784-8 (1974), PA78-6464.

526.Picraux, S. T. and Thomas, G. J., "Correlation of ion channeling and electron microscopy results in the evaluation of heteroepitaxial silicon", J. Appl. Phys., 44, 594-602 (1973), PA73-18040.

527.Picraux, S. T. and Vook, F. L., "Ionization and thermal dependence of implantation disorder in silicon", 2nd International Conference on Ion Implantation in Semiconductors, 1971, p. 1, PA75-25376.

528.Picraux, S. T., Weisenberger, W. H. and Vook, F. L., "Low temperature channeling measurements of ion implantation lattice disorder in single crystal silicon", Radiat. Eff., 7, 1017 (1971), PA74-36491.

529.Pihl, C. F., Bieber, R. L. and Schwuttke, G. H., "Precision lattice parameter studies of ion-implanted silicon", Phys. Status Solidi A, 17, 359-69 (1973), PA76-43778.

530.Pistryak, V. M., Tetelbaum, D. I., Kozlov, V. F., Vasilev, V. K., Zorin, E. I., Pavlov, P. V. and Fogel, Y. M., "Influence of amorphization on the distribution of boron introduced into silicon by ion bombardment", Sov. Phys.-Semicond., 7, 1325-7 (1974), PA77-73271.

531.Pritchard, W. M., "Surface damage of silicon by heavy ions", Trans. Am. Nucl. Soc., 19, 66-7 (1974), PA74-25425.

532.Pronko, P. P., Mitchell, J. B. and Schewchun, J., "Nitrogen implantation of p-silicon at cryogenic temperatures", 2nd International Conference on Ion Implantation in Semiconductors, 1971, p. 495-8, EEA75-13609.

533.Punzel, J., "Electron microscopic investigations on depth of penetration of lattice discrders in ion-bombarded silicon", Wiss. Z Tech. Univ. Dresden, 20, 369-70 (1971), German, EEA75-20090.

534.Reddi, V. G. K. and Sansbury, J. D., "Channeling of phosphorous ions in silicon", Appl. Phys. Lett., 20, 30-1 (1972), EEA75-10109.

535.Reddi, V. G. K. and Sansbury, J. D., "Channeling and dechanneling of ion-implanted phosphorus in silicon", J. Appl. Phys., 44, 2951-63 (1973), PA76-54974.

536.Rimini, E., Luguijjo, E. and Mayer, J. W., "Channeling in Si overlaid with Al and Au films", Phys. Rev. B, 6, 718-28 (1972), PA75-54094.

537.Romanov, S. I. and Smirnov, L. S., "Formation of disordered layers in ion bombardment of crystals", Sov. Phys.-Semicond., 6, 1410-1 (1973), NSA27-26640.

538.Roosild, S., Dolan, R. and Buchanan, B., "Annealing studies of damage introduced by high energy ion implantations of silicon", IEEE Trans. Nucl. Sci., NS-16, 33-6 (1969), EEA73-14014.

539.Ryssel, H. and Kranz, H., "The electrical activation of implanted arsenic in silicon", Appl. Phys., 7, 11-14 (1975), EEA78-19624.

540.Ryssel, H. and Kranz, H., "The influence of quenching on the electrical activation of implanted arsenic in silicon", Phys. Status Solidi A, 28, K25-7 (1975), EEA78-19679.

541.Ryssel, H., Muller, H. and Schmid, K., "Damage dependent electrical activation of boron implanted silicon", Appl. Phys., 3, 321-4 (1974), PA77-37761.

542.Ryssel, H., Muller, H., Schmid, K. and Ruge, I., "Boron doping profiles and annealing behavior of amorphous implanted silicon layers", Proceedings of the 3rd International Conference on Ion Implantation in Semiconductors and Other Materials, 1972, p. 215-24, Publ. 1973, CA81-178847.

543.Schuch, R., "Blocking of a α-particles transmitted through germanium and silicon crystals", Z. Phys., 272, 61-6 (1975), German, PA78-25428.

544.Schulson, E. M. and Marsden, D. A., "A rapid assessment of mercury ion implantation damage in boron-doped silicon using the scanning electron microscope", 9th Annual Conference of the Microbeam Analysis Society (Summaries), 1974, 29A/6 pp., PA77-82208.

545.Schulz, M., "Determination of deep trap levels caused by ion implanted impurities in silicon", Conference on Metal-Semiconductor Contacts, 1974, p. 226-33, EEA78-8474.

546.Schulz, M., "Determination of deep trap levels in silicon using ion-implantation and CV-measurements", Appl. Phys., 4, 225-36 (1974), EEA77-36210.

547.Schultz, M., Goetzberger, A., Franz, I. and Langheinrich, W., "Controlled gold doping of silicon by using ion implantation", Appl. Phys., 3, 275-80 (1974), PA77-37760.

548.Schwarz, G., Trapp, M., Schmiko, R., Butzke, G. and Rogge, K., "Concentration profiles of implanted boron ions in silicon from measurements with the ion microprobe", Phys. Status Solidi A, 17, 653-8 (1973), PA76-52640.

549.Schwettmann, F. N., "Enhanced diffusion during the implantation of arsenic in silicon", Appl. Phys. Lett., 22, 570-2 (1973), PA76-44541.

550. Schwettman, F. N., "Characterization of incomplete activation of high-dose boron implants in silicon", J. Appl. Phys., 45, 1918-20 (1974), PA77-61312.
551. Schwuttke, G. H. and Brack, K., "X-ray interferometry of volume changes in C^+ implanted silicon", Z. Naturforsch. A, 28A, 654-6 (1973), PA76-43817.
552. Schwuttke, G. H. and Brack, K., "Subsurface silicon carbide formation in silicon after high energy carbon (+) ion implantation", Proceedings of the 3rd International Conference on Silicon Carbide, 1973, p. 626-33, Publ. 1974, CA81-160354.
553. Schwuttke, G. H., Brack, K. and Gorey, E. F., "Resistivity and annealing properties of implanted Si: H^+", Radiat. Eff., 6, 103-6 (1970), PA74-21441.
554. Schwuttke, G. H., Brack, K., Gorey, E. F. and others, "Formation and annealing of isolation regions in silicon through Si^+ bombardment", Phys. Status Solidi A, 14, 107-13 (1972), NSA27-6821.
555. Seidel, T. E., "Distribution of boron implanted silicon", 2nd International Conference on Ion Implantation in Semiconductors, 1971, p. 47-57, PA75-25427.
556. Seidel, T. E. and MacRae, A. U., "The isothermal annealing of boron implanted silicon", Radiat. Eff., 7, 1-6 (1971), PA74-36490.
557. Seirmarco, J. A., Joy, R. C. and Kennan, W. A., "Ion implanted silicon sheet resistance curves", Presented at Electrochemical Society Meeting (Abstracts only), San Francisco, Calif., May 1974, EEA77-26518.
558. Seshan, K. and Washburn, J., "The nature and habit planes of defects in P^+ ion-implanted silicon", Phys. Status Solidi A, 26, 345-52 (1974), PA78-6379.
559. Seshan, K. and Washburn, J., "Some new results in the characterization of defects in phosphorus ion-implanted silicon", Radiat. Eff., 26, 31-8 (1975).
560. Shannon, J. M., Ford, R. A. and Gard, G. A., "Annealing characteristics of highly doped ion implanted phosphorous layers in silicon", Radiat. Eff., 6, 217-24 (1970), PA74-21456.
561. Shannon, J. M., Tree, R. and Gand, G. A., "Electrical characteristics of ion implanted boron layers in silicon", Canad. J. Phys., 48, 229-35 (1970), PA73-25857.
562. Shimizu, K., Kawakatsu, H., Ishikawa, Y. and Kanaya, K., "Implantation of aluminum and antimony into silicon targets by argon ion bombardment", Bull. Electrotech. Lab., 33, 151-9 (1969), PA73-59230.
563. Shutov, Y. N., Zorin, E. I., Pavlov, P. V. and Tetelbaum, D. I., "The electrical activity of boron and phosphorus implanted in silicon", Izv. VUZ Fiz., no. 4, 118-9 (1975), Russian, EEA78-28567.
564. Sigmon, T. W., Chu, W. K., Muller, H. and Mayer, J. W., "Analysis of arsenic range distributions in silicon", Appl. Phys., 5, 347-50 (1975), PA78-30232.

565. Sigmon, T. W. and Gibbons, J. F., "Hot electron studies of lattice damage created in silicon by implantation of 2.8 MeV protons", Radiat. Eff., 6, 95-101 (1970), PA74-21394.
566. Sigurd, D. and Domeij, B., "Critical angles of Sb and Bi implanted Si", Phys. Lett. A, 36A, 81-2 (1971), PA74-69149.
567. Skakun, N. A., Dikii, N. P. and Matyash, P. P., "Some characteristics of the distribution of radiation defects resulting from the implantation of boron ions in silicon", Sov. Phys.-Semicond., 8, 854-5 (1975), PA78-33202.
568. Skakun, N. A., Dikii, N. P., Matyash, P. P. and Korol, V. M., "Determination of the location of interstitial sodium atoms in a silicon lattice", Sov. Phys.-Solid State, 15, 123-5 (1973), EEA76-35245.
569. Skakun, N. A., Grigorev, A. N., Dikii, N. P., Matyash, P. P. and Nikolaichuk, L. I., "Investigation into the profile of radiation damage in Si using the elastic scattering of ^{4}He ions", Pis'ma V Zh. Tekh. Fiz., 1, 203-7 (1975), Russian, PA78-72497.
570. Skakun, N. A., Vasilev, V. K., Dikii, N. P., Zorin, E. I., Matyash, P. P., Pavlov, P. V. and Tetelbaum, D. I., "Position of the boron atom in the silicon lattice during ion implantation and subsequent annealing", Sov. Phys.-Solid State, 16, 666-8 (1974), PA78-21541.
571. Smith, B. J., "Projected ion range data for silicon planar technology", Report AERE-R 6660, UKAEA, Harwell, Berks England, February 1971, 5 pp., PA74-61725.
572. Smith, B. J., Stephen, J. and Hammersley, P. J., "Doping centres in partially annealed carbon implanted silicon", Radiat. Eff., 26, 17-22 (1975).
573. Sone, K. and Fukuzawa, F., "Back-scattering energy spectra near the critical angle for proton channeling", Mem. Fac. Eng. Kyoto Univ., 34, 315-24 (1972), PA76-20774.
574. Stephen, I. M., "Crystal damage in ion-implanted silicon", In: Semiconductor silicon, Electrochemical Society, New York, p. 433-44 (1969).
575. Smith, B. J., Hinder, G. W., Marshall, D. C. and Wittam, E. M., "Some observations on high energy nitrogen implantation in silicon", 2nd International Conference on Ion Implantation in Semiconductors, 1971, p. 489-94, EEA75-13608.
576. Stein, H. J., "Electrically active defect annealing in neutron and in ion-damaged Si", Radiat. Eff., 6, 175-80 (1970), PA74-21396.
577. Stein, H. J., "Localized mode of substitutional carbon in ion-implanted silicon", 2nd International Conference on Ion Implantation in Semiconductors, 1971, p. 2-6, PA75-25377.
578. Stein, H. J., "Bonding and thermal stability of implanted hydrogen in silicon", J. Electron. Mater., 4, 159-74 (1975), PA78-38387.

579. Stein, H. J., Vook, F. L. and Borders, J. A., "Depth distribution of vacancies in 400 keV O^+ ion-implanted silicon", Appl. Phys. Lett., 16, 106-8 (1970), PA73-30911.

580. Stein, H. J., Vook, F. L., Brice, D. K., Borders, J. A. and Picraux, S. T., "Infrared studies of the crystallinity of ion-implanted Si", Radiat. Eff., 6, 19-26 (1970), PA74-21437.

581. Stroud, P. T., Lindsay, H. M. and Perkins, J. G., "Some preliminary studies of the structure of ion bombarded thin films", Vacuum, 23 (4), 125 (1973), PA76-72112.

582. Stumpfi, W. and Kalbitzer, S., "Low energy boron and phosphorous implants in silicon. I. Electrical sheet measurements", Radiat. Eff., 6, 205-10 (1970), PA74-21454.

583. Szmid, Z. and Wieteska, K., "α-particle implantation damage in silicon", Phys. Status Solidi A, 23, K3-5 (1974), PA77-52793.

584. Takahashi,, R. and Oshima, M., "Ion bombardment on semiconductors by applying exploding wire technique", Annu. Rep. Eng. Res. Inst. Fac. Eng. Univ. Tokyo, 29, 229-36 (1970), Japanese, EEA74-10449.

585. Tamura, M., "Secondary defects in phosphorus-implanted silicon", Appl. Phys. Lett., 23, 651-3 (1973), PA77-28127.

586. Tamura, M., Ikeda, T. and Tokuyama, T., "Crystal defects and electrical properties in ion-implanted silicon", 2nd International Conference on Ion Implantation in Semiconductors, 1970, p. 96-102, PA75-25380.

587. Tamura, M., Yoshihiro, N. and Ikeda, T., "Tertiary defects in phosphorus-implanted silicon", Appl. Phys. Lett., 27, 427-8 (1975), CA83-186546.

588. Tandon, J. L., Bilger, H. R. and Nicolet, M. A., "Excess noise spectral analysis of boron-implanted layers in silicon", 1973 SWIEEECO Record of Technical Papers, p. 561-8, EEA76-29061.

589. teKatt, E. H. and Schwuttke, G. H., "X-ray double-crystal-diffractometer investigations of implanted silicon: D^+ and N^+", In: Advan. X-Ray Analy., vol. 15, Heinrich, K. F. J., Bennett, C. S., Newkirk, J. B. and Ruad, C. O., (Eds.), Plenum Press, New York, p. 504-15 (1972), NSA27-29371.

590. Tombrello, T. A. and Caywood, J. M., "Molecular-ion effects in backscattering from aligned silicon crystals", Phys. Rev. B, 8, 3065-70 (1973), PA77-20342.

591. Tsai, J. C. C., Morabito, J. M. and Lewis, R. K., "Arsenic implanted and implanted-diffused profiles in silicon using secondary ion emission and differential resistance", Proceedings of the 2nd International Conference on Ion Implantation in Semiconductors and Other Materials, 1972, p. 87-98, Publ. 1973, CA81-178839.

592. Tsuchimoto, T. and Tokyuama, T., "Enhanced diffusion of boron and phosphorous in silicon during hot substrate ion implantation", Radiat. Eff., 6, 121-9 (1970), PA74-21443.

593. Tu, K. N., Chaudhari, P., Lal, K., Crowder, B. L. and Tan, S. I., "X-ray topographic determination of the absence of lateral strains in ion-implanted silicon", J. Appl. Phys., 43, 4262-3 (1972), PA75-80704.
594. Urli, N. B. and Gislon, R., "The presence of deep levels in silicon implanted with channeled low energy phosphorus ions", 2nd International Conference on Ion Implantation in Semiconductors, 1971, p. 466-72, PA75-25870.
595. Uspenskaya, G. I., Genkin, V. M. and Tetelbaum, D. I., "Change of the interplanar distances in deep layers of silicon during bombardment with medium-energy ions", Sov. Phys.-Crystallogr., 18, 224-5 (1973), NSA30-992.
596. Vaidyanathan, K. V., "Local- and defect-mode infrared absorption of carbon ions implanted in silicon", J. Appl. Phys., 44, 583-6 (1973), NSA27-26648.
597. Vaidyanathan, K. V., Chatterjee, P. K. and Streetman, B. G., "Comments on 'evidence for electronic stopping in ion implantation: shallower profile of lighter isotope ^{10}B in Si'", Appl. Phys. Lett., 27, 648-9 (1975).
598. Vasilev, V. K., Dikii, N. P., Zorin, E. I., Matyash, P. P., Pavlov, P. V., Skakun, N. A. and Tetelbaum, D. I., "Degree of substitution of crystal lattice sites by impurity atoms and disordering during the implantation of boron in silicon", Fiz. Tverd. Tela, 16, 1064-7 (1974), Russian, CA81-7008.
599. Vasilev, V. K., Zorin, E. I., Pavlov, P. V. and Tetelbaum, D. I., "Investigation of the distribution of phosphorous implanted in silicon by ion bombardment using an 'arbitrary' orientation of crystals relative to the ion beam", Fiz. Tekh. Poluprov., 3, 1877 (1969), Russian, PA73-30916.
600. Vavilov, V. S., Isaev, N. U., Mukashev, B. N. and Spitsyn, A. V., "Radiation effects in n-type silicon irradiated with 10-MeV protons", Sov. Phys.-Semicond., 6, 907-11 (1972), NSA27-16405.
601. Vikhrev, B. I., Gerasimenko, N. N., Dvurechenskii, A. V. and Smirnov, L. S., "Interaction of hydrogen atoms in silicon with defects introduced by ion bombardment", Sov. Phys.-Semicond., 8, 87-12 (1975), PA78-33160.
602. Vook, F. L., Brice, D. K., Borders, J. A. and Picraux, S. T., "Infrared studies of crystallinity of ion-implanted Si", Radiat. Eff., 6, 19-26 (1970), PA74-21437.
603. Vook, F. L. and Stein, H. J., "Evidence for vacancy motion in low temperature ion-implanted Si", Radiat. Eff., 6, 11-7 (1970), PA74-21436.
604. Westmoreland, J. E., "Channeling effect analysis of lattice disorder in boron implanted silicon", Thesis, California Inst. Technol., Pasadena, 1971, 142 pp., Order No. 72-489, PA75-68036.
605. Westmoreland, J. E., Mayer, J. W., Eisen, F. H. and Welch, B., "Analysis of disorder distribution in boron implanted silicon", Radiat. Eff., 6, 161-74 (1970), PA74-21395.

606.Weyer, G., Andersen, J. U., Deutch, B. I., Golovchenko, J. A. and Nylandsted-Larsen, A., "Direct comparison of Mossbauer and channeling studies of implanted ^{119}Sn in silicon single crystals", Radiat. Eff., 24, 117-21 (1975), PA78-37282.
607.Wu, W. K. and Washburn, J., "Identification of interstitial- and vacancy-type dislocation loops in ion-implanted silicon", J. Appl. Phys., 45, 1085-90 (1974), PA77-56743.
608.Wu, W. K. and Washburn, J., "Boron precipitation in ion implanted silicon", Proc. Annu. Meet., Electron Microsc. Soc. Am., 1975, p. 256-7, CA83-186545.
609.Yudin, V. V., Kurinny, V. I., Akimov, Y. S. and Karatsyuba, A. P., "Spatial distribution of defects produced by boron ion implantation of silicon", Radiat. Eff., 24, 59-61 (1975), PA78-29297.
610.Ziegler, J. F., Crowder, B. L., Cole, G. W., Baglin, J. E. E. and Masters, B. J., "Boron atom distributions in ion-implanted silicon by the (n, ^{4}He) nuclear reaction", Appl. Phys. Lett., 21, 16-17 (1972), PA75-56743.
611.Zorin, E., Pavlov, P. V., Tetelbaum, D. I. and Khokhlov, A. F., "Hall study of silicon layers doped with phosphorous and boron by ion implantation", Phys. Status Solidi A, 6, 337-44 (1971), PA74-69623.
612.Zorin, E. I., Pavlov, P. V., Tetelbaum, D. I. and Khokhlov, A. F., "Positions of the energy levels of ion-implanted impurities in silicon", Sov. Phys.-Semicond., 7, 1340-1 (1974), PA77-69993.

b. Germanium

613.Akimchenko, I. P., Galkin, V. V., Krasnopevtsev, V. V., Krasheninnikov, V. S. and Milyutin, Y. V., "Optical absorption in germanium subjected to bombardment with oxygen ions", Fiz. & Tekh. Poluprov., 5, 2348-50 (1971), Russian, PA75-22766.
614.Arifov, U. A., Aliev, A. A. and Turmashev, E., "The effect of ion-bombardment-produced crystal lattice radiation damage on the angular and energy distributions of scattered ions", Radio Eng. & Electron Phys., 17, 276-8 (1972), PA76-11663.
615.Bjorkquist, K., Sigurd, S., Fladda, G. and Bjarnholt, G., "Lattice location of heavy group III and V elements implanted into Ge", Radiat. Eff., 6, 141-8 (1970), PA74-21432.
616.Brice, D. K., "Ion implantation deposition of energy into electronic processes in silicon and germanium targets", Report SLA-73-230, Sandia Labs., Albuquerque, New Mexico, 1973, NSA28-6368.
617.Campbell, A. B., Mitchell, J. B., Shewchun, J. and Thompson, D. A., "Nitrogen implanted germanium: damage, lattice location, and electrical properties", Can. J. Phys., 53, 303-9 (1975), EEA78-15159.

618. Graczyk, J. F. and Chaudhari, P., "A scanning electron diffraction study of vapor-deposited and ion implanted thin films of Ge. I", Phys. Status Solidi B, 58, 163-79 (1973), PA76-58189.
619. Gusev, V. M., Guseva, M. J., Ionova, E. S., Mansurova, A. N. and Starinin, C. V., "Electrical properties and structure of boron implanted germanium", Phys. Status Solidi A, 21, 413-8 (1974), PA77-28431.
620. Gusev, V. M., Guseva, M. I., Ionova, E. S., Mansurova, A. N. and Starinin, C. V., "Investigation of the Hall effect and structure defects in germanium irradiated with 30-keV Ga^+ ions", Sov. Phys.-Semicond., 7, 1435-8 (1974), EEA77-33884.
621. Gusev, V. M., Guseva, M. I. and Starinin, C. V., "Amorphization of germanium crystals during ion bombardment", Radiat. Eff., 15, 251-4 (1972), PA75-80685.
622. Guseva, M. I. and Mansurova, A. N., "Radiation-enhanced diffusion of boron in germanium during ion implantation", Radiat. Eff., 20, 207-10 (1973), PA77-29021.
623. Herzer, H. and Kalbitzer, S., "Electrical properties of ion implanted germanium", 2nd International Conference on Ion Implantation in Semiconductors, 1971, p. 307-14, EEA75-13603.
624. Holmen, G., Buren, A. and Hogber, P., "Radiation damage in Ge produced and removed by energetic Ge ions", Radiat. Eff., 24, 51-8 (1975), PA78-29296.
625. Holmen, G., Hogberg, P. and Buren, A., "Radiation damage in Ge produced by noble gas ions investigated by the secondary electron emission method", Radiat. Eff., 24, 39-44 (1975), PA78-29294.
626. Holmen, G., Peterstrom, S. and Buren, A., "Radiation enhanced annealing of radiation damage in Ge", Radiat. Eff., 24, 45-50 (1975), PA78-29295.
627. Itoh, T. and Ohdomari, I., "Anneal behavior of aluminum implanted germanium", Jap. J. Appl. Phys., 10, 1002-6 (1971), PA74-61686.
628. Johansson, N. G. E., Sigurd, D. and Bjorkquist, K., "Diffusion effects in ion implanted germanium", Radiat. Eff., 6, 257-62 (1970), PA74-21461.
629. Jones, K. C. and Stevens, F. E. C., "Electrical characteristics of p type germanium layers doped by ion implantation", Electron. Lett., 5, 499-500 (1959), EEA73-10335.
630. Jones, K. W. and Kraner, H. W., "Energy lost to ionization by 254-eV ^{73}Ge atoms stopping in Ge", Phys. Rev. A, 11, 1347-53 (1975), PA78-54230.
631. Lebedev, S. Y. and Omelyanovskaya, N. M., "Structural studies of the near-surface layer of Ge crystals at their bombardment with accelerated ions at various temperatures", Radiat. Eff., 23, 1-5 (1974), PA78-6465.
632. Litovchenko, V. G. and Lysenko, V. S., "Investigation of radiation-type centres generated by bombardment of a 'real' germanium surface with low-energy He^+ ions", Fiz. Tekh. Poluprovodnikov, 4, 72-9 (1970), Russian, EEA73-20992.

633.Ohdomari, I. and Itoh, T., "Carrier generation in germanium implanted with aluminum at room temperature", Jap. J. Appl. Phys., 11, 1709-12 (1972), EEA76-680.
634.Oshiyama, T., Kudo, H. and Mannami, M., "A diffusion model of planar channeling of fast charged particles. Backscattered energy spectra and diffusion constant", J. Phys. Soc. Jap., 36, 1107-13 (1974), PA77-40820.
635.Parsons, J. R. and Hoelke, C. W., "Observation of crystal lattice planes in neutron and ion bombarded Ge", IEEE Trans. Nucl. Sci., NS-16, 37-42 (1969), EEA73-14015.
636.Pavlov, P. V. and Popov, Y. S., "Surface properties of germanium after irradiation with ions of moderate energies", Fiz. Tekh. Poluprovodnikov, 3, 546-9 (1969), Russian, EEA73-1068.
637.Primak, W. and Dayal, Y., "Effect of ion bombardment on the electrical conductivity of germanium", Phys. Rev., 175, 1033-8 (1968), EEA72-2709.
638.Sigurd, D., Fladda, G., Eriksson, L. and Bjorkquist, K., "Lattice disorder in germanium by boron ion bombardment", Radiat. Eff., 3, 45-7 (1970), PA73-61938.
639.Tinsley, A. W., Jones, K. C., Stevens, P. R. C., George, G. G. and Gunnersen, E. M., "Electrical and atom site location properties of ion implanted germanium", European Conference on Ion Implantation, 1970, p. 187-91, PA74-18774.
640.Wood, G., Anderson, G. W., Davey, G. E., Comas, J., Saks, N. S. and Lucke, W. H., "Optical and electrical properties of boron-implanted amorphous germanium thin films", J. Appl. Phys., 45, 4528-33 (1974), PA78-25786.
641.Zwangobani, E. and Macdonald, R. J., "Measurements of the channeling effect in sputtering", Radiat. Eff., 20, 81-8 (1973), PA77-17150.

c. Silicon and Germanium

642.Arifov, U. A. and Aliyev, A. A., "Influence of ion bombardment induced radiation damages in the crystal lattice on the angular regularities in ion scattering", Radiat. Eff., 19, 231-4 (1973), PA77-9932.
643.Baeri, P., Campisano, S. U., Foti, G., Rimini, E. and Davies, J. A., "Channelling effect study of deuteron-induced damage in Si and Ge crystals", Appl. Phys. Lett., 26, 424-6 (1975), PA78-54225.
644.Beyer, W., Stuke, J. and Wagner, H., "Influence of ion implantation on electrical properties of amorphous Ge and Si", Phys. Status Solidi A, 30, 231-40 (1975), EEA78-32956.
645.Bogh, E., Hogild, P. and Stensgaard, I., "Spatial distribution of defects in ion bombarded silicon and germamium", Radiat. Eff., 7, 115-21 (1971), PA74-36492.

646. Campisano, S. U., Foti, G., Grasso, F. and Rimini, E., "Beam-induced lattice disorder in channeling experiments on Si and Ge", Appl. Phys. Lett., 21, 425-7 (1972), PA76-1656.
647. Feldman, L. C. and Appleton, B. R., "Multiple scattering and planar dechanneling in silicon and germanium", Phys. Rev. B, 8, 935-51 (1973), PA76-52655.
648. Johansson, N. G. E., Mayer, J. W. and Marsh, O. J., "Technique used in Hall-effect analysis of ion implanted Si and Ge", Solid-State Electron., 13, 317-35 (1970), PA73-34557.
649. Krautel, H., "Study of the annealing behaviour of high dose implants in silicon and germanium crystals", Radiat. Eff., 24, 255-62 (1975), PA78-42918.
650. Kudo, H., Oshiyama, T. and Mannami, M., "Determination of defect distribution in single crystals by channelling of fast charged particles", J. Phys. Soc. Jap., 36, 214-20 (1974), PA77-16527.
651. Lin, W. K., Olson, H. G. and Powers, D., "Alpha-particle stopping cross section of silicon and germanium", J. Appl. Phys., 44, 3631-4 (1973), PA73-64078.
652. Marcus, G. H., Olson, N. T. and Ryan, J. L., "Radiation damage studies in Si and Ge using channeling", Trans. Am. Nucl. Soc., 15, 731-2 (1972), PA76-24163.
653. Swanson, M. L., Parsons, J. R. and Hoelke, C. W., "Damaged regions in neutron-irradiated and ion-bombarded Ge and Si", 5th International Conference on Electron and Ion Beam Science and Technology, 1972, p. 359-66, PA75-47384.

d. Silicon Carbide

654. Addamiano, A., Anderson, G. W., Comas, J., Hughes, H. L. and Lucke, W. H., "Ion implantation effects of nitrogen, boron, and aluminum in hexagonal silicon carbide", J. Electrochem. Soc., 119, 1355-62 (1972), EEA76-725.
655. Addamiano, A., Comas, J. and Lucke, W. H., "Ion implantation in 6H-SiC", Presented at International Conference on Silicon Carbide (abstracts only), Miami Beach, Fla., September 1973, PA77-9928.
656. Akimchenko, I. P., Vavilov, V. S., Galkin, V. V., Ivanov, V. S., Krasnopevtsev, V. V. and Milyutin, Y. V., "Synthesis of silicon carbide by implantation of Si ions in diamond", Sov. Phys.-Semicond., 6, 1039-41 (1972), EEA76-4550.
657. Akimchenko, I. P., Vavilov, V. S., Krasnopevtsev, V. V., Milyutin, Y. V. and Loi, C. K., "Photoluminescence of β-SiC synthesised by the ion implantation method", Sov. Phys.-Semicond., 7, 659-60 (1973), PA77-28908.
658. Borders, J. A., Picraux, S. T. and Beezhold, W., "Formation of SiC in silicon by ion implantation", Appl. Phys. Lett., 18, 509-11 (1971), PA74-52300.

659.Choyke, W. J. and Patrick, L., "Intrinsic luminescent defects in 15R SiC induced by ion bombardment and annealing", Proceedings of the International Conference on Radiation Damage and Defects in Semiconductors, 1972, p. 218-22, Publ. 1973, PA76-34430.
660.Hart, R. R., Dunlop, H. L. and Marsh, O. J., "Disorder produced in SiC by ion bombardment", Proceedings of the International Conference on Radiation Effects in Semiconductors, 1972, p. 405-10, PA75-47391.
661.Makarov, V. V. and Petrov, N. N., "Influence of ion bombardment on the cathodoluminescence of silicon carbide", Sov. Phys. Solid State, 8, 1272-6 (1966), EEA70-4096.
662.Marsh, O. J., "Ion implantation in SiC", Presented at International Conference on Silicon Carbide, Miami Beach, Florida, 1973, EEA77-5913.
663.Matzke, L. and Koniger, M., "On ion implantation in silicon carbide", Phys. Status Solidi A, 1, 469-78 (1970), PA74-12734.
664.Patrick, L. and Choyke, W. J., "Localized vibrational modes of a persistent defect in ion-implanted SiC", J. Phys. Chem. Solids, 34, 565-7 (1973), NSA27-23788.
665.Rothenmund, W. and Fritzsche, C. R., "Optical absorption and electrical conductivity of SiC films produced by ion implantation", J. Electrochem. Soc., 121, 586-8 (1974), PA77-45473.

3. Group III-V Semiconductors
a. Gallium Arsenide

666.Abdalla, M. I., Palmer, J. F. and Desfeux, C., "Transport properties of O^+ implanted in GaAs", Conference on Gallium Arsenide and Related Compounds, 1974, p. 342-54, Publ. 1975, EEA78-28727.
667.Ambridge, T., Heckingbottom, R., Bell, E. C., Sealy, B. J., Stephens, K. G. and Surridge, R. K., "Effect of dual implants into GaAs", Electron. Lett., 11, 314-15 (1975), EEA78-37388.
668.Anderson, W. J., Douglass, C. A., III, and Park, Y. S., "Electroreflectance measurements of melt-doped and ion-implanted GaAs", J. Appl. Phys., 46, 3870-5 (1975).
669.Arnold, G. W., "Enhanced diffusion in ion-bombarded GaAs (luminescence)", Proceedings of the 2nd International Conference on Ion Implantation in Semiconductors, 1971, p. 151-6, PA75-25446.
670.Bell, E. C., Galccum, A. E., Hemment, P. L. F. and Sealy, B. J., "Heat treatment of ion implanted GaAs", Radiat. Eff., 22, 253-8 (1974), EEA78-568.
671.Benson, R. B., Jr., Littlejohn, M. A., Pao, P. S. and Sarin, H. K., "Some second-phase structures in gallium arsenide annealed after implantation with zinc", Appl. Phys. Lett., 27, 69-71 (1975), EEA78-41457.
672.Bicknell, R., Hemment, P. L. F., Bell, E. C. and Tansey, J. E., "Electrical and structure sensitive measurements on ion implanted GaAs", Phys. Status Solidi A, 12, K9-12 (1972), PA75-60821.

673. Borders, J. A., "Near band edge optical absorption produced by ion implantation in GaAs", Appl. Phys. Lett., 18, 16-18 (1971), PA74-37558.
674. Bottiger, J. and Whitton, J. L., "Directional effects as observed by the channelling effect technique in ion induced radiation damage in GaAs", Radiat. Eff., 19, 201-2 (1973), PA76-79390.
675. Carter, G., Grant, W. A., Haskell, J. D. and Stephens, G. A., "Radiation damage by implanted ions in GaAs and GaP", Radiat. Eff., 6, 277-84 (1970), PA74-21386.
676. Chatterjee, P. K., Vaidyanathan, K. V., McLevige, M. V. and Streetman, B. G., "Photoluminescence from Be-implanted GaAs", Appl. Phys. Lett., 27, 567-9 (1975).
677. Davies, D. E., Kennedy, J. K. and Ludington, C. E., "Comparison of group IV and VI doping by implantation in GaAs", J. Electrochem. Soc., 122, 1374-7 (1975).
678. Davies, D. E., Kennedy, J. K. and Yang, A. C., "Compensation from implantation in GaAs", Appl. Phys. Lett., 23, 615-6 (1973), PA77-17175.
679. Davies, D. E., Roosild, S. and Lowe, L., "The role of elevated temperatures in the implantation of GaAs", Solid-State Electron., 18, 733-6 (1975), EEA78-28651.
680. Deveaud, B., Palmier, J. F., Favennec, P. N. and Kar, R. K., "Photovoltaic effects on O^+ implanted GaAs", Solid State Commun., 17, 1253-6 (1975).
681. Donnelly, J. P., Lindley, W. T. and Hurwitz, C. E., "Silicon- and selenium-ion-implanted GaAs reproducibly annealed at temperatures up to 950°C", Appl. Phys. Lett., 27, 41-3 (1975), EEA78-32995.
682. Eisen, F. H., Harris, J. S., Welch, B., Pashley, R. D., Sigurd, D. and Mayer, J. W., "Properties of tellurium implanted gallium arsenide", Ion Implantation Semicond. Other Mater., Proc. Int. Conf., 3rd, 1972, p. 631-40, Publ. 1973, CA81-178843.
683. Euthymiou, P. C., Nomicos, C. D. and Philadelpheus, A. T., "Behaviour of magnetophotoconductivity of semi-insulating GaAs upon α-particle bombardment", J. Appl. Phys., 44, 521-2 (1973), NSA27-13854.
684. Favennec, P. N., "Implantation of zinc into GaAs at 1 MeV", Proceedings of the 2nd International Conference on Ion Implantation in Semiconductors, 1971, p. 174-81, EEA75-13599.
685. Favennec, P. N., Boulet, J. P., Haridon, H. L. and Rao, E. V. K., "Investigation of defect diffusion in oxygen-implanted GaAs by scanning electron microscope", Conference on Lattice Defects in Semiconductors, 1974, p. 481-6, Publ. 1975, PA78-81256.
686. Favennec, P. N., Pelous, G. P., Binet, M. and Baudet, P., "Compensation of gallium arsenide by oxygen implantation", Conference on Ion Implantation Semicond. Other Mater., 3rd, 1972, p. 621-30, Publ. 1973, CA81-178842.

687. Finstad, T. G., Andersen, S. L. and Olsen, T., "Annealing characteristics and lattice site location of 40 keV Sn implantations in GaAs", Phys. Status Solidi A, 25, 515-21 (1974), EEA78-5264.
688. Fritzsche, C. R. and Rothemund, W., "Sputtering during ion implantation into gallium arsenide", Appl. Phys., 7, 39-44 (1975), EEA78-19625.
689. Gamo, K., Aoki, K., Masuda, K. and Namba, S., "Diffusion of defects in low temperature ion implanted GaAs", Jap. J. Appl. Phys., 10, 1118-9 (1971), PA74-65152.
690. Gavrilov, A. A., Kachurin, G. A., Pridachin, N. B. and Smirnov, L. S., "Diffusion of zinc and cadmium in gallium arsenide irradiated with arsenic ions", Sov. Phys.-Semicond., 8, 1455-6 (1975), EEA78-33042.
691. Gibbons, J. F. and Tremain, R. E., Jr., "The effects of degeneracy on doping efficiency for type implants in GaAs", Appl. Phys. Lett., 26, 199-201 (1975), EEA78-19621.
692. Grob, J. J., Ghitescu, A. and Siffert, P., "Lattice disorder produced in gallium arsenide by cadmium implantation", Ion Implantation Semiconductors and Other Materials, Proc. Int. Conf., 3rd, 1972, p. 611-19, Publ. 1973, CA81-178841.
693. Guivarch, A., Favennec, P. N. and Pelous, G. P., "Annealing behaviour of disorder density in 1 MeV zinc implanted gallium arsenide between 20°C and 1000°C", Proceedings of the International Conference on Radiation Damage and Defects in Semiconductors, 1972, p. 429-37, Publ. 1973, PA76-33578.
694. Guivarch, A. and L'Haridon, H., "Study of defects produced in GaAs by the implantation of Zn^{+} ions (1 MeV) using channeling", J. Phys., 119-24 (1973), French, NSA29-30169.
695. Hanazawa, T., Yamaguchi, J., Gamo, K. and Itoh, N., "Annealing of expansion in ion-implanted GaAs", Jap. J. Appl. Phys., 13, 1487-8 (1974), PA77-76400.
696. Harris, J. S., "The effects of dose rate and implantation temperature on lattice damage and electrical activity in ion-implanted GaAs", Proceedings of the 2nd International Conference on Ion Implantation in Semiconductors, 1971, p. 157-67, PA75-25372.
697. Harris, J. S., Eisen, F. H., Welch, B. and others, "Influence of implantation temperature and surface protection on tellurium implantation in GaAs", Appl. Phys. Lett., 21, 601-3 (1972), NSA27-11199.
698. Harris, J. S., Harris, J. M. and Marcus, H. L., "Fluorine ion implantation profiles in gallium arsenide as determined by Auger electron spectroscopy", Appl. Phys. Lett., 21, 598-60 (1972), EEA76-8512.
699. Hemment, P. L. F., Sealy, B. J. and Stephens, K. G., "Encapsulation of ion implanted gallium arsenide", Ion Implantation Semicond.: Sci. Technol., Proc. Int., 4th, 1974, p. 27-34, Publ. 1975, CA83-171611.

700.Hunsperger, R. G. and Marsh, O. J., "Electrical properties of Cd, Zn, and S ion-implanted layers in GaAs", Radiat. Eff., 6, 263-8 (1970), PA74-21449.

701.Hunsperger, R. G., Wilson, R. G., and Jamba, D. M., "Mg and Be ion implanted GaAs", J. Appl. Phys., 43, 1318 (1972), EEA75-13543.

702.Hunsperger, R. G. and Wolf, E. D., "Anneal behavior of Cd ion implanted GaAs", J. Electrochem. Soc., 118, 1847-51 (1971), EEA75-7407.

703.Hunsperger, R. G., Wolf, E. D., Shifrin, G. A., Marsh, O. J. and Jamba, D. M., "Measurement of lattice damage caused by ion-implantation doping of semiconductors", Proceedings of the International Conference on Radiation Effects in Semiconductors, 1972, p. 393-8, PA75-47374.

704.Ilic, G., Ewan, G. T. and Whitton, J. L., "Cadmium implanted gallium arsenide - a study of residual bombardment disorder and atom location", Radiat. Eff., 18, 47-9 (1973), PA76-60755.

705.Itoh, T. and Kasahara, J., "Annealing behaviour of unimplanted and zinc-implanted GaAs", J. Appl. Phys., 45 (11), 4915-19 (1974), PA78-22489.

706.Itoh, T. and Kushiro, Y., "Arsenic and cadmium implantations into n-type gallium arsenide (Diode fabrication)", J. Appl. Phys., 42, 5120-4 (1971), EEA75-4574.

707.Itoh, T. and Kushiro, Y., "The effects of arsenic ion implantation in GaAs", Proceedings of the 2nd International Conference on Ion Implantation in Semiconductors, 1971, p. 168-73, EEA75-13598.

708.Kachare, A. H., Spitzer, W. G., Euler, F. K. and Kahan, A., "Infrared reflection of ion-implanted GaAs", J. Appl. Phys., 45, 2938-46 (1974), PA77-76243.

709.Kato, Y., Shiraki, Y., Shimada, T. and Komatsubara, K. F., "Electrical conduction owing to defect centres produced by ion implantation in gallium arsenide", Proceedings of the International Conference on Radiation Damage and Defects in Semiconductors, 1972, p. 348-54, Publ. 1973, PA76-33906.

710.Kato, Y., Shiraki, Y., Shimada, T. and Komatsubara, K. F., "Electrical conductivity of disordered layers in GaAs crystal produced by ion implantation", J. Appl. Phys., 45, 1044-9 (1974), EEA77-30094.

711.Lin, M. S., Gamo, K., Masuda, K. and Namba, S., "Dose dependence of photoluminescence degradation in Te ion-implanted GaAs", Jap. J. Appl. Phys., 12, 1092-3 (1973), PA76-64074.

712.Lin, M. S., Takai, M., Gamo, K., Masuda, K. and Namba, S., "Photoluminescence and channeling study on annealing behavior of Te ion-implanted GaAs", Presented at Electrochemical Society Meeting, San Francisco, Calif., May 1974, PA77-49610.

713.Littlejohn, M. A., Hauser, J. R. and Monteith, L. K., "The electrical properties of 60 keV zinc ions implanted into semi-insulating gallium arsenide", Radiat. Eff., 10, 185-90 (1971), PA74-83372.

714. Mamontov, A. P., Okunev, V. D., Gamon, V. I. and Zakharov, B. G., "Distribution of radiation defects in gallium arsenide irradiated with deuterons. (Luminescence study)", Sov. Phys.-Semicond., 6, 747-9 (1972), PA76-1654.

715. Muller, H., Gyulai, J., Mayer, J. W., Eisen, F. H. and Welch, B., "Anodic oxidation and profile determination of ion implanted semi-insulating gallium arsenide", Proceedings of the 4th International Conference on Ion Implantation in Semiconductors, 1974, p. 19-25, Publ. 1975, CA83-171610.

716. Pashley, R. D. and Welch, B. M., "Tellurium-implanted n^+ layers in GaAs", Solid-State Electron., 18, 977-81 (1975), EEA78-41497.

717. Picraux, S. T., "Vaporization of ion-implanted gallium arsenide", Ion Implantation Semicond. Other Mater., Proc. Int. Conf., 3rd, 1972, p. 641-54, Publ. 1973, CA81-178844.

718. Pridachin, N. B., Kachurin, G. A. and Smirnov, L. S., "Mechanism of radiation-accelerated diffusion in ion-irradiated gallium arsenide", Sov. Phys.-Semicond., 8, 1294 (1975), PA78-50435.

719. Pritchard, W. M., "Annealing of heavy ion irradiation damage in gallium arsenide", Trans. Amer. Nucl. Soc., 15, 732-3 (1972), NSA27-6806.

720. Pruniaux, B. R., North, J. C. and Miller, G. L., "Compensation of N-type GaAs by proton bombardment (semi-i.g.f.e.t.)", Proceedings of the 2nd International Conference on Ion Implantation in Semiconductors, 1971, p. 212-21, EEA75-13600.

721. Sansbury, J. D., "Properties of ion implanted silicon, sulfur, and carbon in gallium arsenide", Thesis, Stanford Univ., Calif., 1971, 115 pp., Order No. 71-19756, PA75-64986.

722. Sansbury, J. D. and Gibbons, J. F., "Properties of ion implanted silicon, sulphur, and carbon in gallium arsenide", Radiat. Eff., 6, 269-76 (1970), PA74-21450.

723. Sealy, B. J., "An examination of tellurium ion-implanted GaAs by transmission electron microscopy", J. Mater. Sci., 10, 683-91 (1975), EEA78-19652.

724. Sealy, B. J. and Surridge, R. K., "A new thin film encapsulant for ion-implanted GaAs", Thin Solid Films, 26, L19-22 (1975), EEA78-24230.

725. Sell, D. D. and MacRae, A. U., "Optical detection of surface damage in GaAs induced by argon ion implantation", J. Appl. Phys., 41, 4929-32 (1970), PA73-78366.

726. Skolnik, L. H., Spitzer, W. G., Kahan, A., Euler, F. and Hunsperger, R. G., "Localized vibrational mode absorption of ion-implanted silicon in GaAs", J. Appl. Phys., 43, 2146-50 (1972), PA75-40664.

727. Skolnik, L. H., Spitzer, W. G., Kahan, A. and Hunsperger, R. G., "Infrared localized-vibrational-mode absorption of ion-implanted aluminum and phosphorous in gallium arsenide", J. Appl. Phys., 42, 5223-9 (1971), PA75-13474.

728.Takai, M., Gamo, K., Masuda, K. and Namba, S., "Effects of implantation temperature on lattice location of tellurium implanted in gallium arsenide", Jap. J. Appl. Phys., 12, 1926-30 (1973), PA77-16512.

729.Takai, M., Gamo, K., Masuda, K. and Namba, S., "Lattice site location of cadmium and tellurium implanted in gallium arsenide", Jap. J. Appl. Phys., 14, 1935-42 (1975).

730.Tansey, J. E., "Electrical properties of implanted layers in gallium arsenide", Thesis, Univ. Surrey, England, 1973, PA77-75864.

731.Tinsley, A. W., Grant, W. A., Carter, G. and Nobes, M. J., "The retention of Bi ions implanted in GaAs", Proceedings of the 2nd International Conference on Ion Implantation in Semiconductors, 1971, p. 199-204, PA75-25424.

732.Tinsley, A. W., Stephens, G. A., Nobes, M. J. and Grant, W. A., "Dose rate effects in indium implanted GaAs", Radiat. Eff., 23, 165-9 (1974), PA78-9752.

733.Tunkasiri, T. and Lewis, D., "Damage in gallium arsenide crystals produced by ion implantation, abrasion and ball-milling", J. Mater. Sci., 10, 1367-74 (1975), EEA78-41485.

734.Tunkasiri, T. and Lewis, D., "Twinning in annealed tellurium-implanted gallium arsenide crystals", Radiat. Eff., 27, 109-10 (1975).

735.Weisenberger, W. H., Picraux, S. T. and Vook, F. L., "Low temperature channeling measurements of ion implantation lattice disorder in GaAs", Radiat. Eff., 9, 121-6 (1971), PA74-52231.

736.Weisenberger, W. H., Picraux, S. T. and Vook, F. L., "Low temperature channeling measurements of ion implantation lattice disorder in GaAs", Proceedings of the International Conference on Radiation Effects in Semiconductors, 1972, p. 377-81, PA75-47373.

737.Westmoreland, J. E., "Marsh, O. J. and Hunsperger, R. G., "Lattice disorder produced in GaAs by 60 keV Cd ions and 70 keV Zn ions", Radiat. Eff., 5, 245-9 (1970), PA74-12707.

738.Whitton, J. L. and Bellavance, G. R., "Ion implantation of sulphur into GaAs, GaP, and Ge monocrystals", Radiat. Eff., 9, 127-32 (1971), PA74-52299.

739.Whitton, J. L. and Carter, G., "The implantation profiles of energetic heavy ions in GaAs, GaP and Ge", Proceedings of the Conference on Atomic Collision Phenomena in Solids, 1969, p. 615-32, PA73-75716.

740.Wolf, E. D. and Hunsperger, R. G., "Measurement of ion implantation lattice damage (111) GaAs using the SEM", Appl. Phys. Lett., 16, 526-9 (1970), PA73-53538.

741.Woodcock, J. M., Shannon, J. M. and Clark, D. J., "Electrical and cathodoluminescence measurements on ion implanted donor layers in GaAs", Solid-State Electron., 18, 267-75 (1975), EEA78-15185.

742.Yuba, Y., Gamo, K., Masuda, K. and Namba, S., "Hall effect measurements of Zn implanted GaAs", Jap. J. Appl. Phys., 13, 641-4 (1974), PA77-44939.

743.Zelevinskaya, V. M. and Kachurin, G. A., "Behavior of germanium implanted in GaAs by ion bombardment", Fiz. Tekh. Poluprov., 5, 1150-3 (1971), Russian, PA74-59023.
744.Zelevinskaya, V. M., Kachurin, G. A., Bogomyakov, N. P., Azikov, B. S. and Shirokov, L. L., "Capture of impurity atoms by vacancies in implanted layers", Sov. Phys.-Semicond., 8, 164-6 (1974), PA77-75648.
745.Zelevinskaya, V. M., Kachurin, G. A., Pridachin, N. B. and Smirnov, L. S., "Doping of gallium arsenide by implantation of zinc ions", Fiz. Tekh. Poluprov., 4, 1784-8 (1970), Russian, PA73-78974.
746.Zelevinskaya, V. M., Kachurin, G. A. and Smirnov, L. S., "Properties of GaAs doped by incorporation of tellurium ions", Inorg. Mater., 9, 1171-3 (1973), PA77-60790.

b. Gallium Phosphide

747.Carter, G. and Whitton, J. L., "Ion implantation damage and annealing in GaP", Radiat. Eff., 15, 143-8 (1972), PA75-80706.
748.Davey, J. E., Pankey, T., Malmberg, P. R. and Lucke, W. H., "Ion-implantation induced optical absorption edge shifts in GaP", Appl. Phys. Lett., 17, 323-5 (1970), PA74-5828.
749.Feldman, L. C., Augustyniak, W. M. and Merz, J. L., "Implantation of Bi into GaP. II. Channeling studies", Radiat. Eff., 6, 293-300 (1970), PA74-21452.
750.Haskell, J. D., Grant, W. A., Stephens, G. A. and Whitton, J. L., "The influence of various parameters on radiation damage in GaP", Proceedings of the 2nd International Conference on Ion Implantation in Semiconductors, 1971, p. 193-8, PA75-25373.
751.Hemenger, P. M. and Dobbs, B. C., "Zinc ion implantation of sulfur-doped GaP", Appl. Phys. Lett., 23, 462-4 (1973), PA77-9924.
752.Inada, T. and Ohmuki, Y., "Magnesium and zinc ion implantation into sulfur-doped GaP", Appl. Phys. Lett., 25, 228-30 (1974), PA77-79876.
753.Merz, J. L., Feldman, L. C., Mingay, D. W. and Augustyniak, W. M., "Implantation of Bi into GaP 111. Hot-implant behaviour (Luminescence)", Proceedings of the 2nd International Conference on Ion Implantation in Semiconductors, 1971, p. 182-92, PA75-25432.
754.Merz, J. C., Feldman, L. S. and Sadowski, E. A., "Ion implantation of bismuth into GaP: I. Photoluminescence", Radiat. Eff., 6, 285-92 (1970), PA74-21451.
755.Naguib, H. M., Grant, W. A. and Carter, G., "On the damage induced in GaP single crystals by Te ion bombardment", Radiat. Eff., 18, 279-81 (1973), PA76-60751.
756.Thompson, D. A., Johar, S. S. and Shevchun, J., "Nitrogen implantation into GaP: damage and nitrogen location studies", J. Electron. Mater., 4, 195-207 (1975), EEA78-19668.

757. Wemple, S. H., North, J. C. and Dishman, J. M., "Optical and channeling studies of ion-bombarded GaP", J. Appl. Phys., 45, 1578-89 (1974), PA77-61307.
758. Williams, T. G. and Palmer, D. W., "The annealing behaviour of gallium phosphide in the region 110-500K after 300 keV neon irradiation", Proceedings of the 2nd International Conference on Ion Implantation in Semiconductors, 1971, p. 205-11, EEA75-25374.

c. Others

759. Gonda, S. I. and Makita, Y., "A new luminescence line due to nitrogen implanted into $Al_xGa_{1-x}As$ (x=0.37)", Appl. Phys. Lett., 27, 392-4 (1975).
760. Gonda, S., Makita, Y. and Maekawa, S., "Promotion of radiative recombination in $GaAs_{1-x}P_x$ by N-ion implantation", IEEE Trans. Electron Devices, ED-22, 712-15 (1975), EEA78-41465.
761. Gonda, S., Makita, Y., Maekawa, S., Tanoue, H. and Tsurushima, T., "Photoluminescence of indirect-band-gap $GaAs_{1-x}P_x$ (x=0.52) implanted with nitrogen ions", Jap. J. Appl. Phys., 13, 1483-4 (1974), PA77-76296.
762. Guseva, M. I., Zotova, N. V., Koval, A. V. and Nasledov, D. N., "Behavior of group IV elements introduced into indium arsenide by ion implantation", Sov. Phys.-Semicond., 8, 34-36 (1974), NSA30-27550.
763. Hvalgard, J. O., Finstad, T. and Henriksen, T., "The channeling effect technique and its applications within the semiconductor field", Phys. Norv., 6, 203 (1972), EEA76-25978.
764. Itoh, T. and Oana, U., "Zinc ion implantation into $GaAs_{0.62}P_{0.38}$", Appl. Phys. Lett., 24, 320-2 (1974), PA77-49577.
765. Lezheiko, L. V., Pridachin, N. B. and Smirnov, L. S., "Conduction in disordered regions produced by bombardment of semiconductors with heavy particles", Sov. Phys.-Semicond., 6, 775-7 (1972), EEA76-706.
766. Makita, Y. and Gonda, S., "Annealing temperature dependence of photoluminescence in n-implanted $GaAs_{1-x}P_x$ (x=0.36)", Jap. J. Appl. Phys., 13, 565 (1974).
767. Makita, Y. and Gonda, S. I., "Photoluminescence of boron-implanted $Al_xGa_{1-x}As$ (x=0.37)", Appl. Phys. Lett., 27, 333-4 (1975).
768. Makita, Y., Gonda, S. I., Tanoue, H., Tsurushima, T. and Maekawa, S., "Hot implantation of nitrogen ions into $GaAs_{1-x}P_x$ (x=0.36)", Jap. J. Appl. Phys., 13, 563-4 (1974), EEA77-22897.
769. Saxena, A. N., Weisburg, L. R., Mann, W. B. and Schima, F. J., "Implantation of $^{14}N^+$ into monocrystalline GaN films", Int. J. Appl. Radiat. & Isot., 26, 33-4 (1975), EEA78-28610.
770. Wilkins, M. A. and Dearnaley, G., "The distribution of ^{32}P channelled into indium antimonida", European Conference on Ion Implantation, 1970, p. 193-7, PA74-18781.

4. Group II-VI Semiconductors
a. Cadmium Telluride

771. Agrinskaya, N. V., Arkadeva, E. N., Guseva, M. I. and others, "Some aspects of ion doping of CdTe", Sov. Phys.-Semicond., 6, 407-9 (1972), NSA27-16408.
772. Agrinskaya, N. V., Arkadeva, E. N., Guseva, M. I., Maslova, L. V. and Matveev, O. A., "Ion implantation into CdTe", Proceedings of the International Symposium on Cadmium Telluride a Material for Gamma-Ray Detectors, 1971, 16/5 pp., EEA75-20108.
773. Akutagawa, W., Turnbull, D., Chu, W. K. and Mayer, J. W., "Channeling and electrical investigation of Au doped CdTe", Solid State Commun., 15, 1919-22 (1974), PA78-26333.
774. Gettings, M., "Ion implantation of cadmium telluride", Thesis, Univ. Surrey, England, 1973, EEA77-38775.
775. Gettings, M. and Stephens, K. G., "Annealing behaviour and heavy ion bombardment of high and low resistivity cadmium telluride", Proceedings of the International Symposium on Cadmium Telluride a Material for Gamma-Ray Detectors, 1971, 17 bis/7 pp., EEA75-20111.
776. Gettings, M. and Stephens, K. G., "Investigation of trapping levels in heavy ion implanted cadmium telluride", Radiat. Eff., 18, 275-7 (1973), PA76-60757.
777. Gettings, M. and Stephens, K. G., "Electrical activity and radiation damage in ion implanted cadmium telluride", Radiat. Eff., 22, 53-62 (1974), NSA30-32928.
778. Gettings, M. and Stephens, K. G., "Characterisation of single crystals of cadmium telluride for ion implantation studies", J. Cryst. Growth, 22, 50-2 (1974), PA77-45479.
779. Kachurin, G. A., Gorodetskii, A. E., Loburets, Y. V. and Smirnov, L. S., "Bombardment of thin CdTe films with indium ions", Fiz. Tverdogo Tela, 9, 494-7 (1967), Russian, EEA70-12545.
780. Kachurin, G. A., Gorodetskii, A. E., Zelevinskaya, V. M. and Smirnov, L. S., "Effect of ion bombardment on thin cadmium telluride films", Fiz. Tekh. Poluprovodnikov, 1, 1427-9 (1967), Russian, EEA71-3325.
781. Meyer, O. and Lang, E., "Lattice disorder in CdTe produced by ion implantation", Proceedings of the International Symposium on Cadmium Telluride a Material for Gamma-Ray Detectors, 1971, 17/9 pp., EEA75-20109.
782. Toyoda, K. and Namba, S., "Properties of $Hg_{1-x}Cd_xTe$ formed by Hg ion bombardment on CdTe", Sci. Pap. Inst. Phys. Chem. Res., 65, 26-33 (1971), EEA71-33974.

b. Cadmium Sulphide

783. Armitage, S. A., "Damage studies in ion-implanted CdS", European Conference on Ion Implanation, 1970, 138-42, EEA71-10525.

784.Armitage, S. A., "Radiation-enhanced diffusion of ion-implanted bismuth in cadmium sulphide", J. Phys. D, 8, 2034-42 (1975).
785.Bhar, T. N., Linder, J. S. and Stone, J. L., "Electrical and optical properties of Bi ion activated CdS thin films", 1973 SWIEEECO Record of Technical Papers, p. 569-76, EEA76-25933.
786.Eldridge, G., Chernow, F. and Ruse, G., "Further studies of bismuth-implanted cadmium sulfide", J. Appl. Phys., 44, 3858-61 (1973), EEA76-38196.
787.Eldridge, G., Govind, P. K., Nieman, D. A. and Chernow, F., "Radiation damage studies of bismuth ion-implanted CdS", European Conference on Ion Implantation, 1970, p. 143-7, EEA74-10526.
788.Fewster, R. H. and Bryant, F. J., "The parameters of ytterbium implantation in CdS, ZnS and ZnSe", Philos. Mag., 28, 1225-39 (1973), PA77-17131.
789.Govind, P. K. and Fraiker, F. J., "Electron microscope study of radiation damage in bismuth ion-implanted CdS", J. Appl. Phys., 42, 2476-81 (1971), EEA74-22485.
790.Hou, S. L. and Marley, J. A., Jr., "Photoelectric properties of ion-implanted CdS", Appl. Phys. Lett., 16, 467-9 (1970), EEA73-27749.
791.Hutchby, J. A., "Optical reflection studies of lattice disorder in iodine implanted CdS", Radiat. Eff., 16, 189-92 (1972), NSA27-21520.
792.Shiraki, Y., Shimada, T. and Komatsubava, K. F., "Ion-implantation of nitrogen into n-type cadmium sulfide", 2nd International Conference on Ion Implantation in Semiconductors, 1971, p. 235-40, PA75-15774.
793.Tell, B., "Ion implantation of sodium lithium and neon in cadmium sulphide", Appl. Phys. Lett., 17, 315-8 (1970), PA74-5857.

c. Others

794.Adachi, S. and Machi, Y., "Energy level study of phosphorus-ion-implanted ZnSe", Jap. J. Appl. Phys., 14, 1599-600 (1975).
795.Bontemps, A., Ligeon, E. and Danielou, R., "Channelling studies of ion implantation induced lattice defects in zinc telluride", Radiat. Eff., 22, 195-204 (1974), PA78-1172.
796.Bontemps, A., Ligeon, E. and Fontenille, J., "Energy loss and projected range of alpha particles in zinc telluride", Radiat. Eff., 21, 181-84 (1974), NSA30-31113.
797.Brown, M. R., Cox, A. F. J., Shand, W. A. and Wilhamo, J. M., "Ion implantation of zinc sulphide thin films", Report SRDE-70046, Signals Res. & Dev. Establ., Christchurch, England, 1970, 7 pp., EEA75-9948.
798.Brown, M. R., Cox, A. F. J., Shand, W. A. and Williams, J. M., "Ion implantation doping of zinc sulphide thin films", Solid State Commun., 9, 37-40 (1971), PA74-33324.

799. Bryant, F. J., Goodwin, G. K. and Hagston, W. E., "Atomic displacement effects on the cathodoluminescence of ZnSe implanted with Yb ions", Proc. Phys. Soc., 337, 21-47 (1974), NSA29-27403.
800. Chernow, F., "Ion-implanted epitaxially grown zinc selenide", Report NASA-CR-143086, Photocond. Semicond. Devices Lab., Univ. Colorado, Boulder, Colo., 1975, 49 pp., CA83-187294.
801. Cox, A. F. J., "Cathodoluminescence studies of transition metal ion implanted II-VI compounds", European Conference on Ion Implantation, 1970, p. 158-61, PA74-19415.
802. Georgobiani, A. N., Kotljarevsky, M. B., Zlobin, V. N., Todual, P. A., Generalov, Y. P. and Dementev, B. P., "The ion implantation of ZnS. (Photoluminescence spectral)", Mater. Res. Bull., 8, 893-7 (1973), PA76-72817.
803. Khan, M. K., "Electrical and optical properties of chlorine ion-implanted zinc telluride (Light emitting diodes)", Thesis, Univ. Wyoming, Laramie, 1974, 179 pp., Order No. 74-23695, PA78-38241.
804. Kwietriak, M. and Wardzynski, W., "Luminescence of ZnTe implanted by Al ions", Phys. Status Solidi A, 31, K47-52 (1975).
805. Langguta, G., Lang, E. and Meyer, G., "Lattice disorder and out-diffusion in ion implanted InSb and CdTe", 2nd International Conference on Ion Implantation in Semiconductors, 1971, p. 228-34, PA75-25370.
806. Marine, J., "Properties of ion implanted ZnTe", European Conference on Ion Implantation, 1970, p. 153-7, PA74-19154.
807. Norris, C. B., Barnes, C. E. and Beezhold, W., "Depth-resolved cathodoluminescence in undamaged and ion-implanted GaAs, ZnS, and CdS", J. Appl. Phys., 44, 3209-21 (1973), EEA76-32172.
808. Olley, J. A., Williams, P. M. and Yoffe, A. D., "Radiation damage II-VI semiconductors", European Conference on Ion Implantation, 1970, p. 148-52, EEA74-10527.
809. Segawa, Y., Lyu, H. N. and Namba, S., "Formation of $Hg_{1-x}Cd_xTe$ by ion implantation", Jap. J. Appl. Phys., 13, 911-12 (1974), PA77-57706.
810. Shin, B. K., Park, Y. S. and Look, D. C., "Electrical characteristics of Al-implanted ZnSe", Appl. Phys. Lett., 24, 435-6 (1974), PA77-57085.
811. Takagi, T., Yamada, I., Sasaki, A. and Ishibashi, T., "Electroluminescence in Mn-implanted ZnS thin films", IEEE Conference Record of 1972 Conference on Display Devices, p. 51-4, EEA76-23766.
812. Wang, C. C., Tao, T. F. and Sunier, J. W., "Proton bombardment and isochronal annealing of p-type $Pb_{0.76}Sn_{0.24}Te$", J. Appl. Phys., 45, 3981-7 (1974), EEA78-8445.
813. Williams, P. M. and Yoffe, A. D., "Physical properties of radiation induced and growth defects in II-VI semiconductors", Proceedings of the International Conference on Radiation Effects in Semiconductors, 1972, p. 399-404, PA75-47372.

5. Dielectrics

814. Akimenko, I. P., Galkin, V. V., Krasnopevtsev, V. V., Krasheninnikov, V. S., Milyutin, Y. V. and Spitsyn, A. V., "Optical and chemical property changes of silica and germanium dioxide as a result of structural transformations caused by ion bombardment", Mikroelectronika, no. 2, 166-72 (1973), Russian, CA80-32210.

815. Arnold, G. W., "Ion-implantation effects in noncrystalline SiO_2", IEEE Trans. Nucl. Sci., NS-20, 220-3 (1973), EEA77-22839.

816. Arnold, G. W., "High-density damage in Al_2O_3 by low-energy proton implantation (CTR material)", Presented at 1st Topical Meeting on the Technology of Controlled Nuclear Fusion, San Diego, California, April 1974, PA77-69316.

817. Arnold, G. W., Krefft, G. B. and Norris, C. B., "Atomic displacement and ionization effects on the optical absorption and structural properties of ion-implanted Al_2O_3", Appl. Phys. Lett., 25, 540-2 (1974), PA78-21531.

818. Bach, H., "Partial disintegration and change of concentration profiles at ion bombardment Na silicate glass surfaces", Radiat. Eff., 22, 73-8 (1974), PA77-72499.

819. Bauer, L. O., MacPherson, M. R., Robinson, A. T. and Dill, H. G., "Properties of silicon implanted with boron ions through thermal silicon dioxide", Solid-State Electron., 16, 289-300 (1973), EEA76-16330.

820. Bayly, A. R., "Optical properties of ion bombarded silica glass", Radiat. Eff., 18, 111 (1973).

821. Bayly, A. R., "Ellipsometric analysis of refractive index profiles produced by ion implantation in silica glass", J. Phys. D., 6, 1115 (1973).

822. Beezhold, W. and EerNisse, E. P., "Ion implantation into insulators: charge-removal studies using ion-induced characteristic X-rays", Appl. Phys. Lett., 21, 592-5 (1972), PA76-11669.

823. Brown, F. and Mackintosh, W. D., "The use of Rutherford backscattering to study the behavior of ion-implanted atoms during anodic oxidation of aluminum: Ar, Kr, Xe, K, Rb, Cs, Cl, Br, and I", J. Electrochem. Soc., 120, 1096-102 (1973), PA76-68511.

824. Chen, L. I., Pickar, K. A. and Sze, S. M., "Carrier transport and storage effects in Au ion implanted SiO_2 structures", Solid-State Electron., 15, 979-86 (1972), EEA75-36209.

825. Chou, N. J. and Crowder, B. L., "Effect of O^+ and Ne^+ implantation on the surface characteristics of thermally oxidized silicon", J. Appl. Phys., 41, 1731-8 (1970), PA73-47814.

826. Chu, W. K., Lugujjo, E., Mayer, J. W. and Sigmon, T. W., "Line-shape extraction analysis of silicon oxide layers on silicon by channeling effect measurements", Thin Solid Films, 19, 329-31 (1973), PA77-20346.

827. Clinard, F. W., Jr. and Sanders, W. M., "Axial channeling in ThO_2", J. Appl. Phys., 43, 4937-42 (1972), PA76-17883.

828. Comer, J. J., Bergeron, C. and Lowe, L. F., "Ion damage in synthetic quartz", Mater. Res. Bull., 8, 505-14 (1973), PA76-60748.
829. Dexler, R. J., Watelski, S. B. and Picraux, S. T., "Epitaxial silicon layers grown on ion-implanted silicon nitride layers", Appl. Phys. Lett., 23, 455-77 (1973), PA77-10541.
830. Donovan, R. P. and Simons, M., "Radiation hardening of thermal oxides on silicon via ion implantation", IEEE Trans. Nuclear Sci., NS-16, 203-6 (1969), EEA73-14168.
831. Donovan, R. P. and Simons, M., "Radiation hardening of thermal oxides on silicon by displacement damage", J. Appl. Phys., 43, 2897-9 (1972), PA75-50689.
832. Dunning, K. L., Hubler, G. K., Comas, J., Lucke, W. H. and Hughes, H. L., "Depth profiles of aluminium and sodium near surfaces: nuclear resonance method", Thin Solid Films, 19, 145-56 (1973), EEA77-2487.
833. EerNisse, E. P., "Compaction of ion-implanted fused silica", J. Appl. Phys., 45, 167-174 (1974), NSA30-976.
834. EerNisse, E. P. and Norris, C. B., "Introduction rates and annealing of defects in ion-implanted SiO_2 layers on Si", J. Appl. Phys., 45, 5196-205 (1974), PA78-26303.
835. Fahrner, W., "Contributions of oxygen, silicon, and hydrogen to the interface states of an Si-SiO_2 interface", J. Electrochem. Soc., 121, 784-7 (1974), EEA77-30225.
836. Fahrner, W. and Goetzberger, A., "Surface states induced by ion implantation", Proceedings of the 2nd International Conference on Ion Implantation in Semiconductors, 1971, p. 373-82, EEA75-13662.
837. Fahrner, W. and Goetzberger, A., "Properties of a single-level surface state induced by Be implantation into Si-SiO_2 interfaces", J. Appl. Phys., 44, 725-7 (1973), EEA76-12510.
838. Fritzsche, C. R., Axmann, A. and Seelewind, H., "Effect of ion implantation on electrical properties of SiO_2 films at high fields", European Conference on Ion Implantation, 1970, p. 123-6, EEA74-10099.
839. Fritzsche, C. R. and Rothemund, W., "Ion implantation and annealing effects in SiO_2 layers on silicon studied by optical measurements", J. Electrochem. Soc., 119, 1243-8 (1972), EEA75-39866.
840. Fritzsche, C. R. and Rothemund, W., "Thermal oxidation of silicon after ion implantation", J. Electrochem. Soc., 120, 1603-5 (1973), EEA77-2467.
841. Fuschillo, N., Lalevic, B. and Annamalai, N. K., "Atomic implantation of Nb in amorphous Nb_2O_5 by r.f. sputtering", J. Vac. Sci. & Technol., 12, 52-5 (1975), EEA78-24212.
842. Harari, E. and Royce, B. S. H., "Oxide charge trapping induced by ion implantation in SiO_2", IEEE Trans. Nucl. Sci., NS-20, 288-92 (1973), EEA77-22930.

843. Harris, J. S., Eisen, F. H., Welch, B., Haskell, J. D., Pashley, R. D. and Mayer, J. W., "Influence of implantation temperature and surface protection on tellurium implantation in GaAs", Appl. Phys. Lett., 21, 601-3 (1972), EEA76-8513.

844. Hartemann, P. and Morizot, M., "Increase in surface-acoustic-wave velocity produced by ion implantation in quartz", Electron. Lett., 9, 497-8 (1973), PA77-6425.

845. Hickmott, T. W., "Effect of bombardment by glass-forming ions on thermally stimulated ionic conductivity of sodium in SiO_2", Phys. Rev. Lett., 32, 65-7 (1974), EEA77-10074.

846. Hickmott, T. W., "Formation of passivating glass films by ion implantation of glass-forming components", IBM Tech. Disclosure Bull., 16, 3432 (1974), EEA77-30154.

847. Johnson, N. M., "Electron trapping in ion-implanted silicon dioxide films on silicon", Thesis, Princeton Univ., N. J., 1974, 83 pp., Order No. 74-25956, PA78-25774.

848. Johnson, N. M., Johnson, W. C. and Lampert, M. A., "Electron trapping in aluminum-implanted silicon dioxide film on silicon", J. Appl. Phys., 46, 1216-22 (1975), EEA78-24241.

849. Kellner, W. and Goetzberger, A., "Current gain recovery in silicon nitride passivated planar transistors by hydrogen implantation", IEEE Trans. Electron Devices, ED-22, 531-3 (1975), EEA78-28770.

850. Koch, F. B., Meek, R. L. and McCaughan, D. V., "Implantation of argon into SiO_2 films due to backsputter cleaning (prior to metallization)", J. Electrochem. Soc., 121, 558-62 (1974), EEA77-23143.

851. Kratschmer, W., "Effects of heavy ion radiation on quartz glass", Proceedings of the International Conference on Nuclear Photography and Solid State Track Detectors, Vol. I, 1972, p. 73-9, PA77-18649.

852. Kushner, R. A., McCaughan, D. V., Murphy, V. T. and Heilig, J. A., "Mobilization of sodium in silicon dioxide films by ion bombardment", Phys. Rev. B, 10, 2632-41 (1974), CA81-178712.

853. Li, S. P., "Ion-implantation effect on time-dependent breakdown in SiO_2", J. Appl. Phys., 46, 4833-34 (1975).

854. McCaughan, D. V., Kushner, R. A. and Murphy, V. T., "Ion bombardment effects in SiO_2 and phosphosilicate glass passivated SiO_2 films on silicon", Presented at Electrochemical Society Meeting, Chicago, Ill., May 1973, EEA76-23333.

855. McCaughan, D. V. and Murphy, V. T., "Low energy ion bombardment effects in SiO_2 (C-V degradation and bias-temperature stability)", IEEE Trans. Nucl. Sic., NS-19, 249-55 (1972), EEA76-12500.

856. McCaughan, D. V. and Murphy, V. T., "Low-energy ion bombardment of silicon dioxide films on silicon", J. Appl. Phys., 44, 2008-17 (1973), EEA76-23377.

857.McCaughan, D. V. and Murphy, V. T., "Low-energy ion bombardment of silicon dioxide films on silicon: II, Inert ambient annealing of degradation in MOS devices", J. Appl. Phys., 44, 3182-90 (1973).
858.Moline, R. A., Buckley, R. R., Haszko, S. E. and MacRae, A. U., "Tapered windows in SiO_2 by ion implantation", IEEE Trans. Electron Devices, ED-20, 840 (1973), EEA76-35201.
859.Monfret, A. and Bernard, J., "Chemical and electrical behaviour of ion implanted SiO_2 films", Proceedings of the 2nd International Conference on Ion Implanation in Semiconductors, Physics and Technology, Fundamental and Applied Aspects, 1971, p. 389-96, EEA75-13604.
860.Naguib, H. M., Singleton, J. F., Grant, W. A. and Carter, G., "Lattice disorder in alumina single crystals produced by ion bombardment", J. Mater. Sci., 8, 1633-40 (1973), PA77-1076.
861.Okuyama, Y. and Kamoshida, M., "Masking effects of silicon dioxide and KMER against boron-11 ion implantation", Denki Kagaku, 41, 482-7 (1973), CA80-53378.
862.Pashley, R. D., Mayer, J. W., Eisen, F. H. and Welch, B., "Comparison of encapsulating dielectrics on ion-implanted GaAs", Presented at Electrochemical Society Meeting, San Francisco, California, May 1974, EEA77-26347.
863.Perkins, J. G., "Ion implantation of thin metal and dielectric films", Thesis, Univ. Surrey, Guildford, England, EEA75-16004.
864.Primak, W., "Threshold for radiation effects in silica", Phys. Rev. B, 6, 4846-51 (1972), NSA27-8896.
865.Primak, W., "Ion bombardment of insulators", J. Nucl. Mater., 53, 238-45 (1974), PA77-79075.
866.Prince, J. L. and Schwettmann, F. N., "Diffusion of boron from implanted sources under oxidizing conditions", J. Electrochem. Soc., 121, 705-10 (1974), EEA77-30177.
867.Pringle, J. P. S., "Range profiles for ions implanted into anodic tantalum oxide", J. Electrochem. Soc., 121, 45-55 (1974), PA77-24500.
868.Schimko, R., Richter, C. E., Rogge, K., Schwarz, G. and Trapp, M., "Implanted arsenic and boron concentration profiles in SiO_2 layers", Phys. Status Solidi A, 28, 87-93 (1975), EEA78-19676.
869.Schmid, K., Ryssel, H., Muller, H., Wiedeburg, K. H. and Betz, H., "Properties of Al_2O_3 and SiO_2 surface layers on InSb, investigated by backscattering techniques", Thin Solid Films, 16, S11-12 (1973), EEA76-25986.
870.Schwettmann, F. N., Dexter, R. J. and Cole, D. F., "Etch rate characterization of boron-implanted thermally grown SiO_2", J. Electrochem. Soc., 120, 1566-70 (1973), EEA77-2462.
871.Sixt, G. and Goetzberger, A., "Control of positive surface charge in Si-SiO_2 interfaces by use of implanted Cs ions", Appl. Phys. Lett., 19, 478-9 (1971), EEA75-4627.

872. Sproul, M. E. and Nassibian, A. G., "Effect of O^+ implantation on silicon-silicon dioxide interface properties", Solid-State Electron., 17, 577-82 (1974), EEA77-22970.
873. Stroud, P. T. and Goode, G. C., "The trapping efficiency of low energy gold ions incident on evaporated silicon oxide", Thin Solid Films, 9, 297-300 (1972), PA76-12783.
874. Stroud, P. T., Lindsay, H. M. and Perkins, J. G., "Some preliminary studies of the structure of ion-bombarded thin films", Vacuum, 23, 125-30 (1973), PA76-72112.
875. Tokyuama, T., Toshida, I. and Warabisako, T., "Silicon-silicon dioxide interface states induced by implantation of various ion species", Proceedings of the 3rd International Conference on Ion Implantation in Semiconductors and Other Materials, 1972, p. 159-68, Publ. 1973, CA81-178838.
876. Truche, R., Dupuy, M. and Lafeuille, D., "Study of defects produced by ion implantation of silicon. III", J. Phys. Colloq., no. 5, 93-98 (1973), French, NSA30-19431.
877. Vologdin, E. N., Zhukova, G. A. and Mordkovich, V. N., "Irradiation of oxide-coated silicon with low-energy charged particles", Sov. Phys.-Semicond., 8, 1143-6 (1973), NSA27-26641.
878. Wang, S., Russell, T. and Royce, B. S. H., "Annealing studies of aluminum(+) implanted silicon dioxide thin films", Report AD-785856/6GA, Solid State Mater. Lab., Univ. Princeton, N. J., 1974, 26 pp., CA82-179540.
879. Zaininger, K. H., "Insulation of semiconductor components", Patent German 2425942, Publ. January 1975, CA83-20750.
880. Zhagata, L. A., Freiberga, L. A., Eimanis, I. A., Eglitis, I. E., Kalnynya, R. P. and Feltyn, I. A., "Changes in low-temperature SiO_2 film properties in RF plasma", Latv. PSR Zinat. Akad. Vestis Fiz. Teh. Ser., no. 3, 18-24 (1973), Russian, EEA76-35256.
881. Zorin, E. I., Murel, A. V., Pavlov, P. V., Tulovchikov, V. S. and Shitova, E. V., "Change in the properties of the silicon-silica phase boundary during bombardment with medium-energy ions", Mikroelektronika, no. 6, 122-8 (1973), Russian, CA80-42168.
882. Zuther, G., Prandtke, H. and Schmidt, M., "Dependence of trap density in SiO_2 thin films on the oxygen ion implantation dosis", Phys. Status Solidi A, 20, K123-5 (1973), EEA77-10079.

VI. Device Applications

1. General

883. Balazs, A., Hermann, L. and Gyulai, J., "Determination of junction depth in implanted silicon by 'pulled' anodization and capacitance-voltage measurements", Phys. Status Solidi A, 29, K105-8 (1975), EEA78-33035.

884.Beale, J. R. A., "Application of ion implantation to semiconductor devices", European Conference on Ion Implantation, 1970, p. 81-7, PA74-18773.
885.Comas, J., "Ion-implantation technology and device applications", IEEE Trans. Parts, Hybrids & Packag., PHP-10, 234-9 (1974), EEA78-11451.
886.Crowder, B. L., "Applications of ion implantation for new device concepts", J. Vac. Sci. & Technol., 8, S71-5 (1971), EEA75-4580.
887.Dill, H. G., Finnila, R. M., Leupp, A. M. and Toombs, T. N., "The impact of ion implantation on silicon device and circuit technology", Solid State Technol., 15 (12), 27-35 (1972), EEA76-12648.
888.Fan, G. J. and Hung, R., "Circuit fabrication using an ion-beam writing or selectively ion-implantation process", IBM Tech. Disclosure Bull., 18, 282 (1975), EEA78-41677.
889.Glotin, P., Bernard, J. and Monfret, A., "Implications of ion implantation technology on ion implanted active devices in silicon", Radiat. Eff., 7, 65-72 (1971), PA74-37199.
890.Goetzberger, A., "Application of ion implantation to semiconductor devices", 6th International Congress on Microelectronics, 1974, 7 pp., German, PA78-51135.
891.Kraus, L., "The use of recombination for testing of radiation damaged silicon devices", 1973 8th Annual Meeting of the IEEE Industry Applications Society, p. 635, EEA74-10270.
892.Ku, S. M. and Pillus, C. A., "Method of manufacturing a semiconductor device", Patent USA 3887994, Publ. June 1975, EEA78-41503.
893.Lange, L. N. and Hambleton, K. S., "The application of ion implantation to semiconductor devices", In: Festkorper Problems IX; Advances in Solid State Physics, Madelung, O., (Ed.), Pergamon Press, New York, p. 316-37 (1969), PA73-53853.
894.Lee, D. H. and Mayer, J. W., "Ion-implanted semiconductor devices", Proc. IEEE, 62, 1243-57 (1974), EEA77-34057.
895.Lepselter, M. P., "I2-impact on device fabrication", J. Electrochem. Soc., 120, 101C (1973), EEA76-23151.
896.Lloyd, W. W. and Dexter, R. J., "Ion implantation and conventional epitaxy to produce dielectrically isolated silicon layers", Patent USA 3855009, Publ. December 1974, CA82-79559.
897.MacRae, A. U., "Recent advances in ion implanted junction-device technology", 2nd International Conference on Ion Implantation in Semiconductors, 1971, p. 329-34, EEA75-13690.
898.MacRae, A. U., "Device fabrication by ion implantation", Radiat. Eff., 7, 59-63 (1971), PA74-37201.
899.Marley, J. A., Jr., "Semiconductor device", Patent German 2450903, Publ. April 1975, CA83-70302.
900.Parry, P. D. and Leiby, R. A., "Silicon wafer heating during ion implantation", Presented at 1973 International Electron Devices Meeting Supplement, Washington, D. C., USA, December 1973, PA77-33629.

901. Pleass, C. M., "Semiconductor device production - A challenge to ion implantation", Electron and Ion Beam Science and Technology Fifth International Conference, 1972, p. 225-34, EEA76-8405.

902. Portnoy, W. M. and Leedy, L. M., "Tungsten masking against boron implantation", Electron. Lett., 9, 129-30 (1973), EEA76-19828.

903. Reddi, V. G. K., "Ion implantation technology for silicon device processing", 1973 IEEE International Convention and Exposition, Vol. VI, 56.2/2 pp., EEA77-2494.

904. Reddi, V. G. K. and Yu, A. Y. C., "Ion implantation for silicon device fabrication", Solid State Technol., 15 (10), 35-41 (1972), EEA76-4639.

905. Saczuk, K., "The application of ion implantation technique to the manufacturing of silicon semiconductor devices", Elektronika, no. 11, 441-7 (1972), Polish, EEA76-8395.

906. Stephen, J., "Ion implantation in semiconductor device technology", Radio & Electron. Eng., 42, 265-83 (1972), EEA75-28412.

907. Tokuyama, T., "Application of ion implantation", Oyo Buturi, 43, 1015 (1974), Japanese, EEA78-15415.

908. Tsuchimoto, T., "Semiconductor device having a semiconductor circuit element in an isolated semiconductor region", Patent USA 3773566, Publ. November 1973, CA80-53757.

909. Wittkower, A., "Wafer handling for ion implantation", Electron. Packag. & Prod., 131 (5), 95 (1973).

2. P-N Junctions

910. Ashburn, P. and Morgan, D. V., "The role of radiation damage on the current-voltage characteristics of p-n junctions", Solid-State Electron., 17, 689-98 (1974), EEA77-30241.

911. Bogardus, E. H. and Poponiak, M. R., "Defects in As-implanted p-n junctions", Appl. Phys. Lett., 23, 553-5 (1973), EEA77-5918.

912. Chamberlain, N. G. and Lamb, D. R., "Electron emission from shallow ion-implanted P-N junctions in α-SiC", Int. J. Electron., 32, 565-8 (1972), PA75-40299.

913. Choi, W. E., "A new formation method of silicon p^+-n-n^+ junctions by VDH-implanter", J. Korean Inst. Electr. Eng., 22 (5), 5-11 (1973), Korean, EEA77-18774.

914. Donnelly, J. P. and Harman, T. G., "p-n junction $PbS_{1-x}Se_x$ photodiodes fabricated by Se^+ ion implantation", Solid-State Electron., 18, 288-90 (1975), EEA78-15620.

915. Faulkner, K. R. and Tedkill, A., "Ion implanted p-n junctions in $GaAs_{0.6}P_{0.4}$", 2nd International Conference on Ion Implantation in Semiconductors, 1971, p. 221-7, EEA75-13689.

916. Gislon, R., Urli, N. B. and Spoglia, U., "Study of the characteristics of implanted p-n junctions by C(V,I) measurements", European Conference on Ion Implantation, 1970, p. 92-6, PA74-19185.

917.Gusev, V. M., Guseva, M. I., Kurinnyy, V. I., Titov, V. V., Tsyplenkov, V. S., Baranova, E. K. and Streltsov, L. P., "An investigation of the properties of the oxide film, produced by atomic oxygen bombardment of silicon, and its use as a protection for p-n junctions", Radio Eng. & Electron. Phys., 16, 1357-61 (1971), PA75-51176.

918.Hou, S. L., Beck, K. and Marley, J. A., Jr., "Type conversion and p-n junction formation in ion-implanted ZnTe", Appl. Phys. Lett., 14, 151-4 (1969), EEA72-11263.

919.Kircher, C. J., "Comparison of leakage currents in ion-implanted and diffused p-n junction", J. Appl. Phys., 46, 2167-74 (1975), PA78-66023.

920.Lamouroux, B., "Semiconductor junctions formed by ion implantation or diffusion followed by laser beam irradiated and an anneal at relatively low temperatures", Patent French 2236271, Publ. January 1975, CA83-89841.

921.Mamontov, A. P., Nichipurenko, B. A., Okunev, V. D. and Presnov, V. A., "Insulation of p-n junctions in proton-irradiated gallium arsenide", Sov. Phys.-Semicond., 6 (4), 617-8 (1972), NSA27-16411.

922.Marine, J. and Rodot, H., "p-n junction formation in ion-implanted ZnTe", Appl. Phys. Lett., 17, 352-4 (1970), PA74-5289.

923.Marsh, O. J. and Dunlap, H. L., "Ion-implanted junctions and conducting layers in SiC", Radiat. Eff., 6, 301-12 (1970), PA74-21459.

924.Michel, A. E., Fang, F. F. and Pan, E. S., "Annealing properties of ion-implanted p-n junctions in silicon", J. Appl. Phys., 45, 2991-6 (1974), PA77-76394.

925.Mirsagatov, S. A., Morozkin, V. V., Rubinov, V. M. and Chirva, V. P., "p-n junction obtained in silicon carbide by lithium ion bombardment", Izv. Akad. Nauk Uzb. SSR, Ser. Fiz.-Mat. Nauk, 17 (2), 37-40 (1973), CA79-98406.

926.Park, Y. S. and Chung, C. H., "Type conversion and p-n junction formation in lithium-ion-implanted ZnS", Appl. Phys. Lett., 18, 99-101 (1971), PA74-28549.

927.Park, Y. S. and Shin, B. K., "Ion-implanted ZnSe p-n junction devices", Oyo Buturi, 43 (Suppl.), 508-12 (1974), EEA78-2822.

928.Park, Y. S. and Shin, B. K., "Injection electroluminescence in phosphorous-ion-implanted ZnSe p-n junction diodes", J. Appl. Phys., 45, 1444-6 (1974), EEA77-30683.

929.Ralia, S., Bartashyus, I., Zanevichyus, D. and Pranevichyus, L., "The doping of silicon by ion implantation, (p-n junction characteristics)", Izv. VUZ Fiz., no. 9, 140-1 (1971), Russian, EEA75-1103.

930.Woodcock, J. M. and Clark, D. J., "The ion implantation of donors for n^+-p junctions in GaAs", Gallium Arsenide and Related Compounds, 1974, p. 331-41, Publ. 1975, EEA78-28674.

931.Yudin, V. V., Kurinny, V. I., Akimov, Y. S. and Karatsyuba, A. P., "Generation centre distribution in boron-implanted silicon p-n junctions", Radiat. Eff., 22, 209-10 (1974), EEA78-2753.

3. Diodes

a. Photodiodes

932. Donnelly, J. P. and Harman, T. C., "As$^+$-ion implanted lead telluride p-n junction photodiodes", Solid-State Electron., 18, 1144-6 (1975).

933. Donnelly, J. P., Harman, T. C., Foyt, A. G. and Lindley, W. T., "PbTe photodiodes fabricated by Sb$^+$ ion implantation", J. Nonmet., 1, 123-8 (1973), EEA77-14390.

934. Donnelly, J. P., Harman, T. C., Foyt, A. G. and Lindley, W. T., "PbS photodiodes fabricated by Sb$^+$ ion implantation", Solid-State Electron., 16, 529-34 (1973), NSA27-26634.

935. Donnelly, J. P. and Holloway, H., "Photodiodes fabricated in epitaxial PbTe by Sb$^+$ ion implantation", Appl. Phys. Lett., 23, 682-3 (1973), EEA77-14870.

936. Hsieh, C. M., Mathews, J. R., Seidel, H. D., Pickar, K. A. and Drum, C. M., "Ion-implantation-damage gettering effect in silicon photodiode array camera target", Appl. Phys. Lett., 22, 238-40 (1973), EEA76-16645.

937. Hurwitz, C. E. and Donnelly, J. P., "Planar InSb photodiodes fabricated by Be and Mg ion implantation", Solid-State Electron., 18, 753-7 (1975).

938. McNally, P. J., "Ion implantation in InAs and InSb (and the fabrication of mosaic I.R. detectors)", Radiat. Eff., 6, 149-53 (1970), PA74-21433.

939. Peterman, D. A., Dexter, R. J., Echols, R. W. and Morgan, I. H., "Ion implanted light sensing diodes", Proceedings of the 2nd International Conference on Ion Implantation, 1971, p. 404-9, EEA75-13304.

940. Pickar, K. A., Dalton, J. V., Seidel, H. D. and Mathews, J. R., "Electrical properties of silicon diode array camera targets made by boron ion implantation", Appl. Phys. Lett., 19, 43-4 (1971), PA74-59063.

941. Soderman, D. A., Linden, K. J. and Kennedy, C. A., "Ion implanted PbTe and PbSnTe photodiode infrared detectors", Presented at 1973 International Electron Devices Meeting Supplement, Washington, D. C., December 1973, EEA77-19146.

b. LED

942. Hunsperger, R. G. and Marsh, O. J., "$Ga_{1-x}Al_xAs$ produced by Al^+ ion implantation of GaAs: (electroluminescent diode formation)", Appl. Phys. Lett., 19, 327-9 (1971), EEA75-7395.

943. Itoh, T. and Oana, Y., "Ga(AsP) light-emitting diode formed by ion implantation", J. Appl. Phys., 44, 4982-7 (1973), EEA77-14891.

944. Ono, Y., Saito, K., Shiraki, Y. and Shimada, T., "$GaAs_{1-x}P_x$ light emitting diodes produced by Zn ion implantation. I. Dose dependence of properties of implanted diodes", Jap. J. Appl. Phys., 14, 1489-94 (1975).
945. Shiraki, Y., Shimada, T., Ono, Y. and Komatsubara, K. F., "$GaAs_{1-x}P_x$ light emitting diodes produced by Zn ion implantation. II. Photoluminescence of p-type layers formed by ion implantation and diffusion", Jap. J. Appl. Phys., 14, 1495-500 (1975).
946. Takagi, T., Yamada, I., Sasaki, A. and Ishibashi, T., "Mn-implanted ZnS thin-film electroluminescent device", IEEE Trans. Electron Devices, ED-20, 1110-14 (1973), EEA77-543.

c. Detectors

947. Feuerstein, A. and Kalbitzer, S., "The fabrication of radial position sensitive detectors by ion implantation", Nucl. Instrum. & Methods, 109, 601 (1973), EEA76-30652.
948. Garin, A., Oria, M. and Waast, B., "Charged particle detectors made by implantation", Radiat. Eff., 10, 43-8 (1971), PA74-71710.
949. Herzer, H., Kalbitzer, S., Ponpon, J. P., Stuck, R. and Sittert, P., "Ion implanted high-purity germanium detectors", Nucl. Instrum. & Methods, 101, 31-7 (1972), EEA75-25615.
950. Howes, J. H., "Ion implanted p-n junctions in near intrinsic n-type silicon for nuclear particle detectors", 2nd International Conference on Ion Implantation in Semiconductors, 1971, p. 414-9, EEA75-14776.
951. Howes, J. H. and Krill, G., "The characteristics of large area ion implanted p-n junctions for nuclear radiation detectors", European Conference on Ion Implantation, 1970, p. 97-101, PA74-17877.
952. Kostka, A. and Kalbitzer, S., "Integrated dE-E detector system made by ion implantation", Appl. Phys. Lett., 23, 704-5 (1973), EEA77-16660.
953. Nakamoto, S., Aoki, T. and Kawabata, K., "Damage effects of silicon surface barrier detectors with protons, alphas, and nitrogen ions", Mem. Chubu Inst. Technol., 9, 87-98 (1973), NSA31-634.
954. Ponpon, J. P., Grob, J. J., Stuck, R., Burger, P. and Siffert, P., "Boron implanted contacts on high purity germanium (application to nuclear counters)", 2nd International Conference on Ion Implantation in Semiconductors, 1971, p. 420-9, PA75-25883.
955. Ponpon, J. P., Stuck, R., Siffert, P., Herzer, H. and Kalbitzer, S., "Further studies on implanted high purity germanium detectors", IEEE Trans. Nucl. Sci., NS-19, 281-88 (1972), EEA75-25585.
956. Stoll, H., Yariv, A., Hunsperger, R. G. and Tangonan, G. L., "Proton-implanted optical waveguide detectors in GaAs", Appl. Phys. Lett., 23, 664-5 (1973), EEA77-14869.

957. Zulliger, H. R., Drummond, W. E. and Middleman, L. M., "Properties of ion-implanted silicon detectors", IEEE Trans. Nucl. Sci., NS-19, 306-11 (1972), EEA75-25587.

d. IMPATT

958. Fong, T. T., Ying, R. S. and Lee, D. H., "Ion implanted X-band IMPATT/TRAPATT back-to-back diodes", Proc. IEEE, 61, 1044-5 (1973), EEA76-28832.
959. Lee, D. H., "Low-frequency noise of ion-implanted double-drift IMPATT diodes", Proc. IEEE, 61, 666-70 (1973), EEA76-19315.
960. Lee, D. H. and Ying, R. S., "Ion-implanted complementary IMPATT diodes for D-band", Proc. IEEE, 62, 1297-8 (1974), EEA77-34028.
961. Morris, B. L., Paulnack, C. L., Seidel, T. E., Dunn, C. N. and Smith, L. J., "Improved fabrication and noise performance of silicon double-drift millimeter-wave IMPATT diodes", 1973 International Electron Devices Meeting Technical Digest, p. 486-8, EEA77-14481.
962. Murphy, R. A., Lindley, W. T., Peterson, D. F., Foyt, A. G., Wolfe, C. M., Hurwitz, C. E. and Donnelly, J. P., "Proton-guarded GaAs IMPATT diodes", Proceedings of the 4th International Symposium on Gallium Arsenide and Related Compounds, 1972, p. 224-30, Publ. 1973, EEA76-22931.
963. Niehaus, W. C., Seidel, T. E. and Iglesias, D. E., "Double-drift IMPATT diodes near 100 GHz", IEEE Trans. Electron Devices, ED-20, 765-71 (1973), EEA76-34822.

e. Schottky

964. Knepper, R. W., "Reducing series resistance of a Schottky barrier diode clamp for NPN transistor", IBM Tech. Disclosure Bull., 17, 1609-10 (1974), EEA78-11618.
965. Putney, Z. C., "Self-aligned ion implanted guard ring for high junction breakdown voltage", IBM Tech. Disclosure Bull., 17, 2359-60 (1975), EEA78-19907.
966. Shannon, J. M., "Reducing the effective height of a Schottky barrier using low-energy ion implantation", Appl. Phys. Lett., 24, 369-71 (1974), EEA77-26388.
967. Shannon, J. M., "Increasing the effective height of a Schottky barrier using low-energy ion implantation", Appl. Phys. Lett., 25, 75-7 (1974), PA77-72864.
968. Taylor, P. D. and Morgan, D. V., "Role of radiation damage on the contact resistance of GaAs Schottky barriers", Electron. Lett., 10, 428-9 (1974), EEA78-2767.

f. Others

969. Akimov, Y. S., Gusev, V. M., Kurinnyy, V. I. and Ryzhikov, I. V., "Investigation of the injection characteristics of symmetrical pin-diodes prepared by the method of ion implantation", Radio Eng. & Electron. Phys., 16, 1714-22 (1971), EEA75-36231.
970. Barnoski, M. K., Hunsperger, R. G. and Lee, A., "Ion-implanted GaAs injection laser", Appl. Phys. Lett., 24, 627 (1974), EEA77-34351.
971. Barnoski, M. K., Hunsperger, R. G., Wilson, R. G. and Tangonan, G., "Proton-implanted GaP optical waveguide", J. Appl. Phys., 44, 1925-6 (1973), EEA76-20759.
972. Gibbons, P. E., "The application of ion implantation to avalanche multiplication devices", Proceedings of the 2nd International Conference on Ion Implantation in Semiconductors, 1971, p. 410-13, EEA75-13708.
973. Glew, R. W. and Bryant, F. J., "Cathodoluminescence and ion implantation of cadmium sulphide/cuprous sulphide solar cells", Thin Solid Films, 29, 269-74 (1975).
974. Howes, M. J., Morgan, D. V. and Ashburn, P., "The small signal admittance of carbon implanted p-n diodes", Solid-State Electron., 18, 491-7 (1975), PA78-46739.
975. Hunsperger, R. G. and Hirsch, N., "Ion implanted gallium arsenide semiconductor devices fabricated in semi-insulating gallium arsenide substrates", Patent USA 3914784, Publ. October 1975, CA83-211935.
976. Lee, D. H., Berenz, J. J. and Bernick, R. L., "Ion-implanted n^+ contacts for K band GaAs Gunn-effect diodes", Electron. Lett., 11, 189-91 (1975), EEA78-24291.
977. Lee, D. H., Ying, R. S. and Jamba, D. M., "Ion-implanted double-drift K_a-band diodes", Proc. IEEE, 62, 1025-6 (1974), PA77-57712.
978. Mizutani, T. and Kurumada, K., "GaAs planar Gunn digital devices by sulphur-ion implantation", Electron. Lett., 11, 638-9 (1975).
979. Monteith, L. K., Littlejohn, M. A., Nauser, J. R. and Hendricks, H. D., "Indium-zinc implanted gallium arsenide diodes", Radiat. Eff., 16, 133-4 (1972), EEA76-8647.
980. Pickar, K. A. and Dalton, J. V., "Life time effects in ion-implanted silicon (junction diode)", Radiat. Eff., 6, 89-94 (1970), PA74-21440.
981. Rosinski, W., "The current-voltage characteristics of boron implanted silicon diodes", Electron Technol., 5 (1), 95-102 (1972), PA75-71969.
982. Rosinski, W. and Saczuk, K., "Ion implanted silicon diodes with steep C-V dependence", Electron Technol., 6 (3-4), 63-72 (1973), EEA77-34016.
983. Saczuk, K., Kunicki, J. and Kyrlow, J., "Ion implanted silicon planar diodes", Elektronika, no. 12, 494-6 (1972), Polish, EEA76-12578.

984.Shin, B. K. and Park, Y. S., "Switching and memory effects in phosphorus-ion-implanted ZnSe devices", Proc. IEEE, 62, 538-40 (1974), EEA77-22981.
985.Vaidyanathan, K. V. and Walker, G. H., "The effect of Be^+ ion implanted exponential and uniform impurity profiles on the electrical characteristics of GaAs solar cells", Record of the 10th IEEE Photovoltaic Specialists Conference, 1973, p. 31-3, Publ. 1974, EEA77-40601.
986.Valette, S., Labrunie, G. and Lizet, J., "Optical waveguides in ion-implanted ZnTe", J. Appl. Phys., 46, 2731-3 (1975).
987.Wei, D. T. Y., Lee, W. W. and Bloom, L. R., "Quartz optical waveguide by ion implantation", Appl. Phys. Lett., 22, 5-7 (1973), PA76-17889.

7. Bipolar Transistors

988.Archer, J. A., "Improved microwave transistor structure", Electron. Lett., 8, 499 (1972).
989.Archer, J. A., "Low-noise implanted-base microwave transistors", Solid-State Electron., 17, 387-93 (1974), EEA77-18794.
990.Assemat, J. L., "Application of ion implantation to n-p-n transistors", 2nd International Conference on Ion Implantation in Semiconductors, 1971, p. 351-6, EEA75-16739.
991.Barnoski, M. K. and Loper, D. D., "Ion-implanted bipolar transistor carrier concentration profiles", Solid-State Electron., 16, 433-40 (1973), NSA27-26632.
992.Barnoski, M. K. and Loper, D. D., "Microwave characteristics of ion-implanted bipolar transistors", Solid-State Electron., 16, 441-51 (1973), NSA27-26633.
993.Douglas, E. C. and Wu, C. P., "Semiconductor device fabrication by ion implantation", Patent German 2450070, Publ. April 1975, CA83-70301.
994.Edwards, R., Payne, R. S., Gibbon, C. F., Olmstead, M. L., Moline, R. A., Kushler, D. L., Weber, G. R. and Kraetsch, R. T., "Oxide isolation technology featuring ion implantation and partially self-registered emitters", J. Electrochem. Soc., 120, 101C (1973), EEA76-23498.
995.Evans, W. J., Tretola, A. R., Payne, R. S., Olmstead, M. L. and Speeny, D. V., "Oxide-isolated monolithic technology and applications", IEEE J. Solid-State Circuits, SC-8, 373-80 (1973), EEA76-35362.
996.Fair, R. B., "Cooperative effects between arsenic and boron in silicon during simultaneous diffusions from ion implanted and chemical source predepositions", Solid-State Electron., 17, 17-24 (1974), EEA77-10234.
997.Feng, B. C. and Feng, G. C., "Ion-implanted emitters", IBM Tech. Disclosure Bull., 17, 2942-3 (1975), EEA78-19805.

998. Graul, J., Kaiser, H. and Wilhelm, W. J., "Bipolar high-speed low-power gates with double implanted transistors", IEEE J. Solid-State Circuits, SC-10, 201-4 (1975), EEA78-33209.

999. Graul, J. and Murrmann, H., "Relations between ion implantation parameters and electrical characteristics of double implanted integrated bipolar transistors", 6th International Congress on Microelectronics, 1974, 6 pp., EEA78-24338.

1000. Heinrich, H., Hastings, L. and Rozenbergs, J., "Simultaneous diffusion of ion-predeposited arsenic and boron in silicon", J. Appl. Phys., 45, 4670-1 (1974), PA78-19086.

1001. Hung, R. and Lo, D. C., "Double-ion implanted bipolar transistor", IBM Tech. Disclosure Bull., 27, 1536-7 (1974), EEA78-8553.

1002. Kellner, W. and Goetzberger, A., "Current gain recovery in silicon nitride passivated planar transistors by hydrogen implantations", IEEE Trans. Electron Devices, ED-22, 531-4 (1975), CA83-140842.

1003. Koji, T., "Ion-implanted-base-transistor h_e dispersion and low frequency noise performance", NEC Res. & Dev., 36, 37-41 (1975), EEA78-28779.

1004. Kostka, A. and Worner, K., "Ion implanted linear super high frequency circuits", 6th International Congress on Microelectronics, 1974, 8 pp., German, EEA78-24526.

1005. Ludvik, S. and Froess, P., "Comparison of noise parameters of diffused and ion-implanted microwave transistors", Electron. Lett., 10, 40-1 (1974), EEA77-14486.

1006. Lutz, R. W., MacDougall, J. D., Hanson, J. W. and Tkal, O., "New developments in consumer integrated circuits", Microelectronics, 4 (4), 3-9 (1973), EEA76-29288.

1007. MacRae, A. U., "Invited: bipolar transistors made by ion implantation", Oyo Buturi, 43 (Suppl.), 483 (1974), EEA78-2856.

1008. Masters, B. J., Ku, S. and Schwenker, R. O., "Bipolar transistor structures", IBM Tech. Disclosure Bull., 17, 2944 (1975), EEA78-19806.

1009. Morizot, M., Dubee, H. and Cornette, A., "Microwave transistors fabricated by ion-implantation: selection of doping impurities and prototype realization", 2nd International Conference on Ion Implantation in Semiconductors, 1971, p. 345-50, EEA75-14027.

1010. Nicholas, K. H., Ford, R. A. and Daniel, P. J., "Reduced gain of ion-implanted transistors", Appl. Phys. Lett., 26, 320-2 (1975), EEA78-24317.

1011. Ohmura, Y., Abe, T., Mimura, S., Konaka, M., Kanazawa, M., Ohtsubo, H. and Fujinuma, K., "Enhanced diffusion of substrate impurities into epitaxial layers in Si by proton irradiation (application to non microwave transistors)", 2nd International Conference on Ion Implantation in Semiconductors, 1971, p. 335-9, PA75-25902.

1012.Pavlov, P. V., Volodko, V. G., Zorin, Y. I., Tetelbaum, O. I., Tulovchikov, V. S. and Chirinskaya, T. Y., "Features of concentration profiles during ion implantation and their use in the design of varactors and bipolar transistors", Izv. VUZ Radioelectron., 14, 1353-64 (1971), Russian, EEA75-13540.

1013.Payne, R. S. and Scavuzzo, R. J., "Bipolar transistors and method of manufacture", Patent USA 3856578, Publ. December 1974.

1014.Payne, R. S., Scavuzzo, R. J., Olson, K. H., Nacci, J. M. and Moline, R. A., "Fully ion-implanted bipolar transistors", IEEE Trans. Electron Devices, ED-21, 273-8 (1974), EEA77-14492.

1015.Polata, B., Marley, J. A., Jr. and Kerr, J. T., "Bipolar transistor structure having ion implanted region and method", Patent USA 3855007, Publ. December 1974.

1016.Pruniaux, B., Assemat, J. L., Delandre, M., Dumetz, A., Gabillot, P. and Lagorsse, J. M., "Design and performances of double implanted microwave bipolar transistors", 1973 European Microwave Conference, Vol. I, A11.5/4 pp., EEA77-6008.

1017.Prussin, S., "Role of sequential annealing, oxidation, and diffusion upon defect generation in ion-implanted silicon surfaces. (Bipolar IC processing)", J. Appl. Phys., 45, 1635-42 (1974), PA77-61308.

1018.Schmid, K., Kranz, H., Ryssel, H., Muller, W. and Dathe, J., "Ion-implanted Ge transistor: Basic experiments", Phys. Status Solidi A, 23, 523-30 (1974), PA77-57713.

1019.Shackle, P. W. and Payne, R. S., "The fabrication of bipolar transistors using electron lithography, ion implantation and nickel masked gold metallization", J. Vac. Sci. & Technol., 10, 1090-3 (1973), EEA77-23018.

1020.Sigmon, T. W., "Characteristics of high performance microwave transistors fabricated by ion implantation", 1973 International Electron Devices Meeting Technical Digest, p. 387-9, EEA74-14506.

1021.Stannon, J. M. and Kerr, J. A., "Transistor ion implantation", Patent UK 1332931, October 1973, CA81-55740.

1022.Tansley, T. L., "Base retardation in implanted-diffused transistors", Microelectronics, 6 (3), 16-20 (1975), EEA78-28778.

1023.Tsuchimoto, T., Itoh, K. and Kumagaya, F., "Characteristics of high dose As implanted Si and UHF power transistor with As implanted emitter", 6th International Congress on Microelectronics, 1974, 2 pp., EEA78-24337.

1024.Ueno, S., Nakatsuka, M., Sando, S. and Irie, T., "8 GHz ion implanted PNP microwave transistor", NEC Res. & Dev., no. 30, 47-53 (1973), EEA77-9573.

5. Field-Effect Transistors

a. MOSFET

1025.Baruch, P., Borel, J. and Monnier, J., "Doping of semiconductors", Patent German 2500728, Publ. July 1975, CA83-171668.

1026.Berger, J., "Nonequilibrium phenomena in ion-implanted MOS capacitors", Appl. Phys. Lett., 24, 497-9 (1974), EEA77-30207.
1027.Berger, J., "Deep-channel MOS transistor", IEEE Trans. Electron Devices, ED-22, 314-19 (1975), EEA78-19831.
1028.Bernard, J., "Technology and characteristics of an ion implanted MOSFET", Proceedings of the International Conference on Advanced Microelectronics, 1970, 2 pp., EEA74-1614.
1029.Bernard, J., "Improved MOS technology using ion implantation", Onde Electr., 54, 15-22 (1974), French, EEA77-14598.
1030.Bernard, J., Borel, J., Glotin, P. and Lacour, J., "Integrated circuit containing metal-oxide-silicon transistors obtained by ion implantation", Patent UK 1347782, Publ. February 1974, CA81-55736.
1031.Bernard, J. and Ceng, L., "Speed improvements of ion implanted self aligned gate MOS transistors", 2nd International Conference on Ion Implantation in Semiconductors, 1971, p. 383-8, EEA75-13794.
1032.Bobenrieth, A., Cahen, O. and Lyon-Caen, R., "Fast, low power switching integrated circuits made by electron beam pattern generation and ion implantation", Electron and Ion Beam Science and Technology Fifth International Conference, 1972, p. 265-80, EEA76-8751.
1033.Bobenrieth, A., Tung, P. N. and Arnodo, C., "Junction field effect transistors fabricated by ion implantation", 2nd International Conference on Ion Implantation in Semiconductors, 1971, p. 340-4, EEA75-13793.
1034.Bower, R. W., Dill, H. G., Aubuchon, K. G. and Thompson, S. A., "MOS field effect transistors formed by gate masked ion implantation", IEEE Trans. Electron Devices, ED-15, 757-61 (1968), EEA72-5033.
1035.Brack, K., Kroell, K. and Gansauge, P., "Lateral nonuniform doping of semiconductor structures by ion implantation", IBM Tech. Disclosure Bull., 16, 3287-8 (1974), EEA77-30153.
1036.Buckley, J. and Drew, B., "What's all this about ion implantation?", New Electron., 7 (2), 33, 39 (1974), EEA77-18859.
1037.Christie, K. H. and Johnson, W. S., "n-channel MOSFET field protection by nonmasked boron ion implantation", 1973 International Electron Devices Meeting Technical Digest, p. 464-6, EEA77-14561.
1038.Crawford, B., "Implanted depletion loads boost MOS array performance", Electronics, 45 (9), 85-90 (1972), EEA75-20202.
1039.Dennard, R. H., Gaensslen, F. H., Yu, H. N., Rideout, V. L., Bassous, E. and LeBlanc, A. R., "Ion implanted MOSFET's with very small physical dimensions", IEEE J. Solid-State Circuits, SC-9, 256-68 (1974), EEA77-38839.
1040.Dill, H. G., Bower, R. W. and Toombs, T. N., "Ion-implanted M.O.S. technology", Radiat. Eff., 7, 45-7 (1971), PA74-36541.

1041.Dill, H. G. and Coppen, P. J., "Ion-implanted MOS LSI circuits", 1971 IEEE International Convention Digest, 1971, p. 190-1, EEA74-26902.
1042.Dill, H. G., Toombs, T. N. and Bauer, L. O., "Recent advances in ion implanted MOS technology", 2nd International Conference on Ion Implantation in Semiconductors", 1971, p. 315-28, PA75-25878.
1043.Edwards, J. R. and Marr, G., "Depletion-mode IGFET made by deep ion implantation", IEEE Trans. Electron Devices, ED-20, 283-9 (1973), EEA76-16443.
1044.Fang, F., Hatzakis, M. and Ting, C. H., "Electron beam fabrication of ion implanted high performance FET circuits", J. Vac. Sci. & Technol., 10, 1082-5 (1973), EEA77-23147.
1045.Fang, F. F. and Rupprecht, H. S., "High performance MOS integrated circuit using the ion implantation technique", IEEE J. Solid-State Circuits, SC-10, 205-11 (1975), EEA78-33210.
1046.Forbes, L., "N-channel ion-implanted enhancement/depletion MOSFET's", IEEE J. Solid-State Circuits, SC-8, 184-5 (1973), ibid pp. 226-230, EEA76-15993.
1047.Geipel, H. J. and Shasteen, R. B., "Low-leakage implanted source", IBM Tech. Disclosure Bull., 18, 337 (1975), EEA78-37541.
1048.Gerger, J., "Ion implantation induced potential barrier in MOS capacitors", Presented at 1973 International Electron Devices Meeting Supplement, Washington, D. C., December 1973, EEA77-18762.
1049.Goetzberger, A., "Ion implantation in MOS-structures", Oyo Buturi, 43 (Suppl.), 289-94 (1974), EEA78-2776.
1050.Goetzberger, A. and Schulz, M., "Fundamentals of MOS technology", Meeting of the German Physical Society on Advances in Solid State Physics, 1973, p. 309-36, EEA76-29216.
1051.Hofflinge, B., Bigall, K. D., Zimmer, G. and Krimmel, E. F., "Ion implanted low-voltage MOS circuits", Siemens Forsch.- & Entwicklungsber., 1, 362-8 (1972), EEA76-8739.
1052.Hoffman, G., "Applications of ion-implanted depletion-mode MOS devices", WESCON, 1971, 28/3 4 pp., EEA75-1185.
1053.Hoffman, G., "Suitability of ion-implanted MOS/LSI for automotive applications", IEEE Vehicular Technology 22nd Annual Conference, 1971, 3 pp., EEA75-15569.
1054.Hoffman, G., "Applications for ion implantation depletion MOS devices", Mundo Electron., no. 32, 57-61 (1974), Spanish, EEA78-648.
1055.Hornung, A., Kasprzak, L. A. and Nagarajan, A., "Ion implanted polycrystalline silicon FET gates", IBM Tech. Disclosure Bull., 16, 2901 (1974), EEA77-23051.
1056.Hswe, M., Palmer, R. B., Shopbell, M. L. and Mai, C. C., "Characteristics of P-channel MOS field effect transistors with ion-implanted channels", Solid-State Electron., 15, 1237-43 (1972), EEA75-39955.

1057. Huang, C. and Van der Ziel, A., "Thermal noise in ion-implanted MOSFET's", Solid-State Electron., 18, 509-10 (1975), EEA78-19851.

1058. Huber, R. J. and Smith, K. F., "Method for manufacturing metal oxide semiconductor integrated circuit of reduced size", Patent USA 3874937, Publ. April 1975, EEA78-37667.

1059. Huber, R. J., Smith, K. F., Hill, D. R., Fordemwalt, J. N., Hanson, J. W. and Dobelle, W. H., "Simplified n-channel process achieves high performance", Electronics, 47 (5), 117-22 (1974), EEA77-14584.

1060. Hung, R., "High performance ion-implanted MOSFET technology", International Conference on Large Scale Integrated Circuits, 1974, p. 34-42, EEA78-11743.

1061. Hunsperger, R. G. and Hirsch, N., "GaAs field-effect transistors with ion-implanted channels", Electron. Lett., 9, 577-8 (1973), EEA77-10254.

1062. Hunsperger, R. G. and Hirsch, N., "Ion-implanted microwave field-effect transistors in GaAs", Solid-State Electron., 18, 349-53 (1975), EEA78-11668.

1063. Johnson, W. S., "Method for producing improved transistor devices", Patent USA 3873372, Publ. March 1975.

1064. Johnson, W. S. and Ku, S. M., "Method for manufacturing ion-implanted insulated-gate field-effect semiconductor transistor devices", Patent USA 3852120, Publ. December 1974.

1065. Kamoshida, M., "Threshold voltage and 'gain' term of ion-implanted enhancement-mode n-channel MOS transistors", Appl. Phys. Lett., 22, 404-5 (1973), EEA76-23448.

1066. Kamoshida, M., "Electrical characteristics of boron-implanted n-channel MOS transistors", Solid-State Electron., 17, 621-6 (1974), EEA77-23068.

1067. Kamoshida, M. and Kudoh, O., "Surface depletion region width dependence of threshold voltage shift of ion-implanted MOS transistor", Appl. Phys. Lett., 24, 501-3 (1974), EEA77-30325.

1068. Kozlov, Y. G. and Ozhogin, M. A., "An MOS-transistor with built-in channel obtained by ion implantation", Izv. VUZ Radioelektron., 17 (12), 98-100 (1974), Russian, EEA78-15360.

1069. Kudoh, O., Nakamura, K. and Kamoshida, M., "Implant dose profile dependence of electrical characteristics of ion-implemented MOS transistors", J. Appl. Phys., 45, 4514-19 (1974), EEA78-11659.

1070. Lecrosnier, D. P. and Pelous, G. P., "Ion-implanted FET for power applications", IEEE Trans. Electron Devices, ED-21, 113-18 (1974), EEA74-14522.

1071. Lepselter, M. P., MacRae, A. U. and MacDonald, R. W., "SB-IGFET, II: an ion implanted IGFET using Schottky barriers", Proc. IEEE, 57, 812-13 (1969), EEA72-17741.

1072. MacDougall, J. and Manchester, K., "Ion implantation offers a bagful of benefits for MOS", Electronics, 43 (13), 86-90 (1970), PA73-59219.

1073. MacPherson, M. R., "The adjustment of MOS transistor threshold voltage by ion implantation", Appl. Phys. Lett., 18, 502-4 (1971), PA74-52968.

1074. MacPherson, M. R., "Threshold shift calculations for ion implanted M.O.S. devices", Solid-State Electron., 15, 1319-26 (1972), EEA75-819.

1075. Mai, C. C., Hswe, M. and Palmer, R. B., "Ion implantation combined with silicon-gate technology", IEEE Trans. Electron Devices, ED-19, 1219-21 (1972), EEA75-35392.

1076. Manchester, K. E., MacDougall, J. D., Tkal, O. and Chu, T. W., "Monolithic varactor tuned RF amplifier IC using ion implantation", Digest of Technical Papers of the 1973 IEEE International Solid State Circuits Conference, p. 186-7, EEA73-22988.

1077. Masuhara, T. and Etoh, J., "Low-level currents in ion-implanted MOSFET", IEEE Trans. Electron Devices, ED-21, 799-807 (1974), EEA78-11651.

1078. McCaughan, D. V. and Murphy, V. T., "Low-energy ion bombardment of silicon dioxide films on silicon. II. Inert ambient annealing of degradation in MOS devices", J. Appl. Phys., 44, 3182-90 (1973), EEA76-35251.

1079. Moline, R. A., Gibson, W. C. and Heck, L. D., "An ion-implanted Schottky-barrier gate field-effect transistor", IEEE Trans. Electron Devices, ED-20, 317-20 (1973), EEA73-16444.

1080. Morehead, F. F., "A general calculation of the redistribution of ion implanted profiles in MOS and other processing", Electrochemical Society Fall Meeting (Extended abstracts), 1974, p. 474-6, EEA78-37444.

1081. Nakamura, K. and Kamoshida, M., "An investigation of stability of p-channel ion-implanted MOS transistors by BT treatments", Jap. J. Appl. Phys., 12, 1635-6 (1973), EEA74-380.

1082. Nakamura, K. and Kamoshida, M., "Electrical characteristics of ion-implanted p-channel MOS transistors", J. Appl. Phys., 45, 334-40 (1974), EEA77-23053.

1083. Nakamura, K. and Kamoshida, M., "Annealing characteristics of ion-implanted p-channel MOS transistors", J. Appl. Phys., 45, 4262-7 (1974), PA78-19084.

1084. Nakamura, K., Kudoh, O. and Kamoshida, M., "Noise characteristics of ion-implanted MOS transistors", J. Appl. Phys., 46, 3189-94 (1975), EEA78-37545.

1085. Onnasch, D. and Reimerdes, H. P., "Radiation effects in n-channel MOSFET's", Solid-State Electron., 17, 663 (1974).

1086. Palmer, R. B., Mai, C. C. and Hswe, M., "The effect of oxide thickness on threshold voltage of boron ion implanted MOSFET", J. Electrochem. Soc., 120, 999-1001 (1973), EEA73-32275.

1087. Pan, E. and Fang, F. F., "Lateral spread of ion-implanted impurities in silicon", J. Appl. Phys., 45, 2801-3 (1974), PA77-67735.

1088.Peressini, P. P. and Johnson, W. S., "Threshold adjustment of n-channel enhancement mode FETs by ion implantation", 1973 International Electron Devices Meeting Technical Digest, p. 67-8, EEA74-14562.
1089.Perkins, C. W., "Radiation effects in modified oxide insulators in MOS structure", IEEE Trans. Nucl. Sci., NS-15, 176-81 (1968), PA72-17420.
1090.Pomper, M. and Tihanyi, J., "Ion-implanted ESFI MOS devices with short switching times", IEEE J. Solid-State Circuits, SC-9, 250-6 (1974), EEA77-38468.
1091.Pond, R. J., "Planar indium antimonide N-channel MOSFET by proton bombardment", Thesis, Case Western Reserve Univ., Cleveland, Ohio, 1971, 67 pp., Order No. 72-87, EEA75-28526.
1092.Pruniaux, B. R., North, J. C. and Miller, G. L., "Compensation of N-type GaAs by proton bombardment (semi-i.g.f.e.t.)", Proceedings of the 2nd International Conference on Ion Implantation, 1971, p. 212-21, EEA75-13600.
1093.Rideout, V. L., Gaensslen, F. H. and LeBlanc, A., "Device design considerations for ion implanted n-channel MOSFET's", IBM J. Res. & Dev., 19, 50-9 (1975), EEA78-19840.
1094.Runge, H., "Threshold voltage shift of p-channel MOS-transistors by implantation of donors", Appl. Phys., 8, 43-6 (1975), German, EEA78-41605.
1095.Runge, H., "Modification of the theory of threshold voltage shift of MOS transistors by ion implantation", Phys. Status Solidi A, 30, 147-53 (1975), German, PA78-62115.
1096.Saczuk, K., "The application of ion implantation in MOS-FET technology", Elektronika, no. 8, 334-8 (1971), Polish, EEA75-1159.
1097.Saczuk, K., "Ion implantation in the processing of digital IC elements", Elektronika, 14, 233-8 (1973), Polish, EEA73-35350.
1098.Sansbury, J. D., "MOS field threshold increase by phosphorus-implanted field", IEEE Trans. Electron Devices, ED-20, 473-6 (1973), PA73-33998.
1099.Sasaki, N., "Characteristics of enhancement/depletion (E/D) gate MOSFET fabricated using ion implantation", Solid-State Electron., 18, 777-84 (1975), EEA78-28813.
1100.Sauvage, J. A. and Evans, S. A., "Modeling of depletion load devices made by ion implantation", 1973 International Electron Devices Meeting Technical Digest, p. 61-3, EEA77-14538.
1101.Schemmert, W. and Hofflinger, B., "Optimisation of depletion-mode transistors for an M.O.S. p-channel technology", Electron. Lett., 9, 555-6 (1973), EEA77-6070.
1102.Schemmert, W. and Zimmer, G., "Threshold-voltage sensitivity of ion-implanted M.O.S. transistors due to process variations", Electron. Lett., 10, 151-2 (1974), EEA77-23041.
1103.Shannon, J. M., "Ion-implanted high-frequency MOS transistors", Philips Tech. Rev., 31, 267-70 (1970), EEA74-34098.

1104.Shopbell, M. L., "Ion implantation in MOS processing", Western Electronic Show and Convention (Vol. 16), 1972, 19/3 4 pp., EEA76-8708.
1105.Sigmon, T. W., "Stabilization of MOS structures by boron ion implantation", IEEE Proc., 63, 1619-20 (1975).
1106.Sigmon, T. W. and Swanson, R., "MOS threshold shifting by ion implantation", Solid-State Electron., 16, 1217-32 (1973), EEA77-6039.
1107.Suzuki, S., Nagahashi, Y., Tanaka, T., Muta, H. and Okabayashi, H., "A static RAM with normally-off-type Schottky barrier FETs", Digest of Technical Papers of the 1973 IEEE International Solid State Circuits Conference, p. 36-7, EEA76-23036.
1108.Teszner, S., Lecrosnier, D. P. and Pelous, G. P., "Ion-implantation field-effect semiconductor devices", Patent USA 3767982, Publ. October 1973.
1109.Tokyuama, T., "Invited: ion implantation into MOS structures", Oyo Buturi, 43 (Suppl.), 499-503 (1974), EEA78-2780.
1110.Verjans, J. and Van Overstraeten, R., "Electrical characteristics of boron-implanted n-channel MOS transistors for use in logic circuits", IEEE Trans. Electron Devices, ED-22, 862-8 (1975), EEA78-41606.
1111.Verjans, J. and Van Overstraeten, R., "Measurement of electrical impurity profile of implanted ions, using the pulsed MOS C-V technique", Solid-State Electron., 18, 911-6 (1975), EEA78-41496.
1112.Wada, K. and Nakano, M., "Substrate bias characteristics of $^{31}P^{+}$ implanted N-channel MOSFET", Jap. J. Appl. Phys., 13, 1673-4 (1974), EEA78-647.
1113.Wang, K. L. and Gray, P. V., "Fabrication and characterization of germanium ion-implanted IGFET's", IEEE Trans. Electron Devices, ED-22, 353-5 (1975), EEA78-19833.
1114.Wang, P. P. and Spencer, O. S., "Threshold voltage characteristics of double-boron-implanted enhancement-mode MOSFET's", IBM J. Res. & Develop., 19, 530-38 (1975).
1115.Warabisako, T., Yoshida, I. and Tokuyama, T., "Properties of MOS structures prepared on substrates having ion-implanted impurity distribution profiles", Proceedings of the 4th Conference on Solid State Devices, 1972, p. 181-6, Publ. 1973, CA80-31736.
1116.Welch, B. M., Elsen, F. H. and Higgins, J. A., "Gallium arsenide field-effect transistors by ion implantation", J. Appl. Phys., 45, 3685-7 (1974), EEA78-646.
1117.Yoshida, I., Musuhara, T., Kubo, M. and Tokuyama, T., "Device design of an ion implanted high voltage MOSFET", Oyo Buturi, 44 (Suppl.), 249-55 (1975), EEA78-41618.
1118.Zenker, W. and Mittenbacher, J., "The influence of the implantation of ions on the technology of MOS components", Nachrichtentechnik, 22, 369-70 (1972), German, EEA75-39949.

1119. Zohta, Y., "Rapid determination of semiconductor doping profiles in MOS structures", Solid-State Electron., 16, 124-6 (1973), NSA27-11202.

1120. Zorin, E. I., Pavlov, P. V., Popov, Y. S., Tulovchikov, V. S. and Shitova, E. V., "The influence of ion bombardment of MOS-structure properties", Proceedings of the International Conference on the Physics and Chemistry of Semiconductor Heterojunctions and Layer Structures, 1970, p. V231-42, Publ. 1971, EEA74-26753.

b. CMOSFET

1121. Barile, C. A., Elie, G. T. and Kluge, H. C., II, "Fabrication of complementary F.E.T.'s with ion-implanted source/drains", IBM Tech. Disclosure Bull., 17, 2874-5 (1975), EEA78-19844.

1122. Charles, F., "The charge-depletion complementary pair MOS", Electron. & Microelectron. Ind., no. 168, 62-3 (1973), French, EEA76-26143.

1123. Coppen, P. J., Aubuchon, K. G., Bauer, L. O. and Moyer, N. E., "A complementary MOS 1.2 volt watch circuit using ion implantation", Solid-State Electron., 15, 165-75 (1972), EEA75-12782.

1124. Coppen, P. J., Bauer, L. O. and Dill, H. G., "Ion implanted CMOS technology", WESCON, 1971, 28/1 4 pp., EEA75-1184.

1125. Douglas, E. C., "Ion implantation for threshold control in COSMOS circuits", IEEE Trans. Electron Devices, ED-21, 324-31 (1974), EEA77-30379.

1126. Douglas, E. C. and Dingwall, A. G. F., "Surface doping using ion implantation for optimum guard layer design in COS/MOS structures", IEEE Trans. Electron Devices, ED-22, 849-61 (1975).

1127. Finnila, R. M., Toombs, T. N. and Dill, H. G., "High density ion-implanted CMOS technology", Electron. Packag. & Prod., 13 (1), 83 (1973).

1128. George, J. and Chruma, J., "The application of ion implantation to CMOS", Solid State Technol., 16 (11), 43-6 (1973), EEA77-10301.

1129. Sopira, M. M., Malmberg, P. R. and Meiksin, Z. H., "Fabrication of integrated CMOS transistors using electron lithography and ion implantation", J. Vac. Sci. & Technol., 10, 1086-9 (1973), EEA77-23148.

1130. Swanson, R. M. and Meindl, J. D., "Ion-implanted complementary MOS transistors in low-voltage circuits", IEEE J. Solid-State Circuits, SC-7, 146-53 (1972), EEA75-13759.

c. MNOSFET

1131. Kendall, E. J. M. and Haslett, J. W., "On the removal of the memory properties of MNOS devices", IEEE Trans. Electron Devices, ED-19, 287-8 (1972), EEA75-10267.

1132. Krick, P. J., "MNOS array structure with self-aligned ion implantation", IBM Tech. Disclosure Bull., 17, 1809-10 (1974), EEA78-11732.

1133. Krick, P. J., "The implanted stepped-oxide MNOSFET", IEEE Trans. Electron Devices, ED-22, 62-3 (1975), EEA78-15351.

1134. Maloney, C. R. C., "A simple experiment to compare the switching speeds of an implant auto-registered and a diffused M.I.S.T.", European Conference on Ion Implantation, 1970, p. 107-9, PA74-19197.

1135. Nakamura, K. and Kamoshida, M., "Electrical characteristics of $^{11}B^{+}$-implanted p-channel MNOS transistors", Solid-State Electron., 18, 1085-8 (1975).

d. CCD

1136. Ablassmeier, U., "Improved charge transfer in p-channel charge coupled devices with implanted gaps", 1972 European Solid State Devices Research Conference, p. 175-6, Publ. 1973, EEA76-32279.

1137. Amelio, G. F. and Hosack, H. H., "Semiconductor device having self-aligned implanted barriers with narrow gaps between electrodes", Patent USA 3911560, Publ. October 1975, CA83-211932.

1138. Amelio, G. F., Kim, C. K. and Salsbury, P. J., "Self-aligned implanted barrier two-phase charge coupled devices", Patent USA 3853634, Publ. December 1974, CA82-118486.

1139. French, B. T., "Designer guide to CCD's", EDN, 18 (2), 34 (1973).

1140. Knauer, K. and Pfeiderer, H. J., "Experimental results of charge coupled devices with implanted gaps", Arch. Elektron. & Uebertragungstech., 28, 383-7 (1974), EEA78-641.

1141. Kosonocky, W. F. and Carnes, J. E., "Design and performance of two-phase charge-coupled devices with overlapping polysilicon and aluminum gates", 1973 International Electron Devices Meeting Technical Digest, p. 123-5, EEA77-14544.

1142. Krambeck, R. H., Walden, R. H. and Pickar, K. A., "Doped surface two-phase CCD", Bell Syst. Tech. J., 51, 1849 (1972).

1143. Lees, A. W. and Ryan, W. D., "A simple method of a buried channel charge coupled device", Solid-State Electron., 17, 1163-9 (1974), EEA78-655.

1144. Shimizu, S., Iwamatsu, S. and Ono, M., "Charge-coupled device with buried channels under electrode gaps", Appl. Phys. Lett., 22, 286-7 (1973), EEA76-19923.

6. Resistors

1145. Anon., "Neon-ion implantation for high resistors in IC's", Bedriff. & Tech., 26, 2333-5 (1971), Dutch, EEA75-9566.

1146. Bilger, H. R., Tandon, J. L. and Nicolet, M. A., "Excess noise measurements in ion-implanted silicon resistors", Solid-State Electron., 17, 599-605 (1974), EEA77-23157.

1147.Davies, D. E., Kennedy, J. K. and Lowe, L. F., "Low-resistivity n-type layers in $InAs_xP_{1-x}$ by ion implantation", Electron. Lett., 11, 462-3 (1975), EEA78-37390.

1148.Hanson, J. W., Huber, R. J. and Fordemwalt, J. N., "Ion implanted n-type resistors on high resistivity substrates", J. Vac. Sci. & Technol., 10, 944-7 (1973), EEA77-23144.

1149.Ho, I. T., Ku, S. M., Maley, G. A., Poponiak, M. R., Schwenker, R. P. and Yeh, T. H., "Programmable memory circuits", IBM Tech. Disclosure Bull., 17, 3279 (1975), EEA78-24473.

1150.Koji, T., "Popcorn noise and generation-recombination noise observed in ion-implanted silicon resistors", Electron. Lett., 11, 185-6 (1975), EEA78-24451.

1151.MacDougall, J. D., Manchester, K. E. and Roughan, P. E., "High value implanted resistors for microcircuits", Proc. IEEE, 57, 1538-42 (1969).

1152.MacIver, B. A., "Comparison of boron and neon damage effects in boron ion-implanted resistors", Electron. Lett., 11, 484-5 (1975).

1153.Nicholas, K. H. and Ford, R. A., "New techniques for improving high value ion implanted resistors", 2nd International Conference on Ion Implantation in Semiconductors, 1971, p. 357-61, PA75-25879.

1154.Nicholas, K. H. and Ford, R. A., "Implanted resistors with properties enhanced by damage", 1973 International Electron Devices Meeting Technical Digest, p. 51-3, EEA77-14611.

1155.Oosthoek, D. P., den Boer, J. A. and Hofker, W. K., "The thermal properties of high value gallium and boron implanted resistors in silicon", European Conference on Ion Implantation, 1970, p. 88-91, PA74-19180.

1156.Perloff, D. S., Kerr, J. T. and Marley, J. A., "Ion-implanted resistors and method", Patent USA 3829890, Publ. August 1974.

1157.Rosendal, K., "Ion-implanted planar resistors", Radiat. Eff., 7, 95-100 (1971), PA74-36542.

1158.Runge, H. and Krimmel, E. F., "Resistors and diodes produced by Al-implantation in silicon", Solid-State Electron., 18, 149-50 (1975), PA78-16100.

1159.Seidel, T. E. and Gibson, W. C., "Buried-guarded layer ion-implanted resistors", IEEE Trans. Electron Devices, ED-20, 744-8 (1973), EEA76-29314.

1160.Wiedeburg, K. H., Betz, H. and Kranz, H., "Hall effect and sheet resistivity measurements on ion-bombarded indium antimonide", Phys. Status Solidi A, 31, K69-K71 (1975), CA83-200761.

7. Other Devices

1161.Addamiano, A., Lucke, W. H. and Comas, J., "Phosphor preparation by ion implantation", J. Lumin., 6, 143-5 (1973), PA76-30956.

1162. Akasaka, Y., Horie, K., Nomura, K. and Kawazu, S., "Anomalous enhancement of etching rate of silicon nitride with HF etchant by ion implantation", Oyo Buturi, 43 (Suppl.), 493-8 (1974), EEA78-2745.
1163. Bernard, J., Guernet, G., Montier, M. and Peccoud, L., "Ion implanted thermoresistive device for cryogenic temperatures", Proceedings of the 2nd International Conference on Ion Implantation in Semiconductors, 1971, p. 397-403, PA75-25886.
1164. Brown, W. L., "New ion implantation areas", 2nd International Conference on Ion Implantation in Semiconductors, 1971, p. 430-8, PA75-25417.
1165. Cembali, F., Galloni, R. and Zignani, F., "A technique to obtain deep penetrating ohmic contacts for electrical measurements on ion implanted silicon", J. Phys. E, 7, 698-700 (1974), PA77-70425.
1166. Cerny, F., "Some orientational results from the up-to-date work in the field of the ion implantation technology", Sdelovaci Tech., 20, 251-2 (1972), Czech, PA75-71970.
1167. Coleman, M. G., "Method of making isolated semiconductor devices", Patent USA 3707765, Publ. January 1973, EEA77-5777.
1168. Eklund, K. M., Holmen, G. and Peterstrom, S., "Ion implantation in silicon films on sapphire", Appl. Phys. Lett., 24, 283-4 (1974), PA77-45437.
1169. Feuerstein, A. and Kalbitzer, S., "Ohmic contacts on n^+- and p^+-Si by ion implantation", Appl. Phys. Lett., 22, 19-20 (1973), EEA76-12491.
1170. Freeman, J. H., Gard, G. A. and Mazey, D. J., "Formation of insulating layers by the use of reactive ion beams", European Conference on Ion Implantation, 1970, p. 74-80, PA74-18641.
1171. Furukawa, S., Inoue, O. and Ishiwara, H., "Experimental simulation on the beam heating effects in ion implantation techniques", Jap. J. Appl. Phys., 12, 1075-6 (1973), PA76-64798.
1172. Gansauge, P. and Jansen, H. G., "Method of contacting semiconductor zones", IBM Tech. Disclosure Bull., 16, 3625 (1974), EEA78-538.
1173. Gnap, A. K., Dushchenko, V. K. and Dubrovin, Y. V., "Ion implantation as a method of producing semiconductor capacitors with given volt-Farad characteristics", Radiotekhnika, Kharkov, no. 31, 97-103 (1974), Russian, EEA78-28745.
1174. Hammer, W. N. and Ziegler, J. F., "Ion implantation sweeping with accurate energy feedback control", IBM Tech. Disclosure Bull., 16, 3075 (1974), EEA77-24420.
1175. Hughes Aircraft Co., "Semiconductor manufacture by implantation", Patent UK 1205288, Publ. September 1970.
1176. Johnson, W. S., "Multiple masking technique in ion implantation", IBM Tech. Disclosure Bull., 15, 660-1 (1972), EEA75-39812.

1177.Kelson, G., Stellrecht, H. H. and Perloff, D. S., "A monolithic 10-b digital-to-analog converter using ion implantation", IEEE J. Solid-State Circuits, SC-8, 396-403 (1973), EEA77-5652.
1178.Lee, H. Y., "Dopants implantation by vacuum discharge and heating method", Oyo Buturi, 43 (Suppl.), 422-3 (1974), EEA78-2742.
1179.MacIver, B. A., "Characteristics of ion-implanted hybrid voltage-variable capacitors", Electron. Lett., 9, 210-12 (1973), EEA76-26233.
1180.McCaughan, D. V., "Effects of radiation due to ion implantation and plasma methods in processing (ICs)", Electrochemical Society Fall Meeting (Extended abstracts), 1974, p. 381-3, EEA78-37435.
1181.Moline, R. A. and Foxhall, G. F., "Ion-implanted hyperabrupt junction voltage variable capacitors", IEEE Trans. Electron Devices, 19, 267-73 (1972), EEA75-9823.
1182.Moline, R. A., Lieberman, R., Simpson, J. and MacRae, A. U., "The growth of high quality epitaxial silicon over ion implanted buried arsenic layers", J. Electrochem. Soc., 121, 1362-6 (1974), EEA78-551.
1183.Nishi, H., Sakurai, T., Akamatsu, T. and Furuya, T., "Migration of Mo atoms across Mo-Si interface induced by Ar^+ ion bombardment", Appl. Phys. Lett., 25, 337-9 (1974), PA78-7465.
1184.Parry, P. D. and Leiby, R. A., "Silicon wafer heating during ion implantation", Presented at 1973 International Electron Devices Meeting Supplement, Washington, D. C., December 1973, EEA77-18726.
1185.Prussin, S. and Fern, A., "Adaptation of ion implantation for integrated circuits", J. Electrochem. Soc., 122, 830-2 (1975), EEA78-33172.
1186.Rosinski, W., "The use of ion implantation for the preparation fo metal-semiconductor contacts", Z. Elektr. Inf.- & Energietech., 2, 232-4 (1972), German, EEA76-8547.
1187.Rosinski, W., "Varicaps with steep C-V characteristic", Elektronika, 14, 273-6 (1973), EEA76-35407.
1188.Rosinski, W. and Komicz, T., "Metal-semiconductor contacts prepared by ion implantation", Electron Technol., 7 (1-2), 13-18 (1974), EEA78-8432.
1189.Stein, H. J., "Ge-nitride, Ge-oxide, and Ge-oxynitride formation by ion implantation", J. Electrochem. Soc., 121, 1073-6 (1974), PA77-76396.

VII. Equipment

1. Accelerators

1190.Allison, R. W., Jr., "Particle accelerator producing a uniformly expanded particle beam of uniform cross-sectioned density", Patent USA 3845312, Publ. October 1974, NSA31-12182.

1191. Armstrong, D. D., Brendt, P. J., Erkkila, B. H., Lunsford, J. S., Shipley, J. P., Stokes, R. H. and Sutton, J., "Experimental study of spiral resonators for acceleration of low velocity ions", Part. Accel., 6, 175-82 (1975), EEA78-30653.

1192. Bahrami, H., Chernow, F., Denison, D. and Eldridge, G., "An improved high beam current ion implanter", European Conference on Ion Implantation, 1970, p. 26-8, PA74-17898.

1193. Bannenberg, J. G. and Boerboom, A. J., "Design of a 320 kV ion accelerator for multiply charged heavy ions", IEEE Trans. Nucl. Sci., NS-22, 1675-8 (1975), EEA78-43695.

1194. Barton, M. Q., Culwick, B. B., Curtiss, J. A., Dabrowski, J. J., Frankel, R. S., Harrison, W. E., Martin, F. R., Smith, J. D., Warkentien, R. J. and Weitman, I., "Use of a general-purpose-time-shared computer in accelerator control", Report BNL-18832, Brookhaven National Lab., Upton, N. Y., (1974), 8 pp., NSA31-6876.

1195. Bennett, J. R. J., "Heavy ion accelerators", Report RHEL/R-252, Rutherford High Energy Lab., Chilton, England, 1972, p. 55-72, NSA28-6488.

1196. Bensussan, A., "Spread-out particle beam irradiator", Patent French 2158724, Publ. October 1971, French, NSA29-21660.

1197. Blaha, A., "Ion machines for microtechnology", Slaboproudy Obz., 34, 216-35 (1973), Czech, EEA76-29295.

1198. Bobbitt, J., "Applications of microprocessors in accelerators", IEEE Trans. Nucl. Sci., NS-22, 1065-8 (1975), EEA78-43645.

1199. Brewer, G. R. and Buckey, C. R., "Acceleration system for implanting ions in specimen", Patent USA 3569757, Publ. March 1971, PA74-36561.

1200. Bromley, D. A., "The development of electrostatic accelerators", Nucl. Instrum. & Methods, 122, 1-34 (1974), PA78-14451.

1201. Cecil, J. N. and Bostrom, N. A., "Particle accelerators for ion implantation", Western Electronic Show and Convention, 1971, 28/4, 6 pp., PA74-82974.

1202. Chaumont, J., Lalu, F., Salome, M. and Seide, C., "Progress report on the ion implantation facility at Orsay", Report CONF-730618, Laboratoire Rene Bernas, Orsay, France, 1973, p. 271-75, NSA30-33106.

1203. Cracknell, P. J., Gettings, M. and Stephens, K. G., "Use of low-energy accelerators for ion implantation", Nucl. Instrum. and Methods, 92, 465-69 (1971), PA74-40435.

1204. Creedon, J. M. and Smith, I. D., "Method of generating very intense positive ion beams", Phys. Rev. Lett., 35, 91-4 (1975), EEA78-39286.

1205. Doyle, N. J., Bogdon, J. M. and Chayke, W. J., "A combination vacuum furnace-sample chamber for ion implantation", Rev. Sci. Instrum., 45, 847-8 (1974), PA77-70429.

1206. Duggan, J. L., "Industrial applications of small accelerators", IEEE Trans. Nucl. Sci., NS-22, 1216-28 (1975), EEA78-43659.

1207.Eckstein, W. and Verbeek, H., "Bombardon: an apparatus for investigating the influence of 5-20 keV ions on metal targets", Vacuum, 23 (5), 159-62 (1973), EEA77-1178.
1208.Flatte, S. M., Seiden, A. and Bangerter, R. O., "A method of tuning charged particle beam transport systems", Nucl. Instrum. & Methods, 120, 205-10 (1974), EEA78-6678.
1209.Flerov, G. N. and Barashenkov, V. S., "Practical applications for heavy ion beams", Usp. Fiz. Nauk, 114, 351-73 (1974), Russian, PA78-5343.
1210.Freeman, J. H., "Implantation machines", European Conference on Ion Implantation, 1970, p. 1-18, PA74-17897.
1211.Freeman, J. H., Caldecourt, L. R., Done, K. C. W. and Francis, R. J., "An industrial scale ion implantation facility", European Conference on Ion Implantation, 1970, p. 36-40, PA74-17900.
1212.Freeman, J. M. and Hooton, B. W., "Study of heavy ion beams from a tandem generator", Nucl. Instrum. Methods, 111, 501-7 (1973), NSA28-27946.
1213.Gerasimov, E. I., Solmyshkov, A. I. and Tarvid, G. V., "Aberrations in a straight ion-optical accelerator", Sov. Phys.-Tech. Phys., 19, 131-3 (1974), PA77-77148.
1214.Grunder, H. A., "Status and outlook for heavy-ion accelerator systems", IEEE Trans. Nucl. Sci., NS-22, 1621-5 (1975), EEA78-43688.
1215.Hammer, W. N. and Ziegler, J. F., "Ion implantation sweeping with accurate energy feedback control", IBM Tech. Disclosure Bull., 16, 3075 (1974), PA77-43380.
1216.Herbert, B. K., "Accurate current integrator for ion implanting", SERT J., 8, 209-10 (1974), EEA78-5266.
1217.Hill, A. R., "A fine-focused ion machine with applications to implantation, microscopy and machining", European Conference on Ion Implantation, 1970, p. 29-32, PA74-17899.
1218.Hofmann, E. G., Offermann, B. P. and Tauber, M., "High power accelerator for industrial applications", International Nuclear Industries Fair Technical Meeting No. 8/8, 1972, 13 pp., German, NSA29-21665.
1219.Humphries, S., Jr., "High power electrostatic ion accelerators", Report LPS-136, Cornell Univ., Ithaca, N. Y., Lab. of Plasma Studies, 1973, 19 pp., NSA29-10806.
1220.Jensen, K. and Veje, E., "Construction of a Wien filter heavy ion accelerator", Nucl. Instrum. & Methods, 122, 511-15 (1974), PA78-17077.
1221.King, W. J., "Ion accelerator and ion species selector", Patent USA 3786359, Publ. January 1974, NSA29-21659.
1222.Kiyozumi, K., "Apparatus for vapor deposition and ion implantation", Patent USA 3847114, Publ. November 1974.
1223.Kleinfelder, W. J., "Ion implantation equipment for semiconductor manufacturing", Electron and Ion Beam Science and Technology Fifth International Conference, 1972, p. 241-53, EEA76-8407.

1224.Komarov, V. L., Nalivaiko, G. A., Razgulyaev, I. I., Solnyshkov, A. I. and Tsepakin, S. G., "Ion-beam apparatus", Instrum. and Exp. Tech., 5, 1422-4 (1970), PA74-59053.
1225.Kura, M. and Oshima, K., "Linear particle accelerator", Patent Japan 73-278, Publ. January 1973, NSA29-10807.
1226.Logachev, I. I., "Ion acceleration with rotation of the electron beam in the magnetic field", Kratk. Soobshch. Fiz., no. 8, 27-32 (1973), Russian, NSA29-21664.
1227.MacPherson, M. R., "Production machine requirements for ion implantation processing in the semiconductor industry", Electron and Ion Beam Science and Technology Fifth International Conference, 1972, p. 235-40, EEA76-8406.
1228.McGinty, G. K., Goldsmith, B. J. and Thomas, R. A., "A 200 kV ion implantation equipment at M.R.L., Salfords", European Conference on Ion Implantation, 1970, p. 41-5, PA74-17901.
1229.Nelson, R. S., "Application of cyclotrons in materials science", Proceedings of the 6th International Cyclotrons Conference, 1972, p. 664-76, EEA76-14427.
1230.Nowicki, S., "Double electrostatic quadrupole doublet as an exit lens of narrow beam ion implantation machine", Electron Technol., 6, 3-29 (1973), EEA77-27898.
1231.Ota, S., Nakamura, O. and Anazawa, N., "Stabilizing system of accelerated charged particles", Patent Japan 73-279, Publ. January 1973, Japanese, NSA29-8444.
1232.Pettersson, B. G., "Radiation protection considerations on the design and operation of particle accelerators", Organisation for Economic Cooperation and Development, Paris, France, 1974, NSA30-7479.
1233.Plusch, S. P., "A design and evaluation of an ion implantation system", Report AD-712011, Air Force Inst. Technol., Wright-Patterson AFB, Ohio, 76 pp. (1970), PA74-36543.
1234.Ramler, W. J., "DC accelerators", Technical Developments and Prospects of Sterlization of Ionizing Radiation, 1974, p. 115-35, PA78-11630.
1235.Ribarov, S. P., "A simple linear accelerator for doping semiconductor by ion implantation technology", C. R. Acad. Bulg. Sci., 28, 461-4 (1975), EEA78-39287.
1236.Robertson, G. I., "Rotating scan for ion implantation", J. Electrochem. Soc., 122, 796-800 (1975), PA78-62552.
1237.Rodde, A. F., Jackson, J. H., McCallum, J. G., Mostek, P. J., Robertson, G. I., Weissman, B. and Williams, N., "ERC Mark I ion implantation system", Presented at Electrochemical Society Meeting, San Francisco, Calif., May 1974, PA77-53623.
1238.Ryding, G., "Ion implantation: higher beam current for higher throughput", Electron. Packag. & Prod., 15 (3), 67-74 (1975).
1239.Ryding, G. and Wittkower, A. B., "Industrial ion implantation machines", IEEE Trans. Manuf. Technol., MFT-4, 21-31 (1975), EEA78-37564.

1240.Sidenius, G. and Holck, O., "Universal range ion accelerator and separator", Report CONF-730618, Niels Bohr Inst., Copenhagen, 1973, p. 199-204, NSA30-33048.

1241.Strojnik, A., "Recent developments in high-voltage accelerators and generators", In: High Voltage Electron Microscopy, Swann, P. R., (Ed.), Academic Press, New York, p. 85-92 (1974), NSA30-16555.

1242.Talada, D. D., "Evaluation of a low energy ion implantation system", Report GE-EE-74-36, Air Force Inst. Tech., Wright-Patterson AFB, Ohio, 1974, 77 pp., EEA78-24234.

1243.Vukanic, J. and Terzic, I., "Experimental investigations of Li^+ ion gun characteristics", Nucl. Instrum. Methods, 111, 117-24 (1973), NSA28-22183.

1244.Weber, C., "The use of computers in electron- and ion-gun design", Comput. Phys. Commun., 5, 44-7 (1973), EEA76-16646.

1245.Wegner, H. E., "Advances in electrostatic accelerators", IEEE Trans. Nucl. Sci., NS-22, 946-55 (1975), EEA78-43629.

1246.Wittkower, A. B., Rose, P. H. and Ryding, G., "Advances in ion implantation production equipment", Solid State Technol., 18 (12), 41-5 (1975).

1247.Yada, K., "An ion implanter", Oyo Buturi, 40, 408-14 (1971), Japanese, EEA75-16643.

2. Ion Sources

1248.Abdelaziz, M. E., Morsey, M. W., Hassan, Z. M. and Hanna, N. N., "Proposed multiply charged ion source", IEEE Trans. Nucl. Sci., NS-19, 132-6 (1974), EEA75-21314.

1249.Abdrashitova, D. K., Babaev, V. G. and Guseva, M. B., "Efficient ion source", Sov. Phys.-Tech. Phys., 19, 264-6 (1974), EEA78-10107.

1250.Aldcroft, D., Burcham, J., Cole, H. C., Cowlin, M., Sheffield, J. and Speth, E., "A high power high efficiency ion source", 3rd International Symposium on Toroidal Plasma Confinement, 1973, B24/1 pp., PA77-30542.

1251.Ananin, O. B., Bykovskii, Y. A., Degryarenko, N. N., Kozyrev, Y. P., Silnov, S. M. and Sharkov, B. Y., "Production of C and Al nuclei in a laser source of multiply-charged ions", JETP Lett., 16, 384-7 (1972), EEA76-14319.

1252.Apard, P., Bliman, S., Geller, R., Jacquot, B. and Jacquot, C., "Production of a beam of nitrogen ions, N_{14}^{7+}", C. R. Hebd. Seances Acad. Sci. B, 275, 621-4 (1972), French, EEA76-2407.

1253.Arianer, J., Baron, E., Brient, M., Cabrespoine, A., Liebe, A., Serafini, A. and Ton-That, T., "Multiply charged ion source", Nucl. Instrum. & Methods, 124, 157-60 (1975), PA78-35374.

1254.Arianer, J., Baron, E. and Ton-That, T., "Progress report (1973) on the confined heavy ion source", Report IPNO-73-04, Inst. de Physique Nucleaire, Orsay, France, 1973, 17 pp., French, NSA30-24841.
1255.Aubert, J., Gautherin, G. and Lejeune, C., "Discharge from a new source of ions: the triplasmatron", C. R. Hebd. Seances Acad. Sci. B, 275, 537-40 (1972), French, EEA76-2405.
1256.Avakov, A. S., Belykh, S. F., Evstifeev, V. V. and Evtukhov, R. N., "A source of alkali-metal ions", Instrum. & Exp. Tech., 15, 1002-3 (1972), EEA76-21431.
1257.Axmann, A., "Source materials for ion implantation", Appl. Phys. Lett., 23, 645-8 (1973), PA77-21197.
1258.Axmann, A., "Ionizable materials to produce ions for implantation (semiconductor doping)", Solid State Technol., 17 (11), 36-9, 46 (1974), PA78-16097.
1259.Beaver, E. M., "An improved electron impact ion source power supply", Mass Spectrosc., 22, 203-7 (1974), EEA78-30616.
1260.Bechtold, V., Bruckmann, H., Finken, D., Friedrich, L., Hamdi, K. and Seitz, E., "The energy dependence of charge exchange reactions applied successfully to Lamb-shift-type ion sources", Proceedings of the 3rd International Polarization Phenomena in Nuclear Reactions Symposium, 1970, p. 839-41, Publ. 1971, PA77-18572.
1261.Camplan, J., "Ion sources for ion implantation", European Conference on Ion Implantation, 1970, p. 19-25, PA74-17546.
1262.Chapman, K. R., Rowton, L. and Trimble, S., "Use of a spike cathode with a duoplasmatron ion source", Nucl. Instrum. & Methods, 106, 253-7 (1973), EEA76-14326.
1263.Chintalpudi, S., Clark, D. J., Gavin, B., Grunder, H., Holley, W. R., Richter, R. M., Selph, F. and Zajec, E., "Heavy ion source development at Berkeley", Report LBL-3024, California Univ., Berkeley, (Lawrence Berkeley Lab.), 1974, 19 pp., NSA31-1330.
1264.Ciegg, T. B., Bissinger, G. A., Haeberli, W. and Quin, P. A., "Lamb-shift polarized ion sources for tandem accelerators at Wisconsin and Triangle Universities nuclear laboratory", Proceedings of the 3rd International Polarization Phenomena in Nuclear Reactions Symposium, 1970, p. 835-6, Publ. 1971, PA77-18570.
1265.Darling, R. L. and Davis, R. H., "Analysis of high-charge state output from a Penning ion source", Rev. Sci. Instrum., 44, 375-8 (1973), EEA76-21441.
1266.Davis, R. C., Jernigan, T. C., Morgan, O. B., Stewart, L. D. and Stirling, W. L., "DuoPIGatron. II. Ion source", Rev. Sci. Instrum., 46, 576-81 (1975), EEA78-34979.
1267.DeFries, R. M., "Elemental arsenic/silicon compound for ion implant source", IBM Tech. Disclosure Bull., 18, 421 (1975), EEA78-37403.

1268.Demikhanov, R. A., Kursanov, Y. V. and Skripal, L. P., "A beam-plasma high-intensity ion source", Sov. Phys.-Tech. Phys., 18, 684-5 (1973), EEA77-12001.
1269.Denimal, J., Gasser, H. and Ricaud, C., "Performance of a hot-cathode heavy-ion source installed in a 2 MV Van de Graaff accelerator", Nucl. Instrum. & Methods, 114, 93-9 (1974), EEA77-12084.
1270.Donnally, B. L., "Lamb-shift sources", Proceedings of the 3rd International Polarization Phenomena in Nuclear Reactions Symposium, 1970, p. 295-316, Publ. 1971, PA77-11569.
1271.Doucas, G. and Hyder, H. R. M., "The brightness and emittance of the negative ion beam emitted by a Middleton sputter source", Nucl. Instrum. & Methods, 119, 413-21 (1974), NSA31-4086.
1272.Eck, J. S. and Elliott, D. O., "Direct extraction of a $^{12}C^-$ beam from a diode ion source", Nucl. Instrum. & Methods, 121, 411-12 (1974), PA78-14479.
1273.Ezoe, H. and Itoh, Y., "A combination ion source and surface ionization", Mass Spectrosc., 20, 357-62 (1972), EEA76-21677.
1274.Fitch, R. K., Ghander, A. M., Rushton, G. J. and Singh, R., "Design and operating characteristics of a low pressure ion source", Jap. J. Appl. Phys., 13 (Suppl. 2, pt. 1), 411-14 (1974), PA78-35368.
1275.Fitch, R. K., O'Shea, K. R. and Rushton, G. J., "Multiple beam and wide beam ion sources", J. Phys. E, 5, 622-3 (1972), EEA75-29521.
1276.Fowler, R. G., Sidlow, R., Soulby, G. and West, N. D., "The proton duoplasmatron ion source for the preinjector of the new injector for Nimrod", Report RL-74-048, Rutherford High Energy Lab., Chilton, Berks., England, 1974, 8 pp., PA77-65511.
1277.Franks, J. and Ghander, A. M., "A saddle field ion source of spherical configuration for etching and thinning applications", Vacuum, 24, 489-91 (1974), EEA78-34081.
1278.Franzen, J., Kuper, H. and Riepe, W., "Automatic ion current control of a direct inlet system", Int. J. Mass Spectrom. & Ion Phys., 10, 353-7 (1973), EEA76-17922.
1279.Freeman, J. H. and Sidenius, G., "The technology and chemistry of heavy ion sources", Nucl. Instrum. & Methods, 107, 477-92 (1973), EEA76-21436.
1280.Fremiot, C. and Grando, J., "Ion source employing a microwave resonant cavity", Patent USA 3778656, Publ. December 1973, NSA29-16443.
1281.Gessinn, G., "Characteristics of a positive HF ion-source immersed in an auxiliary magnetic field", Ind. At. & Spat., 18 (4), 8-18 (1974), EEA78-3996.
1282.Ghander, A. M. and Fitch, R. K., "A double beam ion source", Vacuum, 23 (8), 269-70 (1973), EEA77-3856.
1283.Glavish, H. F., "Atomic beam sources", Proceedings of the 3rd International Polarization Phenomena in Nuclear Reactions Symposium, 1970, p. 267-95, Publ. 1971, PA77-11568.

1284.Glavish, H. F., MacKinnon, B. A., Walker, I. J., Whineray, S. and Titterton, E. W., "The Australian National University atomic beam polarized negative ion source", Proceedings of the 3rd International Polarization Phenomena in Nuclear Reactions Symposium, 1970, p. 817-18, Publ. 1971, PA77-18568.
1285.Green, T. S., "Gas utilisation in ion sources", Nucl. Instrum. & Methods, 115, 615-17 (1974), PA77-35381.
1286.Green, T. S. and Goble, C., "Power and particle flux balance in a P.I.G. discharge ion source. I. Ion containment", Nucl. Instrum. & Methods, 116, 157-63 (1974), PA77-35384.
1287.Green, T. S. and Goble, C., "Power and particle flux balance in a P.I.G. discharge ion source. II. Ionisation", Nucl. Instrum. & Methods, 116, 165-8 (1974), PA77-35385.
1288.Grigoryan, V. G., Gubarev, I. A., Latishev, L. A. and Obukhov, V. A., "The influence of the ions' preliminary acceleration on the extraction of the ionic beam from the Penning discharge plasma surface", 3rd International Conference on Gas Discharges, 1974, p. 145-8, PA77-74109.
1289.Gruebler, W., Konig, V. and Schmelzbach, P. A., "Source of polarized negative hydrogen ions for the ETH tandem accelerator", Proceedings of the 3rd International Polarization Phenomena in Nuclear Reactions Symposium, 1970, p. 819-20, Publ. 1971, PA77-18569.
1290.Gwozdz, P. S. and Koehler, J. S., "A simple ion source for implantation doping of semiconductors", Rev. Sci. Instrum., 41, 1677-8 (1970), PA74-5287.
1291.Heinick, E., Helleboid, J. M. and Weber, A., "Preliminary results with a Penning ion source on a 5.5 MV Van de Graaff accelerator", Nucl. Instrum. & Methods, 124, 301-3 (1975), EEA78-17228.
1292.Herman, A. and Vago, G. Y., "Metal ion source", Finommechanika, 11, 233-7 (1972), Hungarian, EEA75-40863.
1293.Hertenberger, G., Roch, K. and Winter, H., "Calculation of charge state densities in electron beam multiply charged ion sources", Z. Naturforsch. A, 28A, 1687-96 (1973), German, PA77-5151.
1294.Hudson, E. D., Mallory, M. L. and Lord, R. S., "An ion source for high-intensity metal ions", Nucl. Instrum. & Methods, 115, 311-12 (1974), PA77-26587.
1295.Irons, F. E., Peacock, N. J. and Pease, R. S., "Plasma produced by laser irradiation of solid targets as a source of highly stripped ions", Sov. J. Quantum Electron., 2, 13-7 (1972), EEA76-2415.
1296.Isler, R. C., "A stabilized electron bombardment ion source", Rev. Sci. Instrum., 45, 308 (1974), EEA77-27864.
1297.Ivanov, B. I., "One possible way of generating high current ion beams", Zh. Eskp. & Teor. Fiz. Pisma, 20, 170-3 (1974), Russian, PA78-525.

1298.Jensen, K. and Weje, E., "A simple, cheap, universal ion source", Nucl. Instrum. & Methods, 115, 307-8 (1974), PA77-26586.
1299.Khirnyi, Y. M. and Kochemasova, L. N., "A source of doubly charged helium ions for an electrostatic accelerator", Instrum. & Exp. Tech., 17, 1269-71 (1974), EEA78-26215.
1300.Kirschner, H., "α-Sources for channeling experiments", Nucl. Instrum. & Methods, 112, 555-7 (1973), German, EEA77-7716.
1301.Ko, W. C. and Winnard, J. R., "Hollow cathode discharge ion source", IBM Tech. Disclosure Bull., 27, 1379 (1974), PA78-14803.
1302.Lang, I., Nefedev, O. K., Tretyakov, Y. P. and Fefilov, B. V., "Automatic measurement and handling of charge spectra of a source of multiply charged ions", Instrum. & Exp. Tech., 16, 763-7 (1973), PA77-14950.
1303.Lebedev, S. Y. and Panin, S. D., "Source of metallic ions", Instrum. & Exp. Tech., 16, 868-9 (1973), PA77-18559.
1304.Lejeune, C., "Theoretical and experimental study of the duoplasmatron ion source. I. Model of the duoplasmatron discharge", Nucl. Instrum. & Methods, 116, 417-28 (1974), EEA77-27853.
1305.Lejeune, C., "Theoretical and experimental study of the duoplasmatron ion source. II. Emissive properties of the source", Nucl. Instrum. & Methods, 116, 429-43 (1974), EEA77-27854.
1306.Luches, A., Provenzano, L. and De Michele, G., "A lithium ion source for cyclotron external injection", Rev. Sci. Instrum., 45, 1028-9 (1974), PA77-80978.
1307.Mattes, H. G., "A simple, high current gold ion source", Rev. Sci. Instrum., 45, 1030-2 (1974), PA77-80979.
1308.Mayhall, D. J. T. and Robinson, W. P., "Interfacing a source utilizing low perveance ion beam extraction to a beam transport system suitable for ion implantation (semiconductor doping)", Solid State Technol., 17 (11), 40-3 (1974), PA78-16098.
1309.Middleton, R., "A survey of negative ion sources for tandem accelerators", Nucl. Instrum. & Methods, 122, 35-43 (1974), PA78-14481.
1310.Middleton, R. and Adams, C. T., "A close to universal negative ion source", Nucl. Instrum. & Methods, 118, 329-36 (1974), EEA77-32033.
1311.Misra, S. N. and Gupta, S. K., "Pressure-yield characteristics of RF ion-sources", Nucl. Instrum. & Methods, 122, 303-6 (1974), EEA78-13265.
1312.Mochizuki, T. and Arikawa, T., "Measurements of beam emittance in low energy ion source", Mass Spectrosc., 20, 281-6 (1972), EEA76-21674.
1313.Mochizuki, T. and Arikawa, T., "Ion beam source for low energy ion-neutral crossed-beam study", Jap. J. Appl. Phys., 12, 1-6 (1973), EEA76-14318.
1314.Moller, W. and Kamke, D., "Properties of a high current long life lithium ion source", Nucl. Instrum. & Methods, 105, 173-7 (1972), EEA76-10278.

1315.Nation, J. A., "A proposed high-current nanosecond source of energetic ions", J. Plasma Phys., 13, 361-6 (1975), EEA78-21697.
1316.Norbeck, E. and York, R. C., "Beams of Be^+ from an RF ion source", Nucl. Instrum. & Methods, 118, 327-8 (1974), EEA77-35125.
1317.Paleev, V. I. and Zandberg, E. Y., "A source of negative ions for vapor substances", Instrum. & Exp. Tech., 17, 489-91 (1974), EEA78-10104.
1318.Pasyuk, A. S. and Tretyakov, Y. P., "Ion source for the production of multiply charged ions from solid materials", Report CONF-720948, Joint Inst. for Nuclear Research, Dubna, USSR, 1972, p. 512-18, NSA29-10827.
1319.Patel, M. H. and Chandraker, K., "A duoplasmatron ion source for ion implantation", 10th International Conference on Phenomena in Ionized Gases, 1971, p. 136, EEA75-14700.
1320.Paty, L., "An ion source using ionisation and desorption in a strong electrical field", Cesk. Cas. Fis. A, 24, 182-3 (1974), Czech, PA78-137.
1321.Penebre, N. A., "Low-power ion source", IBM Tech. Disclosure Bull., 14, 1475 (1971), EEA75-8264.
1322.Philipp, G., Scheib, U. and Hofmann, A., "He^- and Li^- ions from a modified charge exchange source", Nucl. Instrum. & Methods, 115, 507-8 (1974), EEA77-20617.
1323.Piotrowski, A., Raiko, V. I. and Tyrroff, H., "Ion source for radioactive elements", Instrum. & Exp. Tech., 15 (2), 323-6 (1972), EEA76-2409.
1324.Poitevin, J. M., "Study of a high frequency source of H^+ ions: optimal current", C. R. Hebd. Seances Acad. Sci. B, 275, 561-4 (1972), French, EEA76-2406.
1325.Prelac, K. and Sluyters, T., "Formation of negative hydrogen ions in direct extraction sources", Rev. Sci. Instrum., 44, 1451-63 (1973), PA77-35386.
1326.Rushton, G. J. and Fitch, R. K., "The twin anode electrostatic ion gun with thermionic electron injection", J. Phys. E, 7, 313-15 (1974), PA77-30540.
1327.Schlagheck, W. and Moeller, W., "Preparation and properties of a thermionic sodium ion source", Nucl. Instrum. & Methods, 120, 189-91 (1974), NSA31-4085.
1328.Seliger, R. L. and Ward, J. W., "Low perveance ion beam extraction from an electron bombardment discharge", Presented at Electrochemical Society Meeting, San Francisco, Calif., May 1974, PA77-47408.
1329.Shimizu, K., Kawakatsu, H. and Kanaya, K., "Metal ion source for ion implantation system", Jap. J. Appl. Phys., 11, 1192-8 (1972), EEA75-33523.
1330.Sluyters, T. and Prelec, K., "A hollow discharge duoplasmatron as a negative hydrogen ion source", Nucl. Instrum. & Methods, 113, 299-301 (1973), PA77-8441.

1331.Stix, T. H., "Ion containment sources", IEEE Trans. Nucl. Sci., NS-19, 150-5 (1972), EEA75-21317.
1332.Swaminathan, S. and Venkatasubramanian, V. S., "Characteristics of electrostatic charged particle oscillator ion sources", Pramana, 2 (1), 46-50 (1974), NSA29-27454.
1333.Tauth, T., "On the equations describing a quasi-neutral heavy ion plasma confined by a magnetic field: Application to ion sources", Nucl. Instrum. & Methods, 107, 293-300 (1973), French, EEA76-17936.
1334.Thompson, E., "The design of low divergence ion source extraction systems", Part. Accel., 4 (2), 69-73 (1972), EEA76-14327.
1335.Trotel, J., "A calutron-type ion source for use in an ion implantation machine", European Conference on Ion Implantation, 1970, p. 33-5, PA74-17548.
1336.United Kingdom Atomic Energy Authority, "Improvements in or relating to ion sources", Patent UK 1310397, Publ. March 1973, NSA28-16204.
1337.Wilson, R. G. and Jamba, D. M., "Comparison of sources of boron, phosphorus, and arsenic ions", Appl. Phys. Lett., 22, 176-9 (1973), EEA76-17917.
1338.Winter, H., "Generation of intense beams of multiply charged ions", Vakuum-Tech., 23 (3), 70-77 (1974), German, NSA30-33078.
1339.Winter, H. and Wolf, B. H., "Analysis of the duoplasmatron-type discharge as a source of multiply charged heavy ions", Plasma Phys., 16, 791-811 (1974), PA77-78828.
1340.Wittmaack, K., "Beam formation in a triode ion gun", Nucl. Instrum. & Methods, 118, 99-113 (1974), EEA77-32028.
1341.Wittmaack, K. and Schulz, F., "Influence of source parameters on the properties of an ion beam", J. Vac. Sci. & Technol., 10, 918-21 (1973), EEA77-24421.

3. Beam Focusing

1342.Bondarev, B. I., Kushin, V. V., Murin, B. P., Solovev, L. Y. and Fedotov, A. P., "Focusing of superconducting solenoids in high-energy linear proton accelerators", Sov. At. Energy, 34, 172-74 (1973), NSA31-1325.
1343.Carrico, J. P., "On the possibility of a mass-filter ion-emission microscope", J. Phys. E, 8, 18-20 (1975), EEA78-13282.
1344.Chetvertkov, V. L., "Focusing of an astigmatic beam by a doublet and a symmetric triplet of quadrupole lenses", Sov. Phys.-Tech. Phys., 19, 1-2 (1974), NSA31-9313.
1345.Cookson, J. A. and Pilling, F. D., "The use of focused ion beams for analysis", Thin Solid Films, 19, 381-5 (1973), PA77-18555.
1346.Coupland, J. R. and Green, T. S., "Ion-beam deflection in extraction electrodes", Nucl. Instrum. & Methods, 125, 197-9 (1975), EEA78-21693.

1347.Derbenev, Y. S. and Skrinskii, A. N., "Phase effects of colliding beam interactions in the strong-focusing storage rings", Proceedings of the Third All-Union Conference on the Charged Particle Accelerators, 1973, p. 386-89, Russian, NSA30-16635.
1348.Dymnikov, A. D. and Lenkov, D. I., "Isofrequency periodic focusing structure with maximum stability region", Proceedings of the Third All-Union Conference on the Charged Particle Accelerators, 1973, p. 151-55, Russian, NSA30-16615.
1349.Flatte, S. M. and Seiden, A., "A method of tuning charged-particle beam transport systems", Nucl. Instrum. & Methods, 120, 205-10 (1974), NSA31-9314.
1350.Gabovich, M. D., Naida, A. P., Protsenko, I. M., Simonenko, L. S. and Soloshenko, I. A., "Gas focusing of a negative ion beam", Sov. Phys.-Tech. Phys., 19, 546-7 (1974), EEA78-13283.
1351.Gitomer, S. J. and Krishman, C. K., "Numerical simulation of direct energy conversion (from ion beam)", IEEE Trans. Plasma Sci., PS-2, 277-82 (1974), PA78-20468.
1352.Glukhikh, V. A., Gusev, O. A., Komar, E. G., Kuznetsov, V. S., Svinin, M. P. and Fidelskaya, R. P., "Beam focusing in an accelerator with inductive energy storage", Sov. Phys.-Tech. Phys., 19, 366-7 (1974), NSA31-9312.
1353.Greenaway, A. H. and Michelson, D., "The Hooke and Jeeves method applied to the study of the focusing properties of charged particles in an axially symmetric magnetic field", J. Phys. D, 7, 793-804 (1974), PA77-39773.
1354.Heighway, E. A., "Focussing the dipole magnets with large pole gap to bending radius ratios", Nucl. Instrum. & Methods, 123, 413-9 (1975), PA78-27413.
1355.Hofmann, I., "Focusing an electron ring in the presence of a squirrel cage and a conducting cylinder", Report IPP-0-16, Max-Planck-Institut fuer Plasmaphysik, Garching Muenchen (F.R. Germany), 1973, 17 pp., NSA30-10514.
1356.Joy, T. and Tait, N. R. S., "Calculation to first order of the effects of field errors in a 90° double-focusing magnet", Nucl. Instrum. & Methods, 124, 485-9 (1975), PA78-31290.
1357.Kobayashi, M., "Design of the focusing system for KEK linear accelerator", Report KEK-73-4, National Lab. for High Energy Physics, Oho, Ibaraki, Japan, 1973, 42 pp., NSA30-10550.
1358.Malov, A. F., Suzdalev, V. A., Zaboev, A. I. and Fedoseev, E. P., "Nonparaxial focusing of beams of charged particles in the median plane of an axisymmetric magnetic field for an arbitrary range of variation of particle masses", Sov. Phys.-Dokl., 18, 725-6 (1974), NSA31-1341.
1359.Matsuda, H., "Double focusing mass spectrometers of second order", J. Mass Spectrom. Ion Phys., 14, 219-33 (1974), NSA30-26993.

1360.Menat, M., Chavet, I. and Kanter, M., "The lens effect of the extraction electrode aperture on high-intensity ion beams in electromagnetic isotope separators", Nucl. Instrum. & Methods, 118, 135-40 (1974), EEA77-32073.

1361.Novikov, I. A. and Saulit, V. R., "A rigorous mathematical analysis for the focusing of a plane ion beam with a large angle of divergence. I", Sov. Phys.-Tech. Phys., 18, 851-6 (1974), PA77-31105.

1362.Novikov, I. A. and Saulit, V. R., "A rigorous mathematical analysis for the focusing of a plane ion beam with a large angle of divergence. II", Sov. Phys.-Tech. Phys., 18, 857-63 (1974), PA77-35921.

1363.Ryabtsov, A. V. and Danilin, V. G., "Focusing system of the Erevan synchrotron linotron-injector", Report KFTI-73-26, 10th Final Seminar of the Accelerator Department of the KFTI AN Ukrainian SSR; Kharkov, Ukrainian SSR, 1973, p. 28-30, Russian, NSA30-33077.

1364.Shulga, V. L., "Ion beam focusing by the atomic chains of a crystal lattice", Radiat. Eff., 26, 61-6 (1975).

1365.Springer, R. W., Haas, T. W., Grant, J. T. and Hooker, M. P., "An easy method to accurately align ion-bombardment guns for depth profiling in Auger electron spectroscopy", Rev. Sci. Instrum., 45, 1113-14 (1974), EEA78-13278.

1366.Stein, R., "Space and velocity focusing in time-of-flight mass spectrometers", J. Mass Spectrom. Ion Phys., 14, 205-18 (1974), NSA30-26992.

1367.Vlasov, A. D., "High frequency quadrupole focusing at average energies in a proton linear accelerator", Tr. Radiotekh. Inst., no. 11, 44-54 (1972), Russian, NSA30-16606.

1368.Zajdin, A. G., "Some parameter estimates of the linear accelerator with axial increasing focusing field", Report KFTI-73-26, 10th Final Seminar of the Accelerator Department of the KFTI AN Ukrainian SSR; Kharkov, Ukrainian SSR, 1973, p. 31-2, Russian, NSA30-33088.

4. Beam Profiling

1369.Dalglish, R. L. and Kelly, J. C., "Fast response absolute ion beam monitor", Report CONF-720948, Univ. of New South Wales, Kensington, Australia, 1972, p. 183-89, NSA29-10812.

1370.EerNisse, E. P., Peterson, G. D. and Schueler, D. G., "Ion beam profile monitor (for accelerators)", Rev. Sci. Instrum., 46, 266-8 (1975), EEA78-26241.

1371.Hollis, M. J., "An ion beam monitor for nanoampere currents", Nucl. Instrum. & Methods, 103, 337-9 (1972), EEA75-40897.

1372.Kistner, O. C., "Beam profile measuring device and focusing aid", Nucl. Instrum. & Methods, 105, 603 (1972), EEA76-10263.

1373.Martin, F. W., "Acceleration of doubly charged C, N, and O ions at a 3 megavolt van de Graaff", Nucl. Instrum. & Methods, 124, 329-33 (1975), EEA78-17259.
1374.Popescu, D., "The measurement and the integration of particle accelerator beam currents", Stud. & Cercet. Fiz., 26, 681-7 (1974), Rumanian, PA78-17075.
1375.Skoromnyi, G. M. and Revutskii, E. I., "Measurement of the equilibrium charge distribution in He^+ and Be^+ ion beams", Report KHFTI-73-13, 1973, p. 60-65, Russian, NSA31-1316.
1376.Vishnyakov, V. V., Zhuravlev, N. I., Komarov, V. I., Kosarev, G. E., Nguyen, M. S. and Sinaev, A. N., "Instrument for measuring the spatial profile of beams", Instrum. & Exp. Tech., 16, 1632-5 (1973), NSA30-13319.
1377.Wittmaack, K. and Schulz, F., "A telefocus ion gun with variable beam profile", Electron and Ion Beam Science and Technology Fifth International Conference, 1972, p. 181-95, EEA76-10294.
1378.Zhiltsova, L. Y., Larin, A. A. and Shafranov, M. D., "Fabrication of a luminescent screen for measurement of the profiles of charged-particle beams", Instrum. Exp. Tech., 16, 1831-2 (1973), NSA30-13320.

5. Others

1379.Armour, D. G., Grant, W. A., Cavaleru, A. O. R. and Morley, C. M., "Apparatus for the study of low energy ion trapping and gas release in solid targets", Vacuum, 22, 457-62 (1972), PA76-28713.
1380.Artemov, V. M., Iremashvili, D. V. and Shimko, A. I., "Gun for cleaning surfaces of solids by means of ion bombardment", Instrum. & Exp. Tech., 16, 288-90 (1973), EEA77-1171.
1381.Averill, P., Hofmann, A., Little, R., Mieras, H., Paterson, I. M., Strauch, K., Voss, G. A. and Winick, H., "Latest development with colliding beams in CEA", Proceedings of the Third All-Union Conference on the Charged Particle Accelerators, 1973, p. 341-8, 409, NSA30-16642.
1382.Boroffka, H., Krimmel, E. and Runge, H., "Ion implantation device", Patent German 2222736, Publ. November 1973, CA80-42207.
1383.Boroffka, H. and Runge, H., "Method for fast-switching the ion beam in an ion-implantation facility", Vacuum, 23, 447-9 (1973), PA77-21215.
1384.Colligan, J. S., Freeman, J. H., Grant, W. A., Nobes, M. J. and Lewis, G. W., "Versatile isotope separator system for ion implantation and ion-surface interaction studies", Report CONF-730618, Univ. of Salford, England, 1973, p. 276-96, NSA30-33107.

1385. Ewald, H., Hinckel, P., Muenzenberg, G., Nickel, F., Guettner, K. and Wollnik, H., "Velocity filter for the separation of projectiles and reaction products behind the target of a heavy ion accelerator", Report CONF-730618, Physikalisches Institut, Giessen, Germany, 1973, p. 220-5, NSA30-33104.
1386. Forrester, A. T., "Analysis of a magnetic double sheath for ion beam acceleration and deceleration", J. Appl. Phys., 46, 2051-6 (1975).
1387. Hlavka, J., "Ion implantation. Measurement of ion pressure", Phys. Lett. A, 45, 407-8 (1973), CA80-42156.
1388. Ishimaru, H. and Yagi, N., "Pulsed gas shutter", J. Vac. Soc. Jap., 17, 126-30 (1974), Japanese, PA77-77157.
1389. Jacobwitz, L., "Goniometer for ultrahigh vacuum ion implantation studies", Rev. Sci. Instrum., 41, 127-8 (1970), PA73-21830.
1390. Kachurin, G. A., "Installation for ionic doping", Istrum. & Exp. Tech., 14, 1525-6 (1971), PA75-60812.
1391. Khalin, N. F., Golovnya, V. Y., Khvastunov, V. M. and Shevchenko, N. G., "A device for automatic alignment of targets", Instrum. & Exp. Tech., 16, 1323-5 (1973), EEA77-40021.
1392. Krueger, F. R., Persson, B. I. and Kankeleit, E., "Velocity separation of heavy ions by high frequency deflection", Report CONF-730618, Technische Hochschule, Darmstadt, Germany, 1973, p. 268-70, NSA30-33105.
1393. Ku, T. H. and Karol, P. J., "Preparation of powdered targets", Nucl. Instrum. & Methods, 121, 537-40 (1974), PA78-23501.
1394. Marcou, J., "Transient radiation of a particle beam in a slow wave accelerating structure", Nucl. Instrum. & Methods, 123, 205-15 (1975), PA78-20019.
1395. Masters, B. J., "Radiotracer measurements of sputtered contamination incurred during ion implantation processing", IEEE Trans. Nucl. Sci., NS-20, 1032-4 (1973), PA76-57020.
1396. Ohnesorge, W. F. and Butler, H. M., "Recent trends in particle accelerator radiation safety", Report CONF-741040-6, Oak Ridge National Lab., Tennessee, 1974, 6 pp., NSA31-12173.
1397. Pegoraro, A. and Quirci, G., "Goniometer for implantation and channeling experiments", Nucl. Instrum. & Methods, 103, 417-20 (1972), PA75-74788.
1398. Schmid, K., Muller, H. and Ryssel, H., "A universal goniometer for channeling experiments", Nucl. Instrum. & Methods, 99, 121-4 (1972), PA75-25283.
1399. Sharma, J. K. N., "Vacuum systems for ion implantation equipment", Solid State Technol., 17 (12), 42-5 (1974), PA78-26332.
1400. Templeton, L. F., "Vacuum lock design; the key to continuous ion implantation", Solid State Technol., 18 (12), 46-8 (1975).
1401. Tokuyama, T. and Tsuchimaoto, T., "A semiconductor ion injector", Oyo Buturi, 40, 415-21 (1971), Japanese, EEA75-16644.
1402. Venevtsev, I. T., Skoromnyi, G. M. and Revutskii, E. I., "Pulse cascade transformer", Report KHFTI-73-13, 1973, p. 69-71, Russian, NSA31-1322.

1403. Western Electric Co., "Target positioning for ion implantation", Patent UK 1378667, Publ. August 1972, EEA78-15248.
1404. Whitmell, D. S., Kennedy, W. A. D., Mazey, D. J. and Nelson, R. S., "A heavy-ion accelerator-electron microscope link for the direct observation of ion irradiation effects", Radiat. Eff., 22, 163-8 (1974), PA78-139.
1405. Wolfenden, A., "Ion bombardment chamber", Report ORNL-TM-4156, Fusion Technology Studies Progress Report, 1973, p. 71-80, NSA28-3290.
1406. Zlobin, V. G., Kolyada, V. M. and Legkov, B. E., "Chamber for studying channeling and shadow effects", Instrum. & Exp. Tech., 17, 967-8 (1974), EEA78-17217.

G. ETCHING

I. Bibliographies

1. Hawkins, D. T., "Ion milling (ion-beam etching), 1954-1975: A bibliography", J. Vac. Sci. Technol., 12, 1389-98 (1975).

II. Books

2. Anon., "ASTM Standards, Part 31", American Society for Testing and Materials, 1916 Race Street, Philadelphia, Pennsylvania, E3, E340, E407, 1972.
3. Hollahan, J. R. and Bell, A. T., "Techniques and applications of plasma chemistry", John Wiley & Sons, New York, 1974.
4. Holmes, P. J., "The electrochemistry of semiconductors", Academic Press, New York, 1962.
5. Kehl, G. L., "The principles of metallographic laboratory practice", McGraw-Hill Book Co., New York, p. 60-79 (1949).
6. Tegart, W. J., McG., "The electrolytic and chemical polishing of metals in research and industry", Pergamon Press, New York, 1959.

III. Review Articles

7. Denzer, D., "Etching and pickling", Jahrbuch Oberflachentechnik, 1973, 29, 88 (1973), German, MA6-07-57 0246.
8. Dubee, A., "Application of ion etching in microelectronics", Microelectron. & Reliab., 13, 455-62 (1974), French, EEA78-10933.
9. Kolotyrkin, Y. M., Nepodenko, L. M. and Yagupolskaya, L. N., "Potentiostatic method of investigation in physical metallurgy", Ind. Lab., 38, 357 (1972), MA6-06-21 0296.
10. Laukmanis, L. A., "Gas etching of semiconductors", In: Semiconductors and their applications in electronics. V. Izdalel'stro Zinatne, Riga, USSR, p. 43-61 (1971), Russian, EEA75-13577.
11. Litovchenko, V. G., "Processing of semiconductor plates by chemical etching", Poluprovodn. Tekh. & Mikroelektron., no. 19, 3-16 (1975), Russian, EEA78-28647.
12. Norgate, P. and Hammond, V. J., "Ion beam etching", Phys. Technol., 5, 186-203 (1974), CA83-140264.
13. Rossmann, C., "Methods of cleaning metal articles with aqueous solutions", Jahrbuch Oberflachentechnik, 1970, 26, 38 (1970), German, MA3-05-57 0177.
14. Schatt, W. and Worch, H., "The development of metallographic structures by (chemical) solutions", Neue Hutte, 14, 685 (1969), German, MA3-05-21 0238.

15. Thomas, R. S. and Hollahan, J. R., "Use of chemically reactive gas plasmas in preparing specimens for scanning electron microscopy and electron probe microanalysis", Scanning Electron Microsc., 7 (Pt. 1-4), 83-92 (1974), CA81-142552.
16. Valko, A., "Ion-etching and its application in semiconductor technology. I", Hiradstech. Ipari Kut. Intez. Kozl., 14 (3), 5-32 (1974), Hungarian, PA78-33965.
17. Valko, A., "Ion-etching and its application in semiconductor technology. II", Hiradstech. Ipari Kut. Intez. Kozl., 14 (4), 5-27 (1974), Hungarian, EEA78-19627.

IV. Equipment

18. Appels, J. A., Verschuren, M. A. and Nijdam-Paffen, M. M. M., "Apparatus for etching semiconductors", Patent German 2248185, Publ. April 1973, CA79-11276.
19. Avila, A. J., Sauer, H. A., Miller, T. J. and Jaeger, R. E., "Freeze drying of spent plating and etching baths to recover metals", Plating, 60, 239-41 (1973), MA6-08-58 0612.
20. Bernstein, T. and Labuda, E. F., "A spectroscopic monitor for sputter-etching processes", J. Vac. Sci. Technol., 10, 264 (1973).
21. Bicknell, R. W., "A simple rotating jet-thinning apparatus for producing taper sections and electron microscope specimens from silicon and compound semiconductors", J. Phys. D, 6, 1991-7 (1973), PA77-998.
22. Bondarenko, S. D., Perevoshchikov, V. A. and Gusev, V. K., "Chemical polishing of silicon plates with the etchant moving at high speed", Zhur. Priklad. Khim., 43, 1487-91 (1970), Russian, MA3-12-57 0475.
23. Bosch, R., "Devices for etching semiconductor disks", Patent UK 1300333, Publ. December 1972.
24. Cantagrel, M. and Dubee, A., "Technology of erosion by ionic bombardment", AVIRES 2, Colloq. Int. Appl. Sci. Tech. Vide Revetements Etats Surface, C. R., 2nd, 1972, p. 69-74, French, CA81-112550.
25. Chernyaev, V. N., Obichkin, Y. G. and Semenov, E. I., "Continuous control of the thickness and rate of deposition and etching of dielectric and semiconductor films", Prib. Tekh. Eksp., no. 6, 191-2 (1974), Russian, CA82-164142.
26. Collins, J. C. and Pavone, P. J., "Automatic process end-point detection system", IBM Tech. Disclosure Bull., 27, 1342-3 (1974), EEA78-11699.
27. Crockett, C. G., "Glow-discharge ion gun for etching", Vacuum, 23 (1), 11-13 (1973), CA79-46776.
28. Debiec, R. P., "A photoetch facility for high volume thin film pattern generation", Plating, 57, 142-5 (1970), EEA73-27269.

29.Decker, G., "Etching machines and etching procedures for fabrication of printed circuits", Proceedings of the EIPC Conference on Printed Circuit Manufacturing, 1970, p. 27-55, German, EEA74-4319.
30.Dhaliwal, R. S. and Herring, R. B., "Multiple sensor, end-point etch detect system", IBM Tech. Disclosure Bull., 17, 1946-7 (1974), EEA78-15407.
31.Edel, W. A. and Weigel, W., "Dark etch station", IBM Tech. Disclosure Bull., 17, 2933-4 (1975), EEA78-24161.
32.Fitch, R. K., O'Shea, K. R. and Rushton, G. J., "Developments of an electrostatic ion gun for specimen thinning", Electron. Microsc., Proc. Eur. Congr. Electron Microsc., 5th, 1972, p. 296-7, CA79-10958.
33.Fitch, R. K. and Rushton, G. J., "Low-pressure ion source", J. Vac. Sci. Technol., 9, 379-82 (1972), MA5-08-21 0363.
34.Goffredo, D. L. and Dunkelberger, J. A., Jr., "Etchant rinse apparatus", Patent USA 3871914, Publ. March 1975, CA83-47951.
35.Harvilchuck, J. M., "Direct reading metal etch rate monitor", IBM Tech. Disclosure Bull., 12, 1545-6 (1970), EEA73-28830.
36.Havas, H. and Lechaton, J. S., "End-point detection method and apparatus for sputter etching", Patent USA 3664942, Publ. May 1972, CA77-40623.
37.Hoekstra, J. P. and Dill, F. H., "Method to establish the end point during the delineation of oxides on semiconductor surfaces and apparatus therefor", Patent USA 3874959, Publ. April 1975.
38.Kubodera, K., "Argon ion source for ion etching", Jap. J. Appl. Phys., 14, 1637-8 (1975), CA83-187032.
39.Lacal, R., "Capacity of equipment for etching and chemical milling", Galvano, 41, 120-2 (1972), French, MA5-12-57 0309.
40.Lacal, R., "Performance characteristics of etching and chemical milling plants: Conveyorized spray machine", Galvano, 41, 933-8 (1972), French, MA6-05-53 0151.
41.Lendle, E., "Effluent plants for small and medium-size works", Metalloberflache, 27, 25-27 (1973), German, MA6-06-58 0502.
42.Lendle, E., "Use of regulating and dosing equipment for alkaline etching", Chem.-Anlagen Verfahren, no. 5, 65, 67-8, 70, 72 (1974), German, CA83-31743.
43.Marukawa, G., Oikawa, T., Otsuka, R. and Kagaya, N., "Photoresist etching bath", Patent German 1814074, Publ. July 1969, CA71-86622.
44.Moritz, H., "Automated metal etch system for IC-manufacture", Solid State Technol., 18 (8), 54-7 (1975), EEA78-41693.
45.Pick, U., "A simple etching method for thin film patterns", J. Phys. E, 4, 925-6 (1971), EEA75-424.
46.Read, B. H. and Black, R. E., "Electrolytic etch apparatus", Patent USA 3703458, Publ. November 1972, CA78-21679.
47.Reed, B. H. and Black, R. E., "Electrolytic etch apparatus", Patent USA 3865704, Publ. February 1975, CA83-87341.
48.Reichart, P. C. and Schomaker, D. F., "Tunnel diode peak current trimmer", IBM Tech. Disclosure Bull., 12, 1421 (1970), EEA73-28193.

49. Rode, D. L. and Johnson, W. A., "Crystal etch monitoring by internal reflection interferometry", Rev. Sci. Instrum., 41, 672-5 (1970), MA3-10-63 0266.
50. Schmidt, R., "Apparatus and method for improving and accelerating chemical processes in liquids and baths", Patent German 2353936, Publ. May 1975, CA83-105421.
51. Siegel, B., "Continuous process for electrolytic deburring and polishing of metals", Jahrbuch Oberflachentechnik, 1970, 26, 73-9 (1970), German, MA3-05-57 0182.
52. Terentyev, Y. P., Kovalev, L. K., Suvorov, V. N. and Makarov, V. I., "An apparatus for the ion and plasma sputtering", Sov. J. Opt. Technol., 39, 268-70 (1972), EEA76-16308.
53. Werbicki, J., Jr., "A modern approach to metallographic preparation", Met. Finish, 70 (11), 40-3 (1972), MA6-03-21 055.
54. Yoshida, K. and Nagaia, S., "Sputter etching: apparatus and application to the preparation of thin foils for electron microscopy", J. Vac. Soc. Jap., 16, 281-7 (1973), Japanese, PA77-11620.

V. Techniques
1. Ion-beam

55. Alameddine, O. and Bohg, A., "Jet etching arrangement", IBM Tech. Disclosure Bull., 15, 7 (1972).
56. Anon., "Improved ion etching process for integrated circuits", Int. Elektron. Rundsch., 26 (9), 223 (1972), German, EEA75-39176.
57. Bach, H., "Ion-beam etching. Application to electron-microscopic investigations and solid-state analysis. 2", Schott Inform., no. 4, 6-21 (1972), German, CA79-46674.
58. Cantagrel, M., "High resolution patterns engraved by ion etching", IEEE Trans. Electron Devices, ED-22, 483-6 (1975), CA83-186980.
59. Cantagrel, M. and Marchal, M., "Argon ion etching in a reactive gas", J. Mater. Sci., 8, 1711-16 (1973).
60. Dimigen, H. and Hubsch, H., "Generation of deep structures by ion beam etching", Presented at Electrochemical Society Meeting, San Francisco, Calif., May 1974, EEA77-26358.
61. Dimigen, H. and Luthje, H., "An investigation of ion etching", Phillips Tech. Rev., 35, 199-208 (1975).
62. Gindin, I. A., Starodubov, Y. D. and Khaimovich, P. A., "The development of the grain boundary of metal at low temperatures by an ionic-etching technique", Zavod. Lab., 39, 452-3 (1973), Russian, MA6-10-21 0539.
63. Gloersen, P. G., "Ion-beam etching", J. Vac. Sci. & Technol., 12, 28-35 (1975), EEA78-24421.
64. Kasahara, O. and Naraoka, K., "Ion impact etching of semiconductor devices", Patent Japan 73-85440, Publ. November 1973, CA80-88796.
65. Kubodera, K., "An argon ion source for ion etching", Jap. J. Appl. Phys., 14, 1637-38 (1975).

66. Kynaston, D., "(Ion-beam etching does) away with sliced-up specimens", New Scientist, 45, 256-7 (1970), MA3-07-21 0397.
67. Lavoie, F. J., "Machining with ions", Mach. Des., 42 (17), 98-101 (1970), MA3-12-53 0365.
68. Lelakowska, K., "The application of ion etching technique", Elektronika, no. 5, 190-1 (1971), Polish, EEA74-25562.
69. Luthje, H. and Dimigen, H., "Making deep microstructure by ion etching", 6th International Congress on Microelectronics, 1974, 9 pp., German, EEA78-24436.
70. Norgate, P. and Hammond, V. J., "Ion beam etching", Phys. Technol., 5, 186-203 (1974), PA78-6656.
71. Sanwald, R. C. and Gould, D. J., "Ion etching metallographic samples", Metallography, 7, 73-6 (1974), CA81-174449.
72. Schleich, F. and Bach, H., "Physical fundamentals and a new design for ion-etching and sputtering", Presented at Electrochemical Society Meeting, San Francisco, Calif., May 1974, PA77-53627.
73. Schmidt, P. H. and Spencer, E. C., "Ionic etching across a mask", Patent French 2111511, Publ. July 1972, CA78-117640.
74. Smith, H. I., Melngailis, J., Williamson, R. C. and Brogan, W. T., "Ion beam etching of surface gratings (for acoustic surface wave filters)", J. Vac. Sci. & Technol., 10, 1127 (1973), EEA77-24907.
75. Springer, C., Milsch, B. and Gysler, A., "Ion beam etching: a method for the preparation of thin foils", Prakt. Metallogr., 11, 311-22 (1974), PA77-73303.
76. Ward, J. W., "High-vacuum glow-discharge etcher and ion-thinner", Microstructure, 2, 11 (1971).
77. Yasuda, H., "Figuration of wedge-shaped surface by ion etching", Jap. J. Appl. Phys., 12, 780 (1973), MA7-01-12 0081.
78. Yasuda, H., "Ion milling by beam scanning techniques - effect of beam diameter on milled surface shape", Jap. J. Appl. Phys., Suppl. 2 (Pt. 1), 431-4 (1974), EEA78-21711.
79. Zorll, U., "Ion etching. Structure specific erosion process for plastic surfaces", Metalloberflaeche, 29, 337-42 (1975), German, CA83-164886.

2. Sputter

80. Anon., "An improved sputter-etching process", Microelectron. & Reliab., 12, 175 (1973), EEA76-26130.
81. Auyang, R. P. and Koenig, H. R., "Forming a planar surface on a thin-film device", Patent French 2119930, Publ. September 1972, CA78-90165.
82. Bauer, H. J. and Bogardus, E. H., "Controlled RF-sputter etching using atomic absorption spectroscopy", J. Vac. Sci. & Technol., 11, 1144-7 (1974), EEA78-24479.
83. Convertini, U. and Dimigen, H., "Etched structures by ion sputtering", Patent German 2258297, Publ. June 1974, CA81-70146.

84.Davidse, P. D., "RF sputter etching - a universal etch", Thin Film Dielectrics, 1968, p. 130-42, Publ. 1969, EEA73-9898.
85.Dimigen, H. and Luethje, H., "Influence of the target material on sputter etching processes", Thin Solid Films, 27, 155-63 (1975), CA83-124015.
86.Fritz, L. L., "Sputter etch removal rates of insulators, semiconductors, and conductors", Solid State Technol., 14 (12), 43-8 (1971), EEA75-27763.
87.Glawischnig, H., Hoerschelmann, K. and Weidlich, H., "Sputter etching - a new technique in the production of highly reliable metallization systems for microwave transistors", Siemens Forsch. - & Entwicklungsber., 3, 384-9 (1974), German, EEA78-5321.
88.Hanak, J. J. and Russell, J. P., "Sputter etching technique for recording holograms or other fine-detail relief patterns in hard, durable materials", Patent USA 3733258, Publ. May 1973, CA79-37143.
89.Hanfmann, A. M., "Sputtering: The user's viewpoint", IEEE Trans. Ind. Appl., 1A-8, 578-83 (1972), EEA75-35473.
90.Harkins, P. A., Jr., "Sputter etching process", Patent USA 3676317, Publ. July 1972, CA77-94550.
91.Holland, L. and Priestland, C. R. D., "Influence of sputtering and transport mechanisms on target etching and thin-film growth in rf systems. I. Target processes", Vacuum, 22 (4), 133-41 (1972), CA77-93825.
92.Holland, L. and Priestland, C. R. D., "Influence of sputtering and transport mechanisms on target etching and thin-film growth in rf systems. II. Transport and deposition processes", Vacuum, 22 (4), 143-9 (1972), CA77-93824.
93.Hosokawa, N., Matsuzaki, R. and Asamaki, T., "RF sputter-etching by fluoro-chloro-hydrocarbon gases", Jap. J. Appl. Phys., Suppl. 2 (Pt. 1), 435-8 (1974), PA78-37585.
94.Resneua, J. C., "Application of cathodic sputtering and ionic erosion techniques to high-frequency devices", Colloq. Int. Pulverisation Cathodique Ses Appl., C. R., 1st, 1973, p. 315-20, Publ. 1974, French, CA81-70077.
95.Ricker, T. and Schwing, C., "Thin film structures with a gap width comparable to the film thickness fabricated by sputter etching", Solid State Technol., 17 (10), 69-72 (1974), EEA78-5352.
96.Tisoue, T. C. and Cruzan, P. D., "Low-voltage triode sputtering with a confined plasma. V. Application to backsputter definition", J. Vac. Sci. & Technol., 12, 677-88 (1975), EEA78-33173.
97.Vossen, J. L., "R.F. sputtering processes", Report C70-5.2, American Society for Metals, Metals Park, Ohio, 1970, 13 pp., MA4-02-57 0037.
98.Vossen, J. L., "Sputtering technique for coating the inside walls of through-holes in substrates", J. Vac. Sci. Technol., 11, 875-7 (1974), CA81-143210.

99. Vossen, J. L. and Davidson, E. B., "The interaction of photoresists with metals and oxides during RF sputter-etching", J. Electrochem. Soc., 119, 1708-14 (1972), EEA76-8732.
100. Vossen, J. L., Jr., "Radio-frequency sputter etching and photoresist contamination", Patent USA 3692655, Publ. September 1972, CA78-22571.
101. Weissmantel, C., Fiedler, O., Reisse, G., Erler, H. J., Scheit, U., Rost, M. and Herberger, J., "New developments in ion beam sputtering and etching techniques", Jap. J. Appl. Phys., Suppl. 2 (Pt. 1), 509-12 (1974), PA78-38354.
102. Wilkinson, K. D., "RF sputter etching of aluminum, silicon dioxide, and photoresist", Report AD-757878, Air Force Institute of Technology, Patterson, Ohio, 1972, 129 pp., CA79-46782.

3. Plasma

103. Abe, H., "A new undercutting phenomenon in plasma etching", Jap. J. Appl. Phys., 14, 1825-26 (1975).
104. Devaney, J. R. and Sheble, A. M., III, "Plasma etching PROMS and other problems", Solid State Technol., 17 (12), 46-50 (1974), EEA78-11737.
105. Jones, W. K., "Plasma etching as applied to failure analysis", Proceedings of the 12th Annual Reliability Physics Conference, 1974, p. 43-8.
106. Sheble, A. M., Teel, T. P. and Devaney, J. R., "Plasma etching for SEM and EMP examination of microelectronic devices", 9th Annual Conference of the Microbeam Analysis Society, 1974, 31 A/4 pp., EEA78-660.

4. Photo-etching

107. Bertelsen, B. I. and Reimer, A. E., "Photoetch method", Patent USA 3476561, Publ. November 1969, EEA73-13007.
108. Bonnel, M., "Photo-etching techniques", Rev. Tech. Thomson-CSF, 4, 833-56 (1972), French, EEA76-8761.
109. Campbell, D. E. and Harris, R. A., "Mask registration continuous photoetching of strip material", Patent USA 3671243, Publ. June 1972, CA77-145438.
110. Crumley, J. A., Kerstetter, H. A., Hildebrand, J. R. and Shaak, R. N., "Photoetching a continuously moving metallized substrate", Patent German 2145849, Publ. March 1972, CA77-81192.
111. Fischer, W., "Photoetching", Patent German 2143737, Publ. March 1973, CA78-142460.
112. Haddad, M. M. and Schmeckenbecher, A. F., "Activation of photoetched patterns for electroless plating", IBM Tech. Disclosure Bull., 16, 3579 (1974), EEA77-38151.
113. Hayashi, F., Teshima, H. and Tarui, Y., "Photo etching techniques. IV", Bull. Electrotech. Lab., 33, 645-55 (1969), Japanese, EEA73-16635.

114.Hayashi, V. F., Teshima, H. and Tarui, Y., "Photo-etching techniques", Bull. Electrotech. Lab., 36, 134-44 (1972), Japanese, PA75-71301.
115.Konstantouros, E. and Stoll, K., "Glass masks for photoetching", Patent German 2007484, Publ. March 1973, CA78-142459.
116.Kump, H. J., "Two-step photoetching method for semiconductors", Patent USA 3767493, Publ. October 1973.
117.Takagi, T. and Tawara, J., "Thin coils manufactured by photo-etching process", JEE, no. 82, 50-4 (1973), EEA77-18439.
118.Terada, T., "Photoetching", Patent Japan 73-39338, Publ. November 1973, CA80-150259.
119.Vaccari, J. A., "Photo-etching makes precision metal parts", Mater. Eng., 72, 28-30 (1970), MA4-04-53 0132.

5. Chemical

120.Astrina, V. I., Rudakova, T. E., Moiseev, Y. V., Vlasov, S. V. and Zaikov, G. E., "Formation of a rough surface on poly(ethylene terephthalate) films by chemical etching", Plast. Massy, no. 9, 17-8 (1975), Russian, CA83-207192.
121.Bickford, H. G. and Peredy, S. K., "Method of wet etch control", IBM Tech. Disclosure Bull., 17, 3527 (1975), PA78-61931.
122.Buickaen, V., "A new technique of electrochemical thinning for (electron microscope) specimens", Electrochim. Metallorum, 4, 262-6 (1969), Italian, MA3-05-21 0228.
123.Chwalow, M. L. E., "Etch completion indication", IBM Tech. Disclosure Bull., 15, 606-7 (1972), EEA75-39811.
124.Grilletto, C., "Emission spectrographic analysis of boron in 40% ammonium fluoride solution", Appl. Spectrosc., 28, 369-71 (1974), CA81-98978.
125.Henninges, K. and Schulze, B., "Treating semiconductor elements of circular outline", Patent USA 3607478, Publ. September 1971, CA77-67749.
126.Jones, V. V., "Imaging process including etching alcohol-soluble polyamide with aqueous solutions", Patent USA 3645732, Publ. February 1972, CA76-106470.
127.Khatsernov, M. A., "Preparation of diagonal cut sections of surface layers by an etching method", Zavod. Lab., 38, 823-4 (1972), Russian, CA77-144773.
128.Liu, C. J. and Sauer, H., "Resistance/capacitance methods for determining oxide etch end point", IBM Tech. Disclosure Bull., 16, 2706-7 (1974), EEA77-18570.
129.Merenyi, C. and Hasko, F., "Reaction mechanism of the etching of epoxy based materials by chromic acid-sulfuric acid solutions", Oberflaeche Surf., 16 (3), 49-53 (1975), German, CA83-132386.
130.Morris, R. E., "New developments in ammoniacal etchants. I", Electron. Manufacturer, 17 (3), 24-27 (1972), MA5-12-58 0619.

131.Nakabayashi, M., "Etchant for selective etching of nickel or copper layers", Patent Japan 74-03851, Publ. January 1974, CA81-53612.
132.Porter, R. A., "Masking silicon nitride during phosphoric acid etching in semiconductor device manufacture", Patent USA 3657030, Publ. April 1972, CA76-160067.
133.Protzman, T. F.. "Etching layers in photographic systems without use of a resist", Patent German 2146779, Publ. March 1972, CA77-107735.
134.Reagan, B. M. and Stinson, J. P., "Stabilized etch solution", IBM Tech. Disclosure Bull., 27, 1539 (1974), EEA78-8617.
135.Rossi, D. H., "Etch-back or chemical cleaning - a necessary evil?", Solid State Technol., 12 (11), 40-2 (1969), EEA73-16800.
136.Rudakova, T. E., Moiseev, Y. V., Astrina, V. I., Razumova, L. L., Vlasov, S. V. and Zaikov, G. E., "Mechanism of rough surface formation on poly(ethylene terephthalate) films obtained by chemical etching in aqueous alkali solutions", Vysokomol. Soedin., Ser. A, 17, 1797-801 (1975), Russian, CA83-180041.
137.Saatchyan, R. G., Lyubimov, A. A., Moksyakov, A. I. and Frantsuzov, V. F., "Etching solution", Patent USSR 478795, Publ. July 1975, CA83-183052.
138.Shitikov, G., "Determination of the nitric acid concentration in an etching silicon", Poligrafiya, no. 2, 35 (1974), Russian, CA81-19223.
139.Stiehr, W., "Chemical machining. III. A special electrochemical process for chipless production of metal articles", Galvanotechnik, 63, 337-42 (1972), German, MA6-01-53 0003.

6. Electrochemical

140.Arrowsmith, H. W. and Rencken, D. G., "Electrolytic polishing and etching of thin foils", Proceedings, Second Annual Technical Meeting, International Metallographic Society, Inc., 1970, p. 21-7, MA3-07-21 0425.
141.Bloor, D. W., "Electrochemical principles of etching", Metals Australia, 4, 279-82 (1972), MA6-04-32 0293.
142.Hanazono, M. and Asai, O., "Integrated circuits by electrolytic etching", Patent German 2164490, Publ. June 1972, CA77-81197.
143.Johnson, M. J., "Use of the potentiostat for developing etchants", Photo Chem. Mach.-Photo Chem. Etching, 7, 25-9 (1972), CA77-13197.

7. Others

144.Associated Semiconductor Manufacturers, Ltd., "Methods of etching semiconductor body surfaces", Patent UK 1273150, Publ. May 1972.
145.Gabrail, S. I., "Method of etch subdividing semiconductor wafers", Patent USA 3762973, Publ. October 1973.

146. Glendinning, W. B., "Selective etching out of defects in a semiconductor junction", Patent USA 3518131, Publ. June 1970, CA73-60498.
147. Gohs, H., "Spray-etching in the electronics industry", Electron Ind., 2, 189-91 (1971), German, EEA75-3934.
148. Janakiev, N., "Specimen preparation for the scanning electron microscope by successive hot etching", Praktische Metallographie, 8, 543-6 (1971), MA5-05-21 0140.
149. Jurezok, R., "Pickling and etching", Jahrbuch Oberflachentechnik, 1970, 26, 52-5 (1970), German, MA3-05-57 0179.
150. Kolf, G. and Laessig, W., "Vibration etching, a method of evaluation of photoresist characteristics", Polym. Eng. Sic., 11, 492 (1971).
151. Kolot, V. Y., Rybalko, V. F., Palatnik, L. S., Fogel, Y. M. and Chernyakova, L. E., "Use of secondary ionic ion emission to investigate the superficial film formed on gold during electrolytic etching", Prot. Metals, 5, 375-6 (1969), MA4-04-34 0217.
152. Kuznetsov, F. A. and Kokovin, G. A., "Use of thermodynamic calculations for optimization of the growth and etching of crystals in the gas phase", Protsessy Rosta Krist. Plenok Poluprov., 73-90 (1970), Russian, CA77-67061.
153. Legrand-Bonnyns, E. and Ponslet, A., "Determination of the orientation of grains in a polycrystal from the structure revealed by thermal etching", Metallography, 6, 295-31 (1973), French, MA7-01-21 0036.
154. Reynolds, J. H., "(Etching) method of making a semiconductor device having tunnel oxide contacts", Patent USA 3755026, Publ. August 1973.
155. Scherber, W., "Etching semiconductor material", Patent German 2346564, Publ. April 1975, CA83-52105.
156. Schwartz, J. W., "Re-etching the apertures of a color cathode-ray tube shadow mask", Patent USA 3732133, Publ. May 1973, CA79-11292.
157. Taylor, N. J., "The role of Auger electron spectroscopy in surface elemental analysis", Vacuum, 19, 575-8 (1969), MA3-06-23 0302.
158. Vasilevskaya, L. S. and Malkov, N. V., "Conditions for etching epitaxial layers of semiconducting materials of specified thickness with a view to analysis", Zavod. Lab., 37, 1044-5 (1971), Russian, MA5-03-23 0142.
159. Yoshihiko, W., Inoue, I. and Kusano, Y., "Process (etching) of producing semiconductor devices", Patent USA 3807038, Publ. April 1974.

VI. Semiconductors
1. Group IV
a. Silicon

160. Abe, H., "Semiconductor sloped etching using Freon gas plasma", Patent UK 1398019, Publ. June 1975, CA83-156868.
161. Abe, H., Sonobe, Y. and Enomoto, T., "Etching characteristics of silicon and its compounds by gas plasma", Jap. J. Appl. Phys., 12, 154-5 (1973), MA6-09-21 0492.
162. Bean, K. E. and Gleim, P. S., "Forming circuit components within a substrate", Patent USA 3793712, Publ. February 1974, CA80-114049.
163. Bean, K. E., Yeakley, R. L. and Powell, T. K., "Orientation dependent etching and deposition of silicon", Presented at Electrochemical Society Meeting, San Francisco, Calif., May 1974, PA77-45467.
164. Beck, R. W. and Yoder, D. J., "Etching silicon wafers", Patent USA 3813311, Publ. May 1974, CA81-31158.
165. Bellin, P. H. and Zwicker, W. K., "Observation of surface defects in electrolytically etched silicon by infrared microscopy", J. Appl. Phys., 42, 1216-21 (1971), MA4-09-21 0518.
166. Bendik, N. T. and Garnyk, V. S., "Effect of last diffusing impurities on the etch characteristics of dislocation silicon", Sov. Phys. Solid-State, 13, 1464 (1971).
167. Berzina, R., Klotins, E., Nesaule, V., Eglitis, I. and Eimanis, I., "Silicon surface layer study after etching in a high-temperature gas-discharge plasma", Latv. PSR Zinat. Akad. Vestis, Fiz. Teh. Zinat. Ser., no. 3, 25-8 (1973), Russian, CA79-98206.
168. Blokhina, A. P. and Izidtsnov, S. O., "Mechanism of the etching of silicon and the generation of excess carriers", Elektrokhimiya, 6, 1409-12 (1970), Russian, MA4-01-34 0034.
169. Bohg, A., "Ethylene diamine-pyrocatechol-water mixture shows etching anomaly in boron-doped silicon", J. Electrochem. Soc., 118, 401-2 (1971), EEA74-18557.
170. Boltsov, Y. P., Prokhorov, V. I. and Soron, L. M., "Preparation of thin silicon films for transmission electron microscopy by photolithography and dynamic etching", J. Physics E (Sci. Instruments), 3, 702-4 (1970), MA4-02-21 0067.
171. Briska, M., "Etching the rim of Si wafers", IBM Tech. Disclosure Bull., 18, 443 (1975), EEA78-37639.
172. Brooks, C. G., "Process for deep etching a silicon wafer", Patent USA 3499805, Publ. March 1970, EEA73-21106.
173. Church, C. L. and Smith, J. W., II, "Polycrystalline etch", Patent USA 3909325, Publ. September 1975, CA83-201124.
174. Claes, H. A. D. and Gelling, W. G., "Etchants for silicon and germanium printed circuit boards", Patent German 2445882, Publ. April 1975, CA83-52103.
175. d'Aragona, F. S., "Dislocation etch for (100) planes in silicon", J. Electrochem. Soc., 119, 948 (1972), MA5-11-21 0485.

176.Dismukes, J. P. and Ulmer, R., "Gas-phase etching of silicon with chlorine", J. Electrochem. Soc., 118, 634-6 (1971), EEA74-18561.
177.Duke, W. G., "Radio-frequency sputter etching in an argon plasma using silicon as a target", Report AD-742436, Air Force Institute of Technology, Patterson, Ohio, 1971, 127 pp., CA77-119560.
178.Dumin, D. J. and Keen, R., "Anisotropic etching of (100) silicon-on-sapphire", Presented at Electrochemical Society Meeting, Chicago, Ill., May 1973, EEA76-23290.
179.Freyer, J., "Fabrication of thin silicon films for microwave semiconductor transit time devices", J. Electrochem. Soc., 122, 1238-40 (1975), CA83-156704.
180.Fuji Photo Film Co., Ltd., "Etching a silicon semiconductor device", Patent UK 1337625, Publ. November 1973, CA80-75736.
181.Fuji Photo Film Co., Ltd., "Photosensitive etching solution for silicon semiconductor devices", Patent UK 1337626, Publ. November 1973, CA80-88794.
182.Gibbons, J. F., Hechtl, E. O. and Tsurushima, T., "Ion-bombardment-enhanced etching of silicon", Appl. Phys. Lett., 15, 117-9 (1969), MA3-01-21 0008.
183.Glendinning, W. B. and Pharo, W. B., "Continuously etching a silicon substrate", Patent USA 3773578, Publ. November 1973, CA80-53768.
184.Gulyamov, M. M., Dobrovensky, V. V., Savelev, B. N. and Chudakov, V. S., "Effect of crystallization conditions on the anomalous etching of silicon crystals", Izvest. Akad. Nauk SSSR, Neorg. Materialy, 7, 402-5 (1971), Russian, MA4-08-12 1103.
185.Hayashi, F., Tauri, Y. and Teshima, H., "The preferential etching of silicon with a water-amine solution", Bull. Electrotech. Lab., 36, 27-33 (1972), Japanese, EEA75-36276.
186.Hays, R. G. and Rhee, C., "Etch stop for a potassium hydroxide anisotropic etch", Patent USA 3721593, Publ. February 1972, CA78-129754.
187.Hill, M. J., "Electron microscopy of electrolytically etched silicon surfaces", J. Electrochem. Soc., 120, 142-5 (1973), MA6-06-21 0301.
188.Hines, R. E., "A simple cure for a form of window staining which occurs during the etching of silicon integrated circuits", Microelectron. & Reliab., 11, 537 (1972), EEA76-8744.
189.Hoffmeister, W. and Zugel, M., "Determination of the etch rate of silicon in buffered HF using a ^{31}Si tracer method", Internat. J. Appl. Radiation Isotopes, 20, 139-40 (1969), EEA72-6489.
190.Iiaw, H. M. and Faust, J. W., Jr., "The effect of ultrasonic agitation on the etching of silicon", Presented at Electrochemical Society Meeting, Chicago, Ill., May 1973, EEA76-23287.
191.Irbe, I., "Anomalies observed in some discrete device and integrated circuit substrates", J. Electrochem. Soc., 120, 838-9 (1973), CA79-59080.

192. Izidinov, S. O. and Blokhina, A. P., "Induction period in the etching of silicon", Elektrokhimiya, 7, 1221-4 (1971), Russian, MA5-01-34 0043.
193. Izidinov, S. O. and Blokhina, A. P., "Effect of etching time on the kinetics of silicon photo-passivation in alkaline solutions", Elektrokhimiya, 9, 1759-64 (1973), Russian, CA80-77577.
194. Izidinov, S. O. and Sinyushin, M. N., "Contribution of the photochemical decomposition of nitric acid to the acceleration of the silicon etching process under the action of light in nitric acid with the addition of hydrogen fluroide", Elektrokhimiya, 9, 1486-9 (1973), Russian, CA80-42165.
195. Kamins, T. I., "Diffusion in thin silicon films formed by electrochemical etching", J. Electrochem. Soc., 121, 286-8 (1974), PA77-29009.
196. Kamper, M., "A new striation etch of silicon", J. Electrochem. Soc., 117, 261-2 (1970), MA3-07-21 0378.
197. Kapitonova, L. M., Lebedev, A. A. and Mamadalimov, A. T., "Effect of surface treatment on the photoconductivity spectra of p-silicon with zinc", Fiz. Tekh. Poluprov., 8, 529-33 (1974), Russian, CA81-7459.
198. Karpel, N. G., Fedorchuk, O. K. and Kulyas, S. L., "Spectrochemical method for solving analytical problems during epitaxial growth", Zavod. Lab., 38, 671-4 (1972), Russian, CA77-134674.
199. Keller, W. and Muehlbauer, A., "Etching depressions, a new defect in a dislocation-free silicon", Inst. Phys. Conf. Ser., 1975, p. 538-43, CA83-69325.
200. Kendall, D. L., "On etching very narrow grooves in silicon", Appl. Phys. Lett., 26, 195-8 (1975), EEA78-19620.
201. Khandelwal, D. D., "Mesa etching without overhang for semiconductor devices", Patent USA 3894895, Publ. July 1975, CA83-187375.
202. Kondaurov, V. P., Nikolaeva, E. P., Panteleev, V. I. and Khukhryanskii, Y. P., "Low-temperature vapor etching of silicon with liquid epitaxy", Sb. Tr. Po. Poluprovodn. Materialam, Priboram I Ikh Primeneniyu, 1973, p. 44-7, Publ. 1974, Russian, CA83-51934.
203. Kragness, R. C. and Waggener, H. A., "Precision etching of semiconductors", Patent USA 3765969, Publ. October 1973, CA80-20702.
204. Kraus, F., "Thin etching of silicon wafers", Patent German 1521975, Publ. November 1972, CA78-77295.
205. Kuchinsky, V. A., Volynchikova, L. F. and Reznikov, B. S., "Thermodynamic study of the etching of silicon in water vapour", Doklady Akad. Nauk Beloruss, SSR, 15, 525-27 (1971), Russian, MA4-12-15 1890.
206. Kujer, T. J. M., Giling, L. J. and Bloem, J., "Gas phase etching of silicon with HCl (surface measurement techniques)", J. Cryst. Growth, 22, 29-33 (1974), PA77-44788.
207. Lawrence, J., "Preferential etching of silicon", Patent USA 3759768, Publ. September 1973.

208. Lee, D. B., "Anisotropic etching of silicon", J. Appl. Phys., 40, 4569-74 (1969), EEA73-10361.
209. Levin, E. R., Dismukes, J. P. and Coutts, M. D., "Electron microscope characterization of defects on gaseous-etched silicon surfaces", J. Electrochem. Soc., 118, 1171-5 (1971), MA4-12-34 0800.
210. Licentia, "Etching semiconductor wafers", Patent UK 1212250, Publ. November 1970, CA74-36219.
211. Maeda, K. and Takayama, O., "The effect of vapour etching on 'diffuse-up' of buried impurities into epitaxial layer", Jap. J. Appl. Phys., 13, 1903-4 (1974), EEA78-5257.
212. Mayer, K. R., "Simplication of Kaemper's striation etch for silicon", J. Electrochem. Soc., 120, 1780-2 (1973), CA80-64924.
213. McKelvey, A. F., "Selective silicon groove etching using a tantalum oxide mask formed at room temperatures", Patent USA 3591465, Publ. July 1971, CA75-81719.
214. Meek, R. L., "Fabrication of integrated semiconductor devices by electrochemical etching", Patent USA 3661741, Publ. May 1972, CA77-26369.
215. Mikhailov, Y. A., Novikov, V. V., Noskov, A. I. and Umanskaya, S. N., "Etchant", Patent USSR 448253, Publ. October 1974, CA82-164243.
216. Moeller, R., Stelzer, H., Kleinert, M. and Fabian, L., "Etching substrates", Patent German (East) 105008, Publ. April 1974, CA81-178952.
217. Mueller, A., "Solvent mixture with nitric acid and hydrofluoric acid for wet chemical etching of silicon", Patent USA 3592773, Publ. July 1971, CA77-67748.
218. Muraoka, H., Ohashi, T. and Sumitomo, Y., "Controlled preferential etching technology", Presented at Electrochemical Society Meeting, Chicago, Ill., May 1973, EEA76-23315.
219. Muraoka, H., Ohashi, T., Yasui, T. and Fujii, A., "Photosensitive semiconductor device", Patent UK 1314648, Publ. April 1973, CA79-47067.
220. Muto, S. Y., "Etching process", Patent German 2449731, Publ. June 1975, CA83-156855.
221. Okuyama, H. and Miyashita, Y., "Contamination on chemically etched silicon (111) surfaces", Shinku, 15, 445-9 (1972), Japanese, CA78-116776.
222. Petrova, A. G. and Rakov, A. V., "Interaction of a silicon surface with chemically-active electrolytes", Fizika i Tekh Poluprov., 4, 697-701 (1970), Russian, MA3-08-34 0448.
223. Price, J. B., "Anisotropic etching of silicon with potassium hydroxide-water-isopropyl alcohol", In: Semicond. Silicon, Huff, H. R., (Ed.), Electrochem. Soc., Inc.; Princeton, N. J., p. 339-53 (1973), CA82-24600.
224. Pugacz-Muraszkiewicz, I. J., "Detection of discontinuities in passivating layers on silicon by NaOH anisotropic etch", IBM J. Res. & Dev., 16, 523-9 (1972), EEA76-8516.

225.Raetzel, C., Schild, S. and Schlotterer, H., "Etching properties of different silicon films", Extended Abstracts of Electrochemical Society Fall Meeting, 1974, p. 336-8, EEA78-37677.
226.Rai-Choudhury, P., "Sulfur hexafluoride as an etchant for silicon", J. Electrochem. Soc., 118, 266-9 (1971), EEA74-18550.
227.Raider, S. I., Flitsch, R. and Palmer, M. J., "Oxide growth on etched silicon in air at room temperature", J. Electrochem. Soc., 122, 413-18 (1975), CA82-164079.
228.Retajczyk, T. F., Jr., "Etching materials capable of being etched with hydrofluoric acid", Patent USA 3676240, Publ. July 1972, CA77-119843.
229.Riedl, W., "Etching conditions of silicon substrates in the region of SiO_2- and Si_3N_4-unstability, prior to epitaxial growth", Prace Inst. Tele-i Radiotech., 10 (4), 3-24 (1967), EEA70-18812.
230.Roman, W. C. and Wilson, L. R., "Delineating semiconductor junctions", Patent USA 3830665, Publ. August 1974, CA81-128764.
231.Saidov, M. S. and Shukurov, I., "Solution and etching of Si in tin solutions", Inorg. Mater., 8, 1825-7 (1972), EEA76-35203.
232.Sasano, A. and Shimomoto, Y., "Gas for sputter-etching silicon and silicon compounds", Patent Japan 74-43844, Publ. April 1974, CA81-143314.
233.Schink, N. and Wigoureux, S., "Etching of silicon disks", Patent German 2141235, Publ. March 1973, CA78-129738.
234.Schwabe, U., "Localized etching of silicon crystals", Patent German 2359511, Publ. June 1975, CA83-124971.
235.Secco, A. F., "Dislocation etch for (100) planes in silicon", J. Electrochem. Soc., 119, 948-51 (1972), CA77-53926.
236.Shachnev, V. I. and Afanasovich, V. F., "High-temperature etching of silicon single crystals by gaseous hydrogen chloride", Elektron. Tekh. Nauch.-Tekh., Sb. Mater., no. 1, 89-92 (1972), Russian, CA79-119231.
237.Sheherbinina, R. N., Prushinskii, A. K. and Kimarskii, V. I., "Etching of silicon", Patent USSR 352539, Publ. August 1973, CA80-150267.
238.Shcherbinina, R. N., Prushinskii, A. K. and Kimarskii, V. I., "Anisotropic etching of silicon", Aktiviruemye Protsessy Tekhnol. Mikroelektron., no. 1, 21-6 (1975), Russian, CA83-171604.
239.Sloan, B. J., Jr., Martin, B. M., Clevenger, L. H. and Dunn, R. S., "Integrated-circuit process utilizing orientation-dependent silicon etch", Patent USA 3768150, Publ. October 1973.
240.Smeltzer, R. K. and Bean, K. E., "Etching silicon", Patent German 2460653, Publ. July 1975, CA83-171667.
241.Spencer, E. G., Lenzo, P. V. and Schmidt, P. H., "Ion etching of thin windows in silicon", J. Vac. Sci. & Technol., 11, 863-4 (1974), EEA78-10543.
242.Stinson, L. and Neville, R. C., "Silicon etching effects with sulfur hexafluoride", Presented at Electrochemical Society Meeting, Chicago, Ill., May 1973, EEA76-23291.

243. Stoller, A. I., "The etching of deep vertical-walled patterns in silicon", RCA Rev., 31, 271-5 (1970), EEA74-4824.
244. Stoller, A. I. and Opresko, S. T., "Etching semiconductor bodies", Patent USA 3677848, Publ. July 1972, CA77-106774.
245. Stoller, A. I., Speers, R. F. and Opresko, S., "A new technique for etch thinning silicon wafers", RCA Rev., 31, 265-70 (1970), EEA74-4823.
246. Sugawara, K., "Facets formed by hydrogen chloride vapor etching on silicon surfaces through windows in SiO_2 and Si_3N_4 masks", J. Electrochem. Soc., 118, 110-14 (1971), EEA74-14810.
247. Swann, R. C. G., Eldridge, P. G. and Penton, J. I., "Manufacture of semiconductor components using the dry etching technique", Patent German 2213037, Publ. October 1972, CA78-49766.
248. Takeda, H. and Watanabe, S., "Chemical etching solution for use in semiconductor device preparation", Patent Japan 75-09268, Publ. April 1975, CA83-124980.
249. Takigawa, M., Satoh, T. and Shohno, K., "Selective removal of BP from Si substrate", J. Electrochem. Soc., 122, 824-5 (1975), EEA78-33020.
250. Tanaka, T., Yatsuo, T. and Yokota, T., "Silicon etching solutions containing gum arabic as an etching retardation agent", Patent Japan 75-23982, Publ. March 1975, CA83-89822.
251. Theunissen, M. J. J., Appels, J. A. and Verkuylen, W. H. C. G., "Application of preferential electrochemical etching of silicon to semiconductor device technology", J. Electrochem. Soc., 117, 959-65 (1970), MA3-12-63 0312.
252. Tothne, M. K., "Electrochemical method for etching thin silicon layers", Hiki, 10 (2), 45-51 (1970), Hungarian, EEA73-37851.
253. Tsang, W. T. and Wang, S., "Preferentially etched diffraction gratings in silicon", J. Appl. Phys., 46, 2163-6 (1975), CA83-50590.
254. Tseng, C. C., Botez, D. and Wang, S., "Optical bends and rings fabricated by preferential etching", Appl. Phys. Lett., 26, 699-701 (1975), PA78-72697.
255. Ura, M. and Suzuki, T., "High-purity hydrogen bromide for vapor-phase etching semiconductor substrates", Patent Japan 73-59094, Publ. August 1973, CA80-85238.
256. van Dijk, H. J. A. and de Jonge, J., "Preparation of thin silicon crystals by electrochemical thinning of epitaxially grown structures", J. Electrochem. Soc., 117, 553-54 (1970), MA3-10-63 0265.
257. Vasilev, A. D., Gorbach, T. Y., Kotova, N. V. and Svechnikov, S. V., "Method of silicon directional microprofiling in semiconductor electronics", Dopov. Akad. Nauk UkRSR. Ser. A, no. 4, 367-71 (1975), Russian, EEA78-41464.
258. Vasilev, V. and Khinkov, V., "Effect of the surface on the photovoltage appearing in a semiconductor structure during point scanning", Izv. Fiz. Inst. ANEB (At. Nauchnoeksp. Baza), Bulg. Akad. Nauk., 23, 15-24 (1973), Bulgarian, CA79-130467.

259.Voronkov, V. V., Savelev, B. N., Voronkova, G. I., Kobrovenskii, V. V., Gulyamov, M. M. and Filler, A. S., "Nature of the anomalous etching of silicon containing hydrogen", Inorg. Mater., 10, 686-8 (1974), PA78-65821.
260.Voronov, I. N., Lainer, L. V., Nikitin, V. M., Nosova, E. A. and Smirnov, B. V., "Etching reagent", Patent USSR 369651, Publ. February 1973, CA79-58814.
261.Waggener, H. A., "Electrochemically controlled thinning of silicon", Bell Syst. Tech. J., 49, 473-5 (1970), MA3-08-63 0220.
262.Watanabe, S. and Takeda, H., "Chemical etching solution for use in semiconductor device preparation", Patent Japan 75-09269, Publ. April 1975, CA83-124981.
263.Weirauch, D. F., "Correlation of the anisotropic etching of single-crystal silicon spheres and wafers", J. Appl. Phys., 46, 1478-83 (1975), EEA78-24177.
264.Wen, C. P. and Weller, K. P., "Preferential electrochemical etching of P^+ silicon in an aqueous $HF-H_2SO_4$ electrolyte", J. Electrochem. Soc., 119, 547-8 (1972), EEA75-24202.
265.Wenzel, C., "Slip and bowing control by advanced etching techniques", Semiconductor Prod. Solid State Technol., 10 (8), 40-4 (1967), EEA70-23159.
266.Worledge, P. L., "Etching silicon substrates", Patent UK 1254981, Publ. November 1971, CA76-29064.
267.Yamamoto, Y., Shinada, K., Itoh, T. and Yada, K., "Damage in the surface region of silicon produced by sputter-etching", Jap. J. Appl. Phys., 13, 551-2 (1974), CA80-138316.
268.Yasuda, H., "Application of ion-milling techniques to microparabolic surface formation", J. Appl. Phys., 45, 484-6 (1974), PA77-44787.
269.Zavalskii, Y. P., Shaposhnik, K. I., Yatsenko, N. G., Rudoi, V. V. and Naidenova, T. D., "Mechanism of gas etching of the silicon surface by hydrochloric acid vapors", Tr. Voronezh. Univ., no. 74, 77-83 (1971), Russian, CA77-67134.
270.Zavalskii, Y. P., Yatsenko, N. G., Shaposhnik, K. I., Naidenova, T. D. and Pshestanchik, V. R., "Etching of silicon by anhydrous hydrogen chloride", Tr. Voronezh. Univ., no. 74, 100-5 (1971), Russian, CA77-67133.
271.Zwicker, W. K. and Kurtz, S. K., "Anisotropic etching of silicon using electrochemical displacement reactions", In: Semicond. Silicon, Huff, H. R., (Ed.), Electrochem. Soc., Inc.; Princeton, N. J., p. 315-26 (1973), CA82-24639.

b. Germanium

272.Aleksandrov, L. N., Sidorov, Y. G. and Krivorotov, E. A., "Etching of germanium with germanium tetraiodide at low pressures", Izvest. Akad. Nauk SSSR, Neorg. Materialy, 6, 1056-60 (1970), Russian, MA3-11-21 0631.

273. Avdienko, K. L., Vasileva, L. L. and Repinskii, S. M., "Crystal-chemical principles of the reaction of germanium with hydrogen sulfide", Izv. Akad. Nauk SSSR, Neorg. Mater., 9, 1481-5 (1973), Russian, CA79-150251.
274. Baltramiejunas, R., Sakalas, A., Storasta, J. and Vaitkus, J., "Photoconductivity in germanium during neodymium laser excitation", Fiz. Tekh. Poluprov., 6, 760-2 (1972), Russian, CA77-53810.
275. Beckmann, K. H., "On the chemical composition of surface films produced on germanium in different etchants", Surface Sci., 5, 187-96 (1966), EEA70-526.
276. Belyi, V. I., Kuznetsov, F. A. and Pfeifer, J., "The chemical etching of germanium monocrystal producing a characteristic fine-grained layer", Proc. Second Colloq. on Thin Films, Budapest, 1967, p. 181-7, Publ. 1968, MA3-02-21 0078.
277. Belyi, V. I. and Kuznetsov, F. A., "Polishing etching of germanium single crystals by gaseous hydrogen halides", In: Protsessy Sin. Rosta Krist. Plenok Poluprov. Mater., Simp., Mironov, K. E., (Ed.), Novosibirsk, USSR, "Nauka", Sib. Otd., p. 227-32 (1971), Russian, CA77-40032.
278. Campisi, G. J. and Frankl, D. R., "Effects of etching and oxidation on the thermal conductivity of germanium. Comments", Phys. Rev. B, 10, 2644-6 (1974), CA82-22492.
279. Chu, T. L. and Kelm, R. W., "The etching of germanium with water vapor and hydrogen sulfide", J. Electrochem. Soc., 116, 1261-3 (1969), EEA73-4310.
280. Dudnik, E. P., Gavrilenko, O. V. and Nefedov, V. E., "Chemical etchings in high-purity germanium preparation", Nauch. tr. n. i. proekt. in t redkomet. prom-sti, no. 51, 148-51 (1974), Russian, CA83-119630.
281. Ehman, M. F., Faust, J. W., Jr. and White, W. B., "The influence of the complexing agent concentration on the etch rate of germanium", J. Electrochem. Soc., 118, 1443-7 (1971), MA5-01-34 0069.
282. Eletskaya, G. N. and Ryabchenko, L. N., "Photoetching of silicon and germanium", Vop. Mikroelektron., Mater. Semin. "Elem. Ustroistva Mikroelektron", 1970, p. 92-6, Publ. 1971, Russian, CA80-139456.
283. Erdman, W. C. and Schmidt, P. F., "Etching solutions for germanium and silicon", Patent German 2245809, Publ. April 1973, CA78-166025.
284. Grigorev, V. E. and Malkin, G. M., "Polishing etching of Ge by gaseous iodine", Inorg. Mater., 10, 1184-7 (1974), PA78-85647.
285. Grigorev, V. E., Malkin, G. M. and Tumanov, A. A., "Iodine transport of germanium in a flow-type system", Izv. Akad. Nauk SSSR, Neorg. Mater., 9, 1313-15 (1973), Russian, CA79-140364.
286. Inoue, M., "Etching of germanium with water vapor", Jap. J. Appl. Phys., 11, 1147-52 (1972), CA77-93864.

287.Kawada, J., "Studies on after-etching for germanium-indium alloyed junction elements. III. After-etching techniques for germanium-indium alloyed junction elements and the relability of their electric characteristics", J. Metal Finishing Soc. Japan, 20, 400-4 (1969), Japanese, MA3-03-63 0051.
288.Kopylova, S. V., Klimenko, E. A. and Klimenko, A. G., "Preparation of single crystals of germanium with a thickness of 0.5 micron", Pribory i Tekhn. Eksper., no. 5, 174 (1969), Russian, MA3-03-21 0155.
289.Kotosonov, N. V., Nikolskaya, I. P. and Pitanov, V. S., "Depth of the impaired layer and the etching rate of mechanically polished germanium with various crystallographic orientations", Radiofiz. Mikroelektron., 56-9 (1970), Russian, CA77-80432.
290.Kurbatov, B. S. and Kurov, G. A., "Some characteristics of the etching of germanium with iodine vapor", Sov. Phys. Crystallogr., 14, 286-7 (1969), MA3-05-21 0257.
291.Kuznetsov, F. A. and Belyi, V. I., "Etching of germanium single crystals by gaseous HBr", Growth of Crystals, 8, 141-45 (1969), MA4-02-12 0302.
292.Litvinova, I. Y. and Smirnov, Y. M., "Distribution of antimony in dislocation-free germanium single crystals", Izvest. Akad. Nauk SSSR, Neorg. Materialy, 6, 115-16 (1970), Russian, MA3-12-12 0733.
293.Malkin, G. M., Grigorev, V. E. and Tumanov, A. A., "Etching and growth of Ge in a flow system with cyclic diffusion mass exchange", Inorg. Mater., 9, 1269-70 (1973), EEA77-30160.
294.Meer, W. and Armer, G., "Etchant for use in making making mesa transistors", Patent German 1521968, Publ. July 1974, CA81-143316.
295.Nikolskaya, N. P. and Pitanov, V. S., "Exposure of germanium n-n$^+$ junctions by chemical etching", Prib. Tekh. Eksp., no. 2, 238-9 (1973), Russian, CA79-36376.
296.Pfeifer, J. and Haraszthy, E. S., "Investigation of germanium-silicon layer structures by the etching rate method", Thin Solid Films, 20, S31-S32 (1974), CA80-125972.
297.Russell, G. J. and Haneman, D., "Vacuum thermal etching of germanium and silicon surfaces", J. Electrochem. Soc., 114, 398-403 (1967), EEA70-13950.
298.Shershel, V. A., Litvinova, I. Y. and Bessnova, N. V., "Shape of the crystallization front of a n-type germanium", Izvest. Akad. Nauk SSSR, Neorg. Materialy, 6, 113-14 (1970), Russian, MA3-12-12 0732.
299.Shurygin, P. M. and Tinkova, S. M., "Kinetics of gas etching of Ge by hydrogen chloride with controlled convection", Inorg. Mater., 10, 332-5 (1974), PA78-81319.
300.Sullivan, M. V. and Kolb, G. A., "Direct photoetching of evaporated germanium and its use in mask fabrication", Electrochem. Technol., 6, 430-4 (1968), EEA72-10630.

301.Svetashova, E. S., Konorov, P. P. and Tarantov, Y. A., "Etching rate and state of the surface of germanium in alkaline solutions of hydrogen peroxide", Izvest. Akad. Nauk SSSR, Neorg. Materialy, 7, 738-40 (1971), Russian, MA4-10-34 0634.
302.Ugai, Y. A. and Kirichenko, I. V., "(Etch) dissolution of the (work) damaged layer on germanium", Izvest. Akad. Nauk SSSR, Neorg. Materialy, 5, 1875-78 (1969), Russian, MA3-06-21 0309.
303.Ugai, Y. A. and Kirichenko, I. V., "Effect of the ageing of CP-8 when etching germanium crystals", Izvest. Akad. Nauk SSSR, Neorg. Materialy, 6, 980-81 (1970), Russian, MA3-11-21 0629.
304.Ugai, Y. A., Kirichenko, I. V., Kurbanov, K. R., Shkadova, Z. I. and Makhonina, G. N., "Revelation of the zone of elastic deformations by cyclical etching. (Germanium)", Izvest. Akad. Nauk SSSR, Neorg. Materialy, 9, 311-12 (1973), Russian, MA6-07-22 0584.
305.Volchenko, A. G., Lyubalin, M. D. and Tretyakov, V. N., "Anodic dissolution of germanium crystals in order to remove heterogeneity due to impurities", Izv. Akad. Nauk SSSR Ser. Fiz., 37, 2302-5 (1973), Russian, PA77-10549.
306.Wheeler, D. J. and Haneman, D., "Chemical etching of partial splits in Ge and Si", J. Appl. Phys., 40, 3047-48 (1969), MA3-01-21 0016.

c. Silicon Carbide

307.Bartlett, R. W. and Barlow, M., "Surface polarity and etching of beta silicon carbide", J. Electrochem. Soc., 117, 1436-7 (1970), MA4-05-34 0291.
308.Faust, J. W., Jr. and Liaw, H. M., "Tables of etchants for silicon carbide", Silicon Carbide, Proc. Int. Conf., 3rd, 1973, p. 657-65, Marshall, R. C., (Ed.), Univ. S. C. Press, Columbia, S. C., 1974, CA81-177949.
309.Jennings, V. J., "Etching of silicon carbide", Silicon Carbide, Proc. Int. Conf., 2nd, 1968, p. S199-S210, Publ. 1969, CA77-10865.
310.Modrak, E., Swiderski, I. and Niemyski, T., "Etching of α-SiC crystals in hydrogen and argon flow", J. Cryst. Growth, 22, 181-7 (1974), PA77-56998.
311.Parshina, M. P. and Smirnov, G. V., "Etching reagent for semiconductors", Patent USSR 385353, Publ. May 1973, CA79-150787.
312.Sakai, K., Tooi, A., Fukai, S. and Tsujimoto, Y., "Etching silicon carbide", Patent Japan 74-36120, Publ. September 1974, CA83-52025.
313.Sakai, K., Tooi, A., Fukai, S. and Tsujimoto, Y., "Etching silicon carbide by active sputtering", Patent Japan 74-37919, Publ. October 1974, CA83-20708.
314.Von Muench, W. and Pfaffeneder, I., "Thermal oxidation and electrolytic etching of silicon carbide", J. Electrochem. Soc., 122, 642-3 (1975), CA83-51186.

2. Group III-V
a. Gallium Arsenide

315.Arora, B. M., Karulkar, V. T. and Guha, S., "A new mesa-etching technique for gallium arsenide", J. Mater. Sci., 8, 611-3 (1973), EEA76-23350.
316.Bagratishvili, G. D., Dzhanelidze, R. B., Kurdiani, N. L. and Saksaganskii, O. V., "Gas etching of gallium arsenide", Soobshch. Akad. Nauk Gruz. SSR, 79, 85-8 (1975), Russian, CA83-172013.
317.Bakin, N. N., Dedkov, V. D. and Porokhovnichenko, L. I., "Polishing gas etching of gallium arsenide", Izv. Akad. Nauk SSSR, Neorg. Mater., 8, 1364-7 (1972), Russian, CA77-131638.
318.Belyakov, L. V., Goryachev, D. N., Mizerov, M. N. and Portnoi, E. L., "Characteristics of diffraction lattices prepared by photoetching the surface of semiconductors", Zh. Tekh. Fiz., 44, 1331-3 (1974), Russian, CA82-131220.
319.Bhat, R., Baliga, B. J. and Ghandhi, S. K., "Vapor-phase etching and polishing of gallium arsenide using hydrogen chloride gas", J. Electrochem. Soc., 122, 1378-82 (1975), CA83-211397.
320.Butter, E., Gunther, R., Jacobs, B., Jacobs, K., Streubel, P. and Zehe, A., "Etching of natural cleaved faces of (Ga,Al)As/GaAs-heterostructures", Krist. & Tech., 8, 1021-8 (1973), German, PA77-3652.
321.Butter, E., Gunther, R., Jacobs, B., Jacobs, K., Streubel, P. and Zehe, A., "Investigation of etching in natural cleavage faces of (Ga,Al)As/GaAs heterostructures", Krist. & Tech., 8, 1021-8 (1973), German, PA77-20489.
322.Comerford, L. and Zory, P., "Selectively etched diffraction gratings in GaAs", Appl. Phys. Lett., 25, 208-10 (1974), PA77-79875.
323.DiLorenzo, J. V., "In situ etching of gallium arsenide during vapor phase growth of epitaxial gallium arsenide", Patent USA 3808072, Publ. April 1974, CA83-106722.
324.DiLorenzo, J. V., "An in-situ etch for the CVD growth of GaAs: the 'He-etch'", Gallium Arsenide and Related Compounds, 1974, p. 362-8, Publ. 1975, EEA78-28676.
325.Dorskaya, E. N. and Veits, V. V., "Solution for etching gallium arsenide", Patent USSR 439036, Publ. August 1974, CA82-50673.
326.Emelyanov, A. V., Timofeev, V. N. and Shorshorov, M. K., "Kinetic principles of gallium arsenide dissolution in bromine-methanol solutions", Fiz. Khim. Obrab. Mater., no. 4, 39-43 (1972), Russian, CA77-144762.
327.Etienne, D., Chevrier, J. and Bougnot, G., "Chemical evaporation of gallium arsenide under flow of hydrogen and hydrogen chloride", Mater. Res. Bull., 8, 951-60 (1973), French, CA79-108978.
328.Faktor, M. M., Fiddyment, D. G. and Taylor, M. R., "Electrochemically etched tunnels in gallium arsenide", J. Electrochem. Soc., 122, 1566-7 (1975).

329.Gannon, J. J. and Nuese, C. J., "A chemical etchant for the selective removal of GaAs through SiO_2 masks", J. Electrochem. Soc., 121, 1215-19 (1974), EEA78-544.
330.Hartnagel, H. and Weiss, B. L., "Etching polishing of gallium arsenide", J. Mater. Sci., 8, 1061-3 (1973), CA79-98346.
331.Hirao, M. and Nakashima, H., "Vapor etching effect on the impurity profile of gallium arsenide epitaxial layers", Proc. Conf. Solid State Devices, 2nd, 1970, p. 46-52, Publ. 1971, CA82-78882.
332.Iida, S. and Ito, K., "Selective etching of gallium arsenide crystals in H_2SO_4-H_2O_2-H_2O system", J. Electrochem. Soc., 118, 768-71 (1971), EEA74-22498.
333.Ishibashi, Y. and Yamaguchi, M., "Anisotropy in etching and deposition of selective epitaxial growth of gallium arsenide-II", Japan. J. Appl. Physics, 10, 525-6 (1971), MA4-12-12 1790.
334.Ito, K. and Iida, S., "Selective epitaxial deposition of semiconductor compounds", Patent USA 3752714, Publ. August 1973, CA79-98485.
335.Ivonin, I. V., Krasilnikova, L. M. and Yakubenya, M. P., "Electron-microscopic study of the morphology of a gallium arsenide etched surface", Izv. Vyssh. Ucheb. Zaved., Fiz., 15 (5), 116-19 (1972), Russian, CA77-67094.
336.Kuhn-Kuhnenfeld, F., "Selective photoetching of gallium arsenide", J. Electrochem. Soc., 119, 1063-8 (1972), EEA75-36144.
337.Lin, C., Chow, L. and Miller, K. J., "Vapor phase etching of GaAs in the H_2-H_2O flow system", J. Electrochem. Soc., 117, 407-9 (1970), EEA73-21035.
338.Maehama, T., Toyama, N. and Yoshida, S., "Annealing effects on lattice defects in GaAs (Etching)", Oyo Buturi, 43, 44-51 (1974), Japanese, PA77-75531.
339.Nuese, C. J. and Gannon, J. J., "Electrolytic removal of p-type GaAs substrates from thin, n-type semiconductor layers", J. Electrochem. Soc., 117, 1094-7 (1970), MA4-02-63 0043.
340.Noyaki, T. and Saito, T., "A new gas-etching method for vapour growth of gallium arsenide", Jap. J. Appl. Phys., 11, 110-1 (1972), MA5-08-63 0091.
341.Polgar, K., "Removing of GaAs substrates beneath GaP epitaxial layers by chemical etching", Koezl. Magy. Tud. Akad. Musz. Fiz. Kut. Intez., no. 0-16, 55-7 (1975), Hungarian, PA78-77765.
342.Rode, D. L., Schwartz, B. and DiLorenzo, J. V., "Electrolytic etching and electron mobility of GaAs for FET's", Solid-State Electron., 17, 1119-23 (1974), PA87-82414.
343.Saitoh, T., Matsubara, S. and Minagawa, S., "Effects of illumination on preferential etching of N-type GaAs in a CrO_3-HF-$AgNO_3$ solution", J. Electrochem. Soc., 122, 670-4 (1975), EEA78-24196.
344.Sato, Y. and Uchida, M., "Surface contamination of GaAs during the photoetching process", Rev. Elec. Commun. Lab., 22, 674-77 (1974).

345.Schwartz, B., "Preliminary results on the oxidation of GaAs and GaP during chemical etching", J. Electrochem. Soc., 118, 657-8 (1971), EEA74-22497.
346.Shaikh, S. A. and Chou, H. K., "Vapor phase etching of GaAs in $HCl-AsH_3-H_2$", Presented at Electrochemical Society Meeting, San Francisco, Calif., May 1974, PA77-53619.
347.Shifrin, S. S., Osvenskii, V. B., Milvidskii, M. G., Morgulis, L. M. and Grishina, S. P., "Nature of small etch pits in tellurium-alloyed GaAs single crystals", Sov. Phys.-Crystallogr., 18, 818-20 (1974), PA77-82325.
348.Stirland, D. J., "Anisotropic etching behaviour of gallium arsenide junctions", J. Mater. Sci., 9, 969-80 (1974), EEA77-30228.
349.Tarui, Y., Komiya, Y. and Harada, Y., "Preferential etching and etched profile of GaAs", J. Electrochem. Soc., 118, 118-22 (1971), MA4-08-34 0507.
350.Thrush, E. J., "A method for selective substrate removal from thin p-type gallium arsenide layers", J. Phys. E, 7, 493-5 (1974), PA77-49615.
351.Voigt, R. and Neels, H., "Morphology of the dissolution and etching of gallium arsenide", In: Khim. Svyaz Poluprov. Polumetallakh, Sirota, N. N., (Ed.), "Nauka i Tekhnika": Minsk, USSR, p. 404-12 (1972), Russian, CA79-10696.
352.Voigt, R. and Neels, H., "Kinetics of gallium arsenide dissolution", In: Kinet. Mekh. Krist., Sirota, N. N., (Ed.), "Nauka i Tekhnika": Minsk, USSR, p. 136-46 (1973), Russian, CA81-17375.
353.Vozmilova, L. N., "Anisotropic dissolution of gallium arsenide", Izvest. Akad. Nauk SSSR, Neorg. Materialy, 7, 1056-8 (1971), Russian, MA4-11-34 0721.
354.Vozmilova, L. N. and Maslennikova, V. T., "Anisotropic etchants for GaAs", Izv. Akad. Nauk SSSR Neorg. Mater., 8, 1303-5 (1972), Russian, PA75-74660.
355.Yamamoto, A. and Yano, S., "Anodic dissolution of n-type gallium arsenide under illumination", J. Electrochem. Soc., 122, 260-7 (1975), CA82-148754.

b. Gallium Phosphide

356.Dixon, R. W., Hackett, W. H., Jr. and McGahan, T. E., "Preferential etching in gallium phosphide", Patent USA 3791948, Publ. February 1974, CA80-88792.
357.Kawachi, M., "Etching of GaP", Patent Japan 73-44198, Publ. June 1973, CA79-109626.
358.Kocsis, S. and Lendvay, E., "The thermal etching of GaP", Krist. & Tech., 9, 1131-40 (1974), PA78-18515.
359.Meek, R. L. and Schumaker, N. E., "Anodic dissolution and selective etching of gallium phosphide", J. Electrochem. Soc., 119, 1148-52 (1972), PA75-75386.

360.Michel, C., "Effect of substrate microdefects due to thermal etching on the near-junction properties of GaP:Zn,O grown by liquid phase epitaxy", J. Electrochem. Soc., 122, 678-82 (1975), EEA78-24198.
361.Ogirima, M., Hirao, M., Takeda, Y., Honma, K. and Kurata, K., "Etching of gallium phosphide", Patent Japan 74-23800, Publ. March 1974, CA81-128761.
362.Plauger, L. R., "Control chemical etching of gallium phosphide", J. Electrochem. Soc., 121, 455-7 (1974), CA80-113955.

c. Others

363.Becker, R., "Nonblocking contacts on indium phosphide", Solid-State Electron., 16, 1241-9 (1973), PA77-10206.
364.Bontschewa-Mladenowa, Z., Karbanov, S. and Martschewa, N., "Metallographic investigation of the $Ag_2Te_xSe_{1-x}$ system", Monatsh. Chem., 103, 1158-62 (1972), German, MA6-08-31 2351.
365.Chane, J. P., "Semiconductor devices", Patent USA 3887404, Publ. June 1975, CA83-107228.
366.Dyment, J. C., Logan, R. A. and Schwartz, B., "Method for selectively etching $Al_xGa_{1-x}As$ multilayer structures", Patent USA 3801391, Publ. April 1974.
367.Fischer, C. and Heasell, E., "The formation of thermal etch-patterns during diffusion in indium antimonide", Surface Sci., 30, 592-600 (1972), MA6-02-13 0095.
368.Kobayashi, T. K. and Sugiyama, K., "Etching of $Al_xGa_{1-x}As$ in alkaline solution", Jap. J. Appl. Phys., 12, 619-20 (1973), EEA76-23352.
369.Lendvay, E., Kuhn, A., Chevy, A. and Ceva, T., "Dislocation etching of GaSe single crystals", J. Mat. Sci., 6, 305-8 (1971), MA4-12-12 1763.
370.Mazurkevich, Y. S., Zozulya, N. I., Kostyuk, L. S. and Zozulya, Y. I., "Etching and certain properties of (111) planes of indium antimonide", Izv. Akad. Nauk SSSR, Neorg. Mater., 11, 611-16 (1975), Russian, CA83-88875.
371.Morimoto, Y., "Few characteristics of epitaxial gallium nitride-etching and thermal decomposition", J. Electrochem. Soc., 121, 1383-4 (1974), CA81-160185.
372.Olsen, G. H. and Ettenberg, M., "Universal stain/etchant for interfaces in III-V compounds", J. Appl. Phys., 45, 5112-14 (1974), PA78-22491.
373.Orlova, G. M., Sveshnikova, L. L. and Nikiforovskaya, L. P., "The chemical etching kinetics of indium arsenide in acid ferric chloride solutions", Zhur. Priklad. Khim., 43, 784-8 (1970), Russian, MA3-11-34 0608.
374.Pak, K., Nishinaga, T. and Uchiyama, S., "Thermal etching effect of indium phosphide substrate in LPE (liquid phase epitaxial) saturation process", Jap. J. Appl. Phys., 14, 1613-14 (1975), CA83-186465.

375. Pankove, J. I., "Electrolytic etching of GaN", J. Electrochem. Soc., 119, 1118-19 (1972), PA75-71365.
376. Plauger, L. R., "Chemical attack on semiconductor devices", Patent French 2196604, Publ. March 1974, CA81-70133.
377. Portnoi, K. I., Gribkov, V. N., Shehetanov, B. V., Umantsev, E. I. and Silaev, V. A., "Mechanism of the growth and etching of aluminum nitride crystal whiskers", Kristallografiya, 18, 599-604 (1973), Russian, CA79-58486.
378. Rosin, H., "On chemical etching of structural defects in epitaxial $GaAs_{1-x}P_x$ and bulk GaAs", Phys. Status Solidi A, 25, K5-7 (1974), PA77-82324.
379. Straumanis, M. E. and Hu, L., "The anodic dissolution reaction of InSb: Etch patterns, electron number, anodic disintegration and film formation", J. Electrochem. Soc., 118, 433-6 (1971), MA4-09-34 0579.
380. Tuck, B. and Baker, A. J., "Chemical etching of (111) and (100) surfaces of InP", J. Mater. Sci., 8, 1559-66 (1973), PA77-1172.

3. Group II-VI

381. Bala, W., Lozykowski, H. and Sekulski, J., "Electroluminescence of zinc telluride (ZnTe) diodes", Bull. Acad. Pul. Sci., Ser. Sci., Math., Astron. Phys., 21, 1047-53 (1973), CA80-64887.
382. Boinikh, N. M. and Indenbaum, G. V., "Dislocation etch pits on different crystallographic planes of mercury selenide single crystals", Fizika i Khim. Obrabot. Mat., no. 5, 152-5 (1969), Russian, MA3-01-21 0044.
383. Guthnick, E., "Chemical polishing and etching of semiconductors, especially zinc selenide", Patent German (East) 93310, Publ. June 1973, CA79-150788.
384. Indenbaum, G. V., Baronenkova, R. P. and Boinykh, N. M., "Etching of dislocations in cadmium telluride single crystals", Fizika i Khim. Obrabot. Mat., no. 2, 91-6 (1971), Russian, MA4-09-21 0502.
385. Munir, Z. A., Seacrist, L. S. and Hirth, J. P., "Morphology of thermally etched basal surfaces of cadmium selenide", Surface Sci., 28, 357-72 (1971), MA5-11-12 1327.
386. Oczkowski, H. and Poplawski, Z., "Orientation of ZnSe crystals by chemical etching technique", J. Cryst. Growth, 23, 154-6 (1974), EEA77-36221.
387. Orlova, G. M., Blinov, L. N., Belyakova, N. V. and Kozharina, T. P., "Kinetics of the chemical etching of lead chalcogenides", Zh. Prikl. Khim., 48, 1945-9 (1975), Russian, CA83-184194.
388. Pedos, S. I. and Boinykh, N. M., "Monitoring the uniformity of the composition of crystals of cadmium mercury telluride ($Cd_xHg_{1-x}Te$) solid solutions using etching", Zavod. Lab., 40, 825 (1974), Russian, CA82-163224.

389. Pickhardt, V. Y. and Smith, D. L., "Chemomechanical polishing of cadmium sulfide", J. Electrochem. Soc., 121, 1064-6 (1974), CA81-112726.
390. Preier, H. and Pfeiffer, H., "Etching solution for revealing p-n junctions in PbS and $PbS_{1-x}Se_x$", J. Electrochem. Soc., 121, 595-6 (1974), EEA77-22892.
391. Sangwal, K., "Selective etchants for lead sulphide single crystals", Sov. Phys.-Solid State, 15, 943-4 (1973), PA77-28305.
392. Strehlow, W. H., "Chemical polishing of II-VI compounds", J. Appl. Phys., 40, 2928-32 (1969), MA3-01-21 0012.
393. Tokumaru, Y., Okada, Y. and Izuka, T., "Etch patterns and crystallographic polarity in hexagonal ZnS", Jap. J. Appl. Phys., 14, 1241-2 (1975), PA78-76982.
394. Vul, S. P. and Pakhomova, T. V., "Selective etchant for $CdSnAs_2$", Sov. Phys.-Crystallogr., 17, 1110 (1973), EEA77-330.

VII. Dielectrics

1. Silicon Dioxide

395. Alcorn, G. E. and Feeley, J. D., "Plasma etching via holes in sputtered quartz", IBM Tech. Disclosure Bull., 17, 2701-2 (1975), EEA78-19861.
396. Erdman, W. C. and Garbarini, V. C., "Oxide etchant", Patent USA 3860464, Publ. January 1975, CA82-118484.
397. Haken, R. A., Baker, I. M. and Beynon, J. D. E., "An investigation into the dependence of the chemically-etched edge profiles of silicon dioxide films on etchant concentration and temperature", Thin Solid Films, 18, S3-S6 (1973), EEA77-334.
398. Ham, E. J. and Soden, R. R., "Etching silicon oxide to produce a tapered edge thereon", Patent USA 3839111, Publ. October 1974, CA81-178950.
399. Hamazaki, N. and Suzuki, N., "Selective etching of silicon oxide", Patent Japan 74-47242, Publ. December 1974, CA83-20819.
400. Heinecke, R. A. H., "Control of relative etch rates of SiO_2 and Si in plasma etching", Solid-State Electron., 18, 1146-7 (1975).
401. Hooker, C. E. L. and Tomes, D. W., "Silicon dioxide etch rate control by controlled additions of phosphorus pentoxide and boric oxide", Patent USA 3913126, Publ. October 1975, CA83-211933.
402. Kato, H., "Local etching of the oxide coating on a metal layer in a semiconductor circuit", Patent Japan 73-67139, Publ. September 1973, CA80-53755.
403. Lawrence, J., "Controlled etching of silicon dioxide in buffered hydrofluoric acid", Proc. Symp. Oxide Electrolyte Interfaces, 1972, p. 171-80, Alwitt, R. S., (Ed.), Electrochem. Soc.,; Princeton, N. J., 1973, CA83-89699.
404. Lukae, P., "Sectioning technique for silica and its application to penetration studies", Radiochem. Radioanal. Lett., 18, 203-9 (1974), CA81-178212.

405.MacRae, A. U. and Moline, R. A., "Silicon dioxide tapered films", Patent USA 3769109, Publ. October 1973, CA80-31785.
406.Philips Electronic & Associated Industries, Ltd., "NH_4F/HF-based etchant for selectively etching patterns in thin SiO_2 layers", Patent UK 1320560, Publ. June 1973.
407.Schaefer, H. G., Ting, C. Y. and Vivalda, V. J., "Silicon anode plate", IBM Tech. Disclosure Bull., 17, 2911 (1975), EEA78-24160.
408.Schwettmann, F. N., Dexter, R. J. and Cole, D. F., "Etch rate characterization of boron-implanted thermally grown SiO_2", J. Electrochem. Soc., 120, 1566-70 (1973), EEA77-2462.
409.Tamaka, S. and Nakagawa, T., "Treating silicon dioxide film by etching", Patent Japan 74-24123, Publ. June 1974, CA82-79557.
410.Zagata, L., Kalnina, R., Feltins, I., Freiberga, L., Eimanis, N. and Eglitis, I., "Effect of plasma treatment on changes in low-temperature silicon dioxide film properties", Latv. PSR Zinat. Akad. Vestis, Fiz. Teh. Zinat. Ser., no. 3, 18-24 (1973), Russian, CA79-98204.

2. Silicon Nitride

411.Akasaka, Y., Horie, K., Nomura, K. and Kawazu, S., "Anomalous enhancement of etching rate of silicon nitride with HF etchant by ion implantation", Oyo Buturi, 43 (Suppl.), 493-8 (1974), PA78-3657.
412.Bertsch, L. and Schaefer, H., "Etching silicon nitride layers in semiconductors", Patent German 1546045, Publ. November 1974, CA82-164227.
413.Edel, W. A. and Gall, R. H., "Etching process for silicon nitride", IBM Tech. Disclosure Bull., 16, 3231 (1974), EEA77-30382.
414.Godber, G. A., "Selectively etching silicon nitride", Patent USA 3871931, Publ. March 1975, CA83-36594.
415.Kooi, E., "Rapidly etching silicon nitride in the manufacture of semiconductor devices", Patent USA 3598669, Publ. August 1971, CA77-67754.
416.Miller, G. J., Scully, R. J., White, C. D. and Wirtz, L. H., "Controlling etching conditions of silicon nitride Si_3N_4", IBM Tech. Disclosure Bull., 15, 677-8 (1972), EEA75-39206.
417.Murrmann, H., Wildmann, D. and Schleicher, L., "Passivated semiconductor elements", Patent German 2149898, Publ. April 1973, CA79-24845.
418.Ormond, D. W., Jr., "Nonstaining silicon nitride etch", IBM Tech. Disclosure Bull., 17, 1836-7 (1974), EEA78-11657.
419.Panousis, P. T. and Waggener, H. A., "Etching silicon nitride", Patent USA 3715249, Publ. February 1973, CA78-103641.
420.Pridans, J. V. and Yogodzinski, S. G., "Process for maintaining H_3PO_4 concentration", IBM Tech. Disclosure Bull., 16, 3540-1 (1974), EEA78-511.

421. Soldatov, V. S., Zavyalov, Y. P., Medvedev, K. S., Nikolaenko, O. G. and Uskova, Z. A., "Infrared absorption and rate of etching silicon nitride layers obtained by an ion-plasma method", Tr. Mosk. Energ. Inst., no. 142, 71-5 (1972), Russian, CA80-31559.
422. Swann, R. C. G., Eldridge, P. G. and Penton, J. I., "Gas etching of silicon nitride layers", Patent German 2213036, Publ. September 1972, CA77-170383.
423. Toyoda, H. and Hirano, H., "Removal of silicon nitride films", Patent Japan 74-62513, Publ. June 1974, CA81-178919.
424. Voskoiboinikov, V. V., Rabinovich, L. N. and Repinskii, S. M., "Solution of silicon nitride films in phosphoric acid", Inorg. Mater., 9, 1629-30 (1973), PA77-67728.
425. Watari, Y., Inoue, I. and Kusano, Y., "Semiconductor devices", Patent USA 3807038, Publ. April 1974, CA81-7619.
426. Wohlheiter, V. D., "Silicon nitride and aluminum oxide ledge removal techniques for integrated circuit devices", J. Electrochem. Soc., 120, 90C-91C (1973), EEA76-23490.

3. Silicon Dioxide and Silicon Nitride

427. Abe, H., "Semiconductor", Patent USA 3880684, Publ. April 1975, CA83-107204.
428. Caffrey, R. E., Dumbri, A. C. and Tauber, R. N., "Forming windows in composite dielectric layers", Patent USA 3808069, Publ. April 1974, CA81-7620.
429. Harrap, V., "Equal etch rates of silicon nitride and silicon dioxide utilizing hydrofluoric acid dilution and temperature dependence", In: Semicond. Silicon, Huff, H. R., (Ed.), Electrochem. Soc., Inc.; Princeton, N. J., p. 354-62 (1973), CA82-67177.
430. ITT Industries, Inc., "Plasma etching and polymerization", Patent Neth. 74-06222, Publ. November 1974, CA82-179703.
431. Magdo, I. E. and Magdo, S., "Preparing smallest openings in integrated circuits", Patent German 2451486, Publ. July 1975, CA83-140945.
432. Squillace, A. S., Martin, A. E. and Rudmann, J. J., "Silicon nitride-silicon oxide etchant", Patent USA 3811974, Publ. May 1974, CA81-18374.
433. Squillace, A. S., Martin, A. E. and Rudmann, J. J., "Silicon nitride-silicon oxide etchant", Patent USA 3859222, Publ. January 1975.
434. Suzuki, G., Nagata, M., Tomotsu, T. and Ono, H., "Light-sensitive etchant and its use in preparing a semiconductor structure", Patent German 2352349, Publ. May 1974, CA81-43028.

4. Others

435.Ameen, J. G., McBride, D. G. and Philips, G. C., "Etching of high alumina ceramics to promote copper adhesion", J. Electrochem. Soc., 120, 1518-22 (1973), PA77-7291.

436.Buiko, L. D., Dudich, V. M., Kaloshkin, E. P., Koleshko, V. M. and Lashitskii, E. K., "Interference measurement of the etching rate of dielectric films", Prib. Tekh. Eksp., no. 2, 240-1 (1973), Russian, CA79-24616.

437.Cole, H. S., Jr., "Selective etching of metal oxides of tin or indium", Patent USA 3837944, Publ. September 1974, CA82-132163.

438.Danilov, A. V., Ryabov, A. N. and Krasheninnikov, V. A., "Etching of aluminum oxide single crystals in a vanadium pentoxide melt", Zh. Prikl. Khim., 45, 871-2 (1972), Russian, CA77-53315.

439.Feuersanger, A. E. and Harris, L. M., "Photolithographic etching method for nickel oxide", Patent USA 3677847, Publ. July 1972, CA77-119803.

440.Freeman, J. L., Jr. and Santoro, C. J., "Selective etching of glass layers on aluminum", Patent German 2153336, Publ. May 1972, CA77-67738.

441.Green, J. M., "The etch rate and deterioration of magnesium aluminum spinel in hydrogen (Substrate for epitaxial deposition of silicon)", J. Electrochem. Soc., 119, 1765-8 (1972), EEA76-8535.

442.Gregor, L. V., Nufer, R. W., Pugliese, F. G. and Spaulding, E. C., "Removing chemically resistant dielectric coating for an integrated circuit module", IBM Tech. Disclosure Bull., 15, 1096-7 (1972), EEA76-8729.

443.Growe, W. J. and Ritchie, K., "Glassy reliefs by photolithography", Patent French 2128739, Publ. November 1972, CA78-130612.

444.Ishitani, S., "Corrosion treatment of tin oxide thin film", Patent Japan 74-126595, Publ. December 1974, CA82-179587.

445.Kern, W., "A technique for measuring etch rates of dielectric films", RCA Rev., 29, 557-65 (1968), EEA72-10849.

446.Manasevit, H. M., "Gas-phase etching of sapphire. III", J. Electrochem. Soc., 121, 293-7 (1974), PA77-28303.

447.Muraoka, H., Tsutsumi, H., Ohashi, T., Sumitomo, Y. and Yasui, T., "Semiconductor integrated circuit insulated by dielectric material", Patent British 1304643, Publ. January 1973, CA78-116894.

448.Okamoto, F., "Chemical etchants for the fabrication of thin film patterns of silver on acid- and base-sensitive oxides (acoustic surface wave devices)", Jap. J. Appl. Phys., 13, 383-4 (1974), EEA77-17121.

449.Ondracek, G. and Spieler, K., "Gas etching: A new method of producing contrast in the preparation of metallographic specimens. I. Oxides, oxide cermets, and other composites", Praktische Metallographie, 10, 324-41 (1973), MA6-10-21 0533.

450.Sato, J. and Tanabe, T., "Etching insulating films", Patent USA 3709749, Publ. January 1973, CA78-64125.
451.Schmidt, W. A. and Davey, J. E., "Preparation of smooth, crystalline, damage-free sapphire surfaces by gas etching", Nat. Bur. Stand., Spec. Publ., no. 348, 259-65 (1972), CA77-157274.
452.Tanimura, S., Azuma, K., Yasutake, A. and Miyamoto, K., "Etching tin oxide", Patent Japan 74-01137, Publ. January 1974, CA81-70131.
453.Tenney, A. S. and Ghezzo, M., "Etch rates of doped oxides in solutions of buffered HF", J. Electrochem. Soc., 120, 1091-5 (1973), EEA76-35212.
454.Vakulenko, L. N., Postavit, N. A. and Terekhova, G. V., "Etching of iron oxide films", Dielektr. Poluprovodn., 5, 80-2 (1974), Russian, CA81-144201.
455.Wirtz, L. H., "Soft glass etchant", IBM Tech. Disclosure Bull., 16, 3599 (1974), EEA77-38858.
456.Zatulina, O. S., Lomakina, T. P., Dyakova, A. P., Bashavnikov, V. A. and Kharin, A. N., "Etching films of two-layer dielectrics", Izv. Sev.-Kavk. Nauchn. Tsentr. Vyssh. Shk., Ser. Tekh. Nauk, no. 1, 96 (1973), Russian, CA81-160935.

VIII. Metals

1. Copper and Copper Alloys

457.Ammann, H., "Effect of acid permanganate solution during etching bimetal and trimetal plates", Fachh. Chemigr., Lithogr., Tiefdruck, no. 1, 33-7 (1972), German, CA77-54850.
458.Annapolskii, V. D., Komkova, K. P. and Chernova, A. A., "Solution for electrochemical etching of copper", Patent USSR 375820, Publ. March 1973, CA79-38027.
459.Arkharov, V. L., Gaisinskaya, A. M. and Antonova, Z. F., "Chemical analysis of thin surface layers of copper alloys", Dop. Akad. Nauk Ukrain. RSR, no. 3, 264-5 (1972), Ukrainian, MA6-07-23 0518.
460.Bakaev, V. V., Korolev, G. V., Kucherenko, V. I. and Flerov, V. N., "Anodic etching of copper in aqueous sodium chloride solutions", Izv. Vyssh. Uchebn. Zaved., Khim. Khim. Tekhnol., 18, 277-80 (1975), Russian, CA83-67575.
461.Getman, I. T. and Vdovenko, I. D., "Solution for etching copper", Patent USSR 378565, Publ. April 1973, CA79-128708.
462.Grigoreva, E. V., Lyubimova, M. K. and Krivitskii, V. A., "Solution for etching copper and its alloys", Patent USSR 432244, Publ. June 1974, CA82-47017.
463.Grunwald, J. J., Slominski, L. J., Landau, A. and Shah, D. G., "Composition for dissolving metallic copper", Patent USA 3779842, Publ. December 1973, CA80-86667.
464.Haas, R., "Etching copper and copper alloys", Patent German 2216269, Publ. October 1973, CA80-51609.

465. Iida, F. and Kobayashi, T., "Treating of copper and copper alloys before plating", Patent Japan 75-20015, Publ. July 1975, CA83-210780.
466. Inoue, H., Ando, H. and Takano, M., "Soft etchant for copper", Patent Japan 75-30753, Publ. March 1975, CA83-83815.
467. Ito, Y., "Chemical polishing solution for copper and its alloys", Patent Japan 74-104840, Publ. October 1974, CA82-144015.
468. Katz, G., "Ion-etch characteristics of epitaxial copper on sapphire", J. Mat. Sci., 5, 736-40 (1970), MA4-04-12 0472.
469. Kucherenko, V. I. and Flerov, V. N., "Etching of a copper foil in solutions of ferric chloride", Zhur. Priklad, Khim., 43, 1475-8 (1970), Russian, MA3-12-34 0685.
470. Kucherenko, V. I., Sigal, L. N. and Flerov, V. N., "Mechanism of the etching of copper in ferric chloride", Elektrokhimiya, 6, 1694-8 (1970), Russian, MA4-04-34 0200.
471. Kuznetsova, O. M., Klyueva, T. M. and Presnyakov, I. I., "Etching of a copper foil in ammonium persulfate solutions", Obmen Opytom Radioprom., no. 11, 47-8 (1972), Russian, CA80-40058.
472. Lang, A. M., "A new approach to multilayer circuitry", Proceedings of the 5th Technical EIPC Conference on Flexible Circuitry, 1970, p. 88-101, EEA74-621.
473. Mance, A., "Potentiostatic etching and polishing of copper and its alloys", Metallography, 4, 287-96 (1971), MA5-01-57-0002.
474. Mance, A., Perovic, V. and Mihajilovic, A., "Potentiostatic-controlled etching and polishing of copper and its alloys", Metallography, 6, 123-30 (1973), MA6-09-21 0455.
475. Marlow, E. F., Morris, R. E. and Leeds, J. M., "Ammoniacal solution for copper etching", Patent German 2252327, Publ. June 1973, CA79-84804.
476. Matsumoto, A. and Fujimoto, O., "Etching copper and copper alloys having a resist film of a noble metal", Patent Japan 72-47502, Publ. November 1972, CA80-138503.
477. Matsumoto, A. and Fujimoto, O., "Copper-ammonia complex etching solutions for copper and its alloys", Patent Japan 73-27939, Publ. April 1973, CA79-47100.
478. Meek, J. L. and Vieane, E. L., "Copper etching control system for printed-circuit manufacture", Patent USA 3775202, Publ. November 1973, CA80-53718.
479. Missel, L., "High quality, low-cost etching of copper with cupric chloride-sodium chloride", Electronics Manufacturer, 17, 32-5 (1972), MA5-12-57 0361.
480. Missel, L. and Murphy, F. D., "Steady-state etching of copper. Cupric chloride-sodium chloride baths", Metal Finish, 67 (12), 47-52 (1969), MA3-05-57 0201.
481. Mitsubishi Gas Chemicals, Inc., "Chemical dissolution of copper", Patent French 2167327, Publ. September 1973, CA80-53724.
482. Okabe, M., Shibazaki, Y. and Watanabe, C., "Etching solution for copper", Patent Japan 72-43727, Publ. November 1972, CA79-128713.

483. Okabe, M., Shibazaki, Y. and Watanabe, C., "Solution for chemically dissolving copper", Patent Japan 72-43729, Publ. November 1972, CA79-128712.
484. Osawa, K. and Hirasa, K., "Copper etching solution", Patent Japan 74-00138, Publ. January 1974, CA80-138405.
485. Parikh, G. D. and Willard, W. C., "Electrolytic regenerative etching and metal recovery", Patent USA 3784455, Publ. January 1973, CA80-90374.
486. Politychi, A., "Two-layer printed wirings on mylar foil", J. Electrochem. Soc., 119, 1017-8 (1972), MA5-12-58 0671.
487. Popkova, E. G. and Predvoditelev, A. A., "Selective etching at dislocations on metals (Cu)", Sov. Phys. Crystallogr., 17, 528-33 (1972), MA6-04-21 0204.
488. Roelants, G. J., "Continuous etching copper with regeneration of the etching solution", Patent German 2306445, Publ. August 1973, CA79-139322.
489. Roms, Y. G., Finik, G. I., Donner, A. Y. and Melnikov, V. I., "Electrolytic etching of articles of copper and its alloys", Patent USSR 378537, Publ. April 1973, CA79-99845.
490. Ruff, C. W., "Processing of spent copper etching solutions for electroless copper plating", Patent German 2244307, Publ. March 1974, CA81-15988.
491. Sheshadri, B. S. and Setty, T. H. V., "Anodic dissolution of copper single crystal (100), (110), and (111) faces in acid copper sulfate bath in presence of thioglycolic acid", Indian J. Chem., 11, 149-53 (1973), CA79-12774.
492. Shibasaki, Y. and Oyama, K., "Composition for etching copper with reduced sideways etching", Patent USA 3773577, Publ. November 1973, CA80-53794.
493. Shiga, S., "Solutions for chemical dissolution treatment of metallic materials", Patent USA 3756957, Publ. September 1973, CA79-149005.
494. Tirrell, C. E., "Electrolytic reclamation of spent etching fluids", Patent USA 3761369, Publ. September 1973, CA79-132340.
495. Winter, A. T., "Etching studies of dislocation microstructures in crystals of copper fatigued at low constant plastic strain amplitude", Phil. Mag., 28, 57-64 (1973), MA7-11-13 0949.
496. Zhitomirsky, A. N. and Finik, G. I., "Intensification of the chemical etching of copper", Zhur. Priklad. Khim., 44, 2562-63 (1971), Russian, MA5-07-21 0300.
497. Zubekhin, V. P. and Maksimov, V. P., "Development of etch figures on copper", Izv. Voronezh. Gos. Pedagog. Inst., no. 131, 77-80 (1973), Russian, CA81-142198.

2. Aluminum and Aluminum Alloys

498. Agatsuma, T., Kikuchi, A., Nakada, K. and Tomozawa, A., "Beveling aluminium in multilayer metal circuitry", J. Electrochem. Soc., 122, 825-9 (1975), EEA78-33171.

499.Alwan, A. S. and Ruscetta, R. A., "Porous aluminum foil for electrolytic capacitors", Patent USA 3666642, Publ. May 1972, CA77-67721.
500.Alwitt, R. S., "Electrolytic etching of aluminum foil", Patent USA 3779877, Publ. December 1973, CA80-90375.
501.Aoyama, Y., "Alkali etching of aluminum", Patent Japan 73-102042, Publ. December 1973, CA80-136699.
502.Baumgarten, J. and Ernst, M., "An anodic etching technique for orientation determination of coarse-grained cast 99.5% aluminium by optical methods", Praktische Metallographie, 8, 702-10 (1971), MA5-08-21 0332.
503.Bell, G. and Haid, S., "Acid etching agent for aluminum layers for electronics", Patent German 2142784, Publ. March 1973, CA78-129736.
504.Bertsch, L. and Eberhardt, H. P., "Aqueous basic etch solution for preparing separate semiconductor contacts", Patent German 1546032, Publ. August 1973, CA79-150817.
505.Diversey-France, S. A., "Compositions for chemically attacking aluminum and its alloys", Patent French 2169883, Publ. October 1973, CA80-73536.
506.Dunn, C. G., Bolon, R. B., Alwan, A. S. and Stirling, A. W., "A scanning electron microscope study of etched aluminum foil for electrolytic capacitors", J. Electrochem. Soc., 118, 381-90 (1971), EEA74-18137.
507.Durney, L., "Advances in finishing aluminum. Cleaning and etching aluminum", Mod. Met., 29, 69-72 (1973), MA6-08-57 0316.
508.Ehman, M. F. and Faust, J. W., Jr., "Effect of ultrasonic cavitation on the etching processes on single crystals of aluminum", J. Appl. Phys., 41, 3169-71 (1970), MA4-02-21 0137.
509.Engelhardt, R., "Etching behaviour of ultra-pure aluminium single crystals", Praktische Metallographie, 8, 102-13 (1971), MA4-07-21 0361.
510.Franz, E. C., "Process factors affecting finish of aluminum alloy extrusions (by etching, lacquering or anodizing)", Metal Progr., 97, 80-2 (1970), MA3-10-57 0369.
511.Gsell, C., Fremiot, M. and Champier, G., "Correlation between the points of emergence of dislocations and etch figures on aluminium", Praktische Metallographie, 7, 39-43 (1970), MA3-06-21 0277.
512.Hines, R. E., "An improved technique for etching aluminium interconnect patterns on integrated circuits", Microelectron. & Rel., 10, 59 (1971), EEA74-14812.
513.Holt, R. T. and Perry, M., "The role of solute atoms in geometric etch-pitting of aluminium", J. Inst. Met., 101, 151-2 (1973), PA77-3651.
514.Hunt, P. A., "Etchant for aluminum", Patent Neth. 73-03505, Publ. September 1974, CA83-64729.

515.Hunter, M. S., "Improvements in the etching of hard aluminum electrolytic capacitor foil", J. Electrochem. Soc., 117, 1215-8 (1970), MA4-02-63 0044.
516.Ichiki, M., Kasahara, N. and Nakaide, K., "Pretreatment for chromated zinc alloy using strong alkaline solution to selectively dissolve out aluminum from surface layers", Patent Japan 74-43174, Publ. November 1974, CA80-144009.
517.Ishida, Y., "Etching the surface of aluminum or its alloy", Patent Japan 73-43248, Publ. December 1973, CA81-28702.
518.Ishida, Y., "Etching reagent for aluminum or aluminum alloy", Patent Japan 73-43252, Publ. December 1973, CA81-28701.
519.Jackson, N. F., "Tunnel etching of aluminum", Electrocomponent Sci. Technol., 2, 33-44 (1975), CA83-154472.
520.Kape, J. M., "Chemical etching of aluminium in caustic soda based solutions", Trans. Inst. Metal Finishing, 48, 43-50 (1970), MA3-10-57 0375.
521.Kape, J. M., "Etching as a pre-treatment for anodizing", Trans. Inst. Metal Finishing, 49, 22-8 (1971), MA4-09-57 0277.
522.Kape, J. M., "Pretreatment of aluminum by cleaning and etching", Prod. Finish., 26 (8), 12-16 (1973), CA79-117586.
523.Knopf, W. H., "Aluminum etchant studies", Report BDX-613-1105 (Rev.), Bendix Corp., Kansas City, Mo., 1974, 15 pp., CA82-174308.
524.Kosmynina, M. T. and Batashev, K. P., "Electrochemical etching of aluminum in solutions of sodium chloride containing different additives", Zh. Prikl. Khim., 48, 1293-6 (1975), Russian, CA83-105259.
525.Kumai, H., "Etching of metal films in fabricating semiconductor devices", Patent Japan 73-56537, Publ. August 1973, CA79-130632.
526.Kurihara, K., "Etching aluminium foils", Patent Japan 74-40537, Publ. November 1974, CA83-17703.
527.Markova, N. E. and Mamasyan, T. D., "Etch for aluminum foil A-99", Zavod. Lab., 39, 1214-15 (1973), Russian, CA80-51391.
528.McClanahan, R. M. and Machacek, J. F., "Chemical engraving of aluminum", Patent French 2157946, Publ. July 1973, CA80-73560.
529.Milos, S. and Wyrwa, W., "Influence of etching on the surface quality of (anodized) aluminium sheet", Rudy i Metale Niezelazne, 15, 106-9 (1970), Polish, MA3-08-57 0307.
530.Moll, K., "Aqueous concentrate for the preparation and/or regeneration of cleaning and/or etching solutions for aluminum and aluminum alloys", Patent German 1771685, Publ. December 1973, CA80-86668.
531.Moravec, M. and Kudrman, J., "Revealing of etch pits on polycrystalline aluminium", Praktische Metallographie, 9, 642-50 (1972), MA6-05-21 0252.
532.Muehlhaeusser, M. and Roth, C., "Aluminum electrodes for electrolytic capacitors", Patent German 2255309, Publ. May 1974, CA81-83535.
533.Nagata, I., "Etching of an aluminum foil", Patent Japan 73-06701, Publ. February 1973, CA80-18576.

534. Naraoka, K. and Doda, K., "Etching solution for vacuum evaporated aluminum film for use in integrated circuits", Patent Japan 72-39822, Publ. October 1972, CA79-128716.
535. Naraoka, K., Doda, K. and Nishimura, T., "Etching aluminum with alkaline ferricyanide or permanganate solution", Patent Japan 73-89139, Publ. November 1973, CA80-138406.
536. Ohta, T., "Coating of aluminum", Patent Japan 75-33939, Publ. April 1975, CA83-165916.
537. Ohta, T., "Coating of aluminum", Patent Japan 75-33940, Publ. April 1975, CA83-165915.
538. Oikawa, T., Fujimoto, K., Ohtsuka, Y. and Arai, K., "Etching solution for aluminum and its alloys", Patent Japan 72-39823, Publ. October 1972, CA79-128715.
539. Radu, I., "Study of the microstructures of aluminium alloys by thermal etching", Studii Si Cercetari De Metalurgie-1971, 1, 197-204 (1972), Romanian, MA6-09-21 0462.
540. Robison, G. T., "Specification anodizing", Prod. Finish., 34, 28-35 (1970), MA3-12-57 0476.
541. Santarini, G. and Boos, J. Y., "The formation of etch-pits bounded by high-index crystallographic planes, during galvanostatic anodic etching of aluminium single crystals in hydrochloric acid", Scr. Metall., 8, 1121-5 (1974), PA78-3658.
542. Sato, T., Namikata, T. and Hiyama, S., "Electrolytic etching of high purity aluminum characterized by cylindrical pit configuration", Fujitsu Sci. Tech. J., 6, 91-116 (1970), EEA74-7211.
543. Sheasby, P. G. and Smith, A. M., "Electrolytic etching of aluminum", Patent Austrian 324064, Publ. August 1975, CA83-199480.
544. Shenoi, B. A., Narasimhan, K. R., Srinivasan, V. and Anantharaman, A., "Etching of aluminum foils for electrolytic capacitors", Patent Indian 100313, Publ. July 1967, CA81-57609.
545. Shibasaki, Y., Hasegawa, Y. and Amano, S., "Electrolytically etching aluminum articles", Patent Japan 75-16305, Publ. June 1975, CA83-185506.
546. Smith, A. W., "Surface oxide on etched aluminum", J. Electrochem. Soc., 120, 1551-7 (1973), PA77-6558.
547. Springe, G. and Schwall, E., "Pickling and etching of aluminium and its alloys to obtain decorative effects", Metalloberflache, 23, 321-5 (1969), German, MA3-04-57 0119
548. Syska, P., "Etching an aluminum coating for production of semiconductor components", Patent German 2137280, Publ. October 1972, CA77-170382.
549. Ullmann, P. and Clausnitzer, P., "Etching agent for structuring aluminum layers in semiconductor device manufacture", Patent German (East) 90681, Publ. June 1972, CA77-132423.
550. Weber, K. E. and Johnston, G. R., "Characterization of surfaces for adhesive bonding", SAMPE Q, 6, 16-21 (1974), CA82-114912.

551. Wyon, G., "Relationship between structural defects in pure aluminium and etch and thermal patterns", Metaux-Corrosion-Ind., 45, 178-96 (1970), French, MA3-12-21 0664.
552. Wyon, G., "Relationship between structural defects in pure aluminium and etch and thermal patterns. (II)", Metaux-Corrosion-Ind., 45, 197-235 (1970), French, MA3-12-21 0665.
553. Wyon, G., "Relationship between structural defects in pure aluminium and etch and thermal patterns. (III)", Metaux-Corrosion Ind., 45, 308-42 (1970), French, MA4-05-12 0602.

3. Titanium and Titanium Alloys

554. Allen, K. W., Alsalum, H. S. and Wake, W. C., "Bonding of titanium alloys", J. Adhes., 6, 153-64 (1974), CA81-174450.
555. Arndt, R., "Investigations into the etching behaviour of titanium alloys", Prakt. Metallogr., 9, 281-2 (1972), English, German, PA75-50687.
556. Burton, A. C. and Pierpont, G. C., "Bath for treating titanium", Patent USA 3753855, Publ. August 1973, CA79-128707.
557. Fukumoto, I. and Taniguchi, Z., "Etching solution for titanium metal", Patent Japan 72-45256, Publ. November 1972, CA80-40292.
558. Gladkii, I. N., "Chemical and electrochemical properties of titanium in fluorine-containing solutions", Zh. Prikl. Khim., 46, 2197-201 (1973), Russian, CA80-86545.
559. Horn, E., "Investigations of the etching behaviour of titanium alloys", Prakt. Metallogr., 9, 270-80 (1972), PA75-50686.
560. Kurdenkova, T. L., Ronzhin, N. N. and Fridman, G. I., "Mechanism of dissolution of titanium in solutions of binary mixtures of hydrofluoric acid with some mineral acids", Tekhnol. Legk. Splavov, Nauch.-Tekh. Byul., Vses. Inst. Legk. Splavov, no. 3, 67-73 (1973), Russian, CA81-42124.
561. Ling, F. W. and Starke, E. A., Jr., "Thermal etching of beta-titanium-vanadium alloys", Metallography, 5, 399-407 (1972), MA6-08-21 0404.
562. Roni, J. C., "Titanium chemical milling etchant", Patent USA 3844859, Publ. October 1974, CA82-20891.
563. Rozenfeld, I. L., Babkin, Y. A. and Alekseeva, E. I., "Etching (and electrochemical behaviour) of high-strength titanium alloys without hydrogen absorption", Zashchita Metallov, 6, 410-5 (1970), Russian, MA3-11-34 0597.
564. Shuford, D. M., "Chemical milling close-tolerance titanium tank segments", Plating, 57, 605-9 (1970), MA3-10-53 0304.
565. Smirnova, S. A., "Etchant for the removal of titanium films", Patent USSR 391613, Publ. July 1973, CA81-15982.
566. Smith, W. D., "New chemical etchant for beta-titanium alloys", Metallography, 4, 371 (1971), MA5-01-21 0007.
567. Usova, V. V., Tomashov, N. D., Modestova, V. N. and Sinyavskii, V. S., "Solution for etching titanium alloys", Patent USSR 425992, Publ. April 1974, CA81-174964.

568.Yamada, S., "Treating titanium and its alloys with oxalic acid-sodium fluoride solutions prior to joining with adhesives", Patent Japan 73-48336, Publ. July 1973, CA79-117828.

4. Gold

569.Bowman, R. R., "Semiconductor devices", Patent USA 3819433, Publ. June 1974, CA81-70145.
570.Bowman, R. R., "Etchant for gold", Patent USA 3879303, Publ. April 1975, CA83-89838.
571.Doda, K. and Sasano, A., "Etching solutions for gold foils", Patent Japan 73-25639, Publ. April 1973, CA75-69657.
572.Lantz, L. L., "Solution control for hybrid microcircuit etching", Report BDX-613-1012, Bendix Corp., Kansas City, Mo., 1973, 14 pp., CA81-7584.
573.Loya, A., "Methods for making semiconductor devices", Patent USA 3863333, Publ. February 1975, EEA78-37666.
574.MacArthur, D. M. and Skurkiss, P. K., "Gold etchant", Patent USA 3816317, Publ. June 1974, CA81-143333.
575.Naraoka, K. and Maeda, M., "Etching multilayer metallic films", Patent Japan 73-08706, Publ. March 1973, CA80-53770.
576.Ricker, T. and Schwing, C., "Thin film structures with a gap width comparable to the film thickness fabricated by sputter etching", Solid State Technol., 17 (10), 69-72 (1974), CA82-25025.
577.Sorokin, I. N., Semyachko, G. Y. and Lavrishchev, V. P., "Anodic etching of gold films in hydrohalic acid solutions", Sb. Nauch. Tr. Probl. Mikroelektron., Mosk. Inst. Elektron. Tekh., no. 11, 113-19 (1972), Russian, CA80-9695.
578.Yahalom, J., "Selective electrolytic etching of gold, platinum, and rhodium thin layers of thin-layer circuits", Patent German 2351664, Publ. May 1974, CA81-20246.

5. Silver

579.Bogoyavlenskii, A. F., Aleksandrov, Y. I. and Sverdlov, A. I., "Electroplating of silver on the surface of anodized aluminum alloys", Elektrokhim. Osazhdenie Primen. Pokrytii Dragotsennymi Rek. Metal., 26-7 (1972), Russian, CA80-103046.
580.Chen, C. C. and Hendrickson, A. A., "New dislocation etchant for silver", J. Appl. Phys., 42, 5375-8 (1971), MA5-05-21 0194.
581.Chen, C. C. and Hendrickson, A. A., "On the dislocation etchants for silver", J. Inst. Metals, 100, 255-6 (1972), MA5-11-21 0450.
582.Makarova, L. E., "Electrolyte for electroetching of copper, silver, and alloys based on them", Patent USSR 479988, Publ. August 1975, CA83-154737.
583.Okamoto, F., "Chemical etchants for the fabrication of thin film patterns of silver on acid- and base-sensitive oxides", Jap. J. Appl. Phys., 13, 383-4 (1974), CA80-139464.

584. Okamoto, F., "Etching solution for silver", Patent USA 3860423, Publ. January 1975, CA83-46966.
585. Salasoo, N., "A chemical polish-etch for silver", J. Australian Inst. Metals, 15, 188 (1970), MA4-03-21 0153.
586. Sorokin, I. N., Semyachko, G. Y. and Lavrishchev, V. P., "Anodic dissolution of thin silver films", Izv. Vyssh. Ucheb. Zaved., Khim. Khim. Tekhnol., 16, 593-5 (1973), Russian, CA79-26380.

6. Chromium

587. Ammann, H., "Etching solution for offset printing plates", Patent German 2240585, Publ. February 1974, CA80-139557.
588. Janus, A. R., "Chemical etch rate studies on sputtered chromium films", J. Electrochem. Soc., 119, 392-6 (1972), EEA75-19405.
589. Kolmakov, O. A., Belevich, G. M., Rasnetsova, B. E. and Kolmakova, G. Y., "Etching of chromium", Patent USSR 370280, Publ. February 1973, CA79-22868.
590. Kolmakov, O. A., Belevich, G. M., Rasnetsova, B. E., Kolmakova, G. Y. and Kegels, S. L., "Solution for selective etching of chromium films", Patent USSR 397993, Publ. September 1973, CA81-53601.
591. Kolmakova, G. Y., Rasnetsova, B. E., Belevich, G. M. and Korshunova, V. A., "Selective chemical etching of chromium films", Patent USSR 366227, Publ. January 1973, CA79-11273.
592. Konishi, S. and Tadagoshi, M., "Chromium plating from low concentration chromium (VI) oxide baths. XIV. Change in cracks of chromium deposits by electroetching", Kinzoku Hyomen Gijutsu, 25, 151-6 (1974), Japanese, CA82-65966.
593. Mueller, B., Muller-Uri, H. and Hirsch, V., "Etch medium for metals", Patent German (East) 101494, Publ. November 1973, CA81-70134.
594. Skarvinso, E. R., "Etching agents for chromium film masks used in semiconductor manufacture", Patent German 2143522, Publ. March 1972, CA77-11218.
595. Yarovoi, A. A., Revzin, G. E., Gutrov, K. I., Zelepukhina, N. L. and Poleshchuk, Y. I., "Etching of chromium layers", Patent USSR 424333, Publ. April 1974, CA83-125770.

7. Tungsten

596. Kern, W. and Shaw, J. M., "Electrochemical delineation of tungsten films for microelectronic devices", J. Electrochem. Soc., 118, 1699-704 (1971), MA5-02-34 0111.
597. Lea, C. and Mee, C. H. B., "Thermal etching of (100), (110) and (113) oriented faces of tungsten single crystals", Surface Sci., 25, 332-6 (1971), MA4-10-12 1438.
598. MacArthur, D. M. and Skurkiss, P. K., "Mild acid etch for tungsten", Patent USA 3841931, Publ. October 1974, CA82-20886.

599.Makarova, I. E. and Ragozin, Y. I., "Polishing and etching tungsten single crystals in an alkaline tripotassium hexacyanoferrate solution", Patent USSR 473513, Publ. June 1975, CA83-124565.

600.Senzaki, K. and Iizuka, M., "Etching technique for an emitter of field emission microscope", Bull. Electrotech. Lab. (Denki Shikensho IHO)", 36, 644-5 (1972), Japanese, MA6-06-21 0298.

601.Shankoff, T. A. and Chandross, E. A., "High resolution tungsten patterning using buffered, mildly basic etching solutions", J. Electrochem. Soc., 122, 294-8 (1975), EEA78-19863.

602.Tamura, Y., Fujii, T. and Ohba, Y., "Electrolytic and chemical etching of dislocation pits and light figures on pure tungsten", J. Japan Inst. Metals, 35, 1027-31 (1971), Japanese, MA5-08-21 0355.

603.Tamura, Y., Fujii, T. and Ohba, Y., "Dislocation etch-pits and light figures of pure tungsten by electrolytic and chemical etching", Trans. Nat. Research Inst. Metals (Japan), 14, 173-7 (1972), MA6-06-13 0518.

604.Wilson, V. C. and Danko, J. C., "Development of a stable faceted tungsten emitter surface", IEEE Conference Record of the 1969 Thermionic Conversion Specialist Conference, p. 46-52, EEA74-924.

8. Molybdenum

605.Emelyanova, L. I. and Mikhailov, S. K., "Visualizing the dislocation structure of molybdenum by etching in a high-temperature argon flow", Phys. Metals Metallogr., 26, 158 (1968), MA3-09-21 0542.

606.Putilova, I. N., Maslova, G. M. and Suponitskaya, I. I., "Dissolution of molybdenum in 2-N nitric acid and the effect of chromates on the size of the nitric oxide bubbles so formed", Zhur. Priklad. Khim., 43, 1272-4 (1970), Russian, MA3-11-35 1210.

607.Savitsky, E. M., Burkhanov, G. S., Lapshina, L. G., Ottenberg, E. V. and Kataev, R. S., "Study of the structure of single-crystal and polycrystalline molybdenum by colour etching", Izvest. Akad. Nauk SSSR, Metally, no. 3, 169 (1972), Russian, MA5-10-21 0443.

608.Tartakovskaya, F. M., Melnikova, I. A., Kozlova, I. R. and Makarenko, N. V., "(Etching of polycrystalline refractory metals by selective dissolution). I. Selective etching of molybdenum", Zhur. Prikad, Khim., 43, 538-44 (1970), Russian, MA3-11-34 0604.

609.Wlodyka, J., Przybyslawski, A. and Jakubowski, Z., "Molybdenum recovery from waste etching solutions", Rudy i Metale Niezelazne, 14, 636-9 (1969), Polish, MA3-06-43 0092.

9. Nickel

610.Bartosiewicz, L., "Preferential electrolytic etching to reveal the gamma-prime phase in nickel alloys", Praktische Metallographie, 10, 450-61 (1973), MA7-01-22 0081.
611.Eppensteiner, F. W., "Nickel etch composition", Patent USA 3804768, Publ. April 1974, CA81-110233.
612.Kralina, A. A. and Sazonova, V. A., "Relation between the thermal conditions of solidification and the micro-etch patterns of nickel single crystals", Izvest. Akad. Nauk SSSR, Metally, no. 5, 111-6 (1972), Russian, MA6-05-12 0596.
613.Popkova, E. G. and Predvoditelev, A. A., "Selective etching of dislocations in metals (Ni)", Sov. Phys.-Crystallogr., 18, 647-50 (1974), PA77-53003.
614.Schatt, W. and Worch, H., "Dislocation etching of nickel single crystals by potentiostatic electrolysis", Praktische Metallographie, 7, 483-93 (1970), MA4-01-21 0023.

10. Tantalum

615.Kisilenko, N. I. and Parusnikov, V. N., "Electrochemical etching, polishing, and oxidation of tantalum wire", Elektron. Teknika. Nauchno-Tekhn. Sbornik, Materialy, no. 1, 51-7 (1972), Russian, MA6-07-57 0259.
616.Okabayashi, H., "Etching tantalum film", Patent Japan 74-01138, Publ. January 1974, CA81-95366.
617.Shenoi, B. A., Narasimhan, K. R. and Lakshminarasimhan, V., "Etching tantalum anodes", Patent Indian 103971, Publ. August 1968, CA82-9330.

11. Brass

618.Greene, N. D. and Teterin, G. A., "Development of brass etchants by electrochemical techniques", Corrosion Sci., 12, 57-63 (1972), MA5-09-21 0375.
619.Severdenko, V. P. and Boiko, E. I., "Etching of brass in solutions of nitric acid and the effect of ultrasound on the process", Izvest. Akad. Nauk Beloruss, SSR, no. 3, 113-16 (1970), Russian, MA4-02-57 0031.
620.Vidal, N., Arnstein, G. M. and Castro, G., "An electrochemical etchant for observation of dislocations in 70-30 brass", Metall. Trans., 2, 2951-2 (1971), MA5-02-21 0066.

12. Others

621.Braun, Z. I., "Solution for chemical etching of metals", Patent USSR 404907, Publ. October 1973, CA81-67742.

622.Bulenkov, N. A., Ruslanova, V. V., Sakharov, V. V. and Sidorova, M. V., "Revealing dislocations in magnesium silicide alloys by etching", Izvest. Akad. Nauk SSSR, Neorg. Materialy, 6, 1333-4 (1970), Russian, MA3-11-21 0635.
623.Campbell, J. F., Jr., Engvall, A. E. and Lewis, A. E., "Protecting against electrochemical effects during metal etching", Patent USA 3847690, Publ. November 1974, CA82-105471.
624.Castellano, R. N. and Schmidt, P. H., "Chemical polish and etch for lithium, sodium and potassium", J. Electrochem. Soc., 18, 653-4 (1971), MA4-09-21 0539.
625.Chapek, A., Paufler, P. and Shultse, G. E. R., "Revelation of dislocation networks in the intermetallic compound magnesium dizincide by layer-by-layer etching", Kristallografiya, 15, 1256-8 (1970), Russian, MA4-04-21 0215.
626.Chough, E., "Process for etching metals employing ultrasonic vibration", Patent USA 3767941, Publ. October 1973.
627.Dickson, J. L., Dedo, A. and Craig, G. B., "Etching dislocations in zirconium", J. Nuclear Mat., 38, 116-7 (1971), MA4-08-21 0400.
628.Gindin, I. A., Starodubov, Y. D. and Khaimovich, P. A., "On revealing the grain boundaries in metals at low temperatures", Ind. Lab., 39, 616-17 (1973), PA77-21260.
629.Goffredo, D., "The interrelationship of metal property and etching characteristics", Presented at Semiconductor Integrated Circuit Processing and Production Conference, Anaheim, California, February 1971, EEA74-21565.
630.Hill, J. H., Morris, C. J. and Frazer, J. W., "Cathodic etching to prepare metals for oxygen determination", Analyst, 95, 215-8 (1970), MA3-07-23 0371.
631.Hirokawa, T., Honda, K. and Shibuya, T., "Formation of etch hillocks in white tin", J. Cryst. Growth, 24-25, 484-7 (1974), PA78-6653.
632.Johnson, K. A., "Alternating-current electropolishing and electroetching of metals", Proceedings, Second Annual Technical Meeting, International Metallographic Society, Inc., 1970, p. 145-61, MA3-07-21 0436.
633.Jones, R. W. and Thompson, W. B., "Etching metallic layers", Patent German 2247057, Publ. April 1973, CA79-151681.
634.Kiellarson, P. E. and Holmberg, E. G., "Electrolytic etching of patterns on stainless steel", Patent German 1962469, Publ. October 1970, CA74-36949.
635.Mueller, J., "Chemical erosion or metal etching of workpieces", Patent German 2063222, Publ. July 1972, CA77-92015.
636.Nishikawa, H., "Dislocation loops in the surface layer of cadmium crystals", Jap. J. Appl. Phys., 10, 276 (1971), MA4-09-13 0836.
637.Petrov, L. N., Svist, E. I., Savitskaya, O. P. and Goyan, E. B., "Effect of etching of zinc alloys in nitric acid on their strength", Sov. Mater. Sic., 4, 543-4 (1972), MA5-05-31 1105.

638. Piotrowski, T. and Accinno, D. J., "The preparation of ruthenium for optical metallography", Metallography, 4, 521-32 (1971), MA5-07-21 0308.
639. Plessey Co. Ltd., "Making mounting structure for microelectronic devices by etching away different metals in multi-layer structures", Patent UK 1182422, Publ. February 1970, EEA73-30361.
640. Popkova, E. G., Matveeva, G. S. and Predvoditelev, A. A., "On the mechanism of selective etching at dislocations in zinc", Sov. Phys. Crystallogr., 14, 40-4 (1969), MA3-03-21 0158.
641. Popkova, E. G. and Predvoditelev, A. A., "Selective etching on dislocations in metals", Sov. Phys. Crystallogr., 15, 72-7 (1970), MA3-12-21 0702.
642. Rashkov, N. and Gancheva, M., "Reagents for the colour etching of the phases in the layer obtained as a result of the surface nitriding of manganese", Godish. na Vissh. Khim. - Tekhnol. Inst. Sofia, 15, 459-69 (1972), Bulgarian, MA6-09-21 0475.
643. Redpath, D. L. and Joshi, K. C., "Metallographic preparation of aluminum and gold", Microstructures, 2, 21-2 (1971), MA5-01-21 0063.
644. Romankiw, L. T., "Elimination of undercut in an anodically active metal during chemical etching", Patent USA 3853715, Publ. December 1974, CA82-117873.
645. Schatt, W. and Worch, H., "Potentiostatic polishing and dislocation etching", Prakt. Metallogr., 12, 355-65 (1975), English, German, CA83-185324.
646. Seredenko, V. N., Datsko, O. I., Baryakhtar, F. G. and Kushnerev, V. I., "Electrolyte for exposing the structure and for electropolishing of aluminum, cobalt and copper", Zavod. Lab., 39, 1217-18 (1973), Russian, CA80-33115.
647. Slimak, A. and Piotrowski, A., "Theoretical and practical aspects of the selective potentiostatic etching of alloys", Przeglad Odlewnict., 22, 440-45 (1972), Polish, MA7-01-23 0047.
648. Stiehr, W., "Shape etching. Special process for the noncutting production of metal parts. II", Galvanotechnik, 63, 246-54 (1972), German, CA78-36156.
649. Stolyarova, I. N. and Timonova, M. A., "Solution for etching of magnesium alloys", Patent USSR 395523, Publ. August 1973, CA81-28690.
650. Tousek, J., "The preparation of platinum and palladium foils by electrolytic polishing method", Praktische Metallographie, 7, 202-4 (1970), MA3-09-21 0528.
651. Vaghari, D. J., "Effect of temperature on the delineation of dislocations in bismuth", Indian J. Pure Appl. Physics, 10, 84 (1972), MA6-03-21 0131.
652. Verhaege, M. and De Wilde, R., "New method for electrolytic polishing and etching of lead and lead alloys", Prakt. Metallogr., 10, 220-6 (1973), CA79-37973.

653. Yang, Y. S., "Schoone and Fishione's method of electrolytic thinning of metal film specimens for electron microscopy", Ts'ai Liao K'o Hsueh, 5 (2), 76-9 (1974), Chinese, CA81-139468.

H. ISOLATION TECHNIQUES

I. Bibliographies

1. Agajanian, A. H., "A bibliography on semiconductor device isolation techniques", Solid State Technol., 18 (4), 61-5 (1975), EEA78-24498.

II. Review Articles

2. Anon., "Technical status quo of semiconductor devices", JEE, no. 32, 20-6 (1969), EEA73-10649.
3. Bean, K. E. and Runyan, W. R., "Dielectric isolation - review current and future", Presented at the Fall Meeting of the Electrochemical Society, New York, October 1974.
4. Bell, C. G., Bhandarkar, D. R., Feuchi, D. L., Rege, S. L. and Seiwiorek, D. L., "Large scale integration - a designer's viewpoint", Report PB-221-013/6, Carnegie-Mellon University, Pittsburgh, Pa., 1972, 74 pp..
5. Chang, F. Y., "Bipolar device isolation techniques", Presented at 1974 Modern Engineering and Technology Seminar, Taipei, Republic of China, July 1974.
6. Davies, L. W., "Integrated circuits", In: Information, computers, machines and man, Karbowiak, A. E. and Huey, R. M., (Eds.), Wiley, New York, p. 161-7 (1971), EEA75-6968.
7. Kile, C., Jr., "Semiconductor integrated circuits - a review of fabrication processes and component characteristics", In: Integrated circuit technology, Longo, T. A., (Ed.), Boston Technical Publishers, Inc., Central Square, Cambridge, Mass., p. 9-17 (1970).
8. Lepselter, M. P., "Beam-lead technology", Bell Syst. Tech. J., 45, 233-54 (1966), EEA69-11534.
9. Magdo, I., "Dielectric isolation for semiconductor devices", Thesis, University of Budapest, Hungary, 1975, 109 pp..
10. Soderman, L., "Bipolar LSI components with new structures can be made", Eltek. Aktuell Elektron., 16 (12), 20-1 (1973), Swedish, EEA77-14588.
11. Stoller, A. L., "Relationship between the performance of a linear amplifier microcircuit and the isolation technique used in its construction", RCA Rev., 29, 507-24 (1968), EEA72-10546.

III. Junction Isolation

1. Conventional Junctions

12. Berger, H. H., Kroell, K. E. and Wiedmann, S. K., "Masking process for base and isolation diffusion", IBM Tech. Disclosure Bull., 14, 2531 (1972).

13.Chang, A. W., Chang, F. Y., Pak, M. S. and Walsh, T. V., "Diffused trench isolation with p-channels", IBM Tech. Disclosure Bull., 18, 380-1 (1975), EEA78-37633.
14.Donald, R. G., "Complementary transistors in integrated circuits", Solid-State Electron., 13, 815-24 (1970), EEA73-30366.
15.Glinski, V. J., "A three mask bipolar integrated circuit structure", IEEE Trans. Electron Devices, ED-19, 182-90 (1972), EEA75-6956.
16.Kooi, E. and Van Gils, J. A. A., "Semiconductor device", Patent USA 3911471, Publ. October 1975, CA83-201130.
17.Lehning, H., "Current hogging logic (CHL) - A new bipolar logic for LSI", IEEE J. Solid-State Circuits, SC-9, 228 (1974).
18.Lin, H. C., Ho, J. C., Iyer, R. R. and Kwong, K., "Complementary MOS-bipolar transistor structure", IEEE Trans. Electron Devices, ED-16, 945-51 (1969), EEA73-10679.
19.McLouski, R. M., "Fabrication considerations for complementary bipolar circuits", Presented at the Electrochemical Society Meeting, Chicago, Illinois, May 1973, EEA76-23488.
20.Ogawa, T., Miyata, K., Yagi, H. and Sasaki, T., "Integrated semiconductor device having functional regions isolated by pn junctions therebetween", Patent USA 3795846, Publ. March 1974.
21.Plessey Co. Ltd., "Isolation region between adjacent circuit components of an integrated circuit", Patent UK 1208266, Published October 1970, EEA74-9813.
22.Vernikov, M. A., Goikhman, A. N., Nikonov, A. S., Press, F. P. and Rzhanov, G. V., "Planar silicon p-n-p transistor with an epitaxial base formed by isolating etching", In: Poluprovodn. Prib. Mater., Radautsan, S. I., (Ed.), "Shtiintsa", Kishinev, USSR, p. 64-8 (1973), Russian, CA82-67172.
23.Vora, M. and Chang, J. J., "PIN isolation for monolithic integrated circuits", Presented at the International Electron Devices Meeting, Washington, D. C., October 1967.

2. Collector Diffusion (CDI)

24.Anon., "Breakthrough in integrated circuits", Wireless World", 77 (1433), 526 (1971), EEA75-430.
25.Bruchez, J. A., "Application of new isolation technique to digital integrated circuits", International Conference on Microelectronics, 1969, p. 15-16, EEA72-23148.
26.Bruchez, J., "CDI and competitive technologies", Microelectronics, 5 (4), 45-54 (1974), EEA78-28863.
27.Eckl, D. J., Konkle, K. H. and Richardson, F. K., "High performance ECL gates made by CDI and conventional methods", Presented at the International Electron Devices Meeting, Washington, D. C., October 1969.
28.Forshaw, P., "The advantages of c.d.i. over m.o.s. for engineers designing electronic control systems", New Electron., 6 (24), 36-8, 41, 44-5 (1973), EEA77-14596.

29. Grundy, D., "CDI - its capability and application", Microelectronics, 6 (3), 29-40 (1975), EEA78-28864.
30. Grundy, D. L., Brunchez, J. and Down, B., "Collector diffusion isolation packs many functions on a chip", Electronics, 45 (14), 96-104 (1972), EEA75-27734.
31. Grundy, D. L., Bruchez, J. and Down, B., "Collector diffusion isolation: the bipolar LSI", Microelectronics, 4 (3), 10-30 (1972), EEA76-26150.
32. Jenkins, J. A. and Forshaw, P., "Collector diffusion isolation", Electron., no. 22, 21 (1973), EEA76-16468.
33. Krebs, P., "CDI, the bipolar LSI technology with total systems integration capability", Microelectron. & Reliab., 14, 385-7 (1975).
34. Montgomery, R. J., "L.S.I. circuits with total system capabilities", 6th International Congress on Microelectronics, 1974, 9 pp., EEA78-24516.
35. Montgomery, R. J., "Total system integration using c.d.i.", New Electron., 8 (4), 58-9 (1975), EEA78-24487.
36. Murphy, B. T. and Glinski, V. J., "Transistor-transistor logic with high packing density and optimum performance at high inverse gain", IEEE J. Solid-State Circuits, SC-3, 261-7 (1968), EEA72-2283.
37. Murphy, B. T., Glinski, V. J., Gary, P. A. and Pedersen, R. A., "Collector diffusion isolated integrated circuits", Proc. IEEE, 57, 1523-7 (1969), EEA73-521.
38. Nystrom, E., "Rifa develop their own bipolar LSI", Eltek. Aktuell Elektron., 16 (12), 22-3 (1973), EEA77-14589.
39. Rolfe, D., "Cheaper LSI custom circuits", Elektron. Anz., 5, 205-6 (1973), German, EEA77-14587.
40. Rolfe, D., "Towards less costly integrated circuits", Rev. Polytech., no. 3, 313-5 (1974), French, EEA77-23155.

3. Base Diffusion (BDI)

41. DeFalco, J. A., "Coming up fast from behind-denser bipolar devices", Electronics, 44 (15), 76-9 (1971), EEA74-34061.
42. Murphy, B. T., Neville, S. M. and Pedersen, R. A., "Simplified bipolar technology and its application to systems", IEEE J. Solid-State Circuits, SC-5, 7-14 (1970), EEA73-14276.
43. Senhouse, L. S., Jr., Kushler, D. L. and Murphy, B. T., "Base diffusion isolation (BDI) for transistors", Presented at the International Electron Devices Meeting, Washington, D. C., October 1969, EEA73-21307.
44. Senhouse, L. S., Jr., Kushler, D. L. and Murphy, B. T., "Base diffusion isolated transistors for low power integrated circuits", IEEE Trans. Electron Devices, ED-18, 355-8 (1971), EEA74-18767.

4. Epitaxial Isolation

45. Baker, T. H. and Ghafghaichi, M., "Diffused isolation and etched isolation structure", IBM Tech. Disclosure Bull., 15, 1772 (1972).
46. Darwish, M. and Taubenest, R., "C-MOS and complementary isolated bipolar transistor monolithic integration process", J. Electrochem. Soc., 121, 1119-22 (1974), EEA77-38886.
47. Davidsohn, U. S. and Ajamie, A. J., "Method of manufacturing high-speed isolated integrated circuit", Patent USA 3876480, Publ. April 1975.
48. Demizu, K., "Double epitaxial isolation method for the fabrication of integrated circuit transistors", J. Electrochem. Soc., 118, 1627-30 (1971), EEA75-423.
49. Demizu, K. and Yukawa, K., "Double epitaxial isolation method for the fabrication of integrated circuit transistors", In: Semiconductor silicon, Electrochem. Soc., Inc., New York, p. 243-8 (1969), EEA73-9599.
50. Dhaka, V. A., "Development of a high-performance silicon monolithic circuits", Presented at the International Electron Devices Meeting, Washington, D. C., October 1970.
51. Iwamatsu, S., Meguro, S. and Shimizu, S., "A new isolation structure for high density LSI", 1973 International Electron Devices Meeting Technical Digest, p. 244-7, EEA77-14617.

5. Self-Isolation

52. Makimoto, T., Sugawara, K. and Maki, M., "Self-isolated transistor structures of bipolar integrated circuits", Jap. J. Appl. Phys., 9, 782-92 (1970), EEA73-37054.
53. Vora, M. B., "A self-isolation scheme for monolithic integrated circuits", Presented at the 17th International Electron Devices Meeting, Washington, D. C., October 1971, EEA75-12783.
54. Vora, M. B., "A self-isolation scheme for integrated circuits", IBM J. Res. & Develop., 15, 430-5 (1971), EEA75-9573.
55. Vora, M. B., "Making diffused resistors with self-isolation", IBM Tech. Disclosure Bull., 14, 2307 (1972).

6. Others

56. Agraz-Guerena, J., Gary, P. A., Lepselter, M. P. and Pedersen, R. A., "GIMIC - a high-yield high-performance technology for MSI-LSI", 1973 IEEE International Convention and Exposition, Vol. VI, 30.3/4 pp., EEA77-2613.
57. Ashar, K., "Emitter diffusion isolated semiconductor structure", Patent USA 3787253, Publ. January 1974, CA80-75747.
58. Barron, M. B. and Butler, W. J., "Gate-diffusion isolation for JFET depletion-mode bucket brigade circuit", Patent USA 3825996, Publ. July 1974.

59.Panousis, P. T. and Pritchett, R. L., "GIMIC-0 a low cost non-epitaxial bipolar LSI technology suitable for application to TTL circuits", Presented at the International Electron Devices Meeting, Washington, D. C., December 1974.
60.Radiotechnique-Complelec, "Monolithic semiconductor device array prepared by isolation diffusion", Patent French 2228299, Publ. November 1974, CA83-20806.

IV. Dielectric Isolation
1. Oxide Process
a. LOCOS

61.Anon., "Phillips development of local oxidation technique make possible the production of thick oxide layers", Bedriff & Tech., 26, 1462-3 (1971), Dutch, EEA74-36453.
62.Anon., "LOCOS: a new production technology for integrated circuits", Toute Electron., no. 359, 27-8 (1971), French, EEA75-3952.
63.Anon., "LOCOS: a new I.C. technology", Microelectron. & Rel., 10, 471-2 (1971), EEA75-6965.
64.Appels, J. A., Kooi, E., Paffen, M. M., Schatorje, J. J. H. and Verkuylen, W. H. C. G., "Local oxidation of silicon and its application in semiconductor-device technology", Philips Res. Rep., 25, 118-32 (1970), EEA73-28102.
65.Appels, J. A. and Paffen, M. M., "Local oxidation of silicon: New technological aspects", Philips Res. Rep., 26, 157-65 (1971), EEA74-25566.
66.Brandt, B. B. M., Steinmaier, W. and Strachan, A. J., "LOCMOS, a new technology for complementary MOS circuits", Philips Tech. Rev., 34, 19-23 (1974), EEA77-26525.
67.Brandt, B. B. M., Steinmaier, W. and Strachan, A. J., "LOCMOS, a new CMOS technology", Onde Electr., 54, 27-30 (1974), French, EEA77-14600.
68.Hayasaka, A., Noda, H., Suzuki, M. and Kondo, H., "Semiconductor devices by local oxidation of a silicon surface", Patent German 2365056, Publ. April 1975, CA83-52096.
69.Kooi, E., "Processes for compact LOCOS integrated circuits", Presented at the 17th International Electron Devices Meeting, Washington, D. C., October 1971, EEA75-12752.
70. Kooi, E., "MOS effects and silicon device technology", Oyo Buturi, 43 (Suppl.), 427-36 (1974), EEA78-2883.
71.Kooi, E. and Appels, J. A., "Selective oxidation of silicon and its device applications", Presented at the Electrochemical Society Meeting, Chicago, Illinois, May 1973, EEA76-23152. Also: Semiconductor Silicon 1973, p. 860.
72.Kooi, E., van Lierop, J. G., Verkuijlen, W. H. C. G. and de Werdt, R., "LOCOS devices", Philips Res. Rep., 26, 166-80 (1971), EEA74-25567.

73.Kooi, E., DeWerdt, R. and Nijdam-Paffen, M. M. M., "Semiconductor device", Patent USA 3852104, Publ. December 1974, CA82-79561.
74.Kurz, B., Barron, M. B. and Butler, W. J., "Dielectric strip isolation for JFET or MESFET depletion-mode bucket-brigade circuit", Patent USA 3784847, Publ. January 1974.
75.Kurz, B., Barron, M. B. and Butler, W. J., "Dielectric strip isolation for JFET or MESFET depletion-mode bucket-brigade circuit", Patent USA 3825995, Publ. July 1974.
76.Philips Electronic & Assoc. Ind. Ltd., "Semiconductor device manufacture", Patent UK 1208575, Publ. October 1970, EEA74-9820.
77.Philips Electronic & Assoc. Ind. Ltd., "Semiconductor device manufacture", Patent UK 1208576, Publ. October 1970, EEA74-9821.
78.Philips Electronic & Assoc. Ind. Ltd., "Semiconductor device manufacture", Patent UK 1208577, Publ. October 1970, EEA74-9822.
79.Philips Electronic & Assoc. Ind. Ltd., "Semiconductor device production", Patent UK 1208578, Publ. October 1970, EEA74-9823.
80.Schiereck, F. C., Drost, A. and May, P., "LOCMOS LSI circuitry, an example of its application in a minicomputer", International Conference on Large Scale Integrated Circuits, 1974, p. 43-51, EEA78-11744.

b. ISOPLANAR

81.Baker, W. D., Herndon, W. H., Longo, T. A. and Peltzer, D. L., "The impact of isoplanar technology on high-density read/write memories", 1973 IEEE International Convention and Exposition, Vol. VI, 30.1/8 pp., EEA77-6092.
82.Baker, W. D., Herndon, W. H., Longo, T. A. and Peltzer, D. L., "Oxide isolation brings high density to production bipolar memories", Electronics, 46 (7), 65-70 (1973), EEA76-16467.
83.Baker, W. D., Herndon, W. H., Longo, T. A. and Peltzer, D. L., "Isoplanar techniques of higher density, faster memories", New Electron., 7 (4), 58-9 (1974), EEA77-23153.
84.Dhaka, V. A., Muschinske, J. E. and Owens, W. K., "Subnanosecond ECL gate circuit using Isoplanar II", Digest of Technical Papers of the 1973 IEEE International Solid State Circuits Conference, p. 172-3, EEA76-23048.
85.Dhaka, V. A., Muschinske, J. E. and Owens, W. K., "Subnanosecond emitter-coupled logic gate circuit using Isoplanar II", IEEE J. Solid-State Circuits, SC-8, 368-72 (1973), EEA76-35361.
86.Kamins, T. I., "A new dielectric isolation technique for bipolar integrated circuits using thin single-crystal silicon films", Proc. IEEE, 60, 915-16 (1972), EEA75-27754.
87.Peltzer, D., "Isoplanar processing (memory circuits)", Western Electronic Show and Convention, Vol. 16, 1972, 19/2 4 pp., EEA76-8753.
88.Peltzer, D. and Herndon, B., "Isolation method shrinks bipolar cells for fast, dense memories", Electronics, 41 (5), 52-5 (1971), EEA74-15578.

89. Seeds, R. B. and Luce, R. L., "High density LSI with isoplanar MOS", 1973 IEEE International Convention and Exposition, Vol. VI, 23.3/2 pp., EEA77-2611.

c. ROI (Recessed Oxide Isolation)

90. Aboaf, J. A. and Broadie, R. W., "Dielectric isolation", IBM Tech. Disclosure Bull., 18, 3295-6 (1976).
91. Antipov, I., "Isolation of N and P regions in recessed oxide semiconductor structure", IBM Tech. Disclosure Bull., 27, 1545-6 (1974), EEA78-8554.
92. Cass, E. E., Gopalakrishna, Y. R., Kwap, T. W. and Magdo, I. E., "Recessed oxide integrated circuit structure utilizable for mixed and off-chip input", IBM Tech. Disclosure Bull., 16, 617 (1973).
93. Chang, A. W., Chang, F. Y., Pak, M. S. and Walsh, T. V., "Diffused trench isolation with p channels", IBM Tech. Disclosure Bull., 18, 380-1 (1975), EEA78-37633.
94. Dennard, R. H., Rideout, V. L. and Walker, E. J., "Reducing sidewall conduction in recessed oxide field-effect-transistor arrays", Patent USA 3899363, Publ. August 1975, CA83-156878.
95. Klepner, S. P., "Base isolation in recessed oxide insulation", IBM Tech. Disclosure Bull., 17, 431 (1974).
96. Klepner, S. P., Mangone, J. and Marvel, R., "Recessed oxidation method", IBM Tech. Disclosure Bull., 16, 402 (1973).
97. Lechaton, J. S. and Smith, D. C., "Dielectric isolation using a high-resputtering deposition", IBM Tech. Disclosure Bull., 18, 382-3 (1975), EEA78-37634.
98. Lehman, H. S., "Oxidation resistant mask layer and process for producing recessed oxide region in a silicon body", Patent USA 3874919, Publ. April 1975, EEA78-37668.
99. Magdo, I. and Magdo, S., "Complementary MOS transistors", IBM Tech. Disclosure Bull., 15, 1769 (1972).
100. Magdo, I. and Magdo, S., "Fabricating complementary transistors", IBM Tech. Disclosure Bull., 15, 2294 (1972).
101. Magdo, I. and Magdo, S., "Dielectric isolation for high density semiconductor circuits", Patent UK 1323850, Publ. November 1973.
102. Magdo, I. and Magdo, S., "Method of forming dielectric isolation for high density pedestal semiconductor devices", Patent USA 3796613, Publ. March 1974.
103. Magdo, I. E. and Magdo, S., "Dielectrically isolated Schottky barrier structure and method of forming the same", Patent USA 38583231, Publ. December 1974, EEA78-24506.
104. Magdo, I. E. and Magdo, S., "Method of fabricating integrated circuit device structure with complementary elements utilizing selective thermal oxidation and selective epitaxial deposition", Patent USA 3861968, Publ. January 1975.
105. Magdo, S., Magdo, I. and Hatzakis, M., "Dielectrically isolated transistor", IBM Tech. Disclosure Bull., 13, 3238 (1971).

106. Rideout, V. L., "Thermal oxide isolation process for high-density memories", IBM Tech. Disclosure Bull., 17, 3041-3 (1975),

d. PLANOX

107. Morandi, F., "The M.O.S. planox process", Presented at the International Electron Devices Meeting, Washington, D. C., October 1969, EEA73-21347.
108. Morandi, F., "Planox process smoothes path to greater MOS density", Electronics, 44 (26), 44-8 (1971), EEA75-7529.
109. Morandi, F. and Bladowski, R., "A technology for MOS components with low threshold voltages", Int. Elektron. Rundsch., 25, 289-93 (1971), German EEA75-7539.
110. Toombs, T. N., Finnila, R. M., Dill, H. G. and Bauer, L. O., "High density ion-implanted C-MOS technology", Presented at the 17th International Electron Devices Meeting, Washington, D. C., October 1971, EEA75-13787.

e. OXIM

111. Edwards, R., Payne, R. S., Gibbon, C. F., Olmstead, M. L., Moline, R. A., Kishler, D. L., Weber, G. R. and Kraetsch, R. T., "Oxide isolation technology featuring ion implantation and partially self-registered emitters", Presented at the Electrochemical Society Meeting, Chicago, Illinois, May 1973, EEA76-23498; Also: Semiconductor Silicon 1973, p. 905.
112. Evans, W. J., Tretola, A. R., Payne, R. S., Olmstead, M. L. and Speeney, D. V., "Oxide isolated ion-implanted bipolar transistors for high packing density and low power-delay product", Digest of Technical Papers of the 1973 IEEE International Solid State Circuits Conference, p. 174-5, EEA76-23503.
113. Evans, W. J., Tretola, A. R., Payne, R. S., Olmstead, M. L. and Speeney, D. V., "Oxide-isolated monolithic technology and applications", IEEE J. Solid-State Circuits, SC-8, 373-80 (1973), EEA76-35362.

f. POLYPLANAR

114. Dell'Oca, C. J. and Nonnenmacher, H. G., "Dielectric isolation process for single crystal silicon film", Presented at the Fall Meeting of the Electrochemical Society, New York, October 1974.
115. Harris, R. E., "Using a masking layer for producing field-effect devices having oxide isolation", Patent USA 3698966, Publ. October 1972, CA77-157930.
116. Sanders, T. J. and Morcom, W. R., "Polysilicon-filled notch produces flat, well-isolated bipolar memory", Electronics, 46 (8), 117-20 (1973), EEA76-19956.

117. Sanders, T. J., Morcom, W. R. and Kim, C. S., "An improved dielectric-junction combination isolation technique for integrated circuits", 1973 International Electron Devices Meeting Technical Digest, p. 38-40, EEA77-14609.

g. IPOS (Isolation by Porous Oxidized Silicon)

118. Aboaf, J. A., Broadie, R. W. and Kemlage, B. M., "Polysilicon mask for porous anodic etching", IBM Tech. Disclosure Bull., 18, 1439 (1975).
119. Anon., "Process speeds up chip isolation", Electronics, 47 (22), 45 (1974).
120. Badami, A. V. and Broadie, R. W., "Dielectric isolation of silicon devices", IBM Tech. Disclosure Bull., 18, 116-17 (1975), EEA78-37333.
121. Beyer, K. D. and Poponiak, M. R., "Forming deep SiO_2 layers in monocrystalline silicon", IBM Tech. Disclosure Bull., 17, 2885-6 (1975), EEA78-19645.
122. Beyer, K. D. and Poponiak, M. R., "Doping impurities into porous silicon from a nonaqueous solution", IBM Tech. Disclosure Bull., 18, 97 (1975), EEA78-37397.
123. DeRobertis, I. J. and Poponiak, M. R., "Cell for porous applications", IBM Tech. Disclosure Bull., 18, 122-4 (1975), EEA78-37398.
124. Hamano, K., Sakamoto, M. and Yamanaka, H., "Oxidation process of porous silicon", Jap. J. Appl. Phys., 14, 413-14 (1975), EEA78-19662.
125. Koji, T. and Konno, S., "New isolation technique for a low frequency low noise monolithic linear integrated circuit", Presented at the International Electron Devices Meeting, Washington, D. C., December 1974.
126. Ku, S. M., Poponiak, M. R., Schwenker, R. O. and Yeh, T. H., "Preventing inversion of base junctions in large-scale integration devices", IBM Tech. Disclosure Bull., 17, 2329 (1975).
127. Nakajima, S., Watanabe, Y., Yokoyama, T. and Kato, K., "IPOS scheme: A new isolation technique for integrated circuits", Oyo Buturi, 44 (Suppl.), 303-9 (1975), EEA78-41688.
128. Pogge, H. B. and Poponiak, M. R., "Method of fabricating semiconductor devices embodying dielectric isolation", Patent USA 3919060, Publ. November 1975.
129. Poponiak, M. R. and Schwenker, R. O., "High sheet resistivity resistors by the porous silicon technique", IBM Tech. Disclosure Bull., 18, 125-6 (1975), EEA78-37630.
130. Poponiak, M. R., Schwenker, R. O. and Yeh, T. H., "Dielectric isolation of silicon regions", IBM Tech. Disclosure Bull., 27, 1331-2 (1974), EEA78-11726.
131. Watanabe, Y., Arita, Y. and Yokoyama, T., "Porous silicon and its application", Oyo Buturi, 43, 1046-9 (1974), Japanese, EEA78-15231.

132. Watanabe, Y., Arita, Y., Yokoyama, T. and Igarashi, Y., "Formation and properties of porous silicon and its application", J. Electrochem. Soc., 122, 1351-5 (1975).
133. Watanabe, Y., Arita, Y., Yokoyama, T. and Igarashi, Y., "A novel isolation technique by oxidation of porous silicon", Fall Meeting of the Electrochemical Society, (Extended Abstracts), 1973, p. 296.
134. Watanabe, Y. and Sakai, T., "Application of a thick anode film to semiconductor devices", Rev. Elec. Commun. Lab., 19, 899-903 (1971), EEA75-10313.
135. Watanabe, Y. and Sakai, T., "Semiconductor device and method of producing the same", Patent USA 3640806, Publ. February 1972.
136. Yashiro, T., Saito, K. and Suzuki, T., "Isolation layer formation by enhanced anodization of N-type silicon", Electrochemical Society Fall Meeting (Extended abstracts), 1974, p. 351-2, EEA78-37680.

h. Others

137. Archer, J. A., "Improved microwave-transistor structure", Electron. Lett., 8, 499-500 (1972), EEA75-40199.
138. Berglund, C. N., Clemens, J. T. and Nicollian, E. H., "Undercut isolation - a technique for closely spaced and self-aligned metallization patterns for MOS integrated circuits", J. Electrochem. Soc., 120, 1255-60 (1973), EEA77-2605.
139. Brack, K., Gorey, E. F. and Schwuttke, G. H., "Dielectric isolation in semiconductor bodies", Patent British 1274726, Publ. May 1972, CA77-40611.
140. Bratter, R. L. and Gaind, A. K., "Controlling dielectric isolation of a semiconductor device", Patent USA 3886000, Publ. May 1975, CA83-107226.
141. Chappelow, R. E., Chiu, T. L., Li, P. C. and Lin, P. T., "Dielectric isolation structure for integrated circuit device", IBM Tech. Disclosure Bull., 16, 31 (1973).
142. Cohen, R. A. and Mountain, R. W., "Si_3N_4-masked thermally oxidized post-diffused mesa process (SIMTOP)", IEEE Trans. Electron Devices, ED-18, 54-7 (1971), EEA74-10480.
143. Coleman, M. G., "Method of making isolated semiconductor devices", Patent USA 3707765, Publ. January 1973, EEA77-5777.
144. Cosand, A. E., "Very high speed low power bipolar integrated circuit process", Int. Electron Devices Meeting Tech. Dig., 1973, p. 35-37, EEA76-055954.
145. Davidsohn, U. S. and Lee, F., "Dielectric isolated integrated circuit substrate processes", Proc. IEEE, 57, 1532-7 (1969), EEA73-455.
146. Frankovsky, F., Protschka, H. and Zapport, F., "Techniques for investigating integrated circuit dielectric isolation media", IEEE Trans. Nuclear Sci., NS-15, 140-53 (1968), EEA72-9206.

147. Hatcher, O. W., Jr., "Diffused isolated resistor for integrated circuit", Patent USA 3432792, Publ. March 1969, EEA72-17162.
148. Komiya, H., Abe, H., Sonobe, Y. and Matsui, H., "The application of gas plasma to the fabrication of MOS-LSI", 1973 International Electron Devices Meeting Technical Digest, p. 459-62, EEA77-14620.
149. Lee, F. H., "Dielectrically isolated saturating circuits", IEEE Trans. Electron Devices, ED-15, 645-50 (1968), EEA71-28962.
150. Lepselter, M. P., "Semiconductor integrated circuitry with improved isolation between active and passive elements", Patent USA 3411048, Publ. November 1968, EEA72-10650.
151. Marshall, S., Zeto, R. J. and Thornton, C. G., "Dry pressure local oxidation of silicon for IC isolation", J. Electrochem. Soc., 122, 1411-12 (1975).
152. Maxwell, D. A., Beeson, R. H. and Allison, D. F., "The minimization of parasitics in integrated circuits by dielectric isolation", IEEE Trans. Electron Devices, ED-12, 20-4 (1965), EEA70-10093.
153. Moriyama, H. and Tachi, S., "Semiconductor device", Patent USA 3891469, Publ. June 1975, CA83-171674.
154. Neamen, D., Shedd, W. and Buchanan, B., "Effects of ionizing radiation of dielectrically isolated junction field effect transistors", IEEE Trans. Nucl. Sci., NS-19, 400-5 (1972), EEA76-12644.
155. Panousis, P. T. and Schneider, M., "High-pressure steam apparatus for the accelerated oxidation of silicon", Presented at the Electrochemical Society Meeting, Chicago, Illinois, May 1973, EEA76-23487.
156. Powell, R. J., Ligenza, J. R. and Schenider, M. S., "Selective oxidation of silicon in low temperature high pressure steam", IEEE Trans. Electron Devices, ED-21, 636 (1974).
157. Rakowski, D. W., "Oxide, boron isolation", IBM Tech. Disclosure Bull., 15, 3782 (1973).
158. Skaggs, F. L., Harrap, V. and Bean, K. E., "Forming dielectrically isolated silicon diode array vidicon target", Patent USA 3810796, Publ. May 1974, CA81-18373.
159. Sikina, T. V., Joyce, B. D., Formigoni, N. P., Roe, C. C. and Schliesing, F., "A noval oxide isolation process", Presented at the International Electron Devices Meeting, Washington, D. C., October 1966.
160. Skouson, G. W., "Isolation of components on silicon wafer using SiO", Report NASA-CR-580, Westinghouse Defense and Space Center, Baltimore, Md., 1969.
161. Wagner, S. and Povilonis, E. I., "Diffusion of gallium through silicon dioxide films into silicon", J. Electrochem. Soc., 121, 1487 (1974).
162. Zeto, R. J., Thornton, C. G., Hryckowian, E. and Bosco, C. D., "Low temperature thermal oxidation of silicon by dry oxygen pressure above 1 Atm", J. Electrochem. Soc., 122, 1409-10 (1975).

2. Air Gap

163. Allison, D. F., "Semiconductor isolation method utilizing anisotropic etching and differential thermal oxidation", Patent USA 3796612, Publ. March 1974.
164. Allison, D. F., "Semiconductor structure with dielectric and air isolation", Patent USA 3815223, Publ. June 1974, CA81-43022.
165. Allison, D. F., Youmans, A. P. and Wong, T. H., "New etchant puts dielectric isolation in the groove", Electronics, 42 (10), 112-5 (1968), EEA72-12802.
166. Anon., "New semiconductor memory device", Bell Lab. Record, 48 (5), 156 (1970).
167. Bean, K., Crabbe, J., Malone, F. and Hoffman, J., "An improved radiation hardened JFET array produced via innovative dielectric isolation and device processing", Presented at the Electrochemical Society Meeting, Chicago, Illinois, May 1973, EEA76-23458; Also: Semiconductor Silicon 1973, p. 880.
168. Bean, K. E. and Gleim, P. S., "Semiconductor devices, as dielectrically isolated integrated circuits", Patent S. Africa 69-06663, Publ. April 1970, CA73-92757.
169. Clarke, R. N. and Hejmanovwski, R. E., "Electrically isolated semiconductor devices", Patent USA 3875657, Publ. April 1975.
170. Clausen, H. and Rusert, R. B., "Isolation problems get an airing", Electronics, 41 (9), 75-80 (1968), EEA71-15891.
171. Coughlin, J. B., Gelsing, R. J. H., Jochems, P. J. W. and van der Laak, H. J. M., "A monolithic silicon wideband amplifier from DC to 1 GHz", IEEE J. Solid-State Circuits, SC-8, 414-9 (1973), EEA77-5442.
172. Declercq, M., "Application of the anisotropic etching of silicon to the development of complementary structures", Presented at the International Electron Devices Meeting, Washington, D. C., December 1974.
173. Declercq, M. J., Gerzberg, L. and Meindl, J. D., "Optimization of the hydrazine-water solution for anisotropic etching of silicon in integrated circuit technology", J. Electrochem. Soc., 122, 545-52 (1975), EEA78-24477.
174. Formigoni, N. P., "Fabrication of semiconductor integrated circuits", Patent USA 3411200, Publ. November 1968, EEA72-6212.
175. Holmes, F. E. and Salama, C. A. T., "V groove M.O.S. transistor technology", Electron. Lett., 9, 457-8 (1973), EEA77-2598.
176. Holmes, F. E. and Salama, C. A. T., "VMOS - a new MOS technology", Solid-State Electron., 17, 791-7 (1974), CA81-83485.
177. Holmes, F. E. and Salama, C. A. T., "A V-groove oxide isolated bipolar bucket brigade shift register", Solid-State Electron., 17, 1193-200 (1974), EEA78-684.
178. Kuhn, G. L., "Method for filling grooves and moats used on semiconductor devices", Patent USA 3892608, Publ. July 1975.

179. Lawrence, J., May, J. L. and Gordon, J. R., "Isolation of semiconductor devices", Patent Canada 962372, Publ. February 1975, CA83-20796.
180. Mok, T. D. and Salama, C. A. T., "V-shaped notched-channel field-effect transistor", Electron. Lett., 10, 478-80 (1974), EEA78-2870.
181. Mudge, J. and Taft, K., "V-ATE memory scores a new high in combining speed and bit density", Electronics, 45 (15), 65-9 (1972), EEA75-27735.
182. Rodgers, T. J. and Meindl, J. D., "Epitaxial V-groove bipolar integrated circuit process", IEEE Trans. Electron Devices, ED-20, 226-32 (1973), EEA76-16470.
183. Rodgers, T. J. and Meindl, J. D., "VMOS: high-speed TTL compatible MOS logic", IEEE J. Solid-State Circuits, SC-9, 239 (1974).
184. Ronchi, R. and Vernes, P., "Variable sputtering rates for planarization of V groove m.o.s. transistors", IBM Tech. Disclosure Bull., 17, 2349-50 (1975), EEA78-19906.
185. Rosvold, W. C. and Legat, W. H., "Air gap isolated microcircuits (Beam lead devices)", Presented at the International Electron Devices Meeting, Washington, D. C., October 1967.
186. Ruggiero, E. M., "Thermal displays using air-isolated integrated circuits and methods of making", Patent USA 3762038, Publ. October 1973.
187. Schaible, P. M. and Zielinski, L. B., "Forming sidewall dielectric isolation of integrated circuit devices", IBM Tech. Disclosure Bull., 17, 2893-4 (1975), EEA78-19918.
188. Stoller, A. L. and Schlip, W. J., Jr., "Isolation techniques for fabricating integrated circuits. II. Dielectric refill techniques and decal air isolation", RCA Rev., 29, 485-506 (1968), EEA72-10644.
189. Waggener, H. A., Kragness, R. C. and Tyler, A. L., "Anisotropic etching for forming isolation slots in silicon beam leaded integrated circuits", Presented at the International Electron Devices Meeting, Washington, D. C., October 1967.

3. Polysilicon

190. Gaind, A. K., Kozul, A. A. and Palagonia, A. M., "Doped polysilicon isolation process", IBM Tech. Disclosure Bull., 16, 2871 (1974), EEA77-23133.
191. Halsor, J. L. and Lin, H. C., "A method for polycrystalline silicon delineation applicable to a double-diffused MOS transistor", Solid State Technol., 18 (4), 71 (1975), EEA78-24362.
192. Iwamatsu, S., Meguro, S. and Shimizu, S., "A new isolation structure for high density LSI", 1973 International Electron Devices Meeting Technical Digest, p. 244-7, EEA77-14617.
193. Kamins, T. I., "Deformation occurring during the deposition of polycrystalline-silicon films", J. Electrochem. Soc., 121, 681-4 (1974), PA77-57701.

194.Kamins, T. I., "Obtaining high-quality thick films of polycrystalline silicon from dielectric isolation", Patent USA 3862852, Publ. January 1975, CA82-179666.
195.Kesavan, R., Singh, M. N. and Murthy, B. S., "Double poly process of dielectric isolation for integrated circuits", Indian J. Technol., 9 (2), 44-6 (1971), EEA74-33360.
196.King, F. D., Shewchun, J., Thompson, D. A., Barber, H. D. and Pieczonka, W. A., "Polycrystalline silicon resistors for integrated circuits", Solid-State Electron., 16, 701-8 (1973), EEA76-23501.
197.Kobayashi, I., "A new technology for high-power IC", IEEE Trans. Electron Devices, ED-18, 45-9 (1971), EEA74-6936.
198.Kobayashi, I., "Technology for monolithic high-power integrated circuits using polycrystalline Si for collector and isolation walls", IEEE Trans. Electron Devices, ED-20, 399-404 (1973), EEA76-19957.
199.Magdo, I. and Magdo, S., "Fabricating complementary MOS transistors", IBM Tech. Disclosure Bull., 15, 1767 (1972).
200.Naiman, M. L., "Polycrystalline silicon isolation integrated circuits", Presented at the Electrochemical Society Meeting, San Francisco, Calif., May 1974, EEA77-23142.
201.Stein, L., "Providing polycrystalline silicon regions in monolithic integrated circuits", Patent USA 3730765, Publ. May 1973, CA79-36435.
202.Sumitomo, Y., Niwa, K., Sawazaki, H. and Sakai, K., "A new technology for dielectric isolation", Presented at the Electrochemical Society Meeting, Chicago, Illinois, May 1973, EEA76-23497; Also: Semiconductor Silicon 1973, p. 893.

4. Glass and Ceramics

203.Price, W. L. and Fordenwalt, J. N., "A glass dielectric isolation technique", Presented at the International Electron Devices Meeting, Washington, D. C., October 1966.
204.Ramsey, T. H. and Smith, T., "A process for ceramic dielectric isolation for integrated circuits", Semiconductor Prod. Solid State Technol., 10 (4), 35-40 (1967), EEA70-14818.
205.Ramsey, T. H. and Smith, T., "A process for ceramic dielectric isolation for integrated circuits", In: Integrated Circuit Technology, Lango, T. A., (Ed.), Boston Technical Publishers, Inc., Central Square, Cambridge, Mass., p. 97-102 (1970).
206.Ramsey, T. H., Jr., van Tassel, J. H. and Abraham, R. P., "Method for the fabrication of integrated circuits", Patent USA 3453722, Publ. July 1969, EEA73-9572.

5. Others

207. Amick, J. A. and Fisher, A. W., "Isolation techniques for fabricating integrated circuits. I. Laminate substrates", RCA Rev., 29, 475-84 (1968), EEA72-10643.

208. Azarewicz, J. L. and Wrobel, T. F., "Radiation-stimulated failure mechanism in a dielectrically isolated integrated circuit", IEEE Trans. Parts, Hybrids & Packag., PHP-10, 159-64 (1974), EEA78-2907.

209. Bean, K. E., "Double isolation for electronic devices", Patent USA 3769562, Publ. October 1973, CA80-65532.

210. Bean, K. E., "Dielectric isolation", Patent USA 3884733, Publ. May 1975, CA83-107213.

211. Bingham, K. C. and Cuttell, W. R., "Dielectric isolation layers for use in large scale integrated systems", Proceedings Seventeenth Electronic Components Conference, 1967, p. 371-6, EEA71-10970.

212. Brack, K., Ghafghaichi, M. and Wimpfheimer, H., "Integrated-circuit structure electrically isolated by a combination of amorphous silicon walls and isolating pn junction", Off. Gaz., 918 (1), 105 (1974).

213. Broadie, R. W. and Kemlage, B. M., "Dielectric isolation", IBM Tech. Disclosure Bull., 16, 2869 (1974), EEA74-23132.

214. Cook, C. and Scherrer, R. E., "Anodic aluminum isolation (A^2I)", International Electron Devices Meeting, Technical Digest, p. 266-9 (1974).

215. Dash, S., "Semiconductor structure having islands isolated by moats", Patent USA 3901737, Publ. August 1975, CA83-171684.

216. Frescura, B. L., Roger, R. and Schroeder, J., "Mesa isolation - an isolation technique for integrated circuits", Presented at the International Electron Devices Meeting, Washington, D. C., October 1966.

217. Hitachi, Ltd., "Semiconductor device isolation", Patent UK 1362512, Publ. August 1974, CA81-178927.

218. Maddox, E. D., "Current steering chip upgrades performance of D-A converters", Electronics, 47 (7), 125 (1974).

219. McWilliams, D., Fa, C., Larchian, G. and Maxwell, O., "A new dielectric isolation technique for high quality silicon integrated circuits", Presented at the Spring Meeting of the Electrochem. Soc., Toronto, Canada, July 1964.

220. McWilliams, D. A., Fa, C. H., Larchian, G. A. and Maxwell, O. F., Jr., "Electrically isolated semiconductor devices on a common crystalline substrate", Patent Canada 942893, Publ. February 1974, CA82-67310.

221. Oberlin, D. W., "Complementary bipolar transistors for monolithic structures", Presented at the International Electron Devices Meeting, Washington, D. C., October 1968.

222.Payne, M. I. and Husni, S. Y., "Dielectrically isolated integrated circuits using metallic film pockets", Presented at the International Electron Devices Meeting, Washington, D. C., December 1973.
223.Potter, M. D., "Isolated fixed and variable threshold field effect transistor fabrication", Patent USA 3900352, Publ. August 1975, CA83-171682.
224.Sawazaki, H., Sakai, K., Tsutsumi, H., Sumitomo, Y. and others, "Method for manufacturing a semiconductor integrated circuit isolated through dielectric material", Patent USA 3826699, Publ. July 1974.
225.Stehlin, R. A., Dexter, R. J., Kendall, D. L. and Pankratz, J. M., "Dielectrically isolated semiconductor structures", Patent USA 3897274, Publ. July 1975, CA83-124984.
226.Tokyo Shibaura Electric Co., "Method for manufacturing a semiconductor integrated circuit isolated by dielectric material", Patent UK 1304643, Publ. January 1973.
227.Witteles, A. A. and Eisenberg, M. L., "A method for an accurate prediction of degradation characteristics and quality assurance of a 709 amplifier", IEEE Trans. Nucl. Sci., NS-17, 173-7 (1970), EEA74-21346.

V. Other Techniques

228.Bae, M. S., "Low pressure low temperature lamination of handle wafers for dielectrically isolated integrated circuits", Report TN-1116, RCA, Princeton, N. J., 1975, 2 pp., EEA78-37672.
229.Bean, K. E., Gleim, P. S. and Runyan, W. R., "Semiconductor device isolation using silicon carbide", Patent USA 3838441, Publ. September 1974, CA81-143310.
230.Crosthwait, D. L., "Characterization of semiconductor processes", Proceedings of the 4th Annual Scanning Electron Microscope Symposium and Workshop on Forensic Applicationsof the Scanning Electron Microscope, 1971, p. 425-32, EEA74-26555.
231.Davies, E. D., "Passivation and isolation in semiconductor devices", Patent USA 535062, Publ. December 1974, CA83-171675.
232.Katsumura, H., Nakagawa, T. and Takabayashi, M., "Isolation in integrated circuits using preferential epitaxial growth technique", J. Inst. Elect. Commun. Engrs., 49, 683-8 (1966), Japanese, EEA69-7025.
233.Lin, H. C. and Jones, W. N., "Computer analysis of the double-diffused MOS transistors and integrated circuits", Presented at the 17th International Electron Devices Meeting, Washington, D. C., October 1971, EEA75-13774.
234.Naymik, D. A., "Silicon mosaic for integrated devices", IEEE Trans. Electron Devices, ED-12, 497-9 (1965), EEA69-2686.
235.Tsuchimoto, T., "Method for fabricating semiconductor device having semiconductor circuit element in isolated semiconductor region", Patent USA 3773566, Publ. November 1973.

236. Upadhyayula, L. C., Narayan, S. Y. and Douglas, E. C., "Fabrication of 3-terminal-transferred-electron logic devices by proton bombardment for device isolation", Electron. Lett., 11, 201-2 (1975), EEA78-33105.
237. Wiedmann, S. K., "High density bipolar memories and logic", 1973 IEEE International Convention and Exposition, Vol. VI, 30.4/2 pp., EEA76-2614.

I. DEVICE FABRICATION

I. Field-Effect Transistors

1. MOSFET

1. Abe, H., "The application of gas plasma to the fabrication of MOS LSI", Oyo Buturi, 44 (Suppl.), 287-95 (1975), EEA78-41686.
2. Adamic, J. W., Jr., "Increasing the field-inversion voltage of metal-oxide-on-silicon integrated circuits", Patent USA 3753806, Publ. August 1973, CA79-130633.
3. Alcott, G. J., "Self-aligned metal-oxide-semiconductor transistors", Patent USA 3910804, Publ. October 1975, CA83-201126.
4. Arita, S., "Method for manufacturing a MOS integrated circuit", Patent USA 3865650, Publ. February 1975, EEA78-33237.
5. Armstrong, W. E. and Toliver, D. L., "A scanning electron microscope investigation of glass flow in MOS integrated circuit fabrication", J. Electrochem. Soc., 121, 307-10 (1974), EEA77-14594.
6. Atalla, M. M., "Insulated-gate field effect transistor with electrostatic protection means", Patent USA 3413497, Publ. November 1968, EEA72-9602.
7. Bazin, B. and Albert, J., "Method of fabrication of insulated-gate field-effect transistors", Patent USA 3832248, Publ. August 1974, CA83-36597.
8. Beasom, J. D., "A process for simultaneous fabrication of vertical NPN and PNP's Nch, and Pch MOS devices", 1973 International Electron Devices Meeting Technical Digest, p. 41-3, EEA77-14610.
9. Becke, H. W. and White, J. P., "Gallium arsenide insulated gate field effect transistors", Proceedings International Symposium on Gallium Arsenide", 1966, p. 219-27, Publ. 1967, PA70-21048.
10. Beer, A. F., "Insulated gate field effect transistor with plural overlapped gates", Patent USA 3436623, Publ. April 1969, EEA73-4551.
11. Bell, M. S., "Flow line production of metal oxide silicon transistors", Compon. Technol., 4 (1), 14-17 (1969), EEA72-23829.
12. Bernacki, S. F. and Smith, H. I., "Fabrication of silicon MOS devices using X-ray lithography", IEEE Trans. Electron Devices, ED-22, 421-8 (1975), EEA78-24462.
13. Bogardus, E. H., Peressino, P. P. and Reith, T. M., "Silicon field-effect transistors", Patent French 2235483, Publ. January 1975, CA83-89840.
14. Brand, W. L. and Kashkooli, F., "Metal-oxide-semiconductor structure with increased field threshold", Patent USA 3787251, Publ. January 1974, CA80-75753.

15. Brown, M. A. C. S., Letchford, A., Millard, P. A. and Windle, D. J., "Metal-oxide-semiconductor transistors (MOST's)", R. R. E. Newsletter Res. Rev., no. 7, 3 pp., 1968, EEA72-13493.
16. Burt, D. J., "Permanent-channel insulated-gate field-effect transistor", Patent UK 1327597, Publ. August 1973, CA79-150806.
17. Carballo, R. A. and Ing, P. W., "Subtractive process for improving FET gate oxide formations", IBM Tech. Disclosure Bull., 17, 3598-9 (1975), EEA78-33155.
18. Coppen, P. J., "Metal-insulator-semiconductor integrated devices with high inversion voltage to threshold voltage ratios", Patent USA 3892609, Publ. July 1975, CA83-124978.
19. Curtis, T. E., "Metal-oxide-semiconductor transistors", Patent UK 1347849, Publ. February 1974, CA81-7616.
20. Dawson, R. H., Ditrick, N. H. and Mitchell, M. M., "Method of fabricating insulated-gate field-effect devices", Patent USA 3455020, Publ. July 1969, EEA73-10704.
21. Declerck, G. J., Hattori, T., May, G. A., Beaudouin, J. and Meindl, J. D., "Some effect of 'trichloroethylene oxidation' on the characteristics of MOS devices", J. Electrochem. Soc., 122, 436-9 (1975), EEA78-24184.
22. De LaMoneda, F. H., "Fabrication of small contacts to source and drain of IGFET's", IBM Tech. Disclosure Bull., 17, 2361-2 (1975), EEA78-19908.
23. De LaMoneda, F. H., "Three mask self-aligned IGFET fabrication process", IBM Tech. Disclosure Bull., 17, 2363-5 (1975), EEA78-19909.
24. Delivorias, P., "Development of p-channel enhancement MOS triodes", RCA Rev., 29, 630-9 (1968), EEA72-11332.
25. Dennard, R. H. and Spampinato, D. P., "Process for the formation of self-aligned silicon and aluminum gates", Patent USA 3834959, Publ. September 1974.
26. Diemer, G. and van der Maesen, F., "Insulated gate field-effect transistor with widening current path between source and drain", Patent USA 3436619, Publ. April 1969, EEA73-7479.
27. Diemer, G. and van der Maesen, F., "Tapered insulated gate field-effect transistor", Patent USA 3436620, Publ. April 1969, EEA73-4550.
28. Dingwall, A., "Integrated circuit components having a field-effect transistor with insulated gate electrode", Patent German 2436486, Publ. February 1975, CA83-20800.
29. Dunlop, L. J. and Lubart, N. D., "Notched channel IGFET", IBM Tech. Disclosure Bull., 18, 70 (1975), EEA78-37539.
30. Eriaky, G., "Some aspects of construction and fabrication of high frequency MOS transistors", Hiki, 10, 39-44 (1970), Hungarian, EEA72-38098.
31. Everist, J. A., "Processing techniques for low-power high-performance MOS", Des. Electron., 8 (7-8), 48-55 (1971), EEA72-18790.

32.Feldman, C. and Plachy, R., "Vacuum deposited silicon devices on fused silica substrates", J. Electrochem. Soc., 121, 685-8 (1974), PA77-57702.

33.Fouts, D. P., Hornung, A. and Kasprzak, L. A., "In situ clean-up oxidation of FET channel areas", IBM Tech. Disclosure Bull., 16, 3238 (1974), EEA77-30337.

34.Goodman, F. R., Jr., Heald, D. L., Neville, R. C. and Skalnik, J. G., "Teaching integrated-circuit fabrication to engineering undergraduates", Microelectronics, 6 (1), 25-32 (1974), EEA78-24480.

35.Gosney, W. M. and Buehler, M. G., "Metal-insulator-semiconductor structures having reduced junction capacitance", Patent USA 3852119, Publ. December 1974, CA82-79553.

36.Handy, R. M., "MOS semiconductor assembly", Patent German 2435892, Publ. February 1975, CA83-20802.

37.Hapke, M. and Helms, G., "A test structure for controlling the process of manufacturing MOS circuits", Nachrichtentech. Elektron., 23, 393-4 (1973), German, EEA77-2607.

38.Harbison, D. R., Hartman, J. A., McDade, P. J. and Shoemake, G. E., "Nickel-chromium + gold metallization for MOS devices", 1973 International Electron Devices Meeting Technical Digest, p. 363-6, EEA77-14557.

39.Hayashi, Y. and Tarui, Y., "High-frequency insulated-gate field-effect transistor with protective diodes", Patent USA 3798514, Publ. March 1974.

40.Heiman, F. P. and Zaininger, K. H., "Semiconductor device fabrication", Patent USA 3411199, Publ. November 1968, EEA72-6560.

41.Heyerdahl, N. E., "Method of making insulated-gate field-effect transistor", Patent USA 3775262, Publ. November 1973.

42.Hilbourne, R. A., "Insulated gate field effect transistor with channel portions of different conductivity", Patent USA 3430112, Publ. February 1969, EEA73-1210.

43.Hitachi, Ltd., "Field-effect transistor", Patent UK 1322933, Publ. July 1973, CA79-98469.

44.Hofstein, S. R., "Insulated gate field effect transistor", Patent USA 3434021, Publ. March 1969, EEA73-7476.

45.Hubar, R. J., Smith, K. F., Fordemwalt, J. N. and Hanson, J. W., "Method for fabricating MOS devices with a multiplicity of thresholds on a single semiconductor substrate", Patent USA 3868274, Publ. February 1975.

46.Hughes Aircraft Co., "Insulated gate field effect transistor", Patent UK 1139623, Publ. January 1969, EEA72-9603.

47.Ilieva, M., Kamenova, M. and Popova, A., "MOS structures with 'chlorine' grown oxide and tungsten-gold metallization", Thin Solid Films, 30, 281-6 (1975).

48.Imabun, Y., Okuda, M. and Yoshida, K., "Preparation of variable capacitance MOS diode and its characteristics", Bull. Univ. Osaka Prefecture A, 18 (1), 89-86 (1969), EEA73-25043.

49. Imabun, Y., Okuda, M. and Yoshida, K., "MOS diodes with thin oxide layers", Elec. Eng. Jap., 91, 159-68 (1971), EEA75-24353.

50. Kaiser, R., "Methods of producing field-effect transistors having insulated control electrodes", Patent USA 3818582, Publ. June 1974.

51. Kakizaki, K., Kurakami, O. and Inoue, Y., "Method of producing insulated-gate field-effect transistors", Patent USA 3764413, Publ. October 1973, CA79-150798.

52. Kaplan, L. H., Montillo, F. J. and Pastorius, D. L., "Fabrication of self-aligned IGFET devices with continuous oxide in thin and thick regions", IBM Tech. Disclosure Bull., 14, 1568-9 (1971), EEA75-7544.

53. Kerwin, R. E., Klein, D. L. and Sarace, J. C., "Method for making MIS structures", Patent USA 3475234, Publ. October 1969, EEA73-10700.

54. Kim, M. J., "MOSFET fabrication problems", Solid-State Electron., 12, 557-71 (1969), EEA72-23835.

55. Kircher, C. J. and Yu, H. N., "Fabricating a gate field-effect transistor", IBM Tech. Disclosure Bull., 13, 646-8 (1970), EEA74-1609.

56. Kmeta, W. J. and Nadan, J. S., "Theory and practice of MOS processing techniques - 1972", IEEE Trans. Manuf. Technol., MFT-2, 37-49 (1973), EEA77-23052.

57. Komiya, H., Abe, H., Sonobe, Y. and Matsui, H., "The application of gas plasma to the fabrication of MOS-LSI", 1973 International Electron Devices Meeting Technical Digest, p. 459-62, EEA77-14620.

58. Kooi, E., "Insulated-gate field-effect transistor", Patent USA 3752711, Publ. August 1973, CA79-98487.

59. Kump, H. J., Lambert, L. M. and Quinn, R. M., "High-performance field-effect transistor", Patent USA 3814992, Publ. June 1974, CA81-31159.

60. Lealerc, P., "Mosaic of MOS-type semiconductor elements", Patent USA 3775616, Publ. November 1973.

61. Lin, H. C., "Planar graded channel MOS transistor", Patent USA 3883372, Publ. May 1975, CA83-107214.

62. Madani, N. and Nichols, K. G., "The use of a carbon-dioxide laser for the deposition of MOS diodes", Israel J. Technol., 9, 245-8 (1971), EEA74-37304.

63. Middelhoek, J. and Holleman, J., "The prevention of parasitic MOS channels by 'Silox' diffusion", 1972 European Solid State Devices Research Conference, p. 201, Publ. 1973, EEA76-32200.

64. Mullard Ltd., "Insulated gate FET", Patent UK 1145092, Publ. March 1969, EEA72-20443.

65. Mullard Ltd., "Insulated gate field effect transistor", Patent UK 1153428, Publ. May 1969, EEA73-1213.

66. Mullard Ltd., "Insulated gate FET", Patent UK 1153463, Publ. May 1969, EEA73-1214.

67. Norris, P. E. and Shaw, J. M., "Method of making MOS transistors", Patent USA 3766637, Publ. October 1973.

68. Ohwada, A., "Silicon n-channel normally off MOSFET", Toshiba Rev., no. 39, 44-8 (1968), EEA72-13496.
69. Okada, T., Yamagishi, S., Kamoshida, M., Satake, T. and Kishi, S., "Method of making an ohmic contact with a semiconductor substrate", Patent USA 3887993, Publ. June 1975, EEA78-41632.
70. Okumura, T., Tsuchitani, A. and Hasegawa, T., "Insulated-gate field-effect transistor with punch-through effect element", Patent USA 3764864, Publ. October 1973.
71. Parmer, W. F., "Field-effect transistors", Patent USA 3436281, Publ. April 1969, EEA73-1219.
72. Pasztor, G., "New fabrication methods and structures in the field of MOS devices", Hiradstech. Ipari Kut. Intez. Kozl., 13 (3), 5-30 (1973), Hungarian, EEA77-18806.
73. Philco-Ford Corp., "MIS transistor", Patent UK 1153252, Publ. May 1969, EEA70-1212.
74. Pickar, K. A. and Thibault, L. R., "The fabrication of IGFET's using electron beam technology", J. Vac. Sci. & Technol., 10, 1074-7 (1973), EEA77-23057.
75. Platt, S., "Read-write to read-only memory conversion", IBM Tech. Disclosure Bull., 17, 3543 (1975), EEA78-28859.
76. Postolache, C. and Bulucea, C., "The planar/MOS progress for obtaining metal-oxide-semiconductor (MOS) structures of minimum bulk and surface defect densities", Stud. & Cercet. Fiz., 26, 591-604 (1974), Rumanian, PA78-18690.
77. Sakurani, S. and Fujikawa, N., "Electrodes for n-type metal oxide semiconductors", Patent Japan 74-54892, Publ. May 1974, CA82-37938.
78. Sato, S. and Yamaguchi, T., "Insulated gate field effect transistors", Patent UK 1400780, Publ. July 1975, CA83-156875.
79. Seiter, G., "Development of metal oxide silicon transistors", Internat. Elektronische Rundschau, 20, 147-8 (1966), German, EEA70-9117.
80. Shappir, J., Margalit, S. and Kidron, I., "p-channel MOS transistor in indium antimonide", IEEE Trans. Electron Devices, ED-22, 960-1 (1975).
81. Shimakura, K. and Tsunemitsu, H., "Insulated-gate field-effect semiconductor device having low and stable gate threshold voltage", Patent USA 3806778, Publ. April 1974.
82. Sihvonen, Y. T., Parker, S. G. and Boyd, D. R., "Printable insulated-gate field-effect transistors", J. Electrochem. Soc., 114, 96-102 (1967), EEA70-9101.
83. Soderman, L., "Mechanism of MOS structures. III. MOS production processes are many and varied", Eltek. Aktuell Elektron., 17 (2), 30-2, 34 (1974), Swedish, EEA77-26468.
84. Steinmaier, W., "IGFET devices having out-diffused regions", Patent USA 3767487, Publ. October 1973, CA80-31783.
85. Takahashi, S., "Method of producing insulated-gate field-effect semiconductor device having a channel stopper region", Patent USA 3825455, Publ. July 1974.

86.Takahata, K. and Shiba, H., "Method of fabricating a semiconductor device", Patent USA 3864217, Publ. February 1975, EEA78-37339.
87.Tarui, Y., Hayashi, Y. and Sekigawa, T., "Diffusion self-aligned MOST. New approach for high speed devices", Proceedings of the 1st Conference on Solid State Devices, 1969, p. 105-10, Publ. 1970, CA82-67199.
88.Tarui, Y., Hayashi, Y. and Tanaka, K., "Fabrication of the MOST-IC", Bull. Electrotech. Lab., 31, 35-46 (1967), Japanese, EEA71-8591.
89.Terman, L. M., "Aluminum-silicon self-aligned gate process with field shield", IBM Tech. Disclosure Bull., 15, 2852-4 (1973), EEA76-19930.
90.Tomes, D. W., "Improving metal oxide silicon transistors", Compon. Technol., 4 (1), 12-13 (1969), EEA72-23828.
91.Trindade, D. C., "Prediction of failure rate of field-effect transistors", IBM Tech. Disclosure Bull., 16, 3387 (1974), EEA77-30339.
92.Wang, R. C., "Insulated gate field-effect transistor", Patent USA 3886583, Publ. May 1975, CA83-107223.
93.Wilenken, R. N. and Pearce, L. B., "Metal-silicon field-effect transistor", Patent German 2415736, Publ. October 1974, CA82-67303.
94.Yasuda, Y., Hirabayashi, K. and Moriya, T., "Growth and properties of polycrystalline silicon films as applied to MOS-IC technology", Oyo Buturi, 43 (Suppl.), 400-7 (1974), EEA78-5370.
95.Yu, H. N., "Fabrication of self aligned high-frequency MOSFET or MISFET", IBM Tech. Disclosure Bull., 14, 244-5 (1971), EEA74-34085.

2. CMOSFET

96.Antipov, I. and Edel, W. A., "Fabricating complementary field-effect transistor devices", IBM Tech. Disclosure Bull., 27, 1327-8 (1974), EEA78-11654.
97.Athanas, T. G., "Development of COS/MOS technology", Solid State Technol., 17 (6), 54-9 (1974), EEA77-34103.
98.Beer, A., "Complementary m.o.s. l.s.i. technology", New Electron., 8 (4), 74-5 (1975), EEA78-24491.
99.Beyer, K. D. and Yeh, T. H., "Fabricating complementary field-effect transistor devices", IBM Tech. Disclosure Bull., 16, 2876-7 (1974), EEA77-23049.
100.Carbajal, B. G., Gosney, W. M. and Hall, L. H., "Complementary insulated gate field effect devices", Patent USA 3673679, Publ. July 1972, EEA75-858.
101.Declereq, M. J., "A new C-MOS technology using etching of silicon", IEEE J. Solid-State Circuits, SC-10, 191-7 (1975), EEA78-33207.

102. Gaensslen, F. H. and Terman, L. M., "Complementary MOSFET device structure", Patent USA 3840888, Publ. October 1974, CA81-178945.
103. Gosney, W. M. and Hall, L. H., "The extension of self-registered gate and doped-oxide diffusion technology to the fabrication of complementary MOS transistors", IEEE Trans. Electron Devices, ED-20, 469-73 (1973), EEA76-19927.
104. Gregor, L. V., "Producing complementary IGFET devices", IBM Tech. Disclosure Bull., 12, 2062-3 (1970), EEA73-35231.
105. Herman, D. S., Schuster, M. A. and Oehler, H. G., "Multilayer metallization process for CMOSFET devices", Extended Abstracts of the Spring Meeting of the Electrochemical Society, 1971, p. 167-9, EEA75-4600.
106. Holle, E. and Nochel, J., "COS/MOS technology", Electronik, 20 (4), 111-16 (1971), German, EEA74-17485.
107. Kim, M. J., "Complementary MIS FET devices and method of fabrication", Patent UK 3646665, Publ. March 1972, EEA75-36315.
108. Klein, T., "Technology and performance of integrated complementary MOS circuits", IEEE J. Solid-State Circuits, SC-4, 122-30 (1969), EEA73-1198.
109. Shieh, J. M., "Fabrication of complementary MOST by triple diffusion", AEU, 6 (3), 28-9 (1973), EEA78-2867.
110. Sopira, M. M., Malmberg, P. R. and Meiksin, Z. H., "Fabrication of integrated CMOS transistors using electron lithography and ion implantation", J. Vac. Sci. & Technol., 10, 1086-9 (1973), EEA77-23148.
111. Tihanyi, J., "Production of thin layer complementary channel MOS circuits", Patent USA 3859716, Publ. January 1975, EEA78-28817.
112. Wang, R. C., "Method of producing low-threshold complementary insulated-gate field-effect devices", Patent USA 3759763, Publ. September 1973.
113. Westinghouse Electric Corp., "Monolithic complementary MOSFET device", Patent UK 1370255, Publ. October 1974, EEA78-11738.

3. MNOSFET

114. Abbink, H. C. and Goodwin, N. W., "Method of forming a stable MNOS IGFET", Patent USA 3800411, Publ. April 1974, CA80-138491.
115. Artamonov, V. V., Tikhov, S. V., Orlov, O. A. and Eingorin, M. Y., "Memory phenomenon in MNOS structures with silicon nitride film prepared by reactive sputtering", Radiotekh. Electron., 19, 1323-5 (1974), Russian, CA81-143199.
116. Kenyon, R. A. and Yun, B. H., "Density improvement in MNOS array by overlapping gate electrodes", IBM Tech. Disclosure Bull., 18, 68-9 (1975), EEA78-37628.
117. Kim, M. J., "Method of making a metal-insulator-silicon field-effect transistor (MISFET) memory device and the product", Patent USA 3853496, Publ. December 1974.

118. Krick, P. J., "MNOS memory array fabricated on an insulating substrate", IBM Tech. Disclosure Bull., 15, 466-7 (1972), EEA75-39102.
119. Mroczek, W. and Scherber, W., "Method of manufacturing an (Si_3N_4) planar device", Patent USA 3798062, Publ. March 1974.

II. Charge-Coupled Devices

120. Ablassmeier, U., "Improved charge transfer in p-channel charge coupled devices with implanted gaps", 1972 European Solid State Devices Research Conference, p. 175-6, Publ. 1973, EEA76-32279.
121. Amelio, G. F., Krambeck, R. H. and Pickar, K. A., "Charge-coupled devices employing nonuniform concentrations of immobile charge along the information channel", Patent USA 3796932, Publ. March 1974.
122. Amelio, G. F. and Salsbury, P. J., "Unipolar charge-coupled semiconductor structure", Patent USA 3836409, Publ. September 1974, CA81-143303.
123. Ando, T. and Hirata, Y., "Charge transfer device", Patent USA 3848328, Publ. November 1974.
124. Anon., "Two-phase self-aligned MNOS charge transfer device", Kozl. Magy. Tud. Akad. Musz. Fiz. Kut. Intez., no. 0-14, 31-8 (1974), Hungarian, EEA78-2873.
125. Arnett, P. C., Heller, L. G. and Stapper, C. J., "Single-phase charge-coupled semiconductor device", Patent USA 3796933, Publ. March 1974, CA80-126041.
126. Baker, I. M., Beynon, J. D. E. and Copeland, M. A., "Charge-coupled devices with submicron electrode separations", Electron. Lett., 9, 48-9 (1973), EEA76-12638.
127. Berglund, C. N., Clemens, J. T. and Nicollian, E. H., "Undercut isolation - a technique for closely spaced and self-aligned metallization patterns for MOS integrated circuits", J. Electrochem. Soc., 120, 1255-60 (1973), EEA77-2605.
128. Berglund, C. N. and Strain, R. J., "Fabrication and performance considerations of charge-transfer dynamic shift registers", Bell Syst. Tech. J., 51, 655-703 (1972), EEA75-19122.
129. Bertram, W. J., Kosicki, B. B., Morris, F. J., Sealer, D. A., Sequin, C. H., Shankoff, T. A. and Tompsett, M. F., "3-phase charge coupled devices using 3 levels of polysilicon", Presented at 1973 IEDM Supplement, Washington, D. C., December, EEA77-14565.
130. Bertram, W. J., Mohsen, A. M., Morris, F. J., Sealer, D. A., Sequin, C. H. and Tompsett, M. F., "A three-level metallization three-phase CCD", IEEE Trans. Electron Devices, ED-21, 758 (1974).
131. Beynon, J. D. E., Haken, R. A. and Baker, I. M., "Self-aligned gaps - an alternative technique for closely spaced metallization patterns", J. Electrochem. Soc., 121, 1124 (1974), EEA77-38887.
132. Borel, J., Lacour, J. and Merckel, G., "Charge-coupled device and method of fabrication of the device", Patent USA 3829884, Publ. August 1974.

133. Bower, R. W., Zimmerman, T. A. and Mohsen, A. M., "A high density overlapping gate charge coupled device array", 1973 IEDM Technical Digest, p. 30-2, EEA77-14607.
134. Bower, R. W., Zimmerman, T. A. and Mohsen, A. M., "The two-phase offset gate CCD", IEEE Trans. Electron Devices, ED-22, 70-2 (1975), EEA78-15352.
135. Bower, R. W., Zimmerman, T. A. and Mohsen, A. M., "Performance characteristics of the offset gate charge-coupled device", IEEE Trans. Electron Devices, ED-22, 72-4 (1975), EEA78-15353.
136. Browne, V. A. and Perkins, K. D., "Buried channel CCD's with submicron electrode spacings", International Conference on Technology and Applications of Charge Coupled Devices, 1974, p. 100-5, EEA78-11684.
137. Buss, D. D. and Reinberg, A. R., "Charge-coupled devices", Patent USA 3914127, Publ. October 1975, CA83-211934.
138. Chan, Y. T., "Fabrication and operation of CCD structure with silicon layers grown on sapphire substrates", 1973 IEDM Technical Digest, p. 469-72, EEA77-14563.
139. Collins, D. R., Shortes, S. R., McMahon, W. R., Bracken, R. C. and Penn, T. C., "Charge-coupled devices fabricated using aluminum-anodized aluminum-aluminum double-level metallization", J. Electrochem. Soc., 120, 521-6 (1973), EEA76-23459.
140. Erb, D. M., Kotyczka, W., Su, S. C., Wang, C. and Clough, G., "An overlapping electrode buried channel CCD", 1973 IEDM Technical Digest, p. 24-5, EEA77-14537.
141. Fang, T. F. and Liu, H. K., "Fabrication and measurements of a silicon charge-coupled device", Proceedings of 1973 IEEE Southeast-Con., p. T-6/1-6, EEA76-29284.
142. Forbes, L. and Yeargan, J. R., "N-channel charge coupled device", Patent USA 3909925, Publ. October 1975, CA83-211929.
143. Gajda, W. T. J., "Electrode materials for charge coupled devices", Report N74-15435/2, Notre Dame University, Indiana, 1973, 31 pp..
144. Gelberger, P. P. and Salama, C. A. T., "Moat-etched two-phase charge-coupled devices", Solid-State Electron., 17, 301-5 (1974), EEA77-18816.
145. Goser, K. and Stein, K. U., "Charge-shift device in two-phase technology", Patent German 2341179, Publ. March 1975, CA83-52099.
146. Green, R. D. and Heimbigner, G. L., "Charge coupled semiconductor device", Patent UK 1376900, Publ. December 1974, CA83-20830.
147. Green, R. D. and Heimbigner, G. L., "Method of manufacturing control electrodes for charge coupled circuits and the like", Patent USA 3859717, Publ. January 1975, EEA78-28818.
148. Haken, R. A., Beynon, D. E. and Roberts, P. C. T., "Charge-coupled structures with self-aligned sub-micron gaps", IEEE Trans. Electron Devices, ED-22, 289-93 (1975), EEA78-19828.
149. Hashimoto, N. and Saito, N., "N-channel narrow gap charge-coupled devices using electron-beam lithography", Thin Solid Films, 27, 89-93 (197), EEA78-33159.

150. Ibrahim, A. and Sellars, L., "4096-bit charge coupled device serial memory array", 1973 IEDM Technical Digest, p. 141-3, EEA77-14614.

151. Itoh, Y., Sunami, H. and Kamigaki, Y., "Charge-coupled semiconductor device provided with biasing charges", Patent USA 3852801, Publ. December 1974.

152. Jaggers, K. A., "Bonding and interconnections for CCD's", Electron. Lett., 11, 161 (1975), EEA78-19884.

153. Kim, C. K. and Snow, E. H., "p-channel charge-coupled devices with resistive gate structure", Appl. Phys. Lett., 20, 514-6 (1972), EEA75-28505.

154. Kosonocky, W. F. and Carnes, J. E., "Two-phase charge-coupled devices", Report N73-16187, RCA, Princeton, N. J., 1973, 106 pp..

155. Kosonocky, W. F. and Carnes, J. E., "Design and performance of two-phase charge-coupled devices with overlapping polysilicon and aluminum gates", 1973 IEDM Technical Digest, p. 123-5, EEA77-14544.

156. Krambeck, R. H., Retajczyk, T. F., Silversmith, D. J. and Strain, R. J., "A 4160 bit C4D serial memory", Presented at 1973 IEDM Supplement, Washington, D. C., December, EEA77-14621.

157. Krambeck, R. H., Strain, R. J., Smith, G. E. and Pickar, K. A., "Conductively connected charge-coupled device", IEEE Trans. Electron Devices, ED-21, 70-2 (1974), EEA77-6032.

158. Krambeck, R. H., Walden, R. H. and Pickar, K. A., "Implanted-barrier two-phase charge-coupled device", Appl. Phys. Lett., 19, 520-2 (1971), EEA75-10305.

159. Krambeck, R. H., Walden, R. H. and Pickar, K. A., "A doped surface two-phase CCD", Bell Syst. Tech. J., 51, 1849-66 (1972), EEA75-39085.

160. Mavor, J., "Charge-coupled devices. II", Wireless World, 81, 13-17 (1975), EEA78-8583.

161. McCaughan, D. V., "The use of plasmas in CCD 'processing'", International Conference on Technology and Applications of Charge Coupled Devices, 1974, p. 281-9, EEA78-15375.

162. Mifune, T., Ochi, S., Kano, Y. and Shibita, A., "Improvement on structure of charge coupled devices", Proceedings of the 4th Conference on Solid State Devices, 1972, p. 207-11, Publ. 1973, CA81-143204.

163. Morris, F. J. and Shankoff, T. A., "Charge coupled device processing", Electrochemical Society Fall Meeting (Extended abstracts), 1974, p. 479-83, EEA78-37558.

164. Parks, C. M. and Salama, C. A. T., "V.M.O.S. conductively coupled charge-coupled device", Electron. Lett., 9, 593-4 (1973), EEA77-10255.

165. Parks, C. M. and Salama, C. A. T., "V-groove (VMOS) conductively connected charge coupled devices", Solid-State Electron., 18, 1061-7 (1975).

166. Schroder, D. K., "A two-phase germanium charge-coupled device", Appl. Phys. Lett., 25, 747-9 (1974), EEA78-11647.

167.Schuermeyer, F. L., Belt, R. A., Young, C. R. and Blasingame, J. M., "New structures for charge-coupled devices", Proc. IEEE, 61, 1444-5 (1972), EEA76-4665.
168.Shimizu, S., Iwamatsu, S. and Ono, M., "Charge-coupled device with buried channels under electrode gaps", Appl. Phys. Lett., 22, 286-7 (1973), EEA76-19923.
169.Theunissen, M. J. J. and Esser, L. J. M., "PCCD technology and performance", International Conference on Technology and Applications of Charge Coupled Devices, 1974, p. 106-13, EEA78-11685.
170.Troutman, R. R., "Self-aligning structure for charge-coupled and bucket-brigade devices", Patent USA 3810795, Publ. May 1974, CA81-31157.
171.Walsh, L. R., "Charge-coupled semiconductor device", Patent German 2439726, Publ. March 1975, CA83-52093.
172.Wolf, E. D., "Small geometry charge coupled device", Patent USA 3873371, Publ. March 1975.

III. Schottky Diodes

1. Surface Preparation

173.Adams, A. C. and Pruniaux, B. R., "Gallium arsenide surface film evaluation by ellipsometry and its effect on Schottky barriers", J. Electrochem. Soc., 120, 408-14 (1973), PA76-39239.
174.Baitinger, U., Illi, M. and Najmann, K., "Manufacturing high-frequency diodes", IBM Tech. Disclosure Bull., 13, 3159 (1971).
175.Broom, R. F., "Fabrication of metal-semiconductor diodes", IBM Tech. Disclosure Bull., 15, 2161 (1972).
176.Crowell, C. R. and Sze, S. M., "Surface barrier semiconductor translating device", Patent USA 3349297, Publ. October 1967.
177.DiLorenzo, J. V., "Technique for the fabrication of a millimeter wave beam lead Schottky barrier device", Patent USA 3762945, Publ. October 1973.
178.Goldberg, Y. A., Nasledov, D. N. and Tsarenkov, B. V., "Procedure for fabricating surface-barrier structures by chemical deposition of metals on the surface of a semiconductor", Instrum. & Exp. Tech., 14, 899-901 (1971), PA75-25892.
179.Goradia, C. P., "Surface states and the gold-n-silicon surface barrier", Thesis, Univ. of Oklahoma, 1968, 143 pp., Order No. 68-13264, EEA72-9560.
180.Gutknecht, P. and Strutt, M. J. O., "Aluminium Schottky barriers on sputter-etched silicon", Electron. Lett., 7, 298-9 (1971), EEA74-22632.
181.Gutknecht, P. and Strutt, M. J. O., "Thermally oxidized mesa Schottky barrier diodes", IEEE Trans. Electron Devices, ED-21, 172-3 (1974), EEA77-5972.
182.Harreis, H., "Deviation from Schottky's model for the capacity of metal contacts on clean semiconductor surfaces", Z. Phys., 243, 254-65 (1971), German, PA74-33727.

183.Heer, J. and Trinchera, J. A., "Surface gallium arsenide Schottky-type diode and method of fabricating same", Patent USA 3562606, Publ. February 1971, EEA74-15540.
184.Heime, K., Konig, U., Kohn, E. and Wortmann, A., "Very low resistance Ni-AuGe-Ni contacts to n-GaAs", Solid-State Electron., 17, 835-7 (1974), PA77-60860.
185.Hsieh, C. M., "Contact hole opening for fabricating a Schottky barrier diode", IBM Tech. Disclosure Bull., 17, 3287 (1975), EEA78-24310.
186.Iizuka, M., Fujiwara, S., Kano, G., Hasegawa, H., Teramoto, I. and Iwasa, H., "Method of making a Schottky barrier device", Patent USA 3752702, Publ. August 1973.
187.Itoh, Y., "Effects of silicon surface conditions on nickel-silicon contacts", Jap. J. Appl. Phys., 9, 926-30 (1970), EEA73-38022.
188.Jager, H., "Breakdown phenomena in Schottky diodes", Solid-State Electron., 12, 85-8 (1969), German, PA72-18634.
189.Kniepkamp, H., "Method of producing Schottky contacts", Patent USA 3813762, Publ. June 1974.
190.Lawerence, J. E. and Khidr, M. S., "Interface films in Si-Al systems", The Annual Reliability Physics Symposium, 1970, p. 78-9, EEA73-31337.
191.Lepoff, J. H. and Morris, R. A., "Schottky-barrier diodes structured for better high-frequency performance", Hewlett-Packard J., 24 (10), 2-6 (1973), EEA76-35294.
192.McCusker, J. H., "Method for fabricating thin film transistors", Patent USA 3298863, Publ. January 1967.
193.Murrmann, H., "Method of producing Schottky diodes", Patent USA 3846192, Publ. November 1974.
194.Narusawa, T., Komiya, S. and Hiraki, A., "Diffuse interface in Si (substrate) - Au (evaporated film) system", Appl. Phys. Lett., 22, 389-90 (1973), PA76-38751.
195.Nishitani, K., Ishihara, O. and Mitsui, S., "Control of the surface shape of GaAs Schottky barrier diodes by mesa-etching", Jap. J. Appl. Phys., 13, 1323-4 (1974), EEA77-36288.
196.Nobel, D. and Kock, H. G., "Schottky contact devices and method of manufacture", Patent USA 3775200, Publ. November 1973.
197.Okada, T. and Nakamura, M., "Semiconductor device with Schottky barrier formed on (100) plane of GaAs", Patent USA 3457473, Publ. July 1969.
198.Patwari, A. M. and Hartnagel, H. L., "Effect of surface properties on N-type gallium arsenide-nickel and gallium arsenide-aluminum Schottky diodes", Phys. Status Solidi A, 26, 469-75 (1974), CA82-79484.
199.Pawlowski, M. K., "Metal semiconductor diodes", Przeglad Elektron., 7, 365-75 (1966), Polish, PA70-7986.
200.Rosinski, W., "The use of ion implantation for the preparation of metal-semiconductor contacts", Z. Elektr. Inf. - Energietch., 2, 232-4 (1972), German, PA76-15646.

201.Saltich, J. L. and Rutledge, J. L., "Method for making a hot carrier pn-diode", Patent USA 3550260, Publ. December 1970.
202.Sato, Y., Uchida, M., Ishibashi, Y. and Araki, T., "Chip-type planar Schottky barrier diodes fabricated from selectively grown GaAs", Rev. Electr. Commun. Lab., 23, 535-40 (1975), EEA78-41690.
203.Schley, P. and Sperling, R., "The Schottky barrier - principles and technology", Wiss. Z. Tech. Hochsch. Ilmenau, 16, 121-5 (1970), German, EEA73-35201.
204.Shiraki, Y., Shimada, T. and Komatsubara, K. F., "Ion implantation of nitrogen into cadmium sulfide (Diode formation, photoelectric properties and EPR studies)", J. Appl. Phys., 710-8 (1972), EEA75-10127.
205.Sperling, R. and Schley, P., "Some studies of Schottky contacts", Z. Elektr. Inf. - & Energietch., 2, 237-8 (1972), German, PA76-15254.
206.Tove, P. A., Berg, S., Anderson, L. P. and Hyder, A., "Properties of ohmic and rectifying Schottky barriers fabricated by sputtering", Presented at 1970 Conference on Electron Device Techniques, New York, September, PA74-13213.
207.Weller, K. P., Dreeben, A. B., Davis, H. L. and Anderson, W. M., "Fabrication and performance of GaAs p^+n junction and Schottky barrier millimeter IMPATTs", IEEE Trans. Electron Devices, ED-21, 25-31 (1974), EEA77-5444.
208.Wetterau, L. C., Jr. and Owen, R. B., "Random whisker contact method for semiconductor devices", Patent USA 3500144, Publ. March 1970.
209.Wrixon, G. T. and Pease, R. F. W., "Schottky barrier diodes fabricated on epitaxial GaAs using electron beam lithography", Conference on Gallium Arsenide and Related Compounds, 1974, p. 55-60, Publ. 1975, EEA78-28763.
210.Giddings, J. J., Greer, S. E., Procter, R. W. and Schnitzel, R. H., "Control of silicon content in Schottky barrier diode metallurgy", IBM Tech. Disclosure Bull., 16, 615-6 (1973).
211.Heurle, F. M., Esaki, L. and Seki, H., "Schottky-barrier diode formed by sputter-deposition processes", Patent USA 3451912, Publ. June 1969, EEA73-10607.
212.IBM Corp., "Schottky barrier diode", Patent UK 1151643, Publ. May 1969, EEA73-1165.
213.Inoue, M., Kano, G., Suita, S., Matsuno, J. and Takayanagi, S., "Method for manufacturing semiconductor devices", Patent USA 3515583, Publ. June 1970.
214.Inoue, M., Kano, G., Suita, S., Matsuno, J. and Takayanagi, S., "Method of manufacturing semiconductor device", Patent USA 3519479, Publ. July 1970.
215.Kim, H. B., Sweeney, G. G. and Heng, T. M. S., "Analysis of metal-GaAs Schottky barrier diodes by secondary ion mass spectrometry", Conference on Gallium Arsenide and Related Compounds, 1974, p. 307-19, Publ. 1975, EEA78-28766.

216.Kniepkamp, H., "Thick Schottky-barrier contacts", Patent USA 3816270, Publ. June 1974, CA81-43023.
217.Kohler, R., "Vapour-deposited thin-film diodes - a comparison", Nachrichtentechnik, 20, 103-5 (1970), German, EEA73-21223.

2. Metal Deposition

218.Altman, C., Chapman, S. G. and Satya, A. V. S., "Metal-semiconductor junction devices", Patent UK 1401554, Publ. July 1975, CA83-156876.
219.Baker, D., "The metal-semiconductor contact in silicon devices", Conference on Metal-Semiconductor Contacts, 1974, p. 51-6, EEA78-8503.
220.Calviello, J. A., "Tantalum-gallium arsenide Schottky barrier semiconductor device", Patent USA 3886580, Publ. May 1975.
221.Chuss, J. T., "Metallization of selected regions of surfaces and products so formed", Patent USA 3415679, Publ. December 1968.
222.Coleman, D. J., Jr., Wisseman, W. R. and Shaw, D. W., "Reaction rates for Pt on GaAs", Appl. Phys. Lett., 24, 355-7 (1974), PA77-53604.
223.Cunningham, J. A. and Orr, C. D., "Sputtering apparatus for forming ohmic contacts for semiconductor devices", Patent USA 3844924, Publ. October 1974.
224.Daughton, J. M., "Vertical Schottky pad", IBM Tech. Disclosure Bull., 15, 261 (1972).
225.Dorbeck, F. H., "Electrochemically deposited Schottky contacts on GaAs", Solid-State Electron., 9, 1135-7 (1966), EEA70-2643.
226.Driver, M. C. and Geisler, M. J., "Method for making the self-aligned gate contact of a semiconductor device", Patent USA 3866310, Publ. February 1975, EEA78-33161.
227.Esaki, L. and d'Heurle, F., "Schottky diode", IBM Tech. Disclosure Bull., 11, 19 (1968), EEA71-28870.
228.Fujiwara, S., Matsuoka, S., Kano, G. and Yaegashi, Y., "Ultra-high-frequency Schottky barrier diodes with thermal and mechanical stability", Patent Japan 73-20948, Publ. June 1973, CA80-75730.
229.Korwin-Pawlowski, M. L. and Heasell, E. L., "Metal-InSb surface barriers", Conference on Metal-Semiconductor Contacts, 1974, p. 255-62, EEA78-8513.
230.Krikshtopaitis, I. B., "Deposition of gold films on germanium and silicon single crystals", Instrum. & Exp. Tech., 17, 1783 (1975), PA78-69604.
231.Kumar, V., "Reaction of sputtered Pt films on GaAs", J. Phys. & Chem. Solids, 36, 535-41 (1975), EEA78-24311.
232.Ladd, G. O., Jr. and Feucht, D. L., "GaAs Schottky barrier diodes by the fluxed wetting of tin and indium", Solid-State Electron., 13, 485-90 (1970), EEA73-24981.

233.Legat, W. H. and Tiefert, K. H., "Application of sputtering in the fabrication of semiconductor devices", Solid State Technol., 13 (12), 54-7 (1970), EEA74-12767.
234.McColl, M., Camber, W. A. and Miller, M. F., "Electron beam fabrication of submicrometer diameter mixer diodes for millimeter and submillimeter wavelengths", Proc. IEEE, 61, 1446-7 (1972), EEA76-4628.
235.Prakash, O., "Fabrication of gold-n-type gallium arsenide Schottky barrier varactor diodes", Indian J. Pure & Appl. Phys., 12, 7-12 (1974), PA78-3750.
236.Scholl, R. F. and Bleha, W. P., Jr., "Cadmium sulphide thin film sustained conductivity device with cermet Schottky contacts", Patent USA 3805128, Publ. April 1974, CA81-7632.
237.Scobey, I. H., Wallace, C. A. and Ward, R. C. C., "A topotactic reaction between nickel and gallium arsenide (Schottky barrier diode fabrication)", J. Appl. Crystallogr., 6 (Pt. 6), 425-8 (1973), EEA77-2542.
238.Siemens, A. G., "Schottky contacts", Patent UK 1346950, Publ. February 1974, CA81-7613.
239.Siemens, A. G., "Production of Schottky contacts", Patent UK 1348811, Publ. March 1974, CA81-7623.
240.Sinha, A. K., "Metallization scheme for n-GaAs Schottky diodes incorporating sintered contacts and a W diffusion barrier", Appl. Phys. Lett., 26, 171-3 (1975), EEA78-19768.
241.Slynko, V. V., Nikonyuk, E. S. and Matlak, V. V., "Deposition of contacts on semiconductor crystal surfaces", Pribory Tekh. Eksper., no. 3, 203 (1969), Russian, EEA73-24914.
242.Somogyi, K., "Contacts and sample preparation on semiconductor layers for van der Pauw measurements", Kozl. Magy. Tud. Akad. Musz. Fiz. Kut. Intez., no. 0-9, 47-59 (1973), PA77-20683.
243.Sugano, T., Chou, H. K., Yoshida, M. and Nishi, T., "Chemical deposition of Mo on Si", J. Math. Phys. Cambridge, 7, 1028-38 (1968), PA72-6010.
244.Sugano, T., Koshiga, F., Morino, A., Chou, H. K., Yoshida, M., Mishima, K., Nishi, T., Obunai, T. and Matsuda, S., "Metal to silicon contact: Conduction theory, fabrication technologies characteristics and applications", J. Fac. Eng. Univ. Tokyo B, 31, 125-98 (1971), EEA75-13697.
245.Sugano, T., Morino, A., Koshiga, F., Mishima, K., Nishi, T. and Matsuda, S., "Characteristics of contacts between silicon and a few kinds of metals", J. Inst. Elect. Commun. Engrs., 50, 1045-52 (1967), Japanese, EEA71-18189.
246.Tabuchi, S., "Developmental process of Schottky barrier diode", JEE, no. 8, 23-30 (1966), EEA74-20795.
247.Univ. Southern California, "Method of making junction diodes", Patent USA 3857169, Publ. December 1974, EEA78-24316.
248.Van Steenbergen, T. W. and DeWaal, J. A. G., "Schottky junction in a cavity", Patent USA 3777228, Publ. December 1973.

249.Wronski, C. R., Abeles, B., Daniel, R. E. and Arie, Y., "Granular metal-semiconductor Schottky barriers", J. Appl. Phys., 45, 295-9 (1974), PA77-44993.

3. Edge Protection (Guard Rings)

250.Cowpland, M. C. J. and Boothroyd, A. R., "Integrated-circuit Schottky polarity guard", Electron. Lett., 7, 753-4 (1971), EEA75-4671.
251.Goldman, R. L., "Guard ring structure for microwave Schottky diode", Patent USA 3820235, Publ. June 1974.
252.Kano, G., Fujiwara, S., Iizuka, M., Hasegawa, H. and Sawaki, T., "Improvements of reverse I-V characteristics in epitaxial planar Schottky barrier diodes", Appl. Phys. Lett., 15, 138-9 (1969), PA73-4731.
253.Kressel, H. and Hawrylo, F. Z., "Semiconductor component with Schottky barrier layer", Patent German 2311170, Publ. September 1973, CA79-140852.
254.Lepselter, M. P. and Sze, S. M., "Silicon Schottky barrier diode with near-ideal I-V characteristics", Bell Syst. Tech. J., 47, 195-208 (1968), PA71-40525.
255.Rhee, C., Saltich, J. and Zwernemann, R., "Moat-etched Schottky barrier diode displaying near ideal I-V characteristics", Solid-State Electron., 15, 1181-6 (1972), EEA75-39911.
256.Saltich, J. L. and Clark, L. E., "Use of a double diffused guard ring to obtain near ideal I-V characteristics in Schottky barrier diodes", Solid-State Electron., 13, 857-63 (1970), EEA73-31355.
257.Saxena, A. N., "Forward current-voltage characteristics of Schottky barriers on n-type silicon", Surface Sci., 13, 151 (1969), PA72-43082.
258.Saxena, A. N., "Schottky barrier diode", Patent USA 3694719, Publ. September 1972.
259.Shao, T. F., "Schottky barrier diode", Patent USA 3742317, Publ. June 1973.
260.Wolfe, C. M. and Lindley, W. T., "Epitaxially grown guard rings for GaAs diodes", J. Electrochem. Soc., 116, 276-8 (1969), PA72-30943.
261.Wood, R. A., "Schottky barrier alloy-junction guard ring", IBM Tech. Disclosure Bull., 12, 1057 (1969).
262.Zettler, R. A. and Cowley, A. M., "p-n junction Schottky barrier hybrid diode", IEEE Trans. Electron Devices, ED-16, 58-63 (1969), EEA72-11283.
263.Zoroglu, D. S. and Clark, L. E., "Design considerations for high-voltage overlay annular diodes", IEEE Trans. Electron Devices, ED-19, 4-8 (1972), EEA75-4657.
264.Zwernemann, R., "Schottky diode with minimal metal-guard ring overlap", Patent USA 3821772, Publ. June 1974.
265.Zwernemann, R., "Method of making a high current Schottky barrier device", Patent USA 3891479, Publ. June 1975.

266.Zwernemann, R., "High voltage Schottky barrier device", Patent USA 3907617, Publ. September 1975, CA83-201114.

4. Heat Treatment and Silicide Formation

267.Anantha, N. G., Doo, V. Y. and Seto, D. K., "Chromium deposition from dicumene-chromium to form metal-semiconductor devices", J. Electrochem. Soc., 118, 163-5 (1971), PA74-28552.
268.Anderson, R. M. and Reith, T. M., "Microstructural and electrical properties of thin PtSi films and their relationships to deposition parameters", J. Electrochem. Soc., 122, 1337-47 (1975).
269.Andrews, J. M. and Koch, F. B., "Formation of NiSi and current transport across the NiSi-Si interface", Solid-State Electron., 14, 901-8 (1971), EEA75-1024.
270.Andrews, J. M. and Lepselter, M. P., "Reverse current-voltage characteristics of metal-silicide Schottky diodes", Solid-State Electron., 13, 1011-23 (1970), EEA73-35186.
271.Axelrod, N. N., "Group III-V Schottky barrier diodes", Patent USA 3675316, Publ. July 1972.
272.Beck, U., Neumann, H. G. and Becherer, G., "Phase formation in Ni/Si layers", Krist. & Tech., 8, 1125-9 (1973), German, PA77-37037.
273.Blattner, R. J., Evans, C. A., Jr., Lau, S. S., Mayer, J. W. and Ullrich, B. M., "Effect of oxidzing ambients on platinum silicide formation. II. Auger and backscattering analysis", J. Electrochem. Soc., 122, 1732-6 (1975).
274.Borders, J. A. and Sweet, J. N., "Ion backscattering analysis of tungsten films on heavily doped SiGe", J. Appl. Phys., 43, 3803-8 (1972), PA75-71951.
275.Bourdon, B., Hilaire, D., Ripoche, G., Magnabal, J. and Benoit, J., "Low-temperature formation of platinum and palladium silicides", Colloq. Int. Pulverisation Cathodique Ses Appl., C. R., 1st, 1973, p. 211-15, Publ. 1974, French, CA81-83504.
276.Bower, R. W. and Mayer, J. W., "Growth kinetics observed in the formation of metal silicides on silicon", Appl. Phys. Lett., 20, 359-61 (1972), EEA75-24187.
277.Bower, R. W., Sigurd, D. and Scott, R. E., "Formation kinetics and structure of Pd_2Si films on Si", Solid-State Electron., 16, 1461-71 (1973), EEA77-10126.
278.Chiang, Y. S., "Method of making a metal silicide-silicon Schottky barrier", Patent USA 3841904, Publ. October 1974.
279.Chino, K., "Behavior of Al-Si Schottky barrier diodes under heat treatment", Solid-State Electron., 16, 119-21 (1973), EEA76-8665.
280.Christou, A., "Electromigration-thermomigration in Al-Si and Au-Si", 5th Spring Meeting of the Metallurgical Society of AIME, 1973, p. 149-50, PA77-28243.

281.Coe, D. J., Rhoderick, E. H., Gerzon, P. H. and Tinsley, A. W., "Silicide formation in Rh-Si Schottky barrier diodes", Conference on Metal-Semiconductor Contacts, 1974, p. 74-83, PA78-15727.
282.Chu, W. K., Krautle, H., Mayer, J. W., Muller, H., Nicolet, M. A. and Tu, K. N., "Identification of the dominant diffusing species in silicide formation", Appl. Phys. Lett., 25, 454-7 (1974), EEA78-5241.
283.Chu, W. K., Lau, S. S., Mayer, J. W., Muller, H. and Tu, K. N., "Implanted noble gas atoms as diffusion markers in silicide formation", Thin Solid Films, 25, 393-402 (1975), PA78-42017.
284.Danyluk, S. and McGuire, G. E., "Platinum silicide formation: Electron spectroscopy of the platinum-platinum silicide interface", J. Appl. Phys., 45, 5141-4 (1974), PA78-26302.
285.Day, H. M., Christou, A. and Weisenberger, W. H., "Reliability studies of the PtSi-Ta-Au metallization system for microwave power transistors", J. Electrochem. Soc., 121, 92C-3 (1974), EEA77-26452.
286.Drake, W. H. and Kelland, G. M., "A Ti-Si Schottky diode for use in shallow diffused integrated circuit processes", 1972 European Solid-State Devices Research Conferences, 1973, p. 209.
287.Hiraki, A., Nicolet, M. A. and Mayer, J. W., "Low-temperature migration of silicon in thin layers of gold and platinum", Appl. Phys. Lett., 18, 178-82 (1971), EEA75-10171.
288.Hosack, H. H., "Platinum silicide - aluminum Schottky diode characteristics", Appl. Phys. Lett., 21, 256-7 (1972), PA75-75131.
289.Hosack, H. H., "Electrical and mechanical features of the platinum silicide-aluminum reaction", J. Appl. Phys., 44, 3476-85 (1973), PA76-64403.
290.Hutchins, G. A. and Shepela, A., "The growth and transformation of Pd_2Si on (111), (110) and (100) Si", Thin Solid Films, 18, 343-63 (1973), PA77-3666.
291.Jenkins, R. T. and Wilson, G. H., "Schottky-barrier diode process and devices", Patent USA 3623925, Publ. November 1971.
292.Kawamura, T., Shinoda, D. and Muta, H., "Oriented growth of the interfacial PtSi layer or between Pt and Si", Appl. Phys. Lett., 11, 101-3 (1967).
293.Kingzett, T. J. and Ladas, C. A., "Effect of oxidizing ambients on platinum silicide formation. I. Electron microprobe analysis", J. Electrochem. Soc., 122, 1729-32 (1975).
294.Kircher, C. J., "Metallurgical properties and electrical characteristics of palladium silicide-silicon contacts", Solid-State Electron., 14, 507-13 (1971).
295.Kircher, C. J., Mayer, J. W., Tu, K. N. and Ziegler, J. F., "Analysis of formation of hafnium silicide on silicon", Appl. Phys. Lett., 22, 81 (1973).
296.Koos, V. and Neumann, H. G., "Stresses in thin nickel silicide films", Phys. Status Solidi A, 29, K115-16 (1975), CA83-88846.

297.Krautle, H., Nicolet, M. A. and Mayer, J. W., "Silicide formation at low temperatures by metal-SiO_2 interaction", Phys. Status Solidi A, 20, K33-6 (1973), PA77-6571.
298.Ladas, C. and Kingzett, T., "The effect of oxidizing ambients on platinum silicide formation", Electrochemical Society Fall Meeting (Extended abstracts), 1974, p. 437-9, EEA78-37437.
299.Lau, S. S., Chu, W. K., Mayer, J. W. and Tu, K. N., "Evaluation of glancing angle X-ray diffraction and MeV ^{4}He backscattering analyses of silicide formation", Thin Solid Films, 23, 205-13 (1974).
300.Lepley, B., Ravelet, S. and Renard, P., "Heat-treated Au-CdS and Cu-CdS thin-film Schottky diodes", Phys. Status Solidi A, 22, K191-94 (1974).
301.Liau, Z. L., Campisano, S. U., Canali, C., Lau, S. S. and Mayer, J. W., "Kinetics of the initial stages of Si transport through Pd-silicide for epitaxial growth", J. Electrochem. Soc., 122, 1696-9 (1975).
302.Locker, L. D. and Capio, C. D., "Reaction kinetics of tungsten thin films on silicon (100) surfaces", J. Appl. Phys., 44, 4366-9 (1973).
303.Magdo, I. E., "Integratable Schottky barrier structure", Patent USA 3900344, Publ. August 1975, CA83-156879.
304.Mayer, J. W., Mitchell, I. V. and Nicolet, M. A., "Analysis of contact formation and surface layers on semiconductors", Proceedings of the 2nd International Conference on Ion Implantation in Semiconductors, 1971, p. 274-86, PA75-25877.
305.McCaldin, J. O., "Atom movements occurring at solid metal-semiconductor interfaces", J. Vac. Sci. & Technol., 11, 990-5 (1974), PA78-46764.
306.Mhaskar, P. A. and Guha, S., "On the effect of heat treatment on chromium-silicon Schottky barriers", Jap. J. Appl. Phys., 10, 277-8 (1971), PA74-26257.
307.Morabito, J. M. and Rand, M. J., "Composition profiles of CVD platinum and platinum silicide by Auger electron spectroscopy and secondary ion mass spectrometry", Thin Solid Films, 22, 293-303 (1974), PA77-67137.
308.Muta, H. and Shinoda, D., "Solid-solid reactions in Pt-Si systems (contacts)", J. Appl. Phys., 43, 2913-5 (1972), PA75-50715.
309.Nakamura, K., Nicolet, M. A., Mayer, J. W., Blattner, R. J. and Evans, C. A., Jr., "Interaction of Al layers with polycrystalline Si", J. Appl. Phys., 46, 4678-84 (1975).
310.Nickl, J. J. and Koukoussas, J. D., "Transport reactions and single-crystal growth of silicides of group IVA-VIA metals", J. Less-Common Metals, 23, 73-81 (1971).
311.Ogawa, M., Shinoda, D., Kawamura, N., Nozaki, T. and Asanabe, S., "GaAs alloy type Schottky barrier contacts", Proceedings of the Third International Symposium on Gallium Arsenide and Related Compounds, 1970, p. 268-75, Publ. 1971, PA76-2184.

312.Poate, J. M. and Tisone, T. C., "Kinetics and mechanism of platinum silicide formation on silicon", Appl. Phys. Lett., 24, 391-3 (1974), PA77-57691.
313.Rand, M. J. and Roberts, J. F., "Observations on the formation and etching of platinum silicide", Appl. Phys. Lett., 24, 49-51 (1974), PA77-37115.
314.Reith, T. M. and Revitz, M., "Al/PtSi Schottky barrier diodes with a diffusion barrier", IBM Tech. Disclosure Bull., 16, 3586 (1974), EEA77-38820.
315.Richards, B. P., Scobey, I. H. and Wallace, C. A., "Crystallographic aspects of thin PtSi films on Si substrates", J. Appl. Crystallogr., 7 (Pt. 2), 275-80 (1974), PA77-32833.
316.Robinson, G. Y., "Palladium silicide formation observed by Auger electron spectroscopy", Appl. Phys. Lett., 25, 158-60 (1974), PA77-76391.
317.Saltich, J. L. and Terry, L. E., "Effects of pre- and postannealing treatments on silicon Schottky barrier diodes", Proc. IEEE, 58, 492-4 (1970), EEA73-24979.
318.Schinella, R. D. and Herndon, W. H., "Methods for forming metal/metal silicide semiconductor device interconnect system", Patent USA 3777364, Publ. December 1973.
319.Shinoda, D., "Physico-chemical studies of platinum-silicon interfaces", Rev. Phys. Chem. Jap., 42, 125-34 (1972), EEA76-35258.
320.Shinoda, D. and Ogawa, M., "GaAs alloy type Schottky barrier junctions", Oyo Buturi, 40, 432-6 (1971), Japanese.
321.Signetics Corp., "Semiconductor device with a Schottky-barrier diode, and containing platinum-nickel silicide", Patent German 2264322, Publ. July 1973, CA79-98477.
322.Sinha, A. K., "Electrical characteristics and thermal stability of platinum silicide to silicon ohmic contacts metalized with tungsten", J. Electrochem. Soc., 120, 90C (1973), PA76-38889.
323.Sinha, A. K., "Catastrophic Au-Si eutectic formation in certain W/Au metallizations on oxidized silicon wafers", Thin Solid Films, 20, 115-22 (1974), EEA77-10305.
324.Sinha, A. K., "Metal-compound semiconductor interactions", Electrochemical Society Spring Meeting (Extended abstracts), 1975, p. 269-71, PA78-85593.
325.Sinha, A. K., Haszko, S. E. and Sheng, T. T., "Thermal stability of PtSi films on polysilicon layers", J. Electrochem. Soc., 122, 1714-18 (1975).
326.Sinha, A. K., Marcus, R. B., Sheng, T. T. and Haszko, S. E., "Thermal stability of thin PtSi films on silicon substrates", J. Appl. Phys., 43, 3637-43 (1972), PA75-71334.
327.Sinha, A. K., Read, M. H. and Smith, T. E., "Silicide formation in tungsten and other refractory metalizations on PtSi on silicon", J. Electrochem. Soc., 120, 1775-8 (1973), EEA77-10154.
328.Sinha, A. K. and Smith, T. E., "Kinetics of WSi_2 formation in the thin-film system W/PtSi/Si", J. Appl. Phys., 44, 3465-9 (1973), EEA76-35207.

329. Sinha, A. K., Venkatachalam, G. and Bhatia, C. S., "Nucleation and growth of silicon on tungsten and molybdenum in a field emission microscope", J. Appl. Phys., 46, 4685-8 (1975).
330. Sundstrom, K. E., Petersson, S. and Tove, P. A., "Studies of formation of silicides and their barrier heights to silicon", Phys. Status Solidi A, 20, 653-68 (1973), PA77-17183.
331. Thomas, S. and Terry, L. E., "Auger spectroscopy analysis of palladium silicide films", Appl. Phys. Lett., 26, 433-35 (1975).
332. Tu, K. N., "Silicide formation by metal-silicon interactions", Electrochemical Society Spring Meeting (Extended abstracts), 1975, p. 267-8, PA78-85592.
333. Tu, K. N., "Selective growth of metal-rich silicide of near-noble metals", Appl. Phys. Lett., 27, 221-4 (1975), PA78-86388.
334. Tu, K. N., Ziegler, J. F. and Kircher, C. J., "Formation of vanadium silicides by the interaction of V with bare and oxidised Si wafers", Appl. Phys. Lett., 23, 493-5 (1973), PA77-24854.
335. van Gurp, G. J. and Langereis, C., "Structure and growth kinetics of cobalt silicide on Si", Electrochemical Society Spring Meeting, (Extended abstracts), 1975, p. 272-3, PA78-85594.
336. Ziegler, J. F., Mayer, J. W., Kircher, C. J. and Tu, K. N., "Kinetics of the formation of hafnium silicides on silicon", J. Appl. Phys., 44, 3851-7 (1973), EEA76-38195.

IV. Ohmic Contacts

1. Group IV Semiconductors

337. Berner, W. E., "Connecting electrical conductors to a semiconductor body", Patent UK 1327207, Publ. August 1973, CA79-150809.
338. Black, J. R., "Etch pit formation in silicon at aluminium-silicon contacts because of transport of silicon in aluminium by momentum exchange with conducting electrons", Ohmic Contacts to Semiconductors, 1968, p. 311-23, Publ. 1969, EEA73-10556.
339. Bower, R. W., "Characteristics of aluminum-titanium electrical contacts on silicon", Appl. Phys. Lett., 23, 99-101 (1973).
340. Cembali, F., Galloni, R. and Zignani, F., "Technique for obtaining deep penetrating ohmic contacts for electrical measurements on ion implanted silicon", J. Phys. E, 7, 698-700 (1974), CA81-112723.
341. Collins, R. H., Grochowski, E. G. and North, W. D., "Silicon process technology in monolithic memory (metallurgical contact)", IBM J. Res. & Develop., 16, 6 (1972).
342. Day, H. M., Christou, A., Weisenberger, W. H. and Hirvonen, J. K., "Relations between the Ta-Pt-Ta-Au metallization and PtSi ohmic contacts", J. Electrochem. Soc., 122, 769-72 (1975).
343. Engeler, W. E. and Cordes, L. F., "Method for making high-temperature low ohmic contact to silicon", Patent USA 3794516, Publ. February 1974.

344. Feuerstein, A. and Kalbitzer, S., "Ohmic contacts on n^+- and p^+-Si by ion implantation", Appl. Phys. Lett., 22, 19-20 (1973), PA76-18214.
345. Freyer, J. and Harith, W., "Non-injecting ohmic contacts to n-type (1-3Ω cm) silicon", Arch. Elektron. & Uebertragungstech., 28, 482-3 (1974), EEA78-2766.
346. Hall, E. L. and Philofsky, B. M., "Ohmic contacts", Patent USA 3900598, Publ. August 1975, CA83-171679.
347. Hoare, R. A., "Metal-silicon contact resistivity", Electrochemical Society Fall Meeting (Extended abstracts), 1970, p. 461-4, EEA74-11242.
348. Holm-Kennedy, J. W. and Ku, T. P. C., "Preparation of ohmic contacts to n-Ge by diffusion from As-doped SiO_2", J. Electrochem. Soc., 120, 1602-3 (1973), EEA77-2466.
349. Hooper, R. C., Cunningham, J. A. and Harper, J. G., "Electrical contacts to silicon", Solid-State Electron., 8, 831-3 (1965), EEA70-12960.
350. Ibragimov, R. S., "The creation and electrical properties of ohmic contacts on p-Si", Instrum. & Exp. Tech., 15, 909-10 (1972), EEA76-16349.
351. International Standard Electric Corp., "Ohmic contacts", Patent UK 1120693, Publ. July 1968, EEA72-4975.
352. Krysiewicz, J. and Olejarczyk, B., "Apparatus and instrumentation for the technique of manufacturing aluminium contacts and connections for transistors and integrated circuits", Pr. Przem. Inst. Electron., 13, 47-65 (1972), Polish, EEA76-16195.
353. Kubetskii, G. A., Boltovskii, V. V., Niskov, V. Y. and Rumennik, V. I., "Electrical properties of a nickel-silicon ohmic contact", Fiz. & Tekh. Poluprov., 4, 2216-7 (1970), Russian, PA74-19178.
354. Lepselter, M. P., "Semiconductor contacts and protective coatings for planar devices", Patent USA 3287612, Publ. November 1966.
355. Lepselter, M. P., "Method of fabricating semiconductor contacts", Patent USA 3442701, Publ. May 1969, EEA73-4433.
356. Lepselter, M. P. and Andrews, J. M., "Ohmic contacts to silicon", Ohmic Contacts to Semiconductors, 1968, p. 159-86, Publ. 1969, EEA73-10549.
357. Lisichenko, V. I., Petrichenko, N. N. and Pogorelyi, V. A., "Tin contacts to silicon", Instrum. & Exp. Tech., 17, 1512-13 (1974), PA78-58813.
358. Mamedova, G. A. and Bekirov, M. Y., "Ohmic contacts to Ge-Si solid solutions", Izv. Akad. Nauk SSR Ser. Fiz.-Tekh. & Mat. Nauk, no. 3, 3-5 (1974), Russian, PA78-10073.
359. Matsura, E., Matsui, K. and Hasiguti, R. R., "Technique for ohmic connecting leads to silicon", J. Appl. Phys., 33, 1610-11 (1962).
360. May, J. E., "Tungsten-aluminium ohmic contact for n-type silicon", Ohmic Contacts to Semiconductors, 1968, p. 234-42, Publ. 1969, EEA73-10553.

361. Mei, L. and Greene, J. E., "RF-sputtered amorphous Si/crystalline Si junctions", J. Vac. Sci. & Technol., 11, 145-8 (1974), PA77-75954.
362. Meyer, D. E., "Improving aluminium-silicon ohmic contact without high temperature alloy", Ohmic Contacts to Semiconductors, 1968, p. 227-33, Publ. 1969, EEA73-10552.
363. Northrop, D. C. and Puddy, D. C., "Ohmic contacts between evaporated aluminium and n-type silicon", Nucl. Instrum. & Meth., 94, 557-9 (1971), EEA74-30744.
364. Roberts, F. M. and Wilkinson, E. L. G., "The effects of alloying material on regrowth-layer structure in silicon power devices", J. Mater. Sci., 6, 189-99 (1971), EEA74-22616.
365. Sebestyen, T., Hartnagel, H. L. and Herron, L. H., "Thin-phase epitaxy for good semiconductor metal ohmic contacts", IEEE Trans. Electron Devices, ED-22, 1073-7 (1975).
366. Sello, H., "Ohmic contacts and integrated circuits", Ohmic Contacts to Semiconductors, 1968, p. 277-98, Publ. 1969, EEA73-10555.
367. Shadrin, V. S., "A method of connecting ohmic contacts to silicon devices", Instrum. Exper. Tech., no. 4, 1010-11 (1966), PA70-26671.
368. Shevchenko, V. Y. and Skripkin, V. A., "Forming of metal-semiconductors contacts", Pribory Tekh. Eksper., no. 3, 202 (1969), Russian, EEA73-24913.
369. Shinoda, D., "Ohmic contacts to silicon using evaporated metal silicides", Ohmic Contacts to Semiconductors, 1968, p. 200-13, Publ. 1969, EEA73-10551.
370. Sinha, A. K., "Electrical characteristics and thermal stability of platinum silicide to silicon ohmic contacts metalized with tungsten", J. Electrochem. Soc., 120, 90C (1973), EEA76-23382.
371. Terry, L. E. and Wilson, R. W., "Metallization systems for silicon integrated circuits", Proc. IEEE, 57, 1580-6 (1969).
372. Ting, C. Y. and Chen, C. Y., "A study of the contacts of a diffused resistor", Solid-State Electron., 14, 433-8 (1971).
373. Trimbuch, R. and Tippmann, H., "The ohmic metal-semiconductor contact", Z. Elektr. Inf. & Energietech., 2, 236-7 (1972), German, EEA76-8617.
374. Voskoboinikov, V. V. and Sinitsa, S. P., "Ohmic contacts on silicon", Instrum. Exper. Tech., no. 4, 953-4 (1967), PA71-49286.
375. Wolter, A. R., "Metal-silicon junctions formed with metal ions", Ohmic Contacts to Semiconductors, 1968, p. 214-26, Publ. 1969, EEA73-10602.

2. Group III-V Semiconductors

376. Akimova, A. N., Kostylev, S. A., Lisichenko, V. I., Pogorelyi, V. A. and Privalov, V. N., "Ohmic contacts on gallium arsenide obtained by the condensation of a metallic plasma", Instrum. & Exp. Tech., 15, 907-8 (1972), EEA76-16348.

377. Basterfield, J., Josh, M. J. and Burgess, M. R., "Structural features of alloyed contacts to GaAs", Acta Electron., 15, 5 pp. (1972), EEA75-16661.

378. Blakeslee, A. E., Hange, P. S. and Lanza, C., "Microwave oscillations in bulk semiconductors", Report unnumbered, Contract No. DA28-043-AMC-01550(E), IBM T. J. Watson Research Center, Yorktown Heights, New York, 1969.

379. Brady, D. P., Knight, S., Lawleg, K. L. and Uenohara, M., "Recent results with epitaxial GaAs Gunn effect oscillators", IEEE Proc., 54, 1497-8 (1966).

380. Braslau, N., Gunn, J. B. and Staples, J. L., "Metal-semiconductor contacts for GaAs bulk effect devices", Solid-State Electron., 10, 381-3 (1967), EEA70-14845.

381. Collins, N. E. and Halt, I. E., "Ohmic contact for Group III-V p-type semiconductors", Patent USA 3769694, Publ. November 1973.

382. Collins, N. E. and Halt, I. E., "Ohmic contact for Group III-V p-type semiconductors", Patent USA Reissue 27879, Publ. January 1974, CA80-75743.

383. Colliver, D. J., Hilsum, C., Morgan, J. R., Rees, H. D. and Taylor, B. C., "The technology and performance of InP three-level oscillators", 1970 Symposium on Gallium Arsenide, p. 140-7.

384. Copeland, J. A. and Knight, S., "Bulk negative resistance devices (contacts)", In: Semiconductor and Semimetals, Vol. 7, Willardson, R. K. and Beer, A. C., (Eds.), Academic Press, N. Y., p. 61-7 (197[illegible]).

385. Cox, R. H. and Strack, H., "Ohmic contacts for GaAs devices", Solid-State Electron., 10, 1213-8 (1967), EEA71-5905.

386. Dale, J. R. and Josh, M. J., "Alloys for GaAs devices", Solid-State Electron., 7, 177-81 (1964).

387. Drobny, V., "Ohmic contacts for GaAs and InP of the n-type", Elektrotech. Cas., 25, 399-403 (1974), Slovak, EEA77-33977.

388. Dumin, D. J. and Pearson, G. L., "Properties of GaAs diodes between 4.2 and 300°K", J. Appl. Phys., 36, 3418 (1965).

389. Duraev, V. P., Alferov, Z. I., Andreev, V. M., Borodulin, V. I., Garbuzov, D. Z., Morozov, E. P., Portnoi, E. L., Pak, G. T. and Shveikin, V. I., "Ohmic contact of a semiconductor device", Patent USSR 325908, Publ. April 1974, CA81-143347.

390. Duraev, V. P., Kubetskii, G. A., Pugach, M. K. and Shvetkin, V. I., "Ohmic contacts to gallium arsenide", Instrum. Exper. Tech., no. 2, 469-71 (1968), EEA72-4973.

391. Edwards, W. D., Hartman, W. A. and Torrens, A. B., "Special contact resistance of ohmic contacts to gallium arsenide", Solid-State Electron., 15, 387-92 (1972), PA75-29715.

392. Fomin, N. G., Vorobev, Y. V. and Tretyak, O. V., "Obtaining ohmic contacts based on semi-insulating GaAs", Instrum. & Exp. Tech., 17, 887-8 (1974), PA78-33967.
393. Fremunt, R., Kortan, J. and Janouskova, O., "Investigation of Au-Ni ohmic contacts on n-type GaP", J. Phys. D, 6, L95-7 (1973), PA76-46892.
394. Fremunt, R. and Svoboda, V., "Ohmic contacts on gallium arsenide", Cesk. Casopis Fys. A., 20, 42-53 (1970), Czech, EEA73-21176.
395. Gopen, H. J. and Yu, A. Y. C., "Ohmic contacts to epitaxial pGaAs", Solid-State Electron., 14, 515-7 (1971).
396. Gyulai, J., Mayer, J. W., Rodriguez, V., Yu, A. Y. C. and Gopen, H. J., "Alloying behavior of Au and Au-Ge on GaAs", J. Appl. Phys., 42, 3578-85 (1971), PA74-65787.
397. Iglitsyn, M. I., Kolat, E. V. and Fedotova, E. F., "Investigation of the rate of etching and formation of an ohmic contact with gallium phosphide", Sov. Phys. Semicond., 4, 2078 (1971).
398. Ing, D. W., McAvoy, B. R. and Ure, R. W., "Alloying contacts to gallium arsenide by hot hydrogen and HCl gases", Solid-State Electron., 11, 469-71 (1968), PA71-27156.
399. Ito, K. and Inoue, M., "Ohmic contacts to Group III-V compound semiconductors", Patent Japan 73-84741, Publ. November 1973, CA80-88797.
400. Jaros, M. and Hartnagel, H. L., "An understanding of ohmic contact formation with Ge doping of n-GaAs", Solid-State Electron., 18, 1029-30 (1975), EEA78-41499.
401. Ketchow, D. R., "Germanium-doped GaAs for p-type ohmic contacts", J. Electrochem. Soc., 121, 1237-9 (1974), EEA78-583.
402. Klohn, K. L. and Wandinger, L., "Variation of contact resistance of metal GaAs contacts with impurity concentration and its device implication", J. Electrochem. Soc., 116, 507-8 (1969), PA72-30953.
403. Komashko, V. A., Maslovskii, F. N. and Tuz, V. F., "A low-resistance contact with gallium arsenide", Poluprovodn. Tekh. & Mikroelektron., no. 11, 61-3 (1973), Russian, PA76-52899.
404. Krukovskaya, L. P., Golubev, L. V. and Shmartsev, Y. V., "Nonrectifying junction contacts for GaSb", Instrum. & Exp. Tech., 17, 268-9 (1974), PA78-18676.
405. Kusaka, M. and Okazaki, S., "Electrical properties of metal-aluminum antimonide interface", Rep. Res. Lab. Surf. Sci. Okayama Univ., 3, 171-6 (1970), EEA75-10188.
406. Mills, H. T. and Hartnagel, H. L., "Ideal ohmic contacts to InP", Electron. Lett., 11, 621-3 (1975).
407. Nakatsuka, H., Domenico, A. J. and Pearson, G. L., "Improved ohmic contacts to n-type GaP devices", Solid-State Electron., 14, 849-85 (1971).
408. Ohhata, K., Ogawa, M. and Ishikawa, M., "Material for forming an ohmic contact on n-type gallium arsenide", Patent Japan 75-63877, Publ. May 1975, CA83-187373.
409. Paola, C. R., "Metallic contacts for gallium arsenide", Solid-State Electron., 13, 1189-97 (1970), EEA73-35168.

410.Pounds, R. S., Saifi, M. A. and Hahn, W. C., Jr., "Formation of ohmic contacts to III-V semiconductors, using a laser beam", Solid-State Electron., 17, 245-9 (1974), EEA77-22905.
411.Prakash, O., "A new technique for fabrication of ohmic contacts to GaAs devices", J. Inst. Electron. & Telecommun. Eng., 20 (3-4), 153-8 (1974), EEA78-15229.
412.Pruniaux, B. R., "Transport properties of the gold germanium gallium arsenide metal semiconductor system", J. Appl. Phys., 42, 3575-7 (1971).
413.Rideout, V. L., "Ohmic contact technique for N-type GaAs, GaAsP, and GaAlAs", IBM Tech. Disclosure Bull., 16, 3070-1 (1974), PA77-45441.
414.Robinson, G. Y., "Metal-semiconductor technology for Ni/Au-Ge on n-GaAs", 1973 International Electron Devices Meeting Technical Digest, p. 253-5, EEA77-14406.
415.Robinson, G. Y., "Metallurgical and electrical properties of alloyed Ni/Au-Ge films on n-type GaAs", Solid-State Electron., 18, 331-42 (1975), PA78-25759.
416.Ruppel, W., "The photoconductor-metal contact", In: Semiconductors and Semimetals, Vol. 6, Willardson, R. K. and Beer, A. C., (Ed.), Academic Press, New York, p. 315-45 (1970).
417.Salow, H. and Grobe, E., "Ohmic contacts for Gunn elements", Z. Angew. Phys., 25, 137-40 (1968), German, PA71-49282.
418.Schiavone, L. M. and Pritchard, A. A., "Ohmic contacts for moderately resistive p-type InP", J. Appl. Phys., 46, 452 (1975), EEA78-15301.
419.Schwartz, B. and Sarace, J. C., "Low-temperature alloy contacts to GaAs using metal halide fluxes", Solid-State Electron., 9, 859-62 (1966).
420.Sebestyen, T., Hartnagel, H. and Herron, L. H., "New method for producing ideal metal-semiconductor ohmic contacts", Electron. Lett., 10, 372-3 (1974), PA77-75960.
421.Shih, K. K. and Blum, J. M., "Contact resistances of Au-Ge-Ni, Au-Zn and Al to III-V compounds", Solid-State Electron., 15, 1177-80 (1972), EEA75-39871.
422.Sinha, A. K., Smith, T. E. and Levinstein, H. J., "Sintered ohmic contacts to n- and p-type GaAs", IEEE Trans. Electron Devices, ED-22, 218-24 (1975), EEA78-19758.
423.Staples, J. L., "Method for providing electrical contacts to a wafer of GaAs", Patent USA 3386867, Publ. June 1968.
424.Strazynska, H. and Wojcik, I., "Linear contacts for gallium antimonide", Arch. Elektrotech., 22, 279-86 (1973), Polish, EEA78-19854.
425.van Welzenis, R. G., "Avalanche and microwave emission in n-type indium antimonide at liquid nitrogen temperature", Thesis, Tech. U. Eindhoven, Netherlands, 1972, 25 pp..
426.Vogel, K. and Engle, A., "The formation and properties of ohmic contacts on GaSb", Krist. & Tech., 8, 1083-92 (1973), German, PA77-20682.

427.Walker, G. H. and Conway, E. J., "Electrical contacts to ion cleaned n-type gallium arsenide", IEEE Trans. Parts, Hybrids & Packag., PHP-8, 49-50 (1972), PA76-15250.
428.Yu, A. Y. C., Gopen, H. J. and Waits, R. K., "Contacting technology for gallium arsenide", Report AD3755911, Fairchild Camera and Instrument Corporation, Palo Alto, California, 1970.

3. Group II-VI Semiconductors

429.Allan, D. D. M., Hoy, A. J. and Reid, M. A., "Ohmic contacts to cadmium sulphide films", Solid-State Electron., 16, 951-4 (1973), EEA76-29209.
430.Boer, K. W. and Hall, R. B., "Multilayer ohmic contacts to CdS", J. Appl. Phys., 37, 4739-42 (1966), PA70-11503.
431.Janus, H. M., "Formation of ohmic contacts to ZnO", Rev. Sci. Inst., 41, 1099-100 (1970).
432.Jenkins, R., Mead, C. A. and McCaldin, J., "Ohmic contact to zinc sulfide devices", Patent USA 3780427, Publ. December 1973.
433.Kalinnikov, V. T., Aminov, T. G., Vigileva, E. S. and Shapsheva, N. P., "Ohmic contacts to $CdCr_2Se_4$ single crystals", Instrum. & Exp. Tech., 18, 265-6 (1975), PA78-86394.
434.Kaufman, R. G. and Dowbor, P., "Mechanism of formation on Ohmic contacts in ZnSe, ZnS, and mixed crystals $ZnSe_xSe_{1-x}$", J. Appl. Phys., 45, 4487-90 (1974), PA78-19085.
435.Kroger, F. A., Diemer, G. and Klasens, H. A., "Nature of an ohmic metal-semiconductor contact", Phys. Rev., 103, (1956).
436.Mach, R., Treptow, H. and Ludwig, W., "Physical properties of the Au-ZnSe metal-semiconductor contact", Phys. Status Solidi A, 25, 267-73 (1974).
437.Nakai, J., Mitsusada, K. and Yasuoka, A., "Characteristics of the contact between vacuum-deposited cadmium selenide films and gold films", Jap. J. Appl. Phys., 3, 490 (1964).
438.Shin, B. K., Look, D. C. and Park, Y. S., "Ohmic contacts to Al-implanted ZnSe", J. Electrochem. Soc., 122, 450-1 (1975), PA78-47447.
439.Tupenevich, P. A., Kononenko, V. K. and Lyakhovich, A. K., "Formation of low-resistance ohmic contacts for p-type zinc telluride", Instrum. & Exp. Tech., 17, 1509-11 (1974), EEA78-28613.
440.Yamaguchi, M., "Ohmic contacts to CdS", Jap. J. Appl. Phys., 13, 1325-6 (1974), EEA77-36246.
441.Zold, T., "Ohmic contacts to photoconducting CdS single crystal samples using pure indium", Rev. Sci. Instrum., 44, 1251-5 (1973), EEA77-359.

4. All Other Semiconductors

442.Barrus, C. A., "Pulse electroplating of high-resistance materials, poorly contacted devices, and extremely small areas", J. Electrochem. Soc., 118, 833-4 (1971).
443.Gambino, R. J. and Weiser, K., "Low resistance contacts for amorphous chalcogenide semiconductors", IBM Tech. Disclosure Bull., 15, 577-9 (1972), PA75-75121.
444.Goodman, A. M. and Mehl, W. W., "Research on metallic contacts to semiconductors", Report AD-289205, RCA, David Sarnoff Res. Center, 1962.
445.Goriunova, O. F., "The dynamic properties of small area ohmic contacts with semiconductors", Radio Engineering and Electronic Physics, 10, 960-3 (1965).
446.Henisch, H. K., "Contacts between chalcogenide glasses, metals and semiconductors", Report AD-711221, Pennsylvania State Univ., University Park, 1970, 7 pp., EEA74-10564.
447.Kahng, D. and Ryder, R. M., "Small area semiconductor device", Patent USA 3360851, Publ. January 1968.
448.Mitchell, J. P. and Denure, D. G., "Electrical contacts on photoconductive Sb_2S_3 films", Thin Solid Films, 16, 285-96 (1973), PA76-46893.

5. Reliability

449.Baker, D., "The metal-semiconductor contact in silicon devices", Conference on Metal-Semiconductor Contacts, 1974, p. 51-6, PA78-15724.
450.Black, J. R., "Etch pit formation in silicon at aluminium-silicon contacts because of transport of silicon in aluminium by momentum exchange with conducting electrons", Ohmic Contacts to Semiconductors, 1968, p. 311-23, Publ. 1969, EEA73-10556.
451.Brantley, W. A., Schartz, B., Keramidas, V. G., Kammlott, G. W. and Sinha, A. K., "Gallium migration through contact metallization on GaP", J. Electrochem. Soc., 122, 434-5 (1975), PA78-46748.
452.Chino, K., Morita, K., Sano, K., Matsuki, T. and Ida, M., "Reliability of semiconductor devices for 20 GHz digital radio-relay system", Rev. Electr. Commun. Lab., 22, 633-43 (1974), EEA78-8387.
453.McCarthy, J., "Failure of aluminium contacts to silicon in shallow diffused transistors", Microelectronics & Reliability, 9, 8 pp. (1970), EEA73-21297.
454.Ohata, K. and Ogawa, M., "Degradation of gold-germanium ohmic contact to n-GaAs", Proceedings of the 12th Annual Reliability Physics Conference, 1974, p. 278-83.
455.Oliver, C. B. and Bower, D. E., "Theory of failure of semiconductor contacts by electromigration", The Annual Reliability Physics Symposium, 1970, p. 33-5, EEA73-27480.

456.Tibol, G. J., "Schottky barrier semiconductor device", Patent USA 3476984, Publ. November 1969.

J. RADIATION DAMAGE

I. Books

1.Chaffin, R. J., "Microwave semiconductor devices: fundamentals and radiation effects", Wiley-Interscience, New York, 1973, EEA77-9978.
2.Corbett, J. W., "Electron radiation damage in semiconductors and metals", Academic Press, New York, 1966.
3.Corbett, J. W. and Watkins, G. D., "Radiation effects in semiconductors", Gordon and Breach, New York, 1971.
4.Hasiguti, R. R., (Ed.), "Lattice defects in semiconductors", The Pennsylvania State University Press, University Park, London, Pa., 1968.
5.Larin, F., "Radiation effects in semiconductor devices", Wiley, New York, 1968.
6.Olesen, H. L., "Radiation effects on electronic systems", Plenum Press, New York, 1966.
7.Vavilov, V. S., "Effects of radiation on semiconductors", Plenum Press, New York, 1965.

II. Symposia

8.Radiation damage in semiconductors: 7th International Conference on the Physics of Semiconductors, Academic Press, New York, 1964.
9.1965 Paris International Conference on Radiation Effects in Semiconductors, Book: Radiation Damage in Semiconductors, Dunod, Paris, France, 1965.
10.The University of Michigan Engineering Summer Conferences, Notes: Radiation Effects in Semiconductors, University of Michigan, 1967.
11.Symposium on Radiation Effects on Semiconductor Components, Book: Action des Rayonnements sur les Composants a Semiconducteurs, Journees D'electrouiques, Toulouse, France, 1967.
12.Vook, F. L., (Ed.), "Radiation effects in semiconductors: Proceedings of the Conference on Radiation Effects in Semiconductors, Plenum Publishing Co., New York, 1968.
13.International Conference on Radiation Effects in Semiconductors, held at Albany, August 1970, published by Gordon & Breach, New York, N. Y., 1971.
14.Proceedings of the International Conference on Radiation Damage and Defects in Semiconductors, held at Reading, Berks, England, July 1972, published by Institute of Physics, London, England, 1973.
15.Lattice Defects in Semiconductors, conference held at Freiburg, Germany, July 1974, published by Institute of Physics, London, England, 1975.

16.Consultant Symposium on the Physics of Irradiation Produced Voids, held at Harwell, Oxen, England, September 1974, published by UKAEA, Oxen, England, 1975.
17.Annual Conference on Nuclear and Space Radiation Effects, Proceedings Published in December Issues of IEEE Transactions on Nuclear Science.

III. Review Articles

18.Decuyper, J. C., "Review of the behaviour of irradiated transistors and integrated circuits under irradiation", Report CEA BIB 166, Antomique Grenoble, France, 1970, 42 pp., EEA74-22555.
19.Eisen, F. H., "III-V compounds review. (Radiation damage)", International Conference on Radiation Effects in Semiconductors, 1970, p. 273-80, Publ. 1971, PA75-47365.
20.Frank, W., "Radiation damage in metals and semiconductors", In: Ion Implantation Semicond. Other Mater., Proc. Int. Conf., 3rd, 1972, p. 3-16; Crowder, B. L., (Ed.), Plenum, New York, N. Y., 1973, CA81-177961.
21.Holmes-Siedle, A. G., "Radiation sensitivity and amorphous materials - present and future", Rep. Prog. Phys., 37, 699-769 (1974), PA77-69822.
22.Holmes-Siedle, A. G., "Radiation-induced ionization effects in solids - a review of research problems", Proc. IEEE, 62, 1198-209 (1974), PA77-66929.
23.Kravitz, B. L., "Radiation-hardened IC's", Solid State Technol., 13 (1), 57-60 (1970), EEA73-20150.
24.MacKay, J. W. and Klontz, E. E., "Review of radiation effects in germanium", International Conference on Radiation Effects in Semiconductors, 1970, p. 41-50, Publ. 1971, PA75-47375.
25.Miller, G. L., "A brief review of recent advances in compound semiconductors for radiation detectors", IEEE Trans. Nucl. Sci., NS-19, 251-9 (1972), EEA75-17605.
26.Ricketts, L. W., "Radiation effects on microelectronic components and circuits. I", Solid State Technol., 15 (4), 50-5 (1972), EEA75-27768.
27.Rzewuski, H., "Simple radiation defects in semiconductors and their temperature transformations", Electron. Technol., 7 (1-2), 39-50 (1974), PA78-15469.
28.Saidoh, M. and Townsend, P. D., "Mechanism of defect formation", Radiat. Eff., 27, 1-12 (1975).
29.Vook, F. L., "Important unanswered questions - 1970 (Radiation effects in semiconductors)", International Conference on Radiation Effects in Semiconductors, 1970, p. 1-6, Publ. 1971, PA75-47878.

IV. Materials

1. General

30. Banbury, P. C. and Haddad, I. N., "Energy dependence of anisotropy of defect production in electron irradiated diamond-type crystals. II. Theoretical model", Phil. Mag., 14, 841-6 (1966), EEA70-515.

31. Bertolotti, M., De Pasquale, F., Marietti, P., Sette, D. and Vitali, G., "Laser damage on semiconductor surfaces", J. Appl. Phys., 38, 4088-9 (1967), EEA71-3326.

32. Birnbaum, M., "Semiconductor surface damage produced by ruby lasers", J. Appl. Phys., 36, 3688 (1965).

33. Boesch, H. E., Jr., "A computer-aided data reduction technique for radiation-induced conductivity experiments", Report HDL-TR-1545, Harry Diamond Labs., Washington, D. C., 1971, 33 pp., EEA75-16365.

34. Bolotov, V. V., Vasilev, A. V. and Smirnov, L. S., "Dependence of the rate of accumulation of radiation defects in semiconductors on the radiation intensity", Sov. Phys.-Semicond., 7, 1421-3 (1974), PA77-66939.

35. Bourgoin, J. C., "Radioinduced defects in semiconductors", J. Phys. Colloq., no. 5, 49-60 (1973), French, NSA30-982.

36. Bulgakov, Y. V. and Kumakhov, M. A., "Spatial distribution of radiation defects in materials irradiated with beams of monoenergetic particles", Fiz. Tekh. Poluprovodnikov, 2, 1603-7 (1968), Russian, EEA72-6476.

37. Compton, W. D., "An experimental investigation of radiation effects in semiconductors", Report NASA-CR-98835, Illinois Univ., Urbana, Ill., 1968, 16 pp., EEA73-14106.

38. Conrad, E. E., "Considerations in establishing a standard for neutron displacement energy effects in semiconductors", IEEE Trans. Nucl. Sci., NS-18, 200-11 (1971), PA75-19692.

39. Corbett, J. W. and Bourgoin, J. C., "Defect creation in semiconductors", Report AD-777982, State Univ. of New York, Albany, 1973, 355 pp., NSA30-32872.

40. Epshtein, E. M., "Conductivity electron heating by infrared radiation", Izv. VUZ Radiofiz., 13, 1398-402 (1970), Russian, EEA74-4851.

41. Flanagan, T. M., "Carrier scattering from defects in neutron-bombarded semiconductors", IEEE Trans. Nuclear Sci., NS-15 42-6 (1968), EEA72-11187.

42. Gerlach, E., "Scattering of electrons in semiconductors by a charged dislocation", Phys. Status Solidi B, 62, K43-5 (1974), PA77-32928.

43. Grinberg, A. A., Mekhtiev, R. F., Ryvkin, S. M., Salamanov, V. M. and Yaroshetskii, I. D., "Absorption of laser radiation and damage in semiconductors", Soviet Phys. Solid State, 9, 1085-90 (1967), EEA70-20410.

44.Hench, L. L. and Daughenbaugh, G. A., "Radiation effects in semiconducting glasses", J. Nucl. Materials, 25, 58-63 (1968), EEA71-10527.
45.Holmes, R. R., Wilson, D. K. and Blair, R. R., "Neutron damage failure rates for large populations", IEEE Trans. Nucl. Sci., NS-19, 414-17 (1972), EEA76-12241.
46.Iskanderova, Z. A., Kiv, A. E., Malkin, A. A. and Yanchuk, V. A., "Subthreshold impact mechanisms in the formation of radiation defects in semiconductors", Sov. Phys.-Semicond., 7, 1172-4 (1974), EEA77-56768.
47.Kalma, A. H., "Irradiation-produced band-tailing in semiconductors", IEEE Trans. Nucl. Sci., NS-20, 224-28 (1973), NSA30-943.
48.Karpenko, V. P., Kashererininov, P. G., Matveev, O. A. and Tomasov, A. A., "Determination of the parameters of impurity centres in semiconducting materials by the method of charge reversal with the aid of electromagnetic radiation", Sov. Phys.-Semicond., 7, 1268-72 (1974), PA77-72863.
49.Klein, C. A., "Bandgap dependence and related features of radiation ionization energies in semiconductors", J. Appl. Phys., 39, 2029-38 (1968), EEA71-13243.
50.Kobayashi, M., "Stored charge estimation by electron-beam irradiation", Rev. Electr. Commun. Lab., 22, 695-703 (1974), EEA78-7877.
51.Kryachko, V. V. and Synorov, V. F., "Beta spectrograph for irradiating semiconductors in the range 0.1 to 1 keV", Instrum. Exper. Tech., 6, 1346-8 (1966), EEA71-8222.
52.Kyser, D. F. and Wittry, D. B., "Spatial distribution of excess carriers in electron-beam excited semiconductors", Proc. IEEE, 55, 733-4 (1967), EEA70-18821.
53.Landsberg, P. T., "Radiative decay in compound semiconductors", Solid-State Electron., 10, 513-37 (1967), EEA70-16509.
54.Lezheiko, L. V., Pridachin, N. B. and Smirnov, L. S., "Conduction in disordered regions produced by bombardment of semiconductors with heavy particles", Sov. Phys.-Semicond., 6, 775-7 (1972), EEA76-706.
55.Litvinov, V. L., Sadikov, V. P., Smilga, V. P. and Ukhin, N. A., "Asymmetry of the damage regions generated by high-energy particles in semiconductors", Sov. Phys.-Semicond., 6, 460-1 (1972), PA75-74759.
56.Matsko, P., "Effect of neutron irradiation on the change of certain physical parameters of Te-Se-Bi and Te-Sb-Si systems", Fyz. Casopis, 17, 51-60 (1967), Russian, EEA71-15751.
57.Munakata, C., "On the voltage induced by an electron beam in a bulk semiconductor crystal", Jap. J. Appl. Phys., 5, 745-63 (1966).
58.Munakata, C., "Detection of resistivity variation in a semiconductor pellet with an electron beam", Microelectronics and Reliability, 6, 27-33 (1967).

59. Norris, C. B., "Simulation of transient space-charge flow in a semiconductor under ionizing irradiation", J. Appl. Phys., 43, 3769-73 (1972), PA75-71954.
60. Norris, D. I. R. and Brown, L. M., "The stability of voids in materials under irradiation", Scr. Metall., 8, 1045-7 (1974), PA78-1154.
61. Schmidt, P. F., McCaughan, D. V. and Kushner, R. A., "Problems in the analysis of semiconductor device materials exposed to ionizing radiation", Proc. IEEE, 62, 1220-23 (1974), NSA34-6179.
62. Shipatov, E. T. and Konovov, B. A., "Investigation of the channelling of protons in single crystals of ionic compounds and semiconductors", Izv. VUZ Fiz., no. 9, 52-6 (1968), Russian, EEA72-13503.
63. Stoneham, A. M., "The theory of defects in irradiated semiconductors", International Conference on Radiation Effects in Semiconductors, 1970, p. 7-13, Publ. 1971, PA75-47364.
64. Vasilevskaya, V. N., Glinchuk, K. D., Linnik, L. F., Litovchenko, N. M. and Lyadskaya, O. N., "Destruction of laser-irradiated semiconductors", Poluprovodn. Tekh. & Mikroelektron., no. 8, 44-50 (1972), Russian, EEA76-16251.
65. Vitovskii, N. A., Emtsev, V. V., Mashovets, T. V. and Morozov, Y. G., "Kinetics of the interaction between point radiation defects and impurity atoms in semiconductors", Sov. Phys.-Semicond., 8, 1489-90 (1975), PA78-68950.
66. Wiedersich, H., "On the theory of void formation during irradiation", Radiat. Eff., 12, 111-25 (1972), PA75-25365.
67. Wolf, E. D., "Low-energy electron beam interactions with ion-implanted semiconductor surfaces", Record IEEE Ninth Annual Symposium on Electron, Ion, and Laser Beam Technology, 1967, p. 135-44, EEA71-15743.
68. Zander, K., "Interaction of radiation with condensed matter with special regard to semiconductors", Nachrichtentech. Z. (NTZ), 26, 305-11 (1973), German, EEA76-32191.

2. Group IV Semiconductors

a. Silicon

69. Abou-el-Fotouh, F. A. and Newman, R. C., "Electron irradiation damage in silicon containing carbon and diffused ^{18}O", Solid State Commun., 15, 1409-11 (1974), PA77-82203.
70. Abroyan, I. A. and Tsekhnovicher, L. A., "Influence of electron bombardment of the radiation conductivity of silicon", Fiz. Tekh. Poluprovodinkov, 2, 1221-3 (1968), Russian, EEA72-700.
71. Abuladze, M. N., Gerasimov, A. B., Litovchenko, V. G., Melkadze, T. E. and Shillo, A. G., "Kinetics of formation and annealing of radiation defects on the surface of silicon", Sov. Phys.-Semicond., 8, 508-9 (1974), PA78-15709.

72. Akimchenko, I. P., Bulgakov, Y. V. and Kolomenskaya, T. I., "Energy spectra of radiation defects generated in Si by irradiation with deuterons", Fiz. Tekh. Poluprovodnikov, 3, 157-9 (1969), Russian, EEA72-15442.
73. Alexander, D. R. and Antinone, R. J., "MSI testing at dose rates from 10^9 to 10^{12} rad(Si)/S", IEEE Trans. Nucl. Sci., NS-18, 439-42 (1971), EEA75-12739.
74. Arslambekov, V. A., Gorbunova, K. M. and Smushchenko, V. Y., "Effects of electron bombardment on some electro-physical and physicochemical properties of silicon surfaces", Radiats. Defekty Poluprov., 6-8 (1972), Russian, CA79-59114.
75. Azimov, K. S., Gorodetskii, S. M., Grigoreva, G. M., Kreinin, L. B. and Landsman, A. P., "Influence of disordered regions on the recombination in proton-irradiated p-type silicon", Sov. Phys.-Semicond., 7, 1021-5 (1974), EEA77-22761.
76. Barnes, C. E., "Thermal and infection annealing of neutron-irradiated p-type silicon between 76 and 300K", IEEE Trans. Nucl. Sci., NS-16, 28-32 (1969), EEA73-14013.
77. Barnes, C. E., "Absorption measurements in neutron irradiated silicon", International Conference on Radiation Effects in Semiconductors, 1970, p. 203-9, Publ. 1971, PA75-47389.
78. Barrett, M. J., "Electron damage coefficients in P-type silicon", IEEE Trans. Nuclear Sci., NS-14, 82-6 (1967), EEA71-15753.
79. Baruch, P., Bouvatier, M., Picard, C., Epelboin, Y., Monnier, J. and Netange, B., "Irradiation-enhanced diffusion in silicon. Kinetic study and impurity distribution", J. Phys. Colloq., no. 5, 131-6 (1973), French, CA81-55211.
80. Baruch, P., Castaing, C., Monnier, J. and Blanchard, B., "Redistribution of boron in silicon after high-temperature proton irradiation", Appl. Phys. Lett., 26, 77-80 (1975), PA78-33197.
81. Baruch, P., Castaing, C., Monnier, J. and Blanchard, B., "Redistribution of boron in silicon through high-temperature proton irradiation", Lattice Defects in Semiconductors, 1974, p. 453-60, Publ. 1975, PA78-81109.
82. Bass, R. F., "Influence of impurities on carrier removal and annealing in neutron-irradiated silicon", IEEE Trans. Nucl. Sci., NS-14, 78-81 (1967), EEA71-15752.
83. Bass, R. F. and Curtis, O. L., Jr., "Annealing of 10 MeV electron damage in silicon", IEEE Trans. Nucl. Sci., NS-15, 47-54 (1968), EEA72-11193.
84. Batavin, V. V., Kuznetsov, Y. N., Perezhogin, G. A., Pantuev, V. S. and Popova, G. V., "Application of radiation doping in fabrication of uniform standard silicon specimens", Ind. Lab., 38, 1041-2 (1972), EEA76-25966.
85. Bean, A. R., Morrison, S. R., Newman, R. C. and Smith, R. S., "Electron irradiation damage in silicon containing high concentration of boron", J. Phys. C, 5, 379-400 (1972), PA75-22178.

86.Beatty, M. E. and Hill, G. F., "Annealing of 22-, 40-, and 158-MeV proton damage in n- and p-type silicon", Report NASA-TN-D-5119, NASA, Langley Res. Centre, Langley Station, Va., 1969, 26 pp., EEA73-21095.

87.Belitskaya, S. G. and Ordynskaya, V. V., "Changes in some surface properties of silicon single crystals under the influence of illumination", Fiz. Tekh. Poluprovodnikov, 1, 1526-30 (1967), EEA71-5872.

88.Bielle-Daspet, D., "Recombination centres in n- and p-type silicon irradiated by very high energy protons", Proceedings of the International Conference on Radiation Damage and Defects in Semiconductors, 1972, p. 129-39, Publ. 1973, EEA76-19816.

89.Bielle-Daspet, D. and Pouget, M., "Influence of injection rate on carrier lifetime degradation in p-type silicon irradiated by 4.5 MeV electrons", Nucl. Instrum. & Meth., 93, 413-15 (1971), EEA74-26603.

90.Bortnik, M. V., Tkachev, V. D. and Yukhnevich, A. V., "Radiative recombination at radiation defects in silicon", Sov. Phys. Semicond., 1, 290-4 (1967), EEA70-16493.

91.Brandt, W. and Cheng, L. J., "Position escape from annihilation centers in electron-irradiated Si crystals", Phys. Lett. A, 50A, 439-40 (1975), PA78-25416.

92.Brelot, A., "Selective trapping of vacancies (3MeV electron irradiation)", Proceedings of the International Conference on Radiation Damage and Defects in Semiconductors, 1972, p. 191-201, Publ. 1973, EEA76-19818.

93.Brelot, A. and Charlemagne, J., "Infrared studies of low temperature electron irradiated silicon containing germanium oxygen and carbon", International Conference on Radiation Effects in Semiconductors, 1970, p. 161-9, Publ. 1971, PA75-48192.

94.Brower, K. L., "Structure of multiple-vacancy (oxygen) centers in irradiated silicon", International Conference on Radiation Effects in Semiconductors, 1970, p. 189-95, Publ. 1971, PA75-48283.

95.Brozel, M. R., Newman, R. C. and Totterdell, D. H. J., "Interstitial defects involving carbon in irradiated silicon", J. Phys. C, 8, 243-8 (1975), PA78-33154.

96.Brucker, G. J., "Electrical studies of electron-irradiated lithium-containing n-type silicon", Phys. Rev., 183, 712-24 (1969), EEA73-1082.

97.Brucker, G. J., "Annealing of electron bombardment damage in lithium-containing silicon", IEEE Trans. Nucl. Sci., NS-17, 144-9 (1970), EEA74-22463.

98.Brughas, S., Grigonis, A. and Pranevichyus, L., "Examination of the state of a silicon surface after ionic bombardment", Fiz. & Khim. Obrab. Mater., no. 3, 120-22 (1974), Russian, PA77-82776.

99.Buehler, M. G., "Design curves for predicting fast-neutron-induced resistivity changes in silicon", Proc. IEEE, 56, 1741-3 (1968), EEA72-825.

100. Bulgakov, Y. V. and Razumovskii, S. M., "Electrical conductivity in silicon irradiated with fast helium ions", Radiats. Defekty Poluprov., 211-13 (1972), Russian, CA79-59113.
101. Bulgakov, Y. V. and Shulga, V. I., "Application of the impact parameter method to the channeled ion energy distribution calculation", Phys. Lett. A, 48, 429-30 (1974), NSA31-2241.
102. Carter, J. R., Jr., "Defect impurity relationships in electron-damaged silicon", IEEE Trans. Nuclear Sci., NS-13, 24-32 (1966), EEA70-12564.
103. Carter, J. R., Jr., "Radiation damage in lithium doped silicon", IEEE Trans. Nucl. Sci., NS-14, 110-15 (1967), EEA71-15756.
104. Carter, J. R., Jr., "Effect of electron irradiation on lithium-doped silicon", J. Phys. Chem. Solids, 31, 2405-16 (1970), EEA74-7580.
105. Carter, J. R., Jr. and Downing, R. G., "Annealing of radiation damage in lithium doped silicon", Conference Record of the 6th Photovoltaics Specialists Conference, 1967, p. 145-56, EEA71-13245.
106. Chen, C. S., "Piezospectroscopic studies of radiation-induced infrared absorption bands in silicon", Thesis, Rensselaer Polytech. Inst., Troy, N. Y., 1971, 128 pp., Order No. 72-10849. PA75-68650.
107. Chen, C. S. and Corelli, J. C., "Piezospectroscopic study (78K) of radiation-produced absorption hands in Si", International Conference on Radiation Effects in Semiconductors, 1970, p. 197-201, Publ. 1971, PA75-48193.
108. Chen, C. S. and Corelli, J. C., "Spectroscopy study of copper impurities in silicon (before and after neutron irradiation)", J. Appl. Phys., 44, 5622-3 (1973), PA77-36912.
109. Chen, C. S. and Corelli, J. C., "Absorption coefficient of neutron-irradiated silicon doped with lithium or copper", Report 766768/6, Div. Nucl. Eng., Rensselaer Polytech. Inst., Troy, N. Y., 1973, 71 pp., CA80-53616.
110. Cheng, L. J. and Lori, J., "Electrical properties of gamma-irradiated p-type silicon", Phys. Rev. B., 1, 1558-62 (1970), EEA73-24843.
111. Cheng, L. J. and Swanson, M. L., "Photoconductivity in neutron-irradiated p-type Si", J. Appl. Phys., 41, 2627-35 (1970), EEA73-28021.
112. Cherki, M. and Kalma, A. H., "Studies of defects in irradiated silicon; effect of uniaxial stress on the photoconductivity spectra", Solid State Commun., 7, pap. 5.4 (June 1969), EEA72-23540.
113. Colclaser, R. A., Southward, H. D. and Baca, J. P., "Evaluation of radiation damaged epitaxial silicon", Proc. IEEE, 62, 1232-35 (1974), NSA31-6177.
114. Colwell, J. F., "Correlation of proton damage of silicon with neutron and electron damage", Radiat. Eff., 24, 239-45 (1975), PA78-41820.

115. Corelli, J. C., Young, R. C. and Chen, C. S., "A study of the divacancy in irradiated silicon using infrared spectroscopy and infrared photoconductivity measurements", IEEE Trans. Nucl. Sci., NS-17, 128-36 (1970), EEA74-22458.
116. Curtis, O. L., Jr., "Effects of oxygen and dopant on lifetime in neutron-irradiated silicon", IEEE Trans. Nucl. Sci., NS-13, 33-40 (1966), EEA70-12563.
117. Curtis, O. L., Jr. and Bass, R. F., "Radiation effects in bulk lithium- and aluminum-doped silicon", Presented at 3rd Annual Conference on Effects of Lithium Doping on Silicon Solar Cells, Pasadena, California, April 1970, NSA30-32882.
118. Curtis, O. L., Jr. and Germano, C. A., "Injection-level studies in neutron-irradiated silicon", IEEE Trans. Nucl. Sci., NS-14, 68-77 (1967), EEA71-15750.
119. DeAngelis, H. M., Diebold, J. W., Komm, D. S. and Nikula, J. V., "Charge-state effects on annealing of electron-irradiated silicon", Report AFCRL-TR-74-0239, Air Force Cambridge Res. Labs., Bedford, Mass., 1974, 29 pp., EEA78-28725.
120. Doi, K., Izui, K., Ohtsu, H. and Tomimitsu, H., "Energy loss of 1000 keV electrons in thin silicon crystals as measured by high voltage electron microscopy", Radiat. Eff., 26, 129-34 (1975).
121. Drain, D., Pigneret, J., Samueli, J. J. and Sarazin, A., "Neutron induced displacement number in silicon", Solid State Electron., 9, 1081-7 (1966), EEA70-2216.
122. Dressler, L., "X-ray examination of silicon monocrystals bombarded with ions", Phys. Status Solidi A, 11, K65-7 (1972), PA75-47386.
123. Dubovoi, V. K., Konozenko, I. D. and Semenyuk, A. K., "Bulk-gradient γ-ray-induced e.m.f. in p-type Si", Fiz. Tekh. Poluprovodnikov, 4, 1575-6 (1970), Russian, EEA74-1320.
124. Eritsyan, G. N. and Melkonyan, R. A., "Study of infrared absorption in silicon single crystals, irradiated with 50 MeV electrons", Izv. Akad. Nauk Arm. SSR Fiz., 9, 260-3 (1974), Russian, PA77-82198.
125. Fenimore, E., Mortka, T. and Corelli, J. C., "Radiation-induced extrinsic photoconductivity in Li-doped Si", J. Appl. Phys., 43, 1062-73 (1972), PA75-25958.
126. Foll, H., "High-voltage electron microscope studies of low-temperature radiation damage in silicon", Lattice Defects in Semiconductors, 1974, p. 233-9, Publ. 1975, PA78-81118.
127. Galushka, A. P., Dolgolenko, A. P., Dudnik, V. P., Zholkevich, G. A., Marchenko, V. I. and Smorodin, V. E., "Conductivity switching in silicon single crystals with clusters of defects", Radiats. Defekty Poluprov., 31-3 (1972), Russian, CA79-59116.
128. Gerasimenko, N. N., Zaitsev, B. A., Panov, V. I., Smirnov, L. S. and Tishkovskii, E. G., "Changes in the relative rates of generation of radiation defects in silicon", Sov. Phys.-Semicond., 7, 960-61 (1974), NSA30-13285.

129.Gorodetskii, S. M. and Kreinin, L. B., "A possible mechanism for the formation of complex radiation defects in silicon caused by electron bombardment", Dokl. Akad. Nauk., 181, 46-8 (1968), Russian, EEA72-11206.
130.Gorodetskii, S. M. and Kreinin, L. B., "Influence of disordered regions on carrier recombination in irradiated p-type silicon", Sov. Phys.-Semicond., 8, 428-30 (1974), EEA78-5219.
131.Graff, K. and Pieper, H., "Carrier lifetime measurements on electron-irradiated silicon crystals", Phys. Status Solidi A, 30, 593-9 (1975), EEA78-41426.
132.Gregory, B. L., "Minority carrier recombination in neutron irradiated silicon", IEEE Trans. Nucl. Sci., NS-16, 53-62 (1969), EEA73-14017.
133.Grishaev, I. A., Morokhovskii, V. L. and Shramenko, B. I., "Anomalous electron and positron scattering by silicon monocrystal at 1 GeV energy", Report KFTI-73-26, KFTI AN Ukrainian SSR, Kharkov, Ukrainian SSR, 1973, p. 56-67, Russian, NSA30-32892.
134.Gwozdz, P. S. and Koehler, J. S., "Changes in a.c. conductivity of silicon with electron irradiation at 0.5 K", Phys. Rev. B, 6, 4571-4 (1972), EEA76-4450.
135.Haddad, I. N. and Banbury, P. C., "Energy dependence of anisotropy of defect production in electron irradiated diamond-type crystals. I. Experimental measurements on n and p-type silicon", Phil. Mag., 14, 829-40 (1966), EEA70-514.
136.Hemment, P. L. F., "Defect introduction rates and defects identification studies in low-energy electron irradiated silicon", Nucl. Instrum. & Meth., 92, 471-6 (1971), EEA74-18599.
137.Henderson, H. T., "Bulk 'on', 'off', and negative resistance modes in neutron irradiated silicon", Presented at International Electron Devices Meeting, October 1969, Washington, D. C., EEA70-21165.
138.Herrmann, H. A. and Herzer, H., "Doping of silicon by neutron irradiation", J. Electrochem. Soc., 122 (11), 1568-9 (1975).
139.Holmes, R. R., "Carrier removal in neutron irradiated silicon", IEEE Trans. Nucl. Sci., NS-17, 137-43 (1970), EEA74-22459.
140.Isaev, N. U., Karsybaev, M. S., Kazdaev, K. R. and Tkachev, V. D., "Annealing of radiation defects in n-type silicon irradiated with 14-MeV neutrons", Vestn. Belorus, un-ta, ser. 1, no. 2, 81-2 (1974), Russian, CA82-179577.
141.Isaev, N. U., Karsybaev, M. S. and Tkachev, V. D., "Radiation defects in silicon after irradiation with 14 MeV neutrons and annealing", Sov. Phys.-Semicond., 9, 778-9 (1975).
142.Johnson, E. S., "Recombination luminescence in irradiated silicon: effects of thermal annealing and lithium impurity", Thesis, Univ. Illinois, Urbana, 1971, 194 pp., Order No. 72-6967, PA75-48229.

143.Johnson, E. S. and Compton, W. D., "Recombination luminescence in irradiated silicon - effects of thermal annealing and lithium impurity", International Conference on Radiation Effects in Semiconductors, 1970, p. 219-22, Publ. 1971, PA75-48231.
144.Johnson, E. S., Compton, W. D., Noonan, J. R. and Streetman, B. G., "Recombination luminescence from electron irradiated Li-diffused Si", J. Appl. Phys., 44, 5411-18 (1973), PA77-37660.
145.Jones, C. E. and Compton, W. D., "Recombination luminescence in irradiated silicon-effects of uniaxial stress and temperature variations", International Conference on Radiation Effects in Semiconductors, 1970, p. 213-18, Publ. 1971, PA75-48230.
146.Jones, C. E., Johnson, E. S., Compton, W. D., Noonan, J. R. and Streetman, B. G., "Temperature, stress, and annealing effects on the luminescence from electron irradiated silicon", J. Appl. Phys., 44, 5402-10 (1973), PA77-37659.
147.Jones, C. E., Streetman, B. G. and Noonan, J. R., "Reply to "comments on 'luminescence from electron-irradiated silicon'"", J. Appl. Phys., 45, 4655 (1974), NSA30-30099.
148.Kao, K. C. and Pal, N., "Electric conduction in float-zone silicon after prolonged gamma irradiation", Electron. Lett., 4, 571-2 (1968), EEA72-13324.
149.Kawamoto, H., "Lifetime degradation in neutron irradiated silicon", Thesis, Univ. California, Berkeley, 1970, 151 pp., Order No. 71-798, EEA74-26779.
150.Kawamoto, H. and Oldham, W. G., "Annealing of lifetime damage in neutron-irradiated silicon", Appl. Phys. Lett., 16, 346-8 (1970), EEA73-24782.
151.Kazanskii, A. G. and Koshelev, O. G., "Anisotropy of electron scattering in n-type Si", Sov. Phys.-Semicond., 6, 826-8 (1972), EEA76-709.
152.Kimerling, L. C. and Drevinsky, P. J., "Carrier removal effects in neutron-irradiated lithium-doped silicon", IEEE Trans. Nucl. Sci., NS-18 (6), 60-8 (1971), EEA75-10116.
153.Kimerling, L. C., Drevinsky, P. J. and Chen, C. S., "Defect interactions in neutron-irradiated, lithium-doped silicon", Proceedings of the International Conference on Radiation Damage and Defects in Semiconductors, 1972, p. 182-90, Publ. 1973, EEA76-19817.
154.Klimkova, O. A. and Niyazova, O. R., "Radiation-accelerated diffusion of gold in silicon", Fiz. Tver. Tela, 12, 2199-200 (1970), Russian, EEA74-1343.
155.Koifman, A. I. and Niyazova, O. R., "Broadening of the diffusion profiles in silicon under the action of ionizing radiation", Sov. Phys.-Semicond., 6, 656-7 (1972), PA75-84355.
156.Kolomenskaya, T. I., Razumovskii, S. M. and Vintovkin, S. I., "Temperature dependence of the threshold for the formation of radiations defects in silicon bombarded with electrons", Fiz. Tekh. Poluprovodnikov, 1, 789-91 (1967), Russian, EEA70-20406.

157.Konopleva, R. F. and Rubinova, E. E., "Slow process in the kinetics of the decay of the photocurrent in n-type Si irradiated with 660 MeV protons", Sov. Phys.-Semicond., 7, 956-7 (1974), EEA77-18621.
158.Konozenko, I. D., Semenyuk, A. K. and Khivrich, V. I., "Radiation defects in Si of high purity", International Conference on Radiation Effects in Semiconductors, 1970, p. 249-55, Publ. 1971, PA75-47937.
159.Kosmach, V. F., Kulakov, V. M., Ostroumov, V. I. and Petukhov, A. M., "Calculation of the number of defects generated by bombardment of silicon with fast neutrons", Sov. Phys.-Semicond., 6, 364-6 (1972), PA75-71458.
160.Koval, Y. P., Mordkovich, V. N., Temper, E. M. and Kharchenko, V. A., "Optical properties of neutron-irradiated silicon", Sov. Phys.-Semicond., 6 (7), 1152-5 (1973), NSA27-26636.
161.Kozlov, I. P., "Effect of lithium on the formation of radiation defects in silicon", Mater. Resp. Konf. Molodykh Uch. Fiz., 2nd 1972, p. 85-6, Russian. Also Bogush, A. A., (Ed.), Akad. Nauk Beloruss, SSR, Inst. Fiz.: Minsk, USSR, CA82-67175.
162.Laithwaite, K., Newman, R. C. and Totterdell, D. H., "Interstitial defects involving boron in irradiated silicon", J. Phys. C., 8, 236-42 (1975), PA78-33153.
163.Leadon, R. E., "Model for short-term annealing of neutron damage in p-type silicon", IEEE Trans. Nucl. Sci., NS-17, 110-17 (1970), EEA74-22455.
164.Levchuk, L. V., Galushka, A. P. and Konozenko, I. D., "Effect of irradiation by electrons of energies 1.2 MeV on the electrophysical properties of p-Si, grown in a hydrogen atmosphere", Ukr. Fiz. Zh., 17, 851-4 (1972), Russian, PA75-54201.
165.Littlejohn, M. A. and Lade, R. W., "Influence of Co^{60} gamma irradiation on the surface and bulk recombination rates in silicon", IEEE Trans. Nucl. Sci., NS-14, 305-18 (1967), EEA71-15736.
166.Loggins, C. D., Jr., "The effect of electron irradiation damage on the resistivity and piezoresistivity of n-type silicon", Thesis, North Carolina State Univ., Raleigh, 1970, 147 pp., Order No. 71-10311, EEA74-33986.
167.Lotkova, E. N., "Investigation of the infrared absorption spectrum of neutron-irradiated silicon", In: Electrical and optical properties of semiconductors, Consultants Bur., New York, 1968, p. 87-125, EEA72-6508.
168.Magee, T. J., Hermann, A. M. and Deck, R. J., "Observation of defect annealing in neutron-irradiated silicon by space-charge-limited current", Phys. Rev. B, 5, 3364-6 (1972), PA75-33020.
169.Mallon, C. E. and Naber, J. A., "Short-term anneal of 30-MeV electron damage in high-purity n-type silicon", IEEE Trans. Nucl. Sci., NS-17, 123-7 (1970), EEA74-22457.

170. McKeighen, R. E. and Koehler, J. S., "Electron irradiation effects in silicon at liquid helium temperatures using ac hopping conductivity", International Conference on Radiation Effects in Semiconductors, 1970, p. 141-6, Publ. 1971, PA75-47936.
171. Mordkovich, V. N., Solovev, S. P., Temper, E. M. and Kharchenko, V. A., "Resistivity of silicon subjected to neutron bombardment and annealing", Sov. Phys.-Semicond., 8, 139-40 (1974), PA77-75860.
172. Moyer, N. E., "X-ray lattice parameter studies of electron-irradiated silicon", Proceedings of the Santa Fe Conference on Radiation Effects in Semiconductors, 1967, p. 444-51, Publ. 1968, PA74-23856.
173. Murr, L. E. and Szilva, W. A., "Laser-induced fracture in silicon", J. Mater. Sci., 10, 1536-48 (1975).
174. Naber, J. A., Horiye, H. and Passenheim, B. C., "Lithium - an impurity of interest in radiation effects of silicon", International Conference on Radiation Effects in Semiconductors, 1970, p. 233-8, Publ. 1971, PA75-47385.
175. Naber, J. A., Mallon, C. E. and Leadon, R. E., "Charge-state dependence of defects produced in electron-irradiated silicon", Radiat. Damage Defects Semicond., Proc. Int. Conf., 1972, p. 26-33, Publ. 1973, CA79-36382.
176. Nakashima, K., "Illumination induced annealing in electron-irradiated p-type silicon", Phys. Lett. A, 42A, 533-4 (1973), EEA76-12460.
177. Nakashima, K. and Inuishi, Y., "Interaction between radiation defects and impurities in p-type Si", Technol. Rep. Osaka Univ., 18, 415-23 (1968), EEA72-15476.
178. Nakashima, K. and Inuishi, Y., "Studies of recombination centers in gamma-irradiated p-type silicon", J. Phys. Soc. Japan, 29, 1500-12 (1970), EEA74-7578.
179. Narasimha Rao, D. V. G. L., "Laser-induced resistivity changes in silicon", J. Appl. Phys., 39, 4853 (1968), EEA71-28836.
180. Natarajan, K. and Heasell, E. L., "Defect centers in silicon produced by room temperature electron irradiation. I. Optical and photoconductivity studies", Phys. Status Solidi A, 28, 603-8 (1975), PA78-46172.
181. Natarajan, K. and Heasell, E. L., "Defect centers in silicon produced by room temperature electron irradiation. II. Thermally stimulated conductivity studies", Phys. Status Solidi A, 29, 207-12 (1975), PA78-58362.
182. Natarajan, K. and Heasell, E. L., "Defect centers in silicon produced by room temperature electron irradiation. III. Hall coefficient studies", Phys. Status Solidi A, 29, 415-20 (1975), PA78-61729.
183. Nelson, D. G., "Radiation-enhanced diffusion of boron in silicon", Thesis, Stanford Univ., Calif., 1970, 91 pp., Order No. 70-22188, EEA74-22554.

184.Newman, R. C. and Bean, A. R., "Irradiation damage in carbon doped silicon irradiated at low temperatures by 2 MeV electrons (IR spectra)", International Conference on Radiation Effects in Semiconductors, 1970, p. 155-9, Publ. 1971, PA75-48191.
185.Norris, C. B., "Investigation of static electron irradiation effects in bulk Si and thin Si films at energies far below threshold", J. Appl. Phys., 43, 4060-7 (1972), PA75-80655.
186.Ohmura, Y., Abe, T., Mimura, S., Konaka, M., Kanazawa, M., Ohtsubo, H. and Fujinuma, K., "Enhanced diffusion of substrate impurities into epitaxial layers in Si by proton irradiation (application to non microwave transistors)", Proceedings of the 2nd International Conference on Ion Implantation in Semiconductors, Physics and Technology, Fundamental and Applied Aspects, 1971, p. 335-9, EEA75-14026.
187.Ohmura, Y., Zohta, Y. and Kanazawa, M., "Electrical properties of n-type Si layers doped with proton bombardment induced shallow donors", Solid State Commun., 11, 263-6 (1972), EEA75-32541.
188.Othmer, S. and Curtis, O. L., Jr., "Scanning electron microscope measurements of diffusion length in neutron-irradiated silicon", IEEE Trans. Nucl. Sci., NS-20 (6), 204-8 (1973), PA77-41082.
189.Panov, V. I., Smirnov, L. S. and Tishkovskii, E. G., "Interaction between thermal and radiation defects in silicon", Sov. Phys.-Semicond., 9 (8), 1041-2 (1975).
190.Petrov, V. V. and Tkachev, V. D., "Influence of the trapping effects on the photoconductivity spectra of n-type silicon irradiated with fast electrons", Sov. Phys.-Semicond., 8, 916-17 (1975), EEA78-19564.
191.Poblenz, F. W., Taulbee, C. D. and Walker, R. L., "Application of silicon damage to neutron exposure measurement", IEEE Trans. Nucl. Sci., NS-14, 147-52 (1967), EEA71-15503.
192.Price, J. C. and Kiss, A. E., "Activation energy of defects in neutron irradiated silicon", IEEE Trans. Nucl. Sci., NS-15, 3-5 (1968), EEA71-28826.
193.Reuschel, K., Schnoeller, M., Muehlbauer, A., Spenke, E. and Keller, W., "Homogeneously doping silicon single crystal with n-conducting and adjustable dopant concentrating by neutron bombardment", Patent German 2362264, Publ. June 1975, CA83-124974.
194.Reuschel, K., Schnoeller, M., Muehlbauer, A. and Spenke, E., "Homogeneously doped silicon single crystal by neutron irradiation", Patent German 2362320, Publ. June 1975, CA83-124973.
195.Schwegler, E. C., "Calculation of reverse annealing in neutron damaged p-type silicon", IEEE Trans. Nucl. Sci., NS-18 (6), 69-77 (1971), EEA75-10117.
196.Seryapin, V. G., Seryapina, N. V., Smirnov, L. S., Romanov, S. I. and Obodnikov, V. I., "Redistribution of boron in silicon as a result of irradiation with fast electrons", Sov. Phys.-Semicond., 7, 128-9 (1973), EEA76-35244.

197. Sigmon, T. W., Gibbons, J. F. and Norris, C. B., Jr., "Observation of localized radiation damage in silicon", Appl. Phys. Lett., 14, 90-2 (1969), EEA72-15439.
198. Sirota, N. N., Ivanov, G. M. and Korshunov, F. P., "Effects of disordered regions on the rate of change of resistivity in proton-irradiated silicon", Radiats. Defekty Poluprov., 262-3 (1972), Russian, CA79-59077.
199. Sivo, L. L. and Klontz, E. E., "Studies of radiation damage in degenerate silicon irradiated at low temperatures", Phys. Rev., 178, 1264-73 (1969), EEA72-13346.
200. Sprague, J. A. and Rudee, M. L., "The symmetry of defect clusters in neutron-irradiated silicon", Phys. Status Solidi, 36, K83-5 (1969), MA3-05-16 0118.
201. Srour, J. R., "Short-term annealing in electron-irradiated p-type silicon", IEEE Trans. Nucl. Sci., NS-17, 118-22 (1970), EEA74-22456.
202. Srour, J. R., "Stable-damage comparisons for neutron-irradiated silicon", IEEE Trans. Nucl. Sci., NS-20, 190-5 (1973), PA77-41080.
203. Srour, J. R. and Curtis, O. L., Jr., "Increased radiation tolerance in heat-treated aluminum-doped Czochralski grown silicon", Presented at Conference Record of the 8th IEEE Photovoltaic Specialists Conference, Seattle, Wash., August 1970, EEA74-7615.
204. Stannard, J., "Hall effect studies of irradiated Si(Li) at cryogenic temperatures", Presented at the 3rd Annual Conference on Effects of Lithium Doping on Silicon Solar Cells, Pasadena, Calif., April 1970, NSA30-32880.
205. Stein, H. J., "Comparison of neutron and gamma-ray damage in n-type silicon", J. Appl. Phys., 37, 3382-4 (1966), EEA70-521.
206. Stein, H. J., "Electrical studies of neutron-irradiated n-type Si: defect structure and annealing", Phys. Rev., 163, 801-8 (1967), EEA71-3322.
207. Stein, H. J., "Electrical properties of neutron-irradiated silicon at 76°K: Hall effect and electrical conductivity", IEEE Trans. Nucl. Sci., NS-15, 69-76 (1968), EEA72-11195.
208. Stein, H. J., "Defects in silicon: concepts and correlations", International Conference on Radiation Effects in Semiconductors, 1970, p. 125-40, Publ. 1971, PA75-47387.
209. Stein, H. J. and Vook, F. L., "Electrical studies of electron-irradiated n-type Si: Impurity and irradiation-temperature dependence", Phys. Rev., 163, 790-800 (1967), EEA71-3321.
210. Stevens, P. R. C., "Calculations of electron-radiation damage rate as function of depth in silicon", J. Appl. Phys., 39, 1561-3 (1968), EEA71-13240.
211. Thuong Quat, V. and Nicolet, M. A., "Effect of 14 MeV neutrons on space-charge-limited current of electrons in high purity silicon", IEEE Trans. Nucl. Sci., NS-16, 69-80 (1969), EEA73-14018.

212. Thuong Quat, V. and Nicolet, M. A., "Electron trapping in neutron-irradiated silicon studied by space-charge-limited current", J. Appl. Phys., 43, 2755-62 (1972), PA75-51168.
213. Tkachev, V. D. and Yanchenko, A. M., "Mechanism of the thermally stimulated conductivity of silicon irradiated with fast electrons", Sov. Phys.-Semicond., 9, 582-4 (1975).
214. Ukhin, N. A., "Model of disordered regions in silicon produced by fast neutrons", Sov. Phys.-Semicond., 6, 804-6 (1972), EEA76-708.
215. Urli, N. B., "Annealing of γ-ray irradiated lithium compensated p-type silicon", International Conference on Radiation Effects in Semiconductors, 1970, p. 239-43, Publ. 1971, PA75-47390.
216. Usami, A. and Tokuda, Y., "Effect of oxygen and copper on the defect cluster in neutron-irradiated p-type silicon", J. Appl. Phys., 45, 2823-31 (1974), NSA30-13288.
217. van Lint, V. A. J. and Leadon, R. E., "Implications of cluster model of neutron effects in silicon", Lattice Defects on Semiconductors, 1974, p. 227-32, Publ. 1975, PA78-81127.
218. Vavilov, V. S., Mukashev, B. N. and Spitsyn, A. V., "The influence of oxygen and acceptor impurities on the annealing of radiation defects in p-type silicon", Proceedings of the International Conference on Radiation Damage and Defects in Semiconductors, 1972, p. 284-8, Publ. 1973, EEA76-19820.
219. Vavilov, V. S., Spitsyn, A. V., Kolodin, L. G., Mukashev, B. N., Nussupov. K. H., Takibaev, Z. S. and Inishtikbaev, K. B., "On the interaction of the primary radiation defects with impurites in p-type silicon at 80-200K", Lattice Defects in Semiconductors, 1974, p. 185-9, Publ. 1975, PA78-81114.
220. Vedam, K., "Optical properties of silicon and the effect of irradiation on them", Presented at the 3rd Annual Conference on Effects of Lithium Doping on Silicon Solar Cells, Pasadena, Calif., April 1970, NSA30-32885.
221. Vyatskin, A. Y. and Khramov, V. Y., "Interaction of electrons of initial energies up to 4 keV with thin silicon films", Sov. Phys.-Solid State, 16, 2400 (1975), PA78-81140.
222. Walker, J., "Comments on 'luminescence from electron-irradiated silicon'", J. Appl. Phys., 45, 4653-54 (1974), NSA30-30101.
223. Walker, J. W. and Sah, C. T., "Properties of 1 MeV electron-irradiated defect centres in p-type silicon", Phys. Status Solidi A, 11, 513-22 (1972), PA75-54199.
224. Walker, J. W. and Sah, C. T., "Spherical-square-well defect-potential model for 1-MeV electron irradiated defects in silicon", Phys. Rev. B, 8, 5597-603 (1973), PA77-48929.
225. Walker, R. L., Berman, B. L. and Bloom, S. D., "Coherent bremsstrahlung produced by relativistic positrons and electrons on a single crystal of silicon", Phys. Rev. A, 11, 736-42 (1975), PA78-46225.

226.Whorl, C. A. and Evans, A. W., "Development of high-temperature, high radiation-resistant silicon semiconductors", Report NASA-CR-124074, Solitron Devices, Inc., Riviera Beach, Fla., 1972, 65 pp., NSA30-6770.
227.Wikner, E. G. and Horiye, H., "Study of n- and p-type silicon exposed to highly energetic radiation", IEEE Trans. Nucl. Sci., NS-13, 18-23 (1966), EEA70-12557.
228.Wilsey, N. D., Statler, R. I. and Faraday, B. J., "Defect clusters in electron-irradiated silicon", IEEE Trans. Nucl. Sci., NS-15, 55-60 (1968), EEA72-11194.
229.Wong, E. H. and Streetman, B. G., "Isothermal annealing of the 0.97 -eV luminescence in electron-irradiated Si", J. Appl. Phys., 42, 5882-3 (1971), PA75-10428.
230.Yamaguchi, M. and Nagai, O., "Effects of impurities on gamma-irradiated silicon crystal examined by photovoltaic effect of p-n junction diode", Jap. J. Appl. Phys., 11, 1016-23 (1972), EEA75-32517.
231.Young, R. C. and Corelli, J. C., "Photoconductivity studies of radiation-induced defects in silicon", Phys. Rev. B, 5, 1455-67 (1972), PA75-25959.
232.Zaikovskaya, M. A., Kio, A. E., Niyazova, O. R. and Starodubtsev, S. V., "Changes induced in silicon by irradiation with 'sub-threshold' energy electrons", Fiz. Tekh. Poluprovodnikov, 1, 1131-7 (1967), Russian, EEA71-1204.
233.Zotova, L. G. and Khiznichenko, L. P., "Radiation hardening of silicon", Izv. Akad. Nauk Uzb. SSR, Ser. Fiz.-Mat. Nauk, no. 6, 46-48 (1973), Russian, NSA29-30181.

b. Germanium

234.Abdullaev, A., Emstev, V. V., Korchazhkina, R. L. and Mashovets, T. V., "Formation of γ-radiation defects in phosphorus-doped germanium", Fiz. & Tekh. Poluprov., 5, 2229-31 (1971), Russian, PA75-15741.
235.Abdurakhmanova, S. N., Dostkhodzhaev, T. and Mashovets, T. V., "Influence of the type of group V dopant on the annealing of γ-radiation defects in germanium", Sov. Phys.-Semicond., 7, 1229-30 (1974), NSA30-27553.
236.Abiev, A. K. and Ukhin, N. A., "Low temperature irradiation of germanium with 28 MeV electrons", Fiz. Tekh. Poluprovodnikov, 4, 484-7 (1970), Russian, EEA73-27996.
237.Albany, H. J. and Vandevyuer, M., "Effects of electron irradiation on the thermal conductivity of n- and p-type germanium", Phys. Rev., 160, 633-5 (1967), EEA71-1200.
238.Atakova, E. M., "Recombination centres in neutron-irradiated germanium", Fiz. & Tekh. Poluprovodnikov, 5, 593-5 (1971), Russian, EEA74-22471.

239.Baranskii, P. I., Peller, E. V. and Shershel, V. A., "Anistropy of the Hall mobility of current carriers in γ-irradiated n-germanium with an oxygen impurity", Ukr. Fiz. Zh., 17, 834-5 (1975), Russian, CA79-140801.

240.Baranskii, P. I. and Semenyak, A. K., "Changes in the resistivity gradients in n- and p-type germanium under the influence of γ-irradiation", Sov. Phys. Semicond., 1, 961-4 (1968), EEA71-1205.

241.Basman, A. R., Gerasimov, A. B., Gogotishvili, M. K., Dolidze, N. D., Kasparyan, K. I., Kakhidze, N. G. and Konovalenko, B. M., "Influence of donor impurities on the kinetics of annealing of radiation defects in germanium", Sov. Phys.-Semicond., 7, 922-4 (1974), PA77-32894.

242.Basman, A. R., Gerasimov, A. B., Kakhidze, N. G., Konovalenko, B. M. and Tsertsvadze, A. A., "Model of quasistable light-sensitive defects in germanium irradiated at 77K", Sov. Phys.-Semicond., 7, 903-5 (1974), PA77-32892.

243.Bertolotti, M., Stagni, L., Vitali, G. and Muzii, L., "Role of surface treatment in laser damage of germanium", J. Appl. Phys., 42, 5893-5 (1971), PA75-12647.

244.Bolotov, V. V., Vasilev, A. V. and Smirnov, L. S., "Effects of irradiation conditions on radiation defect accumulation processes in semiconductors", Radiats. Defekty Poluprov., 239-40 (1972), Russian, CA79-59078.

245.Bourgoin, J. and Mollot, F., "The vacancy-interstitial pair in electron irradiated germanium", International Conference on Radiation Effects in Semiconductors, 1970, p. 63-8, Publ. 1971, PA75-47379.

246.Brandt, E. H. and Guth, W., "Electrical breakdown in germanium exposed to infrared radiation", Phys. Status Solidi, 31, 607-15 (1969), EEA72-13356.

247.Bronder, T. and Jakschik, J., "Anisotropic Kα X-ray production by 350 keV electrons in germanium single crystals", Radiat. Eff., 13, 209-13 (1972), PA75-48208.

248.Chen, Y. and MacKay, J. W., "Orientation and temperature dependence of electron damage in n-type germanium", Phil. Mag., 19, 357-67 (1969), EEA69-6497.

249.Cleland, J. W. and Paciesas, W. S., "Impurity dependence of defect introduction and annealing in electron irradiated n-type germanium", IEEE Trans. Nucl. Sci., NS-16, 19-22 (1969), EEA73-14011.

250.Den Ouden, G., "Small-angle scattering of x-rays by neutron-irradiated germanium", J. Appl. Phys., 39, 4509-17 (1968), EEA71-28827.

251.Dobrego, V. P. and Ermolaev, O. P., "Hopping conductivity of n-type germanium irradiated with Co^{60} γ-rays", Sov. Phys.-Semicond., 9, 606-7 (1975).

252.Emtsev, V. V., Korkhashkina, R. L. and Mashovets, R. V., "Vacancy-donor complexes in γ-irradiated phosphorus-doped germanium", Phys. Status Solidi A, 10, 43-8 (1972), PA75-29704.

253.Evseev, V. A., Konopleva, R. F. and Yuferev, A. A., "Recombination of n- and p-type germanium irradiated with 660 MeV protons", Fiz. Tekh. Poluprovodnikov, 4, 1549-53 (1970), Russian, EEA74-1317.
254.Flanagan, T. M. and Klontz, E. E., "Bombardment-produced defects in p-type germanium at low temperatures", Phys. Rev., 167, 789-800 (1968), EEA71-13246.
255.Fukouka, N. and Saito, H., "The defects produced by electron irradiation and annealed at about 360K in n-type germanium", Jap. J. Appl. Phys., 13, 1524-32 (1974), PA77-82201.
256.Fukuoka, N., Saito, H. and Tatsumi, Y., "Annealing of defects formed in germanium after irradiations with particles of various energies", J. Phys. Soc. Jap., 36, 793-99 (1974), PA77-32581.
257.Fukuoka, N., Saito, H. and Tatsumi, Y., "Annealing of the E_C-0.2 eV defect in n-type germanium", Lattice Defects in Semiconductors, 1974, p. 206-11, Publ. 1975, PA78-81116.
258.Gerasimov, A. B., Gogotishvili, M. K., Kasparyan, K. I., Kakhidze, N. G. and Eristavi, G. L., "Absorption edge in copper-doped germanium irradiated at 77°K", Radiats. Defekty Poluprov., 114-16 (1972), Russian, CA79-59076.
259.Germano, C. A. and Curtis, O. L., Jr., "Effect of injection level on carrier lifetime in neutron-irradiated germanium", IEEE Trans. Nucl. Sci., NS-13, 47-52 (1966), EEA70-12569.
260.Golubev, N. F., Latyshev, A. V., Polklonskii, N. A. and Stelmakh, V. G., "Change in carrier concentration and mobility in irradiated n-type germanium", Radiats. Defkty Poluprov., 120-1 (1972), Russian, CA79-59118.
261.Gotwais, J. K., "Dilation of electron irradiated germanium", Thesis, Purdue Univ., Lafayette, Ind., 1971, 150 pp., Order No. 72-1868, PA75-68016.
262.Hirata, M., "Migration of divacancy-like defects in germanium bombarded with electrons at 1.5 MeV at 77K", Lattice Defects in Semiconductors, 1974, p. 164-71, Publ. 1975, PA78-81251.
263.Hirata, M. and Fukumoto, M., "Possibility of di-vacancy migration at 0°C in irradiated germanium", Jap. J. Appl. Phys., 13, 8-16 (1974), PA77-28235.
264.Hirata, M., Fukumoto, M. and Miyamoto, Y., "Recovery spectra of N-type germanium bombarded with electron of 1.5 MeV at 77K", Jap. J. Appl. Phys., 13, 1165-6 (1974), EEA77-33878.
265.Hyatt, W. D. and Koehler, J. S., "Low temperature annealing of electron irradiated germanium", International Conference on Radiation Effects in Semiconductors, 1970, p. 59-61, Publ. 1971, PA75-47378.
266.Ito, K., Oka, M. and Saito, H., "Impurity dependence of the annealings of gamma-ray induced defects in p-type germanium (minority carrier lifetime)", Jap. J. Appl. Phys., 12, 1475-85 (1973), EEA77-250.

267.Jaworowski, A., Merski, K., Rzewuski, H. and Werner, Z., "Annealing of Hall-mobility changes in electron irradiated germanium", J. Appl. Phys., 38, 4745-8 (1967), EEA71-3314.
268.Jaworowski, A. and Rzewuski, H., "Defect transformations in n-type germanium irradiated with electron at 140K", Lattice Defects in Semiconductors, 1974, p. 221-6, Publ. 1975, PA78-81117.
269.Konopleva, R. F. and Novikov, S. R., "Formation of defects in lithium-doped germanium due to bombardment with fast neutrons", Sov. Phys.-Semicond., 7 (7), 906-8 (1974), EEA77-18619.
270.Konopleva, R. F., Novikov, S. R. and Rubinova, E. E., "The processes of formation and annealing of radiation defects in germanium, irradiated by high energy protons", International Conference on Radiation Effects in Semiconductors, 1970, p. 353-8, Publ. 1971, PA75-47383.
271.Konopleva, R. F., Novikov, S. R., Rubinova, E. E., Sadikov, V. P. and Ukhin, N. A., "Annealing processes in n-type Ge irradiated with high-energy protons", Fiz. Tekh. Poluprovodnikov, 4, 1517-23 (1970), Russian, EEA74-1314.
272.Konopleva, R. F. and Yuferev, A. A., "On disordered regions in fast-neutron compensated germanium", Phys. Status Solidi A, 21, 49-55 (1974), EEA77-10019.
273.Konorov, P. P., Romanov, O. V., Andreev, A. D. and Belyakov, V. S., "Use of electrochemical dissolution and electrolyte field defects in the study of radiation effects in germanium", Radiats. Defekty Poluprov., 132-3 (1972), Russian, CA79-59117.
274.Kreutz, E. W., Pagnia, H. and Waidelich, W., "Changes of surface conductivity in Ge during X-irradiation", Z. Angew. Phys., 30, 145-50 (1970), EEA74-4868.
275.Kreutz, E. W. and Waidelich, W., "X-ray induced conductivity changes in Ge", Phys. Lett., 31, 117-18 (1970), EEA73-24763.
276.Krynicki, J., Rzewuski, H. and Werner, Z., "Transformation of simple defects in radiation converted p-type Ge at 200K", Radiat. Eff., 21, 193-5 (1974), PA77-69801.
277.Krynko, Y. N., Koval, I. F. and Melnik, P. V., "Mean free path of electrons without energy losses in Ge", Ukr. Fiz. Zh., 20, 508-10 (1975), Russian, PA78-37285.
278.Latyshev, A. V., Stelmakh, V. F., Tkachev, V. D. and Urenev, V. I., "Negative differential resistance of high-purity germanium irradiated with fast neutrons", Fiz. Tekh. Poluprovodnikov, 4, 1529-35 (1970), Russian, EEA74-1315.
279.Laurence, G. and Albany, H. J., "Phonon scattering and induced-energy levels in electron-irradiated Sb-doped Ge in the n to p-type conversion region", International Conference on Radiation Effects in Semiconductors, 1970, p. 119-24, Publ. 1971, PA75-47766.
280.Lehar, A. F. and Whitehouse, J. E., "Low-temperature electron irradiation and annealing of n-type germanium", Lattice Defects in Semiconductors, 1974, p. 198-205, Publ. 1975, PA78-81115.

281. Lifshitz, T. M. and Musatov, A. L., "Infrared-controlled field emission from germanium", JETP Lett., 3, 84-6 (1966), EEA71-3341.
282. MacDonald, P. J. and Palmer, D. W., "Difference in defect production and annealing processes in boron- and carbon-irradiated germanium", Lattice Defects in Semiconductors, 1974, p. 504-12, Publ. 1975, PA78-81134.
283. Mashovets, T. V., Emtsev, V. V. and Abdurakhmanova, S. N., "Model of the process of formation with γ-radiation defects in germanium doped with group V impurities", Sov. Phys.-Semicond., 8, 59-64 (1974), PA77-75538.
284. Mashovets, T. V., Ryvkin, S. M. and Khansevarov, R. Y., "Influence of thermal defects on the radiation stability of germanium", Soviet Phys. Solid State, 9, 407-10 (1967), EEA70-12542.
285. Matula, R. A. and Klontz, E. E., "Defects induced by electron irradiation of p-type germanium", International Conference on Radiation Effects in Semiconductors, 1970, p. 55-8, Publ. 1971, PA75-47377.
286. McCarter, E. R., "A reversible nuclear radiation approach to generation-recombination noise studies in Ge-single crystals", Thesis, Oklahoma State Univ., 1967, 80 pp., Order No. 68-8455, EEA70-2724.
287. Meese, J. M., "Low-temperature recovery of irradiation defects in n-type germanium", Phys. Rev. B, 9, 4373-91 (1974), CA81-55600.
288. Meese, J. M. and MacKay, J. W., "Impurity dependence of the low temperature annealing in n-type germanium", International Conference on Radiation Effects in Semiconductors, 1970, p. 51-4, Publ. 1971, PA75-47376.
289. Mitchell, J. B., Foti, G., Howe, L. M., Davies, J. A., Campisano, S. U. and Rimini, E., "A channeling investigation of proton and deuteron damage in germanium", Radiat. Eff., 26, 193-6 (1975).
290. Parsons, J. R. and Hoelke, C. W., "Observations of crystal lattice planes in neutron and ion bombarded Ge", IEEE Trans. Nucl. Sci., NS-16, 37-42 (1969), EEA73-14015.
291. Roop, R. M., "Effects of electron irradiation of p-type germanium at liquid helium temperatures using ac hopping conductivity", Thesis, Univ. of Illinois, Urbana, 1973, 308 pp., Order No. 74-12164, NSA30-21826.
292. Saito, H., Fukuoka, N. and Tatsumi, Y., "Annealing of point defects in p-type germanium after electron irradiation at liquid nitrogen temperature", International Conference on Radiation Effects in Semiconductors, 1970, p. 81-5, Publ. 1971, PA75-47381.
293. Saito, H., Fukuoka, N., Tatsumi, Y., Yoshida, H. and Nakagawa, M., "Radiation damage studies using the low temperature irradiation facility. VII. Neutron irradiation of p-type germanium at low temperature", Annu. Rep. Res. Reactor Inst. Kyoto Univ., no. 4, 123-5 (1971), PA75-29168.

294.Saito, H., Hirata, M., Fukuoka, N., Tatsumi, Y., Tagawa, S. and Sasuga, N., "Annealing stage near 4.2°K in neutron irradiated germanium", Jap. J. Appl. Phys., 14, 139-40 (1975), CA82-178430.

295.Semenyuk, A. K. M., Konozenko, I. D. and Varentsov, M. D., "Changes in the carrier mobility in γ-irradiated n-type Ge with a layered distribution of Sb impurities", Fiz. Tekh. Poluprovodnikov, 2, 1393 (1968), Russian, EEA72-2691.

296.Shimotomai, M. and Hasiguti, R. R., "The 220K defect in electron irradiated p-type germanium", International Conference on Radiation Effects in Semiconductors, 1970, p. 87-9, Publ. 1971, PA75-47928.

297.Sinishchuk, I. K. and Reznikov, M. Y., "Interaction of lithium with disordered regions formed by fast neutrons in n-type germanium", Sov. Phys.-Semicond., 7, 1131-32 (1974), NSA30-27546.

298.Smirnov, G. A., Tkachev, V. D. and Urenev, V. I., "Interaction of vacancies with lithium atoms in germanium during irradiation with γ quanta", Sov. Phys.-Semicond., 7, 1484-5 (1974), PA77-66910.

299.Smirnov, L. S., Stas, V. F. and Khainovskaya, V. V., "Irradiation and annealing of germanium crystals with high dislocation densities", Fiz. Tekh. Poluprovodnikov, 3, 1760-5 (1969), Russian, EEA73-17425.

300.Srour, J. R. and Curtis, O. L., Jr., "Recombination parameters in low-resistivity gamma-irradiated n-type germanium", International Conference on Radiation Effects in Semiconductors, 1970, p. 113-18, Publ. 1971, PA75-47929.

301.Stas, V. F. and Smirnov, L. S., "Irradiation and annealing of p-type germanium", Fiz. Tekh. Poluprovodnikov, 4, 276-81 (1970), EEA73-24735.

302.Stein, H. J., "Light-sensitive defect formation by electron and neutron irradiation on n-type and p-type germanium near 80K", J. Appl. Phys., 43, 138-44 (1972), PA75-12749.

303.Tomimasu, T., Mikado, T. and Yamazaki, T., "Anomalous energy loss and straggling of 24.8 MeV electrons through thin single crystals of germanium", J. Phys. Soc. Jap., 37, 286 (1974), PA77-72503.

304.Ukhin, N. A. and Abiev, A. K., "Thermal stability of disordered regions in n-type germanium", Fiz. Tekh. Poluprovodnikov, 4, 1152-5 (1970), EEA73-35033.

305.Ukhin, N. A. and Abiev, A. K., "Producing disordered regions in germanium by 28 MeV electrons", International Conference on Radiation Effects in Semiconductors, 1970, p. 351, Publ. 1971, PA75-47931.

306.Vasilev, A. V., Rybakova, L. A. and Smirnov, L. S., "Radiation-defect annihilation centers", Sov. Phys.-Semicond., 6, 631-2 (1972), PA75-83978.

307. Vavilov, V. S., Kazanskii, A. G., Koshelev, O. G. and Abaturov, M. A., "Influence of microwave radiation on the conductivity of compensated p-type Ge at liquid He temperatures", Sov. Phys.-Semicond., 7, 688-9 (1973), EEA77-10026.
308. Vavilov, V. S., Klyava, Y. G., Koshelev, O. G. and Tyapkina, N. D., "Influence of temperature and microwave radiation on the conductivity of beryllium-doped germanium at liquid helium temperatures", Fiz. Tekh. Poluprovodnikov, 4, 1683-8 (1970), Russian, EEA74-1330.
309. Vavilov, V. S., Spitsyn, A. V. and Tchukitchev, M. V., "Electron irradiation of lithium doped germanium at cold temperatures", International Conference on Radiation Effects in Semiconductors, 1970, p. 75-9, Publ. 1971, PA75-47927.
310. Vitovskii, N. A., Emtsev, V. V., Kotina, I. M. and Mashovets, T. V., "A feature of radiation-defect formation in germanium", Fiz. & Tekh. Poluprov., 5, 2231-3 (1971), Russian, PA75-15742.
311. Vitovskii, N. A., Mashovets, T. V. and Maksimov, M. M., "Reduction, by heat treatment, of the rate of formation of radiation defects in germanium", Fiz. Tekh. Poluprovodnikov, 2, 125-7 (1968), Russian, EEA71-15748.
312. Wada, T., Fukuoka, Y. and Arizumi, T., "Negative resistance due to impact ionization in irradiated germanium", J. Phys. Soc. Japan, 25, 165-73 (1968), EEA71-28828.
313. Wada, T., Fukuoka, Y. and Arizumi, T., "Recombination domain instability in high energy electron irradiated germanium", Jap. J. Appl. Phys., 11, 1009-15 (1972), PA75-57208.
314. Whitehouse, J. E., "The kinetics of the annealing of irradiation damage at low temperatures in n-type germanium", International Conference on Radiation Effects in Semiconductors, 1970, p. 69-73, Publ. 1971, PA75-47380.
315. Wolf, D., "14 MeV-neutron irradiation of p-type germanium at 77K", International Conference on Radiation Effects in Semiconductors, 1970, p. 345-9, Publ. 1971, PA75-47930.
316. Yoshihiro, K., Tokumoto, M. and Yamanouchi, C., "Absorption of sub-mm radiation in antimony doped germanium in the metallic type impurity conduction region", J. Phys. Soc. Jap., 37, 1175 (1974), EEA78-5218.

c. Silicon and Germanium

317. Abdurakhmanova, S. N., Vasileva, E. D., Emtsev, V. V. and Mashovets, T. V., "Process of formation of radiation defects by bombardment of silicon-doped germanium with γ rays", Sov. Phys.-Semicond., 8, 1139-40 (1975), PA78-46209.
318. Astakhov, V. M., Golobokov, Y. N., Stas, V. F. and Smirnov, L. S., "Annealing of electron-irradiated germanium and silicon", Sov. Phys.-Semicond., 7, 945-46 (1974), NSA30-13284.

319.Baruch, P., Abel, F., Cohen, C., Bruneaux, M., Palmer, D. W. and Pabst, H., "Study of atomic displacements in He^+ irradiated silicon and a silicon-germanium alloy, through Rutherford scattering", International Conference on Radiation Effects in Semiconductors, 1970, p. 147-53, Publ. 1971, PA75-47388.
320.Campisano, S. U., Foti, G., Grasso, F., Quercia, I. F. and Rimini, E., "Depth dependence of angular dips. Measurements and calculations for protons and deuterons in silicon and germanium", At. Collisions Solids, Proc. Conf., 4th, 1971, p. 259-67, Andersen, S., (Ed.), Gordon and Breach, New York, 1972, CA82-66427.
321.Curtis, O. L., Jr. and Srour, J. R., "Recombination within disordered regions: influence of barrier height on recombination rate and injection level effects", IEEE Trans. Nucl. Sci., NS-20, 196-203 (1973), PA77-41081.
322.Fedotov, Y. A., Gruzdeva, G. A. and Uvarov, E. F., "Effect of neutron irradiation on the current-voltage characteristic of an n-Ge-n-Si isotypic heterojunction", Fiz. Tekh. Poluprovodnikov, 5, 966-7 (1971), Russian, EEA74-26764.
323.Kolyada, V. M., Zlobin, V. G. and Zaporozhets, I. A., "Investigation of proton radiation damage of silicon and germanium single crystals by the shadow effect", Sov. Phys.-Semicond., 6, 1565-7 (1973), CA80-64921.
324.Macko, P., "Influence of neutron radiation on the characteristic quantities of the alloyed semiconductor GeSi", Fyz. Casopis, 20, 36-47 (1970), EEA73-24788.
325.Messenger, G. C., "A two level model for lifetime reduction processes in neutron irradiated silicon and germanium", IEEE Trans. Nucl. Sci., NS-14, 88-102 (1967), EEA71-15754.
326.Oswald, R. B., Jr., "Fracture of silicon and germanium induced by pulsed electron irradiation", IEEE Trans. Nucl. Sci., NS-13, 63-9 (1966), EEA70-12568.
327.Pavlov, P. V., "Change in the structure of germanium and silicon during ion bombardment and subsequent annealing", Sov. Phys.-Cryst., 18, 506-9 (1973), NSA30-16434.
328.Starostin, K. L., "Temperature dependence of the rate of electron removal in n-type Ge and n-type Si during fast-neutron irradiation", Fiz. Tekh. Poluprovodnikov, 4, 1823-4 (1970), Russian, EEA74-1339.
329.Starostin, K. L., Poretskii, L. B., Mordkovich, V. N. and Tuturov, Y. F., "Irradiation of n-type Ge and n-type Si with fast neutrons at various temperatures in the range 78-330 K", Fiz. Tekh. Poluprovodnikov, 1, 1840-4 (1967), Russian, EEA71-13244.
330.Stein, H. J., "Neutron- and proton-induced defects in SiGe alloys: optical absorption", J. Appl. Phys., 45, 1954-61 (1974), PA77-61194.

331.Zdanovich, A. G., Kryuk, V. I., Mints, R. I. and Milman, I. I., "Exo-electron emission of silicon and germanium after bombardment by protons of low energy", Phys. Status Solidi A, 8, K27-9 (1971), PA75-4318.

d. Silicon Carbide

332.Allan, C. G., "A profile analysis of X-ray diffraction peaks of fast neutron irradiated β-silicon carbide", Presented at International Conference on Silicon Carbide (abstracts only), Miami Beach, Fla., September 1973, PA77-24496.
333.Barinov, Y. V., Bulgakov, Y. V., Iglitzyn, M. I., Ilyin, M. A., Kossaganova, M. G., Pavlov, N. M., Reifmann, M. B., Sakharov, B. A. and Solomatin, V. N., "The effect of neutron and alpha-particle irradiation upon the physical properties of α-SiC", International Conference on Radiation Effects in Semiconductors, 1970, p. 331-5, Publ. 1971, PA75-48294.
334.Geiczy, I. I., Nesterov, A. A. and Smirnov, L. S., "Luminescence of irradiated β-SiC", International Conference on Radiation Effects in Semiconductors, 1970, p. 327-30, Publ. 1971, PA75-48232.
335.Hewette, D. M., II, Fitts, R. B., Pollock, C. B., Federer, J. I. and Scott, J. L., "Effects of neutron irradiation on loose and bonded inert particles coated with pryolytic carbon and silicon carbide", Report ORNL-TM-4551, Oak Ridge National Lab., Tenn., 1974, 46 pp., NSA30-27542.
336.Marakov, V. V., "Spatial distribution of the density of generation of electron-hole pairs during bombardment of silicon carbide with electrons up to 20 keV energy", Sov. Phys.-Semicond., 9, 722-4 (1975).
337.McFadden, R. G., "Irradiation-induced structure changes in reaction bonded SiC", Report AECL-4583, Atomic Energy Canada Ltd., Chalk River, Ontario, 1975, 19 pp., PA78-68967.
338.Patrick, L. and Choyke, W. J., "Photoluminescence of radiation defects in ion-implanted 6H SiC", Phys. Rev. B, 5, 3253-9 (1972), PA75-33309.
339.Stevens, R., "Neutron irradiation damage in SiC whiskers", Phil. Mag., 25, 523-8 (1972), PA75-39674.

3. Group III-V Semiconductors

a. Gallium Arsenide

340.Baramidze, N. V. and Kurdiani, N. I., "Franz-Keldysh effect in gallium arsenide irradiated by fast neutrons", Fiz. Tekh. Poluprov., 7, 611-12 (1973), Russian, CA79-36373.
341.Bogdankevich, O. V., Kalendin, V. V. and Kryukova, I. V., "Effect of intense fast-electron beam irradiation on the electrophysical and optical properties of gallium arsenide", Radiats. Defekty Poluprov., 138 (1972), Russian, CA79-59119.

342. Borghi, L., de Stefano, P. and Mascheretti, P., "Photoconductivity of neutron-irradiated gallium arsenide", J. Appl. Phys., 41, 4665-8 (1970), EEA74-1013.
343. Brailovskii, E. Y. and Konozenko, I. D., "Defects introduced into n-type GaAs by irradiation with CO^{60} quanta", Fiz. Tekh. Poluprovodnikov, 2, 1620-22 (1968), Russian, EEA72-6477.
344. Brudnyi, V. N. and Krivov, M. A., "Low-temperature irradiation and annealing of gallium arsenide", Izv. VUZ Fiz., no. 6, 147-9 (1974), Russian, PA77-67721.
345. Dresner, J., "Mobility in epitaxial GaAs under 1-MeV electron irradiation", J. Appl. Phys., 45, 4118-19 (1974), NSA30-32922.
346. Euler, F., Bouthillette, L., Kahan, A. and Stein, H. J., "Neutron displacement damage in heavily doped GaAs (i.r. optical reflectivity studies)", Proceedings of the International Conference on Radiation Damage and Defects in Semiconductors, 1972, p. 113-20, Publ. 1973, EEA76-19815.
347. Gorain, R. A., Kachurin, G. A., Kravchenko, A. F., Prints, V. Y. and Skok, E. M., "Investigation of the energy of deep levels and of the carrier scattering mechanism in epitaxial films of n-type GaAs irradiated with fast electrons", Sov. Phys.-Semicond., 8, 978-81 (1975), EEA78-19592.
348. Grimshaw, J. A., "Defect studies on electron-irradiated GaAs by electrical measurements", Proceedings of the International Conference on Radiation Damage and Defects in Semiconductors, 1972, p. 355-63, Publ. 1973, EEA76-19822.
349. Gulyaeva, A. S., Krasyuk, B. A., Maslov, V. N. and Sakharov, B. A., "Change in the photoluminescence of gallium arsenide single crystals in regions damaged by a laser beam", Sov. Phys.-Dokl., 17, 780-2 (1973), EEA76-16879.
350. Harada, H., Honda, T. and Fujimoto, M., "Electrical properties of proton and helium ion bombardment N-type GaAs", Electr. Commun. Lab. Tech. J., 24, 381-91 (1975), Japanese, EEA78-32963.
351. Kahan, A., Bouthillette, L. and DeAngelis, H. M., "Electron irradiation of heavily doped GaAs:Si and GaAs:Te (IR reflectivity)", International Conference on Radiation Effects in Semiconductors, 1970, p. 281-5, Publ. 1971, PA75-47917.
352. Kalma, A. H., Berger, R. A. and Cesena, R. A., "Electron irradiation of GaAs", Proceedings of the International Conference on Radiation Damage and Defects in Semiconductors, 1972, p. 364-70, Publ. 1973, EEA76-19823.
353. Lang, D. V. and Kimerling, L. C., "A new technique for defect spectroscopy in semiconductors: application to 1 MeV electron-irradiated n-GaAs", International Conference on Lattice Defects in Semiconductors, 1974, p. 581-8, Publ. 1975, EEA78-41545.
354. Martinelli, R. U. and Wang, C. C., "Electron-beam penetration in gallium arsenide", J. Appl. Phys., 44, 3350-1 (1973), CA79-98390.

355. Mattauch, R. J. and Healy, M. P., "Electron irradiated bulk n-type GaAs", IEEE Trans. Nucl. Sci., NS-22, 825-8 (1975), EEA78-32964.
356. McNichols, J. L. and Berg, N., "Neutron-induced metallic spike zones in GaAs", IEEE Trans. Nucl. Sci., NS-18 (6), 21-30 (1971), EEA75-10115.
357. Mirianashvili, S. M. and Nanobashvili, D. I., "Possibility of transmutation doping of gallium arsenide", Fiz. Tekh. Poluprovodnikov, 4, 1879-83 (1970), Russian, EEA74-7558.
358. O'Brien, J. K. and Corelli, J. C., "Photoconductivity of chromium-compensated GaAs after irradiation by 0.8-MeV electrons", J. Appl. Phys., 44, 1921-2 (1973), EEA76-19708.
359. Papeiko, A. S., Radovskii, E. E., Stelmakh, V. F. and Tkachev, V. D., "Radiation damage in copper-doped gallium arsenide", Fiz. Tekh. Poluprovodnikov, 4, 142-4 (1970), EEA73-20997.
360. Reynolds, R. A., "Impact ionization breakdown of n-type epitaxial GaAs at liquid helium temperatures", Solid-State Electron., 11, 385-90 (1968), EEA71-15755.
361. Riemann, V., "Ultrared sensitive ultrasound absorption in GaAs", Telefunken Ztg., 39, 370-2 (1966), EEA71-8206.
362. Sakurai, T., Bamba, Y. and Furuya, T., "Effects of proton bombardment to n-type GaAs", Fijitsu Sci. & Tech. J., 11 (2), 71-80 (1975), EEA78-37392.
363. Smith, J. L., "Surface damage of GaAs from 0.694- and 1.06-μ laser radiation", J. Appl. Phys., 43, 3399-402 (1972), PA75-61574.
364. Vaidyanathan, K. V., "A study of infrared absorption in electron- and neutron-irradiated gallium arsenide", Thesis, Univ. Waterloo, Ontario, Canada, 1971, PA75-72332.
365. Vilkotskii, V. A., Domanevskii, D. S. and Lomako, V. M., "Radiation change of (charge carrier) life in gallium arsenide", Radiats. Defekty Poluprov., 147-9 (1972), Russian, CA79-59121.
366. Zuleeg, R. and Lehovec, K., "Radiation effect on GaAs interface", Report Sci-1 AFCRL-TR-74-0495, McDonnell Douglas Astronautics Co., Huntington Beach, Calif., 1974, 21 pp., EEA78-41634.

b. Gallium Phosphide

367. Bayliss, C. R. and Kirk, D. L., "Departure from stoichiometry in (100) surfaces of GaP, induced by bombardment with 2kV electrons", Thin Solid Films, 29, L35-38 (1975).
368. Brailovskii, E. Y., Eritsvan, G. N. and Tartachnik, V. P., "Radiation-induced defects in gallium phosphide during irradiation with 50MeV electrons", Fiz. Tekh. Poluprovodn., 9, 1805-7 (1975), Russian, CA83-200405.
369. Geist, V. and Flagmeyer, R., "Influence of the crystal lattice on the angular distribution of X-rays emitted from a GaP single crystal by fast proton bombardment", Phys. Status Solidi A, 26, K1-K3 (1974).

370.Morrison, S. R., Newman, R. C. and Thompson, F., "A radiation damage centre in gallium phosphide containing boron (Local mode absorption)", J. Phys. C, 5, L46-9 (1972), PA75-29167.
371.Vilkotskii, V. A. and Pivorarov, V. Y., "Cathodoluminescence of gallium phosphide, containing residual and radiation defects", Zh. Prikl. Spektrosk., 19, 1055-9 (1973), Russian, PA77-41438.

c. Indium Antimonide

372.Abdullaev, A., Lang, I. G., Mashovets, T. V. and Vitovskii, N. A., "Subthreshold defects in p-InSb", International Conference on Lattice Defects in Semiconductors, 1974, p. 346-52, Publ. 1975, PA78-81088.
373.Guenzer, C. S. and Manning, I., "Calculated displacement damage by neutrons in InSb", IEEE Trans. Nucl. Sci., NS-21 (6), 26-9 (1974), PA78-41816.
374.Guseinov, E. K., Nasledov, D. N., Popov, Y. G. and Schleifstein, M., "Some features of the photomagnetic effect and photoconductivity of n-type InSb associated with electron heating by radiation", Fiz. Tekh. Poluprovodnikov, 4, 1689-96 (1970), Russian, EEA74-1002.
375.Katsiashvili, R. G. and Kurdianai, N. I., "Annealing of radiation defects in plastically deformed InSb crystals", Sov. Phys.-Semicond., 7, 1141-3 (1974), PA77-56781.
376.Kim, C. K., "Proton enhanced diffusion of cadmium in indium antimonide", Thesis, Stanford Univ., Calif., 1974, 71 pp., Order No. 75-6873, EEA78-33053.
377.Kreutz, E. W., "X-ray induced changes of transport phenomena in InSb", Z. Angew. Phys., 27, 244-51 (1969), EEA72-23764.
378.Kreutz, E. W., Pagnia, H. and Waidelich, W., "X-ray surface photoconductivity in n- and p-type InSb", Phys. Status Solidi, 31, K147-50 (1969), EEA72-13150.
379.Kurdiani, N. I., "Annealing of radiation defects and electron mobility in indium antimonide irradiated with fast neutrons", Fiz. Tekh. Poluprovodnikov, 3, 1683-6 (1969), Russian, EEA73-13979.
380.Vystavkin, A. N., Galpern, Y. S. and Gubankov, V. N., "Absorption of millimetre and submillimetre radiation in n-type InSb at helium temperatures", Fiz. Tekh. Poluprovodnikov, 1, 1735-7 (1967), Russian, EEA71-10585.
381.Yamada, K. and Sekiguchi, T., "Mechanism of resistivity decrease of N-type InSb by millimeter microwave irradiation", J. Phys. Soc. Jap., 37, 95-103 (1974), EEA77-30098.

d. Others

382.Abrikosov, N. K., Kolokoltsev, V. N. and Skudnova, E. V., "Action of electron irradiation on undoped gallium antimonide", Fiz. & Khim. Obrab. Mater., no. 6, 21-4 (1974), Russian, PA78-46213.

383. Andreev, V. M., Blinov, V. N., Nosenko, V. A., Rezvanov, O. G. and Rodionova, V. N., "Behaviour of a p-Al_xGa_{1-x}-n-GaAs heterojunction during irradiation with α particles", Sov. Phys.-Semicond., 8, 798-9 (1974), EEA78-13327.

384. Kekelidze, N. P. and Kekelidze, G. P., "Electrical and optical properties of InAs and InP compounds and their solid solutions of InP_xAs_{1-x} irradiated with fast neutrons and γ rays", Proceedings of the International Conference on Radiation Damage and Defects in Semiconductors, 1972, p. 387-93, Publ. 1973, EEA76-19824.

385. Poujade, A. M. and Albany, H. J., "n-to p-type conversion by electron irradiation of tellurium doped gallium antimonide: study by thermal conductivity", C.R. Acad. Sci. Ser. B., 270, 840-3 (1970), EEA73-24733.

386. Sokolowski, D. H., "Electrical properties of irradiated GaSb", Thesis, St. Louis Univ., Mo., 1974, 128 pp., Order No. 74-24145, EEA78-19602.

387. Thommen, K., "Effect of low temperature electron-irradiation on the electrical properties of undoped GaSb", IEEE Trans. Nucl. Sci., NS-14, 40-5 (1967), EEA71-13247.

4. Group II-IV Semiconductors
   a. Cadmium Sulphide

388. Ashkin, A., Tell, B. and Dziedzic, J. M., "Laser induced refractive index inhomogeneities and absorption saturation effects in CdS", IEEE J. Quantum Electron., QE-3, 400-6 (1967), EEA71-3342.

389. Bakun, F. I., Britov, A. D., Zarubin, P. V., Kurbatov, L. N., Mashchenko, V. E. and Mochalkin, N. N., "Conductivity of CdS induced by intensive bombardment", Fiz. Tekh. Poluprovodnikov, 3, 147-50 (1969), Russian, EEA72-15441.

390. Baltateanu, N. and Spinulescu, I., "Effect of ionizing radiation on the crystalline structure and electric resistivity of cadmium sulphide thin films", Radiat. Eff., 20, 149-52 (1973), EEA77-14327.

391. Brodin, M. S., Borts, A. A. and Krupa, N. N., "Observation of residual conductance generated by ruby laser radiation in CdS-type crystals", Sov. Phys.-Semicond., 7, 277-8 (1973), EEA76-35166.

392. Bryant, F. J. and Goodwin, G. K., "Thulium as a monitor of lattice damage in cadmium sulphide", J. Phys. D, 7, 683-94 (1974), PA77-28121.

393. Dyakin, V. V., Korsunskaya, N. E., Markevich, I. V. and Sheinkman, M. K., "Effects of changes in local centers in CdS single crystals caused by the ionizing action of γ rays", Sov. Phys.-Semicond., 8, 281 (1974), PA77-72724.

394. Elsby, C. N., "Radiation damage effects in electron irradiated cadmium sulphide platelets at low temperature (Luminescence properties)", Thesis, Air Force Inst. Technol., Wright-Patterson, AFB, Ohio, 1971, 139 pp., Order No. 72-12903, PA75-68703.

395. Galushka, A. P., Ermolovich, I. B., Korsunskaya, N. E., Konozenko, I. D. and Sheinkman, M. K., "Effect of γ and fast-neutron radiation on the electrical properties of cadmium sulfide single crystals", Sov. Phys. Solid State, 8, 831-7 (1966), EEA70-4100.
396. Gibbons, D. J., "Electron bombardment conductivity of CdS", J. Phys. D, 7, 429-32 (1974), EEA77-10037.
397. Kopas, N. F., Galushka, A. P. and Konozenko, I. D., "Preparing single crystals of CdS with radiation-stable sensitivity to ionizing radiation", Inorg. Mater., 9, 490-3 (1973), PA77-33615.
398. Look, D. C., "High-temperature annealing in electron-irradiated CdS", J. Appl. Phys., 45, 492-3 (1974), PA77-44564.
399. Matsuura, K., Honjo, T. and Tsurumi, I., "A new photoconduction band in CdS crystal produced by electron irradiation", J. Phys. Soc. Jap., 38, 591 (1975), EEA78-19580.
400. Ratsch, U., "Changes of the excitonic optical properties of cadmium sulphide introduced by an electron beam", Phys. Status Solidi B, 69, 459-65 (1975).
401. Wruck, D., "Extrinsic photoconductivity of fast-electron-irradiated undoped CdS single crystals", Phys. Status Solidi, 37, 165-80 (1970), EEA73-24554.
402. Yoshida, T., Oka, T. and Kitagawa, M., "Photoconductivity of CdS crystals irradiated with fast electrons", Appl. Phys. Lett., 21, 1-2 (1972), PA75-57254.

b. Cadmium Telluride

403. Abramov, A. A., Vavilov, V. S. and Vodopianov, L. K., "Radiation effects in CdTe induced by thermal neutrons and 1 MeV electrons", International Conference on Radiation Effects in Semiconductors, 1970, p. 323-6, Publ. 1971, PA75-47892.
404. Barnes, C. E. and Kikuchi, C., "The effect of thermal neutron-induced damage on photoluminescence in CdTe", Radiat. Eff., 26, 105-12 (1975).
405. Bryant, F. J. and Totterdell, D. H. J., "Electron radiation damage and the edge emission of cadmium telluride", International Conference on Radiation Effects in Semiconductors, 1970, p. 311-16, Publ. 1971, PA75-48254.
406. Caillot, M., "Photoconductivity studies of native defects and defects created by electron-irradiation in n-type CdTe", Proceedings of the International Symposium on Cadmium Telluride a Material for Gamma-Ray Detectors, 1971, 13/5 pp., PA75-36537.
407. Caillot, M., "Low temperature annealings of the defect created by 1.5 MeV electron irradiation in n-type CdTe (Photoconductivity)", Phys. Lett. A, 38A, 2-4 (1972), PA75-15739.
408. Caillot, M., "Low-temperature irradiation and annealing of n-type cadmium telluride", Lattice Defects in Semiconductors, 1974, p. 280-7, Publ. 1975, PA78-81120.

409.Matsuura, K. and Tsurumi, I., "An annealing stage of electron radiation damages in In-doped n-type CdTe crystals", Phys. Status Solidi A, 28, K175-8 (1975), EEA78-24127.
410.Mayer, J. W., Zanio, K. and Martini, M., "CdTe and trapping analyses", Proceedings of the International Symposium on Cadmium Telluride a Material for Gamma Ray Detectors, 1971, 20/14 pp., EEA75-21555.
411.Taguchi, T., Shirafuji, J. and Inuishi, Y., "Lattice defects and photoconductive lifetime in CdTe single crystals grown from solution", Lattice Defects in Semiconductors, 1974, p. 272-9, Publ. 1975, PA78-81087.
412.Urli, N. B., "On the properties of some defect centers in CdTe", II-VI Semiconducting Compounds International Conference, 1967, p. 1335-47, EEA72-23773.
413.Zanio, K., "Characterization of semi-insulating CdTe by nuclear methods", Proceedings of the International Symposium on Cadmium Telluride a Material for Gamma-Ray Detectors, 1971, 21/12 pp., EEA75-21556.

c. Others

414.Adrianov, A. A. and Kodzhespirov, F. F., "Photoconductivity of $Zn_xCd_{1-x}S$ single crystals irradiated with fast neutrons", Sov. Phys.-Semicond., 8, 1330-1 (1975), EEA78-24134.
415.Arora, B. M., Ahlburn, B. T. and Compton, W. D., "Radiation-induced laser action in CdSe", IEEE Trans. Nucl. Sci., NS-18, 40-4 (1971), PA75-17638.
416.Baranova, R. K. and Slovokhotov, L. I., "Influence of high-energy gamma radiation on the thermoelectric properties of PbSe and GeTe alloys", In: Semiconductor solar energy converters, Consultants Bureau, London, England, p. 55-7, Russian, EEA73-26809.
417.Bryant, F. J. and Baker, A. T. J., "Threshold energy for tellurium displacement in zinc telluride", J. Phys. C, 5, 2283-8 (1972), PA75-68018.
418.Bryant, F. J. and Manning, P. S., "Selenium displacement and the infrared cathodoluminescence of zinc selenide", J. Phys. C, 5, 1914-20 (1972), PA75-61723.
419.Bryant, F. J. and Manning, P. S., "Radiation damage and decay characteristics of zinc selenide emission bands and their equivalence to zinc sulphide emission bands", J. Phys. & Chem. Solids, 35, 97-101 (1974), PA77-25180.
420.Dutchak, Y. I., Mikolaichuk, A. G. and Kogut, A. M., "Effect of x-radiation on the conductivity of thin α-HgS films", Ukrayin Fiz. Zh., 11, 978-81 (1966), Ukrainian, EEA70-500.
421.Harper, H. T., Colwell, J. F., Green, B. A., Leadon, R. E. and Naber, J. A., "Radiation effects and material studies in PbSnTe (electrical and optical properties)", Report RT-8040-002, Intelcom Rad. Tech., San Diego, Calif., 1974, 129 pp., PA78-61730.

422.Howard, C. J. and Sabine, T. M., "X-ray diffraction profiles from neutron-irradiated magnesium oxide", J. Phys. C, 7, 3453-66 (1974), PA77-82205.
423.Joshi, N. V. and Ballestero, R., "Study of the traps created in the single crystal of CdSe by the irradiation of Mo Kα X-ray quantum. (Photoconductivity)", Phys. Lett. A, 43A, 307-8 (1973), EEA76-19754.
424.Khosla, R. P., Fischer, J. R. and Burkey, B. C., "Impact ionization, breakdown, and photoinduced switching in CdSe", Phys. Rev. B, 7, 2551-64 (1973), EEA76-19762.
425.Kiyak, B. R., Pavlenko, A. A. and Proskura, A. I., "Switching and memory effects in polycrystalline and single-crystal zinc sulphide with radiation defects", Sov. Phys.-Semicond., 8, 1469-70 (1975), EEA78-37375.
426.Loferski, J. J., "Radiation damage in II-VI semiconductors", Postepy Fiz., 25, 77-99 (1974), Polish, NSA30-986.
427.Mallon, C. E., Green, B. A., Leadon, R. E. and Naber, J. A., "Study of the effects of radiation on the electrical and optical properties of HgCdTe", Report INTEL-RT-8027-008, Intelcom Rad. Tech., San Diego, Calif., 1974, 87 pp., EEA78-32979.
428.Mallon, C. E., Naber, J. A., Colwell, J. F. and Green, B. A., "Effects of electron radiation on the electrical and optical properties of HgCdTe", IEEE Trans. Nucl. Sci., NS-20 (6), 214-19 (1973), EEA77-22768.
429.Shepherd, F. D., Jr., "Radiation effects on the spectral response of HgCdTe", IEEE Trans. Nucl. Sci., NS-21 (6), 34-9 (1974), EEA78-19578.
430.Tao, T. F., Wang, C. C., Anderson, D., Simons, D. and Schoolar, R. B., "Proton bombardments on $Pb_{1-x}Sn_xTe$ and their isochronal annealing", 1973 Conference on Electron Device Techniques, p. 30-3, EEA77-5916.
431.Wang, C. C., Tao, T. F. and Sumer, J. W., "Proton bombardment and isochronal annealing of p-type $Pb_{0.76}Sn_{0.24}Te$", J. Appl. Phys., 45, 3981-87 (1974), NSA30-27570.
432.Watkins, G. D., "Irradiation effects in II-VI compounds", International Conference on Radiation Effects in Semiconductors, 1970, p. 301-9, Publ. 1971, PA75-47366.
433.Williams, P. M. and Yoffe, A. D., "Physical properties of radiation induced and growth defects in II-VI semiconductors", International Conference on Radiation Effects in Semiconductors, 1970, p. 399-404, Publ. 1971, PA75-47372.

5. Dielectrics

a. Silicon Dioxide

434.Barabash, L. I., Gorban, A. P., Litovchenko, V. G., Litovchenko, P. G. and Chala, S. V., "Use of p-Si/SiO_2 systems in nuclear radiation detectors", Poluprovodn. Tekh. & Mikroelektron., no. 13, 82-5 (1973), Russian, EEA77-12145.

435. Bates, J. B., Hendricks, R. W. and Shaffer, L. B., "Neutron irradiation effects and structure of noncrystalline SiO_2", J. Chem. Phys., 61, 4163-76 (1974), NSA31-9278.

436. Donovan, R. P., Simons, M. and Monteith, L. K., "Radiation hardening of thermal oxides in silicon via ion implantation", IEEE Trans. Nucl. Sci., NS-16, 203-6 (1969), EEA73-14168.

437. Emms, C. G., Holmes-Siedle, A. G., Groombridge, I. and Bosnell, J. R., "Gamma and vacuum ultraviolet irradiations of ion implanted SiO_2 for MOS dielectrics", IEEE Trans. Nucl. Sci., NS-21 (6), 159-66 (1974), EEA78-19703.

438. Fanet, J. M. and Poirier, R., "Charge storage in SiO_2 under low-energy electron bombardment", Appl. Phys. Lett., 25, 183-85 (1974), NSA30-32921.

439. Fitzgerald, D. J. and Grove, A. S., "Radiation-induced increase in surface recombination velocity of thermally oxidized silicon structures", Proc. IEEE, 54, 1601-2 (1966), EEA70-6218.

440. Fotchenkov, A. A. and Kolodieva, S. V., "Some features of the action of penetrating radiation on the electrical properties of a quartz crystal", Sov. Phys.-Crystallogr., 18, 366-9 (1973), EEA77-5857.

441. Fowkes, F. M., Kahlke, W. E. and Butler, S. R., "Radiation effects of MOS gate insulators", Report HDL-TR-200-1, Lehigh Univ., Bethlehem, Pa., 1974, 45 pp., NSA31-6164.

442. Fowkes, F. M. and Witherell, F. E., "Sodium mobility in irradiated SiO_2", IEEE Trans. Nucl. Sci., NS-21 (6), 67-72 (1974), PA78-41996.

443. Ginther, R. J., Kirk, R. D., Sigel, G. H., Jr. and Friebele, E. J., "Effects of oxidation and reduction on the radiation damage in SiO_2 and GeO_2", Am. Ceram. Soc. Bull., 52, 704 (1973), PA77-32576.

444. Gwyn, C. W., "Model for radiation-induced charge trapping and annealing in the oxide layer of MOS device", J. Appl. Phys., 40, 4886-92 (1969), PA73-18354.

445. Harai, E., Wang, S. and Royce, B. S. H., "Low temperature irradiation effects in SiO_2-insulated MIS devices", J. Appl. Phys., 46, 1310-17 (1975), EEA78-24243.

446. Hickmott, T. W., "Parallel between surface states at the Si-SiO_2 interface and the B_2 center in irradiated SiO_2", J. Vac. Sci. & Technol., 9, 311-15 (1972), EEA75-20142.

447. Hughes, H. L., "Radiation-induced perturbations of the electrical properties of the silicon-silicon dioxide interface", IEEE Trans. Nuclear Sci., NS-16, 195-202 (1969), EEA73-14167.

448. Hughes, H. L., "Dependence of MOS device radiation-sensitivity on oxide impurities", IEEE Trans. Nucl. Sci., NS-19, 256-63 (1972), EEA76-12501.

449. Kassabov, J. D., Stoev, I. G. and Kaschieva, S. B., "Influence of γ-radiation on the density of the surface states at the Si-SiO_2 interface", C. R. Acad. Bulg. Sci., 26, 19-22 (1973), EEA76-29187.

450.Kassabov, J., Velchev, N. and Kaschieva, S., "Influence of radiation on the surface states and hole mobility at the Si-SiO_2 interface", Surface Sci., 21, 366-76 (1970), EEA73-35172.
451.Kropman, D. I., Niitra, A. E. and Vykhma, K. A., "Influence of activated SiO_2 film on Si surface", Poluprovodn. Tekh. & Mikroelektron., no. 13, 85-91 (1973), Russian, EEA77-12146.
452.Li, S. P. and Maserjian, J., "Electron radiation effects on time-dependence dielectric breakdown on SiO_2 films", Solid-State Electron., 18, 287-8 (1975), PA78-29751.
453.McCaughan, D. V. and Kushner, R. A., "Degradation of oxide films due to radiation effects in exposure to plasmas in sputter deposition and backsputtering", Proc. IEEE, 62, 1236-41 (1974), NSA31-6180.
454.Mitchell, J. P., "A study of radiation effects on SiO_2 and Al_2O_3 layers using thermoluminescence glow curve techniques", IEEE Trans. Nucl. Sci., NS-15, 154-67 (1968), EEA72-10846.
455.Mordkovich, V. N., Zhukova, G. A. and Temper, E. M., "Radiation damage in dielectric-semiconductor structures", Fiz. Tekh. Poluprovodnikov, 4, 1583-5 (1970), Russian, EEA74-1494.
456.Norris, C. B. and EerNisse, E. P., "Ionization dilatation effects in fused silica from 2 to 18-keV electron irradiation", J. Appl. Phys., 45, 3876-82 (1974), NSA30-27573.
457.Pepper, M., "Low energy electron irradiation of the Si-SiO_2 interface", Thin Solid Films, 14, S7-10 (1972), EEA76-12530.
458.Powell, R. J. and Derbenwick, G. F., "Vacuum ultraviolet radiation effects in SiO_2", IEEE Trans. Nucl. Sci., NS-18, 99-105 (1971), EEA75-10274.
459.Revesz, A. G., "Defect structure and irradiation behaviour of noncrystalline SiO_2", IEEE Trans. Nucl. Sci., NS-18, 113-6 (1971), EEA75-10181.
460.Revesz, A. G., "Defect structure and irradiation behaviour of noncrystalline SiO_2", 7th International Symposium on the Reactivity of Solids, Preprints, 1972, p. 3.2/1-8, EEA76-4596.
461.Royce, B. S. H., Russell, T. J. and Pandolfi, T., "Vacuum UV absorption of Al^+ implanted SiO_2 films on sapphire", 1974 International Conference on Colour Centres in Ionic Crystals, 181/2 pp., PA78-26206.
462.Sarrabayrouse, G. and Buxo, J., "Negative charges storage in thermal SiO_2 under electron irradiation", C. R. Hebd. Seances Acad. Sci. B, 280 (4), 69-71 (1975), French, EEA78-24347.
463.Seung, P. L. and Maserjian, J., "Electron radiation effects on time-dependent dielectric breakdown in SiO_2 films", Solid-State Electron., 18, 287 (1975).
464.Sigsbee, R. A. and Wilson, R. H., "Electron irradiation dilation in silica", Appl. Phys. Lett., 23, 541-2 (1973), CA80-31195.
465.Sigsbee, R. A., Wilson, R. H. and Ellis, G. W., "Electron irradiation compaction of SiO_2", Electrochemical Society Fall Meeting, 1974, p. 386-7, EEA78-37436.

466.Simons, M., "Short-term charge annealing in electron-irradiated silicon dioxide", IEEE Trans. Nucl. Sci., NS-18, 106-12 (1971), EEA75-10275.
467.Simons, M. and Hughes, H. L., "Rapid annealing of pulse-radiation-induced space charge in silicon dioxide", J. Appl. Phys., 42, 3000-1 (1971), EEA74-26056.
468.Simons, M., Jr., "A study of charge storage in silicon dioxide irradiated with low energy electrons", Thesis, Duke Univ., Durham, N. C., 87 pp., Order No. 69-3150, EEA73-10536.
469.Smushchenko, V. Y., Arslambekov, V. A. and Gorbunova, K. M., "Influence of electron and ion irradiation on some surface properties of oxidized silicon", Sov. Phys.-Semicond., 8, 1429-30 (1975), PA78-69229.
470.Sosnowski, T. P., "Ion beam milling of MOS structures", Electrochemical Society Fall Meeting (Extended abstracts), 1974, p. 390-1, EEA78-37681.
471.Srour, J. R., Curtis, O. L., Jr. and Chiu, K. Y., "Charge transport studies in SiO_2: processing effects and implications for radiation hardening", IEEE Trans. Nucl. Sci., NS-21 (6), 73-80 (1974), EEA78-19702.
472.Thomas, S., "Electron-irradiation effect in the Auger analysis of SiO_2", J. Appl. Phys., 45, 161-6 (1974), PA77-37721.
473.Ting, C. H., Leone, R., Scoggan, G. and Kurylo, A., "Thermal annealing characteristics of electron-irradiated MOS capacitors", Presented at Electrochemical Society Meeting, San Francisco, California, May 1974, EEA77-26409.
474.Toscano-Rico, A. and Pfister, J. C., "Hall effect measurements of hole mobility in an inversion layer at the Si-SiO_2 interface", Phys. Status Solidi A, 5, 209-18 (1971), EEA74-22737.
475.Voland, G. and Pagnia, H., "Influence of X-irradiation on the silicon-silicon dioxide interface of MOS structures", Appl. Phys., 3, 77-80 (1974), EEA77-10139.
476.Wardly, G. A., "A study of electron bombardment induced conductivity in thermally grown silicon dioxide", Thesis, Princeton Univ., N. J., 1968, 275 pp., Order No. 69-2569, EEA73-10538.

b. Others

477.Andrews, J. M. and Mitchell, J. P., "Electrical charge storage in gamma irradiated insulating films", Annual Reliability Physics Symposium, 1970, p. 70, EEA73-28228.
478.Anon., "Radiation induced electrical current and voltage in dielectric structures", Report AFCRL-TR-74-0582, Air Force Cambridge Res. Labs., Hanscom AFB, Mass., 1974, 41 pp., EEA78-41544.
479.Arkhipov, Y. V., Morachevskii, N. V., Morozov, V. V. and Faizullov, F. S., "Energy balance and dynamics of damage in transparent dielectrics by laser radiation", Sov. Phys.-Solid State, 14, 1510 (1972), EEA76-5103.

480.Battye, F. L., Liesegang, J., Leckey, R. C. G. and Jenkin, J. G., "Attenuation of low energy electrons in Al_2O_3 as determined by photoelectron spectroscopy", Phys. Lett. A, 49A, 155-6 (1974), PA77-79084.
481.Boling, N. L. and Dube, G., "Laser-induced inclusion damage at surfaces of transparent dielectrics", Appl. Phys. Lett., 23, 658-60 (1973).
482.Gibbons, D. J., "Trapped space charges in electron-bombarded insulators", J. Phys. D, 7, 433-42 (1974), EEA77-10038.
483.Kronenberg, S., Lux, R. A., Nilson, K. L., Jurczyk, G. K. and Pfeffer, R. L., "Gamma ray induced charge buildup in insulators", Report ECOM-4252, Army Electronics Labs., Fort Monmouth, N. J., 1974, 20 pp., EEA78-28886.
484.Mitchell, D. F. and Sewell, P. B., "Quantitative determination of oxygen in thin oxide films on metals by electron-excited X-ray emission", Thin Solid Films, 23, 109-25 (1974), PA77-72660.
485.Mordkovich, V. N., Zhukova, G. A. and Temper, E. M., "Radiation damage in dielectric semiconductor structures", Fiz. Tekh. Poluprov., 4, 1583-5 (1970), Russian, EEA74-1494.
486.Murr, L. E., "Fossilization of electrical corona streamers from silicon nitride crystals in the electron microscope", Phil. Mag., 25, 721-8 (1972), PA75-39220.
487.Pavlov, P. V., Genkina, N. A., Shitova, E. V. and Tetelbaum, D. I., "The role of radiation damage in the crystallization kinetics of thin amorphous dielectric layers", Phys. Status Solidi A, 29, 303-7 (1975), PA78-54233.
488.Revesz, A. G., "Some aspects of structure sensitive properties of semiconductor-insulator interface systems", J. Vacuum Sci. Technol., 6, 596 (1969), EEA73-10531.
489.Schmidt, P. F., Rand, M. J., Mitchell, J. P. and Ashner, J. D., "Radiation-insensitive silicon oxynitride films for use in silicon devices", IEEE Trans. Nucl. Sci., NS-16, 211-19 (1969), EEA73-14169.
490.Teodorescu, G., Jitianu, G. and Ciomirtan, D., "Polymorphic transformations of CdS and α-Al_2O_3 under electron irradiation and differential thermal analyses", Rev. Roum. Phys., 17, 71-80 (1972).
491.Westbrook, R. D., "Germanium crystals for radiation detectors", Presented at Second National Conference on Crystal Growth, Princeton, N. J., July-August 1972, EEA76-8549.
492.Zhorov, G. A., "Connection between degree of blackness of total normal radiation and the electric resistance in oxides", High Temp., 10, 597-9 (1973), EEA76-19698.
493.Zieborak, J. and Zak, J., "Effect of irradiation of Si_3N_4 layers with low energy electrons", Electron Technol., 4, 221-6 (1971), PA75-1561.

V. Devices

1. General

494. Aiken, J. G., Buehler, M. G., Crabbe, J. S. and Matzen, W. T., "Investigation of radiation effects in semiconductors", Report AFCRL-69-0164, Texas Instrument Inc., Dallas, Texas, 1969, 41 pp., EEA73-28004.
495. Alexander, D. R. and Antinone, R. J., "MSI testing of dose rates from 10^9 to 10^{12} rad (Si)/S", IEEE Trans. Nucl. Sci., NS-18 (6), 439-42 (1971), PA75-22599.
496. Azarewicz, J. L. and Wrobel, T. F., "Radiation-stimulated failure mechanism in a dielectrically isolated integrated circuit", IEEE Trans. Parts, Hybrids & Packag., PHP-10, 159-64 (1974), EEA78-2907.
497. Bauerlein, R., "Radiation damage in semiconductor elements on space travel missions", Umschau Wiss Tech., 69, 805-6 (1969), German, EEA73-25056.
498. Bauerlein, R. and Uhl, D., "Method of improving the radiation resistance of silicon transistors with a silicon oxide coating", Patent USA 3829961, Publ. August 1974.
499. Bates, D. J., Silzars, A. and Ballonoff, A., "Some reliability aspects of electron bombarded semiconductor power devices", Solid State Technol., 17 (7), 31-5 (1974), EEA77-39031.
500. Buchanan, B., Dolan, R. and Roosild, S., "Comparison of the neutron radiation tolerance of bipolar and junction field effect transistors", Proc. IEEE, 55, 2188-9 (1967), EEA71-10668.
501. Cassidy, E. T., "Nuclear radiation testing of diodes, silicon controlled rectifiers, transistors and integrated circuits", Report PA-TR-4692, Picatinny Arsenal, Dover, N. J., 1974, 76 pp., EEA78-41405.
502. Cates, H., Darling, R., Davidsohn, U., Normington, P. and Christensen, B., "Semiconducting neutron hardness assurance obtained using an irradiate-anneal screening technique", The Annual Reliability Physics Symposium, 1970, p. 71-2, EEA73-27338.
503. Chauhan, A. S. and Maheshwari, L. K., "X-ray radiation effects on semiconductor devices", Indian J. Pure & Appl. Phys., 13, 226-8 (1975), EEA78-24301.
504. Cochrun, B. L., Salmon, P. C., Clark, J. and Greiling, P. T., "Investigation and design of electronic circuitry related to noise and radiation effects for solid state devices and circuits", Report AD-776064/8, Northeastern Univ., Boston, Mass., 1973, 42 pp., NSA30-24100.
505. Dennehy, W. J., Holmes-Siedle, A. G. and Zaininger, K. H., "Digital logic for radiation environments: a comparison of metal-oxide-semiconductor and bipolar technologies", RCA Rev., 30 (4), 668-708 (1969), EEA73-21332.

506.Frankovsky, F., Protschka, H., Boyd, G. and Boczar, P., "Analysis of radiation-induced coupling effects between integrated circuit elements", Report IBM-68-521, IBM, Owego, N. Y., 1968, 80 pp., EEA72-9221.

507.Gaillard, R., "Photocurrents induced by a pulse of radiation in semiconductor devices", Thesis, Univ. Lyon, France, 1973, French, PA77-3776.

508.Gregory, B. L., "Radiation effects in devices", Proceedings of the International Conference on Radiation Damage and Defects in Semiconductors, 1972, p. 302-14, Publ. 1973, EEA76-19901.

509.Gregory, B. L., "Radiation hardening technology (semiconductor devices)", Electrochemical Society Fall Meeting, 1974, p. 399-402, EEA78-33135.

510.Gregory, B. L. and Gwyn, C. W., "Applications of neutron damage models to semiconductor device studies", IEEE Trans. Nucl. Sci., NS-17, 325-34 (1970), EEA73-22800.

511.Gregory, B. L. and Gwyn, C. W., "Radiation effects on semiconductor devices", Proc. IEEE, 62, 1264-73 (1974), NSA31-6182.

512.Gregory, B. L. and Sander, H. H., "Injection dependence of transient annealing in neutron-irradiated silicon devices", IEEE Trans. Nucl. Sci., NS-14, 116-26 (1967), EEA71-15757.

513.Gregory, B. L. and Sander, H. H., "Transient annealing of defects in irradiated silicon devices", Proc. IEEE, 58, 1328-41 (1970), EEA73-37967.

514.Gwyn, C. W., Scharfetter, D. L. and Wirth, J., "The analysis of radiation effects in semiconductor junction devices", IEEE Trans. Nucl. Sci., NS-14, 153-69 (1967), EEA71-15792.

515.Hauser, J. R., "(Semiconductor) device radiation physics", Extended Abstracts of Electrochemical Society Fall Meeting, 1974, p. 397-8, EEA78-32926.

516.Horiye, H., "Radiation-induced latchup in monolithic integrated microcircuits", Radiat. Eff., 1, 196-206 (1969), EEA73-13018.

517.Hughes, H. L., "A survey of radiation-induced perturbations in metal-insulator-semiconductor structures", 9th Annual Proceedings Reliability Physics, 1971, p. 33-9, PA75-6874.

518.Jeppson, K., "Can (COS)MOS replace bipolar circuits in space?", Eltek. Aktuell Elektron., 16 (10), 32-4 (1973), Swedish, EEA77-2072.

519.Klippenstein, E., Hanks, C. L. and Hamman, D. J., "Effects of nuclear radiation as part of the temperature-vacuum-power environment", IEEE Trans. Nucl. Sci., NS-14, 195-9 (1967), EEA71-15846.

520.Kraus, L., "The use of recombination for testing of radiation damaged silicon devices", 1973 8th Annual Meeting of the IEEE Industry Applications Society, p. 635, EEA77-10270.

521.Leavy, J. F. and Poll, R. A., "Radiation-induced integrated circuit latchup", IEEE Trans. Nucl. Sci., NS-16, 96-103 (1969), EEA73-12986.

522.Litvinov, R. O., "Influence of strong effects on characteristics of semiconductor devices, sensitive to surface states", Poluprovodn. Tekh. & Mikroelektron., no. 7, 3-17 (1972), Russian, EEA76-16196.
523.Maier, R. J., "Radiation induced surface recombination in oxide passivated transistor", IEEE Trans. Nucl. Sci., NS-14, 252-9 (1967), PA71-27193.
524.Maier, R. J., "System to measure semiconductor characteristics over decades of time after a radiation pulse", IEEE Trans. Nucl. Sci., NS-19 (1), 679-81 (1972), PA75-29727.
525.Marek, A., Jaecklin, A. A. and Cornu, J., "Instantaneous temperature profiles inside semiconductor power devices. II", IEEE Trans. Electron Devices, ED-21, 54-60 (1974), EEA77-5775.
526.Messenger, G. C., "Comments on 'comparison of the neutron radiation tolerance of bipolar and junction field effect transistors'", Proc. IEEE, 56, 1246 (1968), EEA71-26000.
527.Mitchell, J. P. and Wilson, D. K., "Surface effects of radiation on semiconductor devices", Bell Syst. Tech. J., 46, 1-80 (1967), EEA70-10822.
528.Nichols, D. K. and Nielsen, R. L., "A comparison of the ionizing radiation response of microcircuits manufactured by five different technologies", IEEE Trans. Nuclear Sci., NS-16, 181-5 (1969), EEA73-12988.
529.Palkuti, L. J., Thatcher, R. K. and Elliott, K. E., "A study of the effect of space radiation on silicon integrated circuits", Report NASA-CR-95679, Battelle Memorial Inst., Columbus, Ohio, 1968, 318 pp., EEA72-9222.
530.Perkins, C. W. and Marshall, R. W., "Radiation effects on monolithic silicon integrated circuits", IEEE Trans. Nucl. Sci., NS-13, 330-8 (1966), EEA70-11072.
531.Ramakrishan, S. S., Veer Raj Urs, A. S., Monteiro, C. and Patwardhan, P. K., "Effect of nuclear radiation on the electrical characteristics of semiconductor devices", Report BARC-697, Bhabha Atomic Res. Centre, Bombay, India, 1973, 13 pp., EEA77-38830.
532.Rand, M. J. and Schmidt, P. F., "Radiation-resistant coatings for semiconductor devices", Patent USA 3765935, Publ. October 1973.
533.Raymond, J. P., "MSI/LSI radiation response, characterization and testing", IEEE Trans. Nucl. Sci., NS-21 (6), 308-14 (1974), EEA78-19897.
534.Ricketts, L. W., "Radiation effects on microelectronic components and circuits", 1971 International Microelectronics Symposium, pap. 5.7/17 pp., EEA75-12772.
535.Roberts, J. S. and Cresswell, M. W., "Selective low level irradiation to improve blocking voltage yield of junctioned semiconductors", Patent USA 3852612, Publ. December 1974, NSA31-11444.

536.Roberts, J. S. and Cresswell, M. W., "Selective irradiation of gated semiconductor devices to control gate sensitivity", Patent USA 3840887, Publ. October 1974, CA81-178937.
537.Sander, H. H. and Gregory, B. L., "Transient annealing in semiconductor devices following pulsed neutron irradiation", IEEE Trans. Nucl. Sci., NS-13, 53-62 (1966), EEA70-10825.
538.Schmidt, P. F., McCaughan, D. V. and Kushner, R. A., "Problems in the analysis of semiconductor device materials exposed to ionizing radiation", Proc. IEEE, 62, 1222-5 (1974), PA77-66931.
539.Silzars, A., Bates, D. J. and Ballonoff, A., "Electron bombarded semiconductor devices", Proc. IEEE, 62, 1119-58 (1974), EEA77-34231.
540.Snow, E. H., Grove, A. S. and Fitzgerald, D. J., "Effects of ionizing radiation on oxidized silicon surfaces and planar devices", Proc. IEEE, 55, 1968-85 (1967), EEA70-20773.
541.Snow, E. H., Tremere, D. A., Yu, A. Y. C. and Albus, H. P., "Study of radiation effects on novel semiconductor devices", Report AFCRL-69-0259, Fairchild Camera Instrument Corp., Mountain View, Calif., 1969, 47 pp., EEA73-21340.
542.Spratt, J. P., Schnable, G. L. and Standeven, J. D., "Impact of the radiation environment on integrated-circuit technology", IEEE J. Solid State Circuits, SC-5, 14-23 (1970), EEA73-13012.
543.Strahan, V. H., Williams, R. A., Krueger, A. F., Nichols, D. K. and Chesavage, A. J., "Radiation effect on thin film integrated circuit elements", Report ECOM-02431-F, Autonetics, Anaheim, Calif., 1967, 45 pp., EEA72-9224.
544.Tarka, J. E., "Mechanisms of radiation initiated semiconductor device failures", Thesis, Univ. Pennsylvania, Phila., 1969, 252 pp., Order No. 70-16219, EEA74-26554.
545.Tianien, O. J. A., Tuomi, T. O., Jauho, P. and Tunkelo, E., "Radiation damage in semiconductors and semiconductor devices", 4th Nordic Solid State Conference, 1970, p. 35-6, EEA73-37944.
546.Vail, P. J. and Mullis, J. L., "Variations of neutron degradation across wafers", IEEE Trans. Nucl. Sci., NS-20, 377-82 (1973), EEA77-23129.
547.Van Lint, V. A. J., "Simulation for radiation effects in electronics", Proc. IEEE, 62, 1192-7 (1974), EEA77-33371.

2. P-N Junctions

548.Ammerlan, C. A. J., "Recovery behavior of Si(Li) junctions after gamma irradiation", J. Appl. Phys., 39, 6023-8 (1968), EEA72-6536.
549.Ashburn, P. and Morgan, D. V., "Role of radiation damage on the current-voltage characteristics of p-n junctions", Solid-State Electron., 17, 689-98 (1974), NSA30-18789.
550.Aukerman, L. W., Millea, M. F. and McColl, M., "Effects of radiation damage on the behavior of GaAs p-n junctions", IEEE Trans. Nucl. Sci., NS-13, 174-80 (1966), EEA70-10842.

551.Balodis, Y. K. and Puritis, T. Y., "Spectral distribution of mesoplasma radiation in silicon p-n junction at pulsed and constant reverse biased conditions", Latv. PSR Zinat. Akad. Vestis Fiz. Teh. Ser., no. 3, 52-7 (1974), Russian, EEA77-38794.

552.Buchler, M. G., "The space-charge region of neutron-irradiated silicon p-n junctions", IEEE Trans. Nucl. Sci., NS-17, 341-7 (1970), EEA74-22801.

553.Czaja, W., "Response of Si and GaP p-n junctions to a 5-40 keV electron beam", J. Appl. Phys., 37, 4236-48 (1966).

554.Darley, H. M. and Christopher, J. E., "Proton-irradiation effects on In-Ge:As tunnel junctions", Phys. Rev. B, 9, 3447-53 (1974), EEA77-33989.

555.Domanevskii, D. S., Litvinov, V. L., Lomako, V. M., Smilga, V. P., Tkachev, V. D. and Ukhin, N. A., "Radiation-induced changes in the current-voltage characteristics of heavily doped p-n junctions in gallium arsenide", Fiz. Tekh. Poluprovodirkov, 4, 1425-31 (1970), Russian, EEA74-1526.

556.Foyt, A. G., Lindley, W. T., Wolfe, C. M. and Donnelly, J. P., "Isolation of junction devices in GaAs using proton bombardment", Solid State Electron., 12, 209-14 (1969), EEA72-13445.

557.Goben, C. A., "Nuclear radiation effects on silicon p-n junctions", Report COO-1624-20, Langley Res. Center, Langley Station, Va., 1969, 43 pp., EEA73-28159.

558.Goben, C. A., Irani, C. H. and Johnson, P. E., "Neutron influence and electric field strength dependencies of the rate of volume damage introduction in silicon p-n junctions", IEEE Trans. Nucl. Sci., NS-16, 43-52 (1969), EEA73-14016.

559.Hoff, P. and Everhart, T. E., "Carrier profiles and collection efficiency in Gaussian p-n junctions under electron beam bombardment", IEEE Trans. Electron Devices, ED-17, 458-65 (1970), EEA73-24789.

560.Karpenko, V. P., Kashershininov, P. G., Matveev, O. A., Prokofev, S. V. and Ryvkin, S. M., "CdTe crystals for γ-spectrometers", Proceedings of the International Symposium on Cadmium Telluride a Material for Gamma-Ray Detectors, 1971, 3/8 pp., EEA75-21479.

561.Karpenko, V. P., Kashershininov, P. G., Matveev, O. A. and Tomasov, A. A., "Determination of the parameters of impurity centres in semiconducting materials by the method of charge reversal with the aid of electromagnetic radiation", Sov. Phys.-Semicond., 7, 1268-72 (1974), EEA77-36261.

562.Klimkova, O. A. and Niyazova, O. R., "Influence of x-ray irradiation on some parameters on p-n junctions in silicon", In: Physics of p-n junctions and semiconductor devices (2nd edition), Ryokin, S. M. and Schmartsev, Y. V., (Eds.), Consultants Bureau, London, England, 1971, p. 314-16, EEA74-37236.

563.Lee, W. I., "Instability in the reverse I-V characteristics of silicon p-n junctions induced by low-energy electron irradiation", J. Korean Phys. Soc., 1, 1-9 (1968), EEA69-13437.

564.Lomako, V. M., Novoselov, A. M. and Tkachev, V. D., "Effect of electron irradiation on current-carrier lifetime in gallium arsenide p-n junctions", Radiats. Defekty Poluprov., 163-4 (1972), Russian, CA79-71476.
565.Pilipovich, V. A., Ivlev, G. D., Morgun, Y. F., Nechaev, N. V., Osinskii, V. I. and Peshko, A. Y., "Formation of p-n junctions in GaAs by means of laser radiation", Zh. Prikl. Spektrosk., 22, 431-7 (1975), Russian, PA78-62574.
566.Sirota, N. N. and Korshunov, F. P., "Investigation of the effect of γ radiation on the parameters and structure of silicon p-n junctions", Fiz. Tekh. Poluprovodnikov, 2, 1671-7 (1968), Russian, EEA72-7869.
567.Trilisskii, V. A. and Shostaks, P. B., "Determination of the recombination centre parameters of an irradiated p-n junction", Latv. PSR Zinat. Akad. Vestis Fiz. Tehn. Ser., no. 3, 8-14 (1972), Russian, PA75-61382.
568.Webb, J. E., "Radiation resistant silicon semiconductor devices", Patent USA 3490965, Publ. January 1970, EEA73-21170.

3. Bipolar Transistors

569.Andre, B., Buxo, J., Esteve, D., Martinot, H. and Simonne, J., "Planar-transistor stability under x-ray irradiation", Electron. Lett., 2, 423-5 (1966), French, EEA70-2685.
570.Arimura, I., "Investigations of an irradiate-anneal technique for neutron hardness assurance of power transistors", IEEE Trans. Nucl. Sci., NS-20, 370-6 (1973), EEA77-23013.
571.Arimura, I. and Rosenberg, C., "Anomalous damage mechanism in pnp silicon transistors due to thermal neutrons", IEEE Trans. Nucl. Sci., NS-20, 274-9 (1973), EEA77-23009.
572.Baeuerlein, R. and Uhl, D., "Improving the radiation resistance of silicon transistors with silicon dioxide surface layers", Patent German 2235069, Publ. February 1974, CA80-150256.
573.Baltateanu, N. and Niculescu, A., "Intense ionizing radiation effects on silicon semiconductor devices", Revue Roum. Phys., 16, 841-53 (1971), EEA75-7518.
574.Bartholomew, C. Y., "Analysis of transistor I_{cba} excess currents by gamma irradiation", IEEE Trans. Electron Devices, ED-16, 153-4 (1969), EEA69-113081
575.Bereisa, J., Jr. and Goben, C. A., "Neutron dependence of neutral base region recombination", IEEE Trans. Nucl. Sci., NS-17, 317-24 (1970), EEA74-22703.
576.Berggren, C. C. and Honnold, V. R., "Transistor design effects on radiation resistance", Report NASA-CR-1167, Hughes Aircraft Co., Fullerton, Calif., 1968, 97 pp., EEA72-11317.
577.Butsens, G. V., Matveev, I. G. and Trilisskii, V. A., "Irradiation effect on current components of gate-controlled bipolar transistor", Latv. PSR Zinat. Akad. Vestis Fiz. Teh. Ser., no. 1, 30-5 (1974), Russian, EEA77-26454.

578. Chott, J. R. and Goben, C. A., "Annealing characteristics of neutron irradiated silicon transistors", IEEE Trans. Nucl. Sci., NS-14, 134-46 (1967), EEA71-14845.
579. Chow, M. C., Azarewicz, J. L. and Goben, C. A., "Recombination statistics for neutron bombarded silicon transistors", IEEE Trans. Nucl. Sci., NS-15, 88-94 (1968), EEA72-11346.
580. Clark, L. E. and Saltich, J. L., "Optimized profiles for neutron and photocurrent hardened power transistors", IEEE Trans. Nucl. Sci., NS-21, 149-51 (1974), EEA78-19798.
581. Coppage, F. N., "Experimental neutron damage equivalences utilizing device parameters", IEEE Trans. Nucl. Sci., NS-20, 349-53 (1973), EEA77-23011.
582. Emelyanova, T. S., Kozhukhova, E. A. and Shakhovtsov, V. I., "Barrier capacitance of diffusion, silicon p-n-transistors irradiated with fast neutrons", Poluprov. Tekh. Mikroelektron., no. 14, 55-64 (1974), Russian, NSA31-637.
583. Emmert, R. R., Berggren, C. C., Peffley, W. M. and Honnold, V. R., "Transient high energy radiation response in the thin-film transistor", IEEE Trans. Nucl. Sci., NS-13, 287-94 (1966), EEA70-10863.
584. Frank, M. and Taulbee, C. D., "Factors influencing prediction of transistor current gain in neutron radiation", IEEE Trans. Nucl. Sci., NS-14, 127-33 (1967), EEA71-15843.
585. Fitzgerald, D. J. and Snow, E. H., "Comparison of surface and bulk effects of nuclear reactor radiation on planar devices", IEEE Trans. Electron Devices, ED-15, 160-3 (1968), EEA71-15749.
586. George, W. L., Clark, L. E. and Sheinkman, M. K., "The influence of neutron radiation effects on the design of high-voltage power transistors", IEEE Trans. Electron Devices, ED-17, 804-9 (1970), EEA73-38074.
587. Goben, C. A., Smits, F. M. and Wirth, J. L., "Neutron radiation damage in silicon transistors", IEEE Trans. Nucl. Sci., NS-15, 14-29 (1968), EEA71-23347.
588. Kazakevich, V. I., Babikova, Y. F. and Kozhemyakin, A. A., "The effect of electron bombardment on second breakdown in transsistors", Radio Eng. & Electron. Phys., 16, 664-7 (1971), EEA75-13700.
589. Kobayashi, M., "Stored-charge estimate by electron-beam irradiation", Electron. & Commun. Jap., 56, 94-100 (1973), EEA77-36299.
590. Lipman, J. A., "Neutron hardness assurance: rework of linear integrated circuits", Thesis, Southern Methodist Univ., Dallas, Tex., 1974, 76 pp., Order No. 75-1660, EEA78-28872.
591. McKenzie, J. M. and Witt, L. J., "Measurement of the neutron energy dependence of base current degradation of a silicon bipolar transistor", IEEE Trans. Nucl. Sci., NS-20, 341-8 (1973), EEA77-23010.

592.Niehaus, D., Frank, M., Stoll, E. and Swick, E., "Neutron hardness assurance for power transistors", IEEE Trans. Nucl. Sci., NS-17, 154-9 (1970), EEA74-22700.
593.Pease, R. L., Coppage, F. N. and Graham, E. D., "Dependence of ionizing radiation induced h_e degradation on emitter periphery", IEEE Trans. Nucl. Sci., NS-21, 41-2 (1974), EEA77-26446.
594.Pease, R. L., Galloway, K. F. and Stehlin, R. A., "Gamma radiation effects on integrated injection logic cells", IEEE Trans. Electron Devices, ED-22, 348-51 (1975), EEA78-19892.
595.Prince, J. L. and Stehlin, R. A., "Effect of Co^{60} gamma radiation on noise parameters of bipolar transistors", IEEE Trans. Nucl. Sci., NS-18, 404-9 (1971), EEA75-13722.
596.Raicu, B., "The influence of the biasing conditions and of the power dissipated in transistors during irradiation on the permanent radiation damage", Revue Roum. Phys., 17, 795-9 (1972), EEA75-39963.
597.Raymond, J. P. and Johnson, R. E., "Detailed lumped-model analysis of transistor ionizing radiation effects", IEEE Trans. Nucl. Sci., NS-13, 95-104 (1966), EEA70-10865.
598.Reich, B., Hakim, E. B. and Hunter, E. T., "The effects of neutron radiation on second breakdown and thermal behavior of silicon transistors", IEEE Trans. Nucl. Sci., NS-15, 108-13 (1968), EEA72-11298.
599.Samueli, J. J., "The effects of nuclear radiation on junction transistors", Action des Rayonnements Nucleaires Sur les Composants Electroniques, 1965, p. M1-M25, French, EEA70-2668.
600.Sanga, M. M. and Oldham, W. G., "Measurement of neutral base lifetime in neutron-irradiated transistors", IEEE Trans. Nucl. Sci., NS-20, 266-73 (1973), EEA77-23008.
601.Stanley, A. G., "Electron irradiation of NPN bipolar transistors with silicon nitride passivation", IEEE Trans. Nucl. Sci., NS-15, 168-75 (1968), EEA72-11299.
602.Stehlin, R. A., Devries, D., Bean, K., Hoffman, J. and Robbins, C., "Bipolar medium-scale integrated circuit hardening", Report TI-03-73-88-VOL-1, Texas Instruments, Inc., Dallas, 1974, 161 pp., NSA31-11426.
603.Tausch, H. J., Antinone, R. J. and Bailey, R. A., "A simple transistor neutron hardness screen using scattering parameters", IEEE Trans. Nucl. Sci., NS-20, 361-9 (1973), EEA77-23012.
604.Tiainen, O. J. A., "Radiation resistance of CdSe thin film transistors", Phys. Status Solidi A, 2, 231-3 (1970), EEA74-4934.
605.Yamaguchi, M. and Nagai, O., "Application of radiation damage to radiation dosimetry", Jap. J. Appl. Phys., 10, 1638 (1971), EEA75-2479.

4. Field-Effect Transistors
a. JFET

606.Deme, S., Pellionisz, P. and Szlavik, F., "Radiation damage of junction and MOS FET type transistors", Meres Es Automat., 17, 85-9 (1969), Hungarian, EEA72-15565.
607.George, W. L., "Optimization of the neutron radiation tolerance of junction field effect transistors", IEEE Trans. Nucl. Sci., NS-16, 81-6 (1969), EEA73-14316.
608.Ginell, W. S., Zuleeg, R., McNichols, J. L., Notthoff, J. K. and Lehovec, K., "Transient response of epitaxial GaAs JFET structures to ionizing radiation", IEEE Trans. Nucl. Sci., NS-20, 171-9 (1973), NSA29-29599.
609.Gregory, B. L., Naik, S. S. and Oldham, W. G., "Neutron produced trapping centres in junction field effect transistors", IEEE Trans. Nucl. Sci., NS-18, 50-9 (1971), PA75-19712.
610.Krishman, I. N. and Chen, T. M., "Effects of electron bombardment on the noise in junction gate field effect transistors", Solid-State Electron., 16, 1233-40 (1973), EEA77-2570.
611.Long, D. M., "JFET transient radiation response", IEEE Trans. Nucl. Sci., NS-18, 332-9 (1971), PA75-19713.
612.Long, D. M., "Transient radiation response of JFETs and MISFETs at cryogenic temperatures", IEEE TRans. Nucl. Sci., NS-21, 119-23 (1974), EEA78-19336.
613.Long, D. M. and Swant, D. H., "Burnout of junction field effect transistors", IEEE Trans. Nucl. Sci., NS-20, 149-57 (1973), NSA29-29596.
614.McNichols, J. L. and Ginell, W. S., "Predicted effects of neutron irradiation on GaAs junction field effect transistors", IEEE Trans. Nucl. Sci., NS-17, 52-4 (1970), EEA73-25047.
615.Miller, D. J. and Ryan, R. D., "Electron and neutron damage in n- and p-channel junction field effect transistors", Radiat. Eff., 21, 99-104 (1974), PA77-67275.
616.Naik, S. S. and Oldham, W. G., "Neutron radiation effects in junction field-effect transistors", IEEE Trans. Nucl. Sci., NS-18 (5), 9-17 (1971), EEA75-7538.
617.Neamen, D., Sheed, W. and Buchanan, B., "Permanent ionizing radiation effects in dielectrically bounded field effect transistors", IEEE Trans. Nucl. Sci., NS-20, 158-65 (1973), NSA29-29597.
618.Notthoff, J. K., "Radiation responses of matched silicon junction field-effect transistors", IEEE Trans. Nucl. Sci., NS-18, 397-403 (1971), PA75-22593.
619.Sheed, W., Buchanan, B. and Dolan, R., "Radiation effects on junction field effect transistors", IEEE Trans. Nucl. Sci., NS-16, 87-95 (1969), EEA73-14317.
620.Stenin, V. Y., "Effect of radiation exposure on field-effect transistor parameters (review)", Instrum. Exper. Tech., no. 2, 273-87 (1969), EEA73-21331.

621. Wang, K. H., Van Der Ziel, A. and Chenette, E. R., "Neutron-induced noise in junction-field-effect transistors", IEEE Trans. Electron Devices, ED-22, 591-3 (1975), EEA78-28794.
622. Zuleeg, R. and Behle, A. F., "Radiation tolerance of GaAs broadband amplifier", Report AFCRL-TR-73-0666, McDonnell Douglas Astronautics Co., Huntington Beach, Calif., 1973, 42 pp., NSA31-627.
623. Zuleeg, R. and Notthoff, J. K., "Behavior of GaAs J-FET's in radiation environments", Proceedings of the National Aerospace Electronics Conference, 1973, p. 167, EEA77-2573.

b. MOSFET

624. Andre, B., Buxo, J., Esteve, D. and Martinot, H., "Effects of ionizing radiation on m.o.s. devices", Solid-State Electron., 12, 123-31 (1969), EEA72-11328.
625. Aubuchon, K. G., "Radiation hardening of P-MOS devices by optimization of the thermal SiO_2 gate insulator", IEEE Trans. Nucl. Sci., NS-18, 117-25 (1971), PA75-19734.
626. Aubuchon, K. G. and Bereisa, J., "Performance of hardened P-MOS devices in severe neutron environments", IEEE Trans. Nucl. Sci., NS-19, 299-304 (1972), EEA76-12642.
627. Baeuerlein, R., "Effect of ionizing radiation on current transport through the channel of a MOS-field effect-transistor", Report AED-CONF-74-089-047, Siemens, A. G., Erlanger (F. R. Germany), 1974, 6 pp., German, NSA31-6167.
628. Barry, A. L. and Page, D. F., "Radiation hardening of MOS transistors for low ionizing dose levels", IEEE Trans. Nucl. Sci., NS-13, 255-61 (1966), PA70-21043.
629. Buxo, J., Esteve, D. and Therez, F., "On the mechanisms responsible for ionizing radiation induced degradation in P-substrate MOS transistors", C. R. Hebd. Sean. Acad. Sci. B, 273, 436-9 (1971), French, EEA75-1156.
630. Chadsey, W. L., Galloway, K. F. and Pease, R. L., "Comment on low-energy electron-beam energy deposition", IEEE Trans. Electron Devices, ED-22, 201-2 (1975), EEA78-15404.
631. Chapeau, A., Buxo, J. and Sarrabayrouse, G., "Influence of oxide type on the irradiation induced degradation of MOS structures", C. R. Hebd. Seances Acad. Sci. B, 276, 459-62 (1973), French, EEA76-19857.
632. Chiwaki, M. and Tomimasu, T., "Radiation damage of m.o.s. f.e.t.s. by x-ray and electron irradiations", Bull. Electrotech. Lab., 35, 47-54 (1971), Japanese, EEA74-30819.
633. Churchill, J. N., Collins, T. W. and Homstrom, F. E., "Electron irradiation effects in MOS systems", IEEE Trans. Electron Devices, ED-21, 768-77 (1974), EEA78-11649.
634. Collins, D. R. and Sah, C. T., "Effects of x-ray irradiation on the characteristics of metal-oxide-silicon structures", Appl. Phys. Lett., 8, 124-6 (1966), EEA70-2628.

635. Courat, J. P. and Sixou, P., "On the modifications to the properties of a metal-oxide-semiconductor structure due to electron irradiation; effect of a continuous polarising voltage applied during irradiation", C. R. Acad. Sci. B., 268, 1778-81 (1969), French, EEA73-1123.
636. Courat, J. P. and Sixou, P., "Modifications in the surface density states of a metal-oxide-semiconductor structure submitted to an electron radiation", C. R. Acad. Sci. B., 269, 82-5 (1969), French, EEA73-4409.
637. Danchenko, V., "Radiation hardening of metal-oxide-semiconductor devices by boron", Patent USA 3799813, Publ. March 1974. CA80-138484.
638. Danchenko, V., "Radiation hardening of MOS devices by boron", Patent USA 3882530, Publ. May 1975.
639. Danchenko, V. and Brashears, S. S., "Investigation of radiation damage in MOSFET's using bias-temperature treatments", IEEE Trans. Nucl. Sci., NS-15, 182-6 (1968), EEA72-11326.
640. Danchenko, V., Desai, U. D. and Brashears, S. S., "Characteristics of thermal annealing of radiation damage in MOSFET's", J. Appl. Phys., 39, 2417-24 (1968), EEA71-15862.
641. Danchenko, V., Desai, U. D., Killiany, J. M. and Brashears, S. S., "Effects of electric fields on annealing of radiation damage in MOSFET's", IEEE Trans. Electron Devices, ED-15, 751-6 (1968), EEA71-5032.
642. Dennehy, W. J., Brucker, G. J. and Holmes-Siedle, A. G., "A radiation-induced instability in silicon MOS transistors", IEEE Trans. Nucl. Sci., NS-13, 273-81 (1966), EEA70-10862.
643. Elliatiolglu, M. R. and Sencer, O., "Effects of X-ray irradiation to the surface-state density in a metal-oxide-semiconductor structure", Tech. J., Ankara Nucl. Res. Cent., 1 (1), 23-25 (1974), NSA30-24783.
644. Esteve, D. and Buxo, J., "Degradation mechanism of MOS structures and transistors under ionising radiation", Electron. Lett., 6, 198-200 (1970), EEA73-25044.
645. Esteve, D. and Buxo, J., "On the behaviour of currents going through MOS structures under ionizing radiations", Solid-State Electron., 14, 257-60 (1971), PA74-33731.
646. Farmer, J. W., "X-ray induced currents and space-charge buildup in MOS capacitors", Thesis, Kansas State Univ., Manhatten, Kansas, 1974, 109 pp., Order No. 74-25599, PA78-37841.
647. Farmer, J. W. and Lee, R. S., "Photocurrents and photoconductive yield in MOS structures during X-irradiation", J. Appl. Phys., 46, 2710-15 (1975), EEA78-33073.
648. Ferber, R. R., "Preprogrammed non-volatile metal-oxide-semiconductor memory by use of ionizing radiation", IEEE Trans. Nucl. Sci., NS-18 (6), 138-40 (1971), PA75-19736.
649. Galloway, K. F., "Simple model for the slope change of C-V curves of irradiated MOS capacitors", J. Appl. Phys., 45, 964-5 (1974), PA77-53153.

650.Gaw, E. T. and Oldham, W. G., "Properties of heavily irradiated MOSFETs", IEEE Trans. Nucl. Sci., NS-21 (6), 124-9 (1974), PA78-42304.
651.Giralt, G., Andre, B., Simmone, J. and Esteve, D., "Behaviour of M.O.S. structures under x-ray irradiation", Electron. Lett., 2, 209-10 (1966), EEA70-6809.
652.Girii, V. A., Kondrachuk, A. V., Kronyushin, S. I., Shakhovtsov, V. I. and Litvinov, R. O., "The effect of irradiation on the carrier mobility in inversion layers of MOS structures", Phys. Status Solidi A, 22, 357-62 (1974), EEA77-18753.
653.Glotin, P., "Radiation effect on MOS devices", Onde Elect., 47, 950-3 (1967), French, EEA71-8223.
654.Gordon, F., Jr. and Wannemacher, H. E., Jr., "The effects of space radiation on MOSFET devices and some application implications of those effects", IEEE Trans. Nucl. Sci., NS-13, 262-72 (1966), EEA70-10873.
655.Grosewald, P., Kerr, D. R. and Powlus, R. A., "Reducing radiation damage in insulated-gate field-effect transistors", IBM Tech. Disclosure Bull., 14, 811-12 (1971), EEA75-1163.
656.Habing, D. H. and Shafer, B. D., "Modeling and radiation effects study of an LSI/MOS logic system", IEEE Trans. Nucl. Sci., NS-18, 263-72 (1971), EEA75-9477.
657.Holmes-Siedle, A. G. and Zaininger, K. H., "Designing MOS systems for radiation environments", Solid State Technol., 12 (5), 40-9 (1969), EEA72-20438.
658.Holmstrom, F. E., Collins, T. W. and Churchill, J. N., "Measured radiation effects in MOS capacitors with a proposed new model", Appl. Phys. Lett., 24, 464-6 (1974), PA77-60861.
659.Huber, E. E., Jr., Cohen, M. S. and Smith, D. O., "Electron beam detection of charge storage in MOS capacitors", Appl. Phys. Lett., 16, 147-8 (1970), EEA73-17657.
660.Hughes, H. L., "A survey of radiation-induced perturbations in metal-insulator-semiconductor structures", Proceedings of the 9th Annual Reliability Physics Conference, 1971, p. 33-9, EEA75-4650.
661.Hughes, H. L. and Giroux, R. A., "Space radiation versus M.O.S. devices and circuits", 12th East Coast Conference on Aerospace and Navigational Electronics, 1965, pap. 1.3.1, EEA70-22904.
662.Kjar, R. A. and Bell, J. E., "Characteristics of MOS circuits for radiation-hardened aerospace systems", IEEE Trans. Nucl. Sci., NS-18, 258-62 (1971), EEA75-10279.
663.Kjar, R. A., Peel, J. L. and Wrigley, C. Y., "Effects of metallic doping on ionization damage in m.o.s.f.e.t.'s", IEEE Trans. Nucl. Sci., NS-16, 207-10 (1969), EEA73-14318.
664.Kreutz, E. W., "X-radiation effects on MOS devices at RT", Z. Angew. Phys., 32, 280-6 (1971), PA75-13296.
665.Kreutz, E. W., "Photoemission studies at RT of X-irradiated MOS structures", Phys. Status Solidi A, 13, 557-67 (1972), EEA76-8592.

666.Kreutz, E. W., Pagnia, H. and Waidelich, W., "Effects of X-radiation on MOS devices at T=79K", Phys. Status Solidi A, 5, 691-9 (1971), EEA74-59077.
667.Lange, R. O., "Space radiation challenges the MOSFET", 1968 IEEE Electromagnetic Compatibility Symposium Record, p. 360-80, EEA72-819.
668.Lindmayer, J., "Radiation and oxide-metal interactions in MOS", IEEE Trans. Nucl. Sci., NS-18, 91-8 (1971), PA75-19731.
669.Lindmayer, J. and Noble, W. P., Jr., "Radiation resistant MOS devices", IEEE Trans. Electron Devices, ED-15, 637-40 (1968), EEA71-28912.
670.Lisovskii, I. P. and Litvinov, R. O., "Effect of ionising radiation on m.o.s. structures", Poluprovodn. Tekh. & Mikroelectron., no. 18, 67-80 (1974), Russian, PA78-25769.
671.Litvinov, R. O., Litovchenko, V. G. and Radeske, C., "Effect of surface impurities on radiation-induced changes in MOS structures", Phys. Status Solidi A, 22, 293-8 (1974), PA77-33034.
672.Long, D. M., "A radiation effects large signal equivalent circuit for M.O.S. transistors", IEEE Trans. Nucl. Sci., NS-14, 210-16 (1967), EEA71-15861.
673.Long, D. M. and Baer, R., "A new radiation-hard M.O.S. transistor", IEEE Trans. Nucl. Sci., NS-14, 303-4 (1967), EEA71-15860.
674.Long, D. M. and Flescher, H. L., "Transient response of M.O.S. transistors and integrated circuits to ionizing radiation", IEEE Trans. Nucl. Sci., NS-13, 295-9 (1966), EEA70-10864.
675.Ma, T. P. and Barker, R. C., "Effect of gamma-ray irradiation on the surface states of MOS tunnel junctions", J. Appl. Phys., 45, 317-21 (1974), PA77-45002.
676.Ma, T. P., Scoggan, G. and Leone, R., "Comparison of the interface-state generation by 25-keV electron beam irradiation in p-type and n-type MOS capacitors", Appl. Phys. Lett., 27, 61-3 (1975), PA78-81521.
677.MacDonald, N. C. and Everhart, T. E., "Selective electron-beam irradiation of metal-oxide-semiconductor structures", J. Appl. Phys., 39, 2433-47 (1968), EEA71-15857.
678.Matsukawa, T., Shimizu, R., Harada, K. and Kato, T., "Investigation of kilovolt electron energy dissipation in solids (by beam-induced currents in MOS capacitor)", J. Appl. Phys., 45, 733-40 (1974), PA77-52794.
679.Matsukawa, T., Shimizu, R., Murata, K. and Harada, K., "Determination of kilovolt electron energy dissipation in depth in solids (m.o.s. structure)", Scanning Electron Microscopy, 1973, p. 277-84, EEA76-38235.
680.Mattauch, R. J. and Lade, R. W., "The effects of Co^{60} gamma radiation on MOS diodes", IEEE Trans. Nucl. Sci., NS-14, 52-7 (1967), PA71-22641.
681.Mitchell, J. P., "Dose rate dependence of surface radiation damage effects in planar silicon devices", IEEE Northesat Electronics Research and Engineering Meeting, 1966, p. 32-3, EEA70-22901.

682.Mitchell, J. P., "Radiation-induced space charge buildup in M.O.S. structures", IEEE Trans. Electron Devices, ED-14, 764-74 (1967), EEA71-5963.
683.Newman, P. A. and Wannemacher, H. E., "Anomalous radiation effect in p-channel MOSFET's under electron irradiation", Proc. IEEE, 55, 562 (1967), EEA70-17036.
684.Newman, P. A. and Wannemacher, H. E., "Radiation induced peaking effect in open geometry p-channel MOSFET's", IEEE Trans. Nucl. Sci., NS-14, 299-302 (1967), EEA71-15866.
685.Nielsen, R. L. and Nichols, D. K., "Total dose effects of ionizing radiation on MOS structures at 90°K", IEEE Trans. Nucl. Sci., NS-20 (6), 319-25 (1973), PA77-45001.
686.Oldham, W. G., "Study of bulk trapping effects in radiation-resistance MOS devices", Report AFCRL-TR-74-0224, Univ. California, Berkeley, 1974, 32 pp., EEA78-24365.
687.Onnasch, D. and Reimerdes, H. P., "Radiation effects in n-channel MOSFETs", Solid-State Electron., 17, 663-6 (1974), EEA77-30341.
688.Peel, J. L. and Eden, R. C., "Study of ionizing radiation damage in MOS structures using internal photoemission", IEEE Trans. Nucl. Sci., NS-18, 84-90 (1971), EEA75-10272.
689.Peel, J. L. and Kjar, R. A., "Photoemission studies of interface barrier energies of irradiated MOS structures", Appl. Phys. Lett., 17, 3-5 (1970), EEA73-31311.
690.Pfeifer, R. F. and Sah, C. T., "A distortion analysis of the M.O.S. transistor", Semiconductor Prod. Solid State Technol., 9 (11), 39-44 (1966), EEA70-2660.
691.Philipp, L. D. and Lauritzen, P. O., "Irradiation damage of MOSFET's operating in the common source mode", IEEE Trans. Nucl. Sci., NS-14, 284-92 (1967), EEA71-15864.
692.Popov, V. D., "Effect of reactor radiation on MOS-structures", Izv. VUZ Radioelektron., 14, 458-60 (1971), Russian, EEA74-18692.
693.Raymond, P. and Pocock, D. N., "Transient ionizing radiation effects in MOS/LSI", IEEE Trans. Nucl. Sci., NS-18, 288-94 (1971), PA75-19737.
694.Revesz, A. G., Zaininger, K. H. and Evans, R. J., "Reduction of radiation sensitivity in M.O.S. structures by aluminum doping of silicon dioxide", J. Electrochem. Soc., 116, 1146-8 (1969), EEA73-4418.
695.Rouanet, J. C., Giraud, J. L. and de Lafond, Y. G., "Effects of 600 MeV protons on MOS transistors", IEEE Trans. Nucl. Sci., NS-16, 106-8 (1969), EEA73-17663.
696.Saminadayar, K. and Pfister, J. C., "The distribution of interface states in electron irradiated MOS structures and the effects of heat treatment with applied field", Solid State Commun., 8, 2135-9 (1970), French, EEA72-33699.
697.Saminadayar, K. and Pfister, J. C., "Bias-temperature treatment on MOS capacitors irradiated with 1 MeV electrons", Radiat. Eff., 20, 55-64 (1973), PA77-16828.

698. Share, S., Epstein, A. S., Kumar, V., Dahlke, W. E. and Haller, W., "Effects of ionizing radiation on thin-oxide (20-1500 Å) MOS capacitors", J. Appl. Phys., 45, 4894-8 (1974), PA78-21961.
699. Share, S. and Martin, R. A., "Effects of ionizing radiation on short-channel thin-oxide (200-A) MOSFETs", IEEE Trans. Electron Devices, ED-22, 619-20 (1975), EEA78-28798.
700. Singh, B. R. and Srivastava, R. S., "Effect of radiations on M-O-S devices", Int. J. Electron., 34, 785-93 (1973), EEA76-26094.
701. Snow, E. H., Albus, H. P., Cagnina, S. F., Molozzi, A. R. and Fitzgerald, D. J., "Radiation study of MOS structures", Report AFCRL-68-0321 Fairchild Semiconductor Corp., Mountain View, Calif., 1968, 34 pp., EEA72-13499.
702. Stanley, A. G., "Comparison of MOS and metal-nitride-semiconductor insulated gate field-effect transistors under electron irradiation", IEEE Trans. Nucl. Sci., NS-13, 248-54 (1966), EEA70-10871.
703. Stanley, A. G., "A model for shifts in the gate turn-on voltage of insulated-gate field-effect devices induced by ionizing radiation", IEEE Trans. Electron Devices, ED-14, 134-8 (1967), EEA70-10874.
704. Stanley, A. G., "Effect of electron irradiation on carrier mobilities in inversion layers of insulated gate field effect transistors", IEEE Trans. Nucl. Sci., NS-14, 266-75 (1967), EEA71-15876.
705. Therez, F., Buxo, J. and Esteve, D., "On the singular behaviour of p-substrate MOS transistors under the action of an ionizing radiation flux", C. R. Hebd. Sean. Acad. Sci. B., 272, 889-92 (1971), French, EEA74-22725.
706. Thomas, A. G., Butler, S. R., Goldstein, J. I. and Parry, P. D., "Electron beam irradiation effects in thick-oxide MOS capacitors", IEEE Trans. Nucl. Sci., NS-21 (4), 14-19 (1974), EEA78-577.
707. Verhoeven, J. L., Saminadayar, K. and Pfister, J. C., "Illumination with ultraviolet light of MOS capacitors", Phys. Status Solidi A, 24, 441-48 (1974).
708. Wegener, H. A. R. and Doig, M. B., "Radiation resistant MNOS memories", IEEE Trans. Nucl. Sci., NS-19, 291-8 (1972), EEA76-12641.
709. Witteles, A. A., "Neutron radiation effects on MOSFET's: theory and experiment", IEEE Trans. Nuclear Sci., NS-15, 126-32 (1968), EEA72-11325.
710. Yamaguchi, M. and Nagai, O., "Surface effects of radiation on MOS diodes", Jap. J. Appl. Phys., 10, 1733 (1971), EEA75-7545.
711. Zaininger, K. H. and Waxman, A. S., "Radiation resistance of Al_2O_3 MOS devices", IEEE Trans. Electron Devices, ED-16, 333-8 (1969), EEA72-17738.

c. CMOSFET

712.Brucker, G. J., "Transient and steady-state radiation response of CMOS/SOS devices", IEEE Trans. Nucl. Sci., NS-21, 201-7 (1974), EEA78-19896.

713.Burghard, R. A. and Gwyn, C. W., "Radiation failure modes in CMOS integrated circuits", IEEE Trans. Nucl. Sci., NS-20, 300-6 (1973), EEA77-23125.

714.Cohen, S., "Total-dose radiation-hardened COS/MOS integrated circuits", Report PRRL-75-CR-16, RCA Solid State Div., Somerville, N. J., 1975, 29 pp., EEA78-41706.

715.Cohen, S. and Hughes, H., "SEM irradiation for hardness assurance screening and process definition", IEEE Trans. Nucl. Sci., NS-21, 387-9 (1974), EEA78-19901.

716.Dennehy, W. J., Holmes-Siedle, A. G. and Leopold, W. F., "Transient radiation response of complementary-symmetry MOS integrated circuits", IEEE Trans. Nucl. Sci., NS-16, 114-19 (1969), EEA73-12987.

717.Gregory, B. L. and Shafer, B. D., "Latch-up in CMOS integrated circuits", IEEE Trans. Nucl. Sci., NS-20, 293-99 (1973), NSA29-29579.

718.Habing, D. H. and Shafer, B. D., "Room temperature annealing of ionization-induced damage in CMOS circuits", IEEE Trans. Nucl. Sci., NS-20, 307-14 (1973), EEA77-23126.

719.Hughes, H. L., "Radiation-hardened complementary metal-oxide-semiconductor devices and circuits", Patent USA 3862930, Publ. January 1975, CA82-179663.

720.King, E. E., Nelson, G. P. and Hughes, H. L., "The effects of ionizing radiation on various CMOS integrated circuit structures", IEEE Trans. Nucl. Sci., NS-19, 264-70 (1972), EEA75-12673.

721.Pease, R. L., "Combined effect of ionizing radiation and bias-temperature stressing on CMOS inverters", Report SLA-73-1122, Sandia Labs., Albuquerque, N. Mex., 1974, 26 pp., NSA29-26926.

722.Peel, J. L. and Kinoshita, G., "Radiation-hardened complementary MOS using SiO_2 gate insulators", IEEE Trans. Nucl. Sci., NS-19, 271-4 (1972), EEA76-12674.

723.Poch, W. and Holmes-Siedle, A. G., "Permanent radiation effects in complementary-symmetry M.O.S. integrated circuits", IEEE Trans. Nucl. Sci., NS-16, 227-32 (1969), EEA73-12989.

724.Poch, W. J. and Holmes-Siedle, A. G., "The long-term effects of radiation on complementary MOS logic networks", IEEE Trans. Nucl. Sci., NS-17, 33-40 (1970), EEA74-22733.

725.Schlesier, K. M., "Radiation hardening of CMOS/SOS integrated circuits", IEEE Trans. Nucl. Sci., NS-21, 152-8 (1974), EEA78-19893.

726.Simons, M., "Rapid annealing in irradiated CMOS transistors", IEEE Trans. Nucl. Sci., NS-21, 172-8 (1974), EEA78-19838.

727.Simons, M., "Transient surface damage", Report RTI-43U-812-FR, Research Triangle Inst., Durham, N. C., 1974, 93 pp., NSA31-6166.

d. MISFET

728. Abe, H., Inoue, Y., Morita, H. and Enomoto, T., "The influence of electron beam evaporation on MIS devices", J. Vac. Soc. Jap., 16 (2), 69-90 (1973), Japanese, EEA76-29198.
729. Ashner, J. D., "Initial and radiation-induced charge distribution in silicon oxynitride m.i.s. structures", IEEE Trans. Electron Devices, ED-18, 384-6 (1971), EEA74-18690.
730. Brucker, G. J., "Interaction of nuclear environment with MNOS memory device", IEEE Trans. Nucl. Sci., NS-21, 186-92 (1974), EEA78-19894.
731. Castagne, R., "Capacitance and parallel conductance of MIS structures bombarded with low energy electrons", C. R. Acad. Sci. B., 266, 937-9 (1968), French, PA71-27211.
732. Castagne, R., "Determination of the density of surface states for M-I-S structures before and after irradiation with low energy electrons", C. R. Acad. Sci. B, 267, 1000-2 (1968), French, EEA72-4963.
733. Cook, K. B., Jr., Davidson, J. L., Gibson, J. D. and Patterson, R. B., III, "Radiation-hardened MIS devices", Patent USA 3886584, Publ. May 1975, CA83-107219.
734. Cricchi, J. R. and Barbe, D. F., "Compensation of radiation effects by charge transport in metal-nitride-oxide-semiconductor structures", Appl. Phys. Lett., 19, 49-51 (1971), EEA74-34014.
735. Cricchi, J. R. and Barbe, D. F., "Electronic recovery from radiation effects in CMNOS structures", IEEE Trans. Nucl. Sci., NS-18, 126-30 (1971), PA75-19735.
736. Dennehy, W. J., Holmes-Siedle, A. G. and Zaininger, K. H., "Process techniques and radiation effects in metal-insulator semiconductor structures", IEEE Trans. Nucl. Sci., NS-14, 276-83 (1967), PA71-27209.
737. Esteve, D. and Buxo, J., "The shielding metallic interlayer technique hardens MIS devices against ionizing radiation effects", Thin Solid Films, 10, 155-8 (1972), EEA75-16755.
738. Esteve, D., Buxo, J., Martinez, A. and Chapeau, A., "Suggestions for hardening MIS devices against radiation", Thin Solid Films, 9, 471-5 (1972), PA75-19738.
739. Grannemann, W. W., "Ionizing radiation effects on GaAsP MIS capacitor structures", IEEE Trans. Nucl. Sci., NS-20, 166-70 (1973), PA77-41122.
740. Giriy, V. A., Kondrachuk, A. V. and Shahovtsov, V. I., "Current transfer through dielectric in irradiated MIS-structures", Solid State Commun., 12, 249-52 (1973), EEA76-16340.
741. Harai, E. and Royce, B. S. H., "The effects of electron and hole trapping on the radiation hardness of Al_2O_3 MIS devices", IEEE Trans. Nucl. Sci., NS-20 (6), 280-7 (1973), PA77-41123.

742.Holmes-Siedle, A. G., Dennehy, W. J. and Zaininger, K. H., "The interrelation of process techniques and space radiation effects in metal-insulator-semiconductor structures", Report NASA-CR-95678, RCA, David Sarnoff Research Center, Princeton, N. J., 1967, 81 pp., PA72-19924.
743.Holmes-Siedle, A. G. and Zaininger, K. H., "The physics of failure of MIS devices under radiation", IEEE Trans. Reliab., R-17, 34-44 (1966), EEA71-16164.
744.Hughes, H. L., "Irradiation effects in the thin film insulators in MIS type structures", J. Vac. Sci. & Technol., 9, 40 (1972), PA75-40315.
745.Kim, M. J., Doyle, J. T., Patterson, W. J. and Pajak, R. W., "Electrical and irradiation performance of MNOS arrays", Jap. J. Appl. Phys., 13, 1837-46 (1974), EEA78-5358.
746.Kuhlmann, G. J. and Grannemann, W. W., "Effects of ionizing radiation of gallium arsenide phosphide metal-insulator semiconductor structures", Report EE-213(73), New Mexico Univ., Albuquerque, 1973, 109 pp., NSA30-24095.
747.Lodi, R. and Dodson, W., "Operation and radiation hardness of fully decoded 16 x 16 MNOS-REPROM integrated circuit", Report SLA-73-7053, Sandia Labs., Albuquerque, N. Mex., (nd), 11 pp., NSA30-6771.
748.Micheletti, F. B. and Kolondra, F., "Relaxation phenomena associated with radiation-induced trapped charge in Al_2O_3 MOS devices", IEEE Trans. Nucl. Sci., NS-18, 131-7 (1971), EEA75-10277.
749.Perkins, C. W., "Radiation effects in metal-insulator-semiconductor capacitors as determined from conductance measurements", Appl. Phys. Lett., 12, 153-5 (1968), EEA71-18308.
750.Perkins, C. W., Aubuchon, K. G. and Dill, H. G., "Radiation effects and electrical stability of metal-nitride-oxide-silicon structures", Appl. Phys. Lett., 12, 385-7 (1968), EEA71-20579.
751.Plotnikov, A. F., Seleznev, V. N., Shubin, V. E. and Ferchev, F. P., "Redistribution of the voltage in a metal-silicon nitride-silicon dioxide-silicon structure under the influence of laser radiation", Sov. J. Quantum Electron., 4, 1054-5 (1975), PA78-46769.
752.Potemkin, V. V. and Kashkarov, P. K., "Influence of neutron bombardment on the low-frequency noise properties of m.i.s. triodes", Radio Eng. & Electron. Phys., 17, 1598-9 (1972), EEA76-38279.
753.Stanley, A. G. and Wegener, H. A. R., "Effects of electron irradiation of metal-nitride-semiconductor insulated gate field-effect transistors", Proc. IEEE, 54, 784-5 (1966), EEA70-2674.
754.Thibault, L. R. and Pickar, K. A., "Electron irradiation of MIS capacitors - annealing and bias-temperature stress studies", Electron and Ion Beam Science and Technology Fifth International Conference, 1972, p. 148-61, EEA76-8627.
755.Wagemann, H. G. and Brauning, D., "MIS-devices under the influence of ionizing irradiation", Nachrichtentech. Z. (NTZ), 26, 254-60 (1973), German, EEA76-29274.

756.Zaininger, K. H. and Holmes-Siedle, A. G., "A survey of radiation effects in metal-insulator-semiconductor devices", RCA Rev., 28, 208-40 (1967), EEA71-1277.

e. CCD

757.Antcliffe, G. A., Roberts, C. G., Barton, J. B., Pharr, K. K., Mueller, D. W. and Collins, D. R., "Operation of CCD's with stationary and moving electron-beam input", IEEE Trans. Electron Devices, ED-22, 857-61 (1975).
758.Barbe, D. F., Killiany, J. M. and Hughes, H. L., "Effects of gamma radiation on charge-coupled devices", Appl. Phys. Lett., 23, 400-2 (1973), EEA77-2563.
759.Killiany, J. M., Baker, W. D., Saks, N. S. and Barbe, D. F., "Effects of ionizing radiation on charge-coupled device structures", IEEE Trans. Nucl. Sci., NS-21, 193-200 (1974), EEA78-19895.
760.Killiany, J. M., Saks, N. S., Baker, W. D. and Barbe, D. E., "Effects of radiation on buried-channel CCD's with doped polysilicon gates and undoped polysilicon interelectrode isolation", Appl. Phys. Lett., 24, 506-8 (1974), EEA77-30326.
761.Williams, E. W., Bosnell, J. R., Windle, D., Machin, D., Hitchcock, A. and Wankling, J. L., "Radiation hardness of aluminium gate CCD's", International Conference on Technology and Applications of Charge Coupled Devices, 1974, p. 123-32, EEA78-11687.

f. SOS

762.Kjar, R. A. and Kinoshita, G., "Transient photocurrents in SOS structures", IEEE Trans. Nucl. Sci., NS-20, 315-18 (1973), EEA77-23127.
763.Kjar, R. A. and Peel, J., "Radiation induced leakage current in n-channel SOS transistors", IEEE Trans. Nucl. Sci., NS-21, 208-10 (1974), EEA78-19839.
764.Neamen, D., Shedd, W. and Buchanan, B., "Radiation induced charge trapping at the silicon sapphire substrate interface", IEEE Trans. Nucl. Sci., NS-21, 211-16 (1974), EEA78-19705.
765.Pontius, D. H., Budenstein, P. P. and Smith, W. B., "Effect of ionizing radiation on second breakdown", Solid-State Electron., 16, 1073-7 (1973), EEA76-35289.
766.Stein, H. J., "Effects of electron and neutron irradiation on the electrical properties of Si on MgO.Al_2O_3 spinel", Solid-State Electron., 16, 1391-7 (1973), EEA77-5836.

5. Diodes

a. IMPATT and TRAPATT

767. Anderson, W. A., "The performance of IMPATT diodes under the influence of gamma radiation", Proc. IEEE, 57, 852-3 (1969), EEA72-17713.
768. Anderson, W. A., "The performance of IMPATT diodes under the influence of high-intensity gamma and neutron radiation", Proc. IEEE, 57, 1441-2 (1969), EEA73-1164.
769. Anderson, W. A., "A radiation pulse interrupts IMPATT diode oscillation", Proc. IEEE, 58, 807 (1970), EEA73-24992.
770. Borrego, J. M., Gutmann, R. J., Geipel, H. J. and Ghandhi, S. K., "Operation of TRAPATT oscillators under transient ionizing radiation conditions", IEEE Trans. Nucl. Sci., NS-20, 144-48 (1973), NSA29-29595.
771. Chaffin, R. J., "The effect of neutron radiation on an IMPATT diode", IEEE Trans. Microwave Theory Tech., MTT-17, 119-20 (1969), EEA72-11286.
772. Chaffin, R. J., EerNisse, E. P. and Hood, J. A., "The effect of neutron irradiation on the TRAPATT diode", Proc. IEEE, 57, 2063-4 (1969), EEA73-10998.
773. Geipel, H. J., Jr., "Transient ionizing radiation effects in TRAPATT diode oscillators", Thesis, Rensselaer Polytechnic Inst., Troy, N. Y., 1974, 143 pp., Order No. 74-26075, NSA31-6173.
774. Groves, I. S., Spreight, J. D., Leigh, P., McIntrye, N., O'Hara, S. and Hemment, P., "Proton isolated GaAs IMPATT diodes", 4th European Microwave Conference, 1974, p. 163-7, EEA78-2837.
775. Gutmann, R. J., Borrego, J. M., Cottrell, P. E. and Ghandhi, S. K., "The performance of injection locked IMPATT oscillators under transient ionizing radiation", IEEE Trans. Nucl. Sci., NS-20, 388-90 (1973), EEA77-23000.
776. Gutmann, R. J., Borrego, J. M. and Ghandhi, S. K., "Radiation effects in transit-time microwave diodes", Proc. IEEE, 62, 1258-66 (1974), EEA77-34026.
777. Wilson, D. K., Lee, H. S. and Noffke, H., "Radiation effects on silicon avalanche (IMPATT) diodes", IEEE Trans. Nucl. Sci., NS-15, 114-25 (1968), EEA72-11280.

b. LED

778. Barnes, C. E., "Radiation effects in electroluminescent diodes", IEEE Trans. Nucl. Sci., NS-18 (6), 322-31 (1971), PA75-20116.
779. Barnes, C. E., "Neutron damage in SiC light-emitting diodes", Appl. Phys. Lett., 20, 86-8 (1972), PA75-19708.
780. Barnes, C. E., "Neutron damage in GaP light-emitting diodes", Appl. Phys. Lett., 20, 110-12 (1972), PA75-20094.

781.Polimadei, R. A., Share, S., Epstein, A. S. and Gowins, G. E., "Radiation susceptibility of high power gallium arsenide infrared light-emitting diodes", Report AD-785661/OGA, Harry Diamonds Lab., Washington, D. C., 1973, 15 pp., CA82-179541.
782.Polimadei, R. A., Share, S., Epstein, A. S., Lynch, R. J. and Sullivan, D., "Performance of $Ga_{1-x}Al_xAs$ light emitting diodes in radiation environments", IEEE Trans. Nucl. Sci., NS-21, 96-102 (1974), PA78-42814.

c. Detectors

783.Bakirov, M. Y. and Azizov, T. S., "Effect of irradiation of accelerated electrons on rectifying properties of silicon photocells", Radiotekhnika i Elektronika, 12, 1859-60 (1967), Russian, EEA71-15829.
784.Blake, K., Epstein, A. S., Polimadei, R. A., Share, S. and Snyder, J. F., "Radiation effects in light pipes and light pipe isolators", Appl. Opt., 13, 1272-74 (1974), NSA30-24104.
785.Coleman, J. A., Love, D. P., Trainor, J. H. and Williams, D. J., "Effects of damage by 0.8 MeV - 5.0 MeV protons in silicon surface-barrier detectors", IEEE Trans. Nucl. Sci., NS-15, 363-72 (1968), EEA71-28474.
786.Eberhardt, J. E., Ryan, R. D. and Tavendale, A. J., "Nuclear radiation detectors from high purity epitaxial n-GaAs", Proceedings of the International Symposium on Cadmium Telluride a Material for Gamma-Ray Detectors, 1971, 29/9 pp., EEA75-21486.
787.Goulding, F. S. and Pehl, R. H., "A survey of radiation damage in semiconductor detectors", IEEE Trans. Nucl. Sci., NS-19, 91-9 (1972), PA75-27874.
788.Grube, R., Fretwurst, E. and Lindstrom, G., "Radiation damage effects from 2 MeV protons in silicon surface barrier detectors", Nucl. Instrum. & Meth., 101, 97-105 (1972), EEA75-25625.
789.Gruhn, C. R., Kuo, T., Maggiore, C., Preedom, B., Samuelson, L. and Chander, J., "Lithium-drifted germanium for charged particle spectroscopy", IEEE Trans. Nucl. Sci., NS-15, 337-46 (1968), EEA71-28456.
790.Johnson, R. T., Jr., "Beta- and gamma-induced conductivity in semiconductors: application to CdS neutron detectors", J. Appl. Phys., 41, 4981-91 (1970), EEA74-1357.
791.Kaganskaya, M. E., Fedoseeva, O. P. and Sinitsyn, B. I., "Influence of neutron irradiation upon the characteristics of fully depleted silicon surface barrier detectors", Sov. At. Energy, 36, 399-401 (1974), NSA31-8784.
792.Karpenko, V. P., Kashernininov, P. G. and Matveev, O. A., "Effect of charge exchange between deep centers near a p-n junction on the properties of germanium", Sov. Phys.-Semicond., 6, 93-5 (1972), EEA75-32624.
793.Kinch, M. A., "Compensated silicon-impurity conduction bolometer", J. Appl. Phys., 42, 5861-3 (1971), EEA75-7276.

794.Kraner, H. W., Chasman, C. and Jones, K. W., "Effects produced by fast neutron bombardment of Ge(Li) gamma ray detectors", Nucl. Instrum. & Meth., 62, 173-83 (1968), EEA71-23040.
795.Lambert, K. P. and Van de Voorde, M., "High radiation dose luminescent and optical dosimetry systems", Int. J. Appl. Radiat. & Isot., 25 (2), 69-79 (1974), PA77-42878.
796.Lazutkin, I. I., Sinitsyn, B. I., Fedoseeva, O. P. and Tsypin, S. G., "The stability of silicon detector surfaces to radiation of neutrons of varying speed", At. Energ., 31, 636-7 (1971), Russian, PA75-38376.
797.Lebedev, S. Y. and Taurbaev, T. I., "Short-circuit currents and (open-circuit) e.m.f.'s of silicon and selenium photocells coated with a layer of U^{235} and irradiated with neutrons", Fiz. Tekh. Poluprovodinkov, 4, 584-7 (1970), EEA73-27761.
798.Liu, Y. M. and Coleman, J. A., "Radiation damage effects by electrons, protons, and neutrons in Si(Li) detectors", IEEE Trans. Nucl. Sci., NS-19, 346-52 (1972), EEA75-25593.
799.Llacer, J. and Kraner, H. W., "Neutron damage and annealing in high purity germanium radiation detectors", Nucl. Instrum. & Meth., 98, 467-75 (1972), EEA75-14756.
800.Loferski, J. J. and Crisman, E. E., "Investigation of the effects in sub-threshold high energy electrons on the properties of silicon photovoltaic cells", Report NASA-CR-109851, Brown Univ., Providence, R. I., 1970, 13 pp., EEA74-4618.
801.Nakamoto, S., Aoki, T., Kawabata, K., Norisawa, K. and Sakisaka, M., "Damage effects in silicon surface barrier detectors with 0.5 to 2 MeV protons", Jap. J. Appl. Phys., 13, 524-28 (1974), NSA30-32285.
802.Nakamoto, S., Aoki, T., Kawabata, K., Norisawa, K. and Sakisaka, M., "Damage effects in silicon surface barrier detectors by 0.5-1.5 MeV electrons", Jap. J. Appl. Phys., 12, 1493-9 (1973), EEA77-1216.
803.Saunders, J. E., "Application of the logarithmic response of silicon diodes to monitoring beam symmetry and electron energy", Phys. Med. Biol., 19, 373-78 (1974), NSA30-13321.
804.Seregina, Y. A., Semenova, N. N. and Kuzminov, B. D., "Effect of neutron irradiation on operation of fission-fragment surface barrier detectors", At. Energ., 32, 70-2 (1972), Russian, PA75-27873.
805.Solovev, S. M., Tybin, L. I. and Eismont, V. P., "Effect of neutron radiation on the spectrometric characteristics of Si(Li) detectors", Sov. J. At. Energy, 33, 1065-66 (1972), NSA30-6776.
806.Stelson, P. H., Dickens, J. K., Raman, S. and Trammell, R. C., "Deterioration of large Ge(Li) diodes caused by fast neutrons", Nucl. Instrum. & Meth., 98, 481-4 (1972), EEA75-14758.
807.Tang, M. D. and Wu, S. S., "Effects of gamma radiation on semiconductor surface barrier detectors", Nucl. Sci. J., 8, 27-40 (1971), NSA30-18788.

808.Triboulet, R., Margaing, Y., Cornet, A. and Siffert, P., "Undoped high-resistivity cadmium telluride for nuclear radiation detectors", J. Appl. Phys., 45, 2759-65 (1974), EEA77-33948.
809.Zaletin, V. M., Protasov. I. I., Matveev, O. A., Skorokhodov, P. I. and Khusainov, A. K., "Gallium arsenide single crystal films for spectrometric detectors of x-ray and soft γ-radiation", At. Energ., 39, 68-9 (1975), Russian, CA83-185190.
810.Zhits, M. M., Klimkova, O. A., Niyazova, O. R. and Popov, N. V., "Radiation-enhanced diffusion-defect production ratio in γ-irradiated silicon detectors", Phys. Status Solidi A, 10, K23-6 (1972), EEA75-16758.

d. Solar Cells

811.Arndt, R. A. and Westerlund, L. H., "A model for non-penetrating proton damage to silicon solar cells", COMSAT Tech. Rev., 1, 117-37 (1971), EEA75-15489.
812.Berman, P. A., "Status of lithium-doped solar cell investigations at the jet propulsion laboratory", Conference Record of the 8th IEEE Photovoltaic Specialists Conference, 1970, p. 229-39, EEA74-9335.
813.Brucker, G. J., Faith, T. J. and Holmes-Siedle, A. G., "Action of lithium in radiation hardened silicon solar cells", Report NASA-CR-96277, RCA, Princeton, N. J., 1968, 47 pp., EEA72-12377.
814.Brucker, G. J., Faith, T. and Holmes-Siedle, A. G., "Injection-level effects in irradiated, lithium-containing solar cells", Conference Record of the 7th Photovoltaic Specialists Conference, 1968, p. 146-54, EEA72-14450.
815.Brucker, G. J. and Markow, B., "Neutron damage in silicon solar cells", Conference Record of the Sixth Photovoltaics Specialists Conference, 1967, p. 53-63, EEA71-10644.
816.Bryant, F. J. and Glew, R. W., "Effect of oxygen on unencapsulated cadmium sulfide solar cells", Thin Solid Films, 23, S53-S56 (1974), NSA30-30103.
817.Carter, J. R., Jr. and Downing, R. G., "Behaviour of lithium in irradiated solar cells", Conference Record of the 8th IEEE Photovoltaic Specialists Conference, 1970, p. 240-6, EEA74-9336.
818.Crisman, E. E. and Loferski, J. J., "Effects of subthreshold high-energy electrons on the properties of silicon photovoltaic cells", Presented at the 3rd Annual Conference on Effects of Lithium Doping on Silicon Solar Cells, Pasadena, Calif., April 1970, NSA30-32886.
819.Crowther, D. L., Lodi, E. A., DePangher, J. and Andrew, A., "An analysis of non-uniform proton irradiation damage in silicon solar cells", IEEE Trans. Nucl. Sci., NS-13, 37-46 (1966), EEA70-6444.

820.Curtin, D. J. and Meulenberg, A., "Statistical analysis of one MeV electron irradiation of silicon solar cells", Conference Record of the 8th IEEE Photovoltaic Specialists Conference, 1970, p. 193-200, EEA74-9330.

821.Faith, T. J., "Damage and recovery characteristics of lithium-containing solar cells", IEEE Trans. Nucl. Sci., NS-18, 371-7 (1971), PA75-21013.

822.Faith, T. J., Brucker, G. J. and Holmes-Siedle, A. G., "Recovery rate and capacitance measurements on irradiated lithium-containing solar cells", IEEE Trans. Nucl. Sci., NS-15, 61-8 (1968), EEA72-11256.

823.Faith, T. J., Brucker, G. J., Holmes-Siedle, A. G. and Wysocki, J., "Long term stability of lithium-doped solar cells irradiated with electrons, protons, or neutrons", Conference Record of the 7th Photovoltaic Specialists Conference, 1968, p. 131-45, EEA72-12359.

824.Faith, T. J., Corra, J. P. and Holmes-Siedle, A. G., "Room temperature stability and performance of lithium-containing solar cells: an evaluation", Conference Record of the 8th IEEE Photovoltaic Specialist Conference, 1970, p. 247-59, EEA74-6414.

825.Fang, P. H., "Gamma irradiation of lithium silicon solar cells", Conference Record of the 7th Photovoltaic Specialists Conference, 1968, p. 113-16, EEA74-12357.

826.Faraday, B. J., "Analysis of radiation damage in silicon by thermal annealing solar cells", Radiat. Eff., 4, 293-301 (1970), EEA73-39739.

827.Ghosh, S. and Sargent, G. A., "The observation of damage regions produced by neutron irradiation in lithium-doped silicon solar cells", J. Mater. Sci., 7, 153-8 (1972), PA75-22179.

828.Ghosh, S. and Sargent, G. A., "Defect clusters in neutron-irradiated silicon solar cells (electron microscope studies)", Proceedings of the International Conference on Radiation Damage and Defects, 1972, p. 165-9, Publ. 1973, EEA76-22587.

829.Goldhammer, L. J. and Anspaugh, B. E., "Electron spectrum irradiations of silicon solar cells", Conference Record of the 8th IEEE Photovoltaic Specialist Conference, 1970, p. 201-8, EEA74-9331.

830.Grimshaw, J. A., "Reverse bias capacity effects and defect energy levels in electron irradiated GaAs p^+-n junctions", Radiat. Eff., 12, 87-95 (1972), PA75-29775.

831.Matsuoka, Y. and Usami, A., "Laser damage of copper-doped silicon solar cells", J. Nucl. Sci. & Technol., 11 (3), 112-13 (1974), PA77-77691.

832.Nashiyama, I., Teranishi, E. and Kageyama, M., "Protons and deuteron irradiation damage in silicon solar cells", Jap. J. Appl. Phys., 10, 1564-74 (1971), EEA75-3338.

833.Passenheim, B. C., Naber, J. A. and Berger, R. A., "Production and annealing of defects in lithium-diffused silicon after irradiation with 30-MeV electrons and fission neutrons at 300K", Conference Record of the 8th IEEE Photovoltaic Specialists Conference, 1970, p. 260-6, EEA74-9337.
834.Reynard, D. L. and Orvis, D. B., "Beta irradiation of lithium doped solar cells", Conference Record of the 7th Photovoltaic Specialists Conference, 1968, p. 108-12, EEA74-12356.
835.Ross, B., "Semiconductor device having improved resistance to radiation damage", Patent USA 3462311, Publ. August 1969, EEA73-10712.
836.Stanley, A. G., "Comparison of low-energy proton damage in ion-implanted and diffused silicon solar cells", Proc. IEEE, 59, 321-2 (1971), EEA74-14448.
837.Statler, R. L., "An evaluation by solar simulation of radiation damage in silicon solar cells", IEEE Trans. Electron Devices, ED-14, 31-6 (1967), EEA70-9137.
838.Statler, R. L., "Radiation effects in silicon solar cells", Radiat. Eff., 4, 305-11 (1970), EEA73-39740.
839.Stofel, E. and Joslin, D., "Low-energy proton irradiation of solar cell back contacts", Conference Record of the 8th IEEE Photovoltaic Specialists Conference, 1970, p. 209-13, EEA74-9332.
840.Stofel, E. J., Steward, T. B. and Ornelas, J. R., "Neutron damage to silicon solar cells", IEEE Trans. Nucl. Sci., NS-16, 97-105 (1969), EEA73-19482.
841.Usami, A., "Effects of impurities on the radiation damage and annealing behavior of Si solar cells", Jap. J. Appl. Phys., 9, 1063-9 (1970), EEA73-39737.
842.Usami, A., "Effects of junction depth on the radiation damage of silicon solar cells", Electron. & Commun. Jap., 54 (5), 113-20 (1972), EEA75-34671.
843.Waddel, R. C., "Solar cell data from the radiation damage experiment on satellite ATS-1", Report NASA-TM-X-65610, NASA, Greenbelt, Md., 1971, 298 pp., PA75-27400.
844.Weingart, J. M., "Defect structure and behavior in electron irradiated lithium diffused silicon solar cells", Conference Record of the 8th IEEE Photovoltaic Specialists Conference, 1970, p. 267-75, EEA74-9338.
845.Wilsey, N. D., "The effect of boron concentration on radiation damage in silicon solar cells", Record of the Ninth IEEE Photovoltaic Specialists Conference, 1972, p. 338-41, EEA76-7361.
846.Wilsey, N. D. and Lambert, R. J., "Low temperature irradiations of silicon solar cells with 1 MeV electrons", Conference Record of the 8th IEEE Photovoltaic Specialists Conference, 1970, p. 169-72, EEA74-6410.
847.Wohlleben, K., "Damage coefficient of Si-n/p solar cells irradiated with electrons and protons", Z. Angew. Phys., 28, 183-9 (1970), German, EEA73-16031.

848. Wysocki, J. J., "Role of lithium in damage and recovery of irradiated silicon solar cells", IEEE Trans. Nucl. Sci., NS-14, 103-9 (1967), EEA71-15832.

e. Lasers

849. Arnold, G. W., "Luminescence in intrinsic and annealed electron-irradiated GaAs:Cd", Phys. Rev., 183, 777-83 (1969), EEA73-949.
850. Barnes, C. E., "Effects of Co^{60} gamma irradiation on epitaxial GaAs laser diodes", Phys. Rev. B., 1, 4735-47 (1970), EEA73-27795.
851. Barnes, C. E., "Increased radiation hardness of GaAs laser diodes at high current densities", J. Appl. Phys., 45, 3485-89 (1974), NSA30-24105.
852. Bogdankevich, O. V., Vavilov, V. S., Danilychev, V. A., Kalendin, V. V. and Kryukova, I. V., "Influence of irradiation on the principal characteristics of a gallium arsenide laser excited with an electron beam", Fiz. Tekh. Poluprovodnikov, 4, 1209-15 (1970), Russian, EEA73-37698.
853. Compton, W. D. and Ahlburn, B. T., "Radiation effects in semi-conducting laser materials (Final report 1 Dec. 68-30 Nov. 69)", Report ARL-70-0076, Illinois Univ., Urbana, 1970, 44 pp., EEA74-10351.
854. Donnelly, J. P., Calawa, A. R., Harman, T. C., Foyt, A. G. and Lindley, W. T., "$Pb_{1-x}Sn_xTe$ photovoltaic diodes and diode lasers produced by proton bombardment", Solid-State Electron., 15, 403-7 (1972), EEA75-16431.
855. Kawabe, M., Masugama, A., Aoki, K., Masuda, K. and Namba, S., "Abnormal laser emission from electron beam excited GaAs", Jap. J. Appl. Phys., 9, 850-1 (1970), EEA73-37700.
856. Noel, B. W., Barnes, C. E. and Southward, H. D., "Neutron irradiation effects on diffused GaAs laser diodes", IEEE Trans. Nucl. Sci., NS-18, 378-86 (1971), PA75-21098.
857. Schroeder, J. O., Noel, B. W. and Southward, H. D., "Radiation damage induced time delay in GaAs lasers", IEEE Trans. Nucl. Sci., NS-20, 261-65 (1973), NSA29-29575.

f. Schottky

858. Chino, K., Okamoto, H., Ariyoshi, H. and Mizushima, Y., "Stress effect of gold-doped and gamma-irradiated Schottky-barrier diodes", Jap. J. Appl. Phys., 8, 502-3 (1969), EEA72-11288.
859. Darley, H. M., "Radiation damage in metal-semiconductor tunnel junctions", Thesis, Univ. of Kentucky, Lexington, 1973, 206 pp., Order No. 74-19611, NSA30-32919.
860. DeAngelis, H. M., Diebold, J. W. and Kimerling, L. C., "Specific observations of radiation effects in semiconductors by junction-capacitance techniques", Proceedings of the International Conference on Radiation Damage and Defects in Semiconductors, 1972, p. 295-301, Publ. 1973, PA76-33926.

861.Greilling, P. T. and Clark, J., "Transient radiation effects in GaAs IMPATT diodes", Proc. IEEE, 60, 753-4 (1972), EEA75-24318.
862.Guetin, P. and Schreder, G., "Effects of an ion bombardment on the characteristics of a metal/n-GaAs tunnel contact", J. Appl. Phys., 43, 549-61 (1972), EEA75-10226.
863.Mora, N. A., "Effects of radiation damage on the zero-bias anomaly in gallium arsenide Schottky-barrier tunnel junctions", Thesis, Brown Univ., Providence, R. I., 1971, 209 pp., Order No. 72-8175, EEA75-36251.
864.Mora, N. A., Loferski, J. J. and Bermon, S., "Effects of neutron irradiation on the electrical properties of heavily doped p-type GaAs and the Mott transistion", Proceedings of the International Conference on Radiation Damage and Defects in Semiconductors, 1972, p. 103-12, Publ. 1973, PA76-33823.
865.Neaman, D. A., "Radiation effects on gallium arsenide phosphide and gallium phosphide Schottky barrier diodes", Thesis, Univ. New Mexico, Albuquerque, 1971, 242 pp., Order No. 72-13796, EEA75-36256.
866.Neamen, D. A. and Grannemann, W. W., "Fast neutron effects on GaAsP Schottky barrier diodes and Hall effect devices", IEEE Trans. Nucl. Sci., NS-19, 215-9 (1972), EEA76-12651.
867.Shah, A. B., "Optical and beta radiation on silicon Schottky barriers", Thesis, Univ. Missouri, Columbia, 1971, 99 pp., Order No. 71-22939, EEA75-32640.
868.Schnurr, R. H., "The effect of transient ionizing radiation upon Schottky barrier diodes", Thesis, Univ. of New Mexico, 1968, 120 pp., Order No. 68-13082, EEA72-13461.
869.Taylor, P. D. and Morgan, D. V., "Role of radiation damage on the contact resistance of GaAs Schottky barriers", Electron. Lett., 10, 428-9 (1974), PA78-6894.
870.Taylor, P. D. and Morgan, D. V., "Radiation effects on GaAs Schottky diodes", Radiat. Eff., 27, 113-14 (1975).
871.Vasilev, A. M., Golovner, T. M., Zakharchuk, O. V. and Fedoseeva, O. P., "Spectral characteristic of surface-barrier radiation detectors", Fiz. Tekh. Poluprovodikov, 3, 1276 (1969), Russian, PA73-15235.
872.Yamashita, S., Miyazaki, E. and Miyaji, K., "Oscillation observed in silicon surface barrier diodes under pulse electron bombardment", Jap. J. Appl. Phys., 2, 806-7 (1963).
873.Yu, A. Y. C. and Snow, E. H., "Radiation effects on silicon Schottky barriers", IEEE Trans. Nucl. Sci., NS-16, 220-6 (1969), EEA73-14237.

g. Others

874.Afanasev, V. N. and Tuturov, Y. F., "L-shaped current characteristic of semiconductor diodes irradiated by fast neutrons", Radiotekhnika i Elektronika, 12, 1851-3 (1967), Russian, EEA71-15797.

875.Been, J. F., "Effects of nuclear radiation on high-reliability silicon power diode I-change in I-5 design characteristics", Report NASA-TN-D-4620, NASA, Lewis Research Center, Cleveland, Ohio, 1968, 37 pp., EEA72-4997.
876.Been, J. F., "Effects of nuclear radiation on a high-reliability silicon power diode. 4. Analysis of reverse bias characteristics", Report NASA-TN-D-7423, NASA, Cleveland, Ohio, 1973, 36 pp., NSA30-9648.
877.Berg, N. and Dropkin, H., "Neutron displacement effects in epitaxial Gunn diodes", IEEE Tran. Nucl. Sci., NS-17, 233-8 (1970), EEA74-22581.
878.Berman, L. S., Kreinin, L. B. and Nasredinov, F. S., "Investigation of the capacitive properties of silicon n^+-p diodes irradiated with fast electrons", Sov. Phys.-Semicond., 6, 252-6 (1972), PA75-71978.
879.Brehm, G. E., "Effects of gamma radiation on gallium arsenide crystals and Gunn diodes", Thesis, Stanford Univ., Calif., 1970, 86 pp., Order No. 71-12860, EEA74-34009.
880.Brudnyi, W. N., Vilisov, A. A., Vyatkin, A. P., Krivov, M. A. and Malyanov, S. V., "Influence of electron irradiation on the parameters of gallium arsenide pulse diodes", Izv. Vuz Fiz., 4, 109-13 (1970), Russian, EEA73-31349.
881.Chaffin, R. J., "Permanent neutron damage in PIN microwave diode switches", IEEE Trans. Nucl. Sci., NS-18, 429-35 (1971), EEA75-12926.
882.Cohn, N. S. and Petree, M. C., "Anomalous second breakdown failure levels in germanium gold-bonded diodes", IEEE Trans. Nucl. Sci., NS-21, 20-22 (1974), NSA31-3569.
883.Crabbe, J. S. and Ashley, K. L., "Capacitance voltage characteristics of neutron irradiated n^+pp^+ and p^+nn^+ junctions", IEEE Trans. Nucl. Sci., NS-18, 410-9 (1971), EEA75-13669.
884.Fairfield, J. M. and Schwuttke, G. H., "Silicon diodes made by laser irradiation", Solid-State Electron., 11, 1175-6 (1968), EEA72-11261.
885.Haas, G. A., "Analysis of physical parameters in electron-beam-irradiated semiconductor diodes", Report NRL-7366, Naval Res. Lab., Washington, D. C., 1972, 32 pp., EEA78-15589.
886.Johnson, J. W. and Fales, C. L., "Proton-radiation damage in Gunn oscillators", Report NASA-TN-D-7348, NASA, Greenbelt, Md., 1973, 28 pp., NSA30-3412.
887.Korshunov, V. D. and Kostyuchenko, E. A., "Action of neutrons on silicon diodes", Izv. VUZ Fiz., no. 6, 15-19 (1968), Russian, EEA72-4980.
888.Krag, W. E., "Electron beam induced degradation in EBS diodes", Report ESD-TR-74-210, Massachusetts Inst. of Tech., Lexington, Mass., 1974, 58 pp., NSA31-6165.
889.Makarov, V. V., "Use of surface-barrier diodes to investigate the interaction of ionizing radiation with a semiconductor", Fiz. Elektron. Trudy LPI, no. 328, 140-3 (1973), Russian, PA78-72905.

890.Marcus, G. H. and Bruemmer, H. P., "Radiation damage in GaAs Gunn diodes", IEEE Trans. Nucl. Sci., NS-17, 230-2 (1970), EEA74-22580.
891.Morgan, D. V. and Ashbrun, P., "The electrical activity of neutron damage centres in silicon diodes", Radiat. Eff., 22, 35-8 (1974), EEA77-36282.
892.Nakagiri, M. and Tsujide, T., "The effects of X-ray irradiation on MAS diodes", Jap. J. Appl. Phys., 11, 1054 (1972), EEA75-28519.
893.Nikolaevskii, I. F. and Shurenkov, V. V., "Some properties of electron-irradiated silicon diodes", Sov. Phys.-Semicond., 7, 1509-11 (1974), EEA78-618.
894.Schnurr, R. H. and Southward, H. D., "Radiation effects upon gallium arsenide devices", IEEE Trans. Nucl. Sci., NS-15, 306-10 (1968), EEA72-11180.
895.Schuttler, R., "The measurement of a fast neutron flux using a p-i-n diode", Action des Rayonnements Nucleaires Sur les Composants Electroniques, 1965, p. L1-L13, French, EEA70-4666.
896.Sigfridsson, B. and Leman, G., "Some aspects on the theory of transient radiation induced diffusion currents in semiconductor diodes", IEEE Trans. Nucl. Sci., NS-14, 179-86 (1967), EEA71-15796.
897.Tarneja, K. S. and Bartko, J., "Tailoring reverse recovery time and forward voltage drop characteristics of a diode by irradiation and annealing", Patent USA 3888701, Publ. March 1973, EEA78-41566.
898.Tarneja, K. S. and Mareinko, F. V., "Irradiation for fast recovery of high power junction diodes", Patent USA 3809582, Publ. May 1974, CA81-18388.
899.Taylor, G., True, R. and Wagner, H., "Performance of electron beam semiconductor amplifiers", Presented at 17th International Electron Devices Meeting, Washington, D. C., October 1971, EEA75-12595.
900.Tobolski, J., Chambers, E. S. and Kinder, J. D., "Tunnel diode switching induced by transient ionizing radiation", IEEE Trans. Nucl. Sci., NS-14, 221-2 (1967), EEA71-15804.
901.Wada, T., Fukuoka, Y. and Arizumi, T., "A new gradually graded P-i-N structure produced by electron irradiation in germanium. I. Formation", Jap. J. Appl. Phys., 11, 343-50 (1972), PA75-25888.
902.Wada, T., Fukuoka, Y. and Arizumi, T., "A new gradually graded P-i-N structure produced by electron irradiation in germanium. II. Space charge limited current", Jap. J. Appl. Phys., 11, 351-7 (1972), PA75-25889.
903.Wada, T., Nakamura, T., Maruyama, A., Fukuoka, Y. and Arizumi, T., "Decrease of built-in voltage in tunnel diodes caused by irradiation", Jap. J. Appl. Phys., 6, 71-6 (1967), EEA70-12988.
904.Wagemann, H. G., Spencker, A., Baunig, D., Roncin, J. C. and Pelous, G., "Irradiation test and preselection of maverick diodes for the Project Symphonic satellite", Microelectron. & Reliab., 12, 467-72 (1973), EEA77-14467.

905.Walker, J. W. and Sah, C. T., "Characteristics of 1.0 MeV electron-irradiated surface-controlled silicon junction diodes", Radiat. Eff., 20, 187-95 (1973), PA77-28480.
906.Wilson, D. K., "Majority and minority carrier trapping in neutron-irradiated silicon diodes by capacitance recovery", IEEE Trans. Nucl. Sci., NS-15, 77-83 (1968), EEA72-11257.
907.Wilson, W. E., "Neutron irradiation effects in LSA oscillators", Proc. IEEE, 59, 1633-4 (1971), EEA74-37179.
908.Wilson, W. E., "Effects of gamma irradiation on LSA oscillator performance", Report SLA-73-5429, Sandia Lab., Albuquerque, N. Mex., 1973, 11 pp., CA80-53615.
909.Wilson, W. E. and Gregory, B. L., "Neutron induced degradation and failure in high power LSA diode oscillators", Proceedings of the 3rd Biennial Cornell Electrical Engineering Conference on High Frequency Generation and Amplification Devices and Application, 1971, p. 247-56, EEA75-13712.

6. Other Devices

910.Chu, C. K., Bartko, J. and Felice, P. E., "Irradiation for fast switching thyristors", Patent USA 3881963, Publ. May 1975, CA83-89835.
911.Fossum, J. G., Sander, H. H. and Gerwin, H. J., "The effects of ionizing radiation on diffused resistors", IEEE Trans. Nucl. Sci., NS-21, 315-22 (1974), EEA78-19898.
912.McKenzie, J. M., "Statistical distribution of neutron semiconductor device degradation", IEEE Trans. Nucl. Sci., NS-21, 23-7 (1974), EEA78-5204.
913.Ovshinsky, S. R., Evans, E. J., Nelson, D. I. and Fritzsche, H., "Radiation hardness of Ovonic devices", IEEE Trans. Nucl. Sci., NS-15, 311-21 (1968), EEA72-11339.
914.Parfeneva, L. S., Solominskii, M. S. and Yureva, G. A., "Radiation stability of n-Si Hall probes", Fiz. Tekh. Poluprovodnikov, 1, 312-13 (1967), Russian, EEA70-11060.
915.Pocok, D. N. and Krebs, M. G., "Radiation hardening of linear microcircuits by terminal photocompensation", Report AFWL-TR-73-110, Northrop Res. Technol. Centre, Hawthorne, Calif., 1974, 124 pp., EEA78-19924.
916.Ribarov, S., "Phenomenologic theory of gamma irradiation", Izv. Fiz. Inst. ANEB, 24, 87-94 (1973), EEA77-29898.
917.Ribarov, S. and Gospodinov, M., "Changes in the alternating current (AC) conductivity of black silicon carbide varistors at irradiation with reactor neutron radiation", Izv. Fiz. Inst. ANEB, 24, 95-100 (1973), EEA77-29899.
918.Ribarov, S., Gospodinov, M. and Krezhov, K., "Effect of the binding substance on the variations in the electric characteristics of varistors from black silicon carbide at irradiation with reactor neutron radiation", Izv. Fiz. Inst. ANEB, Bulg. Akad. Nauk., 24, 81-86 (1973), Bulgarian, NSA30-431.

919.Ribarov, S. R. and Krezhov, K. A., "Effect of adsorbed gases on gamma-radiation stability of nonlinear semiconductor resistors of black silicon carbide", C. R. Acad. Bulg. Sci., 27, 1343-6 (1974), PA78-46207.

920.Ruwe, V. W., "The effect of neutron radiation on unijunction transistors and silicon control rectifiers", Proceedings of the 8th Annual IEEE Region III Convention, 1969, p. 121-4, EEA73-35223.

921.Scott, L., "Investigation of radiation hardened circuit elements", Report TRW-06415-6003-R000, TRW Systems, Redondo Beach, Calif., 1968, 33 pp., EEA72-13500.

922.Silzars, A., Bates, D. J., Long, J. A. and Roberts, L. A., "Design and performance of R.F. bandpass electron bombarded semiconductor amplifiers", Presented at the 17th International Electron Devices Meeting, Washington, D. C., October 1971, EEA75-12596.

923.Sinclair, R. E., "Nuclear EMP: semiconductor circuit failure and protection implementation schemes", Report PEM-21, California Univ., Livermore, (nd), 16 pp., NSA31-11431.

924.Smith, G. D., "Performance of silicon controlled rectifiers in radiation environment", IEEE Trans. Nucl. Sci., NS-13, 341-9 (1966), EEA70-10892.

925.Srour, J. R., "Effects of ionizing radiation on thin-film semiconductor devices", IEEE Trans. Nucl. Sci., NS-18, 359-70 (1971), EEA75-13760.

926.Tesca, D. M., Peden, J. C. and Beilfuss, J. W., "Characteristics and applications of metal oxide varistors for EMP hardening", Report PEM-43, California Univ., Livermore, Calif., 1974, 21 pp., NSA31-8782.

927.Wilson, D. K. and Lee, H. S., "Permanent radiation damage effects in narrow-base pnpn devices", IEEE Trans. Nucl. Sci., NS-14, 15-32 (1967), EEA71-5953.

K. DEFECTS

I. Books

1. Crawford, J. H., Jr. and Slifkin, L. M., "Point defects in solids, Vol. 1, general and ionic crystals", Plenum Press, New York, 1972.
2. Flynn, C. P., "Point defects and diffusion", Clarendon Press, Oxford, England, 1972.
3. Hannay, N. B., (Ed.), "Treatise on solid state chemistry, Vol. 2, defects in solids", Plenum Press, New York, 1975.
4. Henderson, B., "Defects in crystalline solids", Crane, Russak & Company, Inc., New York, 1972.
5. Loretto, M. H. and Smallman, R. E., "Defect analysis in electron microscopy", Halsted Press, New York, 1976.
6. Matare, H. F., "Defect electronics in semiconductors", Wiley-Interscience, New York, 1971.
7. Milnes, A. G., "Deep impurities in semiconductors", Wiley, New York, 1973.

II. Review Articles

8. Adcock, W. A., "Silicon devices and the future", In: Semiconductor silicon, Electrochem. Soc., Inc., New York, p. 36-55 (1969), EEA73-10408.
9. Anon., "Silicon single crystals of extreme perfection", Microelectron. & Reliab., 13, 149 (1974), EEA77-30191.
10. Cheng, L. J. and Corbett, J. W., "Defect creation in electronic materials", Proc. IEEE, 62, 1210-16 (1974), PA77-66930.
11. Corbett, J. W. and Bourgoin, J. C., "Charge-state effects in displacement damage", IEEE Trans. Nucl. Sci., NS-18, 11-20 (1971), PA75-19660.
12. Dale, J. R., "Defects in silicon - their origin and effect on the properties of silicon planar devices", In: Semiconductor silicon, Electrochem. Soc., Inc., New York, p. 622-31 (1969), EEA73-10439.
13. Feltyn, I. A., "Physics of imperfections in semiconductors", In: Semiconductors and their applications in electronics IV, Zinatne Publications, Riga, USSR, p. 5-18 (1970), Russian, EEA74-4839.
14. Frank, W., "The nature of interstitials in silicon and germanium", Lattice Defects in Semiconductors, 1974, p. 23-43, Publ. 1975, PA78-81081.
15. Hansen, W. L. and Haller, E. E., "A view of the present status and future prospects of high purity germanium", IEEE Trans. Nucl. Sci., NS-21, 251-9 (1974), PA77-53607.

16. John, H. F., Faust, J. W., Jr. and Stickler, R., "Microinhomogeneity problems in silicon", IEEE Trans. Pts Mater. & Packag., PMP-2, 51-8 (1966), EEA70-2247.
17. Petroff, P., "Structural defects in III-V compound semiconductors", Lattice Defects in Semiconductors, 1974, p. 73-90, Publ. 1975, PA78-81100.
18. Seeger, A., "Investigation of point defects in silicon and germanium by non-irradiation techniques", International Conference on Radiation Effects in Semiconductors, 1970, p. 29-38, Publ. 1971, PA75-47399.
19. Sirtl, E., "Incorporation mechanisms of point defects during preparation of crystalline semiconductors", In: Semiconductor silicon 1973, Huff, H. R., (Ed.), Electrochem. Soc., Inc., Princeton, N. J., p. 54-74 (1973), CA81-177965.
20. Smith, F. T. J., "Deviations from stoichiometry in II-VI compounds", Am. Ceram. Soc. Bull., 51, 338 (1972), PA75-50657.
21. Stafiescu, D., "The study of the defects of monocrystalline silicon by infrared transmission", Stud. & Cercet. Fiz., 25, 991-9 (1973), Romanian, EEA77-10125.

III. Line Defects

1. Dislocations

22. Abokarov, D. A., Banyulis, E. Y., Bagduev, G. B., Ivanov, G. A., Polikhronidi, N. G. and Chernobai, V. I., "Acceptor action of dislocation in pure tellurium", Sov. Phys.-Semicond., 7, 402-5 (1973), EEA77-260.
23. Abrahams, M. S., Blanc, J. and Buiocchi, C. J., "Like-sign asymmetric dislocations in zinc-blende structure", Appl. Phys. Lett., 21, 185-6 (1972), PA75-71478.
24. Abrahams, M. S., Weisberg, L. R., Buiocchi, C. J. and Blanc, J., "Dislocation morphology in graded heterojunctions: $GaAs_{1-x}P_x$", J. Mater. Sci., 4, 223-35 (1969), EEA73-1163.
25. Afanasev, V. F., "Influence of dislocations on the excess noise of surface-barrier gold-silicon junctions", Fiz. Tekh. Poluprovodnikov, 4, 125-8 (1970), Russian, EEA73-21205.
26. Afanasovich, V. F., Vasilevskaya, V. N., Boltovets, N. S., Konakova, R. V., Maltseva, I. A., Rybalka, V. V., Suvorova, L. M. and Khomenko, L. A., "Effect of heat treatment on distribution of dislocations and manufacture of defect free silicon p^+np^+ diodes", Poluprovodn. Tekh. & Mikroelektron., no. 15, 42-4 (1974), Russian, PA78-4007.
27. Aharoni, H. and Bar-lev, A., "The role of crystal defects in transistor operation at high power levels", Microelectronics, 6 (3), 11-15 (1975).
28. Aharoni, H., Bar-lev, A. and Margalit, S., "Defects in epitaxial layers of silicon-germanium grown on silicon substrates", J. Cryst. Growth, 17, 254-60 (1972), EEA76-4535.

29. Akerman, K., Brafman, M., Kruszewska, O. and Szterk, L., "Detection of dislocations of structural defects in metal and semiconductor mono-crystals", Przeglad Elektron., 7, 376-86 (1966), Polish, EEA70-12533.
30. Aleksandrov, L. N. and Entin, I. A., "On the interaction of a screw dislocation with interfaces during films growth", Phys. Status Solidi A, 27, 665-72 (1975), PA78-37611.
31. Anchev, V., "Effect of dislocations on electrical characteristics of germanium semiconductor devices", Tekh. Misul, 11 (1), 33-40 (1974), Bulgarian, EEA77-26213.
32. Antonov, P. I., Bakholdin, S. I. and Krymov, V. M., "Effect of grain formation on the dislocation structure of shaped germanium monocrystals", Izv. Akad. Nauk SSSR Ser. Fiz., 37, 2328-33 (1973), Russian, EEA77-5884.
33. Aris, F. C. and Lewis, T. J., "Charge trapping phenomena in anthracene crystals and the influence of crystal deformation", J. Chem. Soc. Faraday Trans. I, 69 (pt. 10), 1872-86 (1973), EEA77-5810.
34. Armigliato, A. and Servidori, M., "Dislocation interactions in silicon by combined X-ray topography electron microscopy observations", Phys. Status Solidi A, 25, K1-3 (1974), PA77-82180.
35. Aseev, A. L., Golobokov, Y. N. and Stenin, S. I., "Dislocation processes during plastic deformation of Si and Ge in the range 0.50 to 0.95 of the melting temperature", Phys. Status Solidi A, 28, 355-64 (1975), PA78-41779.
36. Avdonin, N. A., Vakhrameev, S. S., Milvidskii, M. G., Osvenskii, V. B., Sakharov, B. A., Smirnov, V. A. and Shchelkin, Y. F., "Influence of the temperature field and the thermal stress field upon the formation of the dislocation structure in gallium-arsenide single crystals grown by the Czochralski method", Sov. Phys.-Dokl., 16, 772-5 (1972), PA75-32520.
37. Baitinger, U., Arndt, J. and Schnepf, D., "The electrical properties of plastically bent p-type indium antimonide between 50 and 200°K", J. Mater. Sci., 4, 396-404 (1969), EEA73-4312.
38. Baitinger, U., Schnepf, D. and Arndt, J., "The effect of dislocations on the electrical properties of indium antimonide", J. Phys., 30, 87-92 (1969), French, EEA72-9538.
39. Barth, W. and Elsasser, K., "Polarization of the infrared absorption of dislocations in germanium", Phys. Status Solidi, 48, K147-9 (1971), PA75-10381.
40. Black, J. F. and Jungbluth, E. D., "Decorated dislocations and sub-surface defects induced in GaAs by the in-diffusion of zinc", J. Electrochem. Soc., 114, 188-92 (1967), EEA70-10433.
41. Blinowski, A. and Kossowska-Piekarska, M., "The influence of edge dislocations density in Ge on the distribution of the wetted area diameters", Przeglad Elektron., 9, 268-72 (1968), Polish, EEA71-25936.

42. Borle, W. N. and Bagai, R. K., "On process induced defects in silicon wafer during device manufacture", J. Inst. Telecommun. Eng., 19 (2), 81-2 (1973), EEA76-29053.
43. Brotherton, S. D. and Rogers, T. L., "Influence of dislocations on gold diffusion into thin silicon slices", Solid-State Electron., 15, 853-60 (1972), PA75-60833.
44. Buch, F. and Ahlquist, C. N., "Influence of dislocations on electrical conductivity of CdTe (Photoplastic effect)", J. Appl. Phys., 45, 1756-61 (1974), PA77-60826.
45. Campbell, D. R., Tu, K. N. and Schwenker, R. O., "Diffusion of arsenic along dislocations in epitaxial silicon films", Thin Solid Films, 25, 213-20 (1975), CA82-132173.
46. Carlsson, L., "Photoplastic behavior of CdTe", J. Appl. Phys., 43, 2529-36 (1972), PA75-50742.
47. Cavallini, A. and Gondi, P., "Different electrical effects of dislocations in Ge depending on the deformation temperature", Lett. Nuovo, 10 (3), 115-19 (1974), EEA77-26243.
48. Cavallini, A., Gondi, P. and Schintu, F., "Introduction of dislocations of different character depending on the deformation temperature in Ge", Phys. Status Solidi A, 27, 523-31 (1975), PA78-33174.
49. Chaudhri, G. and Feltham, P., "Creep of germanium", J. Mater. Sci., 7, 1161-7 (1972), PA75-84040.
50. Choi, S. K. and Mihara, M., "Impurity effects on the dislocation velocity in gallium arsenide", J. Phys. Soc. Jap., 32, 1154 (1972), PA75-68028.
51. Collins, C. V. and Miles, M. H., "Electrical effects of screw dislocations in germanium", J. Appl. Phys., 42, 5644-8 (1971), PA75-10064.
52. Conti, M. and Corda, G., "Influence of slip plane dislocations on electrical performances of N-P-N planar transistors", IEEE Trans. Electron Devices, ED-18, 1148-50 (1971), EEA75-1142.
53. Cseh, I. and Podor, B., "Photoelectric and galvanomagnetic measurements on p-type germanium with dislocations", Acta Phys. Acad. Sci. Hung., 34, 225-9 (1973), EEA77-5825.
54. Danilchuk, L. N., "Intensity contrast from screw dislocations lying in a reflecting plane", Sov. Phys.-Solid State, 15, 1424-5 (1974), PA77-36907.
55. Das, B. N. and Weinstein, M., "Crystal imperfections in vapor-grown CdS", II-VI Semiconducting Compounds 1967 International Conference, p. 147-66, EEA72-20251.
56. Das, G., "Defects in silicon devices", Proceedings of the 3rd International Conference on High Voltage Electron Microscopy, 1974, p. 277-84.
57. Eltenberg, M., "Effects of dislocation density on the properties of liquid phase epitaxial GaAs", J. Appl. Phys., 45, 901-6 (1974), PA77-53008.

58.Eremenko, V. G., Nikitenko, V. I. and Yakimov, E. B., "Investigation of the nature of the diode effects in dislocations in silicon", Zh. Eksp. & Teor. Fiz., 67, 1148-60 (1974), Russian, EEA78-600.
59.Erofeeva, S. A. and Osipyan, Y. A., "Mobility of screw dislocations in InSb crystals", Sov. Phys.-Solid State, 16, 2076-7 (1975), PA78-61714.
60.Euskirchen, J. and Schoch, H. P., "Influence of dislocations on the etching behaviour of thermally grown SiO_2 films", Telefunken-Rohre, no. 47, 123-6 (1967), German, EEA71-28810.
61.Faivre, G. and Saada, G., "Dislocations in piezoelectric semiconductors, Phys. Status Solidi B, 52, 127-40 (1972), PA75-60805.
62.Gai, P. L. and Howie, A., "Dissociation of dislocations in GaP", Philos. Mag., 30, 939-43 (1974), PA78-12764.
63.Gavrilov, G. M. and Evdokimov, V. I., "Dislocation density in germanium films as influenced by the conditions of their chemical deposition", Izv. Akad. Nauk SSSR Neorg. Mater., 8, 33-6 (1972), Russian, PA75-29035.
64.George, A. and Champier, G., "Double cross-slip in silicon", Philos. Mag., 31, 961-7 (1975), PA78-46196.
65.George, A. and Champier, G., "Meaning of dislocation velocities measured in doped silicon", Phys. Status Solidi A, 29, K79-82 (1975), PA78-54213.
66.Gezci, S. and Woods, J., "Dislocation etch pits in zinc selenide", J. Mater. Sci., 7, 603-8 (1972), PA75-50677.
67.Ghezzi, C. and Servidori, M., "X-ray study of heterogeneous nucleation of dislocations in P-diffused silicon", J. Mater. Sci., 9, 1797-802 (1974), PA78-9729.
68.Gimelfarb, F. A., Govorkov, A. V., Grishma, S. P., Milvidskii, M. G., Fistul, V. I. and Shifrin, S. S., "Microcathodoluminescence study of dislocation decoration in doped gallium arsenide", Sov. Phys.-Crystallogr., 19, 692-3 (1975), PA78-65596.
69.Glaenzer, R. H. and Jordan, A. G., "The effects of contamination on the electrical properties of edge dislocations in silicon", Phil. Mag., 18, 717-23 (1968), EEA72-2704.
70.Glaenzer, R. H. and Jordan, A. G., "The electrical properties of dislocations in silicon. I. The effects on carrier lifetime", Solid-State Electron., 12, 247-58 (1969), EEA72-13369.
71.Glaenzer, R. H. and Jordan, A. G., "The electrical properties of dislocations in silicon. II. The effects on conductivity", Solid-State Electron., 12, 259-66 (1969), EEA72-13370.
72.Goklaney, S. M., Porter, W. A. and Parker, D. L., "Dislocation creation in silicon after thermal oxidation and phosphorous diffusion", J. Inst. Eng. Electron. & Telecommun. Eng. Div., 55, 19-22 (1974), PA78-62560.
73.Golobokov, Y. N. and Stenin, S. I., "On the role of dynamic properties of dislocations during the growth process of the semiconductor crystals", Ukr. Fiz. Zh., 16, 1462-6 (1971), Russian, PA75-6400.

74.Gomez, A., Cockayne, D. J. H., Hirsch, P. B. and Vitek, V., "Dissociation of near-screw dislocations in germanium and silicon", Philos. Mag., 31, 105-13 (1975), PA78-33169.
75.Grazhulis, V. A., Mukhina, V. Y., Osipya, Y. A. and Shevchenko, S. A., "Investigation of electric conductivity and Hall effect in silicon single crystals with dislocations", Zh. Eksp. & Teor. Fiz., 68, 2149-58 (1975), Russian, EEA78-32958.
76.Grenning, D. A. and Herzog, A. H., "Dislocations and their relation to irregularities in zinc-diffused GaAsP p-n junctions", J. Appl. Phys., 39, 2783-90 (1968), EEA71-23273.
77.Grienauer, H. S. and Mayer, K. R., "Depth distribution of dislocations in phosphorus-diffused silicon", Lattice Defects in Semiconductors, 1974, p. 550-6, Publ. 1975, EEA78-41510.
78.Hartnagel, H. and Weiss, B. L., "Dislocations in GaAs produced by device fabrication", Solid-State Electron., 17, 799-803 (1974), PA77-61320.
79.Hasegawa, F. and Ito, H., "Degradation of a Gunn diode by dislocations induced during thermocompression bonding", Appl. Phys. Lett., 21, 107-8 (1972), EEA75-36179.
80.Haworth, W. L., Mattas, R. and Bimbaum, H. K., "Internal friction in germanium at low temperatures", J. Appl. Phys., 43, 3658-62 (1972), PA75-68068.
81.Holonyak, N., Jr., "Junction effects in compound semiconductors (Final report, 15 September 1964 - 14 September 1967)", Report AFCRL-69-0009, Illinois Univ., Urbana, 1969, 15 pp., EEA73-28072.
82.Hu, S. M., "Effects of surface films and film edges on dislocation movement in silicon", J. Appl. Phys., 46, 1465-9 (1975), EEA78-24174.
83.Hu, S. M., "On indentation dislocation rosettes in silicon", J. Appl. Phys., 46, 1470-2 (1975), EEA78-24175.
84.Hu, S. M. and Patrick, W. J., "Effect of oxygen on dislocation movement in silicon", J. Appl. Phys., 46, 1869-74 (1975), PA78-58057.
85.Huntley, F. A. and Willoughby, A. F. W., "The effect of dislocation density on the diffusion of gold in thin silicon slices", J. Electrochem. Soc., 120, 414-22 (1973), EEA76-23260.
86.Inoue, T., Watanabe, J., Miura, T. and Yamamoto, M., "On the formation of striation-type substructure in copper crystals", Trans. Jap. Inst. Met., 14, 285-91 (1973), PA77-1063.
87.Jastrzebska, M. and Figielski, T., "Investigations of photoelectric phenomena on p-type germanium with dislocations", Phys. Status Solidi, 32, 791-6 (1969), EEA72-17515.
88.Johannessen, J. S., "Thickness dependence of the electrical transport properties of germanium films. (Dislocation model)", Phys. Status Solidi A, 11, 469-78 (1972), PA75-54192.
89.Joshi, M. L. and Dash, S., "Dislocation-induced deviation of phosphorous-diffusion profiles in silicon", IBM J. Res. Developm., 10, 446-54 (1966), EEA70-10437.

90. Kamieniecki, E., "Effect of high electric field on photoconductivity in plastically deformed n-Ge", Phys. Status Solidi A, 10, K81-4 (1972), PA75-29766.
91. Kamieniecki, E. and Elsasser, K., "Polarization of photoconductivity related to edge dislocations in n-type germanium", Phys. Status Solidi B, 56, K25-7 (1973), EEA76-16240.
92. Kastner, S. and Hesse, J., "The influence of dislocations on the diffusion of gold in silicon", Phys. Status Solidi A, 25, 261-7 (1974), PA77-79187.
93. Kawakami, T., Takahashi, T., Kawabe, H. and Yamada, T., "Dislocation structure in the surface layer of scratch-deformed Si single crystals", J. Jap. Inst. Met., 38, 186-91 (1974), Japanese, PA77-44539.
94. Khukhryanskii, Y. P., Shchevlev, M. I. and Popovich, B. N., "Spiral etch pits on germanium", Fiz. Tverdogo Tela, 9, 1042-6 (1967), Russian, EEA70-13974.
95. Kirscht, F. G., Richter, H., Baehr, R., Hansch, C. and Brocke, H. D., "Dislocation formation in scratches on single crystal silicon surfaces", Krist. Tech., 10, 1235-44 (1975), German, CA83-200412.
96. Kishino, S. and Iida, S., "Plane defect at the influence and dislocations in epitaxially grown GaAs", J. Electrochem. Soc., 119, 1113-18 (1972), PA75-71488.
97. Kishino, S. and Ogirima, M., "Screw-type dislocations due to lattice mismatch in GaP:N/GaP epitaxy", Philos. Mag., 31, 1239-51 (1975), PA78-61940.
98. Kochurov, V. P., Yanushkevich, V. A., Malov, Y. V. and Pirogova, L. B., "Stresses produced in silicon by the network of Lomer dislocations", Fiz. & Khim. Obrab. Mater., no. 1, 64-70 (1974), Russian, PA77-52760.
99. Koyama, H., Wittry, D. B. and Kato, T., "Propagation of residual dislocations during silicon epitaxy", J. Appl. Phys., 46, 2899-902 (1975), PA78-72718.
100. Krylow, J., "Electric properties of dislocations in p-type germanium", Phys. Status Solidi, 32, 589-96 (1969), EEA72-17646.
101. Krylow, J. and Auleytner, J., "Electric properties of dislocations in n-type germanium", Phys. Status Solidi, 32, 581-8 (1969), EEA72-17645.
102. Kuczynski, G. C., Iyer, K. R. and Allen, C. W., "Effect of light on the mobility of dislocations in germanium", J. Appl. Phys., 43, 1337-41 (1972), PA75-32521.
103. Labusch, R. and Schettler, R., "On the electronic states at dislocations in germanium", Phys. Status Solidi A, 9, 455-68 (1972), PA75-25857.
104. Lake, P. B. and Meyrick, G., "Dislocation nodes observed in Ge single crystals recovery annealed after dual slip deformation", Metall. Trans., 5, 1761-7 (1974), PA78-18314.

105.Lawrence, J. E., "Behavior of dislocations in silicon semiconductor devices: diffusion, electrical", J. Nucl. Energy, 115, 860-5 (1968), EEA71-25897.
106.Mader, S. and Blakeslee, A. E., "Extended dislocations in $GaAs_{0.7}P_{0.3}$", Appl. Phys. Lett., 25, 367-7 (1974), PA78-15474.
107.Mader, S. and Blakeslee, A. E., "On dislocations in $GaAs_{1-x}P_x$", IBM J. Res. & Dev., 19, 151-62 (1975), EEA78-28611.
108.Maksimov, S. K. and Akulova, G. V., "Two types of dislocation network initiated in silicon by the diffusion of phosphorus", Sov. Phys.-Solid State, 15, 2150-1 (1974), PA77-75533.
109.Maksimov, S. K., Lukyanchuk, T. I. and Myshlyaev, M. M., "Determination of dislocation loop nature by the residual contrast at g.b=0 and s=0", Phys. Status Solidi A, 24, 409-14 (1974), PA77-69816.
110.Malov, Y. V. and Rozhanskii, V. N., "Study of dislocation structure of silicon deformed in the easy slip stage", Fiz. Tverdogo Tela, 9, 1028-37 (1967), Russian, EEA70-13967.
111.Maltseva, V. A., Platonov, B. D., Sofronova, S. Y. and Bekker, Y. M., "Effect of mechanical stress on the shapes of dislocation etch pits in silicon", Sov. Phys.-Solid State, 14, 1053-4 (1972), PA75-80553.
112.Martin, J. C., Blasquez, G., Cacqueray, A. D., Brebisson, M. D. and Schiller, C., "The effect of crystal dislocations on burst noise in silicon bipolar transistors", Solid-State Electron., 15, 739-44 (1972), French, EEA75-24346.
113.Maslova, L. A. and Tsivinskii, S. V., "Growth of shaped germanium crystals having low dislocation density by the Bridgman method", Sov. Phys.-Crystallogr., 18, 565 (1974), EEA77-26382.
114.Matsui, J., "Interaction of slip dislocations in silicon crystals having low dislocation densities", Phys. Status Solidi A, 27, 605-13 (1975), PA78-37244.
115.Matsui, J., "Cross-slip dislocations and their multiplication in (001)-oriented silicon wafers", J. Electrochem. Soc., 122, 977-83 (1975), PA78-72478.
116.Matsui, J. and Kawamura, T., "Dislocation generation along swirls in dislocation-free silicon crystals", Presented at Electrochemical Society Meeting, Chicago, Ill., May 1973, EEA76-23300.
117.Matthews, J. W., "Misfit dislocations and the alignment of epitaxial islands", Surface Sci., 31, 241-56 (1971), PA75-56697.
118.Matthews, J. W., "Defects associated with the accommodation of misfit between crystals", J. Vac. Sci. & Technol., 12, 126-33 (1975), PA78-54430.
119.Matthews, J. W., "Some features of the behavior of misfit dislocations during diffusion", Thin Solid Films, 25, 199-211 (1975), PA78-33337.
120.Matthews, J. W. and Blakeslee, A. E., "Defects in epitaxial multilayers. I. Misfit dislocations", J. Cryst. Growth, 27, 118-25 (1974), EEA78-11481.

121. Matthews, J. W. and Blakeslee, A. E., "Defects in epitaxial multilayers. II. Dislocation pile-ups, threading dislocations, slip lines, and cracks", J. Cryst. Growth, 29, 273-80 (1975), PA78-85679.
122. Mettler, K., "On the effect of dislocations on the diffusion of copper in annealed gallium arsenide", Solid-State Electron., 12, 519-26 (1969), German, EEA72-23760.
123. Meyer, M., Miles, M. H. and Ninomiya, T., "Some electrical and optical effects of dislocations in semiconductors", J. Appl. Phys., 38, 4481-5 (1967), EEA71-3282.
124. Miles, M. H., "Extrinsic photoconductivity from edge dislocations in germanium", J. Appl. Phys., 40, 2720-4 (1969), EEA72-23535.
125. Milshtein, S. K. and Nikitenko, V. I., "Influence of single dislocations on local electrical properties of silicon and germanium", Sov. Phys.-Semicond., 6, 1344-9 (1973), EEA76-16271.
126. Milshtein, S. K. and Nikitenko, V. I., "Diode effect at dislocations in silicon", Sov. Phys.-Solid State, 16, 346-7 (1974), PA78-6893.
127. Mitsuhashi, H., Komura, H. and Chikawa, J., "Dislocation effects on the luminescence of CdS crystals", II-VI Semiconducting Compounds 1967 International Conference, p. 179-89, EEA72-20003.
128. Mroczkowski, R. S., Witt, A. F. and Gatos, H. C., "Effects of back-melting on the dislocation density in single crystals: GaSb", J. Electrochem. Soc., 115, 545-7 (1968), EEA71-20520.
129. Muhlbauer, A. and Sirtl, E., "Lamellar growth phenomena in (111)-oriented dislocation-free float zoned silicon single crystals", Phys. Status Solidi A, 23, 555-65 (1974), EEA77-30193.
130. Myshlyaev, M. M., Khodos, I. I., Maksimov, S. K. and Kislov, N. A., "Dislocation networks in deformed Ge", Izv. Akad. Nauk SSSR Ser. Fiz., 38, 2303-21 (1974), Russian, PA78-6421.
131. Nes, E. and Solberg, J. K., "The dragging of precipitate particles by climbing dislocations in silicon", J. Appl. Phys., 44, 488-9 (1973), EEA76-12450.
132. Neubert, D., Kos, J. and Hahn, D., "Problems of life time doping dislocations in silicon", Presented at 1972 European Solid State Devices Research Conference, Lancaster, England, September 1972, EEA76-32108.
133. Neubert, D. and Kos, H. J., "Edge dislocations in silicon as one dimensional surfaces", Jap. J. Appl. Phys., suppl. 2 (pt. 2), 401-4 (1974), PA78-37241.
134. Neumark, G. F., "New model for interface charge-carrier mobility: the role of misfit dislocations", Phys. Rev. Lett., 21, 1252-6 (1968), EEA72-758.
135. Nicolau, I. F., "Junction delineation and dislocation revealing in silicon by the HIO_4-HF-H_2O system", Solid-State Electron., 12, 446-8 (1969), EEA72-15529.

136.Nishida, M., "Distribution of surface state density related to diffusion induced dislocations near the junction formed by diffusion of phosphorus in silicon", Jap. J. Appl. Phys., 11, 673-7 (1972), PA75-44038.
137.Nishida, M., "Effects of diffusion-induced dislocations on the excess low-frequency noise", IEEE Trans. Electron Devices, ED-20, 221-6 (1973), EEA76-16406.
138.Nishida, K., Matsui, J. and Nakajima, M., "Avalanche multiplication enhancement along dislocation lines in a thermally diffused junction", Jap. J. Appl. Phys., 11, 1228-9 (1972), PA75-65027.
139.Noack, J., "Measurement of minority carrier lifetime in silicon of low dislocation density", Phys. Status Solidi, 32, K17-19 (1969), EEA72-17642.
140.Nosov, Y. G. and Antonov, P. I., "Formation of structural defects in shaped indium antimonide crystals", Izv. Akad. Nauk SWSR Ser. Fiz., 37, 2334-9 (1973), Russian, PA77-13573.
141.Novikov, N. N. and Khimenko, M. V., "The structure of dislocation half-loops in germanium, issuing from microhardness impressions", Izv. VUZ Fiz., no. 2, 124-5 (1972), Russian, PA75-36032.
142.Obukhov, L. V., Prutskov, E. G. and Yanushkevich, V. A., "Microstructure and dislocation movement in Si under the action of a laser", Fiz. & Khim. Obrab. Mater., no. 2, 34-8 (1975), Russian, PA78-72475.
143.Oki, S. and Futagami, K., "Mechanical behavior of dislocations in Si single crystals observed by topography with scanning Soller slit assembly", Jap. J. Appl. Phys., 13, 605-10 (1974), PA77-44538.
144.Olsen, G. H., Abrahams, M. S., Buiocchi, C. J. and Zamerowski, T. J., "Reduction of dislocation densities in heteroepitaxial III-V VPE semiconductors", J. Appl. Phys., 46, 1643-6 (1975), EEA78-24178.
145.Osipyan, Y. A. and Fedyaev, Y. I., "Donor action of dislocations in InSb", JETP Lett., 9, 21-2 (1969), EEA72-15477.
146.Osipyan, Y. A. and Shevchenko, S. A., "Anisotropy of hole mobility in plastically deformed p-germanium", JETP Lett., 18, 153-4 (1973), EEA77-5829.
147.Osvenskii, V. B., Proshko, G. P. and Sizov, S. M., "Parameters of injection lasers made of gallium arsenide crystals with different dislocation densities", Kvantovaya Elektron., no. 2, 94-6 (1971), Russian, PA75-17641.
148.Osvenskii, V. B. and Shifrin, S. S., "Relationships governing the multiplication of dislocations in semiconductors", Sov. Phys.-Solid State, 16, 165-6 (1974), PA77-75534.
149.Osvenskii, V. B., Shifrin, S. S. and Milvidskii, M. G., "Mechanism of dislocation multiplication in semiconductors at high temperatures", Izv. Akad. Nauk SSSR Ser. Fiz., 37, 2357-61 (1973), Russian, PA77-10552.

150.Panteleev, V. A. and Lainer, L. V., "Influence of dislocations on the properties of p-n functions in silicon", Izv. VUZ Fiz., no. 5, 145-7 (1969), Russian, EEA73-14197.
151.Parekh, P. C., Gavel, E. and Lyn, V., "Observations of EE dislocations using Sirtl etchant and their influence on transistor parameters", Solid-State Electron., 13, 707-10 (1970), EEA73-24854.
152.Pataki, G., "Recombination of hot electrons on dislocations and the possibility of negative differential conductivity", Phys. Status Solidi A, 8, 411-22 (1971), PA75-9990.
153.Pavlov, P. V. and Dobrokhotov, E. V., "Dislocation model of diffusion in heavily doped layers of plastically deformed silicon", Sov. Phys.-Solid State, 16, 2307-9 (1975), PA78-81244.
154.Petukhov, B. V. and Pokrovskii, V. L., "Influence of charged impurities on the mobility of dislocations in crystals with high Peierls barriers", Fiz. Tver. Tela, 13, 3679-82 (1971), Russian, PA75-29177.
155.Podor, B., "Effect of dislocations on galvanomagnetic properties of n-type Ge", Acta Phys. Hungar., 23, 393-405 (1967), EEA71-13212.
156.Podor, B., "On the dislocation scattering in silicon-on-insulator films", Phys. Status Solidi A, 2, K197-200 (1970), EEA74-7666.
157.Podor, B., "Effect of dislocations on the electrical properties of germanium and silicon", Hiradastechnika, 23 (1), 24-31 (1972), Hungarian, PA75-29697.
158.Porter, W. A., Drew, D. D. and Linder, J. S., "Thermal conditions when dislocations are generated in silicon", J. Appl. Phys., 43, 1477-9 (1972), PA75-32526.
159.Porter, W. A. and Parker, D. L., "Gravitational effects on process-induced dislocations in silicon", Report NASA-CR-120477, Texas A&M Univ., College Station, 1974, 27 pp., EEA78-15249.
160.Queisser, H. J., "Observations and properties of lattice defects in silicon", In: Semiconductor silicon, Electrochem. Soc., Inc., New York, p. 585-95 (1969), EEA73-10436.
161.Ragimov, I. A. and Rozhanskii, V. N., "Use of the diffraction contrast of radial displacements in the determination of the type of dislocation loop", Sov. Phys.-Solid State, 15, 1112-13 (1973), PA77-28108.
162.Ravi, K. V., "Dislocation sources in silicon", Metall. Trans., 4, 681-9 (1973), PA77-9921.
163.Ray, I. L. F. and Cockayne, D. J. H., "The dissociation of dislocations in silicon", Proc. R. Soc. A, 325, 543-54 (1971), PA75-3781.
164.Rosin, H., "On the method and results of a dislocation structure analysis of Ga(As, P) by etching", Phys. Status Solidi A, 29, K5-9 (1975), PA78-54209.
165.Rozgonyi, G. A., Petroff, P. M. and Panish, M. B., "Elimination of dislocations in heteroepitaxial layers by the controlled introduction of interfacial misfit dislocations", Appl. Phys. Lett., 24, 251-4 (1974), PA77-45436.

166.Rozgonyi, G. A., Petroff, P. M. and Panish, M. B., "Control of lattice parameters and dislocations in the system: $Ga_{1-x}Al_x As_{1-y}P_y$/GaAs", J. Cryst. Growth, 27, 106-17 (1974), PA78-30239.
167.Rozgonyi, G. A., Von Neida, A. R., Iizuka, T. and Haszko, S. E., "Defect studies of GaP crystals pulled from non-stoichiometric melts: dislocation and 'saucer' etch pits (photoluminescence and electroluminescence efficiencies)", J. Appl. Phys., 43, 3141-5 (1972), PA75-50595.
168.Rozhanskii, V. N., Arsenteva, I. P., Zakharov, N. D., Govorkov, V. G. and Voinova, N. N., "Pseudodislocational interactions in the system germanium-arsenic", Phys. Status Solidi B, 51, 215-23 (1972), PA75-39697.
169.Sakharov, B. A., Turovskii, B. M. and Frolov, A. V., "Effect of dislocations on lifetime of minority carriers in silicon", Soviet Phys. Dokl., 13, 600-2 (1968), EEA72-9526.
170.Sargent, C. M. and Purdy, G. R., "Epitaxial dislocations", Philos. Mag., 32, 27-36 (1975), PA78-76995.
171.Schaumburg, H., "Velocities of screw- and 60°-dislocations in germanium", Phil. Mag., 25, 1429-58 (1972), German, PA75-47415.
172.Schroder, W. and Wolf, E., "Investigation of the generation of dislocations during the crucible free growth of silicon single crystals", Krist. & Tech., 9, 275-80 (1974), German, PA77-48632.
173.Schroter, W., "Recombination of charge carriers at dislocations in germanium", Phys. Status Solidi A, 19, 159-70 (1973), EEA76-38133.
174.Schwuttke, G. H., "Semiconductor structures with minimum crystallographic defects", Patent USA 3821038, Publ. June 1974, CA81-70139.
175.Scoble, W. R., Jr. and Weissmann, S., "Fatigue induced microstructure in semiconductor crystals", Cryst. Lattice Defects, 4, 123-36 (1973), CA79-97817.
176.Sechenov, D. A., Zakharov, A. G. and Svetichnyi, A. M., "Metal-dielectric-semiconductor structure based on silicon containing dislocations", Izv. VUZ Fiz., no. 2, 163-5 (1972), Russian, PA75-36573.
177.Sharaki, H., "Silicon wafer annealing effect on loop defect generation", Jap. J. Appl. Phys., 13, 1514-23 (1974), EEA78-560.
178.Sheikhet. E. G., Vervyka, A. V., Trainin, A. L. and Veselovskaya, N. V., "Investigation of the degree of imperfection in epitaxial germanium films", Izv. Akad. Nauk SSSR Ser. Fiz., 37, 2306-9 (1973), Russian, PA77-10061.
179.Shulpina, I. L., Polchkova, N. D. and Olkhovikova, T. I., "X-ray topographic study of dislocations introduced into indium antimonide", Izv. Akad. Nauk SSSR Ser. Fiz., 37, 2340-4 (1973), Russian, EEA77-5886.
180.Siethoff, H., "Helical dislocation distribution in silicon-germanium single crystals", J. Appl. Phys., 42, 5201-2 (1971), PA75-6331.

181.Skalicky, P. and Wolfgang, E., "Dislocation loops in antimony telluride", Phys. Status Solidi A, 8, 463-75 (1971), PA75-12771.
182.Smirnov, B. I., "Mean free path of dislocations during the formation of slip bands in LiF crystals", Fiz. Tverdogo Tela, 9, 417-21 (1967), Russian, EEA70-12577.
183.Smirnov, Y. M., "Distribution of dislocations in germanium single crystals (effect of growth parameters)", Izv. Akad. Nauk SSSR Ser. Fiz., 36, 561-2 (1972), Russian, PA75-47414.
184.Sterkhov, V. A., Panteleev, V. A. and Pavlov, P. V., "Diffusion and electrotransport of indium and silver along dislocations in silicon", Fiz. Tverdogo Tela, 9, 681-3 (1967), Russian, EEA70-12510.
185.Sterkhov, V. A. and Pavlov, P. V., "Self-diffusion of germanium along 'intrinsic' dislocations in an external electric field", Fiz. Tverdogo Tela, 9, 643-6 (1967), Russian, EEA70-12548.
186.Stirland, D. J. and Ogden, R., "Dislocation etch-memory effect in gallium arsenide", Phys. Status Solidi A, 17, K1-K4 (1973), CA79-24281.
187.Stringfellow, G. B., Lindquist, P. F., Cass, T. R. and Burmeister, R. A., "Dislocations in vapor phase epitaxial GaP", J. Electron. Mater., 3, 497-515 (1974), PA77-57710.
188.Svoboda, M. and Klier, E., "Influence of deformation on the conductivity and photoconductivity of p-type CdTe", Czech. J. Phys. B, 22, 725-36 (1972), PA75-61264.
189.Tamura, M. and Sunami, H., "Generation of dislocations induced by chemical vapour deposited Si_3N_4 films on silicon", Jap. J. Appl. Phys., 11, 1097-105 (1972), PA75-68031.
190.Tandon, K. N. and Bilello, J. C., "Anisotropy of etching behavior in bcc tungsten and the problem of edge-screw dislocation distributions", Philos. Mag., 30, 75-84 (1974), CA82-10126.
191.Tikhonov, L. V. and Kharkova, G. V., "Special features of the movement of dislocations during the cyclic heat treatment of alloyed monocrystalline Ge", Metallofizika, Kiev, no. 53, 75-81 (1974), Russian, PA78-21509.
192.Timasheva, T. P., Sorokin, L. M., Rychkova, S. V. and Pakhomovskii, Y. A., "Mechanism of defect formation in high antimony monocrystals after high temperature annealing", Izv. Akad. Nauk SSSR Ser. Fiz., 37, 2349-52 (1973), Russian, PA77-10551.
193.Tregubova, A. S., Mokhov, E. N. and Shulpina, I. L., "Structural imperfections of diffusion-doped SiC(6H) layers", Sov. Phys.-Solid State, 14, 3061-3 (1973), EEA76-32193.
194.Turovskii, B. M., "Influence of oxygen on the dislocation structure of silicon single crystals", Sov. Phys.-Solid State, 16, 177-8 (1974), PA77-79055.
195.Turovskii, B. M. and Lainer, L. V., "Behavior of silicon during high-temperature heat treatment", Inorg. Mater., 9, 1661-3 (1973), EEA77-33943.

196.Van Der Merwe, J. H. and Van Der Berg, N. G., "Misfit dislocation energy in epitaxial overgrowths in finite thickness", Surface Sci., 32, 1-15 (1972), PA75-60798.
197.Vandorpe, M., Peccoud, L., Bourret, A., Bresse, J. F. and La Feuille, D., "Study of faults caused by the diffusion of phosphorus in (111) silicon and the correlation with the electron characteristics of diffused diodes", Proceedings of the 7th International Congress on Electron Microscopy, 1970, p. 603-4, French, EEA74-30686.
198.van Weeren, J. H. P., Struikmans, R., Koopmans, G. and Blok, J., "Low-field magnetoresistance and magnetoconductivity in n-type Ge with dislocations", Phys. Status Solidi, 27, 225-35 (1968), EEA71-1b165.
199.Vanyukov, A. V., Baronenkova, R. P., Krotov, I. I. and Indenbaum, G. V., "Growth of CdTe crystals from the vapour phase and their dislocation structure", Sov. Phys.-Crystallogr., 19, 635-7 (1975), PA78-66558.
200.Varker, C. J. and Ravi, K. V., "Electrical piping and related structural imperfection in bi-polar transistors", Spring Meeting of the Electrochemical Society, Extended Abstracts, 1971, p. 155-6, EEA75-1151.
201.Vaulin, Y. D., Migal, N. N., Migal, V. P. and Stenin, S. I., "Influence of crystallization conditions on formation of dislocation structure of germanium homoepitaxial layers produced by gas-transport technique", Phys. Status Solidi A, 15, 697-710 (1973).
202.Verner, V. D., Maksimov, S. K., Myshlyaev, M. M. and Piskunov, D. I., "Possible mechanism of the change in the shape of Frank dislocation loops", Fiz. Tverd. Tela, 16, 3039-42 (1974), Russian, CA82-163225.
203.Vozmilova, L. N., Maslennikova, V. T. and Stupina, N. M., "Exposure of dislocations in GaAs in the (111) As plane (by etching)", Inorg. Mater., 9, 1454-5 (1973), PA77-67114.
204.Wagner, R., "Electronic states of screw dislocations in germanium", Phys. Status Solidi A, 24, 575-90 (1974), German.
205.Walson, R. P. and Birnbaum, H. K., "Dislocation etch pitting in high-purity niobium", Report Number AD-772617/7GA, Dep. Metall. Min. Eng., Univ. Illinois, Urbana, 1973, 11 pp., CA81-7002.
206.Watanabe, M., Muraoka, H. and Yonezawa, T., "Perfect crystal device technology (diffusion induced dislocations in Si)", Oyo Buturi, 44 (suppl.), 169-77 (1975), PA78-81992.
207.Wu, W. K. and Washburn, J., "Identification of interstitial- and vacancy-type dislocation loops in ion-implanted silicon", J. Appl. Phys., 45, 1085-90 (1974), PA77-56743.
208.Yudin, V. V. and Karatsyuba, A. P., "Investigation of the kinetics of negative annealing of silicon films doped with boron ions", Sov. Phys.-Semicond., 7, 508-9 (1973), PA77-20333.

209. Zakharov, N. D., Neumann, W. and Rozhanskii, V. N., "Mechanism of the formation of dislocation loops on precipitates of phosphorus in germanium", Sov. Phys.-Solid State, 16, 1800 (1974), PA78-58064.
210. Zakharov, N. D., Rozhanskii, V. N. and Korchazhkina, P. L., "Defects appearing during the decay of a solid solution of phosphorus in germanium", Sov. Phys.-Solid State, 16, 926-9 (1974), PA78-18321.
211. Zavadskii, V. A., Kuznetsov, V. I. and Mokritskii, V. A., "Decreasing the dislocation density in epitaxial layers of GaP", Inorg. Mater., 9, 1286-7 (1973), EEA77-30162.

IV. Plane Defects

1. Stacking Faults

212. Barson, F., Hess, M. S. and Roy, M. M., "Diffusion pipes in silicon NPN structures", J. Electrochem. Soc., 116, 304-7 (1969), EEA72-15552.
213. Chikawa, J. and Austerman, S. B., "Stacking faults in BeO crystals", J. Appl. Crystallogr., 7 (pt. 3), 394-5 (1974), PA77-56758.
214. Conti, M., Corda, G., Matteucci, R. and Ghezzi, C., "Oxidation stacking faults in epitaxial silicon crystals", J. Mater. Sci., 10, 705-13 (1975), EEA78-19653.
215. Cullis, A. G., Maher, D. M. and Hsieh, C. M., "Observations of defect generation at stacking faults in boron diffused and oxidized silicon films", Proceedings of the 31st Annual Meeting of the Electron Microscopy Society of America, 1973, p. 128-9, EEA77-2493.
216. Dhananjai, P. and Krishma, P., "Influence of stacking faults on the growth of polytype structures. II. Silicon carbide polytypes", Philos. Mag., 32, 1113-32 (1975).
217. Gunther, H., "On a dynamic singularity in the continuum theory of stacking faults", Monatsber. Deut. Akad. Wiss. Berlin, 13, 296-302 (1971), PA75-50680.
218. Hu, S. M., "Formation of stacking faults and enhanced diffusion in the oxidation of silicon", J. Appl. Phys., 45, 1567-73 (1974), PA77-61306.
219. Hu, S. M., "Anomalous temperature effect of oxidation stacking faults in silicon", Appl. Phys. Lett., 27, 165-7 (1975), CA83-124305.
220. Jenkins, M. and Ravi, K. V., "The anisotropy of stacking fault growth in silicon", Presented at Electrochemical Society Meeting, San Francisco, Calif., May 1974, EEA77-22879.
221. Katz, L. E., "Relationship between process-induced defects and soft P-N junctions in silicon devices", J. Electrochem. Soc., 121, 969-72 (1974), EEA77-34019.
222. Laister, D. and Jenkins, G. M., "Stacking-faults in tellurium-doped gallium arsenide", J. Mater. Sci., 3, 584-9 (1968), PA72-23822.

223.Lawrence, J. E., "Stacking faults in annealed silicon surfaces", J. Appl. Phys., 40, 360-5 (1969).
224.Matsui, J. and Kawamura, T., "Spotty defects in oxidized floating-zoned dislocation-free silicon crystals", Jap. J. Appl. Phys., 11, 197-205 (1972), PA75-19104.
225.Ogden, R., "The nature of stacking fault pyramids in (100) epitaxial silicon", Phys. Status Solidi A, 28, K97-100 (1975), PA78-46504.
226.Ogden, R., Bradley, R. R. and Watts, B. E., "Stacking fault structures in carbon-contaminated low-temperature epitaxial silicon", Phys. Status Solidi A, 26, 135-46 (1974).
227.Palatnik, L. S., Vinogradov, V. E. and Boiko, B. T., "Certain type of structural imperfections in cadmium telluride films", Kristallografiya, 19, 1123-4 (1974), Russian, CA82-24539.
228.Pandey, D. and Krishna, P., "Influence of stacking faults on the growth of polytype structures. II. Silicon carbide polytypes", Philos. Mag., 31, 1133-48 (1975), PA78-54218.
229.Pomerantz, D., "A cause and cure of stacking faults in silicon epitaxial layers", J. Appl. Phys., 38, 5020-6 (1967), EEA71-5841.
230.Pomerantz, D., "Control of stacking faults produced by oxidation of silicon substrates", Sixth Annual Reliability Physics Symposium Proceedings, 1967, p. 120-6, Publ. 1968, EEA71-23215.
231.Prussin, S., "Generation of stacking faults and dislocation loops in silicon wafers (Surface damage)", J. Appl. Phys., 43, 733-5 (1972), PA75-25181.
232.Prussin, S., "Generation of stacking faults and prismatic dislocation loops in device-processed silicon wafers", J. Appl. Phys., 43, 2850-6 (1972), PA75-61179.
233.Ravi, K. V., "Generation of dislocation and stacking faults at surface heterogeneities in silicon (Bipolar transistors)", J. Appl. Phys., 43, 1785-92 (1972), EEA75-16730.
234.Ravi, K. V., "On the annihilation of oxidation induced stacking faults in silicon", Philos. Mag., 30, 1081-90 (1974), PA78-10547.
235.Ravi, K. V. and Varker, C. J., "Oxidation-induced stacking faults in silicon. I. Nucleation phenomenon", J. Appl. Phys., 45, 263-71 (1974), PA77-44548.
236.Ravi, K. V., Varker, C. J. and Volk, C. E., "Electrically active stacking faults in silicon", J. Electrochem. Soc., 120, 533-41 (1973), EEA76-23415.
237.Schwuttke, G. H., "Damage profiles in silicon and their impact of device reliability", Report DAHC15-72-C-0274, IBM Corp., N. Y., 1974, 66 pp., EEA78-24231.
238.Schwuttke, G. H., Brack, K. and Hearn, E. W., "The influence of stacking faults on leakage currents of f.e.t. devices", Microelectron. & Reliab., 10, 467-70 (1971), EEA75-7548.
239.Seto, D. K., Barson, F. and Duncan, B. F., "Identification of crystal defects causing diffusion pipes", Presented at Electrochemical Society Meeting, Chicago, Ill., May 1973, EEA76-23438.

240. Shih, K. K., Woolhouse, G. R., Blakeslee, A. E. and Blum, J. M., "Surface imperfections in GaAs-$Al_xGa_{1-x}As$ DH LPE layers", Gallium Arsenide and Related Compounds, 1974, p. 165-73, Publ. 1975, EEA78-28669.
241. Shiraki, H., "Elimination of stacking faults in silicon wafers by HCl added dry O_2 oxidation", Jap. J. Appl. Phys., 14, 747-52 (1975), EEA78-28632.
242. Stevens, R., "Defects in silicon carbide", J. Mater. Sci., 7, 517-21 (1972), PA75-47423.
243. Sugita, Y., Aoshima, T., Yoneda, K. and Yoshinaka, A., "The influence of oxidation-Sirtl etch condition on the stacking fault generation in (111) silicon wafers", J. Electron. Mater., 4, 175-89 (1975), PA78-38388.
244. Tempel, A. and Seifert, W., "Demonstration of stacking faults in GaN epitaxial layers by means of electron diffraction", Krist. & Tech., 10, 741-6 (1975), German, PA78-85358.
245. Tice, W. K. and Huang, T. C., "Circular stacking faults in silicon", Appl. Phys. Lett., 24, 157-8 (1974), PA77-44546.
246. Toncheva, L. T., "Surface defects of silicon reoxidized in water vapour", C. R. Acad. Bulg. Sci., 25, 479-82 (1972), EEA75-32507.
247. Varker, C. J. and Ravi, K. V., "Oxidation-induced stacking faults in silicon. II. Electrical effects in PN diodes", J. Appl. Phys., 45, 272-87 (1974), PA77-44549.
248. Vieweg-Gutberlet, F. G. and Wilhelm, F., "Investigations on 'doping stacking fault' pyramids", Solid-State Electron., 16, 191-5 (1973), EEA76-12467.

2. Twin Boundaries

249. Abrahams, M. S. and Buiocci, C. J., "Twins and stacking faults in vapor grown GaAs", J. Phys. Chem. Solids, 28, 927-30 (1967), EEA70-18815.
250. Aivazova, A. S., Gorin, S. N., Nikolaeva, L. G., Sidyakin, V. G. and Shmatko, C. G., "Nonuniformity of platy single crystals of β-SiC", Inorg. Mater., 9, 1201-4 (1973), PA77-60513.
251. Borle, W. N., Tata, S., Chatterjee, S. N. and Agarwal, S. K., "Growth twins in silicon crystals pulled from the melt", Indian J. Pure & Appl. Phys., 10, 315-17 (1972), PA75-74688.
252. Datsenko, L. I., Bletskan, N. I., Berezenko, L. E., Kislovskii, E. N., Prokopenko, I. V., and Veselkova, A. A., "Study of the structural perfection of a complex twinning group in Si", Metallofizika, Kiev, no. 53, 50-5 (1974), Russian, PA78-21518.
253. Dyer, L. D., "Causes of twin formation in the growth of silicon crystals", In: Semiconductor silicon, Electrochem. Soc., Inc., New York, p. 132-45 (1969), EEA73-10413.
254. Phillips, V. A., "Lattice resolution observations on the structure of twin boundaries, faults and dislocations in epitaxial silicon", Acta Met., 20, 1143-56 (1972), PA75-74779.

255.Siethoff, H., "Coherent twin boundaries in indium arsenide", Phys. Status Solidi A, 9, 227-35 (1972), German, PA75-19106.
256.Stevens, R., "Twin morphology in silicon carbide whiskers", J. Mater. Sci., 7 (6), 723-6 (1972), PA75-53581.
257.Stoemenos, J. and Vincent, R., "Twinning faults in epitaxial films of germanium telluride and GeTe-SnTe alloys", Phys. Status Solidi A, 11, 545-58 (1972), PA75-53574.
258.Tomita, T. and Ishiwata, M., "The inclusion found in solution-grown β-SiC", Presented at International Conference on Silicon Carbide, Miami Beach, Fla., September 1973, PA77-10538.
259.Varma, C. K. R., "Deformation twinning in diamond and the identification of the twinning plane", Scripta Met., 6, 383-6 (1972), PA75-50684.

V. Point Defects

1. Vacancies and Interstitials

260.Agrinskaya, N. V., Arkadeva, E. N. and Matveev, O. A., "Donor-cadmium vacancy associations in cadmium telluride", Proceedings of the International Symposium on Cadmium Telluride a Material for Gamma Ray Detectors, 1971, 9/5 pp., EEA75-20101.
261.Baranova, E. K., Gusev, V. M. and Streltsov, L. N., "Distribution of defects with depth in silicon irradiated with 80 keV ions", Sov. Phys.-Semicond., 7, 1239-40 (1974), EEA77-30199.
262.Bolotov, V. V., Vasilev, A. V. and Smirnov, L. S., "Migration energy of simplest defects in germanium and silicon", Sov. Phys.-Semicond., 8, 332-4 (1974), PA78-6601.
263.Bourgoin, J. C. and Corbett, J. W., "A new mechanism for interstitial migration", Phys. Lett. A, 38A, 135-7 (1972), PA75-19091.
264.Brodin, M. S., Goer, D. B. and Matsko, M. G., "Defect complexes in ZnTe", Sov. Phys.-Semicond., 7, 681-2 (1973), PA77-17115.
265.Chern, S. S. and Kroger, F. A., "The relative magnitude of the diffusion constants of cadmium vacancies", Phys. Status Solidi A, 25, 215-22 (1974), PA77-79186.
266.Chiang, S. S. Y., "Lattice defects in annealed gallium arsenide single crystals", Thesis, Stanford Univ., Calif., 1974, 76 pp., Order No. 75-6818, EEA78-33052.
267.Chiang, S. Y. and Pearson, G. L., "Properties of vacancy defects in GaAs single crystals", J. Appl. Phys., 46, 2986-91 (1975), PA78-76897.
268.de Kock, A. J. R., "The elimination of vacancy-cluster formation in dislocation-free silicon crystals", J. Electrochem. Soc., 118, 1851-6 (1971), PA75-9357.
269.de Kock, A. J. R., "Microdefects in dislocation-free silicon crystals", Thesis, Univ. Nijmegen, Netherlands, 1973, PA77-3495.
270.deKock, A. J. R., "Microdefects in dislocation-free silicon and germanium crystals", Acta Electron., 16, 303-11 (1973), EEA76-38178.

271.de Kock, A. J. R., "Vacancy clusters in dislocation-free silicon and germanium", Philips Tech. Rev., 34, 244-54 (1974), PA78-33970.

272.Emtzev, V. V., Mashovets, I. V., Maximov, M. and Vitovskii, N. A., "On the donor-vacancy type complexes in germanium", International Conference on Radiation Effects in Semiconductors, 1970, p. 105-11, Publ. 1971, PA75-47382.

273.Hansen, W. L. and Haller, E. E., "High-purity germanium-observations on the nature of acceptors", IEEE Trans. Nucl. Sci., NS-19, 260-4 (1972), EEA76-16611.

274.Hashimoto, F. and Kamiura, Y., "Migration energy of quenched-in defect in germanium", J. Phys. Soc. Jap., 33, 271 (1972), PA75-71536.

275.Hashimoto, F. and Kamiura, Y., "Annealing of quenched intrinsic-type germanium", Jap. J. Appl. Phys., 13, 762-72 (1974), PA77-52754.

276.Hasiguti, R. R. and Motomiya, S., "Estimation of the true formation energy of a vacancy in germanium", International Conference on Radiation Effects in Semiconductors, 1970, p. 39, Publ. 1971, PA75-47400.

277.Hoschl, P. and Kubalkova, S., "Electrical behaviour of In in CdTe crystals", Czech. J. Phys. B, 22, 530-3 (1972), PA75-47889.

278.Hu, S. M., "Interaction potential, correlation factor, vacancy mobility, and activation energy of impurity diffusion in diamond lattice", Phys. Status Solidi B, 60, 595-604 (1973), CA80-31216.

279.Jordan, A. S., Von Neida, A., Caruso, R. and Kim, C. K., "Determination of the solidus and gallium and phosphorus vacancy concentrations in GaP", Oyo Buturi, 43 (suppl.), 100 (1974), PA78-1153.

280.Kapitonova, L. M., Kostina, L. S., Lebedev, A. A. and Makhkamov, S., "Investigation of the photon capture cross sections of thermal-defect levels in n-type Si", Sov. Phys.-Semicond., 8, 766-7 (1974), PA78-25750.

281.Kholodar, G. A. and Vinetskii, V. L., "Electrical properties of silicon with divacancies", Phys. Status Solidi A, 30, 47-51 (1975), EEA78-32954.

282.Larkins, F. P., "Electronic structure of the isolated vacancy in silicon", International Conference on Radiation Effects in Semiconductors, 1970, p. 19-21, Publ. 1971, PA75-47935.

283.Lawrence, J. E., "Co-relation of silicon material characteristics and device performance", Presented at Electrochemical Society Meeting, Chicago, Ill., May 1973, EEA78-23318.

284.Littlejohn, M. A. and Loggins, C. D., Jr., "Influence of oxygen-vacancy complex (A center) on piezoresistance of n-type silicon", J. Appl. Phys., 43, 1074-84 (1972), PA75-25861.

285.Magee, T. J. and Morriss, R. H., "Diffraction contrast of small vacancy clusters in silicon", Proceedings of the 28th Annual Meeting of the Electron Microscopy Society of America, 1970, p. 514-15, PA75-6486.

286.Mashovets, T. V. and Emtsev, V. V., "Point defects in germanium", Lattice Defects in Semiconductors, 1974, p. 103-25, Publ. 1975, PA78-81082.
287.Medvedev, S. A., Maksimovskii, S. N., Kiseleva, K. V., Klevkov, Y. V. and Sentyurina, N. N., "Nature of point defects in undoped CdTe", Inorg. Mater., 9, 321-4 (1973), PA77-12943.
288.Morozova, N. K. and Nikitenko, V. A., "Thermodynamic analysis of intrinsic defects in single crystals of ZnS with deviations from stoichiometry", Inorg. Mater., 9, 1383-6 (1973), PA77-69796.
289.Morozova, N. K., Nikitenko, V. A. and Malov, M. M., "Precision measurements of the density of ZnS single crystals with deviations from stoichiometry", Inorg. Mater., 9, 1379-82 (1973), PA77-69795.
290.Narasimhan, K. L. and Guha, S., "Effect of vacancies on electrical and optical properties of amorphous gallium arsenide films", J. Non-Cryst. Solids, 16, 143-7 (1974), PA78-6921.
291.Panov, V. I., Smirnov, L. S. and Tishkovskii, E. G., "Competition between the capture and the transfer of vacancies between complexes of defects", Sov. Phys.-Semicond., 6, 580-3 (1972), PA75-80694.
292.Panteleev, V. A. and Okulich, V. I., "Equilibrium configuration of vacancies in Si and Ge", Sov. Phys.-Solid State, 14, 523-4 (1972), PA75-71467.
293.Panteleev, V. A., Okulich, V. I., Shchukin, R. N. and Shurganov, V. V., "Model of a semivacancy pair in silicon", Sov. Phys.-Semicond., 8, 1170 (1975), PA78-46168.
294.Schneider, M., "Native point defects in PbSe", Lattice Defects in Semiconductors, 1974, p. 353-8, Publ. 1975, PA78-81089.
295.Schroder, W. and Jegerlehner, K., "Defect structure in dislocation-free silicon single crystals from a crucible-free zone melting technique", Krist. & Tech., 9, 265-73 (1974), German, PA77-48628.
296.Sharaki, H., "Silicon wafer annealing effect on loop defect generation", Jap. J. Appl. Phys., 13, 1514-23 (1974), PA77-82818.
297.Shima, K., "A mechanism for shrinkage of voids in CdSe single crystal", Jap. J. Appl. Phys., 14, 1081-2 (1975), PA78-72463.
298.Sizova, N. L., Kliya, M. O., Rozhanskii, V. N. and Urusovskaya, A. A., "Spatial decoration of defects in CsI crystals", Sov. Phys.-Solid State, 14, 743-6 (1972), PA75-80695.
299.Van Vechten, J. A., "The entropy of neutral and ionized vacancies in Si and Ge", Lattice Defects in Semiconductors, 1974, p. 212-20, Publ. 1975, PA78-81084.
300.Van Vechten, J. A., "Divacancy binding enthalpy in semiconductors", Phys. Rev. B, 11, 3910-17 (1975), PA78-69076.
301.Van Vechten, J. A., "Effect of reconstruction of a semiconductor surface on the crystal growth", Appl. Phys. Lett., 26, 593-6 (1975), EEA78-32993.

302. Verner, V. D. and Michugovskii, D. K., "Energy of formation of metal vacancies in gallium arsenide and indium antimonide", Sov. Phys.-Semicond., 7, 1344 (1974), PA77-69803.
303. Vieweg-Gutberlet, F. G., "Recent investigations on swirls and comparison to striations", Presented at Electrochemical Society Meeting, Chicago, Ill., May 1973, EEA76-23299.
304. Voronkov, V. V., "Formation of vacancy pores during the cooling of germanium and silicon", Sov. Phys.-Crystallogr., 19, 137-42 (1974), PA78-6382.
305. Voronkov, V. V., Voronkova, G. I. and Iglitsyn, M. I., "Mobility and concentration of vacancies in Ge and Si", Sov. Phys.-Semicond., 6, 14-20 (1972), PA75-64659.
306. Yoshida, M., Arai, E., Nakamura, H. and Terunuma, Y., "Excess vacancy generation mechanism at phosphorus diffusion into silicon", J. Appl. Phys., 45, 1498-506 (1974), EEA77-30166.
307. Zolotukhin, A. A., Kovalenko, A. K. and Milevskii, L. S., "Interaction of vacancies with interstitial oxygen and iron atoms in silicon (Acceptor levels)", Fiz. Tver. Tela, 13, 3119-21 (1971), Russian, PA75-6485.

2. Impurities

308. Arakelov, A. G., Bibik, V. F., Dadykin, A. A. and Titov, V. A., "On accumulation of alkali impurities on structure defects near silicon surface", Ukr. Fiz. Zh., 20, 30-4 (1975), Russian, PA78-25509.
309. Brelot, A., "Investigation of the rapid interstitial diffusion in silicon of substitutional impurities ejected from their site by irradiation", Presented by French Soc. Metallurgy Autumn Meeting, Paris, France, September-October 1971, French, PA75-15780.
310. Chern, S. S. and Kroger, F. A., "The defect structure of CdTe: Self-diffusion data", J. Solid State Chem., 14, 44-51 (1975), PA78-68948.
311. Chern, S. S., Vydyanath, H. R. and Kroger, F. A., "The defect structure of CdTe: Hall data", J. Solid State Chem., 14, 33-43 (1975), PA78-68947.
312. Dankovskii, Y. V., Dobrokhotov, G. A., Mescheryakov, R. P., Osovskii, M. I., Tybasov, A. G. and Tronov, G. I., "Carbon as an impurity in semiconducting silicon", Inorg. Mater., 9, 1451-3 (1973), PA77-67723.
313. de Kock, A. J. R., Roksnoer, P. J. and Boonen, P. G. T., "Formation and elimination of growth striations in dislocation-free silicon crystals", J. Cryst. Growth, 28, 125-37 (1975), EEA78-11491.
314. Foll, H., Kolbesen, B. O. and Frank, W., "The nature of swirls and its significance for understanding point defects in silicon", Phys. Status Solidi A, 29, K83-7 (1975), PA78-58054.

315.Hall, R. N., "Interactions between quenched-in defects in high-purity germanium", Lattice Defects in Semiconductors, 1974, p. 109-7, Publ. 1975, PA78-81083.
316.Huntley, F. A., "The solubility and diffusion of copper in germanium", Phil. Mag., 26, 1047-52 (1972), PA75-77613.
317.Hutchinson, P. W. and Bastow, B. D., "The nature of inclusions in heavily tellurium-doped gallium arsenide", J. Mater. Sci., 9, 1483-92 (1974), PA78-1164.
318.Hutchinson, P. W. and Dobson, P. S., "The nature of defects in n^+ gallium arsenide", Philos. Mag., 30, 65-73 (1974), PA77-79043.
319.Jindal, B. K., Karelin, V. V. and Tiller, W. A., "Impurity striations in Czochralski grown Al doped Si single crystals", Presented at Second National Conference on Crystal Growth, Princeton, N. J., July-August 1972, EEA76-8565.
320.Kung, J. K. and Spitzer, W. G., "Si-defect concentrations in heavily Si-doped GaAs: Changes induced by annealing", J. Appl. Phys., 45, 4477-86 (1974), EEA78-8447.
321.Maehama, T., Toyama, N. and Yoshida, S., "Annealing effects on lattice defects in GaAs crystal", Rec. Electr. & Commun. Eng. Conversazione Tohoku Univ., 42 (2), 77-86 (1973), PA77-6412.
322.Ravi, K. V., "The heterogeneous precipitation of silicon oxides in silicon", J. Electrochem. Soc., 121, 1090-8 (1974), PA77-75630.
323.Tovstyuk, K. D., Baran, S. R. and Rakin, G. V., "The equilibrium states of the impurities in the dynamic lattice", Phys. Status Solidi B, 51, 197-204 (1972), PA75-40028.
324.Vieweg-Gutberlet, F. G., "Local distribution of oxygen in silicon single crystals and swirl formation", Lattice Defects in Semiconductors, 1974, p. 544, Publ. 1975, EEA78-41509.

3. Others

325.Altshuller, V. M., Belousova, E. A., Kruglov, I. I., Pavlichenko, V. I., Ryzhikov, I. V. and Rychkova, S. V., "Influence of structure defects on the electrical properties and luminescence of α-SiC", Sov. Phys.-Solid State, 14, 247-8 (1972), PA75-71774.
326.Buehler, M. G., "Thermally stimulated measurements: The characterization of defects in silicon p-n junctions", J. Electrochem. Soc., 120, 97C (1973), EEA76-23411.
327.Emtsev, V. V., Klinger, M. I. and Mashovets, T. V., "A possible mechanism for the scattering of charge carriers in germanium containing point defects", Zh. Eskp. & Teor. Fiz. Pis'ma, 19, 575-9 (1974), Russian, PA77-75934.
328.Forbes, L. and Vaughn, C. K., "Deep-level defects in red $GaAs_{1-x}P_x$ light-emitting diodes", Proc. IEEE, 62, 534-5 (1974), PA77-41435.
329.Hashimoto, F., "Effect of specimen size on the formation energy of thermal defect in germanium", J. Phys. Soc. Jap., 32, 865 (1972), PA75-36018.

330.Komatsu, R., "Effects of microdefect elimination on silicon P-N junction characteristics", Jap. J. Appl. Phys., 14, 419-20 (1975), EEA78-19713.
331.Konorova, L. F., "Thermal defects in silicon (Formation, destruction and effect on electrical properties)", Sov. Phys.-Solid State, 14, 1610-11 (1972), EEA76-8487.
332.Konorova, L. F., "Defects developing in silicon during annealing", Sov. Phys.-Solid State, 16, 348-9 (1974), PA78-3523.
333.Leskoschek, W., Feichtinger, H. and Vidrich, G., "Thermally induced defects in n-type and p-type silicon", Phys. Status Solidi A, 20, 601-10 (1973), EEA77-10018.
334.Melikyan, E. G., "Negative differential resistance in characteristics of locally deformed germanium diodes", Sov. Phys.-Semicond., 7, 1375-6 (1974), PA77-72862.
335.Melikyan, E. G. and Mkhitaryan, Z. O., "Dependence of the characteristics of anisotropically deformed germanium diodes on the crystallographic orientation of the base", Sov. Phys.-Semicond., 7, 1338-9 (1974), EEA77-36258.
336.Soloveva, E. V., Karataev, V. V., Milvidskii, M. G., Osvenskii, V. G. and Stolyarov, O. G., "Effect of the type and concentration of structural point defects on the electrophysical properties of undoped gallium arsenide", Sov. Phys.-Solid State, 14, 441-4 (1972), PA75-71905.
337.Stolarski, E., "Contactless measurement of thickness and damage of semiconductor slices", Arch. Elektrotech., 23, 375-86 (1974), Polish, EEA78-5201.
338.Vaipolin, A. A. and Rud, Y. V., "Influence of deviations from stoichiometry on the structures of CdTe crystals", Inorg. Mater., 10, 470-1 (1974), PA78-81077.

VI. Mechanical Damage

339.Ahlquist, C. N. and Carlsson, L., "The influence of light on crack propagation in cadmium telluride", Philos. Mag., 28, 733-8 (1973), PA77-6456.
340.Alekhin, V. P., Artemov, A. S., Mazur, A. I., Ragimov, I. A., Gusev, O. V. and Stankova, A. I., "Effect of abrasive treatment on the nature of structural damage and the depth of the damage layer of silicon single crystals", Fiz. Khim. Obrab. Mater., no. 5, 76-85 (1974), Russian, CA82-24606.
341.Bariss, V. O., Gailitis, L. A., Feltyn, I. A. and Ertelis, U. Y., "Generation of dislocations by macro-abrasions of surface of germanium", In: Semiconductors and their applications in electronics. V., Izdalel'stro 'Zinatne', Riga, USSR, p. 63-86 (1971), Russian, EEA75-13578.
342.Buck, T. M. and Meek, R. L., "Crystallographic damage to silicon by typical slicing, lapping, and polishing operations", Presented at Symposium on Silicon Device Processing, Gaithersburg, Md., June 1970, EEA73-35135.

343.Grigorev, O. N., Gridneva, I. V., Kryukova, I. V., Milman, Y. V., Trefilov, V. I. and Chugunova, S. I., "Effect of machining and thermal treatment on the degree of perfection of the crystal structure of GaAs single crystals", Fiz. & Khim. Obrab. Mater., no. 3, 117-20 (1975), Russian, PA78-86392.
344.Gutsche, H. W., "Surface damage in silicon", Sixth Annual Microelectronics Symposium, 1967, Pap. No. F3-1, EEA71-5840.
345.Huff, H. R. and Bracken, R. C., "Mechanical preparation of silicon slices and dislocation generation", In: Semiconductor silicon, Electrochem. Soc., Inc., New York, p. 610-21 (1969), EEA73-10438.
346.Kawakami, T., Hashimoto, T., Kawabe, H. and Yamada, T., "Lattice defects in surface layer of Si single crystals produced by lapping", Bull. Jap. Soc. Precis. Eng., 5 (4), 117-18 (1971), PA75-36270.
347.Khokhar, R. U. and Haneman, D., "Atomic mismatch on closure of controlled partial splits in silicon", J. Appl. Phys., 43, 317-19 (1972), PA75-19095.
348.Kirvalidze, I. D. and Dzhagarov, A. G., "Effect of mechanical polishing on the anisotropy of silicon microhardness", Elektron. Ionnye Protsessy Tverd. Telakh, 7, 128-30 (1974), Russian, CA83-124385.
349.Konstantinovic, J., Zivanovic, M. and Davidovic, M., "The effect of temperature and pressure on the mosaic spread in a Ge mónocrystal", Nucl. Instrum. & Methods, 65, 233-4 (1968), PA72-5307.
350.Knudsen, J. F., "Depth of saw and lap damage in germanium", In: Advances in X-ray analysis, Vol. 7, Muller, W. M. and others, (Eds.), Plenum Press, New York, p. 159-73 (1964).
351.Mayer, A., "Detection of damage on silicon surfaces: origin and propagation of defects", RCA Rev., 31, 414-30 (1970), EEA74-4866.
352.Olsen, G. H., Abrahams, M. S. and Zamerowski, T. J., "Asymmetric cracking in III-V compounds", Presented at Electrochemical Society Meeting, San Francisco, Calif., May 1974, EEA77-26344.
353.Parker, S. G. and Guildi, R. L., "Damage and defects in $Pb_{1-x}Sn_xTe$ crystals", Presented at Electrochemical Society Meeting, San Francisco, Calif., May 1974, PA77-48625.
354.Pavilainen, V. S. and Timofeev, V. M., "Influence of annealing on the depth of the disturbed layer in single-crystal silicon after grinding", Fiz. & Khim. Obrab. Mater., no. 4, 80-3 (1973), EEA77-2452.
355.Raheja, K., "Depth of damage induced by silicon wafer sawing", Electrochemical Society Fall Meeting (Extended abstracts), 1970, p. 395-8, EEA74-7674.
356.Schumaker, N. E. and Rozgonyi, G. A., "Dicing induced damage in GaP electroluminescent diodes", J. Electrochem. Soc., 119, 1233-7 (1972), EEA75-39622.

357. Sunada, J., "Lattice distortion induced by scratching on the surface of silicon single crystal wafer", Jap. J. Appl. Phys., 13, 1944-54 (1974), PA78-21713.
358. Trantina, G. G. and Johnson, C. A., "Subcritical crack growth in boron-doped SiC", J. Am. Ceram. Soc., 58, 344-45 (1975).
359. Yamagishi, S., "Deformation of single crystal wafers of silicon caused by lapping", Jap. J. Appl. Phys., 10, 589-92 (1971), EEA74-21598.

L. STRUCTURE ANALYSIS

I. Bibliographies

1. Der Boghosian, S., "A report guide to radiographic testing literature, Vol. 3", Report No. AD-676835, Army Materials and Mechanics Research Center, Watertown, Mass., 1968.
2. Haas, T. W., Dooley, G. J. and Grant, J. T., "A bibliograpby of low energy electron diffraction and Auger electron spectroscopy", In: Progress in surface science, Vol. 1, Part 2, Davidson, S. G., (Ed.), Pergamon Press, New York, p. 155-236 (1971), EEA74-35050.

II. Books

3. Anderson, C. A., "Microprobe analysis", Wiley-Interscience, New York, 1973.
4. Auleytner, J., "X-ray methods in the study of defects in single crystals", Pergamon Press, New York, 1966.
5. Azaroff, L., "X-ray spectroscopy", McGraw-Hill Book Co., New York, 1974.
6. Beeston, B. E. P., Horne, R. W. and Markham, R., "Electron diffraction and optical diffraction techniques", North-Holland, Amsterdam, Netherlands, 1972, PA77-24361.
7. Berger, M., "Neutron radiography", American Elsevier, New York, 1965.
8. Birks, L. S., "Electron probe microanalysis", Wiley-Interscience, New York, 1971.
9. Candano, R. and Verbist, J., (Eds.), "Electron spectroscopy", Elsevier Scientific Publishing Co., New York, 1975.
10. DeSoete, D. and others, "Neutron activation analysis", Wiley-Interscience, New York, 1972.
11. Emsley, J. W., Feeney, J. and Sutcliffe, L. H., "High resolution nuclear magnetic resonance spectroscopy", Pergamon Press, New York, 1965.
12. Ertl, G. and Kupper, J., "Low energy electrons and surface chemistry", Verlag Chemie, Weinheim, Germany, 1974.
13. Geschwind, S., "Electron paramagnetic resonance", Plenum Publishing Co., New York, 1972.
14. Goldstein, J. I. and Yakowitz, H., (Eds.), "Practical scanning electron microscopy: electron and ion microprobe analysis", Plenum Press, New York, 1975.
15. Harrick, N. J., "Internal reflection spectroscopy", Interscience Publishers, New York, 1967.
16. Head, A. K., Humble, P., Clearbrough, L. M. and Morton, A. J., "Computed electron micrographs and defect identification", North-Holland, Amsterdam, Netherlands, 1973, PA77-32555.

17. Hirsch, P. B., Howie, A., Nicholson, R. B., Pashley, D. W. and Whelan, M. J., "Electron microscopy of thin crystals", Plenum Press, New York, 1965.
18. Kane, P. F. and Larrabee, G. B., "Characterization of solid surfaces", Plenum Press, New York, 1974.
19. Lancaster, G., "Electron spin resonance in semiconductors", Plenum Publishing Co., New York, 1967.
20. Murr, L. E., "Electron optical applications in material science", McGraw-Hill Book Co., New York, 1970.
21. Newman, R. C., "Infrared studies of crystal defects", Barnes & Noble, New York, 1973.
22. Pendry, J. B., "Low energy electron diffraction", Academic Press, New York, 1974.
23. Phillips, V. A., "Modern metallographic techniques and their applications", Wiley-Interscience, New York, 1971.
24. Reed, S. J. B., "Electron microprobe analysis", Cambridge University Press, New York, 1975.
25. Rymer, T. B., "Electron diffraction", Halsted Press, London, England, 1974.
26. Wells, O. C., "Scanning electron microscopy", McGraw-Hill Book Co., New York, 1974.
27. Wilson, R. G., "Ion mass spectra", Wiley-Interscience, New York, 1974.
28. Wright, H. C., "Infrared techniques", Clarendon Press, Oxford, England, 1973.

III. Review Articles

29. Berger, H., "Neutron radiography", In: Research techniques in nondestructive testing, Sharpe, R. S., (Ed.), Academic Press, New York, p. 269-314 (1970).
30. Booker, G. R., "Scanning electron microscopy: applications", In: Modern diffraction and imaging techniques in material science, North-Holland, Amsterdam, Netherlands, p. 597-611 (1970).
31. Buck, T. M., Meek, R. L., Poate, J. M. and Seidel, T. E., "A review of some recent applications of MeV ^{4}He backscattering to problems in semiconductor technology. I. Technique and typical applications", Presented at Electrochemical Society Meeting, Chicago, Ill., May 1973, EEA76-23338.
32. Cowley, J. M., "Crystal structure determination by electron diffraction", Progr. Mat. Sci., 13, 269-321 (1967).
33. Chu, W. K., Mayer, J. W., Nicolet, M. A., Buck, T. M., Amsel, G. and Eisen, F., "Principles and applications of ion beam techniques for the analysis of solids and thin films", Thin Solid Films, 17, 1-41 (1973), EEA76-32197.
34. Donnadille, B., "Neptune. (Radiography application)", 3rd International Symposium on Non-Destructive Inspection Methods, 1974, p. 131-4, French, EEA77-37503.

35.Estrup, P. J., "Low energy electron diffraction", In: Modern diffraction and imaging techniques in material science, North-Holland, Amsterdam, Netherlands, p. 377-406 (1970).

36.Gallon, T. E. and Matthew, J. A. D., "Auger electron spectroscopy and its application to surface studies", Rev. Phys. Technol., 3, 31-64 (1972), PA75-48357.

37.Hawkesworth, M. R. and Walker, J., "Radiography with neutrons", J. Mat. Sci., 4, 817-835 (1969), PA73-11621.

38.Johnson, W. C., Stein, D. F. and Joshi, A., "Auger electron spectroscopy - a review", Can. J. Spectrosc., 17, 88-92 (1972), PA75-77481.

39.Lancaster, G., "Electron paramagnetic resonance", J. Mat. Sci., 2, 489-95 (1967).

40.Lang, A. R., "Recent applications of x-ray topography", In: Modern diffraction and imaging techniques in material science, North-Holland, Amsterdam, Netherlands, p. 521-52 (1970), PA71-75048.

41.Likhachev, V. A. and Volkova, N. V., "Applications of lasers to analysis and processing of materials", Ind. Lab., 35, 583-9 (1969), PA73-32505.

42.Tompsett, M. F., "Review: scanning high-energy electron diffraction (SHEED) in materials science", J. Mater. Sci., 7, 1069-79 (1972), PA75-74710.

43.Ugai, Y. A., Kirichenko, I. V. and Kurbanov, K. R., "The structure of the disturbed layer in Ge, Si, and GaAs crystals", Izv. Akad. Nauk SSSR Neorg. Mater., 8, 209-12 (1972), Russian, PA75-36010.

44.Vaisfeld, N. M., "Electron microscopic method of glass investigation: Techniques and results. A review", J. Non Cryst. Solids, 6, 283-93 (1971), PA72-15634.

IV. Electron Beam Techniques

1. Scanning Electron Microscope

a. SEM

45.Abrahams, M. S. and Buiocchi, C. J., "Cross sectional electron microscopy of silicon on sapphire", Appl. Phys. Lett., 27, 325-7 (1975).

46.Alferov, Z. I., Andreev, V. M., Murygin, V. I. and Stremin, V. I., "Study of heterojunction and p-n junctions in the aluminum arsenide-gallium arsenide system by means of a scanning electron microscope/microanalyser", Fiz. Tekh. Poluprov., 3, 1470-7 (1969), Russian, MA3-03-63 0044.

47.Balk, L. J. and Kubalek, E., "Use of phase sensitive - (lock-in) - amplification with scanning electron microscopy", Presented at Joint Session on Electron Microscopy, Liege, Belgium, September 1973, PA77-11624.

48. Beaulieu, R. P., Cox, C. D. and Black, T. M., "A SEM surface voltage measurement technique", Proceedings of the 10th Annual Conference on Reliability Physics, 1972, p. 32-5, EEA76-6082.

49. Belykh, L. A., Kutolin, S. A. and Ivanova, N. M., "Visualization of fixed charges and p-n junction heterogeneity in silicon by electron microscope using transistion metals", Elektron. Tekh. 12, no. 4, 68-70 (1971), Russian, PA75-4244.

50. Beyer, H., "Practical interference microscopical surface measurements particularly in semiconductor technology and microelectronics", Jena Rev., 19, 315-23 (1974), EEA78-19651.

51. Boltovets, N. S., Denisyuk, V. A. and Konakova, R. V., "Avalanche breakdown in silicon planar diodes studied by a scanning electron microscope", Mikroelektronika (Akad. Nauk SSSR), 3, 561-2 (1974), Russian, CA82-79479.

52. Bresse, J. F., "Quantitative use of induced current in scanning electron microscopy to study silicon devices", Rev. Phys. Appl., 9, 354 (1974), EEA77-22709.

53. Bunton, G. V., "Observations on silicon carbide devices using the SEM", Solid-State Electron., 12, 977-79 (1969).

54. Burgess, M. R., "Scanning electron microscopy and its application to semiconductor device assessment", Acta Electron., 18, 15-25 (1975), EEA78-17239.

55. Chaudhari, P. and Laibowitz, R. B., "Direct observation of switching phenomena in memory-type materials by electron microscopy", Thin Solid Films, 12, 239-42 (1972), EEA76-717.

56. Chirat, R. and Fontaine, G., "Observations on semiconductors using the induced mode (of the scanning electron microscopy)", Rev. Phys. Appl., 9, 373-6 (1974), French, EEA77-22710.

57. Clarke, D. R., "Observation of crystal defects using the scanning electron microscope", Phil. Mag., 24, 973-9 (1971), PA75-19037.

58. Clarke, D. R. and Howie, A., "Calculations of lattice defect images for scanning electron microscopy", Phil. Mag., 24, 959-71 (1971), PA75-19036.

59. Cline, J. E., Morris, J. M. and Schwartz, S., "Scanning electron mirror microscopy and SEM of integrated circuits", IEEE Trans. Electron Devices, ED-16, 371-75 (1969).

60. Cox, S. B., "Applications of scanning electron mirror microscopy to electronics", Proc. 4th Annual SEM Symposium, Part 1, 1971, p. 433-40.

61. Cox, R. H., Crosthwait, D. L., Jr. and Dobrott, R. D., "The application of the SEM to the development of high-reliability semiconductor products", IEEE Trans. Electron Devices, ED-16, 376-80 (1969).

62. Crosthwait, D. L., "Characterization of semiconductor processes", Proceedings of the 4th Annual Scanning Electron Microscope Symposium and Workshop on Forensic Applications of the Scanning Electron Microscope, 1971, p. 425-32, EEA74-26555.

63.Crosthwait, D. L., Jr., Ghate, P. B. and Smith, D. M., "Screening of metallization step coverage on integrated circuits", Proceedings of the 11th Annual Reliability Physics Conference, 1973, p. 254-60, EEA77-6091.
64.Crosthwait, D. L. and Ivy, F. W., "Voltage contrast methods for semiconductor device failure analysis", Scanning Electron Microscopy, 1974, p. 935-40, EEA78-686.
65.Dascalu, D. and Brezeanu, G., "Bulk breakdown in mesa IMPATT structures", Rev. Roum. Phys., 19, 571-3 (1974), EEA78-5302.
66.Davies, I. G., Hughes, K. A., Sulway, D. V. and Thornton, P. R., "The direct observation of electrical leakage paths due to crystal defects by use of the SEM", Solid-State Electron., 9, 275-9 (1966).
67.Day, H. and Christou, A., "Scanning electron microscopy study of gold-refractory thin film reliability", Proceedings of the 31st Annual Meeting of the Electron Microscopy Society of America, 1973, p. 126-7, EEA77-2555.
68.de Kock, A. J. R., Ferris, S. D., Kimerling, L. C. and Leamy, H. J., "Investigation of defects and striations in as-grown Si crystals by SEM using Schottky diodes", Appl. Phys. Lett., 27, 313-5 (1975).
69.Devaney, J. R., "Application of SEM to integrated circuit failure", Solid State Technol., 13 (3), 73-7 (1970).
70.Dobrott, R. D., "Applications of the scanning electron microscope on active semiconductor devices", Proceedings of the 28th Annual Meeting of the Electron Microscopy Society of America (condensed papers), 1970, p. 378-9, EEA75-4646.
71.Donovan, T. M. and Heinemann, K., "High-resolution electron microscope observation of voids in amorphous Ge", Phys. Rev. Lett., 27, 1794-6 (1971), PA75-9522.
72.Eberhardt, W., Kalus, C. and Schleicher, L., "Investigation of semiconductor devices with the scanning electron microscope", Siemens Forsch. & Entwicklungsber., 1, 315-21 (1972), German, EEA75-39194.
73.Everhart, T. E., "Some electron device applications of the scanning electron microscope", Spring Meeting of the Electrochemical Society (Extended abstracts), 1971, p. 246-7, EEA75-5467.
74.Everhart, T. E., Gonzales, A. J., Hoff, P. H. and MacDonald, N. C., "Applications of SEM to semiconductor devices", Proceedings of the 6th International Conference on Electron Microscopy, Vol. 1, 1966, p. 201-2.
75.Everhart, T. E., Wells, O. C. and Matta, R. K., "Evaluation of passivated integrated circuits using SEM", J. Electrochem. Soc., 111, 929-36 (1964).
76.Flutie, R. E., "Proper SEM inspection of IC step coverage using the NASA Goddard S-311-P12-A specification as criteria", Scanning Electron Microscopy, 1974, p. 963-70, EEA78-689.
77.Franke, W., Heimann, R. and Lacmann, R., "The dissolution forms of single crystal spheres. III. Dissolution forms of Ge and Si", J. Cryst. Growth, 28, 145-50 (1975), PA78-25560.

78. Galstyan, V. G. and Nosikov, S. V., "Special cases of observing the built-in electric local fields of semiconducting objects in a scanning electron microscope", Izv. Akad. Nauk SSSR Ser. Fiz., 38, 1413-6 (1974), Russian, EEA77-36243.
79. Galstyan, V. G., Nosikov, S. V., Press, F. P. and Pastushkov, V. V., "Scanning electron microscopy of defects in dielectric films", Phys. Status Solidi A, 12, 381-7 (1972), PA75-71964.
80. Gaylord, J. W., "Microplasma observations in silicon junctions using a SEM", J. Electrochem. Soc., 113, 753-4 (1966).
81. Gimeltarv, F. A., Govorkov, A. V. and Fistul, V. L., "Investigation of nonuniform semiconductors in a scanning electron microscope-microanalyser operating in a cathodoluminescent mode", Izv. Akad. Nauk SSSR Ser. Fiz., 38, 1409-12 (1974), Russian, EEA77-37491.
82. Gonzales, A. J., "Failure analysis applications of the scanning electron microscope", Proceedings of the 11th Annual Reliability Physics Conference, 1973, p. 179-84, EEA77-5788.
83. Gopinath, A., Sanger, C. C. and Paracchini, C., "Quantitative voltage and temperature distribution studies in GaAs transverse Gunn diodes using an SEM", Proc. 4th Annual SEM Symposium, Part 1, 1971, p. 457-64, PA74-59070.
84. Gratze, S. C., "Observations on epitaxial gallium arsenide varactor diodes using the SEM", Microelectron. & Reliab., 9, 71 (1970).
85. Green, D., "A new method of examination of semiconductor oxides with a scanning electron microscope", Record of the IEEE 8th Annual Symposium on Electron, Ion and Laser Beam Technology, 1966, p. 375-84.
86. Green, D. and Nathanson, H. C., "Observation of inversion layers under insulated-gate electrodes using a SEM", Proc. IEEE, 53, 183-4 (1965).
87. Green, D., Sandor, J. E., O'Keeffe, T. W. and Matta, R. K., "Reversible changes in transistor characteristics caused by SEM examination", Appl. Phys. Lett., 6, 3-4 (1965).
88. Gubenko, A. Y., Butylkina, N. A., Lukyanov, A. E. and Spivak, G. V., "Investigation into diode structures in a scanning electron microscope", Izv. Akad. Nauk SSSR Ser. Fiz., 38, 2403-8 (1974), Russian, PA78-21938.
89. Hackett, W. H., Jr., McGahan, T. E., Dixon, R. W. and Kammlott, G. W., "A scanning electron microscope investigation of etching phenomena in GaP electroluminescent diodes", J. Electrochem. Soc., 119, 973-6 (1972), PA75-71972.
90. Hackett, W. H., Jr., Saul, R. H., Dixon, R. W. and Kammlott, G. W., "Scanning electron microscope characterization of GaP reemitting diodes", J. Appl. Phys., 43, 2857 (1972), PA75-51389.
91. Hersener, J. and Ricker, T., "In-situ SEM observation of electromigration damage mechanisms in thin Al films", Presented at Joint Session on Electron Microscopy, Liege, Belgium, September 1973, PA77-9998.

92.Higuchi, H. and Maki, M., "Observations of channels of MOS-FET using the SEM", Jap. J. Appl. Phys., 4, 1021-22 (1965).
93.Higuchi, H. and Tamura, H., "Measurement of the lifetime of the minority carriers in semiconductors with a SEM", Jap. J. Appl. Phys., 4, 316-17 (1965).
94.Hoare, R. D. and Ahmed, H., "Retarding field and low voltage electron microscopy", Conference on Scanning Electron Microscopy: Systems and Applications 1973, p. 64-7, PA77-15016.
95.Hunter, D. R., Booker, G. R., Paxman, D. H. and Burgess, M., "Use of the SEM for measuring minority carrier lifetimes and diffusion lengths in semiconductor devices", Conference on Scanning Electron Microscopy: Systems and Applications 1973, p. 208-13, EEA77-14460.
96.Ishii, M. and Nakamura, G., "Electron microscope observation of silicon film at elevated temperature", Mitsubishi Denki Lab. Rep., 8, 125-34 (1967), EEA72-6494.
97.Johansen, H., "Failure analysis of integrated circuits using a scanning electron microscope (SEM)", Proceedings of the 5th Colloquium on Microwave Communication, Vol. V, 1974, p. SM-12/201-10, EEA77-34108.
98.Kalmar, G. and Stefaniay, V., "Testing and failure analysis of semiconductors with scanning electron microscope", Hiradastechnika, 26 (2), 50-5 (1975), PA78-51104.
99.Kandilarov, B., Pashov, N., Simov, S., Balabanov, S., Kantardieva, E. and Nikolova, E., "Heads of pin-like whiskers in CdS layers", Phys. Status Solidi A, 9, K95-7 (1972), PA75-25216.
100.Kimura, H., Higuchi, H., Maki, M. and Tamura, H., "Observation of semiconductor elements using SEM", J. Electron Microscopy, 15, 21-25 (1966).
101.Kleinholenz, K., "Electron microscope investigation of defects in ion irradiated silicon, Z. Naturforsch., 24, 912-17 (1969), German, PA69-44683.
102.Komatsu, E., Higuchi, Y. and Niina, T., "A p-n junction in silicon whiskers grown by VLS methods", Appl. Phys. Lett., 10, 42-43 (1967).
103.Korob, E. B., Mirgalovskaya, M. S., Raukhman, M. R. and Sorokina, N. G., "Use of scanning electron microscopy to study p-n junctions in single crystals of InSb", Inorg. Mater., 8, 1854-8 (1972), EEA76-35280.
104.Levin, E. R., "Minority-carrier diffusion lengths in GaAsGe by scanning electron microscopy", Proceedings of the 31st Annual Meeting of the Electron Microscopy Society of America, 1973, p. 190-1, EEA77-5940.
105.Linnell, R. A., "A nondestructive method for examining compositional and doping variations in GaAs-GaAlAs heterostructure material using a scanning electron microscope", J. Electrochem. Soc., 122, 832-5 (1975), PA78-62557.

106. Lipman, J. A., Bruncke, W. C., Crosthwait, D. L., Galloway, K. F. and Pease, R. I., "Use of a scanning electron microscope for screening bipolar surface effects", IEEE Trans. Nucl. Sci., NS-21, 383-6 (1974), EEA78-19800.
107. Lukianof, G. V., "Scanning electron microscope technique for measuring microcircuit chip area voltages in presence of signal sample nonuniformities", IBM Tech. Disclosure Bull., 17, 3275-6 (1975), EEA78-24472.
108. MacDonald, N. C. and Everhart, T. E., "Direct measurement of the depletion layer width variation vs. applied bias for a p-n junction", Appl. Phys. Lett., 7, 267-9 (1965).
109. MacDonald, N. C. and Everhart, T. E., "SEM investigation of planar diffused p-n junction profiles near the edge of a diffusion mask", J. Appl. Phys., 38, 3685-92 (1967).
110. MacKintosh, I. M., "Application of the SEM to solid-state devices", Proc. IEEE, 53, 370-7 (1965).
111. Magee, T. J., Morriss, R. H. and Melvin, C. D., "Diffraction contrast from small spherical voids in silicon", Phys. Status Solidi A, 13, 107-17 (1972), PA75-77571.
112. Matta, R. K., "Application of the SEM to semiconductors", Solid State Technol., 12 (1), 34-42 (1969).
113. Mazey, D. J. and Nelson, R. S., "Electron microscope investigation of damage structures in gallium arsenide bombarded with neon ions", Radiat. Eff., 1, 229-39 (1969), PA73-7905.
114. McAvoy, B. R., Green, D., Ing, D. W. and Ure, R. W., Jr., "Study of inhomogeneities in GaAs using a SEM", Appl. Phys. Lett., 14, 16-18 (1969).
115. Meieran, E. S. and Cass, T. R., "The applications of scanning and transmission electron microscopy in the semiconductor industry", Proceedings of the Fifth International Materials Symposium-the structure and properties of materials-techniques and application of electron microscopy, 1971, p. 1027-63, Publ. 1972, EEA75-32712.
116. Morgulis, L. M., Osvenskii, V. B. and Shifrin, S. S., "Study of the structure of crystals of semiconductors with a scanning electron microscope", Nauch. Tr., Nauch.-Issled. Proekt. Inst. Redkometal. Prom., no. 46, 53-6 (1973), Russian, CA80-31442.
117. Munir, Z. A., Seacrist, L. S. and Hirth, J. P., "Morphology of thermally etched basal surfaces of cadmium selenide", Surface Sci., 28, 357-72 (1971), PA75-12645.
118. Narayanan, G. H. and Kachare, A. H., "An electron microscopy study of the effects of annealing on the defect structure of heavily silicon-doped gallium arsenide", Phys. Status Solidi A, 26, 657-69 (1974), PA78-29263.
119. Nicolas, D. P., "Rule of SEM in microcircuit failure analysis", Scanning Electron Microscopy 1974, p. 955-62, EEA78-688.
120. Owen, G. and Nixon, W. C., "A large field deflection system for scanning electron beam technology", Presented at Electrochemical Society Meeting, San Francisco, Calif., May 1974, PA77-51660.

121.Pasiecznik, J., Jr. and Frey, J., "A high performance, low cost digitally driven SEM system for materials studies and microfabrication", J. Vac. Sci. & Technol., 10, 1012-15 (1973), PA77-42926.

122.Pease, D. E. and Dao, J., "The scanning electron microscope as an analytical tool", IEEE Trans. Nucl. Sci., NS-21, 788-93 (1974), EEA77-27878.

123.Pease, R. L. and Galloway, K. F., "Degradation of bipolar transistor electrical parameters during SEM evaluation", Microelectron. & Reliab., 13, 549-50 (1974), EEA78-11635.

124.Piwczyk, B. and Siu, W., "Specialized scanning electron microscopy voltage contrast techniques for LSI failure analysis", Proceedings of 12th Annual Reliability Physics Conference, 1974, p. 49-53.

125.Pokorny, J. and Laichter, V., "Measuring the diffusion length of charge carriers in semiconductors with the aid of a scanning electron microscope", Slaboproudy Obz., 36, 213-18 (1975), Czech, PA78-66026.

126.Poltavtsev, Y. G., Zakharov, V. P., Pozdnyakova, V. M. and Protas, I. M., "Structure of amorphous GaAs", Sov. Phys.-Crystallogr., 17, 165-7 (1972), PA75-71433.

127.Rai, K. N. and Srivastava, O. N., "Electron microscopic study of ZnS, CdS and HgS crystals in relation to their polytypic growth", Indian J. Pure & Appl. Phys., 10, 108-10 (1972), PA75-60769.

128.Ress, E. and Ress, G., "SEM-investigation of junction for failure analysis of semiconductor devices", Presented at Joint Session on Electron Microscopy, Liege, Belgium, September 1973, PA77-8529.

129.Saparin, Q. V., Spivak, G. V. and Stepanov, S. S., "Observation of the p-n junction at small constant and alternating bias with the SEM", Proceedings of the 6th International Congress on Electron Microscopy, Vol. 1, 1966, p. 609-10.

130.Schwuttke, G. H. and Brandis, E. K., "SEM of twin structures in silicon web dendrite crystals", J. Electrochem. Soc., 115, 669-71 (1968).

131.Shaw, G. G. and Benson, D. K., "Technique for visualizing subgrain structure at fatigue crack tips by ion etching and scanning electron microscopy", Proceedings of the 31st Annual Meeting of the Electron Microscopy Society of America, 1973, p. 86-7, PA77-1538.

132.Sie, C. H., "Scanning electron microscopy techniques for amorphous semiconductor device studies", Proceedings of the 5th Annual Scanning Electron Microscope Symposium Part I and Part II, Workshop on Biological Specimen Preparation for Scanning Electron Microscopy, 1972, p. 193-6, EEA75-24268.

133.Simov, S. B., Kamadjiev, P. R., Gospodinov, M. M. and Gantcheva, V. F., "Observations of hollow cadmium telluride crystals with a scanning electron microscope", J. Cryst. Growth, 26, 294-300 (1974), PA78-26266.

134. Spivak, G. V., Saparin, G. V., Bykov, M. V., Sedov, N. N. and Komolova, L. K., "Determination of the depth of p-n junctions using a SEM", Sov. Phys. Semicond., 3, 1304-6 (1970).
135. Srivastva, M. M. and Srivastava, O. N., "Observation of dislocation loops in gadolinium thin single crystal", Jap. J. Appl. Phys., 11, 1388-9 (1972), PA75-71489.
136. Stevens, R., "Electron microscopy of fracture chips of SiC", Proceedings of the 28th Annual Meeting of the Electron Microscopy Society of America, 1970, p. 466-7, PA75-9691.
137. Styris, D. L. and Thomas, M. T., "Microstructure of dissociated gold-silicon alloy by sputter etching and scanning electron microscopy", Mater. Sci. Eng., 14, 295-7 (1974), PA77-41578.
138. Sulway, D. V., Hughes, K. A., Evans, W. A. and Thornton, P., "Direct observation of MOS transistor channel pinch-off by SEM", Appl. Phys. Lett., 8, 296-8 (1966).
139. Sulway, D. V., Kyaw, H. and Thronton, P. R., "Direct observation of electrical faults in planar transistors made in epitaxially grown silicon", Solid-State Electron., 10, 545-53 (1967).
140. Takagi, M., Honjo, G., Ishii, F. and Nakanura, M., "Structure of vacuum evaporated CdSe films", NHK Tech. J., 24 (3), 61-5 (1972), Japanese, PA75-71327.
141. Thomas, R. T., "The scanning electron microscope and its role in solid state technology", Marconi Rev., 34, 271-85 (1971), EEA75-4558.
142. Thomas, R. T. and Dix, C., "The application of the conductive mode of scanning electron microscope to solid-state devices", Microelectronics, 3 (11), 32-6 (1971), EEA75-3944.
143. Thornhill, J. W. and Mackintosh, I. M., "Application of SEM to semiconductor device structures", Microelectron. & Reliab., 4, 97-100 (1965).
144. Thornton, P. R., "SEM in device diagnostics and reliability physics", IEEE Trans. Electron Devices, ED-16, 360-71 (1969).
145. Titchmarsh, J. M., Joy, D. C., Woolf, R. J. and Booker, G. R., "Diffraction contrast micrograph images from crystallographic defects in transmission and reflection using the SEM", Proceedings of the 5th European Congress on Electron Microscopy, 1972, p. 494-5, PA75-80621.
146. Toropov, A. I. and Pchelyakov, O. P., "Electron microscopic study of germanium surfaces subjected to ionic etching", Fiz. Khim. Obrab. Mater., no. 4, 126-8 (1975), Russian, CA83-186531.
147. van Essen, C. G. and Verhoeven, J. D., "Rotation between micrographs from the scanning electron microscope and electron channelling patterns", J. Phys. E, 7, 768-70 (1974), PA77-68812.
148. Varker, C. J., "Method for the investigation of thin films on a semiconductor substrate in a scanning electron microscope", Patent USA 3772520, Publ. November 1973.

149.Voronin, V. A., Stewart, W. D., Jr., Marshall, G. W., Wagner, J. B., Jr., Macksey, H. M. and Holonyak, N., Jr., "A scanning electron microscope study of $In_{1-x}Ga_xP$", J. Electrochem. Soc., 121, 1657-61 (1974), EEA78-8454.
150.Yang, D. Y., Johnson, W. C. and Lampert, M. A., "Scanning electron micrographs of self-quenched breakdown regions in $Al-SiO_2-(100)$ Si structures", Appl. Phys. Lett., 25, 140-2 (1974), EEA77-38778.
151.Zakharov, V. P. and Denisyuk, V. A., "Defectoscopy of metal-dielectric-semiconductor (MDS) integrated circuits using a scanning electron microscope", Sov. J. Nondestr. Test., 9, 427-31 (1973), EEA77-34105.

b. Cathodoluminescence

152.Aven, M., Devine, J. Z., Bolon, R. B. and Ludwig, G. W., "Scanning electron microscopy and cathodoluminescence of $ZnSe_xTe_{1-x}$ p-n junctions", J. Appl. Phys., 43, 4136-42 (1972), PA75-81118.
153.Balk, L. J. and Kubalek, E., "Cathodoluminescence studies of semiconductors in scanning electron microscopy", Presented at Joint Session on Electron Microscopy, Liege, Belgium, September 1973, EEA77-5680.
154.Davidson, S. M., Iqbal, M. Z. and Northrop, D. C., "SEM cathodoluminescent studies of plastically deformed gallium phosphide", Phys. Status Solidi A, 29, 571-8 (1975), PA78-61711.
155.Gopinath, A. and Savasee, D. M. E., "General physical studies on semiconductors using a scanning electron microscope", Rev. Phys. Appl., 9, 347-53 (1974), EEA77-22708.
156.Lidgard, G., "A scanning electron microscope study of the cathodoluminescence in GaP LED material", Solid State Electron., 15, 159-64 (1972), PA75-22813.
157.Shaw, D. A., Sulway, D. V., Wayte, R. C. and Thornton, P. R., "Simultaneous observation of surface detail and cathodoluminescence in the SEM", J. Appl. Phys., 38, 887-9 (1967).

c. Energy Dispersive Analysis

158.Barbi, N. C., Sandborg, A. O., Russ, J. C. and Soderquist, C. A., "Light element analysis on SEM using a windowless energy dispersive X-ray spectrometer", Proceedings of 7th Annual Symposium on Scanning Electron Microscopy, 1974, p. 151-6.
159.Christou, A., "SEM failure analysis techniques for thin films and microwave power transistor metalization", Proceedings of 7th Annual Symposium on Scanning Electron Microscopy, 1974, p. 971-8, EEA78-636.
160.Gehrke, R. J. and Davies, R. C., "Spectrum fitting technique for energy dispersive X-ray analysis of oxides and silicates with electron microbeam excitation", Anal. Chem., 47, 1537-41 (1975).

161.Waldl, E., Wolfermann, H., Rusovic, N. and Warlimont, H., "High-accuracy, empirical method of quantitative electron microprobe analysis using an energy dispersive system", Anal. Chem., 47, 1017-19 (1975).
162.Yakowitz, H., "X-ray microanalysis in SEM", Proceedings of 7th Annual Symposium on Scanning Electron Microscopy, 1974, p. 1029-40.

d. Electron Beam Induced Current

163.Alferov, Z. I., Andreev, V. M., Korolkov, V. I. and Stremin, V. I., "Study of high-voltage p-n junctions in gallium arsenide and $Al_xGa_{1-x}As$ by recording the current induced with an electron probe", Fiz. Tekh. Poluprov., 4, 1311-15 (1970), Russian, MA3-12-22 1033.
164.Bottoms, W. R., Guterman, D. and Roitman, P., "Contrast mechanisms in electron beam images of interface structures", J. Vac. Sci. & Technol., 12, 134-9 (1975), EEA78-28696.
165.Chang, T. H. P. and Nixon, W. C., "Electron beam induced potential contrast on unbiased planar transistors", Solid-State Electron., 10, 701-4 (1967).
166.Gates, J. L. and Griffith, O. K., "Inverted silicon island edge observation in SOS/CMOS transistors by SEM EBIC methods", Appl. Phys. Lett., 27, 43-5 (1975), EEA78-33442.
167.Gopinath, A. and De-Monts de-Savasse, T., "Electron-beam induced conductivity, and its use in SEM semiconductor material and device studies", Proceedings of the 7th International Congress on Electron Microscopy, 1970, p. 593-4, EEA74-34010.
168.Kajimura, T. and Nakamura, M., "Observation of doping profiles in Gunn diodes with a scanning electron microscope using the β-conductivity", J. Phys. D, 6, 1597-605 (1973), EEA76-35273.
169.Kato, T., Koyama, H., Shimizu, R. and Matsukawa, T., "SEM observation and contrast mechanism of stacking faults in an epitaxial silicon layer", J. Appl. Phys., 45, 3732-7 (1974), PA78-6663.
170.Kato, T., Matsukawa, T., Koyama, H. and Fujikawa, K., "Scanning electron microscopy of charging effect on silicon", J. Appl. Phys., 46, 2288-92 (1975), PA78-62090.
171.Kato, T., Matsukawa, T. and Shimizu, R., "Scanning electron microscope investigation of electric activity of stacking faults in silicon epitaxial layer", Appl. Phys. Lett., 26, 415-16 (1975), EEA78-28602.
172.Kawado, S., Hayafuji, Y. and Adachi, T., "Observation of lattice defects in silicon by scanning electron microscopy utilizing beam induced current generated in Schottky barriers", Jap. J. Appl. Phys., 14, 407-8 (1975), PA78-41596.

173. Munakata, C., "Application of electron-beam-induced conductivity to measurement of resistivity, electric field and potential distribution in GaAs", Electron and Ion Beam Science and Technology, Third International Conference, 1969, p. 580-9.
174. Schick, J. D., "Junction depth measurements in a scanning electron microscope", Presented at Electrochemical Society Meeting, San Francisco, Calif., May 1974, EEA77-26410.
175. Schick, J. D., "Failure analysis of integrated circuits with SEM beam induced currents", Scanning Electron Microscopy 1974, p. 949-54, EEA78-687.
176. Schick, J. D. and Wilson, H. R., "Determining doping levels in silicon devices", IBM Tech. Disclosure Bull., 17, 2299-300 (1975), EEA78-24155.
177. van Opdorp, C., Peters, R. C. and Klerk, M., "Use of Schottky-diode collectors for SEM determination of bulk diffusion lengths", Appl. Phys. Lett., 24, 125-6 (1974), EEA77-22764.

2. Transmission Electron Microscope

178. Abrahams, M. S., Buiocci, C. J. and Williams, B. F., "Microstructural effects on the minority carrier diffusion length in epitaxial GaAs", Appl. Phys. Lett., 18, 220-3 (1971), EEA74-18510.
179. Aseev, A. L., Vasin, O. I., Stenin, S. I., Soldatenko, N. N. and Tkhorik, Y. A., "Investigation of germanium films and Ge-Si interface structure by transmission electron microscopy", Thin Solid Films, 30, 73-82 (1975).
180. Berezhkova, G. V., Gerasimov, Y. M., Govorkov, V. G. and Kozlovskaya, E. P., "High-voltage electron microscopy in study of dislocation structure of α-aluminum oxide crystals", Izv. Akad. Nauk SSSR, Ser. Fiz., 38, 2382-8 (1974), Russian, CA82-78932.
181. Bernewitz, L. I., Kolbesen, B. O., Mayer, K. R. and Schuh, G. E., "TEM observation of dislocation loops correlated with individual swirl defects in as-grown silicon", Appl. Phys. Lett., 25, 277-9 (1974), PA78-1156.
182. Boitsov, Y. P., Prokhorov, V. I. and Sorokin, L. M., "Preparation of thin silicon films for TEM by photolithography and dynamic etching", J. Phys. E, 3, 702-4 (1970), MA4-02-21 0067.
183. Chang, S. C. and Shimaoka, G., "Crystallinity and epitaxy of evaporated germanium films on sodium chloride substrates", Jap. J. Appl. Phys., suppl. 2, 601-4 (1974), EEA78-19664.
184. Chase, B. D. and Holt, D. B., "Jet polishing of semiconductors. II. Electrochemically formed tunnels in GaP", J. Electrochem. Soc., 119, 314-17 (1972), EEA75-20059.
185. Chase, B. D. and Holt, D. B., "Transmission electron microscope observation on GaP electroluminescent diode materials", J. Mater. Sci., 7, 265-78 (1972), PA75-29992.
186. Chen, L. J., Seshan, K. and Thomas, G., "Contrast and resolution of small dislocation loops in high-voltage electron microscopy", Phys. Status Solidi A, 28, 309-19 (1975), PA78-41778.

187. Cullis, A. G. and Katz, L. E., "Electron microscope study of electrically active defects in silicon", Presented at Electrochemical Society Meeting, San Francisco, Calif., May 1974, PA77-44552.

188. Darlington, E. H. and Valdre, U., "Imaging of weak Lorentz objects (p-n junctions) by high voltage Fresnel TEM and STEM", J. Phys. E, 8, 321-4 (1975), EEA78-19712.

189. Das, G., "Precipitation of copper in silicon", J. Appl. Phys., 44, 4459-67 (1973), PA77-20306.

190. Das, G. and O'Neil, N. A., "Preparation of large-area electron-transparent samples from silicon devices", IBM J. Res. & Dev., 18, 76-9 (1974), PA77-41503.

191. Dudko, G. V., Myshlyaev, M. M., Severin, N. V. and Cherednichenko, D. I., "An electron-microscopic investigation of defects in silicon after electron-beam thermal treatment", Izv. Akad. Nauk SSSR Ser. Fiz., 36, 1371-4 (1972), Russian, PA75-64656.

192. Emmanuel, A. and Pollock, H. M., "Growth of polycrystalline silicon films grain size", J. Electrochem. Soc., 120, 1586-91 (1973), EEA77-2465.

193. English, C. A., Griffiths, V. W. and Venables, J. A., "The applications of channel plates in transmission and scanning electron microscopes", Acta Electron., 16, 43-57 (1973), EEA76-16640.

194. Evans, A. G. and Sharp, J. V., "Transmission electron microscopy of silicon nitride", Proceedings of the Fifth International Materials Symposium - the structure and properties of materials-techniques and application of electron microscopy, 1971, p. 1141-54, Publ. 1972, PA75-60998.

195. Faber, E. S., Tauber, R. N. and Broyde, B., "TEM deposition of silicon by electron irradiation of tetramethylsilane", J. Appl. Phys., 40, 2958-64 (1969), PA72-44628.

196. Garnak, A. E., Khashimov, F. R. and Chuprova, L. N., "Method of thinning monocrystalline plates of CdSe for the electron-microscopic investigation by transillumination", Ind. Lab., 40, 547 (1974), PA78-33388.

197. Grienauer, H. S., Kolbesen, B. O. and Mayer, K. R., "Direct TEM observation of dislocation loops at individual swirl defects in as-grown silicon", Lattice Defects in Semiconductors, 1974, p. 531-7, Publ. 1975, PA78-81102.

198. Hashimoto, N. and Koga, Y., "Observation of vanadium silicides on silicon by TEM", J. Appl. Phys., 39, 5798-800 (1968), MA2-04-21 0194.

199. Hill, M. J., "Ion irradiation damage of silicon in the transmission electron microscope", Phys. Status Solidi A, 8, 199-204 (1971), PA76-6477.

200. Hold, D. B., "Study of defects in epitaxial films of ZnTe by TEM", J. Mater. Sci., 4, 935-43 (1969).

201. Iwanaga, H., Shibata, N. and Suzuki, K., "Transmission electron microscopic studies on ZnO crystals", Jap. J. Appl. Pbys., 13, 185-6 (1974), PA77-28101.

202.Liliental, Z., Wnuk, B. and Torun, J., "Transmission electron microscopy of some semiconductor epitaxial layers", Presented at Proceedings of the 5th Czechoslovakia Conference on Electronics and Vacuum Physics, Brno, Czechoslovakia, October 1972, EEA76-8575.

203.Markova, T. I., "Preparation of gallium arsenide samples for TEM", Zavod. Lab., 35, 1095-6 (1969), Russian, MA3-04-21 0206.

204.Merli, P. G., Missitoli, G. F. and Pozzi, G., "Observation of p-n junction in transmission electron microscopy by out-of-focus technique", Phys. Status Solidi A, 16, K89-91 (1973), EEA76-23417.

205.Meyer, S., Aseev, A. L., Klimenko, A. G., Klimenko, F. A. and Stenin, S. I., "Formation of dislocation structure in silicon layers on nonorienting substrates", Krist. Tech., 10, 247-57 (1975), CA82-118311.

206.Morgulis, L. M., Osvenskii, V. B. and Milvidskii, M. G., "Electron-microscope investigation of tellurium-doped gallium arsenide single crystals", Sov. Phys.-Solid State, 16, 142-3 (1974), PA77-79054.

207.Narayanan, G. H. and Kachare, A. H., "Electron microscopy study of the effects of annealing on the defect structure of heavily silicon-doped gallium arsenide", Phys. Status Solidi A, 26, 657-69 (1974), CA82-66428.

208.Nes, E. and Solberg, J. K., "In situ transmission electron microscope investigation of the annealing of copper precipitate colonies in silicon", J. Appl. Phys., 44, 486-7 (1973), EEA76-12449.

209.Newbury, D. M. and Kirk, D. L., "The effect of annealing treatment on the microstructure of layers of zinc selenide deposited onto (100) germanium surfaces", Thin Solid Films, 22, 323-30 (1974), PA77-67126.

210.Orlova, A., "Determination of the dislocation density in subgrain boundaries and the disorientation of sub-grains by means of TEM", Cesk. Casopis Fys. A., 18, 11-16 (1968), Czech, PA73-11709.

211.Osiecki, R. and Thomas, G., "High voltage electron microscopy of interfacial defects in GaAs (device processing)", Mater. Sci. Eng., 10, 53-61 (1972), EEA76-28462.

212.Osvenskii, V. B., Morgulis, D. M. and Milvidskii, M. G., "Electron microscopic investigation of single crystals of gallium arsenide doped with n-type impurities", Izv. Akad. Nauk SSSR Ser. Fiz., 38, 1447-51 (1974), Russian, PA77-72480.

213.Pankratz, J. M., Sprague, J. A. and Rudee, M. L., "Investigation of neutron-irradiation damage in silicon by TEM", J. Appl. Phys., 39, 101-6 (1968).

214.Petroff, P. M. and DeKock, A. J. R., "Characterization of swirl defects in floating-zone silicon crystals", J. Cryst. Growth, 30, 117-24 (1975), CA83-200416.

215. Phillips, V. A., "Direct resolution of crystal defects", Proceedings of the 28th Annual Meeting of the Electron Microscopy Society of America, 1970, p. 508-9, PA75-6491.
216. Phillips, V. A. and Hugo, J. A., "Observations of (111) atomic planes in silicon", Appl. Phys. Lett., 13, 67-68 (1968).
217. Phillips, V. A. and Hugo, J. A., "The resolution of lattice planes and lattice defects in semiconductors including observations on boundaries", Acta Metall., 18, 123-35 (1970), PA73-18075.
218. Prokhorov, V. I. and Sorokin, L. M., "Preparation of thin crystals of silicon and layers deposited on its surface for transmission electron microscopy", Prib. Tekh. Eksp., no. 3, 220-2 (1973), Russian, CA79-119190.
219. Rai-Choudhury, P., "Gallium doped epitaxial silicon", J. Cryst. Growth, 8, 165-71 (1971), PA74-21435.
220. Rhines, W. C., Stevenson, D. A., Barrett, C. R. and Anderson, L. J., "Diffusion of zinc into gallium arsenide inside the transmission electron microscope", Proc., Electron Microsc. Soc. Amer., 31, 124-5 (1973), CA80-88678.
221. Sato, H., Shinozaki, S. and Yessik, M., "Direct observation and identification of long-period structures of SiC by transmission electron microscopy", J. Appl. Phys., 45, 1630-4 (1974), PA77-60509.
222. Sato, H., Shinozaki, S. and Yessik, M., "Transmission electron microscopy on growth characteristics of SiC crystals during reaction sintering", Presented at International Conference on Silicon Carbide, Miami Beach, Fla., September 1973, PA77-25420.
223. Sealy, B. J., "An examination of tellurium ion-implanted GaAs by transmission electron microscopy", J. Mater. Sci., 10, 683-91 (1975), PA78-37254.
224. Sealy, B. J. and Crocker, A. J., "Electron microscopy of epitaxial PbTe films", Thin Solid Films, 26, 135-46 (1975), EEA78-19690.
225. Sigurd, D., Arnal, H. J., Ottaviani, G. and Mayer, J. W., Crystallization of Ge and Si in metal films. II", J. Appl. Phys., 45, 1740-5 (1974), PA77-61310.
226. Sorokin, L. M. and Prokhorov, V. I., "Investigation of the structure defects generated by the diffusion of phosphorous in silicon", Fiz. Tver. Tela, 12, 3431-6 (1970), Russian, PA74-21419.
227. Sugano, T. and Kawai, N. J., "Direct observation of Si-SiO_2 interface by transmission electron microscope", Annu. Rep. Eng. Res. Inst. Fac. Eng. Univ. Tokyo, 33, 147-52 (1974), Japanese, PA78-37473.
228. Szilva, W. A., Murr, L. E. and Sattler, M. L., "A technique for selected-area observations in silicon wafers by transmission electron microscopy. (Polishing technique)", J. Mater. Sci., 9, 859-61 (1974), PA77-60387.

229. Tisone, T. C. and Lau, S., "An electron microscope study of interdiffusion and compound formation in Ta-Au thin films", Proceedings of the 31st Annual Meeting of the Electron Microscopy Society of America, New Orleans, La., August 1973, EEA77-2595.

230. Tsong, I. S. T. and Barber, D. J., "Development of the surface topography on silica glass due to ion bombardment. (ion thinning for transmission electron microscopy)", J. Mater. Sci., 7, 687-93 (1972), PA75-50462.

3. Microprobe

231. Alferov, Z. I., Andreev, V. M., Konnikov, S. G., Kolyskin, V. I., Larionov, V. R. and Selovanova, G. N., "Investigation of multiple layer heterojunction structures by means of an electron-beam microprobe-analyzer", Krist. & Tech., 8, 1037-43 (1973), PA77-20672.

232. Alferov, Z. I., Konnikov, S. G., Misurnyj, V. A., Tret'Jakov, D. M. and Godlinnik, T. B., "Electron-beam microprobe analysis of epitaxial $Ga_xIn_{1-x}P$ solid solutions", Krist. & Tech., 8, 1029-35 (1973), PA77-3657.

233. Beaman, D. R. and Isasi, J. A., "Electron beam microanalysis. I. The fundamentals and applications", Mater. Res. & Stand., 11 (11), 8-35, 48, 8-68 (1971), PA75-16712.

234. Beaman, D. R. and Isasi, J. A., "Electron beam microanalysis. II. Experimental considerations and qualitative analysis", Mater. Res. & Stand., 11 (12), 12-31, 51-6 (1971), PA75-37068.

235. Beatty, R. P. and Gerhard, J., "Instrumental technique for analysis of laminated metallization for silicon integrated circuits", In: Ohmic contacts to semiconductors, Electrochem. Soc., Inc., New York, p. 324-31 (1969), MA3-04-21 0209.

236. Bilenko, D. I., Novikova, E. A. and Smolyar, V. A., "Determination of the depth of occurrence of p-n junctions with the aid of an electronic probe", Izv. VUZ Fiz., no. 8, 16-20 (1973), Russian, EEA77-360.

237. Blakeslee, A. E., Segmuller, A. and Kuptsis, J. D., "Determination of alloy composition in semiconductor superlattice crystals", Presented at Second National Conference on Crystal Growth, Princeton, N. J., July-August 1972, EEA76-8568.

238. Bottoms, W. R. and Guterman, D., "Electron beam probe studies of semiconductor-insulator interface", J. Vac. Sci. & Technol., 19, 965-71 (1974), PA78-50683.

239. Brummer, O. and Roder, A., "Depiction of crystal disturbances in silicon with the electron beam probe", Exp. Tech. Phys., 19, 489-93 (1971), German, PA75-29706.

240. Brummer, O. and Roder, A., "Measurements of the diffusion length of minority charge carriers in silicon p-n junctions by electron beam microprobe", Wiss. Z. Tech. Univ. Dresden, 20, 371-3 (1971), German, PA75-44133.

241. Cocca, F. J. and Carroll, K. G., "Electron microprobe analysis of impurity heterogeneities in thermally grown silicon oxide", IEEE Trans. Electron Devices, ED-15, 962-6 (1968), EEA72-4966.

242. Colby, J. W., "Failure analysis using the electron microprobe", Proceedings of the 11th Annual Reliability Physics Conference, 1973, p. 189-93, EEA77-5790.

243. DiGiacomo, G., "Microprobe technique for determination of thickness and phosphorous concentration of gate oxide phosphosilicate glass in FET devices", J. Electrochem. Soc., 121, 419-22 (1974), EEA77-23054.

244. Fiermans, L. and Vennik, J., "Direct observations of imperfections in silicon by means of the Castaing probe", C. R. Acad. Sci. B, 264, 638-41 (1967), French, EEA70-16435.

245. Guthrie, J. W., "Identification of AgCl as a surface contaminant on hybrid microcircuit capacitors using ion microprobe techniques", J. Electrochem. Soc., 121, 1617-20 (1974), EEA78-8625.

246. Kim, C. K., "Analysis of $GaAs_xP_{1-x}$ grown from Ga-GaAs GaP ternary system: Quantitative correction and solid solubility isothermos", Proc. 4th National Conf. Electron Microprobe Analysis, 1969, p. 21.

247. Kormany, T., "Investigation of semiconductors by the electron beam microanalyser", In: Annual of the research institute for telecommunication 1973, TIK: Budapest, Hungary, p. 311-27 (1974), EEA78-15247.

248. Kormany, T. and Nagy, G., "The solution of semiconductor problems with the electron beam X-ray microprobe", Krist. & Tech., 8, 1201-8 (1973), German, PA77-6667.

249. Kormany, T. and Nagy, G., "Failure-analysis of microwave devices by electron-probe", Proceedings of the 5th Colloquium on Microwave Communication, Vol. V, 1974, p. SM-13/211-20, EEA77-34035.

250. Legrand, C. and Trillat, E. C., "Use of electron microprobe to study ion implantation", Electron and Ion Beam Science and Technology, Fourth International Conference, 1970, p. 545-52.

251. Luby, S. and Schilder, J., "Electron microprobe compositional analysis of sputtered tantalum-aluminium films", Czech. J. Phys. B, B25, 91-100 (1975), EEA78-11714.

252. Mitchell, D. F. and Sewell, P. B., "Quantitative determination of oxygen in thin oxide films on metals by electron-excited X-ray emission", Thin Solid Films, 23, 109-25 (1974), PA77-72660.

253. Nealey, C. C., Laakso, C. W. and Hagon, P. J., "Planar silicon device analysis with the electron microprobe microanalyzer", Proceedings of the Symposium on the Electron Microprobe, 1966, p. 748-83.

254. Noda, M., Maruno, S. and Yamada, T., "Formation of negative resistance path in amorphous germanium thin films", Jap. J. Appl. Phys., 9, 1424-5 (1970), PA74-26241.

255. Rado, W. G., Johnson, W. J. and Crawley, R. L., "The composition of $Ga_{1-x}Al_xAs$ grown from Ga-rich solution as a function of growth temperature and solution composition", J. Electrochem. Soc., 119, 652-3 (1972), PA75-47286.
256. Sawicka, E., "The investigations of the black plague", 3rd Symposium on the Reliability of Electronics, 1973, p. 359-64, EEA77-6018.
257. Selezneva, M. A., Kupriyanova, T. A. and Ditsman, S. A., "Determination of small diffusion lengths in semiconductors by an electron probe method taking into account the effect of the surface recombination rate and sample size", Izv. Akad. Nauk SSSR, Ser. Fiz., 38, 2328-32 (1974), Russian, CA82-67152.
258. Smith, J. P. and Kraus, H., "Electron microprobe characterization of the stoichiometry and homogeneity of mixed-crystal semiconductor materials", Appl. Spectrosc., 24, 580-2 (1970), PA74-5173.
259. Stave, J., Becherer, G. and Stigge, H. J., "Electron microprobe determination of the doping profile in pnp-silicon semiconductors", Krist. & Tech., 8, 1063-72 (1973), German, EEA77-2508.
260. Stave, J., Becherer, G. and Stigge, H. J., "On the determination of the doping profile in pnp-type silicon semiconductors using the electron beam microprobe", Krist. & Tech., 8, 1063-72 (1973), German, EEA77-10159.
261. Wittry, D. B., "Applications of scanning electron microprobes to solid state electronics", Proc. 5th National Conf. Electron Probe Analysis, 1970, p. 20.

4. Auger Electron Spectroscopy

262. Amelio, G. F., "Band structures of silicon by characteristic Auger electron spectrum analysis", Surface Sci., 22, 301-18 (1970), MA4-03-15 0300.
263. Bassett, P. J., Gallon, T. E. and Prutton, M., "A high energy resolution Auger electron spectrometer using concentric hemispheres", J. Phys. E., 5, 1008-13 (1972), PA75-74610.
264. Bauer, E., "On the nature of annealed semiconductor surfaces", Phys. Lett., 26, 530-1 (1968), EEA71-15734.
265. Brillson, L. J., "Surface photovoltage and Auger spectroscopy studies of (1120) CdS surface", J. Vac. Sci. & Technol., 12, 249-52 (1975), PA78-62074.
266. Card, H. C. and Singer, K. E., "In-depth Auger analysis of aluminum-silicon interfacial reactions", Thin Solid Films, 28, 265-8 (1974), CA83-140836.
267. Chang, C. C. and Quintana, G., "Silicon diffusion through thin tungsten films on silicon, studies with Auger spectroscopy", J. Electron Spectrosc. & Relat. Phenom., 2, 363-76 (1973), EEA77-18857.
268. Charig, J. M. and Skinner, D. K., "Auger electron spectroscopy of nickel deposits on the silicon (111) surface", Surface Sci., 19, 283-90 (1970), MA3-07-12 0693.

269. Chou, N. J., Hammer, R., Van der Meulen, Y. J. and Cahill, J., "Auger and ellipsometric study of phosphorus segregation in oxidised degenerate silica", Appl. Phys. Lett., 24, 200-2 (1974), PA77-41497.
270. Gonzales, A. J., "Failure analysis applications of Auger electron spectroscopy", Proceedings of the 11th Annual Reliability Physics Conference, 1973, p. 185-8, EEA77-5789.
271. Grant, J. T., "Combined low energy electron diffraction and Auger electron spectroscopy studies of InAs surfaces", J. Vac. Sci. & Technol., 9, 911 (1972), PA75-50449.
272. Harman, R., Liday, J. and Vesely, M., "Auger electron spectroscopy of silicon surfaces", Proc. Int. Conf. Solid Surf., 2nd, 1974, p. 823-6, CA82-177696.
273. Harris, L. A., "Miscellaneous topics in Auger electron spectroscopy", J. Vac. Sci. & Technol., 11, 23-8 (1974), PA77-79828.
274. Inoue, T., Horiuchi, S., Yoshii, S. and Tanuma, Y., "Auger studies of diffused impurity profiles in SiO_2/Si and poly Si/SiO_2/Si structures", Oyo Buturi, 44 (Suppl.), 329-37 (1975), PA78-81995.
275. Ludeke, R., Esaki, L. and Chang, L. L., "$Ga_{1-x}Al_xAs$ superlattices profiled by Auger electron spectroscopy", Appl. Phys. Lett., 24, 417-19 (1974), PA77-57692.
276. Lyo, K. and Komiya, S., "Detection of sodium on silicon and silicon dioxide surfaces by cylindrical mirror Auger analyzer", Shinku, 16 (1), 41-4 (1973), Japanese, CA79-24766.
277. Moon, R. L. and James, L. W., "Auger spectra of HCl vapor-etched n^+ GaAs (100) substrates", J. Electrochem. Soc., 120, 581-3 (1973), EEA76-23345.
278. Morabito, J. M., "Three-dimensional elemental analysis with Auger electron spectroscopy and secondary ion mass spectrometry", Presented at Electrochemical Society Meeting, San Francisco, Calif., May 1974, EEA79-26529.
279. Nishijima, M. and Murtotani, T., "Combined Auger electron spectroscopy and electron impact desorption studies of silicon surfaces", Surface Sci., 32, 459-65 (1972), PA75-64556.
280. Powell, B. D., Woodruff, D. P. and Griffiths, B. W., "A scanning Auger electron microscope for surface studies", J. Phys. E, 8, 548-52 (1975), EEA78-34989.
281. Ranke, W. and Jacobi, K., "Electron stimulated oxidation of GaAs, studies by quantitative Auger electron spectroscopy", Surf. Sci., 47, 525-42 (1975), PA78-29517.
282. Ridgway, J. W. T. and Haneman, D., "Silicon (111)-7 structures obtained by cleavage at room temperature", Appl. Phys. Lett., 17, 130-1 (1970).
283. Robinson, G. Y. and Jarvis, N. L., "Auger electron spectroscopy and sputter etching of Ni/Au-Ge on n-GaAs", Appl. Phys. Lett., 21, 507-10 (1972), EEA76-4568.

284.Thomas, J. H., III and Morabilo, J. M., "Detectability limits for boron and phosphorus in silicon by Auger electron spectroscopy (AES)", Surf. Sci., 41, 629-33 (1974), EEA77-14400.
285.Thomas, R. E. and Haas, G. A., "Auger examination of contaminants in thin-film metallizations", J. Appl. Phys., 46, 963-5 (1975), EEA78-24171.
286.Uebbing, J. J., "Auger electron spectroscopy of contaminated gallium arsenide surfaces", J. Vac. Sci. Technol., 7, 81-3 (1970), MA3-12-23 0694.
287.Uebbing, J. J., "Use of Auger electron microscopy in determining the effect of carbon and other surface contaminants on GaAs-Cs-O photocathodes", J. Appl. Phys., 41, 802-4 (1970).
288.Uebbing, J. J. and Taylor, N. J., "Auger electron spectroscopy of clean gallium arsenide", J. Appl. Phys., 41, 804-8 (1970), PA73-27648.
289.Vlachova, B., "Auger spectroscopy of silicon", Proceedings of the 5th Czechoslovakia Conference on Electronics and Vacuum Physics, 1972, Ic-2/2 pp., PA76-15062.
290.Waldrop, J. R. and Harris, J. S., "Potential profiling across semiconductor junctions by Auger electron spectroscopy in the scanning electron microscope", J. Appl. Phys., 46, 5214-17 (1975).

5. Low Energy Electron Diffraction

291.Auer, P. P. and Monch, W., "On structural and electronic properties of cleaved silicon surfaces", Jap. J. Appl. Phys., (Suppl. 2), 397-400 (1974), PA78-37800.
292.Bauer, E., "On the nature of annealed semiconductor surfaces", Phys. Lett., 26A, 530-1 (1968).
293.Bauerle, F., Monch, W. and Henzler, M., "Correlation of electronic surface properties and surface structure on cleaved silicon surfaces", J. Appl. Phys., 43, 3917-19 (1972), PA75-77800.
294.Bellina, J. J., Jr., "An ordered boron structure after deposition on Si(111)", Jap. J. Appl. Phys., 13, 1659-60 (1974), PA78-1261.
295.Carron, G. J., Walford, L. K. and Schoeffel, J. A., "Investigation of thermal stresses in silicon by source image distortion", J. Appl. Phys., 40, 2372-5 (1969), PA72-35164.
296.Chang, C. C., "LEED studies of thin-film silicon overgrowths on α alumina", In: The structure and chemistry of solid surfaces, Somorjai, G. A., (Ed.), John Wiley and Sons, New York, Pap. No. 77, 1969, MA4-04-12 0577.
297.Duke, C. B., "Determination of the structure and properties of solid surfaces by electron diffraction and emission", In: Aspects of the study of surfaces, Vol. 27, Prigogine, I. and Rice, S. A., (Eds.), Wiley, New York, p. 1-209 (1974), PA78-69091.

298. Florio, J. V., "The determination of surface structures by low energy electron diffraction (LEED) - the Si(111) surface", Presented by American Crystallographic Association summer meeting, Ames, Iowa, August 1971, PA75-3502.
299. Florio, J. V. and Robertson, W. D., "Phase transformations of the silicon (111) surface", Surface Sci., 22, 459-64 (1970), PA74-5069.
300. Gallon, T. E., Prutton, M. and Wray, L., "Combined low energy electron diffraction and Auger electron spectroscopy studies of InAs surfaces", J. Vac. Sci. & Technol., 9, 911-14 (1972), PA75-50464.
301. Haidinger, W. and Barnes, S. C., "LEED investigations of clean and Au-stabilized Si surfaces", Surface Sci., 20, 313-25 (1970), PA73-55703.
302. Holloway, P. H. and Hudson, J. B., "A versatile substrate design for LEED and AES studies in uhv", Rev. Sci. Instrum., 43, 1045-7 (1972), PA75-60617.
303. Kolaczkiewica, J., Koziol, C. and Mroz, S., "An investigation of impurities segregation to the (001) nickel surface during thermal treatment. Work function changes and Auger electron spectroscopy using the LEED camera", Acta Phys. Pol. A, A41, 783-4 (1972), PA75-53554.
304. Margoninski, Y., "Investigation of the surface reaction of germanium with oxygen by LEED and conductivity measurements", J. Vac. Sci. & Technol., 9, 920-3 (1972), PA75-50448.
305. Miyata, N. and Hanawa, T., "LEED study of the growth of silver films on GaP surfaces", Jap. J. Appl. Phys., (Suppl. 2), 593-6 (1974), PA78-37600.
306. Nesterenko, B. A., Borodkin, A. D. and Snitko, O. V., "Atom vibrations on clean and doped (111) silicon surface and on clean (110) GaAs surface", Surface Sci., 32, 576-82 (1972), PA75-71318.
307. Palmberg, P. W. and Peria, W. T., "LEED studies on Ge and Na-covered Ge", Surface Sci., 6, 57-97 (1967).
308. Ridgway, J. W. T. and Haneman, D., "Correlation of LEED surface structures and surface tear marks on cleaved Si surfaces", Surf. Sci., 18, 441-5 (1969), PA73-31064.
309. Russell, G. J. and Haneman, D., "Oxygen adsorption on vacuum cleaved GaAs at liquid nitrogen temperatures by LEED", Surface Sci., 27, 362-6 (1971), PA75-3561.
310. Takeda, K., Hanawa, T. and Shimojo, T., "Growth of Au film on GaAs(110) observed by LEED and AES", Jap. J. Appl. Phys., (Suppl. 2), 589-92 (1974), PA78-37599.
311. Takeishi, Y., Hara, H., Sato, T. and Sasaki, I., "Characteristic properties of Si 311 surfaces", Jap. J. Appl. Phys., 7, 679 (1968), EEA71-23243.
312. Thomas, R. N. and Francombe, M. H., "A LEED study of the homoepitaxial growth of thick silicon films", Appl. Phys. Lett., 11, 108-10 (1967).

313. Tracy, J. C. and Burkstrand, J. M., "Recent advances in surface characterization using LEED and Auger electron spectroscopy", Crit. Rev. Solid State Sci., 4, 381-93 (1974), PA78-37477.
314. van Bommel, A. J. and Meyer, F., "A LEED study of the PH_3 adsorption on the silicon (111) surface", Surface Sci., 8, 381-98 (1967).
315. van Bommel, A. J. and Meyer, F., "LEED study of a nickel-induced surface structure on silicon (111)", Surface Sci., 8, 467-72 (1967).

6. High Energy Electron Diffraction

316. Baehr, R. and Kressner, F. H., "Surface evaluation of polished silicon discs", Krist. & Tech., 10, 443-54 (1975), German, PA78-76980.
317. Colella, R. and Menadue, J. F., "Comparison of experimental and n-beam calculated intensities for glancing incidence high-energy electron diffraction", Acta Crystallogr. A, A28, 16-22 (1972), PA75-9238.
318. Henderson, R. C., "Silicon cleaning with hydrogen peroxide solutions: a high energy electron diffraction and Auger electron spectroscopy study", J. Electrochem. Soc., 119, 772-5 (1972), EEA75-28402.
319. Henderson, R. C. and Helm, R. F., "High-energy electron diffraction and Auger electron spectroscopy study of silicon epitaxy by silane pyrolysis", J. Vac. Sci. & Technol., 9, 164 (1972), PA75-39523.
320. Henderson, R. C. and Helm, R. F., "Silicon homoepitaxial thin films via silane pyrolysis: a HEED and Auger electron spectroscopy study", Surface Sci., 30, 310-34 (1972), PA75-39528.
321. Krause, G. O., "Characterization of Si surfaces by RHEED", In: Semiconductor silicon, Electrochem. Soc., Inc., New York, p. 574-84 (1969).
322. Laermans, C., Michiels, L. and deBock, A., "Reflection electron diffraction (RHEED) study of oblique growth orientations in CdS films", Presented at Joint Session on Electron Microscopy, Liege, Belgium, September 1973, EEA77-5915.

7. Electron Spectroscopy for Chemical Analysis

323. Clarke, R. A., Tapping, R. L., Hopper, M. A. and Young, L., "An ESCA study of the oxide at the Si-SiO_2 interface", J. Electrochem. Soc., 122, 1347-50 (1975).
324. Flitsch, R. and Raider, S., "ESCA studies of dielectric films", Presented at Electrochemical Society Meeting, Chicago, Ill., May 1973, EEA76-23266.
325. Hedman, J., "ESCA applied to solid-state physics", At. Energy Rev., 12, 763-85 (1974).

326. Klasson, M., Berndtsson, A., Hedman, J., Nilsson, R., Nyholm, R. and Nordling, C., "Electron escape depth in silicon", J. Electron Spectrosc. & Relat. Phenom., 3, 427-34 (1974), PA77-67688.
327. Leonhardt, G., Berndtsson, A., Hedman, J., Klasson, M., Nilsson, R. and Nordling, C., "ESCA studies of some $A^{III}-B^{V}$ compounds with Ga and As", Phys. Status Solidi B, 60, 241-8 (1973).
328. Pietrokowsky, P., Wood, J. K. and Scheffer, L. K., "ESCA applied to semiconductor materials and processing technology", Presented at Electrochemical Society Meeting, Chicago, Ill., May 1973, EEA76-23337.
329. Raider, S. I., Flitsch, R. and Palmer, M. J., "Oxide growth on etched silicon in air at room temperature", J. Electrochem. Soc., 122, 413-18 (1975), EEA78-24182.

8. Others

330. Calverly, A. and Wight, D. R., "A scanning electron beam study of the relationship between the cathodoluminescence and the electroluminescence efficiencies of GaP diodes", Solid-State Electron., 13, 382-5 (1970), PA73-34839.
331. Casey, H. C., "Investigation of inhomogeneities in GaAs with electron-beam excitation", J. Electrochem. Soc., 114, 153-8 (1967).
332. Casey, H. C., Jr. and Kaiser, R. H., "Analysis of n-type GaAs with electron beam excited radiative recombination", J. Electrochem. Soc., 114, 149-53 (1967).
333. Czaja, W., "Detection of partial dislocations in silicon with the scanning electron beam technique", J. Appl. Phys., 37, 918-19 (1966).
334. Devaux, J. and Henoc, P., "Microscopic and electron diffraction study of silicon strips, diffused into phosphorous", Ann. Telecomns., 25, 88-92 (1970), French, PA73-51050.
335. Engelke, H., "Contactless short/open test method using two electron beams", Presented at Electrochemical Society Meeting, San Francisco, Calif., May 1974, EEA77-26504.
336. English, F. L., "Inhomogeneities in the resistivity of single crystal cadmium sulfide", II-IV Semiconducting Compounds International Conference, 1967, p. 190-8, EEA71-20554.
337. Heydenreich, J., "Study of charge distribution in insulator layers with a reflection electron microscope", Electron Microsc., Proc. Eur. Congr. Electron Microsc., 5th, 1972, p. 490-1, German, CA79-10955.
338. Holt, D. B. and Chase, B. D., "Scanning electron beam excited charge collection micrography of GaAs lasers", J. Mater. Sci., 3, 178-82 (1968).
339. Igras, E., "The electron mirror microscope and its applications for semiconductor studies", Electronic Equip. News, 12, 18-23 (1970), EEA73-37991.

340.Igras, E. and Krylow, J., "Electron mirror microscopic investigation of silicon surface layers with implanted impurities", 3rd Symposium on the Reliability of Electronics, 1973, p. 135-43, PA77-10205.

341.Kreutle, M. and Meyer-Ehmsen, G., "Determination of structure potentials and absorption potentials of silicon from electron diffraction intensities", Phys. Status Solidi A, 8, 111-18 (1971), PA75-3729.

342.Lovyagin, P. N., Fedina, L. I., Aleksandrov, L. N. and Krivorotov, E. A., "Electron microscopic investigation of the initial stages in the growth of epitaxial silicon films on the real surface of silicon and germanium during ionic sputtering", Izv. Akad. Nauk SSSR Ser. Fiz., 38, 1528-32 (1974), EEA77-36217.

343.Mathews, J. F., "Microscopy of internal crystal imperfections in Si p-n junction diodes by the use of electron beams", Appl. Phys. Lett., 3, 206-7 (1963).

344.Moss, S. C. and Graczyk, J. F., "Evidence of voids within the as-deposted structure of glassy silicon", Phys. Rev. Lett., 23, 167-71 (1969), PA73-14814.

345.Munakata, C., "Measurement of the homogeneity of a semiconductor with an electron beam", Jap. J. Appl. Phys., 4, 815 (1965).

346.Munakata, C., "Measurement of minority carrier lifetime with a non-ohmic contact and an electron beam", Microelectron. & Reliab., 5, 267-70 (1966).

347.Nagasirna, N., "Structure analysis of thermal oxide films of silicon by electron diffraction and infra-red absorption", Jap. J. Appl. Phys., 9, 875-88 (1970).

348.Orlov, P. N., Ukrainskii, Y. M., Galkin, B. D. and Skvortsov, K. F., "Quality of a gallium arsenide surface layer after abrasive finishing", Izv. Vyssh. Ucheb. Zuved., Mashinostr., no. 4, 158-61 (1973), Russian, CA79-24291.

349.Pchelyakov, O. P. and Stenin, S. I., "Application of high resolution replicas and diffraction electron microscopy to an investigation into the vaporization and growth of Si and Ge", Izv. Akad. Nauk SSSR Ser. Fiz., 38, 1518-21 (1974), Russian, PA77-73257.

350.Pirouz, P. and Boswarva, I. M., "Pseudo-Kikuchi pattern contrast from tilted crystals", Conference on Scanning Electron Microscopy: Systems and Applications 1973, p. 328-31, PA77-24375.

351.Prokhorov, V. I. and Sorokin, L. M., "Obtaining thin Si crystals and layers deposited on a Si surface for transilluminational electron microscopy", Instrum. & Exp. Tech., 16, 918-20 (1973), PA77-24359.

352.Schroder, B., "High resolution electron energy-loss spectroscopy of amorphous germanium and silicon films", AIP Conf. Proc., no. 20, 114-19 (1974), PA78-1170.

353.Sebeson, J. M., Hindermann, D. K. and Harrod, W. L., "Noncontact testing of interconnections in film integrated circuits using an electron beam", Proceedings of the 11th Annual Reliability Physics Conference 1973, p. 138-45, EEA77-6065.
354.Shimamoto, N., Ohtsuki, Y. H. and Fukamachi, T., "Exchange effects on Kikuchi band and abnormal absorption in electron diffraction", J. Phys. Soc. Jap., 33, 459-61 (1972), PA75-64631.
355.Shimaoka, G. and Chang, S. C., "Structure of silicon films evaporated onto a clean NaCl substrate", J. Vac. Sci. & Technol., 9, 235-8 (1972), PA75-39524.
356.Sittig, R. and Zimmermann, W., "Doping inhomogeneities in silicon observed by a high-resolution electroreflectance technique", Phys. Status Solidi A, 12, 663-70 (1972).
357.Soychak, F. J. and Warley, S. D., "E-beam or X-ray semiconductor in-line defect detection", IBM Tech. Disclosure Bull., 18, 339 (1975), PA78-69601.
358.Swaminathan, V., Narayanan, G. H. and Copley, S. M., "Diffuse streaks in the electron diffraction patterns from CdTe single crystals", J. Appl. Phys., 46, 3716-19 (1975).
359.Taheri, E. H. Z. and Hogarth, C. A., "Observations of the behaviour of some thin film cathodes in the scattering electron microscope", J. Non-Cryst. Solids, 15, 386-94 (1974), EEA77-38790.
360.Warminski, T., "Measurements of the electric field distribution on semiconductor surfaces using the electron mirror microscope (E.M.M.)", Phys. Status Solidi A, 8, K17-20 (1971), EEA75-4637.
361.Wittry, D. B. and Kyser, D. F., "Measurement of diffusion lengths in direct-gap semiconductors by electron beam excitation", J. Appl. Phys., 38, 375-82 (1967).
362.Yoshida, M. and Hirota, S., "Reflection electron diffraction of gold-diffused silicon", Jap. J. Appl. Phys., 11, 917-18 (1972), PA75-50463.

V. Ion Techniques
1. Probe

363.Anon., "Implantation and analysis by means of ionic beams", Ned. Tijdschr. Natuurk., 38 (19), 30-2 (1973), Dutch, EEA76-19684.
364.Colby, J. W., "Failure analysis using the ion microprobe", Proceedings of the 11th Annual Reliability Physics Conference, 1973, p. 194-7, EEA77-5791.
365.Colby, J. W. and Katz, L. E., "Determination of boron segregation coefficient in silicon by ion microprobe analysis", Presented at Electrochemical Society Meeting, San Francisco, Calif., May 1974, PA77-49596.
366.Colby, J. W., Miller, P. and Bridges, A. D., "Ion microprobe measurement of boron profile redistribution during phosphorus diffusion", Presented at Electrochemical Society Meeting, San Francisco, Calif., May 1974, EEA77-22883.

367. Gerber, R. M. and Dzimianski, J. W., "Use of an ion microprobe in semiconductor failure analysis", J. Vac. Sci. & Technol., 10, 1072-3 (1973), PA77-42954.
368. Liebl, H., "Ion probe microanalysis", Rev. Sci. Instrum., 8, 797-808 (1975).
369. Nagai, K. and Yamaguchi, M., "On improvements of the measurement accuracy in diffusion profiles of doped elements in semiconductors by IMMA", Jap. J. Appl. Phys., 13 (Suppl. 2), 395-8 (1974), EEA78-19663.

2. Field Ion

370. Block, J. H., Ernst, L. and Ernst, N., "Short note on 'observations of Si in the field of microscope'", Jap. J. Appl. Phys., 14, 1813-16 (1975).
371. Janssen, A. P. and Jones, J. P., "A study of the growth of germanium, and silicon, on tungsten, by field emission and field-ion microscopy", Surf. Sci., 41, 257-76 (1974), PA77-13110.
372. Mueller, E. W. and Tsong, T. T., "Field ion microscopy, field ionization, and field evaporation", Progr. Surface Sci., 4 (Pt. 1), 139 pp. (1973), CA80-138014.
373. Nishikawa, O. and Utsumi, T., "Field ion microscopic study of electric contact phenomena", J. Appl. Phys., 45, 672-6 (1974), EEA77-26400.
374. Van Oostrom, A., "Field ion microscopy", Acta Electron., 16, 59-71 (1973), French, EEA76-16641.

3. Backscattering

375. Allen, C. R. and Thomas, C. R., "Detection of phosphorus in heavily diffused silicon by He^+ backscattering", Electron. Lett., 9, 475-7 (1973), EEA77-2449.
376. Borders, J. A. and Picraux, S. T., "Characterization of silicon metallization systems using energetic ion backscattering", Proc. IEEE, 62, 1224-31 (1974), NSA31-5564.
377. Buck, T. M. and Wheatley, G. H., "Studies of solid surfaces with 100 keV $^4He^+$ and H^+ ion beams", Surface Sci., 33, 35-55 (1972), PA75-77485.
378. Caywood, J. M., "Measurement of the solubility of germanium in aluminum utilizing MeV He^+ backscattering", Metall. Trans., 4, 735-43 (1973), EEA77-2606.
379. Geist, V., Flagmeyer, R. and Otto, G., "Study of $A^{III}B^V$ semiconductor crystals using Rutherford backscattering of energetic protons", Krist. Tech., 10, 1115-33 (1975), German, CA83-200414.
380. Harrington, W. L., Honig, R. E., Goodman, A. M. and Williams, R., "Low-energy ion-scattering spectroscope (ISS) of the SiO_2/Si interface", Appl. Phys. Lett., 27, 644-45 (1975).

381.Meek, R. L. and Gibson, W. M., "Determination of light impurities on silicon by 90° forward ion scattering through thin targets", Nucl. Instrum. & Meth., 98, 375-7 (1972), PA75-16714.
382.Sachse, G. W., Miller, W. E. and Gross, C., "An investigation of RF sputter etched silicon surfaces using helium ion backscatter", Solid-State Electron., 18, 431-5 (1975), EEA78-19687.
383.Sigurd, D., "Metal-semiconductor interactions studied by ion backscattering techniques", Conference on Metal-Semiconductor Contacts, 1974, p. 141-59, PA78-18680.
384.Tsukamoto, K., Akasaka, Y. and Horie, K., "Microanalysis using backscattering method", J. Vac. Soc. Jap., 18, 15-22 (1975), PA78-58821.
385.Wood, D. R. and Morgan, D. V., "The characterization of GaAs surfaces", J. Electrochem. Soc., 122, 773-7 (1975), PA78-62548.

4. Mass Spectrometry

386.Assour, J. M., "Identification of chemical constituents of defects in silicon", J. Electrochem. Soc., 119, 1270-2 (1972), EEA75-39815.
387.Ban, V. S., "Mass spectrometric studies of vapor-phase crystal growth. II. GaN", J. Electrochem. Soc., 119, 761-5 (1972), PA75-53606.
388.Ban, V. S., "CVD of silicon studied by mass spectrometry", Presented at Electrochemical Society Meeting, San Francisco, Calif., May 1974, PA77-53622.
389.Brantley, W. A., Schartz, B., Keramidas, V. G., Kammlott, G. W. and Sinha, A. K., "Gallium migration through contact metallizations on GaP", J. Electrochem. Soc., 122, 434-5 (1975), EEA78-19711.
390.Clegg, J. B. and Millett, E. J., "Analysis of semiconductor surfaces by spark source mass spectrometry", Acta Electron., 18, 27-32 (1975), PA78-30231.
391.Colligon, J. S., "Surface compositional analysis using low energy ion bombardment induced emission processes", Vacuum, 24, 373-88 (1974), EEA78-11526.
392.Didenko, P. I., Litovchenko, V. G., Marchenko, R. I. and Romanova, G. F., "Study of Si-Al_2O_3 and Si-SiO_2-Al_2O_3 structures by mass spectrometry of the secondary ions", Poluprovodn. Tekh. & Mikroelektron., no. 18, 90-2 (1974), Russian, EEA78-11571.
393.Duhamel, A. P., "The analysis of thin films at Au/Pd/Ti utilizing an ion-beam surface mass analyzer", Microstructural Science, Vol. I. Proceedings of the 5th Annual Technical Meeting of the International Metallographic Society, 1972, p. 121-3, Publ. 1974, EEA78-11709.
394.Farrow. R. F., "The kinetics of silicon deposition on silicon by pyrolysis of silane. A mass spectrometric investigation by molecular beam sampling", J. Electrochem. Soc., 121, 899-907 (1974), PA77-70419.

395.Frantsuzov, A. A. and Makrushin, N. I., "Mass-spectrometer study of the initial stage of silicon oxidation", Sov. Phys. Tech. Phys., 20, 372-4 (1975).
396.Hall, R. N., "Chemical impurities and lattice defects in high-purity germanium", IEEE Trans. Nucl. Sci., NS-21, 260-72 (1974), CA80-13-325.
397.Kacerovsky, P., "Mass spectrometry of the surface of epitaxially grown GaP monocrystals", Elektrotech. Cas., 25, 386-9 (1974), Czech, EEA77-33927.
398.Lin, S. S., "Mass spectrometric analysis of vapors in oxidation of Si_3N_4 compacts", J. Amer. Ceram. Soc., 58, 160 (1975).
399.Lin, S. S., "Mass spectrometric analysis of vapor in chemical vapor deposition of alumina", J. Electrochem. Soc., 122, 1405-8 (1975).
400.Oda, S., Hirata, S., Kamada, H. and Kudo, K., "Determination of impurities in semiconductor materials by spark source mass spectrometry", Annu. Rep. Eng. Res. Inst. Fac. Eng. Univ. Tokyo, 30, 277-82 (1971), Japanese, PA75-16691.
401.Pikus, G. Y. and Shnyukov, V. F., "Secondary emission method of investigation of structural defects in Ge crystals irradiated by inert gas ions", Izv. Akad. Nauk SSSR Ser. Fiz., 37, 2609-12 (1973), Russian, PA77-20331.
402.Schulz, F., Wittmaack, K. and Maul, J., "Implications in the use of secondary ion mass spectrometry to investigate impurity concentration profiles in solids", Radiat. Eff., 18, 211-15 (1973), EEA76-35241.
403.Wittmaack, K., Maul, J. and Schulz, F., "A versatile in-depth analyzer", Presented at Electrochemical Society Meeting, San Francisco, Calif., May 1974, EEA77-26361.

VI. X-Ray Techniques
1. Topography

404.Abe, T., Abe, Y. and Chikawa, J., "The structure and origin of growth striations in dislocation-free Si crystals", Presented at Electrochemical Society Meeting, Chicago, Ill., May 1973, EEA76-23297.
405.Altshuler, V. M., Fokin, A. S., Pavlenko, Y. S. and Yudina, N. I., "X-ray diffraction investigation of dislocations at the film-substrate interface of autoepitaxial silicon films", Fiz. Tver. Tela, 13, 2827-9 (1971), Russian, PA75-1184.
406.Ando, Y., Kato, N. and Patel, J. R., "X-ray diffraction topographs of silicon crystals with superposed oxide film. III. Intensity distribution", J. Appl. Phys., 44, 4405-12 (1973), PA77-20501.
407.Arnaud, M., "The contribution of X-ray topography in characterizing silicon substrates and diffusion: correlations", Revue Tech. Thomson-CSF, 4, 528-44 (1972), French, EEA75-39819.

408. Authier, A., "X-ray topography as a tool in crystal growth studies", J. Cryst. Growth, 13-14, 34-8 (1972), PA75-47216.
409. Authier, A. and Patel, J. R., "X-ray topographic determination of the intrinsic or extrinsic nature of stacking faults", Phys. Status Solidi A, 27, 213-22 (1975), CA82-78935.
410. Authier, A., Simon, D. and Senes, A., "Correlation between lattice defects and electrical properties of silicon Zener diodes", Phys. Status Solidi A, 10, 233-9 (1972), PA75-33032.
411. Berger, H., "X-ray investigation of crystal perfection of GaAs single crystals and epitaxial layers", Kristall & Tech., 7, 403-18 (1972), German, PA75-74767.
412. Blet, G., "X-ray topography study of defects due to ion implantation", C. R. Hebd. Sean. Acad. Sci. B, 273, 983-4 (1971), French, PA75-9527.
413. Brown, A. S. and Springthorpe, A. J., "Defects in epitaxial gallium phosphide layers (p-n junctions)", Phys. Status Solidi A, 7, 495-502 (1971), PA75-4248.
414. Chikawa, J. and Fujimoto, I., "Video display technique for X-ray diffraction topography", NHK Tech. Monogr., no. 23, 3-20 (1974), EEA78-6679.
415. Chikawa, J., Fujimoto, I. and Abe, T., "X-ray topographic observation of moving dislocations in silicon crystals", Appl. Phys. Lett., 21, 295-8 (1972), PA75-77585.
416. Chikawa, J., Fujimoto, I., Endo, S. and Mase, K., "X-ray television topography for quick inspection of Si crystals", Presented at Electrochemical Society Meeting, Chicago, Ill., May 1973, EEA76-23321.
417. Chiki, Z., Shulpina, I. L. and Belokurova, I. N., "Growth defects in crystals of Ge-Si alloys", Izv. Akad. Nauk SSSR Ser. Fiz., 37, 2345-8 (1973), Russian, PA77-13575.
418. Ciszek, T. F., "Copper decoration and X-ray topography of point defects in dislocation-free silicon crystals grown under various conditions", Presented at Electrochemical Society Meeting, Chicago, Ill., May 1973, EEA76-23302.
419. Dashevsky, M. Y., Isaakyan, V. A. and Khatsernov, M. A., "The X-ray topographic investigation of defect distribution of dendritic silicon crystals along their cross section", Krist. & Tech., 8, 1453-5 (1973), PA77-32840.
420. DeKock, A. J. R., Roksnoer, P. J. and Boonen, P. G. T., "Microdefects in swirl-free silicon crystals", In: Semicond. silicon, Huff, H. R., (Ed.), Electrochem. Soc., Inc.: Princeton, N. J., p. 83-94 (1973), CA81-160355.
421. Fabre, E. and Schiller, C., "Optical and crystallographical characterization of gallium arsenide", Acta Electron., 15, 13 pp. (1972), EEA75-16617.
422. Fabre, E. and Schiller, C., "Correlation between the presence of dislocations and vacancies in n-type gallium arsenide. (Photoluminescence)", Solid State Commun., 10, 81-4 (1972), PA75-16607.

423. Fairfield, J. M. and Schwuttke, G. H., "The influence of crystallographic defects on device performance", J. Electrochem. Soc., 113, 1229-31 (1966).
424. Fuchs, E., "Two new analytical methods for materials research and component development", Elektroniker, 12 (9), EL9-EL12 (1973), German, PA77-18706.
425. Gajda, W. J., Jr., "Topographic examination of ion-implanted silicon", Proceedings of the National Electronics Conference, 1970, p. 229-33, EEA75-1038.
426. George, A., Escaravage, S., Schroter, W. and Champier, G., "Measurement of the speeds of displacement of dislocations in silicon by X-ray topography", Acta Crystallogr. A, A28, S161 (1972), French, PA76-15058.
427. Gerward, L., "Lattice defects and dynamical diffraction of X-rays in silicon", 4th Nordic Solid-State Conf., 1970, p. 10-13, PA73-75150.
428. Gerward, L., "Dynamical images of dislocations in X-ray diffraction topographs", Phys. Status Solidi A, 2, 143-8 (1970), German, PA71-5281.
429. Ghezzi, C. and Servidori, M., "X-ray study of heterogeneous nucleation of dislocations in phosphorus-diffused silicon", J. Mater. Sci., 9, 1797-802 (1974), CA82-66423.
430. Grieco, M. J., "Investigation of imperfections in silicon substrates using copper displacement technique", J. Electrochem. Soc., 121, 289-93 (1974), EEA77-14382.
431. Grienauer, H., "High-temperature x-ray topography of silicon", In: Semicond. silicon, Huff, H. R., (Ed.), Electrochem. Soc., Inc.: Princeton, N. J., p. 459-68 (1973), CA82-24538.
432. Grienauer, H. and Speck, W., "Non-destructive testing of crystals by X-ray topography", Siemens Rev., 39, 181-5 (1972), EEA75-23489.
433. Halliwell, M. A. G., Childs, J. B. and O'Hara, S., "Reflection X-ray topography of compound semiconductors", Proceedings of the 4th International Symposium on Gallium Arsenide and Related Compounds, 1972, p. 98-105, Publ. 1973, EEA76-23372.
434. Howard, J. K. and Smith, P. J., "Analysis of defect distribution in transistor structures with reflection and transmission X-ray topography", IBM J. Res. & Develop., 15, 123-31 (1971), EEA74-25564.
435. Itoh, N., Morikawa, Y., Nakau, T. and Kuroda, H., "X-ray topographic studies of strains in silicon implanted with ions at high doses", Jap. J. Appl. Phys., 14, 2069-70 (1975).
436. Jegerlehner, K. and Mohr, U., "X-ray topography of bent semiconductor slices", Kristall & Tech., 7, 303-12 (1972), German, PA75-39601.
437. Kannan, V. C., "Direct observation of motion of dislocations in deformed silicon by X-ray topography", Report UCRL-18673 (PB 183570), Univ. California, Berkeley, 1969.

438. Kawado, S. and Maruyama, T., "X-ray study of small dislocation loops in thermally oxidized silicon", J. Appl. Crystallogr., 5 (Pt. 4), 281-5 (1972), PA75-64669.
439. Kawamura, T. and Matsui, J., "X-ray diffusion and topographic study of silicon crystals nearly free from dislocations", 8th International Congress of Crystallography, 1969, p. 7-24, PA73-58816.
440. Kishino, S. and Iida, S., "Plane defect at the influence and dislocations in epitaxially grown GaAs", J. Electrochem. Soc., 119, 1113-18 (1972), EEA75-36149.
441. Kishino, S., Nakashima, H., Ito, R. and Nakada, O., "X-ray topographic study of dark-spot defects in gallium arsenide-gallium aluminum arsenide ($Ga_{1-x}Al_xAs$) double-heterostructure wafers", Appl. Phys. Lett., 27, 207-9 (1975), CA83-124306.
442. Kishino, S., Ogirima, M. and Kurata, K., "A cross-batch pattern in $GaAs_{1-x}P_x$ epitaxially grown on GaAs substrate", J. Electrochem. Soc., 119, 617-22 (1972), PA75-47110.
443. Kiyosawa, T., Igaki, K. and Ohashi, N., "Vapor phase crystal growth of zinc selenide under controlled partial pressure and its crystal structure", Trans. Jap. Inst. Met., 13, 248-54 (1972), PA75-64594.
444. Kozaki, S., Hashizume, H. and Kobra, K., "High-resolution video display of X-ray topographs with the divergent Laue method", Jap. J. Appl. Phys., 11, 1514 (1972), PA75-83922.
445. Lefeld-Sosnowska, M., Zielinska-Rohozinska, E. and Gronkowski, J., "X-ray diffraction investigations of defects generated in Si single crystals by thermal oxidation", Phys. Status Solidi A, 29, 535-43 (1975), PA78-62564.
446. Mader. S., Blakeslee, A. E. and Angilello, J., "The interpretation of dislocation contrast in X-ray topographs of $GaAs_{1-x}P_x$", J. Appl. Phys., 45, 4730-4 (1974), PA78-25396.
447. McFarlane, S. H., III and Wang, C. C., "Lang topographic studies of III-V heteroepitaxial films grown on sapphire and spinel", J. Appl. Phys., 43, 1724-32 (1972), PA75-32368.
448. Meleshko, L. I. and Sosnina, E. I., "X-ray topographic photography of single crystals using a conical cassette", Metallofizika, Kiev, no. 32, 92-6 (1971), Russian, PA75-43545.
449. Milevskii, L. S. and Smolskii, I. L., "Change in the mobility of dislocations during their emergence to the surface in crystals with high Peierls barriers", Fiz. Tverd. Tela, 16, 1028-31 (1974), Russian, CA81-7009.
450. Milvidskii, M. G., Osvenskii, V. B., Novikov, A. G., Fomin, V. G. and Grishina, S. P., "Influence of heat treatment on the structural perfection of single crystals of gallium arsenide alloyed with tellurium", Sov. Phys.-Crystallogr., 18, 519-20 (1974), PA77-48759.
451. Naukkarinen, K. and Tuomi, T. O., "X-ray topographic study of defects in selenium and tellurium single crystals", J. Appl. Phys., 40, 3054C (1969), PA72-44709.

452.Naukkarinen, K., Tuomi, T. O. and Korhonen, U., "Dislocation networks in vapour-grown selenium single crystals as seen by X-ray topography", Phys. Status Solidi A, 11, 327-34 (1972), PA75-50676.
453.Nishida, K., Matsui, J. and Nakajima, M., "Dislocation-free junctions formed by diffusing gallium and boron into Si", Jap. J. Appl. Phys., 14, 713-14 (1975), PA78-54964.
454.Patel, J. R. and Authier, A., "X-ray topography of defects produced after heat treatment of dislocation-free silicon containing oxygen", J. Appl. Phys., 46, 118-25 (1975), PA78-29269.
455.Patel, J. R. and Freeland, P. E., "Burgers vector of dislocations generated for dislocation velocity measurements in semiconductors", J. Appl. Phys., 41, 2814-17 (1970), PA73-58805.
456.Pierron, E. D., et al, "Defects studies of semiconductor materials", Met. Trans., 1, 639-42 (1970).
457.Posen, H. and Bruce, J. A., "Observation of slip distribution in alpha silicon carbide crystals", Presented at International Conference on Silicon Carbide, Miami Beach, Fla., September 1973, PA77-24482.
458.Renninger, M., "X-ray topographic studies on the lattice distortions of single crystals of silicon due to scratching", J. Appl. Crystallogr., 5 (Pt. 3), 163-9 (1972), PA75-43626.
459.Rozgonyi, G. A. and Haszko, S. E., "Reflection x-ray topography of GaAs and GaP cleavage faces", J. Electrochem. Soc., 117 (12), 1562-67 (1970), PA74-18624.
460.Rupprecht, H. and Schwuttke, G. H., "Distribution of boron-induced defects in shallow diffused surface layers of silicon", J. Appl. Phys., 37, 2862-6 (1966).
461.Scheihem, E. G. and Scott, V. Y., "X-ray examination of the regularity of the structure of plane single crystals of germanium", Bull. Acad. Sci. Phys. Ser., 33, 2016-19 (1969), PA73-65242.
462.Schiller, C., "Interface analysis by X-ray diffraction topography", Solid-State Electron., 13, 1163-6 (1970), EEA73-35167.
463.Schuhmann, B. and Schulz, M., "X-ray topographic investigations on highly doped gallium arsenide single crystals", Krist. & Tech., 9, 1029-39 (1974), German, PA77-82184.
464.Schwuttke, G. H., "In-process control of structural defects in semiconductor manufacturing", Symposium on Test Methods and Measurements of Semiconductor Devices, 1967, Pap. No. 501, EEA71-11011.
465.Schwuttke, G. H., "Damage profiles in silicon and their impact on device reliability", Report 008850, IBM Corp., East Fishkill Lab., Hopewell Junction, N. Y., 1975, 62 pp., CA83-171575.
466.Schwuttke, G. H. and Fairfield, J. M., "Dislocation in silicon due to localized diffusion", J. Appl. Phys., 37, 4394-6 (1966).
467.Schwuttke, G. H. and Howard, J. K., "X-ray stress topography of thin films on germanium and silicon", J. Appl. Phys., 39, 1581-91 (1968).

468.Segmuller, A., "X-ray diffraction topography of germanium wafers", IBM J. Res. Developm., 12, 448-57 (1968), EEA72-4908.
469.Shulpina, I. L., Polchkova, N. D. and Olkhovikova, T. I., "X-ray topographic study of dislocations introduced into indium antimonide", Izv. Akad. Nauk SWSR Ser. Fiz., 37, 2340-4 (1973), Russian, PA77-13574.
470.Toncheva, L. T. and Vassilev, I. S., "X-ray topographic studies of defects in silicon during thermal oxidation", C. R. Acad. Bulg. Sci., 27, 1045-8 (1974), PA78-18307.
471.Toncheva, L., Vassilev, I. S., Kassabov, J. and Nadjakov, E., "X-ray topographic study of the imperfection due to boron diffusion in silicon", C. R. Acad. Bulg. Sci., 23, 41-4 (1970), EEA73-27994.
472.Vassilev, I. S. and Toncheva, L. T., "X-ray diffraction topography of defects in silicon induced by localized boron diffusion", C. R. Acad. Bulg. Sci., 25, 731-4 (1972), PA75-77567.
473.Verma, A. R., "X-ray topographic studies of dislocations in silicon carbide crystals", Presented at International Conference on Silicon Carbide, Miami Beach, Fla., September 1973, PA77-24481.
474.Walford, L. K. and Carron, G. J., "Analytical descriptions of observed stress patterns of X-ray topographs of silicon", J. Appl. Phys., 39, 5802-4 (1968), MA2-04-12 0452.
475.Wang, P., "In process measurement of structural defects in silicon by X-ray topography", Solid State Technol., 12 (1), 25-33 (1969), EEA72-12807.
476.Wattenberg, U., "Comparative investigation of X-ray topographic methods for illustrating crystal defects", J. Appl. Crystallogr., 5 (Pt. 3), 169-77 (1972), German, PA75-47405.
477.Wonsiewicz, B. C. and Patel, J. R., "Computer simulation of X-ray topographs of stacking faults in silicon", J. Appl. Crystallogr., 8 (Pt. 1), 67-8 (1975), PA78-18325.
478.Zeyfang, R., "X-ray diffraction study of imperfections in epitaxial silicon on sapphire", Thin Solid Films, 6, 321-8 (1970), PA74-28283.

2. Borrmann Effect

479.Borodina, T. I., Katznelson, A. A. and Kissin, V. I., "Multiple-beam Borrmann effect in perfect crystals of germanium and silicon", Phys. Status Solidi A, 3, 105-10 (1970), PA74-5169.
480.Feldman, R. and Post, B., "Absorption and excitation in multiple Borrmann diffraction", Phys. Status Solidi A, 12, 273-6 (1972), PA75-60736.
481.Haber, G. and Post, B., "Absorption effects in perfect germanium crystals", Presented at American Crystallographic Association Summer Meeting, Ames, Iowa, August 1971, PA75-3711.
482.Huang, T. C., "Comment on 'multiple-beam Borrmann effect in perfect crystals of germanium and silicon' by T. I. Borodina et al", Phys. Status Solidi A, 10, K149-50 (1972), PA75-39602.

483. Iveronova, V. I., Katsnelson, A. A., Borodina, T. I. and Sapkova, I. G., "Some peculiarities of the multiple-beam Borrmann effect in germanium", Phys. Status Solidi A, 11, 39-44 (1972), PA75-47330.
484. Kyutt, R. N. and Efimov, O. N., "Temperature dependence of the anomalous transmission of X-rays in silicon", Fizika Tverdogo Tela, 11, 283-9 (1969), Russian, MA2-06-32 0374.
485. Melnichuk, I. V., Nikulitsa, V. G. and Rarenko, I. M., "Anomalous transmission of X rays in CdSb single crystals", Phys. Status Solidi A, 25, K173-K176 (1974).
486. Melnichuk, I. V., Nikulitsa, V. G. and Rarenko, I. M., "Determination of the disturbed surface layer thickness of CdSb by the method of integral characteristics of X-ray anomalous transmission", Ukr. Fiz. Zh., 20, 115-18 (1975), Russian, PA78-25554.
487. Miller, D. C. and Witt, A. F., "Analysis of growth related strain in Czochralski grown indium antimonide using X-ray anomalous transmission topography", J. Cryst. Growth, 29, 19-28 (1975), EEA78-33022.
488. Novikova, L. P., Stafeev, V. I. and Milevskii, L. S., "Investigation of structure defects in indium arsenide by the method of anomalous X-ray transmission", Fiz. & Tekh. Poluprov., 5, 2164-7 (1971), Russian, PA75-12770.
489. Sheikhet, E. G. and Efimov, O. N., "Photoelectric absorption and diffuse scattering in the anomalous transmission of X-rays in Ge crystals", Sov. Phys.-Solid State, 14, 625-7 (1972), PA75-77584.
490. Sheikhet, E. G., Efimov, O. N. and Veselovskaya, N. V., "Anomalous transmission of X-rays by silicon single crystals containing dislocations", Sov. Phys.-Solid State, 14, 223-4 (1972), PA75-71496.
491. Thomas, J. E., Stuber, A. B. and Dearlove, J. W., "X-ray study of lithium precipitation in germanium", J. Appl. Phys., 43, 251-3 (1972), PA75-9813.
492. Umeno, M., "An effect of simultaneous reflection on anomalous transmission patterns of germanium", Phys. Status Solidi A, 11, 501-11 (1972), PA75-53662.
493. Wenzl, H. F., "Investigation of lithium precipitates in germanium crystals by anomalous X-ray transmission", Z. Naturforsch. A, 26A, 495-501 (1971), PA75-19093.

3. Orientation Determination

494. Hashizume, H. and Kohra, K., "Effects of lattice distortion on equal inclination fringes in the asymmetric diffraction of X-rays", Phys. Status Solidi A, 13, K9-11 (1972), PA75-77532.
495. Keppler, U., Meier, M., Neifeind, A. and Rohmer, C., "X-ray crystal orientation determination for semiconductor single crystals", Krist. & Tech., 10, 79-84 (1975), EEA78-11506.

496.Korolev, V. E., Litvinov, Y. M. and Malyukov, B. A., "Determination of the crystallographic polarity of gallium phosphide single crystals", Izv. Sib. Otd. Akad. Nauk SSSR, Ser. Khim. Nauk, no. 4, 118-21 (1975), Russian, CA83-171126.
497.Lambert, V. L., "X-ray examination of epitaxial PbSnTe and PbTe", J. Appl. Phys., 46, 2304-5 (1975), PA78-61938.
498.Persson, E. G., "The Laue case of X-ray diffraction in nearly perfect germanium crystals: the 111 and 440 reflections", Phys. Scripta, 5, 145-9 (1972), PA75-71403.
499.Piotrowski, T., "X-ray method for determining orientation of GaAs crystals for quantum electronics purposes", Arch. Elektrotech, 17, 677-80 (1968), EEA72-11203.
500.Posen, H., "An atlas of Laue patterns for SiC polytypes", Presented at International Conference on Silicon Carbide, Miami Beach, Fla., September 1973, PA77-24451.
501.Scheidegger, R. and Galetti, G., "The orientation of single crystals by the Laue reflection method", Brown Boveri Rev., 53, 48-60 (1966), EEA70-10419.
502.Wang, C. C. and Cullen, G. W., "Epitaxial orientation effects of III-V semiconductors grown on dielectric oxides", Electrochemical Society Fall Meeting (Extended abstracts), 1974, p. 334-5, PA78-69616.

4. Stress Analysis

503.Bakutis, J., Valacka, K. and Venisloviene, R., "Internal macrostresses in CdTe thin films", Liet. Fiz. Rinkinys, 13, 415-20 (1973).
504.Cohen, B. G. and Focht, M. W., "X-ray measurement of elastic strain and annealing in semiconductors", Solid State Electron., 13, 105-12 (1970), EEA73-14058.
505.Ishida, K., Matsui, J., Kamejima, T. and Sakuma, I., "X-ray study of $Al_xGa_{1-x}As$ epitaxial layers", Phys. Status Solidi A, 31, 255-62 (1975), PA78-81327.
506.Kamins, T. I. and Meieran, E. S., "X-ray measurements of stress in thin single-crystal silicon films", J. Appl. Phys., 44, 5064-6 (1973), PA77-29006.
507.Louzon, T. J. and Spencer, T. H., "X-ray diffraction stress measurements in thin films (on substrates)", Solid State Technol., 18, 25-8 (1975), EEA78-37599.
508.Nagai, H., "Structure of vapour-deposited $Ga_xIn_{1-x}As$ crystals", J. Appl. Phys., 45, 3789-94 (1974), PA78-6664.
509.Rozgonyi, G. A. and Ciesielka, T. J., "X-ray determination of stresses in thin films and substrates by automatic Bragg angle control", Presented at Electrochemical Society Meeting, Chicago, Ill., May 1973, EEA76-23336.
510.Zeyfang, R., "Residual stresses in thin single crystals bonded to an amorphous substrate: silicon-integrated circuits", J. Appl. Phys., 42, 1182-5 (1971), PA74-30837.

5. Others

511. Anon., "X-ray at line speed", Tool. & Prod., 39 (3), 46-7 (1973), EEA77-9303.
512. Armigliato, A., Bergamini, P. and Morettini, L., "Thin films and bulk specimen X-ray microanalysis in the electron microscope", 9th Annual Conference of the Microbeam Analysis Society, 1974, 15A/3 pp., PA77-77294.
513. Auleytner, J., Furmanik, Z. and Krylow, J., "Application of the moving-slit X-ray automonochromatization method in structural studies of planar diodes and an attempt to correlate electrical properties with lattice defects", Acta Phys. Pol. A, A41, 507-21 (1972), PA75-47944.
514. Baldin, S. A. and Ioannesyants, L. M., "X-ray analyzer with a Ge(Li)-detector", Instrum. & Exp. Tech., 14, 424-6 (1971), PA75-24089.
515. Beezhold, W., "Ion-induced characteristic X-ray microanalysis", Presented at Electrochemical Society Meeting, Chicago, Ill., May 1973, EEA76-23320.
516. Belugina, N. V., Bublik, V. T., Gorelik, S. S. and Shvartsman, V. L., "Precision determination of the lattice constant of germanium using a widely-divergent x-ray beam", Zavod. Lab., 36, 306-7 (1970), Russian, MA3-08-21 0493.
517. Belyatskaya, N. S., Grishina, S. P., Lopatin, E. V., Milvidskii, M. G., Ostvenskii, V. B. and Fomin, V. G., "Structural properties of single crystals of gallium arsenide heavily doped with donor impurities", Sov. Phys.-Crystallogr., 17, 126-31 (1972), PA75-71432.
518. Blokhin, M. A., Zoinmer, G., Volkov, V. F. and Monastyrsky, L. M., "X-ray L-spectra of germanium", Fiz. Tver. Tela, 11 (1), 17-21 (1969), Russian, MA2-05-32 0340.
519. Boehm, J. M., Prager, P. R. and Barnea, Z., "The effect of surface damage on the intensity of X-rays diffracted by ground spherical single crystals", Acta Crystallogr. A, A30 (Pt. 3), 335-7 (1974), PA77-44786.
520. Bonefacic, A., "X-ray study of diffuse streak diffraction pattern from silicon crystals", Acta Cryst., A25 (5), 589-94 (1969), MA2-12-12 1629.
521. Brodsky, M. H., Title, R. S. and Weiser, K., "Structural, optical and electrical properties of amorphous silicon films", Phys. Rev. B, 1, 2632-41 (1970), PA73-47761.
522. Burge, R. E. and Smart, J. E., "Inelastic electron scattering and incoherent X-ray scattering in silicon at small angles", Phil Mag., 17, 1285-8 (1968), MA2-07-15 0626.
523. Chikawa, J. and Oguro, Y., "Quick observation of X-ray diffraction images for inspecting semiconductor devices", J. Soc. Instrum. & Contr. Eng., 11, 148-54 (1972), Japanese, EEA75-16612.
524. Daweritz, L. and Dornics, M., "The influence of stoichiometry on the structure of vacuum deposited CdSe films", Phys. Status Solidi A, 20, K37-9 (1973), PA77-6572.

525.Deslattes, R. D., "Optical and x-ray interferometry of a silicon lattice spacing", Appl. Phys. Lett., 15, 386-8 (1969).
526.Donaghey, L. F. and Bissinger, R. H., "Characterization of GaAs substrates and epitaxial $GaAs_{1-x}P_x$ layers by divergent X-ray beam diffraction", J. Electron. Mater., 4, 131-58 (1975), PA78-38386.
527.Dove, D. B., Hence, L. L., Gould, K. W. and Loehman, R. F., "Structure, properties and radiation sensitivity of electrically bistable materials (Annual Report, 6 Feb 1970 - 5 Feb 1971)", Report TR-2, Univ. Florida, Gainesville, 1971, 58 pp., PA75-1516.
528.Bublik, V. T., Gorelik, S. S. and Dubrovina, A. N., "Strongly doped germanium and silicon", Sov. Phys. Solid State, 10, 2247-9 (1969), MA2-06-21 0280.
529.Ejder, E., "Growth and morphology of GaN", J. Cryst. Growth, 22, 44-6 (1974), PA77-45478.
530.Fripp, A. L., Stermer, R. L. and Catlin, A., "Structure of silicon films deposited on oxidized silicon wafers", J. Electrochem. Soc., 117, 1569-71 (1970), PA74-15949.
531.Hashizume, H., Ishida, H. and Kohra, K., "Improvement in the visibility of crystal Moire fringes with the use of a parallel X-ray beam", Phys. Status Solidi A, 12, 453-63 (1972), PA75-71404.
532.Houska, C. R., "Nondestructive atomic diffusion studies using X-ray diffraction - experimental considerations", High Temp. - High Pressures, 4, 417-29 (1972), EEA76-27662.
533.Jakimavicius, J. and Batarunas, J., "Use of structural amplitudes of dispersions of X rays for studying the zoned structure in Si and CdS", Liet. Fiz. Rinkinys., 13, 731-43 (1973).
534.Jennings, L. D., "Measurement of the integrated X-ray intensities of Ge and Si", J. Appl. Phys., 40, 5038-44 (1969).
535.Kikuta, S. and Kohra, K., "Variation with thickness in the profile of Laue case diffraction curve of X-rays from a thin silicon crystal", J. Phys. Soc. Japan, 25, 924 (1968), MA2-03-21 0152.
536.Lefeld-Sosnowska, M. and Malgrange, C., "Observation of oscillations in rocking curves of the Laue reflected and refracted beams from thin silicon single crystals", Phys. Status Solidi, 30, K23-5 (1968), MA2-03-21 0136.
537.Lisoivan, V. I. and Dikovskaya, R. R., "Local precision determination of lattice constants of a silicon single crystal", Instrum. Exper. Tech., 4, 992-4 (1969), PA73-50499.
538.Malyukov, B. A., Ukransky, Y. M. and Korolev, V. E., "X-ray spectral determination of the thickness of the oxide film on silicon discs", Zavod. Lab., 34, 678-80 (1968), Russian, MA2-07-22 0503.
539.Novikov, N. N. and Khimenko, M. V., "X-ray investigation of hole injections in germanium dislocations", Metallofizika, Kiev, no. 48, 93-9 (1973), Russian, PA77-32569.
540.Olkhovikova, T. L. and Shulpina, I. L., "An X-ray investigation of the defects in indium antimonide single crystals", Izv. Akad. Nauk SSSR Ser. Fiz., 36, 563-6 (1972), Russian, PA75-50675.

541.Oyama, Y., "Solid solution in the ternary system, Si_3N_4-AlN-Al_2O_3", Jap. J. Appl. Phys., 11, 760-1 (1972), PA75-43823.
542.Park, Y. H. and Lau, B. W., "X-ray studies on the thermal expansion of silicon single crystal", J. Korean Phys. Soc., 4, 9-12 (1971), PA75-47757.
543.Patel, J. R., "X-ray diffuse scattering from silicon containing oxygen clusters", J. Appl. Cyrstallogr., 8 (Pt. 2), 186-91 (1975), PA78-41745.
544.Pruett, H. D. and Lien, S. Y., "X-ray imaging technique for observing liquid-encapsulation Czochralski crystal growth", Presented at Electrochemical Society Meeting, Chicago, Ill., May 1973, EEA76-23269.
545.Rek, Z., "X-ray study of the effect of hydrostatic pressure on the dislocation structure of silicon single crystals", Acta Phys. Pol. A, A41, 635-40 (1972), PA75-47419.
546.Renner, O., "Density measurements of thin germanium films by total reflection of X-rays", Czech. J. Phys. B, 22 (10), 1007-16 (1972), PA75-74630.
547.Schiller, C. and Jarrousse, B., "Anomalous X-ray scattering by semiconductors of the zinc-blende structure", J. Appl. Crystallogr., 1, 117-18 (1968), French.
548.Sheikhet, E. G. and Shkot, V. Y., "X-ray diffraction study of the structural perfection of plane germanium single crystal", Izvest. Akad. Nauk Fiz., 33, 2016-19 (1969), Russian, MA3-06-12 0533.
549.Shulpina, I. L. and Kovev, E. K., "X-ray diffraction study of defects formed in silicon during deep diffusion of phosphorus", Fiz. Tverdogo Tela., 9, 1008-14 (1967), Russian, EEA70-13966.
550.Sonomura, H., Nanmori, T., Saito, S. and Miyauchi, T., "Composition determination of $Al_xGa_{1-x}P$ alloys using diffraction X-ray intensities", J. Appl. Phys., 45, 5109-11 (1974), PA78-22490.
551.Uno, R., Okano, T. and Yukino, K., "Electron distribution in gallium arsenide as revealed by the X-ray diffraction", J. Phys. Soc. Japan, 28, 437-42 (1970), PA73-30872.
552.Willoughby, A. F. W., Driscoll, C. M. H. and Bellamy, B. A., "A study of non-stoichiometry in gallium arsenide by precision lattice parameter measurements", J. Mater. Sci., 6, 1389-96 (1971), PA75-6434.

VII. Infrared Techniques

553.Baimlet, M., "Application of i.r. radiation to nondestructive testing of materials and components", Jemna Mech. & Opt., 19, 132-7 (1974), Czech, EEA77-33508.
554.Baker, J. A., "Determination of parts per billion of oxygen in silicon", Solid-State Electron., 13, 1431-4 (1970), EEA74-1416.
555.Baranova, E. K., "Infrared absorption-spectroscopy study of the formation of defects in Si irradiated with ions of different masses", Sov. Phys.-Semicond., 9, 763-4 (1975).

556.Black, J. F., Lanning, E. and Perkowitz, S., "Infrared techniques for semiconductor characterization", Infrared Phys., 10, 125-39 (1970), PA73-72055.
557.Bobrova, V. S., Vavilov, V. S. and Galkin, G. N., "Reflection of light in the infra-red region from silicon with a high concentration of non-equilibrium carriers", Fiz. Tver. Tela, 12, 1232-35 (1970), Russian, MA3-08-32 0524.
558.Chen, C. S. and Corelli, J. C., "Infrared spectroscopy of divacancy-associated radiation-induced absorption hands in silicon", Phys. Rev. B, 5, 1505-17 (1972), PA75-22777.
559.Cohen, B. G., "Infrared microscopy - some applications to solid state devices", Solid State Technol., 17 (7), 36-40 (1974), EEA77-38866.
560.Demeshina, A. I., Korohazhkina, R. L., and Kuznetsova, N. N., "Long-wavelength infrared absorption in n-type Ge due to the interaction of impurities", Sov. Phys. Semicond., 4, 363-4 (1970), MA3-11-32 0755.
561.Fischer, B. and Marschall, N., "Resonant excitation of plasmons in GaAs", Proc. of 10th Int. Conf. on the Phys. of Semicond., 1970, p. 845-7, PA74-59319.
562.Fitzpatrick, J. R. and Conn, G. K. T., "Non-destructive measurement of carrier concentration in epitaxial gallium arsenide", Brit. J. Appl. Phys., 2, 1407-9 (1969), MA3-04-22 0198.
563.Gardner, E. E. and Schumann, P. A., "Measurement of diffused impurity concentration", IBM Tech. Disclosure Bull., 11, 300 (1968), EEA72-705.
564.Gauster, W. B. and Bushnell, J. C., "Laser-induced infrared absorption in silicon", J. Appl. Phys., 41, 3850-3 (1970), PA73-69129.
565.Genkin, G. M. and Zilberberg, V. V., "Raman scattering of light in semiconductors", Fiz. Tekh. Poluprov., 3, 1506-11 (1969), Russian, MA3-03-32 0133.
566.Griffin, D. D., "Infrared techniques for measuring temperature and related phenomena of microcircuits", Appl. Opt., 7, 1749-56 (1968).
567.Gross, C. and Gaetano, G., "Comparison of infrared and activation analysis results in determining the oxygen and carbon content in silicon", J. Electrochem. Soc., 19, 926-9 (1972), EEA75-36139.
568.Hall, R. N., "Chemical impurites and lattice defects in high-purity germanium", IEEE Trans. Nucl. Sci., NS-21, 260-72 (1974), EEA77-26315.
569.Keenan, W. A. and Poponiak, M. R., "Infrared television microscope study of processed semiconductor materials", Presented at Electrochemical Society Meeting, Chicago, Ill., May 1973, EEA76-23325.

570.Kolbesen, B. O., "Simultaneous determination of the total content of boron and phosphorus in high-resistivity silicon by i.r. spectroscopy at low temperatures", Appl. Phys. Lett., 27, 353-5 (1975).
571.Leung, P. C., Skolnik, L. H., Allred, W. P. and Spitzer, W. G., "Infrared absorption study of Li-diffused Mg-doped GaAs", J. Appl. Phys., 43, 4096-104 (1972), PA75-81360.
572.Lihl, F., Tomiser, J., Skalicky, P. and Kuster, M., "Observation of dislocations in silicon by stress-birefringence", Z. Angew. Phys., 32, 287-91 (1971), German, PA75-15763.
573.Pajot, B., "Improvement in the detection of oxygen in silicon by infra-red absorption", Solid-State Electron., 12, 923-5 (1969), EEA73-10393.
574.Palik, E. D., "Magneto-optical studies of semiconductors in the far infrared", Presented at 1972 Spring Meeting of the Optical Society of America, New York, April 1971, PA75-36541.
575.Pasechnik, Y. A., Snitko, O. V., Getsko, O. M., Romanenko, V. F. and Sergeev, O. T., "Study of SiO_2-SiC dielectric-semiconductor structure by infra-red spectroscopic techniques", Poluprovodn. Tekh. & Mikroelektron., no. 18, 35-9 (1974), Russian, EEA78-11567.
576.Peterman, D. and Workman, W., "Infrared radiometry of semiconductor devices", Microelectron. & Reliab., 6, 307-15 (1967).
577.Pliskin, W., "Use of infrared spectroscopy for the characterization of dielectric films on silicon", Presented at Electrochemical Society Meeting, Chicago, Ill., May 1973, EEA76-23326.
578.Schulz, M. and Sarodnik, R., "Defect structure treatments on GaAs single crystals with the aid of IR microscopy", Krist. & Tech., 8, 1413-20 (1973), German, PA77-32568.
579.Riney, T. O., "Residual thermoelastic stress in bonded silicon wafers", J. Appl. Phys., 32, 454-60 (1961).
580.Schanda, J. and Sass, G., "On some problems of IR spectroscopy of semiconductors", Kozl. Magy. Tnd. Akad. Musz, Fis. Kat. Intez., 3, 33-44 (1971), Hungarian, PA74-59311.
581.Sedgwick, T. O. and Smith, J. E., Jr., "Raman scattering spectroscopy applicated to the study of the chemical vapor deposition systems", Electrochemical Society Fall Meeting (Extended abstracts), 1974, p. 520-2, PA78-73321.
582.Sher, A. H., "Improved infrared response technique for detecting defects and impurities in germanium and silicon phi-n diodes", Report NBS-SP-400-13, Nat. Bur. Stand., Washington, D. C., 1975, 20 pp., EEA78-37502.
583.Sher, A. H. and Keery, W. J., "Improved infrared-response technique for determining impurity and defect levels in semiconductors", Appl. Phys. Lett., 20, 120-2 (1972), EEA75-10196.
584.Skolnik, L. H., "Application of infrared active localized vibrational modes to the investigation of impurity incorporation in semiconductors", Thesis, Univ. Southern California, Los Angeles, 202 pp., Order No. 72-6109, PA75-47433.

585. Stahescu, D., "The study of the defects of monocrystalline silicon by infrared transmission", Stud. & Cercet Fiz., 25, 991-9 (1973), Romanian, PA77-21211.
586. Sunshine, R. A., "Infrared observation of current distributions in large area power transistors", 1973 8th Annual Meeting of the IEEE Industry Applications Society, p. 637-40, EEA77-10238.
587. Sunshine, R. A. and Goldsmith, N., "Infrared transmission microscopy utilizing a high-resolution video display", RCA Rev., 33, 383-92 (1972), EEA75-36165.
588. Szep, I. C. and Barsony, I., "Determination of the oxygen content in thin silicon slices", Hiki, 9, 5-12 (1969), Hungarian, EEA73-4289.
589. Taksau, S. I. and Matsushita, Y., "Induced defects in crystals through mechanical processing", Oyo Buturi, 44 (Suppl.), 259-68 (1975), EEA78-41489.
590. Tkachev, V. D. and Lappo, M. T., "Investigation of divacancies in silicon. (IR absorption and photoconductivity)", Presented at International Conference on Radiation Effects in Semiconductors, Albany, New York, August 1970, PA75-48194.

VIII. Neutron Techniques

591. Barcus, G. D., "Neutron radiography", 28th Annual Technical Conference Transactions of the American Society for Quality Control, 1974, p. 376-81, EEA78-83.
592. Bereznai, T., DeCorte, F. and Hoste, J., "Neutron activation analysis of semiconductor silicon", Radiochem. & Radioanal. Lett., 17, 279-28 (1974), PA78-19092.
593. Boganch, Y., Kvitter, P. and Sabo, E., "Non-destructive neutron activation analysis for routine testing of impurities in semiconductor silicon", Atommaya Energiya, 24, 421-4 (1968), Russian, EEA71-23218.
594. Bujdoso, E., Miskei, M. and Ormos, G., "Determination of some trace impurities in gallium and gallium arsenide by neutron activation analysis", Proc. Physico-Chem. Methods in Chem. Analysis, Volume 2, 1968, p. 262-9, MA2-01-23 0055.
595. Chalmeton, V., "Neutron radiography", Acta Electron., 16, 73-84 (1973), EEA76-16642.
596. Doi, K., Minakawa, N., Motohashi, H. and Masaki, N., "A trail of neutron diffraction topography", J. Appl. Crystallogr., 4 (Pt. 6), 528-30 (1971), PA75-3710.
597. Eichhorn, F., "Neutron and X-ray diffraction investigations of silicon implanted by phosphorus ions", Phys. Status Solidi A, 30, 155-62 (1975), EEA78-33036.
598. Eichhorn, F., Kosmowski, M., Schopf, H. G. and Schulz, G. E., "Dynamical neutron diffraction from germanium crystals containing randomly distributed dislocations of low density", Acta Cryst., A25 (Suppl.), 204 (1969), PA73-58815.

599.Hewitt, J. S., Aydogdu, M. K., Blumenauer, G. R. and Robitaille, H. A., "Novel techniques in neutron radiography with a 35 MeV linear electron accelerator", Non-Destr. Test., 7, 315-23 (1974), EEA78-2226.

600.Jaskolska, H., Walis, L. and Rowinska, L., "A study of purity of semiconductor silicon plates, and the determination of concentration profiles of gallium and phosphorus by neutron activation and autoradiography", Radiochem. & Radioanal. Lett., 20, 229-39 (1975), EEA78-19684.

601.Kikuta, S., Kohra, K., Minakawa, N. and Doi, K., "An observation of neutron Pendellosung fringes in a wedge-shaped silicon single crystal", J. Phys. Soc. Jap., 31, 954-5 (1971), PA75-9429.

602.Kim, C. K. and Voris, S. S., "Halogen impurities in CdS and ZnSe single crystals by thermal neutron activation", J. Phys. & Chem. Solids, 35, 1381-4 (1974), EEA77-36224.

603.Krasivina, L. E., Zverev, B. P., Simakhin, Y. F., Usmanova, M. M. and Zhumaev, N., "A neutron-activation-analysis investigation of the distributions of antimony and phosphorus introduced into silicon by ionic implantation", Ind. Lab., 40, 1444-5 (1974), EEA78-28620.

604.Kudo, K., "Application of radioactiviation analysis for the determination of trace amounts of elements in semiconductor materials", Oyo Buturi, 44 (Suppl.), 319-27 (1975), EEA78-41492.

605.Larrabee, G. B., "Neutron activation analysis of semiconductors", Presented at 10th National Meeting of the Society for Applied Spectroscopy, St. Louis, Mo., October 1971, PA75-16715.

606.Lorch, E., "Conventional and elastic neutron diffraction from vitreous silica", J. Phys. C., 3, 1314-22 (1970), PA73-55684.

607.Lunde, G., "Analysis of surface and bulk impurities in silicon single crystal slices by neutron activation", Solid State Technol., 13 (1), 61-5 (1970), EEA73-21070.

608.Martin, J. A., "Neutron activation and autoradiography of silicon", In: Semiconductor silicon, Electrochem. Soc., Inc., New York, p. 547-57 (1969).

609.Michalec, R., Sedlakova, L., Chalupa, B., Cech, J. and others, "Neutron diffraction by flexural vibrating single crystals of Si, Ge, and SiO_2 in Bragg case", Phys. Status Solidi A, 23, 667-73 (1974).

610.Nagy, L. G., "Nondestructive and destructive neutron activation investigation of impurities in semiconductor purity silicon", Proc. Physico-Chem. Methods in Chem. Analysis, Budapest, Volume 2, p. 341-8 (1968), MA2-01-23 0059.

611.Nagy, L. G., Schiff, E., Szokolyi, L. and Nyeste, G., "Study of the efficiency of radiochemical separation methods used in the neutron activation analysis of high purity germanium", Proc. Physico-Chem. Methods in Chem. Analysis, Volume 2, 1968, p. 446-54, MA2-01-23 0062.

612. Nilsson, G. and Nelin, G., "Study of the homology between silicon and germanium by thermal-neutron spectrometry", Phys. Rev. B, 6, 3777-86 (1972), PA75-84190.
613. Nunes, A. C., "A neutron diffraction search for noncentrosymmetric thermal oscillations in germanium and silicon", Acta Cryst., A25 (Suppl.), 32 (1969), PA73-23507.
614. Perezhogin, G. A. and Meshcheryakov, V. G., "Application of neutron activation analysis to the investigation of dopants distribution in epitaxial silicon films", Ind. Lab., 35, 979-80 (1970), MA2-12-21 0660.
615. Schmidt, P. F. and Adda, L. P., "Interface-state generation by gold diffusion through SiO_2 films on silicon: MOS and neutron-activation results", J. Appl. Phys., 45, 1826-33 (1974), EEA77-30219.
616. Wolfstirn, K. B., "Survey analysis of impurities in gallium arsenide by instrumental neutron activation", J. Phys. Chem. Solids, 31, 601-12 (1970), MA3-10-23 0515.

IX. Laser Techniques

617. Black, J. F. and Sherman, B., "Investigation of semiconductor wafers using a 3.39 micron laser-raster microscope", Messtechnik, 79, 61-5 (1971), German, EEA74-18596.
618. Black, J. F., Sherman, B. and Fowler, V., "Examination of semiconductor wafers with a scanned laser infrared microscope", In: Semiconductor silicon, Electrochem. Soc., Inc., New York, p. 693-705 (1969), PA73-18623.
619. Harris, W. J. and Woods, D. C., "Thermal stress studies using optical holographic interferometry (NDT)", Mater. Eval., 32 (3), 50-6 (1974), EEA77-35853.
620. Mohaupt, G. and Patzmann, G., "A few examples of applying laser microspectral analysis in semiconductor engineering", Jena Rev., 19, 252-6 (1974), EEA78-8463.
621. Oswald, D. R. and Munro, D. F., "A laser scan technique for electronic materials surface evaluation", J. Electron. Mater., 3, 225-42 (1974), PA77-27020.
622. Sbar, N. L., "The application of laser scanning to detect detection in unexposed high-resolution photographic plates", 1973 IEEE/OSA Conference on Laser Engineering and Applications Digest of Technical Papers, p. 58, EEA77-10308.
623. Seman, J. A. and Wiley, J. D., "A scanning device for measurement of doping homogeneity in GaP", J. Appl. Phys., 43, 4646-50 (1972), EEA76-4510.
624. Sugawara, K., Yoshimi, T., Okuyama, H. and Shirasu, T., "In situ monitoring of film deposition using He-Ne laser system. I. Measurements of CVD insulating film at 6328 A", J. Electrochem. Soc., 121, 1233-5 (1974), EEA78-547.

625.Sugawara, K., Yoshimi, T. and Okuyama, H., "In situ monitoring of film deposition using He-Ne laser system. II. Measurements of polycrystalline silicon film in the infrared", J. Electrochem. Soc., 121, 1235-7 (1974), EEA78-548.
626.Williams, R. and Woods, M. H., "Laser-scanning photoemission measurements of the silicon-silicon dioxide interface", J. Appl. Phys., 43, 4142-7 (1972), PA75-81119.

X. Etching Techniques

627.Angilello, J., Potemski, R. M. and Woolhouse, G. R., "Etch pits and dislocations in (100) GaAs wafers", J. Appl. Phys., 46, 2315-16 (1975), PA78-61703.
628.Antonov, P. I., Grigorev, N. S. and Stepanov, A. V., "Deflection of the orientation of profiled germanium crystals, drawn from the melt", Izv. Akad. Nauk SSSR, Ser. Fiz., 36, 486-93 (1972), Russian, PA75-47425.
629.Assour, J. M., "Identification of chemical constituents of defects in silicon", J. Electrochem. Soc., 119, 1270-2 (1972), PA75-77913.
630.Boinykh, N. M., Sokolov, A. M., Indenbaum, G. V. and Vanyukov, A. V., "Structure of cadmium mercury telluride solid solution crystals obtained from a melt", Izv. Akad. Nauk SSSR, Met., no. 5, 240-2 (1974), Russian, CA82-24550.
631.D'Aragona, F. S., "Dislocation etch for (100) planes in silicon", J. Electrochem. Soc., 119, 948-51 (1972), PA75-71495.
632.deKock, A. J. R. and Boonen, P. G. T., "A new method for revealing striations in high-resistive floating-zone silicon crystals", J. Electrochem. Soc., 119, 1241-3 (1972), PA75-74787.
633.Dolivo-Dobrovolskaya, G. I., Lyubalin, M. and Mokievskii, V. A., "Symmetry and form of etch pits on germanium crystals", Izv. Akad. Nauk SSSR Ser. Fiz., 36, 567-9 (1972), Russian, PA75-50446.
634.Dorling, G. W., Bull, R. K., Durrani, S. A., Fremlin, J. H. and Khan, H. A., "Anisotropic etching of charged-particle tracks in crystals", Radiat. Eff., 23, 141-3 (1974), PA78-6480.
635.Faust, J. W., Jr., Tung, Y. and Liaw, H. M., "Etch pits on pure polytypes of silicon carbide", Silicon Carbide, Proc. Int. Conf., 3rd, 1973, p. 215-21, Publ. 1974, CA81-160348.
636.Hayashi, F., Tarui, Y. and Teshima, H., "The preferential etching of silicon with a water-amine solution", Bull. Electrotech. Lab., 36, 27-33 (1972), Japanese, PA75-71316.
637.Hockey, B. J. and Mitchell, J. W., "Etch pit studies of dislocation arrangements resulting from the deformation of single crystals of copper 7.5 at.% aluminium alloys", Phil. Mag., 26, 409-23 (1972), PA75-64664.
638.Huber, A. and Linh, N. T., "Metallographic display of crystal defects in InP", J. Cryst. Growth, 29, 80-4 (1975), French, PA78-61895.

639. Iizuka, T., "Etching studies of anomalous defects in p-type GaP crystals", J. Electrochem. Soc., 121, 1085-90 (1974), PA77-75528.
640. Ilin, Y. L. and Saenko, I. V., "Macrostructure of gallium phosphide crystals studied by a chemical etching method", Izv. Leningr. Elektrotekhn. In-Ta, no. 167, 83-5 (1975), Russian, CA83-200407.
641. Ivonin, I. V., Krasilnikova, L. M. and Yakubenya, M. P., "An electron-microscope study of the morphology of the etched surface of gallium arsenide", Izv. VUZ Fiz., no. 5, 116-9 (1972), Russian, PA75-56575.
642. Johnston, W. D., Jr., Callahan, W. M. and Miller, B. I., "Observation of dark-line degradation sites in AlGaAs/GaAs DH laser material by etching and phase-contrast microscopy", J. Appl. Phys., 45, 505-7 (1974), PA77-51606.
643. Karpel, N. G., Kulyas, S. L., Fedorchuk, O. K. and Stochik, V. V., "Spectrochemical determination of the impurities in semiconductor epitaxial structures", Ind. Lab., 37, 536-40 (1971), PA75-29194.
644. Keller, W. and Muhlbauer, A., "Etching depressions, a new defect in dislocation-free silicon", Lattice Defects in Semiconductors, 1974, p. 538-43, Publ. 1975, PA78-81074.
645. Kocsis, S. and Lendvay, E., "Thermal etching of gallium phosphide", Krist. Tech., 9, 1131-40 (1974), CA81-142153.
646. Lizuka, T., "Etching studies of impurity precipitates in pulled GaP crystals", J. Electrochem. Soc., 118, 1190-4 (1971), EEA74-30661.
647. Mikosza, A. G. and Lawn, B. R., "Section-and-etch study of Hertzian fracture mechanics", J. Appl. Phys., 42, 5540-5 (1971), PA75-9699.
648. Miller, J. W., Nagata, N. and Vreeland, T., Jr., "Observations of dislocations in zinc by chemical etching", Mater. Sci. Eng., 13, 121-4 (1974), CA79-150241.
649. Nishida, M., "Kinetic study on the first-order network of etched Ge surfaces", Jap. J. Appl. Phys., 11 (6), 835-9 (1972), PA75-50447.
650. Oczkowski, H. and Poplawski, Z., "Orientation of ZnSe crystals by chemical etching technique", J. Cryst. Growth, 23, 154-6 (1974), PA77-73262.
651. Ohishi, M. and Ohmori, K., "Etch pits and optical images of zinc selenide single crystals", Oyo Buturi, 43, 1108-10 (1974), Japanese, CA82-132180.
652. Osipyan, Y. A., Smirnova, I. S. and Strukova, G. K., "Detection of dislocations in cadmium sulfide by the method of selective etching", Fiz. Khim. Obrab. Mater., no. 6, 145-7 (1974), Russian, CA82-178427.
653. Patel, A. R. and Mathai, K. J., "Etch hillocks on carbon faces of α SiC (Dislocation sites)", J. Phys. D, 5, 390-3 (1972), PA75-29005.

654. Pfeifer, J. and Haraszthy, E. S., "The investigation of Ge-Si layer structures by means of the etching rate method (ERM)", Thin Solid Films, 20, S31-2 (1974), PA77-28306.
655. Platonov, B. D., Maltseva, V. A., Krasyuk, B. A. and Bekker, Y. M., "Effect of the thermophysical properties of materials on the nature of the etching of dislocations", Inzh.-Fiz. Zh., 27, 744 (1974), Russian, CA82-66426.
656. Porter, W. A., Gupta, A. and Swindle, D. L., "Use of Sirtl etch for silicon-slice evaluation", J. Electrochem. Soc., 120, 589-90 (1973), EEA76-23349.
657. Richter, H. and Schulz, M., "Dislocation detection on the (100) face of gallium arsenide single crystals", Krist. Tech., 9 (9), 1041-50 (1974), German, CA81-160335.
658. Rosin, H., "Chemical etching of structural defects in epitaxial $GaAs_{1-x}P_x$ and bulk GaAs", Phys. Status Solidi A, 25, K5-K7 (1974).
659. Rozsa, E. and Stefaniay, V., "Comparative investigations on silicon single crystals", Hiradstech. Ipari Kut. Kozl., 11 (2), 33-45 (1971), Hungarian, EEA75-1009.
660. Secco d'Aragona, F., "Annealing behavior and etching phenomena of microdefects in dislocation-free float-zone silicon", Phys. Status Solidi A, 7, 577-82 (1971), PA75-3569.
661. Simanovskii, A. A., "Effect of crystallization conditions on the morphology of zinc selenide crystals", Rost Krist., 9, 197-201 (1972), Russian, CA79-58498.
662. Tokumaru, Y., Okada, Y. and Iizuka, T., "Etch patterns and crystallographic polarity in hexagonal zinc sulfide", Jap. J. Appl. Phys., 14, 1241-2 (1975), CA83-124308.
663. Ugai, Y. A., Kirichenko, I. V. and Kurbanov, K. R., "Identification of defects in germanium, using a chemical etchant", Izv. Akad. Nauk SSSR, Neorg. Mater., 10, 1921-5 (1974), Russian, CA82-78927.
664. Vieweg-Gutberlet, F., "Method of investigation of inhomogeneous dopant distributions in silicon single crystals", Solid-State Electron., 12, 9 pp. (1969), German, EEA73-4331.
665. Winter, A. T., "Etching studies of dislocation microstructures in crystals of copper fatigued at low constant plastic strain amplitude", Philos. Mag., 28, 57-64 (1973), PA77-1527.

XI. Electron Paramagnetic Resonance

666. Alexander, H., Kenn, M., Nordhofen, B., Weber, E. and Sander, W., "EPR spectra in silicon with dislocations", International Conference on Lattice Defects in Semiconductors, 1974, p. 433-8, Publ. 1975.
667. Ammerlaan, C. A. J. and Watkins, G. D., "Electron-paramagnetic-resonance detection of optically induced divacancy alignment in silicon", Phys. Rev. B, 5, 3988-93 (1972), PA75-40764.

668. Bauer, C. F. and Whitmore, D. H., "Characterization of paramagnetic hole centers in Al_2O_3 and Al_2O_3Fe by optical and EPR absorption", J. Solid State Chem., 11, 38-52 (1974).
669. Broude, S. V., Grazhulis, V. A., Kveder, V. V. and Osipyan, Y. A., "Investigation of properties of the dislocation EPR spectra in silicon", Zh. Eksp. & Teor. Fiz., 66, 1469-78 (1974), Russian, PA77-73014.
670. Brower, K. L., "EPR of a Jahn-Teller distorted (111) carbon interstitially in irradiated silicon", Phys. Rev. B, 9, 2607-17 (1974), PA77-61035.
671. Caplan, P. J., AuCoin, T. R., Helbert, J. N., Poindexter, E. H., Wagner, B. E. and Zeto, R. J., "MNR and ESR of defects and tagged defects in silicon oxide systems", Electrochemical Society Fall Meeting (Extended abstracts), 1974, p. 523-4, EEA78-37450.
672. Cullis, P. R. and Marko, J. R., "Electron paramagnetic resonance properties of n-type silicon in the intermediate impurity-concentration range", Phys. Rev. B, 11, 4184-200 (1975), PA78-66244.
673. Daly, D. F. and Noffke, H. E., "An EPR study of fast neutron radiation damage in silicon", International Conference on Radiation Effects in Semiconductors, 1970, p. 179-87, Publ. 1971, PA75-48282.
674. de Wit, J. G. and Ammerlaan, C. A. J., "Electron paramagnetic resonance on divacancies in phosphorus-implanted silicon", Proceedings of the 2nd International Conference on Ion Implantation in Semiconductors, Physics and Technology, Fundamental and Applied Aspects, 1971, p. 39-46, PA75-26265.
675. de Wit, J. G., Ammerlaan, C. A. J. and Sieverts, E. G., "ENDOR study of the divacancy in silicon", International Conference on Lattice Defects in Semiconductors, 1974, p. 178-84, Publ. 1975.
676. Ditina, Z. Z. and Strakhov, L. P., "Investigation of the surface of cadmium selenide by the EPR method", Sov. Phys. Solid State, 9, 2000-03 (1968), MA1-07-15 0698.
677. Emelyanenko, O. V., Ilmenkov, G. V., Kolchanova, N. M., Nasledov, D. N., Smetannikova, Y. S. and Yarmarkin, V. K., "Anomalous properties of A^3B^5 crystals containing transition metal atoms", Presented at 11th International Conference on Physics of Semiconductors, Warsaw, Poland, July 1972, PA75-68405.
678. Galli, A., "On the central EPR resonance in reactor-irradiated α-Al_2O_3", Lett. Nuovo Cimento, 11, 547-52 (1974), PA78-10217.
679. Ginodman, V. B., Gladkov, P. S., Zhurkin, B. G., "Mikhaelevich, V. G. and Penin, N. A., "EPR of phosphorus in compensated n-type silicon under uniaxial compression", Fiz. Tekh. Poluprov., 4, 1736-9 (1970), Russian, MA4-02-15 0187.
680. Goldstein, B., "Li-defect interactions in electron-irradiated n-type silicon by EPR measurements", International Conference on Radiation Effects in Semiconductors, 1970, p. 223-31, PA75-48293.

681. Gorelkinskii, Y. V., Sigle, V. O. and Takibaev, Z. S., "EPR of conduction electrons produced in silicon by hydrogen ion implantation", Phys. Status Solidi A, 22, K55-K57 (1974), CA81-18291.
682. Gothe, K. H., "EPR studies on the interface between semiconductor and insulator", Wiss. Z. Tech. Hochsch., Ilmenau, 19 (3-4), 101-5 (1973), German, CA80-53620.
683. Haneman, D., "Review of electron paramagnetic resonance investigations of semiconductor surfaces", International Conference on Solid Surfaces, 2nd, Proc., 1974, p. 371-80.
684. Hester, R. K., Sher, S., Soest, J. F. and Weisz, G., "Nuclear-magnetic-resonance detection of charge defects in gallium arsenide", Phys. Rev. B, 10, 4262-73 (1974), PA78-29255.
685. Hiraki, A., "EPR in electron bombarded phosphorus-doped n-type germanium - an attempt to detect the presence of bombardment-induced interstitial phosphorus", International Conference on Radiation Effects in Semiconductors, 1970, p. 91-5, Publ. 1971, PA75-48279.
686. Ho, L. T., "Time dependence of the excitation spectrum of neutral magnesium donors in silicon", Phys. Status Solidi A, 28, K73-K75 (1975).
687. Holm, R. T. and Furdvna, J. K., "Observation of helicon-excited electron paramagnetic resonance in a high-mobility semiconductor", Solid State Commun., 15, 1459-62 (1974).
688. Inoue, M. and Huang, C. Y., "EPR studies of Mn^{24} in $Cd_{1-x}Mg_xTe$ mixed single crystals", J. Phys. Soc. Jap., 32, 763-70 (1972), PA75-26261.
689. Izumi, T. and Matsumori, T., "Paramagnetic centers created in $Si-SiO_2$ structure by ion implantation", Jap. J. Appl. Phys., 14, 1067-8 (1975), EEA78-37414.
690. Kamimura, H., "Knight shift and MNR line shape in metallic region of correlated impurity band electrons", Phil. Mag., 29, 65-71 (1974).
691. Kerr, R. K. and Schwerdtfeger, C. F., "The effect of sulphur vacancies on the EPR spectra of intrinsic and cobalt doped $CdIn_2S_4$", J. Phys. & Chem. Solids, 33, 1795-7 (1972), PA75-68735.
692. Khusidman, M. B. and Neshpor, V. S., "A study of certain semiconducting compounds of boron by the EPR method", Poroshkovaya Met., 9, 67-8 (1970), Russian, MA4-02-15 0211.
693. Kushnir, A. S., Nosachev, B. G., Omelchenko, S. A., Shtambur, I. V. and Yakunin, A. Y., "ESR structural study of single-crystal ZnSe", Sov. Phys. Solid State, 17, 953-4 (1975).
694. Lee, Y. H. and Brosious, P. R., "New EPR spectra in neutron-irradiated silicon. II", Radiat. Eff., 22, 169-72 (1974), PA78-1512.
695. Lee, Y. H. and Corbett, J. W., "EPR study of defects in neutron-irradiated silicon: quenched-in alignment under (110)-uniaxial stress", Phys. Rev. B, 9, 4351-61 (1974), NSA30-16466.

696.Lee, Y. H. and Corbett, J. W., "EPR evidence of the self-interstitials in neutron-irradiated silicon", Solid State Commun., 15, 1781-84 (1974).

697.Mann, H. M., Gemmell, D. S. and Marshall, S. A., "Study and evaluation of CdTe by use of channeling and electron paramagnetic resonance measurements", Proceedings of the International Symposium on Cadmium Telluride a Material for Gamma-ray Detectors, 1971, 15/8 pp., EEA75-20107.

698.Miida, R., Nakada, M., Kimura, H., Komiya, H. and Ibuki, S., "Photosensitive EPR in zinc selenide silicon (-doped) crystals", J. Phys. Soc. Japan, 29, 246 (1970), PA73-72497.

699.Murakami, K., Masuda, K., Gamo, K. and Namba, S., "ESR studies on defects and amorphous phase in silicon produced by ion implantation", Jap. J. Appl. Phys., 12 (9), 1307-16 (1973), PA76-79913.

700.Olekseeva, M. F., Slinko, E. I., Tovstyuk, K. D. and Handozhko, A. G., "Nuclear magnetic resonance in PbTe", Phys. Status Solidi A, 21, 759-62 (1974), PA77-28712.

701.Pavlov, N. M., Iglitsyn, M. I., Kosaganova, M. G., Solomatin, V. N. and Barinov, Y. V., "EPR spectra of irradiated n-type silicon carbide", Fiz. Tver. Tela, 13, 2814-17 (1971), Russian, PA75-1729.

702.Petrova, A. G., Rakov, A. V., Yarygin, V. P., Ivanov, E. N. and Pashintsev, Y. I., "EPR spectra of Cu^{2+} and Mn^{2+} adsorbed on the surface of silicon", Fiz. Tekh. Poluprov., 5, 1440-3 (1971), Russian, PA74-59446.

703.Raudorf, T. W., "Electron spin resonance in degenerate semiconductors", Thesis, McGill Univ., Montreal, Canada, 1971, PA75-48273.

704.Schmidt, U., Weber, E., Alexander, H. and Sander, W., "On the magnetic properties of dislocations in silicon", Solid State Commun., 14, 735-9 (1974).

705.Seqsa, K. H. and Spenke, S., "EPR on manganese-doped GaAs epitaxial layers", Phys. Status Solidi A, 27, 129-34 (1975).

706.Shiota, I., Miyamoto, N. and Nishizawa, J., "Photoinduced paramagnetic defects on silicon surface", Proc. Int. Conf. Solid Surf., 2nd, 1974, p. 417-20, CA82-177800.

707.Sobolev, E. V. and Shcnerbakova, M. J., "EPR spectra of the diamond and their bearing on the optical properties", Presented at Conference on Magnetic Resonance and Related Phenomena, Bucharest, Romania, September 1970, PA75-26274.

708.Title, R. S., Brodsky, M. H. and Crowder, B. L., "EPR studies in amorphous silicon", Proc. of 10th Int. Conf. on the Phys. of Semicond., 1970, p. 794-98, PA74-59447.

709.Toyoda, Y. and Hayashi, Y., "Bolometric detection of EPR in phosphorous-doped silicon at low temperature", J. Phys. Soc. Japan, 29, 247-48 (1970), PA73-72491.

710.Urban, P. and Sperlich, G., "Exchange interaction of S state ions with charge carriers in SnTe", Solid State Commun., 16, 927-30 (1975).
711.Vollmer, H. and Geist, D., "Electron paramagnetic resonance (EPR) of phosphorus in silicon-rich silicon-germanium alloys", Phys. Status Solidi B, 62, 367-76 (1974).
712.Wang, S., Shirafuji, J. and Inuishi, Y., "Saturation of ESR in neutron-irradiated silicon", Technol. Rep. Osaka Univ., 23, 469-74 (1973), PA77-13347.
713.Watkins, G. D., "A microscopic view of radiation damage in semiconductors using e.p.r. as a probe", IEEE Trans. Nucl. Sci., NS-16, 13-18 (1969), EEA73-14010.
714.Watkins, G. D., "EPR studies of the lattice vacancy and low-temperature damage processes in silicon", Lattice Defects in Semiconductors, 1974, p. 1-22, Publ. 1975, PA78-81113.
715.Watkins, G. D., "EPR observation of close Frenkel pairs in irradiated ZnSe", Phys. Rev. Lett., 33, 223-25 (1974), NSA30-27544.
716.Watkins, G. D., "Isotope effects in the EPR spectrum of a tin-vacancy pair in silicon", Solid State Commun., 17, 1205-8 (1975).

XII. Synergism

717.Alessandrini, E. I., Campbell, D. R. and Tu, K. N., "Surface reactions of MOS structures", J. Appl. Phys., 45, 4888-93 (1974), EEA78-11456.
718.Anderson, R. M., "The precipitation of Si in AlCuSi thin films", Proceedings of the 31st Annual Meeting of the Electron Microscopy Society of America, 1973, p. 34-5, EEA77-2609.
719.Arthur, J. R., "Surface stoichiometry and structure of GaAs and GaP", J. Vac. Sci. & Technol., 9, 910 (1972), PA75-50445.
720.Baker, J. F. C. and Ogden, R., "Crystal imperfections in silicon epitaxial layers grown on ion-implanted substrates", J. Mater. Sci., 10, 1259-61 (1975), PA78-65839.
721.Bell, E. C., Glaccum, A. E., Hemment, P. L. F. and Sealy, B. J., "Heat treatment of ion implanted GaAs", Radiat. Eff., 22, 253-8 (1974).
722.Berenbaum, L., Thorpe, W. R. and DiGiacomo, G., "Electromigration behavior of Al-Cu-Si thin films", Presented at Electrochemical Society Meeting, San Francisco, Calif., May 1974, EEA77-26500.
723.Cullis, A. G. and Katz, L. E., "Electron microscope study of electrically active impurity precipitate defects in silicon", Philos. Mag., 30, 1419-43 (1974), PA78-19091.
724.Dolivo Dobrovolskaya, G. I., Mokievskii, V. A., Antonov, P. I. and Sachkov, G. V., "Morphological features of germanium obtained by the Stepanov method", Izv. Akad. Nauk SSSR Ser. Fiz., 36 (3), 507-13 (1972), Russian, PA75-47328.

725. Ettenberg, M., McFarlane, S. H. and Gilbert, S. L., "Solution growth and characterization of GaAs and $Al_xGa_{1-x}As$ on GaP (dislocation etching and X-ray examination)", Proceedings of the 4th International Symposium on Gallium Arsenide and Related Compounds, 1972, p. 29-36, Publ. 1973, EEA76-23367.
726. Faulkner, D. and Stevens, R., "Characterisation of polycrystalline silicon carbide by scanning and transmission electron microscopy", Proceedings of the 28th Annual Meeting of the Electron Microscopy Society of America, 1970, p. 400-1, PA75-9697.
727. Fontana, P. V., Decosterd, J. P. and Wegmann, L., "Investigations of carbon residues on surfaces of silicon integrated circuits", J. Electrochem. Soc., 121, 146-50 (1974), EEA77-10297.
728. Herfert, R. E., "The use of advanced analytical techniques for characterization of the chemical, physical, and crystallographic nature of a surface", Proceedings of the 31st Annual Meeting of the Electron Microscopy Society of America, 1973, p. 132-3, EEA77-3094.
729. Horiuchi, S., "Investigation of polycrystalline silicon layers by electron microscopy and X-ray diffraction", Solid-State Electron., 18, 1111-3 (1975).
730. John, H. F., Faust, J. W., Jr. and Stickler, R., "Microinhomogeneity problems in silicon", IEEE Trans. Pts. Materials Packaging, PMP-2, 51-8 (1966), EEA70-2247.
731. Klimenko, A. P. and Tkhorik, Y. A., "Film properties of germanium obtained by evaporation in vacuum", Poluprovodn. Tekh. & Mikroelektron., no. 17, 50-60 (1974), Russian, EEA78-2751.
732. Leamy, H. J. and Mogab, C. J., "Epitaxial growth of β SiC on Si: Growth defects", Proceedings of the 31st Annual Meeting of the Electron Microscopy Society of America, 1973, p. 36-7, EEA77-2491.
733. Ogirima, M., "Vapor-phase epitaxial growth and defects of gallium phosphide", Jap. J. Appl. Phys., 11, 281-7 (1972), PA75-29032.
734. Palatnik, L. S., Vinogradov, V. E. and Boiko, B. T., "Type of structural imperfection in CdTe films", Sov. Phys.-Crystallogr., 19, 698-9 (1975), PA78-65840.
735. Ram, U. S., Dubey, M. and Singh, G., "On the growth mechanism of silicon carbide polytypes", Presented at International Conference on Silicon Carbide, Miami Beach, Fla., September 1973, PA77-24448.
736. Ramsey, J. N., "The synergism of complementary analytical techniques in solving S/C device problems", 1974 Digests of the Intermag. Conference, 17-5/1 pp., PA78-16102.
737. Ravi, K. V. and Varker, C. J., "Growth 'striations' and 'swirls' in float-zone silicon single crystals", Presented at Electrochemical Society Meeting, Chicago, Ill., May 1973, EEA76-23301.
738. Sandulova, A. V., Mocharnyuk, G. F., Sheherbai, K. S., Varshava, S. S. and Skoropad, R. N., "Morphology of GaAs crystals grown from the gas phase", Izv. Akad. Nauk SSSR Neorg. Mater., 8, 945-6 (1972), Russian, PA75-56656.

739. Servidori, M. and Armigliato, A., "Electron microscopy of silicon monophosphide precipitates in P-diffused silicon", J. Mater. Sci., 10, 306-13 (1975), PA78-25409.
740. Sheble, A. M., Teel, T. P. and Devaney, J. R., "Plasma etching for SEM and EMP examination of microelectronic devices", 9th Annual Conference of the Microbeam Analysis Society, 1974, 31A/4 pp., EEA78-660.
741. Shifrin, S. S., Osvenskii, V. B., Milvidskii, M. G., Morgulis, L. M. and Grishina, S. P., "Nature of small etch pits in tellurium-doped gallium arsenide single crystals", Kristallografiya, 18, 1299-302 (1973), Russian, CA80-64564.
742. Todd, C. J., Ashwell, G. W. B., Speight, J. D. and Heckingbottom, R., "Thermally induced processes at Au-GaAs interfaces: an assessment by Rutherford backscattering, SEM, and depth profiling Auger electron spectroscopy", Conference on Metal-Semiconductor Contacts, 1974, p. 171-83, PA78-19093.

XIII. Others

743. Beyer, H., "Some practical aspects of the interference-microscopic measurement of surfaces with particular application to semiconductors and microelectronics", Feingeraete Tech., 23, 416-22 (1974), German, EEA78-2665.
744. Bogenschutz, A. F. and George, U., "The analytical investigation of doped silicon dioxide coats and doping gas mixture", Feinwerktech. & Messtech., 83, 114-16 (1975), PA78-62540.
745. Danilov, G. M., Medvedev, Y. V. and Petrov, A. S., "Installation for contactless measurement of the nonuniformity of the parameters of semiconductor materials and dielectrics", Instrum. & Exp. Tech., 16, 923-5 (1973), EEA77-11925.
746. Dzhafarov, T. D., Demakov, Y. P. and Rumyantseva, A. V., "On a possibility of determining nonuniformities in $Al_xGa_{1-x}As$ solid solution compositions", Krist. & Tech., 9, 1399-402 (1972), EEA78-19672.
747. Ganapolskii, E. M., "State of chromium atom impurities in gallium arsenide", Sov. Phys. Solid State, 16, 1868-72 (1975), PA78-62282.
748. Ghatalia, A. K. and Thomas, D. R., "Semiconductor process defect monitor", IBM Tech. Disclosure Bull., 17, 2577-8 (1975), EEA78-19637.
749. Haller, E. E. and Hansen, W. L., "Impurities in high-purity germanium as determined by Fourier transform spectroscopy", IEEE Trans. Nucl. Sci., NS-21, 279-86 (1974), PA77-53609.
750. Heinrichs, J., "Surface plasmons at metal-vacuum and metal-dielectric interfaces", Solid State Commun., 12, 167-70 (1973), EEA76-16339.
751. Hwang, C. J., Haszko, S. E. and Bergh, A. A., "UV microprobe technique for measurement of minority-carrier diffusion length in GaP p-n junction material", J. Appl. Phys., 42, 5117-19 (1971).

752.Jones, C. L. and Willoughby, A. F. W., "The complete base profile shape in a pushed-out diffused transistor analyzed by radiotracer methods", Appl. Phys. Lett., 25, 114-16 (1974), PA77-67718.
753.Knolle, W. R. and Retajczyk, T. F., Jr., "Monitoring sodium contamination in silicon devices and processing materials by flame emission spectrometry", J. Electrochem. Soc., 120, 1106-11 (1973), EEA76-35213.
754.Latshaw, G. L., "A Mossbauer effect study of iron implanted into silicon, diamond, and germanium", Thesis, Stanford Univ., Calif., 1971, 110 pp., Order No. 72-5941, PA75-48152.
755.Lightowlers, E. C., North, J. C., Jordan, A. S., Derick, L. and Merz, J. L., "Nuclear microanalysis of oxygen concentration in liquid-phase epitaxial gallium phosphide", J. Appl. Phys., 44, 4758-68 (1973), PA77-18777.
756.Lightowlers, E. C., North, J. C. and Lorimor, O. G., "Nitrogen concentration in GaP measured by optical absorption and by proton-induced nuclear reactions", J. Appl. Phys., 45, 2191-200 (1974), EEA77-30173.
757.Lindholm, A. W. and Vanat, W. F., "Pyrolytic oxide deposit inspection process", IBM Tech. Disclosure Bull., 17, 2592 (1975), EEA78-19540.
758.Lutt, B. D., Milyavskii, Y. S., Maksimov, D. E., Rudnevskii, N. K. and Shevchenko, V. Y., "Determination of departures from stoichiometric composition of semiconductor compounds by chemico-analytical methods", Izv. Akad. Nauk SSSR Neorg. Mater., 8, 439-45 (1972), Russian, PA75-43503.
759.Meier, A. A., Fistul, V. I., Yureva, I. M., Aronov, A. S. and Vikk, G. A., "Optical scanning for composition uniformity in $GaAs_{1-x}P_x$ solid solutions", Ind. Lab., 39, 746-9 (1973), PA77-26708.
760.Milvidskii, M. G., Osvenskii, V. B., Proshko, G. P. and Kholodnyi, L. P., "Nature of defects in gallium arsenide heavily doped with tellurium", Sov. Phys.-Semicond., 6, 194-7 (1972), PA75-71926.
761.Nagasima, N. and Kubota, N., "Structures of Si films chemically vapor-deposited on amorphous SiO_2 substrates", Jap. J. Appl. Phys., 14, 1105-12 (1975), EEA78-37415.
762.Papkov, V. S. and Surovikov, M. V., "Internal stresses in heteroepitaxial silicon coatings on sapphire", Izv. Akad. Nauk SSSR Ser. Fiz., 37, 2371-6 (1973), Russian, PA77-13108.
763.Patrick, W. J. and Patzner, E. J., "Detection of surface defects on silicon wafers by scattered light measurements", In: Semicond. silicon, Huff, H. R., (Ed.), Electrochem. Soc., Inc.: Princeton, N. J., p. 482-90 (1973), CA82-24536.
764.Piotrowski, T., "Optical methods of orienting GaAs crystals for quantum electronics purposes", Arch. Elekrotech, 17, 673-6 (1968), EEA72-11202.

765. Protopopescu, M., Aloman, A., Dobre, A. and Visa, I., "Contributions to the study of macro and microstructures of semiconductor materials produced in S.R.R.", Bul. Inst. Politeh. 'Gheorghe Gheorghiu-Dej'Bucuresti, 33 (4), 117-37 (1971), PA75-61257.
766. Schade, H., Nuese, C. J. and Gannon, J. J., "Direct evidence for generation of defect centers during forward-bias degradation of $GaAs_{1-x}P_x$ electroluminescent diodes", J. Appl. Phys., 42, 5072-5 (1971), EEA75-4406.
767. Schmidt, M. and Schmidt, L., "Optical measurements of stress distributions in GaAs at low temperature", Appl. Phys., 8, 47-9 (1975), PA78-77759.
768. Seidowski, E., Klimkiv, A. V. and Bleeck, P., "Morphology of InAs crystal needles from the gas phase", Krist. & Tech., 10, 361-3 (1975), German, PA78-73279.
769. Spry, R. J. and Henes, J. D., "Mossbauer-type photoluminescence spectra of irradiated germanium", International Conference on Radiation Effects in Semiconductors, 1970, p. 99-104, Publ. 1971, PA75-48243.
770. Stelmak, J., "Defects detection in a metal-oxide-silicon structure", IBM Tech. Disclosure Bull., 15, 177-8 (1972), EEA75-36292.
771. Stolarski, E., "Contactless measurement of thickness and damage of semiconductor slices", Arch. Elektrotech., 23, 375-86 (1974), Polish, EEA78-5201.
772. Taguchi, T., Shirafuji, J. and Inuishi, Y., "Photoluminescence study on lattice defects on CdTe", Technol. Rep. Osaka Univ., 23, 195-203 (1973), EEA76-28616.
773. Tihanyi, J., "Investigation of surface phenomena on semiconductor devices by a light spot scanning method", Symposium on Test Methods and Measurements of Semiconductor Devices, 1967, Pap. No. 508, EEA71-10573.
774. Tong, A. H. and Schumann, P. A., "Junction depth measurement by the scanning light spot technique", In: Semiconductor silicon, Electrochem. Soc., Inc., New York, p. 706-720 (1969).
775. Turovskii, B. M. and Lainer, L. V., "Method of evaluating the quality of silicon single crystals containing small-angle boundaries", Ind. Lab., 40, 854-5 (1974), EEA78-28616.
776. Vecsernyes, L., "Problems of spectrochemical determination of metal impurities in semiconductor silicon", Symposium on Test Methods and Measurements of Semiconductor Devices, 1967, Pap. No. 516, EEA71-10538.
777. Vedam, K. and So, S. S., "Characterization of real surfaces by ellipsometry", Surface Sci., 29, 379-95 (1972), PA75-25187.
778. Virtmanis, A. S., Feltyn, I. A. and Freiberga, L. A., "Influence of local defects on the C-V characteristics of MIS-structures", Latv. PSR Zinat. Akad. Vestis Fiz. Tehn. Ser., no. 3, 23-6 (1972), Russian, EEA75-32593.
779. Zuev, V. A., Gorban, A. P., Peikov, P. K., Gavrilenko, V. I., Romanyuk, B. N. and Sukach, G. A., "Study of thermochemical destruction of silicon surface by different electrophysical methods", Ukr. Fiz. Zh., 19, 88-93 (1974), Russian, PA77-20491.

M. ELECTRICAL AND OPTICAL PROPERTIES OF MIS DEVICES

I. Bibliographies

1.Schlegel, E. S., "A bibliography of metal-insulator-semiconductor studies", IEEE Trans. Electron Devices, ED-14 (11), 728-49 (1967), EEA71-5868.
2.Schlegel, E. S., "Additional bibliography of metal-insulator-semiconductor studies", IEEE Trans. Electron Devices, ED-15 (12), 951-4 (1968), EEA72-5034.
3.Storm, H. F., "Field-effect transistor (FET) bibliography", IEEE Trans. Electron Devices, ED-14, 710-16 (1967), EEA71-3441.
4.Storm, H. F., "Field-effect transistor (FET) bibliography", IEEE Trans. Electron Devices, ED-16, 957-63 (1969), EEA73-10681.

II. Books

5.Bailey, F. J., "Introduction to semiconductor devices", George Allen & Unwin Ltd., London, England, 1972, EEA75-654.
6.Cobbold, R. S., "Theory and applications of field-effect transistors", Wiley-Interscience, New York, 1970.
7.Greenway, D. L. and Harbeke, G., "Optical properties and band structure of semiconductors", Pergamon Press, New York, 1968.
8.Jowett, C. E., "Semiconductor devices: testing and evaluation", Business Books, London, England, 1974, EEA77-33854.
9.Kane, P. F. and Larrabee, G. B., "Characterization of semiconductor materials", McGraw-Hill Book Co., New York, 1970.
10.Moss, T. S., Burrell, G. J. and Ellis, B., "Semiconductor opto-electronics, John Wiley & Sons, Halsted Press division, New York, 1973.
11.O'Dwyer, J. J., "The theory of electrical conduction and breakdown in solid dielectrics", Clarendon Press, Oxford, England, 1973.
12.Paul, R., "Field-effect transistors: Physical principles and properties", VEB Verlag Technik, Berlin, Germany, 1972, German.
13.Runyan, W. R., "Semiconductor measurements and instrumentation", McGraw-Hill Book Co., New York, 1975.

III. Review Articles

14.Appels, J. A., Kalter, H. and Kooi, E., "Some problems of MOS technology", Philips Tech. Rev., 31, 225-36 (1970), EEA74-34095.
15.Dabrowski, G., "Active components in electronics", Elektroniker, 14 (3), EL20-6 (1975), German, EEA78-24348.
16.Flietner, H., "Electronics of semiconductor surfaces", Z. Elektr. Inf. - & Energietch., 2, 210-11 (1972), German, EEA76-8606.

17. Garcia-Moliner, F., "Semiconductor surfaces", Electron. & Fis. Apl., 17, 202 (1974), PA78-3737.
18. Goetzberger, A., "Interface states in Si-SiO_2-interfaces", International Conference on Technology and Applications of Charge Coupled Devices, 1974, p. 47-58, PA78-25776.
19. Goetzberger, A. and Schulz, M., "Fundamentals of MOS technology", Meeting of the German Physical Society on Advances in Solid State Physics, 1973, p. 309-36, EEA76-29216.
20. Gray, P. V., "The silicon-silicon dioxide system", Proc. IEEE, 57 (9), 1543-51 (1969), EEA73-4422.
21. Grosvalet, J., "Physical functioning principles of metal-oxide-semiconductor devices", Onde Elect., 47, 937-43 (1967), French, EEA71-8273.
22. Komiya, Y. and Tarui, Y., "Survey reports on the characteristics of the SiO_2 and the devices of MOS structure", Bull. Electrotech. Lab., 31 (4), 162-73 (1967), Japanese, PA71-18148.
23. Kooi, E., "MOS effects and silicon device technology", Oyo Buturi, 43 (Suppl.), 427-36 (1974), PA78-6907.
24. Litovchenko, V. G., "Some physical and technological problems in semiconductor/dielectric structures", Poluprovodn. Tekh. & Mikroelektron., no. 8, 3-14 (1972), Russian, EEA76-16321.
25. Nicollian, E. H., "Insulator films and their interfaces in semiconductor devices", J. Vac. Sci. & Technol., 9, 12 (1972), EEA75-20141.
26. Osburn, C. M., "Dielectric breakdown in silicon dioxide films", Electrochemical Society Fall Meeting (Extended abstracts), 1974, p. 495-6, EEA78-37560.
27. Paul, R., "MIS transistors: present status and progress", Nachrichtentechnik, 20, 1-9 (1970), German, EEA73-14320.
28. Revesz, A. G. and Zaininger, K. H., "The Si-SiO_2 solid-solid interface system", RCA Rev., 29, 22-76 (1968).
29. Roitsin, A. B., "Theory of deep centers in semiconductors (review)", Sov. Phys.-Semicond., 8, 1-17 (1974), PA77-75815.
30. Seidel, T. E., Meek, R. L., Poate, J. M. and Cullis, A. G., "A review of diffusion and ion damage gettering in silicon", Electrochemical Society Fall Meeting (Extended abstracts), 1974, p. 535-6, EEA78-37453.
31. Stiles, P. J., "Properties of carriers at the Si-SiO_2 interface in MOSFET structures", J. Vac. Sci. & Technol., 11, 958-61 (1974), EEA78-24255.
32. Sugano, T., Hoh, K., Sakaki, H., Iizuka, T., Hirai, K., Kuroiwa, K. and Kakemoto, K., "Quantum state and electron transport at silicon-silicon dioxide interface and MIS device technology", J. Fac. Eng. Univ. Tokyo B, 32, 155-371 (1973), EEA77-18747.
33. Uemura, Y., "Landau levels and electronic properties of semiconductor interfaces", Jap. J. Appl. Phys., 13, Suppl. 2 (Pt. 2), 17-24 (1974), PA78-37828.
34. van der Meulen, Y. J., "Some recent studies of very thin SiO_2 films", J. Vac. Sci. & Technol., 11, 985-99 (1974), EEA78-24258.

35. Zaininger, K. H. and Heiman, F. P., "The C-V technique as an analytical tool. I", Solid State Technol., 13 (5), 49-56 (1970), EEA73-38002.

IV. Theory

36. Abram, R. A., "Localized states in inverted silicon-silicon dioxide interfaces", J. Phys. C., 6, L379-81 (1973), EEA77-344.
37. Ando, T. and Uemura, Y., "Theory of oscillatory g factor in an MOS inversion layer under strong magnetic fields", J. Phys. Soc. Jap., 37, 1044-52 (1974), PA77-79412.
38. Avakyants, G. M. and Arutyunyan, V. M., "On double injection theory in compensated semiconductors with deep acceptor centres and electron traps", Izv. Akad. Nauk Arm. SSR Fiz., 8, 429-37 (1973), Russian, EEA77-26399.
39. Baccarani, G. and Severi, M., "On the determination of the interface-state density in the presence of statistical fluctuations of the surface potential", Alta Freq., 43, 56E-60 (1974), PA77-44998.
40. Baccarani, G. and Severi, M., "On the accuracy of the theoretical high-frequency semiconductor capacitance for inverted MOS-structures", IEEE Trans. Electron Devices, ED-21, 122-5 (1974), PA77-24830.
41. Barrie, R. and Cheung, C. Y., "The distribution function for impurity states in semiconductors", Canad. J. Phys., 44, 2231-40 (1966), EEA70-6209.
42. Boudry, M. R., "Theoretical origins of N_{ss} peaks observed in Gray-Brown MOS studies", Appl. Phys. Lett., 22, 530-1 (1973), EEA76-25992.
43. Edwards, J. R., "Investigations of electron scattering mechanisms at the silicon-silicon dioxide interface", Thesis, Univ. Illinois, Urbana, 1970, 149 pp., Order No. 70-20957, EEA74-22624.
44. Garcia, N. and Falicov, L. M., "Localized states of p-type inversion layers in semiconductor field-effect transistors", Phys. Rev. B, 11, 728-31 (1975), EEA78-15269.
45. Gupta, H. M. and van Overstraeten, R. J., "Theory of thermal dielectric relaxation and direct determination of trap parameters", J. Phys. C, 7, 3560-72 (1974), EEA78-586.
46. Hess, K. and Sah, C. T., "Dipole scattering at the $Si-SiO_2$ interface", Surf. Sci., 47, 650-4 (1975), EEA78-15281.
47. Iizuka, T. and Sugano, T., "Calculations of energy levels of oxygen and silicon vacancies at the $Si-SiO_2$ interface", Jap. J. Appl. Phys., 12, 73-9 (1973), EEA76-12513.
48. Krambeck, R. H., "Numerical calculation of impurity redistribution during thermal oxidation of semiconductors", J. Electrochem. Soc., 121, 588-91 (1974), EEA77-22890.
49. Krowne, C. M. and Holm-Kennedy, J. W., "Energy relaxation of electrons in the (100) n-channel of a Si-mosfet. I. Bulk phonon treatment", Surf. Sci., 46, 197-321 (1974), EEA78-19739.

50.Krowne, C. M. and Holm-Kennedy, J. W., "Energy relaxation of electrons in the (100) n-channel of a Si-mosfet. II. Surface phonon treatment", Surf. Sci., 46, 232-50 (1974), EEA78-19740.
51.Ning, T. H. and Sah, C. T., "Theory of scattering of electrons in a nondegenerate-semiconductor-surface inversion layer by surface-oxide charges", Phys. Rev. B, 6, 4605-13 (1972), EEA76-8588.
52.Noti, G., "Normalization of wave functions in an MIS structure", Solid-State Electron., 15, 723-5 (1972), EEA75-24368.
53.Pepper, M., Pollitt, S. and Adkins, C. J., "The spatial extent of localized state wavefunctions in silicon inversion layers", J. Phys. C, 7, L273-7 (1974), PA77-70072.
54.Rutledge, J. L. and Armstrong, W. E., "Effective surface mobility theory", Solid-State Electron., 15, 215-19 (1972), EEA75-13652.
55.Williams, R., "Image force interactions at the interface between an insulator and a semiconductor", J. Appl. Phys., 45, 1239-42 (1974), PA77-57172.
56.Wu, C. Y. and Thomas, G., "Theory of the electron mobility in inversion layers on oxidized silicon surface at room temperature", Phys. Rev. B, 6, 4581-8 (1972), EEA76-4580.
57.Wu, C. Y. and Thomas, G., "Two-dimensional electron-lattice scattering in thermally oxidized silicon surface-inversion layers", Phys. Rev. B, 9, 1724-32 (1974), EEA77-22754.
58.Zakharov, A. K. and Neizvestnyi, I. G., "Thermal generation and recombination of carriers at non-equilibrium variations of the inversion layer", Phys. Status Solidi A, 30, 419-27 (1975), PA78-72911.

V. Electrical Properties

1. Dielectric Breakdown

59.Altchek, L. and Klein, N., "Influence of the law of electrical conductivity on thermal switching and breakdown", IEEE Trans. Electron. Devices, ED-20, 801-11 (1973).
60.Bardhan, A. R., Srivastava, P. C., Bhattacharya, I. B. and Bhattacharya, D. L., "Electrical breakdown in very thin Al_2O_3 films", Int. J. Electron., 39, 343-51 (1975), EEA78-41522.
61.Bhattacharyya, A., Chaudhari, P. K. and Joshi, M. L., "Degradation in thin dielectric films", Presented at Electrochemical Society Meeting, San Francisco, Calif., May 1974, PA77-45251.
62.Carnes, J. E. and Duffy, M. T., "Self-healing breakdown measurements of pyrolytic aluminium oxide films on silicon", J. Appl. Phys., 42, 4350-6 (1971), EEA75-1076.
63.Chou, N. J. and Eldridge, J. M., "Effects of material and processing parameters on the dielectric strength of thermally grown SiO_2 films", J. Electrochem. Soc., 117, 1287-93 (1970), EEA74-7656.
64.DiStefano, T. H., "Dielectric breakdown induced by sodium in MOS structures", J. Appl. Phys., 44, 527-8 (1973), EEA76-12509.

65. Garbarino, P. L. and Sandison, R. D., "Nondestructive location of oxide breakdowns on MOSFET structures", J. Electrochem. Soc., 120, 834-5 (1973), EEA76-29271.
66. Gould, R. D. and Hogarth, C. A., "Current voltage characteristics, dielectric breakdown and potential distribution measurement in Au-SiO_x-Au thin film diodes and triodes", Int. J. Electron., 37, 157-75 (1974), EEA78-2770.
67. Grabowski, W., "The physical mechanism of breakdown and measurement methods of breakdown voltages in MOS transistors", Elektronika, no. 12, 497-500 (1972), Polish, EEA76-12639.
68. Green, J. M., Osburn, C. M. and Sedgwick, T. O., "The influence of silicon heat treatments on the minority carrier generation and the dielectric breakdown in MOS structures", J. Electron. Mater., 3, 579-99 (1974), EEA77-30230.
69. Hilbiber, D. F. and Lucas, R. C., "Nondestructive determination of MOSFET gate breakdown voltage", Intervia, 14, 402-3 (1967), EEA70-20829.
70. Inagaki, T. and Nishimura, Y., "Dielectric breakdown voltage characteristics of evaporated silicon oxide films", Fijitsu Sci. Tech. J., 5, 103-22 (1969), EEA73-13535.
71. Klein, N., "The mechanism of self-healing electrical breakdown in MOS structures", IEEE Trans. Electron Devices, ED-13, 788-805 (1966), PA70-21060.
72. Laverty, S. J. and Ryan, W. D., "Electrical breakdown of MOS structures and its dependence upon the oxidation process", Internat. J. Electronics, 26, 471-6 (1969), EEA73-10683.
73. Lenzlinger, M., "Gate protection of MIS devices", IEEE Trans. Electron Devices, ED-18, 249-57 (1971), EEA74-13255.
74. Leuing, A. F., "MOS gate insulators for single crystal CdS", Solid-State Electron., 17, 1213-14 (1974), EEA78-596.
75. Linholm, L. W. and Plachy, R. F., "Electrostatic gate protection using an arc gap device", Proceedings of the 11th Annual Reliability Physics Conference, 1973, p. 198-202, EEA77-6044.
76. McCaughan, D. V. and Heilig, J. A., Jr., "Dielectric strength and interface-state behaviour of oxygen plasma-grown SiO_2 films annealed at high temperature", Int. J. Electron., 34, 737-40 (1973), EEA76-26000.
77. Mitros, J. and Tworek, E., "The effect of condensation heat of gate electrode metal upon stability and dielectric strength of MOS structures", Electron Technol., 6 (1-2), 75-81 (1973), PA77-53150.
78. Morgan, D. V., Howes, M. J., Pollard, R. D. and Waters, D. G. P., "Electroforming and dielectric breakdown in thin aluminium oxide films", Thin Solid Films, 15, 123-31 (1973), Russian, EEA76-19880.

79.Nekraso, M. M., Kletchenkov, I. I., Kizema, I. G., Karplyuk, A. I. and Chuchko, A. A., "Investigation of the influence of some technological factors on the electrical strength of the oxide film gate in m.o.s.-structures", Dielektr. & Poluprovod, no. 6, 72-5 (1974), Russian, PA78-50689.

80.Nishimura, Y., Inagaki, T. and Sasaki, H., "Dielectric breakdown voltage characteristics of evaporated silicon oxide films", Electronics Commun., 52 (3), 116-24 (1969), EEA73-13543.

81.Osburn, C. M., "Measuring breakdown voltage distribution of an array of MOSFET's", IBM Tech. Disclosure Bull., 14, 502-3 (1971), EEA74-37309.

82.Osburn, C. M., "Dielectric breakdown in SiO_2 films on silicon", Presented at Spring Meeting of the Electrochemical Society, Washington, D. C., May 1971, EEA75-4606.

83.Osburn, C. M., "Dielectric breakdown properties of SiO_2 films grown in halogen and hydrogen-containing environments", J. Electrochem. Soc., 121, 809-15 (1974), EEA77-30226.

84.Osburn, C. M. and Bassous, E., "Improved dielectric reliability of SiO_2 films with polycrystalline silicon electrodes", J. Electrochem. Soc., 122, 89-92 (1975), EEA78-8484.

85.Osburn, C. M. and Chou, N. J., "Accelerated dielectric breakdown of silicon dioxide films", J. Electrochem. Soc., 120, 1377-84 (1973), EEA77-342.

86.Osburn, C. M. and Ormond, D. W., "Dielectric breakdown in silicon dioxide films on silicon. I. Measurement and interpretation", J. Electrochem. Soc., 119, 591-7 (1972), EEA75-23671.

87.Osburn, C. M. and Ormond, D. W., "Dielectric breakdown in silicon dioxide films on silicon. II. Influence of processing and materials", J. Electrochem. Soc., 119, 597-603 (1972), EEA75-23672.

88.Osburn, C. M. and Ormond, D. W., "Sodium-induced barrier-height lowering and dielectric breakdown on SiO_2 films on silicon", J. Electrochem. Soc., 121, 1195-8 (1974), EEA78-580.

89.Osburn, C. M. and Ormond, D. W., "The effect of silicon wafer imperfections on minority carrier generation and dielectric breakdown in MOS structures", J. Electrochem. Soc., 121, 1229-33 (1974), EEA78-582.

90.Osburn, C. M. and Raider, S. I., "The effect of mobile sodium ions on field enhancement dielectric breakdown in SiO_2 films on silicon", J. Electrochem. Soc., 120, 1369-76 (1973), EEA77-341.

91.Osburn, C. M. and Weitzman, E. J., "Electrical conduction and dielectric breakdown in silicon dioxide films on silicon", J. Electrochem. Soc., 119, 603-9 (1972), EEA75-24275.

92.Perelygin, A. I., Sadofev, Y. G., Vellikok, L. F., Komarov, I. N. and Nyukhitkov, V. M., "Electrical strength of thin dielectric films in MDS structures", Izv. VUZ Fiz., no. 9, 100-3 (1974), Russian, PA78-10077.

93.Raider, S. I., "Time-dependent breakdown of silicon dioxide films", Appl. Phys. Lett., 23, 34-6 (1973), EEA76-29186.

94.Ridley, B. K., "Mechanism of electrical breakdown in SiO_2 films", J. Appl. Phys., 46, 998 (1975).

95.Seung, P. L., "A measurement technique of time-dependent dielectric breakdown in MOS capacitors", Microelectron. & Reliab., 13, 209-14 (1974), EEA77-30231.
96.Svensson, C. and Shumka, A., "Time dependent breakdown in silicon dioxide films", Int. J. Electron., 38, 69-80 (1975), EEA78-8479.
97.Tsuchida, N. and Ueda, M., "The influence of cathode metal on dielectric breakdown strength of evaporated SiO thin films", Electr. Eng. Jap., 93 (4), 1-7 (1973), EEA78-11535.
98.Vromen, B. H., "Nitrogen-silicon reaction and its influence on the dielectric strength of thermal silicon dioxide", Appl. Phys. Lett., 27, 152-4 (1975), PA78-86387.
99.Wang, P. and van Buren, N., "SEM observation of self-healing breakdown in thermally grown SiO_2 films", J. Electrochem. Soc., 117, 127-8 (1970), EEA73-16986.
100.Wilmsen, C. W., Vasbinder, G. C. and Chan, Y. K., "Electrical conduction through thermal and anodic oxides in InSb", J. Vac. Sci. & Technol., 12, 56-9 (1975), PA78-50685.
101.Worthing, F. L., "DC dielectric breakdown of amorphous SiO_2 films at room temperature", J. Electrochem. Soc., 115, 88-92 (1968).
102.Yang, D. Y., Johnson, W. C. and Lampert, M. A., "Scanning electron micrographs of self-quenched breakdown regions in Al-SiO_2-(100)Si structures", Appl. Phys. Lett., 25, 140-2 (1974), EEA77-38778.
103.Zalar, S. M., "A microphase mechanism of subintrinsic dielectric breakdown in MIS structures", Presented at Electrochemical Society Spring Meeting, Toronto, Canada, May 1975.

2. Dielectric Properties

104.Burkhardt, P. J., "Dielectric relaxation in thermally grown SiO_2 films", IEEE Trans. Electron Devices, ED-13, 268-75 (1966).
105.Butlin, R. S., "Microwave integrated circuits", Thesis, Univ. Warwick, Coventry, England, 1973, EEA77-38875.
106.Danilina, T. I., Motoshkin, V. V., Vedernikov, V. A. and Mukhacheva, N. S., "The moisture-stability of SiO_2 films", Izv. VUZ Fiz., no. 7, 125-7 (1972), Russian, EEA75-35703.
107.Hartl, M., "Insulating layers for field-effect transistors", Nachrichtentech., 23, 501-6 (1970), German, EEA74-4933.
108.Kosenko, V. Y., Onipko, A. F., Svechnikov, S. V. and Stepanchuk, V. P., "Some characteristics of SiO_2 films, obtained by sputtering of quartz in a high-frequency discharge", Poluprovodn. Tekh. & Mikroelektron., no. 17, 29-33 (1974), Russian, EEA78-2958.
109.Kriegler, R. J. and Bartnikas, R., "Dielectric relaxation in Si-SiO_2-Cr structures", IEEE Trans. Electron Devices, ED-17, 1010-11 (1970), EEA74-1497.

110. Marciniak, W., "Low frequency dielectric relaxation in SiO_2 layers on Si substrates", Bull. Acad. Pol. Sci. Ser. Sci. Tech., 18 (11), 93-9 (1970), EEA74-15499.
111. Petruchuk, I. I. and Sazonova, E. V., "Relaxation polarization in SiO_2 and surface potential stability in the metal-dielectric-semiconductor system", Elektron. Tekh., 12, no. 4, 66-8 (1971), Russian, EEA75-4630.
112. Sedgwick, T. O. and Krongelb, S., "The effect of water on the dielectric constant of vapour deposited silica films", J. Electrochem. Soc., 118, 349-51 (1971), EEA74-18696.
113. Texas Instruments Inc., "Integrated circuit production", Patent UK 1162500, Publ. August 1969, EEA72-23888.

3. Junction Breakdown

114. Antognetti, P. and Bulucca, C., "Deep depletion and avalanche studies of the miniature avalanche MOS diode", Proceedings of the International Conference on the Physics and Chemistry of Semiconductor Heterojunctions and Layer Structures, 1970, p. V-55-63, Publ. 1971, EEA74-26897.
115. Armstrong, G. A. and Magowan, J. A., "The distribution of mobile carriers in the pinch-off region of an insulated-gate field-effect transistor and its influence on device breakdown", Solid-State Electron., 14, 723-33 (1971), EEA74-30835.
116. Bakowski, M. and Hansson, B., "Influence of bevel angle and surface charge on the breakdown voltage of negatively beveled diffused p-n junctions", Solid-State Electron., 18, 651-7 (1975), EEA78-28712.
117. Bateman, I. M., Armstrong, G. A. and Magowan, J. A., "Drain voltage limitations of MOS transistors", Solid-State Electron., 17, 539-50 (1974), EEA77-23066.
118. de Graff, H. C., "Gate-controlled surface breakdown in silicon p-n junctions", Philips Res. Rep., 25, 21-32 (1970), EEA73-21329.
119. Dunn, P. J. and Mellor, P. J. T., "The degradation of MOS transistors resulting from junction avalanche breakdown", Microelectron. & Rel., 11, 369-76 (1972), EEA75-39946.
120. Matsumoto, K. and Haneta, Y., "Effect of thin oxide film on breakdown voltage of silicon N^+P junction", Jap. J. Appl. Phys., 13, 367-8 (1974), EEA77-14528.
121. Muller, R. G., "Gate-enhanced vs channel-current induced breakdown for floating gate avalanche injection", Solid-State Electron., 17, 503-5 (1974), EEA77-23064.
122. Nekrasov, M. M., Kletchenkov, I. I., Kizema, I. G., Karplyuk, A. I. and Chuchko, A. A., "On the question of increasing breakdown voltage in planar p-n junctions", Dielektr. & Poluprovod, no. 6, 68-71 (1974), Russian, EEA78-24286.
123. Neugebauer, C. A., Burgess, J. F., Joynson, R. E. and Mundy, J. L., "Channel shortening in MOS transistors during junction walkout", Appl. Phys. Lett., 19, 287-9 (1971), EEA75-1155.

124.Nicollian, E. H. and Berglund, C. N., "Creation of interface states by avalanche effect", Sixth Annual Reliability Physics Symposium Proceedings, 1967, p. 66-7, Publ. 1968, EEA71-23300.
125.Tsujide, T., "New phenomena on drain-to-source breakdown voltage of n-channel FET", Appl. Phys. Lett., 25, 230-3 (1974), EEA77-38837.
126.Wanlass, F. M., "Diode with sharp reverse-bias breakdown characteristic", Patent USA 3423606, Publ. January 1969, EEA72-15573.

4. Noise Effects

127.Bell, D. A., "1/f noise", Radio Electronic Engr., 38, 199-200 (1969), EEA73-17669.
128.Berz, F., "Theory of low frequency noise in Si MOST's", Solid-State Electron., 13, 631-47 (1970), EEA73-25051.
129.Bhat, T. A. R., "Current noise in Si-MOSFET's", Arch. Elektron. & Ubertragungstech., 26, 343-8 (1972), EEA75-28506.
130.Cheng, Y. C., Kriegler, R. J., Haslett, J. W., Kendall, E. J. M. and Scholz, F. J., "Low-frequency noise in HCl processed IGFET's", Proc. IEEE, 62, 859-60 (1974), EEA77-26486.
131.Christensson, S. and Lundstrom, I., "Low frequency noise in MOS transistors: II experiments", Solid-State Electron., 11, 813-20 (1968), EEA71-26023.
132.Christensson, S., Lundstrom, I. and Svensson, C., "Low frequency noise in MOS transistors: I Theory", Solid-State Electron., 11, 797-812 (1968), EEA71-26022.
133.Churchill, M. J. and Lauritzen, P. O., "Carrier density fluctuation noise in silicon junction field effect transistors at low temperatures", Solid-State Electron., 14, 985-93 (1971), EEA75-1171.
134.Das, M. B., "FET Noise sources and their effects on amplifier performance at low frequencies", IEEE Trans. Electron Devices, ED-19, 338-48 (1972), EEA75-10269.
135.Das, M. B. and Moore, J. M., "Measurements and interpretation of low-frequency noise in FET's", IEEE Trans. Electron Devices, ED-21, 247-57 (1974), EEA77-14525.
136.Dill, H. G., "Noise contribution of the offset-gate IGFET with an additional gate electrode", Electron. Lett., 3, 341-2 (1967), EEA71-1275.
137.Elliott, D. A., "Low temperature noise in MOST amplifiers", Solid-State Electron., 14, 1041-7 (1971), EEA75-1173.
138.Flinn, I., Bew, G. and Berz, F., "Low frequency noise in MOS field effect transistors", Solid-State Electron., 10, 833-45 (1967), PA70-38717.
139.Fu, H. S., "Surface 1/f noise on oxidized silicon", Thesis, Univ. Illinois, Urbana, 1971, 161 pp., Order No. 71-21123, EEA75-28524.

140.Fu, H. S. and Sah, C. T., "Theory and experiments on surface 1/f noise", IEEE Trans. Electron Devices, ED-19, 273-85 (1972), EEA75-10266.
141.Halladay, H. E. and van der Ziel, A., "Test of the thermal-noise hypothesis in MOSFET's", Electron. Lett., 4, 336-7 (1968), EEA71-26025.
142.Halladay, H. E. and van der Ziel, A., "On the high frequency excess noise and equivalent circuit representation of the MOSFET with n-type channel", Solid-State Electron., 12, 161-76 (1969), EEA72-20439.
143.Haslett, J. W. and Kendall, E. J. M., "Gate flicker noise in MOSFET's", Proc. IEEE, 60, 1111-12 (1972), EEA75-36303.
144.Haslett, J. W. and Trofimenkoff, F. N., "Gate noise in MOSFET's at moderately high frequencies", Solid-State Electron., 14, 239-45 (1971), EEA71-18815.
145.Haslett, J. W. and Trofimenkoff, F. N., "Effects of the substrate on surface state noise in silicon MOSFET's", Solid-State Electron., 15, 117-31 (1972), EEA75-7551.
146.Hawkins, R. J. and Bloodworth, G. G., "Two components of 1/f noise in MOS transistors", Solid-State Electron., 14, 929-32 (1971), EEA75-1167.
147.Hawkins, R. J., Mansour, I. R. M. and Bloodworth, G. G., "The spectrum of current noise in MOST's at very low frequencies", Brit. J. Appl. Phys. (J. Phys. D), 2, 1059-62 (1969), EEA72-20426.
148.Hielscher, F. H. and End, J. H., III, "Effects of oxygen-reduction treatment on the threshold voltage and low-frequency noise of MOS transistors", Appl. Phys. Lett., 24, 27-9 (1974), EEA77-18804.
149.Hsu, S. T., "Surface state related 1/f noise in MOS transistors", Solid-State Electron., 13, 1451-9 (1970), EEA74-1612.
150.Hsu, S. T., Fitzgerald, D. J. and Grove, A. S., "Surface-state related 1/f noise in p-n junctions and MOS transistors", Appl. Phys. Lett., 12, 287-9 (1968), EEA71-20570.
151.Huang, C. and van der Ziel, A., "A new type of flicker noise in microwave MOSFETs", Solid-State Electron., 18, 885-6 (1975), EEA78-37549.
152.Jones, B. L. and Hurlston, R. E., "VHF noise due to surface states in MOS devices", Proc. IEEE, 58, 152-3 (1970), EEA73-17667.
153.Kakovkin, V. V., Zaitsev, B. D. and Shvedov, Y. Y., "Measurement of very low frequency noise in MOS transistors by a frequency conversion method", Izv. VUZ Radioelektron., 14, 67-71 (1971), Russian, EEA74-15568.
154.Katto, H., "Low frequency noise in MOSFET", Oyo Buturi, 43, 1164-7 (1974), Japanese, EEA78-24360.
155.Katto, H., Kamigaki, Y. and Itoh, Y., "MOSFET's with reduced low frequency 1/f noise", Oyo Buturi, 44 (Suppl.), 243-8 (1975), EEA78-41617.
156.Kirk, E. W. and van der Ziel, A., "Induced gate noise in MOS FET's", Solid-State Electron., 14, 945-8 (1971), EEA75-1169.

157.Klaassen, F. M., "Noise of MOSFET's at very high frequencies", Conference on Physical Aspects of Noise in Electronic Devices, 1968, p. 91-4, EEA72-20446.
158.Klaassen, F. M., "A computation of the high-frequency noise quantities of a MOSFET", Philips Res. Rep., 24, 559-71 (1969), EEA73-17666.
159.Klaassen, F. M., "On the geometrical dependence of 1/f noise in MOS transistors", Philips Res. Rep., 25, 171-4 (1970), EEA73-31419
160.Klaassen, F. M., "On the substrate current noise in MOS transistors beyond pinchoff", Proc. IEEE, 59, 331-2 (1971), EEA74-13262.
161.Klaassen, F. M., "Characterization of low 1/f noise in MOS transistors", IEEE Trans. Electron Devices, ED-18, 887-91 (1971), EEA74-34082.
162.Klaassen, F. M. and Prins, J., "Thermal noise of MOS transistors", Philips Res. Rep., 22, 505-14 (1967), EEA71-3436.
163.Klaassen, F. M. and Prins, J., "Noise of VHF and UHF MOS tetrodes", Philips Res. Rep., 23, 478-84 (1968), EEA72-20435.
164.Kobayashi, I., Nakahara, M. and Atsumi, M., "An investigation of the relation between the interface state and 1/f noise behaviors in silicon MOS transistor with the negative bias-heat treatment", Proc. IEEE, 61, 1145-6 (1973), EEA76-32276.
165.Lee, S. J. and Van Der Ziel, A., "Flicker noise compensation in high-impedance MOSFET circuits", Solid-State Electron., 16, 1301-2 (1973), EEA77-2572.
166.Leuenberger, F., "1/f noise of MOS transistors", Helv. Phys. Acta, 41, 448-50 (1968), German, EEA72-2808.
167.Leuenberger, F., "Absence of 1/f noise in M.O.S. transistors operated in saturation", Electron. Lett., 7, 561 (1971), EEA75-1158.
168.Leuenberger, F., "Charge pumping and low-frequency noise in MOS structures", Phys. Status Solidi A, 8, 545-50 (1971), EEA75-7550.
169.Leupp, A. and Strutt, M. J. O., "Noise behaviour of the MOSFET at VHF and UHF", Electron. Lett., 4, 313-14 (1968), EEA71-26027.
170.Leupp, A. and Strutt, M. J. O., "High-frequency FET noise parameters and approximation of the optimum source admittance", IEEE Trans. Electron Devices, ED-16, 428-31 (1969), EEA72-17739.
171.Leventhal, E. A., "Derivation of 1/f noise in silicon inversion layers from carrier motion in a surface band", Solid-State Electron., 11, 621-7 (1968), EEA71-23369.
172.Mansour, I. R. M., Hawkins, R. J. and Bloodworth, G. G., "Measurement of current noise in MOS transistors from 5 x 10^{-5} to 1 Hz", Radio Electronic Engr., 35, 212-6 (1968), EEA71-15856.
173.Mantena, N. R. and Lucas, R. C., "Experimental study of flicker noise in MIS field-effect transistors", Electron. Lett., 5, 607-8 (1969), EEA73-10675.
174.Mavor, J., "Low-noise operation of MOS transistors in common-gate connection", Electron. Lett., 3, 406-7 (1967), EEA71-1272.

175. Mavor, J., "Noise parameters for metal-oxide-semiconductor transistor", Proc. Instn. Elect. Engrs., 113, 1463-7 (1966), EEA70-1055.
176. Mavor, J. and Reed, K. B., "Equivalent two-port thermal-noise representation of metal-oxide semiconductor transistors", IEEE Trans. Electron Devices, ED-14, 111-12 (1967), EEA70-9105.
177. Mochkin, V. S., Andreevskii, B. M. and Peresvetov, N. A., "Low-frequency noise of MOS-transistors with a built-in channel", Izv. VUZ Radioelektron., 18 (3), 10-16 (1975), Russian, EEA78-24355.
178. Mochkin, V. S. and Shvedov, Y. Y., "Investigation of low-frequency noise in MOS transistors", Izv. VUZ Radioelektron., 16 (7), 27-36 (1973), Russian, EEA76-35326.
179. Nakahara, M., "Anomalous low-frequency noise in MOS transistors at low temperatures", Electron. & Commun. Jap., 55 (6), 99-105 (1972), EEA76-29268.
180. Nakahara, M., Iwasaw, H. and Yosutaka, K., "Anomalous low-frequency noise enhancement beyond pinch-off in silicon n-channel MOS transistors", Proc. IEEE, 57, 2177-8 (1969), EEA73-14323.
181. Naryshkin, A. K. and Gerasimova, T. I., "Experimental investigation of the noise of MOS transistors", Telecommun. & Radio Eng. Pt. 2, 27 (10), 124-5 (1972), EEA76-38280.
182. Paul, R., "Thermal noise of MOS transistors", Nachrichtentechnik, 17, 458-66 (1967), EEA71-13322.
183. Paul, R., "Noise and noise factor of MOS transistors in the low and medium frequency ranges", Nachrichtentechnik, 19, 67-70 (1969), German, EEA72-9600.
184. Paul, R., "Physical causes of noise in MIS transistors. I", Nachrichtentechnik, 20, 259-62 (1970), German, EEA73-35236.
185. Paul, R., "Physical causes of noise in MIS transistors. II. Pt. III", Nachrichtentechnik, 20, 289-94 (1970), German, EEA73-38109.
186. Potemkin, V. V. and Kashkarov, P. K., "The noise of field triodes with an insulated gate at low frequencies", Radio Eng. & Electron Phys., 17, 518-20 (1972), EEA76-8695.
187. Rao, P. S. and van der Ziel, A., "Noise and y-parameters in MOS FET's", Solid-State Electron., 14, 939-44 (1971), EEA75-1168.
188. Rigaud, D., Nougier, J. P., Touboul, A. and Lecoy, G., "Analysis of the thermal noise behaviour in j.g.f.e.t. at low temperature", Physica, 56, 161-7 (1971), EEA75-1164.
189. Rogers, C. G., "Low frequency noise in MOST's at cryogenic temperatures", Solid-State Electron., 11, 1099-1104 (1968), EEA72-5042.
190. Ronen, R. S., "Low-frequency 1/f noise in MOSFET's", RCA Rev., 34, 280-307 (1973), EEA77-381.
191. Sah, C. T. and Hielscher, F. H., "Evicence of the surface origin of the 1/f noise", Phys. Rev. Lett., 17, 956-8 (1966), EEA70-2617.
192. Shoji, M., "Analysis of high-frequency thermal noise of enhancement mode MOS field-effect transistors", IEEE Trans. Electron Devices, ED-13, 520-4 (1966), PA70-1979.

193. Soares, R. A., "Noise in MOS transistors", Des. Electron., 8 (7-8), 25-30 (1971), EEA74-22726.
194. Strutt, M. J. O., "Noise and flicker in field effect transistors", Scienta Elec., 16 (3), 86-103 (1970), German, EEA74-15574.
195. Takagi, K., "Noise in MOS field-effect tetrodes", Rec. Elec. & Commun. Eng. Conversazione Tohoku Univ., 40 (2), 53-9 (1971), Japanese, EEA75-16753.
196. Takagi, K., "Drain noise in field-effect transistors", Bull. Kyushu Inst. Technol. (Sci. & Technol.), no. 27, 125-32 (1973), Japanese, EEA77-10252.
197. Takagi, K. and Van der Ziel, A., "Non-thermal noise in MOSFET's and MOS tetrodes", Solid-State Electron., 12, 907-13 (1969), EEA73-10691.
198. Tanaka, T., Nagano, K. and Nameki, N., "The 1/f noise MOS transistors", Japan J. Appl. Phys., 8, 1020-6 (1969), EEA73-4523.
199. Van der Ziel, A., "Noise in junction-and MOSFET's at high temperatures", Solid-State Electron., 12, 861-6 (1969), EEA73-10689.
200. van der Ziel, A., "Limiting flicker noise in MOSFETs", Solid-State Electron., 18, 1031 (1975), EEA78-41624.
201. Yau, L. D. and Sah, C. T., "Theory and experiments of low-frequency generation-recombination noise in MOS transistors", IEEE Trans. Electron Devices, ED-16, 170-7 (1969), EEA72-11330.
202. Yau, L. D. and Sah, C. T., "Geometrical dependences of the low-frequency generation-recombination noise in MOS transistors", Solid-State Electron., 12, 903-5 (1969), EEA73-10690.
203. Yau, L. D. and Sah, C. T., "On the 'excess white noise' in MOS transistors", Solid-State Electron., 12, 927-36 (1969), EEA73-10692.
204. Zaitsev, B. D., Shverov, E. E. and Zaika, V. V., "Dependence of low-frequency noise of MOS transistors on operating conditions and temperature", Izv. VUZ Radioelektron., 13, 1159-60 (1970), Russian, EEA74-4930.
205. Zalud, V., "Thermal noise of MOS transistors", Sdelovaci Tech., 16, 436-7 (1968), Czech., EEA73-1182.

5. Temperature Effects

206. Baland, B. and Pinard, P., "Influence of temperature on MOS structures in the inversion region", Thin Solid Films, 10, 449-53 (1972), French, EEA75-24288.
207. Barker, R. W. J. and Hart, B. L., "Determination of zero-thermal-drift operating point in m.o.s.f.e.t.s.", Des. Electron., 9, 36-7 (1971), EEA75-1157.
208. Brindha, T. M. and Suryan, G., "Temperature dependence of the I-V characteristics in Ag-SiO_2-SiC (P-type) (MOS) structures", Presented at Nuclear Physics and Solid State Physics Symposium, Roorkee, India, December 1969, EEA73-24916.
209. Cheng, Y. C., "On the temperature dependence of MOS threshold", Phys. Lett., 29A, 444-5 (1969), EEA73-7475.

210.Cobbold, R. S. C., "Temperature effects in MOS transistors", Electron. Lett., 2, 190-1 (1966), EEA70-6811.
211.Craig, G. R. and Sesnic, S. S., "Thermal effects in JFET and MOSFET devices at cryogenic temperatures", 1970 SWIEECO Record of Technical Papers of the 22nd Southwestern IEEE Conference and Exhibition, 1970, p. 241-5, EEA73-35245.
212.Fedotov, Y. A., Chernyavskii, I. A., Turinov, V. I. and Koshkin, V. S., "The effect of thermal treatment on the properties of MOS structures", Latv. PSR Zinat. Akad. Vestis Fiz. Tehn. Ser., no. 1, 57-9 (1971), Russian, EEA74-22617.
213.Gardner, C. G., Jr. and Bostian, C. W., "Pinch-off voltage shifts in MOSFET's at helium temperatures", Proc. IEEE, 56, 2082 (1968), EEA72-2828.
214.Goetzberger, A., "Behavior of MOS inversion layers at low temperature", IEEE Trans. Electron Devices, ED-14, 787-9 (1967), EEA71-6030.
215.Goetzberger, A. and Irvin, J. C., "Low-temperature hysteresis effects in metal-oxide-silicon capacitors caused by surface-state trapping", IEEE Trans. Electron Devices, ED-15, 1009-14 (1968), EEA72-7928.
216.Goetzberger, A. and Nicollian, E. H., "Temperature dependence of inversion-layer frequency response in silicon", Bell Syst. Tech. J., 46, 513-22 (1967), PA70-38719.
217.Gorban, A. P., Litovchenko, V. G. and Peikov, P. K., "Effect of low-temperature treatment, on electrical and recombination characteristics of silicon/silicon-dioxide systems", Poluprovodn. Tekh. & Mikroelektron., no. 10, 7-22 (1972), Russian, EEA76-16325.
218.Green, R. R., "MOSFET operation at 4.2°K", Rev. Opt., 39, 1495-7 (1968), EEA73-4524.
219.Hasegawa, K., Morita, H., Enomoto, T. and Fukuwatari, H., "The influence of heat-treatment on the Si-SiO_2 interface states in the poly Si-SiO_2-Si system", Presented at Electrochemical Society Meeting, San Francisco, California, May 1974, EEA77-26406.
220.Iida, K., "The effects of heat treatment on the interface characteristics in the Si-Al_2O_3 and Si-SiO_2-Al_2O_3 systems", Jap. J. Appl. Phys., 11, 288-95 (1972), EEA75-13648.
221.Kano, K. and Kunig, H., "Effect of thermal instability on ultrahigh frequency performance of insulated-gate field-effect transistors", Solid-State Electron., 12, 719-29 (1969), PA73-15245.
222.Kasabov, I., Ilieva, M. and Kovacheva, I., "Effect of thermal-field treatment on the parameters of MOS structures", Izv. Fiz. Inst. ANEB, 24, 67-70 (1974), PA77-60862.
223.Kassabov, J. D., Ilieva, M. N. and Kovacheva, I. S., "Temperature investigations of the system Au/W-SiO_2-Si", C. R. Acad. Bulg. Sci., 26, 23-5 (1973), EEA76-29188.
224.Kirschner, N., "The thermal resistance of integrated MOS transistors", Microelectron. & Reliab., 14, 37-9 (1975), EEA78-15363.

225. Kolenstov, K., Peikov, P., Smirnov, N. and Dimov, V., "Influence of thermal treatment on system Si-pyrolytic SiO_2 in nitrogen atmosphere", C. R. Acad. Bulg. Sci., 23, 1477-80 (1970), EEA74-26700.
226. Kolodziejski, J. F., "Some results of testing m.o.s. transistors at elevated temperatures", Electron. Lett., 8, 418-19 (1972), EEA75-32676.
227. Kolodziejski, J. F., "Bias-temperature treatment and avalanching of p-n junctions in MOS transistors", Electron Technol., 6 (1-2), 135-42 (1973), EEA77-26467.
228. Kriegler, R. J. and Devenyi, T. F., "Temperature-bias aging of (HCl) MOS structures", Proceedings of the 11th Annual Reliability Physics Conference, 1973, p. 153-8, EEA77-6041.
229. Mak, M. M. and Hampshire, M. J., "MOS characteristics at low temperatures", J. Phys. D, 7, L191-5 (1974), EEA78-2772.
230. Nannoni, R., "Trial interpretation of the variations of the complex capacity of a metal/oxide/semiconductor (MOS) structure as a function of the frequency, the temperature and the direct potential difference", C. R. Acad. Sci. B, 265, 1259-62 (1967), French, PA71-18142.
231. Nassibian, A. G. and Sproul, M. E., "Temperature dependence of gold-doped MOSFET devices", Proc. Inst. Radio Electron. Eng. Aust., 32, 239-47 (1971), EEA74-26886.
232. Nassibian, A. G. and Sproul, M. E., "The influence of gold on (110) substrates and the temperature dependence of MOS systems", J. Phys. D, 8, 1507-12 (1975), EEA78-41527.
233. Nathanson, H. C., Jund, C. and Grosvalet, J., "Temperature dependence of apparent threshold voltage of silicon MOS transistors at cryogenic temperatures", IEEE Trans. Electron Devices, ED-15, 362-8 (1968), EEA71-23368.
234. Paul, R., "The temperature behaviour of MOS transistors", Nachrichtentechnik, 18, 292-6 (1968), German, EEA72-809.
235. Pelikan, L., "Temperature stabilisation of circuits with m.o.s. transistors", Sdelovaci Tech., 19, 284-6 (1971), Czech, EEA75-1179.
236. Peralta, L. and Thuillier, J. C., "Thermally stimulated currents from MOS structure on tellurium", Phys. Status Solidi A, 22, 473-7 (1974), PA77-45006.
237. Peykov, P. H., Kaschieva, S. B. and Matev, I. M., "Influence of thermal treatment on MOS structures with damaged Si-SiO_2 interface", C. R. Acad. Bulg. Sci., 27, 1477-80 (1974), EEA78-28680.
238. Rai, B. P., Singh, K. and Srivastava, R. S., "Temperature dependence of MOS devices", Phys. Status Solidi A, 28, 407-10 (1975), PA78-46767.
239. Reynolds, F. H., "The response of the threshold voltages of the transistors in simple MOS circuits to tests at elevated temperatures", Proceedings of the 9th Annual Reliability Physics Conference, 1971, p. 46-56, EEA72-3997.

240. Reynolds, F. H., "Thermally accelerated aging of semiconductor components", Proc. IEEE, 62, 212-22 (1974), EEA77-9972.
241. Reynolds, F. H., "Response of an m.o.s. integrated circuit to tests at elevated ambient temperatures", Proc. Inst. Electr. Eng., 121, 913-19 (1974), EEA78-2917.
242. Rogers, C. G., "MOST's at cryogenic temperatures", Solid-State Electron., 11, 1079-91 (1968), EEA72-818.
243. Schumacher, F. and Stoev, I., "Measurement of the thermal resistance of MIS varactors", Arch. Elektron. & Uebertragungstech., 28 (1), 12-14 (1974), EEA77-10249.
244. Sesnic, S. S. and Craig, G. R., "Thermal effects in JFET and MOSFET devices at cryogenic temperatures", IEEE Trans. Electron Devices, ED-19, 933-42 (1972), EEA75-28512.
245. Sesnic, S. S., Craig, G. R., Champion, J. W. and LesKovar, B., "Investigations of germanium FET and MOSFET at LHe temperatures", 21st Annual Southwestern IEEE Conference and Exhibition, 1969, 8 pp., EEA73-4544.
246. Singh, B. R. and Srivastava, R. S., "Effect of temperature on M-O-S capacitances", Int. J. Electron., 32, 347-51 (1972), EEA75-13762.
247. Skvortsov, I. M. and Nikolaeva, V. V., "Interaction of Si and SiO_2 during heat treatment of masked silicon plates in hydrogen", Izv. Akad. Nauk SSSR Neorg. Mater., 8 (9), 1587-90 (1972), Russian, EEA76-8519.
248. Sproul, M. E. and Nassibian, A. G., "Temperature dependence of turn-on voltage of gold-doped MOSFETs", IEEE Trans. Electron Devices, ED-22, 8-14 (1975), EEA78-5330.
249. Stillger, J., "Temperature dependence of the drain current of MOS-transistors", Z. Angew. Phys., 25, 177-82 (1968), German, EEA71-28903.
250. Sugano, T., Sakaki, H. and Hoh, K., "Effect of heat treatments on interface mobility of carriers in MOS structures", Annu. Rep. Eng. Res. Inst. Fac. Eng. Univ. Tokyo, 29, 143-51 (1970), PA74-16445.
251. Thuillier, J. C., "MOS capacitor on tellurium at low temperature", Phys. Status Solidi A, 6, 91-7 (1971), EEA74-65831.
252. Tokuda, A. R. and Lauritzen, P. O., "MOSFET thresholds at 4.2K induced by cooling bias", IEEE Trans. Electron Devices, ED-21, 606-7 (1974), EEA78-2872.
253. Urban, S., "Aging of semiconductor-oxide structures at elevated temperature", Phys. Status Solidi A, 23, 641-8 (1974), PA77-57177.
254. Urban, S., "Aging of semiconductor-oxide structures in an electric field", Phys. Status Solidi A, 26, 237-42 (1974), EEA78-2781.
255. Vadasz, L. and Grove, A. S., "Temperature dependence of MOS transistor characteristics below saturation", IEEE Trans. Electron Devices, ED-13, 863-6 (1966), EEA70-10884.

256.Valkov, S. A., "Temperature stabilization of MOS transistor gain", Proc. IEEE, 59, 1374-5 (1971), EEA74-34100.
257.Viswanathan, C. R. and Ogura, S., "Effect of heating under bias on photoelectric threshold in MOS structure", Appl. Phys. Lett., 12, 220-3 (1968), EEA71-15859.
258.Wang, R. and Dunkley, J., "Threshold voltage variations with temperature in MOS transistors", IEEE Trans. Electron Devices, ED-18, 386-8 (1971), EEA74-18794.
259.Wei, L. S. and Simmons, J. G., "Effect of temperature and voltage sweep rate of C-V characteristics of MIS capacitors", Solid-State Electron., 17, 1021-8 (1974), EEA78-591.
260.Yau, L. D. and Sah, C. T., "Temperature and field dependence of the generation-recombination noise and the thermal emission rates at the gold acceptor center in silicon", Solid-State Electron., 13, 1213-18 (1970), PA73-68877.
261.Zuleeg, R. and Lehovec, K., "Temperature dependence of the saturation current of MOST's", IEEE Trans. Electron Devices, ED-15, 987-9 (1968), EEA72-9597.

6. Frequency Dependence

262.Arandjelovic, V., "High-frequency MOST structures", Proc. IEEE, 58, 143-4 (1970), EEA73-14325.
263.Asami, Y., Matsubara, T. and Taki, Y., "Realization of negative-resistance and its applications on very high frequency oscillations by using MOS type FET", Technol. Rep. Seikei Univ., no. 11, 909-19 (1971), Japanese, EEA75-7552.
264.Balakirev, M. V. and Bogachev, V. M., "Frequency and transient characteristics of metal-oxide-semiconductor transistors and synthesis of their equivalent circuits", Radio Eng. & Electron. Phys., 16, 1884-94 (1971), EEA75-39952.
265.Beneking, H. and Kasugai, H., "Gigahertz p-channel enhancement silicon MOSFET with nonoverlapping gate", IEEE J. Solid State Circuits, SC-5, 328-30 (1970), EEA74-5116.
266.Burns, J. R., "High-frequency characteristics of the insulated-gate field-effect transistor", RCA Rev., 28, 385-418 (1967), EEA71-3438.
267.Calzolari, P. U., "Frequency limitations and optimal doping distributions for a field-effect transistor", Alta Frequenza, 36, 752-4 (1967), EEA71-1273.
268.Castagne, R., Hesto, P. and Vapaille, A., "The very low frequency response of the inversion layer of an MIS structure", C. R. Acad. Sci. B, 268, 1578-81 (1969), French, PA72-47147.
269.Das, M. B., "High-frequency network properties of MOS transistors including the substrate resistivity effects", IEEE Trans. Electron. Devices, ED-16, 1049-69 (1969), EEA73-10682.
270.Eaton, D. H. and Sah, C. T., "Frequency response of Si-SiO_2 interface states on thin oxide MOS capacitors", Phys. Status Solidi A, 12, 95-109 (1972), EEA75-32594.

271. Green, M. A., Temple, V. A. K. and Shewchun, J., "Frequency response of the current multiplication process in MIS tunnel diodes", Solid-State Electron., 18, 745-52 (1975), EEA78-28758.
272. Hopkins, R. S., Jr., "High frequency Y-parameters of the MOS transistor", Thesis, Rutgers State Univ., New Jersey, 1970, 88 pp., Order No. 71-3063, EEA74-30838.
273. Huneck, M., "High frequency properties of an n-channel MOSFET of the reduced-density mode. I", Nachrichtentechnik, 16, 451-5 (1966), German, EEA70-9106.
274. Johnson, H., "A high-frequency representation of the MOS transistor", Proc. IEEE, 54, 1970-1 (1966), EEA70-14607.
275. Katsube, T. and Adachi, Y., "Interface states of MOS diodes with thin SiO_2 films", Jap. J. Appl. Phys., 12, 320-1 (1973), EEA76-19864.
276. Katto, H. and Sah, C. T., "Frequency response of the surface state admittance in weakly inverted thin SiO_2-Si MOS capacitors", Phys. Status Solidi A, 13, 417-28 (1972), EEA76-4581.
277. McNutt, M. J., "Theory and experiments on high frequency metal-oxide-semiconductor capacitance", Thesis, Univ. Illinois, Urbana, 1974, 99 pp., Order No. 75-11694, EEA78-41543.
278. Mitchell, M. M., Dawson, R. and Ditrick, N. H., "High-frequency characteristics of the MOS tetrode", IEEE Northeast Electronics Research and Engineering Meeting, 1966, p. 132-3, EEA70-22911.
279. Mogensen, P. and Chan, W., "MOS multiplexer switches can do well at high frequencies", Electronics, 41 (23), 152-6 (1968), EEA72-799.
280. Nachmanson, R. C., Jorkow, W. G. and Dobrowolski, D. D., "Study of the metal insulator semiconductor structure (MIS) within a large frequency range", Nachrichtentechnik, 22 (2), 37-8 (1972), German, EEA75-13649.
281. Nagai, K., Hayashi, Y. and Tarui, Y., "A high-frequency equivalent circuit of MOS FET", Bull. Electrotech. Lab., 36, 34-40 (1972), Japanese, EEA75-36277.
282. Nakhmanson, R. S., "Frequency dependence of the photo-EMF of strongly inverted Ge and Si MIS structures. I. Theory", Solid-State Electron., 18, 617-26 (1975), PA78-62116.
283. Nakhmanson, R. S. and Erkov, V. G., "Measurement of the capacitance and conductance of a metal-insulator-semiconductor structure at sonic and subsonic frequencies", Instrum. & Exp. Tech., no. 3, 858-60 (1970), EEA74-18800.
284. Nakhmanson, R. S., Ovsyuk, Z. S. and Popov, L. K., "Frequency dependence of photo-EMF of strongly inverted Ge and Si MIS structures. II. Experiments", Solid-State Electron., 18, 627-34 (1975), PA78-62117.
285. Nuzillat, G. and Sixou, P., "Effect of the polarizing field on the radio-frequency spectrum of metal-anodic oxide-silicon structures", C. R. Acad. Sci. B, 266 (4), 218-21 (1968), French, PA71-18143.

286.Nuzillat, G. and Sixou, P., "Study of high density of surface states in MOS structures by means of the complex capacity variations as a function of frequency", J. Phys., 30, 71-81 (1969), French, PA72-21528.
287.Paul, R., "High frequency parameters of MOS transistors", Proceedings of the Second International Congress on Microelectronics, 1966, p. 53-74, Publ. 1967, German, EEA71-13324.
288.Paul, R., "High frequency properties of MOS transistors", Symposium on Test Methods and Measurements of Semiconductor Devices, 1967, Pap. No. 201, PA71-18146.
289.Paul, R., "Experimental high-frequency behaviour of MOS transistors", Nachrichtentechnik, 17, 255-60 (1967), German, EEA71-8272.
290.Paul, R., "Contributions to the direct-current and low-frequency behaviour of the MOS transistor", Nachrichtentechnik, 17, 87-92 (1967), German, EEA70-20848.
291.Pierret, R. F., "Frequency variation of the MOS-transistor V_D=0 admittance", Solid-State Electron., 11, 253-60 (1968), EEA71-8277.
292.Saks, N. S., "Low frequency conductance and capacitance measurements on MOS capacitors in weak inversion", Solid-State Electron., 18, 737-44 (1975), EEA78-33096.
293.Schumacher, F., "Influence of the C(V) transition characteristic from C_{max} to C_{min} on frequency multiplication with MIS punch through varactors", Arch. Elektron. & Uebertragungstech., 29 (2), 94-6 (1975), EEA78-11083.
294.Sixou, P., Danas, P. and Courat, J. P., "On the study of irradiated MOS structures at very low frequency", Solid State Electron., 15, 133-5 (1972), EEA75-7465.
295.Soares, R., "MOS transistors at high frequencies", Des. Electron., 8 (9), 42-8 (1971), EEA74-22728.
296.Temple, V. and Shewchun, J., "Exact frequency dependent complex admittance of the MOS diode including surface states, Shockly-Read-Hall (SRH) impurity effects, and low temperature dopant impurity response", Solid-State Electron., 16, 93-113 (1973), EEA76-8698.
297.Van Nielen, J. A., "A simple and accurate approximation to the high-frequency characteristics of insulated-gate field-effect transistors", Solid-State Electron., 12, 826-9 (1969), EEA73-4528.
298.Zohta, Y., "Frequency dependence of $\Delta V/\Delta(C^{-2})$ of MOS capacitors", Solid-State Electron., 17, 1299-1309 (1974), PA77-82493.

7. Switching Behavior

299.Ahrenholz, P., "A study of the switching properties of metal-silicon oxynitride - metal structures", Z. Elektr. Inf. & Energietech., 2, 226-7 (1972), German, EEA76-8615.

300.Bellmann, S., Landgrat-Dietz, D. and Moschwitzer, A., "On switching times of MOS transistors in integrated arrangements", Nachrichtentechnik, 20, 251-4 (1970), German, EEA73-35235.
301.Janicke, O., "On the switching behaviour of MOS transistors", Nachrichtentechnik, 22, 265-9 (1972), German, EEA75-36302.
302.Kawashiro, S., Nakahara, K. and Tarui, Y., "Switching characteristics of the MOS transistor", Bull. Electrotech. Lab., 31, 114-22 (1967), Japanese, EEA71-8279.
303.Kravchenko, A. B., Plotnikov, A. F., Popov, Y. M., Seleznev, V. N. and Shubin, V. E., "Illumination induced switching in metal-insulator-semiconductor structures", Sov. Phys.-Semicond., 8, 522-3 (1974), PA78-15735.
304.Laverty, S. J., "Conduction and switching phenomena in thermally grown silicon dioxide films", Int. J. Electron., 30, 165-74 (1971), EEA74-10649.
305.Morgan, D. V. and Howes, M. J., "Threshold and memory switching in silicon oxide films", Thin Solid Films, 20, S7-S9 (1974), EEA77-10082.
306.Nakahara, K., Kawashiro, S. and Tarui, Y., "Switching characteristics of the MOS transistor. II", Bull. Electrotech. Lab., 33, 775-60 (1969), Japanese, EEA73-17661.
307.Paul, R., "Switching behaviour of MOS transistors with ohmic load", Nachrichtentechnik, 17, 143-7 (1967), German, EEA71-1271.
308.Saffir, O. S., "Getting more speed from MOS", Electronics, 42 (4), 106-9 (1969), EEA72-9595.
309.Shatzkes, W., Av-Ron, M. and Anderson, R. M., "On the nature of conduction and switching in SiO_2", J. Appl. Phys., 45, 2065-77 (1974), EEA77-30222.
310.Wilmsen, C. W. and Allender, M. C., "Conductance switching of thermally grown SiO_2 (MOS structures)", J. Appl. Phys., 45, 1912-14 (1974), EEA77-30221.

8. Charge Transport

311.Ferris-Prabhu, A. V., "Time dependence of charge transport in MIS memory transistors", Appl. Phys. Lett., 20, 149-50 (1972), EEA75-13747.
312.Ferris-Prabhu, A. V., "Charge transfer in layered insulators", Solid-State Electron., 16, 1086-7 (1973), EEA76-35265.
313.Gnadinger, A. P. and Rosenzweig, W., "Polarization and charge motion in metal-Al_2O_3-SiO_2-Si structures", J. Electrochem. Soc., 121, 700-5 (1974), EEA77-30224.
314.Hofstein, S. R., "Space-charge-limited ionic currents in silicon dioxide films", Appl. Phys. Lett., 10, 291-3 (1967), EEA70-26717.
315.Hofstein, S. R., "Tritium-labeled field-induced proton transport in SiO_2 films", Appl. Phys. Lett., 11, 95-6 (1967), EEA70-38718.

316.Hughes, R. C., "Charge-carrier transport phenomena in amorphous SiO_2 - direct measurement of the drift mobility and lifetime", Phys. Rev. Lett., 30, 1333-36 (1973).
317.Hughes, R. C., "Hole mobility and transport in thin SiO_2 films (m.o.s. structure)", Appl. Phys. Lett., 26, 436-8 (1975), PA78-62109.
318.Kalnynya, R., Pentyush, E., Feltyn, I. and Freiberg, L., "Charge motion in MOS-structures on germanium with pyrolytic SiO_2 films", Latv. PSR Zinat. Akad. Vestis Fiz. Tehn. Ser., no. 5, 70-5 (1968), Russian, PA72-28544.
319.Komiya, Y., Tarui, Y. and Nagai, K., "Transient ion current in silicon oxides of the MOS structure. I", Bull. Electrotech. Lab., 30 (1), 34-44 (1966), Japanese, EEA70-2251.
320.Kuhn, M. and Silversmith, D. J., "Ionic contamination and transport of mobile ions in MOS structures", J. Electrochem. Soc., 118, 966-70 (1971), EEA74-30738.
321.Leadon, R. E. and Snowden, D. P., "Charge transport and polarization effects in dielectrics of MSI/LSI devices", Report RT-8046-002 HDL-TR-064-1, Intelcom Rad. Tech., San Diego, Calif., 1974, 58 pp., EEA78-33238.
322.Moeen-Ziai, B., "Electrical conduction in rf sputtered silicon dioxide", Thin Solid Films, 22, S31-S36 (1974).
323.Ohwada, A., "Charge transport phenomena in MOS system", Japan J. Appl. Phys., 6, 65-70 (1967), PA70-24217.
324.O'Reilly, T. J. and DeLucia, J., "Injection current flow through thin insulator films", Solid-State Electron., 18, 965-8 (1975), EEA78-41535.
325.Popova, L. I., Antov, B. Z. and Vitanov, P. K., "Charge migration on the oxide surface in MOS structures", Bulg. J. Phys., 2, 41-50 (1975), EEA78-41516.
326.Powell, R. J., "Hole transport and trapping in SiO_2 films", Presented at the Electrochemical Society Spring Meeting, Toronto, Canada, May 1975.
327.Powell, R. J. and Berglund, C. N., "Photoinjection studies of charge distribution in oxides of MOS structures", J. Appl. Phys., 42, 4390-7 (1971), EEA75-1077.
328.Reimer, H., "Results of investigation of charge transport in SiO_2", 3rd Symposium on the Reliability of Electronics, 1973, p. 343, EEA77-5942.
329.Schlegel, E. S., "Influence of p-n junctions on the mobile charge density in MOS transistors", J. Appl. Phys., 42, 425-6 (1971), EEA74-10651.
330.Srour, J. R., Curtis, O. L., Jr. and Chiu, K. Y., "Charge transport studies in SiO_2: processing effects and implications for radiation hardening", IEEE Trans. Nucl. Sci., NS-21, 73-80 (1974), PA78-42326.
331.Todd, D. A., "An investigation of charge transport mechanisms in evaporated oxide films on etched silicon surfaces", Thesis, **Oklahoma**, Norman, 48 pp., Order No. 69-1996, EEA73-10537.

332.Verwey, J. F., "Hole currents in thermally grown SiO_2", J. Appl. Phys., 43, 2273-87 (1972).
333.Verwey, J. F., "Avalanche-injected hole current in SiO_2", Appl. Phys. Lett., 21 (9), 417-19 (1972), EEA76-662.
334.Verwey, J. F., "Time dependence of injection currents in SiO_2", Philips Res. Rep., 28, 66-74 (1973), EEA76-25944.
335.Verwey, J. F., "Nonavalanche injection of hot carriers into SiO_2", J. Appl. Phys., 44, 2681-7 (1973), EEA76-26001.
336.Verwey, J. F. and de Maagt, B. J., "Avalanche-injected electron currents in SiO_2 at high injection densities", Solid-State Electron., 17, 963-71 (1974), EEA78-8578.
337.Weinberg, Z. A., "High-field transport in silicon dioxide films on silicon induced by corona charging", Thesis, Princeton Univ., Princeton, N. J., 1974, 124 pp., Order No. 75-23250, CA83-211753.
338.Zirinsky, S., "Charge transfer properties of MNOS structures as influenced by processing parameters", J. Electron. Mater., 4, 591-624 (1975), PA78-58420.

9. Instabilities

339.Barile, C. A., Dockerty, R. C. and Nagarajan, A., "Insulator charge and V_b stability of SNOS capacitors", J. Electrochem. Soc., 121, 907-11 (1974), EEA77-34054.
340.Breed, D. J., "Non-ionic room temperature instabilities in MOS devices", Solid-State Electron., 17, 1229-43 (1974), PA77-82491.
341.Cadambanguidi, R., Viswanathan, R. and Ogura, S., "Interface phenomena in integrated circuit oxides", Report UCLA 70-80, Univ. California, Los Angeles, 1970, 79 pp., EEA75-9593.
342.Cerny, A., "Stability problems in MOS transistors", Proceedings Second International Congress on Microelectronics, 1966, p. 75-84, Publ. 1967, German, EEA71-13330.
343.Cerny, A. and Benes, O., "Charge behaviour within a SiO_2 layer from the viewpoint of MOS transistor stability", XII International Scientific Colloquium, 1967, p. 75-80, Publ. 1968, German, EEA73-10699.
344.Chang, C. Y., "Method of stabilizing a silicon-base MOS device with zinc", Patent USA 3873384, Publ. March 1975.
345.Chaudhari, P. K. and Beyar, D. C., "Threshold voltage instability in MNOS structures", Presented at Electrochemical Society Meeting (abstracts only), San Francisco, Calif., May 1974, EEA77-26475.
346.Chen, P. C. Y., "Interface instability of r.f. sputtered silicon nitride films on silicon", Thin Solid Films, 21, 245-53 (1974), PA77-45010.
347.Curry, J. J. and Nigh, H. E., "Bias instability in two-layer m.i.s. structures", Annual Reliability Physics Symposium, 1970, p. 8, EEA73-28226.
348.Dimov, V. I. and Greorgiev, V. K., "Stabilized surfaces of MOS structures", C. R. Acad. Bulg. Sci., 25, 909-12 (1972), EEA75-853.

349. Dockerty, R. C., Barile, C. A., Nagarajan, A. and Zalar, S. M., "Improved V_t stability of SNOS FETs by oxygen annealing", Proceedings of the 11th Annual Reliability Physics Conference 1973, p. 159-62, EEA77-6042.

350. Erb, D. M., Dill, H. G. and Toombs, T. N., "Electron gate currents and threshold stability in the n-channel stacked gate MOS tetrode", IEEE Trans. Electron Devices, ED-18, 105-9 (1971), EEA74-7775.

351. Eversteyn, F. C. and Peek, H. L., "Preparation and stability of enhancement n-channel MOS transistors with high electron mobility", Philips Res. Rep., 24, 15-33 (1969), EEA72-17740.

352. Faggin, F., Forsythe, D. D. and Klein, T., "Room temperature threshold shift observed on silicon gate MOS devices", The Annual Reliability Physics Symposium (presentation abstracts), 1970, p. 11-12, EEA73-28227.

353. Fujimoto, S., Maekawa, M., Inoue, Y. and Haneta, Y., "Effects of diffusion on instabilities of poly Si-Si_3N_4-SiO_2-Si structures", Presented at Spring Meeting of the Electrochemical Society, Washington, D. C., May 1971, EEA75-4644.

354. Gad, H., "Stability measurements in MISFET's", Arch. Tech. Mess. & Messtech. Prax., no. 470, 53-4 (1975), German, EEA78-33143.

355. Gnadinger, A. P. and Rosenzweig, W., "Basic instability mechanisms in metal-Al_2O_3-SiO_2-Si structures", J. Electrochem. Soc., 120, 92C (1973), EEA76-23883.

356. Hara, H., Okamoto, Y. and Oknuma, H., "A new instability in MOS transistor caused by hot electron and hole injection from drain avalanche plasma into gate oxide", Jap. J. Appl. Phys., 9, 1103-12 (1970), EEA73-38108.

357. Hofstein, S. R., "Stabilization of MOS devices", Solid-State Electron., 10, 657-70 (1967), PA70-38715.

358. Kano, K. and Kunig, H., "Effect of thermal instability on ultrahigh frequency performance of insulated-gate field-effect transistors", Solid-State Electron., 12, 719-29 (1969), EEA73-10693.

359. Kolodziejski, J. F., "The instabilities in MOS transistors due to slow trapping effects", 3rd Symposium on the Reliability of Electronics, 1973, p. 253-61, EEA77-6046.

360. Komiya, Y., Tarui, Y. and Nagal, K., "Bias temperature treatment of MOS structure and charge distribution in silicon oxides", Bull. Electrotech. Lab., 30 (1), 25-33 (1966), Japanese, EEA70-2250.

361. Kriegler, R. J., "Ion stabilities in MOS structures", Proceedings of the 12th Annual Reliability Physics Conference, 1974, p. 250-8.

362. Leighton, W., "Room temperature instabilities in MOS structures with pyrolytic SiO_2 films", Proc. IEEE, 55, 214 (1967), EEA70-22829.

363. Lindmayer, J., "Stability and surface mobility in MOS structures", Proceedings of the International Conference on the Physics and Chemistry of Semiconductor Heterojunctions and Layer Structures, 1970, p. V-39, Publ. 1971, PA74-49149.

364. Marciniak, W., "Physical and chemical reasons for the instability of Mn-SiO_2-Si structures", Electronika, no. 5, 215-20 (1970), Polish, PA70-79024.

365. Miura, Y. and Matukura, Y., "Instabilities of MOS structure", Jap. J. Appl. Phys., 6, 582-7 (1967), EEA70-14546.

366. Morris, J. D. and White, J. J., "Sodium redistribution in SiO_2 films on silicon during immersion in hot electrolytes", Extended Abstracts of the Spring Meeting of the Electrochemical Society, 1971, p. 232-4, EEA72-4607.

367. Muravskii, B. S., Kuznetsov, V. I., Frizen, G. I. and Chernyi, V. N., "Investigation of the kinetics of a surface-barrier instability", Sov. Phys.-Semicond., 6, 1797-802 (1973), EEA76-26017.

368. Nicollian, E. H., "Interface instabilities", Proceedings of the 12th Annual Reliability Physics Conference, 1974, p. 267-72.

369. Nishimatsu, S., Hashimoto, N., Masuhara, T. and Nagata, M., "Stability investigation of n-channel MOS-FET utilizing alumina film for large scale dynamic memory array", Oyo Buturi, 43 (Suppl.), 367-71 (1974), EEA78-2882.

370. Osburn, C. M., "Characterization of thin oxide films for FET applications", J. Solid State Chem., 12, 232-7 (1975), PA78-21986.

371. Rai, B. P., Singh, K. and Srivastava, R. S., "Stability of MOS capacitors prepared from oxide grown in O_2-HCl gas", J. Phys. E, 8, 628-9 (1975), PA78-66037.

372. Reynolds, F. H., "Room-temperature threshold-voltage instabilities in an MOS integrated circuit", Proceedings of the 11th Annual Reliability Physics Conference, 1973, p. 170-6, EEA77-6087.

373. Rossel, P., Martinot, H. and Esteve, D., "Temporal instabilities of MOS transistors. II. Disymmetric behaviour", Solid-State Electron., 13, 425-38 (1970), French, EEA73-21335.

374. Rossel, P., Martinot, H. and Serhal, A., "Temporal instabilities of MOS transistors. I. Linear behaviour", Solid-State Electron., 13, 257-68 (1970), French, EEA73-17670.

375. Schlegel, E. S. and Schnable, G. L., "A negative-ion type instability in MOS devices", J. Electrochem. Soc., 119, 165-8 (1972), EEA75-13765.

376. Shimakura, K. and Tsunemitsu, H., "Insulated-gate field effect semiconductor device having low and stable gate threshold voltage", Patent USA 3806778, Publ. April 1974, CA81-18392.

377. Silversmith, D. J., "Dielectric loss, electrode deterioration, and the sodium problem in MOS structures", J. Electrochem. Soc., 119, 121-2 (1972), EEA75-13647.

378.Simonne, J., "Evolution of the voltage at the boundaries of the inversion layer of a m.o.s. structure in the unstable regime", C. R. Hebd. Seances Acad. Sci. B, 275, 853-6 (1972), French, EEA76-8582.
379.Singh, B. R., Tyagi, B. D. and Marathe, B. R., "Proton-induced instability in MOS devices", Presented at Electrochemical Society Meeting, San Francisco, Calif., May 1974, EEA77-26479.
380.Swaroop, B., "Charge instability due to gate electrode processing in MOS capacitors", Electrochemical Society Fall Meeting (Extended abstracts), 1974, p. 384-5, EEA78-33103.
381.Woods, M. H., "Instabilities in double dielectric structures", Proceedings of the 12th Annual Reliability Physics Conference, 1974, p. 259-66.

10 Surface States

a. General

382.Brotherton, S. D., "The influence of platinum in the Si-SiO_2 system", Solid-State Electron., 13, 1113-15 (1970), PA73-63296.
383.Card, H. C. and Rhoderick, E. H., "Conductance associated with interface states in MOS tunnel structures", Solid-State Electron., 15, 993-8 (1972), EEA75-36210.
384.Chaudhari, P. K., Bhattacharyya, A. and Beyar, D. C., "Surface-state generation during stress aging of HCl thermal Si-SiO_2 interfaces", Presented at Electrochemical Society Meeting, San Francisco, California, May 1974, EEA77-26407.
385.Chou, N. J., "Electrochemical studies in MOS structures", Electrochemical Society Fall Meeting, Extended Abstracts, 1970, p. 424-4, PA74-13231.
386.Cooper, J. A., Jr., Ward, E. R. and Schwartz, R. J., "Surface states and insulator traps at the Si_3N_4 - GaAs interface", Solid-State Electron., 15, 1219-27 (1972), EEA75-39954.
387.Deal, B. E., "The XVII drifts of silicon oxide technology", Presented at Electrochemical Society Meeting, San Francisco, California, May 1974, EEA77-26474.
388.Gibbon, C. F., Pappas, P. N. and Seidel, H. D., "In-process control of interface properties in the diode array camera tube target", Solid-State Electron., 14, 707-21 (1971), EEA74-30405.
389.Goetzberger, A., "Problems of the Si-SiO_2 interface", Verh. Dtsch. Phys. Ges., no. 8, 908 (1972), German, EEA76-19881.
390.Goetzberger, A., Lopez, A. S. and Strain, R. J., "On the formation of surface states during stress aging of thermal Si-SiO_2 interfaces", J. Electrochem. Soc., 120, 90-6 (1973), EEA76-12512.
391.Grosvalet, J., "Method of measurement of semiconductors surface properties - application to the study of M.O.S. devices and reliability", Symposium on Test Methods and Measurements of Semiconductor Devices, 1967, Pap. No. 401, EEA71-10570.

392. Kar, S., "Interface states and potential barriers in metal-(20 to 50 A) SiO_2-Si structures", Thesis, Lehigh Univ., Bethlehem, Pa., 1971, 232 pp., Order No. 71-27720, EEA75-32699.
393. Kar, S. and Dahlke, W. E., "Interface states in MOS structures with 20-40 A thick SiO_2 films on nondegenerate Si", Solid-State Electron., 15, 221-37 (1972), EEA75-13653.
394. Kuper, A. B., "Impurities and interface states in the silicon dioxide/silicon system (Final report)", Report NASA-CR-97738, Case Western Reserve Univ., Cleveland, Ohio, 1968, 47 pp., EEA73-7418.
395. Nakagiri, M., "Surface state generation in MOS structures by applying a high field to the SiO_2 film", Jap. J. Appl. Phys., 13, 1610-17 (1974), EEA78-587.
396. Pieper, H., "Some results of a study of the SiO_2-Si interface using the thermally stimulated charge release", Phys. Status Solidi A, 23, K55-8 (1974), EEA77-26419.
397. Raider, S. I., Gdula, R. A. and Petrak, J. R., "Nitrogen reaction at silicon-silicon dioxide interface", Presented at Electrochemical Society Spring Meeting, Toronto, Canada, May 1975.
398. Semushkina, N. A. and Semushkin, G. B., "Energy spectrum of the surface states in the Si-SiO_2 system", Sov. Phys.-Solid State, 15, 1-5 (1973), EEA76-32216.
399. Taylor, G. W., "Comments on 'Surface-state density by photovoltage measurement-II'", J. Phys. D, 5, L52-4 (1972), EEA75-24279.
400. Uritskii, V. Y., "Effect of different conditions of thermal oxidation of silicon on the density of surface states", Radiotekhnika i Elektronika, 13, 764-5 (1968), Russian, EEA71-25940.

b. Fast Surface States

401. Adamcik, I. and Schrofel, J., "Control of semiconductor-insulator interface characteristics in the MIS system", Proceedings of the 5th Czechoslovakia Conference on Electronics and Vacuum Physics (Extended abstracts), 1972, II.c)-5/3 pp., EEA76-8626.
402. Arnold, E. and Poleshuk, M., "Carrier generation at the silicon-silicon dioxide interface under pulsed conditions", J. Appl. Phys., 46, 3016-18 (1975), CA83-89701.
403. Baccarani, G., Severi, M. and Soncini, G., "A new method for the determination of the interface-state density in the presence of statistical fluctuations of the surface potential (in M.O.S. devices)", Appl. Phys. Lett., 23, 265-7 (1973), EEA76-38215.
404. Barabash, L. N., Litovchenko, P. G., Lastovetskii, V. F. and Chala, S. V., "Effect of electrophysical parameters of the silicon surface on surface-barrier characteristics of detectors", Poluprovodn. Tekh. & Mikroelektron., no. 19, 60-9 (1975), Russian, PA78-54557.

405.Bellavin, B. K. and Uritskiy, V. Y., "The density of surface states at the Si-SiO_2 interface in MOS structures", Radio Eng. & Electron. Phys., 17, 699-700 (1972), EEA76-12526.
406.Bredimas, V. and Thuillier, J. C., "Influence of the recombination dynamics on the response of a continuum of surface states", Phys. Status Solidi A, 10, 149-51 (1972), EEA75-16686.
407.Brews, J. R., "The role of lateral nonuniformities in MOS devices", Electrochemical Society Fall Meeting (Extended abstracts), 1974, p. 497, EEA78-37561.
408.Brillson, L. J., "Observation of extrinsic surface states on (1120) CdS", Surf. Sci., 51, 45-60 (1975), PA78-81505.
409.Broux, G., Van Overstraeten, R. and Declerck, G., "Experimental results on fast surface states and 1/f noise in M.O.S. transistors", Electron. Lett., 11, 97-8 (1975), EEA78-15347.
410.Brugler, J. S. and Jespers, P. G. A., "Charge pumping in MOS devices", IEEE Trans. Electron Devices, ED-16, 297-302 (1969), EEA72-13488.
411.Buchel, M. and Luth, H., "Surface photovoltage spectroscopy of electronic surface states on cleaved germanium (111) surfaces", Surf. Sci., 50, 451-64 (1975), PA78-69226.
412.Caplain, A., Pautrat, J. L. and Pfister, J. C., "Effect of the type of silicon on some fundamental properties of the silicon-silicon oxide interface", Phys. Status Solidi A, 21, 665-75 (1974), CA80-113976.
413.Castagne, R. and Vapaille, A., "On the determination of the interface density of states of the system SiO_2-Si in the forbidden band region", C. R. Acad. Sci. B., 270, 276-8 (1970), French, EEA73-21173.
414.Castagne, R. and Vapaille, A., "Density of states at the Si-SiO_2 interface determined from measurements at very low frequencies: example of results and accuracy", C. R. Acad. Sci. B, 270, 552-5 (1970), French, EEA73-21175.
415.Castro, P. L. and Deal, B. E., "Low-temperature reduction of fast surface states associated with thermally oxidized silicon", J. Electrochem. Soc., 118, 280-6 (1971), EEA74-18695.
416.Costagne, A., Mayer, F. and Vapaille, A., "Effects of crystal orientation on the state density at the interface between anodically grown silicon dioxide and silicon", Surf. Sci., 41, 595-600 (1974), EEA77-14442.
417.Courat, J. P., Nuzillat, G. and Sixou, P., "On the evaluation of surface states densities in metal-oxide-silicon structures", Solid State Commun., 7, 1123-8 (1969), French, PA72-49131.
418.Csabay, O. and Frank, H., "Influence of surface treatment of silicon on the effective impurity charge density in surface states of MOS structures", Solid-State Electron., 16, 985-9 (1973), CA79-98412.

419. Csabay, O., Frank, H., Kamensky, I. and Svacina, J., "Relations between the surface treatment of silicon, the density of surface-state charge and the effective lifetime of minority carriers of metal-insulator-semiconductor structures", Cesk. Cas. Fys., 24, 142-51 (1974), Slovak, CA81-83487.

420. Csabay, O., Harman, R., Harmatha, L. and Malik, K., "Effect of the thickness of a layer removed from silicon surfaces by chemical etching on the effective surface state density of MOS structures", Cesk. Cas. Fis. A, 24, 53-7 (1974), Slovak, PA77-53148.

421. Csabay, O. and Weber, B., "Effect of crystallographic orientation of silicon substrate on the effective surface state charge density of MOS structures", Cesk. Cas. Fis. A, 21, 617-19 (1971), Slovak, EEA75-13748.

422. Das, P., Motamedi, M. E. and Webster, R. T., "Determination of semiconductor surface properties using surface acoustic waves", Appl. Phys. Lett., 27, 120-2 (1975), PA78-81496.

423. Deal, B. E., MacKenna, E. L. and Castro, P. L., "Characteristics of fast surface states associated with SiO_3-Si and Si_3N_4-SiO_2-Si structures", J. Electrochem. Soc., 116, 997-1005 (1969), EEA73-4415.

424. Declerck, G. and Van Overstraeten, R., "Dependence of fast surface states on surface potential", Presented at Eurocon 71 Digest, Lausanne, Switzerland, October 1971, EEA75-4648.

425. Declerck, G., Van Overstraeten, R. and Broux, G., "Measurement of low densities of surface states at the Si-SiO_2-interface", Solid-State Electron., 16, 1451-60 (1973), EEA77-10178.

426. Deuling, H., Klausmann, E. and Goetzberger, A., "Interface states in Si-SiO_2 interfaces", Solid State Electron., 15, 559-71 (1972), EEA75-20147.

427. Dockerty, R. C. and Cowan, M. D., "Lateral annealing effects in p-channel silicon gate FET's having a SiO_2-Si_3N_4 gate dielectric", Presented at Electrochemical Society Meeting, San Francisco, Calif., May 1974, EEA77-26480.

428. Erlaky, G., Valko, A. and Schiller, V., "Investigations of fast surface states in MOS systems", Proceedings of the International Conference on the Physics and Chemistry of Semiconductor Heterojunctions and Layer Structures, 1970, p. V-91-9, Publ. 1971, EEA74-26899.

429. Fahrner, W. and Goetzberger, A., "Energy dependence of electrical properties of interface states in Si-SiO_2 interfaces", Appl. Phys. Lett., 17, 16-18 (1970), EEA73-31312.

430. Fogels, E. A. and Salama, C. A. T., "Characterization of surface states at the Si-SiO_2 interface using the quasi-static technique", J. Electrochem. Soc., 118 (12), 2002-6 (1971), EEA75-7461.

431. Fowler, A. B., "Process for the elimination of interface states in MOS structures", Patent USA 3849204, Publ. November 1974.

432. Henne, M. and Kirpal, A., "A possibility for determining the surface charge density and other parameters of the MOS field effect transistor by means of the substrate control", Nachrichtentechnik, 21, 414-15 (1971), German, EEA75-7549.
433. Hickmott, T. W., "Parallel between surface states at the Si-SiO_2 interface and the B center in irradiated SiO_2", J. Vac. Sci. & Technol., 9, 311-15 (1972), EEA75-20142.
434. Hirose, M., Hiraki, S., Nakashita, T. and Osaka, Y., "Surface states in tunnelable MOS structures", Jap. J. Appl. Phys., 14, 999-1004 (1975), EEA78-33079.
435. Hunter, W. R., "Thin oxide MOS capacitor studies of fast surface states", Thesis, Univ. Illinois, Urbana-Champaign, 1971, 143 pp., Order No. 72-12221, EEA72-36313.
436. Hunter, W. R., Eaton, D. H. and Sah, C. T., "Thin-oxide MOS capacitance studies of fast surface states", Appl. Phys. Lett., 17, 211-13 (1970), EEA74-1438.
437. Iwauchi, S. and Tanaka, T., "Interface properties of Al_2O_3-Ge structure and characteristics of Al_2O_3-Ge MOS transistors", Jap. J. Appl. Phys., 10, 260-5 (1971), PA74-30854.
438. Kalnynya, R. P., Feltyn, I. A. and Freiberg, L. A., "Electrical properties of pyrolytic films of SiO_2 on silicon and germanium", Fiz. Tekh. Poluprovodnikov, 4 (4), 813 (1970), Russian, EEA73-31314.
439. Kamieniecki, E., "Low-temperature photocapacity measurement in MOS structure", Solid-State Electron., 16, 1487-93 (1973), PA77-20696.
440. Kar, S., "Interface charge characteristics of MOS structures with different metals on steam grown oxides", Solid-State Electron., 18, 723-32 (1975), PA78-62120.
441. Kassabov, J., "Investigation of Si-SiO_2 interface by MOST structures", C. R. Acad. Bulg. Sci., 23, 29-32 (1970), EEA73-28112.
442. Kassabov, J. and Ilieva, M., "Rapid technique for determination of surface state density and energy distribution of MOS structures", C. R. Acad. Bulg. Sci., 23, 33-6 (1970), EEA73-24911.
443. Katto, H. and Itoh, Y., "Surface state density evaluation using high-frequency MOS capacitance technique", Phys. Status Solidi A, 21, 627-37 (1974), PA77-28498.
444. Kobayashi, I., Nakahara, M. and Atsumi, M., "Study of the Si-SiO_2 interface state with the negative bias-heat treatment approach", Proc. IEEE, 61, 249-50 (1973), EEA76-8587.
445. Lam, Y. W., "Surface-state density and surface potential in MIS capacitors by surface photovoltage measurements. I", J. Phys. D, 4, 1370-5 (1971), EEA74-30742.
446. Lam, Y. W. and Rhoderick, E. H., "Surface-state density and surface potential in MIS capacitors by surface photovoltage measurements. II", J. Phys. D, 4, 1376-89 (1971), EEA74-30743.

447.Lam, Y. W. and Rhoderick, E. H., "Author's reply (to Comments on 'Surface-state density by photovoltage measurements-II')", J. Phys. D, 5, L54 (1972), EEA75-24280.

448.Leuenberger, F., "Vapour-deposited silicon dioxide for device applications", Thin Solid Films, 22, 245-53 (1974), EEA77-33965.

449.Litovchenko, V. G. and Kovbasyuk, V. P., "The influence of the thermal treatment on the electronic structure of the silicon surface", Ukrayin Fiz. Zh., 13, 773-8 (1968), Russian, EEA71-23279.

450.Ma, T. P., "Surface states in thick oxide MOS tunnel junctions", Thesis, Yale Univ., New Haven, Conn., 1974, 160 pp., Order No. 75-15332, CA83-156709.

451.McNutt, M. J. and Sah, C. T., "Experimental observations of the effects of oxide charge inhomogeneity on fast surface state density from high-frequency MOS capacitance-voltage characteristics", Appl. Phys. Lett., 26, 378-80 (1974), EEA78-24238.

452.Maekawa, M., "Electrical characteristics of polycrystalline silicon-silicon nitride-silicon dioxide-silicon structures", Jap. J. Appl. Phys., 11, 1251-8 (1972), EEA75-36203.

453.Maserjian, J., Petersson, G. and Svensson, C., "Saturation capacitance of thin oxide MOS structures and the effective surface density of states of silicon", Solid-State Electron., 17, 335-9 (1974), PA77-41127.

454.Meyer, F., "Surface-state densities on clean semiconductor surfaces measured by ellipsometry. Comments", Phys. Rev. B, 9, 3622-6 (1974), CA81-18270.

455.Moskal, D. N., Gorban, A. P. and Kropman, D. I., "Some electrical properties of the reactive oxide with silicon", Poluprovodn. Tekh. & Mikroelektron., no. 19, 35-40 (1975), Russian, PA78-54556.

456.Nakagiri, M., "Surface state generation in MOS structures by applying a high field to the SiO_2 film", Jap. J. Appl. Phys., 13, 1610-17 (1974), EEA78-587.

457.Nanonni, R., "Concentration and distribution of surface states in silicon", Phys. Status Solidi, 34, K99-102 (1969), EEA73-7415.

458.Nannoni, R., "Properties of silicon-anodic oxide interfaces", Phys. Status Solidi A, 3, 235-48 (1970), EEA74-4863.

459.Nicollian, E. H., Goetzberger, A. and Lopez, A. D., "Expedient method of obtaining interface state properties from MIS conductance measurements", Solid State Electron., 12, 937-44 (1969), EEA73-10534.

460.Nishimatsu, S. and Ashikawa, M., "Simple method for measuring the interface state density", Rev. Sci. Instrum., 45, 1109-12 (1974), CA81-160933.

461.Ohkura, I., Kajiwara, Y. and Sugioka, Y., "Interface states of Metal-Nitride-Oxide-Silicon (MNOS)", Mitsubishi Denki Lab. Rep., 11, 173-82 (1970), EEA74-26710.

462. Pagnia, H., "Influence of illumination on oxide charges and interface states of InSb MOS devices", Phys. Lett. A, 51A, 389-90 (1975), EEA78-24265.

463. Pandey, K. C. and Phillips, J. C., "Energy bands of reconstructed surface states of cleaved-Si", Phys. Rev. Lett., 34, 1450-3 (1975), PA78-62077.

464. Pautrat, J. L. and Pfister, J. C., "Electronic properties of the Si-SiO_2 interface as a function of oxide growth conditions. I. Surface states", Phys. Status Solidi A, 11, 287-96 (1972), EEA75-24283.

465. Peikov, P. H., "Heterogeneity of the Si-SiO_2 system", C. R. Acad. Bulg. Sci., 27, 321-3 (1974), EEA77-33975.

466. Pepper, M. and Eccleston, W., "The influence of the electrode on the low temperature annealing of interface states in the Si-SiO_2 system", Thin Solid Films, 8, 133-42 (1971), EEA75-1087.

467. Pospelov, V. V., Suris, R. A., Fetisov, Y. A., Fuks, B. I. and Khafizov, R. Z., "A new method of determining the density of the states on the surface of a semiconductor", Zh. Eskp. & Teor. Fiz. Pis'ma, 21, 448-51 (1975), Russian, PA78-72888.

468. Sato, T., Takeishi, Y. and Hara, H., "Effects of crystallographic orientation on mobility, surface state density, and noise in p-type inversion layers on oxidized silicon surfaces", Jap. J. Appl. Phys., 8, 588-98 (1969), EEA72-15564.

469. Saunders, A. F. and Wright, G. T., "Interface states in the silicon/silicon-oxide system observed by thermally stimulated charge release", Electron. Lett., 6, 207-9 (1970), EEA73-24921.

470. Schlegel, E. S., "Effect of aluminum linewidth on the annealing of fast states in MOS structures", IEEE Trans. Electron Devices, ED-19, 839-40 (1972), EEA75-20214.

471. Schroder, D. K. and Guldberg, J., "Interpretation of surface and bulk effects using the pulsed MIS capacitor", Solid-State Electron., 14, 1285-97 (1971), EEA75-4640.

472. Severi, M. and Sanciri, G., "Surface-state density at the (hydrogen-chloride) oxide-silicon interface", Electron. Lett., 8, 402-4 (1972), EEA75-32586.

473. Shirokov, A. A., Boznkov, V. G. and Vyatkin, A. P., "Fast surface states activated by silver ions at the germanium-neutral electrolyte interface", Izv. Vyssh. Uchebn. Zaved., Fiz., 18 (8), 91-6 (1975), Russian, CA83-170013.

474. Simmons, J. G. and Mar, H. A., "Interfacial effects in MIS systems", 1972 Annual Conference on Electrical Insulation and Dielectric Phenomena, p. 67-74, Publ. 1973, EEA76-26025.

475. Singh, B. R. and Srivastava, R. S., "Surface state density at the silicon/silicon dioxide interface in MOS structures", Indian J. Pure & Appl. Phys., 10, 114-16 (1972), EEA75-32684.

476. Singh, B. R., Tyagi, B. D., Chandorkar, A. N. and Marathe, B. R., "Surface-state distribution and ion migration in thermally grown SiO_2 films", Presented at Electrochemical Society Meeting, San Francisco, California, May 1974, EEA77-26408.

477. Singh, R. and Hartnagel, H., "Reduction in surface charge density by new GaAs passivation method", J. Phys. D, 8, L42-3 (1975), PA78-25768.

478. Stoev, I. G. and Antov, B. Z., "Investigation of surface states in the MOS system by gate controlled diode structure", C. R. Acad. Bulg. Sci., 25, 1331-4 (1972), EEA76-12498.

479. Swaroop, B., "The effect of ambient-annealing on dielectric film (SiO_2 and Si_3N_4) deposited on silicon. I. Interface states", 1971 Annual Conference on Electrical Insulation and Dielectric Phenomena, p. 233-41, Publ. 1972, EEA75-32702.

480. Swaroop, B., "Hydrogen annealing effect on silicon-insulator(s) interface states", J. Phys. D, 6, 1090-2 (1973), EEA76-26010.

481. Tanaka, T. and Iwauchi, S., "Effects of interface states on the characteristics of MOS transistors", Jap. J. Appl. Phys., 7, 1473-83 (1968), EEA72-5039.

482. Testemale, E., Seve, G. and Lassabatere, L., "Interfaces in the silicon-anodic Al_2O_3-nickel structures", Thin Solid Films, 26, 243-52 (1975), French, PA78-50687.

483. Tidey, R. J., Stradling, R. A. and Pepper, M., "Peaked structure appearing in the field effect mobility of silicon MOS devices at liquid helium temperatures", J. Phys. C, 7, L353-5 (1974), EEA78-584.

484. Tschopp, L. L., "Scattering of conduction electrons by surface oxide charge at the silicon-silicon oxide interface", Thesis, Univ. Illinois, Urbana-Champaign, 1971, 133 pp., Order No. 72-12416, EEA75-36314.

485. Tsui, D. C. and Allen, S. J., Jr., "Localization and the minimum metallic conductivity in Si inversion layers", Phys. Rev. Lett., 34, 1293-5 (1975), EEA78-24266.

486. Uritskii, V. Y. and Konorov, P. P., "Surface states and built-in charge in silicon-silicon dioxide system", Sov. Phys.-Semicond., 8, 1055-6 (1975), PA78-37840.

487. Ushirokawa, A., Warashina, M. and Yamamoto, K., "Generation mechanism and distribution of Si-SiO_2 interface states with BT treatment", Oyo Buturi, 40, 518-22 (1971), Japanese, EEA75-16685.

488. van Overstraeten, R. J., Declerck, G. and Broux, G. L., "The influence of surface potential fluctuations on the operation of the MOS transistor in weak inversion", IEEE Trans. Electron Devices, ED-20, 1154-8 (1973), EEA77-2568.

489. Venkateswaran, K., "Comments on 'Surface-state spectra from thick-oxide MOS tunnel junction'", Solid-State Electron., 18, 908 (1975), EEA78-37480.

490. Verhoeven, J. L., Saminadayar, K. and Pfister, J. C., "Illumination with ultraviolet light of MOS capacitors", Phys. Status Solidi A, 24, 441-8 (1974), PA77-70074.

491. Virtmanis, A., Feltins, I., Freiberga, L., Eglitis, I. and Eimanis, I., "Effect of plasma treatment on the surface of the silicon-silicon dioxide interface", Latv. PSR Zinat. Akad. Vestis, Fiz. Teh. Zinat. Ser., no. 4, 20-4 (1975), Russian, CA83-156490.

492.Walker, L. G., "Determination of surface state density in tunnel MOS devices from current-voltage characteristic", Solid-State Electron., 17, 763-5 (1974), PA77-57181.

493.Wang, K. L. and Gray, P. V., "Interface properties of germanium MIS structures", Electrochemical Society Fall Meeting (Extended abstracts), 1974, p. 501-3, PA78-72912.

494.Whelan, M. V., "Electrical behavior of defects at a thermally oxidized silicon surface", Report PRRS 1970-6, Eindhoven Univ. Technol., Netherlands, 1970, 93 pp., EEA74-7671.

495.Yeow, Y. T., Lamb, D. R. and Brotherton, S. D., "An investigation of the influence of low-temperature annealing treatments on the interface state density at the Si-SiO_2", J. Phys. D, 8, 1495-506 (1975), EEA78-41526.

496.Ziegler, K. and Klausmann, E., "Static technique for precise measurements of surface potential and interface state density in MOS structures", Appl. Phys. Lett., 26, 400-2 (1974), EEA78-24239.

c. Fixed Charge

497.Badcock, F. R. and Lamb, D. R., "Stability and surface charge in the MOS system", Int. J. Electron., 24, 1-9 (1968).

498.Deal, B. E., Sklar, M., Grove, A. S. and Snow, E. H., "Characteristics of the surface state charge (Q_{ss}) of thermally oxidized silicon", J. Electrochem. Soc., 114, 266-74 (1967).

499.Fowkes, F. M. and Hess, D. W., "Control of fixed charge of Si-SiO_2 interface by oxidation-reduction treatments", Appl. Phys. Lett., 22, 377-9 (1973), EEA76-23373.

500.Hess, D. W. and Deal, B. E., "Effect of nitrogen and oxygen/nitrogen mixture on oxide charges in MOS structures", J. Electrochem. Soc., 122, 1123-7 (1975), EEA78-41478.

501.Joannopoulos, J. D. and Cohen, M. L., "Intrinsic surface states of (110) surfaces of Group IV and III-V semiconductors", Phys. Rev. B, 10, 5075-81 (1974), CA82-148759.

502.Kamins, T. I. and Deal, B. E., "Characteristics of Si-SiO_2 interfaces beneath thin silicon films defined by electrochemical etching", J. Electrochem. Soc., 122, 557-60 (1975), EEA78-24247.

503.Komiya, Y., Tarui, Y. and Mori, K., "Measurements of the positive charge distribution in SiO_2 by channel conductance method and step etching method", Bull. Electrotech. Lab., 36, 104-11 (1972), Japanese, EEA75-36193.

504.Lamb, D. R. and Badcock, F. R., "The effect of ambient, temperature, and cooling on the surface charge at the Si/SiO_2 interface", Int. J. Electron., 24, 11-16 (1968).

505.Laverty, S. J. and Ryan, W. D., "The components of M.O.S. oxide charge and their dependence upon the oxidation process", Int. J. Electron., 26, 369-76 (1969), EEA73-4522.

506. Mirsch, S., "Electronic properties of the boundary between semiconductor and insulator in M.I.S. structures with annealed dielectric", Z. Elektr. Inf.-& Energietech., 2, 219-21 (1972), German, EEA76-8611.

507. Mochkin, V. S. and Gorokhov, V. N., "Electrical characteristics of a MDS structure containing silicon nitride, obtained by reactive cathode sputtering", Izv. VUZ Fiz., no. 2, 66-70 (1974), Russian, EEA77-18731.

508. Montillo, F. and Balk, P., "High-temperature annealing of oxidized silicon surfaces (interface states)", J. Electrochem. Soc., 118, 1463-8 (1971), EEA74-37198.

509. Pautrat, J. L. and Pfister, J. C., "Electronic properties of the $Si\text{-}SiO_2$ interface as a function of oxide growth conditions. II. Fixed charge", Phys. Status Solidi A, 11, 669-75 (1972), EEA75-28448.

510. Peikov, P. C., "Fluctuation of the fixed oxide charge in the MIS structures and the C-V measurements", C. R. Acad. Bulg. Sci., 26, 1311-14 (1973), EEA77-18728.

511. Peikov, P. H., "Heterogeneity of the $Si\text{-}SiO_2$ system", C. R. Acad. Bulg. Sci., 27, 321-3 (1974), PA77-70071.

512. Perez de la Sota, R. and Velasco, G., "Characterization and measurement of the electric charge in $Si\text{-}SiO_2$ interfaces", An. Fis., 67 (3-4), 143-51 (1971), Spanish, EEA74-30733.

513. Przewlocki, H., "Investigation of the influence of technology on $Si\text{-}SiO_2$ system and MOS structure electrical properties", Electron. Technol., 3 (3), 103-22 (1970), EEA74-18220.

514. Richman, P., "Method for selectively establishing regions of different surface charge densities in a silicon wafer", Patent USA 3867196, Publ. February 1975, EEA78-37662.

515. Sasaki, N., "Change of $Si\text{-}SiO_2$ interface change by BT treatment", Jap. J. Appl. Phys., 12, 1458-9 (1973), EEA77-345.

516. Sproul, M. E. and Nassibian, A. G., "Field-effect mobilities of gold-doped MOS structures", J. Appl. Phys., 46, 1255-60 (1975), PA78-46759.

517. Szep, I. C. and Solt, K., "Reduction of positive charge in SiO_2 layers by PbO", Proceedings of the International Conference on the Physics and Chemistry of Semiconductor Heterojunctions and Layer Structures, 1970, p. V-205-11, Publ. 1971, EEA74-26750.

518. Tarui, Y., Komiya, Y. and Teshima, H., "The built-in voltage and the charge distributions in the oxide of MOS structure", Jap. J. Appl. Phys., 6, 191-204 (1967), EEA70-15009.

519. Tschopp, L. L., "Scattering of conduction electrons by surface oxide charge at the silicon-silicon dioxide interface", Thesis, Univ. Illinois, Urbana-Champaign, 1971, 133 pp., Order No. 72-12416, EEA75-36314.

520. Wagner, S. and Povilonis, E. I., "Interface states and fixed charge as a function of small changes in orientation of (111) silicon wafers", J. Electrochem. Soc., 122, 86-8 (1975), PA78-18687.

521. Wienhold, H., "Oxide charges and densities of interface states existing on surfaces of SiO_2-Si/Ge", Phys. Status Solidi A, 1, K49-51 (1970), EEA74-4861.

11. Leakage Currents

522. Barker, R. W. J., "Techniques for minimizing MOSFET's gate leakage current", Proc. IEEE, 59, 283-4 (1971), EEA74-13261.
523. Barker, R. W. J. and Hart, B. L., "Minimization of MOSFET gate leakage current in buffer amplifier circuits", Proc. IEEE, 57, 1766 (1969), EEA73-9409.
524. Biswas, B. N. and Datta, G., "Zero gate leakage current in the enhancement mode MOSFET", Proc. IEEE, 56, 202-3 (1968), EEA71-15867.
525. Gosney, W. M., "Subthreshold drain leakage currents in MOS field-effect transistors", IEEE Trans. Electron Devices, ED-19, 213-19 (1972), EEA75-7537.
526. Ham, W. E., "Leakage currents in silicon on sapphire MOS transistors", Presented at 1973 International Electron Devices Meeting Supplement (abstracts only), Washington, D. C., December, EEA77-18820.
527. Kennedy, E. J., "Gate leakage currents in MOS field-effect transistors", Proc. IEEE, 54, 1098-9 (1966), EEA70-1059.
528. Lin, H. C. and Shiota, P. S., "Insulated gate-field effect transistor structure with reduced current leakage", Patent USA 3440502, Publ. April 1969, EEA73-10703.
529. Nassibian, A. G., "Effect of gold on surface properties and leakage current of MOS transistors", Solid-State Electron., 10, 891-6 (1967), PA70-32827.
530. Negro, V. C., "Comments on 'gate leakage currents in MOS field-effect transistors'", Proc. IEEE, 55, 1540-1 (1967), EEA70-22892.
531. Negro, V. C. and Pannone, L., "Self-heating and gate leakage current in a guarded MOSFET", Proc. IEEE, 60, 342-3 (1972), EEA75-10298.
532. Stuart, R. A. and Eccleston, W., "Leakage currents of m.o.s. devices under surface depletion conditions", Electron. Lett., 8, 225-7 (1972), EEA75-20206.

12. Substrate Properties

533. Bergveld, P., "The influence of substrate bias upon the AC characteristics of MOS transistors", Proc. IEEE, 57, 72-3 (1969), EEA72-13491.
534. Brotherton, S. D., "Dependence of MOS transistor threshold voltage on substrate resistivity", Solid-State Electron., 10, 611-16 (1967), PA70-26695.
535. Das, M. B., "Dependence of the characteristics of MOS-transistors on the substrate resistivity", Solid-State Electron., 11, 305-22 (1968), EEA71-10677.

536. Doronkin, Y. F., Lyakhovich, V. V. and Miroshnichenko, S. I., "Effect of the substrate on transfer admittance of a saturated MOS-transistor", Izv. VUZ Radioelektron., 16, 97-81 (1973), Russian, EEA76-26096.

537. Galkin, V. N., Konstantinov, Y. G. and Shilkov, V. M., "Effect of substrate bias voltage on threshold potential of MOS transistors", Izv. VUZ Radioelektron., 17, 122-3 (1974), Russian, EEA77-23047.

538. Hayashi, Y. and Tarui, Y., "Effect of substrate on MOST's and MOS-IC's, mainly on the dc characteristics", Electron. & Commun. Jap., 54, 93-103 (1972), EEA75-32679.

539. Nakahara, M., "On the enhancement of substrate terminal current caused from impact ionization at Si-SiO_2 interface of silicon MOS transistor", Fujitsu Sci. Tech. J., 5 (3), 51-79 (1969), EEA73-14313.

540. Nakahara, M., Wasawa, H. I. and Yasutake, K., "Anomalous enhancement of substrate terminal current beyond pinch-off in silicon n-channel MOS transistors and its related phenomena", Proc. IEEE, 56, 2088-90 (1968), EEA72-2829.

541. Richman, P., "Complementary MOS field-effect transistors on high-resistivity silicon substrates", Solid-State Electron., 12, 377-83 (1969), EEA72-15569.

542. Ryan, R. D., "Substrate current in silicon p-channel MOS transistors", Proc. IEEE, 57, 1424-5 (1969), EEA73-1205.

13. Other Properties

543. Abdullaev, G. B., Gajiev, N. D. and Talibi, M. A., "Investigation of physical properties of metal-dielectric-semiconductor structures based on selenium and tellurium", Proceedings of the International Conference on the Physics and Chemistry of Semiconductor Heterojunctions and Layer Structures, 1970, p. V-53, Publ. 1971.

544. Antula, J., "Thickness fluctuations and electric field penetration in thin metal-insulator-metal structures", Solid-State Electron., 11, 1063-7 (1968), EEA72-817.

545. Armstrong, G. A. and Magowan, J. A., "Pinch-off in insulated-gate field effect transistors", Solid-State Electron., 14, 760-3 (1971), EEA74-30836.

546. Baikov, V. D. and Sharugin, V. I., "Power considerations in MOS-transistor dynamic elements", Izv. VUZ Radioelektron., 14, 412-19 (1971), Russian, EEA74-18796.

547. Bandali, M. B., "SCL-MOST in the triode model, Solid-State Electron., 15, 1023-6 (1972), EEA75-36309.

548. Bandali, M. B. and Wright, G. T., "Experimental study of surface channel in insulated-gate field-effect transistor", Electron. Lett., 7, 142-4 (1971), EEA74-15566.

549. Bashenov, V. K., Dyakov, V. V. and Presnov, V. A., "Effects of deep traps on MIS-capacitors", Phys. Status Solidi A, 1, K145-9 (1970), EEA74-1611.

550.Baum, G., "Control of substratum of MOS transistors", Nachrichtentechnic, 18, 358-60 (1968), German, EEA72-812.
551.Boesen, G. F. and Jacobs, J. E., "ZnO field-effect transistor", Proc. IEEE, 56, 2094-5 (1968), EEA72-5041.
552.Borawski, R. and Lane, R. Q., "D.C. and R.F. performance of a new epitaxial depletion mode "N" channel MOST", Proceedings of National Electronics Conference, 1966, p. 150-3, PA70-11544.
553.Bouat, J. and Thuillier, J. C., "Surface quantum transport in MOS structure on tellurium", Oyo Buturi, 43 (Suppl.), 327-31 (1974), PA78-6906.
554.Brotherton, S. D., "The influence of platinum in the $Si-SiO_2$ system", Solid-State Electron., 13, 1113-15 (1970), EEA73-35166.
555.Brotherton, S. D., "Electrical properties of gold at the silicon-dielectric interface", J. Appl. Phys., 42, 2085-94 (1971), EEA74-18694.
556.Brotherton, S. D. and Phillips, J. B., "The control of threshold voltage of complementary pair MOS transistors", Conference on Integrated Circuits, 1967, p. 126-36, EEA73-483.
557.Brown, D. M., Cady, W. R., Sprague, J. W. and Salvagni, P. J., "The p-channel refractory metal self-registered MOSFET", IEEE Trans. Electron Devices, ED-18, 931-40 (1971), EEA74-34083.
558.Brumme, R. and Ebest, G., "On the voltage dependence of MOST channel conductance", Nachrichtentechnik, 19, 251-4 (1969), German, EEA72-20434.
559.Cagnina, S. F. and Snow, E. H., "Properties of gold doped MOS structures", J. Electrochem. Soc., 114, 1165-73 (1967), EEA71-8214.
560.Calzolari, P. U., Graffi, S. and Morandi, C., "Field-enhanced carrier generation in MOS capacitors", Solid-State Electron., 17, 1001-11 (1974), EEA78-590.
561.Card, H. C. and Rhoderick, E. H., "Studies of tunnel MOS diodes. II. Thermal equilibrium considerations", J. Phys. D, 4 (10), 1602-11 (1971), EEA74-37247.
562.Chang, C. and Tsao, K. Y., "Electrical properties of diffused zinc on SiO_2-Si MOS structures", Solid-State Electron., 12, 411-15 (1969), PA72-33362.
563.Collins, D. R., "The effect of gold on the properties of the $Si-SiO_2$ system", J. Appl. Phys., 39, 4133-43 (1968), EEA71-49317.
564.Cooper, J. A., Jr. and Schwartz, R. J., "The effect of an energy-dependent capture cross section on data interpretation using the MOS conductance technique", J. Appl. Phys., 44, 5613-14 (1973), PA77-37315.
565.Deal, B. E., Snow, E. H. and Mead, C. A., "Barrier energies in metal-silicon dioxide-silicon structures", J. Phys. Chem. Solids, 27 (11-12), 1873-9 (1966), PA70-7997.
566.DeVilbiss, W. F. and Charlson, E. J., "MOS transistors on semi-insulating gallium arsenide", Proceedings of the 1968 Microelectronics Symposium, Pap. E8-1-E8-5, EEA71-28908.

567. Diener, K. H., "The harmonic distortions of the output current of MOS transistors", Nachrichtentechnik, 19, 106-11 (1969), German, EEA72-11331.
568. Diener, K. H., "On the harmonic voltage distortions of the vanishing point of capacitively loaded MOS transistors", Nachrichtentechnik, 20, 299-301 (1970), German, EEA73-38110.
569. Diener, K. H., "On the harmonic voltage distortions of vanishing points of the inductively loaded MOS transistor", Nachrichtentechnik, 21, 110-12 (1971), German, EEA74-15571.
570. Dobrovolskii, V. N. and Chaika, G. Y., "Electrostatic attenuation of the longitudinal field component in the near-surface region of MOS structures", Poluprovodn. Tekh. & Mikroelektron., no. 10, 34-8 (1972), Russian, EEA76-16327.
571. Dorda, G., "The band splitting in silicon inversion layers", Proceedings of the International Conference on the Physics and Chemistry of Semiconductor Heterojunctions and Layer Structures, 1971, Pap. V-83-90, EEA74-26898.
572. Dorey, A. P. and Maddern, T. S., "The effect of strain on MOS transistors", Solid-State Electron., 12, 185-9 (1969), EEA72-11333.
573. Dudnikov, V. P., Zaitsev, B. D., Perelman, A. A. and Shvedov, E. E., "Time-dependent drift of MOS transistors", Izv. VUZ Radioelektron., 13, 1358-66 (1970), Russian, EEA74-10650.
574. Duffy, M. T. and Revesz, A. G., "Interface properties of Si-(SiO_2)-Al_2O_3 structures", J. Electrochem. Soc., 117, 372-7 (1970), EEA73-21186.
575. Eisele, I. and Dorda, G., "Negative magnetoresistance in n-channel (100) silicon inversion layers", Phys. Rev. Lett., 32, 1360-3 (1974), PA77-57175.
576. Falicov, L. M. and Garcia, N., "Hole cyclotron masses in silicon MOS devices", Solid State Commun., 17, 473-5 (1975), EEA78-41538.
577. Fang, P. H., "Dispersion mechanisms of the MOS impedance", J. Appl. Phys., 39, 2472-3 (1968), EEA71-15858.
578. Gentsch, H. and Pruller, H., "MOS sandwich grid diode for production of high ionization at the solid gas phase boundary", Z. Naturforsch. A, 26A, 1010-16 (1971), German.
579. Gnadinger, A. P. and Talley, H. E., "Channel shape in an insulated gate field-effect transistor", Proc. IEEE, 58, 916-17 (1970), EEA73-35237.
580. Goecke, D., "Field effect transistors. II. (Metal-oxide semiconductor field-effect transistors)", Microtecnic, 21, 491-3 (1967), EEA71-8281.
581. Golder, J., "Continuation of an investigation of the differential charge pumping effect in MOS transistors", Helv. Phys. Acta, 44, 866-83 (1972), German, EEA75-24357.
582. Gorban, A. P., Litovchenko, V. G., Romanyuk, B. N., Pataki, G., Nemeth-Sallay, M. and Lorinczy, A., "On the nature of recombination centres of the Si-SiO_2 interface", Phys. Status Solidi A, 24, K1-5 (1974), PA77-67283.

583.Hart, B. L. and Barker, R. W. J., "Source-follower voltage gain", Int. J. Electron., 31, 537-40 (1971), EEA75-1160.

584.Hayashi, T., "Negative resistance phenomena in metal-oxide-semiconductor (MOS) diodes", J. Inst. Elect. Commun. Engrs. Japan, 48, 909-15 (1965), EEA70-2638.

585.Herman, A. R. and Hancsecz, B. V., "Influence of the metallization on the charges in various MOS-structures", Proceedings of the International Conference on the Physics and Chemistry of Semiconductor Heterojunctions and Layer Structures, 1971, Pap. V-109-15.

586.Hesse, H., "Investigations on p-channel MOS-field-effect tetrodes", Nachrichtentech. Z., 23, 491-4 (1970), German, EEA74-4932.

587.Holm-Kennedy, J. W., "New effect in insulated-gate field-effect transistors: transverse voltage anisotropy at the inverted semiconductor surface", Phys. Rev. Lett., 32, 111-14 (1974), PA77-20693.

588.Ilieva, M. and Stoeva, R., "Low-temperature treatment of the silicon-silicon dioxide-molybdenum-gold and silicon-silicon dioxide-tungsten-gold systems in a hydrogen ambient", Dokl. Bolg. Akad. Nauk, 26, 171-3 (1973), CA79-59075.

589.Kamins, T. I., "Field-effects in polycrystalline-silicon films", Solid-State Electron., 15, 789-99 (1972), EEA75-24251.

590.Kar, S., "Interface states and potential barriers in metal-(20 to 50A) SiO_2-Si structures", Thesis, Lehigh Univ., Bethlehem, Pa., 1971, 232 pp., Order No. 71-27720, EEA75-32699.

591.Kar, S. and Dahlke, W. E., "Metal-dependent interface states in thin MOS structures", Appl. Phys. Lett., 18, 401-3 (1971), EEA74-26699.

592.Kar, S. and Dahlke, W. E., "Potentials and direct current in Si-(20 to 40A) SiO_2-metal structures", Solid-State Electron., 15, 869-75 (1972), EEA75-32695.

593.Kassabov, J., "Investigation of Si-SiO_2 interface by MOST structures", C. R. Acad. Bulg. Sci., 23 (1), 29-32 (1970), EEA73-28112.

594.Kassabov, J. D. and Ilieva, M. N., "Effect of bias-temperature treatments on the system metal-SiO_2-Si", C. R. Acad. Bulg. Sci., 25, 897-900 (1972), EEA76-741.

595.Katayama, Y., Yoshida, I., Koteva, N. and Komatsubara, K. F., "Differential negative resistance of n-type inversion layer in silicon MOS field-effect transistor", Appl. Phys. Lett., 20, 31-3 (1972), EEA75-10260.

596.Kim, T. W. and Grannemann, W. W., "A study of metal insulator semiconductor magnetoresistance, bulk, and surface properties of indium antimonide", Report NASA-CR-118330, Univ. New Mexico, Albuquerque, 1971, 45 pp., EEA75-10301.

597.Kingston, F. E. and Lee, K., "Field effect transistors at 4.2°K", Rev. Sci. Instrum., 39, 599-601 (1968), PA72-27207.

598.Komiya, Y., Tarui, Y. and Kubo, K., "Si MIS characteristics using vapour-grown Al_2O_3 films. II", Bull. Electrotech. Lab., 36, 85-95 (1972), Japanese, EEA75-36191.
599.Komiya, Y., Tarui, Y., Teshima, H. and Yoshida, Y., "The built-in voltage and the charge distributions in the oxide of MOS structure", Bull. Electrotech. Lab., 31 (4), 123-39 (1967), Japanese, EEA71-8278.
600.Koomen, J., "Investigation of the MOST channel conductance in weak inversion", Solid-State Electron., 16, 801-10 (1973), EEA76-29279.
601.Koomen, J., "Interface studies of the MOS-structure by transfer-admittance measurements", Solid-State Electron., 17, 321-8 (1974), PA77-41126.
602.Korchlov, V. P. and Sukharev, Y. G., "Zinc telluride thin film FET's with an isolated gate", Radiotekhnika I Elektronika, 15, 638-9 (1970), Russian, EEA73-25041.
603.Korwin-Pawlowski, M. L. and Heasell, E. L., "Characteristics of MOS capacitors formed on p-type InSb", Phys. Status Solidi A, 24, 649-52 (1974).
604.Kubo, S., "Effects of preoxidation treatment on MOS transistor characteristics", Jap. J. Appl. Phys., 6, 119 (1967), EEA70-13004.
605.Labuda, E. F. and Ryden, W. D., "Neutralizing charge in semiconductor bodies and dielectric coatings induced by cathodic etching", Patent USA 3791952, Publ. February 1974, CA80-88790.
606.Lamb, D. R., Moghal, G. R. and Hawkins, R. J., "The effect of gold doping on the threshold voltage, Hall mobility, gain, and current noise of MOS transistor", Int. J. Electron., 30, 141-7 (1971), PA74-21756.
607.Lassabatere, L., Seve, G. and Mendolia, J., "Study of oxide capture centres in an MOS structure by thermal simulation", Thin Solid Films, 6, 381-95 (1970), French, PA74-28555.
608.Lattin, W. W. and Rutledge, J. L., "Impact ionization current in MOS devices", Solid-State Electron., 16, 1043-6 (1973), EEA76-35263.
609.Lazarus, M. J., "Observation of the Boltzmann limit for an IGFET", IEEE Trans. Electron Devices, ED-22, 365 (1975), EEA78-19835.
610.Learn, A. J., "Effects of MOS metallization geometry and processing on mobile impurities", J. Electrochem. Soc., 122, 1127-32 (1975), EEA78-41479.
611.Leuenberger, F., "Dependence on threshold voltage of silicon p-channel MOSFET's on crystal orientation", Proc. IEEE, 54, 1985-7 (1966), EEA70-14605.
612.Lundstrom, I., Christensson, S. and Svensson, C., "Carrier trapping hysteresis in MOS transistors", Phys. Status Solidi A, 1, 395-407 (1970), EEA74-7784.

613.Lundstrom, I., Shivaraman, S., Svensson, C. and Lundkuist, L., "A hydrogen-sensitive MOS field-effect transistor", Appl. Phys. Lett., 26, 55-7 (1975), EEA78-15341.
614.Maeda, K., Shirai, K. and Nakano, J., "Electrical characteristics of polycrystalline silicon-aluminum oxide-silicon dioxide-silicon structure", Jap. J. Appl. Phys., 13, 1173-4 (1974), EEA77-34055.
615.Mar, H. A. and Simmons, J. G., "Surface-generation statistics and associated thermal currents in metal-oxide-semiconductor structures", Phys. Rev. B, 11, 775-83 (1975), CA82-164082.
616.Mirsch, S., "Electronic properties of the boundary between semiconductor and insulator in M.I.S. structures with annealed dielectric", Z. Elektr. Inf. - & Energietech., 2, 219-21 (1972), German, EEA76-8611.
617.Mochkin, V. S., "An MOS transistor with a negative resistance controlled by gate voltage", Izv. VUZ Radioelektron., 17, 90-2 (1974), Russian, EEA77-23048.
618.Nakhmanson, R. S., Dobrovolski, P. P. and Amineva, S. S., "Anomalous recombination in silicon MIS structures", Phys. Status Solidi A, 9, 699-711 (1972), EEA72-13651.
619.Neizvestnyj, I. G., "Surface capacity and conductivity of germanium M.I.S. structures", Z. Elektr. Inf. - & Energietech., 2, 224 (1972), German, EEA76-8613.
620.Nishimatsu, S., Tokuyama, T. and Matsushita, M., "Characteristics of metal/alumina/silicon dioxide/silicon (MAOS) system", In: Thin film dielectrics, Vratney, F., (Ed.), Electrochem. Soc., Inc., New York, p. 338-54 (1969), EEA73-10541.
621.Nishizawa, J., Koyanagi, M. and Kimura, M., "Impedance measurements to study a semiconductor surface and thin insulating films on its surface", Jap. J. Appl. Phys., 13, Suppl. 2 (Pt. 1), 773-80 (1974), PA78-37827.
622.Pavlov, Y. A., Zherebstov, A. S., Rakov, A. V. and Kolorev, M. A., "Effects of local fields caused by elastic voltages at the boundary of Si with SiO_2 on electrical properties of MOS structures", Radiotekhnika I Elektronika, 14, 1348-9 (1969), Russian, PA73-8573.
623.Pepper, M., Pollitt, S. and Adkins, C. J., "Variable-range hopping in a silicon inversion layer", Phys. Lett. A, 47A, 71-2 (1974), PA77-28496.
624.Pepper, M., Pollitt, S. and Adkins, C. J., "Anderson localisation of holes in a Si inversion layer", Phys. Lett. A, 48A, 113-14 (1974), EEA77-30232.
625.Prince, J. L. and Wortman, J. J., "On the deeply depleted MOS capacitor", Proc. IEEE, 58, 842-4 (1970), EEA73-25049.
626.Rakov, A. V., Kolorev, M. A., Pavlov, E. A., Petrova, A. G. and Shepovalova, G. A., "Investigation of the influence of metal surface impurities in silicon on the properties of metal-oxide-semiconductor structures", Fiz. Tekh. Poluprovodnikov, 2, 526-9 (1968), Russian, EEA71-20575.

627.Richman, P., "The effect of gold doping upon the characteristics of MOS field-effect transistors with applied substrate voltage", Proc. IEEE, 56, 774-5 (1968), EEA71-18221.
628.Richman, P., "Modulation of space-charge-limited current flow in insulated-gate field-effect tetrodes", IEEE Trans. Electron Devices, ED-16, 759-66 (1969), EEA73-1201.
629.Sakaki, H. and Sugano, T., "Negative differential resistance and thermal effect in silicon MOS field-effect transistors", Oyo Buturi, 43 (Suppl.), 314-22 (1974), EEA78-2878.
630.Sato, S. and Sah, C. T., "Measurements of the thermal-emission rates of electrons and holes at the gold centers in silicon using the small-signal-pulsed field effect", J. Appl. Phys., 41, 4175-81 (1970), EEA73-37869.
631.Schemmert, W. and Hofflinger, B., "Optimisation of depletion-mode transistors for an M.O.S. p-channel technology", Electron. Lett., 9, 555-6 (1973), EEA77-6070.
632.Sedgwick, T. O. and Krongelb, S., "Negative surface charge in Ge MOS structures", J. Electrochem. Soc., 117, 1199-200 (1970), EEA73-37997.
633.Shappir, J., "Punch-through phenomena in m.o.s. transistors", European 71 Digest, 2 pp., EEA75-4719.
634.Sjostrand, M. E. and Stiles, P. J., "The channel conductivity in n-type Si inversion layers at very low electron densities", Solid State Commun., 16, 903-7 (1975), PA78-37839.
635.Sproul, M. E. and Nassibian, A. G., "Field-effect mobilities of gold-doped MOS structures", J. Appl. Phys., 46, 1255-60 (1975), EEA78-24242.
636.Stern, F., "Evidence for a mobility edge in inversion layers. II", J. Vac. Sci. & Technol., 11, 962-4 (1974), PA78-46763.
637.Stolarski, E., "The influence of point mechanical stress on the drain current of MOS-FET's", Arch. Elektrotech., 20, 833-5 (1971), Polish, EEA75-10259.
638.Stuart, R. A. and Eccleston, W., "Punchthrough currents in short-channel m.o.s.t. devices", Electron. Lett., 9, 586-8 (1973), EEA77-6028.
639.Sweet, J. N. and Holloway, P. H., "Correlations between interfacial oxides and resistance changes for W contacts on heavily doped Si-Ge", Thin Solid Films, 27, 263-71 (1975), PA78-58418.
640.Takahashi, S. and Haneta, Y., "Electric properties of flat MOS transistors", Denki Kagaku, 41, 498-502 (1973), CA80-42142.
641.Tanaka, T., "Shift of the gate threshold voltage of MOS transistors due to the introduction of shallow impurities", Jap. J. Appl. Phys., 10, 84-91 (1971), EEA74-13257.
642.Tarui, Y., Komiya, Y. and Teshima, H., "The built-in voltage and the charge distributions in the oxide of MOS structure", Jap. J. Appl. Phys., 6, 191-204 (1967), EEA70-8470.
643.Tarui, Y., Teshima, H. and Matsuoka, T., "Si MOS characteristics using vapor grown Al_2O_3 films", Bull. Electrotech. Lab., 33, 576-87 (1969), Japanese, EEA73-17660.

644.Terao, H., Ito, T. and Sakai, Y., "Interface properties of InAs-MIS structures and their application to FET", Electr. Eng. Jap., 94, 127-327 (1974), PA78-50681.
645.Toda, M., "Propagation of surface charge waves in the n-InSb MOS structure in a magnetic field", J. Phys. Soc. Japan, 27, 428-36 (1969), PA72-47150.
646.Toda, M., "Nonlinear effects of surface charge waves in a n-InSb MOS structure in a transverse magnetic field", Jap. J. Appl. Phys., 9, 49-57 (1970), PA73-23736.
647.Tsao, K. Y., "Electrical properties of silicon m.o.s. structures with gold and copper diffused into the interface", Thesis, Lehigh Univ., Bethlehem, Pa., 1974, 202 pp., Order No. 74-21442, EEA78-19744.
648.Ushirokawa, A., Suzuki, E. and Warashina, M., "Avalanche injection effects in MIS structures and realization of N-channel enhancement type MOS FET's", Jap. J. Appl. Phys., 12, 298-407 (1973), EEA76-19932.
649.van Nielen, J. A., "Operation and D.C. behaviour of MOS transistors", Philips Tech. Rev., 31, 209-15 (1970), EEA74-34094.
650.van Nielen, J. A., Theunissen, M. J. J. and Appels, J. A., "MOS transistors in thin monocrystalline silicon layers", Philips Tech. Rev., 31, 271-5 (1970), EEA74-34099.
651.Wei, L. S. and Simmons, J. G., "Transient emission and generation currents in metal-insulator-semiconductor capacitors", Solid-State Electron., 18, 853-7 (1975), EEA78-37475.
652.Weissman, R. H., "The electrical characteristics of insulator-gallium arsenide interfaces", Thesis, Stanford Univ., California, 1971, 184 pp., Order No. 71-19776, EEA75-32607.
653.Whelan, M. V., "Electrical behavior of defects at a thermally oxidized silicon surface", Report PRRS 1970-6, Eindhoven Univ. Technol., Netherlands, 1970, 93 pp., EEA74-7671.
654.Witt, W., Huber, F. and Delivorias, P., "MOS transistors with anodically formed metal oxides as gate insulators", Proc. IEEE, 55, 687 (1967), EEA70-17035.
655.Zerbst, M., "New results on MIS transistors", In: Festkorper problems IX; Advances in solid state physics, Madelung, O., (Ed.), Pergamon Press, New York, p. 300-15 (1969), EEA73-28221.

VI. Characterization

1. Impurity Profiles

a. MIS Technique

656.Adda, L. P. and Clemens, J. T., "A diagnostic technique of the identification of low-level, surface-diffused impurities in silicon", Presented at Electrochemical Society Meeting, San Francisco, Calif., May 1974, PA77-49591.
657.Baccarani, G., Solmi, S. and Soncini, G., "The silicon impurity profile as revealed by high-frequency non-equilibrium MOS C-V characteristics", Alta Freq., 41, 113-15 (1972), EEA75-28504.

658.Beister, G., Schmidt, V., Hottewitzsch, P. and Ehwald, K. E., "Impulse measurement of doping profiles in MOS structures using the second harmonic method", Phys. Status Solidi A, 28, K53-6 (1975), EEA78-19680.

659.Boucher, J. and Rakotondrazafy, C., "Direct determination of doping in a silicon substrate from studies of the temperature variation of a MOS structure", C. R. Hebd. Seances Acad. Sci. B, 277, 683-6 (1973), French, PA77-13569.

660.Brown, D. M., Connery, R. J. and Gray, P. V., "Doping profiles by MOSFET deep depletion C(V)", J. Electrochem. Soc., 122, 121-7 (1975), PA78-26305.

661.Fahrner, W. and Goetzberger, A., "Determination of deep energy levels in Si by MOS techniques", Appl. Phys. Lett., 21 (7), 327-31 (1972), EEA76-736.

662.Jones, J. E., "The characterization of thin heavily doped layers in silicon", Thesis, McMaster Univ., Hamilton, Ontario, Canada, 1974, PA78-66562.

663.Le Bloa, A. and Fortin, B., "Capacitive equivalent circuit of the depletion layer of a MOS transistor used for determination of impurity concentration near insulator-semiconductor interface", Phys. Status Solidi A, 25, 653-62 (1974), EEA78-5336.

664.Margalit, S. and Neugroschel, A., "Redistribution of boron and phosphorous in silicon after two oxidation steps used in MOST fabrication", IEEE Trans. Electron Devices, ED-19, 861-8 (1972), EEA75-24358.

665.Nicollian, E. H., Hanes, M. H. and Brews, J. R., "Using the MIS capacitor for doping profile measurements with minimal interface state error", IEEE Trans. Electron Devices, ED-20, 380-9 (1973), EEA76-16313.

666.Nishi, Y. and Konaka, M., "Determination of impurity distribution in silicon epitaxial film using MOS capacitor", Jap. J. Appl. Phys., 5, 1116 (1966), EEA70-2235.

667.Ogura, S., "Metal-insulator-semiconductor structure for impurity profile measurement", IBM Tech. Disclosure Bull., 18, 1140-1 (1975), EEA78-41467.

668.Pao, H. C., "A simplified MIS method of measuring the semiconductor impurity concentration", Solid-State Electron., 11, 509-12 (1968), EEA71-15714.

669.Rakov, A. V., Korolev, M. A., Pavlov, E. A., Petrova, A. G. and Shepovalova, G. A., "Investigation of the influence of metal surface impurities in silicon on the properties of metal-oxide-semiconductor structures", Fiz. Tekh. Poluprovodnikov, 2, 526-9 (1968), Russian, EEA71-20575.

670.Sadovnichii, A. A., Grotte, A. M., Peresunko, S. P. and Maksyuta, I. I., "Determination of a concentration profile of epitaxial structures using C-V characteristics of MOS varactors", Poluprovodn. Tekh. & Mikroelektron., no. 15, 59-66 (1974), Russian, EEA78-2748.

671.Shannon, J. M., "D.C. measurement of the space charge capacitance and impurity profile beneath the gate of an MOST", Solid-State Electron., 14, 1099-106 (1971), EEA75-1176.
672.Tihanyi, J., "Measurement of substrate impurity profile of MIS field-effect transistors", Siemens Forsch. - & Entwicklungsber., 1, 363-8 (1972), EEA75-39957.
673.van Gelder, W. and Nicollian, E. H., "Silicon impurity distribution as revealed by pulsed MOS C-V measurements", J. Electrochem. Soc., 118, 138-41 (1971), EEA74-15570.
674.Ziegler, K., Klausmann, E. and Kar, S., "Determination of the semiconductor doping profile right up to its surface using the MIS capacitor", Solid-State Electron., 18, 189-98 (1975), PA78-16101.
675.Zohta, Y., "Rapid determination of semiconductor doping profiles in MOS structures", Solid-State Electron., 16, 124-6 (1973), EEA76-8602.

b. Other Techniques

676.Ambridge, T. and Faktor, M. M., "An electrochemical technique for automatic depth profiles of carrier concentration", Gallimm Arsenide and Related Compounds, 1974, p. 320-30, Publ. 1975, PA78-54989.
677.Baroux, B., Dugue, M. and Sev, A., "Design and construction of a profile analyser for measuring impurity distribution in a semiconducting material", Revue Tech. Thomson-CSF, 4, 507-26 (1972), French, EEA75-39818.
678.Cardwell, M. J. and Peart, R. F., "Measurement of carrier-concentration profiles in epitaxial indium phosphide", Electron. Lett., 9, 88-9 (1973), EEA76-16286.
679.Chakrabarti, U. K., "Approximate solution for the redistribution of impurities during second oxidation", IEEE Trans. Electron Devices, ED-22, 566-8 (1975), PA78-62542.
680.Frank, H. and Kordos, P., "Measuring doping profiles on epitaxial GaAs layers", Elcktrotech. Cas., 26, 92-101 (1975), Czech, PA78-29520.
681.Frenzel, M., "A contribution to the determination of penetration depths of dopants in silicon dioxide films on silicon", Krist. & Tech., 8, 965-73 (1973), German, PA77-4050.
682.Gadzhiev, A. R. and Shlimats, I. S., "Determination of the nature of the distribution of impurities, by measuring the DC hopping conductivity", Sov. Phys.-Semicond., 8, 120-1 (1974), PA77-75858.
683.Glover, G. H. and Tantraporn, W., "R-I profiling: A new technique for measuring semiconductor doping profiles", Appl. Phys. Lett., 25, 348-9 (1974), PA78-7466.
684.Glover, G. H. and Tantraporn, W., "Doping profile measurements from avalanche space-charge resistance: A new technique", J. Appl. Phys., 46, 867-74 (1975), PA78-58816.

685. Greene, J. E., Sequeda-Orsorio, F., Streetman, B. G., Noonan, J. R. and Kirkpatrick, C. G., "Measurement of boron impurity profiles in Si using glow discharge optical spectroscopy", Appl. Phys. Lett., 25, 435-8 (1974), PA78-10527.
686. Gupta, D. C., "Diode voltage-capacitance method for measuring resistivity and impurity profile in a silicon epitaxial layer", Solid State Technol., 11 (2), 31-4 (1968), EEA71-13182.
687. Gupta, D. C. and Chan, J. Y., "Direct measurement of impurity distribution in semiconducting materials", J. Appl. Phys., 43, 515-22 (1972), PA75-15775.
688. Inoue, T., Horiuchi, S., Yoshii, S. and Tanuma, Y., "Auger studies of diffused impurity profiles in SiO_2/Si and poly Si/SiO_2/Si structures", Oyo Buturi, 44 (Suppl.), 329-37 (1975), EEA78-41493.
689. Lile, D. L. and Davis, N. M., "Semiconductor profiling using an optical probe", Solid-State Electron., 18, 699-704 (1975), PA78-62569.
690. Lis, D., Rosinski, W. and Stano, A., "On certain measurement method of impurities concentration distribution and epitaxial layers resistivity", Arch. Elektrotech., 17, 889-907 (1968), Polish, EEA72-7811.
691. Mattis, R. L. and Buchler, M. G., "Semiconductor measurement technology: A BASIC program for calculating dopant density profiles from capacitance-voltage data", Report NBS-SP-400-11, Nat. Bur. Stand., Washington, D. C., 1975, 40 pp., EEA78-41505.
692. Mitsuhashi, G., "Impurity profile in silicon epitaxial wafer with buried layers", NEC Res. & Dev., no. 36, 68-74 (1975), PA78-62562.
693. Morris, B. L., "Some device applications of spreading resistance measurements on epitaxial silicon", J. Electrochem. Soc., 121, 422-6 (1974), PA77-45459.
694. Muller, G. L. and Robinson, D. A. H., "Looking inside semiconductor devices", Bell Lab. Rec., 53, 129-35 (1975), EEA78-19537.
695. Nuyts, W. and van Overstraeten, R., "Computer calculations of impurity profiles in silicon. II", Phys. Status Solidi A, 15, 455-72 (1973), EEA76-16300.
696. Schumann, P. A., Schneider, C. P. and Pietrogallo, L. A., "Three-dimensional impurity profile of an epitaxial layer", Solid State Technol., 16 (9), 54-5 (1973), PA77-1450.
697. Sharma, R. S., "Impurity distribution in the diffused layers in silicon", Indian J. Phys., 48, 641-6 (1974), EEA78-8433.
698. Tihanyi, J., "Doping profile measurements on silicon epitaxial layers with field controlled planar diodes", Solid-State Electron., 13, 309-15 (1970), EEA73-17502.
699. Yamashita, A., Aoki, T. and Yamaguchi, M., "A method for determining a GaAs epitaxial layer impurity profile", Jap. J. Appl. Phys., 14, 991-8 (1975), EEA78-33029.

700.Zohta, Y. and Ohmura, Y., "Determination of the spatial distribution of deep centers from capacitance measurements of pn junctions", Appl. Phys. Lett., 21, 117-19 (1972), EEA75-36219.

2. C-V Characteristics

701.Agatsuma, T. and Morita, J., "Use of drain capacitance-voltage characteristics as a process control tool for the threshold voltage of silicon gate MOSFET's", IEEE Trans. Electron Devices, ED-22, 961-2 (1975), EEA78-41609.
702.Antognetti, P., "Experimental observations of the metal-oxide system aluminum oxide on silicon-silicon", Alta Frequenza, 39, 632-4 (1970), Italian, EEA73-72095.
703.Baccarani, G., Severi, M. and Soncini, G., "Electrical characteristics of MOS structures on (111) and (100) oriented n-type silicon as influenced by use of hydrogen chloride during thermal oxidation", J. Electrochem. Soc., 120, 1436-8 (1973), EEA77-343.
704.Baccarani, G., Solmi, S. and Soncini, G., "Errors induced by impurity redistribution on the surface-state measurements in MOS capacitors", Alta Frequenza, 40, 674-8 (1971), EEA74-76780.
705.Betko, J., Dubecky, F., Merinsky, K. and Cervenak, J., "C-V characteristics analysis of MOS structures on high resistivity silicon", Elektrotech. Cas., 26, 365-76 (1975), Slovak, PA78-62111.
706.Blomniek, E. A., Montsevoi, Y. A. and Temper, E. M., "Measurement of capacitance-voltage characteristics of MOS structures by the pulse method", Fiz. Tekh. Poluprovodnikov, 1, 1270-1 (1967), Russian, EEA71-1270.
707.Breed, D. J. and Kramer, R. P., "A new type of charge trapping in MOS systems", Philips Res. Rep., 28, 75-9 (1973), EEA76-26011.
708.Brews, J. R. and Lopez, A. D., "A test for lateral nonuniformities in MOS devices using only capacitance curves", Solid-State Electron., 16, 1267-77 (1973), EEA77-2571.
709.Brodie, D. E., Kawaliw, I. and Tokarsky, R., "Surface potentials in vacuum-deposited MIS structures", Canad. J. Phys., 46, 2207-12 (1968), PA72-5740.
710.Brotherton, S. D. and Burton, P., "The influence of non-uniformly doped substrates on MOS C-V curves", Solid-State Electron., 13, 1591-5 (1970), EEA74-7668.
711.Castagne, R., "Determination of the density of slow states of the capacitance of a metal-insulator-semiconductor by studying the charge while applying a linearly growing voltage", C. R. Acad. Sci. B, 267, 866-9 (1968), French, PA72-10024.
712.Castagne, R. and Vapaille, A., "Effect of spatial fluctuations of the potential at the Si-SiO_2 interface on the capacity-voltage curves of a MOS structure", C. R. Acad. Sci. Ser. B, 270, 1347-50 (1970), French, EEA73-28114.

713. Castagne, R. and Vapaille, A., "C(V) curves and apparent density of states at the Si/SiO_2 interface of MOS structures with (111) surface orientation", Solid State Commun., 9, 1347-51 (1971), French, EEA74-37207.

714. Choe, Y. M. and Murphy, N. S. J., "Reciprocal C-V plots using a pulse technique", Solid-State Electron., 18, 1032-3 (1975), EEA78-1537.

715. Cova, S., Tenconi, S. M. and Dotti, D., "Automated digital measurements of CV curves and derivatives, and of capacitance transients for semiconductor evaluation", IEEE Trans. Instrum. & Meas., IM-22, 347-55 (1973), PA77-5105.

716. Doering, E., "Examinations of metal-aluminium oxide-silicon system", Z. Angew. Phys., 29, 346-9 (1970), German, EEA74-1613.

717. El-Hoshy, A., Brown, D. M. and Engeler, W. E., "Molybdenum-films as partial diffusion masks in MOS processing", Appl. Phys. Lett., 17, 261-3 (1970), EEA74-1490.

718. Faraone, L. and Nassibian, A. G., "Nickel doped MOS capacitors and N-channel MOSFET's", Aust. Telecommun. Res., 7 (2), 32-5 (1973), EEA77-26463.

719. Feltynsh, I. A., Freiberga, L. A., Eglitis, I. E. and Eimanis, I. E., "Investigation of MOS-structures with r.f. glow discharge deposited SiO_2 films", Latv. PSR Zinat. Akad. Vestis Fiz. Tehn. Ser., no. 2, 48-52 (1970), Russian, PA73-51090.

720. Freyberga, L. A., "Evaluation of state density employing capacity of MOS structures", In: Semiconductors and their applications in electronics IV, Zinatne Publications, Riga, USSR, p. 137-55 (1970), Russian, PA73-5639.

721. Gabilli, E., Severi, M., Soncini, G. and Zini, Q., "Electrical properties of m.o.s. capacitors with oxide grown in the presence of HCl", Alta Freq., 41, 854-9 (1972), EEA76-16432.

722. Galloway, K. F., "Ionic clustering as a possible source of error in determining MOS C-V characteristics", Thin Solid Films, 23, S41-4 (1974), PA77-72871.

723. Gnadinger, A. P., "Surface mobility near threshold and other parameters of insulated gate field effect transistors", Thesis, Univ. Kansas, Lawrence, 1970, 158 pp., Order No. 70-25338, EEA74-30840.

724. Gnadinger, A. P. and Talley, H. E., "Surface mobility near the transistor region of an insulated gate field effect transistor", 1970 SWIEECO Record of Technical Papers of the 22nd Southwestern IEEE Conference and Exhibition, 1970, p. 171-5, EEA73-35244.

725. Goetzberger, A., "Ideal MOS curves for silicon", Bell Syst. Tech. J., 45, 1097-122 (1966), EEA70-1983.

726. Goetzberger, A. and Irvin, J. C., "Low-temperature hysteresis effects in metal-oxide-silicon capacitors caused by surface-state trapping", IEEE Trans. Electron Devices, ED-15, 1009-14 (1968), PA72-18644.

727.Goetzberger, A. and Klausmann, E., "A capacitance mapping technique for investigation of localized recombination generation sites in $Si-SiO_2$ interfaces", Proc. IEEE, 58, 799-80 (1970), EEA73-24930.

728.Goklaney, S. M. and Porter, W. A., "The effect of dislocations on the capacitance vs. bias voltage characteristic of MOS capacitors", IEE-IERE Proc. India, 12, 184-7 (1974), EEA78-8689.

729.Goodman, A. M., "A useful modification of the technique for measuring capacitance as a function of voltage", IEEE Trans. Electron Devices, ED-21, 753-7 (1974), PA78-29746.

730.Goodman, A. M., "An investigation of the silicon-sapphire interface using the MIS capacitance method", IEEE Trans. Electron Devices, ED-22, 63-5 (1975), PA78-29747.

731.Gorban, A. P., Litovchenko, V. G., Moskal, D. N. and Peikov, P. K., "Determination of generation-recombination characteristics in MIS-structures by means of non-equilibrium C(V) curves", Ukr. Fiz. Zh., 20, 66-71 (1975), Russian, PA78-25773.

732.Gorban, A. P., Litovchenko, V. G. and Peikov, P. K., "Investigation of the fast surface state spectrum of MIS structures by differential C-U method", Phys. Status Solidi A, 10, 289-96 (1972), EEA75-16687.

733.Gray, P. V. and Brown, D. M., "Freeze-out characteristics of the MOS varactor", Appl. Phys. Lett., 13, 247-8 (1968), PA72-10025.

734.Gupta, D. C. and Anantha, N. G., "Measurement of epitaxial layer resistivity using MOS capacitance method", Proc. IEEE, 55, 1108 (1967), EEA70-20220.

735.Haneta, Y., "C-V characteristics of MOS diodes prepared by SiH_4-NO_2 system", Jap. J. Appl. Phys., 8, 929-34 (1969), PA73-18356.

736.Herman, A. R. and Jancsecz, B. V., "Influence of the metallization on the charges in various MOS-structures", Proceedings of the International Conference on the Physics and Chemistry of Semiconductor Heterojunctions and Layer Structures, 1970, p. V-109-15, Publ. 1971, PA74-52966.

737.Hielscher, F. H. and Preier, H. M., "Non-equilibrium C-V and I-V characteristics of metal insulator-semiconductor capacitors", Solid-State Electron., 12, 527-38 (1969), PA72-44996.

738.Holtz, F., "Determination of the MOS flatband condition by electroreflectance and capacitance measurements", Phys. Status Solidi A, 21, 469-78 (1974), EEA77-14429.

739.Iida, K., "Effects of bias application on C-V characteristics in MA(O)S structures", Jap. J. Appl. Phys., 12, 807-13 (1973), EEA76-29193.

740.Imabun, Y., "Gate-controlled MOS variable capacitors", Electr. Eng. Jap., 93, 123-8 (1973), EEA78-5415.

741.Jakubowski, A. and Jagodzinski, P., "The method for calculating capacitance-voltage characteristics of ideal MIS structures", Electron Technol., 4 (3), 37-44 (1971), EEA75-13643.

742.Jakubowski, A. and Jagodzinski, P., "Effect of series resistance on C-V characteristics of MIS capacitor", Arch. Elektrotech., 24, 209-10 (1975), Polish, EEA78-41512.

743.Kaden, G., "The influence of iron diffusion on the electrical properties of Si-SiO_2-m.o.s. structures", Phys. Status Solidi A, 3, 161-71 (1970), German, EEA74-4862.

744.Kajiwara, Y., Nagai, S. and Hagino, H., "Electrical properties of polycrystalline Si film (in MIS structure)", Presented at Electrochemical Society Meeting, San Francisco, California, May 1974, EEA77-22940.

745.Kamins, T. I. and Deal, B. E., "Characteristics of Si-SiO_2 interfaces beneath thin silicon films defined by electrochemical etching", J. Electrochem. Soc., 122, 557-60 (1975), PA78-46762.

746.Kar, S., "Determination of Si-metal work function differences by MOS capacitance technique", Solid-State Electron., 18, 169-81 (1975), EEA78-8494.

747.Kim, J. C., "Interface properties of InSb MIS structures", IEEE Trans. Parts, Hybrids & Packag., PHP-10, 200-7 (1974), EEA78-11541.

748.Kobayashi, K. and Ohta, K., "Electronic processes in metal-silicon nitride-silicon dioxide-silicon systems", Jap. J. Appl. Phys., 11, 538-45 (1972), EEA75-20139.

749.Korwin-Pawlowski, M. L. and Heasell, E. L., "Characteristics of MOS capacitors formed on p-type InSb", Phys. Status Solidi A, 24, 649-52 (1974), PA77-70075.

750.Kranzer, D. and Gassaway, J. D., "C/V curves for gate-controlled diodes in silicon on sapphire", Electron. Lett., 11, 50-2 (1975), EEA78-8523.

751.Kuhn, M. and Nicollian, E. H., "Nonequilibrium effects in quasi-static MOS measurements", J. Electrochem. Soc., 118, 370-3 (1971), PA74-37235.

752.Laukmanis, L. A. and Pentyush, E. V., "Conductance and susceptance measurements in MOS-structures", Latv. PSR Zinat. Akad. Vestis Fiz. Teh. Ser., no. 6, 19-22 (1973), Russian, PA77-35349.

753.Lehovec, K., "C-V analysis of a partially-depleted semiconducting channel", Appl. Phys. Lett., 26, 82-4 (1975), PA78-33555.

754.Lideikis, T. P. and Shilalnikas, V. I., "On the transient response of a MOS capacitor", Litov. Fiz. Sb., no. 3, 413-20 (1972), Russian, EEA76-16317.

755.Litovchenko, V. G. and Serba, A. A., "On the measurement of high-velocity generation recombination characteristics of MOS structures with the aid of minority carrier C(V) characteristics", Poluprovodn. Tekh. & Mikroelektron., no. 19, 40-2 (1975), Russian, EEA78-28703.

756.Lopez, A. D., "Using the quasistatic method for MOS measurements", Rev. Sci. Instrum., 44, 200-4 (1973), EEA76-16329.

757.Lopez, A. D. and Strain, R. J., "MOS surface potential and the gross nonuniformity", Solid-State Electron., 16, 507-11 (1973), EEA76-16338.

758.Machalica, P., "Measurement of C-V characteristics of MIS structures biased with triangle voltage", Pr. Przem. Inst. Elektron., 15, 31-45 (1974), Polish, PA78-72910.
759.McCaughan, D. V., Murphy, V. T. and Walden, R. H., "Some pitfalls in fast ramp C-V measurements", Solid-State Electron., 16, 1423-7 (1973), PA77-13230.
760.McCaughan, D. V. and Wonsiewicz, B. C., "Effects of dislocations on the properties of metal SiO_2-silicon capacitors", J. Appl. Phys., 45, 4982-5 (1974), PA78-21963.
761.McMillan, L., "MOS C-V techniques for IC process control", Solid State Technol., 15 (9), 47-52 (1972), EEA75-39191.
762.McNutt, M. J. and Sah, C. T., "Effects of spatially inhomogeneous oxide charge distribution on the MOS capacitance-voltage characteristics", J. Appl. Phys., 45, 3916-21 (1974), PA78-6898.
763.Mattauch, R. J. and Lade, R. W., "Surface inversion of MOS diodes", Proc. IEEE, 55, 246 (1967), EEA70-22841.
764.Miller, H., "Surface state studies of thick silicon dioxide layers using MOS structures", Proceedings of the Microelectronics Symposium, 1968, Pap. E1-1-E1-9, EEA71-28958.
765.Nadan, J. S., "Experimental capacitance-voltage characteristics of three-terminal digital MOS varactors", Proc. IEEE, 61, 1512-13 (1973), EEA76-38383.
766.Neugroschel, A., Margalit, S. and Bar-Lev, A., "n-channel enhancement m.o.s. transistor built in germanium-doped silicon", Electron. Lett., 9, 153-4 (1973), EEA76-19926.
767.Neugroschel, A., Margalit, S. and Bar-Lev, A., "Interface properties of oxidized germanium-doped silicon", J. Phys. D, 6, 1606-21 (1973), EEA76-35255.
768.Nicollian, E. H. and Goetzberger, A., "The Si-SiO_2 interface electrical properties as determined by the metal-insulator-silicon conductance technique", Bell Sys. Tech. J., 46, 1055-1133 (1967).
769.Nienerowski, K. and Weber, E. H., "Simultaneous determination of volume trap concentration and interface trap density from C-V characteristics", Phys. Status Solidi A, 7, 407-11 (1971), EEA75-4631.
770.Nishimatsu, S. and Ashikawa, M., "A simple method for measuring the interface state density (Analogue computation)", Rev. Sci. Instrum., 45, 1109-12 (1974), EEA78-11572.
771.Noble, W. P., Jr., "Semiconductor-insulator capacitance measurement", IBM Tech. Disclosure Bull., 13, 3226 (1971), EEA74-30825.
772.Olenski, J. and Machalica, P., "Technique for plotting non-equilibrium C/V curves of an MOS capacitor", Electron. Lett., 11, 232-4 (1975), EEA78-28606.
773.Owen, S. J. T., "Voltage shifts on sputtered SiO_2 films in MIS structures", Thin Films, 2, 19-24 (1971), EEA74-37208.
774.Pauwels, H. J., "Carrier concentration from MIS C-V measurements", IEEE Trans. Electron Devices, ED-20, 332-3 (1973), EEA76-16312.

775. Pavlov, Y. A., Zherebtsov, A. S., Rakov, A. V. and Korolev, M. A., "Effects of local fields caused by elastic voltages at the boundary of Si with SiO_2 on electrical properties of MOS structures", Radiotekhnika I Elektronika, 14, 1348-9 (1969), Russian, EEA73-4525.

776. Pierret, R. F. and Small, D. W., "Effects of lateral surface generation on the MOS-C linear-sweep and C-t transient characteristics", IEEE Trans. Electron Devices, ED-20, 457-8 (1973), EEA76-16314.

777. Pierret, R. F. and Small, D. W., "Stationary room temperature MOS-C deep-depletion characteristics", Solid-State Electron., 18, 79-85 (1975), PA78-25771.

778. Popova, L. I., Vitanov, P. K. and Antov, B. Z., "C-V and I-V characteristics of MIS structures with pyrolytic SiO_2 as dielectric", Thin Solid Films, 23, 15-22 (1974).

779. Preier, H., "Different mechanisms affecting the inversion layer transient response", IEEE Trans. Electron Devices, ED-15, 990-97 (1968), PA72-12172.

780. Przewlocki, H., "Determination of semiconductor surface region resistivity by the MIS structure capacitance measurement method", Arch. Elektrotech., 18, 539-57 (1969), Polish, EEA73-10672.

781. Schipanski, D., "Investigation of C-U-curves at low temperatures in the Ge-SiO_2 system", Wiss. Z. Tech. Hochsch. Ilmenau, 18, 71-8 (1972), German, EEA75-32606.

782. Schipanski, D., "Study of germanium-insulator structures at low temperature by means of the C-V method", Z. Elektr. Inf. - & Energietech., 2, 224-5 (1972), German, EEA76-12540.

783. Schwartz, R. J., Dockerty, R. C. and Thompson, H. W., Jr., "Capacitance voltage measurements on n-type InAs MOS diodes", Solid-State Electron., 14, 115-24 (1971), PA74-33730.

784. Svensson, C., Lundstrom, I. and Christensson, S., "Pulsed C-U measurements on MIS devices", Phys. Status Solidi A, 1, 389-94 (1970), EEA74-7783.

785. Tanguy, J., "Study of the metal-organic monolayer-semiconductor structures", Thin Solid Films, 13, 33-9 (1972), EEA76-4590.

786. Tarui, Y., Komiya, Y. and Nagai, K., "Measurement of the surface states by utilizing MOS structure. II", Bull. Electrotech. Lab., 30 (1), 8-24 (1966), Japanese, EEA70-2252.

787. Temple, V. A. K. and Shewchun, J., "The exact low-frequency capacitance-voltage characteristics of metal-oxide-semiconductor (MOS) and semiconductor-insulator-semiconductor (SIS) structures", IEEE Trans. Electron Devices, ED-18, 335-42 (1971), PA74-24135.

788. Tokutaka, H., Kubo, S. and Akiyama, K., "Effects of aluminum electrode and hydrogen gas during heat treatments on MOS structure", Jap. J. Appl. Phys., 5, 841-2 (1966), EEA70-11542.

789. Tsujide, T., "C/V shift characteristics of metal-Al_2O_3-Si and metal-Al_2O_3-SiO_2-Si structures", Jap. J. Appl. Phys., 11, 62-8 (1972), EEA75-7546.

790.Tyagai, V. A., Petrova, N. A., Shirshov, Y. M., Frolov, O. S. and Snitko, O. V., "Electrophysical properties of silicon/silicon-dioxide/electrolyte structures", Poluprovodn. Tekh. & Mikroelektron., no. 10, 80-8 (1972), Russian, EEA76-16328.
791.Uematsu, S., Kondo, H. and Nakao, Y., "Influence of sputtering on the SiO_2-Si structure", Mitsubishi Denki Lab. Rep., 12, 29-35 (1971), EEA75-16683.
792.Ugai, Y. A., Kolesnikov, V. G., Anokhin, V. Z., Fetisova, S. V. and Bulavin, V. G., "Chemical reactions and electrical properties in heterogeneous systems SiO_2-metal", Inorg. Mater., 9, 153-5 (1973), EEA76-38219.
793.Ushirokawa, A. and Warashina, M., "Simplifications of C-V method for the metal-insulator-semiconductor structure and estimation of internal constants", Electron. & Commun. Jap., 54, 98-106 (1971), Japanese, EEA75-28437.
794.Van Overstraeten, R., Declerck, G. and Broux, G., "Graphical technique to determine the density of surface states at the Si-SiO_2 interface of MOS devices using the quasistatic C-V method", J. Electrochem. Soc., 120, 1785-7 (1973), PA77-13228.
795.Virtmanis, A. S., Zhagata, L. A., Laukmanis, L. A., Feltyn, I. A., Freiberga, L. A. and Teteris, V. A., "Investigation of low-temperature SiO_2 films", Latv. PSR Zinat. Akad. Vestis Fiz. Teh. Ser., no. 4, 3-7 (1973), Russian, EEA77-22950.
796.Werner, W. M., "The work function difference of the MOS-system with aluminium field plates and polycrystalline silicon field plates", Solid-State Electron., 17, 769-75 (1974), PA77-60870.
797.Zelinka, J. and Starosta, K., "Evaluation of the MIS-structure properties on the basis of measuring the reciprocal value of capacitance", Elektrotech. Cas., 25, 35-45 (1972), PA77-41121.
798.Zerbst, M., "Relaxation effects in semiconductor insulator interfaces", Z. Angew. Phys., 22, 30-3 (1966), German, PA70-11545.

3. I-V Characteristics

799.Asmontas, S., Silalnikas, V. and Juciene, V., "The electrical conductivity of Au-SiO_2-Si structure", Litov, Fiz. Sbornik, no. 5-6, 809-17 (1968), Russian, EEA73-4420.
800.Av-Ron, M. and Shatzkes, M., "Current-voltage characteristics of Al-SiO_2-Si structures", Appl. Phys. Lett., 21, 233-5 (1972), EEA75-36190.
801.Batavin, V. V., "Influence of SiO_2 precipitates on current-voltage characteristics of p-n junctions in silicon", Fiz. Tekh. Poluprovodnikov, 4, 760-3 (1970), Russian, EEA73-31341.
802.Biryukov, V. N., "Voltage-current characteristics of MOS transistors", Izv. VUZ Radioelektron., 16 (12), 98-100 (1973), Russian, EEA77-18811.
803.Bonnet, J. and Lassabatere, L., "Pulsed study of injection and conduction in Si-SiO_2-Al structures", Thin Solid Films, 22, 177-87 (1974), French, PA77-60874.

804. de Maagt, B. J. and Verwey, J. F., "Determination of the capture cross-section of hole-trapping centres in SiO_2", Philips Res. Rep., 28, 210-18 (1973), EEA77-2515.

805. DeAtley, E. and Johnston, W. C., "Use of V-I relations in MOS output design", Electro-Technology, 82 (4), 64-70 (1968), EEA72-4460.

806. Dill, H. G., "Study of the different modes of operation of the insulated gated field effect transistor. I", Scientia Electrica, 13 (3), 45-88 (1967), PA70-22637.

807. Dill, H. G., "Study of the different modes of operation of the insulated gated field-effect transistor. II", Scientia Electrica, 13 (4), 89-104 (1967), EEA71-15869.

808. Fa, C. H. and Jew, T. T., "The poly-silicon insulated-gate field-effect transistor", IEEE Trans. Electron Devices, ED-13, 290-1 (1966), EEA70-2658.

809. Forgacs, G., Lorinczy, A. and Nemeth-Sallay, M., "Non-ohmic conduction in MOS systems with thin oxide layers", Phys. Status Solidi A, 20, K113-17 (1973), PA77-16827.

810. Gaman, V. I., Gazarov, V. D. and Reznikov, V. A., "Voltage-current characteristics of a metal-dielectric-semiconductor system", Izv. VUZ Fiz., no. 2, 15-19 (1969), Russian, PA72-44997.

811. Hoeneisen, B. and Mead, C. A., "Current voltage characteristics of small size m.o.s. transistors", IEEE Trans. Electron Devices, ED-19, 382-3 (1972), EEA75-10270.

812. Hsu, S. T., Scott, J. H. and Ham, W. E., "Anomalous current-voltage characteristics in SOS MOS FETs", Presented at 1973 International Electron Devices Meeting Supplement (abstracts only), p. 4, EEA77-14567.

813. Huang, J. S. T., "Characteristics of a depletion-type IGFET", IEEE Trans. Electron Devices, ED-20, 513-14 (1973), EEA76-19929.

814. Ilin, V. N., "Approximation of volt-ampere characteristics of MOS transistors", Pribory Sist. Upravleniya, no. 7, 26-8 (1968), Russian, EEA72-2831.

815. Kappallo, W. K. and Walsh, J. P., "A current voltage technique for obtaining low-frequency C-V characteristics of MOS capacitors", Appl. Phys. Lett., 17, 384-6 (1970), EEA74-7767.

816. Lehovec, K. and Zuleeg, R., "Negative resistance of a modified insulated-gate field-effect transistor", Proc. IEEE, 62, 1163-5 (1974), EEA77-34056.

817. Mills, T. G. and Kroger, F. A., "Electrical conduction at elevated temperatures in thermally grown silicon dioxide films", J. Electrochem. Soc., 120, 1582-6 (1973), EEA77-2430.

818. Mohan Rao, G. R. and Carr, W. N., "MOSFET devices with trapezoidal gates: I-V characteristics and magnetic sensitivity", Solid-State Electron., 16, 483-90 (1973), EEA76-16449.

819. Neugebauer, C. A., Burgess, J. F., Joynson, R. E. and Mundy, J. L., "I-V characteristics of MOS capacitors with polycrystalline silicon field plates", J. Appl. Phys., 43, 5041-4 (1972), EEA76-12504.

820.Ohwada, A., "Silicon n-channel normally-off insulated-gate field-effect transistors", Jap. J. Appl. Phys., 7, 862-9 (1968), PA72-3684.
821.Popova, L. I., Vitanov, P. K. and Antov, B. Z., "C-V and I-V characteristics of MIS structures with pyrolytic SiO_2 as dielectric", Thin Solid Films, 23, 15-22 (1974), EEA77-36263.
822.Przewlocki, H. M. and Marcinak, W., "The triangular voltage sweep method as a tool in studies of mobile charge in MOS structures", Phys. Status Solidi A, 29, 265-74 (1975), EEA78-28700.
823.Sakaki, H. and Sugano, T., "Negative differential resistance and thermal effect in Si MOS transistors at cryogenic temperatures", Annu. Rep. Eng. Res. Inst. Fac. Eng. Univ., Tokyo, 32, 141-6 (1973), Japanese, EEA77-23034.
824.Schwab, H., "Current/voltage characteristics of MOS field-effect transistors in the linear and saturation region", Electron. Lett., 10, 356-8 (1974), German, EEA77-38838.
825.Spinluscu Carnaru, I., Stanescu, C. and Draghici, I., "Metal-oxide-semiconductor transistors", Stud. Cercet. Fiz., 24, 741-52 (1972), Rumanian, EEA76-818.
826.Stoev, I. G., "$I_{gr}(V_g)$-$I_d(V_g)$ characteristics of MOS transistor", C. R. Acad. Bulg. Sci., 27, 897-9 (1974), PA78-29745.
827.Strikha, V. I. and Boozanyova, E. V., "The influence of local electronic states on I-V and C-V characteristics of the metal-thin insulator-semiconductor structures", Proceedings of the International Conference on the Physics and Chemistry of Semiconductor Heterojunctions and Layer Structures, 1970, p. V-189-96, Publ. 1971.
828.Testemale, E., Seve, G. and Lassabatere, L., "Electrical conduction in silicon-alumina-nickel structures", Thin Solid Films, 26, 253-72 (1975), French, PA78-50688.
829.Tsang, P. J., "CVD Ta-doped Al_2O_3 films and its electrical conduction", Presented at Electrochemical Society Meeting, San Francisco, Calif., May 1974, PA77-45465.
830.Weight, G. T., "Current/voltage characteristics, channel pinch-off and field dependence of carrier velocity in silicon insulated-gate field-effect transistors", Electron. Lett., 6, 107-9 (1970), EEA73-17662.

4. Other Techniques

831.Amelio, G. F., "A new method of measuring interface state densities in MIS devices", Surface Sci., 29, 125-43 (1972), EEA75-16754.
832.Anderson, K., "Device manufacturers test problems", Digest of Papers from Semiconductor Memory Test Symposium, 1972, p. 17-26, Publ. 1973, EEA76-26157.

833. Anon., "New method for measuring the surface potential in MOS structures", Kozl. Magy. Tud. Akad. Musz. Fiz. Kut. Intez., no. 0-14, 39-45 (1974), Hungarian, PA78-6902.

834. Bateman, I. M. and Magowan, J. A., "MOST channel-length measurement", Electron. Lett., 6, 669-71 (1970), EEA74-10647.

835. Benes, O. and Cerny, A., "A new method for evaluation of MIS structure insulation films", Tesla Electron., 3 (1), 3-10 (1970), EEA73-75759.

836. Brown, G. A., Lovelace, K. and Hutchins, C., "A process control test for lateral charge spreading susceptibility", Proceedings of the 11th Annual Reliability Physics Conference, 1973, p. 203-7, EEA77-6088.

837. Danklefs, R. and Thornton, H., "Testing complex MOS: the how and why", Electron. Eng., 29 (10), 42-6 (1970), EEA74-6916.

838. Declercq, M. and Jespers, P., "Analysis of interface properties in MOS transistors by means of 'charge pumping' measurements", Rev. HF, 9, 244-53 (1974), EEA78-15368.

839. Dellow, F., "Automatic measurement of MOST threshold voltage and gain factor", Instrum. Pract., 25 (2), 109-10 (1971), EEA74-13256.

840. Flemming, J. P., "Applications of the scanning low energy electron probe (SLEEP) for MOS device evaluation", Solid-State Electron., 14, 1061-6 (1971), EEA75-1174.

841. Grosvalet, J., "Methods of measurement of semiconductors' surface properties-application to the study of MOS devices and reliability", Symposium on Test Methods and Measurements of Semiconductor Devices, 1967, Pap. No. 401, PA71-18147.

842. Hartstein, A. and Fowler, A. B., "High temperature 'variable range hopping' conductivity in silicon inversion layers", J. Phys. C, 8, L249-53 (1975), EEA78-28691.

843. Henne, M. and Kirpal, A., "The threshold voltage of MIS transistors and its measurement", Nachrichtentechnik, 22, 365-6 (1972), German, EEA75-39947.

844. Hess, K., "On the magnetoresistance of silicon surface inversion layers", Phys. Status Solidi A, 31, 159-64 (1975), EEA78-41532.

845. Igarashi, T., Wakabayashi, J. and Kawaji, S., "Hall effect in silicon inversion layers under strong magnetic fields", J. Phys. Soc. Jap., 38, 1549 (1975), PA78-69245.

846. Ikeda, H., "A precise method for measuring low-frequency small-signal conductance parameters of an MOS transistor", Electron. & Commun. Jap., 54, 98-106 (1971), Japanese, EEA75-39935.

847. Ikeda, H., "A measuring method for MOST transconductance and its variation", IEEE Trans. Instrum. & Meas., IM-21, 198-204 (1972), EEA75-32682.

848. Ikeda, H., "An elegant method for measuring MOST drain-source conductance in the saturated current region", IEEE Trans. Instrum. & Meas., IM-21, 234-6 (1972), EEA75-32683.

849. Kamins, T. I. and MacDonald, N. C., "Measurements of the scattering of conduction electrons by localized surface charges", Phys. Rev., 167, 754-8 (1968), EEA71-13234.

850.Kar, S. and Dahlke, W. E., "Metal-dependent interface states in thin MOS structures", Appl. Phys. Lett., 18, 401-3 (1971), PA74-49150.
851.Komiya, Y., Tarui, Y. and Mori, K., "Measurements of the positive charge distribution in SiO_2 by channel conductance method and step etching method", Bull. Electrotech. Lab., 36, 104-11 (1972), Japanese, EEA75-36193.
852.Koomen, J., "The measurement of interface state charge in the MOS system", Solid-State Electron., 14, 571-80 (1971), EEA74-34104.
853.Kuhne, H., "Digital measurements of dynamic mutual conductance in field effect transistors", Radio Fernsehen Elektron., 22, 10-12 (1973), German, EEA76-19934.
854.Lam, Y. W., "Measurements of commercial MIS capacitors by the surface-photovoltage methods", Jap. J. Appl. Phys., 12, 916-23 (1973), EEA76-29195.
855.Lee, J. H. and Satya, A. V. S., "Three-device FET diagnostic testing scheme", IBM Tech. Disclosure Bull., 17, 2905-7 (1975), EEA78-24357.
856.Lewis, E. T., "The application of an analog computer for the determination of charge centers in the gate structure of the variable threshold FET", Solid-State Electron., 14, 475-80 (1971), EEA74-26889.
857.Litvinov, R. O., "The determination of the properties of M.I.S. structures by means of the photo-capacitive effect", Z. Elektr. Inf. - & Energietech., 2, 215-17 (1972), German, EEA76-8609.
858.Longo, H. E. and Zerbst, M., "Surface term analysis in the case of MIS transistors", Z. Angew. Phys., 24, 138-42 (1968), German, PA71-31415.
859.Mar, H. A. and Simmons, J. G., "Determination of bulk-trap parameters in MIS structures", Appl. Phys. Lett., 25, 503-5 (1974), PA78-25760.
860.Marciniak, W. and Ruslk, M., "Simple method of M.O.S.-transistor threshold-voltage measurement", Electron. Lett., 10, 202-4 (1974), EEA77-26466.
861.Mellor, P. J. T., "Techniques for the measurement of source and drain series resistance in m.o.s. transistors", Proc. Inst. Elec. Eng., 118, 1393-8 (1971), EEA75-1165.
862.Mellor, P. J. T., "Double-extrapolation method for m.o.s.t. threshold-voltage", Electron. Lett., 8, 236-7 (1972), EEA75-20207.
863.Nakhmanson, R. S. and Dobrovolskii, P. P., "Investigations of MIS structure inhomogeneities using a scanning mercury probe", Phys. Status Solidi A, 19, 225-41 (1973), EEA76-38222.
864.Parker, W. C. and Forbes, L., "Experimental characterization of gold-doped infrared-sensing MOSFET", IEEE Trans. Electron Devices, ED-22, 916-24 (1975).
865.Rowe, B., "Simple measurement of MOST gain factor", Electron. Lett., 4, 437-8 (1968), EEA71-28902.

866. Rudan, M. and Spadini, G., "Consideration on the deployment of phase detector for differential admittance measurement of MOS device", Alta Freq., 43, 236-40 (1974), Italian, EEA77-31913.
867. Rusek, M., "The analysis of continuous measurement of the transconductance in MIS transistors", Elektronika, 14, 384-90 (1973), Polish, EEA77-2566.
868. Sakaki, H., Hoh, K. and Sugano, T., "Determination of interface-state density and mobility ratio in silicon surface inversion layers (on m.o.s.t.s)", Annu. Rep. Eng. Res. Inst. Fac. Eng. Univ. Tokyo, 30, 193-7 (1971), EEA75-10262.
869. Salama, C. A. T. and Holmes, F., "On the determination of surface recombination velocity from the transient response of MIS structures", Solid-State Electron., 13, 1204-5 (1970), PA73-62394.
870. Sonoda, G., "FET stray and coupling capacitance equalization technique", IBM Tech. Disclosure Bull., 27, 1355 (1974), EEA78-8568.
871. Staples, E. J., Jr., "Detection of Rayleigh surface waves on silicon with the MOS-FET", Thesis, Southern Methodist Univ., Dallas, Tex., 1971, 118 pp., Order No. 72-663, EEA75-28527.
872. Voschenkov, A. M. and Zemel, J. N., "Admittance studies of surface quantisation in (100)-oriented Si metal-oxide-semiconductor field-effect transistors", Phys. Rev. B, 9, 4410-21 (1974), EEA77-34059.
873. Wallinga, H., "A method for the measurement of the turn-on condition in MOS transistors", Solid-State Electron., 14, 1093-8 (1971), EEA75-448.
874. Wohlgemuth, P. and Gall, D., "Measurement of barrier heights in M.I.S. systems", Z. Elektr. Inf. - & Energietech., 2, 217-19 (1972), German, EEA76-8610.
875. Yoshida, B., Miyahara, T., Wada, A. and Yamada, K., "A method for measuring low-frequency small-signal conductance parameters of a MOS-FET", Bull. Nagoya Inst. Technol., 24, 217-23 (1972), Japanese, EEA78-5328.

5. Instrumentation

876. Abend, P., Becker, H., Braunig, D., Bublitz, R., Herdam, G., Spencker, A., Wagemann, H. G. and Wawer, W., "Computer controlled measurement of semiconductor devices", Presented at 1st International Symposium on CAMAC in Real Time Computer Applications, Luxembourg, Belgium, December 1973, French and German, EEA77-14243.
877. Abowitz, G. and Arnold, E., "Simple mercury drop electrode for MOS measurements", Rev. Sci. Instrum., 38, 564-5 (1967), EEA70-24219.
878. Ahlers, H. and Ullrich, K., "An automatically operating method for determining the profiles of imperfections in semiconductor components", Nachrichtentechnik, 17, 323-7 (1967), German, EEA71-3791.

879.Aleinikov, N. M., Likholet, A. N., Mudrov, A. K., "An instrument for monitoring the surface state of semiconductors by the method of contact potential difference (in device manufacture)", Instrum. & Exp. Tech., 17, 1778-80, EEA78-37404.
880.Armstrong, W. E. and Lidke, J. R., "An analog system for measurement of effective surface mobility in MOS devices", IEEE J. Solid-State Circuits, SC-8, 180-2 (1973), EEA76-16440.
881.Astakhov, O. P. and Sgibnev, I. V., "Equipment for measuring the Hall constant of semiconductors at temperatures up to 1000°C", Meas. Tech., 16, 575-6 (1973), PA77-35350.
882.Batavin, V. V., Pantuev, V. S. and Priliprko, V. I., "An apparatus for four-probe specific resistance measurement on epitaxial silicon films", Meas. Tech., 17, 1753-4 (1974), PA78-69609.
883.Battista, M. A., "Test vehicle for semiconductor surfaces", IBM Tech. Disclosure Bull., 13, 1433-4 (1970), EEA74-7779.
884.Brieke, R., Linse, V. and Preusche, H. J., "Instrument with automatic range switching for measuring MOS transistors", Nachrichtentech. Elektron., 24, 141-3 (1974), EEA77-24196.
885.Connelly, R. N., "A tester for MOS integrated circuits", Syst. Technol., no. 15, 6-8 (August 1972), EEA75-35409.
886.Curran, L., "Meeting the MOS/LSI challenge: a special report on testers", Electronics, 44 (10), 68-75 (1971), EEA74-21677.
887.DeBar, D. E., "Capacitance bridge for MOS devices", IBM Tech. Disclosure Bull., 17, 2013-14 (1974), EEA78-16947.
888.Dudnikov, V. P., Zaitsev, B. D. and Shvedov, E. E., "An instrument for time-dependent drift measurements of MOS-transistors", Izv. VUZ Radioelektron., 13 (12), 1470-3 (1970), Russian, EEA74-11271.
889.Gordon, B. J., Stover, H. L. and Harp, R. S., "A new impurity profile plotter for epitaxy and devices", Symposium on Silicon Device Processing (programme and abstracts), 1970, p. 50, EEA73-35130.
890.Hammer, R., "A mercury contact probe for MOS measurements on oxidized silicon", Rev. Sci. Instrum., 41, 292-3 (1970), EEA73-21333.
891.Hammer, R., "A multicontact probe for dielectric strength measurements of an array of MOS capacitors", Rev. Sci. Instrum., 42, 1542 (1971), EEA75-749.
892.Hammer, R., "High temperature multicontact test probe", IBM Tech. Disclosure Bull., 14, 1924-5 (1971), EEA75-10293.
893.Harvilchuck, J. M. and Carruthers, R. A., "Automated charge/time/temperature measurement apparatus", Presented at Electrochemical Society Meeting, San Francisco, Calif., May 1974, EEA77-23056.
894.Hoekstra, J. P., "Probe for MOS measurements", IBM Tech. Disclosure Bull., 13, 2981-2 (1971), EEA74-18798.
895.IBM Corp., "Field effect transistor test apparatus", Patent UK 1380287, Publ. January 1975, EEA78-28816.
896.Janz, G., "Combined transistor-MOSFET-diode tester", Elektron. Int., no. 1-2, 16-17 (1974), German, EEA77-24192.

897. Karlovsky, J., "Semi-automatic bridge for the measurement of the capacity ratio C_{min}/C_{max} of a M.O.S. structure", Z. Elektr. Inf. - & Energietech., 2, 223 (1972), German, EEA76-8612.
898. Kopp, R. J., Williamson, T. A. and Grannemann, W. W., "Instrumentation providing automated data for use in metal insulator-semiconductor interface studies", ISA Trans., 10, 209-15 (1971), EEA75-10183.
899. McKnight, S. R., "Aluminum contacting probe", IBM Tech. Disclosure Bull., 17, 1624-5 (1974), EEA78-11730.
900. Odenweller, R. L., Jr., "An aging and testing facility for high-reliability transistors and diodes", West. Elect. Engr., 11, 20-9 (1967), EEA71-3794.
901. Ormond, D. W., Jr., "High-temperature mulicontact probe", IBM Tech. Disclosure Bull., 14, 568 (1971), EEA74-37310.
902. Pistor, R. L., "Circuit for automating metal-oxide-silicon retention-time measurements", IBM Tech. Disclosure Bull., 17, 1038-9 (1974), EEA78-5331.
903. Pogson, I., "Breakdown voltage tester for semiconductors", Electron. Aust., 36 (10), 52-5, 108 (1975), EEA78-24084.
904. Reimer, H., "Device for determination of surface state parameters in MIS structures", Wiss. Z. Elektrotech., 11, 202-12 (1968), German, EEA72-20441.
905. Roddan, A. C. and Vizir, V., "An instrument for automatic measurement of epitaxial layer thickness", Symposium on Silicon Device Processing (programme and abstracts), 1970, p. 49, EEA73-35129.
906. Ryssel, H., Schmid, K. and Muller, H., "A sample holder for measurement and anodic oxidation of ion implanted silicon", J. Phys. E, 6, 492-4 (1973), PA76-37662.
907. Schmitt, A., "State of the art and prospects of automated semiconductor measurement techniques", 6th International Congress on Microelectronics, 1974, 8 pp., German, EEA78-24527.
908. Schneider, B., "Current stage test of semiconductor", EC-Nyt, no. 42, 2-3 (1975), Danish, EEA78-24085.
909. Shewchun, J., Ghanekar, K. M. and Yager, R., "A computer controlled automatic system for measuring the conductivity and Hall effect in semiconducting samples", Rev. Sci. Instrum., 42, 1797-1807 (1971), EEA75-10169.
910. Smith, K. D., "A minicomputer-aided MOS array tester", Comput. Des., 9 (12), 65-71 (1970), EEA74-12628.
911. Storm, H. F., "MOSFET analyzer", IEEE Trans. Instrum. Meas., IM-19, 2-5 (1970), EEA73-17664.
912. Viswanathan, C. R., Howes, R. and Hinkle, V. O., "MOS-CV test system for IC process control and monitoring", Solid State Technol., 17 (3), 43-6, 57 (1974), EEA77-23156.

VII. Gettering

913. Aubuchon, K. G., Harari, E., Leong, D. H. and Chang, C. P., "Effects of HCl gettering, Cr doping and Al^+ implantation on hardened SiO_2", IEEE Trans. Nucl. Sci., NS-21, 167-71 (1974), EEA78-19631.
914. Buck, T. M., Pickar, K. A., Poate, J. M. and Hsieh, C. M., "Gettering rates of various fast-diffusing metal impurities at ion-damaged layers on silicon", Appl. Phys. Lett., 21, 485-7 (1972), EEA76-4504.
915. Chu, C. K., Johnson, J. E. and Riding, A. J., "HCl gettering for high power silicon devices", Presented at Electrochemical Society Meeting, San Francisco, Calif., May 1974, PA77-45468.
916. Dash, S., "Gettering and dislocations", In: Semiconductor silicon 1973, Electrochem. Soc., Inc., Princeton, N. J., p. 626-38.
917. d'Hervilly, G., Edel, W. A., Jacquemin, J. and Westdorp, W. A., "Method of gettering impurities in semiconductor devices introducing stress centers and devices resulting thereof", Patent USA 3874936, Publ. April 1975, EEA78-37430.
918. Graff, K. and Pieper, H., "Carrier lifetime of heat-treated silicon crystals", J. Electron. Mater., 4, 281-98 (1975).
919. Lawrence, J. E., "On lattice disorders, solute diffusion precipitation, and gettering silicon devices", In: Semiconductor silicon, Electrochem. Soc., Inc., New York, p. 596-609 (1969), EEA73-10437.
920. MacKenna, E. L., "Silicon and silicon dioxide gettering in perspective", Electrochemical Society Fall Meeting (Extended abstracts), 1974, p. 531-4, EEA78-37452.
921. Meek, R. L. and Gibbon, C. F., "Preliminary results of an ion scattering study of phosphosilicate glass gettering", J. Electrochem. Soc., 121, 444-7 (1974), PA77-45461.
922. Meek, R. L., Seidel, T. E. and Cullis, A. G., "Diffusion gettering of Au and Cu in silicon", J. Electrochem. Soc., 122, 786-96 (1975), PA78-62551.
923. Nagasaki, T., Poponiak, M. P. and Yeh, T. H., "Gettering technique and structure", IBM Tech. Disclosure Bull., 17, 3587-8 (1975), PA78-62544.
924. Poponiak, M. R., Keenan, W. A. and Schwenker, R. O., "Gettering of bipolar devices with diffused isolation region", In: Semiconductor silicon 1973, Electrochem. Soc., Inc., Princeton, N. J., p. 201-16.
925. Rozgonyi, G. A., Petroff, P. M. and Read, M. H., "Elimination of oxidation-induced stacking faults by preoxidation gettering of silicon wafers. I. Phosphorus diffusion-induced misfit dislocations", J. Electrochem. Soc., 122, 1725-29 (1975).
926. Schweitzer, S. and Ziffermayer, G., "Apparatus for the gettering of semiconductors", Patent USA 3856472, Publ. December 1974.
927. Schweitzer, S. and Ziffermayer, G., "Getters for semiconductors", Patent German 2203123, Publ. July 1973, CA79-98490.

928.Seidel, T. E. and Meek, R. L., "Ion implantation damage gettering and phosphorus diffusion gettering of copper and gold in silicon", Ion Implantation Semicond. Other Mater., Proc. Int. Conf., 3rd, 1972, p. 305-15, Publ. 1973, CA81-178845.
929.Seidel, T. E., Meek, R. L. and Cullis, A. G., "Direct comparison of ion-damage gettering and phosphorus-diffusion gettering of Au in Si", J. Appl. Phys., 46, 600-9 (1975), PA78-47430.
930.Yoshimura, K., Murase, K., Kobayashi, T. and Aoyama, H., "Ultra-stable MOS input operational amplifier", NEC Res. Dev., no. 34, 72-9 (1974).

VIII. MIS as a Research Tool
1. Tunneling Spectroscopy

931.Bhattacharyya, A. and Chaudhari, P. K., "Dependence of oxide processing on tunneling in MNOS structures", Presented at Electrochemical Society Meeting, Chicago, Ill., May 1973, EEA76-23380.
932.Card, H. C., "Potential barriers to electron tunneling in ultrathin films of SiO_2", Solid State Commun., 14, 1011-14 (1974), EEA77-26422.
933.Clarke, R. A. and Shewchun, J., "Non-equilibrium effects on metal-oxide-semiconductor tunnel currents", Solid-State Electron., 14, 957-73 (1971), EEA75-1170.
934.Fish, L. and Gundjan, A., "Nonlinearity of the MOS tunnelling structure I-V characteristic for submillimetre radiation detection", 1972 European Solid State Devices Research Conference, 1972, p. 178-9, Publ. 1973, EEA76-32471.
935.Green, M. A. and Shewchun, J., "Current multiplication in metal-insulator-semiconductor (MIS) tunnel diodes", Solid-State Electron., 17, 349-65 (1974), EEA77-18754.
936.Guetin, P. and Schreder, G., "Tunnelling in metal-semiconductor contacts under pressure", Philips Tech. Rev., 32, 211-20 (1971), EEA75-32634.
937.Ivanchenko, Y. M. and Medvedev, Y. V., "Excitation of oscillations of charge carrier density in a tunnel junction", Sov. Phys.-JETP, 37, 520-4 (1973), PA78-6909.
938.Kasprzak, L. A., Laibowitz, R. B. and Ohring, M., "Effect of oxide thickness (10-30A) on Si-SiO_2 barrier height in MOS tunnel structures", Presented at Electrochemical Society Meeting, San Francisco, California, May 1974, EEA77-22937.
939.Lewicki, G. and Maserjian, J., "Oscillations in MOS tunnelling", J. Appl. Phys., 46, 3032-9 (1975), EEA78-41523.
940.Ma, T. P. and Barker, R. C., "Surface-state spectra from thick-oxide MOS tunnel junctions", Solid-State Electron., 17, 913-29 (1974), PA78-18691.
941.Maes, H. and Van Overstraeten, R., "Low-field tunnelling current in thin-oxide M.N.O.S. memory transistors", Electron. Lett., 9, 19-21 (1973), EEA76-12636.

942.Maserjian, J., "Tunneling in thin MOS structures", J. Vac. Sci. & Technol., 11, 996-1003 (1974), EEA78-24260.
943.Maserjian, J. and Petersson, G. P., "Tunneling through thin MOS structures: Dependence on energy (E-k)", Appl. Phys. Lett., 25, 50-2 (1974), EEA77-36239.
944.Nicollian, E. H. and Goetzberger, A., "Comment on dispersion mechanisms of the MOS impedance", J. Appl. Phys., 40, 899 (1969), EEA73-4744.
945.Petersson, G. P., Svensson, C. M. and Maserjian, J., "Resonance effects observed at the onset of Fowler-Nordheim tunneling in thin MOS structures", Solid-State Electron., 18, 449-51 (1975), EEA78-24269.
946.Polanco, J. I. and Roberts, G. G., "Thermally assisted tunnelling in dielectric films. II", Phys. Status Solidi A, 13, 603-6 (1972), EEA76-8593.
947.Preier, H., "Contributions of surface states to MOS impedance", Appl. Phys. Lett., 10, 361-3 (1967), PA70-32828.
948.Schein, L. B. and Compton, W. D., "Electron local-mode phonon interaction in metal-insulator-semiconductor tunnel junctions", Appl. Phys. Lett., 17, 236-9 (1970), EEA74-1489.
949.Snow, E. H., "Fowler-Nordheim tunneling in SiO_2 films", Solid State Commun., 5, 813-15 (1967), EEA71-8280.
950.Strassburg, J., "Conduction processes in thin silica films", Z. Elektr. Inf. - & Energietech., 2, 208-9 (1972), German, EEA76-8605.
951.Swaroop, B., "Conduction in composite structures of silicon dioxide-nitride films", Presented at 39th Annual Conference on Electrical Insulation and Dielectric Phenomena, Pocono Manor, Pa., October 1970, EEA74-4452.
952.Temple, V. A. K., Green, M. A. and Shewchun, J., "Equilibrium-to-nonequilibrium transistion in MOS (surface oxide) tunnel diode", J. Appl. Phys., 45, 4934-43 (1974), PA78-21962.
953.Torok, M. I., "Tunneling in forward biased MOS tunnel structures", Acta Phys. & Chem. Szeged, 18, 139-46 (1972), EEA76-19855.
954.Venkateswaran, K., "Comments on 'Surface state spectra from thick-oxide MOS tunnel junction'", Solid-State Electron., 18, 908 (1975), PA78-69252.
955.Walachova, J., "Comments on line shape in tunneling spectroscopy induced by impurity local-mode; application to measurements on In-SiO_2-degenerated pSi junction", Proceedings of the 5th Czechoslovakia Conference on Electronics and Vacuum Physics, 1972, II.b)-15/2 pp., EEA76-8623.

2. Carrier Mobility

956.Baccarani, G., Mazzone, A. M. and Morandi, C., "The diffuse scattering model of effective mobility in the strongly inverted layer of MOS transistors", Solid-State Electron., 17, 785-9 (1974), PA77-60871.

957.Bandali, M. B., "The effects of the field dependence of carrier mobility on the validity of the gradual channel approximation in insulated-gate field-effect transistors", Solid-State Electron., 14, 1325-7 (1971), EEA75-4713.

958.Bar-Lev, A. and Margalit, S., "Changes of mobility along a depletion type MOS transistor channel", Solid-State Electron., 13, 1541-6 (1970), EEA74-7787.

959.Baum, G., "Drift velocity saturation in MOSFET's", Solid-State Electron., 13, 789-98 (1970), German, EEA73-28217.

960.Baum, G. and Beneking, H., "Drift velocity saturation in MOS transistors", IEEE Trans. Electron Devices, ED-17, 481-2 (1970), EEA73-25046.

961.Brews, J. R., "Carrier-density fluctuations and the IGFET mobility near threshold", J. Appl. Phys., 46, 2193-203 (1975), EEA78-33157.

962.Chen, J. T. C. and Muller, R. C., "Carrier mobilities at weakly inverted silicon surfaces", J. Appl. Phys., 45, 828-34 (1974), EEA77-26403.

963.Cheng, Y. C., "The influence of surface on hole mobility in metal-oxide-silicon structure", Jap. J. Appl. Phys., 13, Suppl. 2 (Pt. 2), 363-6 (1974), EEA78-19723.

964.Dobrovolskii, V. N. and Abessenova, L. N., "Mobility distribution of charge carriers in a field-effect MOS-transistor channel", Poluprovodn. Tekh. & Mikroelektron., no. 11, 66-70 (1973), Russian, EEA76-29276.

965.Dobrovolskii, V. N. and Ninidze, G. K., "Gate capacity of a field-effect-MOS-transistor", Poluprovodn. Tekh. & Mikroelektron., no. 11, 71-4 (1973), Russian, EEA76-29277.

966.Fedorovich, Y. V. and Dumish, L. K., "Possible activation of mobile particles on the free surface of an oxide in a metal-oxide-semiconductor system", Sov. Phys.-Semicond., 6, 1954-6 (1973), EEA76-26019.

967.Fowler, A. B., "Substrate bias effects on electron mobility in silicon inversion layers at low temperatures", Phys. Rev. Lett., 34, 15-17 (1975), PA78-18689.

968.Frohman-Bentchkowsky, D., "On the effect of mobility variation on MOS device characteristics", Proc. IEEE, 56, 217-18 (1968), EEA71-10678.

969.Hara, H., Ohuchi, K., Sato, T. and Ohnuma, H., "Measurements of the field-effect and effective mobilities in MOS transistors", Electron. & Commun. Jap., 54 (7), 92-7 (1971), Japanese, EEA75-39934.

970. Hsu, S. T. and Scott, J. H., Jr., "Mobility of current carriers in silicon-on-sapphire (SOS) films", RCA Rev., 36, 240-53 (1975).
971. Hughes, R. C., "Hole mobility and transport in thin SiO_2 films (m.o.s. structure)", Appl. Phys. Lett., 26, 436-8 (1975), EEA78-28678.
972. Kamins, T. I., "Hall mobility in dielectrically isolated single-crystal silicon films defined by electrochemical etching", Solid-State Electron., 17, 667-74 (1974), PA77-57097.
973. Kassabov, J., Velchev, N. and Gancheva, V., "On the effect of gate oxide thickness upon the Hall mobility and other magneto-electrical characteristics in MOST structures", Solid-State Electron., 17, 41-5 (1974), PA77-16831.
974. Molchanov, A. A. and Khodosh, L. S., "Effect of mobility versus transfer field dependence on characteristics of MOS-transistors", Izv. VUZ Radioelectron., 13, 892-3 (1970), Russian, EEA74-1608.
975. Murphy, N. St. J., Berz, F. and Flinn, I., "Carrier mobility in silicon MOST's", Solid-State Electron., 12, 775-86 (1969), EEA73-4527.
976. Oakley, R. E. and Pepper, M., "The influence of dielectric charge on the field effect mobility in MIS inversion layers", Presented at 1972 European Solid State Devices Research Conference, Lancaster, England, September 1972, EEA76-32220.
977. Pals, J. A. and van Heck, W. J. J., "Peaked structure in field-effect mobility of silicon MOS transistors at very low temperatures", Appl. Phys. Lett., 23, 550-2 (1973), PA77-13225.
978. Ribeiro, C. A. and Pfister, J. C., "Hall effect measurements of hole mobility in an inversion layer of a MOS structure at low temperatures", Solid State Commun., 10, 63-9 (1972), EEA75-10189.
979. Rusck, M., "The dependence of MOS-transistor parameters on carrier mobility", Elektronika, no. 7-8, 297-9 (1972), Polish, EEA75-39936.
980. Sakaki, H. and Sugano, T., "Anisotropic channel conductivity of MOSFETs on the (110) surface of silicon", Annu. Rep. Eng. Res. Inst. Fac. Eng. Univ. Tokyo, 30, 183-92 (1971), EEA75-10261.
981. Sugano, T., "Electron transport in silicon surface inversion layer", Oyo Buturi, 43 (Suppl.), 309-13 (1974), PA78-6904.
982. Tarui, Y., Nagai, K. and Hayashi, Y., "The charging carriers of the interface states of the MOS structure and their effect on the static characteristics of the MOS transistor", Electron. & Commun. Jap., 53 (2), 146-53 (1971), EEA74-26866.
983. Tidey, R. J. and Stradling, R. A., "Peaked structure appearing in the field effect mobility of silicon MOS devices at temperatures above 20K", J. Phys. C, 7, L356-61 (1974), EEA78-585.
984. Tolomasov, V. A., Rubtsova, R. A. and Abrosimova, L. N., "Dependence of the mobility of carriers on their density in homoepitaxial antimony-doped silicon films", Sov. Phys.-Semicond., 8, 1298 (1975), EEA78-24119.

985.Toscano-Rico, A. and Pfister, J. C., "Hall effect measurements of hole mobility in an inversion layer on the Si-SiO_2 interface", Phys. Status Solidi A, 5, 209-18 (1971), EEA74-22737.

3. Minority Carrier Lifetime

986.Calzolari, P. U., Graffi, S., Mazzone, A. M. and Morandi, C., "Bulk lifetime determination from current and capacitance transient response of M.O.S. capacitors", Alta Freq., 41, 848-53 (1972), EEA76-16431.
987.Calzolari, P. U., Morandi, C. and Graffi, S., "Effects of non-uniform doping on generation-lifetime measurement in M.O.S. capacitors", Electron. Lett., 9, 273-4 (1973), EEA77-338.
988.Ciszek, T. F. and Schwuttke, G. H., "Growth and characterization of silicon ribbons produced by a capillary action shaping technique", Phys. Status Solidi A, 27, 231-41 (1975), EEA78-11517.
989.Clement, G. and Voos, M., "Minority carrier lifetime measurement in gold doped silicon MOS structures", Surface Sci., 11 (1), 147-8 (1968), EEA71-49315.
990.Grinthal, E. T. and Pierret, R. F., "Comments on 'a linear-sweep MOS-C technique for determining minority carrier lifetimes'", IEEE Trans. Electron Devices, ED-20, 508-9 (1973), EEA76-19859.
991.Guzev, A. A., Kuryshev, G. L. and Sinitsa, S. P., "Investigation of carrier capture on the surface of silicon in metal-oxide-semiconductor structure", Fiz. Tekh. Poluprov., 4, 1456-61 (1970), Russian, EEA74-1493.
992.Heiman, F. P., "On the determination of minority carrier lifetime from the transient response of an MOS capacitor", IEEE Trans. Electron Devices, ED-14, 781-4 (1967), EEA71-6028.
993.Hofstein, S. R., "Minority carrier lifetime determination from inversion layer transient response", IEEE Trans. Electron Devices, ED-14, 785-6 (1967), EEA71-6029.
994.Jones, E. and Barber, H. D., "The measurement in diffused layers of minority carrier recombination generation and transport and of surface properties", Presented at Electrochemical Society Meeting, Chicago, Ill., May 1973, EEA76-23209.
995.Kamins, T. I., "Minority-carrier lifetime in dielectrically isolated-single-crystal silicon films defined by electrochemical etching", Solid-State Electron., 17, 675-81 (1974), EEA77-30074.
996.Kano, Y. and Shibata, A., "On the determination of minority carrier lifetime and surface recombination velocity from the transient response of MOS capacitors", Jap. J. Appl. Phys., 11, 1161-4 (1972), EEA75-32520.
997.Muller, J. and Schiek, B., "Transient responses of a pulsed MIS-capacitor", Solid-State Electron., 13, 1319-32 (1970), EEA73-38113.
998.Pierret, R. F., "A linear-sweep MOS-C technique for determining minority carrier lifetimes", IEEE Trans. Electron Devices, ED-19, 869-73 (1972), EEA75-24359.

999.Pierret, R. F., "The gate-controlled diode S_0 measurement and steady-state lateral current flow in deeply MOS structures", Solid-State Electron., 17, 1257-69 (1974), PA77-82492.

1000.Rakotondrazafy, C. and Boucher, J., "Determination of the minority carrier lifetime in a m.o.s. structure by a frequential study in modulated light", C. R. Hebd. Seances Acad. Sci. B, 275, 553-6 (1972), French, EEA76-740.

1001.Rhodes, C. S. L. and Salama, C. A. T., "Determination of the lifetime from thermal generation and optical injection in a pulsed MOS capacitor", J. Phys. D, 6, 1798-815 (1973), EEA77-347.

1002.Rhodes, C. S. L. and Salama, C. A. T., "A simplified method for measuring lifetime using steady-state back illumination of an MOS capacitor", J. Electrochem. Soc., 121, 1219 (1974), EEA78-581.

1003.Schroder, D. K. and Nathanson, H. C., "On the separation of bulk and surface components of lifetime using the pulsed MOS capacitor", Solid State Electron., 13, 577-82 (1970), EEA73-24851.

1004.Thanh, D. V., Lorinczy, A. and Pataki, G., "Recombination processes in silicon based MOS structures", Magy. Fiz. Foly., 21 (2), 73-102 (1973), Hungarian, EEA76-38221.

1005.Young, D. R. and Osburn, C. M., "Minority carrier generation studies in MOS capacitors on N-type silicon", J. Electrochem. Soc., 120, 1578-81 (1973), PA77-6677.

1006.Young, D. R. and Osburn, C. M., "Minority carrier generation studies in MOS capacitors on n-type silicon. Reply to comments", J. Electrochem. Soc., 121, 850 (1974), CA81-143215.

1007.Zechnall, W., "Investigation of the transient response of a MOS-capacitor with respect to a reliable measurement of minority-carrier-lifetime in silicon", Presented at Electrochemical Society Meeting, Chicago, Ill., May 1973, EEA76-23207.

1008.Zechnall, W., "Minority carrier generation studies in MOS capacitors on n-type silicon", J. Electrochem. Soc., 121, 849 (1974), CA81-143205.

1009.Zechnall, W. and Werner, W. M., "Determination of generation lifetime from the small-signal transient behavior of MOS capacitors", Solid-State Electron., 18, 971-6 (1975), EEA78-41536.

4. Others

1010.Hess, K., Neugroschel, A., Shiue, C. C. and Sah, C. T., "Non-ohmic electron conduction in silicon surface inversion layers at low temperatures", J. Appl. Phys., 46, 1721-7 (1975), EEA78-24246.

1011.Holtz, F., "Determination of the MOS flat-band condition by electroreflectance and capacitance measurements", Phys. Status Solidi A, 21, 469-78 (1974), CA80-113974.

1012.Jakubowski, A. and Jagodzinski, P., "Investigation of the $CdSe-SiO_x$ system properties by the MIS structure capacitance measurement method", Electron Technol., 4 (3), 45-64 (1971), EEA75-13644.
1013.Kar, S., "Determination of Si-metal work function differences by MOS capacitance technique", Solid-State Electron., 18, 169-81 (1975), PA78-15734.
1014.Kassing, R. and Deppe, H. R., "Positive flat band voltages measured on MIS diodes consisting of evaporated CdS films", Thin Solid Films, 13, 27-31 (1972), EEA76-4589.
1015.Krowne, C. M. and Holm-Kennedy, J. W., "Energy relaxation of electrons in the (100) n-channel of a Si-mosfet. I. Bulk phonon treatment", Surf. Sci., 46, 197-321 (1974), PA78-42309.
1016.Krowne, C. M. and Holm-Kennedy, J. W., "Energy relaxation of electrons in the (100) n-channel of a Si-mosfet. II. Surface phonon treatment", Surf. Sci., 46, 232-50 (1974), PA78-42310.
1017.Mar, H. A. and Simmons, J. G., "Determination of the energy distribution of interface traps in MIS systems using non-steady-state techniques", Solid-State Electron., 17, 131-5 (1974), PA77-24837.
1018.Mar, H. A. and Simmons, J. G., "Determination of bulk-trap parameters in MIS structures", Appl. Phys. Lett., 25, 503-5 (1974), EEA78-11527.
1019.Mar, H. A. and Simmons, J. G., "Determination of bulk trap parameters using thermal dielectric relaxation techniques", Solid-State Electron., 17, 1181-5 (1974), EEA78-595.
1020.Peykov, P. H., "Possibility for determination of a photon flux by means of linear MOS-C technique", C. R. Acad. Bulg. Sci., 27, 639-41 (1974), EEA78-814.
1021.Snejdar, V. and Jerhot, J., "On the determination of the electronic properties of the inaccessible thin film semiconductor-solid substance interface by the MIS capacitance method", Thin Solid Films, 26, 341-8 (1975), EEA78-24285.

IX. Optical Properties

1. Photoelectric Effects

1022.Balabanov, S., "Amplification of pulse photoelectric current in a mos-system with CdS", 5th IMEKO Symposium on Photon Detections, 1971, p. 421-5, EEA75-24371.
1023.Bottka, N., "Effect of photocarriers on surface barrier electroreflectance in MOS capacitor configurations", J. Appl. Phys., 44, 5626-8 (1973), PA77-37316.
1024.DiStefano, T. H. and Viggiano, J. M., "Interface imaging by scanning internal photoemission", IBM J. Res. & Dev., 18, 94-9 (1974), EEA77-26396.
1025.Farmer, J. W. and Lee, R. S., "Photocurrents and photoconductive yield in MOS structures during X-irradiation", J. Appl. Phys., 46, 2710-15 (1975), PA78-66036.

1026. Foss, N. A., "Field-enhanced photoelectron emission from metal-oxide-semiconductor structures", J. Appl. Phys., 42, 3762-5 (1971), PA74-73767.
1027. Gorban, A. P., Litovchenko, V. G., Serba, A. A. and Moskal, D. N., "Use of photo-effects in MOS-structures with a nonstationary depletion layer for detection of ionisation", Poluprovodn. Tekh. & Mikroelektron., no. 11, 27-34 (1973), Russian, EEA76-29200.
1028. Inoue, S. and Okamoto, K., "Photoresponse of MOS-FET", Rep. Univ. Electro-Commun., 23, 173-80 (1972), Japanese, EEA77-36471.
1029. Inoue, S. and Okamoto, K., "Photoresponse of gate-controlled MOS diode", Rep. Univ. Electro-Commun., 24, 267-73 (1973), Japanese, EEA77-36472.
1030. Ivanov, Y. L. and Farbshtein, I. I., "Photoelectric properties of a tellurium-based metal-oxide-semiconductor structure", Sov. Phys.-Semicond., 7, 1219-20 (1974), PA77-60873.
1031. Kalnina, R., Feltins, I., Eglitis, I. and Eimanis, I., "Use of the internal photoemission method to study metal-dielectric-semiconductor structures with a plasma-grown silicon dioxide film", Latv. PSR Zinat. Akad. Vestis, Fiz. Teh. Zinat. Ser., no. 5, 113-16 (1974), Russian, CA82-67153.
1032. Kassabov, J., Velchev, N. and Stoev, I., "Photoeffect in MOST structures", 5th IMEKO Symposium on Photon Detectors, 1971, p. 413-20, EEA75-24291.
1033. Kassabov, J. D., Velchev, N. B. and Stoev, I. G., "Physical analysis of the photoeffect in silicon MOST structures", C. R. Acad. Bulg. Sci., 27, 43-5 (1974), PA77-57170.
1034. Komiya, Y., Sakamoto, T., Suzuki, E. and Tarui, Y., "Visible and infrared photoinjection of electrons into insulating layers of MIS memory devices", Oyo Buturi, 43 (Suppl.), 356-61 (1974), EEA78-2880.
1035. Krag, W. E., Phelan, R. J., Jr., Dimmock, J. O., "Photoresponse properties of InSb MOS devices", Solid State Commun., 7 (11), Pap. No. 7.3 (1969), EEA72-39658.
1036. Kreutz, E. W., "Photoelectric spectroscopy of oxide states with MOS structures at T=79K", Appl. Phys., 1, 161-6 (1973), EEA76-23375.
1037. Lakatos, A. I. and Mort, J., "Transient photocurrents and internal photoemission in semiconductor-insulator-metal systems", Solid State Commun., 7 (11), Pap. No. 7.6 (1969), PA72-39656.
1038. Litovchenko, V. G., Gorban, A. P. and Peikov, P. H., "On the nature of surface centres in the semiconductor-dielectric system", C. R. Acad. Bulg. Sci., 27, 1635-7 (1974), PA78-50680.
1039. Nachmanson, R. C. and Luzevic, L. V., "Kinetics of the photo e.m.f. in silicon and germanium M.I.S. structures", Z. Elektr. Inf. - & Energietech., 2, 212-5 (1972), German, EEA76-8608.
1040. Okamoto, H., "Impedance measurement of the illuminated metal-SiO_2-Si diodes", IEEE Trans. Electron Devices, ED-17, 83-5 (1970), EEA73-14314.

1041.Okamoto, K., "Photo responses of MOS channel", Rep. Univ. Electro-Commun., no. 28, 111-14 (1970), Japanese, EEA74-26893.
1042.Okamoto, K. and Mochizuki, M., "Photoresponses of MOS transistor", Proc. IEEE, 59, 1351-2 (1971), EEA74-34101.
1043.Pagnia, H., "Photocurrents in InSb MOS devices", Phys. Lett. A, 48A, 260-2 (1974), PA77-70073.
1044.Pierret, R. F. and Roesner, B. B., "Photoemission from surface centres", Appl. Phys. Lett., 24, 366-9 (1974), PA77-53147.
1045.Pierret, R. F. and Sah, C. T., "Quantitative analysis of the effects of steady state illumination on the MOS-capacitor. II. Experimental results", Solid-State Electron., 13, 289-302 (1970), EEA73-17672.
1046.Plotnikov, A. F. and Vavilov, V. S., "Photoelectric effect in metal-oxide-semiconductor structures", Sov. Phys.-Semicond., 7, 599-603 (1973), EEA77-14446.
1047.Powell, R. J. and Berglund, C. N., "Photoinjection studies of charge distributions in oxides of MOS structures", J. Appl. Phys., 42, 4390-7 (1971).
1048.Sah, C. T. and Fu, H. S., "Transient response of MOS capacitors under localized photoexcitation", IEEE Trans. Electron Devices, ED-21, 202-9 (1974), PA77-28494.
1049.Sebenne, C., Guichar, G., Balkanski, M. and Bolmont, D., "Surface states from photoemission threshold on silicon (111) face", Jap. J. Appl. Phys., 13, Suppl. 2 (Pt. 2), 405-8 (1974), PA78-37801.
1050.Stoev, I., Antov, B. and Velchev, N., "Surface local photoeffect in MOST structures", Izv. Fiz. Inst. ANEB, 24, 59-66 (1973), EEA77-30334.
1051.Tansley, T. L. and Ralph, J. E., "Photoeffects in metal-insulator-gallium arsenide diodes", J. Phys. D. Appl. Phys., 3, 807-11 (1970), EEA73-25040.
1052.Thomas, J. H., III, "Transient photodepopulation measurements of electron trap distributions in thin SiO_2 films on silicon", J. Appl. Phys., 44, 811-14 (1973), EEA76-12319.
1053.Triebswasser, S., "Photosensitive insulated gate field effect transistor", Patent USA 3459944, Publ. August 1969, EEA73-21349.
1054.Viswanathan, C. R., Loo, R. and Ogura, S., "Photoelectric studies of heteroepitaxial semiconductor films (MIS structures)", J. Vac. Sci. & Technol., 9, 230-3 (1972), EEA75-20222.
1055.Viswanathan, C. R. and Ogura, S., "Silicon-oxide interface studies by a photoelectric technique", Proc. IEEE, 57, 1552-7 (1969), EEA73-1206.
1056.Viswanathan, C. R. and Ogura, S., "Band-structure studies on silicon using a metal-oxide-semiconductor junction", Phys. Rev. B, 1, 624-7 (1970), EEA73-24764.
1057.Viswanathan, C. R. and Ogura, S., "Study of the role of impurities in MOS structures and the observation of optical transition in silicon", 30th Annual Conference on Physical Electronics, 1970, Pap. E-4, EEA73-25054.

1058. Viswanathan, C. R. and Ogura, S., "Direct measurement of built-in voltage in MOS structures", J. Appl. Phys., 42, 5865-8 (1971), EEA75-7458.
1059. Zuev, V. A. and Litovchenko, V. G., "Inter-extrinsic transitions in an Si-SiO_2 system", Poluprovodn. Tekh. & Mikroelektron., no. 8, 68-73 (1972), Russian, EEA76-16323.
1060. Zuev, V. A., Litovchenko, V. G., Popov, V. G. and Sukach, G. A., "On photoelectric properties in the system Si-SiO_2", Ukr. Fiz. Zh., 18, 1141-9 (1973), Russian, EEA76-32217.
1061. Zuev, V. A., Litovchenko, V. G., Popov, V. G. and Sukach, G. A., "Photoelectric methods for determining electrical characteristics in metal-dielectric-semiconductor structures", Poluprovodn. Tekh. & Mikroelektron., no. 12, 32-49 (1973), Russian, EEA76-38224.
1062. Zuev, V. A., Sachenko, A. V., Oleinik, V. A., Popov, V. G. and Sukach, G. A., "Condensator photoelectromotive force in the system Si-SiO_2 at high levels of injections", Ukr. Fiz. Zh., 19, 450-7 (1974), Russian, EEA77-18761.

2. Others

1063. Astafev, A. S., Agalarzade, P. S., Grishaeva, N. I., Cherntsov, S. M. and Izidinov, S. O., "Investigation of the interface between a semiconductor and a low-temperature dielectric (SiO_2, 300°C)", Fiz. & Tekh. Poluprov., 5, 1193-6 (1971), Russian, EEA74-30735.
1064. Gavrilenko, V. I., Zuev, V. A. and Litovchenko, V. G., "Quality control procedure for surfaces of semiconductors and semiconductor devices based on the optical effect of electroreflection", Poluprovodn. Tekh. & Mikroelektron., no. 18, 49-53 (1974), Russian, EEA78-11366.
1065. Gorban, A. P., Litovchenko, V. G. and Serba, A. A., "Investigation of the recombination mechanisms in MIS structures by a photocapacitance method", Phys. Status Solidi A, 29, K163-7 (1975), PA78-62114.
1066. Gradinaru, G., "Characterization of Si-SiO_2 system by the surface avalanche method", Autom. & Electron., 17, 263-71 (1973), Rumanian, EEA77-18849.
1067. Gupta, D. C., "Non-destructive determination of carrier concentration in epitaxial silicon using a total internal reflection technique", Solid-State Electron., 13, 543-52 (1970), EEA73-24850.
1068. Hasegawa, H., Furukawa, M. and Yanai, H., "Slow wave propagation along a microstrip line on Si-SiO_2 systems", Proc. IEEE, 59, 297-9 (1971), EEA74-13442.
1069. Hasegawa, H., Furukawa, M. and Yanai, H., "Properties of microstrip line on Si-SiO_2 system", IEEE Trans. Microwave Theory & Tech., MTT-19, 869-8 (1971), EEA74-37463.

1070. Hasegawa, H. and Yanai, H., "Characteristics of parallel-plate waveguide filled with a silicon-silicon oxide system", Electron. & Commun. Jap., 53 (10), 63-73 (1970), EEA75-13932.
1071. Holtz, F., "Determination of the MOS flatband condition by electroreflectance and capacitance measurements", Phys. Status Solidi A, 21, 469-78 (1974), PA77-28497.
1072. Kropman, D. I., Niitra, A. E. and Vykhma, K. A., "Influence of activated SiO_2 film on Si surface", Poluprovodn. Tekh. & Mikroelektron., no. 13, 85-91 (1973), Russian, EEA77-12146.
1073. Lamb, D. R., "Some electrical properties of the silicon-silicon dioxide system", Thin Solid Films, 5, 247-76 (1970), EEA73-35173.
1074. Litovchenko, V. G., Gorban, A. P., Litvinov, R. O., Peikov, P. K., Radeske, V. N. and Serba, A. A., "Differential photocapacitive effect in MOS devices", Poluprovodn. Tekh. & Mikroelektron., no. 19, 29-35 (1975), Russian, EEA78-28702.
1075. Litovchenko, V. G., Gorban, A. P. and Lysenko, V. S., "Some electrical and recombination properties of Si-SiO_2 system", Proceedings of the International Conference on the Physics and Chemistry of Semiconductor Heterojunctions and Layer Structures, 1970, p. I.-296-77, Publ. 1971, EEA74-22629.
1076. Litovchenko, V. G., Litvinov, R. O. and Gorban, A. P., "Measurement of recombination characteristics of MOS structures with the aid of a photocapacitive effect", Poluprovodn. Tekh. & Mikroelektron., no. 10, 30-4 (1972), Russian, EEA76-16326.
1077. Litvinov, R. O., "The determination of the properties of M.I.S. structures by means of the photo-capacitive effect", Z. Elektr. Inf. - & Energietech., 2, 215-17 (1972), German, EEA76-8609.
1078. Maronchuk, Y. E. and Senoshenko, O. V., "Negative photoconductivity of Al-Si_3N_4-n-GaAs metal-insulator-semiconductor structures", Sov. Phys.-Semicond., 8, 895-6 (1975), PA78-33562.
1079. Nachmanson, R. C. and Luxevic, I. V., "Kinetics of the photo e.m.f. in silicon and germanium M.I.S. structures", Z. Elektr. Inf. - & Energietech., 2 212-5 (1972), German, EEA76-8608.
1080. Ning, T. H. and Yu, H. N., "Optically induced injection of hot electrons into SiO_2", J. Appl. Phys., 45, 5373-8 (1974), PA78-25793.
1081. Olofsson, G., "Investigation of optical properties of a GaP MIS structure", Phys. Status Solidi A, 22, 175-84 (1974), PA77-33033.
1082. Rakotondrazafy, C. and Boucher, J., "Influence of continuous illumination on the high frequency C(V) characteristics of a m.o.s. structure", C. R. Hebd. Seances Acad. Sci. B, 278, 599-602 (1974), French, PA77-45000.
1083. Saifi, M. A. and Stolen, R. H., "Far-infrared interference technique for determining epitaxial silicon thickness", J. Appl. Phys., 43, 1171-7 (1972), EEA75-13542.
1084. Sato, K., "Contribution of surface electrons to the infra-red optical properties of silicon", J. Phys. Soc. Japan, 27, 89-97 (1969), MA3-03-32 0103.

1085. Schumann, P. A., Jr., "Infrared interference method of measuring epitaxial layer thickness", J. Electrochem. Soc., 116, 409-13 (1969).

1086. Schumann, P. A., Jr., "Measurement of semiconductor epitaxial layer thickness by infrared reflectance", Symposium on Silicon Device Processing (programme and abstracts), 1970, p. 42-3, EEA73-35125.

1087. Schumann, P. A., Jr. and Schneider, C. P., "Measurement of silicon epitaxial layers less than 1mm thick by infrared interference", J. Appl. Phys., 41, 3532-35 (1970), PA73-61638.

1088. Stern, F., "Quantum and continuum results for inversion layers", Proceedings of the 10th International Conference on the Physics of Semiconductors, 1970, p. 451-8, EEA74-30758.

1089. Walsh, R. J., "Measurement of layer thickness of silicon epitaxial wafers", Semiconductor Products and Solid State Technology, 7 (8), 23-27 (1964).

1090. Weinberg, Z. A., Johnson, W. C. and Lampert, M. A., "Determination of the sign of carrier transported across SiO_2 films on Si", Appl. Phys. Lett., 25, 42-3 (1974), EEA77-33971.

1091. Zuev, V. A., Ilyushin, B. N. and Litovchenko, V. G., "Electroluminescence in gallium arsenide and silicon carbide MOS devices", Poluprovodn. Tekh. & Mikroelektron., no. 15, 29-35 (1974), Russian, PA77-73195.

N. FET TECHNOLOGIES AND APPLICATIONS

I. Books

1.American Micro-Systems, Inc., (Staff), "MOS integrated circuits: theory, fabrication, design and systems applications of MOS LSI", Van Nostrand Reinhold, New York, 1972.

2.Carr, W. N. and Mize, J. P., "MOS-LSI design and applications", McGraw-Hill, New York, 1972.

3.Crawford, R. H., "MOSFET in circuit design", McGraw-Hill Book Co., New York, 1967.

4.Flynn, G., "MOS and CMOS logic IC's", H. W. Sams, Indianapolis, Ind., 1973.

5.Mavor, J., (Ed.), "M.O.S.T. integrated-circuit engineering", Peter Peregrinus, London, England, 1973, EEA77-6078.

6.Melen, R. and Garland, H., "Understanding CMOS integrated circuits", H. W. Sams, Indianapolis, Ind., 1975.

7.Nanavati, R. P., "Semiconductor devices: BJTS, JFETS, MOSFETS and integrated circuits", Intext Educational Publishers, New York, 1975.

8.Richman, P., "MOS field-effect transistors and integrated circuits", John Wiley & Sons, New York, 1973, EEA77-10265.

9.Wallmark, J. T. and Carlstedt, L. G., "Field-effect transistors in integrated circuits", John Wiley & Sons, New York, 1974, EEA77-36330.

II. Review Articles

10.Anon., "Many faces of MOS", Electron. Des., 19 (25), 34-5 (1971), EEA75-16745.

11.Anon., "Field effect transistors", Electra, 52 (2), 14-19 (1972), Dutch, EEA75-13758.

12.Avery, L. R., "Mixed-discipline monolithic integrated circuits - bipolar, MOS and COS/MOS technologies and future trends", Microelectron. & Reliab., 13, 349-61 (1974), EEA78-5362.

13.Beneking, H., "Review of different types of field-effect transistors", Nachrichtentech Z., 23, 485-91 (1970), German, EEA74-4931.

14.Bishop, A. and Jones, P., "A review of the present position regarding complementary metal oxide semiconductors (C-MOS)", Elektroniker, 13 (11), EL11-EL17 (1974), German, EEA78-8606.

15.Borel, J., "Optimisation of enhanced MOS transistor circuits", Proceedings Second International Congress on Microelectronics, 1966, p. 559-78, 1967, French, EEA71-13326.

16.Borel, J., Suat, J. P., Bernard, J. and Garcia, M., "SOS technology for MOS integrated circuits", Onde Electr., 54, 113-20 (1974), EEA77-18862.

17.Brown, D. M., "The self-registered MOSFET-A brief review", Solid State Technol., 15 (4), 33-7 (1972), EEA75-27767.
18.Bruun, G., "Field effect transistors with special reference to MOS transistors", Forsknang, 77, 226-36 (1968), Danish, EEA72-2823.
19.Carlson, R. S., "Proposer for MOS technology", Microelectronics, 16 (2), 5-8 (1974), EEA78-24481.
20.Carlyle, J. S. and Leibbrand, R., "MOS technology introduction and survey. I", Elektronik, 18 (1), 3-6 (1969), German, EEA72-7394.
21.Carlyle, J. S. and Leibbrand, R., "MOS technology introduction and survey. II", Elektronik, 18 (2), 47-52 (1969), German, EEA72-9596.
22.Cassee, P. R., "Technology of recent f.e.t. microelectronic circuits", Scienta Elec., 17 (3), 73-90 (1971), EEA75-3950.
23.Cheroff, G., Critchlow, D. L., Dennard, R. H. and Terman, L. M., "IGFET circuit performance: n-channel versus p-channel", IEEE J. Solid State Circuits, SC-4, 267-71 (1969), EEA73-7472.
24.Corradetti, M., "Why MOS?", Electronic Compon., 9 (9), 7005-7 (1968), EEA71-28906.
25.Crawford, R. H., "Considerations in the limitations of MOSFET IC's", Northeast Electronics Research and Engineering Meeting, 1967, p. 174-5, EEA71-28985.
26.de Troye, N. C., "Low power digital integrated circuits", 6th International Congress on Microelectronics, 1974, 9 pp., EEA78-24515.
27.Durrant, N. F., "The insulated gate transistor: a brief history", Electronic Engng., 42 (507), 70-3 (1970), EEA73-21348.
28.Edwards, R., "The anatomy of IGFET's", Bell Lab. Record, 46, 246-54 (1968), EEA72-796.
29.Erdelyi, J., Ugray, L., Fejes, L. and Kovacs, F., "Development trends of semiconductor type (monolithic) integrated circuits", Hiradastechnika, 25 (1), 20-7 (1974), Hungarian, EEA77-34094.
30.Forte, S., "Seconder for MOS technology", Microelectronics, 6 (2), 14-17 (1974), EEA78-24483.
31.Frohman-Bentchkowsky, D., "Trends in memory technologies", 6th International Congress on Microelectronics, 1974, 3 pp., EEA78-24513.
32.Georgiev, V. K. and Toneva, R. S., "On the physical limitations in MOS technology", C. R. Acad. Bulg. Sci., 26, 1021-3 (1973), EEA77-14583.
33.Goral, A., "Perspective MOS techniques", Elektronika, no. 10, 393-7 (1972), Polish, EEA76-4657.
34.Grannemann, W. W., Southward, H. D., Erteza, A. and Phillips, D. H., "The metal oxide semiconductor integrated circuit technology", J. Sci. & Ind. Res., 31, 299-311 (1972), EEA76-19959.
35.Gschwend, H. G. and Rein, W., "MOS-circuits to the scientists' wish", Elektron. Ind., 4 (9), 184-7 (1973), German, EEA77-14520.
36.Gunzel, G., "The MOS technique", Elektron. Ind., 4 (5-6), 90-3, 95 (1972), German, EEA75-32680.

37.Henkel, D., "Progress in bipolar and unipolar semiconductor technology. I", Radio Fernsehen Elektron., 22, 657-8, 672-3 (1973), German, EEA77-14601.
38.Henkel, D., "Progress in bipolar and unipolar semiconductor technology. II", Radio Fernsehen Elektron., 22, 704-7 (1973), EEA77-14602.
39.Hill, J., "Custom design in m.o.s.", New Electron., 6 (19), 24, 27-8, 33-4 (1973), EEA77-10299.
40.Hill, J., "Custom design in m.o.s. integrated circuits", GEC J. Sci. & Technol., 41 (1), 15-20 (1974), EEA77-34093.
41.Hodges, D. A., "A review and projection of semiconductor components for digital storage", Proc. IEEE, 63, 1136-47 (1975), EEA78-28869.
42.Hoeneisen, B. and Mead, C. A., "Fundamental limitations in microelectronics. I. MOS technology", Solid-State Electron., 15, 819-29 (1972), EEA75-23385.
43.Jahn, R., "The state of the art of integration techniques", Fernmeldetechnik, 15 (1), 22-5 (1975), German, EEA78-24460.
44.Johnson, E. O., "Power-delay energy comparison of bipolar and IGFET digital devices and circuits", IEEE Trans. Electron Devices, ED-22, 1044-5 (1975).
45.Kamins, T. I., "Chemically vapor deposited polycrystalline-silicon films", IEEE Trans. Parts, Hybrids & Packag., PHP-10, 221-9 (1974), EEA78-11450.
46.Leland Seely, J., "A survey of MOS technologies", IEEE International Convention Digest, 1971, p. 184-5, EEA74-22750.
47.Lepselter, M. P., "Silicon integrated circuit technology", Oyo Buturi, 43 (Suppl.), 13-17 (1974), EEA78-681.
48.Leuenberger, F., "Monolithic integrated MOS circuits", Tech. Mitt. PTT, 51, 61-7 (1973), French, EEA76-12677.
49.Kiener, M., "C-MOS technology", Mitt. AGEN, no. 17, 3-19 (1974), German, EEA78-15412.
50.Kroeger, J. and Threewitt, B., "Review the basics of MOS logic", Electron. Des., 22 (6), 98-105 (1974), EEA78-8169.
51.Lacour, J., "Evaluation of the various m.o.s. technologies", Onde Electr., 54, 3-10 (1974), French, EEA77-14597.
52.McDonald, A., "Choosing between MOS processes", New Electron., 6 (4), 52-3 (1973), EEA76-16472.
53.Memelink, O. W., "Integrated circuits - a revolution in electronics", European 71 Digest, 2 pp., EEA75-3962.
54.Miles, J. F., "Metal-oxide-semiconductor transistors and their applications", Mullard Tech. Commun., 9, 54-63 (1966), PA70-1996.
55.Nishida, S., "Advances and problems in MOS devices", JEE, no. 76, 54-6 (1973), EEA76-29272.
56.Oehmichen, J. P., "The MOST transistors", Electronique Industr., no. 100, 23-7 (1967), French, EEA71-3434.
57.Paul, R., "MIS components", Nachrichtentech. Elektron., 23, 282-93 (1973), German, EEA77-2569.

58.Perrin, R. M., "A MOS course for the non-technical professional", Microelectronics, 4 (3), 2-9 (1973), EEA76-26101.
59.Rhodes, J. G. L., "Systems engineering: the circuit designer is not a dying breed", Electron. Eng., 43 (516), 38-40 (1971), EEA74-14821.
60.Rossler, F., "The potentialities in MOS technique", Radio Fernsehen Elektron., 23, 448-51 (1974), German, EEA78-2918.
61.Rossler, F., "Aspects for the design of integrated MIS circuits", Nachrichtentech. Elektron., 25, 94-9 (1975), German, EEA78-24494.
62.Rule, A. F. E., "The MOST physics and technology", In: MOS integrated circuit design, Wolfendale, E., (Ed.), Butterworth, London, England, p. 1-22 (1973), EEA76-19937.
63.Seeley, J. L., "Advances in the state-of-the-art of MOS device technology", Semiconductor Prod. Solid State Technol., 10 (3), 59-62 (1967), EEA70-10876.
64.Sev, A., "Applications of metal oxide with isolated grid semiconductor", Onde Elect., 47, 954-8 (1967), French, EEA71-8274.
65.Shenton, G., "Semiconductor memories - which direction?", New Electron., 8 (4), 44-6, 50, 54 (1975), EEA78-24486.
66.Shenton, G. and Smart, J., "MOS semiconductors: how electronics complement mechanical design", Des. Eng. Mater. & Compon., 58-62 (March 1972), EEA75-13751.
67.Sigournay, N., "MOS packs value into silicon chips", New Sci., 66 (943), 25-7 (1975), EEA78-33226.
68.Soderman, L., "This is how MOS circuits operate", Eltek. Aktuell Elektron., 16 (12), 30-1 (1973), Swedish, EEA77-14521.
69.Sonde, B. S., "MOS FET integrated circuits", Electro-Technol., 15 (3), 73-8 (1971), EEA76-11959.
70.Spinuluscu Carnaru, I. and Stanescu, C., "Metal-oxide-semiconductor transistors", Stud. Cercet. Fiz., 24, 741-52 (1972), Rumanian, EEA75-818.
71.Systems Design Group, "Integrated circuit technology and application. X.", Electronic Compon., 10, 441-20 (1969), EEA72-12790.
72.Tarui, Y., Nagai, K. and Hayashi, Y., "Nonvolatile semiconductor memory", Oyo Buturi, 43, 990-1002 (1974), Japanese, EEA78-15366.
73.Thomas, E. and Callan, J. D., "Slow, but small may win the race", Electronics, 40 (4), 179-82 (1967), EEA70-13106.
74.Torrero, E. A., "Focus on semiconductor memories", Electron. Des., 23 (7), 98-107 (1975), EEA78-24453.
75.Tummers, L. J., "MOS transistors", Philips Tech. Rev., 31, 206-8 (1970), EEA74-34093.
76.Vinsani, M. and Cislaghi, E., "Semiconductor memories", Alta Freq., 42, 430-43 (1973), Italian, EEA77-9283.
77.Wallmark, J. T., Boesenberg, W. A., Ross, E. C., Flatley, D. and Parker, H., "MOS field effect transistor technology", Report NASA-CR-1113, RCA Princeton, N. J., 1968, 67 pp., EEA72-13498.
78.Warner, R. M., Jr., "Comparing MOS and bipolar integrated circuits", IEEE Spectrum, 4 (6), 50-8 (1967), EEA70-17263.

79. Werner, W., "MOS-technologies", Elektro-Anz., 25, 387-9 (1972), German, EEA76-8683.
80. West, J., "Practical circuit design using MOS", Des. Electron., 8 (6), 30-2 (1971), EEA74-15561.
81. Wolf, H. P., "MOS integrated circuits", Tech. Rundsch., 66 (11), 25, 27, 29 (1974), German, EEA77-23159.

III. Technologies

1. Silicon-Gate

82. Abbas, S. A., Barile, C. A. and Dockerty, R. C., "Low-leakage, n-channel silicon gate FET with a self-aligned field shield", 1973 International Electron Devices Meeting Technical Digest, p. 371-3, EEA77-14559.
83. Agatsuma, T. and Morita, J., "Use of drain capacitance-voltage characteristics as a process control tool for the threshold voltage of silicon gate MOSFET's", IEEE Trans. Electron Devices, ED-22, 961 (1975).
84. Antipov, I., "Field effect device", Patent USA 3899373, Publ. May 1974, CA83-171677.
85. Barile, C. A., Dockerty, R. C. and Nagarajan, A., "Method for stabilizing FET devices having silicon gates and composite nitride-oxide gate dielectrics", Patent USA 3793090, Publ. February 1974.
86. Blanchard, R. A., Gargini, P. A., May, G. A. and Melen, R., "High-voltage simultaneous diffusion silicon-gate CMOS", IEEE J. Solid-State Circuits, SC-9, 103-10 (1974), EEA77-23119.
87. Bulucea, C., "Avalanche injection into the oxide in silicon gate-controlled devices. II. Experimental results", Solid-State Electron., 18, 381-91 (1975), EEA78-19849.
88. Chang, C. S. and Jen, T. S., "Complementary field-effect transistors having p-doped silicon gates", Patent USA 3821781, Publ. June 1974.
89. Chang, C. S. and Jen, T. S., "Complementary field-effect transistor having p-doped silicon gate", Patent USA 3865654, Publ. February 1975, CA83-20795.
90. Clemens, J. T. and Labuda, E. F., "Impurity diffusion in SiO_2 layers and related effects on the MOS properties of the Si-gate technology", Presented at Electrochemical Society Meeting, San Francisco, Calif., May 1974, EEA77-26478.
91. Dockerty, R. C., Abbas, S. A. and Barile, C. A., "Low-leakage n- and p-channel silicon-gate FET's with an SiO_2-Si_3N_4-gate insulator", IEEE Trans. Electron Devices, ED-22, 33-9 (1975), EEA78-15349.
92. English, P. D., "The impact of the silicon gate process on the application of MOS circuits", Microelectronics, 3 (9), 39-46 (1970), EEA73-13259.
93. Faggin, F. and Klein, T., "A faster generation of MOS devices with low thresholds is riding the crest of the new wave silicon-gate IC's", Electronics, 42 (20), 88-94 (1969), EEA73-6765.

94. Feldman, C. and Plachy, R., "Vacuum deposited silicon devices on fused silica substates", J. Electrochem. Soc., 121, 685-8 (1974), EEA77-30176.
95. Fujimoto, S., "Silicon-gate, insulated-gate field effect transistor", Patent USA 3798752, Publ. March 1974, CA80-138474.
96. Fujimoto, S., Mackawa, M. and Inoue, Y., "Effects of diffusion on instabilities of poly Si-Si_3N_4-SiO_2-Si structures", Extended Abstracts of the Spring Meeting of the Electrochemical Society, 1971, p. 299-31, EEA75-4644.
97. Harlow, J. E., III, Swann, R. C. G., Penton, J. I. and Bakker, M. B., "Planar silicon gate MOS process", Patent USA 3761327, Publ. September 1973.
98. Hayashi, Y., Nagai, K. and Tarui, Y., "Electrically erasable non-volatile optical memory for visible and infrared light pattern", Oyo Buturi, 43 (Suppl.), 362-6 (1974), EEA78-2881.
99. Horiuchi, S., "Threshold voltage shift of n-channel Si-gate MOSFET's", IEEE Trans. Electron Devices, ED-22, 1038-43 (1975).
100. Iwamatsu, S., Tominaga, Y., Anzai, N., Tomozawa, A. and Agatsuma, T., "Silicon gate P channel MOS technology", Hitachi Hyoron, 55 (4), 37-42 (1973), Japanese, EEA77-6071.
101. Iwamatsu, S., Tominaga, Y., Ito, K. and Hashimoto, H., "Hitachi, Hyoron, 56 (7), 17-22 (1974), Japanese, EEA78-679.
102. Lawrence, J., MacArthur, A. T. P., Wu, S. M., van Tongerloo, E. H. and Laverty, S. J., "MOS Si-gate beam leads", Presented at Electrochemical Society Meeting (abstracts only), Chicago, Ill., May 1973, EEA76-23507.
103. Lin, H. C., Arzoumanian, Y. F., Halsor, J. L., Giuliano, M. N. and Benz, H. F., "Effect of silicon-gate resistance on the frequency response of MOS transistors", IEEE Trans. Electron Devices, ED-22, 255-64 (1975), EEA78-19825.
104. Lin, H. C., Halsor, J. L. and Hayes, P. J., "Shielded silicon gate complementary MOS integrated circuit", IEEE Trans. Electron Devices, ED-19, 1199-207 (1972), EEA75-35391.
105. Maekawa, M., "Electrical characteristics of polycrystalline silicon-silicon nitride-silicon dioxide-silicon structures", Jap. J. Appl. Phys., 11, 1251-8 (1972), EEA75-36203.
106. Moreau, J. P., "Silicon gate MOS technologies applications", Onde Electr., 54 (3), 108-12 (1974), EEA77-18814.
107. O'Dell, A. D., "Silicon gate n-channel m.o.s.", New Electron., 6 (18), 51-2, 57-8 (1973), EEA77-6076.
108. Ohuchi, K., Tanaka, S., Ara, K., Tanuma, Y., Kubota, N. and Horichi, S., "A 1Kb N-silicon-gate static random-access memory with two different depletion-load thresholds", Oyo Buturi, 44 (Suppl.), 211-16 (1975), EEA78-41683.
109. Rai-Choudhury, P. and Hower, L., "Growth and characterization of polycrystalline silicon", J. Electrochem. Soc., 120, 1761-6 (1973), EEA77-10106.
110. Richman, P., "Suppression of parasitic thick-field conduction mechanisms in silicon-gate MOS integrated circuits", Electron. Lett., 7, 12-13 (1971), EEA74-7771.

111. Saraswat, K. C., "High voltage silicon gate MOS integrated circuits", Thesis, Stanford Univ., Calif., 1975, 134 pp., Order No. 75-13593, EEA78-41700.
112. Schlageter, J., "Advances in silicon gate MOS LSI cells", 1972 IEEE Region Six (US Western region) Conference Record, p. 142-4, EEA76-4223.
113. Shewchun, J., Clarke, R. A. and Temple, V. A. K., "Experimentally observed admittance properties of the semiconductor-insulator-semiconductor (SIS) diode", IEEE Trans. Electron Devices, ED-19, 1044-50 (1972), EEA75-28529,
114. Shewchun, J., Temple, V. and Clarke, R. A., "Exact frequency-dependent complex admittance of the semiconductor-insulator-semiconductor (SIS) diode including surface states and Shockley-Read-Hall (SRH) impurity effects", J. Appl. Phys., 43, 3487-99 (1972), EEA75-32687.
115. Turley, A. P., McLouski, R. M., Reid, P. R. and McCann, D. H., "A p-channel silicon gate process with Si_3N_4/SiO_2 gate dielectric", Electrochemical Society Fall Meeting (Extended abstracts), 1974, p. 462-3, EEA78-37557.
116. Vadasz, L. L., Grove, A. S., Rowe, T. A. and Moore, G. E., "Silicon-gate technology", IEEE Spectrum, 6 (10), 28-35 (1969), EEA73-6767.
117. Vittoz, E., Gerber, B. and Leuenberger, F., "Silicon gate CMOS frequency divider for electronic wrist watch", IEEE J. Solid-State Circuits, SC-7, 100-4 (1972), EEA75-12657.
118. Walsh, P. S. and Creighton, O., "A comparison of n and p-channel silicon gate technologies and device parameters", 1972 European Solid State Devices Research Conference, p. 201-2, Publ. 1973, EEA76-32280.
119. Watt, A. S. M. and Elliot, A. B. M., "Nonuniform threshold voltage instabilities in p-channel silicon-gate MOS transistors", Electron. Lett., 11, 559-60 (1975).
120. Yasuda, Y., Hirabayashi, K. and Moriya, T., "Growth and properties of polycrystalline silicon films as applied to MOS-IC technology", Proceedings of the 5th International Conference on Solid State Devices, 1973, p. 400-7, CA81-142563.

2. CMOS

121. Adley, J. M. and Elie, G. T., "Gallium deep pocket for complementary field-effect transistors", IBM Tech. Disclosure Bull., 16, 3590 (1974), EEA77-38846.
122. Altman, L., "Performance is enhanced, chip complexity increases (CMOS technology)", Electronics, 48 (10), 80-2 (1975), EEA78-23707.
123. Altman, L., "Communications subsystems need just one chip apiece. (Special functions in CMOS)", Electronics, 48 (10), 83-5 (1975), EEA78-23708.

124.Anantha, N. G. and Jambotkar, C. G., "Integrated circuits incorporating complementary FET devices", IBM Tech. Disclosure Bull., 16, 3245 (1974), EEA77-30386.
125.Anon., "Focus on CMOS", Electron. Des., 20 (8), 54-61 (1972), EEA75-28507.
126.Anon., "Discussing complementary MOS", Microelectronics, 4 (2), 20-8 (1972), EEA75-35315.
127.Anon., "The combination of HiNIL and CMOS: the best design technique for noisy conditions", Elektronik, 13 (3), 29-30 (1975), Danish, EEA78-24459.
128.Anon., "MOS TAP", Elektro, 1, 242-5 (1975), EEA78-28852.
129.Arthur, F., "CMOS applications", Electron. Equip. News, 19-22 (1975), EEA78-24452.
130.Athanas, T., "Development of low threshold voltage COS/MOS integrated circuits", 1971 IEEE International Convention Digest, 1971, p. 538-9, EEA74-25580.
131.Bishop, A., "C-MOS - past, present and future", New Electron., 7 (4), 70-1, 73, 75 (1974), EEA77-18860.
132.Bishop, A. and Jones, P., "C-MOS - a status report", Microelectron. & Reliab., 13, 363-72 (1974), EEA78-5363.
133.Bishop, R. A., "Complementary MOS technique", Elektroniker, 11, 215-19 (1972), German, EEA76-4656.
134.Bishop, R. A., "Higher packing density and lower power with complementary m.o.s.", Electron. Eng., 44 (536), 61-3 (1972), EEA75-39930.
135.Bishop, R. A., "Complementary m.o.s. offers many advantages to the digital-systems designer", Electron. Eng., 44 (537), 67-70 (1972), EEA75-811.
136.Bishop, R. A. and Carley, D. R., "Fundamentals of COS/MOS integrated circuits", Solid State Technol., 17 (4), 85-9 (1974), EEA77-30401.
137.Borbely, E., "CMOS and noise immunity", Electron, no. 65, 47 (1974), EEA78-8175.
138.Brugger, C., "Complementary MOS integrated circuits with low consumption", Revue Polytech., no. 1293, 829-37 (1971), French, EEA75-4711.
139.Burghard, R. A. and Gwyn, C. W., "Radiation failure modes in CMOS integrated circuits", IEEE Trans. Nucl. Sci., 20, 300-6 (1973), CA80-88666.
140.Daoust, P. J. and Gruodis, A. J., "Small signal receiver circuit for complementary metal-oxide semiconductor circuits", IBM Tech. Disclosure Bull., 16, 3577-8 (1974), EEA77-38884.
141.Della Mussia, J. P., "In 1972 the c.m.o.s. technique will mark the beginning of a new era for integrated circuits", Inter Electron., 26 (29), 32-9 (1971), French, EEA75-3940.
142.Ferrie, R. G., "In-circuit IC tester checks TTL and C-MOS", Electronics, 47 (11), 120-1 (1974), EEA77-26025.
143.Gregory, B. L. and Shafer, B. D., "Latch-up in CMOS integrated circuits", IEEE Trans. Nucl. Sci., NS-20, 293-9 (1973), EEA77-23124.

144. Hampel, D. and Stewart, R. G., "EMP hardened CMOS circuits", Report HDL-TR-192-1, RCA Government Communications Systems, Somerville, N. J., 1973, 35 pp., EEA78-41703.
145. Hayashi, Y. and Tarui, Y., "Complementary MOS device", Bull Electrotech. Lab., 30 (1), 59-68 (1966), Japanese, EEA70-2616.
146. Healy, J., "Testing CMOS: a total-testing concept", Electron, no. 66, 27-8, 31 (1975), EEA78-19886.
147. Herrmann, G., "Testing CMOS components and circuits", Elektronik, 24, 103-6 (1975), German, EEA78-41673.
148. Hnatek, E. R., "Test results from screening 4000 series CMOS IC's", Eval. Eng., 14 (3), 20-1 (1975), EEA78-33167.
149. Hoover, M. V., "An introduction to the characteristics and applications of COS/MOS transistors in linear service", 8th Asilomar Conference on Circuits, Systems and Computers, 1974, p. 319-27, Publ. 1975, EEA78-23668.
150. Johannessen, S., "Construction with CMOS. II. Application examples", Elektro, 87 (9), 15-16, 19-20 (1974), EEA77-34092.
151. Johnson, P. A., "Complementary m.o.s. integrated circuits: properties, circuits and uses of c.m.o.s. with particular reference to hybrid a. & d. circuits", Wireless World, 79, 395-400 (1973), EEA76-29296.
152. Johnson, P. A. and Valeri, A. M., "Complementary MOS integrated circuits", Antenna, 46 (4), 131-40 (1974), Italian, EEA77-38453.
153. Jones, P., "C.m.o.s. technology and applications", New Electron., 8 (8), 32, 35 (1975), EEA78-24493.
154. Kalin, W. F., "Using complementary MOS circuits for control instrumentation", Instrum. & Control Syst., 45, 75-8 (1972), EEA76-362.
155. Kalthoff, M., "Complementary MOS-technique will be dynamically attractive", Elektroniker, 13 (5), EL18-19 (1974), German, EEA77-33721.
156. Lavooy, H., "The use of the CD4047A in COS/MOS timing applications", Polytech. Tijdschr. Elektrotech. Elektron., 30 (2), 35-44 (1975), Dutch, EEA78-14928.
157. Marciniak, W. and Iwanicka, K., "Complementary MOS-transistors in integrated circuits", Elektronika, 14 (8), 352-8 (1973), Polish, EEA76-38306.
158. Merchant, K., "The simulation of n-channel field-effect transistors in the circuit analysis of CMOS circuits", Nachrichtentech. Z. (NTZ), 28, 133-7 (1975), EEA78-15364.
159. Moore, E. E., "Performance characteristics of RCA's low-voltage COS/MOS devices", 1971 IEEE International Convention Digest, p. 188-9, EEA74-22752.
160. Morgan, D. K. and Steudel, G., "The COS/MOS phase locked loop circuit - a multipurpose building block", Elektron. Ind., 4 (7-8), 154-6 (1973), German, EEA77-9667.
161. Nagano, K. and Tanaka, T., "Aluminum oxide complementary MOS transistors", Proceedings of the 1st Conference on Solid State Devices, 1969, p. 132-6, Publ. 1970, CA82-79477.

162.Nguyen-Huu, A., "Advances in CMOS device technology", Comput. Des., 14 (1), 87-92 (1975), EEA78-15344.
163.Polinsky, M. A., Schade, O., Jr. and Keller, J. P., "CMOS-bipolar monolithic integrated-circuit technology", 1973 International Electron Devices Meeting Technical Digest, p. 229-31, EEA77-14615.
164.RCA Corp., "Complementary insulated gate field effect transistors", Patent UK 1265932, Publ. March 1972, EEA75-32704.
165.Richards, B. W. and Murray, L. A., "Sealed-junction beam-lead C-MOS devices", 1971 International Microelectronics Symposium, Pap. No. 6.3, 5 pp., EEA75-12773.
166.Rigby, G. A., "CMOS integrated circuits - a status report", Aust. Electron. Eng., 7 (5), 14-18 (1974), EEA77-38876.
167.Riseman, J., "Forming complementary metal-oxide semiconductor field-effect transistors by simultaneous double diffusion from the solid state", IBM Tech. Disclosure Bull., 16, 2897 (1974), EEA77-23050.
168.Sakamoto, H. and Forbes, L., "Grounded load complementary FET circuits: Sceptre analysis", IEEE J. Solid-State Circuits, SC-8, 282-4 (1973), EEA76-28790.
169.Sanquini, R. L., "Building C-MOS bipolar circuits on monolithic chip enhances IC specs", Electronics, 47 (20), 103-5 (1974), EEA78-676.
170.Santoni, A. and Trolese, G., "Design ideas with COS/MOS", New Electron., 7 (9), 27, 29, 30, 33, 34 (1974), EEA77-26049.
171.Schanis, R. S., "Reduce system noise with CMOS circuits", Electron. Des., 21 (25), 112-15 (1973), EEA77-18279.
172.Schlesier, K. M. and Norris, P. E., "CMOS hardening techniques", IEEE Trans. Nucl. Sci., NS-19, 275-81 (1972), EEA75-12640.
173.Schmidt, B., "Realizing the full low-power potential of the complementary m.o.s. integrated circuit technology", Proceedings of the National Electronics Conference, 1970, p. 265-7, EEA75-450.
174.Shah, R. B., "Overvoltage indicator can be added to C-MOS IC tester", Electronics, 48 (3), 99 (1975), EEA78-11724.
175.Soderman, L., "Is CMOS the ultimate generation of circuits?", Eltek. Aktuell Elektron., 17 (7-8), 48-50 (1974), Swedish, EEA78-678.
176.Soderman, L., "Is CMOS the ultimate circuit family? II", Eltek. Aktuell Elektron., 17 (9), 38-43 (1974), Swedish, EEA78-2509.
177.Stewart, R. G. and Hampel, D., "EMP hardened CMOS circuits", IEEE Trans. Nucl. Sci., NS-21, 332-9 (1974), EEA78-19900.
178.Walker, R., "C-MOS specifications: Don't take them for granted", Electronics, 48 (1), 103-7 (1975), EEA78-8165.
179.Walker, W., "Examining worst-case fan-out of standard C-MOS buffers", Electronics, 47 (13), 125 (1974), EEA77-29737.
180.Wang, R. C. C., "Low threshold complementary IGFETs", Thesis, Arizona State Univ., Tempe, 1971, 159 pp., Order No. 72-4216, EEA75-28522.

181.Wang, R. and DeMassa, T. A., "Silicon-aluminum gate complementary IGFET's", IEEE Trans. Electron Devices, ED-21, 130-1 (1974),
182.West, J., "The CMOS key to portable electronics", New Sci., 66, 550-2 (1975), EEA78-37647.
183.White, M. H. and Cricchi, J. R., "Complementary MOS transistors", Solid-State Electron., 9, 991-1008 (1966), PA70-1982.
184.Wilson, J. P., "ASEA seeks niche in custom CMOS market", New Electron., 8 (14), 29-30 (1975), EEA78-37648.
185.Wolf, H. P., "Technology and design of CMOS circuits. I", Elektron. Anz., 6 (2), 27-30 (1974), German, EEA77-26512.
186.Wolf, H. P., "Technology and design of CMOS circuits. II", Elektron. Anz., 6 (3), 59-65 (1974), German, EEA77-26513.
187.Yates, V., "CMOS circuits save power and are becoming faster", Eltek. Aktuell Elektron., 16 (4), 26-8 (1973), Swedish, EEA76-29311.
188.Yates, V., "Interfacing with CMOS", Electron. Components, 14, 381-4 (1973), EEA76-23020.
189.Yates, V., "The current trend in CMOS technology", Electron. Equip. News, 15 (7), 37-9 (1973), EEA77-5571.

3. MNOS

190.Andrasi, A., Horvath, Z. J., Nemethne-Sallay, M. and Stubnya, G., "Investigation on MNOS-structure. I", Kozl. Magy. Tud. Akad. Musz. Fiz. Kut. Intez., no. 0-14, 13-20 (1974), Hungarian, PA78-6901.
191.Anon., "Charge Storage in MNOS-system", Kozl. Magy. Tud. Akad. Musz. Fiz. Kut. Intez., no. 0-14, 5-12 (1974), Hungarian, PA78-6900.
192.Augusta, B. and Landler, F., "MNOS device characteristics and applications in memories", International Conference on Large Scale Integrated Circuits, 1974, p. 226-32, EEA78-11690.
193.Borisov, B. S., Kandyba, P. Y., Pospelov, V. V. and Svidzinskii, K. K., "New trends in the utilisation of MNOS structures", Izv. VUZ Radioelektron., 16 (4), 44-9 (1973), Russian, EEA76-26097.
194.Brewer, J. E., "MNOS secondary storage", Proceedings of the IEEE 1974 National Aerospace and Electronics Conference, p. 32-6, EEA78-8622.
195.Brown, M. A. C. S., Bounden, J. E. and Vanstone, G. F., "A simple equivalent circuit for the MNOST memory element", Solid State Electron., 15, 707-19 (1972), EEA75-24367.
196.Carlstedt, G., Petersson, G. P. and Jeppson, K. O., "A content-addressable memory cell with MNOS transistors", IEEE J. Solid-State Circuits, SC-8, 338-43 (1973), EEA76-35359.
197.Conti, M. and Vanin, A., "Conduction and biasing mechanisms into MNOS structures for memory", Ind. Ital. Elettrotec. & Elettron., 25, 893-9 (1972), Italian, EEA76-8688.
198.Cricchi, J. R., "MNOS nonvolatile semiconductor storage", 1974 Digests of the Intermag. Conference, 6-5/1 pp., EEA78-8204.

199. Cricchi, J. R., Blaha, F. C. and Fitzpatrick, M. D., "The drain-source protected MNOS memory device and memory endurance", 1973 International Electron Devices Meeting Technical Digest, p. 126-9, EEA77-14545.
200. Endo, N. and Nishi, Y., "A double-diffused MNOS transistor as a new non-volatile memory", 1973 International Electron Devices Meeting Technical Digest, p. 78-81, EEA77-14543.
201. Fowler, A. B., "Eliminating interfacial trapping states in metal-insulator-oxide-semiconductor structures", Patent French 2235482, Publ. January 1975, CA83-89839.
202. Gentil, P., LeGoascoz, V. and Borel, J., "MNOS memory devices", Onde Electr., 52, 427-34 (1972), French, EEA76-4664.
203. Goodwin, N, W., "Metal-nitride-oxide-semiconductor memory device", Patent USA 3803705, Publ. April 1974, CA80-150264.
204. Gordon, N. and Johnson, W. C., "Switching mechanism in thin-oxide MNOS devices", IEEE Trans. Electron Devices, ED-20, 253-6 (1973), EEA76-16441.
205. Gregory, P. J., "MNOS - A solution for non-volatile semiconductor memory", Colloquium Digest on Computer Memories, 1973, p. 2, EEA76-16451.
206. Horninger, K. H., "A fully decoded MNOS store in single-channel technology, using novel storage transistors", Bull. Assoc. Suisse Electr., 64, 1258-63 (1973), German, EEA77-6068.
207. Hsia, Y., "Memory applications of the MNOS", Western Electronic Show and Convention (Vol. 16), 1972, 4/3 5 pp., EEA76-8706.
208. Huang, J. S., "Switching properties of m.n.o.s. memory transistors", Electron. Lett., 8, 469-70 (1972), EEA75-39932.
209. Katsube, T., Adachi, Y. and Ikoma, T., "Trap centers in MNOS memory devices measured by thermally stimulated currents", Jap. J. Appl. Phys., 12, 1633-4 (1973), EEA77-379.
210. Kendall, E. J. M. and Haslett, J. W., "On the removal of the memory properties of MNOS devices", IEEE Trans. Electron Devices, ED-19, 287-8 (1972), EEA75-10267.
211. Kendall, J. T., "Recent progress in MNOS memory technology", Microelectron. & Reliab., 13, 413-14 (1974), EEA78-5367.
212. Kim, M. J., "Charge storage MISFET memory devices", Jap. J. Appl. Phys., 13, 1847-59 (1974), EEA78-5359.
213. Kim, M. J., Doyle, J. T., Patterson, W. J. and Pajak, R. W., "Electrical and irradiation performance of MNOS arrays", Jap. J. Appl. Phys., 13, 1837-46 (1974), EEA78-5358.
214. Kobayashi, K., "Influence of device parameters on memory performance of MNOS structure", Jap. J. Appl. Phys., 11, 555-63 (1972), EEA75-20218.
215. Kobayashi, K. and Ohta, K., "Electronic processes in MNOS system (II) Transitions between the types of C-V, I-V and G-V characteristics", Jap. J. Appl. Phys., 12, 881-7 (1973), EEA76-29194.

216.Koike, S. and Kambara, G., "The optical writing of MNOS memory with Au deep traps", Solid-State Electron., 17, 994-6 (1974), EEA78-8580.

217.Koike, S., Kambara, G., Matsuda, T. and Tsuru, S., "Non-volatile MNOS memory", Natl. Tech. Rep., 20, 217-26 (1974), Japanese, EEA78-649.

218.Lee, H. S., "MNOS varactor bootstrap memory cell", IBM Tech. Disclosure Bull., 17, 3214 (1975), EEA78-28801.

219.Lundkvist, L., Lundstrom, I. and Svensson, C., "Discharge of MNOS structures", Solid-State Electron., 16, 811-23 (1973), EEA76-26016.

220.Lundkvist, L., Lundstrom, I., Svensson, C. and Wallmark, J. T., "Discharge of MNOS structures at applied voltage and/or elevated temperature", 1972 European Solid State Devices Research Conference, p. 233-4, Publ. 1973, EEA76-32283.

221.Lundstrom, K. I. and Svensson, C. M., "Properties of MNOS structures", IEEE Trans. Electron Devices, ED-19, 826-36 (1972), EEA75-20212.

222.Mavor, J., "Reply to comment on metal-nitride-oxide-silicon avalanche-injection memory", Electron. Lett., 9, 349 (1973), EEA77-6025.

223.Mellor, P. J. T. and Dunn, P. J., "Comment on metal-nitride-oxide-silicon avalanche-injection memory", Electron. Lett., 9, 252-3 (1973), EEA77-2565.

224.Oakley, R. E. and Browne, V. A., "The effect of repeated switching cycles on the MNOS memory transistor", 1972 European Solid State Devices Research Conference, p. 233, Publ. 1973, EEA76-32282.

225.Pasch, L., "The MNOS-LAD (metal-nitride-oxide-silicon avalanche-diode) - a new semiconductor storage element", Nachrichtentech. Elektron., 23 (2), 73-6 (1973), German, EEA76-19906.

226.Reid, P. R., McLouski, R. M., Cricchi, J. R. and Blaha, F. C., "The effect of Si_3N_4 deposition parameters on MNOS memory transistor response", J. Electrochem. Soc., 121, 85C (1974), EEA77-23055.

227.Sewell, F. A., Jr., "The light-sensitive MNOS memory transistor", IEEE Trans. Electron Devices, ED-20, 563-72 (1973), EEA76-23735.

228.Simmons, J. G., "Non-steady-state measurements on variable-threshold MNOS transistors", J. Vac. Sci. & Technol., 9, 39 (1972), EEA75-20219.

229.Taylor, G. W. and Simmons, J. G., "Effects of bulk trapping on the memory characteristics of thick-oxide MNOS variable-threshold capacitors", Solid-State Electron., 17, 1-10 (1974), PA77-16829.

230.Uranwala, J. S., Simmons, J. G. and Mar, H. A., "Nonsteady-state techniques for determining the energy distribution of interface traps in MNOS memory devices", Appl. Phys. Lett., 26, 697-9 (1975), EEA78-37537.

231.Wallmark, J. T. and Scott, J. H., "Switching and storage characteristics of MIS memory transistors", RCA Rev., 30 (2), 335-65 (1969), EEA72-23830.
232.Wei, L. S. and Simmons, J. G., "Trapping, emission and generation in MNOS memory devices", Solid-State Electron., 17, 591-8 (1974), PA77-49083.
233.White, M. H. and Cricchi, J. R., "Characterization of thin-oxide MNOS memory transistors", IEEE Trans. Electron Devices, ED-19, 1280-8 (1972), EEA76-814.
234.Wojciechowska, D. and Zak, J., "Surface electric properties of Al-Si_3N_4-Si structures", Electron. Technol., 6 (1-2), 63-73 (1973), PA77-53149.
235.Yanai, H., Kawate, K. and Horibata, Y., "Effect of bias-voltage sweep-rate on C-V characteristics of MNOS diodes", Annu. Rep. Eng. Res. Inst. Fac. Eng. Univ., 33, 167-71 (1974), Japanese, PA78-37826.

4. SOS

236.Allison, J. F., Dumin, D. J., Heiman, F. P., Meuller, C. W. and Robinson, P. H., "Thin-film silicon: preparation, properties, and device applications", Proc. IEEE, 57, 1490-8 (1969), EEA73-1088.
237.Allison, J. F., Heiman, F. P. and Burns, J. R., "Silicon-on-sapphire complementary MOS memory cells", IEEE J. Solid-State Circuits, SC-2, 206-10 (1967), EEA71-13341.
238.Boleky, E. J., "The performance of a complementary MOS transistors on insulating substrates", RCA Rev., 31, 372-95 (1970), EEA74-4935.
239.Boleky, E. J. and Meyer, J. E., "High-performance low-power CMOS memories using silicon-on-sapphire technology", IEEE J. Solid-State Circuits, SC-7, 135-45 (1972), EEA75-13799.
240.Crossley, P. A. and Ham, W. E., "Use of test structures and results of electrical tests for silicon-on-sapphire integrated circuit processes", J. Electron. Mater., 2, 465-83 (1973), EEA77-10298.
241.Cullen, G. W. and Corboy, J. F., Jr., "A comparison of the semiconducting properties of thin films of silicon on sapphire and spinel", J. Electrochem. Soc., 121, 1345-50 (1974), PA78-1357.
242.Cullen, G. W. and Dougherty, F. C., "Silicon on spinel: the interaction between deposition constituents and the substrate surface", J. Cryst. Growth, 17, 230-40 (1972), EEA76-4532.
243.Della-Mussia, J. P., "MOS with TTL speed", Inter Electron., 27 (47), 34-6 (1972), French, EEA75-20215.
244.Dumin, D. J., "The current status of silicon-on-sapphire", Electrochemical Society Fall Meeting (Extended abstracts), 1974, p. 321-2, EEA78-37674.

245.Dumin, D. J. and Robinson, P. H., "Electrically and optically active defects in silicon-on-sapphire films", J. Cryst. Growth, 3-4, Spe., 214-18 (1968), EEA72-20215.
246.Eaton, S. S., "Sapphire brings out the best in C-MOS", Electronics, 48 (12), 115-20 (1975), EEA78-28843.
247.Falckenberg, R., "Method of dividing Mg-Al spinel substrate wafers coated with semiconductor material and provided with semiconductor components", Patent USA 3816906, Publ. June 1974.
248.Farrington, D., "Anomalous current-voltage characteristics of floating bulk MOSFET's (in SOS)", Electrochemical Society Fall Meeting (Extended abstracts), 1974, p. 345-7 (1974), EEA78-37556.
249.Flatly, D. W. and Ham, W. E., "Electrical instabilities in SOS/MOS transistors", Electrochemical Society Fall Meeting (Extended abstracts), 1974, p. 487-9, EEA78-37559.
250.Gehweiler, W. F., Kaiser, H. W. and Pridgen, J. I., "CMOS/SOS correlator and multiplier", Proceedings of the IEEE 1974 National Aerospace and Electronics Conference, p. 252-9, EEA78-5375.
251.Goser, K., Pomper, M. and Splittgerber, H., "High-speed and high-density SOS-circuits", 6th International Congress on Microelectronics, 1974, 4 pp., EEA78-24517.
252.Gottlieb, G. E. and Corboy, J. F., Jr., "The development of a dual rate technique for the growth of silicon-on-sapphire films", J. Cryst. Growth, 17, 261-70 (1972), EEA76-4536.
253.Ham, W. E., "Surface charge effects on the resistivity and Hall coefficient of thin silicon-on-sapphire films", Appl. Phys. Lett., 21, 440-3 (1972), EEA76-739.
254.Hampel, D., McGuire, K. E. and Prost, K. J., "CMOS/SOS serial-parallel multiplier", IEEE J. Solid-State Circuits, SC-10 (5), 307-14 (1975), EEA78-37620.
255.Heiman, F. P., "Thin-film silicon-on-sapphire deep depletion M.O.S. transistors", IEEE Trans. Electron Devices, ED-13, 855-62 (1966), EEA70-10883.
256.Hynecek, J., "Elastoresistance of n-type silicon on sapphire", J. Appl. Phys., 45, 2631-5 (1974), EEA77-33876.
257.Hynecek, J., "Hall effect in silicon on sapphire", J. Appl. Phys., 45, 2806-7 (1974), EEA77-33985.
258.Hynecek, J., "Graphical method for determining the flat band voltage for silicon on sapphire", Solid-State Electron., 18, 119-20 (1975), PA78-15732.
259.Ipri, A. C., "Electrical properties of silicon films on sapphire using the MOS Hall technique", J. Appl. Phys., 43, 2770-5 (1972), EEA75-28439.
260.Johannessen, J. S., "Magnetoresistance in germanium on sapphire", Phys. Status Solidi A, 26, K183-6 (1974).
261.Kaiser, H. W., "CMOS/SOS for high speed signal processing", 1973 IEEE International Convention and Exposition, Vol. VI, 23.2/2 pp., EEA77-2610.
262.King, G., "Silicon-on-sapphire IC's: technology forecast", Circuit Manufac., 15 (9), 48-52 (1975).

263.Kjar, R. A. and Kinoshita, G., "Transient photocurrents in SOS structures", IEEE Trans. Nucl. Sci., NS-20, 315-18 (1973), EEA77-23127.
264.Kranzer, D., "Spatial dependence of the carrier lifetime in thin films of silicon on sapphire", Appl. Phys. Lett., 25, 103-5 (1974), PA77-70083.
265.Kurthy, Z., Solt, G., Szabo, L., Herman, A. and Kurthy, K. J., "Active thin film devices, II", Hiki, 8 (2), 5-22 (1968), Hungarian, EEA72-7960.
266.Lilen, H., "Combining the speed of bipolar integrated circuits with the consumption of the MOS; MOSIC's on sapphire (SOS) appear on the market", Electron. & Microelectron. Ind., no. 156, 41-5 (1972), EEA75-27778.
267.Maev, K. L., Stadnik, A. V., Kosenko, V. Y. and Grigorev, O. N., "Mechanical stresses in silicon films on sapphire", Poluprovodn. Tekh. & Mikroelektron., no. 17, 60-5 (1974), Russian, EEA78-2752.
268.McGreivy, D. J. and Viswanathan, C. R., "The effect of arsenic in improving lifetime in silicon-on-sapphire films", Appl. Phys. Lett., 25, 505-6 (1974), EEA78-11721.
269.Mercier, J., "Physical and electrical investigations on silicon epitaxial layers on sapphire substrates", J. Electrochem. Soc., 117 (5), 666-75 (1970).
270.Micheletti, F. B., "Recent SOS technology advances and applications", Solid State Technol., 18 (8), 39-46 (1975), EEA78-41691.
271.Phillips, D. H., "Silicon-on-sapphire device photoconduction predictions", IEEE Trans. Nucl. Sci., NS-21, 217-20 (1974), EEA78-19706.
272.Popov, Y. A., Boronin, A. T. and Sladkov, A. B., "Problems of designing highly integrated schemes based on silicon-on-sapphire technology", International Symposium on Discrete Systems, Pt. I, 1974, p. 214-21, EEA78-19926.
273.Portlock, J. J., "SOS technology - fabrication problems and their solution", New Electron., 8 (4), 70-1 (1974), EEA78-24489.
274.Rai-Choudhury, P., Schroder, D. K., Heng, T. M. S. and Sweeny, G. G., "Impurity distribution and electrical properties of silicon-on-sapphire films", Electrochemical Society Fall Meeting (Extended abstracts), 1974, p. 339-41, EEA78-37678.
275.Rapp, A. K. and Ross, E. C., "Silicon-on-sapphire substrates overcome MOS limitations", Electronics, 45 (20), 113-16 (1972), EEA75-39931.
276.Ronen, R. S. and Robinson, P. H., "Recent advances in thin-film silicon devices on sapphire substrates", Proc. IEEE, 59, 1506-10 (1971), EEA74-36482.
277.Ronen, R. S. and Strauss, L., "The silicon-on-sapphire MOS tetrode - some small-signal features, LF to UHF", IEEE Trans. Electron Devices, ED-21, 100-9 (1974), EEA77-6034.

278.Sarace, J. C. and Ipri, A. C., "Low threshold deep depletion technology for CMOS/SOS integrated circuits", Electrochemical Society Fall Meeting (Extended abstracts), 1974, p. 484-6, EEA78-37686.

279.Schroder, D. K. and Rai-Choudhury, P., "Silicon-on-sapphire with microsecond carrier lifetimes", Appl. Phys. Lett., 22, 455-7 (1973), EEA76-25960.

280.Smith, W. B., Pontius, D. H. and Budenstein, P. P., "Second breakdown and damage in junction devices", IEEE Trans. Electron Devices, ED-20, 731-44 (1973), EEA76-29233.

281.Somogyi, K., "Contacts and sample preparation on semiconductor layers for van der Pauw measuremenets", Kozl. Magy. Tud. Akad. Musz. Fiz. Kut. Intez., no. 0-9, 47-59 (1973), EEA77-10161.

282.Suat, J. P., et al., "Main characteristics of MIS/SOS technology", International Conference on Large Scale Integrated Circuits, 1974, p. 52-63, EEA78-11745.

283.Sunshine, R. A., "Second-breakdown phenomena in avalanching silicon-on-sapphire diodes", IEEE Trans. Electron Devices, ED-19, 873-85 (1972), PA75-47945.

284.Tango, H., Iwamura, J., Maeguchi, K. and Isobe, M., "N-channel Si-gate MOS devices on sapphire substrate", Oyo Buturi, 44 (Suppl.), 225-31 (1975), EEA78-41685.

285.Vanek, O. and Tvarozek, V., "Silicon integrated circuits on sapphire substrate", Sdelovaci Tech., 23 (2), 63-4 (1975), Slovak, EEA78-19922.

286.Young, D. R., "Metal-oxide-semiconductor field-effect transistor structure", IBM Tech. Disclosure Bull., 17, 1208-9 (1974), EEA78-5333.

287.Zaininger, K. H. and Wang, C. C., "MOS and vertical junction device characteristics of epitaxial silicon on low aluminum-rich spinel", Solid-State Electron., 13, 943-50 (1970), PA73-62395.

5. MAOS

288.Clemens, J. T. and Labuda, E. F., "Vapor transport of phosphorus in integrated circuit fabrication", Extended Abstracts of the Spring Meeting of the Electrochemical Society, 1971, p. 187-8, EEA75-433.

289.Clemens, J. T., Labuda, E. F. and Berglund, C. N., "Aluminum oxide/silicon dioxide, double-insulator, MOS structure", Bell Syst. Tech. J., 54, 687-719 (1975), PA78-62110.

290.DiMaria, D. J., "Effects of interface barrier energies of metal-aluminium oxide-semiconductor (MAS) structures as a function of metal electrode material, charge trapping, and annealing", J. Appl. Phys., 45, 5454-6 (1974), EEA78-11550.

291.Gnadinger, A. P. and Rosenzweig, W., "Polarization and charge motion in metal-Al_2O_3-SiO_2-Si structures", J. Electrochem. Soc., 121, 700-5 (1974), EEA77-30224.

292.Hamano, K. and Ohta, K., "Electrical properties of MAOS structures as revealed by conductance technique", Jap. J. Appl. Phys., 11, 1053 (1972), EEA75-28518.

293.Jund, C., Kervella, B. and Grosvalet, J., "A mechanism of evaluation in MAOS systems", 9th Annual Proceedings Reliability Physics, 1971, p. 9-15, EEA75-4649.

294.Komiya, Y., Tarui, Y. and Kubo, K., "Si MIS characteristics using vapour-grown Al_2O_3 films. II", Bull. Electrotech. Lab., 36, 85-95 (1972), Japanese, EEA75-36191.

295.Kuhn, M. and Ligenza, J. R., "Plasma grown Al_2O_3-SiO_2 MOS structures", Electrochemical Society Fall Meeting, Extended abstracts, 1970, p. 358-61, EEA74-7672.

296.Nagano, K., Iwauchi, S. and Tanaka, T., "Al_2O_3-silicon MOS field effect transistors", Japan. J. Appl. Phys., 8, 277 (1969), EEA72-11324.

297.Tsujide, T., Igarashi, R., Onoda, K., Wada, T. and Nakagiri, M., "Read-only memory using MAS transistors", IEEE J. Solid-State Circuits, SC-5 (5), 203-7 (1970), EEA73-38100.

298.Nakegiri, M. and Wada, T., "Electrical characteristics of metal Al_2O_3-SiO_2-Si (MAOS) transistors", Jap. J. Appl. Phys., 11, 1484-90 (1972), EEA76-817.

299.Norris, P. E. and Micheletti, F. B., "Metal-oxide-semiconductor transistor including a gate insulator layer of aluminum oxide", Patent USA 3735482, Publ. May 1973, CA79-59193.

300.Pappu, R. V. and Boothroyd, A. R., "Memory behavior of metal-plasma-anodized Al_2O_3 and SiO_2 semiconductor (MAOS) capacitors", Appl. Phys. Lett., 22, 72-4 (1973), EEA76-12634.

301.Potter, M. D., "Soft turn-on of MAOS devices", J. Electrochem. Soc., 121 (3), 89C (1974), EEA77-26476.

302.Raymond, R. K. and Das, M. B., "Use of anodic aluminum oxide in MOS structures", 1973 International Electron Devices Meeting Technical Digest, p. 359-62, EEA77-14556.

303.Salama, C. A. T., "Conduction in Al_2O_3 films and charge storage in MAOS structures", J. Electrochem. Soc., 118, 1993-8 (1971), EEA75-7460.

304.Sato, S. and Yamaguchi, T., "Study of charge storage behavior in metal-alumina-silicon dioxide-silicon (MAOS) field effect transistor", Solid-State Electron., 17, 367-75 (1974), EEA77-18817.

305.Tsang, P. J. and Ooo, V. Y., "Threshold voltage and electron mobility of N-channel IGFET's with Al_2O_3/SiO_2 gate insulation", Electrochemical Society Fall Meeting, Extended abstracts, 1970, p. 421-3, EEA74-7796.

306.Tsujide, T. and Iida, K., "The effect of electrode on the C-V hysteresis behaviour of MAS and MAOS structures", Jap. J. Appl. Phys., 11, 600-1 (1972), EEA75-16681.

307.Wada, T., Nakagiri, M., Iwase, K. and Nakanuma, S., "MAS-ROM - the fabrication and memory characteristics", Microelectronics, 4 (2), 8-19 (1972), EEA75-36301.

308.Wada, T., Onoda, K., Ishiguro, H. and Nakanuma, S., "MAS-ROM - electrically reprogrammable ROM with decoder", IEEE J. Solid-State Circuits, SC-7, 375-81 (1972), EEA75-36287.
309.Waxman, A. and Zaininger, K. H., "Al_2O_3-silicon insulated gate field effect transistors", Appl. Phys. Lett., 12, 109-10 (1968), PA71-22638.
310.Yamaguchi, T. and Sato, S., "Circuit array and system of MAOS type fully decoded electrically alterable read only memory-IC", Jap. J. Appl. Phys., 13, 1414-20 (1974), EEA77-38490.

6. FAMOS

311.Abbas, S. A., Barile, C. A., Lane, R. D. and Liu, P. T., "Electrically erasable floating gate FET memory cell", Patent USA 3836992, September 1974.
312.Anantha, N. G., Dockerty, R. C. and Lucy, J. M., Sr., "Electrically erasable floating gate field-effect transistor memory cell", IBM Tech. Disclosure Bull., 17, 2311-13 (1975), EEA78-24356.
313.Berger, J., "Improved MOS transducer-drive circuit (piezoelectric transducers)", IEEE J. Solid-State Circuits, SC-8, 92-3 (1973), EEA76-10104.
314.Chang, C. S. and Gokhale, B. V., "Electrically erasable FAMOS memory cell", IBM Tech. Disclosure Bull., 15, 2922-4 (1973), EEA76-19931.
315.DiStefano, T. H. and Laibowitz, R. B., "Electrically reprogrammable floating avalanche injection MOS (FAMOS) structure", IBM Tech. Disclosure Bull., 15, 3264-6 (1973), EEA76-26100.
316.Dockerty, R. C., "N-channel FAMOS cell with a Schottky electrode", IBM Tech. Disclosure Bull., 16, 3244 (1974), EEA77-30338.
317.Frohman-Bentchkowsky, D., "Memory behavior in a floating-gate avalanche-injection MOS (FAMOS) structure", Appl. Phys. Lett., 18, 332-4 (1971), EEA71-22724.
318.Frohman-Bentchkowsky, D., "A fully decoded 2048-bit electrically programmable FAMOS read-only memory", IEEE J. Solid-State Circuits, SC-6, 301-6 (1971), EEA74-34081.
319.Frohman-Bentchkowsky, D., "FAMOS - a new semiconductor charge storage device", Solid-State Electron., 17, 517-29 (1974), EEA77-23065.
320.Kohler, E., "The floating-gate transistor - a new storage cell to be programmed electrically", Nachrichtentechnik, 22, 366-7 (1972), German, EEA75-39948.
321.Lochner, D., "Programmable semiconductor memories", Funk-Tech, 30, 320-1 (1975), EEA78-33205.
322.Pepper, M., "Memory elements (FAMOS)", Patent USA 3774087, Publ. November 1973.
323.Yu, K. K., Brody, T. P. and Chen, P. C. Y., "Experimental realization of floating-gate-memory thin-film transistor", Proc. IEEE, 63, 826-7 (1975), EEA78-19848.

7. MOS/Bipolar

324. Antipov, I., "NPN N-channel bipolar FET devices", IBM Tech. Disclosure Bull., 17, 2935-6 (1975), EEA78-19845.
325. Beasom, J., "Better bipolar-MOS process yields linear ICs with good AC and DC specs", Electronics, 47 (22), 65-70 (1974), EEA78-677.
326. Beasom, J. D., "Fabrication of MOS devices and complementary bipolar transistor devices in a monolithic substrate", Patent USA 3865649, Publ. February 1975, EEA78-37660.
327. Burt, D. J., "Combined MOS/bipolar transistor integration", International Conference on Microelectronics, 1969, p. 54-5, EEA72-23155.
328. Cheney, G. T. and Marr, G., "BIGFET makes IGFET more versatile", Bell Lab. Rec., 50 (6), 194-8 (1972), EEA75-27733.
329. Crawford, R. H., "MOS-bipolar high-voltage driver circuit", Patent USA 3778688, Publ. December 1973.
330. Everist, J. A., "Technology teamwork - m.o.s. and bipolar", Electron. Engineering, 47 (563), 55-8 (1975), EEA78-8604.
331. Graf, S. and Polinsky, M., "MOS-bipolar monolithic integrated circuit technology", Presented at International Electron Devices Meeting, Washington, D. C., October 1970, EEA74-6934.
332. Lin, H. C., Ho, J. C., Lyer, R. R. and Kwong, K., "Complementary MOS-bipolar transistor structure", IEEE Trans. Electron Devices, ED-16, 945-51 (1969), EEA73-10679.
333. McCaffrey, T. P. and Gray, P. R., "Monolithic MOS/bipolar integrated circuit structure", Patent USA 3838440, Publ. September 1974, CA81-161227.
334. Mackowiak, E. and Montier, M., "Advantages of MOS and bipolar transistors on the same chip - application to a fast complex memory", International Conference on Large Scale Integrated Circuits, 1974, p. 74-88, French, EEA78-11747.
335. Marr, G. and Mowery, G. I., "A depletion-load, p-channel, bipolar-IGFET technology", Bell Syst. Tech. J., 54, 69-79 (1975), EEA78-11722.
336. Polinsky, M. and Graf, S., "MOS-bipolar monolithic integrated circuit technology", IEEE Trans. Electron Devices, ED-20, 239-44 (1973), EEA76-16471.
337. Yu, K. K., Lin, H. C. and Kwong, K., "The MOS-BI: A monolithic MOS-bipolar integrated structure (trigger circuit)", Presented at International Electron Devices Meeting, Washington, D. C., October 1967, EEA72-10647.

8. DMOS

338. Barson, F., "Double-diffused MOS device and process of fabricating", IBM Tech. Disclosure Bull., 14, 2900-1 (1972), EEA75-24361.
339. Beckett, H., "Double diffused MOSFETs", Electron. Equip. News, 15 (9), 56-7 (1974), PA77-25264.

340. Cauge, T. P. and Kocsis, J., "High-voltage, high-frequency double-diffused metal-oxide-semiconductor device", Patent USA 3845495, Publ. October 1974, CA82-37937.

341. Cauge, T. P., Kocsis, J. and Buchanan, J. R., "Forming metal-oxide-semiconductor structure using double diffusion", Patent USA 3909320, Publ. September 1975, CA83-201121.

342. Cauge, T. P., Kocsis, J., Sigs, H. J. and Vendelin, G. D., "Double-diffused MOS transistor achieves microwave gain", Electronics, 44 (4), 99-104 (1971), EEA74-13252.

343. Coppes, A., "D-MOS FET single gate 1 GHz amplifier", New Electron., 6 (22), 66, 69, 71-2 (1973), EEA77-13871.

344. Harris, R. E., "Double diffused MOS transistors", Presented at International Electron Devices Meeting, Washington, D. C., October 1970, EEA72-7946.

345. Kraybill, A. V., Rapshys, W. and Yester, F. R., "Self-aligned double-diffused metal-oxide-semiconductor devices", Patent USA 3863330, Publ. February 1975, CA83-52063.

346. Lilen, H., "DMOS integrated circuits for telephone switching", Electron. & Microelectron. Ind., no. 203, 57-60 (1975), French, EEA78-25530.

347. Lin, H. C. and Jones, W. N., "Computer analysis of the double-diffused MOS transistors and integrated circuits", Presented at 17th International Electron Devices Meeting, Washington, D. C., October 1971, EEA75-13774.

348. McLintock, G. A. and Thomas, R. E., "Effects of process changes of double-diffused MOST characteristics", 1973 International Electron Devices Meeting Technical Digest, p. 367-70, EEA77-14558.

349. McLintock, G. A., Thomas, R. E., Cobbold, R. S. C. and Hogeboom, J. G., "Forward and reverse characteristics of self-aligned double-diffused M.O.S. transistors", Electron. Lett., 8, 463-5 (1972), EEA75-36282.

350. Margalit, S., Gilboa, H., Bar-Lev, A. and Kirdron, I., "Diffused MOS transistor", Presented at 8th Convention of the Electrical and Electronics Engineers, Tel Aviv, Israel, April-May 1973, EEA77-5454.

351. Pocha, M. D., Gonazalez, A. G. and Dutton, R. W., "Threshold voltage controllability in double-diffused-MOS transistors", IEEE Trans. Electron Devices, ED-21, 778-84 (1974), EEA78-11650.

352. Rodgers, T. J., Asai, S., Pocha, D., Dutton, R. W. and Meindl, D., "An experimental and theoretical analysis of double-diffused MOS transistors", IEEE J. Solid-State Circuits, SC-10, 322-31 (1975), EEA78-37622.

353. Sigg, H. J., Vendelin, G. D., Cauge, T. P. and Kocsis, J., "D-MOS transistor for microwave applications", IEEE Trans. Electron Devices, ED-19, 45-53 (1972), EEA75-4867.

9. ESFI MOS

354.Druminski, M., Kuhl, C., Preuss, E. and Schwidefsky, F., "Quality evaluation of thin silicon films by various investigation methods", Electrochemical Society Fall Meeting (Extended abstracts), 1974, p. 329-31, EEA78-37676.

355.Goser, K. and Pomper, M., "Five-transistor memory cells in ESFI MOS technology", IEEE J. Solid-State Circuits, SC-8, 324-6 (1973), EEA76-35356.

356.Goser, K. and Pomper, M., "Static ESFI MOS(SOS) cells for high-density memories", Siemens Forsch. & Entwicklungsber, 4, 220-5 (1975), EEA78-37652.

357.Goser, K., Pomper, M. and Tihanyi, J., "High-density static ESFI MOS memory cells", IEEE J. Solid-State Circuits, SC-9, 234-8 (1974), EEA77-38881.

358.Horninger, K., "A high-speed ESFI SOS programmable logic array with an MNOS version", IEEE J. Solid-State Circuits, SC-10, 331-6 (1975), EEA78-37623.

359.Pomper, M., Horninger, K., Tihanyi, J. and Goser, C., "Fast low-power memory devices using ESFI MOS technology", Onde Electr., 54, 187-91 (1974), EEA77-26524.

360.Pomper, M. and Tihanyi, J., "Ion-implanted ESFI MOS devices with short switching times", IEEE J. Solid-State Circuits, SC-9, 250-6 (1974), EEA77-38468.

361.Raetzel, C., Schild, S. and Schlotterer, H., "Etching properties of different silicon films", Electrochemical Society Fall Meeting (Extended abstracts), 1974, p. 336-8, PA78-69617.

362.Tihanyi, J. and Schlotterer, H., "Influence of the floating substrate potential on the characteristics of ESFI MOS transistors", Solid-State Electron., 18, 309-14 (1975), EEA78-11666.

363.Tihanyi, J. and Schlotterer, H., "Properties of ESFI MOS transistors due to the floating substrate and the finite volume", IEEE Trans. Electron Devices, ED-22, 1017-23 (1975).

10. Non Conventional Dielectrics

364.Alexander, J. H., Chiffick, R. C., Drake, C. F. and Scanlan, I. F., "Polarisable dielectrics for MOST's", Electron. Lett., 6, 813-14 (1970), PA74-9500.

365.Fillard, J. P. and Manifacier, J. C., "SnO_2-semiconductor heterojunction: Electrical properties of a particular MOS structure", Jap. J. Appl. Phys., 9, 1012-3 (1970), EEA73-37999.

366.Hsueh, Y. W., "Metal-TiO_2-SiO_2-Si field effect transistor", Electrochemical Society Fall Meeting (Extended abstracts), 1974, p. 506-7, EEA78-37562.

367.Kurogi, Y., Endo, N. and Sugibuchi, K., "Ferroelectric field effect memory device using bismuth titanate film", Oyo Buturi, 44 (Suppl.), 197-202 (1975), EEA78-41614.

368.Wakefield, J. and Gamble, H. S., "Titanium dioxide as gate insulator for MOS transistors", Electron. Lett., 6, 507-8 (1970), EEA73-35228.

11. Others

369.Abbas, S. A., Dockerty, R. C., Johnson, W. S. and Peressini, P. P., "Short channel field-effect transistor", IBM Tech. Disclosure Bull., 17, 3263 (1975), EEA78-24358.
370.Abe, T., "High-density, high-integration C^2 MOS may rival P MOS in performance", JEE, no. 92, 20-3 (1974), EEA78-2908.
371.Balog, M., Schieber, M., Patai, S. and Michman, M., "Thin films of metal oxides on silicon by chemical vapor deposition with organometallic compounds", J. Cryst. Growth, 17, 298-301 (1972), EEA75-4577.
372.Bazin, B., "SAMNOS technology", Solid-State Electron., 15, 649-51 (1972), EEA75-24366.
373.Berth, M., "Refractory gate m.o.s. technology", Onde Electr., 54, 23-6 (1974), French, PA77-25277.
374.Borel, J., Bernard, J. and Suat, J. P., "A depletion load self-aligned technology", Solid-State Electron., 16, 1377-81 (1973), EEA77-5589.
375.Brown, D. M., Engeler, W. E., Garfinkel, M. and Gray, P. V., "Self-registered molybdenum-gate MOSFET", J. Electrochem. Soc., 115, 874-6 (1968), PA71-49316.
376.Brown, D. M., Engeler, W. E., Tiemann, J. J., Lavoo, N. T., Carlson, R. O. and Connery, R. J., "High frequency MOS digital capacitor", IEEE Trans. Electron Devices, ED-22, 938-44 (1975).
377.Burock, R., DeBolt, J. R. and Parente, R. N., "Manufacturing beam lead, insulated-gate, field effect transistor (IGFET) integrated circuits", Western Electric Eng., 17 (3), 2-16 (1973), EEA76-38315.
378.Charlson, E. J. and Lien, J. C., "An Al p-silicon MOS photovoltaic cell", J. Appl. Phys., 46 (9), 3982-7 (1975).
379.Cunningham, J. A. and Carbajal, B. G., "Metal and oxide systems for MOS integrated circuits", Extended abstracts of the Spring Meeting of the Electrochemical Society, 1971, p. 50-1, EEA74-33357.
380.Curtis, T. and Kernhajian, H., "High-speed molybdenum-gate m.o.s.t. shift register", Electron. Lett., 7, 729-30 (1971), EEA75-4699.
381.Davidsohn, U. S., "Diffusion guarded metal-oxide-silicon field-effect transistors", Patent USA 3868721, Publ. February 1975.
382.Demassa, T. A. and Catalano, G. T., "The inhomogeneous channel FET: ICFET", Solid-State Electron., 16, 847-51 (1973), EEA76-29280.
383.Dill, H. G., "Field effect device with overlapping insulated gates", Patent USA 3454844, Publ. July 1969, EEA73-14331.

384.Dill, H. G. and Thomas, N., "Complementary silicon-insulator-semiconductor field-effect transistor (SISFET) integrated circuit", Patent UK 1363581, Publ. August 1974, CA81-178926.
385.Farzan, B. and Salama, C. A. T., "Depletion MOS power transistors", Electron. Lett., 11, 565-6 (1975).
386.Fowler, A. B. and Young, D. R., "Field-effect transistor storage", IBM Tech. Disclosure Bull., 17, 1834-5 (1974), EEA78-11656.
387.Fu, H. S. and Sah, C. T., "A distributed gate bistable MOS transistor", Solid-State Electron., 14, 799-804 (1971), EEA74-37311.
388.Garnache, R. R. and Kenney, D. M., "Two hot-step FET process", IBM Tech. Disclosure Bull., 14, 1971-2 (1971), EEA75-10294.
389.Holle, E., "MOS-protection circuits", Elektroniker, 14 (4), EL1-2 (1975), German, EEA78-28846.
390.Hughes, W. C., Lemmond, C. Q., Parks, H. G., Ellis, G. W., Possin, G. E. and Wilson, R. H., "A semiconductor nonvolatile electron beam accessed mass memory", Proc. IEEE, 63, 1230-40 (1975), EEA78-28870.
391.Iizuka, H., Sato, T., Masuoka, F., Ohuchi, K., Hara, H., Ishikawa, M. and Takeishi, Y., "Stacked-gate avalanche-injection type metal-oxide-semiconductor (SAMOS) transistor memory", Proceedings of the 4th Conference on Solid State Devices, 1972, p. 158-66, Publ. 1973, CA80-31734.
392.Kano, G., Iwasa, H., Takagi, H. and Teramoto, I., "The lambda diode: a versatile negative-resistance device", Electronics, 48 (13), 105-9 (1975), EEA78-28844.
393.Kaplan, L. H., "Self aligning IGFET with refractory metal gate", IBM Tech. Disclosure Bull., 14, 2299-3000 (1972), EEA75-20216.
394.Kaplan, L. H., Montillo, F. J. and Pastorius, D. L., "Fabrication of self-aligned IGFET devices with continuous oxide in thin and thick regions", IBM Tech. Disclosure Bull., 14, 1568-9 (1971), EEA75-7544.
395.Kassabov, J. D., Georgiev, V. K. and Stoeva, R. D., "MOST and MOS IC with low threshold voltage by means of the W-Au metal system", C. R. Acad. Bulg. Sci., 25, 1617-19 (1972), EEA76-19925.
396.Kovtonyuk, N. K., Morozov, V. A. and Zakharov, I. S., "Photocurrent in high resistivity silicon MOSOM structures", Izv. VUZ Radioelektron., 17 (9), 10-14 (1974), Russian, PA78-18686.
397.Lee, H. S. and Moore, A. J., "Self-aligned field-effect transistor gate process", IBM Tech. Disclosure Bull., 17, 2591 (1975), EEA78-19843.
398.Leung, A. F., "MOS gate insulators for single crystal CdS", Solid-State Electron., 17, 1213-14 (1974), PA77-82490.
399.Lundstrom, K. I., Shivaraman, M. S. and Svensson, C. M., "A hydrogen-sensitive Pd-gate MOS transistor", J. Appl. Phys., 46, 3876-81 (1975).

400. Maeda, K., Shirai, K. and Nakano, J., "Electrical characteristics of polycrystalline silicon-aluminum oxide-silicon dioxide-silicon structure", Jap. J. Appl. Phys., 13, 1173-4 (1974), EEA77-34055.
401. Mai, C. C., Chan, T. C. and Palmer, R. B., "Three-mask self-aligned MOS technology", IEEE Trans. Electron Devices, ED-20, 1162-4 (1973), EEA77-2604.
402. Masuda, H., Masuhara, T. and Nagata, M., "E/D gate MOS transistor", Electron. & Commun. Jap., 57, 113-21 (1974), EEA78-33149.
403. Masuhara, T., Nagata, M. and Hashimoto, N., "A high-performance N-channel MOS-LSI using depletion-type load elements", Digest of Technical Papers of the 1971 International Solid State Circuits Conference, p. 12-13, EEA74-21587.
404. Mohan Rao, G. R. and Carr, W. N., "Magnetic sensitivity of a MAGFET of uniform channel current density", Solid-State Electron., 14, 995-1001 (1971), EEA75-1172.
405. Moline, R. A. and Reutlinger, G. W., "Self-aligned maskless chan stops for IGFET integrated circuits", IEEE Trans. Electron Devices, ED-20, 1129-32 (1973), EEA77-2603.
406. Muller, R. S., "Resistive-gated i.g.f.e.t. structures", Eurocon 71 Digest, 1 pp., EEA75-4718.
407. NCR Corp., "Nonvolatile memory cells", Patent UK 1370870, Publ. October 1974, EEA78-11671.
408. Paul, R., "Double control of MOS transistors", Nachrichtentechnik, 18, 369-74 (1968), German, EEA72-807.
409. Pelikan, L. and Keller, M., "The m.o.s. resonant transistor", Sdelovaci Tech., 22, 243-5 (1974), Czech, EEA78-658.
410. Powell, N. R. and Irwin, J. M., "Flexible high speed FFT and MOS monolithic chips", 8th Asilomar Conference on Circuits, Systems and Computers, 1974, p. 524-8, Publ. 1975, EEA78-23777.
411. Pruniaux, B. R., North, J. C. and Pager, A. V., "A semi-insulated gate gallium-arsenide field-effect transistor", IEEE Trans. Electron Devices, ED-19, 672-4 (1972), EEA75-16747.
412. RCA, "Insulated gate field effect transistors", Patent UK 1380466, Publ. January 1975, EEA78-28819.
413. Ronen, R. S., "Method of simultaneously making a SIGFET and a MOSFET", Patent USA 3837071, Publ. September 1974.
414. Sanchez-Lassise, J. H., "Transition from electrode controlled to bulk controlled current in MIS devices using silicon nitride as an insulator", Thesis, Univ. Arkansas, Fayetteville, 1971, 136 pp., Order No. 71-27696, EEA75-32700.
415. Scheerer, W., "IGFET with subdivided channel", IBM Tech. Disclosure Bull., 12, 2345-6 (1970), EEA73-35232.
416. Shappir, J., Margalit, S. and Kidron, I., "p-channel MOS transistor in indium antimonide", IEEE Trans. Electron Devices, ED-22, 960-1 (1975), EEA78-41608.
417. Singh, B. R., "A study of the polarization phenomenon in lead silicate glass using the MIGOS structure", J. Phys. D, 7, 443-50 (1974), PA77-20692.

418.Sitaram, R. S. and Townsend, W. G., "A voltage controlled variable resistance MOSFET", Int. J. Electron., 38, 253-7 (1975), EEA78-15358.
419.Soderman, L., "Four phase MOS circuits are most dynamic", Eltek. Aktuell Elektron., 17 (5-6), 48-50 (1974), EEA77-36328.
420.Stand. Microsystems, "Threshold voltage control of MOS device", Patent UK 1381435, Publ. January 1975, EEA78-28875.
421.Syrzycki, M., "The dual-gate MOS transistor", Elektronika, no. 10, 405-10 (1971), Polish, EEA75-4700.
422.Takabashi, S., "Insulated-gate field-effect semiconductor device having a channel stopper region", Patent USA 3825455, Publ. July 1974, CA81-178916.
423.Talamonti, L., "Binary capacitors in MOS circuits", Elektronik, 22 (5), 179-81 (1973), German, EEA76-26142.
424.Talamonti, L., "Binary capacitances in MOS circuits", Electron. & Microelectron. Ind., no. 176, 79-82 (1973), French, EEA77-2602.
425.Terman, L. M., "Aluminum-silicon self-aligned gate 1-device cell with narrow word line patch", IBM Tech. Disclosure Bull., 15, 1163-4 (1972), EEA76-8689.
426.Verwey, J. F. and Kramer, R. P., "Atmos - An electrically reprogrammable read-only memory device", IEEE Trans. Electron Devices, ED-21, 631-6 (1974), EEA77-38841.
427.Werner, R. E., "Bidirectional IGFET with symmetrical linear resistance with specific substrate voltage control", Patent USA 3441748, Publ. April 1969, EEA73-9369.
428.White, M. H., Lampe, D. R., Fagan, J. L., Kub, F. J. and Barth, D. A., "A nonvolatile charge-addressed memory (NOVCAM) cell", IEEE J. Solid-state Circuits, SC-10, 281-7 (1975), EEA78-37616.
429.Yamaguchi, T. and Sato, S., "Investigation of continuously variable threshold voltage devices (CVTD)", Oyo Buturi, 44 (Suppl.), 233-42 (1975), EEA78-41616.
430.Yamamoto, T. and Morimoto, M., "Thin-MIS structure Si negative-resistance diode", Appl. Phys. Lett., 20, 269-70 (1972), EEA75-20200.
431.Yamazaki, S., Hatakeyama, K., Kagawa, I. and Yamashita, Y., "Properties of nonvolatile semiconductor memories by using silicon clusters in the gate insulator", 1973 International Electron Devices Meeting Technical Digest, p. 355-8, EEA77-14555.
432.Yu, K. K., Reitboeck, H. J. and Brody, T. P., "A novel analog-memory cell for application in pattern recognition", Presented at 1973 International Electron Devices Meeting Supplement, Washington, D. C., December 1973, EEA77-18821.

IV. Applications

1. Memory

433.Abbott, R. A., Regitz, W. M. and Karp, J. A., "A 4K MOS dynamic random-access memory", IEEE J. Solid-State Circuits, SC-8, 292-8 (1973), EEA76-35352.

434. Allen, C. A. and Lund, D. F., "Dynamic m.o.s. memory array chip", Patent USA 3685027, Publ. August 1972, EEA76-821.
435. Anon., "First Australian designed 'MOS' micro-integrated circuits", Austral. J. Instrum. Control, 25 (3), 18-20 (1969), EEA71-23826.
436. Anon., "A static 4096-bit RAM with a particularly high packing density", Elektronik, 23 (4), 116 (1974), German, EEA77-23039.
437. Anon., "C.m.o.s. r.a.m. applications", New Electron., 7 (24), 75, 78, 81, 85 (1974), EEA78-8620.
438. Askin, H. O., Boss, C. D. and Lee, J. M., "Gated output word decoder", IBM Tech. Disclosure Bull., 27, 1359-60 (1974), EEA78-8569.
439. Baitinger, U., Frantz, H., Haug, W. and Remshardt, R., "IGFET read-only memory", IBM Tech. Disclosure Bull., 14, 1429-30 (1971), EEA75-7541.
440. Bapat, D. and Mrazek, D., "System considerations in dynamic MOSRAM's", Electron. Equip. News, 13 (10), 30-2, 5 (1972), EEA75-13757.
441. Baranyai, A., "MOS-LSI memory systems", Hiradstech. Ipari. Kut. Intez. Kozl., 12 (2), 33-52 (1972), Hungarian, EEA75-39937.
442. Basiladze, S. G., "Design of integrated write and readout circuits using a type of MOS-transistor", Izv. VUZ Radioelektronika, 13, 381-8 (1970), Russian, EEA73-27259.
443. Basiladze, S. G. and Sonin, M. S., "MOSFET's in integrated digital computer modules", Izv. VUZ Radioelektronika, 11, 747-64 (1968), Russian, EEA72-2825.
444. Beer, A. F., "A MOS transistor store with discretionary wiring", Philips Tech. Rev., 31, 286-95 (1970), EEA74-34117.
445. Beer, A. F., Nicholas, K. H. and Swift, P., "An integrated MOST memory using discretionary wiring", International Conference on Microelectronics, 1969, p. 29-30, EEA72-23841.
446. Bially, T., "Audio frequency spectrum analysis using MOS memory elements", IEEE Trans. Audio Electroacoust., AU-18, 201-3 (1970), EEA73-24923.
447. Bizjak, J. F., "A new MOS random access memory", Aust. Electron. Eng., 5 (12), 10-14 (1972), EEA76-11952.
448. Blair, K. A., "Penetrate the mystique of MOS specs", Electron Des., 20 (8), 64-9 (1972), EEA75-28508.
449. Boll, H. J. and Lynch, W. T., "Design of a high-performance 1024-B switched capacitor p-channel IGFET memory chip", IEEE J. Solid-State Circuits, SC-8, 310-18 (1973), EEA76-35355.
450. Boonstra, L., Lambrechtse, C. W. and Salters, R. H. W., "Design considerations for a 4K 1 MOS/BIT RAM", Onde Electr., 54, 175-80 (1974), French, EEA77-26523.
451. Boysel, L. L. and Carter, G. P., "MOS complex array system design", Electro-Technology, 83 (2), 35-7 (1969), EEA72-12803.
452. Carlstedt, G. L. and Svensson, C. M., "MNOS memory transistors in simple memory arrays", IEEE J. Solid-State Circuits, SC-7, 382-8 (1972), EEA75-36288.

453. Chapple, I. K., "The application of MOS memories to replace core storage in large computers", Microelectron. & Reliab., 13, 415 (1974), EEA78-5368.
454. Cohen, M. S. and Moore, J. S., "Physics of the MOS electron-beam memory", J. Appl. Phys., 45, 5335-48 (1974), PA78-25766.
455. Cricchi, J. R., Lancaster, E. and Strull, G., "A large-scale complementary MOS memory", IEEE Trans. Aerospace Electronic Syst., AES-3, Suppl. 677-86 (1967), EEA71-14549.
456. Dailey, J. R. and Kuntzleman, H. C., "IGFET functional memory cells", IBM Tech. Disclosure Bull., 13, 3122-4 (1971), EEA74-18799.
457. Dingwall, A. G. F. and Stricker, R. E., "High density COS/MOS 1024-bit static random access memory", IEEE J. Solid-State Circuits, SC-10, 197-200 (1975), EEA78-33208.
458. Dubois, R., "M.O.S. and their applications in memories. I", Automatisme, 18, 78-84 (1973), French, EEA76-26085.
459. Dubois, R., "MOS and their applications in memories. II", Automatisme, 18, 165-74 (1973), French, EEA76-26086.
460. Dubois, R., "M.O.S. devices and their applications in memories. III", Automatisme, 18, 206-17 (1973), EEA76-29307.
461. Feryszka, R. and Yung, A. K., "Two transistor dynamic memory cell", Report TN-1007, RCA, Princeton, N. J., 1975, 2 pp., EEA78-41704.
462. Fischle, A. S. and Holmstrom, F. E., "Discrete memory elements are formed by evaporating metal pads onto a silicon dioxide coated silicon wafer. The metal pads surround a central metallized circle", IBM Tech. Disclosure Bull., 15, 741-2 (1972), EEA76-816.
463. Fontenay, R., "A 4096-bit random access memory", Electron. & Microelectron. Ind., no. 168, 32-3 (1973), French, EEA76-23021.
464. Foss, R. C. and Harland, R., "Peripheral circuits for one-transistor cell MOS RAM's", IEEE J. Solid-State Circuits, SC-10 (5), 255-61 (1975), EEA78-37612.
465. Foss, R. C. and Monico, J., "MOS memory cells and their engineering implications", International Conference on Large Scale Integrated Circuits, 1974, p. 233-41, EEA78-11691.
466. Friedberg, H., "Properties and applications of electronically programmable MOS-ROMs", Elektron. Ind., 4 (4), 57-60 (1973), German, EEA76-29310.
467. Friedrich, J. H., "A coincident-select MOS storage array", IEEE J. Solid-State Circuits, SC-3, 280-5 (1968), EEA72-2284.
468. Frohman-Bentchkowsky, D., "Silicon gate MOS technology and its memory applications", WESCON 1970 Technical Papers, 7 pp., EEA74-7791.
469. Gargione, F., Keyzer, W. and Noel, G., "A multilayer hybrid thin-film 1024 x 3 COS/MOS memory", New Electron., 6 (13), 25-9 (1973), EEA76-38320.

470. Ginovker, A. S., Gusev, A. A., Konkov, V. P., Kuryshev, G. L., Mishin, A. I., Sinitsa, S. P. and Tepman, B. G., "Integrated circuit with a programming structure employing MOS storage transistors", Radio Eng. & Electron. Phys., 17, 1225-6 (1972), EEA76-26151.
471. Gladu, R. G. and Olson, L. T., "Multigated 'dribble' powered MOS FET storage cell", IBM Tech. Disclosure Bull., 17, 3336 (1975), EEA78-28300.
472. Goser, K., "Low power memory cells in MOS techniques with special processes", Proceedings of the International Conference on Space and Communications, 1971, 1 pp., EEA74-34089.
473. Goser, K., "Low power memories in modified P/MOS techniques and CMOS technique", Nachrichtentech. Z., (NTZ), 26 (1), 9-15 (1973), German, EEA76-11973.
474. Grochowski, E. G., "Rewritable IGFET read-only store", IBM Tech. Disclosure Bull., 14, 263 (1971), EEA74-34089.
475. Hart, B. L. and Baker, R. W. J., "'Zero' decay analogue store using guarded gate MOSFETs", Radio & Electron. Eng., 42, 114-16 (1972), EEA75-15966.
476. Hartmann, H. L. and Lang, H., "Problems of intermediate storage in asynchronously or synchronously operating multiplex devices for digital signals", Nachrichtentech. Z., 23, 620-4 (1970), German, EEA74-8337.
477. Heissing, H. and Mitterer, R., "A 4096-bit MOS memory device with single-transistor cells", Siemens Forsh. & Entwicklungsber., 4, 197-202 (1975), EEA78-41696.
478. Hodge, W. A. and Flamingam, D. H., "What you need to know about testing 4K MOS memories", Tekscope, 6 (5), 3-7 (1974), EEA78-41699.
479. Hoff, M. E., Jr., "Silicon-gate dynamic MOS crams 1024 bits on a chip", Electronics, 43 (16), 68-73 (1970), EEA73-38093.
480. Hoffman, W. K. and Kalter, H. L., "An 8K B random-access memory chip using the one-device FET cell", IEEE J. Solid-State Circuits, SC-8, 298-305 (1973), EEA76-35353.
481. Horninger, K., "New semiconductor storage elements. Electrically re-programmable storage matrix with MNOS transistors", Elektron. Ind., 4 (5/6), 94-5 (1972), German, EEA75-32681.
482. Horninger, K., "Storage arrays with MNOS transistors", Siemens Forsch. & Entwicklungsber., 4, 213-19 (1975), EEA78-37561.
483. Hsieh, J. C. and Sonoda, G., "MOSFET storage array addressing system", IBM Tech. Disclosure Bull., 13, 2383-4 (1971), EEA74-18797.
484. Ieda, N., Yoshimura, H. and Asaoka, T., "High speed MOS memory", Rev. Electr. Commun. Lab., 21, 473-81 (1973), EEA77-18848.
485. Iyer, R. R., "Protective device for MOS integrated circuits", Proc. IEEE, 56, 1223-4 (1968), EEA71-26066.
486. James, R. P., "Three-device MOSFET charge-storage cell", IBM Tech. Disclosure Bull., 14, 1435 (1971), EEA75-7542.
487. Jordan, W. F., "MOS arrays come on strong", Electronics, 41 (26), 54-6 (1968), EEA72-2279.

488.Karp, J. A., "Dynamic refresh memories", 1971 IEEE International Convention Digest, 1971, p. 36-7, EEA74-22715.
489.Kasperkovitz, D. and van Santen, J. G., "Integration density and power dissipation of MOS and bipolar shift registers: a comparison", Microelectron. & Rel., 9, 497-501 (1970), EEA74-6881.
490.Katholing, G., "MOS read only memory (MOS-ROM) GDR 101", Siemens Electron. Components Bull., 7 (2), 33-6 (1972), EEA75-36311.
491.Kim, M. J., "Metal insulator silicon field effect transistor memory", Patent USA 3853496, Publ. December 1974, CA82-67308.
492.King, J. L., "Method for testing MOS memory store device", Patent USA 3813032, Publ. May 1974.
493.Kobayashi, K. and Haneta, Y., "Memory performance of MOS transistors with SiO_2 films prepared by SiH_4-H_2O system", Jap. J. Appl. Phys., 12, 715-20 (1973), EEA76-26006.
494.Kolev, K., Smirnov, N. D. and Peikov, P. H., "Memory effect in MOS structures", C. R. Acad. Bulg. Sci., 27, 461-3 (1974), EEA78-11534.
495.Koo, T. K., "The non-volatility characteristics of MNOS memory FET and EAROM arrays", Proceedings of the IEEE 1974 National Aerospace and Electronics Conference, p. 37-41, EEA78-8623.
496.Lambrechtse, C. W., Salters, R. H. W. and Boonstra, L., "A-4096 bit one-transistor per-bit RAM with internal timing and low dissipation", Digest of Technical Papers of the 1973 IEEE International Solid State Circuits Conference, p. 26-7, 194, EEA76-23033.
497.Lattin, B., "Testing dynamic RAM's", WESCON 1971 Technical Papers, 3/4, 5 pp., EEA75-1183.
498.Lauffer, D. and Lim, P., "A user's look at MOS-RAMS for main frame memory", 1971 IEEE International Convention Digest, 1971, p. 72-3, EEA74-22717.
499.Laughton, W. J., "Molybdenum gates open the door to faster MOS memories", Electronics, 44 (8), 68-73 (1971), EEA74-15562.
500.Lea, R. M., "Design for a high-speed m.o.s. associative memory", Electron. Lett., 8, 391-3 (1972), EEA75-32707.
501.LeBlanc, A. R., "Field-effect transistor memory circuit", IBM Tech. Disclosure Bull., 15, 1292-3 (1972), EEA76-8691.
502.Levine, L. and Myers, W., "Timing: a crucial factor in LSI-MOS main-memory design", Electronics, 48 (14), 107-11 (1975), EEA78-33197.
503.Linford, J., "MOS memories for digital systems", Electron. Equip. News, 13 (9), 42-5 (1972), EEA75-10265.
504.Linton, R. H. and Sonoda, G. Y., "MOSFET associative cell", IBM Tech. Disclosure Bull., 17, 3340-1 (1975), EEA78-28302.
505.Lockwood, G. C., Naber, C. T. and Koo, T. K., "A 1024 bit MNOS electrically-alterable ROM", Western Electronic Show and Convention (Vol. 16), 1972, 4/5 6 pp., EEA76-8752.
506.McConnell, J. W. and Meadows, R. A., "MOS adaptive memory elements as weights in an adaptive pattern classifier", IEEE J. Solid-State Circuits, SC-1, 94-9 (1966), EEA71-1292.

507. McCracken, T. E., "Development of a solid state refresh memory tonal display", Thesis, Univ. Missouri, Columbia, 1974, 323 pp., EEA78-37657.
508. McKenny, V., "Depletion-mode transistors boost speed of MOS random access memory", Electronics, 44 (4), 80-5 (1971), EEA74-13270.
509. Maine, S. G. T., "Will m.o.s. stimulate a new computer generation?", Electron. Eng., 43 (525), 27-30 (1971), EEA75-385.
510. Marvin, C. E., "MOS/LSI throughout", 1971 IEEE International Convention Digest, p. 74-5, EEA74-22718.
511. Masuoka, F., Ishikawa, M., Sato, T. and Iizuka, H., "An electrically programmable and erasable 2048-bit stacked-gated MOS ROM", Presented at 1973 International Electron Devices Meeting Supplement, Washington, D. C., December, EEA77-18819.
512. Matukura, Y., "MOS memory", J. Inst. Electron. & Commun. Eng. Jap., 55, 544-9 (1972), Japanese, EEA75-39945.
513. Miller, H. S., "COS/MOS random access memories", 1971 IEEE International Convention Digest, p. 34-5, EEA74-22714.
514. Mitterer, R. W., "Dynamic 4-kilobit MOS read-write memories", International Conference on Large Scale Integrated Circuits, 1974, p. 253-63, EEA78-11241.
515. Moschwitzer, A. and Pasch, L., "MNOS storage FET's for integrated semiconductor memories", Z. Elek. Inf. - & Energietech., 1 (1), 37-50 (1971), German, EEA75-10300.
516. Moschwitzer, A., Pasch, L. and Sauer, M., "General principles and new structures of components for storing information by means of MIS field electrodes", Nachrichtentechnik, 22, 362-5 (1972), German, EEA75-39962.
517. Mostyn, R. C., "The use of metal oxide silicon transistors in computer circuits", Marconi Rev., 31 (170), 189-205 (1968), EEA72-806.
518. Mundy, J. L., Burgess, J. F., Joynson, R. E. and Neugebauer, C., "Low-cost associative memory", IEEE J. Solid-State Circuits, SC-7, 364-9 (1972), EEA75-36285.
519. Muoio, A. W., "Character/symbol generation by MOS read only memory", Proc. Soc. Inform. Display, 11 (1), 6-15 (1970), EEA73-25697.
520. Namimoto, K., "Microcomputer and its applications", Toshiba Rev., no. 97, 19-23 (1975), EEA78-41698.
521. Noble, W. P., Jr., "Bulk access-surface storage memory cells", IBM Tech. Disclosure Bull., 16, 3170-2 (1974), EEA77-30335.
522. Noda, A., "Leak-compensated analog memory with pair MOSFET's", IEEE J. Solid-State Circuits, SC-5, 75-7 (1970), EEA73-25038.
523. Oliva, I., "MOS shift registers - application and new technology", Ind. Ital. Elettrotec. & Elettron., 26, 290-4 (1973), Italian, EEA76-34916.
524. Pallum, C., "N-channel MOS registers and stores", Electron. & Microelectron. Ind., no. 162, 52-5 (1972), French, EEA76-4235.

525. Patel, S. and Barry, J. N., "Custom-designed m.o.s. arrays for use in digital systems", Electron. Eng., 44 (536), 64-6 (1972), EEA75-39091.
526. Phelan, R. J., Jr. and Dimmock, J. O., "Imaging and storage with a uniform MOS structure", Appl. Phys. Lett., 11, 359-61 (1967), PA71-13863.
527. Pleshko, P. and Terman, L. M., "An investigation of the potential of MOS transistor memories", IEEE Trans. Electronic Comput., EC-15, 423-7 (1966), EEA70-1332.
528. Podraza, G. V., Nissim, S. and Brewer, D., "A low power MOS monolithic memory array", First Annual IEEE Computer Conference, 1967, p. 87-90, EEA71-22148.
529. Pomeranz, J. N., "Complementary metal-oxide-semiconductor J-K flip-flop and shift cell", IBM Tech. Disclosure Bull., 16, 3613-14 (1974), EEA78-366.
530. Popov, V. P., Soldatenko, V. A., Chervan, B. A. and Antonov, M. A., "Signal distortion in a dynamic MOS-transistor shift register", Izv. VUZ Radioelektron., 17 (12), 31-4 (1974), Russian, EEA78-14915.
531. Priel, U., "Semiconductor arrays get bigger and denser", Electronics, 42 (2), 100-2 (1969), EEA73-9577.
532. Proebsting, R. and Green, R., "A TTL compatible 4096-bit n-channel RAM", Digest of Technical Papers of the 1973 IEEE International Solid State Circuits Conference, p. 28-9, EEA76-23034.
533. Raisanen, W., Hillis, D., Marley, J. and Trolsen, G., "MOS-bipolar LSI memory", Proceedings of the International Conference on Advanced Microelectronics, 1970, 2 pp., EEA74-417.
534. Reed, J. A., "MOS dynamic memory array and refreshing system", Patent USA 3858185, Publ. December 1974, EEA78-15420.
535. Reese Brown, J., Jr., "Pattern sensitivity in MOS memories", Digest of Papers from Semiconductor Memory Test Symposium, 1972, p. 33-46, Publ. 1973, EEA76-26159.
536. Regitz, W. M., "Three-transistor-cell 1024-bit 500-ns MOS RAM", IEEE J. Solid-State Circuits, SC-5, 181-6 (1970), EEA73-38099.
537. Regitz, W. M. and Reed, J. A., "MOS memory decoder circuit", Patent USA 3863230, Publ. January 1975, EEA78-33160.
538. Repchick, D. P., "3-device cell MOSFET memory system", IBM Tech, Disclosure Bull., 14, 285-6 (1971), EEA74-34090.
539. Schein, D., "An electronic long-term store for analogue quantities", Tech. Mitt. Aeg-Telefunken, 60, 379-83 (1970), German, EEA74-10657.
540. Seewann, E., "Dynamic MOSFET shift register array clock driver", IBM Tech. Disclosure Bull., 16, 2767-8 (1974), EEA77-22489.
541. Shenton, G., "A systems approach to the design of MOS memory components", Microelectron. & Reliab., 13, 419-24 (1974), EEA78-5369.
542. Shenton, G., "The 4K r.a.m. - a change of direction?", New Electron., 8 (10), 47, 50 (1975), EEA78-28867.

543. Smith, R. L., "The MOSFET approach to small scale integration of large scale circuits", Sixth Annual Microelectronics Symposium, 1067, Pap. No. C6-1, EEA71-5988.
544. Sonoda, G. Y., "I.g.f.e.t. address powered Schmidt storage cell", IBM Tech. Disclosure Bull., 17, 3337 (1975), EEA78-28301.
545. Spampinato, D. P. and Terman, L. M., "High-speed complementary memory cell", IBM Tech. Disclosure Bull., 14, 1723-5 (1971), EEA75-10292.
546. Stein, K. U., Sihling, A. and Doering, E., "Storage array and sense/refresh circuit for single-transistor memory cells", IEEE J. Solid-State Circuits, SC-7, 336-40 (1972), EEA75-36284.
547. Talamonti, L., "Two-level capacitor boosts MOS memory performance", Electronics, 46 (4), 115-17 (1973), EEA76-12635.
548. Tarui, Y., Hayashi, Y. and Nagai, K., "Electrically reprogrammable nonvolatile semiconductor memory", IEEE J. Solid-State Circuits, SC-7, 369-75 (1972), EEA75-36286.
549. Terman, L. M., "MOSFET memory circuits", Proc. IEEE, 59, 1044-58 (1971), EEA74-22735.
550. Tunzi, B. R., "MOS random-access arrays", Electronics, 42 (2), 102-5 (1969), EEA73-9578.
551. Uchida, Y., Endo, N., Saito, S., Konaka, M., Nojima, I., Nishi, Y. and Tamaru, K., "A 1024-bit MNOS RAM using avalanche-tunnel injection", IEEE J. Solid-State Circuits, SC-10 (5), 288-94 (1975), EEA78-37617.
552. Vanstone, G. F., "Metal-nitride-oxide-silicon-capacitor arrays as electrical and optical stores", Electron. Lett., 8, 13-14 (1972), EEA75-9822.
553. Walther, T. R. and McCoy, M. R., "A three transistor MOS memory cell with internal refresh", Microelectronics, 4 (3), 41-3 (1972), EEA76-26102.
554. White, G., "MOS buffer memories offer economies", Electron, no. 2, 12-14 (April 1972), EEA75-24355.
555. Wotruba, G., "A large-scale integrated memory circuit using single-transistor cells with a density of 1600 bit/mm^2 in N-silicon gate technology", Siemens Forsch. & Entwicklungsber, 4, 207-12 (1975), EEA78-37650.
556. Wuttke, G. H., "RAM's in CMOS-technique", Elektron. Anz., 7 (5), 127-32 (1975), EEA78-33200.
557. Wynhoven, R. M., "MOST stores provide more capacity", Electron. Equip. News, 14 (4), 67 (1972), EEA75-32677.
558. Zdravkovic, S. and Sanja, S., "Memory realisations in MOS techniques", Tehnika, 28, 744-744e (1973), Croatian, EEA76-26105.

2. Logic

559. Altman, L., "Logic, memory, I/O chips build computer systems (CMOS)", Electronics, 48 (10), 85-8 (1975), EEA78-23709.
560. Bajenseco, T. I., "Is I^{2}L going to compete with C-MOS?", Electron. & Microelectron. Ind., no. 200, 49-51 (1975), French, EEA78-24456.

561.Baranyai, A., "Static and dynamic properties of MOS logic systems", Hiradastechnika, 23 (6), 161-6 (1972), Hungarian, EEA75-31718.
562.Bilous, O. and Turnbull, J., "An approach to the design of custom LSI logic modules", International Conference on Large Scale Integrated Circuits, 1974, p. 199-206, EEA78-11755.
563.Birchel, R., "CMOS - a logic family with (almost) ideal properties", Bauelem. Elektrotech., 8 (5), 46-60 (1973), German, EEA76-31860.
564.Birchel, R., "The effects of interference on CMOS circuits compared to bipolar logic families", Bauelem. Elektrotech., 10 (3), 46, 48-50, 52, 54-6 (1975), German, EEA78-28842.
565.Bishop, R. A., "COS/MOS technology simplifies logic circuits", Elektron. Ind., 4 (12), 288-9 (1973), German, EEA77-18281.
566.Borel, J., Lacour, J. and Verdonne, M., "MOS transistors integrated circuits", Onde Elec., 47, 944-9 (1967), French, EEA71-8311.
567.Cagliano, D. and Taddei, R., "MOS-D: a program for the placement of integrated logic gates", Proceedings of the 2nd International Computing Symposium, 1972, p. 335-44, EEA75-31825.
568.Cook, P. W., Critchlow, D. L. and Terman, L. M., "Comparison of MOSFET logic circuits", IEEE J. Solid-State Circuits, SC-8, 348-55 (1973), EEA76-34918.
569.Critchlow, D. L., "The n-channel IGFET for logic and memory", 1968 NEREM Record, p. 142-3, EEA72-17742.
570.Critchlow, D. L., Dennard, R. H. and Schuster, S. E., "Design of large-scale integrated logic circuits using MOS devices", Sixth Annual Microelectronics Symposium, 1967, Pap. C5, EEA71-8317.
571.Crook, C., "Comparing the power of C-MOS with TTL", Electronics, 47 (10), 132-3 (1974), EEA77-22487.
572.Dabrowski, G., "Digital integrated circuits - features and concepts. II", Elektron Int., no. 23-24, 415-16 (1974), German, EEA78-8607.
573.Dansky, A. H., "Multiple threshold IGFET ternary circuits", IBM Tech. Disclosure Bull., 27, 1356-7 (1974), EEA78-8186.
574.Ellul, J. P., Copeland, M. A. and Chan, C. H., "MOS capacitor pull-up circuits for high-speed dynamic logic", IEEE J. Solid-State Circuits, SC-10, 298-306 (1975), EEA78-37619.
575.Elmasry, M. I., "An optimal two-phase M.O.S.-LSI logic system", International Electrical, Electronics Conference and Exposition Digest, 1973, p. 136-7, EEA77-9715.
576.Feller, A. and Lombardi, T., "Design, development, fabrication and delivery of register and multiplexer units", Report NASA-CR-123562, RCA, Van Nuys, Calif., 1971, 51 pp., EEA75-35425.
577.Fleishhammer, W., Schneider, G. and Koppe, G., "T^3L achieves C-MOS noise immunity while retaining TTL speed", Electronics, 47 (5), 94-7 (1974), EEA77-18278.

578.Gordeev, B. K., "Equivalent circuits of series- and parallel-connected MOS transistors", Izv. VUZ Radioelektron., 15, 906-10 (1972), Russian, EEA76-366.
579.Grimes, D. W., "Ternary CMOS logic device", IBM Tech. Disclosure Bull., 17, 1145-6 (1974), EEA78-5070.
580.Halligan, J., "CMOS: a new logic type for control systems", Contr. Eng., 19, 68-70 (1972), EEA75-31701.
581.Hammer, W., "Frequency division logic", Eurocon 71 Digest, 2 pp., EEA75-3892.
582.Harrison, L., "CMOS: The tolerant logic", Electron, no. 27, 13-19, 38 (1973), EEA76-35351.
583.Karstad, K., "CMOS for general-purpose logic design", Comput. Des., 12 (5), 99-106 (1973), EEA76-29308.
584.Khodosh, L. S. and Khotyanov, B. M., "Analysis of logic elements of MOS integrated circuits with depletion-type loads", Izv. VUZ Radioelektron., 18 (1), 11-16 (1974), Russian, EEA78-19902.
585.Kostuch, D. J., "Time delay for MOSFET integrated logic", IBM Tech. Disclosure Bull., 13, 519-20 (1970), EEA74-6943.
586.Krebs, P. E. and Croft, G. F., "MOS integrated circuits using gate capacitance storage", Conference on Integrated Circuits, 1967, p. 137-43, EEA73-484.
587.Leuenberger, F., "Monolithic integrated MOS circuits", Tech. Mitt. PTT, 51 (2), 61-7 (1973), French, EEA76-12677.
588.Lilen, H., "Complementary symmetry metal oxide semiconductor: Complementary MOS integrated circuit logic", Electronique Industr., no. 122, 245-50 (1969), EEA72-23144.
589.Liu, T. K., "Synthesis of logic networks with (metal oxide semiconductors) complex cells", Report PB-210208, Univ. Illinois, Urbana, 1972, 232 pp., EEA76-8132.
590.McCluskey, B. R., "Integrated logic circuits using complementary MOS transistors", International Conference on Microelectronics, 1969, p. 56-7, EEA72-23156.
591.Mouftah, H. T. and Jordan, I. B., "Integrated circuits for ternary logic", Proceedings of the 1974 International Symposium on Multi-Valued Logic, p. 285-302, EEA78-5095.
592.Mouftah, H. T. and Jordan, I. B., "Implementation of 3-valued logic with COS MOS integrated circuits", Electron. Lett., 10, 441-2 (1974), EEA78-2505.
593.Muller, K. D., "Reducing the power consumption of CMOS circuits by logic structural means", 6th International Congress on Microelectronics, 1974, 14 pp., German, EEA78-24524.
594.Naumov, Y. Y., "Pulse noise immunity of MOS logic circuits", Izv. VUZ Radioelektron., 14, 1382-4 (1971), Russian, EEA75-12667.
595.Ohta, K., Morimoto, M., Saitoh, M., Fukuda, T., Morino, A., Shimizu, K., Hayashi, Y. and Tarui, Y., "A high-speed logic LSI using diffusion self-aligned enhancement depletion MOS IC", IEEE J. Solid-State Circuits, SC-10, 314-22 (1974), EEA78-37621.
596.Payne, A. J., "Linear and interface circuits", Microelectron. & Reliab., 13, 373-8 (1974), EEA78-5364.

597. Pound, A. E., "Designing MOS for maximum testability", WESCON 1971 Technical Papers, 3/2, 3 pp., EEA75-436.
598. Reynolds, F. H. and Morton, W. D., "Metal-oxide-semiconductor (MOS) integrated circuits. II. Simple logic circuits", Post Off. Elec. Eng. J., 63, Pt. 2, 105-112 (1970), EEA74-13260.
599. Secaze, G., "C-MOS or LPSTTL - which logic to choose?", Electron. & Microelectron. Ind., no. 200, 61-5 (1975), French, EEA78-24457.
600. Shenton, G., "Problems of optimum partitioning the logic in MOS systems", Contr. & Instrum., 4, 41-3 (1972), EEA75-39089.
601. Soderman, L., "How MOS circuits function. V. Good MOS circuits are dynamic", Eltek. Aktuell Elektron., 77 (4), 46-7 (1974), EEA77-26035.
602. Thomas, T., Rombeek, H., Gilmour, W. and Caughey, M., "Logiflex - faster design for integrated circuits", Telesis, 3 (3), 72-8 (1973), EEA77-14605.
603. Verjans, J. R. and Van Overstraeten, R. J., "NENDEP - a simple n-channel MOS technology for logic circuits", IEEE J. Solid-State Circuits, SC-10, 212-18 (1975), EEA78-33211.
604. Wallace, J., "Designing control logic with c.m.o.s.", New Electron., 7 (16), 20-1, 25 (1974), EEA77-38496.
605. Yen, T. T., "C-MOS flip-flop can do more than logic tasks", Electronics, 48 (6), 123-6 (1975), EEA78-14897.

3. Microprocessors

606. Alaspa, A. A. and Dingwall, A. G. F., "COS/MOS parallel processor array", IEEE J. Solid-State Circuits, SC-5, 221-7 (1970), EEA73-38101.
607. Anon., "The Mostek approach to microprocessors", New Electron., 8 (3), 22-3 (1975), EEA78-15413.
608. Anon., "First I^{2}L processor in four-bit design surpasses N-MOS", Electronics, 48 (2), 29-32 (1975).
609. Chrones, C., "Testing microprocessor chip: A large-scale challenge", Electron. Packag. & Prod., 15 (4), 35-42 (1975).
610. Chung, D., "Four-chip microprocessor family reduces system parts counts", Electronics, 48 (5), 87-93 (1975), EEA78-14896.
611. Dieckmann, H. W., "Computer on a semiconductor crystal. The microprocessor, a programmable LSI circuit", Elektrotech. Z. (ETZ) B, 27, 378-80 (1975), German, EEA78-37568.
612. Fox, W. A. and Reyling, G. F., Jr., "A single chip 16-bit microprocessor for general applications", Microelectron. & Reliab., 14, 389-97 (1975).
613. Hoff, M. E., "Microprocessors", 6th International Congress on Microelectronics, 1974, 6 pp., EEA78-24520.
614. King, E. E. and Nelson, G. P., "Radiation testing of an 8-bit CMOS microprocessor", IEEE Trans. Nucl. Sci., NS-22, 2120-1 (1975).
615. Kirk, B. R., "Microprocessors for everyday applications", Microelectron. & Reliab., 14, 377-8 (1975).

616.Martin, D. P. and Berland, K. S., "IC's interface keyboard to microprocessor", Electronics, 48 (5), 83-5 (1975).
617.Mizuno, M., Suzuki, M., Shimakura, K. and Takemae, K., "An N-channel 16-bit variable architecture microprocessor", 6th International Congress on Microelectronics, 1974, 5 pp., EEA78-24521.
618.Morris, J. H., Patel, H. and Schwartz, M., "Scamp microprocessor aims to replace mechanical logic", Electronics, 48 (19), 81-7 (1975).
619.Peterson, J. J., "Design and development of MOS sealer integrated circuit", Report HDL-TR-050-1, Collins Radio Co., Newport Beach, Calif., 1974, 124 pp., EEA78-23776.
620.Rattner, J., Cornet, J. C. and Hoff, M. E., Jr., "Bipolar LSI computing elements usher in new era of digital design", Electronics, 47 (18), 89-96 (1974).
621.Reyling, G., "Performance and control of multiple microprocessor systems", Comput. Des., 13 (3), 81-6 (1974).
622.Schmid, H., "Monolithic processors", Comput. Des., 13 (10), 87-95 (1974).
623.Shima, M. and Faggin, F., "In switch to n-MOS microprocessor gets a 2-μs cycle time", Electronics, 47 (8), 95-100 (1974), EEA77-22481.
624.Shima, M., Faggin, F. and Mazor, S., "N-channel 8-bit single chip microprocessor", ISSCC, Digest of Technical Papers, 1974, p. 56-7.
625.Tarui, T., Namimoto, K. and Takahashi, Y., "Twelve-bit microprocessor nears minicomputer's performance level", Electronics, 47 (6), 111-16 (1974), EEA77-14585.
626.Uebe, F., "n-channel MOS memories - new possibilities for microprocessor memory design", Microelectron. & Reliab., 13, 387-400 (1974), EEA78-5366.
627.Weissberger, A. J., "MOS/LSI microprocessor selection", Electron. Des., 22 (12), 100-4 (1974), EEA78-15403.
628.Young, L., Bennett, T. and Lavell, J., "N-channel MOS technology yields new generation of microprocessors", Electronics, 47 (8), 88-95 (1974), EEA77-22480.

4. Consumer Electronics

a. Clocks and Watches

629.Anon., "Four-digit clock circuit uses one mos chip", Electron. Engineering, 47 (565), 29 (1975), EEA78-14894.
630.Bernstein, H., "Complementary MOS precision timer (CMOS clock circuit)", Elektron. Ind., 4, 291-2 (1973), German, EEA77-18282.
631.Borden, H., Mingione, J. and Nance, P., "P/MOS chip, drives liquid crystal display, for digital alarm clock", Electronics, 45 (3), 66-9 (1972), EEA75-11032.
632.Della-Mussia, J. P., "A complete clock on a MOS/LSI", Inter Electron., 26 (30), 33-4 (December 1971), French, EEA75-6921.

633.den Otter, C. W., Sr., "A completely electronic stopwatch", Elektronik, 24 (8), 73-6 (1975), German, EEA78-37607.
634.Eaton, S. S., "COS/MOS crystal oscillators advance timekeeping technology", 1973 IEEE International Convention and Exposition, Vol. VI, 51.3/3 pp., EEA77-2193.
635.Forrer, M. P., "Survey of circuitry for wristwatches", Proc. IEEE, 60, 1047-54 (1972), EEA75-35317.
636.Kato, H., "CMOS integrated circuits for electronic watches", Natl. Tech. Rep., 21, 46-53 (1975), Japanese, EEA78-33223.
637.Luce, N. A., "C/MOS digital wristwatch features liquid crystal display", Electronics, 45 (8), 93-7 (1972), EEA75-15913.
638.Morgan, D. K., "A digital auto clock using COS/MOS LSI technology", 1971 IEEE Vehicular Technology 22nd Annual Conference, 5 pp., EEA75-12689.
639.Nezik, D. and Rogner, R., "The TN quartz master clock C with integrated MOS frequency dividers", Nachr. Telefonbau & Normalzeit, no. 75, 51-4 (1974), German, EEA77-36055.
640.Raabe, M., "Integrated CMOS circuit with low power consumption for quartz-crystal wristwatches", Elektronik, 22, 424-6 (1973), German, EEA77-9681.
641.Tourki, P., "Quartz oscillator in complementary M.O.S. technology", Electron. Lett., 9, 451-3 (1973), EEA77-5569.
642.Walker, G. M., "Inside electronic watches: a micropower movement", Electronics, 44 (8), 97-100 (1971), EEA74-15563.

b. Radio and Television

643.Austin, W. M., Dean, J. A., Griswold, D. M. and Hart, O. P., "TV applications of MOS transistors", IEEE Trans. Broadcast Televis. Receivers, BTR-12, 68-76 (1966), EEA70-9099.
644.Evans, W., Moeller, C. and Milbourn, E., "Color TV remote control goes all-electronic", Electronics, 43 (11), 102-11 (1970), EEA73-32340.
645.Hanf, H., "Problems in the development of unipolar (MOS) circuits at the VEB Radio Works, Erfurt", Radio Fernsehen Elektron., 20, 413-15 (1971), German, EEA74-36563.
646.Izumi, Y., "MOS IC for TV automatic synchronizing circuit", Nat. Tech. Rep., 17, 512-18 (1971), Japanese, EEA75-10941.
647.Kleinman, H. M., "Application of dual-gate MOS field-effect transistors in practical radio receivers", IEEE Trans. Broadcast Televis. Receivers, BTR-13, 72-81 (1967), EEA74-22906.
648.Nakamura, J., Nakahara, I., Sato, T. and Sasaki, R., "A development design of VHF color television tuner using dual-gate MOSFET", Nat. Tech. Rep., 15, 253-62 (1969), Japanese, EEA73-5082.
649.Schurmann, J., "Ion implanted UHF MOS FET for TV tuner and receiver front end systems", 6th International Congress on Microelectronics, 1974, 12 pp., EEA78-25846.

650.Weaver, S., "Dual-gate MOSFET's in TV IF amplifiers", IEEE Trans. Broadcast Televis. Receivers, BTR-16 (2), 96-110 (1970),
651.Weaver, S., "TV design considerations using high-gain dual-gate MOSFETs", IEEE Trans. Broadcast & Telev. Receivers, BTR-19, 87-98 (1973), EEA76-27418.

c. Calculators

652.Anon., "MOS IC for electronic calculators. I", Mando Electron., no. 41, 91-102 (1975), Spanish, EEA78-37646.
653.Aubry, M., "Use of microelectronics in scientific pocket calculators", Electron. & Microelectron. Ind., no. 200, 29-31 (1975), EEA78-24455.
654.Furuta, Y., "MOS LSI for desk-top calculator", Natl. Tech. Rep., 21, 36-45 (1975), Japanese, EEA78-28865.
655.Kawagoe, H., Sakamoto, T. and Tsuji, N., "High breakdown voltage MOS LSI for desk-top calculator", Hitachi Hyoron, 56 (9), 25-30 (1974), Japanese, EEA78-19888.
656.Kawasaki, T., "MOS LSI for desk-top calculator", Nat. Tech. Rep., 17, 563-71 (1971), Japanese, EEA75-9587.
657.Kubo, M. and Nagata, M., "A flexible approach to design of desk-top calculator MOS LSI", JAEU, 4 (2), 24-7 (1971), EEA75-6923.
658.Tsuji, N. and Kawagoe, H., "Some problems on one chip MOS LSI for desk-top calculator", International Conference on Large Scale Integrated Circuits, 1974, p. 357-66, EEA78-11760.

d. Others

659.Bishop, A. and Woodruff, A., "A digital speedometer using c.m.o.s. I. Speed-measuring circuits using inductive pickup", Wireless World, 80 (1465), 316-20 (1974), EEA77-39957.
660.Bishop, A. and Woodruff, A., "Digital speedometer using c.m.o.s. II. Average-speed indication", Wireless World, 80, 382-5 (1974), EEA77-39958.
661.Caillaud, F., Durand, G., LeCan, C. and LeFloch, G., "C-MOS ICs for remote control", International Conference on Large Scale Integrated Circuits, 1974, p. 295-30, French, EEA78-11758.
662.Carley, D. R. and Bishop, A., "COS/MOS circuits for consumer applications", Electron, no. 39, 21, 23, 25, 31 (1973), EEA77-18283.
663.Chapman, M. J., "MOS circuits in industrial and domestic control", Electron. & Power, 19, 238-40 (1973), EEA76-31864.
664.Debenham, M. J., "MOS in telecommunications", Microelectron. & Reliab., 13, 417 (1974), EEA78-6125.
665.Frandon, P., "C-MOS IC telephone keyboard encoder", International Conference on Large Scale Integrated Circuits, 1974, p. 329-38, French, EEA78-12789.

666.Gheewala, T. R., Melen, R. D. and White, R. L., "A CMOS implantable multielectrode auditory stimulator for the deaf", IEEE J. Solid-State Circuits, SC-10, 472-80 (1975).
667.Grinsberg, E., "Integrated circuits in consumer electronics", 6th International Congress on Microelectronics, 1974, 8 pp., German, EEA78-24522.
668.Hittinger, W. C., "Metal-oxide-semiconductor technology", Sci. Am., 229 (2), 48-57 (1973), EEA76-38314.
669.Horiuchi, S. and Melen, R. D., "A 24 x 6 interfaced-scan MOS image sensor", IEEE J. Solid-State Circuits, SC-8, 286-9 (1973), EEA76-29596.
670.Hughes, C. J. and Jones, W. G., "LSI in telephone exchange peripheral units", International Conference on Microelectronics, 1969, p. 33-4, EEA72-24336.
671.Kubo, M., Nagata, M., Masuhara, T. and Hori, R., "Some design considerations of electrical characteristics of MOS-LSI", Elec. Eng. Jap., 91 (2), 151-9 (1971), EEA75-23368.
672.Link, F. J., "The use of a complementary MOS sequence timer", Comput. Des., 10 (2), 74-5 (1971), EEA74-17481.
673.Mcintosh, B. C. and Emerson, P. G., "Consumer radio goes LSI", Proceedings of the National Electronics Conference, 1970, p. 547-9, EEA75-2191.
674.Marston, R. M., "20 easy-to-build COSMOS burglar alarms. I", Radio-Electron., 46 (4), 33-5, 88-9 (1975), EEA78-28339.
675.Nishiyama, A., Yoshino, H., Yoshida, T. and Yamaguchi, T., "Learning machine using MOS IC weighting elements", Electron. & Commun. Jap., 57, 112-18 (1974), EEA78-28292.
676.Penney, J. D., "MOS and bipolar ICs in consumer applications", Microelectron. & Reliab., 13, 379-86 (1974), EEA78-5365.
677.Plummer, J. D. and Meindl, J. D., "MOS electronics for a portable reading aid for the blind", IEEE J. Solid-State Circuits, SC-7, 111-19 (1972), EEA75-14859.
678.Pshaenich, A., "Audible gas alarm uses CMOS gates to provide time delays and transient protection", Electron. Des., 22 (10), 132 (1974), EEA78-9965.
679.Valkov, S. A., "Hybrid MOS-integral circuit of a multivibrator for electronic telephone dialing service", Elektro Prom.-st. & Priborostr., 8, 166-8 (1973), Bulgarian, EEA77-7185.
680.van Barneveld, E. J., "Large-scale integration in MOS-technique", Ingenieur, 83 (27), ET94-7 (1971), Dutch, EEA74-29749.
681.Vreeland, R. W. and Yaeger, C. L., "A nine channel 'COS/MOS' EEG telemeter for long-term seizure monitoring", Proceedings of the 26th Annual Conference on Engineering in Medicine and Biology, 1973, p. 402, EEA77-12375.
682.Vreeland, R. W. and Yeager, C. L., "Low level 'COS/MOS' multiplexing for simplified EEG telemetry", Biotelemetry, II, 1974, p. 55-7, EEA78-17446.

5. Electronic Instruments

683. Abbott, E. A., Jr., "Electronic switching circuit employing insulated gate field effect transistors", Patent USA 3428832, Publ. February 1969, EEA73-393.

684. Ando, T. and Ishihara, Y., "A self-scanned image sensor employing MOS structure", Proc. IEEE, 58, 247-8 (1970), EEA73-20785.

685. Ando, T., Ishihara, Y. and Akahoshi, T., "Scan generator using MOST integrated circuits", J. Inst. Telev. Eng. Jap., 25, 550-7 (1971), Japanese, EEA75-5267.

686. Andreev, A. and Burzakov, I., "Differential amplifier stage with MOS-transistors in an integrated embodiment", Elektro Prom.-st & Priborostr., 9, 237-8 (1974), Bulgarian, EEA78-11147.

687. Anon., "Putting MOS circuits to work", Electronics, 45 (14), 20-1 (1972), EEA75-13752.

688. Anon., "For construction of communications equipment: MOS/LSI-circuits are the only suitable", Eltek. Aktuell Elektron., 17 (11), 68-71 (1974), EEA78-9404.

689. Anon., "Pulse generator using CMOS", Electron, no. 52, 47 (1974), EEA77-29693.

690. Anon., "COS/MOS digital ICs", Elektor, 1 (2), 241 (1975), EEA78-28851.

691. Anon., "Data transmission using CMOS", Electron. Equip. News, 24-5 (1975), EEA78-25529.

692. Barker, R. W. J. and Hart, B. L., "Offset-voltage temperature-coefficient in dual MOSFET source followers", Electron. Lett., 7 (5-6), 123-5 (1971), EEA74-14788.

693. Barker, R. W. J. and Hart, B. L., "Gate-circuit protection safeguards m.o.s.f.e.t. amplifier", Electron. Eng., 44 (527), 31-2 (1972), EEA75-7528.

694. Barker, R. W. J. and Hart, B. L., "Sub-ohm voltage-variable-resistance using MOSFET's", Int. J. Electron., 38, 247-51 (1975), EEA78-15010.

695. Basiladze, S. G. and Mochkin, V. S., "Design of dynamic inverters using MOS transistors", Izv. VUZ Radioelektronika, 12 (8), 887-95 (1969), Russian, EEA73-12966.

696. Beiner, E. T., Boyarevich, V. Y. and Grube, A. P., "Analysis of MOS-transistor switching circuits by simulation technique", Latv. PSR Zinat. Akad. Vestis Fiz. Teh. Ser., no. 4, 119-22 (1974), Russian, EEA78-5136.

697. Benz, H. F. and Mattauch, R. J., "A linear voltage-tunable distributed null device", IEEE J. Solid-State Circuits, SC-7, 499-503 (1972), EEA76-417.

698. Bilotti, A., "Operation of a MOS transistor as a variable resistor", Proc. IEEE, 54, 1093-4 (1966), EEA70-1060.

699. Bilotti, A., "A distributed MOS attenuator", Proc. IEEE, 55, 562-3 (1967), EEA70-16950.

700. Bishop, R. A., "COS/MOS drive circuits for DSM (dynamic scattering mode) liquid crystal displays", Proceedings of Electronics Meeting 1972 on Display Systems, p. 111-26, EEA76-16715.

701.Borejko, W., "A low level d.c. chopper with the MOSFET", Prace Inst. Elektrotech., 17 (62), 33-61 (1970), Polish, EEA74-4166.
702.Bratt, B., "Improve CRT-display systems with NMOS", Electron. Des., 22 (1), 100-9 (1974), EEA77-20477.
703.Breedlove, P. S., "ECL/MOS for optimum minicomputer systems", Comput. Des., 11 (8), 61-6 (1972), EEA75-35283.
704.Briggmann, D., "Multi-channel oscillator for the 2m band in CMOS technique. I", Funk-Tech., no. 21, 757-8 (1974), German, EEA78-8124.
705.Brown, D. M., Engeler, W. E., Tiemann, J. J., Lavoo, N. T., Carlson, R. O. and Connery, R. J., "High-frequency MOS digital capacitor", IEEE Trans. Electron Devices, ED-21, 938-44 (1975), EEA78-41520.
706.Buhler, W., "Active elements in the new computer generation", Radio TV Electron., 33, 783-7 (1973), German, EEA77-137.
707.Burns, J. R. and Parrillo, L. C., "An integrated MOS transistor amplifier", Northeast Electronics Research and Engineering Meeting, 1967, p. 54-5, EEA71-29212.
708.Callan, J. D., "A metal oxide silicon digital differential analyzer", IEEE Northeast Electronics Research and Engineering Meeting, 1966, p. 76-7, EEA71-23132.
709.Cernoch, M., "Integrated MOS switch", Automatizace, no. 2, 39-45 (1972), Czech, EEA75-16200.
710.Cheney, G. T. and Thomasco, T. F., "IGFET/BIGFET decoder/driver with memory for LED displays", IEEE J. Solid-State Circuits, SC-7, 425-7 (1972), EEA75-35956.
711.Ciarlo, D. R., "MOSFET detector evaluation (for X-ray dosimetry)", IEEE Trans. Nucl. Sci., NS-21, 390-4 (1974), PA78-46823.
712.Ciarlo, D. R., Kalibjian, R., Mayeda, K. and Bostev, T. A., "Metal-oxide-semiconductor X-ray detectors", IEEE Trans. Nucl. Sci., NS-19, 350-5 (1972), EEA75-17611.
713.Claiborne, L. T. and Staples, E. J., "Surface wave filters in integrated circuits", IEEE Trans. Sonics & Ultrasonics, SU-19, 413 (1972), EEA75-27739.
714.Claiborne, L. T., Staples, E. J. and Harris, J. L., "MOSFET ultrasonic surface-wave detectors for programmable matched filters", Appl. Phys. Lett., 19, 58-60 (1971), PA74-65474.
715.Cobbold, R. S. C., "MOS transistor as a 4-terminal device", Electron. Lett., 2, 189-90 (1966), EEA70-6810.
716.Coers, G., "MOSFET network minimizes audio oscillator distortion", Electronics, 45 (1), 85 (1972), EEA75-6885.
717.Corney, A. C., "Simple, medium-precision high-resistance measuring device", Electron. Eng., 42 (510), 46-7 (1970), EEA73-34432.
718.Cowle, B. S. and Sowden, R. J., "MOS multiple-decade counters", Des. Electron., 8 (5), 56-8 (1971), EEA74-14782.
719.Davies, T. K. and Manning, B. C., "An NMR preamplifier with improved signal-to-noise performance", Rev. Sci. Instrum., 44, 1513-14 (1973), EEA77-20832.

720. Defranould, P. H., "n- and p-channel m.o.s.f.e.t.s as Rayleigh-surface-wave detectors", Electron. Lett., 9, 125-6 (1973), EEA76-21996.
721. Dickson, J. F. and Pile, C., "MOS frequency divider circuits for digital frequency synthesisers", International Conference on Microelectronics, 1969, p. 11-12, EEA72-23065.
722. Downey, J. B., Vadasz, L. and Louwerse, A. L., "MOS array composition with standard functional building blocks", Proceedings of the Microelectronics Symposium, 1968, Pap. B6-1-B6-7, EEA71-28904.
723. Dragotinov, A., "MOS-transistorized universal voltmeter", Radio Fernsehen Elektron., 20, 679-81 (1971), German, EEA75-8223.
724. Drapalik, L. F. and Pletscher, W., "Digital techniques. II", Radio TV Electron., 34, 913-5 (1974), German, EEA78-8574.
725. Droszez, J., "Properties of a field effect MOSFET-type transistor as a variable resistor of voltage control", Rozpr. Electrotech., 18, 725-38 (1972), EEA76-8693.
726. Dunlop, J., "m.o.s.ts store analogue voltages for long periods", Electron. Eng., 43 (525), 34-5 (1971), EEA75-1187.
727. Ellis, G. W., Possin, G. E. and Wilson, R. H., "Diode detection of information stored in electron-beam-addressed MOS structure", Appl. Phys. Lett., 24, 419-21 (1974), PA77-57168.
728. English, P., "Digital applications of MOS integrated circuits", Marconi Rev., 31 (170), 171-88 (1968), EEA72-263.
729. Everist, J. A., "Using MOS in equipment", Des. Electron., 8 (5), 51-2, 5 (1971), EEA71-15560.
730. Ewins, A. J., "Wien-bridge audio oscillator", Wireless World, 77 (1425), 104-7 (1971), EEA74-12737.
731. Fadrhons, J., "A cascade amplifier for h.f. applications of the MOSFET transistors KF 521", Sdelovaci Tech., 19 (2), 44-6 (1971), Czech., EEA74-33191.
732. Fischer, J. E., "MOS capacitors for surface barrier electroreflectance measurements", Rev. Sci. Instrum., 42, 872-7 (1971), EEA74-26887.
733. Flamm, P., "Digital voltmeter for simple construction with MOS-LSI modules", Elektronik, 23 (2), 59-63 (1974), German, EEA77-11877.
734. Foltz, J. W., "Silicon gate CMOS in micropower digital systems", WESCON 1971 Technical Papers, 29/4, 10 pp., EEA75-372.
735. Forbes, L. and Wittmer, L. L., "Experimental verification of operation of indium-doped infrared-sensing MOSFET", IEEE Trans. Electron Devices, ED-22 (12), 1100-101 (1975).
736. Forbes, L. and Yeargan, J. R., "Design for silicon infrared sensing MOSFET", IEEE Trans. Electron Devices, ED-21, 459-62 (1974), EEA77-30673.
737. Foss, N. A. and Ward, S. A., "MOS photodiode", Patent USA 3813541, Publ. May 1974.
738. Frantz, H. and Sonoda, G., "MOSFET substrate self-biasing", IBM Tech. Disclosure Bull., 13, 2385-6 (1971), EEA74-17442.

739. Fry, P. W., "A low cost analogue-to-digital converter based on MOS integrated circuits", International Conference on Microelectronics, 1969, p. 60-1, EEA72-23157.

740. Fry, P. W., "A MOST integrated differential amplifier", IEEE J. Solid-State Circuits, SC-4, 166-8 (1969), EEA73-276.

741. Fry, P. W. and Hoey, S. J., "A silicon MOS magnetic field transducer of high sensitivity", IEEE Trans. Electron Devices, ED-16, 35-9 (1969), EEA72-9598.

742. Fukatko, T., "Calculation of the amplification of an amplifier with a m.o.s.f.e.t.", Sdelovaci Tech., 20, 202-3 (1972), Czech, EEA75-32697.

743. Fullagar, D., "A new c.m.o.s. analogue gate technology", New Electron., 7 (24), 72 (1974), EEA78-8619.

744. Funk, R. E., "Low-power digital frequency synthesis application demonstrates unique COS/MOS performance characteristics", WESCON 1971 Technical Papers, 29/3, 16 pp., EEA75-2077.

745. Funk, R. E. and Bishop, A., "COS/MOS simplifies equipment design", New Electron., 6 (9), 24, 27-8 (1973), EEA76-23026.

746. Gad, H., "Nonlinear gate current of the common-source MISFET oscillator with inductive feedback", Nachrichtentech. Z. (NTZ), 28, 59-62 (1975), EEA78-11179.

747. Gantsev, V. A., Gorbunov, V. A. and Nemchinov, V. M., "Amplifying stage using MOS transistors with a nonlinear load resistance", Telecommun. & Radio Eng. Pt. 2, 28 (10), 123-6 (1973), EEA78-5020.

748. Gasquet, H. L., "The channel 6 color beat what can be done about it?", IEEE Trans. Broadcast & Telev. Receivers, BTR-18, 11-17 (1972), EEA75-14487.

749. Gazin, J. F., "The MOS transistor - an analogue switching device. I", Electron. Components, 16 (14), 11-16 (1974), EEA78-642.

750. Gazin, J. F., "The MOS transistor - an analogue switching device. II. Applications", Electron. Components, 16 (15), 31, 33-4 (1974), EEA77-38575.

751. Gazin, J. F., "MOS switches and multiplexers-analogue applications", Toute Electron., no. 395, 21-5 (1975), EEA78-37550.

752. Gentsch, H. and Pruller, H., "MOS sandwich grid diode for production of high field ionization at the solid gas phase boundary", Z. Naturforsch. A, 26A, 1010-16 (1971), German, EEA74-26722.

753. George, J. B. and Adams, W. S., "MOSFET devices in analog computers", Instrum. Control Syst., 40 (9), 119-22 (1967), EEA71-4684.

754. Gibson, E. D., "High-performance MOS/LSI MODEMS", Proceedings of the IEEE 1972 International Conference on Communications, 29-18/6 pp., EEA76-4228.

755. Gillette, G. C. and Crosby, J., "C-MOS teams with liquid crystals to make a reliable digital VOM", Electronics, 46 (25), 99-102 (1973), EEA77-7639.

756. Goller, L., "Field-effect-transistors as voltage controlled linear resistors", Nachrichtentech. Z., 23, 507-11 (1970), German, EEA74-4279.

757.Gordeev, B. K. and Krylov, B. A., "Asymptotic parameters of cascade switching networks using complementary-symmetry MOS-transistors", Izv. VUZ Radioelektron., 17 (12), 10-19 (1974), Russian, EEA78-15359.
758.Goser, K., "MOS transistors as high-speed switches in a selection matrix", Nachrichtentech. Z., 23, 512-21 (1970), EEA74-4354.
759.Green, R. R., "MOSFET operation at 4.2°K", Rev. Opt., 39, 1495-7 (1968), PA73-4746.
760.Gritsenko, O. T., Levental, I. Y. and Arslanov, R. A., "Direct-coupled electrometer amplifier with a MOS transistor", Instrum. & Exp. Tech., 18, 498-9 (1975).
761.Hage, D. and Givens, S., "Switching high-frequency signals with FET integrated circuits", Microelectronics, 5 (4), 32-44 (1975), EEA78-37645.
762.Hart, B. L. and Barker, R. W., "An electrometer MOST", Electron. Compon., 11, 1322 (1970), EEA74-5660.
763.Hart, B. L. and Barker, R. W. J., "A precision-timed ramp generator with zero idling power consumption", Proc. IEEE, 61, 1047-8 (1973), EEA76-25637.
764.Hart, P. A. H. and Klaassen, F. M., "The MOS transistor as a small-signal amplifier", Philips Tech. Rev., 31, 216-24 (1970), EEA74-33186.
765.Hayashi, Y. and Tarui, Y., "Elementary consideration on MOS transistor integrated linear circuits: on the linearity of gain", J. Inst. Elect. Commun. Engrs. Japan, 49, 773-4 (1966), Japanese, EEA70-6803.
766.Hayashi, Y., Tarui, Y. and Komuro, T., "Design and characteristics of the MOST-IC", Bull. Electrotech. Lab., 31 (4), 7-34 (1967), Japanese, EEA71-8310.
767.Hazman, I., "Four quadrant multiplier using m.o.s.f.e.t. differential amplifiers", Electron. Lett., 8, 63-5 (1972), EEA75-9524.
768.Heng, T. M. S. and Nathanson, H. C., "Vertical m.o.s. transistor geometry for power amplification at gigahertz frequencies", Electron. Lett., 10, 490-2 (1974), EEA78-4957.
769.Hernandez, J., "Multi-phase multiplexing in digital MOS systems", Proceedings of the National Electronics Conference, 1972, p. 281-6, EEA76-5722.
770.Hill, L. K., "Analogue switching circuits. II. More inside inf for designers", Electromech. Des., 16 (7), 28-31 (1972), EEA76-4281.
771.Hoffman, C. R., "Positive voltage translation circuit", IBM Tech. Disclosure Bull., 17, 2392-3 (1975), EEA78-19910.
772.Hohne, M., Koch, G. and Meder, M., "Properties and preliminary technical data of a group of 4-bit binary counters in the MOS high volt technique", Radio Fernsehen Elektron., 23, 721-5 (1974), German, EEA78-11232.

773.Hooper, D. E., "An introduction to analogue techniques in MOST integrated circuits", Proc. Instn. Radio Electronics Engrs. Australia, 29 (5), 151-63 (1968), EEA71-26056.
774.Hu, C. and Muller, R. S., "A resistive-gated IGFET tetrode", IEEE Trans. Electron Devices, ED-18, 418-25 (1971), EEA74-22732.
775.Husty, J. and Snitil, F., "A DC amplifier with interrupted zero setting for automatic bridge balancing", Sdelovaci Tech., 20, 17-19 (1972), Czech, EEA75-15873.
776.Ikeda, H., "An MOS transistor rf oscillator suitable for MOS-IC", Proc. IEEE, 56, 1638-9 (1968), EEA71-29219.
777.Ikeda, H., "A variable capacitance circuit with MOS transistors", J. Inst. Televis. Engrs. Japan, 23 (1), 35-43 (1969), Japanese, EEA73-12967.
778.Illin, V. N., "Some problems of information transmission using induced-channel MOS transistor switches", Izv. VUZ Radioelektronika, 13, 636-8 (1970), Russian, EEA73-35587.
779.Jacob, L., "Monolithic optoelectronic arrays", Electron., no. 60, 36-7 (1974), EEA78-5354.
780.Johns, R. H., "FET electrometer", Phys. Teach., 12 (2), 113-14 (1974), PA77-26441.
781.Josephy, R. D., "MOS transistors for power amplification in the h.f. band", Philips Tech. Rev., 31, 251-8 (1970), EEA74-33188.
782.Joynson, R. E., Mundy, J. L., Burgess, J. F. and Neugebauer, C., "Eliminating threshold losses in MOS circuits by bootstrapping using varactor coupling", IEEE J. Solid-State Circuits, SC-7, 217-24 (1972), EEA75-19137.
783.Kalibjian, R., Ciarlo, D., Mayeda, K. and Boster, T., "Time-resolved X-ray detection using MOS-C detectors", IEEE Trans. Nucl. Sci., NS-19, 339-45 (1972), EEA75-25592.
784.Kassabov, J., "Long channel MOS transistor (LOMOST) as a current stabilizer", Solid-State Electron., 13, 161-4 (1970), EEA73-14328.
785.Kasabov, I. D. and Georgiev, V. K., "Microresistor MOS integrated circuits", Elektro Prom.-st & Priborostr., 9 (4), 43-7 (1974), Bulgarian, EEA77-38879.
786.Kerins, J. H., "Low power circuit design using p-channel MOS", 1971 IEEE International Convention Digest, p. 186-7, EEA74-22751.
787.King, N. A. B., "The metal oxide silicon transistor and its advantages in micro-circuit technology", Marconi Rev., 31 (170), 152-70 (1968), EEA72-805.
788.Klein, R., "Applications of the U108 D integrated MOS circuit", Radio Fernsehen Elektron., 24 (2), 44-8 (1975), German, EEA78-15416.
789.Kondratskii, B. A. and Sonin, M. S., "Integrated amplifying stage using MOS transistors", Izv. VUZ Radioelektron., 13, 1261-2 (1970), Russian, EEA74-9695.
790.Koomen, J. and van den Akker, J., "A MOST inverter with improved switching speed", IEEE J. Solid-State Circuits, SC-7, 231-7 (1972), EEA75-19138.

791.Kou, C. K., Sikorski, M. E. and Scheibner, E. J., "MOS transducers with digital measurement capability", Proc. IEEE, 56, 1599-600 (1968), EEA71-28909.
792.Kozlov, Y. G., Ozhogin, M. A. and Filatov, A. S., "MOS-transistor integrated-circuit modulators", Izv. VUZ Radioelektron., 16 (4), 87-9 (1973), Russian, EEA76-45604.
793.Kubelik, I. and Triska, A., "Application of MOS transistor for the compensation of a bridge in measurement of the magneticoresistance of amorphous semiconductors", Cesk. Cas. Fis. A, 24 (1), 58-60 (1974), Czech, EEA77-31912.
794.Kuhne, H., "Circuit application examples with the double MOS field effect transistor SMY 51", Radio Fernsehen Elektron., 22, 572-4 (1973), German, EEA77-2188.
795.Lake, D. W. and Foy, W. R., "MOS revolutionizes modem design", Electron. Engineer, 31 (10), DC5-6, 8, 9, 10 (1972), EEA76-4209.
796.Laudon, L. and Szilagyi, A., "Continuous industrial pH-transmitter with MOS-transistors", Meres Es Automat., 17, 310-13 (1969), Hungarian, EEA73-7473.
797.Lazarus, M. J., Gibson, M. P. G. and Ryall, M., "Operation of dual gate MOS field effect transistors at 77°K and use in rf and video amplifiers", J. Phys. E, 4, 58-60 (1971), EEA74-10652.
798.Liesenfeld, W., "Remote control and data processing with C-MOS", Elektrotechnik, no. 11, 14-15 (1975), German, EEA78-33148.
799.Ma, S. P. F. and Sklar, R. E., "MOSFET momentary switch circuit", Patent USA 3593036, Publ. July 1971, EEA74-36662.
800.McCann, M. R., "Interfacing with MOS circuits", Microelectronics, 4, 2-7 (1972), EEA75-35314.
801.McCreary, J. L. and Gray, P. R., "All-MOS charge redistribution analog-to-digital conversion techniques - Part I", IEEE J. Solid-State Circuits, SC-10, 371-8 (1975).
802.McDonald, A., "MOS in telecommunications", 6th International Congress on Microelectronics, 1974, 8 pp., EEA78-25557.
803.McGrenera, D., "MOS subsystems prise open the v.d.n. market", Electron. Engineering, 45, 44-5 (1973), EEA76-14132.
804.McKeon, E. F. and Carlson, F. M., "Small-signal r.f. amplification of MOS devices", Proceedings of the National Electronics Conference, 1966, p. 88-93, EEA70-10885.
805.McKinley, W. R., "MOS complex arrays-today and tomorrow", IEEE International Convention Record, 1967, Pt. 6, Pap. No. 50-4, EEA71-3452.
806.McVey, E. S. and Fiskeaux, C. D., "A sample-hold circuit using MOSFETs", IEEE Trans. Ind. Electron. & Control Instrum., IECI-21, 262-3 (1974), EEA77-38579.
807.Maine, S. G. T. and Butler, D. S., "Custom design of m.o.s. integrated circuits", New Electron., 6 (18), 39-40, 43, 45-6, 48-9 (1973), EEA77-6075.
808.Marston, R. M., "40 projects using COS/MOS digital ICs", Radio-Electron., 45 (9), 58-60 (1974), EEA78-8199.

809.Marston, R. M., "Using COSMOS digital ICs. II", Radio-Electron., 45 (10), 51-3 (1974), EEA78-8200.
810.Martinot, H., Rossel, P. and Salviac, M., "MOS transistor inverters", Onde Elec., 50, 790-803 (1970), French, EEA74-4079.
811.Matsushita Electric Industrial Co., Ltd., "MOS control device", Patent UK 1121482, Publ. July 1968, EEA72-5045.
812.Miyoshi, D. S. and Cotts, R. M., "Helium cooled radio frequency preamplifier for use in NMR", Rev. Sci. Instrum., 39, 1881-5 (1968), PA72-6439.
813.Montier, M., Gharadjenaghi, F. and Robert, J., "Monolithic display device using nematic liquid crystal and MOS integrated circuit", Onde Electr., 54, 501-4 (1974), French, EEA78-12045.
814.Morita, Y., Takahashi, H., Matayoshi, H. and Fukuta, M., "Si UHF MOS high-power FET", IEEE Trans. Electron Devices, ED-21, 733-4 (1974), EEA78-645.
815.Muller, J., "An upper sideband upconverter using MIS varactors", IEEE J. Solid-State Circuits, SC-7, 43-50 (1972), EEA75-10526.
816.Mrazek, D., "Low voltage offers new jobs for MOS commutators", Electronics, 43 (12), 82-7 (1970), EEA73-30551.
817.Nabettani, H., Ueda, A., Uota, K. and Ishii, M., "MOS-FET timing circuit", Patent USA 3862441, Publ. January 1975, EEA78-28350.
818.Negro, V. C., Cassidy, M. E. and Graveson, R. T., "A guarded insulated gate field effect electrometer", IEEE Trans. Nuclear Sci., NS-14, 135-42 (1967), PA71-5336.
819.Negro, V. C., Watnick, S. and Raft, P., "A temperature-compensated electrometer for environmental measurements", IEEE Trans. Nucl. Sci., NS-21, 805-11 (1974), PA77-51108.
820.Nienhuis, R. J., "A MOS tetrode for the u.h.f. band with a channel 1.5 micrometers long", Philips Tech. Rev., 31, 259-65 (1970), EEA74-34097.
821.Norman, R. H., "Complex MOS circuits applications", IEEE Northeast Electronics Research and Engineering Meeting, 1965, p. 172-3, EEA70-9115.
822.Novik, V. K., Nikonov, A. S., Sopov, O. V., Levina, I. A. and Gavrilova, N. D., "Operation of pyroelectrics using field effect transistors", Radiotekhnika I Elektronika, 15, 642-4 (1970), Russian, EEA73-25042.
823.O'Brien, T. E., "Monolithic level shifter lets MOS, TTL share same network", Electronics, 44 (14), 70-2 (1971), EEA74-29790.
824.Oehmichen, J. P., "MOS transistors utilization", Onde Elec., 47, 959-62 (1967), French, EEA71-8275.
825.Ohmori, Y., Sunazawa, M. and Fujimura, Y., "MOS IC crosspoint switch for space division digital switching networks", IEEE J. Solid-State Circuits, SC-9, 142-7 (1974), EEA77-24031.
826.Ohwada, A., Yamamoto, S., Funjinuma, K. and Narita, T., "Single and dual gate FET's for ultrahigh frequency applications", Proceedings of the 1st Conference on Solid State Devices, 1969, p. 117-24, Publ. 1970, CA82-67174.

827.Ozhogin, M. A., Sonin, M. S. and Yavorovskaya, R. Y., "Differential MOS transistor integrated circuit without resistors", Izv. VUZ Radioelektron., 14, 831-4 (1971), Russian, EEA74-37307.
828.Pacak, M., "MOSFET transistor circuits", Slaboproudy Obzor, 30, 512-16 (1969), Czech, EEA73-12978.
829.Pease, R. F. W., Henderson, R. C. and Dalton, J. V., "IGFET inverter circuits made with electron lithography", J. Vac. Sci. & Technol., 10, 1078-81 (1973), EEA77-23146.
830.Perkins, C. R., "Complex MOS circuit arrays", Proceedings of the 1970 Electronic Components Conference, 1970, p. 475-9, EEA74-4937.
831.Pershenkov, V. S. and Tkachev, B. V., "MOS thyristors", Izv. VUZ Radioelektron., 15, 847-50 (1972), Russian, EEA76-815.
832.Petrov, I. N. and Andreev, D. G., "Digital measurement equipment and MOS-integrated circuits. Problems and prospects", Elektro Prom.-st. & Priborostr., 9, 224-6 (1974), Bulgarian, EEA78-13069.
833.Phelan, R. J., Jr. and Dimmock, J. O., "InSb MOS infrared detector", Appl. Phys. Lett., 10, 55-8 (1967), EEA70-22246.
834.Plessey Co., Ltd., "Circuit arrangements embodying metal oxide semiconductor transistors", Patent UK 1290011, Publ. September 1972.
835.Pollard, J. and Rietsch, J. C., "Transmodulation characteristics of a high-frequency MOS tetrode. II", Toute Electron., no. 353, 21-7 (1971), French, EEA74-22742.
836.Porter, D., "Integrated arrays offer many advantages in o.c.r. systems", Electron. Eng., 44 (529), 35-7 (1972), EEA75-13275.
837.Poujois, R., Baylac, B., Barbier, D. and Ittel, J. M., "Low-level MOS transistor amplifier using storage techniques", Digest of Technical Papers of the 1973 IEEE International Solid State Circuits Conference, p. 152-3, 216-17, EEA75-22985.
838.Preisig, J. O., "Constant input impedance attenuator using Mosfet's", Report TN-967, RCA, Princeton, N. J., 1974, 6 pp., EEA78-2546.
839.Radio Corp. of America, "MOS tetrode", Patent UK 1123037, Publ. August 1968, EEA72-5044.
840.Rao, G. R. M., "MOSFET magnetic sensors and application to integrated circuits", Thesis, Southern Methodist Univ., Dallas, Texas, 1971, 121 pp., Order No. 72-6364, EEA75-31833.
841.Riksaasen, T., "Integrated electronic circuits", Telektronikk, no. 3, 87-94 (1970), Norwegian, EEA74-10975.
842.Rossler, F., "New MOS block elements of radio VEB combine in Erfurt", Radio Fernsehen Elektron., 20, 553-5 (1971), German, EEA75-3948.
843.Rossler, F., "Electrical design of LSI-MOS trigger circuits", Radio Fernsehen Elektron., 23, 377-80 (1974), German, EEA77-38497.
844.Sakai, M., Tanaka, H. and Yoshimura, K., "Monolithic integrated-circuit driver for numerical display tubes", IEEE J. Solid-State Circuits, SC-4, 263-7 (1969), EEA73-4091.

845. Sanchez-Cordoves, J., "MOS transistors in information theory", Metal. & Electr., 37 (426), 172-4 (1973), Spanish, EEA76-25668.
846. Santilli, R. A., "Gain control biasing circuits for field-effect transistors", Patent USA 3443240, Publ. May 1969, EEA73-9422.
847. Saraswat, K. C., Meindl, J. D. and Berger, J., "A high voltage MOS switch", IEEE J. Solid-State Circuits, SC-10, 136-42 (1975), EEA78-19823.
848. Schar, F., "Distributed mosfet amplifier using microstrip", Int. J. Electron., 34, 721-30 (1973), EEA76-25576.
849. Schumann, H., "A multivibrator for long pulse times with integrated field effect components", Elektroniker, 10, 233 (1971), German, EEA75-6889.
850. Secaze, G., "The phase-locked loop in MOS and its applications", Electron. & Microelectron. Ind., no. 180, 71-9 (1973), French, EEA77-18265.
851. Seeger, B., "Some applications of MOS-field effect transistors", Elektron Int., no. 5-7, 78-9 (1974), German, EEA77-23042.
852. Seely, J. L., "MOS TTL interfaces: problems and solutions", Western Electronic Show and Convention, WESCON, 1969, 4 pp., EEA73-28223.
853. Seyfried, P., "Possibilities of the application of MOS stores in electronic instruments", Uerh. Deut. Phys. Ges., no. 10, 800 (1971), German, EEA74-59078.
854. Shenton, G. D., "Impact of MOS LSI technology on instrumentation", Contr. & Instrum., 4 (6), 40-1 (1972), EEA75-27798.
855. Smart, J. M., "Processing analogue signals - the role of m.o.s.", Electron, no. 36, 53-5 (1973), EEA77-9755.
856. Snow, E. H., "Variable threshold insulated gate field effect device", Patent USA 3428875, Publ. February 1969, EEA72-23842.
857. Sobajic, M. V., Veselinovic, D. I. and Danilovic, D. Z., "i.g.f.e.t.'s improve the linearity of bootstrap sweep generators", Electron. Engineering, 45 (543), 63-5 (1973), EEA76-22992.
858. Soderholm, L. G., "Basic MOS chip serves timer-counter", Des. News, 27 (12), 42-3 (1972), EEA75-39090.
859. Stepanenko, I. P. and Zaitsev, B. D., "MOS-transistor as an amplifying component", Izv. VUZ Radioelektronika, 12, 388-407 (1969), Russian, EEA72-19248.
860. Storm, H. F., "Insulated gate field effect transistor used as a voltage-controlled linear resistor", Patent USA 3577019, Publ. May 1971, EEA74-22757.
861. Strachan, A. J. and Townsend, W. G., "A proposed voltage variable resistor MOST", Proc. IEEE, 56, 2096-7 (1968), EEA72-2830.
862. Suarez, R. E., Gray, P. R. and Hodges, D. A., "All-MOS charge redistribution analog-to-digital conversion techniques - Part II", IEEE J. Solid-State Circuits, SC-10, 379-87 (1975).
863. Svec, A., "Evaluation of the working point adjustment of a transistorized amplifier stage with a MOSFET", Sdelovaci Tech., 19 (3), 90 (1971), Slovak, EEA74-33196.

864.Tanigawa, H. and Ando, T., "Influence of the charge pumping effect on an integrated solid-state image sensor using MOST switches", Proc. IEEE, 61, 491-2 (1973), EEA76-20255.
865.Taylor, R. S., "Broadband doubler using MOSFET's", Proc. IEEE, 54, 1584-5 (1966), EEA70-6944.
866.Thiels, R. and Crawford, B., "Transient response of practical MOS single-polarity integrated circuits", Southwestern IEEE Record of Technical Program, 1967, Pap. 22-6-1-22-6-8, EEA71-5999.
867.Townsend, W. G. and Strachan, A. J., "A voltage variable resistor MOST", Microelectron. & Rel., 9, 491-6 (1970), EEA74-7781.
868.Vavilov, V. S., Plotnikov, A. F. and Shubin, V. E., "Possibility of using InSb-based MOS structures as image converters", Fiz. Tekh. Poluprovodnikov, 4, 598-600 (1970), Russian, PA73-51088.
869.Vendelin, G. D., Sigg, H. J., Cauge, T. P. and Kocsis, J., "m.o.s. transistors for microwave receivers", New Electron., 6 (3), 30-1 (1973), EEA76-17621.
870.Vichev, B., "Analog multipliers with m.o.s. transistors", Izv. Inst. Elektron., 6, 143-8 (1972), Russian, EEA76-4288.
871.Volkov, S. A., "Analysis of an MOS-transistor symmetric multivibrator", Izv. VUZ Radioelektron., 17 (7), 78-82 (1974), Russian, EEA78-8127.
872.Wagner, W., "Examples of application of integrated MOS circuits, type U 101D to U 105D", Radio Fernsehen Elektron., 23, 452-4, 459-61 (1974), German, EEA78-2919.
873.Whelan, M. V., "Resistive MOS-gated diode light scanner", 1972 European Solid State Devices Research Conference, p. 208, Publ. 1973, EEA76-32472.
874.Whelan, M. V. and Daverveld, L. A., "New self-scanned structures: resistive-MOS-gated diode arrays", 1973 International Electron Devices Meeting Technical Digest, p. 416-18, EEA77-14886.
875.White, S. A., "MOS/LSI brings complex circuits within reach of any designer", Electronics, 44 (14), 62-6 (1971), EEA74-25375.
876.White, S. A., "An 'adaptive' MOSFET oscillator", 6th Asilomar Conference on Circuits and Systems, 1972, p. 415-18, Publ. 1973, EEA76-23011.
877.Wilenken, R., "Beware of CMOS-switch failure modes", Electron. Des., 23 (6), 68-71 (1975), EEA78-28789.
878.Wolfram, J., "Some applications of MOS - circuits U 105D and U 107D", Radio Fernsehen Elektron., 22, 369-71 (1973), German, EEA77-5661.
879.Wolleson, D., "Analogue-signal commutation using m.o.s.i.c.s.", Electron, no. 2, 15-16 (April 1972), EEA75-24356.
880.Yagi, H., Inoue, M., Naito, T. and Tatsukawa, T., "NMR marginal oscillator with MOS-FET operating at low temperature", Jap. J. Appl. Phys., 12, 1794-8 (1973), PA77-5130.
881.Yoshimura, K., Murase, K., Kobayashi, T. and Aoyama, H., "An ultrastable MOS input operational amplifier", NEC Res. & Dev., no. 34, 72-9 (1974), EEA78-11156.

882.Zandra, W., "COS MOS multivibrators and their applications", Radio Elektron. Schau, 50, 584-6 (1974), German, EEA78-8242.
883.Zarkiewicz, R., "MOSFET - transistors in h.f. circuits", Elektronika, 15 (4), 178-82 (1974), Polish, EEA77-36021.
884.Zemel, J. N., "Ion sensitive field effect transistors and related devices", Anal. Chem., 47 (2), 255A, 258-9, 264-5, 268 (1975), EEA78-28786.
885.Zolotarevskii, V. I., Zaika, V. V., Gerasimenko, S. I., Fabrikov, S. P. and Sakhno, A. I., "Transmission errors of multichannel integrated MOS switches", Izv. VUZ Radioelektron., 14, 1376-9 (1971), Russian, EEA75-13763.

V. Reliability and Cost

886.Anlock, E. S., Prosser, J. F. and Remis, B. R., "Reliability study of an n-channel silicon gate FET with field shield", Proceedings of the 12th Annual Conference on Reliability Physics, 1974, p. 70-3.
887.Banghart, L. E., "System cost savings with MOS integrated circuitry", Proceedings of the National Aerospace Electronics Conference, 1965, p. 292-8, EEA70-14813.
888.Baranyai, A., Kerek, L. and Barsony, I., "Experimental method for non-destructive analysis of MOC IC's", Hiradstech. Ipari Kut. Intez. Kozl., 14, 16-37 (1974), Hungarian, EEA77-38880.
889.Beattie, C. G., Adams, J. D., Carrell, S. L., George, T. D. and Valek, M. H., "Elements of semiconductor device reliability", Proc. IEEE, 62, 149-68 (1974), EEA77-9969.
890.Behera, S. K. and Speer, D. P., "A procedure for the evaluation and failure analysis of m.o.s. memory circuits using the scanning electron microscope in potential contrast mode", Proceedings of the 10th Annual Conference on Reliability Physics 1972, p. 5-11, EEA76-4250.
891.Boysel, L., "Cutting systems costs with MOS", Electronics, 42 (2), 105-7 (1969), EEA73-9579.
892.Brewer, R., "The practical application of IEC recommendations on electronics reliability to the field of semiconductor devices", 3rd Symposium on the Reliability of Electronics, 1973, p. 753-67, EEA77-5798.
893.Cave, K. J. S., "MOST's: Grounds for confidence", Compon. Technol., 3 (2), 18-20 (1968), EEA72-798.
894.Cheney, G. T., Freyman, R. L. and Mannele, A. A., "Reliability of $Al_2O_3SiO_2$ IGFET integrated circuits", Proceedings of the 9th Annual Conference on Reliability Physics, 1971, p. 62-6, EEA75-3982.
895.Child, J. W. M. and de Voil, G. F., "Destruction of m.o.s. device by static", New Electron., 7 (5), 16-17 (1974), EEA77-26483.
896.Colbourne, E. D., Coverley, G. P. and Behera, S. K., "Reliability of MOS LSI circuits", Proc. IEEE, 62, 244-59 (1974), EEA77-9705.

897. DeStafeno, J. J., Lukianoff, G. V. and Touw, T. R., "Testing semiconductor devices by voltage polarity coding", IBM Tech. Disclosure Bull., 17, 2301-2 (1975), EEA78-24465.
898. DiStefano, T. H. and Tu, K. N., "The physics of interface interactions related to reliability of future electronics devices", Report AFCRL-TR-0263, IBM Thomas J. Watson Res. Center, Yorktown Heights, N. Y., 1974, 94 pp., EEA78-24290.
899. Drew, L. C. and Sheahan, J. E., "MOS/LSI procurement: a joint business venture", Western Electronic Show and Convention, WESCON, 1969, 4 pp., EEA73-23884.
900. Durrant, N. F., "MOSFET", Research, 2 (2), 18-19 (1969), EEA72-23832.
901. Feeleston, W. and Pepper, M., "Modes of failure of MOS devices (in integrated circuits)", Microelectron. & Rel., 10, 325-38 (1971), EEA75-4708.
902. Fitzgerald, D. J., Parker, G. H. and Spiegel, P., "Reliability studies of MOS Si-gate arrays", Proceedings of the 9th Annual Reliability Physics Conference, 1971, p. 57-61, EEA75-4724.
903. Forsythe, D. D., "Surface-charge induced failures observed on MOS integrated circuits", Microelectron. & Rel., 8, 339-47 (1969), EEA73-14319.
904. Gajda, J. J., "Techniques in failure analysis of MOS devices", Proceedings of the 12th Annual Conference on Reliability Physics, 1974, p. 30-37.
905. Gear, G., "MOS/LSI failure analysis techniques", Proceedings of the 11th Annual Reliability Physics Conference 1973, p. 127-32, EEA77-6086.
906. Gibbeson, J. R., "Specification D4000: the British Post Office approach to long life and high reliability in MOS LSI", International Conference on Large Scale Integrated Circuits, 1974, p. 167-76, EEA78-11752.
907. Hamiter, L. C., Jr., "Integrated electronics - How reliable are MOS IC's? As good as bipolars, says NASA", Electronics, 42 (13), 106-10 (1969), EEA72-19373.
908. Hamvas, V. T., "Ensure error-free MOS IC operation in noisy environments", Electron. Des., 19, 58-60 (1971), EEA75-3976.
909. Heinze, W. and Ludwig, J., "Reliability in the production of semiconductor devices and how to get it", 3rd Symposium on the Reliability of Electronics, 1973, p. 101-20, EEA77-5796.
910. Kalmar, G. and Pataki, A., "The importance of the X-ray inspection method in the production of reliable semiconductor devices", 3rd Symposium on the Reliability of Electronics, 1973, p. 171-9, EEA77-5598.
911. Katholing, G. K., "MOS technology in large units", Polytech. Tijdschr. Elektrotech. Elektron, 30 (4), 124-30 (1975), Dutch, EEA78-19921.

912.Kern, W., Vossen, J. L. and Schnable, G. L., "Improved reliability of electron devices through optimized coverage of surface topography", Proceedings of the 11th Annual Reliability Physics Conference, 1973, p. 214-23, EEA77-5792.
913.Koschel, H., "Failure analysis, one of the key positions in quality and reliability assurance of semiconductor devices", 3rd Symposium on the Reliability of Electronics, 1973, p. 263-73, EEA77-5797.
914.Kostychev, G. I., "Physical aspects of the reliability of integrated microcircuits employing MIS transistors", Telecommun. & Radio Eng. Pt. 2, 29, 104-8 (1974), EEA78-37656.
915.Kroeger, J. and Threewitt, B., "Heed the limitations of MOS I/O circuitry", Electron Des., 22 (10), 82-8 (1974), EEA78-8172.
916.Lane, C. H., "Electrical overstress failures in silicon devices", Report AD747795, Rome Air Dev. Center, Griffiss AFB, N. Y., 1972, 46 pp., EEA76-23153.
917.Longley, M. R., "Practical considerations and recommendations for semiconductor burn-in", Eval. Eng., 12 (4), 4-5 (1973), EEA77-5773.
918.Luce, R. L., "MOS integrated circuit technologies", 1971 IEEE International Convention Digest, p. 536-7, EEA74-25579.
919.Lucey, R. J. and Firman, A. H., "Process monitoring in transistor production", Solid State Technol., 18, 57-60 (1975), EEA78-24329.
920.McKenna, R., "An MOS quality and reliability program", Microelectron. & Rel., 10, 67-74 (1971), EEA74-18806.
921.Morozowich, M., "Failure analysis of a semiconductor", Solid State Technol., 18 (4), 69 (1975), EEA78-24089.
922.Nichols, E. D., "MOS/LSI reliability prediction", Proceedings of the 1972 Reliability and Maintainability Symposium, p. 474-8, EEA75-15993.
923.Noyce, R. N., "Probable evolution of technique and prices, of semiconductor memories", Onde Elec., 50, 757-62 (1970), French, EEA74-4916.
924.Pinon, M., "Qualification and supervision of custom design MOS LSI production line", International Conference on Large Scale Integrated Circuits, 1974, p. 159-66, French, EEA78-11751.
925.Porter, W. A. and McKee, W. R., "Influence of material quality and process-induced defects on semiconductor device performance and yield", Report NASA-CR-120557, Texas A & M Univ., College Station, 1974, 54 pp., EEA78-37347.
926.Raisanen, W., Hillis, D., Marley, J. and Trolsen, G., "Large scale integrated circuit bipolar memories (MOS)", Onde Elec., 50 (9), 763-7 (1970), French, EEA74-4917.
927.Reynolds, F. H., Parrott, R. W. and Braithwaite, D., "Use of tests at elevated temperatures to accelerate the life of an MOS integrated circuit", Proc. Inst. Elec. Eng., 118, 475-85 (1971), EEA74-18808.

928. Ross, E. C., Goodman, A. M. and Duffy, M. T., "The effects of SiO_2 thickness variation on the operation of direct-tunneling mode m.n.o.s. memory transistors", Annual Reliability Physics Symposium, 1970, p. 6, EEA73-28225.
929. Schick, J. D., "Electrical contact through an SiO_2 overlay by ultrasonic probe", IBM Tech. Disclosure Bull., 17, 3288 (1975), EEA78-24475.
930. Schlegel, E. S., Keen, R. S. and Schnable, G. L., "The effects of insulator surface-ion migration on MOS and bipolar integrated circuits", The Annual Reliability Physics Symposium, 1970, p. 4-5, EEA73-25055.
931. Schnable, G. L., Ewald, H. J. and Schlegel, E. S., "MOS integrated circuit reliability", IEEE Trans. Rel., R-21, 12-19 (1972), EEA75-9569.
932. Schroen, W., "Failure analysis of surface inversion", Proceedings of the 11th Annual Reliability Physics Conference 1973, p. 117-23, EEA77-5786.
933. Schroen, W. H., Aiken, J. G. and Brown, G. A., "Reliability improvement by process control", Proceedings of the 10th Annual Conference on Reliability Physics, 1972, p. 42-8, EEA76-4641.
934. Soderman, L., "This is how MOS circuits work. II. Reliability and economy are closely connected", Eltek. Aktuell Elektron., 17 (1), 44-6 (1974), Swedish, EEA77-18854.
935. Spivak, H. A., Hawkins, L. E. and Nasuti, A. J., "An MOS-LSI memory reliability program for a commercial computer application", Proceedings of the 1971 Annual Symposium on Reliability, p. 331-7, EEA74-18818.
936. Szep, I. C. and Tihanyi, J., "Low-cost fabrication of MOS-IC's", Proceedings of the International Conference on Advanced Microelectronics, 1970, 1 pp., EEA74-1616.
937. Tanaka, T. and Nishi, Y., "The nature and mechanism of degradation in MNOS read write memory", Oyo Buturi, 44 (Suppl.), 203-9 (1975), EEA78-41615.
938. Thornton, P. R., "SEM in device diagnostics and reliability physics", IEEE Trans. Electron Devices, ED-16, 360-71 (1969).
939. Thornton, P. R., Davies, I. G., Shaw, D. A., Sulway, D. V. and Wayte, R. C., "Device failure analysis by SEM", Microelectron. and Reliab., 8, 33-53 (1969).
940. Trindade, D. C., "Time zero determination of FET reliability", Patent USA 3889188, Publ. June 1975, EEA78-41633.
941. Urban, S., "Aging of semiconductor-oxide structures in an electric field", Phys. Status Solidi A, 26, 237-42 (1974), PA78-6908.
942. van Beek, H. W., "MOS/LSI circuit design: designing-in reliability", Proceedings of the 10th Annual Conference on Reliability Physics 1972, p. 36-41, EEA76-4694.
943. Viele, A. A., "Failure analysis technique for locating the fail site in MOSFET (LSI) logic chips with sputtered SiO_2 passivation", Proceedings of the 12th Annual Reliability Physics Conference, 1974, p. 16-21.

944.Vincoff, M. N. and Schnable, G. L., "Reliability of complementary MOS integrated circuits", IEEE Trans. Reliab., R-24 (4), 255-60 (1975), EEA78-37570.

945.Woods, M. H. and Tuska, J. W., "Degradation of M.N.O.S. memory transistor characteristics and failure mechanism model", Proceedings of the 10th Annual Conference on Reliability Physics, 1972, p. 120-5, EEA76-4675.

O. CHARGE-COUPLED DEVICES

I. Bibliographies

1. Agajanian, A. H., "A bibliography on charge-coupled devices", Solid State Technol., 19 (5), 48-54 (1976).
2. Nunn Associates, "International abstracts on charge-coupled devices, 1970-1974", Ovum Ltd., 22 Grays Inn Road, London, England, 1974.

II. Books and Symposia

3. Sequin, C. H. and Tompsett, M. F., "Charge transfer devices", Academic Press, Inc., New York, 1975.
4. Proceedings of the International Conference on Technology and Applications of Charge Coupled Devices; Held at University of Edinburgh, 25-27 September 1974; Published by the University of Edinburgh, Scotland, 1974, EEA78-24364.

III. Review Articles

5. Altman, L., "New m.o.s. technique points way to junctionless devices", Electronics, 43 (10), 112-18 (1970), EEA73-28210.
6. Amelio, G. F., "Physics and applications of charge-coupled devices", New Electron., 6 (22), 19-28 (1973), EEA77-14529.
7. Amelio, G. F., "Charge-coupled devices", Sci. Am., 230 (2), 22-31 (1974), EEA77-30340.
8. Ando, T., Itoh, Y. and Hara, H., "Charge coupled devices", J. Inst. Electron. Commun. Eng. Jap., 55, 539-43 (1972), Japanese, EEA75-39944.
9. Anon., "Charge transfer devices: Principles of operation and basic characteristics", JEE, no. 99, 38-40, 42-4 (1975), EEA78-28805.
10. Beynon, J. D. E., "Charge-coupled devices", Electron. & Power, 19, 188-92 (1973), EEA76-26089.
11. Borsuk, J. and Bordoloi, K., "Charge-coupled devices - a review", Proceedings of 1973 IEEE Southeast-Con., p. T-5/1-7, EEA76-29283.
12. Boyle, W. S. and Smith, G. E., "Charge coupled semiconductor devices", Bell Syst. Tech. J, 49, 587-93 (1970), EEA73-24917.
13. Boyle, W. S. and Smith, G. E., "Charge-coupled devices - a new approach to MIS device structures", IEEE Spectrum, 8 (7), 18-27 (1971), EEA74-30824.
14. Burt, D. J., "Basic operation of the charge coupled device", International Conference on Technology and Applications of Charge Coupled Devices, 1974, p. 1-12, EEA78-11673.

15.Burt, D. J., "Charge-coupled devices and their applications", Electron & Power, 21, 93-7 (1975), EEA78-19815.
16.Burt, D. J., "Charge coupled devices, today and for the future", Microelectron. & Reliab., 14, 379-81 (1975).
17.Collet, M. G. and Esser, L. J. M., "Charge transfer devices", Meeting of the German Physical Society on Advances in Solid State Physics, 1973, p. 337-58, EEA76-29282.
18.French, B. T., "Designer's guide to charge coupled devices", EDN, 18 (2), 34 (1973).
19.Goser, K. and Pfleiderer, H. J., "Charge-transfer devices", Elektronik, 23 (1), 3-8 (1974), German, EEA77-10253.
20.Jacobson, R. A., "Charge coupling - a remarkable new way to manipulate information", Machine Design, 46 (12), 120 (1974).
21.Kosonocky, W. F. and Carnes, J. E., "Basic concepts of charge-coupled devices", RCA Rev., 36, 566-93 (1975).
22.Kosonocky, W. F. and Sauer, D. J., "The ABCs of CCDs", Electron. Des., 23 (8), 58-63 (1975), EEA78-33147.
23.Lilen, H., "Integrated charge transfer devices", Electron. & Microelectron. Ind., no. 154, 63-6 (1972), French, EEA75-24258.
24.Nyblom, A., "The next revolution in electronics: Charge coupled devices", Tek. Tidskr., 105, 17-19 (1975), Swedish, EEA78-15369.
25.Poirer, R., "Charge transfer semiconductor devices", Rev. Tech. Thomson-CSF, 5, 33-48 (1973), French, EEA76-29278.
26.Sachenko, A. V. and Snitko, O. V., "Charge-coupled devices", Poluprovodn. Tekh. & Mikroelektron., no. 15, 3-13 (1974), Russian, EEA78-2885.
27.Sangster, F. L. J., "Bucket brigade and other charge transfer devices", Proceedings of the 5th International Conference on Solid-State Devices, 1973, p. 275-81.
28.Schoppnies, E., "Charge transfer circuits", Radio Fernsehen Electron., 24, 210-14 (1975), German, EEA78-33158.
29.Smith, G. E., "Charge transfer devices", 9th Annual Proceedings Reliability Physics, 1971, p. 29-32, EEA75-4723.
30.Smith, G. E., "Charge coupled devices", 1972 IEEE Region Six (US Western Region) Conference Record, p. 107-9, EEA76-4249.
31.Smith, G. E., "An overview of charge coupled devices", Oyo Buturi, 43, 263 (1974), EEA78-2874.
32.Sugano, T., "Charge coupled device", J. Inst. Electron. & Commun. Eng. Jap., 54, 1268-75 (1971), Japanese, EEA75-4729.
33.Svenson, R., "CCD: A new type of semiconductor component", Elektronik, 12 (5), 23-31 (1974), Danish, EEA77-34051.
34.Theunissen, M. J. J., "Charge transfer devices. I. Physical principles", Onde Electr., 54, 317-24 (1974), EEA77-38851.
35.Tompsett, M. F., "Charge transfer devices", J. Vac. Sci. & Technol., 9, 1166-81 (1972), EEA75-36297.
36.Tompsett, M. F., "CCD technology", International Conference on Technology and Applications of Charge Coupled Devices, 1974, p. 75-83, EEA78-11681.

37. Williams, T., "Charge-coupled devices. I. Introduction, early device structure and operation", Wireless World, 80, 472-6 (1974), EEA78-2886.

IV. Clocking

38. Bharat, R., "The feasibility of slow-wave-structure clocking of high-frequency charge-coupled devices", Report AD-768786/6, Battelle Columbus Laboratories, Ohio, 1973, 20 pp.
39. Brotherton, S. D., "A theoretical analysis of c.c.d. operation with square clock pulses", Solid-State Electron., 17, 341-8 (1974), EEA77-23061.
40. Brotherton, S. D. and Singh, M. P., "Influence of clock waveforms on the performance of charge coupled devices", Solid-State Electron., 17, 981 (1974).
41. Gelberger, P. P. and Salama, C. A. T., "A uniphase charge-coupled device", Proc. IEEE, 60, 721-2 (1972), EEA75-24374.
42. Haken, R. A., Baker, I. M. and Beynon, J. D. E., "A simple driving and frequency characterization technique for two-phase charge-coupled devices", IEEE J. Solid-State Circuits, SC-9, 199-203 (1974), EEA77-30331.
43. Ho, I. T. and Yu, H. N., "Fabrication of high-density charge-coupled devices with self-isolated clock lines", IBM Tech. Disclosure Bull., 15, 1052-3 (1972), EEA76-4662.
44. Kosonocky, W. F., "Two-phase charge-coupled devices", Report TN-931, RCA, Princeton, N. J., 1973, 2 pp., EEA76-32278.
45. Melen, R. D. and Meindl, J. D., "One-phase CCD: a new approach to charge-coupled device clocking", IEEE J. Solid-State Circuits, SC-7, 92-3 (1972), EEA75-10308.
46. Mohsen, A. M., McGill, T. C., Anthony, M. and Mead, C. A., "Push clocks: a new approach to charge-coupled devices clocking", Appl. Phys. Lett., 22, 172-5 (1973), EEA76-15979.
47. Mohsen, A. M., McGill, T. C. and Mead, C. A., "Charge transfer in overlapping gate charge-coupled devices", IEEE J. Solid-State Circuits, SC-8, 191-207 (1973), EEA76-26091.
48. Pfleiderer, H. J. and Knauer, K., "Generating overlapped clock phases for CCD array", Electronics, 48 (3), 97, 99 (1975), EEA78-11201.
49. Vanstone, G. F., Roberts, J. B. G. and Long, A. E., "The measurement of charge residual for CCD transfer using impulse and frequency responses", Solid-State Electron., 17, 889 (1974).

V. Applications

1. General

50. Alfke, P., "Charge coupled device applications", Microelectron. & Reliab., 13, 345-7 (1974), EEA78-5083.
51. Altman, L., "The new concept for memory and imaging charge coupling", Electronics, 44 (13), 50-9 (1971), EEA74-26907.

52. Altman, L., "Charge-coupled devices move in on memories and analog signal processing", Electronics, 47 (16), 91-101 (1974), EEA77-36903.
53. Baker, W. D., Barbe, D. F. and Sumney, L. W., "Charge-coupled device applications", Nav. Res. Rev., 27 (12), 22-45 (1974).
54. Burt, D. J., Coltman, H. D. and Simpson, P. I., "Charge-coupled devices", GEC J. Sci. & Technol., 41, 53-62 (1974), EEA78-2871.
55. Carnes, J. E. and Kosonocky, W. F., "Charge-coupled devices and applications", Solid State Technol., 17 (4), 67-77 (1974).
56. Konig, S., "Principle and applications of charge coupled semiconductor components", Fernseh - & Kino-Tech., 29, 107-10 (1975), German, EEA78-33154.
57. Lilen, H., "CCD (charge-coupled devices) are commercialised from now on", Electron. & Microelectron. Ind., no. 202, 20-1 (1975), EEA78-28790.
58. Mavor, J., "Current status of CCD", Electron, no. 71, 15-17 (1975), EEA78-33153.
59. Moeller, A., "Charge transfer components - new IC for data storage and picture scanning", Int. Elektron. Rundsch., 27, 239-44 (1973), German, EEA77-2855.
60. Boss, M. H., "C.c.d.s. - have they a future?", New Electron., 8 (4), 73 (1975), EEA78-24490.
61. Ross, M. H., "What can CCD do for you?", Microelectron. & Reliab., 14, 363-5 (1975).
62. Smith, G. E., "Charge coupled devices", 6th International Congress on Microelectronics, 1974, 2 pp., EEA78-24367.
63. Theunissen, M. J. J., "Charge transfer devices. II. Applications", Onde Electr., 54, 405-13 (1974), EEA78-5335.
64. Vanstone, G. F., "Future developments in charge coupled devices", International Conference on Technology and Applications of Charge Coupled Devices, 1974, p. 243-55, EEA78-11688.

2. Shift Register/Memory

65. Ablassmeier, U. and Doering, E., "CCD arrays of high storage density in Al-Si-gate-technology", Siemens Forsch.- & Entwicklungsber., 4, 226-30 (1975), German, EEA78-37653.
66. Berglund, C. N., Powell, R. J., Nicollian, E. H. and Clemens, J. T., "Two-phase stepped oxide CCD shift register using undercut isolation", Appl. Phys. Lett., 20, 413-4 (1972), EEA75-23383.
67. Chan, C. H. and Chamberlain, S. G., "A CCD serial to parallel shift register", IEEE J. Solid-State Circuits, SC-8, 388-91 (1973), EEA76-34920.
68. Chan, C. H., Chamberlain, S. G. and Colton, D. R., "A charge-coupled serial to parallel shift register", International Electrical, Electronics Conference and Exposition Digest, 1973, p. 140-1, EEA77-9716.

69. Collins, D. R., Barton, J. B., Buss, D. C., Kmitz, A. R. and Schroeder, J. E., "CCD memory options", Digest of Technical Papers of the 1973 ISSCC, p. 136-7, EEA76-23042.

70. Collins, D. R., Rhines, W. C., Barton, J. B., Shortes, S. R., Brodersen, R. W. and Tasch, A. F., Jr., "Electrical characteristics of 500-bit Al-Al_2-O_3-Al CCD shift registers", Proc. IEEE, 62, 282-4 (1974), EEA77-9706.

71. Cooke, I. J., "MOS charge coupled shift registers for memory applications", Presented at Colloquium Digest on Computer Memories, London, England, February 1973, EEA76-16452.

72. Dessertenne, B. T., Lakhoua, N. and Poirier, R., "Two-phase 128 bits, charge-coupled shift register", Onde Electr., 54, 325-30 (1974), French.

73. Gunsagar, K. C., Guidry, M. R. and Amelio, G. F., "A CCD line addressable random-access memory (LARAM)", IEEE J. Solid-State Circuits, SC-10, 268-73 (1975), EEA78-37614.

74. Gutlove, N., Amelio, G., Green, A. and Wakeman, R., "Charge coupled device memories for synthetic aperture radar", Presented at Proceedings of the National Aerospace Electronics Conference, Dayton, Ohio, May 1973, EEA77-3632.

75. Heller, L. G. and Lee, H. S., "Digital signal transfer in charge-transfer devices", IEEE J. Solid-State Circuits, SC-8, 116-25 (1973), EEA76-16437.

76. Ho, I. T. and Riseman, J., "Reach-through mode operation of single-electrode double-threshold charge-coupled memory cell", IBM Tech. Disclosure Bull., 15, 412-3 (1972), EEA75-39938.

77. Kosonocky, W. F. and Carnes, J. E., "Charge-coupled digital circuits", IEEE J. Solid-State Circuits, SC-6, 314 (1971).

78. Kosonocky, W. F. and Carnes, J. E., "Two-phase charge-coupled devices with overlapping polysilicon and aluminum gates", RCA Rev., 34 (1), 164-202 (1973), EEA76-26104.

79. Kosonocky, W. F. and Carnes, J. E., "Advanced investigation of two-phase charge-coupled devices", Report N73-33722/2, RCA, Princeton, N. J., 1973, 77 pp.

80. Krupp, R. S. and Tomko, L. A., "Switching networks of planar shifting arrays", Bell Syst. Tech. J., 52, 991-1007 (1973), EEA76-36167.

81. Nishizawa, J. and Nonaka, T., "Charge-coupled semiconductor memory device", Patent USA 3829885, Publ. August 1974.

82. Pao, H., Gunsagar, K., Guidry, M. and Amelio, G. F., "Application of CCD in memory systems", Proceedings of the IEEE 1974 National Aerospace and Electronics Conference, 1974, p. 62-70, EEA78-5338.

83. Patrin, N. A., "Fast access-low power charge-transfer device for high-performance memory", IBM Tech. Disclosure Bull., 18, 343-4 (1975), EEA78-37631.

84. Rosenbaum, S. D. and Caves, J. T., "CCD memory arrays with fast access by on-chip decoding", 1974 ISSCC Digest of Technical Papers, p. 210-1.

85.Rosenbaum, S. D. and Caves, J. T., "8192-bit block addressable CCD memory", IEEE J. Solid-State Circuits, SC-10, 273-80 (1975), EEA78-37615.
86.Schneider, P. and Witte, J., "CCD memories in a working memory system", Siemens Forsch. - & Entwicklungsber., 4, 231-7 (1975), EEA78-41625.
87.Sequin, C. H., "Two-dimensional charge-transfer arrays", IEEE J. Solid-State Circuits, SC-9, 134-42 (1974), EEA77-23120.
88.Tchon, W. E. and Huang, J. S. T., "Staggered oxide C4D structure with clocked source repeater", 1973 IEDM Technical Digest, p. 27-8, EEA77-14003.
89.Terman, L. M., "Small area stored charge memory cells", IBM Tech. Disclosure Bull., 15, 1227-9 (1972), EEA76-8690.
90.Thornber, K. K., "Incomplete charge transfer in IGFET bucket-brigade shift registers", IEEE Trans. Electron Devices, ED-18, 941 (1971).
91.Tompsett, M. F., Amelio, G. F. and Smith, G. E., "Charge coupled 8-bit shift register", Appl. Phys. Lett., 17, 111-5 (1970), EEA73-34343.
92.Waaben, S., "Charge-transfer electronics-tandem matrix semiconductor memory selection", Digest of Technical Papers of the 1971 ISSCC, p. 84-5, EEA74-22797.
93.Western Electric Co., "Charge coupled device which prevents trapping of the charge", Patent UK 1376639, Publ. December 1974, EEA78-15372.
94.Western Electric Co., "Charge coupled devices", Patent UK 1376640, Publ. December 1974, EEA78-15373.
95.Western Electric Co., "Reversible two-phase charge coupled storage device", Patent UK 1379141, Publ. January 1975, EEA78-19923.
96.White, M. H., Lampe, D. R., Fagan, J. L., Kub, F. C. and Barth, D. A., "A nonvolatile charge-addressed memory (NOVCAM) cell", IEEE J. Solid-State Circuits, SC-10, 281-8 (1975).
97.Windle, D. J., Williams, E. W., Vanstone, G. F., Bounden, J. E. and Mavor, J., "A serial parallel serial charge coupled array", New Electron., 7, 8 (1974), EEA77-26047.
98.Zaininger, K. H., "Charge-coupled devices for memory applications: a review", International Conference on Technology and Applications of Charge Coupled Devices, 1974, p. 178, EEA78-11739.
99.Zimmerman, T. A. and Miller, C. S., "Charge coupled devices in signal processing systems", Report AD-782574/8, TRW Systems Group, Redondo Beach, California, 1974, 145 pp.

3. Analog Signal Processing

100.Anon., "Distributed floating-gate amplifier (DFGA)", Report ED-AX-53, Fairchild Space & Defence Systems, N. Y., 1974, 84 pp., EEA78-28876.

101.Buss, D. D., Bailey, W. H. and Collins, D. R., "Analysis and applications of analogue CCD circuits", International Symposium on Circuit Theory, 1973, p. 3-7, EEA77-3454.

102.Buss, D. D., Bailey, W. H. and Tasch, A. F., Jr., "Signal processing applications of charge-coupled devices", International Conference on Technology and Applications of Charge Coupled Devices, 1974, p. 179-97, EEA78-12725.

103.MacLennan, D. J., "Charge-coupled devices. III. Signal processing", Wireless World, 81, 61-5 (1975), EEA78-15370.

104.MacLennan, D. J. and Mavor, J., "Novel technique for the linearisation of charge-coupled devices", Electron. Lett., 11, 222-3 (1975), EEA78-33145.

105.Mclennan, D. J., Mavor, J., Vanstone, G. F. and Windle, D. J., "Novel tapping technique for charge-coupled devices", Electron. Lett., 9, 610-11 (1973), EEA77-10257.

106.Matsushima, K., "Analog signal processing with CCDs has many applications", JEE, no. 99, 11-14 (1975), EEA78-28804.

107.Mayer, G. J., "The chirp Z-transform - A CCD implementation", RCA Rev., 36, 759-73 (1975).

108.Meindl, J. D., "Implantable integrated electronics", Biotelemetry. II, Davos, Switzerland, 1974, p. 32-6, EEA78-17439.

109.Philips Ltd., "Charge transfer device has minimal crosstalk", Patent UK 1395025, Publ. May 1975, EEA78-41631.

110.Roberts, J. B. G., "Radar doppler processing using CCD and SAW devices", Electron. Lett., 11, 525-6 (1975).

111.Roberts, J. B. G. and Chesswas, M., "Video integration with CCD delay lines", International Conference on Technology and Applications of Charge Coupled Devices, 1974, p. 214-20, EEA78-13023.

112.Roberts, J. B. G., Chesswas, M. and Eames, R., "Video integration using charge-coupled devices", Electron. Lett., 10, 169-71 (1974), EEA77-22550.

113.Sakaue, T., Iwasawa, M., Horiike, Y. and Matsushima, K., "An improved charge-coupled analog delay line", Oyo Buturi, 44 (Suppl.), 181-7 (1975), EEA78-41613.

114.Sequin, C. H., "Experimental investigation of a linear 500-element 3-phase charge-coupled device", Bell Syst. Tech. J., 53, 581 (1974).

115.Sequin, C. H. and Mohsen, A. M., "Linearity of electrical charge injection into charge-coupled devices", IEEE J. Solid-State Circuits, SC-10, 81-91 (1975), EEA78-19822.

116.Simpson, I., "The charge coupled device as an analogue store", New Electron., 6, 20-1 (1973), EEA76-35327.

117.Takemoto, I., Ohba, S., Kubo, M. and Ashikawa, M., "A wide-band low-noise charge transfer video delay line", IEEE J. Solid-State Circuits, SC-9, 415-22 (1974), EEA78-2906.

118.Tiemann, J. J., Engeler, W. E., Baertsch, R. D. and Brown, D. M., "Charge-transfer devices filter complex communications signals", Electronics, 47 (23), 113-16 (1974), EEA78-3475.

119.Tompsett, M. F., "Surface potential equilibration method of setting charge in charge-coupled devices", IEEE Trans. Electron Devices, ED-22, 305-9 (1975), EEA78-19830.
120.Tompsett, M. F. and Zimany, E. J., Jr., "The use of charge coupled devices for analogue delay", 1972 European Solid State Devices Research Conference, p. 174-5, Publ. 1973, EEA76-31946.
121.Tompsett, M. F. and Zimany, E. J., Jr., "Use of charge-coupled devices for delaying analog signals", IEEE J. Solid-State Circuits, SC-8, 151-7 (1973), EEA76-16439.
122.Wallinga, H., "A comparison of CCD analog input circuit characteristics", International Conference on Technology and Applications of Charge Coupled Devices, 1974, p. 13-22, EEA78-11674.
123.Whatley, M. M., Benning, C. J., Bailey, W. H. and Buss, D. D., "Charge-coupled device applications in radar signal processing", Proceedings of the IEEE 1974 National Aerospace and Electronics Conference, 1974, p. 284-91, EEA78-6273.
124.White, M. H., Lampe, D. R. and Fagan, J. L., "CCD and MNOS devices for programmable analog signal processing and digital nonvolatile memory", 1973 IEDM Technical Digest, p. 130-3, EEA77-14546.
125.White, M. H. and Webb, W. R., "Study of the use of charge-coupled devices in analog signal processing systems", Report AD-783703/2, Westinghouse Defense and Electronic Systems Center, Baltimore, Md., 1974, 186 pp.
126.Whitehouse, H. J. and Means, R. W., "Signal processing using charge coupled devices and surface wave devices", Proceedings of the 1973 IEEE Ultrasonic Symposium, p. 572-7.
127.Williams, T., "Tapping the potential of c.c.d. for radar", New Electron., 7, 37 (1974), EEA77-27509.
128.Yanai, H., Suzuki, N. and Yonei, K., "Threshold logic using charge-coupled devices", Annu. Rep. Eng. Res. Inst. Fac. Eng. Univ. Tokyo, 32, 127-32 (1973), Japanese, EEA77-22478.

4. Filters

129.Bounden, J. E. and Tomlinson, M. J., "c.c.d. Chebyshev filter for radar m.t.i. applications", Electron. Lett., 10, 89-90 (1974), EEA77-24087.
130.Buss, D. D., Collins, D. R., Bailey, W. H. and Reeves, C. R., "Transversal filtering using charge-transfer devices", IEEE J. Solid-State Circuits, SC-8, 138-46 (1973), EEA76-16026.
131.Buss, D. D. and Hewes, C. R., "Tunable filters using charge transfer devices", Report AD-782818/9, Texas Instruments, Inc., Dallas, Texas, 1974, 35 pp.
132.Collins, D. R., Bailey, W. H., Gosney, W. M. and Buss, D. D., "Charge-coupled-device analogue matched filters", Electron. Lett., 8, 328-9 (1972), EEA75-27460.

133. Hartmann, C. S., Claiborne, L. T., Buss, D. D. and Staples, E. J., "Programmable transversal filters using surface waves, charge transfer devices, and conventional digital approaches", International Specialist Seminar on Component Performance and Systems Applications of Surface Acoustic Wave Devices, 1973, p. 102-14, EEA77-7127.
134. Ibrahim, A., Sellars, L., Foxall, T. and Steenaart, W., "CCD's for transversal filter applications", 1974 IEDM Technical Digest, p. 240-3.
135. Lampe, D. R., White, M. H., Fagan, J. L. and Mims, J. H., "Electrically reprogrammable analog transversal filters", Digest of the Technical Papers of the 1973 ISSCC, p. 156-7.
136. MacLennan, D. J. and Mavor, J., "Transversal filtering using charge-coupled devices", International Conference on Technology and Applications of Charge Coupled Devices, 1974, p. 221-8, EEA78-11269.
137. Roberts, J. B. G., Eames, R. and Roche, K. A., "Moving-target-indicator recursive radar filter using bucket-brigade circuits", Electron. Lett., 9, 89-90 (1973), EEA76-17561.
138. Sekula, J. A., Prince, P. R. and Wang, C. S., "Non-recursive matched filters using charge-coupled devices", 1974 IEDM Technical Digest, p. 244-7.
139. Thornber, K. K., "Optimum linear filtering for charge-transfer devices", IEEE J. Solid-State Circuits, SC-9, 285-91 (1974), EEA77-38840.
140. Wupper, H., "A method for the realization of filters with the aid of charge-transfer devices", 7th Asilomar Conference on Circuits, Systems and Computers, 1973, p. 354-8, Publ. 1974, EEA77-29836.

5. Image Sensors

141. Amelio, G. F., "The impact of large CCD image sensing area arrays", Electron. Components, 17 (3), 21-2 (1975), EEA78-11995.
142. Amelio, G. F., "The impact of large CCD image sensing area arrays. II. (distributed gate amplifier application)", Electron. Components, 17 (4), 17-18 (1975), EEA78-15345.
143. Amelio, G. F., Bertram, W. J., Jr. and Tompsett, M. F., "Charge-coupled imaging devices: design consideration", IEEE Trans. Electron Devices, ED-18, 896-92 (1971).
144. Anon., "Charge-coupled device image sensor study", Report N74-17167/9, Texas Instruments, Inc., Dallas, Texas, 1973, 100 pp.
145. Anon., "Charge coupled device image sensor study", Report N74-17168/7, Fairchild Camera and Instrument Corp., Mountain View, California, 1974, 85 pp.
146. Bankowski, W. F., Kumar, V. R. and Tartamella, J. D., "Charge-coupled devices optical sensor storage device", IBM Tech. Disclosure Bull., 16, 2915-16 (1974), EEA77-23352.

147. Barbe, G. F., "Imaging devices using the charge-coupled concept", Proc. IEEE, 63, 38 (1975).
148. Berger, J. L., "Performance comparison of CCD (charge coupled devices) and BBD (bucket brigade devices) working as optical sensor", Onde Electr., 54, 331-42 (1974), French, EEA77-39068.
149. Brown, R. W. and Chamberlain, S. G., "Quantum efficiency of a silicon gate charge-coupled optical imaging array", Phys. Status Solidi A, 20, 675-85 (1973), EEA77-10554.
150. Carnes, J. E. and Kosonocky, W. F., "Sensitivity and resolution of charge-coupled imagers at low light levels", RCA Rev., 33, 607-22 (1972), EEA76-20277.
151. Chamberlain, S. G., Chan, C. H. and Colton, D. R., "Experimental results and analysis on charge coupled optical imaging arrays", 1972 European Solid State Devices Research Conference, p. 176, Publ. 1973, EEA76-32463.
152. Collins, D. R., Barton, J. B., Robinson, D. A. and Williams, C. S., "Aspects of CCD imaging", 7th Asilomar Conference on Circuits, Systems and Computers, 1973, p. 722, Publ. 1974, EEA77-30721.
153. Compton, R. D., "The solid state imaging revolution", Electro-Opt. Syst. Des., 6 (4), 22-31 (1974), EEA77-37357.
154. Doshier, A. J., "Feasibility study of Pb(1-x)Sn_xTe charge coupled devices for infrared imaging applications", Thesis, Naval Post-graduate School, Monterey, California, 1973, 85 pp., Order No. AD-761473.
155. Dyck, R. H. and Jack, M. D., "Low light level performance of a charge-coupled area imaging device", International Conference on Technology and Applications of Charge Coupled Devices, 1974, p. 154-61, EEA78-15650.
156. French, D. E., Sekula, J. A. and Hartman, J. M., "'Off'-focal-plane time delay and integration (OTDI) using charge coupled devices", 1974 IEDM Technical Digest, p. 437-41.
157. Friedrich, O. M., Jr., "Optical data processing and optical communications", Electro-Opt. Syst. Des., 6, 36-9 (1974), PA77-43357.
158. Gray, P. A., "Back surface imaging of thinned C.C.D.", International Conference on Technology and Application of Charge Coupled Devices, 1974, p. 162-7, EEA78-12013.
159. Hamaoui, H., Amelio, G. and Rothstein, J., "Application of an area-imaging charge-coupled device for television cameras", IEEE Trans. Broadcast & Telev. Receivers, BTR-20, 78-85 (1974), EEA77-24140.
160. Hartsell, G. A. and Kmetz, A. R., "Design and performance of a three phase double level metal 160 x 100 element CCD imager", 1974 IEDM Technical Digest, p. 59-62.
161. Kim, C. K., "Two-phase charge coupled linear imaging devices with self-aligned implanted barriers", 1974 IEDM Technical Digest, p. 55-8.

162.Kim, C. K. and Dyck, R. H., "Low light level imaging with buried channel charge-coupled devices", Proc. IEEE, 61, 1146-7 (1973), EEA76-33086.
163.Kohn, E. S., "Fabrication of rear illuminated CCD imager", Report TN-964, RCA, Princeton, N. J., 1974, 3 pp., EEA78-3074.
164.Kohn, E. S. and Schultz, M. L., "Charge-coupled scanned IR imaging sensors", Report AD-780546/8, RCA Laboratories, Princeton, N. J., 1974, 65 pp.
165.Kosonocky, W. F., Carnes, J. E., Kovac, M. G., Levine, P., Shallcross, F. V. and Rodgers, R. L., III, "Control of blooming in charge-coupled imagers", RCA Rev., 35, 3-24 (1974), EEA78-847.
166.Kovac, M. G., "Charge transfer image sensor", Report TN-397, RCA, Princeton, N. J., 1973, 3 pp., EEA77-6326.
167.Lagnado, I., "Signal processing image sensor using charge coupled devices", International Conference on Technology and Applications of Charge Coupled Devices, 1974, p. 198-205, EEA78-12014.
168.Lee, R. N., Scharnhorst, P. K. and Schodar, R. B., "Preparation and properties of PbS-Si heterojunctions for infrared charge coupled imaging", Report AD-775895/6, Naval Ordnance Lab., White Oak, Md., 1974, 24 pp.
169.Leonberger, F. J., McWhorter, A. L. and Harman, T. C., "PbS MIS devices for charge-coupled infrared imaging applications", Appl. Phys. Lett., 26, 704-7 (1975), PA78-72907.
170.Lilen, H., "First commercial photo CCDs", Electron. & Microelectron. Ind., no. 201, 28-31 (1975), French, EEA78-24768.
171.Melen, R., "The tradeoffs in monolithic image sensors: MOS vs CCD", Electronics, 46 (11), 106-11 (1973), EEA76-26501.
172.Melen, R. D. and Meindl, J. D., "A transparent-electrode CCD image sensor", Digest of Technical Papers of the 1973 ISSCC, p. 130-1, 208, EEA76-23712.
173.Melen, R. D. and Meindl, J. D., "A transparent-electrode CCD image sensor for a reading aid for the blind", IEEE J. Solid-State Circuits, SC-9, 41-9 (1974), EEA77-16883.
174.Michon, G. J. and Burke, H. K., "Charge injection imaging", Digest of Technical Papers of the 1973 ISSCC, p. 138-9, EEA76-23715.
175.Michon, G. J. and Burke, H. K., "Operational characteristics of CID imager", 1974 ISSCC Dig. of Tech. Pap., p. 26-27.
176.Nelson, R. D., "Accumulation-mode charge-coupled device", Appl. Phys. Lett., 25, 568-70 (1974), EEA78-11991.
177.Powers, J. P., "Analysis of a multipurpose electro-optical signal processor", Report AD-787475/3ST, Naval Postgraduate School, Monterey, California, 1974, 32 pp.
178.Rodgers, R. L., III, "Beam-scanned silicon targets for camera tubes", 1973 IEEE International Convention and Exposition, Vol. VI, 1.1/5 pp., EEA77-2864.
179.Rodgers, R. L., III, "Charge-coupled imager for 525-line television", RCA Eng., 20, 79 (1974).

180. Rodgers, R. L., III, "A large area CCD imager for TV applications", International Conference on Technology and Applications of Charge Coupled Devices, 1974, p. 153, EEA78-13022.
181. Sander, W. B., "CCD light change monitor for sensing movement etc.", Patent USA 3859518, Publ. January 1975, EEA78-26029.
182. Scharnkorst, P. K., "The infrared extended, gates CCI (charge coupled imager)", Report AD-775896/4, Naval Ordnance Lab., White Oak, Md., 1974, 32 pp., CA81-128680.
183. Schnitzler, A. D., "Prospects of charge-coupled imagers", Presented at 2nd European Electro-Optics Conference, Montreux, Switzerland, April 1974, EEA77-26781.
184. Seib, D. H., "Carrier diffusion degradation of modulation transfer function in charge coupled imagers", IEEE Trans. Electron Devices, ED-21, 210-7 (1974), EEA77-14906.
185. Sequin, C. H., "Blooming suppression in charge coupled area imaging devices", Bell Syst. Tech. J., 51 (8), 1923-6 (1972), EEA76-1069.
186. Sequin, C. H., "Interlacing in charge-coupled imaging devices", IEEE Trans. Electron Devices, ED-20, 535-41 (1973), EEA76-23703.
187. Sequin, C. H., "Experimental investigation of a linear 500-element 3-phase charge-coupled device", Bell Syst. Tech. J., 53, 581-610 (1974), EEA77-26464.
188. Sequin, C. H., Morris, F. J., Shankoff, T. A., Tompsett, M. F. and Zimany, E. J., "Charge-coupled area image sensor using three levels of polysilicon", IEEE Trans. Electron Devices, ED-21, 712 (1974).
189. Sequin, C. H., Sealer, D. A., Bertram, W. J., Buckley, R. R., Morris, F. J., Shankoff, T. A. and Tompsett, M. F., "Charge-coupled image-sensing devices using three levels of polysilicon", 1974 ISSCC Dig. of Tech. Pap., p. 24-25.
190. Sequin, C. H., Sealer, D. A., Bertram, W. J., Jr., Tompsett, M. F., Buckley, R. R., Shankoff, T. A. and McNamara, W. J., "A charge-coupled area image sensor and frame store", IEEE Trans. Electron Devices, ED-20, 244-52 (1973), EEA76-16699.
191. Sequin, C. H., Shankoff, T. A. and Sealer, D. A., "Measurements on a charge-coupled area image sensor with blooming suppression", IEEE Trans. Electron Devices, ED-21, 331-41 (1974), EEA77-30672.
192. Shortes, S. R., Chan, W. W., Rhines, W. C., Barton, J. B. and Collins, D. R., "Development of a thinned, backside-illuminated charge-coupled device imager", 1973 IEDM Technical Digest, p. 415, EEA77-14885.
193. Shortes, S. R., Chan, W. W., Rhines, W. C., Barton, J. B. and Collins, D. R., "Characteristics of thinned backside illuminated charge-coupled device imagers", Appl. Phys. Lett., 24, 565-7 (1974), EEA77-30708.
194. Steckl, A. J., Nelson, R. D., French, B. T. and Schechter, D., "Application of charge-coupled devices to infrared detection and imaging", International Conference on Technology and Applications of Charge Coupled Devices, 1974, p. 256-69, EEA78-12015.

195.Tompsett, M. F., Amelio, G. F., Bertram, W. J., Jr., Buckley, R. R., McNamara, W. J., Mikkelsen, J. C., Jr. and Sealer, D. A., "Charge-coupled imaging devices: experimental results", IEEE Trans. Electron Devices, ED-18, 992-6 (1971), EEA74-36986.

196.Tompsett, M. F., Bertram, W. J., Sealer, D. A. and Sequin, C. H., "Charge coupled image sensing devices", 1972 European Solid State Devices Research Conference, p. 176-7, Publ. 1973, EEA76-32464.

197.Tompsett, M. F., Bertram, W. J., Sealer, D. A. and Sequin, C. H., "Charge-coupling improves its image, challenging video camera tubes", Electronics, 46 (2), 162-9 (1973), EEA76-9003.

198.Tompsett, M. F., Sealer, D. A., Sequin, C. H. and Shankoff, T. A., "Charge coupled image sensing - state of the art", 1973 IEEE Intercon Technical Papers, Pap. 1.4.

199.Watson, H. A., "Charge-coupled linear and area-imaging devices", 1972 SID Symposium. Digest of Technical Papers, p. 710-71, EEA75-32338.

200.Watson, H. A., "Charge-coupling technology leads to compact video cameras", Bell Lab. Rec., 51, 266-71 (1973), EEA77-7474.

201.Weckler, G. P., "Solid state image sensing with photodiode arrays", 1973 IEEE Intercon Technical Papers, Pap. 1.2.

202.Weimer, P. K., Kovac, M. G., Shallcross, F. V. and Pike, W. S., "Self-scanned image sensors based on charge transfer by the bucket-brigade method", IEEE Trans. Electron Devices, ED-18, 996-1006 (1971), EEA74-36987.

203.Weimer, P. K., Pike, W. S., Kovac, M. G. and Shallcross, F. V., "The design and operation of charge-coupled image sensors", Digest of Technical Papers of the 1973 ISSCC, p. 132-3, EEA76-23713.

204.Weimer, P. K., Pike, W. S., Shallcross, F. V. and Kovac, M. G., "Video processing in charge-transfer image sensors by recycling of signal through the sensor", RCA Rev., 35, 341 (1974).

205.White, M. H., Lampe, D. R., Blaha, F. C. and Mack, I. A., "Characterization of surface channel CCD image arrays at low light levels", IEEE J. Solid-State Circuits, SC-9, 1-12 (1974), EEA77-10592.

206.White, M. H., Lampe, D. R., Mack, I. A. and Blaha, F. C., "Characterization of charge-coupled device line and area-array imaging at low light levels", Digest of Technical Papers of the 1973 ISSCC, p. 134-5, EEA76-23714.

207.Willett, E. and Forst, H., "Artillery-launched charge-coupled device", Report AD-786697/3SL, Fairchild Space and Defense Systems, Syosset, N. Y., 1974, 78 pp.

208.Williams, T., "Charge-coupled devices. IV. Imaging applications", Wireless World, 81, 133-8 (1975), EEA78-12010.

209.Woehrn, D., "Solid state imaging devices and charge coupled devices prospects", Onde Electr., 54, 343-8 (1974), French, EEA77-39069.

210. Zaininger, K. H. and Carnes, J. E., "CCD (charge-coupled device) image sensors for solid state television cameras: a review", International Broadcasting Convention, 1974, p. 58-68, EEA77-39801.

VI. Other Charge Transfer Devices

1. Bucket-Brigade

211. Arita, S. and Kayahara, M., "Bucket brigade device (BBD) analog delay device", Natl. Tech. Rep., 21, 54-62 (1975), Japanese, EEA78-33224.
212. Berglund, C. N., "The bipolar transistor bucket-brigade shift register", IEEE J. Solid-State Circuits, SC-7, 180-5 (1972), EEA75-12661.
213. Berglund, C. N. and Boll, H. J., "Performance limitations of the IGFET bucket-brigade shift register", IEEE Trans. Electron Devices, ED-19, 852-60 (1972), EEA75-23283.
214. Brunn, E. and Olesen, O., "Analog capacitance ROM with IGFET bucket-brigade shift register", IEEE J. Solid-State Circuits, SC-10, 55-9 (1975), EEA78-8609.
215. Buss, D. D., Bailey, W. H. and Collins, D. R., "Matched filtering using tapped bucket-brigade delay lines", Electron. Lett., 8, 106-7 (1972), EEA75-12533.
216. Buss, D. D., Bailey, W. H. and Eversole, W. L., "Noise in MOS bucket-brigade devices", IEEE Trans. Electron Devices, ED-22 (11), 977-81 (1975).
217. Butler, W. J., Barron, M. B. and Kurz, B., "Level-shift compensation in m.o.s. bucket-brigade circuits operated in an analogue mode", Electron. Lett., 8, 286-7 (1972), EEA75-23326.
218. Butler, W. J., Barron, M. B. and Puckette, C. M., "Bucket-brigade band-width characteristics", Electron. Lett., 8, 449-50 (1972).
219. Butler, W. J., Barron, M. B. and Puckette, C. M., "Practical considerations for analog operation of bucket-brigade circuits", IEEE J. Solid-State Circuits, SC-8, 157-68 (1973), EEA76-15992.
220. Butler, W. J., Puckette, C. M. and Barron, M. B., "Implementation of a moving-target indicator by bucket-brigade circuits", Electron. Lett., 8, 543-4 (1972), EEA76-1986.
221. Butler, W. J., Puckette, C. M. and Smith, D. A., "Differential mode of operation for bucket-brigade circuits", Electron. Lett., 9, 106-8 (1973), EEA76-17318.
222. Frey, W., "Bucket-brigade device with improved charge transfer", Electron. Lett., 9, 588-9 (1973).
223. Frey, W. and Heinle, W., "A general method of calculating the charge-transfer coefficient of MOS bucket-brigade devices", Arch. Elektron. & Uebertragungstech., 28, 181-2 (1974), EEA77-23037.
224. Hinkle, F. E., "C-MOS decade divider clocks bucket-brigade delay line", Electronics, 48 (6), 117-8 (1975), EEA78-33144.

225.Kovac, M. G., Pike, W. S., Shallcross, F. V. and Weimer, P. K., "Solid state imaging emerges from charge transport", Electronics, 45 (5), 72-7 (1972), EEA75-13755.
226.Kovac, M. G., Shallcross, F. V., Pike, W. S. and Weimer, P. K., "Image sensors based on charge-transfer by integrated bucket-brigades", Presented at 17th IEDM, Washington, D. C., October 1971, EEA75-13344.
227.Lewis, E. T., "Bucket-brigade shift-register operation - exact correlation between experimental data and a computer model", IEEE J. Solid-State Circuits, SC-8, 207-21 (1973), EEA76-25537.
228.Mok, T. D. and Salama, C. A. T., "Logic array using charge-transfer devices", Electron. Lett., 8, 495 (1972).
229.Mok, T. D. and Salama, C. A. T., "m.o.s. bucket-brigade AND-OR logic array", Electron. Lett., 9, 135-6 (1973), EEA76-19442.
230.Mok, T. D. and Salama, C. A. T., "Performance and characterization of logic arrays using charge-transfer semiconductor devices", International Electrical, Electronics Conference and Exposition Digest, 1973, p. 166-7, EEA74-10267.
231.Mok, T. D. and Salama, C. A. T., "A charge-transfer-device logic cell", Solid-State Electron., 17, 1147 (1974).
232.Pike, W. S., Kovac, M. G., Shallcross, F. V. and Weimer, P. K., "An experimental solid-state TV camera using a 32 x 44 element charge-transfer bucket-brigade sensor", RCA Rev., 33, 483-500 (1972), EEA76-9983.
233.Sangster, F. L. J., "M.O.S. integrated bucket-brigade transversal filters", Eurocon 1971 Digest, 2 pp., EEA75-3765.
234.Sangster, F. L. J., "The bucket brigade and other charge transfer devices", Oyo Buturi, 43 (Suppl.), 275-81 (1974), EEA78-2876.
235.Terui, Y., Horiuchi, S. and Yoshino, M., "MOS bucket-brigade device for low voltage operation", Natl. Tech. Rep., 21, 63-72 (1975), Japanese, EEA78-33225.
236.Thornber, K. K., "Incomplete charge transfer in IGFET bucket-brigade shift registers", IEEE Trans. Electron Devices, ED-18, 941-50 (1971).
237.Troutman, R. R., "Method for making self-aligning structure for charge-coupled and bucket brigade devices", Patent USA 3810795, Publ. May 1974.
238.Vogl, N. G., "Bucket-brigade shift register tutorial", Electrochemical Society Fall Meeting (Extended abstracts), 1974, p. 477-8, EEA78-37685.
239.Vogl, N. G., Jr. and Deliduka, B. J., "Half-million pel bucket-brigade optical scanner", 1974 ISSCC Dig. of Tech. Dig., p. 30-31, 220.

2. Surface Charge Transport

240.Baertsch, R. D., Engeler, W. E. and Tiemann, J. J., "A new surface charge analog store", 1973 IEDM Technical Digest, p. 134-7, EEA77-14547.

241.Engeler, W. E., Tiemann, J. J. and Baertsch, R. D., "Surface charge transport in silicon", Appl. Phys. Lett., 17, 469-72 (1970), EEA74-7653.
242.Engeler, W. E., Tiemann, J. J. and Baertsch, R. D., "The surface-charge transistor", IEEE Trans. Electron Devices, ED-18, 1125 (1971).
243.Engeler, W. E., Tiemann, J. J. and Baertsch, R. D., "Surface-charge transport in a multielement charge-transfer structure", J. Appl. Phys., 43, 2277-85 (1972), EEA75-19157.
244.Tiemann, J. J., Baertsch, R. D., Engeler, W. E. and Brown, D. M., "A surface-charge shift register with digital refresh", IEEE J. Solid-State Circuits, SC-8, 146-51 (1973), EEA76-15991.
245.Tiemann, J. J., Engeler, W. E., Baertsch, R. D. and Brown, D. M., "Intracell charge-transfer structures for signal processing", IEEE Trans. Electron Devices, ED-21, 300-8 (1974), EEA77-18807.
246.Tiemann, J. J., Engeler, W. E. and Baertsch, R. D., "A surface-charge correlator", IEEE J. Solid-State Circuits, SC-9, 403-10 (1974), EEA78-2905.

3. Others

247.Bharat, R., "A frequency divider using a charge-coupled device", Report AD-774253/9, Army Electronics Command, Fort Monmouth, N. J., 1973, 15 pp.
248.Board, K., Shannon, J. M. and Gill, A., "c.c.f.e.t.: An active charge-coupled device", Electron. Lett., 11, 452-3 (1975),
249.Esser, L. J. M., "Peristaltic charge-coupled device: a new type of charge-transfer device", Electron. Lett., 8, 620-1 (1972), EEA76-8682.
250.Esser, L. J. M., "Peristaltic charge-coupled device for high-speed charge transfer", 1974 ISSCC Dig. of Tech. Pap., p. 28-29, 219.
251.Esser, L. J. M., Collet, M. G. and van Santen, J. G., "The peristaltic charge coupled device", 1973 IEDM Technical Digest, p. 17-20, EEA77-14535.
252.Fraser, J. C., Alexander, D. H., Finnila, R. M. and Su, S. C., "An extrinsic silicon charge coupled device for detecting infrared radiation", 1974 IEDM Technical Digest, p. 442-5.
253.Goser, K. and Knauer, K., "Nonvolatile CCD memory with MNOS storage capacitors", IEEE J. Solid-State Circuits, SC-9, 148-50 (1974), EEA77-23121.
254.Hughes, A. J., Eccleston, W. and Stuart, R. A., "A CCD on gallium arsenide", International Conference on Technology and Applications of Charge Coupled Devices, 1974, p. 270-3, EEA78-11689.
255.Knauer, K. and Goser, K., "Non-volatile CCD-memory with MNOS capacitors", International Conference on Technology and Applications of Charge Coupled Devices, 1974, p. 274-80, EEA78-11740.
256.Knauer, K., Pomper, M. and Tihanyi, J., "CCD in ESFI MOS technique", Nachrichtentech. Z., 28, 356-7 (1975), German, EEA78-41681.

257. Roosito, S. A., "Schottky barrier infrared detector arrays with charge coupled device readout", Report AD-164744/5, Department of the Air Force, Washington, D. C., 1974, 12 pp.
258. Salama, C. A. T., "Two-phase MNOS charge-coupled device", Electron. Lett., 8, 21-2 (1972), EEA75-10264.
259. Takemoto, I., Ashikawa, M., Kubo, M. and Ohba, S., "A charge transfer device with self-aligned electrodes", 1973 IEDM Technical Digest, p. 473-6, EEA77-14564.
260. Takemoto, I., Sunami, H., Ohba, S., Aoki, M. and Kubo, M., "Bulk charge-transfer device", Oyo Buturi, 43 (Suppl.), 264-8 (1974), EEA78-2804.
261. Tompsett, M. F., "A simple charge regenerator for use with charge-transfer devices and the design of functional logic arrays", IEEE J. Solid-State Circuits, SC-7, 237-42 (1972), EEA75-19139.

VII. Characterization

1. Transfer Efficiency

262. Amelio, G. F., Tompsett, M. F. and Smith, G. E., "Experimental verification of the charge coupled device concept", Bell Syst. Tech. J., 49, 593-600 (1970), EEA73-24918.
263. Barbe, D. F., Killiany, J. M. and Hughes, H. L., "Effects of gamma radiation on charge-coupled devices", Appl. Phys. Lett., 23, 400-2 (1973), EEA77-2563.
264. Barbe, D. F. and Saks, N. S., "A tradeoff analysis of transfer speed versus charge-handling capacity for CCD's", International Conference on Technology and Applications of Charge Coupled Devices, 1974, p. 114-22, EEA78-11686.
265. Berger, J., Brugler, S. and Melen, R., "Measurement of transfer efficiency of charge-coupled devices", IEEE J. Solid-State Circuits, SC-6, 421-2 (1971).
266. Brodersen, R. W., Buss, D. D. and Tasch, A. F., Jr., "Experimental characterization of transfer efficiency in charge coupled devices", IEEE Trans. Electron Devices, ED-22, 40 (1975).
267. Burt, D. J., "Performance limitations of charge coupled devices", International Conference on Technology and Applications of Charge Coupled Devices, 1974, p. 84-91, EEA78-11682.
268. Carnes, J. E., Kosonocky, W. F. and Ramberg, F. G., "Drift-aiding fringing fields in charge-coupled devices", IEEE J. Solid-State Circuits, SC-6, 322-6 (1971), EEA74-34114.
269. Clough, G. A., Kotyczka, W. and Erb, D. M., "Experimental observation of signal degradation in recirculating charge-coupled devices", 1974 IEDM Technical Digest, p. 251-2.
270. Collins, D. R., Rhines, W. C., Barton, J. B., Shortes, S. R., Brodersen, R. W. and Tasch, A. F., Jr., "Electrical characteristics of long CCD shift registers fabricated using Al-Al_2-O_3-Al double level metallization", 1973 IEDM Technical Digest, p. 29, EEA77-14004.

271.Copeland, M. A., Ibraham, A. A., Chamberlain, S. G., Jurgens, W. L., Chan, C. and Caves, J. T., "Floating gate studies and photosensitivity investigation on charge-coupled devices", 1972 ISSCC Digest of the Technical Papers, p. 134-5.
272.Hartnagel, H. L. and Hutchinson, T. J., "Increased signal speed by GaAs CCD's", Arch. Elektron. & Uebertragungstech., 29, 286-8 (1975), EEA78-24346.
273.Killiany, J. M., Saks, N. S., Baker, W. D. and Barbe, D. E., "Effects of radiation on buried-channel CCD's with doped polysilicon gates and undoped polysilicon interelectrode isolation", Appl. Phys. Lett., 24, 506-8 (1974), EEA77-30326.
274.Krambeck, R. H., "Zero loss transfer across gaps in a CCD", Bell Syst. Tech. J., 50, 3169-75 (1971), EEA75-7558.
275.Krambeck, R. H., Retajczyk, T. F. and Yan, L. D., "Transfer efficiency of E-beam fabricated C4D's", 1974 IEDM Technical Digest, p. 248-50.
276.Kubo, M., "The bulk charge transfer device", Oyo Buturi, 44 (Suppl.), 173-9 (1975), EEA78-41612.
277.Levine, P. A., "Measurement of CCD transfer efficiency by use of feedback to increase the effective number of transfers", IEEE J. Solid-State Circuits, SC-8, 104-8 (1973), EEA76-16435.
278.Mohsen, A. M., McGill, T. C. and Mead, C. A., "Charge transfer in overlapping gate charge-coupled devices", IEEE J. Solid-State Circuits, SC-8, 191 (1973).
279.Mohsen, A. M. and Tompsett, M. F., "The effects of bulk traps on the performance of bulk channel charge-coupled devices", IEEE Trans. Electron Devices, ED-21, 701-12 (1974), EEA78-643.
280.Patrin, N. A., "Performance of very high density charge coupled devices", IBM J. Res. & Dev., 17, 241-8 (1973), EEA77-13984.
281.Roberts, P. C. T., Singh, M. P. and Lamb, D. R., "Identification of loss mechanism in 3-phase surface-channel charge-coupled devices", Proc. Inst. Electr. Eng., 122, 1089-91 (1975), EEA78-41619.
282.Singh, M. P., Roberts, P. C. T., Lamb, D. R. and Brotherton, S. D., "Influence of clocking waveform on charge transfer in three-phase C.C.D.'s", International Conference on Technology and Application of Charge Coupled Devices, 1974, p. 39-46, EEA78-11678.
283.Tompsett, M. F., Kosicki, B. B. and Kahng, D., "Measurements of transfer inefficiency of 250-element undercut-isolated charge coupled devices", Bell Syst. Tech. J., 52, 1-7 (1973), EEA76-16433.
284.Williams, E. W., Brookes, R. and Windle, D., "Optimum operating conditions for an aluminium-gate CCD shift register", J. Phys. D, 7, 2063-76 (1974), EEA78-376.
285.Williams, E. W., Vanstone, G. F., Windle, D. J., Brookes, R., Rich, E. C. and Jones, B. K., "Optical measurement of the charge transfer efficiency of a charge-coupled device", J. Phys. D, 7, L4-7 (1974), EEA77-6037.
286.Williams, T., "Spotlight on CCD transfer efficiency", New Electron., 7 (2), 30 (1974), EEA77-18812.

2. Noise Effects

287.Barbe, D. F., "Noise and distortion considerations in charge-coupled devices", Electron. Lett., 8, 207-8 (1972), EEA75-20205.
288.Carnes, J. E. and Kosonocky, W. F., "Noise sources in charge-coupled devices", RCA Rev., 33, 327-43 (1972), EEA75-36304.
289.Carnes, J. E., Kosonocky, W. F. and Levine, P. A., "Measurements of noise in charge-coupled devices", RCA Rev., 34, 553-65 (1973), EEA77-18815.
290.Chai, H. D., Galetto, L. V. and King, J. H., "Noise compensator for charge coupled devices", IBM Tech. Disclosure Bull., 16, 1099-1100 (1973).
291.Emmons, S. P. and Buss, D. D., "Noise measurements on the floating diffusion input for charge-coupled devices", J. Appl. Phys., 45, 5303 (1974).
292.Emmons, S. P., Tasch, A. F., Jr. and Caywood, J. M., "A low-noise CCD input with reduced sensitivity to threshold voltage", 1974 IEDM Technical Digest, p. 233-5.
293.Mohsen, A. M., Tompsett, M. F. and Sequin, C. H., "Noise measurement in charge coupled devices", IEEE Trans. Electron Devices, ED-22, 209-18 (1975), EEA78-19824.
294.Thornber, K. K., "Noise suppression in charge transfer devices", Proc. IEEE, 60, 1113-4 (1972), EEA75-35319.
295.Thornber, K. K. and Tompsett, M. F., "Spectral density of noise generated in charge transfer devices", IEEE Trans. Electron Devices, ED-20, 456 (1973), EEA76-16445.

P. SCHOTTKY DIODES

I. Books

1. Burstein, E. and Lundquist, S., (Eds.), "Tunneling phenomena in solids", Plenum Press, New York, 1969.
2. Duke, C. B., "Tunneling in solids", Academic Press, New York, 1969, p. 102-110.
3. Henisch, H. K., "Rectifying semiconductor contacts", Clarendon Press, Oxford, England, 1957.
4. Milnes, A. G. and Feucht, D. L., "Heterojunctions and metal-semiconductor junctions", Academic Press, New York, 1972 (Schottky diodes: p. 156-200; Ohmic contacts: p. 288-305; Bibliography: p. 306-73.)
5. Schwartz, B., (Ed.), "Ohmic contacts to semiconductors", The Electrochemical Society, New York, 1969.
6. Sze, S. M., "Physics of semiconductor devices", Wiley-Interscience, New York, 1969, Chapter 8.
7. Metal-Semiconductor Contacts, Conference held at Manchester, England, April 1974; Published by Institute of Physics, London, England, 1974.

II. Review Articles

8. Andrews, J. M., "The role of the metal-semiconductor interface in silicon integrated circuit technology", J. Vac. Sci. & Technol., 11, 972-84 (1974), EEA78-24478.
9. Anon., "Schottky diodes - where we stand today", Microwaves, 10 (4), 44-6 (1971).
10. Atalla, M. M., "Metal-semiconductor Schottky barriers, devices and applications", Proceedings of Microelectronics Symposium, 1966, p. 123-57.
11. Cordes, L. F., Garfinkel, M. and Taft, E. A., "Development and fabrication of improved Schottky power diodes", Report NASA CR-134624, Research and Development Center, General Electric Company, Schenectady, N. Y., 1974, 67 pp.
12. Crowell, C. R. and Sze, S. M., "Hot electron transport and electron tunneling in thin film structure", In: Physics of thin films, Vol. 4, Haas, G., (Ed.), Academic Press, New York, p. 325-71 (1967).
13. Daw, A. N. and Mitra, R. N., "Electroless plating on semiconductors - a survey", J. Sci. & Ind. Res., 33, 23-6 (1974), EEA78-2724.
14. Duke, C. B., "Tunneling in semiconductors", J. Vac. Sci. & Technol., 7, 22-7 (1970), PA73-56330.

15. Gossick, B. R., "Metal-semiconductor rectifiers and transistors", Solid State Electron., 6, 445-52 (1963).
16. Henry, R., "Technology of Schottky diodes and relations between physical and electrical parameters", Rev. Tech. Thomson CSF, 2, 475-95 (1970), French, PA73-69177.
17. Hofmeister, E., "One hundred years of crystal rectifiers", Proceedings of the Symposium on: Physics - the Foundation of Technology, 1974, p. 421-52, German, EEA78-15143.
18. Mead, C. A., "Metal-semiconductor surface barriers", Solid-State Electron., 9, 1023-33 (1966), EEA70-2225.
19. Padovani, F. A., "Metal-semiconductor barrier devices", In: Semiconductors and Semimetals, Vol. 7A, Willardson, R. K. and Beer, A. C., (Eds.), Academic Press, New York, p. 75-146 (1971).
20. Rhoderick, E. H., "Transport processes in Schottky diodes", Conference on Metal-Semiconductor Contacts, 1974, p. 3-19, PA78-15720.
21. Rideout, V. L., "A review of the theory and technology of ohmic contacts to group III-V compound semiconductors", Solid-State Electron., 18, 541-50 (1975), PA78-46754.
22. Stratton, R., "Tunneling in Schottky barrier rectifiers", In: Tunneling phenomena in solids, Plenum Press, New York, 1969, Chapter 8.
23. Yu, A. Y. C., "The metal semiconductor contact: an old device with a new future", IEEE Spectrum, 7 (3), 83-9 (1970), EEA73-21253.

III. Theory

1. Diffusion and Thermionic Emission

24. Andrews, J. M., Jr. and Lepselter, M. P., "Reverse current-voltage characteristics of metal-silicide Schottky diodes", Presented at International Electron Devices Meeting, Washington, D. C., October 1968, EEA72-7896.
25. Baccarani, G., Calzolari, P. U. and Graffi, S., "Current transport in MSM devices", J. Appl. Phys., 45, 341-4 (1974), PA77-45013.
26. Bethe, H. A., "Theory of the boundary layer of crystal rectifiers", Report 43-12, MIT Radiation Lab., November 1942.
27. Calcolari, P. U., Graffi, S. and Pierini, G., "A numerical analysis of the Schottky barrier", Alta Freq., 41, 690-5 (1972), EEA76-4623.
28. Chaika, G. Y. and Strikha, V. I., "Effect of electron interaction on the current characteristics and on the equivalent circuit of the contact between metal and semiconductor", Radiotekhnika I Elektronika, 14, 178-80 (1969), Russian, EEA72-13422.
29. Chang, C. Y. and Sze, S. M., "Carrier transport across metal-semiconductor barriers", Solid-State Electron., 13, 727-40 (1970), EEA73-31330.
30. Christov, S. G., "Electron currents across metal-semiconductor interfaces", Phys. Status Solidi A, 15, 665-80 (1973), PA76-24506.
31. Crowell, C. R. and Beguwala, M., "Recombination velocity effects on current diffusion and IMREF in Schottky barriers", Solid-State Electron., 14, 1149-57 (1971), EEA75-1126.

32.Crowell, C. R. and Sze, S. M., "Quantum-mechanical reflection of electrons at metal-semiconductor barriers. Electron transport in semiconductor-metal-semiconductor structures", J. Appl. Phys., 37, 2683-90 (1966), EEA69-14400.
33.Crowell, C. R. and Sze, S. M., "Current transport in metal-semiconductor barriers", Solid-State Electron., 9, 1035-48 (1966), EEA70-2224.
34.El-Gahaly, M., Nigrin, J. and Goud, P. A., "Stationary charge transport in metal-semiconductor-metal (MSM) structures", J. Appl. Phys., 44, 4672-80 (1973), PA77-20680.
35.Fonash, S. J., "Current transport in metal semiconductor contact - a unified approach", Solid-State Electron., 15, 783-7 (1972), EEA75-24287.
36.Fonash, S. J., Pivot, J., Roger, J. A. and Cachard, A., "AC properties of biased MIM structures", J. Appl. Phys., 45, 1223-32 (1974), EEA77-30213.
37.Lindsey, H. A. and DeMassa, T. A., "A study of the metal-semiconductor (n-type) rectifying contact", Report TR-IAFOSR 69-1224TR, Arizona State Univ., Tempe, 1969, 199 pp., EEA73-21196.
38.Morino, A. and Sugano, T., "Forward characteristics of Si Schottky diodes", Jap. J. Appl. Phys., 9, 1484-94 (1970), PA74-24124.
39.Parrott, J. E., "Reformulation of boundary conditions at metal-semiconductor contacts", Conference on Metal-Semiconductor Contacts, 1974, p. 20-6, EEA78-8500.
40.Rhoderick, E. H., "The conduction mechanism in Schottky diodes", 1972 European Solid State Devices Research Conference, p. 208-9, Publ. 1973, PA76-55336.
41.Rhoderick, E. H., "Comments on the conduction mechanism in Schottky diodes", J. Phys. D, 5, 1920-9 (1972), PA75-75128.
42.Rideout, V. L. and Crowell, C. R., "Effects of image force and tunneling on current transport in metal-semiconductor (Schottky barrier) contacts", Solid-State Electron., 13, 993-1009 (1970), EEA73-31332.
43.Schottky, W., Naturwiss, 26, 843 (1938).
44.Schultz, W., "On the theory of rectification of metal-semiconductor contacts", Z. Physik, 138, 596-612 (1954).
45.Stokoe, T. Y. and Parrott, J. E., "Inclusion of carrier temperature effects in a thermionic-diffusion theory of the Schottky barrier", Solid-State Electron., 17, 477-84 (1974), PA77-44997.
46.Stratton, R., "On approximate thermonic and field emission equations", Solid-State Electron., 8, 175 (1965).
47.Strikha, V. I., "Determining the transmission coefficient of the gap in a rectifying metal-semiconductor contact", Radio Eng. & Electron. Phys., 15, 2148-50 (1970), EEA75-4638.
48.Strikha, V. I., "Determination of all physical parameters of Schottky-barrier diode", Poluprovodn. Tekh. & Mikroelektron., no. 7, 93-102 (1972), Russian, EEA76-16391.

49. Strikha, V. I. and Chaika, G. F., "Current-voltage characteristic and equivalent circuit of a metal-semiconductor contact, obtained by allowing for the perturbation of the electron distribution function by the current", Fiz. Tekh. Poluprovodnikov, 3, 601-3 (1969), Russian, PA72-47146.
50. Strikha, V. I. and Kilchitskaya, S. S., "Surface recombination velocity at a contact between a semiconductor and a metal", Sov. Phys. Semicond., 1, 831-6 (1967), EEA70-22827.
51. Strikha, V. I. and Sheka, D. I., "The voltage-current characteristics of a metal semiconductor contact of small area on the assumptions of the diffusion theory", Izv. VUZ Fiz., no. 7, 7-15 (1967), Russian, EEA71-8165.
52. Strikha, V. I. and Sheka, D. I., "Theory of voltage-current characteristic of the contact between metal and thin semiconductor film", Poluprovodn. Tekh. & Mikroelektron., no. 7, 105-10 (1972), Russian, PA76-24508.
53. Vaitkus, R. L., "Numerical solution of an idealized two-carrier metal-semiconductor junction diode", Thesis, Univ. Minnesota, Minneapolis, 1971, 187 pp., Order No. 72-5587, EEA75-32642.
54. Wilkinson, J. M., "Evidence for current diffusion effects in silicon Schottky barriers resulting from measurements on titanium-silicon diodes", Conference on Metal-Semiconductor Contacts, 1974, p. 27-35, PA78-15722.
55. Williams, R., "The effect of barrier recombination on production of hot electrons in a metal by forward bias injection in a Schottky diode", RCA Rev., 30, 306-13 (1969), PA72-44990.

2. Thermionic-field and Field Emission Tunneling

56. Card, H. C., "On the direct currents through interface states in metal-semiconductor contacts", Solid-State Electron., 18, 881-3 (1975), EEA78-37476.
57. Chandorkar, A. N., Khokle, W. S. and Singh, A., "Current-voltage characteristics of tunneling Schottky barriers", Int. J. Electron., 37, 119-25 (1974), PA77-60855.
58. Charbonnier, F. M., Strager, R. W., Swanson, L. W. and Martin, E. E., "Nottingham effect in field and thermionic field emission", Phys. Rev. Lett., 13, 397 (1964).
59. Combescot, R. and Schreder, G., "Tunnelling in metal-semiconductor contacts. I. Influence of the impurities", J. Phys. C, 6, 1363-84 (1973), EEA76-23386.
60. Crowell, C. R. and Rideout, V. L., "Thermionic-field resistance maxima in metal-semiconductor (Schottky) barriers", Appl. Phys. Lett., 14, 85-8 (1969), PA72-33353.
61. Crowell, C. R. and Rideout, V. L., "Normalized thermionic-field (T-F) emission in metal-semiconductor (Schottky) barriers", Solid-State Electron., 12, 89-105 (1969), PA72-15144.
62. Duke, C. B., Rice, M. J. and Steinrisser, F., "Tunneling measurement of electron-plasmon interactions in degenerate semiconductors", Report R-420, Illinois Univ., Urbana, 1969, 45 pp., EEA73-21262.

63.Fowler, R. H. and Nordheim, L., "Electron emission in intense electric fields", Proc. Roy. Soc., A119, 173-81 (1928).
64.Guetin, P. and Schreder, G., "Quantitative aspects of the tunneling resistance in n-GaAs Schottky barriers", J. Appl. Phys., 42, 5689-98 (1971), EEA75-7500.
65.Guinet, C. and Zylbersztejn, A., "Tunneling in metal-InSb contacts", Phys. Lett. A, 33A, 67-8 (1970), EEA74-1504.
66.Gundlach, K. H., "Note on the effect of the band structure on tunneling in metal-semiconductor-metal junctions", Phys. Status Solidi, 30, K135-8 (1968), EEA72-6541.
67.Kalygina, V. M., Stepanov, V. E. and Gaman, V. I., "Current-voltage characteristics of a metal-dielectric contact, allowing for thermionic field emission of electrons", Izv. VUZ Fiz., no. 4, 123-6 (1973), Russian, PA76-44145.
68.Kane, E. O., "Theory of tunneling", J. Appl. Phys., 32, 83 (1961).
69.Mahmond, A. A., "Analysis of small-area metal-semiconductor contacts", 21st Annual Southwestern IEEE Conference and Exhibition, 1969, 8 pp., PA73-4956.
70.Murphy, E. L. and Good, R. H., "Thermionic emission, field emission and the transition region", Phys. Rev., 102, 1464 (1956).
71.Padovani, F. A., "Forward voltage-current characteristics of metal-silicon Schottky barriers", J. Appl. Phys., 38, 891-2 (1967), PA70-26669.
72.Padovani, F. A. and Stratton, R., "Field and thermionic field emission in Schottky barriers", Solid-State Electron., 9, 695 (1966).
73.Padovani, F. A. and Stratton, R., "The accuracy of the WKB approximation for tunneling in metal-semiconductor junctions", Appl. Phys. Lett., 13, 167-9 (1968), PA72-7821.
74.Pellegrini, B., "New quantum and electronic theory of metal-semiconductor contacts", Phys. Rev. B, 7, 5299-312 (1973), PA76-44148.
75.Pellegrini, B., "A detailed analysis of the metal-semiconductor contact", Solid-State Electron., 17, 217-37 (1974), PA77-41118.
76.Rideout, V. L., "Current transport by thermally assisted tunneling in metal-semiconductor (Schottky barrier) contacts", Thesis, Univ. Southern California, Los Angeles, 1970, 196 pp., Order No. 70-23182, PA74-59062.
77.Schreder, G., "Contribution to the study of the tunnel effect in semiconductors", Thesis, Univ. Paris, France, 1972, French, PA76-44150.
78.Soronevich, Y. R., Strikha, V. N. and Sheka, D. N., "Current-voltage characteristic of the Schottky barrier diode allowing for tunneling through the space-charge region", Poluprovodn. Tekh. & Mikroelektron., no. 11, 51-6 (1973), Russian, EEA76-29204.
79.Stratton, R., "Theory of field emission from semiconductors", Phys. Rev., 125, 67 (1962).
80.Stratton, R., "Energy distributions of field emitted electrons", Phys. Rev., 135, A794 (1964).

81. Stratton, R. and Padovani, F. A., "Influence of ellipsoidal energy surfaces on the differential resistance of Schottky barriers", Phys. Rev., 175, 1072-6 (1968), PA72-13904.
82. Strikha, V. I. and Sheka, D. I., "On the influence of image forces and tunnel effect on current-voltage characteristics of the metal-semiconductor contact with the Schottky barrier", Ukr. Fiz. Zh., 18, 592-7 (1973), Russian, PA76-41266.
83. Tantraporn, W., "On the study of metal-semiconductor contacts", IEEE Trans. Electron Devices, ED-19, 331-8 (1972), EEA75-10178.
84. Thornber, K. K., McGill, T. C. and Mead, C. A., "The tunneling time of an electron", J. Appl. Phys., 38, 2384 (1967).
85. Wilson, A. H., "A note on the theory of rectification", Proc. Roy. Soc., A136, 487 (1932).

3. Minority Carrier Injection

86. Andersson, L. P., Hyder, A. and Berg, S., "Minority carrier injection and resistance modulation in silicon surface-barrier diodes", Nucl. Instrum. & Methods, 114, 237-9 (1974), PA77-18608.
87. Calzolari, P. U. and Graffi, S., "Minority carrier transport in metal-semiconductor junctions", Solid-State Electron., 16, 1501-3 (1973), PA77-20687.
88. Card, H. C. and Rhoderick, E. H., "The effect of an interfacial layer on minority carrier injection in forward-biased silicon Schottky diodes", Solid-State Electron., 16, 365-74 (1973), EEA76-16399.
89. Clarke, R. A., Green, M. A. and Shewchun, J., "Contact area dependence of minority-carrier injection in Schottky barrier diodes", J. Appl. Phys., 45, 1442-3 (1974), PA77-60856.
90. Decker, D. R., Dunn, C. N. and Frost, H. B., "The effect of injecting contacts on avalanche diode performance", IEEE Trans. Electron Devices, ED-18, 141-6 (1971), PA74-24120.
91. Green, M. A. and Shewchun, J., "Minority carrier effects upon the small signal and steady-state properties of Schottky diodes", Solid-State Electron., 16, 1141-50 (1973).
92. Jager, H. and Kosak, W., "Modulation effect by intense hole injection in epitaxial silicon Schottky-barrier diodes", Solid-State Electron., 16, 357-61 (1973), EEA76-16398.
93. Kumar, R. C., "Inclusion of minority charge carriers in the solution of Poisson's equation for a metal semiconductor contact in thermal equilibrium", Int. J. Electron., 33, 689-95 (1972), EEA76-4570.
94. Ladbrooke, P. H., "Interfaces and recombination currents in Schottky barrier diodes", Solid-State Electron., 15, 139-40 (1972), EEA75-7507.
95. Martinez, A., Enea, G. and Esteve, D., "The recombination current in Schottky diodes", C. R. Hebd. Sean. Acad. Sci. B, 274, 323-6 (1972), French, PA75-25891.

96.Melchy, M. A., "An introduction to the generalized field theory. II. Forward conduction in Schottky diodes", Int. J. Electron., 29, 525-32 (1970), PA74-5628.
97.Niimi, T., "Analysis of admittance of Schottky barrier diode with minority carrier injections", Trans. Inst. Electron. Commun. Eng. Japan. A, B, C., 53 (4), 27 (1970), EEA74-4900.
98.Scharfetter, D. L., "Minority carrier injection and charge storage in epitaxial Schottky barrier diodes", Solid-State Electron., 8, 299-311 (1965).
99.Stricha, W. I. and Kiltschitzkaja, S. S., "On the surface recombination speed in the metal-semiconductor contact", Nachrichtentecknik, 21, 359-61 (1971), German, EEA75-1083.
100.Viktorovitch, P. and Kamarinos, G., "The chemical theory of the minority carrier injection ration in a real Schottky diode", J. Phys. D, 8, 246-53 (1975), PA78-25753.
101.Yu, A. Y. C. and Snow, E. H., "Minority carrier injection of metal-silicon contacts", Solid-State Electron., 12, 155-60 (1969), PA72-26565.

4. Image Force Lowering and Potential Energy Profile

102.Berg, S. and Andersson, L. P., "Voltage-dependent reverse current in high resistivity silicon surface-barrier diodes", Nucl. Instrum. & Methods, 114, 241-4 (1974), EEA77-10217.
103.Crowell, C. R., "A simplified self-consistent model for image force and interface charge in Schottky barriers", J. Vac. Sci. & Technol., 11, 951-7 (1974), PA78-46751.
104.Demoulin, E. and van de Wiele, F., "Inversion layer at the interface of Schottky diodes", Solid-State Electron., 17, 825-33 (1974), PA77-60859.
105.Harrick, N. J., "Effect of metal-to-semiconductor potential on the semiconductor surface barrier height", Phys. Chem. Solids, 8, 106-8 (1959).
106.Inkson, J. C., "Schottky barriers and plasmons", J. Vac. Sci. & Technol., 11, 943-6 (1974), PA78-46750.
107.Kanev, S. K., Nikolov, P. S. and Doudov, K. P., "Calculation of electrostatic potential distribution in semiconductor contact region during injection into a blocking contact due to illumination", C. R. Acad. Bulg. Sci., 25, 309-12 (1972), PA75-68447.
108.Krupp, H., "Physical models of the static electrification of solids", Proceedings of the Third Conference on Static Electrification, 1971, p. 1-16, PA76-2166.
109.Kumar, R. C., "The penetration of charge into the metal electrode of a metal-semiconductor contact in equilibrium", Int. J. Electron., 29, 365 (1970), EEA73-37993.
110.Lerach, L. and Albrecht, H., "Analytical solution for the avalanche breakdown voltage of p^+-n junctions and Schottky diodes. Anomalous behaviour of low energy bandgap high-mobility semiconductors", Phys. Status Solidi A, 30, 625-36 (1975), PA78-81514.

111.Lubberts, G. and Burkey, B. C., "A method for calculating band profile and capacitance of nonuniformly doped Schottky barriers", Solid-State Electron., 18, 805-10 (1975).
112.Oakes, J. G. and Lee, C. A., "An inductive phase delay in the retarding field region of Schottky barrier BARITT diodes", 1973 International Electron Devices Meeting Technical Digest, p. 479-82, EEA77-14479.
113.Pellegrini, B., "Current-voltage characteristics of silicon metallic-silicide interfaces", Solid-State Electron., 18, 417-26 (1975), PA78-46753.
114.Strikha, V. I. and Sheka, D. I., "On the influence of image forces and tunnel effect on current-voltage characteristics of the metal-semiconductor contact with the Schottky barrier", Ukr. Fiz. Zh., 18, 592-7 (1973), Russian, EEA76-23398.
115.Tereshchenko, A. K., "On the mirror image force effects in forming potential barriers on metal-semiconductor contacts", Poluprovodn. Tekh. & Mikroelektron., no. 12, 85-9 (1973), Russian, PA76-72401.
116.Zyuganov, A. N., "Non-diffusion theory of volt-ampere characteristics in metal-monopolar semiconductor junctions with current limitation by a space charge in any form", Poluprovodn. Tekh. & Mikroelektron., no. 12, 108-19 (1973), Russian, PA76-72403.

5. Richardson Constant and Effective Mass

117.Conley, J. W., Duke, C. B., Mahan, G. D. and Tiemann, J. J., "Electron tunneling in metal-semiconductor barriers", Phys. Rev., 150, 466-9 (1966), PA70-1916.
118.Crowell, C. R., "Richardson constant for thermionic emission in Schottky barrier diodes", Solid-State Electron., 8, 395-9 (1965).
119.Crowell, C. R., "Richardson constant and tunneling effective mass for thermionic and thermionic-field emission in Schottky barrier diodes", Solid-State Electron., 12, 55-9 (1969), EEA72-9573.
120.Padovani, F. A., "The Richardson constant for thermionic emission in Schottky barrier diodes", Solid-State Electron., 12, 135-6 (1969), EEA72-15145.
121.Sroubek, Z., "Electron tunneling and band structure of $SrTiO_3$ and $KTaO_3$", Phys. Rev. B, 2, 3170-5 (1970), PA74-9490.

6. Surface States, Barrier Heights and Electrostatic Screening

122.Anderson, C. L., Crowell, C. R. and Kao, T. W., "Effects of thermal excitation and quantum-mechanical transmission and photothreshold determination of Schottky barrier height", Solid-State Electron., 18, 705-14 (1975), EEA78-28713.

123.Andrews, J. M. and Philips, J. C., "Chemical bonding and structure of metal-semiconductor interfaces", Phys. Rev. Lett., 35, 56-9 (1975), PA78-62103.
124.Bardeen, J., "Surface states and rectification at a metal-semiconductor contact", Phys. Rev., 71, 717 (1947).
125.Barret, C., Castagne, R. and Vapalle, A., "Decrease in the apparent height of a metal-semiconductor barrier due to spatial fluctuations of the interface density of states or of the thickness of the insulating film", C. R. Hebd. Sean. Acad. Sci. B, 273, 23-5 (1971), French, PA74-59060.
126.Cowley, A. M. and Sze, S. M., "Surface states and barrier height of metal-semiconductor systems", J. Appl. Phys. 36, 3212-20 (1965).
127.Crowell, C. R., "Metal-semiconductor interfaces", Surface Sci., 13, 13-16 (1969), PA72-43094.
128.Crowell, C. R. and Roberts, G. I., "Surface state and interface effects on the capacitance voltage relationship in Schottky barriers", J. Appl. Phys., 40, 3726-30 (1969), EEA73-4452.
129.Eimers, G. W. and Stevens, E. H., "A composite model for Schottky diode barrier height", IEEE Trans. Electron Devices, ED-18, 1185-6 (1971), EEA75-4672.
130.Flores, F., Louis, E. and Yndurain, F., "One-electron properties of the metal-semiconductor junction for zinc-blende compounds", J. Phys. C, 6, L465-9 (1973), EEA77-10155.
131.Forlani, F., Minnaja, N. and Sacchi, G., "The potential barrier at the metal-semiconductor contact", Alta Frequenza, 36, 826-34 (1967), Italian, PA71-35924.
132.Geppert, D. U., Cowley, A. M. and Dore, B. V., "Correlation of metal-semiconductor barrier height and metal work function: Effects of surface states", J. Appl. Phys., 37, 2458-67 (1966).
133.Gossick, B. R., "Electrical characteristics of a metal-semiconductor contact. I", Surface Sci., 21, 123-35 (1970), EEA73-35171.
134.Hariu, T. and Shibata, Y., "Control of Schottky barrier height by thin high-doped layer", Proc. IEEE, 63, 1523-4 (1975), EEA78-41529.
135.Heime, K., "The relation between the height of a metal-semiconductor barrier and the electronegativity and atomic number of the metal", Solid-State Electron., 13, 1505-7 (1970), German, PA74-16402.
136.Heine, V., "Electron states at surfaces", Proceedings of the 10th International Conference on the Physics of Semiconductors, 1970, p. 228-32, PA74-59026.
137.Hirabayashi, K., "Dielectric theory of the barrier height at metal-semiconductor and metal-insulator interfaces", Phys. Rev. B, 3, 4023-5 (1971), PA74-52884.
138.Korol, A. N. and Strikha, V. I., "Effects of metal-semiconductor contact imperfections on V-I characteristics", Poluprovodn. Tekh. & Mikroelektron., no. 18, 39-42 (1974), Russian, PA77-25757.

139. Kurtin, S., McGill, T. C. and Mead, C. A., "Fundamental transition in the electronic nature of solids", Phys. Rev. Lett., 22, 1433-6 (1969).
140. Levine, J. D., "Schottky barrier anomalies and interface states", J. Appl. Phys., 42, 3991-9 (1971), PA74-73740.
141. Levine, J. D., "Power law reverse current-voltage characteristic in Schottky barriers", Solid-State Electron., 17, 1083-6 (1974), PA77-82480.
142. McGill, T. C., "Phenomenology of metal-semiconductor electrical barriers", J. Vac. Sci. & Technol., 11, 935-42 (1974), EEA78-24251.
143. McGill, T. C. and Mead, C. A., "Electrical interface barriers", J. Vac. Sci. & Technol., 11, 122-7 (1974), PA77-75962.
144. MacPherson, A. C., Day, H. M., Kajiyama, K., Sakata, S. and Mizushima, Y., "Comments on 'Schottky-barrier devices with low barrier height' (and reply)", Proc. IEEE, 63, 980-1 (1975), EEA78-19779.
145. Mead, C. A. and Spitzer, W. G., "Fermi level position at semiconductor surfaces", Phys. Rev. Lett., 10, 471-2 (1963).
146. Mead, C. A. and Spitzer, W. G., "Fermi level position at metal-semiconductor interfaces", Phys. Rev., A134, 713-16 (1964).
147. Monch, W., "On metal-semiconductor surface barriers", Surface Sci., 21, 443 (1970).
148. Nunes de Oliveria, L. and Leal Ferreira, G. F., "Formation and dissipation of a Schottky barrier in a conducting dielectric", Phys. Rev. B, 11, 2311-15 (1975), PA78-42302.
149. Padovani, F. A., "Graphical determination of barrier height and excess temperature of a Schottky barrier", J. Appl. Phys., 37, 921 (1966).
150. Padovani, F. A., "The Schottky barrier space charge", 20th Annual Southwestern IEEE Conference, 1968, 3 pp., PA73-4677.
151. Peckerar, M., "On the origin of the increase in Schottky barrier height with interfacial oxide thickness", J. Appl. Phys., 45, 4652 (1974), PA78-18677.
152. Penin, N. A., "The blocking layer capacitance of a metal-semiconductor contact taking account of the majority charge carrier density", Radio Engng. Electronic Phys., 11, 1134-6 (1966), PA70-14977.
153. Perlman, S. S., "Barrier height diminution in Schottky diodes due to electrostatic screening", IEEE Trans. Electron Devices, ED-16, 450-4 (1969), EEA72-20362.
154. Phillips, J. C., "Elementary excitations at metal-semiconductor interfaces", Phys. Rev. B, 1, 593-5 (1970), PA73-40813.
155. Phillips, J. C., "Microscopic theory of covalent-ionic transition at metal-semiconductor interfaces", Solid-State Commun., 12, 861-4 (1973), PA76-38892.
156. Phillips, J. C., "Chemical bonding at metal-semiconductor interfaces", J. Vac. Sci. & Technol., 11, 947-50 (1974), EEA78-24253.

157. Radzievskii, I. A. and Strikha, V. I., "Frequency conversion on a V-I characteristic of a contact between metal and semiconductor", Poluprovodn. Tekh. & Mikroelektron., no. 8, 73-80 (1972), Russian, EEA76-16324.
158. Rangarajan, S., "Effective electron-electron interaction near an interface between a metal and a semiconductor", Proceedings of the Nuclear Physics and Solid State Physics Symposium, 1973, p. 17-19, PA78-29744.
159. Rhoderick, E. H., "The physics of Schottky barriers", J. Phys. D, 3, 1153-67 (1970), EEA73-31350.
160. Schrader, L., "The calculation of the depletion layer width of a metal-semiconductor contact including the charge of the majority carriers", Phys. Status Solidi A, 22, K199-201 (1974), PA77-44996.
161. Seitanyan, G. B. and Tkhorik, Y. A., "Role of surface states in the formation of the Schottky barrier at the contact of metal with gallium arsenide", Poluprovodn. Tekh. & Mikroelektron., no. 11, 33-40 (1973), Russian, EEA76-29201.
162. Strikha, V. I., "On the voltage-current characteristics of the metal-semiconductor contact in the presence of a dielectric layer and a local level", Acta Phys. Pol. A, A44, 741-6 (1973), Russian, PA77-13220.
163. Strikha, V. I. and Panichevskaya, V. I., "On the influence of a surface condition on the differential slope of half-logarithmic Schottky diode volt-amp characteristics", Poluprovodn. Tekh. & Mikroelektron., no. 12, 92-7 (1973), Russian, EEA76-38259.
164. Strikha, V. I., Panichevskaya, V. I. and Buzaneva, E. V., "The determination of the energy spectrum of surface electronic states at a metal-semiconductor contact with a Schottky barrier", Izv. VUZ Fiz., no. 11, 40-6 (1973), Russian, PA77-13221.
165. Viktorovitch, P. and Kamarinos, G., "Chemical-thermodynamic theory of a metal-semiconductor contact", Conference on Metal-Semiconductor Contacts, 1974, p. 36-48, PA78-15723.
166. Yndurain, F., "Density of states and barrier height of metal-Si contacts", J. Phys. C, 4, 2849-58 (1971), PA75-10071.

7. Deep Lying Impurities

167. Bleicher, M. and Lange, E., "Schottky-barrier capacitance measurements for deep level impurity determination", Solid-State Electron., 16, 375-80 (1973), PA76-24510.
168. Crowell, C. R. and Nakano, K., "Deep level impurity effects on the frequency dependence of Schottky barrier capacitance", Solid-State Electron., 15, 605-10 (1972), EEA75-24320.
169. DeGryse, R., Gomes, W. P., Cardon, F. and Vennik, J., "On the interpretation of Mott-Schottky plots determined at semiconductor/electrolyte systems", J. Electrochem. Soc., 122, 711-2 (1975).

170. Glover, G. H., "Determination of deep levels in semiconductors from C-V measurements", IEEE Trans. Electron Devices, ED-19, 138-43 (1972), PA75-13268.
171. Hesse, K. and Strack, H., "On the frequency dependence of GaAs Schottky barrier capacitances", Solid-State Electron., 15, 767-74 (1972), EEA75-24325.
172. Korol, A. N., Kitsai, M. Y., Strikha, V. I. and Sheka, D. I., "Effect of deep impurity levels on Schottky barrier diode characteristics", Solid-State Electron., 18, 375-9 (1975), EEA78-19781.
173. Krambeck, R. H., "The effects of traps on the capacitance of selenium Schottky barrier diodes", Thesis, Carnegie Mellon Univ., Pittsburgh, Pa., 1968, 93 pp., Order No. 68-17635, EEA73-10631.
174. Malachowski, M., "Energy distribution of electrons contributing to the current in reverse biased metal-semiconductor junction", Electron. Technol., 6 (3-4), 49-61 (1973), EEA77-33979.
175. Malachowski, M. J., "Energy distribution of electrons to the current in metal-semiconductor system", Acta Phys. Pol. A, A44, 361-72 (1973), PA76-79768.
176. Roberts, G. I., "The effects of deep-lying impurity energy levels on the capacitance-voltage relationship of metal-semiconductor (Schottky) barriers", Thesis, Univ. Southern California, Los Angeles, 1971, 137 pp., Order No. 71-27951, EEA75-32639.
177. Roberts, G. I. and Crowell, C. R., "Capacitance energy level spectroscopy of deep-lying semiconductor impurities using Schottky barriers", J. Appl. Phys., 41, 1767-76 (1970), EEA73-24967.
178. Schrader, L., "The calculation of the depletion layer width of a metal-semiconductor contact with one deep level", Phys. Status Solidi A, 29, 511-15 (1975), PA78-62104.
179. Simmons, J. G., "Theory of metallic contacts on high resistivity Solids. II. Deep traps", J. Phys. & Chem. Solids, 32, 2581-93 (1971), PA74-83419.
180. Srivastava, S. K. and Bhattacharyya, R., "Effect of density gradient of deep traps on Schottky barrier capacitance in thin films", Int. J. Electron., 31, 257-64 (1971), PA74-64827.
181. Wronski, C. R., "Effects of deep centers on n-type GaP Schottky barriers", J. Appl. Phys., 41, 3805-12 (1970), PA73-65746.

8. Photoresponse

182. Crowell, C. R., Kao, T. W., Anderson, C. L. and Rideout, V. L., "A convenient operational equivalent to the Fowler photothreshold plot", Surface Sci., 32, 591-4 (1972), PA75-72087.
183. Dalal, V. L., "Simple model for internal photoemission", J. Appl. Phys., 42, 2274-9 (1971), PA74-53042.

184. Dussel, G. A., Boer, K. W. and Stirn, R. J., "Photoconductor-metal contact at higher densities", Phys. Rev. B, 7, 1443-54 (1973), EEA76-12520.
185. Fowler, R. H., "Sensitivity curves for clean metals at various temperatures", Phys. Rev., 38, 45 (1931).
186. Pinsker, T. N., "Hot electrons in a metal-semiconductor system: Specular reflection from the surface", Fiz. Tekh. Poluprovodnikov, 2, 237-46 (1968), Russian, PA71-35752.

9. IMREF in Barrier Region

187. Ladbrooke, P. H., "IMREF behaviour in metal-semiconductor barriers", J. Phys. D, 6, L37-9 (1973), PA76-30539.

IV. Physical and Electrical Properties

1. Barrier Height

a. Group IV Semiconductors

188. Ahmad, K., "Contact potentials and barrier heights in gold-silicon and aluminum-silicon contacts", J. Appl. Phys., 43, 4706-13 (1972), EEA76-4574.
189. Anderson, L. P., Hyder, A. and Misra, M., "Semiconductor contacts to silicon surface-barrier detectors", Nucl. Instrum. & Methods, 118, 537-9 (1974), EEA77-35197.
190. Andrews, J. M., "Field dependent barrier at an ideal metal-semiconductor interface", Presented at 30th Annual Conference on Physical Electronics, Milwaukee, Wis., March-April 1970, EEA73-24943.
191. Archer, R. J. and Atalla, M. M., "Metal contacts on cleaned silicon surfaces", Conference on Clean Surfaces (vol. 101), 1963, p. 697-708.
192. Archer, R. J. and Yep, T. O., "Dependence of Schottky barrier height on donor concentration", J. Appl. Phys., 41, 303-11 (1970), PA73-31380.
193. Arizumi, T. and Hirose, M., "Transport properties of metal-silicon Schottky barriers", Jap. J. Appl. Phys., 8, 749-54 (1969), PA72-47143.
194. Arizumi, T., Hirose, M. and Altaf, N., "Au-Ag alloy-silicon Schottky barriers", Jap. J. Appl. Phys., 7, 870-4 (1968), PA72-3675.
195. Arizumi, T., Hirose, M. and Altaf, N., "Au-Cu alloy and Ag-Cu alloy-silicon Schottky barriers", Jap. J. Appl. Phys., 8, 1310-13 (1969), PA73-10586.
196. Barret, C., LeBars, P. and Vapaille, A., "Influence of non-uniform thickness of the oxide interfacial layer on the barrier height of a metal-silicon contact", C. R. Hebd. Seances Acad. Sci. B., 280, 133-6 (1975), French, EEA78-24307.

197.Barret, C. and Vapaille, A., "Effect of crystalline orientation on the barrier height of a metal-semiconductor junction", C. R. Hebd. Seances Acad. Sci. B, 276, 271-4 (1973), French, PA76-27310.

198.Basterfield, J., Shannon, J. M. and Gill, A., "The nature of barrier height variations in alloyed Al-Si Schottky barrier diodes", Solid-State Electron., 18, 290-1 (1975), EEA78-15322.

199.Beguwala, M. and Crowell, C. R., "Determination of hafnium-p-type silicon Schottky barrier height", J. Appl. Phys., 45, 2792-4 (1974), PA77-67279.

200.Brook, P. and Whitehead, C. S., "Hyperabrupt junctions in Au-Si Schottky diodes by ion implantation", Electron. Lett., 4, 335-7 (1968), EEA71-25913.

201.Broom, R. F., "Doping dependence of the barrier height of palladium-silicide Schottky diodes", Solid-State Electron., 14, 1087-92 (1971), EEA75-1124.

202.Burton, C. E., Pertney, W. M. and Leedy, H. M., "Anomalous forward characteristics of a metal-thin epitaxial silicon junction", Solid-State Electron., 14, 1071-6 (1971), EEA75-1111.

203.Cowley, A. M., "Titanium-silicon Schottky barrier diodes", Solid-State Electron., 13, 403-14 (1970), EEA73-21258.

204.Crowell, C. R., Saraco, J. C. and Sze, S. M., "W-semiconductor Schottky barrier diodes", Trans. Met. Soc. AIME, 233, 278 (1965).

205.Crowell, C. R., Sze, S. M. and Spitzer, W. G., "Equality of the temperature dependence of the gold-silicon surface barrier and the silicon energy gap in an n-type Si diodes", Appl. Phys. Lett., 4, 91-2 (1964).

206.Curtis, O. and Gossick, B. R., "Transient behavior of gold-germanium surface barriers", IRE Trans. Electron Devices, 3, 163-7 (1956).

207.Donald, D. K., "Experiments on Hg-Si surface barriers", J. Appl. Phys., 34, 1758 (1963).

208.Guinet, C., "New method of determination of a metal-semiconductor barrier height", Appl. Phys. Lett., 25 (10), 600-2 (1974), EEA78-11532.

209.Gutknecht, P. and Strutt, M. J. O., "Schottky barrier height dependence on Si crystal orientation", Appl. Phys. Lett., 21, 405-7 (1972), EEA76-776.

210.Hirose, M., Altaf, N. and Arizumi, T., "Contact properties of metal-silicon Schottky barriers", Jap. J. Appl. Phys., 9, 260-4 (1970), PA73-37525.

211.Hoshino, H. and Okazaki, S., "Au-As alloy-Si Schottky barrier", Jap. J. Appl. Phys., 7, 686-7 (1968), PA71-40522.

212.Itoh, Y., "Silicon surface effect on Schottky barriers", In: Semiconductor silicon, Electrochemical Society, Inc., New York, p. 350-7 (1969), EEA73-10557.

213.Jager, H. and Kosak, W., "Barrier energies of metal-semiconductor contacts of Si and Ge with metals of I and VII side groups", Solid-State Electron., 12, 511 (1969), German, PA72-44991.

214.Kahng, D., "Conduction properties of the Au-N-type-Si Schottky barrier", Solid-State Electron., 6, 281-95 (1963).
215.Kahng, D. and Lepselter, M. P., "Planar epitaxial Si Schottky barrier diodes", Bell. Syst. Tech. J., 44, 1525 (1965).
216.Kano, G., Inoue, M., Matsuno, J. and Takayanagi, S., "Mo-Si Schottky barrier", J. Appl. Phys., 37, 2985 (1966).
217.Komarovskikh, K. F. and Stafeyev, V. I., "An investigation of some alloyed metal-germanium junctions", Radio Engng. Electronic Phys., 11, 1939-46 (1966), EEA70-17010.
218.Landkammer, F. J., "Barrier heights of metal-semiconductor contacts", Solid State Commun., 5, 247-53 (1967), German, PA70-26691.
219.Majewski, Z., "The height of the potential barrier in Schottky junction made by the method of local sputtering", Electron Technol., 3 (4), 51-60 (1970), PA74-80228.
220.Malm, H. L., "Properties of metal surface barriers on high purity germanium", IEEE Trans. Nucl. Sci., NS-22, 140-4 (1975), PA78-62100.
221.Matsushita Electronics Corp., "Molybdenum-silicon Schottky diode", Electronics, 39 (9), 167-8 (1966).
222.Mhaskar, P. A. and Guha, S., "Characteristics of metal silicon Schottky barrier diodes", Presented at Symposium on Electronics, Madras, India, March 1971, EEA74-15539.
223.Munoz-Merino, E., "Metal-polycrystalline silicon barriers", Phys. Status Solidi A, 15, K167-9 (1973), PA76-24507.
224.Narain, J., Borrego, J. M. and Gutmann, R. J., "Effect of chromium thickness of AuCr-nSi Schottky and Au-Cr-n-p^+ Si BARITT diodes", Electron. Lett., 11, 178-9 (1975), EEA78-19772.
225.Noble, W. P., Jr., Braun, I. and Henisch, H. K., "Contact barriers on cleaved germanium surfaces", Solid-State Electron., 10, 45-9 (1967), EEA70-4060.
226.Reith, T. M. and Schick, J. D., "The electrical effect on Schottky barrier diodes of Si crystallization from Al-Si metal films", Appl. Phys. Lett., 25, 524-6 (1974), EEA78-11529.
227.Saltich, J., "Investigation of Schottky barrier heights on n-type silicon as a function of doping density", Ohmic Contacts to Semiconductors, 1968, p. 187-99, Publ. 1969, EEA73-10550.
228.Saxena, A. N., "Hafnium-silicon Schottky barriers: large barrier height on p-type silicon and ohmic behavior on n-type silicon", Appl. Phys. Lett., 19, 71-3 (1971), PA74-65811.
229.Saxena, A. N., Grob, J. J., Hage-Ali, M. and Siffert, P., "Hafnium-silicon surface barriers", Conference on Metal-Semiconductor Contacts, 1974, p. 160-70, PA78-18681.
230.Schley, P. and Sperling, R., "Electrophysical properties of Mo-Si Schottky barriers", Nachrichtentechnik, 20, 452-3 (1970), German, PA74-16439.
231.Severin, P. J. and Poodt, G. J., "Capacitance-voltage measurements with a mercury-silicon diode", J. Electrochem. Soc., 119, 1384-9 (1972), PA76-2182.

232. Smith, B. L. and Rhoderick, E. H., "Schottky barriers on p-type silicon", Solid-State Electron., 14, 71-5 (1971), EEA74-15494.
233. Szydlo, N. and Poirier, R., "Alkaline-metals-silicon Schottky barriers", J. Appl. Phys., 44, 1386-7 (1973), PA76-24505.
234. Szydlo, N., Poirier, R. and Kleefstra, M., "Reverse I-V characteristics of the Na-Si Schottky barrier", Appl. Phys. Lett., 17, 477-8 (1970), PA74-13210.
235. Szydlo, P. N., "Typical current-voltage curve of sodium-silicon N", Revue Tech. Thomson-CSF, 3, 205-24 (1971), PA74-49120.
236. Thanailakis, A., "Metal contacts on ultra-high vacuum cleaved silicon", Conference on Metal-Semiconductor Contacts, 1974, p. 59-66, PA78-15725.
237. Thanailakis, A., "Contacts between simple metals and atomically clean silicon", J. Phys. C, 8, 655-68 (1975), PA78-29742.
238. Thanailakis, A., Chan, D. S. and Northrop, D. C., "Activation energy for the slow states in the aluminium-germanium contact", J. Phys. D, 5, 1930-6 (1972), PA75-75135.
239. Thanailakis, A. and Northrop, D. C., "Slow voltage-induced barrier height changes in metal-germanium surface barriers", J. Phys. D, 4, 1776-84 (1971), EEA75-1082.
240. Thanailakis, A. and Northrop, D. C., "Metal-germanium Schottky barriers", Solid-State Electron., 16, 1383-9 (1973), PA77-13223.
241. Turner, M. J. and Rhoderick, E. H., "Metal-silicon Schottky barriers", Solid-State Electron., 11, 291-300 (1968), PA71-18116.
242. Walter, F. J., Dabbs, J. W. T. and Roberts, L. D., "Large-area germanium surface-barrier counters", Rev. Sci. Inst., 31, 756-62 (1960).
243. Yu, A. Y. C. and Mead, C. A., "Characteristics of aluminum-silicon Schottky barrier diode", Solid-State Electron., 13, 97-104 (1970), EEA73-14217.
244. Yu, A. Y. C. and Snow, E. H., "Surface effects on metal-silicon contacts", J. Appl. Phys., 39, 3008-16 (1968), PA71-40536.

b. Group III-V Semiconductors

245. Archer, R. J., Yep, T. O. and Muray, K., "Control of thin film interface barriers", Report unnumbered, Contract No. F33615-68-C-1054, Hewlett-Packard Company, Palo Alto, California, 1968.
246. Coleman, D. J. and Irvin, J. C., "GaAs Schottky diodes with near-ideal characteristics", Proc. IEEE, 59, 1121-2 (1971), EEA74-22681.
247. Cowley, A. M., "Depletion capacitance and diffusion potential of GaP Schottky barrier diodes", J. Appl. Phys., 37, 3024 (1967).
248. Cowley, M. and Heffner, H., "Gallium phosphide-gold surface barrier", J. Appl. Phys., 35, 255-6 (1964).
249. Furukawa, Y. and Ishibashi, Y., "Transient phenomena in the capacitance of GaAs Schottky barrier diodes", Jap. J. Appl. Phys., 5, 837-8 (1966), PA70-5036.

250. Furukawa, Y. and Ishibashi, Y., "Transient phenomena in capacitance and reverse current in a GaAs Schottky barrier diode", Jap. J. Appl. Phys., 6, 13-20 (1967), PA70-24204.
251. Furukawa, Y. and Ishibashi, Y., "Trapping effects in Au-n-type GaAs Schottky barrier diodes", Jap. J. Appl. Phys., 6, 503-8 (1967), PA70-21018.
252. Goldberg, Y. A., Posse, E. A. and Tsarenkov, B. V., "Ideal GaP surface-barrier diodes", Electron. Lett., 7, 601-2 (1971), EEA75-1094.
253. Hackam, R. and Harrop, P., "Correlation of the Schottky barrier height measurements determined from the capacitance method and the saturation current", Solid-State Electron., 15, 1031-2 (1972), EEA75-36245.
254. Hariu, T. and Shibata, Y., "Optimum Schottky barrier-height for high-efficiency microwave transferred electron diodes", Proc. IEEE, 63, 823-4 (1975), EEA78-19754.
255. Huang, C. I. and Li, S. S., "Reverse I-V characteristics in Au-GaAs Schottky diode in the presence of interfacial layer", Proc. IEEE, 61, 477-8 (1973), EEA75-19907.
256. Ida, M., Sato, Y., Shimada, K. and Uchida, M., "GaAs Schottky barrier reliability", Elec. Commun. Lab. Tech. J, 20, 1761-8 (1971), Japanese, EEA75-16705.
257. Kadrnka, E. and Holzbach, J., "A gallium-arsenide diode with hot carriers", Sdelovaci Tech., 19 (2), 36 (1971), Czech, EEA74-34055.
258. Kajiyama, K., Mizushima, Y. and Sakata, S., "Schottky barrier height of n-$In_xGa_{1-x}As$ diodes", Appl. Phys. Lett., 23, 458-9 (1973), EEA77-5965.
259. Kajiyama, K., Sakata, S. and Ochi, O., "Barrier height of Hf/GaAs diode", J. Appl. Phys., 46, 3221-2 (1975), EEA78-37512.
260. Kalibjian, R. and Mayeda, K., "Photovoltaic and electron-voltaic properties of diffused and Schottky barrier GaAs diodes", Solid-State Electron., 14, 529-40 (1971), PA74-65807.
261. Maksimova, N. K., Vyatkin, A. P., Pronina, I. E. and Zharov, A. A., "Surface barrier gold-gallium arsenide junction", Izv. VUZ Fiz., no. 7, 89-92 (1968), Russian, EEA72-4965.
262. McColl, M., Millea, M. F. and Mead, C. A., "Zero-bias contact resistances of Au-GaAs Schottky barriers", Solid-State Electron., 14, 677-83 (1971), PA74-62159.
263. Millea, M. F., McColl, M. and Mead, C. A., "Schottky barriers on GaAs", Phys. Rev., 177, 1164-72 (1969), PA72-21519.
264. Minden, H. T., "Gallium arsenide dual Schottky barrier diodes", Solid-State Electron., 16, 1185-8 (1973).
265. Nannichi, Y., Nishida, K. and Mizusawa, K., "Determination of the composition of $Ga_xAl_{1-x}As$ from the photoresponse of Schottky barriers", Jap. J. Appl. Phys., 9, 332-3 (1970), EEA73-21038.
266. Neamen, D. A. and Grannemann, W. W., "Experimental determination of the generation rate conversion factor using a pulse of ionizing radiation of arbitrary shape", IEEE Trans. Nucl. Sci., NS-18, 318-21 (1971), PA75-19706.

267. Neamen, D. A. and Grannemann, W. W., "Electrical characteristics of GaAsP Schottky barrier diodes", Solid-State Electron., 14, 1319-23 (1971), EEA75-7504.
268. Ohura, J. and Takeishi, Y., "Electrical properties of metal-GaAs Schottky barrier contacts", Jap. J. Appl. Phys., 9, 458-67 (1970), EEA73-35184.
269. Padovani, F. A., "Thermionic emission of Au-GaAs Schottky barriers", Solid-State Electron., 11, 193-200 (1968), PA71-13931.
270. Padovani, F. A. and Sumner, G. G., "Experimental study of gold-gallium arsenide Schottky barriers", J. Appl. Phys., 36, 3744-7 (1965); ibid 37, 2520 (1966).
271. Pearson, G. L., "Fundamental studies of the metallurgical, electrical, and optical properties of gallium phosphide", Report NASA-CR-95950, Stanford Univ., Solid-State Electronics Lab., Calif., 1968, 14 pp., PA72-35434.
272. Pruniaux, B. R., "Transport properties of the gold germanium gallium arsenide metal semiconductor system", J. Appl. Phys., 42, 3575-7 (1971), PA74-65786.
273. Pruniaux, B. R. and Adams, A. C., "Dependence of barrier height of metal semiconductor contact (Au-GaAs) on thickness of semiconductor surface layer", J. Appl. Phys., 43, 1980-2 (1972), PA75-33030.
274. Rideout, V. L., "Dependence of barrier height on energy gap in Au n-type $GaAs_{1-x}P_x$ Schottky diodes", Solid-State Electron., 17, 1107-8 (1974), PA77-82481.
275. Seiranyan, G. B. and Tkhorik, Y. A., "On the Schottky barrier height of metal-GaAs systems", Phys. Status Solidi A, 13, K115-18 (1972), PA76-8264.
276. Seiranyan, G. B. and Tkhorik, Y. A., "The Schottky barrier height at metal-GaAs (110) contact", Izv. Akad. Nauk. Arm. SSR Fiz., 7, 77-80 (1972), Russian, PA75-54214.
277. Seiranyan, G. B. and Tkhorik, Y. A., "On a rectifying barrier in gallium arsenide Schottky diodes", Poluprovodn. Tekh. & Mikroelektron., no. 12, 79-85 (1973), Russian, PA76-75791.
278. Senechal, R. R. and Basinski, J., "Capacitance measurements on Au-GaAs Schottky barriers", J. Appl. Phys., 39, 4581-9 (1968), PA72-5720.
279. Smith, B. L., "The effect of surface treatment on gallium arsenide Schottky barrier diodes", Solid-State Electron., 11, 502-4 (1968), PA71-27178.
280. Smith, B. L., "Near ideal Au-GaP Schottky diodes", J. Appl. Phys., 40, 4675-6 (1969), EEA73-10585.
281. Smith, B. L., "Au-(n-type) InP Schottky barriers and their use in determining majority carrier concentrations in n-type InP", J. Phys. D, 6, 1358-62 (1973), EEA76-29196.
282. Smith, B. L. and Abbott, M., "Minority carrier diffusion length in liquid epitaxial GaP", Solid-State Electron., 15, 361-70 (1972), EEA75-16650.

283. Sonomura, H., Yamamoto, N. and Miyauchi, T., "Some properties of GaP alloyed diodes", Electron. Commun. Japan, 52, no. 5, (1969), EEA73-20771.
284. Tsarenkov, B. V., Goldberg, Y. A., Izergin, A. P., Posse, E. A., Ravich, V. N., Rafiev, T. Y. and Silvestrova, N. F., "Metal-n-GaP surface-barrier structures", Sov. Phys. Semicond., 6, 610-4 (1972), EEA76-750.
285. Uebbing, J. J. and Bell, R. L., "Cesium-GaAs Schottky barrier height", Appl. Phys. Lett., 11, 357-8 (1967), PA71-13797.
286. van den Dries, J. G. A. M., "Influence of chemical reactions on the barrier height of metal semiconductor junctions", J. Phys. & Chem. Solids, 35, 130-2 (1974), PA77-24826.
287. van den Dries, J. G. A. M. and Post, A. G., "Reverse characteristics of Ag-GaS Schottky barriers", Solid State Commun., 12, 709-11 (1973), PA76-33993.
288. Vogel, K. and Engel, A., "Relation between surface treatment and barrier height in metal-GaSb contacts", Krist. & Tech., 9, 67-76 (1974), German, PA77-13222.
289. Vyatkin, A. P., Maksimova, N. K. and Pekarskii, Y. N., "Gallium arsenide Schottky-barrier pulse diodes", Izv. VUZ Radioelektron., 14, 703-5 (1971), Russian, EEA74-30783.
290. Wei, C. H. and Yee, S. S., "Comments on 'Dependence of barrier height of metal-semiconductor contact (Au-GaAs) on thickness of semiconductor surface layer'", J. Appl. Phys., 45, 971 (1974), PA77-53144.
291. White, H. G. and Logan, R. A., "GaP surface barrier diodes", J. Appl. Phys., 34, 1990-7 (1963).
292. Wronski, C. R., "Current-voltage characteristics of silver-n-type GaP Schottky barriers", RCA Rev., 30, 314-21 (1969), PA72-49119.
293. Zuev, V. A., Seiranyan, G. B., Sukach, G. A. and Tkhorik, Y. A., "Potential barrier height anisotropy on free and metallized GaAs surface", Phys. Status Solidi A, 16, K35-8 (1973), PA76-30534.

c. Group II-VI Semiconductors

294. Akobirova, A. T., Maslova, L. V., Matveev, O. A. and Khusainov, A. K., "Mechanism of rectification at a metal-CdTe contact", Sov. Phys.-Semicond., 8, 1103-4 (1975), PA78-46755.
295. Baker, W. D., "Investigation on ZnTe", Thesis, Carnegie-Mellon Univ., Pittsburgh, Pa., 1971, 30 pp., Order No. 72-8057, PA75-64962.
296. Baker, W. D. and Milnes, A. G., "Schottky barriers on ZnTe", J. Appl. Phys., 43, 5152-3 (1972), PA76-18215.
297. Bujaiti, M., "CdS-metal barriers from photovoltage measurements", Brit. J. Appl. Phys. (J. Phys. D), 1, 581-4 (1968), PA71-40511.
298. Consigny, R. L. and Madigan, J. R., "Anisotropic surface barriers on CdSe", Solid State Commun., 7, 189-93 (1969), PA72-24194.

299. Courteas, E. and Chernov, F., "Contact barriers on insulating CdS", Appl. Phys. Lett., 8, 3 (1966).
300. Goodman, A. M., "Evaporated metallic contacts to conducting cadmium sulfide single crystals", J. Appl. Phys., 35, 573-80 (1964).
301. Guillien, R., Lepley, B. and Ravelet, S., "Comparative study of the metal-semiconductor barrier Au-CdS formed on thin films and single crystals of cadmium sulphide. (Solar cells)", C. R. Hebd. Seances Acad. Sci. B, 278, 19-22 (1974), French, PA77-33031.
302. Kirov, K. I. and Ivanov, N. A., "Study of the injection of holes in the CdS-Au contact", C. R. Acad. Bulg. Sci., 25, 901-4 (1972), PA75-84346.
303. Kohler, R. and Wauer, L., "Some investigations of evaporated Au-CdS-type metal-semiconductor barriers", Solid-State Electron., 14, 581-8 (1971), PA74-65808.
304. Kurz, W. and Heinz, K., "Drift measurements in CdS platelets with a Schottky contact", Phys. Status Solidi A, 13, K49-52 (1972), PA75-77884.
305. Kusaka, M., Kanakura, M. and Okazaki, S., "Interface properties and Schottky barrier on the polar surfaces of CdS and GsP crystals", Jap. J. Appl. Phys., suppl. 2 (Pt. 2), 437-40 (1974), PA78-37822.
306. Kusaka, M., Matsui, T. and Okazaki, S., "Interface properties and Schottky barriers on polar surfaces of CdS", Surf. Sci., 41, 607-10 (1974), PA77-24828.
307. Learn, A. J., Scott-Monck, J. A. and Spriggs, R. S., "Barriers at evaporated metal-polycrystalline CdS interfaces", Appl. Phys. Lett., 8, 144 (1966).
308. McCarthy, S. J. and Yee, S. S., "Large barrier Cu-CdS contacts", Solid-State Electron., 16, 115-16 (1973), PA76-8265.
309. McCarthy, S. J. and Yee, S. S., "Effect of chemisorbed oxygen on the Au, Cu-CdS barrier height", Solid-State Electron., 16, 1435-9 (1973), PA77-20685.
310. Mead, C. A., "Surface barriers on ZnSe and ZnO", Phys. Lett., 18, 218 (1965).
311. Mead, C. A., "Surface states on semiconductor crystals: barriers on the Cd(Se:S) system", Appl. Phys. Lett., 6, 103-4 (1965).
312. Mead, C. A. and Spitzer, W. C., "Photoemission from gold and copper into CdS", Appl. Phys. Lett., 2, 745 (1963).
313. Mehta, R. R. and Sharma, B. S., "Photoconductive gain greater than unity in CdSe films with Schottky barriers at the contacts", J. Appl. Phys., 44, 325-8 (1973), PA76-21107.
314. Nakai, J., Mitsusada, K. and Yasuoka, A., "Contact between vacuum-deposited CdSe films and Au films", Jap. J. Appl. Phys., 3, 490 (1964).
315. Neville, R. C., "Some electronic properties of ZnO and $SrTiO_3$", Thesis, California Inst. Technol., Pasadena, 1971, 105 pp., Order No. 72-420, PA75-71985.

316.Neville, R. C. and Mead, C. A., "Surface barriers on zinc oxide", J. Appl. Phys., 41, 3795-800 (1970).
317.Neville, R. C. and Mead, C. A., "Tunneling currents in zinc oxide", J. Appl. Phys., 41, 5285-90 (1970), EEA74-4889.
318.Scott-Monck, J. A. and Learn, A. J., "Rectifying contacts under evaporated CdS", Proc. IEEE, 56, 68 (1968).
319.Spitzer, W. G. and Mead, C. A., "Barrier-height studies on metal-semiconducting systems", J. Appl. Phys., 34, 3061 (1963).
320.Stirn, R. J., Boer, K. W. and Dussel, G. A., "CdS-metal contact at higher densities", Phys. Rev. B, 7, 1433-43 (1973), PA76-21149.
321.Stirn, R. J., Boer, K. W., Dussel, G. A. and Voss, P., "Work function of metal-CdS contacts at higher current densities", Solid State Commun., 7, 7-8 (1969), EEA72-19961.
322.Touskova, J. and Kuzel, R., "The rectifying contact of gold on etched CdTe single crystals", Phys. Status Solidi A, 10, 91-7 (1972), EEA75-16714.
323.Touskova, J. and Kuzel, R., "The rectifying contacts on CdTe of n-type", Phys. Status Solidi A, 15, 257-66 (1973), PA76-18216.
324.Walpole, J. N., Nill, K. W., Calawa, A. R. and Harman, T. C., "Metal-semiconductor contacts on $Pb_{1-x}Sn_xTe$", Report TR 5, MIT, Cambridge, Mass., 1970, 25 pp., PA74-21754.
325.Zotov, V. V. and Serdvuk, V. V., "Effects resulting from the occurrence of a contact barrier at the boundary between cadmium sulphide and an indium electrode", Elektron. Tekh. 12, no. 4, 31-6 (1971), Russian, EEA75-4629.
326.Zuleeg, R. and Muller, R. S., "Space-charge limited currents and Schottky-emission currents in thin film CdS diodes", Solid-State Electron., 7, 575-82 (1964).

d. All Other Semiconductors

327.Artobolevskaya, E. S., Chenskii, E. V., Gvozdover, R. S. and Petrov, V. I., "Electrical conduction mechanism in the ferroelectric semiconductor SbSI", Sov. Phys. Solid State, 14, 1935-9 (1973), PA76-33955.
328.Assimos, J. A. and Trivich, D., "Photovoltaic properties and barrier heights of single-crystal and polycrystalline Cu_2O-Cu contacts", J. Appl. Phys., 44, 1687-93 (1973), PA76-36663.
329.Carnes, J. E. and Goodman, A. M., "Evaporated metallic contacts to conducting $SrTiO_3$ single crystals", J. Appl. Phys., 38, 3001 (1967).
330.Fonstad, C. G. and Rediker, R. H., "Electrical properties of high-quality stannic oxide crystals", J. Appl. Phys., 42, 2911-8 (1971), PA74-49008.
331.Goodman, A. M., "Metal-semiconductor barrier height measurement by the differential capacitance method - one carrier system", J. Appl. Phys., 34, 329-38 (1963).

332.Lakatos, A. I. and Roberts, G. G., "Double Schottky-barrier capacitance on trigonal selenium", J. Appl. Phys., 39, 5308-9 (1968), EEA72-23798.
333.Lanyon, H. P. D. and Richardson, R. E., Jr., "Rectification at the metal-selenium interface", Phys. Status Solidi A, 7, 411-9 (1971).
334.Neville, R. C. and Mead, C. A., "Surface barrier energies on strontium titanate", J. Appl. Phys., 43, 4657-63 (1972), PA76-5052.
335.Rideout, V. L. and Wemple, S. H., "Optical constants of evaporated Au and Pt films on $KTaO_3$", J. Opt. Soc. Am., 56, 749 (1966).
336.Saxena, A. N., "Multi-plane metal-semiconductor junction device", Patent USA 3450958, Publ. June 1969.
337.Saxena, A. N. and McDad, J. C., "Distributed barrier metal-semiconductor junction device", Patent USA 3450957, Publ. June 1969.
338.Wemple, S. H., Kahng, D. and Braun, H. J., "Surface barrier diodes on semiconducting $KTaO_3$", J. Appl. Phys., 28, 353 (1967).
339.Zolyan, T. S., "Nonlinearity of current-voltage characteristics at a liquid semiconductor-metal interface", Sov. Phys.-Semicond., 6, 1750-1 (1973), PA76-41265.

2. I-V Characteristics and Breakdown Phenomena

340.Anantha, N. G. and Ashar, K. G., "Planar mesa Schottky barrier diode", IBM J. Res. & Develop., 15, 442-5 (1971), PA75-22584.
341.Anderson, W. A. and Milano, R. A., "I-V characteristics for silicon Schottky solar cells", Proc. IEEE, 63, 206-8 (1975), EEA78-14109.
342.Bohg, A., "Localization of breakdown regions in Schottky diodes", IBM Tech. Disclosure Bull., 17, 3309 (1975), EEA78-37511.
343.Card, H. C., "On the direct currents through interface states in metal-semiconductor contacts", Solid-State Electron., 18, 881-3 (1975), PA78-69244.
344.Cervenak, J., "The volt-ampere characteristics of thin-film metal InSb metal structures", Czech. J. Phys. B., 18, 1629-35 (1968), EEA72-7855.
345.Chaika, G. E., "Negative resistance of metal-semiconductor contacts", Sov. Phys. Semicond., 6, 115-9 (1972), EEA75-32604.
346.Chino, K., Suzuki, T. and Mizushima, Y., "Properties of a Schottky-barrier transistor and carrier-coupling devices", Rev. Electr. Commun. Lab., 23, 541-5 (1975), EEA78-41576.
347.Conrad, R. W. and Bebbs, J. F., Jr., "Current oscillations in GaAs Schottky-barrier diodes", Phys. Status Solidi A, 2, K101-3 (1970), EEA74-10583.
348.Glover, G. H., "Charge multiplication in Au-SiC(6H) Schottky junction", J. Appl. Phys., 46, 4842-44 (1975).
349.Hackam, R. and Harrop, P., "Electrical properties of nickel-low-doped n-type gallium arsenide Schottky-barrier diodes", IEEE Trans. Electron Devices, ED-19, 1231-5 (1972), EEA76-777.

350.Hayashi, T. and Saito, N., "Transient current phenomena in reverse biased Al-n type GaAs Schottky diode", Rep. Fac. Eng. Shizuoka Univ., no. 25, 1-9 (1974), Japanese, EEA78-33124.
351.Heime, K., "Schottky diodes", Fernmelde-Ingen., 24, 1-40 (1970), German, EEA73-38038.
352.Iizuka, M., Fujiwara, S., Kano, G., Hasegawa, H., Teramoto, I. and Iwasa, H., "Schottky barrier-type semiconductor device with improved backward breakdown voltage characteristic", Patent USA 3742315, Publ. June 1973.
353.Kano, G., "Avalanche breakdown voltages in punch-through Si epitaxial planar Schottky barrier diodes", Jap. J. Appl. Phys., 8, 463-7 (1969), PA72-26576.
354.Kano, G., "Experiment on the evaluation of metal-semiconductor Schottky barriers", Jap. J. Appl. Phys., 8, 1144-8 (1969), PA73-12158.
355.Kano, G., Iizuka, M., Fujiwara, S., Hasegawa, H. and Sawaki, T., "Schottky barrier semiconductor device having a substantially nonconductive barrier for preventing undesirable reverse-leakage currents and method for making", Patent USA 3763408, Publ. October 1973.
356.Kassing, R. and Kahler, E., "Low-frequency current oscillations in high-resistivity, Au-doped silicon junctions with two Schottky contacts", Phys. Status Solidi A, 12, 209-13 (1972), PA75-61383.
357.Klose, H., Muller, G. and Thiessen, K., "The liquid crystal technique for the observation of electric fields in p-n and Schottky junctions", Phys. Status Solidi A, 16, K97-9 (1973), EEA76-23418.
358.Krempasky, J. and Cervenak, J., "The influence of contacts on the VA-characteristics of thin-layer systems with amorphous semiconductors", Czech. J. Phys. B, 21, 285-94 (1971), EEA74-18685.
359.Kumar, R., Jindal, S. and Bhattacharyya, A. B., "Breakdown and capacitance properties of hyperabrupt epitaxial Schottky barrier diodes", Solid-State Electron., 18, 999-1002 (1975), EEA78-41577.
360.Ladbrooke, P. H., "Reverse current characteristics of some imperfect Schottky barriers", Solid-State Electron., 16, 743-9 (1973), PA76-39301.
361.Lunze, K., Kohler, E., Moschwitzer, A. and Elschner, H., "Solid-state component parts. IX", Nachrichtentechnik, 20, 345-51 (1970), German, EEA73-38044.
362.Martinez, A., Esteve, D., Peyriguer, J. M., Lagorsse, J. M. and Dameme, D., "Electrical characterization of M-Si contacts and technology control", Conference on Metal-Semiconductor Contacts, 1974, p. 67-73, PA78-15726.
363.Melikyan, E. G., "Surface barrier instability of the current in deformed germanium", Sov. Phys.-Semicond., 7, 1116 (1974), PA77-49087.
364.Murarka, S. P., "Forward I-V characteristics of Pt/n-GaAs Schottky barrier contacts", Solid-State Electron., 17, 985-91 (1974), EEA78-8537.

365. Okuto, Y. and Crowell, C. R., "Avalanche breakdown of highly doped passivated Si Schottky barriers", Oyo Buturi, 43 (Suppl.), 390-4 (1974), EEA78-2833.
366. Olson, H. M., "A mechanism for catastrophic failure of avalanche diodes", IEEE Trans. Electron Devices, ED-22, 842-9 (1975), EEA78-41570.
367. Patwari, A. M. and Hartnagel, H. L., "Effect of surface properties on n-type GaAs-Ni and GaAs-Al Schottky diodes", Phys. Status Solidi A, 26, 469-75 (1974), PA78-15718.
368. Peyriguer, J. M., Martinez, A. and Esteve, D., "On the current-voltage characteristics of Schottky diodes", C. R. Hebd. Seances Acad. Sci. B, 277, 627-9 (1973), French, EEA77-10212.
369. Piqueras, J. and Munoz, E., "Negative resistance characteristics in Au-polycrystalline GaAs barriers", Solid-State Electron., 17, 406-8 (1974), EEA77-18781.
370. Rand, M. J., "I-V characteristics of PtSi-Si contacts made from CVD platinum", J. Electrochem. Soc., 122, 811-5 (1975).
371. Rees, H. D., "Injection properties of contacts to InP", Conference on Metal-Semiconductor Contacts, 1974, p. 105-15, PA78-15729.
372. Rusu, A. and Bulucea, C., "Enhanced breakdown voltage in planar metal-overlap laterally diffused (MOLD) Schottky diodes", Appl. Phys. Lett., 27, 620-1 (1975).
373. Siekanowicz, W. W., Huang, H. and Enstrom, R., "Schottky-barrier diodes for electron-beam-semiconductor applications", 1973 International Electron Devices Meeting Technical Digest, p. 520, EEA77-14831.
374. Siekanowicz, W. W., Huang, H., Enstrom, R. E., Martinelli, R. U., Ponczak, S. and Olmstead, J., "Current-gain characteristics of Schottky-barrier and p-n junction electron-beam semiconductor diodes", IEEE Trans. Electron Devices, ED-21, 691-70 (1974), EEA78-788.
375. Sinha, A. K. and Poake, J. M., "Effect of alloying behavior on the electrical characteristics n-GaAs Schottky diodes metallized with W, Au and Pt", Appl. Phys. Lett., 23, 666-8 (1973), PA77-28489.
376. Touskova, J. and Kuzel, R., "Rectifying contacts on n-type CdTe", Phys. Status Solidi A, 15, 257-66 (1973).
377. Tsarenko, B. V., Goldberg, Y. A. and Prosse, E. A., "Mechanism of the flow of the forward current in surface-barrier GaP structure", Sov. Phys.-Semicond., 7, 1546-50 (1974), PA77-82483.
378. Tserng, H. Q., Wisseman, W. R. and Hasty, T. E., "Characterization of GaAs Schottky-Read IMPATT diodes", 1973 International Electron Devices Meeting Technical Digest, p. 483-5, EEA77-14480.
379. Vincent, G., Bois, D. and Pinard, P., "Conductance and capacitance studies in GaP Schottky barriers", J. Appl. Phys., 46, 5173-8 (1975).
380. Vogel, H. and Thomas, P., "Electronic differentiation of current-voltage characteristics of semiconductor diodes with high resistances", Z. Angew. Phys., 27, 277-9 (1969), German.

381. Wilkinson, J. M., "Current diffusion effects in titanium-n silicon Schottky diodes", Solid-State Electron., 17, 583-5 (1974), EEA77-23007.
382. Wilson, J. I. B. and Allen, J. W., "A compensation law for reverse-biased ZnSe Schottky diodes", Solid-State Electron., 18, 759-62 (1975).
383. Wu, S. Y., "Au-SiC Schottky barrier diodes", Solid-State Electron., 17, 683-7 (1974), PA77-57167.
384. Yamashita, A. and Araki, H., "Pre-breakdown phenomena of Au-GaAs Schottky diode", Jap. J. Appl. Phys., 10, 1282 (1971), EEA74-37245.

3. C-V Characteristics

385. Deneuville, A., "Characterization of the interface states at a Ag/Si interface from capacitance measurements", J. Appl. Phys., 45, 3079-84 (1974), PA77-75961.
386. Ghosh, A. K., Feng, T., Morel, D. I., Shaw, R. F. and Rowe, C. A., Jr., "Photovoltaic and rectification properties of Al/Mg phthalocyanine/Al Schottky-barrier cells", J. Appl. Phys., 45, 230-6 (1974), PA77-37322.
387. Glover, G. H., "The C-V characteristics of Schottky barriers on laboratory grown semiconducting diamonds", Solid-State Electron., 16, 973-83 (1973), EEA76-35260.
388. Goodman, A. M., "Metal-semiconductor barrier height measurement by the differential capacitance method - one carrier system", J. Appl. Phys., 34, 329-38 (1963).
389. Goodman, A. M., "Metal-semiconductor barrier height measurement by the differential capacitance method without an ohmic reference contact. One-carrier system", J. Appl. Phys., 36, 1411-4 (1965).
390. Goodman, A. M. and Perkins, D. M., "Metal semiconductor barrier-height measurement by the differential capacitance method. Degenerate one-carrier system", J. Appl. Phys., 35, 3351-3 (1964).
391. Harreis, H. and Heiland, G., "Capacitance of metal contacts on clean semiconductor surfaces (ZnO and Si surface states)", Surface Sci., 24, 643-6 (1971), PA74-24129.
392. Hirose, M., Kido, M., Nakashita, T. and Kurinobu, S., "Interfacial oxide layer in Au-Si Schottky barriers", Oyo Buturi, 43, 798-802 (1974), Japanese, PA78-18678.
393. Huang, C. I. and Li, S. S., "Analyses of transient capacitance experiments for Au-GaAs Schottky barrier diodes in the presence of deep impurities and the interfacial layer", Solid-State Electron., 16, 1481-6 (1973), PA77-20686.
394. Katoda, T. and Sugano, T., "Hall effect, Schottky barrier capacitance, and photoluminescence spectra measurements for GaAs epitaxial layers and their correlation", J. Electrochem. Soc., 121, 1066-73 (1974), PA77-75888.
395. Korwin-Pawlowski, M. L. and Heasell, E. L., "Metal-InSb surface barriers", Conference on Metal-Semiconductor Contacts, 1974, p. 255-62, PA78-18683.

396.Korwin-Pawlowski, M. L. and Heasell, E. L., "The properties of some metal-InSb surface barrier diodes", Solid-State Electron., 18, 849-52 (1975), EEA78-37474.
397.Kusaka, M., Kunisue, H. and Okazaki, S., "The electrical properties of the CdS surface influenced by oxygen", Rep. Res. Lab. Surf. Sci. Okayama Univ., 3, 163-8 (1969), EEA74-26773.
398.Neville, R. C., "Anomalous capacitance of Schottky barriers on strontium titanate", J. Appl. Phys., 46, 350-3 (1975), EEA78-15262.
399.Nishino, T., Yamano, F. and Hamakawa, Y., "A method of measuring differential photovoltage spectrum at Schottky barrier", Jap. J. Appl. Phys., 14, 1885-96 (1975).
400.Pals, J. A., "Properties of Au, Pt, Pd and Rh levels in silicon measured with a constant capacitance technique", Solid-State Electron., 17, 1139-45 (1974), EEA78-594.
401.Pankov, Y. D., "Anomalous barrier capacitance of semiconductors with short mean free path", Fiz. Tekh. Poluprovodnikov, 4, 978-80 (1970), Russian, EEA73-31316.
402.Smith, B. L. and Rhoderick, E. H., "Possible sources of error in the deduction of semiconductor impurity concentrations from Schottky-barrier (C,V) characteristics", Brit. J. Appl. Phys. (J. Phys. D), 2, 465-7 (1969), PA72-30918.
403.Walpole, J. N. and Nill, K. W., "Capacitance-voltage characteristics of metal barriers on p-PbTe and P-InAs: effects of the inversion layer", J. Appl. Phys., 42, 5609-17 (1971), EEA75-7456.

4. Noise Effects

404.Anand, Y., "Low frequency noise in Schottky barrier diodes", Proc. IEEE, 57, 855-6 (1969), EEA72-17714.
405.Anand, Y., "Noise factor dependence on mismatch of low level r.f. signal", Electron. Lett., 7, 111-2 (1971), EEA74-13229.
406.Buchs, J. D., "The noise of microwave Schottky diodes at 70 MHz", Frequenz, 26, 218-23 (1972), German, EEA75-36507.
407.Carruthers, T., "Bias-dependent structure in excess noise in GaAs Schottky tunnel junctions", Appl. Phys. Lett., 18, 35-7 (1971), EEA74-18737.
408.Cooke, H. F., "Low-noise microwave transistors and mixer diodes", Des. Electron., 7 (3), 46-53 (1969), EEA73-14573.
409.Cowley, A. M. and Zettler, R. A., "Shot noise in silicon Schottky barrier diodes", IEEE Trans. Electron Devices, ED-15, 761-9 (1968), EEA72-5001.
410.Haus, H. A., Statz, H. and Pucel, R. A., "Noise measure of metal-semiconductor-metal Schottky-barrier microwave diodes", Electron. Lett., 7, 667-9 (1971), EEA75-1454.
411.Hsu, S. T., "Low-frequency excess noise in metal-silicon Schottky barrier diodes", IEEE Trans. Electron Devices, ED-17, 496-506 (1970), EEA73-28168.

412. Hsu, S. T., "Flicker noise in metal semiconductor Schottky barrier diodes due to multistep tunneling processes", IEEE Trans. Electron Devices, ED-18, 882-7 (1971), PA74-69636.
413. Iizuka, H. and Hitaoka, S., "Low-noise GaAs Schottky-barrier beam-lead mixer diodes", Proc. IEEE, 58, 1372-3 (1970), EEA73-38422.
414. Kitsai, M. E. and Strikha, V. I., "Shot noise in Mo-Si diodes with Schottky barriers", Sov. Phys.-Semicond., 8, 1152-3 (1975), EEA78-19784.
415. Knott, K. F., "Low frequency noise and reliability of microwave Schottky barrier diodes", 3rd Symposium on the Reliability of Electronics, 1973, p. 247-52, EEA77-5990.
416. Minniti, R. J., Jr., Neudeck, G. W. and Anderson, R. M., "Shot noise in platinum-gallium arsenide Schottky barrier diodes", J. Appl. Phys., 42, 1886-91 (1971), PA74-37217.
417. Mircea, A. and Perichon, R., "Excess noise in gallium arsenide avalanche diodes", Acta Electron., 17 (2), 6 pp. (1974), French, EEA77-22995.
418. Neudeck, G. W., Minniti, R. J., Jr. and Anderson, R. M., "The ideality factor and the high-frequency noise of Schottky-barrier-type diodes", IEEE J. Solid-State Circuits, SC-7, 89-90 (1972), EEA75-10220.
419. Neudeck, G. W., "High frequency shot noise in Schottky barrier diodes", Solid-State Electron., 13, 1249-56 (1970), EEA73-35191.
420. Peyriguer, J. M., Martinez, A. and Esteve, D., "On the existence, in Schottky diodes, of correlation laws between the d.c. characteristics and the amplitude of low frequency background noise", C. R. Hebd. Seances Acad. Sci. B, 276, 511-14 (1973), French, EEA76-23423.
421. Strikha, V. I., "Shot noise in Schottky diodes", Fiz. Tekh. Poluprov., 4, 1991-2 (1970), Russian, PA74-7685.
422. Strikha, V. I. and Kitsai, M. Y., "Shot noise in a Schottky-barrier diode assuming surface electron states at the contact", Poluprovodn. Tekh. & Mikroelektron., no. 9, 38-45 (1972), Russian, EEA76-16392.
423. Strikha, V. I. and Kitsai, M. Y., "Shot noise of Schottky-barrier diodes with disturbed distribution function of current carriers", Poluprovodn. Tekh. & Mikroelektron., no. 10, 88-90 (1972), Russian, EEA76-16394.
424. Strikha, V. I. and Kitsai, M. Y., "Shot noise in a Schottky barrier diode during tunneling through a space charge region", Poluprovodn. Tekh. & Mikroelektron., no. 12, 89-92 (1973), Russian, PA76-72402.
425. Swallow. G. H., Oxley, T. H. and Hansom, A. M., "Recent advances in the burnout properties of low noise gallium arsenide Schottky barrier mixer diodes", 1973 European Microwave Conference, Vol. I, A7.6/3 pp., EEA77-5976.

426.Viola, T. J., "High frequency noise in Schottky barrier diodes", Thesis, Univ. Virginia, Charlottesville, 1974, 114 pp., Order No. 74-23250, EEA78-15323.
427.Viola, T. J., Jr. and Mattauch, R. J., "High-frequency noise in Schottky-barrier diodes", Proc. IEEE, 61, 393 (1973), EEA76-12610.
428.Viola, T. J., Jr. and Mattauch, R. J., "Unified theory of high-frequency noise in Schottky barriers", J. Appl. Phys., 44, 2805-8 (1973), PA76-44146.
429.Zaklikiewicz, A., "Results of noise measurements on low frequency range of semiconductor Schottky barrier diodes", Arch. Elektrotech., 17, 215-9 (1968), Polish, EEA71-20627.

5. Photoresponse

430.Bercha, D. M., Nebola, I. I., Radavtsan, S. I., Syrbu, N. N., Kiosev, V. K. and Tezlevan, V. E., "Photoelectric properties of Ni-CdP_2 Schottky barriers", Sov. Phys.-Semicond., 8, 1343-7 (1975), PA78-66031.
431.Bois, D. and Boulou, M., "Photocapacitance studies in high-purity GaAs", Phys. Status Solidi A, 22, 671-5 (1974), PA77-44964.
432.Bucher, H. K., Burkey, B. C., Lubberts, G. and Wolf, E. L., "Photodetection by barrier modulation in Cu-diffused Au/CdS junctions", Appl. Phys. Lett., 23, 617-19 (1973), PA77-16818.
433.Diehl, H., "Theory of photo effects at p-n junctions and metal-semiconductor contacts", Phys. Stat. Solidi, 9, 621-34 (1965), German.
434.Itoh, Y. and Hashimoto, N., "Photoelectric measurement of tungsten silicide and n-type silicon barriers", J. Appl. Phys., 38, 899-900 (1967), PA70-26670.
435.Langmann, U., "Photoluminescence of n-GaAs at transparent Schottky contacts", Appl. Phys., 1, 219-21 (1973), EEA76-25993.
436.Lebedev, A. A., Ovezov, K., Prochukhan, V. D. and Rud, Y. V., "Cu-$ZnSiP_2$ Schottky diodes", Pisma V. Zh. Tekh. Fiz., 1, 195-9 (1975), Russian, PA78-72904.
437.Li, S. S., Lindholm, F. A. and Wang, C. T., "Quantum yield of metal-semiconductor photodiodes", J. Appl. Phys., 43, 4123-9 (1972), PA75-81116.
438.Lubberts, G., Burkey, B. C., Bucher, H. K. and Wolf, E. L., "Photodetection by light-induced barrier modulation in Cu-diffused Au-CdS diodes", J. Appl. Phys., 45, 2180-90 (1974), PA77-60857.
439.Mach, R. and Ludwig, W., "Impact ionization investigations on ZnSe Schottky barrier diodes", Phys. Status Solidi A, 23, 507-16 (1974), EEA77-30237.
440.Makarov, V. V., "Optical absorption of silicon carbide investigated as a surface-barrier diode", Sov. Phys.-Semicond., 6, 1556-7 (1973), EEA76-19891.
441.Padovani, F. A., "Light source control system for barrier height studies", Rev. Sci. Instrum., 39, 772-4 (1968), PA71-35875.

442.Peka, G. P., Strikha, V. I. and Timofeeva, G. K., "Recombinational radiation in GaAs-metal contacts", Ukr. Fiz. Zh., 18 (12), 2063-6 (1973), Russian, PA77-33499.

443.Radautsan, S. I., Molodyan, I. P., Bulyarskii, S. V., Syrbu, N. N. and Tezlevan, V. E., "Photoelectrical properties of Au-$CdIn_2S_4$ surface-barrier diode", Phys. Status Solidi A, 21, 617-22 (1974), EEA77-14430.

444.Radautsan, S. I., Syrbu, N. N. and Kiosev, V. K., "Optical modulation spectra of CdP_2 crystals and Ni-CdP_2 Schottky barriers", Phys. Status Solidi B, 64, 459-65 (1974), PA77-70049.

445.Radautsan, S. I., Syrbu, N. N., Tezlevan, V. E. and Bol, I. B., "Franz-Keldysh effect in Ag-CdP_2 Schottky barriers", Sov. Phys.-Semicond., 8, 564-7 (1974), PA78-25772.

446.Rochon, P. and Fortin, E., "Photovoltaic effect of a Au-Ge barrier in a quantizing magnetic field", Can. J. Phys., 52, 1173-7 (1974), PA77-67277.

447.Ruppel, W., "Fundamentals of interface photoeffects", 6th International IMEKO Symposium on Photon-Detectors, 1974, p. 256-67, PA78-77179.

448.Ryall, M. D. and Allen, J. W., "Exciton emission from forward-biased zinc selenide Schottky diodes", J. Phys. & Chem. Solids, 34, 2137-41 (1973), PA77-25176.

449.Seiranyan, G. B., Tkborik, Y. A., Zuev, V. A. and Sukach, G. A., "Barrier photo-emf of metal-GaAs Schottky diodes under strong excitation conditions", Sov. Phys.-Semicond., 8, 625 (1974), PA78-21958.

450.Spicer, W. E., Gregory, P. E., Chye, P. W., Babalola, I. A. and Sukegawa, T., "Photoemission study of the formation of Schottky barriers", Appl. Phys. Lett., 27, 617-19 (1975).

451.Spitzer, W. G. and Mead, C. A., "Conduction band minima of $Ga(As_{1-x}P_x)$", Phys. Rev., 133, 872 (1964).

452.Stillman, G. E., Wolfe, C. M., Rossi, J. A. and Foyt, A. G., "Unequal electron and hole impact ionization coefficients in GaAs", Presented at 1973 International Electron Devices Meeting, Washington, D. C., December 1973, PA77-32971.

453.Sze, S. M., Crowell, C. R. and Kahng, D., "Photoelectric determination of the image force dielectric constant for hot electrons in Schottky barriers", J. Appl. Phys., 36, 2534-6 (1964).

454.Tansley, T. L., "Temperature dependence of hole diffusion length from spectral response of Au-n^+ GaAs photodiodes", J. Phys. D, 5, 1146-52 (1972), PA75-44178.

455.Tsarenkov, B. V., Goldberg, Y. A., Gusev, G. V. and Ogurtsov, V. I., "Photoelectric properties of Au-n-GaP surface-barrier structures in the ultraviolet part of the spectrum", Sov. Phys.-Semicond., 8, 264-5 (1974), PA77-72866.

456.Tuzzolino, A. J., Hubbard, E. L., Perkins, M. A. and Fan, C. Y., "Photo effects in silicon surface-barrier diodes", J. Appl. Phys., 33, 148-55 (1962).

457.Wolf, E. L., "High gain photodetection by barrier modulation in compensated Schottky junctions", Conference on Metal-Semiconductor Contacts, 1974, p. 91-101, PA78-15728.
458.Zuev, V. A. and Seiranyan, G. B., "Laser-aided photo-excitation as a method for investigating the Schottky barrier", Kvantovaya Elektron., Kiev, no. 8, 80-2 (1974), Russian, PA78-42301.

6. Pressure Sensitivity

459.Boitsov, Y. P. and Prokhorov, V. I., "Resolving power of the method of measuring the resistivity from the leakage resistance of a metal-semiconductor contact", Instrum. & Exp. Tech., 14, 281-4 (1971), EEA74-37195.
460.Cenerini, M., Costato, M. and Reggiani, L., "Pressure dependence of the reverse current in surface barrier", Lett. Nuovo Cimento,
461.Chino, K. and Ariyoshi, H., "Effect of surface polishing on the stress-sensitivity of a Schottky-barrier diode", J. Math. Phys. Cambridge, 7, 1130 (1968), PA72-5725.
462.Chino, K., Ariyoshi, H. and Mizushima, Y., "Induced stress-sensitivity of ion-bombarded silicon-metal contacts", Jap. J. Appl. Phys., 9, 406 (1970), PA73-53865.
463.Chino, K. and Mizushima, Y., "Behavior of the Schottky-barrier diode under uniaxial stress", Jap. J. Appl. Phys., 13, 155-63 (1974), EEA77-14475.
464.Chino, K., Okamoto, H., Ariyoshi, H. and Mizushima, Y., "Stress effect of gold-doped and gamma-irradiated Schottky-barrier diodes", Jap. J. Appl. Phys., 8, 502-3 (1969), PA72-24202.
465.Fonash, S. J., "The effect of stress on metal semiconductor junctions", Solid-State Electron., 16, 253-63 (1973), PA76-21151.
466.Fonash, S. J., "Comments on barrier changes in stressed metal-semiconductor contacts", J. Appl. Phys., 45, 496-8 (1974), PA77-44994.
467.Groth, O., Hok, B. and Tove, P. A., "Stress effect in sputtered metal-semiconductor junctions", Phys. Status Solidi A, 20, 589-94 (1973), PA77-16823.
468.Guetin, P. and Schreder, G., "Band-structure effects in metal-GaSb tunnel contacts under pressure", Phys. Rev. Lett., 27, 326-30 (1971), PA74-65783.
469.Guetin, P. and Schreder, G., "Phonom tunneling spectroscopy in n-Ge Schottky barriers under pressure", Phys. Rev. B, 7, 3697-702 (1973), PA76-36379.
470.Kano, G., Fujiwara, S., Iizuka, M., Hasegawa, H. and Sawaki, T., "Schottky barrier pressure-sensitive semiconductor device with air space around periphery of metal-semiconductor junction", Patent USA 3786320, Publ. January 1974.
471.Kano, G. and Kato, H., "Pressure-sensitive transistor having a Schottky barrier junction formed in the emitter", Patent USA 3757176, Publ. September 1973.

472. Kramer, P. and van Ruyven, L. J., "Position of the band edges of silicon under uniaxial stress", Appl. Phys. Lett., 20, 420-2 (1972), PA75-51167.
473. Mabrouk, S. A. E., "Effects of gold doping on pressure sensitivity and barrier characteristics of the Al-n-Si Schottky barrier structure", Thesis, Case Western Res. Univ., Cleveland, Ohio, 1971, 108 pp., Order No. 72-6314, EEA75-32641.
474. Mora, N. A., Bermon, S. and Pollak, F. H., "Effects of uniaxial compression on the zero-bias anomaly in p-silicon Schottky-barrier tunnel junctions", Phys. Rev. Lett., 28, 225-8 (1972), EEA75-10228.
475. Okamoto, H., Ariyoshi, H. and Mizushima, Y., "Observation on stress effect in Schottky barrier diodes", Solid-State Electron., 12, 441-4 (1969), PA72-33344.
476. Pickering, C., Vyas, M. K. R., Pitt, G. D. and Adams, A. R., "Effect of pressure on the high electric field instabilities in n-type GaAs", J. Phys. C, 8, 129-37 (1975).
477. Pokalyakin, V. I. and Stepanov, G. V., "The investigation of pressure sensitivity of silicon-gold diodes", Presented at 5th Czechoslovakia Conference on Electronics and Vacuum Physics, Brno, Czechoslovakia, October 1972, PA76-15255.
478. Polyakova, A. L., "Pressure sensitivity of Schottky barrier diode as a pressure transducer", Proceedings of the 7th International Congress on Acoustics, 1971, p. 221-4, EEA75-16720.
479. Rideout, V. L. and Crowell, C. R., "Pressure sensitivity of gold-potassium tantalate Schottky barrier diodes", Appl. Phys. Lett., 10, 329-32 (1967), EEA70-17007.
480. Toyama, N. and Shimizu, T., "The stress effect on Schottky barrier diode", Bull. Kyushu Inst. Technol. (Sci. & Technol.), no. 29, 139-46 (1974), Japanese, EEA78-8529.
481. Yelinson, M. I., Pokalyakin, V. I., Polyakova, A. L., Stepanov, G. V. and Shklovskaya-Kordi, V. V., "Current sensitivity of Schottky diodes to pressure", Radiotekhnika I Elektronika, 15, 210-12 (1970), Russian, EEA73-21257.

7. Temperature Effects

482. Balazs, J., Fercenzi, G. and Somogyi, M., "Investigation of thermally stimulated current (TSC) in nGaP-Au Schottky barriers", Kozl. Magy. Tud. Akad. Musz. Fiz. Kut. Intez., no. 0-14, 85-97 (1974), Hungarian, PA78-10074.
483. Banerjee, A. R. and Gardiner, J. G., "Unbalance and temperature effects in Schottky barrier diode mixers", Int. J. Electron., 34, 619-28 (1973), EEA76-25647.
484. Card, H. C., "Thermally-induced changes in barrier heights of aluminium contacts to p- and n-type silicon", Solid State Commun., 16, 87-9 (1975), EEA78-11582.

485.Card, H. C., "One-dimensional analysis of heat treated aluminium-silicon junctions", Conference on Metal-Semiconductor Contacts, 1974, p. 129-37, PA78-18679.
486.Cho, A. Y. and Casey, H. C., Jr., "Properties of Schottky barriers and p-n junctions prepared with GaAs and $Al_xGa_{1-x}As$ molecular beam epitaxial layers", J. Appl. Phys., 45, 1258-63 (1974), PA77-57163.
487.Crowell, C. R. and Sze, S. M., "Temperature dependence of avalance multiplication in semiconductors", Appl. Phys. Lett., 9, 242-4 (1966), EEA70-533.
488.El-Gabaly, M. A. and Goud, P. A., "On the temperature dependence of the small signal admittance of punch-through injection transit-time diodes", 1973 European Microwave Conference, Vol. I, A11.1/4 pp., EEA77-5987.
489.Forlain, F., Minnaja, N. and Sacchi, G., "Contribution to the study of metal contacts of semiconductor real surfaces", Electron. Lett., 3, 196-8 (1967), EEA70-14551.
490.Griffiths, B., Stow, C. D. and Syms, P. H., "An accurate diode thermometer for use in thermal gradient chambers", J. Phys. E, 7, 710-14 (1974), PA77-68640.
491.Hackam, R. and Harrop, P,. "Temperature dependence of the Schottky barrier height in gallium arsenide", Solid State Commun., 11, 669-72 (1972), EEA75-36246.
492.Klose, H., "On the potential barrier of Schottky junctions", Phys. Status Solidi A, 14, 457-62 (1972), PA76-15252.
493.Mach, R., Treptow, H. and Ludwig, W., "Physical properties of the Au-ZnSe metal-semiconductor contact", Phys. Status Solidi A, 25, 567-73 (1974), PA78-10076.
494.Murarka, S. P., "High-temperature stability of AuPt/n-GaAs Schottky barrier diodes", Solid-State Electron., 17, 869-76 (1974), EEA77-30294.
495.Nannichi, Y. and Pearson, G. L., "Gallium phosphide devices for use at high temperatures", Conference Record of the 7th Photovoltaic Specialists Conference, 1968, p. 16-21, EEA74-10611.
496.Nannichi, Y. and Pearson, G. L., "Properties of GaP Schottky barrier diodes at elevated temperatures", Solid-State Electron., 12, 341-8 (1969), PA72-33345.
497.Rideout, V. L. and Crowell, C. R., "Current transport by thermally assisted tunneling in metal-semiconductor contacts", 30th Annual Conference on Physical Electronics, 1970, Pap. C5, EEA73-24942.
498.Sinha, A. K. and Poate, J. M., "Relative thermal stabilities of thin-film contacts to n-GaAs metallized with W, Au and Pt", Jap. J. Appl. Phys., suppl. 2 (pt. 1), 841-4 (1974), EEA78-19717.
499.Smith, M. A. and Carter, J., "Constant-temperature method for evaluating deep-level parameters in Schottky-barrier TSC measurements", J. Phys. D, 8, 254-61 (1975).

500. Tantraporn, W., "Determination of low barrier heights in metal-semiconductor contacts", J. Appl. Phys., 41, 4669-71 (1970), EEA74-1498.
501. Vernon, S. M. and Anderson, W. A., "Temperature effects in Schottky-barrier silicon solar cells", Appl. Phys. Lett., 26, 707-9 (1975), PA78-72901.
502. Wada, O., Yanagisawa, S. and Takanashi, H., "The effect of heat treatment on Al-GaAs Schottky barriers", Jap. J. Appl. Phys., 12, 1814-15 (1973), PA77-6673.
503. Wasse, M. P., Clerk, B. W. and Conlon, R. F. B., "AgSn cathode contact in gallium-arsenide transferred-electron devices", Electron. Lett., 9, 189-90 (1973), EEA76-26029.

8. Contact Resistance

504. Becker, R., "Nonblocking contacts on indium phosphide", Solid-State Electron., 16, 1241-9 (1973), EEA77-5929.
505. Chang, C. Y., Fang, Y. K. and Sze, S. M., "Specific contact resistance of metal-semiconductor barriers", Solid-State Electron., 14, 541-50 (1971), EEA74-34024.
506. Dzhunaidov, S. S., Knigin, P. I., Urmanov, N. A. and Shopen, V. I., "Characteristics of alloyed Au:Sb contacts with cadmium-compensated silicon", Sov. Phys.-Semicond., 8, 384-5 (1974), PA78-6895.
507. Farmer, R. H. and Ishii, T. K., "A new method for calculating the upper cutoff frequency of a Schottky barrier diode", Proceedings of the National Electronics Conference, 1970, p. 191-5, EEA75-1127.
508. Gutai, L. and Mojzes, I., "Determination of semiconductor-metal contact resistance by an angle-dependent geometrical magneto-resistance method", Appl. Phys. Lett., 26, 325-6 (1975), PA78-46746.
509. Gwarek, W., Karatnicki, P. and Morawski, T., "The perturbation method to series resistance measurements of the Schottky diode", Elektronika, 15, 470-1 (1974), Polish, EEA78-15312.
510. Heime, K., Konig, U., Kohn, E. and Wortmann, A., "Very low resistance Ni-AuGe-Ni contacts to n-GaAs", Solid-State Electron., 17, 835-7 (1974), EEA77-30247.
511. Joseph, R. R., Prokop, G. S. and Ting, C. Y., "Current crowding effects at aluminum silicon contacts", The Annual Reliability Physics Symposium, 1970, p. 36-7, EEA73-24104.
512. Nougier, J. P. and Rolland, M., "Metal semiconductor contact: resistivity and noise", Solid-State Electron., 16, 1399-405 (1973), PA77-13224.
513. Shepela, A., "The specific contact resistance of Pd_2Si contacts on n- and p-Si", Solid-State Electron., 16, 477-81 (1973), PA76-27311.
514. Stratton, R. and Padovani, F. A., "Differential resistance peaks of Schottky barrier diodes", Solid-State Electron., 10, 813-21 (1967), EEA70-19154.

515.Tippmann, H. and Trimbuch, R., "Specfic contact resistance of Al-p-Si contacts", Wiss. Z. Tech. Hochsch. Ilmenau, 20, 133-47 (1974), German, PA78-33554.
516.Tsuda, N., "Resistance anomalies in metal-semiconductor tunnel junctions", J. Phys. Soc. Japan, 24, 962 (1968), EEA71-20568.
517.Yu, A. Y. C., "Electron tunneling and contact resistance of metal-silicon contact barriers", Solid-State Electron., 13, 239-47 (1970), EEA73-14178.

9. Ohmic Contacts

518.Almazov, A. B., Kulikova, E. V. and Ryzhikov, I. V., "Determination of a metal-semiconductor contact resistance", Fiz. Tekh. Poloprovodikov, 3, 754-7 (1969), Russian, EEA73-4410.
519.Armantrout, G. A. and Looney, J. C., "Metal-semiconductor contact measurements using a VHF bridge", Solid State Technol., 11 (4), 29-37 (1968), EEA71-18168.
520.Astridge, R. A. and Holt, J., "Alloyed contacts to thick, large area semiconductor devices", J. Mater. Sci., 5, 640-4 (1970), EEA74-1500.
521.Bareikis, V. A. and Liberis, Y. S., "Low frequency current fluctuations in an ohmic point contact for a metal and a p-type germanium", Litov. Fiz. Sb., 13, 881-91 (1973), Russian, EEA77-38792.
522.Berger, H. H., "Contact resistance and contact resistivity", J. Electrochem. Soc., 119, 507-14 (1972), EEA75-24274.
523.Bower, R. W., "Characteristics of aluminum-titanium electrical contacts on silicon", Appl. Phys. Lett., 23, 99-101 (1973), EEA76-32297.
524.Crowell, C. R., "General normalized current voltage characteristics of quasi-ohmic contacts", Ohmic Contacts to Semiconductors, 1968, p. 17-30, Publ. 1969, PA73-21164.
525.Duraev, V. P., Pugach, M. K. and Shveikin, V. I., "Dependence of the resistance of nonrectifying contacts with GaAs on the frequency and density of the current", Instrum. & Exp. Tech., 14, 612-3 (1971), EEA75-10185.
526.Fremunt, R., Misek, J. and Kratena, L., "Some optical properties of Au-Ni contacts on n-type GaP", Phys. Status Solidi A, 20, K41-3 (1973), PA77-6676.
527.Gesellschaft Fur Gleichrichterhan & Elektronik mbH, "Semiconductor contact electrode", Patent UK 2207010, Publ. September 1970, EEA74-10567.
528.Goncharov, V. P. and Okhrimenko, Y. A., "Properties of ohmic contacts with gallium antimonide single crystals that have been grown by the gas-transport reaction method", Instrum. & Exp. Tech., 17, 1189-90 (1974), EEA78-19709.
529.Gu, J., Kitahar, T., Kawakami, K. and Sakaguchi, T., "Ohmic contact and impurity conduction in P-doped CdTe", J. Appl. Phys., 46, 1184-5 (1975), EEA78-19579.

530. Henriksen, T., Andreasen, T., Andersen, S. L. and Olsen, T., "Alloying behavior of metal films on GaAs", Phys. Norv., 7 (2), 100 (1973), PA77-16822.
531. Mey, G. D., "Electrical contacts to evaporated cadmium selenide", Electrocompon. Sci. & Technol., 1, 39-41 (1974), PA78-15717.
532. Nishov, V. Y. and Kubetskii, G. A., "Resistance of ohmic contacts between metals and semiconductor films", Sov. Phys. Semicond., 4, 1553-4 (1971).
533. Pawlikowski, J. M., Kochan, B. and Becla, P., "The ohmic metal - p-type Zn_3As_2 contact", Acta Phys. Pol. A, A45, 803-5 (1974), PA77-53143.
534. Prakash, O., "Exact computation of bulk resistance for evaporated ohmic contacts", J. Inst. Electron. & Telecommun. Eng., 20, 488-91 (1974), PA78-62102.
535. Robinson, G. Y. and Jarvis, N. L., "Auger electron spectroscopy and sputter etching of Ni/Au-Ge on n-GaAs", Appl. Phys. Lett., 21, 507-10 (1972), PA76-4903.
536. Semenov, G. M. and Pyrkov, V. V., "High-speed monitoring of the contact joints to semiconductor power diodes", Defekoskopiya, no. 6, 90-5 (1971), Russian, PA75-25893.
537. Sinha, A. K., "Electrical characteristics and thermal stability of platinum silicide-to-silicon ohmic contacts metalized with tungsten", J. Electrochem. Soc., 120, 1767-71 (1973), EEA77-10153.
538. Sokolov, Y. F. and Elenkrig, B. B., "Resistances of non-rectifying contacts with semiconductors", Sov. Phys.-Semicond., 7, 437-8 (1973), PA76-75793.
539. Wagner, H. P. and Besocke, K. H., "Alternating current method for separating the contact influence from bulk properties of semiconductors", J. Appl. Phys., 40, 2916-22 (1969), EEA72-23799.
540. Weiss, B. L. and Hartnagel, H. L., "Complex distribution effects of thin component ohmic-contact layers on GaAs", Electron. Lett., 11, 263-4 (1975), PA78-62099.
541. Yokoyama, N., Ohkawa, S. and Ishikawa, H., "Effects of the heating rate in alloying of Au-Ge to n-type GaAs on the ohmic properties", Jap. J. Appl. Phys., 14, 1071-2 (1975), EEA78-37467.

V. Schottky Barrier Diode as a Research Tool

1. Impurity Profiling and Deep Lying Impurities

542. Anon., "Determination of the composition and carrier concentration of epitaxial $GaAs_{1-x}P_x$ layers by the method of Schottky barriers", Kozl. Magy. Tud. Akad. Musz. Fiz. Kut. Intez., no. 0-14, 99-108 (1974), Hungarian, PA78-6922.
543. Adams, A. R., "Schottky barrier diodes", J. Phys. E, 5, 958-61 (1972), PA75-75127.
544. Baranowski, J. J. and Higgins, V. J., "Computer-aided technique for the accurate determination of the semiconductor impurity doping profile", Report AD-697127, U. S. Army Electronics Command, Fort Monmouth, New Jersey, 1969.

545. Beguwala, M. and Crowell, C. R., "Experimental determination of properties of deep-level systems in semiconductor junctions by admittance measurements", J. Electrochem. Soc., 120, 89C-90C (1973), EEA76-23426.
546. Bradshaw, A. and Knappett, J. E., "Epitaxial growth of device quality gallium arsenide layers", Solid State Technol., 13 (10), 45-50 (1970).
547. Califano, F. P. and Luciano, A., "An automatic test set for measuring the doping profile of semiconductor epitaxial layers", Rev. Sci. Instrum., 41, 865-9 (1970).
548. Carroll, J. E., "Measurement of donor density of pulsing Schottky-barrier diodes", Electron. Lett., 4, 259-60 (1968), EEA71-20567.
549. Carter, W. E., Gummel, H. K. and Chawala, B. R., "Interpretation of capacitance vs. voltage measurements of p-n junctions", Solid-State Electron., 15, 195-202 (1972).
550. Coerver, L. E., "Note on the interpretation of C-V data in semiconductor junctions", IEEE Trans. Electron Devices, ED-17, 436 (1970), EEA73-21215.
551. Copeland, J. A., "A technique for directly plotting the inverse doping profile of semiconductor wafers", IEEE Trans. Electron Devices, ED-16, 445-9 (1969).
552. Copeland, J. A., "Diode edge effect on doping-profile measurements", IEEE Trans. Electron Devices, ED-17, 404-7 (1970).
553. Dean, R. H., "A practical technique for controlling field profile in thin layers of n-GaAs", IEEE Trans. Electron Devices, ED-19, 1144-8 (1972).
554. Dean, R. H., "Reflection amplification in thin layers of n-GaAs", IEEE Trans. Electron Devices, ED-19, 1148-56 (1972).
555. De Man, H. J. J., "On the calculation of doping profiles from C(V) measurements on two-sided junctions", IEEE Trans. Electron Devices, ED-17, 1087-8 (1970).
556. De Man, H. J. J., "A fast algorithm for the calculation of junction capacitance and its application for impurity profile determination", Solid-State Electron., 15, 177-87 (1972).
557. Dmitruk, N. I., Tereshchenko, A. K., Maevo, O. I., Lyashenko, V. I. and Raskevich, A. M., "Influence of deep levels on the space charge capacitance and the field effect in gallium arsenide", Sov. Phys.-Semicond., 7, 469-73 (1973), PA77-24829.
558. Fabre, E., "Characterization of deep centres in gallium arsenide by photocapacitive methods", C. R. Acad. Sci. Ser. B, 270, 848-51 (1970), French, EEA73-24734.
559. Ferenezi, G. and Balazs, J., "Remarks on the interpretation of TSC measurements in GaP p-n and metal-semiconductor junctions", Conference on Metal-Semiconductor Contacts, 1974, p. 249-54, EEA78-8512.
560. Goldberg, Y. A. and Tsarenkov, B. V., "An investigation of the relaxation of the capacitance of surface-barrier structures and determination of the parameters of impurity atoms in n-type GaAs", Fiz. & Tekh. Poluprov., 5, 1781-6 (1970), Russian, EAA75-4631.

561.Grimmeiss, H. G., "The application of metal-semiconductor barriers to the characterization of deep level impurities", Conference on Metal-Semiconductor Contacts, 1974, p. 187-200, EEA78-8507.

562.Grimshaw, J. A. and Osborne, D. N., "Charge neutrality in semiconductors with implanted impurity profiles", Solid-State Electron., 14, 603-11 (1971).

563.Hamilton, B., "The detection of minority carrier traps by optically induced exchange in n-type GaP Schottky barriers", Conference on Metal-Semiconductor Contacts, 1974, p. 218-25, EEA78-8510.

564.Hamilton, B. and Smith, B. L., "Inadvertent deep centers in n-type GaP from Schottky barrier photocapacitance", Appl. Phys. Lett., 22, 674-6 (1973), EEA76-29109.

565.Heime, K., "Investigations of epitaxial GaAs-layers by means of Schottky contacts", Z. Angew. Phys., 32, 374-81 (1972), German, PA75-25431.

566.Huang, C. I., "Investigation of deep level impurities (oxygen and chromium) in bulk gallium arsenide and gold-gallium arsenide Schottky diodes", Thesis, Univ. Florida, Gainesville, Fla., 1973, 87 pp., Order No. 74-9601, CA81-31084.

567.Hughes, F. D., "The characterization of epitaxial GaAs using Schottky barriers", Acta Electron., 15 (1), 11 pp. (1972), EEA75-16616.

568.Johnson, W. C. and Panousis, P. T., "The influence of Debye length on the C-V measurement of doping profiles", IEEE Trans. Electron Devices, ED-18, 965 (1971).

569.Karpinsky, J., Cap, J., Tomek, J. and Karlovsky, J., "An instrument for automatic measurement and recording of distribution of impurities in semiconductors", Elektrotech. Cas., 25, 404-11 (1974), Czech, EEA77-33929.

570.Kennedy, D. P., Murley, P. C. and Kleinfelder, W., "On the measurement of impurity atom distributions in silicon by the differential capacitance technique", IBM J. Res. Develop., 12, 399-409 (1968).

571.Kennedy, D. P. and O'Brien, R. R., "On the measurement of impurity atom distributions by differential capacitance technique", IBM J. Res. Develop., 13, 212-4 (1969).

572.Klohn, K. L. and Wandinger, L., "Variation of contact resistance of metal-GaAs contacts with impurity concentration and its device implication", J. Electrochem. Soc., 116, 507-8 (1969).

573.Komaka, M., Abe, T. and Sato, K., "Non-destructive determination of impurity concentration in silicon epitaxial layer using metal-silicon Schottky barrier", Jap. J. Appl. Phys., 7, 790-1 (1968), EEA71-25898.

574.Ladbrooke, P. H., "Reverse current characteristics of some imperfect Schottky barriers", Solid-State Electron., 16, 743-9 (1973), EEA76-23359.

575.Lambert, L. M. and Kump, H. J., "The measurement of deep levels using C-V techniques", Phys. Status Solidi A, 21, 233-45 (1974), EEA77-10172.
576.Livingstone, A. W. and Allen, J. W., "Impact ionization of deep impurities in zinc selenide", J. Phys. C, 6, 3491-500 (1973), PA77-10158.
577.Losee, D. L., "Admittance spectroscopy of deep impurity levels: ZnTe Schottky barriers", Appl. Phys. Lett., 21, 54-6 (1972), PA75-61391.
578.Losee, D. L., "Admittance spectroscopy of impurity levels in Schottky barriers", J. Appl. Phys., 46, 2204-14 (1975).
579.Marfang, Y., Lascaray, J. and Triboulet, R., "Study of deep and shallow levels by photocapacitance in Au-n-CdTe Schottky barriers", Conference on Metal-Semiconductor Contacts, 1974, p. 201-9, EEA78-8508.
580.Meyer, N. I. and Guldbrandsen, T., "Method for measuring impurity distributions in semiconductor crystals", Proc. IEEE, 51, 1631-7 (1963).
581.Miller, G. J., "A feedback method for investigating carrier distribution in semiconductors", IEEE Trans. Electron Devices, ED-19, 1103-8 (1972).
582.Pals, J. A., "Properties of Au, Pt, Pd and Rh levels in silicon measured with a constant capacitance technique", Solid-State Electron., 17, 1139-45 (1974).
583.Parker, G. H., "Tunneling in Schottky barriers", Thesis, California Inst. Technol., Pasadena, 1970, 52 pp., Order No. 70-14846, EEA74-26833.
584.Pokorny, J. and Fremunt, R., "Determining the concentration profile of dopants in epitaxial GaAs by the method of differential capacitance in a Schottky diode", Slaboproudy Obz., 33, 461-6 (1972), Czech, PA75-83905.
585.Roberts, G. I. and Crowell, C. R., "Capacitance energy level spectroscopy of deep-lying semiconductor impurities using Schottky barriers", J. Appl. Phys., 41, 1767-76 (1970).
586.Sah, C. T., Rosier, L. L. and Forbes, L., "Low-temperature high-frequency capacitance measurements of deep- and shallow-level impurity center concentrations", Appl. Phys. Lett., 15, 161-4 (1969).
587.Salama, C. A. T. and Holmes, F., "A technique for directly plotting the doping profile of semiconductor wafers", Solid-State Electron., 14, 178-81 (1971), PA74-33710.
588.Schulz, M., "Deep trap levels of ion-implanted germanium in silicon measured by Schottky contact techniques", Appl. Phys. Lett., 23, 31-3 (1973), EEA76-29185.
589.Schulz, M., "Determination of deep trap levels in silicon using ion-implantation and CV-measurements", Appl. Phys., 4, 225-36 (1974), PA77-72720.

590. Smith, B. L., "Au-(n-type) InP Schottky barriers and their use in determining majority carrier concentrations in n-type InP", J. Phys. D, 6, 1358-62 (1973).
591. Smith, B. L., "Deep level energy spectroscopy in GaP using thermal release of trapped charge in Schottky barriers", Conference on Metal-Semiconductor Contacts, 1974, p. 210-17, EEA78-8509.
592. Smith, B. L. and Carter, M. A., "A constant-temperature method for evaluating deep-level parameters in Schottky-barrier TSC measurements", J. Phys. D, 8, 254-61 (1975), PA78-25754.
593. Spiwak, R. R., "Design and construction of a direct-plotting capacitance inverse-doping profiler for semiconductor evaluation", IEEE Trans. Instrum. Meas., IM-18, 197-202 (1969).
594. Sugano, T. and Katoda, T., "Electron mobility and deep traps in GaAs epitaxial layer", Annu. Rep. Eng. Res. Inst. Fac. Eng. Univ. Tokyo, 32, 157-62 (1973), Japanese, EEA77-22767.
595. Tansley, T. L., "AC profiling by Schottky gated cloverleaf", J. Phys. E, 8, 52-4 (1975), EEA78-15139.
596. Tantraporn, W. and Glover, G. H., "Avalanche breakdown criterion from doping profile integral", IEEE Trans. Electron Devices, ED-19, 697-8 (1972), PA75-33025.
597. Teshima, H., Tarui, Y. and Saeki, T., "A measurement of the impurity distributions in Si epitaxial growth layers by the Schottky diode method", Bull. Electrotech. Lab., 33, 639-44 (1969), Japanese, EEA73-17413.
598. Tsarenkov, B. V., Goldberg, Y. A., Posse, E. A. and Shulinskaya, M. M., "Method for the determination of the parameters of impurity centers in a semiconductor from the relaxation of the reverse current in surface-barrier structures as applied to GaAs", Sov. Phys. Semicond., 6, 2003-6 (1973), PA76-50267.
599. Zohta, Y., "Frequency dependence of C and $\Delta V/\Delta(C^{-2})$ of Schottky barriers containing deep impurities", Solid-State Electron., 16, 1029-36 (1973).

2. Tunneling Spectroscopy

600. Abdurakhmanov, K. P., Kotov, B. A., Okuneva, N. M., Plachenova, E. L. and Usov, O. A., "Some electronic properties of heavily doped p-type GaAs", Sov. Phys. Solid State, 14, 2234-6 (1973), PA76-33995.
601. Binnig, G. and Hoenig, H. E., "Energy gap of the superconducting semiconductor $SrTiO_{3-x}$ determined by tunneling", Solid State Commun., 14, 597-601 (1974), PA77-37358.
602. Card, H. C. and Rhoderick, E. H., "Studies of tunnel MOS diodes. I. Interface effects in silicon Schottky diodes", J. Phys. D, 4, 1589-601 (1971).
603. Card, H. C. and Rhoderick, E. H., "Studies of tunnel MOS diodes. II. Thermal equilibrium considerations", J. Phys. D, 4, 1602-11 (1971), PA74-73754.

604.Carruthers, T., "Temperature and bias dependence of zero-bias tunneling structure in GaAs Schottky junctions", J. Appl. Phys., 41, 3870-2 (1970), PA73-68863.
605.Carruthers, T., "Bias and temperature dependence of tunnelling phenomena in GaAs Schottky-barrier junctions", Phys. Rev. B, 10, 3356-74 (1974), PA78-21954.
606.Chen, T. T., Chen, W. H. and Shaw, S. Y., "Electron tunneling in metal-Ge-metal junctions", Phys. Lett. A, 46A, 303-4 (1974), PA77-20684.
607.Christopher, J. E., Darley, H. M., Lehman, G. W. and Tripathi, S. N., "Comparison of experiment and theory for tunneling in In-Ge:P Schottky-barrier junctions", Phys. Rev. B, 11, 754-63 (1975), EEA78-15270.
608.Combescot, R. and Schreder, G., "Tunnelling in metal-semiconductor contacts. I. Influence of the impurities", J. Phys. C, 6, 1363-84 (1973), EEA76-23386.
609.Combescot, R. and Schreder, G., "Tunnelling in metal-semiconductor contacts. II. Influence of the electron-phonon interaction", J. Phys. C, 7, 1318-36 (1974), PA77-44995.
610.Conley, J. W., Duke, C. B., Mahan, G. D. and Tiemann, J. J., "Electron tunneling in metal-semiconductor barriers", Phys. Rev., 150, 466 (1966).
611.Conley, J. W. and Tiemann, J. J., "Experimental aspects of tunneling in metal-semiconductor barriers", J. Appl. Phys., 38, 2880-4 (1967), PA71-35932.
612.Crowell, C. R., "Interpretation of tunnel emission and capacitance measurements in the presence of dielectric film - thickness fluctuations", Appl. Phys. Lett., 8, 328 (1966).
613.Cullen, D. E., "Tunneling spectroscopy in P type silicon", Thesis, Univ. Illinois, Urbana, 1970, 140 pp., Order No. 70-20943, EEA74-34030.
614.Davis, L. C. and Steinrisser, F., "One-electron and phonon-assisted tunneling in n-Ge Schottky barriers", Phys. Rev. B, 1, 614-23 (1970), PA73-44285.
615.Eklund, P. C. and Van Zandt, L. I., "Inelastic electron tunneling through Schottky barriers on vanadium-doped Ti_2O_3", Phys. Rev. B, 11, 784-92 (1975), PA78-33553.
616.Fritzsche, H., "Experimental investigation of the electronic properties of solids", Report AFOSR-71-0327TR, Univ. Chicago, Ill., 1971, 16 pp., PA75-47789.
617.Guetin, P. and Schreder, G., "Phonon structures in metal-semiconductor tunnel junctions", Solid State Commun., 8, 291-4 (1970), EEA73-17617.
618.Guetin, P. and Schreder, G., "Tunneling in metal-semiconductor contacts under pressure", Philips Tech. Rev., 32, 211-20 (1971), PA75-61393.
619.Guetin, P. and Schreder, G., "Tunneling spectroscopy and band-structure effects in n-GaSb under pressure", Phys. Rev. B, 6, 3816-35 (1972), EEA76-749.

620. Inkson, J. C., "Many body effects near metal semiconductor interfaces", Presented at 11th International Conference on Physics of Semiconductors, Warsaw, Poland, July 1972, PA76-2183.
621. Inkson, J. C., "Many-body effects at metal-semiconductor junctions. I. Surface plasmons and the electron-electron screened interaction", J. Phys. C, 5, 2599-610 (1972), PA75-75119.
622. Inkson, J. C., "Many-body effect at metal-semiconductor junctions. II. The self energy and band structure distortion", J. Phys. C, 6, 1350-62 (1973), EEA76-23385.
623. Johnson, K. W. and Olson, D. H., "Electron tunneling into $KTaO_3$ Schottky barrier junctions", Phys. Rev. B, 3, 1244-8 (1971), PA74-28548.
624. Kieu, V. C., Granger, R. and Finck, C., "Electrical properties of indium-lead telluride tunnel junctions", Phys. Status Solidi A, 15, 491-500 (1973), French, EEA76-16181.
625. Losee, D. L. and Wolf, E. L., "Tunneling spectroscopy of CdS Schottky-barrier junctions", Phys. Rev., 187, 925-32 (1969), EEA73-14244.
626. Mahan, G. D., "Many-body theory of tunneling: polarons in Schottky junctions", In: Tunneling phenomena in solids, Plenum Press, New York, p. 305-13 (1969), PA73-21073.
627. Ngai, K. L. and Economou, E. N., "Theory of electron-surface-plasmons interactions in tunneling, low-energy electron diffraction, and photoemission", Phys. Rev. B, 4, 2132-46 (1971), PA74-69484.
628. Ngai, K. L., Economou, E. N. and Cohen, M. H., "Theory of inelastic electron-surface plasmon interactions in metal-semiconductor tunnel junctions", Phys. Rev. Lett., 22, 1375-8 (1969), EEA72-17715.
629. Parker, G. H. and Mead, C. A., "Tunneling in CdTe Schottky barriers", Phys. Rev., 184, 780-7 (1969), EEA73-7446.
630. Schreder, G., "Contribution to the study of the tunnel effect in semiconductors", Thesis, Univ. Paris, France, 1972, French, EEA76-26023.
631. Sood, B. R. and Pak, Y. L., "Zero-bias anomaly in GaAs/Mo tunnel junctions", J. Appl. Phys., 45, 2579-81 (1974), PA77-67278.
632. Sroubek, Z., "Electron tunneling in indium-$SrTiO_3$:Nb Schottky barriers", Solid State Commun., 7, 1561-4 (1969), EEA73-17616.
633. Sroubek, Z. and Kubec, F., "Electron tunneling in indium-TiO_2:Nb Schottky barriers", Solid State Commun., 12, 767-70 (1973), PA76-33994.
634. Steinrisser, F., Davis, L. C. and Duke, C. B., "Electron and phonon tunneling spectroscopy in metal-germanium contacts", Phys. Rev., 176, 912-4 (1968), PA72-13909.
635. Thomas, P. and Queisser, H. J., "Electron-phonon coupling in the barriers of GaAs Schottky diodes", Phys. Rev., 175, 983-4 (1968), PA72-7825.
636. Tsuda, N., "Tunneling anomalies in Nb-GaAs junctions above 9.2K", J. Phys. Soc. Japan, 29, 514 (1970), EEA74-1499.

637.Wolf, E. L. and Losee, D. L., "Zero bias anomalies in metal-semiconductor tunnel junctions", Solid State Commun., 7, 665-8 (1969), PA72-35456.
638.Wolf, E. L. and Losee, D. L., "Spectroscopy of Kondo and spin-flip scattering: high-field tunneling studies of Schottky-barrier junctions", Phys. Rev. B, 2, 3660-87 (1970), PA74-9485.
639.Wolf, E. L., Losee, D. L., Cullen, D. E. and Compton, W. D., "Anomalous metal semiconductor tunneling near the Mott transition", Phys. Rev. Lett., 26, 438-42 (1971), EEA74-15534.

3. Density of States and E-k Relationship

640.Barret, C. and Vapaille, A., "Determination of the density and relaxation of silicon-metal interface states. (Schottky barrier diode)", C. R. Hebd. Seances Acad. Sci. B, 278, 837-9 (1974), French, EEA77-26392.
641.Barret, C. and Vapaille, A., "Determination of the distribution of interface states for a Mo-Si junction", C. R. Hebd. Seances Acad. Sci. B, 279, 665-8 (1974), French, EEA78-11533.
642.Barret, C. and Vapaille, A., "Determination of the density and the relaxation time of silicon-metal interfacial states", Solid-State Electron., 18, 25-7 (1975), PA78-25758.
643.Eastman, D. E. and Freeouf, J. L., "Relation of Schottky barriers to empty surface states on III-V semiconductors", Phys. Rev. Lett., 34, 1624-7 (1975), PA78-62078.
644.Flietner, H., "The E(k) relation for a two-band scheme of semiconductors and the application to the metal-semiconductor contact", Phys. Status Solidi B, 54, 201-8 (1972), PA76-1999.
645.Goldberg, Y. A., Rahev, T. Y., Tsarenkov, B. V. and Yakovlev, Y. P., "Surface-barrier structure of metal and n-type $Ga_{1-x}Al_xAs$, and their energy band diagram", Sov. Phys. Semicond., 6, 398-401 (1972), PA75-75134.
646.Mahan, G. D. and Conley, J. W., "The density of states in metal-semiconductor tunneling", Appl. Phys. Lett., 11, 29 (1967), EEA70-22855.
647.Padovani, F. A. and Stratton, R., "Experimental E-k relationship determination using Schottky barriers", Phys. Rev. Lett., 16, 1202 (1966).
648.Panichevskaya, V. I. and Strikha, V. I., "The effect of the insulator layer thickness on the spectrum of electron surface states in a metal-semiconductor contact with a Schottky barrier", Radio Eng. & Electron. Phys., 18, 1458-9 (1973), EEA78-28709.
649.Panichevskaya, V. I., Strikha, V. I. and Novitskii, V. A., "Influence of metal alloying of silicon surfaces on surface state spectra and current transfer at the metal-semiconductor contact with the Schottky barrier", Poluprovodn. Tekh. & Mikroelektron., no. 18, 30-4 (1974), Russian, PA78-25756.

650.Sah, C. T., Wang, C. T. and Lee, S. H., "Junction edge region thermally stimulated capacitance (TSCAP) of n-Si doped with phosphorus and bismuth", Appl. Phys. Lett., 25, 523-4 (1974), EEA78-11528.

4. Other Applications

651.Bobylev, B. A., Kravchenko, A. F., Loburets, Y. V. and Terekhov, A. S., "Electroreflection from Schottky barriers", Sov. Phys. Semicond., 6, 1511-5 (1973), PA76-33982.
652.Crowell, C. R., Spitzer, W. G., Howarth, L. E. and LaBate, E. E., "Attenuation length measurements of hot electrons in metal films", Phys. Rev., 127, 2006-15 (1962).
653.Glover, G. H., "Ionization rate in GaAs determined from photomultiplication in a Schottky barrier", J. Appl. Phys., 44, 3253-7 (1973).
654.Gutkin, A. A., Dmitriev, M. V. and Nasledov, D. N., "Use of a Schottky barrier in determination of the optical absorption coefficient of a semiconductor", Sov. Phys. Semicond., 6, 1218-9 (1973), PA76-15521.
655.Izuzaki, K., Hariu, T., Ono, S. and Shibata, Y., "Measurement of electron temperature of hot carriers using Schottky barriers", Rep. Res. Inst. Elec. Commun. Tohoku Univ., 22, 1-9 (1970), PA74-30832.
656.Kapitonova, L. M., Kostina, L. S., Lebedev, A. A. and Makhkamov, S., "Investigation of the photon capture cross sections of thermal-defect levels in n-type Si", Sov. Phys.-Semicond., 8, 766-7 (1974), EEA78-11588.
657.Klose, H., Muller, G. and Thiessen, K., "The liquid crystal technique for the observation of electric fields in p-n and Schottky junctions", Phys. Status Solidi A, 16, K97-9 (1973), PA76-38887.
658.Mak, M. K. and Hampshire, M. J., "Surface potential evaluation in MIS structures using Schottky diodes", Solid-State Electron., 18, 907 (1975), EEA78-37479.
659.Mantena, N. R. and Barrera, J. S., "Measurement of Schottky barrier edge capacitance correction", Solid-State Electron., 12, 1000-2 (1969), EEA73-10598.
660.Namordi, M. R. and Thompson, H. W., "Aluminum-silicon Schottky barriers as semiconductor targets for EBS devices", Solid-State Electron., 18, 499-508 (1975).
661.Smith, B. L., "Background energy level spectroscopy in GaP using thermal release of trapped space charge in Schottky barriers", Appl. Phys. Lett., 21, 350-2 (1972), EEA76-737.
662.Thompson, W. A. and von Molnar, S., "Schottky barrier probe tunneling into superconductors", J. Appl. Phys., 41, 5218-23 (1970), EEA74-4387.
663.Toda, M. and Tosima, S., "Magnetocapacitance effect in a metal-$CdCr_2Se_4$ junction", J. Appl. Phys., 43, 1751-6 (1972), PA75-36555.

664. Tyagai, V. A., Popov, V. B. and Smtko, O. V., "Determination of the electric field intensity at a metal-semiconductor contact by the electro-absorption method", Sov. Phys. Semicond., 6, 148-9 (1972), EEA75-32605.
665. van Opdorp, C., Peters, R. C. and Klerk, M., "Use of Schottky-diode collectors for SEM determination of bulk diffusion lengths", Appl. Phys. Lett., 24, 125-6 (1974), PA77-41078.
666. Viscakas, J., Karpinskas, S., Sakalauskas, S. and Smilga, A., "The measurement of potential distribution in semiconductors of high resistance", Litov. Fiz. Sbornik, no. 5-6, 867-73 (1968), Russian, EEA73-4318.
667. Woods, M. H., "The use of a Schottky barrier to determine impact ionization coefficients with application to silicon", Thesis, Princeton Univ., New Jersey, 1971, 100 pp., Order No. 72-2761, PA75-44091.
668. Woods, M. H., Johnson, W. C. and Lampert, M. A., "Use of a Schottky barrier to measure impact ionization coefficients in semiconductors", Solid-State Electron., 16, 381-94 (1973), EEA76-16334.
669. Young, M. L. and Wight, D. R., "Concentration dependence of the minority carrier diffusion length and lifetime in GaP", J. Phys. D, 7, 1824-37 (1974).

VI. Practical Applications
1. Memory

670. Aldrich, R. A., "Low storage Schottky barrier diode transistors", Presented at International Electron Devices Meeting, Washington, D. C., October 1968, EEA72-7964.
671. Bentivoglio, G. L., Forlani, F., Minnaja, N. and Sacchi, G., "Integrated matrices of Schottky diodes for permanent memories", Advance in Electronics - Proceedings of the 15th Electronics Congress, 1968, p. 173-81, Publ. 1969, Italian, EEA73-7441.
672. Bertolucci, B., "CAMAC magnetostrictive readout system using Schottky memories (for use with spark chamber)", IEEE Trans. Nucl. Sci., NS-20, 361-6 (1973), PA76-28676.
673. Breuer, D. R. and Buie, J. L., "Monolithic bipolar semiconductor device employing a cermet for both Schottky barrier and ohmic contact", Patent USA 3737742, Publ. June 1973, CA79-59191.
674. Broom, R. F., Moser, A. and Rideout, A. J., "Vertical Schottky diode-memory device", IBM Tech. Disclosure Bull., 15, 2158 (1972).
675. Chang, A. W., "Transistor-diode clamping and biasing circuit", IBM Tech. Disclosure Bull., 15, 1085-6 (1972), EEA76-8179.
676. Crowder, B. L., Hovel, H. J. and Urgell, J. J., "Silicon Schottky barrier bistable memory element", IBM Tech. Disclosure Bull., 15, 891 (1972).
677. Dhaka, V. A., "Integrated circuits with surface barrier diodes", Patent 3506893, Publ. April 1970.

678.Dorler, J. A., Forneris, J. L. and Swietek, D. J., "Schottky barrier diode read-only memory", Patent USA 3780320, Publ. December 1973.

679.Dorler, J. and Swietek, D., "Integrated Schottky diode and transistor circuit", IBM Tech. Disclosure Bull., 14, 3214-5 (1972), EEA75-27742.

680.Gaensslen, F. H., "Schottky barrier read-only memory", IBM Tech. Disclosure Bull., 14, 252 (1971), EEA74-34116.

681.Guhman, G. F., "Diode-capacitor memory cells supported on a sapphire substrate", IBM Tech. Disclosure Bull., 16, 3166-7 (1974), EEA77-30284.

682.Heald, R. A. and Hodges, D. A., "Design of Schottky-barrier diode clamped transistor layouts", IEEE J. Solid-State Circuits, SC-8, 269-75 (1973), EEA76-29312.

683.Hodges, D. A., Lepselter, M. P., Lynes, D. J., MacDonald, R. W., MacRae, A. U. and Waggener, H. A., "Low-power bipolar transistor memory cells", IEEE J. Solid State Circuits, SC-4, 280-4 (1969), EEA73-4555.

684.Hovel, H. J., "Schottky barrier method sneak path elimination in memory devices", IBM Tech. Disclosure Bull., 14, 356-7 (1971), EEA74-37243.

685.Ingebrigtsen, K. A., Cohen, R. A. and Mountain, R. W., "A Schottky-diode acoustic memory and correlator", Appl. Phys. Lett., 26, 596-8 (1975).

686.Koch, S., "Fixed data memory utilizing Schottky diodes", Patent USA 3774170, Publ. November 1973.

687.Kurz, B., Barron, M. and Butler, W., "Applications of metal-silicon structures in integrated circuits", Eurocon 71 Digest, 1971, 2 pp., EEA75-3877.

688.Landler, P. F., "Dynamic Schottky cell", IBM Tech. Disclosure Bull., 17, 3215-16 (1975), EEA78-28753.

689.Lim, I. C., "A full adder using Schottky-barrier diodes and a tunnel diode", J. Korean Inst. Electron. Eng., 9, 22-8 (1972), Korean, EEA76-19455.

690.Luxton, H. E. G., "Gallium arsenide field effect transistors - their performance and application up to X-band frequencies", 4th European Microwave Conference, 1974, p. 92-6, EEA78-2427.

691.Mette, H. L. and Ahlstrom, E. R., "Magneto-Schottky-diode-integrated circuits", Report ECOM-3527, Army Electronics Command, Fort Monmouth, New Jersey, 1972, 27 pp., EEA75-35415.

692.Mick, J. R., "A Schottky m.s.i. four-bit shifter", New Electron., 7 (17), 42, 44, 49, 52 (1974), EEA78-378.

693.Mick, J. R., "Using Schottky 3-state outputs in bus-organized systems", EDN, 19 (23), 35-39 (1974).

694.Minden, H. T., "Gallium arsenide dual Schottky barrier diodes", Solid-State Electron., 16, 1185-8 (1973), EEA76-35299.

695. Morris, G. E., Hall, G. A., Cook, C. F. and Higgins, V. J., "Investigations of RF induced burnout in microwave mixer diodes: A continuing study", IEEE-S-MTT International Microwave Symposium, 1974, p. 139-41.
696. Nakamura, M., Kodera, H. and Migitaka, M., "Computer study on GaAs Schottky barrier IMPATT diodes", Solid-State Electron., 16, 663-7 (1973), EEA76-22927.
697. Nawata, K., Ikeda, M. and Ishii, Y., "Millimeter-wave GaAs Schottky-barrier IMPATT diodes", IEEE Trans. Electron Devices, ED-21, 128-30 (1974), EEA77-13858.
698. Pierre, L., "The evolution of Schottky diode integrated circuits", Electronique, no. 134, 415-9 (1970), French, EEA73-30380.
699. Platt, S., "Double emitter-bit line Schottky memory cell", IBM Tech. Disclosure Bull., 15, 260 (1972), EEA75-35304.
700. Rattner, J., Cornet, J. C. and Hoff, M. E., Jr., "Bipolar LSI computing elements usher in new era of digital design", Electronics, 47 (18), 89-96 (1974), EEA78-675.
701. Rein, H. M., Kosak, W. and Kotte, W., "Fast integrated digital circuits with Schottky diodes", Wiss. Ber. AEG-Telefunken, 45, 130-41 (1973), German, EEA76-19962.
702. Sugano, T. and Morino, A., "Design and fabrication experiment of Schottky diodes", J. Inst. Elect. Commun. Engrs., 50, 1053-60 (1967), Japanese, EEA71-18188.
703. Suzuki, S., Nagahashi, Y., Tanaka, T., Yamada, T., Muta, H., Okabayashi, H. and Yamada, K., "A static random-acess memory with normally-off-type Schottky barrier FET's", IEEE J. Solid-State Circuits, SC-8, 326-31 (1973), EEA76-35357.
704. Toda, J. K. and Luraya, J. L. P., "Reduction of storage time of a transistor using a Schottky barrier diode", Proc. IEEE, 55, 2064 (1967).
705. Waaben, S., "Electronically-variable semiconductor memory using two diodes per memory cell", 1970 IEEE International Solid State Circuits Conference, p. 46-7, EEA73-24998.
706. Waaben, S., "Charge-transfer electronics tandem matrix semiconductor memory selection", Digest of Technical Papers of the 1971 International Solid-State Circuits Conference, p. 84-5, EEA74-22797.

2. Logic

707. Baldey, R., "Schottky diode clamped TTL", Presented at Colloquium Digest on Very High Speed Bi-polar Logic: ECL-Schottky TTL, London, England, October 1971, EEA75-452.
708. Bertin, C. L. and Williams, T. A., "FET - Schottky diode logic", IBM Tech. Disclosure Bull., 13, 1276 (1970).
709. Cavinaugh, C., "Using Schottky to upgrade performance of your TTL system", EDN, 18 (18), 38 (1973).

710.Chin, W. B., Gruodis, A. J., Kraft, W. R. and Tai, E., "Integrated Schottky diode-transistor design and layout", IBM Tech. Disclosure Bull., 17, 3580-1 (1975), EEA78-28861.
711.Fulkerson, D. E., "Direct-coupled transistor-transistor logic: A new high-performance LSI gate family", IEEE J. Solid-State Circuits, SC-10, 110-17 (1975).
712.Hewlett, F. W., Jr., "Schottky I^2L", IEEE J. Solid-State Circuits, SC-10, 343-8 (1975), EEA78-37625.
713.Hintz, L., "Most Schottky TTL circuits use NPN inputs, but PNP is better - here's why", New Electron., 15, 28 (1973).
714.Kan, D. T., "Shunt-feedback Schottky clamped logic gates", IEEE J. Solid-State Circuits, SC-7, 404-11 (1972), EEA75-35294.
715.Kataoka, S., Hashizume, N., Kawashima, M. and Komamiya, Y., "High field domain functional logic devices with multiple control electrodes", Cornell Electronic Engineering Conference, 4th Bienn., Proc., 1973, p. 225-34.
716.Konishi, Y. and Hoshino, N., "12 GHz low-noise converter", Proceedings of the IEEE 1972 International Conference on Communications, 26-19/4 pp., EEA76-4227.
717.Kurz, B. and Barron, M. B., "Improved Schottky clamped (T^2L) circuits", IEEE J. Solid-State Circuits, SC-7, 175-9 (1972), EEA75-12660.
718.McLay, A., "The kinks and kernels of I.p. Schottky and c.m.o.s.", New Electron., 8 (4), 42-3 (1975), EEA78-24485.
719.Mathev, K. F., "Schottky direct-coupled transistor logic circuit", 6th Asilomar Conference on Circuits and Systems, 1972, p. 410-14, Publ. 1973, EEA76-23049.
720.Miles, T. E., "Schottky TTL vs ECL for high speed logic", Comput. Des., 11, 79-86 (1972), EEA76-4232.
721.Noyee, R. N., Bohn, R. E. and Chua, H. T., "Schottky diodes make IC scene", Electronics, 42 (14), 74-80 (1969), EEA72-9374.
722.Ribenyi, A., "Schottky diode in TTL circuits", Proceedings of the International Conference on the Physics and Chemistry of Semiconductor Heterojunctions and Layer Structures, 1971, p. V349-54, EEA74-26822.
723.Richer, J. W. and Eades, J. D., "An ultra-fast line-driving monolithic logic circuit", Conference on Integrated Circuits, 1967, p. 1-8, EEA73-503.
724.Scheffer, C. W. and Wu, L. L., "Schottky barrier diode clamp fabrication", IBM Tech. Disclosure Bull., 14, 3410 (1972).
725.Shagurin, M. I., Petrov, L. N. and Tatyanin, V. I., "Comparison of TTL and DTL variants employing Schottky barrier diodes", Izv. VUZ Radioelektron., 14, 1365-9 (1971), Russian, EEA75-12666.
726.Suzuki, S., Muta, H., Tanaka, T., Yamada, K., Nagahashi, Y., Okabayashi, H. and Kawamura, N., "Logic circuits with 2μm gate Schottky barrier FETs", Oyo Buturi, 44, 219-24 (1975), EEA78-41684.
727.Van Tuyl, R. L. and Liechti, C. A., "High-speed integrated logic with GaAs MESFET's", IEEE J. Solid-State Circuits, SC-9, 269-76 (1974), EEA77-38469.

728.Yu, A. Y. C. and Snow, E. H., "Fast switching pnp transistor", Patent USA 3571674, Publ. March 1971, EEA74-22722.

3. Microwave Diodes

729.Ahmad, K., "RF burnout in X-band Schottky mixers", J. Appl. Phys., 43, 4826-7 (1972), EEA76-4130.
730.Anand, Y., "RF burnout dependence on variations in barrier capacitance of mixer diodes", Proc. IEEE, 61, 247 (1973), EEA76-7994.
731.Anand, Y. and Doherty, W. E., "Reciprocity in silicon Schottky-barrier diodes", Electron. Lett., 3, 236-7 (1967), PA70-29847.
732.Anand, Y. and Howell, C., "A burnout criterion for Schottky-barrier mixer diodes", Proc. IEEE, 56, 2098 (1968), EEA72-3038.
733.Anand, Y. and Moroney, W. J., "Microwave mixer and detector diodes", Proc. IEEE, 59, 1182-90 (1971), EEA74-27187.
734.Anon., "TESLA semiconductor microwave devices", TESLA Electron., 7 (3), 91-2 (1974), EEA78-8390.
735.Assd. Electrical Ind. Ltd., "Schottky barrier diode", Patent UK 1139495, Publ. January 1969, EEA72-11290.
736.Araki, T. and Hirayama, M., "A quasi-millimeter-wave integrated mixer, ECL-1384", Rev. Electr. Commun. Lab., 20, 788-92 (1972), EEA76-13918.
737.Barber, M. R., "Noise figure and conversion loss of the Schottky barrier mixer diode", IEEE Trans. Microwave Theory Tech., MITT-15, 629-35 (1967), EEA71-9066.
738.Buchs, J. D., "A stripline mixer with subharmonic local oscillator using two Schottky barrier diodes in an antiparallel connection", Arch. Elek. Ubertrag., 25, 52-3 (1971), German, EEA74-22926.
739.Buchs, J. D., "Down-conversion by means of Schottky barrier diodes", Nachrichtentech. Z. (NTZ), 25, 305-11 (1972), German, EEA75-32912.
740.Cerniglia, N. P., Tonner, R. C., Berkovits, G. and Solomo, A. H., "Beam-lead Schottky barrier diodes for low noise integrated microwave mixers", Presented at International Electron Devices Meeting, Washington, D. C., October 1967, EEA72-7891.
741.Cohen, L. D., "Microwave characterization of the properties and performance of GaAs Schottky barrier mixer diodes", Proc. IEEE, 59, 288-9 (1971), EEA74-13238.
742.Coleman, D. J., Jr., "Transit-time oscillations in BARITT diodes", J. Appl. Phys., 43, 1812-8 (1972), EEA75-20468.
743.Cowley, A. M., "Low noise X-band Schottky barrier video detector diode", Presented at International Electron Devices Meeting, Washington, D. C., October 1968.
744.Day, H. M., Cohen, E. D., Gleason, K. R. and Macpherson, A. C., "Microwave micromanipulator", Rev. Sci. Instrum., 41, 1754-6 (1970).

745. Day, H. M., Gleason, K. R. and Macpherson, A. C., "The role of edge capacitance in the design of microwave Schottky barrier detector diodes", Solid-State Electron., 13, 1111-3 (1970).
746. Day, H. M., MacPherson, A. C. and Bradshaw, E. F., "Multiple contact Schottky barrier microwave diode", Proc. IEEE, 54, 1955-6 (1966), EEA70-14554.
747. Dickens, L. E., "A millimeter-wave pumped X-band uncooled parametric amplifier", Proc. IEEE, 60, 328-9 (1972), EEA75-10528.
748. Dickens, L. E., Cotton, J. M., Jr. and Geller, B. D., "A mixer and solid state L.O. for a 60 GHz receiver", 1971 IEEE GMTT International Microwave Symposium Digest of Technical Papers, p. 188-90, EEA75-1875.
749. Dobramysl, D., "Integrated mixer for 18 and 26 GHz", 1971 IEEE GMTT International Microwave Symposium Digest of Technical Papers, p. 18-19, EEA74-36350.
750. Dragone, C., "Performance and stability of Schottky barrier mixers", Bell Syst. Tech. J., 51, 216-95 (1972), EEA76-4119.
751. Edrich, J., "A parametric amplifier for 46 GHz", Proc. IEEE, 59, 1125-6 (1971), EEA74-22990.
752. Elder, H. E., "Active solid-state devices", Bell Syst. Tech. J, 47, 1323-77 (1968), EEA72-687.
753. Engleson, M., "Comments on Schottky barrier mixer diodes", Electronic Commun., 3 (2), 9-10 (1968), EEA72-23847.
754. Fleri, D. A. and Cohen, L. D., "Nonlinear analysis of the Schottky-barrier mixer diode", IEEE Trans. Microwave Theory & Tech., MTT-21, 39-43 (1973), EEA76-7985.
755. Fuchs, G., "Schottky diode mixer in hybrid technique", Nachrichtentechnik, 22, 314-5 (1972), German, EEA75-39076.
756. Gardiner, J. G., "The signal-handling capacity of the square-wave switched ring modulator", Radio & Electron. Eng., 41, 465-70 (1971), EEA75-3844.
757. Gardiner, J. G. and Banerjee, A. R., "Conversions loss stability and gain comparison in Schottky barrier mixers", Electron. Lett., 6, 829-30 (1970), EEA74-10844.
758. Gardiner, J. G. and Yousif, A. M., "Distortion effects in switching diode modulators arising from local-oscillator interference", Proc. Inst. Elec. Eng., 118, 339-42 (1971), EEA74-9739.
759. Genzabella, C. F. and Howell, C. M., "Gallium arsenide Schottky mixer diodes", Proceedings of International Symposium on Gallium Arsenide, 1966, p. 131-7, Publ. 1967, PA70-21019.
760. Gerzon, P. H., Barnes, J. W., Waite, D. W. and Northrop, D. G., "The mechanism of rf spike burn-out in Schottky barrier microwave mixers", Solid-State Electron., 18, 343-8 (1975).
761. Gerzon, P. H. and Waite, D. W., "The R.F. burnout mechanism in silicon Schottky barrier microwave mixers", Presented at Colloquium Microwave Mixers and Mixer Diodes (Digest), London, England, April 1972, EEA75-20488.
762. Glance, B. and Snell, W. W., Jr., "Low-noise, integrated, millimeter-wave receiver", Bell Syst. Tech. J., 53, 1321-28 (1974).

763.Goldwassser, R. E. and Lee, C. A., "Silicon Schottky barrier avalanche diodes", Presented at International Electron Devices Meeting, Washington, D. C., October 1969, EEA73-21265.
764.Henne, W., "The Schottky diode and its applications", Bull. Assoc. Suisse Electr., 63, 1491-7 (1972), German, EEA76-8084.
765.Huang, H. C., Levine, P. A., Gobat, A. R. and Klatskin, J. B., "GaAs IMPATT diodes with improved CW efficiency and noise performance", Presented at 17th International Electron Devices Meeting, Washington, D. C., October 1971, EEA75-13705.
766.Hughes, G. W. and White, R. M., "MIS and Schottky barrier microstrip devices", Proc. IEEE, 61, 1460-1 (1972), EEA76-5397.
767.Hughes, G. W. and White, R. M., "Microwave properties of nonlinear MIS and Schottky-barrier microstrip", IEEE Trans. Electron Devices, ED-22, 945-56 (1975), EEA78-41710.
768.Ida, M., Sato, Y., Uchida, M. and Shimada, K., "GaAs Schottky barrier diodes for ultra high frequency communication systems", Elec. Commun. Lab. Tech. J, 20, 1747-59 (1971), Japanese, EEA75-20700.
769.Ida, M., Sato, Y., Uchida, M. and Shimada, K., "GaAs Schottky barrier diodes for ultra high frequency communication systems", Rev. Electr. Commun. Lab., 21, 800-6 (1973), EEA77-36290.
770.Irvin, J. C., "Schottky barrier devices", In: Microwave semiconductor devices and their circuit applications", McGraw-Hill, New York, p. 349-69 (1969).
771.Irvin, J. C., Coleman, D. J., Johnson, W. A., Tatsuguchi, T., Decker, D. R. and Dunn, C. N., "Fabrication and noise measure of high-power GaAs IMPATT diodes", Presented at International Electron Devices Meeting, Washington, D. C., October 1970, EEA74-10607.
772.Irvin, J. C. and Vanderwal, N. C., "An inside look at Schottky-barrier devices", Bell Lab Record, 47, 57-62 (1969), EEA72-15508.
773.Jager, D. and Rabus, W., "Bias-dependent phase delay of Schottky contact microstrip line", Electron. Lett., 9, 201-3 (1973), EEA76-27025.
774.Jager, D., Rabus, W. and Eickhoff, W., "Bias-dependent small-signal parameters of Schottky contact microstrip lines", Solid-State Electron., 17, 777-83 (1974), EEA77-31119.
775.Jehnicka, J., "Demodulation with respect to minimal deviations of differential gain", Sdelovaci Tech., 20, 42-3 (1972), Czech, EEA75-19115.
776.Johnson, K. M., "X-band integrated circuit mixer with reactively terminated image", IEEE J. Solid-State Circuits, SC-3, 50-9 (1968), EEA71-29238.
777.Kahng, D., "Au-n-type GaAs Schottky barrier and its varactor application", Bell Syst. Tech. J., 42, 215 (1964).
778.Kanerva, H. H. J., Leppavitori, S. and Stubb, T., "A Schottky diode and its behaviour in the gigahertz region", State Inst. Tech. Res. Rep. II, no. 29, 5-15 (1969), EEA73-17595.
779.Kim, C. K. and Armstrong, L. D., "GaAs Schottky-barrier avalanche diodes", Solid-State Electron., 13, 53-6 (1970), EEA73-14251.

780.Kitazume, S., Kobayashi, H. and Ishihara, H., "Up- and down-converters in the 60 GHz frequency range", NEC Res. Dev., no. 34, 62-71 (1974).
781.Klamka, J., "Microwave semiconductor devices", Elektronika, 15, 320-6 (1974), Polish, EEA78-2828.
782.Klamka, J., "World's research achievements and development tendencies in microwave semiconductor devices", Elektronika, 14, 465-71 (1973), Polish, EEA77-14241.
783.Knerr, R. H. and Murray, J. H., "Microwave amplifier using several IMPATT diodes in parallel", IEEE Trans. Microwave Theory & Tech., MTT-22, 569-72 (1974), EEA77-18173.
784.Konishi, Y. and Hoshino, N., "12 GHz low-noise converter", Proceedings of the IEEE 1972 International Conference on Communications, 26-19/4 pp., EEA76-4227.
785.Kuno, H. J. and Tully, J. W., "New solid state components for millimeter wave systems", Microwave J., 14 (11), 34-8 (1971), EEA75-32923.
786.Landkammer, F. J., "The design of Schottky barrier mixers for use at millimeter frequencies", Electronic Commun., 3, 8 (1968), EEA71-26922.
787.Landkammer, F. J., "The design of Schottky barrier mixers for use at millimeter frequencies", Electronic Commun., 3, 8-9 (1968), EEA72-17082.
788.Landkammer, F. J., "A millimeter mixer with Schottky barrier diodes", Northeast Electronic Research and Engineering Meeting Digest, 1967, p. 58-9, Publ. 1968, PA73-8542.
789.Leedy, H. M., Stover, H. L., Bryan, R. P., Garvin, H. L. and Morehead, H. G., "Advanced millimetre-wave mixer diodes, GaAs and silicon, and a broadband low-noise mixer", Proceedings of the 3rd Biennial Cornell Electrical Engineering Conference on High Frequency Generation and Amplification Devices and Application, 1971, p. 451-62, EEA75-14054.
790.Licchti, C. A., "Down-converter using Schottky-barrier diodes", IEEE Trans. Electron Devices, ED-17, 975-83 (1970), EEA74-305.
791.Liu, S. G. and Risko, J. J., "A punch-through PN-junction-Schottky-barrier microwave oscillator", Presented at 17th International Electron Devices Meeting, Washington, D. C., October 1971, EEA75-14016.
792.Macpherson, A. C., Cohen, E. D. and Herndon, M., "Rectification of microwave power using Schottky barrier diodes", Presented at URSI 1969 Spring Meeting, Washington, D. C., April, PA74-24123.
793.Macpherson, A. C., Cohen, E. D. and Herndon, M., "Microwave power rectification with commercial Schottky barrier diodes", J. Microwave Power, 4, 278-85 (1969), EEA74-9231.
794.Macpherson, A. C. and Day, H. M., "Design and fabrication of high burn-out Schottky barrier crystal video diodes", Solid-State Electron., 15, 409-15 (1972), EEA75-17036.

795.Migitaka, M., Nakamura, M., Saito, K. and Sekine, K., "Millimeter-wave GaAs Schottky-barrier IMPATT diodes", Proc. IEEE, 61, 1448-9 (1972), EEA76-4627.
796.Misawa, T., Moline, R. A. and Tretola, A. R., "10 GHz Si Schottky-barrier IMPATT diode with hyperabrupt impurity distribution produced by ion implantation", Solid-State Electron., 15, 189-93 (1972), EEA75-13701.
797.Morino, A. and Sugano, T., "Conversion loss of Schottky diode microwave down-convertors", Electron. & Commun. Jap., 54 (3), 68-74 (1971), Japanese, EEA75-32911.
798.Moroney, W. J., "Low barrier height gallium arsenide microwave Schottky diode using gold-germanium alloy", Patent USA 3711745, Publ. January 1973.
799.Moroney, W. J. and Anand, Y., "Low barrier height gallium arsenide microwave Schottky diodes using gold-germanium alloy", Proceedings of the Third International Symposium on Gallium Arsenide and Related Compounds, 1971, p. 259-67, EEA76-4147.
800.Morris, G. E. and Hall, G. A., "RF burnout of Ku-band mixer diodes", IEEE Trans. Microwave Theory Tech., MTT-22, 745-46 (1974).
801.Nakamura, M., Kodera, H. and Migitaka, M., "Computer study on GaAs Schottky barrier IMPATT diodes", Solid-State Electron., 16, 663-7 (1973), EEA76-22927.
802.Nelson, T. M. and Goldie, M., "Fast-acting varactors for sub-nanosecond power limiting in receiver protectors", IEEE-S-MTT International Microwave Symposium, 1974, p. 176-78.
803.Nio, K. and Kadoo, S., "Experiments with GaAs Schottky barrier IMPATT diodes", NHK Lab. Note, no. 156, 1-8 (1972), EEA76-15894.
804.Ohta, T. and Nakano, H., "Characteristics of reflected-type Schottky-barrier diode mixer", Electron. & Commun. Jap., 54 (11), 101-7 (1971), EEA76-7961.
805.Oxley, T. H., Lord, J. F., Ming, K. J. and Clarke, J., "Image recovery mixers", Proceedings of the 1971 European Microwave Conference, 4 pp., EEA75-1500.
806.Oxley, T. H. and Summers, J. G., "Metal-gallium arsenide diodes as mixers", Proceedings of International Symposium on Gallium Arsenide, 1966, p. 138-50, Publ. 1969, PA70-21020.
807.Oxley, T. H., Summers, J. and Hansom, A., "The development of gallium arsenide Schottky barrier diodes as mixers and detectors", 1969 European Microwave Conference, p. 483-6, Publ. 1970, EEA73-35468.
808.Oxley, T. H. and Swallow, G. H., "Beam lead mixer and detector diodes for microwave integrated circuit applications", Microwave J., 14 (12), 31-4 (1971), EEA75-32925.
809.Peppiatt, H. J. and McDaniel, A. V., Jr., "Intermodulation analysis and design of a Schottky barrier diode mixer", Proc. IEEE, 56, 96 (1968), EEA71-8534.

810.Puntenney, D. G., Vetter, R. J., Weeks, W. L., Ziemer, P. L. and Born, G. S., "Microwave dosimetry using electrochemical effects", J. Microwave Power, 9 (1), 39-45 (1974).
811.Rabus, W., "Harmonic frequency generation along Schottky-contact-microstrip lines", Arch. Elektron. & Uebertragungstech., 28 (1), 1-11 (1974), German, EEA77-9560.
812.Rodgers, J. M. and Pomeroy, R. C., "Q band GaAs Schottky-barrier IMPATT", Electron. Lett., 7, 21 (1971), EEA74-10845.
813.Rusch, W. V. T. and Burrus, C. A., "Planar millimeter-wave epitaxial silicon Schottky-barrier converter diodes", Solid-State Electron., 11, 517-25 (1968), PA71-31408.
814.Sato, K., Yoshida, M. and Matsui, T., "Schottky barrier diode for UHF mixer", Toshiba Rev. (Internat. Ed.), no. 32, 37-43 (1967), EEA71-10629.
815.Sato, Y., Uchida, M., Shimada, K., Ida, M. and Imai, T., "GaAs Schottky barrier diode, ECL-1314", Rev. Elec. Commun. Lab., 18, 638-44 (1970), EEA74-12759.
816.Schneider, M. V. and Snell, W. W., Jr., "Stripline downconverter with subharmonic pump", Bell Syst. Tech. J, 53, 1179-83 (1974).
817.Schneider, M. V. and Wrixon, G. T., "Development and testing of a receiver at 230 GHz", IEEE-S-MTT International Microwave Symposium, 1974, p. 120-22.
818.Shinohara, S., "Millimeter-wave frequency-multiplier circuit using a semiconductor diode", Rep. Fac. Eng. Shizuoka Univ., no. 21, 67-77 (1970), Japanese, EEA74-33317.
819.Sisson, M. J. and Swallow, G. H., "Beam lead diodes for millimetre wavelengths", Presented at Colloquium Digest on Millimetre-Wave Hybrid and Monolithic Integrated Circuits, London, England, April 1973, EEA76-22949.
820.Small, T. and Pye, T., "Recent developments in the field of semiconductor components for UHF and microwaves", Alta Frequenza, 36, 245-7 (1967), Italian, EEA71-11330.
821.Steinbrecher, D. H., "Solid-state millimeter-wave power generation and amplification", Report NASA-CR-86400, MIT, Cambridge, Mass., 1970, 7 pp., PA74-19196.
822.Stover, H. L., Leedy, H. M. and Morehead, H. G., "Solid-state devices and components for mm-wave receiver-transmitter systems", Microwave J., 16 (2), 35-41 (1973), EEA76-21134.
823.Strobel, H. J., "Linear h.f. rectifier for measuring purposes using Schottky diodes", Radio Mentor, 34 (2), 87-9 (1968), German, EEA71-17150.
824.Swallow, G. H., "Some aspects of high burnout gallium arsenide diodes", Presented at Colloquium on Microwave Mixers and Mixer Diodes (Digest), London, England, April 1972, EEA75-20487.
825.Swallow, G. H., Hansom, A. M. and Oxley, T. H., "Some aspects of high-burnout gallium arsenide Schottky diodes", GEC J. Sci. & Technol., 40 (3), 126-31 (1973), EEA77-10214.

826.Sze, S. M., Lepselter, M. P. and MacDonald, R. W., "Metal-semiconductor IMPATT diode", Solid-State Electron., 12, 107-9 (1969), PA72-15117.

827.Tantraporn, W. and Yu, S. P., "Efficiencies of Schottky-barrier GaAs and both complementary structures of Si IMPATT diodes", IEEE Trans. Electron Devices, ED-20, 492-6 (1973), EEA76-19304.

828.Tatsuguchi, I. and Aslaksen, E. W., "Integrated 4-GHz balanced mixer assembly", IEEE J. Solid-State Circuits, SC-3, 21-6 (1968), EEA71-19166.

829.Turner, R. J., "Analog gate and zener diode give 70-dB isolation at 80 MHz", Electronics, 47 (23), 121, 123 (1974).

830.Wacker, R. W. and Mao, S., "Surface oriented GaAs Gunn oscillators and Schottky barrier diodes", Presented at International Electron Devices Meeting, Washington, D. C., October 1967, EEA72-7892.

831.Walker, A. V., "Schottky barrier diodes for small signal applications", New Electron., 8 (6), 32-4 (1975), EEA78-24312.

832.Watanabe, T., Kodera, H. and Migitaka, M., "GaAs 50 GHz Schottky-barrier IMPATT diodes", Electron. Lett., 10, 7-8 (1974), EEA77-9563.

833.Weller, K. P., "Small-signal theory of a transit-time-negative-resistance device utilizing injection from a Schottky barrier", RCA Rev., 32, 372-82 (1971), EEA75-7502.

834.Weller, K. P., Wen, C. P., Young, A. F. and Dreeben, A. B., "Millimeter wave GaAs avalanche oscillator", Proceedings of the 3rd Biennial Cornell Electrical Engineering Conference on High Frequency Generation and Amplification Devices and Application, 1971, p. 191-8, EEA75-14037.

835.Wortmann, A., Heime, K. and Beneking, H., "A new concept for microstrip-integrated GaAs Schottky-diodes", IEEE Trans. Electron. Devices, ED-22, 198-200 (1975), EEA78-14778.

836.Wortmann, A. K. and Kohn, E. E., "A monolithic integrated Schottky diode for microwave mixers", Solid-State Electron., 18, 1095-8 (1975).

837.Wosinski, T. and Jelenski, A., "Investigation of the parameters of Schottky-barrier diodes", Elektronika, 16, 195-8 (1975), Polish, EEA78-28748.

838.Wrixon, G. T., "Low-noise diodes and mixers for the 1 - 2-mm wavelength region", IEEE Trans. Microwave Theory Tech., MTT-22, 1159-65 (1974).

839.Yahara, T., Ogiso, K., Hoshiga, H. and Shirahata, K., "GaAs Schottky barrier diodes", Mitsubishi Denki Lab. Rep., 12, 17-27 (1971), EEA75-16710.

840.Yell, R. W., "Signal-compression performance of Schottky-barrier diodes in microwave mixers", Electron. Lett., 5, 360-1 (1969), EEA73-345.

841.Young, D. T. and Irvin, J. C., "Millimeter frequency conversion using Au-n-type GaAs Schottky barrier epitaxial diodes with novel contacting technique", Proc. IEEE, 53, 2130-1 (1966).

842. Zettler, R. A. and Cowley, A. M., "Hybrid hot carrier diodes", Hewlett-Packard J., 20, 13-20 (1969).

4. Schottky-Gate Field-Effect Transistors

843. Ames, I., d'Heurle, F., Esaki, L., Landauer, R. W. and Seki, H. H., "Field effect transistors utilizing Schottky barrier principle", IBM Tech. Disclosure Bull., 9, 1470-1 (1967).
844. Anantha, N. G., "Semiconductor device embodying field-effect transistors and Schottky barrier diodes", Patent USA 3749987, Publ. July 1973.
845. Anastassiou, A. and Strutt, M. J. O., "Experimental and computed four scattering and four noise parameters of GaAs FET's up to 4 GHz", IEEE Trans. Microwave Theory & Tech., MTT-22, 138-40 (1974), EEA77-9566.
846. Anastassiou, A. and Strutt, M. J. O., "Experimental and calculated four gain and four noise parameters of GaAs-FET's up to 4 GHz", Arch. Elektron. & Uebertragungstech., 28 (1), 37-42 (1974), EEA77-10250.
847. Anon., "New field effect transistor: MESFET", Elektronik, 10 (3), 31-2 (1972), Danish, EEA75-20209.
848. Anon., "GaAs FET promises to end noise - gain tradeoff", Microwaves, 12 (10), 9 (1973).
849. Asai, S., Kurono, H., Takahashi, S., Hirao, M. and Kodera, H., "Single- and dual-gate GaAs Schottky-barrier FET's for microwave frequencies", Oyo Buturi, 43 (Suppl.), 442-7 (1974), EEA78-2884.
850. Asai, S., Murai, F. and Kodera, H., "GaAs dual-gate FET with low noise and wide dynamic range", 19th International Electron Devices Meeting Technical Digest, 1973, p. 64-7.
851. Asai, S., Murai, F. and Kodera, H., "GaAs dual-gate Schottky-barrier FET's for microwave frequencies", IEEE Trans. Electron Devices, ED-22, 897-904 (1975).
852. Bachem, K. H., Erlaki, G. and Markert, W., "Vapor-grown submicron epitaxial gallium arsenide layers of MESFET", Fourth International Conference on Chemical Vapor Deposition, 1973, p. 296-305, CA81-161134.
853. Baechtold, W., "Noise behavior of Schottky barrier gate field-effect transistors at microwave frequencies", IEEE Trans. Electron Devices, ED-18, 97-104 (1971).
854. Baechtold, W., "Noise temperature in silicon in the hot electron region", IEEE Trans. Electron Devices, ED-18, 1186-7 (1971).
855. Baechtold, W., "Noise behavior of GaAs field-effect transistors with short gate lengths", IEEE Trans. Electron Devices, ED-19, 674-80 (1972).
856. Baechtold, W., "X- and Ku-band amplifiers with GaAs Schottky-barriers field-effect transistors", IEEE J. Solid-State Circuits, SC-8, 54-8 (1973), EEA76-7981.

857.Baechtold, W., "Author's reply to 'Comments on 'noise behavior of GaAs field-effect transistors with short gate length''", IEEE Trans. Electron Devices, ED-20, 840-1 (1973), EEA76-35324.
858.Baechtold, W., Daetwyler, K., Forster, T., Mohr, T. O., Walter, W. and Wolf, P., "Si and GaAs 0.5 μm-gate Schottky-barrier field-effect transistors", Electron. Lett., 9, 232-4 (1973), EEA76-25564.
859.Baechtold, W. and Wolf, P., "An improved microwave silicon MESFET", Solid-State Electron., 14, 783-90 (1971).
860.Barrera, J. S. and Archer, R. J., "InP Schottky-gate field-effect transistors", IEEE Trans. Electron Devices, ED-22, 1023-30 (1975).
861.Barron, M., Kurz, B. and Butler, W., "Application of JFET-MESFET devices to bucket-brigade circuits", Presented at 17th International Electron Devices Meeting, Washington, D. C., October 1971, EEA75-13782.
862.Baudet, P. and Binet, M., "GaAs MESFET device", Onde Electr., 54, 31-5 (1974), French, EEA77-14531.
863.Beneking, H., Jahncke, J. and Naumann, J., "Procedures for the measurement of s-parameters and the construction of stripline connected GaAs-MESFET's in the 10 GHz range", Proceedings of the 3rd Biennial Cornell Electrical Engineering Conference on High Frequency Generation and Amplification Devices and Application, 1971, p. 427-36, EEA75-14052.
864.Beneking, H. and Kohn, E., "High speed GaAs-MESFET differential amplifier stage with integrated current source", International Electron Devices Meeting Technical Digest, 1974, p. 292-5.
865.Camisa, R. L., Goel, J. and Druckier, I., "GaAs MESFET linear power-amplifier stage giving 1W", Electron. Lett., 11, 572-3 (1975).
866.Clouser, P. L. and Risser, V. V., "C-band field effect transistor amplifiers", IEEE J. Solid State Circuits, SC-5, 323-7 (1970), EEA74-5115.
867.Davis, R. T., "Low noise GaAs FET's climb in frequency to 8 GHz", Microwaves, 10 (12), 8 (1971).
868.Davis, R. T., "Broader bandwidths at higher power achieved using IMPATT and Gunn diodes", Microwaves, 11 (2), 32 (1972).
869.Davis, R. T. and Hardeman, L., "Look out, foreign R & D is closing the gap", Microwaves, 10 (6), 34-40 (1971).
870.Dmitruk, N. L., Tereshchenko, A. K., Maevo, O. I., Lyashenko, V. I. and Raskevich, A. M., "Influence of deep levels on the space charge capacitance and the field effect in gallium arsenide. (Schottky diode)", Sov. Phys.-Semicond., 7, 469-73 (1973), EEA77-10190.
871.Doerbeck, F. H., "A planar GaAs Schottky barrier field effect transistor with a self-aligned gate", Proceedings of the Third International Symposium on Gallium Arsenide and Related Compounds, 1970, p. 251-8, Publ. 1971, EEA76-4673.
872.Drangeid, K. E., "High-speed field-effect structure", IBM Tech. Disclosure Bull., 11, 332-3 (1968).

873.Drangeid, K. E., "Investigation of highly-doped Schottky-barrier field-effect transistor", Report AD715282 and AD728635, IBM T. J. Watson Research Center, Yorktown, New York, 1970, 17 pp. and 44 pp..

874.Drangeid, K. E., Broom, R. F., Jutzi, W. G., Mohr, T. O., Moser, A. and Sasso, G., "A memory-cell array with normally off-type Schottky-barrier FET's", IEEE J. Solid-State Circuits, SC-7, 277-82 (1972), EEA75-28509.

875.Drangeid, K. E. and Sommerhalder, R., "Dynamic performance of Schottky barrier field-effect transistors", IBM J. Res. Developm., 14, 82-94 (1970), PA73-53904.

876.Drangeid, K. E., Sommerhalder, R. and Walter, W., "High speed GaAs Schottky barrier field-effect transistors", Electron. Lett., 6, 228 (1970).

877.Driver, M. C., Geisler, M. J., Barrett, D. L. and Kim, H. B., "S-band microwave power FET", 19th International Electron Devices Meeting Technical Digest, 1973, p. 393-95.

878.Drukier, I., Camisa, R. L., Jolly, S. T., Huang, H. C. and Narayan, S. Y., "Medium-power GaAs field-effect transistors", Electron. Lett., 11, 104-5 (1975), EEA78-15348.

879.Dumke, W. P., Rideout, V. L. and Woodall, J. M., "GaAs field-effect transistors with self-registered gates", IBM Tech. Disclosure Bull., 14, 1248-9 (1971), EEA75-10286.

880.Fujuta, M., "High-power GaAs FET shatters X-band 1W barrier", JEE, no. 103, 32-5 (1975), EEA78-41610.

881.Graffeuil, J. and Caiminade, J., "Low-frequency noise in GaAs Schottky-gate f.e.t.s.", Electron. Lett., 10, 266-8 (1974), EEA77-30329.

882.Hogeboom, J. G. and Cobbold, R. S. C., "Etched Schottky-barrier MOSFET's", Electron. Lett., 7, 133-4 (1971), EEA74-15565.

883.Honig, W. M., "Gate-controlled source-to-drain resistance of FET's", IEEE Trans. Electron Devices, ED-21, 732-33 (1974).

884.Hower, P. L., Hooper, W. W., Tremere, D. A., Lehrer, W. and Bittmann, C. A., "The Schottky barrier gallium arsenide field-effect transistor", 2nd International Symposium on Gallium Arsenide, 1968, p. 187-94.

885.Huang, H. C., Drukier, I., Camisa, R. L., Narayan, S. Y. and Jolly, S. T., "High-efficiency GaAs MESFET amplifiers", Electron. Lett., 11, 508-9 (1975).

886.Jadus, D. K., "Buried field effect transistor", IBM Tech. Disclosure Bull., 13, 1431-2 (1970), EEA74-7778.

887.Jahneke, J., "The microwave properties of GaAs MESFET's in stripline technique", Nachrichtentech. Z. (NTZ), 26, 193-9 (1973), German, EEA76-29273.

888.Jimenez, J. J., Oliva, J. A. and Septier, A., "Very low noise cryogenic MESFET amplifier", European Microwave Conference, Proc., 1973, 4 pp..

889.Jimenez, J. and Vapaille, A., "Characteristics of a surface barrier gallium arsenide field effect transistor (MESFET) operating at 4.2K", C. R. Hebd. Sean. Acad. Sci. B, 272, 1119-22 (1971), French, PA74-49117.
890.Jutzi, W., "A comparison of MOS- and MES-field-effect transistors with 1 μm channel length for integrated D.C.-coupled circuits", Elektron. Rechenanlagen, 14, 19-27 (1972), German, EEA75-13756.
891.Kim, H. B. and Driver, M. C., "Application of face-growth to self-aligned Schottky barrier gate field effect transistors", Patent USA 3855690, Publ. December 1974.
892.Kniepkamp, H., Gisdakis, S., Kellner, W. and Ristow, D., "Technology and microwave performance of a 1 μm GaAs Schottky-barrier field effect transistor", Siemens Forsch. - & Entwicklungsber., 4, 274-80 (1975), EEA78-41626.
893.Kohn, E., "Normally-off m.e.s.f.e.t. with fast switching behaviour", Electron. Lett., 10, 505 (1974), EEA78-5329.
894.Kohn, E., "V-shaped-gate GaAs m.e.s.f.e.t. for improved high-frequency performance", Electron. Lett., 11, 160 (1975), EEA78-19817.
895.Kohn, E., Wuller, R., Stahlmann, R. and Beneking, H., "High-speed 1 μm GaAs m.e.s.f.e.t.", Electron. Lett., 11, 171-2 (1975), EEA78-19818.
896.Kurtin, S. and Mead, C. A., "GaSe Schottky barrier gate FET", Proc. IEEE, 56, 1594-5 (1968), EEA71-28911.
897.Lepselter, M. P. and Sze, S. M., "SB-IGFET: An insulated-gate field effect transistor using Schottky barrier contacts for source and drain", Proc. IEEE, 56, 1400-2 (1968), PA71-49318.
898.Liechti, C. A., "Performance of dual-gate GaAs MESFET's as gain-controlled low-noise amplifiers and high-speed modulators", IEEE Trans. Microwave Theory & Tech., MTT-23, 461-9 (1975), EEA78-23563.
899.Liechti, C. A. and Tillman, R. L., "Application of GaAs Schottky-gate FETs in microwave amplifiers", Digest of Technical Papers of the 1973 IEEE International Solid State Circuits Conference, p. 74-5, EEA76-22939.
900.Liechti, C. A. and Tillman, R. I., "Design and performance of microwave amplifiers with GaAs Schottky-gate field-effect transistors", IEEE Trans. Microwave Theory & Tech., MTT-22, 510-17 (1974), EEA77-18170.
901.Maeda, M. and Minai, Y., "Microwave variable-gain amplifier with dual-gate GaAs FET", IEEE Trans. Microwave Theory & Tech., MTT-22, 1226-30 (1974).
902.Maeda, M., Takahashi, S. and Kodera, H., "CW-oscillation characteristics of GaAs Schottky-barrier gate field-effect transistors", Proc. IEEE, 63, 320-1 (1975), EEA78-8573.
903.Matino, H., "Microwave GaAs Schottky gate FET", Electron. & Commun. Jap., 56 (7), 90-6 (1973), EEA78-8054.
904.Mead, C. A., "Schottky barrier gate field effect transistor", Proc. IEEE, 54, 307-8 (1966).

905.Middelhoek, S., "Projection masking, thin photoresist layers and interference effects", IBM J. Res. Develop., 14, 117-24 (1970).
906.Middelhoek, S., "Metalization process in fabrication of Schottky barrier FET's", IBM J. Res. Develop., 14, 148-51 (1970).
907.Mohr, T., "Selective metal masking in semiconductor device preparation", IBM Tech. Disclosure Bull., 13, 318-19 (1970), EEA74-7777.
908.Mohr, T. O., "Silicon and silicon-dioxide processing for high-frequency MESFET preparation", IBM J. Res. Develop., 14, 142-7 (1970).
909.Mohr, T. O., "Highly-doped Schottky barrier field-effect transistors", Report AD-740-959, IBM T. J. Watson Research Center, Yorktown Heights, New York, 1971.
910.Nozaki, T., Ogawa, M., Terao, H. and Watanabe, H., "Multi-layer epitaxial technology for the Schottky barrier GaAs field-effect transistor", Conference on Gallium Arsenide and Related Compounds, 1974, p. 46-54, Publ. 1975, EEA78-28820.
911.Ostoich, V., Jeppesen, P. and Slaymaker, N., "Direct modulation of D.H. GaAlAs lasers with GaAs MESFET", Electron. Lett., 11, 515-16 (1975).
912.Pengelly, R. S., "Broad band lumped-element X band GaAs F.E.T. amplifier", Electron. Lett., 11, 58-60 (1975).
913.Philippow, P., "Aluminium-silicon-Schottky-FET -- manufacture and measuring results", Nachrichtentechnik, 22, 372-3 (1972), German, EEA75-39950.
914.Philippow, P. and Kohler, E., "Manufacture of Al-Si Schottky contacts for Schottky FET", Nachrichtentech. Elektron., 23, 428-30 (1973), German, EEA77-10260.
915.Philippow, P. and Kohler, E., "The Al-Si Schottky field effect transistors - a new component to be integrated in semiconductor block circuits. I", Nachrichtentech. Elektron., 24, 105-9 (1974), German, EEA77-18861.
916.Philippow, P. and Kohler, E., "The Al-Si-Schottky field effect transistor - a component to be integrated in semiconductor block circuits", Nachrichtentech. Elektron., 24, 134-7 (1974), EEA77-23058.
917.Rossel, P. and Cabot, J. J., "Output-resistance properties of the GaAs Schottky-gate field-effect transistor in saturation", Electron. Lett., 11, 150-2 (1975), EEA78-19816.
918.Rossel, P., Cabot, J. J. and Graffeuil, J., "Bistable switching on gallium arsenide Schottky gate field-effect transistors", Appl. Phys. Lett., 25, 510-11 (1974), EEA78-11646.
919.Shinoda, D. and Kawamura, N., "Schottky barrier-type field-effect transistor", Patent USA 3767984, Publ. October 1973.
920.Slaymaker, N. A. and Turner, J. A., "Microwave FET amplifiers with centre frequencies between 1 and 11 GHz", European Microwave Conference, Proc., 1973, 3 pp..

921.Statz, H. and Munich, W. V., "Silicon and gallium arsenide field-effect transistors with Schottky-barrier gate", Solid-State Electron., 12, 111-7 (1969), PA72-24205.
922.Sugeta, T., Ida, M. and Uchida, M., "Microwave performance of GaAs-Schottky barrier gate FET's", Electr. Commun. Lab. Tech. J., 24, 323-34 (1975), Japanese, EEA78-33152.
923.Suzuki, S., Nagahashi, Y., Tanaaka, T., Muta, H. and Okabayashi, H., "A static RAM with normally-off-type Schottky barrier FETs", Digest of Technical Papers of the 1973 IEEE International Solid State Circuits Conference, p. 36-7, 196, EEA76-23036.
924.Uchida, M., Ida, M. and Sato, Y., "Fabrication technique for GaAs Schottky-barrier gate FET's", Electr. Commun. Lab. Tech. J., 24, 311-21 (1975), Japanese, EEA78-33151.
925.Wolf, P., "Microwave properties of Schottky-barrier-field-effect transistors", IBM J. Res. Develop., 14, 125-41 (1970).
926.Yu, H. N., "Insulated gate field-effect transistor", IBM Tech. Disclosure Bull., 14, 253-4 (1971), EEA74-34087.

5. Photo Detectors

927.Anand, Y. A., "High RF burnout resistance silicon Schottky barrier mixer diodes", International Electron Devices Meeting Technical Digest, 1974, p. 129.
928.Anderson, W. A. and Delahoy, A. E., "Schottky barrier diodes for solar energy conversion", Proc. IEEE, 61, 1457-8 (1972), EEA76-7291.
929.Anderson, W. A. and Delahoy, A. E., "Advances in Schottky barrier solar cell technology", 8th Intersociety Energy Conversion Engineering Conference Proceeding, Pt. I, 1973, p. 326-30, EEA77-1769.
930.Anon., "New photodetector is fast responding and efficient", Bell Lab. Record, 45, 160 (1967).
931.Archer, R. J. and Cohen, J., "Schottky barrier infrared detector having ultrathin metal layer", Patent USA 3757123, Publ. September 1973.
932.Baertsch, R. D. and Richardson, J. R., "An Ag-GaAs Schottky barrier ultraviolet detector", J. Appl. Phys., 40, 229-35 (1969), PA72-43125.
933.Carr, T. G., Richmond, J. C. and Wagner, R. G., "Position-sensitive Schottky barrier photodiodes: time-dependent signals and background saturation effects", IEEE Trans. Electron Devices, ED-17, 507-13 (1970), EEA73-24562.
934.Chaudhry, R., "Fabrication and testing apparatus for LID Si surface barrier detectors", Report BARC-689, Bhabha Atomic Res. Centre, Bombay, India, 1973, 9 pp., EEA77-20725.
935.Cohen, J., Vilms, J. and Archer, R. J., "Investigation of Schottky barriers for optical detection and cathodic emission", Report AFCRL-69-0287, Hewlett-Packard Co., Palto Alto, Calif., 1969, 44 pp., EEA73-21264.

936. Cowley, A. M., "Low noise X-band Schottky barrier video detector diode", Presented at International Electron Devices Meeting, Washington, D. C., October 1968, EEA72-7897.
937. Dalal, V. L., "Analysis of photoemissive Schottky barrier photodetectors", J. Appl. Phys., 42, 2280-4 (1971), EEA74-22215.
938. Dauchot, J. P. and Van Cakenberghe, J., "Schottky photodiodes and adsorption", Phys. Status Solidi A, 26, K109-11 (1974), PA78-25755.
939. Davis, P. H., "The Schottky barrier silicon photodetector in perspective with other detection devices in the 200 to 1100 nm range", Proceedings of the Technical Program, Electrical Optical Systems Design Conference, 1971, p. 349-56, EAA76-1121.
940. Davis, P. H. and Wendland, P. H., "The Schottky-barrier silicon photodetector in perspective with other detection devices in the 200 nm to 1100 nm range", Proceedings of the 1st European Electro-Optics Markets and Technology Conference, 1972, p. 393-9, Publ. 1973.
941. Fetterman, H. R., Clifton, B. J., Tannenwald, P. E. and Parker, C. D., "Submillimeter detection and mixing using Schottky diodes", Appl. Phys. Lett., 24, 70-2 (1974), EEA77-19136.
942. Fetterman, H. R., Clifton, B. J., Tannenwald, P. E., Parker, C. D. and Penfield, H., "Submillimeter heterodyne detection and harmonic mixing using Schottky diodes", IEEE Trans. Microwave Theory & Tech., MTT-22, 1013-15 (1974), EEA78-15631.
943. Fonasch, S. J., "The role of the interfacial layer in metal-semiconductor solar cells", J. Appl. Phys., 46, 1286-89 (1975).
944. Green, M. A., "Enhancement of Schottky solar cell efficiency above its semiempirical limit", Appl. Phys. Lett., 27, 287-8 (1975).
945. Henry, R. J., "Behavior of silicon Schottky barrier photodiodes at liquid nitrogen temperatures", Report DLR-MITT-70-08, Deutsche Forsch. & Versuchsanstalt Luft & Raumfahrt, Stuttgart, Germany, 1970, 14 pp., EEA74-30444.
946. Hohnke, D. K. and Holloway, H., "Epitaxial PbSe Schottky-barrier diodes for infrared detection", Appl. Phys. Lett., 24, 633-5 (1974), EEA77-34260.
947. Kajiyama, K., Sakata, S. and Mizushima, Y., "Schottky-barrier devices with low barrier height", Proc. IEEE, 62, 1289-90 (1974), EEA77-34027.
948. Kondo, R., Okimura, H. and Sakai, Y., "Electrical properties of semiconductor photodiodes with semitransparent films", Jap. J. Appl. Phys., 10, 1547-54 (1971), EEA75-796.
949. Korwin-Paulowski, M. and Tarchin, D. W., "High frequency photodiode with n-type Au-Si surface barrier", Przeglad Elektron., 9, 261-7 (1968), Polish, PA71-49432.
950. Li, S. S. and Cox, E. R., Jr., "A new high-sensitivity integrated silicon Schottky-barrier phototransistor", Presented at 1973 International Electron Devices Meeting, Washington, D. C., December 1973, EEA77-14868.

951.Lindley, W. T., Phelan, R. J., Jr., Wolfe, C. M. and Foyt, A. G., "GaAs Schottky barrier avalanche photodiodes", Appl. Phys. Lett., 14, 197-9 (1969), PA72-26575.
952.Logothetis, E. M., Holloway, H., Varga, A. J. and Wilkes, E., "Infrared detection by Schottky barriers in epitaxial PbTe", Appl. Phys. Lett., 19, 318-20 (1971), EEA75-7265.
953.Love, J. A. and Sizelove, J. R., "Voltage enhanced spectral response in the Schottky photodiode", Appl. Optics, 7, 718-20 (1968), PA71-31458.
954.Lubberts, G., Buecher, H. K., Burkey, B. C. and Wolf, E. L., "Photodetection by light-induced barrier modulation in Cu-diffused Au-CdS diodes", 19th International Electron Devices Meeting Technical Digest, 1973, p. 92-95.
955.McColl, M., Millea, M. F. and Silver, A. H., "The superconductor-semiconductor Schottky barrier diode detector", Appl. Phys. Lett., 23, 263-4 (1973), EEA76-38541.
956.Marinov, M. G. and Lefterov, D. P., "On the formation of a barrier junction in surface-barrier detectors on the basis of p-type silicon", C. R. Acad. Bulg. Sci., 27, 1485-7 (1974), EEA78-28681.
957.Padovani, F. A. and Sammer, G. C., "Schottky barrier light sensitive storage device formed by random metal particles", Patent USA 3634692, Publ. January 1972, EEA75-20176.
958.Palmer, J. P., "Fiber-optics coupled detector arrays using thin film cadmium sulfide Schottky diodes", International Electron Devices Meeting Technical Digest, 1974, p. 67-8.
959.Peckerar, M. C., "Electron-beam studies of Schottky-barrier detector surfaces", IEEE Trans. Nucl. Sci., NS-20, 383-7 (1973), EEA77-24452.
960.Pyee, M., Clairon, A. and Auvray, J., "Using a Schottky diode in detecting and mixing at 30 THz", Electron. Lett., 10, 500-1 (1974), French, EEA78-5646.
961.Richardson, J. R. and Baertsch, R. D., "Zinc sulfide Schottky barrier ultra-violet detectors", Solid-State Electron., 12, 393-7 (1969), PA72-33347.
962.Richmond, J. C. and Wagner, R. G., "Position-sensitive Schottky barrier photodiodes", 1970 Solid State Sensors Symposium, p. 62-4, EEA74-1062.
963.Schulz, W., "Photo detectors for continuous position measurements", Elektronik, 21, 13-5 (1972), German, EEA75-11068.
964.Sharpless, W. M., "Evaluation of a specially designed GaAs Schottky-barrier-photodiode using 6328-A radiation modulated at 4 GHz", Appl. Optics, 9, 489-94 (1970), EEA73-17189.
965.Shepherd, F. D., Jr. and Yang, A. C., "Silicon Schottky retinas for infrared imaging", 1973 International Electron Devices Meeting Technical Digest, p. 310-13, EEA77-14843.
966.Spratt, J. P. and Schwarz, R. F., "Metal-silicon Schottky diode arrays as infrared vidicon retinae", 1973 International Electron Devices Meeting Technical Digest, p. 306-9, EEA77-19108.

967.Stillman, G. E., Wolfe, C. M., Foyt, A. G. and Lindley, W. T., "Schottky barrier $In_xGa_{1-x}As$ alloy avalanche photodiodes for 1.06 μm", Appl. Phys. Lett., 24, 8-10 (1974), PA77-37758.
968.Stillman, G. E., Wolfe, C. M. and Melngailis, I., "Monolithic integrated $In_xGa_{1-x}As$ Schottky-barrier waveguide photodetector", Appl. Phys. Lett., 25, 36-8 (1974), EEA77-36474.
969.Stirn, R. J. and Yeh, Y. M., "Solar and laser energy conversion with Schottky barrier solar cells", Record of the 10th IEEE Photovoltaic Specialists Conference, 1973, p. 15-24, Publ. 1974, EEA77-40599.
970.Tseng, C. C. and Wang, S., "Integrated grating-type Schottky-barrier photodetector with optical channel waveguide", Appl. Phys. Lett., 26, 632-5 (1975).
971.Turner, R. J., "Schottky diode pair makes an r.f. detector stable", Electronics, 47 (9), 94-5 (1974), EEA77-22463.
972.Vilisov, A. A., Vyatkin, A. P., Maksimova, N. K., Miloserdova, L. I. and Pekarskii, Y. N., "Sensitivity of gallium arsenide detector diodes", Izv. VUZ Radioelektron., 14, 585-7 (1971), Russian, EEA74-27179.
973.Voronov, V. N., Ginzberg, A. D., Naumov, A. I., Pavelev, D. G., Uskova, V. M. and Fedoseev-Svhetsov, L. I., "Some results of the application of Schottky barrier diodes in mixers for radiometers in the wave range 0.9-3.8 mm", Izv. VUZ Radiofiz., 16, 1125-6 (1973), Russian, EEA77-2878.
974.Wang, C., "New grating type gold-n-type silicon Schottky-barrier photodiode for 0.4-1.1 μm photodetection", Thesis, Univ. Florida, Gainesville, Fla., 1973, 88 pp., Order No. 74-9616, CA81-31076.
975.Wang, C. T. and Li, S. S., "A new grating-type gold n-type silicon Schottky barrier photodiode", IEEE Trans. Electron Devices, ED-20, 522-6 (1973).
976.Yamamoto, T. and Takase, Y., "Photoresponse of an Au-Si contact gate field-effect transistor", Electron. & Commun. Jap., 53, 175-81 (1970), EEA75-9913.
977.Yeh, C. and Shabde, S. N., "Light-sensitive metal-semi-insulator-n^+ GaAs diodes", Proc. IEEE, 59, 1636-7 (1971), EEA74-37251.

6. Particle Detectors

978.Baertsch, R. D., "Schottky barrier atomic particle and X-ray detector", Patent USA 3598997, Publ. August 1971.
979.Caywood, J. M., Mead, C. A. and Mayer, J. W., "Influence of carrier diffusion effects on window thickness of semiconductor detectors", Nucl. Instrum. & Methods, 79, 329-32 (1970).
980.Gibbons, P. E., "Metal-semiconductor contacts for solid state radiation detectors", Nucl. Instrum. & Methods, 29, 289-92 (1964).
981.Glasow, P., "Semiconductor radiation detectors in the research satellite Azur", Siemens Rev., 38, 12-4 (1971).
982.Mayer, J. W., "Germanium and silicon surface barrier diodes as alpha particle detectors", J. Appl. Phys., 30, 1937-44 (1959).

983. Muggleton, A. H. F., "Semiconductor devices for gamma ray, X-ray and nuclear radiation detection", J. Phys. E, 5, 390 (1972).

7. Power Rectifiers

984. Bates, D. J., "Semiconductors inside tubes make high-performance RF amplifiers", Electronics, 47 (15), 85-88 (1974).
985. Beilstein, K. E. and Bertin, C. L., "Regulated MOSFET power supply device", IBM Tech. Disclosure Bull., 15, 817-8 (1972), EEA76-249.
986. Bixby, B. and Carlan, A., "Schottky barrier diodes improve high frequency rectifier efficiency", Proceedings of the 1972 Seventh Annual Meeting of the IEEE Industry Application Society, p. 439-46, EEA76-7216.
987. Heymann, H. and Petruzella, J., "The development and fabrication of a power Schottky diodes", Report N70-11411, General Electric Company, Auburn, New York, 1969.
988. Hoeneisen, B. and Mead, C. A., "Power Schottky diode design and comparison with the junction diode", Solid-State Electron., 14, 1225-36 (1971), EEA75-4678.
989. Kim, C., "High-power high-efficiency operation of Read-type IMPATT-diode oscillators", Electron. Lett., 9, 173-4 (1973), EEA76-22914.
990. Lindley, W. T., Krag, W. E., Wolfe, C. M., Sasiela, R. J., Murphy, R. A., Hurwitz, C. E., Kostishack, D. F., Yakutis, A. J. and Foyt, A. G., "Schottky barrier GaAs EBS amplifier", Proceedings of the 4th International Symposium on Gallium Arsenide and Related Compounds, 1972, p. 295-300, Publ. 1973, EEA76-23702.
991. Migitaka, M., Nakamura, M., Saito, K. and Sekine, K., "A high power GaAs Schottky barrier IMPATT diode in 30 GHz band", 1973 European Microwave Conference, Vol. I, A8.5/4 pp., EEA77-5981.
992. Moore, D. W., "Analyzing the principles of Schottky diodes", Electron. Equip. News, 15 (7), 61-4 (1973), EEA77-8861.
993. Page, D. J., "Schottky barrier power rectifier", Patent USA 3652905, Publ. March 1972.
994. Page, D. J., "Theoretical performance of the Schottky barrier power rectifier", Solid-State Electron., 15, 505-15 (1972), EEA75-19055.
995. Polgar, P., Mouyard, A. and Shiner, B., "A high-current metal-semiconductor rectifier", IEEE Trans. Electron Devices, ED-17, 725-31 (1970), EEA73-38040.
996. Wisseman, W. R., Shaw, D. W., Adams, R. L. and Hasty, T. E., "GaAs Schottky-Read diodes for X-band operation", IEEE Trans. Electron Devices, ED-21, 317-23 (1974), EEA77-26441.

VII. Exploratory Devices

1. Surface Barrier Transistors

997.Anantha, N. G., "Fabricating a metal base transistor", IBM Tech. Disclosure Bull., 13, 2149-50 (1970).
998.Angell, J. B. and Keiper, F. P., Jr., "Surface-barrier transistor, (Part III - Circuit Applications)", Proc. IRE, 41, 1709-12 (1953).
999.Anon., "Semiconductor-metal-semiconductor triode", Report unnumbered, Contract No. AF33(615)-3506, Hewlett-Packard Co., Palo Alto, California, 1966.
1000.Antula, J., "A method of investigation of hot electron transport through thin metal films using the metal-base triode configuration", Thin Solid Films, 13, 93-7 (1972), EEA76-4592.
1001.Atalla, M. M. and Sosher, R. W., "Hot-carrier triodes with thin-film metal base", Solid-State Electron., 6, 245 (1963).
1002.Bradley, W. E., "The surface-barrier transistor, (Part I - Principles)", Proc. IRE, 41, 1702-6 (1953).
1003.Cervenak, J., "Space charge limited electric currents in thin-film metal-InSb-metal structures", Czech. J. Phys. B., 19, 1379-88 (1969); PA73-28375.
1004.Geppert, D. V., "A metal base transistor", Proc. IRE, 50, 1527 (1962).
1005.Kansas, R., "Surface-barrier transistor, (Part IV - On the high-frequency performance of transistor)", Proc. IRE, 41, 1712-5 (1953).
1006.Lavine, J. M., "A survey of hot-electron and thin-film transistors", Semiconductor Products and Solid State Technology, 7 (5), 17-23 (1964).
1007.May, G. A., "The Schottky-barrier-collector transistor", Solid-State Electron., 11, 613-9 (1968), PA71-40542.
1008.Mead, C. A., "The tunnel-emission amplifier", Proc. IRE, 48, 359 (1960).
1009.Moll, J. L., "Comparison of hot electron and related amplifiers", IEEE Trans. Electron Devices, ED-10, 299 (1963).
1010.Nadkarni, G. S. and Simmons, J. G., "Determination of the defect nature of MoO_3 films using dielectric-relaxation currents", J. Appl. Phys., 43, 3650-6 (1972), PA75-68453.
1011.Pietoulet, B. and Rousoyre, M., "Mechanism of metal-semiconductor interfacial transistors", Phys. Status Solidi, 9, 369 (1965), French.
1012.Schwarz, R. F. and Walsh, J. F., "Surface-barrier transistor, (Part V - The properties of metal to semiconductor contacts)", Proc. IRE, 41, 1715-20 (1953).
1013.Siemens, A. G., "Metal base transistor", Patent UK 1144147, Publ. March 1969, EEA72-20403.
1014.Spratt, J. P., Schwarz, R. F. and Kane, W. M., "Hot electrons in metal film: injection and collection", Phys. Rev. Lett., 6, 341 (1961).

1015.Sze, S. M. and Gummel, H. K., "Appraisal of semiconductor-metal-semiconductor transistor", Solid-State Electron., 9, 751 (1966).
1016.Tiley, J. W. and Williams, R. A., "Surface-barrier transistors (Part II - Electrochemical techniques for fabrication)", Proc. IRE, 41, 1706-8 (1953).
1017.Zeidenbergs, G., "Multiemitter Schottky barrier transistor with negligible emitter-to-emitter gain", IBM Tech. Disclosure Bull., 14, 1104 (1971).

2. Acoustic Wave Transducers

1018.Bert, A. G. and Kantorowicz, G., "Phase and amplitude storage of an acoustic signal in a Schottky-diode correlator", Electron. Lett., 11, 523-4 (1975).
1019.Chao, C., "Monolithic surface-acoustic-wave phase shifter", Electron. Lett., 9, 49-51 (1973), EEA76-14781.
1020.Melikyan, E. G., "Electroacoustic interaction in a metal-semiconductor contact", Radiotekhnika I Elektronika, 12, 1303-4 (1967), Russian, EEA71-10557.
1021.Takada, S., Kayakawa, H., Ishiguro, T. and Mikoshiba, N., "Depletion-layer transduction of surface waves on piezoelectric semiconductor", Jap. J. Appl. Phys., 11, 24-30 (1972), EEA75-7463.
1022.White, D. L., "The depletion layer and other high-frequency transducers using fundamental modes", In: Physical acoustics, vol. 1, part B, Mason, W. P., (Ed.), Academic Press, New York, 321-52 (1964).

3. Cold Cathode Emitters

1023.Forier, R. P., de Morre, R., Hiernaut, J. P. and van Cakenberghe, J., "Electron emission at metal-semiconductor junctions", Vacuum, 22, 531-34 (1972), PA76-50268.
1024.Itoh, T., Matsuda, I., Hasegawa, K. and Umeoka, K., "Field-induced photoelectron emission from silicon surface-barrier diodes", J. Appl. Phys., 38, 3395-6 (1967), PA70-32939.
1025.Stolte, C. A. and Archer, R. J., "pn-Schottky hybrid cold-cathode diode", Appl. Phys. Lett., 19, 497-8 (1971), PA75-13353.
1026.Stolte, C. A., Vilms, J. and Archer, R. J., "The Schottky barrier cold cathode", Solid-State Electron., 12, 945-54 (1969), EEA73-10027.
1027.Swank, R. K., "Characteristics of a ZnS:Pd:Cs_2O cold cathode", J. Appl. Phys., 41, 778-81 (1970), EEA73-17120.
1028.Williams, R. and Wronski, C. R., "Electron emission from the Schottky barrier structure ZnS:Pt:Cs", Appl. Phys. Lett., 13, 231 (1968).

4. Bistable Switches

1029.Abdullaev, G. B., Allakhverdiev, K. R., Antonov, V. B., Nani, R. K. and Salnev, E. Y., "Switching and memory effect at the contact between a $CdIn_2S_4$ crystal and a metal", Sov. Phys.-Dokl., 16, 588-90 (1972), EEA75-10190.
1030.Anacker, W., Herrell, D. J. and Park, K. C., "Fabrication of switchable resistor/Schottky barrier memory cell", IBM Tech. Disclosure Bull., 14, 2151-3 (1971), EEA75-16709.
1031.Biard, J. R., "Unitary semiconductor high speed switching device utilizing a barrier diode", Patent 3463975, Publ. August 1969.
1032.Chang, L. L. and Esaki, L., "Nonvolatile Schottky diode with barrier height controlled by ferroelectric polarization", IBM Tech. Disclosure Bull., 14, 1250-1 (1971), EEA75-10225.
1033.Hassayer, C., Martinez, A., Pujol, J. E. and Esteve, D., "On a 'memory' effect in n-type silicon Schottky diodes in the presence of metallic impurities", C. R. Hebd. Seances Acad. Sci. B, 276, 351-4 (1973), French, EEA76-19904.
1034.Holm-Kennedy, J. W., Heald, D. L., Williams, G. and Chang, Y. W., "Observation and explanation of multistable nonvolatile memory in silicon FIT diodes", Appl. Phys. Lett., 22, 167-9 (1973), EEA76-16358.
1035.Moser, A., "Bistable switching in metal-semiconductor junctions", Appl. Phys. Lett., 20, 244-5 (1972), EEA75-16703.

5. Luminescent Diodes

1036.Allen, J. W., Livingstone, A. W. and Turvey, K., "Electroluminescence in reverse-biased zinc selenide Schottky diodes", Solid-State Electron., 15, 1363-9 (1972), EEA76-785.
1037.Card, H. C. and Smith, B. L., "Green injection luminescence from forward-biased Au-GaP Schottky barriers", J. Appl. Phys., 42, 5863-5 (1971), EEA75-7285.
1038.Haeri, S. Y. and Rhoderick, E. H., "Electroluminescence from gallium phosphide Schottky diodes", Conference on Metal-Semiconductor Contacts, 1974, p. 84-90, EEA78-8903.
1039.Holloway, H., Weber, W. H., Logothetis, E. M., Varga, A. J. and Young, K. F., "Injection luminescence and laser action in epitixial PbTe diodes", Appl. Phys. Lett., 21, 5-6 (1972).
1040.Livingstone, A. W., Turvey, K. and Allen, J. W., "Electroluminescence in forward-biased zinc selenide Schottky diodes", Solid-State Electron., 16, 351-6 (1973), EEA76-16695.
1041.Nill, K. W., Calawa, A. R., Harman, T. C. and Walpole, J. N., "Laser emission from metal-semiconductor barriers on PbTe and $Pb_{0.8}Sn_{0.2}Te$", Appl. Phys. Lett., 16, 375-7 (1970), EEA73-27786.

6. Gunn Effect Elements

1042. DiLorenzo, J. V., Niehaus, W. C. and Rode, D. L., "Gallium arsenide Schottky barrier avalanche diode array", Patent USA 3896473, Publ. July 1975, CA83-124985.

1043. DiLorenzo, J. V., Niehaus, W. C., Velebir, J. R. and Iglesias, D. E., "Beam lead plated heat sink GaAs IMPATT's", Presented at 1973 International Electron Devices Meeting, Washington, D. C., December 1973, EEA77-18783.

1044. Goto, G., Nakamura, T. and Isobe, T., "Two-dimensional domain dynamics in a planar Schottky-gate Gunn-effect device", IEEE Trans. Electron Devices, ED-22, 120-6 (1975), EEA78-11597.

1045. Heime, K., "Planar Schottky-gate Gunn devices", Electron. Lett., 7, 610-3 (1971), EEA75-1042.

1046. Hubner, R., "Gunn elements and Schottky-barrier diodes", Funk-Tech., no. 13, 477-80 (1972), German, EEA75-32556.

1047. Irvin, J. C., "GaAs IMPATT diodes in perspective", Cornell Electronic Engineering Conference, 4th Bienn., Proc., 1973, p. 287-98.

1048. Shimizu, N., Kumabe, K. and Kanbe, H., "Characteristics of a gallium-arsenide travelling-wave amplifier with Schottky-barrier contacts", Electron. Lett., 9, 29-30 (1973), EEA76-11859.

1049. Sugeta, T., Tanimoto, M., Ikoma, T. and Yanai, H., "Characteristics and applications of a Schottky-barrier-gate Gunn-effect digital-device", IEEE Trans. Electron Devices, ED-21, 504-15 (1974), EEA77-34003.

1050. Takeuchi, M., Higashisaka, A. and Sekido, K., "GaAs bulk effect pulse regenerator with a Schottky barrier control gate", Proc. IEEE, 60, 740-1 (1972), EEA75-23249.

1051. Utsugi, Y., Harada, H. and Kurumada, K., "Threshold condition of Schottky gate-Gunn pulse device", Rev. Electr. Commun. Lab., 23, 278-89 (1975), EEA78-37493.

1052. Yanai, H. and Sugeta, T., "Schottky-gate Gunn effect digital device", Annu. Rep. Eng. Res. Inst. Fac. Eng. Univ. Tokyo, 30, 223-8 (1971), Japanese, EEA75-10212.

1053. Yanai, H., Sugeta, T., Tamaki, Y. and Tanimoto, M., "Signal-processing of a Schottky-gate Gunn-effect digital functional device", Annu. Rep. Eng. Res. Inst. Fac. Eng. Univ. Tokyo, 31, 139-44 (1972), Japanese, EEA76-8101.

7. Microphones and Pressure Sensors

1054. Gummel, H. K. and Rideout, V. L., "Shock-mounting for electromechanical transducer", Patent USA 3414689, Publ. December 1968.

1055. Kahng, D. and Wemple, S. H., "Surface-barrier diode transducer using high dielectric semiconductor material", Patent USA 3443041, Publ. June 1965.

8. Others

1056. Brodsky, M. H., "Contact barriers to semiconductor crystals", IBM Tech. Disclosure Bull., 17, 1814-16 (1974), EEA78-11610.
1057. Brookbanks, D. M., Griffith, I. and White, P. M., "Integral heat sink contacts for CW indium phosphide transferred electron oscillators and amplifiers", Conference on Metal-Semiconductor Contacts, 1974, p. 116-22, EEA78-8521.
1058. Dahlberg, R., "Controllable Schottky diode", Patent USA 3495141, Publ. February 1970, EEA73-31384.
1059. Dorler, J., Forneris, J. L. and Swietek, D. J., "Multijunction Schottky barrier diode", IBM Tech. Disclosure Bull., 16, 628 (1973).
1060. Epple, R. R., "Semiconductor device having a rectifying junction surrounded by a Schottky contact", Patent USA 3760241, Publ. September 1973.
1061. Hogeboom, J. G., "Schottky barrier diodes and their application as Schottky barrier resistors", Thesis, Univ. Toronto, Ontario, 1973, CA81-112716.
1062. Krishna, S., "Schottky barrier plasma thyristor circuit", Patent USA 3829880, Publ. August 1974.
1063. Lepselter, M. P., "Schottky barrier diodes as impedence elements and method of making", Patent USA 3770606, Publ. November 1973.
1064. Normington, P. J. C., "Schottky diode clipper device", Patent USA 3746945, Publ. July 1973.
1065. Soshea, R. W., "Surface barrier semiconductor limiter employing low barrier height metals on silicon", Patent USA 3486086, Publ. December 1969.
1066. Soshea, R. W. and Zettler, R. A., "Hybrid semiconductor device including diffused-junction and Schottky-barrier diodes", Patent USA 3463971, Publ. August 1969.
1067. Telefunken Patentverwertungs GmbH, "Controllable Schottky diode", Patent UK 1173575, Publ. December 1969, EEA73-31381.

Q. MODELING

I. Bibliographies

1.Agajanian, A. H., "A bibliography on semiconductor device modeling", Solid-State Electron., 18, 917-29 (1975), EEA78-41409.

II. Books

2.Carroll, J. E., "Physical models for semiconductor devices", Crane, Russak, New York, 1974.

3.Gray, P. E., DeWitt, D., Boothroyd, A. R. and Gibbons, J. F., "Physical electronics and circuit models of transistors", John Wiley & Sons, Inc.: New York, 1964.

4.Gray, P. E. and Searle, C. L., "Electronic principles - physics, models and circuits", John Wiley & Sons, Inc.: New York, 1969.

5.Hamilton, D. J., Lindholm, F. A. and Marshak, A. H., "Principles and applications of semiconductor device modeling", Holt, Rinehart and Winston, Inc.: New York, 1971.

6.Herskowitz, G. L. and Schilling, R. B., (Eds.), "Semiconductor device modeling for computer-aided design", McGraw-Hill Book Company, Inc.: New York, 1972.

7.Linvill, J. G., "Models of transistors and diodes", McGraw-Hill Book Company, Inc.: New York, 1963.

8.Linvill, J. G. and Gibbons, J. F., "Transistors and active circuits", McGraw-Hill: New York, 1961.

III. Review Articles

9.Arendt, A., "The RCI model - A general model for semiconductor devices", IEEE Trans. Electron Devices, ED-20, 5-12 (1973), EEA76-4413.

10.Berger, H. H., "Models for contacts to planar devices", Solid-State Electron., 15, 145-58 (1972), EEA75-13654.

11.Brown, R. W., "Transistor models for circuit analysis programs", Electronic Engng., 41 (502), 50-5 (1969), EEA73-12735.

12.Brown, R. W., "Device modeling", In: Computer aided design techniques, Wolfendale, E., (Ed.), Butterworth, London, England, p. 80-114 (1970), EEA74-6692.

13.Brown, R. W., "Component models for computer-aided circuit design: II", Electron. Compon., 12, 441-4 (1971), EEA74-17299.

14.Brown, R. W., "Component models for computer-aided circuit design: III. Diodes and FETs", Electron. Compon., 12, 607-10 (1971), EEA74-21178.

15.Butler, E. M., "On-line transistor modeling in a manufacturing environment", IEEE Trans. Circuit Theory, CT-20, 683-7 (1973), EEA77-72.
16.Hart, P. A. H., "Device modeling", Proceedings of the European Semiconductor Device Research Conference on Solid State Devices, 1971, p. 31-43, EEA76-660.
17.Lange, J., "A survey of the present state of microwave transistor modeling and simulation as applied to circuit design", IEEE Trans. Electron Devices, ED-18, 1168-74 (1971), EEA75-4863.
18.Lindholm, F. A., "Generation and selection of large signal dynamic models for bipolar and field-effect transistors", IEEE Proceedings of the International Symposium on Circuit Theory, 1972, p. 385, EEA75-32674.
19.Linvill, J. G., "The evolution of modeling of bipolar semiconductor devices", IEEE Proceedings of the International Symposium on Circuit Theory, 1972, p. 287-91, EEA75-32672.
20.Logan, J., "Modeling for circuit and system design", Proc. IEEE, 60, 78-85 (1972), EEA75-7516.
21.McAffe, L. C., Jr., "Techniques for the optimization of numerical models of semiconductor devices", Thesis, Univ. Michigan, Ann Arbor, 1970, 180 pp., Order No. 71-15233, EEA74-37322.
22.McCalla, W. J., "Choosing the right model", 1973 WESCON Technical Papers, Western Electronic Show and Convention, 14/2-66 pp., EEA77-10240.
23.Peterson, B. R., "Use of optimization techniques for modeling semiconductor device characteristics", Thesis, Utah State Univ., Logan, 1971, 179 pp., Order No. 72-4734, EEA75-28534.
24.Philipp, L. D., "The development and use of scattering parameters for semiconductor junction devices", Thesis, Univ. New Mexico, Albuquerque, 1971, 195 pp., Order No. 72-4011, EEA75-28467.
25.Philipp, L. D. and Boatwright, L. T., "The development and use of scattering parameters for semiconductor junction devices", Report EE-183(71)ONR-005, Univ. New Mexico, Albuquerque, 1971, 198 pp., EEA75-16701.

IV. Computer Analysis

26.Berka, Z., "Analyzing the properties of semiconductor structure mathematical models with the aid of a computer", Slaboproudy Obz., 36, 259-64 (1975), Czech, EEA78-32931.
27.Boerl, J., Merckel, C., Monnier, J., Saintot, P., Vandorpe, J., Stern, M. and Maffei, M., "A connection between technology and models using a computer analysis", Digest of Technical Papers of the 1973 IEEE International Solid State Circuits Conference, 1973, p. 102-3, 205, EEA76-23157.
28.Brown, R. W., "Component models for computer aided circuit design. I", Electron. Compon., 12, 259-60 (1971), EEA74-14631.

29. Bulucea, C., Rusu, A. and Postolache, C., "Surface breakdown in silicon planar junctions - a computer-aided experimental determination of the critical field", Solid-State Electron., 17, 881-8 (1974), EEA78-8575.
30. Calzolari, P. U., Graffi, S. and Picrini, G., "A new program for the DC one-dimensional analysis of semiconductor devices", Alta Freq., 41, 598-60 (1972), EEA76-656.
31. Case, G. R., "Acquisition of computer aided design/analysis models for semiconductors", J. Test. & Eval., 1, 295-300 (1973), EEA76-34782.
32. Caughey, M., Luke, D., McCready, L. and Rombeek, H., "Circuit simulation by computer", Telesis, 2 (6), 17-24 (1972), EEA75-39195.
33. Golembeski, J. J., "Linear circuit models derived via computer optimization", IEEE Trans. Educ., E-12, 162-9 (1969), EEA73-12740.
34. Gummel, H. K., "Computer device modeling", In: Festkorper Problems IX; Solid State Physics, Advances in, Madelung, O., (Ed.), Pergamon Press, New York, p. 338-9 (1969), EEA73-28139.
35. Javed, A. and Goud, P. A., "Computer modeling of a microwave varactor including injection effects", 1969 European Microwave Conference, p. 519, Publ. 1970, EEA73-38430.
36. Jenkins, F. S. and Shi Ping Fan, "TIME - a nonlinear d.c. and time-domain circuit-simulation program", IEEE J. Solid-State Circuits, SC-6, 182-82 (1971), EEA74-25343.
37. Klopfenstein, R. W. and Chung, P. W., "Computer solution of one-dimensional Poisson's equation", IEEE Trans. Electron Devices, ED-22, 329-33 (1975), EEA78-19630.
38. Kurz, G., "Models of transistors for network design by digital computers", Radio Fernsehen Elektron., 20, 384-6 (1970), German, EEA74-34122.
39. Leggett, J. D., "Circuit theory and problems in computer aids for digital integrated electronics", Presented at International Symposium on Circuit Theory 1973, Toronto, Canada, April 1973, EEA77-2581.
40. Lesnicki, A., "Determination of parameter values of the transistor model based on the measured twoport characteristics", Arch. Elektrotech., 23, 325-36 (1974), Polish, EEA78-5206.
41. McAffe, L. C., Jr., "The optimization of numerical models of semiconductor devices", Proceedings of the 8th Annual Allerton Conference on Circuit and System Theory, 1970, p. 625-34, EEA74-37323.
42. Nakamura, T. and Nishizawa, J., "CAD for realization of an impurity distribution in a semiconductor", Electron. & Commun. Jap., 5 (12), 90-6 (1972), PA77-6977.
43. Neudorfer, M. L. and Yee, S. S., "Computer analysis of current instabilities in piezoelectric semiconductors", IEEE Trans. Electron Devices, ED-16, 1069-76 (1969), EEA73-10488.
44. Pocock, D. N. and Raymond, J. P., "Simplified terminal models of complex linear microcircuits for use in computer aided systems analysis", IEEE Trans. Nucl. Sci., NS-18, 212-9 (1971), EEA75-9571.

45.Raymond, J. P. and Krebs, M. G., "Lumped model analysis of semiconductor devices using the NET-2 circuit/system analysis program", IEEE Trans. Nucl. Sci., NS-19, 103-7 (1972), EEA76-12256.
46.Russo, P. M. and Rohrer, R. A., "Computer optimization of the transient response of an ECL gate", IEEE Trans. Circuit Theory, CT-18, 197-9 (1971), EEA74-12745.
47.Seidman, T. I. and Choo, S. C., "Iterative scheme for computer simulation of semiconductor devices", Solid-State Electron., 15, 1229-35 (1972), EEA75-39966.
48.Sergert, J. E., Bowers, J. C. and Nienhaus, H. A., "Computer modeling of semiconductor devices", International Microelectronics Symposium, Paper 2A-5, 11 pp., 1973.
49.Washburn, L. A., "Determination of optimum burn-in time: a composite criterion", IEEE Trans. Rel., R-19, 134-40 (1970), EEA74-18829.
50.Wilson, C. L., "Analytic method for approximating two-dimensional device structures using one-dimensional device analysis programs", 1970 SWIEECO Record of Technical Papers of the 22 Southwestern IEEE Conference and Exhibition, p. 161-5, EEA73-31443.
51.Wright, M. L. and Mantena, N. R., "LSA mode calculations using a simple computer model", Int. J. Electron., 28, 19-27 (1970), EEA73-17695.
52.Yu, S. P. and Tantraporn, W., "A computer simulation scheme for various solid-state devices", IEEE Trans. Electron Devices, ED-22, 515-22 (1975).
53.Zaniolo, C. and McNamee, L. P., "A simple method for determining static parameters of large signal semiconductor diode and transistor models", Solid-State Electron., 15, 1295-302 (1972), EEA76-659.

V. Mathematical Analysis

54.Andrew, R., "Improved formulation of Gummel's algorithm for solving the 2-dimensional current-flow equations in semiconductor devices", Electron. Lett., 8, 536-8 (1972), EEA76-657.
55.Aronov, D. A., Kotov, E. P. and Kotov, Y. P., "The influence of carrier capture in deep traps on the small-signal characteristics of double-injection currents in semiconductors", Phys. Status Solidi A, 30, 97-105 (1975), EEA78-33088.
56.Barth, K., Diener, K. H., Elschner, H., Landgraf-Deitz, D. and Moschwitzer, A., "Application of a new algorithm in the analysis of semiconductor structures", Nachrichtentech. Elektron., 23, 445-7 (1973), German, EEA77-14497.
57.Bereisa, J., Jr. and Goben, C. A., "Determination of graded base parameters in the presence of recombination at low to moderate injection levels", 1970 SWIEECO Record of Technical Papers of the 22 Southwestern IEEE Conference and Exhibition, p. 181-5, EEA73-35257.

58. Blum, A. S., "A two-dimensional modeling of the electron and hole concentrations in semiconductor structures under transient conditions", Thesis, Washington Univ., St. Louis, Mo., 1971, 228 pp., Order No. 71-19840, EEA75-28535.
59. Brantley, W. A., "Calculated elastic constants for stress problems associated with semiconductor devices", J. Appl. Phys., 44, 534-5 (1973), EEA76-12242.
60. David, P., "Spectral analysis of a physical system representable by a stationary linear active electronic network", Ann. Telecommun., 28, (1-2), 55-70 (1973), French, EEA76-15843.
61. di Maio, B. and Tamburello, C., "A method to find the coefficients of a mathematical model of a planar diffused transistor", Alta Frequenza, 38, 606-14 (1969), EEA73-4488.
62. Dierking, W. H., Katz, G. E. and Steele, E. L., "Transient radiation current generator model for semiconductor devices", IEEE Trans. Nuclear Sci., NS-16, 144-52 (1969), EEA73-14343.
63. Domingos, H., "Transient temperature rise in silicon semiconductor devices", IEEE Trans. Electron Devices, ED-22, 20-2 (1975), EEA78-5203.
64. Ezawa, H., Kawaji, S., Kuroda, T. and Nakamura, K., "Electron mobility in a semiconductor inversion layer. Possible contribution from bulk phonons", Surface Sci., 24, 659-62 (1971), EEA74-13220.
65. Forbes, L. and Sah, C. T., "Application of the distributed equilibrium equivalent circuit model to semiconductor junctions", IEEE Trans. Electron Devices, ED-16, 1036-41 (1969), EEA73-10580.
66. Green, M. A. and Shewchun, J., "Efficient numerical solution of the transmission-line equivalent-circuit model of a semiconductor", Electron. Lett., 9, 474-5 (1973), EEA77-2373.
67. Green, M. A. and Shewchun, J., "The general transmission line equivalent circuit model for degenerate and non-degenerate carrier concentrations in semiconductors", Solid-State Electron., 17, 717-23 (1974), EEA77-30021.
68. Green, M. A. and Shewchun, J., "Application of the small-signal transmission line equivalent circuit model to the AC, DC and transient analysis of semiconductor devices", Solid-State Electron., 17, 941-9 (1974), EEA78-8392.
69. Gregory, B. L. and Gwyn, C. W., "Application of neutron damage models to semiconductor device studies", IEEE Trans. Nucl. Sci., NS-17, 325-34 (1970), EEA74-22890.
70. Gummel, H. K., "A self-consistent iterative scheme for one-dimensional steady state transistor calculations", IEEE Trans. Electron Devices, ED-11, 455-65 (1964), EEA68-16000.
71. Hennig, F. and Sah, C. T., "Matrix analysis of distributed semiconductor circuit models", Solid-State Electron., 16, 1081-3 (1973), EEA76-35089.
72. Hockney, R. W., Warriner, R. A. and Reiser, M., "Two-dimensional particle models in semiconductor-device analysis", Electron. Lett., 10, 484-6 (1974), EEA78-2675.

73.Hsu, S. T. and Whittier, R. J., "Characterization of burst noise in silicon devices", Solid-State Electronics, 12, 867-8 (1969), EEA73-10594.
74.Jesshope, C. R., "Numerical solutions to the 2-dimensional time-dependent semiconductor equations", Electron. Lett., 11, 431-3 (1975), EEA78-37343.
75.Kani, K., Kato, H. and Ohtsuki, T., "A nonlinear lumped-network model of semiconductor devices and its application", Electron. & Commun. Jap., 55 (7), 76-83 (1972), EEA76-29059.
76.Kani, K. and Yokota, A., "A nonlinear lumped network model of semiconductor devices with consideration of recombination kinetics", IEEE Trans. Electron Devices, ED-19, 1028-38 (1972), EEA75-32711.
77.Kato, H., Kani, K. and Ohtsuki, T., "A nonlinear lumped network model of semiconductor devices and its application", NEC Res. & Dev., no. 28, 28-39 (1973), EEA76-29240.
78.Kennedy, D. P., "Mathematical simulation of the effects of ionizing radiation on semiconductors", Report AFCRL-69-0086, IBM Corp., Hopewell Junction, N. Y., 1969, 50 pp., EEA73-28126.
79.Konjuchov, A. G., Lebl, L. and Soukrup, M., "Variation of quality and reliability values of electronic component parts in series production", Slaboproudy Obz., 36, 171-6 (1975), EEA78-19544.
80.Kurata, M., "A small-signal calculation for one-dimensional transistors", IEEE Trans. Electron Devices, ED-18, 200-10 (1971), EEA74-13247.
81.Lee, C. M., Lomax, R. J. and Haddad, G. I., "Semiconductor device simulation", IEEE Trans. Microwave Theory & Tech., MTT-22, 160-77 (1974), EEA77-14247.
82.Lindholm, F. A. and Hamilton, D. J., "A systematic modeling theory for solid state devices", Solid-State Electron., 7, 771-84 (1964).
83.McAfee, L. C., Jr., "Equation formulation considerations for efficient numerical modeling of semiconductor phenomena", Solid-State Electron., 18, 799-804 (1975), EEA78-32932.
84.McNutt, M. J. and Sah, C. T., "High frequency space charge layer capacitance of strongly inverted semiconductor surfaces", Solid-State Electron., 17, 377-85 (1974), PA77-41128.
85.Mock, M. S., "On the computation of semiconductor device current characteristics by finite difference methods", J. Eng. Math., 7, 193-205 (1973), EEA76-32066.
86.Neumark, G. F., "New model for interface charge-carrier mobility: the role of misfit dislocations", Phys. Rev. Lett., 21, 1252-6 (1968), PA72-3339.
87.Novash, V. I., Korolev, O. P., Korenyako, V. N. and Savin, S. N., "Mathematical model of the intelligence part of a directional semiconductor-type remote-control relay", Izv. VUZ Energ., no. 4, 29-35 (1972), Russian, EEA75-27997.
88.Ohtsuki, T. and Kani, K., "A unified modeling scheme for semiconductor devices with applications of state-variable analysis", IEEE Trans. Circuit Theory, CT-17, 26-32 (1970), EEA73-17694.

89.Pastzor, G., "The use of the Linvill model for modeling semiconductor devices", Hiradstech. Ipari Kut. Intez. Kozl., 11 (4), 5-17 (1971), Hungarian, EEA75-13668.
90.Pastzor, G., "The use of the Linvill model for modeling semiconductor devices", Hiradstech. Ipari Kut. Intez. Kozl., 11 (3), 34-55 (1971), Hungarian, EEA75-4736.
91.Reiser, M., "Difference methods for the solution of the time-dependent semiconductor flow equations", Electron. Lett., 7, 353-5 (1971), EEA74-26295.
92.Reiser, M., "On the stability of finite difference scheme in transient semiconductor problems", Comput. Methods Appl. Mech. & Eng., 2 (1), 65-8 (1973), EEA76-12254.
93.Rohr, P., Lindholm, F. A. and Allen, K. R., "Questionability of drift-diffusion transport in the analysis of small semiconductor devices", Solid-State Electron., 17, 729-34 (1974), EEA77-30022.
94.Sah, C. T., "Exact large-signal lumped-network representation of a semiconductor", Electron. Lett., 8, 88-91 (1972), EEA75-13807.
95.Sandberg, I. W., "On the solvability of the equations of networks containing semiconductor devices, and reverse-bias breakdown phenomena", IEEE Trans. Circuit Theory, CT-18, no. 4, 475-6 (1971), EEA74-21236.
96.Sandberg, I. W., "Conditions for the existence of a global inverse of semiconductor-device nonlinear-network operators", IEEE Trans. Circuit Theory, CT-19, 34-6 (1972), EEA75-3736.
97.Schelonka, E. P., "Discrete charge models for simulating solid state systems functions", 21st Annual Southwestern IEEE Conference and Exhibition, 1971, p. 385-8, EEA75-7567.
98.Snejdar, V., Jerhot, J. and Berkova, D., "Influence of scattering mechanisms on the transconductance of CdSe thin film transistors", Thin Solid Films, 13, 47-50 (1972), EEA76-4670.
99.Stanclik, J., "Modelling of electronic components for statistical design", Rozpr. Electrotech., 20, 713-23 (1974), Polish, EEA78-24097.
100.Startin, R. A. and Ash, E. A., "Perturbation techniques for semiconductor-device computer experiments", Proc. Inst. Electr. Eng., 119, 1549-54 (1972), EEA76-4417.
101.van Vliet, K. M., "General transport theory of noise in PN junction-like devices. I. Three-dimensional Green's function formulation", Solid-State Electron., 15, 1033-53 (1972), EEA75-39892.
102.Vandorpe, D. and Xuong, N. H., "Mathematical 2-dimensional model of semiconductor devices", Electron. Lett., 7, 47-50 (1971), EEA74-10681.
103.Wernstedt, J. and Bergmann, S., "Application of identification methods in piece good processes in semiconductor production", Proceedings of the 3rd IFAC Symposium on Identification and System Parameter Estimation, Pt. I, 1973, p. 517-21, EEA77-5917.

104.Wilson, P. R., "Integrals involving the Gaussian distribution function", Microelectronics, 6 (1), 11-16 (1974), EEA78-24096.
105.Zombory, L., Kohalmi, Z. and Veszely, G., "Self-consistent modeling of transferred electron devices", Period. Polytech. Elec. Eng., 17 (1), 27-33 (1973), EEA77-5950.
106.Zyuganov, A. N., Pismennyi, Y. G. and Tkhorik, Y. A., "Effect of finite concentration of charge carriers near the cathode on V-I characteristics of monopolar semiconductors with arbitrary compensation", Poluprovodn. Tekh. & Mikroelektron., no. 15, 47-9 (1974), Russian, EEA78-2783.

VI. P-N Junctions

107.Anon., "How accurate is the parabola model of a pn-junction?", Elektron. Ind., 6 (6), 124-7 (1975), German, PA78-66019.
108.Aydarova, V. Y., Kurtsin, V. M. and Nisnevich, Y. D., "Current equations in illuminated six-layer p-n-p-n-p-n structure", Radio Eng. & Electron. Phys., 16, 703-5 (1971), EEA75-13294.
109.Basavaraj, T. N. and Bhattacharyya, A. B., "Depletion-layer characterisation of single-diffused p-n junctions", Solid-State Electron., 17, 765-7 (1974), EEA77-30244.
110.Beister, G., Duckert, K., Hottewitzsch, P. and Klose, H., "Capacitance of p^+-n junctions with deep centres. I. Theory", Phys. Status Solidi A, 19, 479-86 (1973), EEA77-365.
111.Boccassi, G. and Capocaccia, F., "High injection in narrow uniform p-n junctions: a second-approximation model", IEEE Trans. Electron Devices, ED-22, 602-9 (1975), PA78-62089.
112.Buehler, M. G., "The space-charge region of neutral-irradiated silicon p^+n junctions", IEEE Trans. Nucl. Sci., NS-17, 341-7 (1970), EEA74-22801.
113.Choo, S. C., "Analytical approximations for an abrupt pn junction under high-level conditions", Solid-State Electron., 16, 793-9 (1973), EEA76-26048.
114.Cornu, J., "Analytical model for p-n junction under high-injection conditions", Electron. Lett., 7, 509-10 (1971), EEA75-1093.
115.Ferenczi, G., "Determination of relevant parameters of the dopants Zn and Te from C-T measurements in GaP|Zn,O| p-n junctions", J. Appl. Phys., 45, 1480-1 (1974), PA77-60850.
116.Gwyn, C. W., "Calculated small signal characteristics for irradiated pn junctions", IEEE Trans. Nucl. Sci., NS-19, 355-61 (1972), EEA76-12581.
117.Iskender-zade, Z. A., Dzhafarova, E. A., Akhundov, M. R. and Pashabekova, U. S., "High frequency barrier layer capacitance of diffused p-n junctions with deep traps", Izv. Akad. Nauk Az. SSR Ser. Fiz.-Tekh. & Mat. Nauk, no. 4, 3-8 (1974), EEA78-11542.
118.Kennedy, D. P., "The potential and electrical field at the metallurgical boundary of an abrupt p-n semiconductor junction", IEEE Trans. Electron Devices, ED-22, 988-94 (1975).

119. Kuzmicz, W. and Swit, A., "Some problems of the theory of forward biased p-n junction", Electron Technol., 4, 235-55 (1971), EEA75-4652.
120. McMurray, L. R. and Kleiner, C. T., "Adaptation of the p-n junction burnout model to circuit analysis codes", IEEE Trans. Nucl. Sci., NS-19, 76-81 (1972), EEA76-12579.
121. Perdomini, A. and Gislon, R., "A model of the p-n junction with exponential doping profile", Int. J. Electron., 34, 145-59 (1973), EEA76-12582.
122. Smiley, C. F., Yau, L. D. and Sah, C. T., "Application of the transmission line equivalent circuit model to the analysis of the PN junction admittance under d.c. bias", Solid-State Electron., 16, 895-901 (1973), EEA76-29246.
123. Stafeev, V. I. and Agarev, V. N., "Photosensitivity of multilayer structures with very large numbers of p-n junctions", Sov. Phys.-Semicond., 8, 247-9 (1974), PA77-75959.
124. Steele, E. L., "Computer modeling of p-n junction capacitance", Proceedings of 1973 IEEE Southeast-Con., p. T-4/1-5, EEA76-29247.

VII. Diodes

1. General

125. Afanasev, I. J. and Finkel'shtein, E. Y., "Diffusion currents of a three-dimensional planar diode model", Latv. PSR Zinat. Akad. Vestis Fiz. Teh. Ser., no. 3, 26-31 (1974), Russian, EEA77-38815.
126. Afanasev, I., Finkel'shtein, E. and Pantyukh, D., "Dependence of a diffusion current of a semiconductor diode three-dimensional model on the diffusion length of charge carriers and the rate of surface recombination", Latv. PSR Zinat. Akad. Vestis Fiz. Tekh. Ser., no. 6, 48-57 (1970), Russian, EEA74-10685.
127. Alabedra, R. and Lecoy, G., "Ground noise at low frequencies of voltage regulation semiconductor diodes", Onde Elec., 51, 724-9 (1971), French, EEA74-37249.
128. Barna, A. A. and Horelick, D., "A simple diode model including conductivity modulation", IEEE Trans. Circuit Theory, CT-18, 233-40 (1971), EEA74-13224.
129. Charlot, J. J. and Vapaille, A., "Numerical simulation of platinum doped diodes behaviour in forward-reverse switching", Onde Electr., 54, 243-8 (1974), French, EEA77-30020.
130. Chua, L. O. and Tseng, C. W., "A memristive circuit model for p-n junction diodes", Int. J. Circuit Theory & Appl., 2, 367-89 (1974), EEA78-8524.
131. Garver, R. V., "New model for microwave diodes", 1969 European Microwave Conference, 1970, p. 493-6, EEA74-35469.
132. Glover, K. J., Gardiner, J. G. and Howson, D. P., "Double-tuned modulator calculations using an improved diode model", Radio & Electron. Eng., 42, 37-44 (1972), EEA75-9462.

133.Guldberg, J. and Schroder, D. K., "Theoretical and experimental gain of electron-excited silicon targets", IEEE Trans. Electron Devices, ED-18, 1029-35 (1971), EEA74-36893.
134.Hart, B. L., "Diode switching using charge analysis", Wireless World, 77 (1425), 139-42 (1971), EEA74-13277.
135.Karadeniz, M. and Toker, C., "A new high frequency modelling of the step-recovery diodes with continuously varying parameters", Proceedings of the 5th Colloquium on Microwave Communication, Vol. IV, Microwave Circuits and Devices, 1974, p. MT-36/301-9, EEA77-36285.
136.Luque, A. and Dehesa, C., "Two-piece model of hot point generation in reverse biased p-n diodes", Solid-State Electron., 17, 165-77 (1974), PA77-24819.
137.Monnier, J. and Borel, M., "Mathematical models and technological parameters", Presented at the Proceedings of the International Conference on Advanced Microelectronics, Paris, France, April 1970, EEA71-1632.
138.Roulston, D. J. and Varshney, R. C., "Semiquantitative treatment of transients in silicon epitaxial diodes", Electron. Lett., 5, 548-50 (1969), EEA73-7430.
139.Smith, M. J., Jr., "Modelling and analysis of nonideal polyphase diode converters", Simulation, 17, 237-43 (1971), EEA75-22648.
140.Stone, P. S., "A numerical investigation of the steady-state behavior of silicon diodes", Thesis, Univ. Minnesota, Minneapolis, 1971, 150 pp., Order No. 72-5581, EEA75-32627.
141.Tajima, T., "Characteristics of silicon diodes for temperature compensation", Trans. Soc. Instrum. Contr. Eng., 7 (1), 9-13 (1971), Japanese, EEA74-22648.
142.Togatoff, W. W., "On the non-linear theory of transient phenomena in diodes and thyristors", Elektrotech. Cas., 25, 483-97 (1974), Czech, EEA77-36280.
143.Tsikerman, L. Y., Krivonosov, A. I. and Mints, V. A., "Linear models and structural diagrams of semiconductor diodes used as temperature transducers", Meas. Tech., 15, 886-8 (1972), EEA76-17772.
144.Wang, P. P. and Branin, F. H., Jr., "Multi-section network modelling of junction diodes", Modelling and Simulation, Vol. 4, 1973, p. 470-4, EEA78-28746.
145.Wunderlin, W., "Dynamic performance of semiconductor diodes with large signals", Elektroniker, 12 (2), EL1-7 (1973), German, EEA76-19689.
146.Yu, S. P. and Tarntraporn, W., "A computer simulation scheme for various solid-state devices", IEEE Trans. Electron Devices, ED-22, 515-22 (1975), EEA78-28554.

2. IMPATT and TRAPATT

147.Allen, S. G., "Simple TRAPATT circuit model", Electron. Lett., 9, 178-80 (1973), EEA76-22915.

148. Carroll, J. E., "Effects of diode-package cutoff frequency on TRAPATT operation", Electron. Lett., 9, 387-8 (1973), EEA77-2538.
149. Chaffin, R. J., "A proposed circuit model for microstrip-line TRAPATT oscillators", IEEE Trans. Microwave Theory & Tech., MTT-18 (11), 983-5 (1970), EEA74-8031.
150. Conn, D. R., Mitchell, R. H. and Stewart, J. A. C., "State-space analysis of an IMPATT diode lumped model", Proc. IEEE, 58, 809-10 (1970), EEA73-24993.
151. Cottam, M. G., "Analytic theory for silicon double-sided n^+-n-p-p^+ TRAPATT-diode structures", Electron. Lett., 9, 353-4 (1973), EEA77-5967.
152. Cowley, A. M., Fazarinc, Z., Hall, R., Hamilton, S., Yen, C. and Zettler, R., "The singly-tuned IMPATT oscillator", 1970 IEEE International Solid State Circuits Conference, p. 10-11, EEA73-25330.
153. Decker, D. R., "IMPATT diode quasi-static large-signal model", IEEE Trans. Electron Devices, ED-21, 469-79 (1974), EEA77-34021.
154. Doumbia, I., Salmer, G. and Constant, E., "High-frequency limitation on silicon IMPATT diode: velocity modulation", J. Appl. Phys., 46, 1831-3 (1975), EEA78-28754.
155. Hirachi, Y., Nishi, H., Shinoda, M. and Fukukawa, Y., "Millimeter-wave IMPATT diodes with improved efficiency by using ion-implanted ohmic contact", Proc. IEEE, 63, 1367-8 (1975), EEA78-41574.
156. Jerinic, G., "Hybrid computer simulation of IMPATT diodes", Thesis, Univ. Wisconsin, Madison, 1972, 213 pp., Order No. 72-9131, EEA75-36253.
157. Kotsebue, K. L. and Regier, R. D., "TRAPATT-diode model for circuit design", Electron. Lett., 8, 205-6 (1972), EEA75-18951.
158. Kotzebue, K. L. and Regier, R. D., "The design of lumped-element TRAPATT circuits", 1973 IEEE G-MTT International Microwave Symposium, 1973, p. 178-80, EEA77-5468.
159. Kramer, N. B., "Characterization and modelling of IMPATT oscillators", IEEE Trans. Electron Devices, ED-15, 838-46 (1968), EEA73-7444.
160. Kremer, R., "Small-signal models for IMPATT diodes", Arch. Elektron. & Uebertragungstech., 29, 296-304 (1975), German, EEA78-33117.
161. Kremer, R., "Large-signal model for the IMPATT diode", Arch. Elektron. & Uebertragungstech., 29, 361-6 (1975), German, EEA78-41567.
162. Kremer, R., "Large-signal admittances of IMPATT diodes obtained from amplifier measurements", Nachrichtentech. Z. (NTZ), 28, 292-9 (1975), German, EEA78-41573.
163. Kuno, H. J., Fong, T. T. and English, D. L., "Characterization of IMPATT diodes at millimeter-wave frequencies", IEEE Trans. Electron Devices, ED-19, 752-7 (1972), EEA75-20168.

164.Kuvas, R. L., "Noise in IMPATT diode oscillators and amplifier", Proceedings of the 3rd Biennial Cornell Electrical Engineering Conference on High Frequency Generation and Amplification Devices and Application, 1971, p. 145-54, EEA75-14034.
165.Lee, C. A. and Frey, J., "Comparison of plasma formation in n^+p and p^+n TRAPATT diodes", Electron. Lett., 9, 318-20 (1973), EEA77-5966.
166.McCready, W. J. L. and Stewart, J. A. C., "Derivation of a large signal IMPATT diode oscillator lumped model", Int. J. Electron., 32, 69-79 (1972), EEA75-7773.
167.Moffat, J. W. and Mitchell, R. H., "Large-signal lumped modeling and characterization on an IMPATT diode", IEEE Trans. Electron Devices, ED-19 (12), 1297-8 (1972), EEA76-778.
168.Moffat, J. W. and Mitchell, R. H., "Large-signal lumped modeling and characterization of an IMPATT diode", Electron., 35, 465-77 (1973), EEA76-38255.
169.Olson, H. M., "Thermal runaway of IMPATT diodes", IEEE Trans. Electron Devices, ED-22, 165-8 (1975), EEA78-15313.
170.Regier, R. D. and Kotzebue, K. L., "Numerical techniques for TRAPATT circuit analysis and design", 1973 IEEE G-MTT International Microwave Symposium, 1973, p. 47-9, EEA77-5369.
171.Slaymaker, N. A. and Carroll, J. E., "Verification of simple TRAPATT-oscillator model", Proc. Inst. Elec. Eng., 119, 1113-18 (1972), EEA75-32932.
172.Stewart, J. A. C., Conn, D. R. and Mitchell, H. R., "Small signal lumped model for IMPATT diodes", Int. J. Electron., 29, 301-13 (1970), EEA73-38042.
173.Stewart, J. A. C., Conn, D. R. and Mitchell, H. R., "State-space analysis of general IMPATT diode small-signal lumped models", IEEE Trans. Microwave Theory & Tech., MTT-18, 835-42 (1970), EEA74-7716.
174.Tozer, R. C. and Hobson, G. S., "Frequency/temperature relationships of CW IMPATT diodes", Electron. Lett., 8, 74-5 (1972), EEA75-10214.
175.Tozer, R. C. and Hobson, G. S., "Degradation of the power- and frequency-temperature performance of IMPATT diodes at lower frequencies", IEEE Trans. Electron Devices, ED-22, 156-8 (1975), EEA78-24309.

3. Avalanche

176.Culshaw, B. C., "Time-domain model of the device/circuit characteristics of a trapped-plasma mode avalanche-diode oscillator", Electron. Lett., 7, 339-40 (1971), EEA74-27168.
177.Hoefflinger, B., Snapp, C. P. and Stark, L. A., "High-speed computer program for avalanche diode microwave generators with multiple resonances", Proceedings of the 2nd Biennial Cornell Electrical Engineering Conference, 1969, p. 402-11, EEA70-21271.

178. Isobe, T. and Nakamura, T., "Computer application for studying avalanche diodes", Fujitsu Sci. Tech. J., 6 (4), 41-57 (1970), EEA74-15529.
179. Kawarada, K. and Mizushima, Y., "Properties of a new avalanche diode oscillator computed by a large-signal, self-consistent simulation", Rev. Elec. Commun. Lab., 19, 875-85 (1971), EEA75-10229.
180. Maffei, A., "Unidimensional mathematical model of transient operation of avalanche diodes", Thesis, Univ. Lyon, France, 1972, French, EEA76-26072.
181. Majewski, M., "Analytical model of avalanche diode in transit-time mode", Elektronika, 15, 513-18 (1974), Polish, EEA78-11615.
182. Peterson, D. F. and Steinbrecher, D. H., "Circuit model for characterizing the nearly linear behavior of avalanche diodes in amplifier circuits", IEEE Trans. Microwave Theory & Tech., MTT-21, 19-27 (1973), EEA76-8036.
183. Robinson, B. B., "A qualitative model of the avalanche diode", Proc. IEEE, 57, 1783 (1969), EEA73-10626.
184. Schroeder, W. R. and Haddad, G. I., "Comments on simplified model for avalanche-resonance-pumped semiconductor diodes", Electron. Lett., 6, 114-15 (1970), EEA73-17606.
185. Steinbrecher, D. H. and Peterson, D. F., "Small-signal model with frequency-independent elements for the avalanche region of a microwave negative resistance diode", IEEE Trans. Electron Devices, ED-17, 883-91 (1970), EEA73-38419.
186. Vidallon, G., Thomas, J. and Geronimi, S., "Non-linear analysis of the operation of the avalanche diode. Application to frequency multipliers", Electron. & Fis. Apl., 16, 221-5 (1973), French, EEA78-2537.

4. Schottky

187. Alley, G. D. and Talley, H. E., "A theoretical study of the high-frequency performance of a Schottky-barrier field-effect transistor fabricated on a high-resistivity substrate", IEEE Trans. Microwave Theory & Tech., MTT-22, 183-9 (1974), EEA77-14527.
188. Alley, G. D., Talley, H. E. and Wright, G. L., "Two-dimensional distributed theory for a microwave Schottky barrier field effect transistor", 1973 IEEE G-MTT International Microwave Symposium, p. 233-5, EEA77-6045.
189. Demoulin, E. and van de Wiele, F., "Inversion layer at the interface of Schottky diodes", Solid-State Electron., 17, 825-33 (1974), EEA77-30293.
190. Druzhinin, A. V. and Knyazev, A. Y., "Application of a semiconductor model with surface donors to an oxide cathode", Izv. Akad. Nauk SSSR Ser. Fiz., 37, 2590-3 (1973), Russian, PA77-21147.
191. Eimers, G. W. and Stevens, E. H., "A composite model for Schottky diode barrier height", IEEE Trans. Electron Devices, ED-18, 1185-6 (1971), EEA75-4672.

192.Glover, G. H., "The C-V characteristics of Schottky barriers on laboratory grown semiconducting diamonds", Solid-State Electron., 16, 973-83 (1973), EEA76-35260.
193.Green, M. A. and Shewchun, J., "Numerical analysis of semiconductor devices with planar, cylindrical and radial symmetry", Int. J. Electron., 37, 285-8 (1974), EEA78-2668.
194.Hesse, K. and Strack, H., "On the frequency dependence of GaAs Schottky barrier capacitances", Solid-State Electron., 15, 767-74 (1972), EEA75-24325.
195.Kottwitz, A., Leimer, F. and Stotzel, H., "Some remarks on the relaxation behaviour of thin film structures Al-a-GeSe-Al", Phys. Status Solidi A, 24, K49-52 (1974), PA77-67272.
196.Kurumada, K., "Theoretical investigation of Schottky gate Gunn effect digital devices", Electr. Commun. Lab. Tech. J., 22, 1399-411 (1973), Japanese, EEA76-38254.
197.Kurumada, K., "Analysis of Schottky-barrier-gate Gunn effect digital devices", Rev. Electr. Commun. Lab., 21, 807-18 (1973), EEA77-36270.
198.Kurumada, K. and Harada, H., "Analytical approximation theory of Schottky-gate gunn-device characteristics", Electron. & Commun. Jap., 56 (6), 94-100 (1973), EEA78-2797.
199.Lerach, L. and Albrecht, H., "Analytical solution for the avalanche breakdown voltage of p^+-n junctions and Schottky diodes. Anomalous behavior of low energy bandgap high-mobility semiconductors", Phys. Status Solidi A, 30, 625-36 (1975).
200.Oakes, J. G., "The small-signal admittance of the Schottky barrier BARITT diodes", Thesis, Cornell Univ., Ithaca, N. Y., 1974, 122 pp., Order No. 74-26299, EEA78-24315.
201.Posdziech, G., "A model for describing a contact in semiconductor planar technology", Z. Elektr. Inf. - & Energietech., 4, 234-40 (1974), German, PA78-6896.
202.Rhoderick, E. H., "A note on Levine's model of Schottky barriers", J. Appl. Phys., 46, 2809 (1975), EEA78-33076.
203.Vaitkus, R. L., "Numerical solution of an idealized two-carrier metal-semiconductor junction diode", Thesis, Univ. Minnesota, Minneapolis, 1971, 187 pp., Order No. 72-5587, EEA75-32642.
204.Vickers, V. E., "Model of Schottky barrier hot-electron-mode photodetection", Appl. Opt., 10, 2190-2 (1971), PA75-5433.

5. Gunn

205.Cohen, J. and Gilden, M., "A mathematical model of a varactor package in microstrip line", IEEE Trans. Microwave Theory & Tech., MTT-21, 412-13 (1973), EEA76-27050.
206.de Groot, J., "The interaction of a Gunn-device with a microwave circuit", Presented at MOGA 70 8th International Conference on Microwave and Optical Generation and Amplification (Summaries), Amsterdam, Netherlands, September 1970, EEA74-1485.

207. Gunshor, R. L. and Kak, A. C., "Lumped-circuit representation of Gunn diodes in domain mode", IEEE Trans. Electron Devices, ED-19, 765-70 (1972), EEA75-20119.
208. Isobe, T., Nakamura, T. and Goto, G., "Computer simulation of two-dimensional Gunn devices", Fujitsu Sci. & Tech. J., 9 (2), 77-99 (1973), EEA76-32222.
209. Kuznetsov, K. V., Levinstein, M. E. and Shur, M. S., "The investigation of the Gunn diode operation in a resonator using the computer model", Solid-State Electron., 14, 207-20 (1971), EEA74-18677.
210. Levinshtein, M. E. and Shur, M. S., "The investigation of the Gunn diode operation in a resonator using computer model", Presented at MOGA 70 8th International Conference on Microwave and Optical Generation and Amplification (Summaries), Amsterdam, Netherlands, September 1970, EEA74-1484.
211. Malyshev, V. A. and Alekseev, Y. I., "Study of the current in a kinematic model of a Gunn diode", Izv. VUZ Radioelektron., 13, 1027-30 (1970), Russian, EEA74-1818.
212. Ramaswamy, S. and Steele, E. I., "Gunn diode parameters from a small signal analysis", Proceedings of 1973 IEEE Southeast-Conf., p. T-2/1-5, EEA76-29226.
213. Sapelnikov, A. N. and Staroselskii, V. J., "Electronic model of the Gunn diode", Izv. VUZ Radioelektron., 14, 105-6 (1971), Russian, EEA74-15486.
214. Templin, A. S. and Gunshor, R. I., "Analytic model for varactor-tuned waveguide Gunn oscillators", IEEE Trans. Microwave Theory & Tech., MTT-22, 554-6 (1974), EEA77-18172.
215. Vaganov, V. I. and Kuznetsov, U. M., "Gunn-diode modeling", Izv. VUZ Radioelektronika, 13, 409-12 (1970), Russian, EEA75-28353.

6. PIN

216. Aronov, D. A. and Mamatkulov, R., "Diffusive capacitance quality of a p-i-n diode for a large harmonic signal", Radio Eng. & Electron. Phys., 17, 1907-11 (1972), EEA77-2532.
217. Brousseau, M., Brabant, J. C. and Barrau, J., "A relaxation model for oscillations in Si PIN devices", 1972 European Solid State Devices Research Conference, p. 234-6, Publ. 1973, EEA76-32260.
218. Cheung, D. K.-P., "Small signal impedance of P^+ -I-N^+ diode under high level injection", Thesis, Ohio State Univ., Columbus, 1971, 150 pp., Order No. 72-4447, EEA75-28466.
219. Profirescu, M. D., "Computer simulation of switching in microwave PIN diodes", 4th European Microwave Conference, 1974, p. 390-3, EEA78-2435.
220. Schunemann, K. and Muller, J., "Contact layer charge storage in PIN-diodes", Proceedings of the 5th Colloquium on Microwave Communication, Vol. IV, Microwave Circuits and Devices, 1974, p. MT-56/511-20, EEA77-34020.

221. Schunemann, K. and Unger, H. G., "Transient phenomena in pin-diodes", Presented at Proceedings of the 4 Colloquium on Microwave Communication, Budapest, Hungary, April 1970, EEA70-38031.

7. BARITT

222. Christie, J. and Stewart, J. A. C., "Small-signal BARITT noise measures", IEEE Trans. Electron Devices, ED-22, 836-41 (1975), EEA78-41569.
223. El-Gabaly, M. and Nigrin, J., "Design considerations for BARITT diodes", Symposium on Computers, Electronics and Control, 1974, IV-4/9 pp., EEA78-37515.
224. van de Roer, T. G., "Analytical small-signal theory of BARITT diodes", Report TH-74-E-46, Eindhoven Univ. Technol., Netherlands, 1974, 40 pp., EEA78-2836.

8. Others

225. Antongnetti, P., Chiabrera, A. and Ridella, S., "Recombination instabilities in semiconductors doped with deep two-level traps", J. Appl. Phys., 43, 4676-80 (1972), EEA76-4415.
226. Biro, V., "New method for the design of varactor frequency multipliers", Presented at Proceedings of the 4 Colloquium on Microwave Communication, Budapest, Hungary, April 1970, Russian, EEA73-38030.
227. Crabbe, J. S., "Development of a deep trap model for the capacitance characteristics of silicon $p^+ nn^+$ diodes irradiated with fast neutrons", Thesis, Southern Methodist Univ., Dallas, Texas, 1971, 220 pp., Order No. 72-16302, EEA75-36236.
228. Crabbe, J. S. and Ashley, K. L., "Capacitance voltage characteristics of neutron irradiated $n^+ pp^+$ and $p^+ nn^+$ junctions", IEEE Trans. Nucl. Sci., NS-18, 410-19 (1971), EEA75-13669.
229. Jirina, M., "A diode model with a series, current dependent resistance", Slaboproudy Obz., 33, 549-54 (1972), Czech, EEA76-8409.
230. Mueller, R. R. and Nitz, B., "Analogue-computer model for a gallium-arsenide l.s.a. diode", Electron. Lett., 5, 498-9 (1969), EEA73-10706.
231. Pfleiderer, H., "Magnetodiode model", Solid-State Electron., 15, 335-53 (1972), EEA75-13635.
232. Redd, J. C. and Kotzebue, K. L., "A cascaded-impedance-inverter model of wide-band frequency triplers", 1971 IEEE-GMTT International Microwave Symposium Digest of Technical Papers, p. 96-7, EEA74-36466.
233. Schnall, E., "A convenient way to model the handy Zener diode", Electronics, 44 (21), 67-71 (1971), EEA74-37241.
234. Streetman, B. G. and Holonyak, N., Jr., "Current oscillations in deep-level doped semiconductors", IBM J. Res. Developm., 13, 529-32 (1969), EEA73-17685.

VIII. Bipolar Transistors

1. General

235.Anon., "The inside story", Electronics, 43 (4), 72-3 (1970).
236.Arkhanelskaya, I. T., "An approximation of the charge carrier transfer process in mathematical transistor model", Radio Eng. & Electron. Phys., 18, 739-45 (1973), EEA77-34042.
237.Blundell, A. J., "New approach to transistor circuit analysis", Wireless World, 77 (1428), 287-92 (1971), EEA74-22709.
238.Blundell, A. J., "New approach to transistor circuit analysis", Wireless World, 77 (1429), 329-34 (1971), EEA74-30863.
239.Glasford, G. M., "Comparisons of nonlinear distortion models for bipolar transistors", 8th Asilomar Conference on Circuits, Systems and Computers, 1974, p. 351-5, Publ. 1975, EEA78-24101.
240.Gopinath, B. and Mitra, D., "When are transistors passive?", Bell Syst. Tech. J., 50, 2835-47 (1971), EEA75-9294.
241.Hower, P. L. and Govil, P. K., "Comparison of one- and two-dimensional models of transistor thermal instability", IEEE Trans. Electron Devices, ED-21, 617-23 (1974), EEA77-38826.
242.Jensen, R. W., "A class of extended diode and transistor models", Modelling and Simulation, Vol. 4, 1973, p. 479-83, EEA78-28556.
243.Lindholm, F. A., "Network representation of semiconductor device behaviour", 8th Asilomar Conference on Circuits, Systems and Computers, 1974, p. 18, Publ. 1975, EEA78-24340.
244.Rudenko, A. I., "Higher harmonics of the dynamic characteristic of a transistor", Sov. Phys.-Semicond., 8, 534 (1974), EEA78-8556.
245.Stanclik, J., "Modelling method of a bipolar transistor in the linear working region", Arch. Elektrotech., 23, 313-24 (1974), Polish, EEA78-5304.
246.Togatov, V. V., "To the non-linear theory of transient phenomena in transistors", Elektrotech. Cas., 25, 577-89 (1974), Russian, EEA78-2849.

2. Ebers-Moll Model

247.Balim, G. M., "An extension of the Ebers-Moll equations", Radio Eng. & Electron. Phys., 17, 1729-31 (1972), EEA76-38252.
248.Baranyi, A. and Newcomb, R. W., "Multiple DC solution one-transistor circuits", IEEE Trans. Circuit Theory, CT-19, 626-8 (1972), EEA75-38952.
249.Bittner, J. and Muller, W., "Equivalent circuit of the transistor after Ebers and Moll, as a basis for computer aided designs", Nachrichtentech. Elektron., 23, 186-7 (1973), German, EEA76-28796.
250.Coston, W. T., "A transistor companion model including junction capacitances to be used in nonlinear periodic steady-state analysis", IEEE J. Solid-State Circuits, SC-10, 40-6 (1975), EEA78-8549.

251.Ebers, J. J. and Moll, J. L., "Large-signal behavior of junction transistors", Proc. Inst. Radio Engrs., 42, 1761-72 (1954), EEA58-937.
252.Getreu, I., "Modeling the bipolar transistor. I. First-level non-linear model", Electronics, 47 (19), 114-20 (1974), EEA77-38695.
253.Getreu, I., "Modeling the bipolar transistor. III", Electronics, 47 (23), 137-42 (1974), EEA78-2847.
254.Jensen, R. W., "A bipolar transistor model including low current effects", 7th Asilomar Conference on Circuits, Systems and Computers, 1973, p. 668-70, Publ. 1974, EEA77-30317.
255.Jirina, M., Baresova, B. and Stastina, H., "Temperature dependent properties of the parameters in the Ebers-Moll transistor model", Slaboproudy Obzor, 32, 149-56 (1971), Czech, EEA74-18626.
256.Lindholm, F. A. and Hamilton, D. J., "Incorporation of the Early effect in the Ebers-Moll model", Proc. IEEE, 59, 1377-8 (1971), EEA74-34072.
257.Moll, J. L., "Large-signal transient response of junction transistors", Proc. Inst. Radio Engrs., 42, 1773-84 (1954), EEA58-938.
258.Richards, G. A., "D.C. modelling of bipolar transistors", Marconi Rev., 33 (178), 239-45 (1970), EEA73-38126.
259.Solin, N. N. and Chemodanov, A. Y., "Signal generation in transistor oscillators", Izv. VUZ Radioelektron., 15, 868-72 (1972), Russian, EEA76-310.
260.Uhlir, J., "A non-linear static transistor model", Acta Polytech. III, no. 2, 7-17 (1973), Czech, EEA78-32927.
261.Zilka, Z., "Parameters identification in an injection model of a transistor", Slaboproudy Obz., 35, 201-6 (1974), Czech, EEA77-30312.
262.Zilka, Z., "The Ebers-Moll model of the bipolar transistor and its variations", Slaboproudy Obz., 36, 322-7 (1975), Slovak, EEA78-37523.

3. Charge Control Model

263.Aaronson, G. and Schilling, R. B., "A comparison of two transistor models", IEEE Trans. Electron Devices, ED-19, 1-3 (1972), EEA75-4683.
264.Arkhangelskaya, I. T. and Arkhangelskii, A. Y., "A simple method of transfer and storage of charge carriers in a semiconductor with a bipolar transistor model", Izv. VUZ Radioelektron., 17, 104-11 (1974), Russian, EEA77-34040.
265.Arribas de Costa, F. and Melia, J. M., "Method of measurement of the parameters of a model for control of charge", Rev. Telecomun., 28 (109), 3-8 (1972), Spanish, EEA76-35310.
266.Beaufoy, R. and Sparks, J. J., "The junction transistor as a charge-controlled device", ATE J., 13, 310-27 (1957).
267.Beck, F., Geissler, U. and Khakzar, H., "Investigation of validity for the Gummel-Poon-model", Int. Elektron. Rundsch., 28, 147-9 (1974), German, EEA77-36301.

268. Chawla, B. R., "Circuit representation of the integral charge-control model of bipolar transistors", IEEE J. Solid-State Circuits, SC-6, 262-4 (1971), EEA74-26846.

269. Daseking, H. W., "Parameter determination for the modified Gummel-Poon NPN transistor model", 8th Asilomar Conference on Circuits, Systems and Computers, 1974, p. 356-60, Publ. 1975, EEA78-24102.

270. de Graaff, H. C., "A modified charge-control theory for saturated transistors", Philips Res. Rep., 26, 191-215 (1971), EEA74-26850.

271. Getreu, I., "Modeling the bipolar transistor. II", Electronics, 47 (22), 71-5 (1974), EEA78-624.

272. Gummel, H. K., "A charge-control transistor model for network analysis programs", Proc. IEEE, 56, 751 (1968), EEA71-18434.

273. Gummel, H. K., "A charge control relation for bipolar transistors", Bell Syst. Tech. J., 49, 115-20 (1970), EEA73-17622.

274. Gummel, H. K. and Poon, H. C., "An integral charge control model of bipolar transistors", Bell Syst. Tech. J., 49, 827-52 (1970), EEA73-31388.

275. Hart, B. L., "The transistor charge model: a tutorial development", Electron. Compon., 12, 1113-18 (1971), EEA75-1139.

276. Khakzar, H., "Nonlinear distortion calculation of a bipolar transistor using the Gummel-Poon model", Arch. Elektron. & Uebertragungstech., 28, 237-41 (1974), German, EEA77-30297.

277. Louw, W. J., "The transistor as a charge controlled device", Trans. S. Afr. Inst. Elec. Eng., 62 (3), 49-58 (1971), Afrikaans, EEA74-18816.

278. Mar, J., "Time-domain method of measuring transistor parameters", IEEE J. Solid-State Circuits, SC-6, 223-6 (1971), EEA74-25347.

279. Narayanan, S. and Poon, H. C., "An analysis of distortion in bipolar transistors using integral charge control model and Volterra series", IEEE Trans. Circuit Theory, CT-20, 341-51 (1973), EEA76-29254.

280. Persky, G., "Hot carrier effects in the integral charge-control model for bipolar transistors", Bell Sys. Tech. J., 51, 455-60 (1972), EEA75-20177.

281. Poon, H. C., "Modeling of bipolar transistor using integral charge control model with application to third-order distortion studies", IEEE Trans. Electron Devices, ED-19, 719-31 (1972), EEA75-20185.

282. Poon, H. C. and Meckwood, J. C., "Modeling of avalanche effect in integral charge control model", IEEE Trans. Electron Devices, ED-19, 90-7 (1972), EEA75-4673.

283. Scharfetter, D. L., "Wafer processing oriented IC design", Presented at 1974 IEEE International Symposium on Circuits and Systems, San Francisco, Calif., April 1974, EEA78-690.

284. Srivastava, A. and Gupta, M. L., "Switching transients of transistors for a linearly varying input", Indian J. Pure & Appl. Phys., 11, 621-4 (1973), EEA77-18810.

285.Winkel, J. T., "Extended charge-control model for bipolar transistors", IEEE Trans. Electron Devices, ED-20, 389-94 (1973), EEA76-19912.
286.Zimmer, C. R., "An integral charge control model (ICCM): theory and experiment", 1973 WESCON Technical Papers, Western Electronic Show and Convention, 14/4-6 pp., EEA77-10242.

4. Statistical Models

287.Abdullaev, D. A. and Smolyak, A. M., "Storage in multistate elements with independent state outputs by means of reservation within each element", Izv. Akad. Nauk USSR Ser. Tekh. Nauk, no. 5, 6-11 (1971), Russian, EEA75-18997.
288.Alberding, N., "Random walk models of semiconductor noise", J. Appl. Phys., 44, 1911-12 (1973), EEA76-19690.
289.Butler, E. M., "Techniques for statistical d.c. modeling of bipolar transistors", Proceedings of the 1974 IEEE International Symposium on Circuits and Systems, p. 725-9, EEA77-38831.
290.De Man, H., Merten, R. and Grevens, H., "The use of the transistor simulation program SITCAP for statistical modeling of bipolar transistors", Presented at AGARD Conference on Computer Aided Design for Electronic Circuits, Lyngby, Denmark, May 1973, EEA77-5381.
291.Delcaro, L. G. and Zoldan, G. G., "Method for the constrained identification of transistor differential parameters", Tec. Ital., 38 (7-9), 147-52 (1973).
292.Kennedy, D. P., Murley, P. C. and O'Brien, R. R., "A statistical approach to the design of diffused junction transistors", IBM J. Res. Developm., 8, 482-95 (1964), EEA68-9873.
293.Logan, J., "Statistical circuit design: characterization and modelling for statistical design", Bell Syst. Tech. J., 50, 1105-47 (1971), EEA74-26294.
294.Woo, W. K. and Metcalfe, J. E., "Optimal allocation of inspection effort in multistage manufacturing processes", Western Electric Eng., 16 (3), 8-16 (1972), EEA75-35268.

5. Other Models

295.Aaronson, G. and Schilling, R. B., "A mathematical foundation for linear lumped modeling techniques", Solid-State Electron., 18, 301-8 (1975), EEA78-11638.
296.Achuthan, M. K. and Vitto, T. J., "Variation of junction transistor current amplification factor with emitter current", Int. J. Electron., 36, 195-206 (1974), EEA77-10227.
297.Andresciani, V. and Mazzei, P., "Bipolar transistor model for linear and digital active networks", Note Recens. Not., 19, 189-207 (1970), Italian, EEA74-411.

298. Barna, A., "Small-signal modeling of distributed base regions in bipolar transistors", Arch. Elektron. & Uebertragungstech., 28, 242-4 (1974), EEA77-30298.
299. Bass, S. C. and Peak, S. C., "Terminal models of digital gates allowing waveform simulation", International Symposium on Circuit Theory 1973, p. 287-9, EEA77-2582.
300. Beldiman, A., "Methods of assessing bipolar transistors starting from physical and electrical measurements carried out on reference models", Autom. & Electron., 17 (2), 72-80 (1973), Italian, EEA76-29251.
301. Blice, R. D., Munarin, J. A. and Pease, R. L., "Bipolar transistor screening methods for neutron hardness assurance", IEEE Trans. Nucl. Sci., NS-19, 115-20 (1972), EEA76-12612.
302. Bowler, D. J. and Lindholm, F. A., "High current regimes in transistor collector regions", IEEE Trans. Electron Devices, ED-20, 257-63 (1973), EEA76-16407.
303. Callahan, M. J., Jr., "Models for the lateral p-n-p transistor including substrate interaction", IEEE Trans. Electron Devices, ED-19, 122-3 (1972), EEA75-3978.
304. Calzolari, P. U., Ghigi, P. R. and Graffi, S., "Some considerations on the two-dimensional theory of drift transistors", Alta Freq., 41, 866-9 (1972), EEA76-16401.
305. Chiu, T. L. and Ghosh, H. N., "A diffusion model for arsenic in silicon", IBM J. Res. Develop., 15, 472-6 (1971).
306. Chou, S., "Small-signal characteristics of lateral transistors", Solid-State Electron., 15, 27-38 (1972), EEA75-7522.
307. Clark, L. E., "Physical dependencies and modelling of output admittances and gain-voltage relationships in bipolar transistors", Proc. IEEE, 61, 1445-6 (1972), EEA76-4638.
308. Conn, D. R. and Mitchell, R. H., "State-space analysis of two-dimensional current flow in the base region of a junction transistor", Int. J. Electron., 31, 353-7 (1971), EEA74-37305.
309. Cottin, P., "Drift-Transistor Linvill model for cad program", Presented at Proceedings of the International Conference on Advanced Microelectronics, Paris, France, April 1970, EEA74-1595.
310. Curtis, J., "Modelling transistors for improved performance", Microwaves, 14 (2), 33-4, 36, 38, 40-1, 43 (1975), EEA78-24095.
311. De Graaff, H. C., "A.C. small-signal model for bipolar transistors in saturation", Electron. Lett., 7, 73-5 (1971), EEA74-10618.
312. De Graaff, H. C., "Collector models for bipolar transistors", Solid-State Electron., 16, 587-600 (1973), EEA76-19918.
313. de Mello, H. A., "Lumped models of bipolar transistors and evaluation of their parameters (technical report)", Report TR-4825-3, Stanford Univ., California, 1969, 51 pp., EEA74-7755.
314. Doring, H., "Numerical analysis of the large-signal behaviour of bipolar transistors", Radio Fernsehen Elektron., 24, 308-10 (1975), German, EEA78-33131.

315.Dubock, P., "D.C. numerical model for arbitrarily biased bipolar transistors in two dimensions", Electron. Lett., 6, 53-5 (1970), EEA73-14273.
316.Engl, W. L. and Manck, O., "Two-dimensional numerical analysis of bipolar transistor transients", 1973 International Electron Devices Meeting Technical Digest, p. 381-3, EEA77-14505.
317.Esteve, D. and Martinot, H., "Excess surfaces currents in p-n junctions and bipolar transistors", Solid-State Electron., 14, 693-705 (1971), EEA74-34025.
318.Gaur, S. P., "Two dimensional carrier flow in a transistor structure under non-isothermal conditions", Thesis, Univ. Massachusetts, Amherst, 1974, 110 pp., Order No. 75-6030, EEA78-33134.
319.Gaur, S. P., "Quasisaturation-region operation of n-p-n -n power transistors", Electron. Lett., 11, 446-7 (1975), EEA78-37516.
320.Ghosh, H. N., "A distributed model of the junction transistor and its application in the prediction of the emitter-base diode characteristic, base impedance, and pulse response of the device", IEEE Trans. Electron Devices, ED-12, 513-31 (1965), EEA69-4049.
321.Giacoletto, L. J., "A model for d.c. characteristics of bipolar transistors", Int. J. Electron., 36, 719-33 (1974), EEA77-30306.
322.Glasford, G. M., "Low voltage d-c mathematical models for bipolar junction transistors based on the diffusion equations expressed in common emitter form", Proceedings of the National Electronics Conference, 1970, p. 238-43, EEA75-1152.
323.Glasford, G. M., "Nonlinear distortion models for bipolar junction transistors", 7th Asilomar Conference on Circuits, Systems and Computers, 1973, p. 548-52, Publ. 1974, EEA77-30316.
324.Golubev, A. P. and Kremlev, V. Y., "Nonlinear models of a planar transistor, taking two-dimensional current spreading in the base region into consideration", Izv. VUZ Radioelektron., 14, 135-44 (1971), Russian, EEA74-15553.
325.Groendijk, H., "Modeling base crowding in a bipolar transistor", IEEE Trans. Electron Devices, ED-20, 329-30 (1973), EEA76-16408.
326.Gummel, H. K. and Poon, H. C., "A compact bipolar transistor model", 1970 IEEE International Solid State Circuits Conference, p. 78-9, EEA73-25025.
327.Gwyn, C. W., Summers, G. G. and Corbett, W. T., "Modeling of the saturation characteristics of high voltage transistors", IEEE Trans. Nucl. Sci., NS-17, 63-9 (1970), EEA74-22699.
328.Hahn, L. A. and Ashley, K. L., "Quantitative model for epitaxial-transistor collector characteristics", Electron. Lett., 6, 485-7 (1970), EEA73-35207.
329.Hajj, I. N., Roulston, D. J. and Bryant, P. R., "d.c. analysis of transistor circuits via a three-terminal piecewise-linear modeling approach", International Symposium on Circuit Theory 1973, p. 268-71, EEA77-2029.

330.Heimeier, H. H., "A two-dimensional numerical analysis of a silicon N-P-N transistor", IEEE Trans. Electron Devices, ED-20, 708-14 (1973), EEA76-29253.

331.Houselander, L. S., Chow, H. Y. and Spence, R., "Transistor characterization by effective large-signal two-port parameters", IEEE J. Solid-State Circuits, SC-5, 77-9 (1970), EEA73-21285.

332.Jain, A. K., Bhatnagar, S. K. and Srivastava, G. P., "Effects of surface-recombination and geometry in allow junction transistors", Solid-State Electron., 14, 451-6 (1971), EEA74-26854.

333.Jesshope, C. R., Zaluska, E. J. and Kemhadjian, H. A., "Comparison of numerical-solution methods for 2-dimensional bipolar-transistor-analysis algorithm", Electron. Lett., 11, 285-6 (1975), EEA78-28553.

334.Kumar, R., "Collector capacitance and high-level injection effects in bipolar transistors", Thesis, Univ. Rochester, N. Y., 1974, 136 pp., Order No. 74-22600, EEA78-8557.

335.Kutoviy, I. V. and Slepov, L. I., "Structure and circuit models of N-transistor", Sov. Autom. Control, 8, 72-5 (1975), EEA78-41410.

336.Last, J. D., Lucas, D. W. and Sumerling, G. W., "A numerical analysis of the d.c. performance of small geometry lateral transistors", Solid-State Electron., 17, 1111-18 (1974), EEA78-632.

337.Leduc, P., "Unit transistor: d.c. bipolar transistor model whose parameters are the geometrical dimensions of transistors", Presented at AGARD Conference Proceedings No. 130 Computer Aided Design for Electronic Circuits, Lyngby, Denmark, May 1973, French, EEA77-6013.

338.Linder, J. S., Willis, G. W., Jr. and Barnard, H. M., "Multiple emitter transistor large signal model", Presented at 21st Annual Southwestern IEEE Conference and Exhibition, San Antonio, Texas, April 1969, EEA73-4510.

339.Linder, S., Willis, G. W., Jr. and Barnard, H. M., "Applications of linear graph theory to lumped models of solid-state devices", IEEE Trans. Educ., E-13, 83-7 (1970), EEA70-38082.

340.Logan, J., Lindholm, F. A., Rohr, P. and Hamilton, D. J., "Modelling the 'early effect' in bipolar transistors using an empirical but effective parameter", Proc. IEEE, 60, 335-6 (1972), EEA75-10248.

341.Long, E. L., "Current-induced field enhancement of lateral transistors", Presented at International Electron Devices Meeting, Washington, D. C., October 1970, EEA74-10638.

342.Magdo, S., "Theory and operation of space-charge-limited transistors with tranverse injection", IBM J. Res. & Dev., 17, 443-58 (1973), EEA77-18786.

343.Miyata, T., Miura, T. and Mano, K., "Large signal turn-on response of the junction transistors including nonlinear effects", IEEE Trans. Electron Devices, ED-16, 1042-8 (1969), EEA73-10641.

344.Pallottini, L., "Hybrid transistor models", Elettrificazione, no. 3, 139-45 (1973), Italian, EEA76-32263.
345.Pals, J. A., "On the noise of a transistor with d.c. current crowding", Philips Res. Rep., 26, 91-102 (1971), EEA74-22802.
346.Pasztor, G., "The use of the Linvill model for modelling semiconductor devices. III. Transistor models", Hiradstech. Ipari Kut. Intez. Kozl., 12, 5-23 (1972), Hungarian, EEA75-39915.
347.Pasztor, G., "Linvill-model describing semiconductor regions with inhomogeneous doping and recombination", Hiradstech. Ipari Kut. Intez. Kozl., 14 (4), 28-39 (1974), Hungarian, EEA78-19795.
348.Paul, R., "The bipolar transistor model", Nachrichtentech. Elektron., 23, 250-1 (1973), German, EEA76-35309.
349.Poon, H. C. and Gummel, H. K., "Modeling of emitter capacitance", Proc. IEEE, 57, 2181-2 (1969), EEA73-14344.
350.Poon, H. C. and Wilson, C. L., "Calculations of collector current variation in bipolar transistors", Solid-State Electron., 17, 973-5 (1974), EEA78-8555.
351.Raburn, W. D. and Causey, W. H., "Calculations of sustaining voltage in silicon bi-polar transistors", Microelectronics, 6 (3), 4-6 (1975), EEA78-28776.
352.Rey, G., Dupuy, F. and Bailbe, J. P., "A unified approach to the base widening mechanisms in bipolar transistors", Solid-State Electron., 18, 863-6 (1975), EEA78-37525.
353.Rey, G., Lemaire, K. and Bailbe, J. P., "Behaviour of a bipolar transistor in static service: IBIS model", Onde Elect., 50, 503-16 (1970), French, EEA73-38085.
354.Richards, G. A., "The application of optimisation techniques to bi-polar transistor modelling", Proceedings of the Conference on Optimization Techniques in Circuit and Control Applications, 1970, p. 51-6, EEA73-31442.
355.Rieder, K., "Influence of emitter-base contact separation on current amplification in bipolar transistors", Solid-State Electron., 14, 1251-8 (1971), German, EEA75-4691.
356.Rohr, P. and Lindholm, F. A., "Modifications of equivalent circuits for narrow-base transistors", IEEE J. Solid-State Circuits, SC-10, 65-72 (1975), EEA78-8550.
357.Rohr, P., Lindholm, F. A. and Allen, K. R., "Questionability of drift-diffusion transport in analysis of small semiconductor devices", Solid-State Electron., 17, 729-34 (1974).
358.Roulston, D. J., Chamberlain, S. G. and Sehgal, J., "High-level asymptotic variation of transistor base resistance and current gain", Electron. Lett., 7, 438-40 (1971), EEA74-30802.
359.Saraeva, E. A., "Refinement of an equivalent circuit of a multi-emitter transistor operating as a frequency multiplier", Radiotekhnika, Kharkov, no. 19, 71-6 (1971), Russian, EEA75-28502.
360.Schilling, R. B., "A bipolar transistor model for device and circuit design", RCA Rev., 32, 341-71 (1971), EEA75-7517.

361. Seltz, D. and Kidron, I., "A two-dimensional model for the lateral p-n-p transistor", IEEE Trans. Electron Devices, ED-21, 587-93 (1974).
362. Sirsi, R. M., Boothroyd, A. R. and Thomas, R. E., "Determination of a collector junction model for planar bipolar transistors", IEEE Trans. Electron Devices, ED-20, 878-83 (1973), EEA76-38266.
363. Slatter, J. A. G., "Fundamental modelling of cylindrical-geometry bipolar transistors", Electron. Lett., 8, 222-3 (1972), EEA75-20180.
364. Vandorpe, D., "Mathematical study of a bipolar transistor polarized in direct current", Onde Elect., 49, 717-22 (1969), French, EEA73-17649.
365. Verwey, J. F., "On the mechanism of h_{fe} degradation by emitter-base reverse current stress", Microelectron. & Rel., 9, 425-32 (1970), EEA74-1502.
366. Wheatley, C. F., "Observation of the influence of base widening upon beta and f_t", RCA Rev., 32, 247-50 (1971), EEA74-30812.
367. White, A. B., "Basewidth modulation and nonlinear in CADA models", IEEE J. Solid-State Circuits, SC-7, 428-31 (1972), EEA75-36260.
368. Wolszczak, B., "Bipolar transistor model for circuit analysis programs", Elektronika, 309-11 (1972), Polish, EEA75-38910.
369. Zajc, B., "Distribution of minority-carrier concentration and electrical field in double-diffused structures", Elektrotech. Vestn., 41, 152-6 (1974), Slovene, EEA78-24318.
370. Zaluska, E. J., Dubock, P. A. and Kemhadjian, H. A., "Practical 2-dimensional bipolar-transistor-analysis algorithm", Electron. Lett., 9, 599-60 (1973), EEA77-10225.

6. Computer Analysis

371. Aubrey, R. A. and Kraus, H. L., "Nonlinear large-signal modeling of transistor-collector characteristics for computer-aided circuit design", IEEE J. Solid-State Circuits, SC-5, 50-2 (1970), EEA73-12737.
372. Bernstein, S., "Represent transistor beta variations", Electron. Des., 22 (19), 146-8 (1974), EEA78-24320.
373. Brown, R. W., "TRAMP - A new transistor modelling programme", Electron. Components, 14, 745-50 (1973), EEA76-37835.
374. Choma, J., Jr., "Models for computer-aided analysis of large-signal transistor networks", Proceedings of the Fifth Asilomar Conference on Circuits and Systems, 1971, p. 436-40, Publ. 1972, EEA76-803.
375. Choma, J., Jr., "A model for the computer-aided noise analysis of broad-banded bipolar circuits", IEEE J. Solid-State Circuits, SC-9, 429-35 (1974), EEA78-2391.

376.Covello, R. J., "An optimization technique to calculate bipolar transistor parameters", Presented at AGARD Conference Proceedings No. 130 Computer Aided Design for Electronic Circuits, Lyngby, Denmark, May 1973, EEA77-6011.
377.Davies, P. A., "Computer modelling of bipolar transistors", Thesis, Univ. Coll., Swansea, Wales, 1973, EEA77-2554.
378.DeMan, H. J. and Mertens, R., "SITCAP - a simulator of bipolar transistors for computer-aided circuit analysis programs", Digest of Technical Papers of the 1973 IEEE International Solid State Circuits Conference, p. 104-5, 205, EEA76-22895.
379.Drobish, W. E., "Higher order models for computer-aided circuit design", IEEE J. Solid-State Circuits, SC-7, 208-11 (1972), EEA75-15772.
380.Dutton, R. W., "Bipolar transistor modeling of avalanche generation for computer circuit simulation", IEEE Trans. Electron Devices, ED-22, 334-8 (1975).
381.Fossum, J. G., "A new approach to bipolar device modeling for CAD", Proc. IEEE, 60, 756-7 (1972), EEA75-24342.
382.Fossum, J. G., "A bipolar device modeling technique applicable to computer-aided circuit analysis and design", IEEE Trans. Electron Devices, ED-20, 582-93 (1973), EEA76-23432.
383.Fossum, J. G., "Expanding modeling of bipolar devices", 1973 WESCON Technical Papers, Western Electronic Show and Convention, Vol. 17, 14/1-6 pp., EEA77-10239.
384.Ghosh, H. N., de la Moneda, F. H. and Dono, N. R., "Computer-aided transistor design, characterization, and optimization", Solid State Electron., 10, 705-26 (1967), EEA70-19161.
385.Gokhale, B. V., "Numerical solutions for a one-dimensional silicon n-p-n transistor", IEEE Trans. Electron Devices, ED-17, 594-602 (1970).
386.Greenbaum, J. R., "Easy-to-use Hypi program makes possible transition from nonlinear to linear transistor model", Electronics, 46 (2), 174-9 (1973), EEA76-7906.
387.Hachtel, G. D. and Joy, R. C., "A new efficient one-dimensional analysis program for junction device modeling", Proc. IEEE, 60, 86-98 (1972), EEA75-7481.
388.Henderson, J. C., Scarborough, R. J. D. and Silvester, R., "Effect of emitter oxide-window-edge shape in computer-simulation studies of ion-implanted transistors", Electron. Lett., 6, 623-5 (1970), EEA73-38125.
389.Hruby, J., "A method of computing the parameters of a linearized transistor model", Slaboproudy Obzor, 33, 349-53 (1972), Czech, EEA75-36271.
390.Hubig, E. and Franz, R., "Presentation of transistor characteristics by means of a process computer", Mess. & Pruef., no. 1, 32-3 (1974), German, EEA77-18575.
391.Jorke, G., "Computer simulation of electronic structures", Nachrichtentech. Elektron., 23, 442-4 (1973), German, EEA77-14248.

392.Kim, R. H., "A computer-aided design and analysis of heat dissipation in a junction transistor", Modeling and Simulation, Vol. 5, Pt. II, 1974, p. 973, EEA78-41595.

393.Kjaerstad, O. G., "Transistor equivalents", Presented at AGARD Conference Proceedings No. 130 Computer Aided Design for Electronic Circuits, Lyngby, Denmark, May 1973, EEA77-6010.

394.Kuo, Y. L., "Distortion analysis of bipolar transistor circuits", IEEE Trans. Circuit Theory, CT-20, 709-16 (1973), EEA76-37849.

395.Manck, O. and Engl, W. L., "Two-dimensional computer simulation for switching a bipolar transistor out of saturation", IEEE Trans. Electron Devices, ED-22, 339-47 (1975), EEA78-19796.

396.McBride, L. C., "Large-signal modeling of bipolar transistors for computer-aided circuit analysis", Thesis, Stanford Univ., California, 1971, 119 pp., EEA75-35112.

397.McEwan, J. R., "Computer simulation of radiation induced emitter crowding", IEEE Trans. Nucl. Sci., NS-21, 353-8 (1974), EEA78-19799.

398.Moshkin, V. I., "Transit time through a space-charge region of the collector-base junction", Izv. VUZ Radioelektron., 17, 113-15 (1974), Russian, EEA78-8551.

399.Paratov, G. M., Litkens, E. M. and Shulgina, G. S., "Modelling of the relationship between performance characteristics and structural parameters of a transistor using a digital computer", Izv. VUZ Radioelektronika, 13, 701-9 (1970), Russian, EEA73-35215.

400.Peterson, O. G., "Numerical method for the solution of the transient behavior of bipolar semiconductor devices", Thesis, Univ. Pennsylvania, Philadelphia, 1971, 200 pp., Order No. 72-6215, EEA75-32670.

401.Reschke, D., "Simulation of the static and dynamic characteristics of transistors on an analogue computer", XII International Scientific Colloquium, 1967, p. 117-23, Publ. 1968, German, EEA73-10717.

402.Rohrer, R., Fan, S. P. and Claudio, L., "Automated bipolar junction transistor d.c. model parameter determination", IEEE J. Solid-State Circuits, SC-6, 260-2 (1971), EEA74-25350.

403.Simonescu, M., "Establishing of dynamic parameters of the transistor in small signal regime superposed on a statical injection regime", Stud. Cercet. Fiz., 24, 649-64 (1972), Rumanian, EEA76-798.

404.Slotboom, J. W., "Computer-aided two-dimensional analysis of bipolar transistors", IEEE Trans. Electron Devices, ED-20, 669-79 (1973), EEA76-29252.

405.Sokal, N. O., "Modeling transistors and diodes for computer-aided nonlinear analysis", Proceedings of the National Electronics Conference, 1968, p. 729-30, EEA73-16330.

406.Thielking, R. C., "Using improved transistor models in computer aided analysis of an RGB video amplifier", Proceedings of the National Electronics Conference, 1970, p. 490-5, EEA75-108.

407. Venkateswaran, S. and Hasan, M. M., "Characterization of planar transistors using digital computer", J. Inst. Telecommun. Eng., 17, 444-7 (1971), EEA75-24335.
408. Wieder, A. W., Engl, W. L. and Lehning, H., "Computer-aided device modeling and design procedure for current hogging logic (CHL)", IEEE J. Solid-State Circuits, SC-10, 352-9 (1975), EEA78-37627.
409. Wolszczak, B., "Universal model of bipolar transistor for computer-aided circuit analysis programs", Elektronika, 14, 306-8 (1973), EEA76-37836.
410. Wolszczak, B. and Stanclik, J., "Transistor model for computer aided circuit analysis programs", Elektronika, no. 4, 166-8 (1972), Polish, EEA75-24329.
411. Zilka, Z., "The static parameters of a diode model of the transistor", Slaboproudy Obzor, 31, 312-17 (1970), Czech, EEA73-31397.

7. High Frequency Transistors

412. Andresciani, V. and De Leo, R., "Equipment for the determination of microwave transistor S-parameters", Alta Freq., 41, 167-8 (1972), Italian, EEA75-20456.
413. Andresciani, V. and de Sario, M., "A model of the bipolar transistor for high frequencies", Alta Freq., 42, 36-40 (1973), Italian, EEA76-19359.
414. Bailey, R. L., "Large-signal nonlinear analysis of a high-power high-frequency junction transistor", IEEE Trans. Electron Devices, ED-17, 108-19 (1970), EEA73-14278.
415. Baranyi, A. and Radvanyi, A., "Modelling high-frequency transistors for CAD", Hiradastechnika, 23, 358-64 (1972), Hungarian, EEA76-11825.
416. Bardyla, T. I., Radionchyk, K. N. and Rudenko, V. N., "Transistor circuit analysis at high frequencies and large signals", Presented at Proceedings of the 4 Colloquium on Microwave Communication, Budapest, Hungary, April 1970, EEA73-38091.
417. Benjamin, J. A., "New design concepts for microwaves power transistor", Microwave J., 15 (10), 39 (1972), EEA76-8668.
418. Bogacev, V. M., "On the spectrum analysis of the currents of a high-frequency transistor in the segment-linear and segment-parabolic approximations of the characteristics", Elektro Prom. -st. & Priborostr., 7, 223-5 (1972), Bulgarian, EEA75-36259.
419. Chen, P. T., "Design and applications of 2-6.5 GHz transistor amplifiers", Digest of Technical Papers of the 1973 IEEE International Solid State Circuits Conference, p. 76-7, EEA76-22940.
420. Conn, D. R. and Mitchell, R. H., "Small-signal modelling and characterization of microwave transistors", Int. J. Electron., 34, 655-66 (1973), EEA76-26077.
421. Di Maio, B., Prestigiacomo, G. and Tamburello, C., "Identification of transistors at microwaves", Alta Freq., 41, 163-7 (1972), Italian, EEA75-20455.

422. Faldella, E. and Iuculano, G., "Mathematical model with minimum number of independent coefficients for transistors for high frequencies", Electron. Lett., 10, 26-7 (1974), EEA77-10226.
423. Ghosh, H. N., Ashar, K. G., Oberai, A. S. and DeWitt, D., "Design and development of an ultralow-capacitance, high-performance pedestal transistor", IBM J. Res. Develop., 15, 436-41 (1971).
424. Ghosh, H. N., Oberai, A. S., Chang, J. J. and Vora, M. B., "An arsenic emitter structure for high-performance silicon transistors", IBM J. Res. Develop., 15, 457-63 (1971).
425. Harrison, R. G., "Modeling a microwave power transistor", Digest of Technical Papers of the 1971 International Solid State Circuits Conference, p. 36-7, EEA71-23029.
426. Hartmann, K. and Strutt, M. J. O., "Computer simulation of small-signal and noise behavior of microwave bipolar transistors up to 12 GHz", IEEE Trans. Microwave Theory & Tech., MTT-22, 178-83 (1974), EEA77-14493.
427. Kronquist, R. L., Fourrier, J. Y., Pestie, J. P. and Brilman, M. E., "Determination of a microwave transistor model based on an experimental study of its internal structure", Solid-State Electron., 18, 949-63 (1975), EEA78-41590.
428. Lange, J. and Carr, W. N., "An application of device modeling to microwave power transistors", IEEE J. Solid-State Circuits, SC-7, 71-80 (1972), EEA75-10539. See also: Thesis, Southern Methodist Univ., Dallas, Texas, 1971, 118 pp., Order No. 71-19602, EEA75-28789.
429. Macnee, A. B. and Talsky, R. J., "High-frequency transistor modeling for circuit design", IEEE J. Solid-State Circuits, SC-7, 320-2 (1972), EEA75-28533.
430. Meninger, M., "Approximating transfer parameters by a rational fractional function using the least squares method", Slaboproudy Obz., 33, 509-13 (1972), Czech, EEA76-4418.
431. Roulston, D. J., Chamberlain, S. G. and Sehgal, J., "High level, high frequency modelling of double diffused transistors", 1971 International Electrical and Electronics Conference, p. 18-19, EEA72-10251.
432. Slatter, J. A. G., "Transistor modelling at microwave frequencies", Electron. Equip. News, 14 (6), 85-6 (1972), EEA75-39913.
433. Stangerup, P., "Linear transistor models in HF network analysis, adaptation between measurements and analysis by computer", Presented at AGARD Conference Proceedings No. 130 Computer Aided Design for Electronic Circuits, Lyngby, Denmark, May 1973, EEA77-6012.
434. Wahl, A. J., "Distributed theory for microwave bipolar transistors", IEEE Trans. Electron Devices, ED-21, 40-9 (1974), EEA77-5446.

8. Other Bipolar Devices

435.Abbott, N. E., "Modeling and analysis of tuned power amplifiers", Thesis, Univ. Michigan, Ann Arbor, 1971, 289 pp., Order No. 72-14782, EEA75-35251.
436.Aiken, J. G., "The bipolar tetrode as a process control device", J. Electrochem. Soc., 120 (3), 100C (1973), EEA76-23442.
437.Berezin, A. S. and Mochalkina, O. R., "Lateral-structure transistors and thyristors", Izv. VUZ Radioelektron., 15, 811-25 (1972), Russian, EEA76-794.
438.Bridgen, R., Cusdin, A. R. and van Nielen, J. A., "Large signal transistor models for logic circuit simulation", Presented at Colloquium on Non-linear Circuit Analysis, London, England, January 1971, EEA74-9761.
439.Fejes, L. and Beke, I., "Calculation of the operating points and characteristics of the TTL gate", Hiki, 11, 30-56 (1971), Hungarian, EEA74-25485.
440.Ho, C. F., "Transistor simulated four layer device and its applications", JAEU, 4 (3), 22-8 (1971), EEA75-7515.
441.Hodges, D. A. and Heald, R. A., "Design of Schottky-barrier diode-clamped transistor layouts", Digest of Technical Papers of the 1973 IEEE International Solid State Circuits Conference, p. 106-7, EEA76-23444.
442.Hower, P. L., "Optimum design of power transistor switches", IEEE Trans. Electron Devices, ED-20, 426-35 (1973), EEA76-16409.
443.Kamoshida, M., Kani, K., Sato, K. and Okada, T., "Heat transfer analysis of beam-lead transistor chip", IEEE Trans. Electron Devices, ED-17, 863-70 (1970), EEA73-38077.
444.Kani, K., Kamoshida, M. and Okada, T., "Heat transfer analysis of beam-lead transistor", Electron. & Commun. Jap., 53, 157-64 (1970), EEA74-33377.
445.Lee, F. C. Y. and Wilson, T. G., "Analysis and modeling of a family of two-transistor parallel inverters", IEEE Trans. Magnetics, MAG-9, 414-18 (1973).
446.McGhee, J., "A transient model for a three terminal p-n-p-n switch and its use in predicting the gate turn-on process", Int. J. Electron., 35, 73-80 (1973), EEA76-29264.
447.Ollivier, P. M., "Microwave YIG-tuned transistor oscillator amplifier design: application to C band", IEEE J. Solid-State Circuits, SC-7, 54-60 (1972), EEA75-10538.
448.Olmstead, J. Einthoven, W., Ponczak, S. and Kannam, P. J., "High-level current gain in bipolar power transistors", RCA Rev., 32, 221-46 (1971), EEA74-34073.
449.Pfleiderer, H. J., "The influence of the collector-emitter capacitance on the transfer function of the bucket-brigade shift register", Nachrichtentech. Z. (NTZ), 26, 203-10 (1973), German, EEA76-28900.

450.Rein, H. M. and Bruchmann, H., "Analog transistor simulating integrated circuits", Int. Elektron. Rundsch., 25, 227-31 (1971), German, EEA74-37272.
451.Saltich, J. L., Volk, C. E. and Clark, L. E., "Optimum vertical design and evaluation techniques for high-current, high-voltage transistors", IEEE Power Electronics Specialists Conference 1972, p. 74-9, EEA77-375.
452.Saraeva, E. A., "Refinement of an equivalent circuit of a multi-emitter transistor operating as a frequency multiplier", Radiotekhnika, Kharkov, no. 19, 71-6 (1971), Russian, EEA75-28502.
453.Senhouse, L. S., Jr., "One dimensional modelling of TRIM transistors", Presented at 17th International Electron Devices Meeting (abstracts), Washington, D. C., October 1971, EEA75-13736.
454.Spirito, P., "Study of behaviour of a monostable circuit with a transistor in avalanche region", Alta Freq., 41, 927-34 (1972), Italian, EEA76-37952.
455.Thielking, R. C., "Using improved transistor models in computer-aided analysis of an RGB video amplifier", IEEE Trans. Broadcast. & Telev. Receivers, BTR-17, 24-9 (1971), EEA74-14714.
456.Tucker, R. S., "Gain-bandwidth limitations of microwave transistor amplifiers", IEEE Trans. Microwave Theory & Tech., MTT-21, 322-7 (1973), EEA76-22923.
457.Wang, P. P. and Branin, F. H., Jr., "Multi-section model of an N^+ -P-v-N^+ high voltage power switching transistor", IEEE Power Electronics Specialists Conference 1973, p. 80-9, EEA77-376.

IX. Field-Effect Transistors

1. General

458.Arnold, S., "F.E.T. characterisation for microstrip applications", Presented at Colloquium on RF Measurements on Solid-State Devices (digest), London, England, May 1971, EEA74-27205.
459.Biehl, B. L., "Evaluation of f.e.t. models for CASA", IEEE Proceedings of the International Symposium on Circuit Theory, 1972, p. 384, EEA75-32701.
460.Keyes, R. W., "Effect of randomness in the distribution of impurity ions on FET thresholds in integrated electronics", IEEE J. Solid-State Circuits, SC-10, 245-7 (1975), EEA78-33214.
461.Lindholm, F. A., "Unified modeling of field-effect devices", IEEE J. Solid-State Circuits, SC-6, 250-9 (1971), EEA74-26871.
462.Lindholm, F. A. and Gray, P. R., "Large signal and small-signal models for arbitrarily-doped four-terminal field-effect transistors", IEEE Trans. Electron Devices, ED-13, 819-29 (1966).
463.Mihran, T. G., "A five-parameter model for current and cross modulation in field-effect transistors with high drain voltage", IEEE Trans. Electron Devices, ED-22, 982-8 (1975).
464.Miller, D. M. and Meyer, R. G., "Nonlinearity and cross modulation in field-effect transistors", IEEE J. Solid-State Circuits, SC-6, 244-50 (1971), EEA74-26870.

465.Muggli, R. A., "Program gives FET model from specs", Electron. Des., 21 (12), 92-5 (1973), EEA76-35319.
466.Reiser, M., "A two-dimensional numerical FET model for DC, AC, and large-signal analysis", IEEE Trans. Electron Devices, ED-20, 35-45 (1973), EEA76-4658.
467.Robinson, F. N. H., "Understanding the field effect transistor", Electron. Compon., 12, 1231-2 (1971), EEA75-7531.
468.Sewell, F. A., Jr., Wegener, H. A. R. and Lewis, E. T., "Charge storage model for variable threshold FET memory element", Appl. Phys. Lett., 14, 45-7 (1969), EEA70-9593.
469.Snejdar, V., "The electrical conductivity of semiconductor films used in thin-film field-effect transistors", Slaboproudy Obzor, 31, 293-8 (1970), Czech, EEA73-35239.
470.Stenin, V. Y., "The large signal equivalent circuit of a field-effect transistor", Telecommun. & Radio Eng. Pt. 2, 27, 140-2 (1972), EEA73-12650.
471.Wedlock, B. D., "Static large-signal field-effect transistor models", Proc. IEEE, 58, 593-5 (1970), EEA73-21330.

2. JFET

472.Abe, A., Hirao, T., Sugawa, T. and Takayanagi, S., "High frequency cascade JFET fabricated with a differential ion doping technology", 6th International Congress on Microelectronics, 1974, 4 pp., EEA78-24368.
473.Burgess, J. P. and Green, P. F. L., "A computer oriented large-signal JFET model", 1971 International Electrical and Electronics Conference, p. 20-21, EEA75-10302.
474.Catalano, G. T., "An analysis and numerical solution for the electrical characteristics of field effect devices", Solid-State Electron., 18, 583-6 (1975), EEA78-28809.
475.Chiabrera, A. and Frisianai, A. L., "Modeling of junction-gate field effect transistors", Proceedings of the National Electronics Conference, 1968, p. 285-90, EEA73-17674.
476.Chiu, T. L. and Ghosh, H. N., "Characteristics of the junction-gate field effect transistor with short channel length", Solid-State Electron., 14, 1307-17 (1971), EEA75-4712.
477.Compton, J. B., "High-frequency junction f.e.t. characterisation and application", Des. Electron., 7 (6), 36-41 (1970), EEA73-21325.
478.Erratico, P. G., Maloberti, F. and Svelto, V., "JGFET thermal noise in hot electrons regime", Alta Freq., 41, 683-9 (1972), EEA76-4652.
479.Kasser, R., "A new noise equivalent circuit for the junction FET with uncorrelated noise sources", Proc. IEEE, 58, 1171-2 (1970), EEA73-31421.
480.Kennedy, D. P. and O'Brien, R. R., "Computer aided two-dimensional analysis of the junction field-effect transistor", IBM J. Res. Developm., 14, 95-116 (1970), EEA73-28214.

481.Lehovec, K. and Miller, R. S., "Field distribution in junction field-effect transistors at large drain voltages", IEEE Trans. Electron Devices, ED-22, 273-81 (1975), EEA78-19826.
482.Lehovec, K. and Seeley, W. G., "On the validity of the gradual channel approximation for junction field effect transistors with drift velocity saturation", Solid-State Electron., 16, 1047-54 (1973), EEA76-35330.
483.Misra, M. and Prasad, H. C., "Drain-source capacity of junction field effect transistor", Solid-State Electron., 15, 325-7 (1972), EEA75-13770.
484.Regier, F. A., "Junction field-effect transistor analysis as a boundary-value problem", Modelling and Simulation, Vol. 4, 1973, p. 427-9, EEA78-24366.
485.Yamaguchi, K., Toyabe, T. and Kodera, H., "Two-dimensional analysis of triode-like operation of junction gate FET's", IEEE Trans. Electron Devices, ED-22, 1047-9 (1975).

3. MOSFET

a. Two-Dimensional Models

486.Armstrong, G. A., Magowan, J. A. and Ryan, W. D., "Two-dimensional solution of the d.c. characteristics for the MOST", Electron. Lett., 5, 406-7 (1969), EEA73-1196.
487.Armstrong, G. A., Magowan, J. A. and Burt, D. J., "Two-dimensional analysis of double-gate MOS transistor", Electron. Lett., 5, 633-4 (1969), EEA73-10676.
488.de la Moneda, F. H., "Threshold voltage from numerical solution of the two-dimensional MOS transistor", IEEE Trans. Circuit Theory, CT-20, 666-73 (1973), EEA76-37842.
489.Loeb, H. W., Andrew, R. and Love, W., "Application of 2-dimensional solutions of the Shockley-Poisson equation to inversion-layer MOST devices", Electron. Lett., 4, 352-4 (1968), EEA71-26021.
490.Mock, M. S., "A two-dimensional mathematical model of the insulated-gate field-effect transistor", Solid-State Electron., 16, 691-9 (1973), EEA76-19935.
491.van Drope, D., "MOS two-dimensional study", Onde Elec., 51, 837-43 (1971), French, EEA75-4634.

b. Statistical Models

492.Clemens, J. T. and Labuda, E. F., "A statistical analysis of the threshold voltage of a double insulator", J. Electrochem. Soc., 120, 100C (1973), EEA76-23496.
493.Hagan, B. J. and Magowan, J. A., "An improved statistical approach to the prediction of large scale integrated M.O.S.T. circuit performance", International Conference on Computer Aided Design, 1972, p. 257-62, EEA75-19265.

494.Kamins, T. I. and Muller, R. S., "Statistical considerations in MOSFET calculations", Solid-State Electron., 10, 423-31 (1967), EEA70-14603.

495.Mar, H. A. and Simmons, J. G., "Surface-generation statistics and associated thermal currents in metal-oxide-semiconductors structures", Phys. Rev. B, 11, 775-83 (1975), PA78-33560.

496.Morozov, A. A. and Khodosh, L. S., "Initial component parameter data for a statistical analysis of MOS integrated circuits", Izv. VUZ Radioelektron., 14, 1380-2 (1971), Russian, EEA75-13764.

c. Other Models

497.Armstrong, G. A. and Magowan, J. A., "The distribution of mobile carriers in the pinch-off region of an insulated-gate field-effect transistor and its influence on device breakdown", Solid-State Electron., 14, 723-33 (1971), EEA74-30835.

498.Armstrong, G. A. and Magowan, J. A., "Modelling of short-channel m.o.s. transistors", Electron. Lett., 6, 313-15 (1970), EEA73-31416.

499.Armstrong, G. A. and Magowan, J. A., "Derivation of simple expressions for interelectrode capacitances of IGFET's as a function of bias condition", Electron. Lett., 7, 281-3 (1971), EEA74-22730.

500.Baccarani, G. and Mazzone, A. M., "The influence of an energy-dependent relaxation time on the specular and diffuse scattering models of surface conduction in MOS structures", Solid-State Electron., 18, 718-20 (1975), EEA78-28714.

501.Baccarani, G., Mazzone, A. M. and Morandi, C., "The diffuse scattering model of effective mobility in the strongly inverted layer of MOS transistors", Solid-State Electron., 17, 785-9 (1974), EEA77-30343.

502.Baikov, V. D., Narmazinskii, A. N. and Nemchinov, V. M., "Effect of substrate potential on the current characteristics of MOS transistors", Izv. VUZ Radioelektron., 12, 461-70 (1969), Russian, EEA73-1203.

503.Balas, J., "Output response of the MOS transistor TESLA KF 520", Sdelovaci Tech., 16, 372-3 (1968), Czech, EEA73-1181.

504.Barron, M. B., "Low level currents in insulated gate field effect transistors", Solid-State Electron., 15, 293-302 (1972), EEA75-13769.

505.Baum, G., "Calculation of the immobile part of the space charge in a semiconductor of MOS structure", Solid State Electron., 11, 1113-8 (1968), German, PA72-12172.

506.Baum, G. and Beneking, H., "An analysis of the static and dynamic behavior of MOS-FET's with inclusion of the saturation region", Arch. Elekt. Ubertragung., 22 (1), 1-7 (1968), German, PA71-22639.

507.Berman, A. and Kerr, D. R., "Inversion charge redistribution model of the high-frequency MOS capacitance", Solid-State Electron., 17, 735-42 (1974), PA77-57180.

508.Berz, F. and Prior, C. G., "Test of McWhorter's model of low-frequency noise in Si m.o.s.t.s.", Electron. Lett., 6, 595-7 (1970), EEA73-38095.
509.Bierhenke, H. and Goser, K., "Analysis of the small signal behaviour of MOS-transistors using the fast Fourier Transform", Arch. Elektron. & Uebertragungstech., 29 (1), 20-3 (1975), German, EEA78-8562.
510.Boudry, M. R., "Theoretical origins of N_{ss} peaks observed in Gray-Brown MOS studies", Appl. Phys. Lett., 22, 530-1 (1973), EEA76-25992.
511.Breed, D. J., "A new model for the negative voltage instability in MOS devices", Appl. Phys. Lett., 26, 116-18 (1975), EEA78-15342.
512.Brews, J. R., "Surface potential fluctuations generated by interface charge inhomogeneities in MOS devices", J. Appl. Phys., 43, 2306-13 (1972), EEA75-20135.
513.Brews, J. R., "Admittance of an MOS device with interface charge inhomogeneities", J. Appl. Phys., 43, 3451-5 (1972), EEA75-32686.
514.Brews, J. R., "Limitations upon photoinjection studies of charge distributions close to interfaces in MOS capacitors", J. Appl. Phys., 44, 279-84 (1973), EEA76-12508.
515.Brews, J. R., "An improved high-frequency MOS capacitance formula", J. Appl. Phys., 45, 1276-9 (1974), PA77-60863.
516.Brews, J. R., "Error analysis of high-frequency MOS capacitance calculations", Solid-State Electron., 17, 447-56 (1974), PA77-45008.
517.Brews, J. R., "Theory of the carrier-density fluctuations in an IGFET near threshold", J. Appl. Phys., 46, 2181-92 (1975), EEA78-33156.
518.Bulucea, C., "Avalanche injection into the oxide in silicon gate-controlled devices", Solid-State Electron., 18, 363-74 (1975), EEA78-11669.
519.Burns, J. R., "Large-signal transit-time effects in the MOS transistor", RCA Rev., 30, 15-35 (1969), EEA72-17729.
520.Buxo, J., Esteve, D., Enea, G. and Martinez, A., "A 'four parameters' model that fits the degradation curve $\Delta V_g(V_g)$ of MOS transistors under irradiation", Solid-State Electron., 15, 1029-31 (1972), EEA75-36310.
521.Cagnina, S. F. and Snow, E. H., "Properties of gold doped MOS structures", J. Electrochem. Soc., 114, 1165-72 (1967), PA71-13896.
522.Calzolari, P. U., Graffi, S., Mazzone, A. M. and Morandi, C., "Bulk lifetime determination from current and capacitance transient response of M.O.S. capacitors", Alta Freq., 41, 848-53 (1972), EEA76-16431.
523.Chang, L. L. and Moore, J. S., "Generalized current and conductance extrema in metal-insulator-semiconductor tunnel junctions", J. Appl. Phys., 40, 5315-19 (1969), PA73-18351.

524. Cheney, G. T. and Kotch, R. A., "A simple theory for threshold voltage modulation in IGFET's", Proc. IEEE, 56, 887-8 (1968), EEA71-23372.

525. Cheng, Y. C., "Effect of charge inhomogeneities on silicon surface mobility (M.O.S. devices)", J. Appl. Phys., 44, 2425-7 (1973), EEA76-23379.

526. Cherry, E. M., "Small-signal high-frequency response of the insulated-gate field-effect transistor", IEEE Trans. Electron Devices, ED-17, 569-77 (1970), EEA74-4928.

527. Chiu, T. L., "Experiments and theory on the characteristics of metal-oxide-semiconductor transistors in the current saturation range", Thesis, Univ. Illinois, Urbana, USA, 115 pp., Order No. 69-10660, EEA73-21338.

528. Chiu, T. L. and Sah, C. A., "Correlation of experiments with a two-section model theory of the saturation drain conductance of MOS transistors", Solid State Electron., 11, 1149-63 (1968), EEA72-5043.

529. Chou, N. J., "Application of triangular voltage sweep method to mobile charge studies in MOS structures", J. Electrochem. Soc., 118, 601-9 (1971), PA74-37231.

530. Collins, T. W. and Churchill, J. N., "Exact modeling of the transient response of an MOS capacitor", IEEE Trans. Electron Devices, ED-22, 90-101 (1975), PA78-21960.

531. Cooper, J. A., Jr. and Schwartz, R. J., "The effect of an energy-dependent capture cross section on data interpretation using the MOS conductance technique", J. Appl. Phys., 44, 5613-14 (1973), EEA77-18741.

532. Cooper, J. A., Jr. and Schwartz, R. J., "Electrical characteristics of the SiO_2-Si interface near midgap and in weak inversion", Solid-State Electron., 17, 641-54 (1974), EEA77-30240.

533. Critchlow, D. L., Dennard, R. H. and Schuster, S. E., "Design and characteristics of n-channel insulated-gate field-effect transistors", IBM J. Res. & Dev., 17, 430-42 (1973), EEA77-18808.

534. Csabay, O. and Frank, H., "Determination of the influence exerted by the thickness of the oxide layer on the effective charge density of imperfections of MOS structures by means of a simple graphical-numerical method", Nachrichtentech. Elektron., 23, 108-12 (1973), German, EEA76-19869.

535. Das, M. B., "Switching characteristics of MOS and junction-gate field-effect transistors", Proc. Instn. Elect. Engrs., 114, 1223-30 (1967), EEA71-3437.

536. Deuling, H. J., "Influence of acoustic phonons on tunneling through metal-oxide-semiconduction junctions", Phys. Rev. B, 1, 1644-7 (1970), EEA73-25050.

537. Dirilten, H., "On the mathematical models characterizing faulty four-phase MOS logic arrays", IEEE Trans. Comput., C-21, 301-5 (1972), EEA75-6918.

538.Edwards, J. R. and Marr, G., "Depletion-mode IGFET made by deep ion implantation", IEEE Trans. Electron Devices, ED-20, 283-9 (1973), EEA76-16443.
539.El-Mansy, Y. A., "Modelling of insulated-gate field-effect transistors", Thesis, Carleton Univ., Ottawa, Canada, 1974, EEA78-37551.
540.El-Sissi, H. and Cobbold, R. S. C., "Numerical calculation of the ideal C/V characteristics of nonuniformly doped m.o.s. capacitors", Electron. Lett., 9, 594-6 (1973), PA77-16824.
541.Fitchen, F. C., "Limiting amplitude MOSFET small-signal model", Proc. IEEE, 55, 2176-7 (1967), EEA71-10680.
542.Fitchen, F. C. and Ellerbruch, V. G., "Linear operation of the MOSFET complementary pair", IEEE J. Solid-State Circuits", SC-6, 422-3 (1971), EEA75-314.
543.Freestone, H. T., "Theory of the insulated gate field effect transistor", Thesis, Univ. Pennsylvania, 1968, 83 pp., Order No. 69-99, EEA73-10695.
544.Frohman-Bentchkowsky, D. and Grove, A. S., "Conductance of MOS transistors in saturation", IEEE Trans. Electron Devices, ED-16, 108-13 (1969), EEA72-9599.
545.Fu, H. S., "Surface 1/f noise on oxidized silicon", Thesis, Univ. Illinois, Urbana, 161 pp., Order No. 71-21123, EEA75-28524.
546.Gaensslen, F. H. and LeBlanc, A., "Simulation of micron MOS device characteristics", Presented at Third European Solid State Device Research Conference, Munchen, Germany, September 1973.
547.Galkin, V. N. and Shilkov, V. M., "Output resistance of MOS transistors at saturation", Izv. VUZ Radioelektron., 16, 122-3 (1973), Russian, EEA76-16447.
548.Georgin, G., "Some notions about MOS transistors", Onde Electr., 54, 11-14 (1974), French, EEA77-14530.
549.Giraud, J. L., "Mathematical model of a MOS transistor submitted to high energy radiations", Report (unnumbered), Univ. Toulouse, France, February 1970, 122 pp., EEA75-16757.
550.Gnadinger, A. P. and Talley, H. E., "Quantum mechanical calculation of the carrier distribution and the thickness of the inversion layer of a MOS field-effect transistor", Solid-State Electron., 13, 1301-9 (1970), EEA73-35241.
551.Golder, J. and Baldinger, E., "Determination of the relationship between the condition of the boundary layer and the charge pumping effect in MOS transistors", Helv. Phys. Acta, 44, 387-400 (1971), German, PA74-45333.
552.Goser, K., "Channel formation in an insulated gate field-effect transistor (IGFET) and its equivalent circuit", Siemens Forsch. - & Entwicklungsber, 1, 3-9 (1971), EEA75-4715.
553.Goser, K. and Steinhubl, K., "Distribution of gate channel capacitance between source and drain in the equivalent circuit of a MOS transistor", Siemens Forsch. - & Entwicklungsber., 1, 284-6 (1972), German, EEA75-39958.

554.Grimmer, F. and Goser, K., "A model for the turn-off behaviour of MOS transistors", Arch. Elektron. & Ubertragungstech., 26, 197-201 (1972), German, EEA75-24348.
555.Hachtel, G. D. and Mack, M. H., "A graphical study of the current distributions in short-channel IGFETS", Digest of Technical Papers of the 1973 IEEE International Solid State Circuits Conference, p. 110-11, 206, EEA76-23466.
556.Hachtel, G. D., Mack, M. and O'Brien, R. R., "Semiconductor device analysis via finite elements", 8th Asilomar Conference on Circuits, Systems and Computers, 1974, p. 332-8, Publ. 1975, EEA78-24100.
557.Hara, H., "Theoretical analysis of fundamental characteristics of channel-doped MOS transistor", Electron. & Commun. Jap., 55 (10), 99-106 (1972), EEA75-35320.
558.Hart, B. L. and Barker, R. W. J., "First-order theory of MOSFET hybrid-mode operation", Int. J. Electron., 38, 625-30 (1975), EEA78-28799.
559.Hashemian, R. and Khakzar, H., "A unified model for MOS transistors with wide range of channel length", 8th Asilomar Conference on Circuits, Systems and Computers, 1974, p. 529-32, Publ. 1975, EEA78-24372.
560.Haslett, J. W. and Trofimenkoff, F. N., "Small-signal, high-frequency equivalent circuit for the metal-oxide-semiconductor field-effect transistor", Proc. Instn. Elect. Engrs., 116, 699-702 (1969), EEA72-15566.
561.Haslett, J. W. and Trofimenkoff, F. N., "Gate noise in MOSFETs at moderately high frequencies", Solid-State Electron., 14, 239-45 (1971), EEA74-18815.
562.Hemmert, R. S., "Invariance of the Hall effect MOSFET to gate geometry", Solid-State Electron., 17, 1039-43 (1974), EEA78-651.
563.Heydemann, M., "Numerical method of investigation for MOST structures", Electron. Lett., 6, 735-7 (1970), French, EEA74-1602.
564.Hofstein, S. R., "An analysis of deep depletion thin-film MOS transistors", IEEE Trans. Electron Devices, ED-13, 846-55 (1966), EEA70-10881.
565.Huang, J. S. T., "Characteristics of a depletion-type IGFET", IEEE Trans. Electron Devices, E-20, 513-14 (1973), EEA76-19929.
566.Huang, J. S. T. and Taylor, G. W., "Modeling of an ion-implanted silicon-gate depletion-mode IGFET", IEEE Trans. Electron Devices, ED-22, 995-1001 (1975).
567.Hundorf, S., "Two models for interpreting instabilities of MOS structures", Nachrichtentechnik, 20, 444-5 (1970), German, PA74-16449.
568.Iwauchi, S. and Tanaka, T., "The effects of traps in a semiconductor on the characteristics of MOS transistors", Jap. J. Appl. Phys., 7, 1237-46 (1968), EEA71-28905.

569.Jeppson, K. O., "Influence of the channel width on the threshold voltage modulation in m.o.s.f.e.t.s.", Electron. Lett., 11, 297-9 (1975), EEA78-28788.
570.Johnson, E. O., "The insulated-gate field-effect transistor - a bipolar transistor in disguise", RCA Rev., 34, 80-94 (1973), EEA76-26103.
571.Kalinowski, J. J., "A compact MOST model for design analysis", Proc. IEEE, 60, 1000-1 (1972), EEA75-32692.
572.Karakhanyan, E. R., Nosov, Y. R. and Shilin, V. A., "An analysis of the lag of an MOS transistor", Radio Eng. & Electron. Phys., 17, 668-73 (1972), EEA76-8696.
573.Kassabov, J. and Velchev, N., "Magneto-electrical properties of silicon MOST structures", Solid-State Electron., 14, 1159-65 (1971), EEA75-1178.
574.Katto, H. and Itoh, Y., "Analytical expressions for the static MOS transistor characteristics based on the gradual channel model", Solid-State Electron., 17, 1283-92 (1974), EEA78-657.
575.Kennedy, D. P. and Murley, P. C., "Steady state mathematical theory for the insulated gate field effect transistor", IBM J. Res. & Dev., 17, 2-12 (1973), EEA76-26093.
576.Kennedy, D. P. and Phillips, A., Jr., "Source-drain breakdown in an insulated gate, field-effect transistor", 1973 International Electron Devices Meeting Technical Digest, p. 160-3, EEA77-14552.
577.Kiener, M., "MOS transistors on inhomogeneously doped substrates - threshold voltage calculation and optimization of the complementary substrate concentrations", Z. Angew. Math. & Phys., 25, 120-4 (1974), German, PA77-67286.
578.Kotani, M., "MOSFET transient analysis by the hyperbolic function description", Trans. Inst. Electron. Commun. Eng. Japan. A.B.C., 53 (3), 25 (1970), EEA74-4102.
579.Landgraf-Dietz, D., "The equation of characteristics for a MOSFET with any given geometry", Nachrichtentechnik, 18, 69-70 (1968), German, EEA71-13328.
580.Landgraf Dietz, D., "A non-linear RC line used as an analogue model for the field effect transistor with insulated control electrode", Nachrichtentechnik, 19, 337-9 (1969), EEA73-7474.
581.Lattin, W. W., "MOS models: Present status and future needs", 8th Asilomar Conference on Circuits, Systems and Computers, 1974, p. 23-4, Publ. 1975, EEA78-24099.
582.Laur, J. and Jayadevaiah, T. S., "Effect of field-dependent mobility on m.o.s. transistor characteristics", Electron. Lett., 8, 482-3 (1972), EEA75-39933.
583.Laur, J. and Jayadevaiah, T. S., "Carrier concentration in the inversion layer of a MOS field effect transistor", Solid-State Electron., 16, 644-6 (1973), EEA76-19936.
584.Lazarus, M. J., "The short-channel IGFET", IEEE Trans. Electron Devices, ED-22, 351 (1975), EEA78-19832.
585.Lee, H. S., "An analysis of the threshold voltage for short-channel IGFET's", Solid-State Electron., 16, 1407-17 (1973), EEA77-6040.

586.Lo, A. W. and Gibson, J. J., "Simple MOSFET modeling for digital applications", IEEE J. Solid-State Circuits, SC-8, 391-3 (1973), EEA76-35322.
587.Lukes, Z., "A general equivalent circuit of the insulated-gate field-effect transistors", Elektrotech. Casopis, 20, 413-29 (1969), Czech, EEA73-7471.
588.Lunsmann, P. D., "Equivalent circuit for the conductance of an MOSFET", Electron. Lett., 4, 100-1 (1968), EEA71-23371.
589.Madland, G. R., "MOS integrated circuits: the designer's dilemma", Western Electronic Show and Convention, WESCON, 1969, 14 pp., EEA73-23882.
590.Maeda, H., "Transport phenomena in Si MOS inversion layers", Solid State Phys., 6, 639-47 (1971), Japanese, EEA75-10191.
591.Maeda, H., "Electron conductivity-mass anisotropy in silicon MOS inversion layers", J. Phys. Soc. Jap., 32, 575 (1972), EEA75-20217.
592.Mansour, I. R. M., Hawkins, R. J. and Bloodworth, G. G., "Physical model for the current noise spectrum of MOST's", Brit. J. Appl. Phys. (J. Phys. D), 2, 1063-82 (1969), EEA72-20427.
593.Mansour, I. R. M., Talkham, E. A. and Barboor, A. I., "Investigations on the effect of drift-field-dependent mobility of MOST characteristics. I. Q_b constant", IEEE Trans. Electron Devices, ED-19, 899-907 (1972), EEA75-28510.
594.Marciniak, W. and Przewlocki, H. M., "Equilibrium distribution of the uncompensated mobile charge in the dielectric layer of a MOS structure", Phys. Status Solidi A, 24, 359-66 (1974), PA77-67282.
595.Margalit, S., Neugroshel, A. and Bar-Lev, A., "Redistribution of boron in silicon after two oxidation steps used in MOST fabrication", Proceedings of the 7th Convention of Electrical and Electronics Engineers in Israel, 1971, p. 142-54, EEA74-82981.
596.Marinov, J. P. and Kucarov, S. I., "Equivalent circuits and y-parameters of MOS-transistors in discrete embodiment", Elektro Prom.-st. & Proborostr., 7 (2), 45-7 (1972), Bulgarian, EEA75-24354.
597.Martinot, H. and Rossel, P., "Carrier multiplication in the pinchoff region of m.o.s. transistors", Electron. Lett., 7, 118-20 (1971), French, EEA74-15564.
598.Martinot, H., Rossel, P. and Salviac, M., "MOS transistor inverters", Onde Elec., 50, 790-803 (1970), French, EEA74-4079.
599.Masuhara, T., Etoh, J. and Nagata, M., "MOS-FET model including low-current region near threshold", Electron. & Commun. Jap., 56 (8), 125-32 (1973), EEA78-19821.
600.Masuhara, T., Etoh, J. and Nagata, M., "A precise MOSFET model for low-voltage circuits", IEEE Trans. Electron Devices, ED-21, 363-71 (1974), EEA77-26471.

601.Matsumoto, Y. and Uemura, Y., "Scattering mechanism and low temperature mobility of MOS inversion layers", Jap. J. Appl. Phys., 13, Suppl. 2 (Pt. 2), 367-70 (1974), EEA78-19724.
602.Merckel, G., Borel, J. and MacKowiak, E., "Equivalent diagram of the MOS transistor for use in the IMAG I programme", Onde Elec., 49, 84-9 (1969), French, EEA72-7930.
603.Meyer, J. E., "MOS models and circuit simulation", RCA Rev., 32, 42-63 (1971), EEA74-22739.
604.Mitterer, R. and Essl, D., "A program for optimizing digital static MOS-circuits", Elektron. Rechenanlagen, 13 (2), 60-5 (1971), German, EEA74-17298.
605.Mochkin, V. S., "The static characteristics of MOS transistors with an imbedded channel", Radio Eng. & Electron. Phys., 18, 1087-92 (1974), EEA78-650.
606.Mohsen, A. M. and Morris, F. J., "Measurements on depletion-mode field effect transistors and buried channel MOS capacitors for the characterization of bulk transfer charge-coupled devices", Solid-State Electron., 18, 407-16 (1975), EEA78-19850.
607.Molchanov, A. A. and Khodosh, L. S., "Effect of mobility versus transfer field dependence on characteristics of M.O.S.-transistors", Izv. VUZ Radioelectron., 13, 892-3 (1970), Russian, EEA74-1608.
608.Negro, V. C. and Goldstein, L. H., "An analytical expression for MOSFET gate leakage current", Proc. IEEE, 61, 1509-10 (1973), EEA76-38278.
609.Nishida, S., "Advances and problems in MOS devices", JEE, no. 76, 54-6 (1973), EEA76-29272.
610.Owens, A. R. and Perry, M. A., "Temperature coefficient of drift of MOS transistors", Electron. Lett., 2, 309-10 (1966), EEA70-9108.
611.Pacak, M., "Calculating the gain of stages using MOSFET transistors", Slaboproudy Obzor, 30 (3), 110-12 (1969), Czech, EEA72-13495.
612.Panigrahi, G., "Numerical calculation of low-frequency capacitance voltage curves of m.o.s. capacitors with nonconstant profiles", Electron. Lett., 9, 43-4 (1973), EEA76-12637.
613.Pao, H. C. and Sah, C. T., "Effects of diffusion current on characteristics of metal-oxide (insulator)-semiconductor transistors", Solid-State Electron., 9, 927-37 (1966), PA70-1980.
614.Parris, S. R., "MOS IC's: answers to applicational problems", Western Electronic Show and Convention, WESCON, 1969, 8 pp., EEA73-23883.
615.Pershenkov, V. S., Stepanenko, I. P. and Yavorskaya, R. Y., "Differential parameters of MOS-transistors in the region of saturation", Izv. VUZ Radioelektron., 14, 337-43 (1971), Russian, EEA74-18795.
616.Peters, D. W., "Turn-on delay time of MOS transistors", Proc. IEEE, 56, 89-90 (1968), EEA71-8276.

617.Pfeifer, R. F. and Sah, C. T., "A distortion analysis of the MOS transistor", Semiconductor Prod. Solid State Technol., 9 (11), 39-44 (1966), EEA70-2660.

618.Pierret, R. F. and Sah, C. T., "An MOS-oriented investigation of effective mobility theory", Solid-State Electron., 11, 279-90 (1968), PA71-22433.

619.Pierret, R. F. and Sah, C. T., "Quantitative analysis of the effects of steady-state illumination on the MOS-capacitor. I. Theory", Solid-State Electron., 13, 269-88 (1970), EEA73-17671.

620.Pollard, I. J. and Rietch, J. C., "Transmodulation characteristics of a high-frequency MOS tetrode", Toute Electron., no. 352, 19-22 (1971), French, EEA74-26890.

621.Poon, H. C., "Development of a compact process-oriented DC model for short channel IGFETs", 8th Asilomar Conference on Circuits, Systems and Computers, 1974, p. 508-12, Publ. 1975, EEA78-24369.

622.Popa, A., "An injection level dependent theory of the MOS transistor in saturation", IEEE Trans. Electron Devices, ED-19, 774-81 (1972), EEA75-20210.

623.Popov, V. P., Pertsev, A. V., Soldatenko, V. A. and Biryukov, V. N., "Calculation of the switching characteristics of channel transistors with an MOS structure", Radio Eng. & Electron. Phys., 16, 667-73 (1971), EEA75-13768.

624.Powell, M. W. and Price, J. B., "Design and performance of MOSFET circuits using silicon on insulating substrates", 1973 IEEE International Convention and Exposition, Vol. VI, 23.4/3 pp., EEA77-2612.

625.Przewlocki, H., "Investigation of the influence of technology on Si-SiO_2 system and MOS structure electrical properties", Electron. Technol., 3 (3), 103-22 (1970), PA74-41147.

626.Przewlocki, H. M. and Marciniak, W., "Triangular voltage sweep method as a tool in studies of mobile charge in MOS structures", Phys. Status Solidi A, 29, 265-74 (1975), CA83-51951.

627.Rao, P. S., "The effect of the substrate upon the gate and drain noise parameters of MOSFET's", Solid-State Electron., 12, 549-55 (1969), EEA72-23834.

628.Rao, P. S., "A theoretical study of excess noise in MOSFET's", Thesis, Univ. Minnesota, Minneapolis, 1970, 173 pp., Order No. 70-27157, EEA74-26894.

629.Reynolds, F. H., "Simple mathematical model of shift of threshold voltage induced in an m.o.s. transistor by testing at elevated temperatures", Proc. Inst. Electr. Eng., 119, 1683-6 (1972), EEA76-4666.

630.Richman, P., "Theoretical threshold voltages for MOS field effect transistors", Solid-State Electron., 11, 869-76 (1968), EEA71-26024.

631.Rideout, V. L., Gaensslen, F. H. and LeBlanc, A., "Device design considerations for ion implanted MOSFETs", 1973 International Electron Devices Meeting Technical Digest, p. 148-51, EEA77-14549.

632. Roberts, R. and Hamilton, A., "Analysis of MOSFET", Honeywell Comput. J., 4 (3), 32-7 (1970), EEA74-18793.
633. Rogovoskaya, E. T., "Influence of size quantization on the conduction in a metal-oxide-semiconductor system", Sov. Phys.-Semicond., 7, 811-12 (1973), PA77-28503.
634. Rossel, P., "Properties of the real part of the input admittance of a MOS transistor with zero drain-source voltage", C. R. Hebd. Seances Acad. Sci. B, 277, 447-9 (1973), French, EEA77-2564.
635. Rossel, P. and Martinot, H., "Model of a metal-oxide-silicon transistor operating in the saturation region", Electron. Lett., 9, 414-6 (1973), French, EEA77-6027.
636. Ruch, J. G., "Electron dynamics in short channel field effect transistors", IEEE Trans. Electron Devices, ED-19, 652-4 (1972), EEA75-16746.
637. Sah, C. T. and Fu, H. S., "Current and capacitance transient response of MOS capacitor. I. General theory and applications to initially depleted surface without surface states", Phys. Status Solidi A, 11, 297-310 (1972), EEA75-24284.
638. Sah, C. T. and Fu, H, S., "Current and capacitance transient responses of MOS capacitor. II. Recombination centers in the surface space charge layer", Phys. Status Solidi A, 14, 59-70 (1972), EEA76-4583.
639. Sah, C. T., Pierret, R. F. and Tole, A. B., "Exact analytical solution of high frequency lossless MOS capacitance-voltage characteristics and validity of charge analysis", Solid-State Electron., 12, 681-8 (1969), PA73-8568.
640. Sah, C. T., Tole, A. B. and Pierret, R. F., "Error analysis of surface density determination using the MOS capacitance method", Solid-State Electron., 12, 689-709 (1969), PA73-8569.
641. Sauvage, J. A. and Evans, S. A., "Modeling of depletion load devices made by ion implantation", 1973 International Electron Devices Meeting Technical Digest, p. 61-3, EEA77-14538.
642. Schroder, D. K. and Nathanson, H. C., "On the separation of bulk and surface components of lifetime using the pulsed MOS capacitor", Solid-State Electron., 13, 577-82 (1970), EEA73-24851.
643. Schroeder, J. E. and Muller, R. S., "IGFET analysis through numerical solution of Poisson's equation", IEEE Trans. Electron Devices, ED-15, 954-61 (1968), EEA72-5035.
644. Sequin, C. and Baldinger, E., "Interface states in metal-oxide-semiconductor field-effect transistors", Solid-State Electron., 13, 1527-40 (1970), EEA74-7786.
645. Shannon, J. M., "Simple expressions for the effect of substrate doping on the gain of MOST's", Electron. Lett., 5, 181-2 (1969), PA72-28543.
646. Shichman, H. and Hodges, D. A., "Modeling and simulation of insulated-gate field-effect transistor switching circuits", IEEE J. Solid-State Circuits, SC-3 (3), 285-9 (1968), EEA72-2839.
647. Shilin, V. A., "A high-frequency model of an MOS-transistor", Izv. VUZ Radioelektron., 16, 43-8 (1973), Russian, EEA76-26095.

648.Simmons, J. G. and Taylor, G. W., "Theory of non-steady-state interfacial thermal currents in MOS devices, and the direct determination of interfacial trap parameters", Solid-State Electron., 17, 125-30 (1974), EEA77-14437.
649.Simmons, J. G. and Wei, L. S., "Theory of transient emission current in MOS devices and the direct determination interface trap parameters", Solid-State Electron., 17, 117-24 (1974), PA77-24835.
650.Smith, D. P., "An accurate model for the short channel insulated gate field-effect transistor", Thesis, Stanford Univ., Calif., 1971, 97 pp., Order No. 71-23560, EEA75-28525.
651.Staples, E. J., Jr., "Detection of Rayleigh surface waves on silicon with the MOS-FET", Thesis, Southern Methodist Univ., Dallas, Texas, 1971, 118 pp., Order No. 72-663, EEA75-28527.
652.Stepaenko, I. P. and Khudykina, V. M., "Analysis and design of single-pulse MOS-transistor stages, allowing for nonlinear distortions", Izv. VUZ Radioelektron., 16, 67-72 (1973), Russian, EEA76-25667.
653.Sugano, T., "Pulse response of MOS capacitor with guard ring", Annu. Rep. Eng. Res. Inst. Fac. Eng. Univ. Tokyo, 33, 127-34 (1974), Japanese, EEA78-19698.
654.Sugano, T. and Iizuka, T., "Transient response of MOS capacitance to step function voltage", Annu. Rep. Eng. Res. Inst. Fac. Eng. Univ. Tokyo, 32, 151-6 (1973), Japanese, EEA77-23036.
655.Szanto, L., "MOS-transistors model parameter calculation by the aid of optimization strategies", Elektrotech. Cas., 25, 636-45 (1974), Slovak, EEA78-2869.
656.Szanto, L., "A MOS transistor model related to field dependence of carrier mobility and channel length variation", TESLA Electron., 8, 20-5 (1975), EEA78-28814.
657.Talkhan, E. A., Mansour, I. R. M. and Barboor, A. I., "Investigations on the effect of drift-field-dependent mobility on MOST characteristics. II. Q_b field dependent", IEEE Trans. Electron Devices, ED-19, 908-16 (1972), EEA75-28511.
658.Treleaven, D. H. and Trofimenkoff, F. N., "MOSFET equivalent circuit at pinch-off", Proc. IEEE, 54, 1223-4 (1966), EEA70-1069.
659.Troutman, R. R., "Subthreshold slope for insulated gate field-effect transistors", IEEE Trans. Electron Devices, ED-22, 1049-51 (1975).
660.Troutman, R. R. and Chakravarti, S. N., "Subthreshold characteristics of insulated-gate field-effect transistors", IEEE Trans. Circuit Theory, CT-20, 659-65 (1973), EEA76-37841.
661.Tutto, P., "The influence of inhomogeneous distribution of recombination centres on the inversion behaviour of MOS capacitors", Phys. Status Solidi A, 21, 693-702 (1974), German, PA77-28501.
662.van der Ziel, A. and Nabbaum, A., "Normalized characteristic of saturated high-field MOSFET's", Solid-State Electron., 17, 413-14 (1974), EEA77-23062.

663.van Nielen, J. A. and Memelink, O. W., "The influence of the substrate upon the DC characteristics of silicon MOS transistors", Philips Res. Rep., 22, 55-71 (1967), EEA70-9107.

664.van Overstraeten, R. J., Declerck, G. and Broux, G. L., "Inadequacy of the classical theory of the MOS transistor operating in weak inversion", IEEE Trans. Electron Devices, ED-20, 1150-3 (1973), EEA77-6031.

665.Van Overstraeten, R. J., Declerck, G. J. and Muls, P. A., "Theory of the MOS transistor in weak inversion - new method to determine the number of surface states", IEEE Trans. Electron Devices, ED-22, 282-8 (1975), EEA78-19827.

666.Van Overstraeten, R. J., De Man, H. J. and Mertens, R. P., "Transport equations in heavy doped silicon", IEEE Trans. Electron Devices, ED-20, 290-8 (1973), EEA76-16212.

667.Varshney, R. C., "Simple theory for threshold-voltage modulation in short-channel m.o.s. transistors", Electron. Lett., 9, 600-2 (1973), EEA77-10256.

668.Verwey, J. F., Kramer, R. P. and de Maagt, B. J., "Mean free path of hot electrons at the surface of boron-doped silicon", J. Appl. Phys., 46, 2612-19 (1975), PA78-66035.

669.Vlasov, V. A. and Popov, P. I., "A thermal stability criterion for semiconductor devices", Izv. VUZ Radioelektron., 18, 124-7 (1975), Russian, EEA78-19539.

670.Wienhold, H., "Relaxation effects in epitaxial Si-MOS-structures", Phys. Status Solidi, 30, K17-21 (1968), German, PA72-35458.

671.White, M. H., Blaha, F. C. and Herman, D. S., "A study relating m.o.s. processes to a model of the Al-SiO_2-Si system", Symposium on Silicon Device Processing (programme and abstracts), 1970, p. 54-5, EEA73-35177.

672.Wu, S. H. and Anderson, R. L., "MOSFET's in the OK approximation. Static characteristics of MOSFET's in the OK approximation", Solid-State Electron., 17, 1125-37 (1974), EEA78-654.

673.Yamada, E., "Theory of hot electrons and saturation velocities in silicon inversion layers", Oyo Buturi, 43 (Suppl.), 323-6 (1974), PA78-6905.

674.Yau, L. D., "A simple theory to predict the threshold voltage of short-channel IGFET's", Solid-State Electron., 17, 1059-63 (1974), EEA78-652.

675.Yau, L. D., "Simple I/V model for short-channel I.G.F.E.T.s in the triode region", Electron. Lett., 11, 44-5 (1975), EEA78-8566.

676.Young, T. K. and Dutton, R. W., "Extraction and sensitivity of parameters for higher order MOS models", 8th Asilomar Conference on Circuits, Systems and Computers, 1974, p. 518-23, Publ. 1975, EEA78-24371.

677.Zahn, M. E., "Calculation of the turn-on behavior of MOST", Solid-State Electron., 17, 843-54 (1974), EEA77-30344.

678.Zechnall, W., "A model of the capacitance and interface charge buildup in MOS capacitors", Presented at Electrochemical Society Meeting, San Francisco, Calif., May 1974, EEA77-22941.

d. Computer Analysis

679.Barron, M. B., "Computer aided analysis of insulated gate field-effect transistors", Thesis, Stanford Univ., Calif., 1970, 200 pp., Order No. 70-18375, EEA74-22743.
680.Borel, J. and Merckel, G., "A model for MOS transistor utilisation with IMAG II program", Presented at Proceedings of the International Conference on Advanced Microelectronics, Paris, France, April 1970, EEA74-1618.
681.Calzolari, P. U., Graffi, S. and Pierini, G., "An accurate computer analysis of relaxation effects in MOS capacitors", Alta Freq., 43, 396-406 (1974), PA78-6897.
682.Curran, L., "Computers make a big difference in MOS designs", Electronics, 42 (21), 82-92 (1969), EEA73-6486.
683.Dellow, F., "MOST - characteristic simulation using an analogue computer", Des. Electron., 9, 21-3 (1972), EEA75-13749.
684.Drenick, P. E., "Computer-animated algorithms (m.o.s.t. model)", IEEE Trans. Educ., E-17, 130-2 (1974), EEA77-17743.
685.Goser, K., "Channel formation in an i.g.f.e.t. and its equivalent circuit for CAD", 1970 IEEE International Solid State Circuits Conference, p. 98-9, EEA73-25053.
686.Hamilton, A. W., "A new model of the insulated gate field effect transistor for use in computer aided circuit analysis", Proceedings of the 15th Midwest Symposium on Circuit Theory, Vol. I, 1972, v.5/9 pp., EEA76-8705.
687.Jenkins, F. S., Lane, E. R., Lattin, W. W. and Richardson, W. S., "MOS-device modeling for computer implementation", IEEE Trans. Circuit Theory, CT-20, 649-58 (1973), EEA76-37840.
688.Liblich, S. and Nassibian, A. G., "A theory for computer evaluation and comparison of experimental CV curves of MOS capacitors", Solid-State Electron., 17, 636-7 (1974), PA77-49085.
689.McGuffin, R. W., McG.Luke, D. and Ryan, W. D., "Simple charge-controlled transient model of the m.o.s. transistor for computer-aided design and analysis programs", Electron. Lett., 5, 638-40 (1969), EEA73-10677.
690.Merckel, G., Borel, J. and Cupcea, N. Z., "An accurate large-signal m.o.s. transistor for use in computer-aided design", IEEE Trans. Electron Devices, ED-19 (5), 681-90 (1972), EEA75-15775.
691.Rehn, A., "Computer-aided analysis of characteristic values illustrated by a field-effect transistor model", Nachrichtentech. Elektron., 24, 97-101 (1974), German, EEA77-18813.
692.Reiser, M., "Computer-aided design of field-effect transistors", Eurocon 71 Digest, 2 pp., EEA75-4720.

693.Schweikert, D. G., "Computer-generated IGFET layout using a vertically-packed Weinberger arrangement", Digest of Technical Papers of the 1971 International Solid State Circuits Conference, p. 118-19, EEA74-21590.
694.Smith, D. P. and Linvill, J. G., "An accurate short-channel IGFET model for computer-aided circuit design", Digest of Technical Papers of the 1971 International Solid State Circuits Conference, p. 40-1, EEA74-22753.

4. MNOSFET

695.Arnett, P. C., "Spatial charge distribution in MNOS modeling", In: Semiconductor silicon 1973, Electrochem. Soc., Inc., Princeton, N. J., p. 803-13 (1973).
696.Brewer, J. E. and Hadden, D. R., Jr., "Block-oriented random access MNOS memory", AFIPS Natl. Comput. Conf., Expo Conf. Proc., Vol. 43, 1974, p. 837-40.
697.Ferris-Prabhu, A. V., "Maximum tunneling distance in MNOS devices. Theory", Phys. Status Solidi A, 11, 81-6 (1972), EEA75-24281.
698.Ferris-Prabhu, A. V., "Theory of MNOS memory device behavior", IBM J. Res. & Dev., 17, 125-34 (1973), EEA76-35325.
699.Ginovker, A. S., Gritsenko, V. A. and Sinitsa, S. P., "Two-band conduction of amorphous silicon nitride", Phys. Status Solidi A, 26, 489-95 (1974).
700.Svensson, C. and Lundstrom, I., "Theory of the thin-oxide M.N.O.S. memory transistor", Electron. Lett., 6, 645-7 (1970), EEA73-38096.
701.Uchida, Y., "A two dimensional analysis of the write operation of MNOS transistors using avalanche-tunnel injection", Presented at 1973 International Electron Devices Meeting Supplement, Washington, D. C., December 1973, EEA77-14566.
702.Zirinsky, S., "Charge transfer properties of MNOS structures as influenced by processing parameters", J. Electron. Mater., 4, 591-624 (1975).

5. Schottky-Gate FET

703.Alley, G. D., Talley, H. E. and Wright, G. L., "Two dimensional distributed theory for a microwave Schottky barrier field effect transistor", 1973 IEEE G-MTT International Microwave Symposium, Digest of Technical Papers, p. 233-5, EEA77-6045.
704.Asai, S., Murai, F. and Kodera, H., "GaAs dual-gate Schottky-barrier FET's for microwave frequencies", IEEE Trans. Electron Devices, ED-22, 897-904 (1975), EEA78-41607.
705.Baechtold, W., "Noise behaviour of GaAs field-effect transistors with short gate lengths", IEEE Trans. Electron Devices, ED-19, 674-80 (1972), EEA75-17025.

706.Barnes, J. J. and Lomax, R. J., "Two-dimensional finite element simulation of semiconductor devices", Electron. Lett., 10, 341-3 (1974), EEA77-34050.
707.Barnes, J. J. and Lomax, R. J., "Transient 2-dimensional simulation of a submicrometer gate-length MESFET", Electron. Lett., 11, 519-21 (1975).
708.Dawson, R. H., "Equivalent circuit of the Schottky-barrier field-effect transistor at microwave frequencies", IEEE Trans. Microwave Theory & Tech., MTT-23, 499-501 (1975), EEA78-24352.
709.Iversen, S., "Comments on 'Effect of source lead inductance on the noise figure of a GaAs FET'", Proc. IEEE, 63, 983-4 (1975).
710.Jahncke, J., "The microwave properties of GaAs MESFET's in strip-line technique", Nachrichtentech. Z. (NTZ), 26, 193-9 (1973), German, EEA76-29273.
711.Jutzi, W. and Reiser, M., "Threshold voltages of normally off MESFET's", IEEE Trans. Electron Devices, ED-19, 314-22 (1972), EEA75-10268.
712.Reiser, M., "Large-scale numerical simulation in semiconductor device modelling", Comput. Methods Appl. Mech. Eng., 1, 17-38 (1972), EEA75-28531.
713.Vendelin, G. D. and Omori, M., "Circuit model for the GaAs m.e.s.f.e.t. valid to 12 GHz", Electron Lett., 11, 60-1 (1975), EEA78-15346.

6. MISFET

714.Balland, B., Boutard, M. and Pinard, P., "On the parallel conductance of MIS structures", Thin Solid Films, 17, 139-43 (1973), French, EEA76-35266.
715.Braunig, D. and Wagemann, H. G., "Experiments and model of non-equilibrium behavior of MIS varactors using the linear ramp technique", IEEE Trans. Electron Devices, ED-21, 241-7 (1974), EEA77-14524.
716.Brumme, R., "The model of the MIS transistor", Nachrichtentech. Elektron., 23, 252-3 (1973), German, EEA76-35328.
717.Chang, W. H., "MIS array potential calculation", Solid-State Electron., 16, 491-6 (1973), EEA76-16337.
718.Chang, W. H. and Lee, H. S., "Potential distribution of an inhomogeneously doped MIS array", IBM J. Res. & Dev., 18, 47-52 (1974), PA77-41106.
719.De Gryse, R., Renders, L. and Vennik, J., "A solution for the high-frequency C versus V curve of MIS structures", Phys. Status Solidi A, 20, 247-52 (1973), EEA77-2518.
720.De Gryse, R., Valeminck, R. and Vennik, J., "The evaluation of interface parameters in MIS structures. I. Theory", Solid-State Electron., 18, 151-9 (1975), PA78-15733.
721.Esteve, D. and Simonne, J. J., "MIS capacitance in heavy inversion", Electron. Lett., 5, 194-5 (1969), PA72-28542.

722. Ferris-Prabhu, A. V., "Time dependence of charge transport in MIS memory transistors", Appl. Phys. Lett., 20, 149-50 (1972), EEA75-13747.
723. Forlani, F., Ninnanja, N. and Pagiola, E., "Theory of the influence of surface states on the impedance of a semiconductor-insulator interface", Phys. Status Solidi, 17, 657-72 (1966), EEA70-2635.
724. Fquiluz, A., Lee, T. K., Quinn, J. J. and Chiu, K. W., "Interface excitations in metal-insulator-semiconductor structures", Phys. Rev. B, 11, 4989-93 (1975), PA78-77193.
725. Gad, H., "A new approach to the approximation of I-V characteristics of MISFET's", Nachrichtentech. Z. (NTZ), 27, 352-5 (1974), German, EEA77-38849.
726. Gad, H., "Harmonic distortions of the drain current in MISFET's with oscillation circuit loads", Arch. Elektron. & Uebertragungstech., 28, 471-7 (1974), German, EEA78-2866.
727. Guzev, A. A., Kurishev, G. L. and Sinitsa, S. P., "Scattering mechanisms in inversion channels of MIS structures on silicon", Phys. Status Solidi A, 14, 41-50 (1972), EEA76-4582.
728. Junghans, B., "A model for insulator charge in MIS structures", Phys. Status Solidi A, 2, 731-8 (1970), German, EEA74-1508.
729. Longo, H. E., "Current-multiplication of MIS transistor", Z. Angew. Phys., 29, 166-8 (1970), German, EEA73-28218.
730. Longo, H. E., "Drainfield and source to drain resistance of MIS transistor near saturation", Z. Angew. Phys., 28, 333-6 (1970), German, EEA73-21337.
731. Michael, M. and Leonard, W. F., "MIS capacitance and derivative of capacitance, with application to nonparabolic band semiconductors", Solid-State Electron., 17, 71-85 (1974), PA77-16832.
732. Poksheva, J. G., "Analysis and modelling of soft breakdown and related 'kink' phenomena in MIS/SOS transistors", 8th Asilomar Conference on Circuits, Systems and Computers, 1974, p. 513-17, Publ. 1975, EEA78-24370.
733. Reimer, H., "On the problem of determining and separating the influence of surface states and volume states on the impedance of MIS structures", Electron. Technol., 3 (4), 61-9 (1970), EEA74-37302.
734. Rosinski, A. and Ryzko, J., "Physical interpretation of measurements of the m.i.s. structures G-U characteristics", Arch. Elektrotech., 20, 817-21 (1971), Polish, EEA75-10258.
735. Ross, E. C. and Wallmark, J. T., "Theory of the switching behavior of MIS memory transistors", RCA Rev., 30, 366-81 (1969), EEA72-23831.
736. Schroder, J. and Seidel, M., "Circuit variations of dynamically operated MIS structures", Wiss. Z. Tech. Univ. Dres., 21, 535-7 (1972), German, EEA76-4671.
737. Simmons, J. G. and Wei, L. S., "Theory of dynamic charge and capacitance characteristics in MIS systems containing discrete surface traps", Solid-State Electron., 16, 43-52 (1973), EEA76-8697.

738.Simmons, J. G. and Wei, L. S., "Theory of dynamic charge current and capacitance characteristics in MIS systems containing distributed surface traps", Solid-State Electron., 16, 53-66 (1973), EEA76-4667.

739.Sinonne, J. J., "A method to extract interface state parameters from the MIS parallel conductance technique", Solid-State Electron., 16, 121-4 (1973), EEA76-8601.

740.Temple, V. A. K. and Shewchun, J., "An investigation of the effect of two-band model of the barrier on the tunneling characteristics of degenerate MIS diodes", Solid-State Electron., 17, 417-26 (1974), PA77-45007.

741.Thornber, K. K., Kahng, D. and Neppell, C. T., "Bias-temperature-stress studies of charge retention in dual-dielectric charge-storage cells", Bell Syst. Tech. J., 53, 1741-70 (1974), EEA78-8565.

742.Trullemans, C. and Van de Wiele, F., "Lateral carrier flow along m.i.s. structures", European 1971 Digest, 2 pp., EEA75-4647.

743.Ushirokawa, A., Warashina, M. and Nagami, A., "2-dimensional parameter conductance method for estimation of interface states in MIS structures", Jap. J. Appl. Phys., 12, 388-97 (1973), EEA76-19865.

X. Charge-Coupled Devices

744.Berglund, C. N., "Analog performance limitations of charge-transfer dynamic shift registers", IEEE J. Solid-State Circuits, SC-6, 391-4 (1971), EEA75-360.

745.Berglund, C. N. and Thornber, K. K., "A fundamental comparison of incomplete charge transfer in charge transfer devices", Bell Syst. Tech. J., 52, 147-82 (1973), EEA76-19924.

746.Berglund, C. N. and Thronber, K. K., "Incomplete transfer in charge-transfer devices", IEEE J. Solid-State Circuits, SC-8, 108-16 (1973), EEA76-16436.

747.Brotherton, S. D., "A theoretical analysis of c.c.d. operation with square clock pulses", Solid-State Electron., 17, 341-8 (1974), EEA77-23061.

748.Buss, D. D., Holmes, J. D. and Rosser, L. A., "Investigation of semiconductor charge devices", Report AD-772637/5, Texas Instruments, Inc., Dallas, Texas, 1973, 167 pp..

749.Carnes, J. E. and Kosonocky, W. F., "Fast-interface-state losses in charge-coupled devices", Appl. Phys. Lett., 20, 261-3 (1972), EEA75-16742.

750.Carnes, J. E., Kosonocky, W. F. and Ramberg, E. G., "Free charge transfer in charge-coupled devices", IEEE Trans. Electron Devices, ED-19, 798-808 (1972), EEA75-20211.

751.Chamberlain, S. G., Copeland, M. A. and Ibrahim, A. A., "Some general experimental studies on charge-coupled device circuits", Int. J. Electron., 35, 833-46 (1973), EEA77-2177.

752.Chan, C. H. and Chamberlain, S. G., "Numerical methods for the charge transfer analysis of charge-coupled devices", Solid-State Electron., 17, 491-9 (1974), EEA77-23063.
753.Chan, C. H. and Chamberlain, S. G., "Charge transfer analysis in two-phase step oxide charge coupled devices", International Conference on Technology and Applications of Charge Coupled Devices, 1974, p. 29-37, EEA78-11676.
754.Chang, W. H. and Heller, L. G., "Structure dependence of free-charge transfer in charge-coupled devices", IBM J. Res. & Dev., 18, 436-42 (1974), EEA78-8567.
755.Chang, W. H. and Lee, H. S., "Potential distribution of an inhomogeneously doped MIS array", IBM J. Res. & Dev., 18, 47-52 (1974), EEA77-23046.
756.Collet, M. G. and Viegenthart, A. C., "Calculations on potential and charge distributions in the peristaltic charge-coupled device", Philips Res. Rep., 29, 25-44 (1974), EEA77-26484.
757.Copeland, M. A., "A large-signal model for charge coupled devices", 1971 International Electrical and Electronics Conference, Conference digest, p. 22-3, EEA75-10315.
758.Daimon, Y., McGill, T. C. and Mohsen, A. M., "Numerical and analytical study of CCD operation", International Conference on Technology and Applications of Charge Coupled Devices, 1974, p. 38, EEA78-11677.
759.Daimon, Y., Mohsen, A. M. and McGill, T. C., "Final stage of the charge-transfer process in charge-coupled devices", IEEE Trans. Electron Devices, ED-21, 266-72 (1974), EEA77-14526.
760.Daimon, Y., Mohsen, A. M. and McGill, T. C., "Charge transfer in buried-channel charge-coupled devices", 1974 ISSCC Dig. of Tech. Pap., p. 146-7, 245.
761.El-Sissi, H. and Cobbold, R. S. C., "Buried-channel C.C.D. models based on a 1-dimensional analysis", Electron. Lett., 10, 198-9 (1974), EEA77-26465.
762.El-Sissi, H. and Cobbold, R. S. C., "One-dimensional study of buried-channel charge-coupled devices", IEEE Trans. Electron Devices, ED-21, 437-47 (1974), EEA77-30332.
763.El-Sissi, H. and Cobbold, R. S. C., "Potentials and fields in buried-channel CCD's: a two-dimensional analysis and design study", IEEE Trans. Electron Devices, ED-22, 77-90 (1975), EEA78-11652.
764.El-Sissi, H. and Cobbold, R. S. C., "On the decay of small amounts of signal charge in surface-channel CCD's", IEEE Trans. Electron Devices, ED-22, 154-6 (1975), EEA78-11653.
765.El-Sissi, H. and Cobbold, R. S. C., "Influence of signal charge on the final decay characteristics of surface-channel charge-coupled-devices", Solid-State Electron., 18, 867-70 (1975), EEA78-37548.

766.Epifanov, A. A., Frolov, O. S., Krolevets, K. M. and Derkach, Y. P., "Numerical analysis of fringing fields in charge-coupled devices", Poluprovodn. Tekh. & Mikroelektron., no. 18, 42-9 (1974), Russian, EEA78-11665.
767.Fawcett, W. and Vanstone, G. F., "A two-dimensional dynamic particle model of the surface CCD transport process", International Conference on Technology and Applications of Charge Coupled Devices, 1974, p. 22-8, EEA78-11675.
768.Gunsager, K. C., Kim, C. K. and Phillips, J. D., "Performance and operation of buried channel charge coupled devices", 1973 IEDM Technical Digest, p. 21-3, EEA77-14536.
769.Hanneman, H. W. and Esser, L. J. M., "Field and potential distributions in charge-transfer devices", Philips Res. Rep., 30, 56-72 (1975), EEA78-28807.
770.Hara, H., "A theoretical analysis on fundamental performances of charge-coupled devices", Electron. & Commun. Jap., 54, 133-40 (1971), EEA75-32678.
771.Heller, L. G., Chang, W. H. and Lo, A. W., "A model of charge transfer in bucket brigade and charge-coupled devices", IBM J. Res. & Develop., 16, 184-7 (1972), EEA75-28516.
772.Heller, L. G. and Lee, H. S., "Charge-transfer device modeling (shift register)", J. Electrochem. Soc., 120 (3), 100C (1973), EEA76-23457.
773.Jayadevaiah, T. S. and Laur, J., "Simple model for charge-coupled devices", Electron. Lett., 7, 751-3 (1971), EEA75-4726.
774.Joyce, W. B. and Bertram, W. J., "Linearized dispersion relation and Green's function for discrete-charge-transfer devices with incomplete transfer", Bell Syst. Tech. J., 50, 1741-59 (1971), EEA74-37316.
775.Kent, W. H., "Charge distribution in buried-channel charge-coupled devices", Bell Syst. Tech. J., 52, 1009-24 (1973), EEA76-35318.
776.Kim, C. K. and Lenzlinger, M., "Charge transfer in charge-coupled devices", J. Appl. Phys., 42, 3586-94 (1971), EEA74-34092.
777.Lamb, D. R., Singh, M. P., Roberts, P. C. T. and Brotherton, S. D., "Influence of surface states on the performance of three-phase charge-coupled devices", International Conference on Technology and Applications of Charge Coupled Devices, 1974, p. 59-66, EEA78-11679.
778.Lee, H. S. and Heller, L. G., "Charge-control method of charge-coupled device transfer analysis", IEEE Trans. Electron Devices, ED-19, 1270-9 (1972), EEA76-813.
779.Lees, A. W. and Ryan, W. D., "Simple model of a buried channel charge coupled devices", Solid-State Electron., 17, 1163-9 (1974), EEA78-655.
780.McKenna, J. and Schryer, N. L., "The potential in a charge coupled device with no mobile minority carriers and zero plate separation", Bell Syst. Tech. J., 52, 669-96 (1973), EEA76-26087.

781. McKenna, J. and Schryer, N. L., "The potential in a charge-coupled device with no mobile minority carriers", Bell Syst. Tech. J., 52, 1765-93 (1973), EEA77-10251.
782. McKenna, J. and Schryer, N. L., "Analysis of field-aided, charge-coupled device transfer", Bell Syst. Tech. J., 54, 667-85 (1975), EEA78-28787.
783. McKenna, J., Schryer, N. L. and Walden, R. H., "Design considerations for a two-phase, buried-channel, charge-coupled device", Bell Syst. Tech. J., 53, 1581-97 (1974), EEA78-5326.
784. Melen, R. D. and Meindl, J. D., "A transparent-electrode CCD image sensor", Digest of Technical Papers of the 1973 IEEE International Solid State Circuits Conference, p. 130-1, EEA76-23712.
785. Mohsen, A. M., Bower, R., McGill, T. C. and Zimmerman, T., "Overlapping-gate buried-channel charge-coupled devices", Electron. Lett., 9, 396-8 (1973), EEA77-6026.
786. Mohsen, A. M., McGill, T. C., Daimon, Y. and Mead, C. A., "The influence of interface states on incomplete charge transfer in overlapping gate charge-coupled devices", IEEE J. Solid-State Circuits, SC-8, 125-38 (1973), EEA76-16438.
787. Mohsen, A. M., McGill, T. C. and Mead, C. A., "Charge transfer in overlapping gate charge-coupled devices", IEEE J. Solid-State Circuits, SC-8, 191-207 (1973), EEA76-26091.
788. Ong, D. G., "Characterisation and modelling of carrier generation in charge-coupled devices operated at low frequency", Thesis, Purdue Univ., Lafayette, Ind., 1974, 155 pp., Order No. 75-10934, EEA78-41630.
789. Ong, D. G. and Pierret, R. F., "Approximate formula for surface carrier concentration in charge-coupled devices", Electron. Lett., 10, 6-7 (1974), EEA77-10258.
790. Ong, D. G. and Pierret, R. F., "Thermal carrier generation in charge-coupled devices", IEEE Trans. Electron Devices, ED-22, 593-602 (1975), EEA78-28297.
791. Reschke, D., "Possibilities of calculating the charge transport in charge-coupled (CCD) components using a hybrid computer system", Nachrichtentech. Elektron., 23 (3), 89-92 (1973), German, EEA76-19933.
792. Singh, M. P., Roberts, P. C. T. and Lamb, D. R., "Operation of three-phase charge coupled devices with two-level clocking waveforms", Int. J. Electron., 38, 631-9 (1975), EEA78-28800.
793. Strain, R. J., "Properties of an idealized traveling-wave charge-coupled device", IEEE Trans. Electron Devices, ED-19, 1119-80 (1972), EEA75-36289.
794. Strain, R. J. and Schryer, N. L., "A nonlinear diffusion analysis of charge-coupled-device transfer", Bell Syst. Tech. J., 50, 1721-40 (1971), EEA74-37315.
795. Sugano, T. and Iizuka, T., "Effect of carrier velocity saturation on transfer characteristics of charge coupled devices", Annu. Rep. Eng. Res. Inst. Fac. Eng. Univ. Tokyo, 32, 147-50 (1973), Japanese, EEA77-23035.

796.Suzuki, N. and Yanai, H., "Computer analysis of surface-charge transport between transfer electrodes in a charge-coupled device", IEEE Trans. Electron Devices, ED-21, 73-83 (1974), EEA77-6033.
797.Tanigawa, H. and Ando, T., "A nonlinear CCD lumped model and its application to output stage performance analysis", IEEE Trans. Electron Devices, ED-22, 575-81 (1975), EEA78-28793.
798.Tanikawa, H., Shimohashi, A., Tokuhira, S. and Ohtsuki, O., "The operational mode in a buried-channel charge-coupled device", Jap. J. Appl. Phys., 14, 1629-30 (1975).
799.Thornber, K. K., "Operational limitations of charge transfer devices", Bell Syst. Tech. J., 52, 1453-82 (1973), EEA77-11256.
800.Thornber, K. K., "Error rates of digital signals in charge transfer devices", Bell Syst. Tech. J., 52, 1795-809 (1973), EEA77-11258.
801.Thornber, K. K., "Theory of noise in charge-transfer devices", Bell Syst. Tech. J., 53, 1211-62 (1974), EEA78-5054.
802.Tompsett, M. F., "The quantitative effects of interface states on the performance of charge-coupled devices", IEEE Trans. Electron Devices, ED-20, 45-55 (1973), EEA76-4659.
803.Vanstone, G. F., Roberts, J. B. G. and Long, A. E., "The measurement of the charge residual for CCD transfer using impulse and frequency responses", Solid-State Electron., 17, 889-95 (1974).
804.Wallinga, H. T. and Van Ruyven, H. L. M., "Diagnostic analysis of the charge transfer in CCDs", 1974 ISSCC Dig. of Tech. Pap., p. 148-49.
805.Wilson, E. A. and Tchon, W. E., "Calculation of transfer potentials in charge coupled devices with arbitrary voltage and charge boundary conditions using the finite element method", 1973 SWIEEECO Record of Technical Papers, p. 549-55, EEA76-26107.
806.Yanai, H. and Suzuki, N., "Analysis on surface-charge transfer in charge transfer devices", Annu. Rep. Eng. Res. Inst. Fac. Eng. Univ. Tokyo, 31, 151-6 (1972), Japanese, EEA76-8680.

XI. Ohmic Contacts

807.Akopyan, A. A. and Gribnikov, Z. S., "High-frequency electrical conductivity of ambipolar semiconductors with ohmic contacts", Sov. Phys.-Semicond., 8, 1078-84 (1975), PA78-46680.
808.Berger, H. H., "Models for contacts to planar devices", Solid-State Electron., 15, 145-58 (1972), EEA75-13654.
809.Frank, H., "Ohmic contacts on GaAs and GaP", Elektrotech. Cas., 25 (4-6), 294-303 (1974), Czech, PA77-60853.
810.Gossick, B. R., "Electrical characteristics of a metal semiconductor contact. II", Surface Sci., 25, 465-90 (1971), PA74-30852.
811.Gossick, B. R., "Electrical characteristics of a metal-semiconductor contact. III", Surface Sci., 28, 469-88 (1971), EEA75-7466.

812. Keenan, W. A., Schumann, P. A., Jr., Tong, A. H. and Phillips, R. P., "A model for the metal semiconductor contact in the spreading resistance probe", Ohmic Contacts to Semiconductors, 1968, p. 263-76, Publ. 1969, EEA73-10554.

813. Kellner, W., "Planar ohmic contacts to N-type GaAs: determination of contact parameters using the transmission line model", Siemens Forsch. - & Entwicklungsber., 4, 137-40 (1975), PA78-50668.

814. Kordos, P. and Hansak, L., "Knowledge on electric properties of ohmic contacts on GaAs at low temperatures", Elektrotech. Cas., 25, 304-11 (1974), Czech, PA77-60854.

815. Kuczogi, E., "The fritting of stationary contacts as a semiconductor phenomenon", Elektrotech. Z. (ETZ) A, 93, 465-7 (1972), German, EEA75-35590.

816. Murmann, H. and Widmann, D., "Current crowding on metal contacts to planar devices", IEEE Trans. Electron Devices, ED-16, 1022-4 (1969), EEA73-10526.

817. Pellegrini, B. and Salardi, G., "A model of ohmic contacts to semiconductors", Solid-State Electron., 18, 791-8 (1975).

818. Polyakov, N. N. and Konkov, V. L., "The derivation of the formulae for the spreading resistance of a plane contact of circular form", Izv. Vuz. Fiz., no. 9, 100-5 (1970), Russian, PA74-7655.

819. Strikha, V. I., Popova, G. D. and Buzaneva, Y. V., "Principles of manufacture of metal-to-semiconductor ohmic contacts", Poluprovodn. Tekh. & Mikroelektron., no. 19, 82-6 (1975), Russian, EEA78-28740.

820. Strikha, V. I., Popova, G. D. and Buzaneva, Y. V., "Physical principles of manufacture of metal/semiconductor ohmic contacts", Poluprovodn. Tekh. & Mikroelektron., no. 20, 20-35 (1975), Russian, EEA78-37469.

821. Tantraporn, W., "On the study of metal-semiconductor contacts", IEEE Trans. Electron Devices, ED-19, 331-8 (1972), EEA75-10178.

822. Vilms, J. and Wandinger, L., "Theory of contact resistance for one type of ohmic contact", Ohmic Contacts to Semiconductors, 1968, p. 31-47, Publ. 1969, EEA73-10546.

XII. Silicon Dioxide

823. Adachi, H. and Shibata, Y., "Asymmetric two-site hopping model for high-electric-field-excited thermally stimulated current in evaporated silicon oxide films", J. Phys. D, 8, 1120-32 (1975), EEA78-28593.

824. Allen, W. G. and Atkinson, C., "Comparison of models for redistribution of dopants in silicon during thermal oxidation", Solid-State Electron., 16, 1283-7 (1973), PA77-10553.

825. Card, H. C., "Population of interface states in devices with thin insulating layers", Presented at 1972 European Solid State Devices Research Conference, Lancaster, England, September 1972, EEA76-32218.

826. Chakrabarti, U. K., "Approximate solution for the redistribution of impurities during second oxidation", IEEE Trans. Electron Devices, ED-22, 566-8 (1975), EEA78-28792.
827. Eaton, D. H. and Sah, C. T., "Series equivalent circuit representation of SiO_2-Si interface and oxide trap states", Solid-State Electron., 16, 841-6 (1973), EEA76-29208.
828. Georgiev, V. K. and Dimov, V. I., "Properties of surface ions on SiO_2", C. R. Acad. Bulg. Sci., 25, 743-6 (1972), EEA75-39857.
829. Gray, K. E., "Electron transmission through ultra-thin insulating barriers", Solid State Commun., 13, 1787-9 (1973), EEA77-5936.
830. Nagaraj, B. and Suryan, G., "Electrical conduction through thin films of SiO_2 on SiC", Presented at the 1st National Sympsoium on Solid Thin Films, Bangalore, India, April 1971, EEA74-26759.
831. O'Dwyer, J. J., "Two-carrier model for high field conduction in SiO_2", J. Appl. Phys., 44, 5438-40 (1973), EEA77-18653.
832. Ruffa, A. R., "Models for electronic processes in SiO_2", J. Non-Cryst. Solids, 13, 37-54 (1973).
833. Sah, C. T., Ning, T. H. and Tschopp, L. L., "The scattering of electrons by surface oxide charges and by lattice vibrations at the silicon-silicon dioxide interface", Surface Sci., 32, 561-75 (1972), PA75-71317.
834. Sandor, J. E., "A model for the 'fixed' charge at the Si/SiO_2 interface", Proc. IEEE, 57, 1184-6 (1969), EEA73-14175.
835. Schemmel, R. R., Philipp, L. D. and Gordon, R. L., "Modeling of time dependent insulation dielectric properties", Am. Ceram. Soc. Bull., 51, 648, 50 (1972), EEA76-4747.
836. Wojtowicz, P. J., "Two dimensional phase transistions of mobile ions at SiO_2-Si interfaces", RCA Rev., 36, 132-48 (1975), EEA78-28708.

XIII. Thyristors

837. Balenovich, J. D., Gillot, D. M. and Motto, J. W., Jr., "Thyristor high frequency ratings by concurrent testing and computer simulation", 1971 Sixth Annual Meeting of the IEEE Industry and General Applications Group, p. 753-7, EEA75-10254.
838. Beattie, W. C., "Analogue modelling of thyristors", Proc. Inst. Electr. Eng., 120, 786-8 (1973), EEA76-29265.
839. Beattie, W. C. and Monteith, W., "Digital modeling of a thyristor", Proc. Inst. Electr. Eng., 120, 789-90 (1973), EEA76-29266.
840. Beattie, W. C., Monteith, W. and Ho, H. H., "Improved logic model for thyristor", Proc. Inst. Electr. Eng., 121, 1422-3 (1974), EEA78-2863.
841. Kats, B. S., "The equivalent circuit of a low-power thyristor", Telecommun. Radio Eng. Pt. 2, 26 (8), 122-4 (1971), EEA75-24347.
842. Pershenkov, V. S. and Tkachev, B. V., "MOS thyristors", Izv. VUZ Radioelektron., 15, 847-50 (1972), Russian, EEA75-815.

843.Prajoux, R., Thierry, P. and Jalade, J., "Extension of small-signal model for thyristor-rectifier control loops when using an asynchronous delay-type firing circuit", Electron. Lett., 9, 228-9 (1973), EEA76-25845.

844.Silard, A. and Marinescu, V., "Computer-aided experimental investigation of the correct turn-on in thyristor with amplifying gate", Electron. Lett., 11, 419-20 (1975), EEA78-37530.

845.Stepanov, P. and Zur, A., "A digital model for investigation of the switching behaviour of a thyristor with regard to the thermal drift of the constants of the material", Wiss. Z. Tech. Hochsch. Ilmenau, 20, 103-10 (1974), German, EEA78-15308.

846.Timurdzhi, V. G., Voronov, N. A. and Chausov, O. G., "Modelling of heat processes in thyristors of a pulsed convertor", Izv. VUZ Elektromekh., no. 4, 465-9 (1974), Russian, EEA78-298.

XIV. Other Devices

847.Adler, D., "Theory gives shape to amorphous materials", Electronics, 43 (20), 61-72 (1970), EEA74-534.

848.Board, K., "Transient temperature response of solid-state devices", Int. J. Electron., 39, 75-80 (1975), EEA78-32928.

849.Celtekligil, U., "Digital computation of nonlinear differential equations for SCR circuits with the help of fictitious inductances in the equivalent circuits of SCR's", IEEE Trans. Ind. Electron. & Control Instrum., IECI-22, 404-8 (1975), EEA78-37534.

850.Choma, J., Jr., "Frequency domain stability criteria for large signal tuned common base amplifiers", IEEE Proceedings of the International Sympsoium on Circuit Theory, 1972, p. 303-7, EEA75-31652.

851.Chua, L. O., "Modelling of three terminal devices: a black box approach", IEEE Trans. Circuit Theory, CT-19, 555-62 (1972), EEA75-38945.

852.Cohen, M. H., Neale, R. G. and Paskin, A., "A model for an amorphous semiconductor memory device", J. Non-Cryst. Solids, 8-10, 885-91 (1972), EEA75-32577.

853.Cooper, D., Williams, R. and Durnya, F., "Development, characterization and application of a new geometry inverter high power switch", Conference Record of 1969 4th Annual Meeting of the IEEE Industry and General Applications Group, p. 499-506, EEA73-12433.

854.Dixit, P. N., Lal, P. and Srivastava, S. K., "Tunnel-induced impact ionization in the metal-thin-insulator-semiconductor-metal system", Phys. Status Solidi A, 30, 337-43 (1975), EEA78-33091.

855.Driscoll, M. M., "Two-stage self-limiting series mode type quartz-crystal oscillator exhibiting improved short-term frequency stability", IEEE Trans. Instrum. & Meas., IM-22, 130-8 (1973), EEA76-25626.

856. Enea, G. and Esteve, D., "Numerical analysis of the behaviour of p-π-n diodes and p-n-π-n thyristors in direct polarization", C. R. Hebd. Seances Acad. Sci. B, 278, 323-6 (1974), French, EEA77-18777.
857. Ghoshtagore, R. N., "Model of doped-oxide-source diffusion in silicon", Solid-State Electron., 17, 1065-73 (1974), EEA78-571.
858. Goldenblum, A., "Band-filling process in AlGaAs-GaAs heterojunctions", Rom. Sci. Abstr., 9, 947 (1972).
859. Jackson, D. P. and Morgan, D. V., "Computer modelling of collision processes in solids", Contemp. Phys., 15, 25-48 (1974), PA77-28128.
860. Jaydevaiah, T. S. and Busmundrud, O., "Amorphous-crystalline silicon junctions", Electron. Lett., 8, 75-7 (1972), EEA75-10198.
861. Korb, F., "Thermal rating of separately cooled semiconductors for line commutated convertors", Elektrotech. Z. (ETZ) A, 92, 100-7 (1971), German, EEA74-16988.
862. Langer, P. H. and Goldstein, J. I., "Impurity redistribution during silicon epitaxial growth and semiconductor device processing. (Computer model)", J. Electrochem. Soc., 121, 563-71 (1974), PA77-45471.
863. Lee, D. H., "Double injection in cylindrical germanium structures", J. Appl. Phys., 44, 4708-16 (1973), PA77-20681.
864. Molchanov, A. A., Svechnikov, S. V., Stepanchuk, V. P. and Sharadkin, A. M., "The small signal theory of optoelectronic circuits", Poluprovodn. Tekh. & Mikroelektron., no. 12, 12-24 (1973), Russian, EEA76-38533.
865. Morris, B. L., "Some device applications of spreading resistance measurements on epitaxial silicon", J. Electrochem. Soc., 120 (3), 91C (1973), EEA76-23156.
866. Newland, F. J., "Switching and memory effects in glassy semiconductors", Inst. Eng. Aust. Elec. Eng. Trans., EE-8, 32-6 (1972), EEA75-39843.
867. Ollins, R. I. and Ratner, S. J., "Computer-aided design and optimization of a broad-band high-frequency monolithic amplifier", IEEE J. Solid-State Circuits, SC-7, 487-92 (1972), EEA76-277.
868. Owen, H. A., Jr., Wilson, T. G., Feng, S. Y. M. and Lee, F. C. Y., "A computer-aided design procedure for flyback set-up DC-to-DC converters", IEEE Trans. Magn., MAG-8, 289-91 (1972), EEA76-22526.
869. Palfy, M. and Bohonyey, F., "Investigations of the encasement of semi-conductor power elements based on electrical analogy", Elektrotechnika, 63, 413-15 (1970), Hungarian, EEA74-15584.
870. Parry, P. D. and Leiby, R. A., "Silicon wafer heating during ion implantation", Presented at 1973 International Electron Devices Meeting Supplement, Washington, D. C., December 1973, EEA77-18726.
871. Patterson, J. T., Lomax, R. J. and Haddad, G. I., "Quenched-domain mode admittance model for transferred-electron devices", IEEE Trans. Electron Devices, ED-21, 654-6 (1974), EEA77-38810.

872.Pocock, D. N. and Krebs, M. G., "Terminal modeling and photo-compensation of complex microcircuits", IEEE Trans. Nucl. Sci., NS-19, 86-93 (1972), EEA76-12657.
873.Popov, V. P. and Bickart, T. A., "RC transmission line with non-linear resistance large-signal response computation (m.o.s.t. model)", IEEE Trans. Circuits & Syst., CAS-21, 666-71 (1974), EEA77-30018.
874.Quast, W., "Small-signal admittance of the insulator n-type-gallium-arsenide interface region", Electron. Lett., 8, 419-21 (1972), EEA75-32587.
875.Redshaw, S. C. and Rushton, K. R., "Heat conduction with melting simulated on a resistance-capacitance network", J. Phys. D, 3, 1727-36 (1970), EEA74-6996.
876.Robrock, R. B., "A lumped model for characterizing single and multiple domain propagation in bulk GaAs", IEEE Trans. Electron Devices, ED-17, 93-102 (1970), EEA73-14127.
877.Robrock, R. B., "Extension of the lumped bulk device model to incorporate the process of domain dissolution", IEEE Trans. Electron Devices, ED-17, 103-7 (1970), EEA73-14128.
878.Robrock, R. B., "Effects of device parameters on domain dynamics in bulk GaAs", Proc. IEEE, 58, 804-5 (1970), EEA73-24900.
879.Rohr, P., Lindholm, F. A. and Allen, K. R., "Questionability of drift-diffusion transport in the analysis of small semiconductor devices", Solid-State Electron., 17, 729-34 (1974).
880.Rusu, A. and Bulucea, C., "Numerical calculation of electrical potential distribution in two-dimensional semiconductor structures", Rev. Roum. Sci. Tech. Ser. Electrotech. & Energ., 18, 463-78 (1973), EEA77-5973.
881.Stewart, J. A. C. and Wakefield, J., "Simulation of Pt-n-p^+ silicon punchthrough device", Electron. Lett., 8, 378-9 (1972), EEA75-32706.
882.Stutz, T. O. J. and Muller, A., "Controlled stability of flip-flop circuit", IEEE J. Solid-State Circuits, SC-7, 211-12 (1972), EEA75-15917.
883.Szedon, J. R. and Stelmak, J. P., "Effects of organic and in-organic dielectric films on semiconductor devices", IEEE Trans. Elect. Insulation, EI-5, 3-10 (1970), EEA73-17557.
884.Tandon, J. L., Bilger, H. R. and Nicolet, M. A., "Excess noise spectral analysis of boron-implanted layers in silicon", 1973 SWIEEECO Record of Technical Papers, p. 561-8, EEA76-29061.
885.van Laar, J., "The physical model of negative electron affinity photoemitters", Acta Electron., 16, 215-27 (1973), EEA76-26512.
886.Yarbrough, R. B., "Circuit models for transformers", IEEE Trans. Educ., E-12, 181-8 (1969), EEA73-15925.

XV. Integrated Circuits

1. MOS IC

887. Basiladze, S. G., "Calculation of the parameters of a memory subsystem using one type of MOS-transistor", Izv. VUZ Radio-elektronika, 13, 3-11 (1970), Russian, EEA73-25039.

888. Bazin, B., Crawford, R. H. and Sevin, L. J., "Theoretical effects of overlap capacitance upon MOS inverter output", Proc. IEEE, 56, 1088-9 (1968), EEA71-23395.

889. Carr, W. N. and Hong, S. T., "The MAGFET: A new integrated circuit building block", 1970 SWIEECO Record of Technical Papers of the 22 Southwestern IEEE Conference and Exhibition, p. 166-70, EEA73-35256.

890. Cupcea, N., "Some aspects of the integrated MOS transistor theory", Bul. Inst. Politeh. 'Gheorghe Gheorghiu-Dej' Bucuresti, 36, 135-51 (1974), EEA78-5327.

891. Habing, D. H. and Shafer, B. D., "Modeling and radiation effects study of an LSI/MOS logic system", IEEE Trans. Nucl. Sci., NS-18, 263-72 (1971), EEA75-9477.

892. Hara, H., "A design theory of MOS inverters", Trans. Inst-Electron, Commun. Eng. Japan. A,B,C, 53 (4), 29-31 (1970), EEA74-4084.

893. Hayashi, Y., Ezaki, J. and Tarui, Y., "Design of ED-MOS buffer", Electron. & Commun. Jap., 55 (10), 107-14 (1972), EEA76-38304.

894. Hayashi, Y., Tarui, Y. and Hashimoto, S., "Design theory of ED-MOS-IC (enhancement-depletion mode)", Electron. & Commun. Jap., 55 (7), 84-92 (1972), EEA76-32288.

895. Jaeger, R. C., Lin, H. C. and Linholm, L. W., "Comments on 'an optimized output stage for MOS integrated circuits' (and reply)", IEEE J. Solid-State Circuits, SC-10, 185 (1975), EEA78-19891.

896. Kotani, M., "Pulse response analysis of a complementary MOSFET inverter by the hyperbolic functional description", Proc. IEEE, 58, 1967-8 (1970), EEA74-4165.

897. Nazaryan, K. K. and Pryanishnikov, V. A., "Calculation of interference from the source in a parallel converter using a MOSFET", Izv. VUZ Priborostr., 16, 112-17 (1973), Russian, EEA76-25648.

898. Petrenko, A. I., Savatev, V. A., Tetelbaum, A. Y., Abakumov, V. G., Shelpuk, G. D., Takachenko, Y. I., Lukashenko, I. B., Zubenko, U. M. and Popudribko, T. A., "Design of morphology for MOS-LSI circuits allowing for interaction between human operator and computer", Izv. VUZ Radioelektron., 17, 81-92 (1974), Russian, EEA77-36329.

899. Sieber, P. and Wolff, D., "Using TELEMOS to produce custom-built MOS circuits", Elektronik, 23, 429-32 (1974), German, EEA78-2900.

900. Szanto, L., "A transistor model for the circuit analysis of MOS type integrated circuits", Slaboproudy Obz., 36, 265-9 (1975), Czech, EEA78-33229.

901.Young, T. K. and Dutton, R. W., "MSINC - a Modulator Simulator for Integrated Nonlinear Circuits with MOS model example", Proceedings of the 1974 IEEE International Symposium on Circuits and Systems, p. 720-4, EEA77-38370.

2. Other IC's

902.Beale, J. R. A., Bridgen, R. and Cusdin, A. R., "Modelling of transistors for fast integrated logic circuits", Electron. Lett., 6, 301-3 (1970), EEA73-31391.
903.Bhattacharyya, A. B. and Gupta, M. L., "Single-section lumped models for integrated circuit resistors", Solid-State Electron., 16, 1506-9 (1973), EEA77-10303.
904.Burgomistrov, V. A. and Mischenko, V. P., "Modelling of high-frequency semiconductor integrated microcircuits with the help of low-frequency models", Telecommun. & Radio Eng., Pt. 2, 29 (7), 132-4 (1974), EEA78-37345.
905.Choma, J., "A process-oriented model for the simulation of base pushout in integrated bipolar devices", IEEE Trans. Electron Devices, ED-22, 1079-86 (1975).
906.Cannon, D. L., "Semiconductor integrated circuit component models", Report THEMIS-RE-2-70, Cullen Coll. Engng., Houston, Texas, 1970, 39 pp., EEA74-22803.
907.Dickhaut, R. H., "Simplified microcircuit modeling", IEEE Trans. Nucl. Sci., NS-18, 227-36 (1971), EEA75-9572.
908.Fetcher, B. C., "An electric integrated circuit reliability model", Report NASA-CR-125062, California Inst. Technol., Pasadena, 1972, 84 pp., EEA76-841.
909.Fulton, A. W., "Modelling the collector diffusion isolation integrated circuit transistor", Thesis, Stanford Univ., California, 1971, 119 pp., Order No. 72-11552, EEA75-35413.
910.Glasford, G. M., "Comparisons of incremental high frequency circuit models for electron devices", Proceedings of the National Electronics Conference, 1968, p. 279-84, EEA73-17698.
911.Greenbaum, J. R., Horen, R., Andrews, J. L. and Peden, J. C., "Radiation models for digital integrated circuits", IEEE Trans. Nucl. Sci., NS-19, 96-102 (1972), EEA76-12659,
912.Hilbourne, R. A., "The design of M.O.S. integrated circuits", AGARD Lecture Series No. 75 Custom Design for Large Scale Integration (LSI), 1975, p. 3/1-20, EEA78-19929.
913.Hoertt, D. J., "Microcircuits: the brain and the brawn", IEEE Trans. Ind. Electron. & Control Instrum., IECI-22, 159-63 (1975), EEA78-19856.
914.Holt, A. G. J. and Bowron, P., "An alternative derivation of the small-signal low-frequency admittance matrix for a semiconductor integrated circuit", IEEE Trans. Electron Devices, ED-16, 966-8 (1969), EEA73-9583.

915. Idleman, T. E., Jenkins, F. S., McCalla, W. J. and Pederson, D. O., "SLIC - a simulator for linear integrated circuits", IEEE J. Solid-State Circuits, SC-6, 188-203 (1971), EEA74-25344.
916. Katoh, H., "A lump-element approach to microwave integrated circuits", Electron. & Commun. Jap., 56 (6), 47-53 (1973), EEA78-2418.
917. Kidron, I., "Integrated circuit model for lateral PNP transistors including isolation junction interactions", Int. J. Electron., 31, 421-40 (1971), EEA75-419.
918. Kleiner, C., Nelson, J., Vassallo, F. and Heaton, E., "Integrated circuit model development for EMP", IEEE Trans. Nucl. Sci., NS-21, 323-31 (1974), EEA78-19899.
919. Kobylarz, T. J., Herman, F. H., Graf, A. J. and Calella, P. C., "Statistic for improvement of IC process control", 7th Asilomar Conference on Circuits, Systems and Computers, 1973, p. 143-7, Publ. 1974, EEA77-30408.
920. Kostychev, G. I., "Physical approach to estimating the reliability of integrated circuits", Izv. VUZ Radioelektron., 17, 86-90 (1974), Russian, EEA77-30350.
921. Leidich, N. A. J., "Basic circuit concepts for monolithic linear integrated circuits", Funk-Tech., 30, 278-81, 284 (1975), German, EEA78-33203.
922. Lindholm, F. A., "Integrated-circuit transistor and diode models for network-analysis programs", IEEE Trans. Circuit Theory, CT-18, 122-8 (1971), EEA74-12638.
923. Lindholm, F. A., Director, S. W. and Bowler, D. L., "Assessing model adequacy and selecting model complexity in integrated-circuit simulation", IEEE J. Solid-State Circuits, SC-6, 213-22 (1971), EEA74-25346.
924. Lindsted, R. D. and DiCicco, D. A., "Analytical and experimental thermal analysis of multiple heat sources in integrated semiconductor chips", IBM J. Res. & Develop., 16, 303-6 (1972), EEA75-35394.
925. Madzy, T. M., "Static discharge modeling techniques for evaluation of integrated (FET) circuit destruction", Conference on Electromagnetic Compatibility 1975, p. 134-9, EEA78-28877.
926. Makimoto, T., "Modelling of transistors in integrated circuits and its application", Electron. & Commun. Jap., 53 (10), 143-9 (1970), EEA75-10318.
927. Muehldorf, F. I., "Fault clustering: modeling and observation on experimental LSI chips", IEEE J. Solid-State Circuits, SC-10, 237-44 (1975), EEA78-33213.
928. Nichols, J. S., Alexander, D. R. and Hittinger, G. N., "Characterization and modeling of the 709 integrated circuit operational amplifier in an ionizing radiation environment", IEEE Trans. Nucl. Sci., NS-16, 172-6 (1969), EEA73-12852.
929. Orton, D. W. R., "Large scale integration design", Electron. Equip. News, 16 (2), 44-6 (1974), EEA77-38878.

930.Pederson, D. O. and Solomon, J. E., "The need and use of macromodels in IC subsystem design", Presented at the 1974 IEEE International Symposium on Circuits and Systems, San Francisco, Calif., April 1974, EEA78-668.

931.Petrenko, A. I., "An automated system for the design of integrated circuits", Izv. VUZ Radioelektron., 16, 15-22 (1973), Russian, EEA76-32290.

932.Pilling, D. J. and Skalnik, J. G., "A circuit model for predicting transient delays in LSI logic circuits", 6th Asilomar Conference on Circuits and Systems, 1972, p. 424-8, Publ. 1973, EEA76-23050.

933.Plaud, B., "A failure rate prediction model for LSI devices", International Conference on Large Scale Integrated Circuits, 1974, p. 141-9, EEA78-11750.

934.Roche, M., "Integrated components models", Presented at Proceedings of the International Conference on Advanced Microelectronics, Paris, France, April 1970, EEA74-1594.

935.Snethen, T. J. and Jessep, D. C., "The circuit failure modelling challenge for LSI", International Symposium on Circuit Theory 1973, p. 425-30, EEA77-2580.

936.Spence, H. W., "Computer aids for LSI design. Perspective 1974", Onde Electr., 55, 135-9 (1975), EEA78-24496.

937.Stapper, C. H., Jr., "On a composite model to the IC yield problem", IEEE J. Solid-State Circuits, SC-10, 537-9 (1975).

938.Steele, E. L., "Variable current gain model for semiconductor circuit analysis", Proceedings of the National Electronics Conference, 1968, p. 731-5, EEA73-16331.

939.Szporko, J., "Large scale integration problems in electronics", Elektronika, 14, 428-33 (1973), Polish, EEA77-2600.

940.Tozoni, O. V. and Dianov, V. I., "Calculating the electromagnetic interactions between integrated circuit conductors with a crystalline substrate", Kibern. & Vychisl. Tekh., no. 26, 3-15 (1974), Russian, EEA78-11706.

941.Ustyuzhaninov, V. N., "Systems approach to the analysis of characteristics of large scale integrated circuits using uniform structures", Izv. VUZ Radioelektron., 17 (9), 3-9 (1974), Russian, EEA78-8611.

942.Wilkinson, D. R. and Silva, L. F., "A general controlled source (in integrated circuit form)", Proceedings of the 15th Midwest Symposium on Circuit Theory, Vol. II, 1972, X.3/10 pp., EEA76-7939.

3. Computer-Aided Design

943.Andreeva, A. N., Pelovska, E. I., Barzakov, I. S., Petrova, R. S. and Zarkov, I. P., "A MOS transistor model for computer-aided design of integrated circuits", C. R. Acad. Bulg. Sci., 27, 1191-4 (1974), EEA78-24450.

944.Bowers, J. C., "Terminal modeling of linear and digital integrated circuits for CAD applications", Proceedings of the 15th Midwest Symposium on Circuit Theory, Vol. I, 1972, iv.4/10 pp., EEA76-8134.
945.Brown, R. W., "Component models for computer aided circuit design. IV. Integrated circuits", Electron. Compon., 12, 858-68 (1971), EEA74-29471.
946.Capocaccia, E., "A batch-processing approach to computer aided design of MOS large scale integrated circuits", Proceedings of the 2nd International Computing Symposium 1972, p. 324-34, EEA75-27772.
947.Critchlow, D. L. and Schuster, S. E., "Computer-aided design of large-scale integrated MOS logic circuits", Proceedings 1968 Microelectronics Symposium, 1968, Pap. B8-1-B8-9, EEA71-28974.
948.Cutler, A. B., "Computer-aided design for MOS polycell large scale integrated circuits", Presented at the Semiconductor Integrated and Production Conference and Exhibition (abstracts), Anaheim, Calif., February 1971, EEA74-21578.
949.Ebertin, M. A., "The design of MOS/LSI through the use of computer aided design techniques", Proceedings of the International Conference on Advanced Microelectronics, 1970, 2 pp., EEA74-414.
950.Ebertin, M. A., "Large scale design of MOS integrated circuits using computers (CAD)", Onde Elec., 50, 737-45 (1970), French, EEA74-4175.
951.Engl. W. L., Pernards, P. and Milynski, D. A., "Computer-aided topological design for integrated circuits", IEEE Trans. Circuit Theory, CT-20, 717-25 (1973), EEA76-37850.
952.Forte, M., "Computer aided design in the realization of LSI systems", Proceedings of the International Conference on Advanced Microelectronics, 1970, 1 pp., EEA74-412.
953.Forte, S. S., "The application of computer aided design to MOS integrated circuits", Advance in Electronics - proceedings of the 15th Electronics Congress, 1968, p. 3-11, Italian, EEA73-6493.
954.Fossum, J. G., "Systematic computer-aided three-dimensional modeling of integrated bipolar devices", Thesis, Univ. Arizona, Tucson, 1971, 200 pp., Order No. 72-2169, EEA75-27771.
955.Fossum, J. G. and Hamilton, D. J., "Systematic computer-aided multidimensional modeling of integrated bipolar devices", Solid-State Electron., 16, 1-14 (1973), EEA76-8410.
956.Frohman-Bentchkowsky, D. and Vadasz, L., "Computer-aided design and characterization of digital MOS integrated circuits", IEEE J. Solid-State Circuits, SC-4, 57-64 (1969), EEA72-12804.
957.Greenbaum, J. R., "Digital IC models for computer-aided design. Part 1. TTL NAND gate", Electronics, 46 (25), 121-5 (1973). See also: ibid 46 (26), 107-112 (1973); ibid, 47 (2), 98-101 (1974); ibid, 47 (5), 125 (1974), EEA77-18574.

958.Greenbaum, J. R. and Miller, W. A., "Digital-IC models for computer-aided design. IV. AND-OR-INVERT gate", Electronics, 47 (3), 124-6 (1974), EEA77-13975.
959.Hamilton, A. W., "A new technique for the modeling of digital logic circuitry for computer aided circuit analysis", Proceedings of the 15th Midwest Symposium on Circuit Theory, Vol. I, 1972, iv.3/10 pp., EEA76-8133.
960.Hesse, W. C. A. and Schoppnies, E., "Integrated circuits by computer for the computer", Radio Fernsehen Elektron., 21, 193-6 (1972), German, EEA75-23124.
961.Kennedy, D. P., "Computer aided design - fact or fable?", 1973 International Electron Devices Meeting Technical Digest, p. 6-9, EEA77-14577.
962.Kozemchak, E. B., "Computer analysis of digital integrated circuits by macromodelling", IEEE Proceedings of the International Symposium on Circuit Theory, 1972, p. 379, EEA75-31750.
963.Lin, H. L. and Jones, W. N., "Computer analysis of the double-diffused MOS transistor for integrated circuits", IEEE Trans. Electron Devices, ED-20, 275-83 (1973), EEA76-16442.
964.Lindholm, F. A., "Device modeling for computer-aided design and analysis of integrated circuits", Solid-State Electron., 12, 831-40 (1969), EEA73-9234.
965.Mandl, W., "Computer-aided design and MOS memories", Western Electronic Show and Convention, 1970, 5 pp., EEA74-6706.
966.Rabbat, N. B., "A computer-aided approach for the prediction of LSI transient circuit performance", 7th Asilomar Conference on Circuits, Systems and Computers, 1973, p. 136-42, Publ. 1974, EEA77-30407.
967.Rabbat, N. B., Ryan, W. D. and Hossain, S. Q. A. M. A., "A computer modeling approach for LSI digital structures", IEEE Trans. Electron Devices, ED-22, 523-31 (1975), EEA78-28855.
968.Rosinski, A. T., "Nonlinear DC models of planar transistor adapted for design and analysis of integrated circuits on digital computers", Arch. Elektrotech., 21, 201-9 (1972), Polish, EEA75-31808.
969.Rusakov, S. G. and Khodash, L. S., "Computer-aided analysis of static states of MOS bipolar integrated structures", Izv. VUZ Radioelektron., 16, 115-8 (1973), Russian, EEA76-32291.
970.Schiotz, H., "Digital circuit and system development by computer", Elektro, 86 (19), 39-42 (1973), Norwegian, EEA77-10296.
971.Spence, H. W., "Computer aids for LSI design", International Conference on Large Scale Integrated Circuits, 1974, p. 188-98, EEA78-11754.
972.Tabarnyi, V. G., Logvinenko, N. P. and Kubasov, G. F., "Analysis of the operation of dynamic MOS-LSI circuits using a computer-aided design", Izv. VUZ Radioelektron., 17 (6), 93-7 (1974), Russian, EEA77-34095.

973.Treble, D. P., "Dimensional checking of M.O.S. L.S.I. layouts", International Conference on Computer-Aided Design, 1972, p. 7-11, EEA75-19268.
974.Tsuboi, Y., Tsuchiya, R. and Asakawa, T., "Computer-aided design of MOS LSI", Proc. Inst. Radio Electron. Eng. Aust., 32 (6), 196-203 (1971), EEA74-29475.
975.Van Lier, M. C. and Otten, R. H. J. M., "Automatic IC layout: The model and technology", IEEE Trans. Circuits & Syst., CAS-22, 845-54 (1975), EEA78-41680.
976.Vladimirescu, A., "Calculator-aided design of MOS integrated circuits", IEEE J. Solid-State Circuits, SC-10, 151-61 (1975), EEA78-19889.
977.Wieder, A. W., Engl, W. L. and Lehning, H., "Computer-aided device modeling and design procedure for current hogging logic (CHL)", IEEE J. Solid-State Circuits, SC-10, 352-60 (1975), EEA78-37627.

XVI. Second Breakdown

978.Asakawa, T. and Tsubouchi, N., "Second breakdown in MOS transistors", IEEE Trans. Electron Devices, ED-13, 811-12 (1966), EEA70-9103.
979.Bayzdrenko, A. A. and Gusev, V. A., "On the problem of second breakdown in transistors", Radio Eng. & Electron. Phys., 18, 1297-8 (1973), EEA78-5316.
980.Dean, E. B., Lutzky, M., Petree, M. C. and Shaklan, F. R., "A diode model which predicts second breakdown", 8th Asilomar Conference on Circuits, Systems and Computers, 1974, p. 361-6, Publ. 1975, EEA78-24103.
981.Chabanne, J. P., "The second breakdown phenomenon in power transistors", Electron. & Microelectron. Ind., no. 192, 123-31 (1974), French, EEA78-5307.
982.Cohn, N. S. and Petree, M. C., "Anomalous second breakdown failure levels in germanium gold-bonded diodes", IEEE Trans. Nucl. Sci., NS-21, 20-2 (1974), EEA78-616.
983.Demizu, K. and Yamamoto, Y., "Second breakdown of IC structured power transistors", IEEE Trans. Electron Devices, ED-22, 352-3 (1975), EEA78-24461.
984.Gaur, S. P., Navon, D. H. and Teerlinck, R. W., "Transistor design and thermal stability", IEEE Trans. Electron Devices, ED-20, 527-34 (1973), EEA76-23431.
985.Hower, P. L., "Quantitative models for second breakdown in transistors", IEEE Power Electronics Specialists Conference 1973, p. 67-73, EEA77-374.
986.Krishna, S. and Gray, P., "Second breakdown: hot spots of hot cylinders", Proc. IEEE, 62, 1182-3 (1974), EEA77-34041.
987.Popescu, C., "The second breakdown in reverse biased transistor as an electrothermal switching", IEEE Trans. Electron Devices, ED-21, 428-36 (1974), EEA77-30303.

988. Raburn, W. D., "An investigation of current mode second breakdown", Report BER-177-93, Univ. Alabama, University, 1974, 34 pp., EEA78-24098.
989. Ramanan, K. V., "A new physical model for second breakdown in semiconductor devices", Presented at Symposium on Electronics (electron devices) abstracts only, Hyderabad, India, February 1972, EEA75-13707.
990. Roman, G., "A model for computation of second breakdown in transistors", Solid State Electron., 13, 961-80 (1970), EEA73-35220.
991. Smith, W. B., Pontius, D. H. and Budenstein, P. P., "Second breakdown and damage in junction devices", IEEE Trans. Electron Devices, ED-20, 731-44 (1973), EEA76-29233.
992. Wang, P. P., "Thermal instability and secondary breakdown in power transistors", IEEE Trans. Aerosp. & Electron. Syst., AES-7, 1195-200 (1971), EEA75-16708.

XVII. Miscellaneous

993. Cariolaro, G. L., "Noise in a three pole device", Alta Frequenza, 37, 975-7 (1968), Italian, EEA73-7490.
994. Hobson, G. S., "A model to demonstrate the properties of electronic velocity distribution functions", Int. J. Elec. Eng. Educ., 9, 404-10 (1972), EEA75-13808.
995. Horst, R. L., "Forehearth temperature control improvement by application of solid-state instruments", Conference Record of 1969 4th Annual Meeting of the IEEE Industry and General Applications Group, p. 57-63, EEA73-10716.
996. Oiyak, V. D., Krasnobaev, A. M., Romanovskii, V. F. and Ostrenko, V. S., "Determination of the thermal resistance of the pellet valve by electro-thermal analogy", Elektroteknika, 41 (2), 30-2 (1970), Russian, EEA73-23226.
997. Oren, R., "Discussion of various views on popcorn noise", IEEE Trans. Electron Devices, ED-18, 1194-5 (1971), EEA75-4656.
998. Stefankiewicz, Z., "Transistor distance protection with a new form of characteristics in the R-X coordinate system", Przegl. Elektrotech., 47, 362-7 (1971), Polish, EEA74-37321.
999. Vandorpe, D. and Monnier, J., "Two-dimensional analysis of planar diffusion", Electron. Lett., 8, 554-5 (1972), EEA76-4412.

R. PACKAGING

I. Bibliographies

1.Agajanian, A. H., "A bibliography on electronic packaging. II", Solid State Technol., 18 (10), 56-63 (1975).
2.Anon., "Packaged circuits", DDC Report AD-752400, 1972.

II. Books

3.Allen, B. M., "Soldering handbook", Drake Publishers, Inc., New York, 1972.
4.Bruins, P. F., (Ed.), "Packaging with plastics", Gordon and Breach, New York, 1975.
5.Hamer, D. W. and Biggers, J. V., "Thick film hybrid microcircuit technology", Wiley-Interscience, New York, 1973.
6.Harper, C. A., "Handbook of electronic packaging", McGraw-Hill Book Co., New York, 1969.
7.Hnatek, E. R., "User's handbook of integrated circuits", Wiley-Interscience, New York, 1973.
8.Jowett, C. E., "Electronic engineering processes", Business Books, London, England, 1972.
9.Miller L. F., "Thick film technology and chip joining", Gordon and Breach, New York, 1972.
10.Planer, G. V. and Phillips, S., "Thick film circuits", Butterworth, London, England, 1972.
11.Ranney, M. W., "Microencapsulation technology", Noyes Development Corporation, Park Ridge, N. J., 1969.
12.Rikoski, R. A., "Hybrid microelectronic circuits: the thick film", Wiley-Interscience, New York, 1973, EEA77-14632.
13.Scott, A. W., "Cooling of electronic equipment", John Wiley & Sons, New York, 1974.
14.Seely, J. H. and Chu, R. C., "Heat transfer in microelectronic equipment", Marcel Dekker, Inc., New York, 1972.
15.Sideris, G., "Microelectronic packaging interconnection and assembly of integrated circuits", McGraw-Hill Book Co., New York, 1968.
16.Szilard, J., "Sealing and potting compounds", Noyes Press, Park Ridge, N. J., 1972.

III. Review Articles

17.Anderson, J. C., "Applications of thin films in microelectronics", Thin Solid Films, 12, 1-15 (1972), EEA75-39211.
18.Anon., "Laser-beam trimming system for hybrid circuits", Elektron. Ind., 5 (1-2), 14-15 (1974), German, PA77-43356.

19. Anon., "Forecast of 1980 hybrid issues", Circuits Manuf., 15 (3), 52, 4 (1975), EEA78-33242.
20. Axelrod, N. N., "Applications of thin film dielectrics in microelectronics", Phys. Electron. Ceram., Proc. Electron. Phenomena Ceram. Conf., 1969, p. 313-30, Publ. 1971, CA77-157488.
21. Bailey, D. C., "A plastic dual-in-line approach for thick film hybrids", International Electronic Circuit Packaging Symposium, 1969, 4 pp., EEA73-27360.
22. Basseches, H. and Gerstenberg, D., "Status of thin film integrated circuit technology", Thin Solid Films, 12, 295-303 (1972), EEA75-848.
23. Bilous, O. and Rymaszewski, E. I., "Medium density monolithic logic technology", International Electronic Circuit Packaging Symposium, 1969, 7 pp., EEA73-27288.
24. Blanks, H. S., "A review of new methods and attitudes in reliability engineering", Microelectron. & Reliab., 12, 301-19 (1973), EEA77-9231.
25. Boswell, D., "The interaction of hybrids and semiconductors", New Electron., 6 (17), 17, 20 (1973), EEA77-2618.
26. Boswell, D., "Thin and thick film for professional and military applications", New Electron., 8 (6), 54-5 (1975), EEA78-24439.
27. Boswell, M. D., "Encapsulation of hybrid integrated circuits. A review of materials, processes and structures", International Symposium on Materials for Electronic Components, 1975, p. 133-41, EEA78-33252.
28. Broda, R., "Tips on selecting an integrated circuit package", Insulation, 14 (10), 54-8 (1968), EEA72-302.
29. Buchoff, L. S., "Selection of encapsulation methods and materials", Spring Meeting of the Electrochemical Society (extended abstracts), 1971, p. 85-6, EEA74-36548.
30. Buhler, W., "Technology of semiconductors", Elektro-Monteur, 25 (4), 103-10 (1974), German, EEA77-36159.
31. Cherniack, G. B. and Arnold, W. R., "An advanced technique for complex electronic components based on SLT technology", Presented at Electronic Components Conference, Washington, D. C., May 1969, EEA72-23129.
32. Clough, W. L. and Collins, J. M., "A review of plastic packaging for hybrid thick-film circuits", Microelectronics, 3 (11), 4-15 (1971), EEA75-3987.
33. Cometta, M., "Thick film circuits technology", Elettrificazione, no. 5, 236-40 (1975), Italian, EEA78-33182.
34. Corradetti, M., "Packages: their impact on cost and design of MOS-LSI circuits", Presented at International Conference on Advanced Microelectronics, Paris, France, April 1970, EEA74-393.
35. Deal, B. E., "Current concepts in the passivation and encapsulation of semiconductor devices", Proceedings of the 10th Electrical Insulation Conference, 1971, p. 63-8, EEA75-13534.

36.DeCoursey, D. T., "Thick film microcircuitry and packaging technology - state of the art", Electron. Mfr., 16 (6), 19-22 (1971), EEA74-33370.
37.Dodd, R., "Sizing-up the package in the age of implosion", Electron. Engineering, 45 (548), 49-51 (1973), EEA77-29.
38.DuBois, J. H., "A review of the uses and properties of inorganic plastics", Insul./Circuits, 20 (11), 31-2 (1974), EEA78-2931.
39.Dunbridge, B., "LSI technology overview", Communications Systems and Technology Conference, 1974, p. 22-8, EEA77-34111.
40.Dyson, A. F., "Thick film - some of the why's and wherefore's", New Electron., 7 (10), 108, 110-11 (1974), EEA77-30368.
41.Fehr, G. K., "Microcircuit packaging and assembly - state of the art", Solid State Technol., 13 (8), 41-7 (1970), EEA74-354.
42.Fehr, G. K., "Microcircuit packages and assembly techniques. I", Insul./Circuits, 17 (11), 37-40 (1971), EEA75-9585.
43.Fehr, G. K., "Microcircuit packages and assembly techniques. II", Insul./Circuits, 17 (12), 16-20 (1971), EEA75-12744.
44.Flock, W. M. and Guthrie, E. A., "The LID revisited", Solid State Technol., 17 (10), 42 (1974), EEA78-5342.
45.Frieser, R. G., "A review of solder glasses", J. Electrochem. Sci. & Technol., 2 (3), 163-99 (1975).
46.Furman, T. T., "Integrated micro circuit production a PhD project in engineering design", Engineering Design - a key to the management of successful enterprises, 1974, p. 42-62, EEA78-2928.
47.Gantz, R., "Choice consideration of application of discrete and integrated semiconductor devices for hybrid circuit", Pr. Przem. Inst. Elektron., 13 (3-4), 21-33 (1972), Polish, EEA78-33248.
48.Gray, P., "Electro-thermal interactions in integrated circuit design", International Symposium on Circuit Theory 1973, p. 82-5, EEA77-6058.
49.Gregor, L. V., "Thin film processes for microelectronic application", Am. Ceram. Soc. Bull., 51, 246-7 (1972), EEA76-4703.
50.Hawkes, P. L., "The technology of tantalum film microcircuits", Trans. Inst. Metal Finishing, 47 (Pt. 3), 114-20 (1969), EEA73-3532.
51.Hicks, J., "Multilayer thick films", New Electron., 8 (6), 72 (1975), EEA78-24440.
52.Holmes, P. J. and Corkhill, J. R., "Thick-film integrated circuits. I", Electronic Compon., 10 (10), 1171-6 (1969), EEA73-6766.
53.Lazar, M., "Packaging design: managing the packaging of electronic systems", Electromech. Des., 16 (3), 24-9 (1972), EEA75-31898.
54.Manfield, H. G., "Packaging techniques for electronic systems", In: Microcircuits and their applications, Iliffe Books, London, England, p. 101-12 (1969), EEA72-14978.
55.Manfield, H. G., "Electronic circuit encapsulation", BEAMA Electrical Insulation Conference, 1970, 13 pp., EEA74-4230.
56.Mann, T., "Why go thick film?", Electron. Components, 16 (2), 30-1 (1974), EEA77-10289.

57.Miller, L. F., "A survey of chip joining techniques", Solid State Technol., 12 (8), 47-52 (1969), EEA73-3527.
58.Miller, L. F., "A survey of chip joining techniques. II", Solid State Technol., 12 (9), 33-41 (1969), EEA73-6739.
59.Myers, T. R., "Flip-chip microcircuit bonding systems", Proceedings of the Electronic Components Conference, 1969, p. 131-44, EEA72-23131.
60.Nicasie, M. J., "A classification service for packaging literature", Open, 6 (2), 19-23 (1974), Dutch, EEA77-17830.
61.Philofsky, E., "Review of bonding and interconnecting methods in microelectronics", J. Vac. Sci. & Technol., 12, 36 (1975), EEA78-24542.
62.Philofsky, E. and Hall, E., "A review of the limitations of aluminum thin films on semiconductor devices", IEEE Trans. Parts, Hybrids & Packag., PHP-11, 281-9 (1975).
63.Pletscher, W., "Electronic technology on ceramics, hybrid circuits in thick film technique", Radio TV Electron., 34 (1), 33-5 (1974), German, EEA77-14630.
64.Pricer, W. D., "Integrated circuits without an integrated circuit theory - an overview", International Symposium on Circuit Theory 1973, p. 78-81, EEA77-6057.
65.Reddig, H., "What does thick film technology offer?", Sprechsaal Keram., Glas, Baustoffe, 106 (21), 828, 830, 831, 833-35 (1973).
66.Reissing, T. C., "The whole story of thick film circuits", Radio Mentor Electron., 39, 508-10 (1973), German, EEA77-14580.
67.Rogers, B. C., "IC packages: which one to use?", Solid State Technol., 11 (9), 45-6 (1968), EEA72-2273.
68.Rose, A. S., "The promises and problems of microelectronic interconnection processes", IEEE International Convention Digest, 1970, p. 324, EEA73-37076.
69.Rostky, G., "Focus on packaging", Electron. Des., 19 (16), 36-47 (1971), EEA74-36546.
70.Scrupski, S. E., "Plastic ceramic duel stirs up new design concepts for LSI packages. I", Electronics, 44 (8), 75-86 (1971), EEA74-17571.
71.Scrupski, S. E., "System designers eye multichip LSI packages. II", Electronics, 44 (9), 65-8 (1971), EEA74-17645.
72.Seely, J. H., "Microelectronic packaging: definitions and challenges", Insul./Circuits, 18 (12), 31-4 (1972), EEA76-8759.
73.Siemund, H., "Modern packaging techniques used by Grundig AG", Grundig Tech. Inf., 22, 547-9 (1975), German, EEA78-36594.
74.Smalcelj, I., "Survey of some processes in assembly of semiconductor elements", Elektrotechnica, 71 (3), 193-5 (1971), Croatian, EEA74-33940.
75.Socolovsky, A., "Course: Packaging with ICs. II", Electron. Eng., 31 (4), 32-3, 37-8, 40, 42-3 (1972), EEA75-27954.
76.Socolovsky, A. and Ruth, S. B., "Packaging with integrated circuits", Electron. Eng., 31 (2), 51-2 (1972), EEA75-19242.

77.Spandorfer, L. M., "Large scale integration - an appraisal", In: Computers, advances in, Academic Press, New York, p. 179-238 (1968), EEA73-20104.
78.Sparks, M., "The packaging of electronic components", Bell Lab. Record, 47, 102-7 (1969), EEA72-17219.
79.Stephen, S., "Problems in the mechanical construction of modules", Prace Przemysl. Inst, Telekomun., 20 (68), 59-66 (1970), Polish, EEA74-29901.
80.Stuhlbarg, S. M., "What is needed to get started packaging hybrid integrated circuits", IEEE International Convention Digest, 1970, p. 422-3, EEA73-37102.
81.Szalay, M., "Development trends of printed circuit boards", Finommech. & Mikrotech., 13, 267-74 (1974), Hungarian, EEA78-8608.
82.Tapp, C. M. and Wiley, T. A., "Review of hybrid microelectronics", Report SLA-73-1063, Sandia Labs, Albuquerque, N. Mex., 1974, 65 pp., EEA78-37696.
83.Wang, C. H., "Fine line technique for microelectronic devices", Proceedings of the 1970 Electronic Components Conference, p. 570-5, EEA74-402.
84.Wheatley, C. F., Jr., "Integrated solid-state devices", Conference on Automotive Electronics, 1974, p. 38-66, EEA77-30347.
85.Zawadzka-Ananjew, D. and Gandurska, J., "A survey of some methods of hybrid microcircuit protection", Elektronika, 15 (2), 59-63 (1974), Polish, EEA77-30418.

IV. Chip Bonding

86.Ailawadhi, M. L. and Kakar, A. K., "Torque and thermal cycling as methods of testing reliability of reflow-soldered chip-to-substrate joints", Proceedings of the 11th Annual Reliability Physics Conference 1973, p. 77-82, EEA77-6114.
87.Anon., "Semiconductor chip bonding method", Patent USA 3667848, Publ. June 1972.
88.Baker, T. H., Ghafghaichi, M. and Totta, P. A., "Semiconductor integrated-circuit chip structure protected against impact damage from other chips during chip handling", Patent USA 3781609, Publ. December 1973.
89.Bankes, R. B., Large, D. M., Naumann, H. K. and Sweitzer, S. E., "Mass soldering of leads to semiconductor chips", Tech. Dig., no. 33, 9-10 (1974), EEA77-23170.
90.Bowman, R. and Bond, R., "Spider bond packaging", Electron. Packag. Prod., 6 (6), 15-16, 18 (1970), EEA74-6918.
91.Brownewell, D., "Evaluation of solder bump flip chips in thick film solder reflow hybrids", Insul./Circuits, 21 (5), 33-6 (1975), EEA78-28885.
92.Buchoff, L. S., "Stretching the chip bonding alternatives with reliable, low cost elastomeric connectors", Insul./Circuits, 21 (12), 23-4 (1975).

93. Bull, D. N., "Advances in low temperature die bonding techniques", Solid State Technol., 17 (9), 60-5 (1974), EEA78-2927.
94. Cass, E. E., Gustafson, R. M. and Young, P. M., "Semiconductor chip having connecting pads arranged in a nonorthogonal array", Patent USA 3795845, Publ. March 1974.
95. Dalbo, J. M., Mele, J. J. and Prasad, C., "Metallurgical vacuum furnace fluxless chip joining (by solder reflow)", IBM Tech. Disclosure Bull., 17, 2880 (1975), EEA78-19945.
96. David, R. F. S., "Advances in epoxy die-attach", Solid State Technol., 18 (9), 40-4 (1975).
97. Dionne, E., Graff, W. S., Riseman, J. and Siwy, M. J., Jr., "Method for modified solder reflow chip joining", IBM Tech. Disclosure Bull., 16, 2900 (1974), EEA77-23167.
98. Drop, J. G., "Integrated circuit personalization at the module level", IBM Tech. Disclosure Bull., 27, 1344 (1974), EEA78-11700.
99. Duffek, E. F., Funk, E. J., Jankowski, A. S., Lane, J. C., Lehner, W. L., Oliver, F. F. and Schneider, M., "Structure for simultaneously attaching a plurality of semiconductor dice to their respective package leads", Patent USA 3765590, Publ. October 1973.
100. Early, R. C., "Thick film/flip chips - a systems approach", IEEE Trans. Manuf. Technol., MFT-4, 2-7 (1975).
101. Eger, H., "Method of separating semiconductor chips from a semiconductor substrate", Patent USA 3668774, Publ. June 1972.
102. Forlani, F., "Method for mounting integrated-circuit chips on interconnection supports", Patent USA 3785044, Publ. January 1974.
103. Geilker, C. D., "Mounting IC flat-packs", Radio-Electron., 42 (6), 64 (1971), EEA75-4041.
104. Goldman, L. S., "Self alignment capability of controlled collapse chip joining", 22nd Electronic Components Conference, 1972, p. 332-39.
105. Goldman, L. S., Herdzik, R. J., Koopman, N. G. and Wild, R., "Lead-indium for joining a semiconductor chip to a substrate", IBM Tech. Disclosure Bull., 16, 3610-11 (1974), EEA78-692.
106. Gorski, D. A., Zappella, P. I. and Asch, A. E., "Flip-chip header for IR detector arrays operating at cryogenic temperatures", IEEE Trans. Parts, Hybrids & Packag., PHP-11, 312-4 (1975).
107. Grossman, S. E., "Film-carrier technique automates the packaging of IC chips", Electronics, 47 (10), 89-95 (1974), EEA77-23164.
108. Hall, R. N., "Semiconductor chip bonder", Patent USA 3617682, Publ. November 1971.
109. Hegarty, B. A. and Trevail, L. H., "High heat dissipation solder-reflow flip-chip transistor", Patent USA 3823469, Publ. July 1974.
110. Helda, R. W., "Spider bonding technique for IC's", International Electronic Circuit Packaging Symposium, 1969, 2 pp., EEA73-27358.

111.Herdzik, R. J., Jeannotte, D. A., Peterson, G. W. and Sullivan, M. J., "Improved semiconductor chip to substrate solder bond using a locally dispersed ternary intermetallic compound", Patent USA 3839727, Publ. October 1974.
112.Herrman, E. P., "Silicon chip separation to permit chip butting", Report TN-1110, RCA, Princeton, New Jersey, 1975, 2 pp., EEA78-37671.
113.Hugle, W. B., Bambera, J. L. and Pedrotti, D. B., "Flip-chip assembly", Solid State Technol., 12 (8), 62-7 et seq. (1969), EEA73-3528.
114.Johnson, K. I., Scott, M. H., Battarbee, W. and Griggs, R., "An investigation into flip chip bonding", Microelectron. & Rel., 10, 451-9 (1971), EEA75-6963.
115.Johnson, K. I., Scott, M. H., Battarbee, W. and Griggs, R., "The feasibility of ultrasonic welding flip chips to thick film conductors", Proceedings of the Conference on Hybrid Microelectronics, 1973, p. 71-4, EEA77-408.
116.Kulesza, F. W., "Where epoxy die bonding for microelectronics stands today", Insul./Circuits, 20 (12), 31-3 (1974), EEA78-7865.
117.Leinkram, C. Z., "Process for bonding chip devices to hybrid circuitry", Patent USA 3680196, Publ. August 1972.
118.Makhijani, M. P., Pastmann, H. H., Solis, M., Jr. and Wilk, R. H., "Waffled surface dicing pallet", IBM Tech. Disclosure Bull., 27, 1378 (1974), EEA78-8613.
119.Marinace, J. C., "Process of bonding chips in a substrate recess by epitaxial growth of the bonding material", Patent USA 3433686, Publ. March 1969.
120.Masson, J. E. and Le Pape, B., "Die bonder", IBM Tech. Disclosure Bull., 16, 3621-2 (1974), EEA78-91.
121.Matcovich, T. J., "Microcircuit interconnection techniques and applications", Am. Ceram. Soc. Bull., 51, 359 (1972), EEA75-27728.
122.McNutt, R. D., "Chip removal/replacement process", IBM Tech. Disclosure Bull., 16, 2971-2 (1974), EEA77-26533.
123.Miller, L. F., "Controlled collapse reflow chip joining", IBM J. Res. Developm., 13, 239-50 (1969), EEA72-17153.
124.Miller, L. F., "A critique of chip-joining techniques", Solid State Technol., 13 (4), 50-62 (1970), EEA73-33998.
125.Miller, L. F., "Flip chip interconnection systems", IEEE International Convention Digest, 1970, p. 330-1, EEA73-37078.
126.Palmer, G. G., "Method of embedding semiconductor chip within a dielectric layer flush with surface", Patent USA 3615946, Publ. October 1971.
127.Pietrucha, B. M. and Reiss, E. M., "Epoxy bonding for integrated circuits", Electron. Prod. Methods & Equip., 4 (6), 22-5 (1975), EEA78-41709.
128.Salzer, J. M., "Evaluating the economic factors of automated chip bonding", Insul./Circuits, 21 (2), 29-31 (1975), EEA78-15390.
129.Sello, H., "Use of aluminum bumps in multichip technology", Am. Ceram. Soc. Bull., 51, 360 (1972), EEA75-27786.

130.Siegel, M., "Flip chips are alive and well", International Electronic Circuit Packaging Symposium, 1969, 4 pp., EEA73-27286.
131.Slemmons, J. W., Salmassy, S. and Bester, M. H., "Flip-chip bonding with formed projections", International Electronic Circuit Packaging Symposium, 1969, 9 pp., EEA73-27287.
132.Texas Instruments, "Mounting for high-speed integrated circuits", Patent UK 1376939, Publ. December 1974, EEA78-19951.
133.Tognelli, C., "Notching die and compound die", IBM Tech. Disclosure Bull., 17, 2838-9 (1975), EEA78-19944.
134.Waite, G. C., "Semiconductor chip attachment with small bump flip chips", IEEE Trans. Manuf. Technol., MFT-4, 8-13 (1975).
135.Youmans, A. P., Rose, R. E. and Greenman, W. F., "A multichip package utilizing In-Cu flip-chip bonding", Proc. IEEE, 57, 1599-605 (1969), EEA73-459.

V. Beam-Lead Devices

136.Acosta, J. and Segal, H., "A beam lead packaging concept and its reliability", 22nd Electronic Components Conference, 1972, p. 201-6.
137.Adams, J. R., "Analysis and development of a thermocompression bond schedule for beam lead devices", IEEE Trans. Parts, Hybrids & Packag., PHP-8, 22-6 (1972), EEA75-39209.
138.Athanas, T. G. and Anastasio, A. A., "Making beam leads for semiconductor devices", Patent USA 3765970, Publ. October 1973, CA80-20731.
139.Bachner, F. J., "Aluminium interconnection and beam leads on polyimide-coated copper substrates", IEEE Trans. Parts, Hybrids & Packag., PHP-8, 30-4 (1972), EEA75-27730.
140.Bachner, F. J., "Flexible, beam-leaded substrates utilizing aluminum interconnections and beam leads", Am. Ceram. Soc. Bull., 51, 360 (1972), EEA75-27787.
141.Bachner, F. J., Cohen, R. A. and McMahon, R. E., "Rigid and non-rigid beam-lead substrates", Solid State Technol., 13 (8), 62-6 (1970), EEA74-357.
142.Bachner, F. J., Cohen, R. A., Mountain, R. and McMahon, R. E., "A beam-lead substrate package for a six-stage TTL shift register", Microelectron. & Reliab., 8, 307-12 (1969), EEA73-13078.
143.Boore, F. J. and Sutter, D. M., "Differential pressure test: a quantitative stress test method for bonded beam-leaded devices", Presented at Proceedings of the 21st Electronic Components Conference, Washington, D. C., May 1971, EEA74-29847.
144.Brown, L. W., "Beam lead testing, handling, and failure analysis", Report BDX-613-975 (Rev.), Bendix Corp., Kansas City, Mo., 1974, 64 pp., NSA31-2715.
145.Cabezas, M., "Cost/effectiveness of beam leads in hybrid circuits", Insul./Circuits, 20 (12), 27-9 (1974), EEA78-8624.
146.Chen, A. Y., "How not to destroy beam leads in testing IC chips", Electronics, 45 (9), 119 (1972), EEA75-19238.

147. Condra, L. W. and Gambale, R. A., "Thin film thermocouple (beam lead bonding), Tech. Dig., no. 32, 17-18 (1973), EEA77-18872.
148. Crishal, J. M. and Rice, E. J., "Beam leads and method of fabrication", Patent USA 3620932, Publ. November 1971.
149. Culver, H. E., Schilling, H. and Thun, R. E., "Advanced thin-film metallurgy for beam-leaded integrated circuits", J. Vac. Sci. & Technol., 10, 170-5 (1973), EEA76-23482.
150. Duffek, E. F., Jr. and Blech, I. A., "Method of fabrication of solder reflow interconnections for facedown bonding of semiconductor devices", Patent USA 3480412, Publ. November 1969.
151. Eleftherion, M. P., "Beam lead interconnection systems", IEEE International Convention Digest, 1970, p. 326-7, EEA73-37077.
152. Fuller, C. R., "Method of forming beam leads on semiconductor devices and integrated circuits", Patent USA 3653999, Publ. April 1972.
153. Gelsing, R. J. and Van Steensel, K., "A multilayer interconnection system with gold beam leads", Microelectron. & Reliab., 8, 325-9 (1969), EEA73-13014.
154. Gregory, P. L., "Beam lead semiconductor package", Patent USA 3857993, Publ. December 1974, EEA78-15427.
155. Hamer, C. A. and Loro, A., "Fabrication of beam leads", Patent USA 3760238, Publ. September 1973.
156. Harris, R. E. and Cullen, D. F., "Process for laser scribing beam lead semiconductor wafers", Patent USA 3824678, Publ. July 1974.
157. Harris, R. E., Nash, M. T. and Cullen, D. F., "Fabrication of beam-leaded MOS/LSI", 1971 International Microelectronics Symposium, Pap. no. 6.6, 4 pp., EEA72-12775.
158. Hayasaka, T. and Hattori, S., "On the mechanism of thermocompression bonding of gold beam leads", Rev. Electr. Commun. Lab., 23, 344-52 (1975), EEA78-33163.
159. Herring, C. R., Sattler, T. J. and Monahan, J. J., "Sequential bonder", Tech. Dig., no. 36, 17-18 (1974), EEA78-11779.
160. Howland, F. L., "Beam-lead and wire thermocompression bonding", Proceedings of the 21st Electronic Components Conference, 1971, p. 285-8, EEA74-33376.
161. Howland, F. L., "Beam-lead silicon integrated circuits and hybrid circuit assembly", Am. Ceram. Soc. Bull., 51, 360 (1972), EEA75-27785.
162. Keys, L. K., "Beam lead devices in hybrid microcircuits - techniques and applications", Insul./Circuits, 18 (2), 17-24 (1972), EEA75-19279.
163. Labuda, E. F., Herb, G. K., Ryden, W. D., Fritzinger, L. B. and Szabo, J. M., Jr., "The uniform gold process - a metal pattern generation technique for integrated circuits with a beam lead metallization system", Presented at Electrochemical Society Meeting, San Francisco, Calif., May 1974, EEA77-26517.

164. Legat, W. H. and Tielert, K. H., "Application of sputtering in the fabrication of semiconductor devices", Solid State Technol., 13 (12), 54-7 (1970), EEA74-12767.
165. Lepselter, M. P., "Semiconductive device including beam leads", Patent USA 3426252, Publ. February 1969.
166. Lesk, A. and Black, J. R., "Lead bonding techniques and physics of failure consideration", Presented at the Annual Reliability Physics Symposium (presentation abstracts), Las Vegas, Nev., April 1970, EEA73-27297.
167. Lohrmann, G., "Integrated circuits", Tech. Rundsch., 65 (30), 35-9 (1973), German, EEA77-6053.
168. Murray, L. A. and Mao, R. A., "Reliability of COS/MOS beam lead devices", 22nd Electronic Components Conference, 1972, p. 226-8.
169. Nash, M. T., "Beam lead plating process", Patent USA 3699010, Publ. October 1972.
170. Rose, A. S. and Greig, W. J., "Packaging of beam lead devices", Presented at 21st Electronic Components Conference, Washington, D. C., May 1971, EEA74-29846.
171. Sard, R., Okinaka, Y. and Waggener, H. A., "Electroless gold beam lead plating", J. Electrochem. Soc., 121, 62-6 (1974), EEA77-14592.
172. Shubiak, M. T. and Wadsworth, D. J., "Tradeoffs in packaging beam lead integrated circuit systems", Presented at IEEE International Computer Group Conference (Memories, Terminals and Peripherals), Washington, D. C., June 1970, EEA74-405.
173. Snyder, M. G., "Advanced semiconductor packaging through beam leads", International Electronic Circuit Packaging Symposium, 1969, 2 pp., EEA73-27357.
174. Sutter, D. M. and Yatsko, M. A., "Hybrid integrated circuit package for beam-leaded devices", 22nd Electronic Components Conference, 1972, p. 168.
175. Sylvania Electric Products Inc., "Integrated circuit", Patent UK 1143506, Publ. February 1969, EEA71-19366.
176. Tsurumiya, K., Hirano, Y. and Takahashi, A., "Microwave beam-lead transistor", Fujitsu Sci. & Tech. J., 10 (2), 105-24 (1974), EEA77-36300.
177. Tuzi, T., Awane, K. and Matunami, M., "Semiconductor device", Patent USA 3771219, Publ. November 1973, CA80-53767.
178. Wiley, T. A., "Early production history on hybrid microcircuits containing beam-lead semiconductors", Report BDX-613-1116-Rev., Bendix Corp., Kansas City, Mo., 1974, 29 pp., EEA78-37697.
179. Youngberg, D. A. and Lenhardt, B. W., "Rework technique removes beam-lead devices intact for post-mortem tests", Insul./Circuits, 21 (12), 21-2 (1975).
180. Zeien, R. H., Fenster, H. and Breig, W. J., "Screen printing for beam lead device bonding", International Microelectronics Symposium, 1971, 4-5, 5 pp., EEA75-12804.

VI. Soldering Techniques

181.Andrade, A. D. and Lampe, T., "Inspection techniques for resistance reflow soldered flatpack or ribbon leads", Insul./Circuits, 21 (1), 32-4 (1975), EEA78-10931.
182.Barrington, J., Larry, J. R., Hicks, W. T., Cole, G. R., Anders, F. J. and Hoffman, L. C., "Solderable and non-solderable microcircuit conductors: a status report", Microelectronics, 4 (4), 10-14 (1973), EEA76-26163.
183.Bemier, D., "Soldering hybrid microcircuits with paste solder", Insul./Circuits, 21 (12), 27-8 (1975).
184.Boynton, K. G., "Oil in wave soldering", Electron. Prod. Methods & Equip., 2 (8), 60, 62, 64 (1973), EEA77-13633.
185.Browne, L. T., "Soldering method and apparatus for ceramic circuits", Patent USA 3797100, Publ. March 1974.
186.Browne, L. T., "Reflow soldering of hybrid circuits", Insul./ Circuits, 21 (9), 27-9 (1975), EEA78-41711.
187.Bruder, S. I. and Faure, L. H., "Solder bucket pads contactor", IBM Tech. Disclosure Bull., 17, 1635-6 (1974), EEA78-11701.
188.DeBoskey, W. R., "Indium-lead solder for component mounting", IBM Tech. Disclosure Bull., 17, 1141 (1974), EEA78-5384.
189.Fenner, M. W., "Soldering chemicals in the electronics industry", Electron. Mfr., 17 (4), 19-20 (1972), EEA75-23432.
190.Finn, J. B., "Compatibility of glassy materials for monolithic packaging", Proceedings of the Electronic Components Conference, 1969, p. 322-8, EEA72-23212.
191.Harden, E. W., Mazzeo, N. J. and McCaffrey, P. M., "Solder transfer timing", IBM Tech. Disclosure Bull., 16, 3074 (1974), EEA77-23168.
192.Holley, J. H., "Solder sealing of large complex hybrid microcircuits", Proceedings of the Electronic Components Conference, 1970, p. 5-11, EEA74-499.
193.Kager, A., "Have present soldering techniques ground to a halt?", Elektron. J., 9 (11), 54, 56-7 (1974), German, EEA78-4794.
194.Kageyama, K., Yoshikawa, T. and Minatoguchi, T., "Soldering of hybrid integrated circuit", Anritsu Tech. Bull., no. 30, 127-31 (1973), Japanese, EEA78-11777.
195.Klose, J., "Experiences with light beam soldering", Finommech. & Mikrotech., 12, 358-61 (1973), Hungarian, EEA77-17839.
196.Kunert, A. and Oesterhelt, G., "Solder for contacting a semiconductor body and method for its production", Patent USA 3607240, Publ. September 1971.
197.Langan, J. P., "Plated coatings can aid soldering in electronic packaging", Insul./Circuits, 20 (9), 32 (1974), EEA77-38152.
198.Langan, J. P., "Important considerations for hand soldering (electronic assemblies)", Insul./Circuits, 21 (9), 20-1 (1975), EEA78-40481.

199.Langan, J. P., "Solder performs speed production and provides uniform, reliable connections", Insul./Circuits, 21 (4), 35-6 (1975), EEA78-23271.
200.Levitin, I. B. and Rozov, A. F., "Pulsed i.r. heating for soldering electronic components", Svetotekhnika, no. 8, 12-14 (1974), Russian, EEA77-38153.
201.McGougan, C. J., "Flow-solder wave profiles", Electron. Equip. News, 15 (5), 42-4 (1973), EEA77-173.
202.Maronna, G., "Welding and soldering techniques for joining small electronic components", Finommech. & Mikrotech., 12, 337-40 (1973), Hungarian, EEA77-17838.
203.Murray, J., "Examining soldering techniques", Electron. Equip. News, 15 (5), 31-3 (1973), EEA74-175.
204.Noth, R. W. and Sommer, G. C., Jr., "Solder transfer technique", IBM Tech. Disclosure Bull., 17, 2590 (1975), EEA78-19935.
205.Olson, H. C. and Knauss, G. L., "RF hybrid microcircuit soldering techniques", Report SLA-73-1073, Sandia Labs., Albuquerque, N. Mex., 1974, 44 pp., EEA78-41716.
206.Rabmovich, E. M., "Solder glasses and glass-crystalline cements", Izv. Akad. Nauk SSSR Neorg. Mater., 7, 545-60 (1971).
207.Schmidtke, H., "Soldering technique in electronics", Elektroniker, 11 (2), 49-58 (1972), German, EEA75-23431.
208.Smith, R. L. and Hewett, R. D., "Soldering with active organic-acid flux", Electron. Packag. & Prod., 15 (11), 57-61 (1975).
209.Stoicheva, V. and Karastoyanova, E., "Glasses for soldering tungsten and molybdenum", Stroit. Mater. Silikat. Prom., 14 (8-9), 28-30 (1973).
210.Sylvester, J. A., "Hand or manual soldering", Insul./Circuits, 21 (11), 76-84 (1975).
211.Thwaites, C. J., "The attainment of reliability in modern soldering techniques for electronic assemblies", Int. Metall. Rev., 17, 149-74 (1972), EEA76-81.
212.Tomlin, J. E., "Inter/nepcon - the metallurgy of soldering microconnections", Electronic Engng., 41 (500), 26-7 (1969), EEA73-9556.
213.Woeltle, R. and Deyl, V., "Soft-soldering contact in semiconductor device", Patent German 2340423, Publ. February 1975, CA83-20799.
214.Wozniakowski, B., "Rosin solder flux", Elektronika, 14, 483-8 (1973), Polish, EEA77-13632.

VII. Bonding Techniques

1. Thermocompression

215.Brantley, W. A. and Harrison, D. A., "Localized plastic deformation of GaP and GaAs generated by thermocompression bonding", J. Electrochem. Soc., 120, 1281-4 (1973), EEA77-2458.
216.Davies, R. W. and Ellington, T. S., "Composite thermocompression bonding thermode", Tech. Dig., no. 34, 9-10 (1974), EEA77-36313.

217. Fujiwara, S., "Thermocompression-bonded semiconductor device", Patent USA 3729807, Publ. May 1973, CA79-11274.
218. Higgins, J. J., Berejka, A. J. and Spenadel, L., "Bonding glass with thermoplastic sealing compositions", Patent USA 3654005, Publ. April 1972.
219. Holloway, P. H. and Long, R. L., Jr., "On chemical cleaning for thermocompression bonding", IEEE Trans. Parts, Hybrids & Packag., PHP-11, 83-8 (1975), EEA78-32166.
220. Intrator, A. M. and Adams, N., "Process of making thermoelectrostatic bonded semiconductor devices", Patent USA 3781978, Publ. January 1974.
221. Kraft, W., "Method of manufacturing a thermocompression contact", Patent USA 3843428, Publ. October 1974.
222. Parris, S. R., "MOS IC's: answers to applicational problems", Western Electronic Show and Convention, 1969, 8 pp., EEA73-23883.
223. Ramsey, T. H., "Metallurgical behavior of gold wire in thermal compression", Solid State Technol., 16 (10), 43-7 (1973), EEA77-10322.

2. Ultrasonic

224. Harman, G. G., "Microelectronic ultrasonic bonding", Report NBS-SP-400-2, Nat. Bur. Stand., Washington, D. C., 1974, 102 pp., EEA77-36338.
225. Kashiwabara, M., Hattori, S., Aoki, T. and Salto, Y., "Deformation characteristics and bonding process of fine aluminum wire in ultrasonic bonding", Rev. Elec. Commun. Lab., 19, 791-7 (1971), EEA75-9590.
226. Khadpe-Kamat, S., "Study of ultrasonic bonding of microcircuit interconnection", Thesis, Drexel Univ., Philadelphia, Pa., 1974, 127 pp., Order No. 74-23505, EEA78-15429.
227. Matcovich, T. J., Coren, R. I., Kelemen, D. G. and Galli, R. J., "Ultrasonic bonding of integrated circuits to thick film pedestals on ceramic substrates", Proceedings of the Electronic Components Conference, 1969, p. 254-75, EEA72-23135.
228. Perkins, J. P., "Ultrasonic assembly techniques", Electron. Prod. Methods & Equip., 2 (8), 66-8, 108 (1973), EEA77-13634.
229. Sakamoto, Y., Nishizuka, H. and Kawai, K., "Vibration element for supersonic bonding", Patent USA 3750926, Publ. August 1973.
230. Wirsing, C. E., "An ultrasonic bond, monometallic, gold interconnection technique for integrated packages", Solid State Technol., 16 (10), 48-50 (1973), EEA77-10323.

3. Others

231. Becker, P., "Problems in acceptance testing of semiconductor components. III. Influence of packaging material and bonding technique on the reliability", Radio Mentor Electron., 39, 444-7 (1973), German, EEA77-5162.

232.Benson, H. L. and Fewell, L. K., "Wire bonding to palladium-silver", International Microelectronics Symposium, 1971, 5 pp., EEA75-12771.
233.Chang, W. H., Kiseda, J. R., Rifkin, A. A. and Staats, R. W., "Personalizing prepacked semiconductor devices", IBM Tech. Disclosure Bull., 17, 1950-1 (1974), EEA78-15428.
234.Clark, R. J. and Loukianoff, M. A., "Adhesion of polyimide to integrated circuit substrates", IBM Tech. Disclosure Bull., 16, 2812 (1974), PA77-49975.
235.Cranston, B. H., "Methods for bonding leads and testing bond strength", Patent USA 3634930, Publ. January 1972.
236.Cranston, B. H., "Explosive bonding of electrical interconnections", IEEE Trans. Parts, Hybrids & Packag., PHP-8, 27-32 (1972), EEA75-39178.
237.Davis, D. O., "Factors in high reliability wire bonding", Presented at the Annual Reliability Physics Symposium (presentation abstracts), Las Vegas, Nev., April 1970, EEA73-28108.
238.Dehaine, G. and Kurzweil, K., "Tape automated bonding moving into production", Solid State Technol., 18 (10), 46-52 (1975).
239.Foote, R. S., "Bonding system for semiconductor device", Patent USA 3593412, Publ. July 1971.
240.Fowler, W. H., "Multiple position nest for bonding", Tech. Dig., no. 36, 9-10 (1974), EEA78-11778.
241.Gillis, T. B. and Lash, R. E., "Low temperature processing for hi-rel multi-chip hybrids", 24th Electronic Components Conference, 1974, p. 181-5, EEA77-34119.
242.Grechus, G. K. and Svitak, J. J., "Lead frame bonding", Tech. Dig., no. 38, 13-14 (1975), EEA78-28884.
243.Hamer, C. A., Loro, A. and Payne, J. R., "Bonding method for multiple chip arrays", Patent USA 3859723, Publ. January 1975, EEA78-24544.
244.Hartleroad, R. J. and Grabowski, J. P., "Multiple magnetic alignment of semiconductor devices for bonding", Patent USA 3868764, Publ. March 1975, EEA78-37338.
245.Hartleroad, R. J. and Grabowski, J. P., "Laminated template for semiconductor device bonding", Patent USA 3868765, Publ. March 1975, EEA78-41400.
246.Hartleroad, R. J., Krajewski, J. C. and Rayon, R. G., "Magnetic alignment for semiconductor device bonding", Patent USA 3887997, Publ. June 1975, EEA78-41733.
247.Herbst, D. L. and Slawsen, W., "Problems associated with wire-bonding in thick film circuits", Presented at Semiconductor Integrated Circuit Processing and Production Conference (abstracts), Anaheim, Calif., February 1971, EEA74-21611.
248.Hicks, R. E., "Substrate bonding bumps for large-scale arrays", Patent USA 3647533, Publ. March 1972.
249.Husni, S. Y., "Method of connecting semiconductor device to terminals of package", Patent USA 3668770, Publ. June 1972.

250.Kageyama, K., Yoshikawa, T. and Nishimura, Y., "Wire bonding for thin-film integrated circuit", Anritsu Tech. Bull., no. 30, 121-6 (1973), Japanese, EEA78-11776.
251.Kimball, J., "Epoxy device bonding and die handling techniques for hybrid microcircuits", Solid State Technol., 16 (10), 55-8 (1973), EEA77-10318.
252.Kurz, F. W. A., "Bonding or impregnating composition", Patent USA 3707386, Publ. December 1972.
253.Larkin, M., "Lead bonding and reliability", New Electron., 6 (3), 24 (1973), EEA73-16461.
254.Law, J. M., Strickler, A. A. and Von Kaenel, W., "Automatic pin insertion and bonding to a metalized pad on a (ceramic) substrate surface", Patent USA 3736651, Publ. June 1973.
255.Lind, R. S., "Method of sealing ceramic to nonmetallic using indium alloy seal", Patent USA 3747173, Publ. July 1973.
256.Ludwig, D. P., "Compliant bonding - a production method", International Microelectronics Symposium, 1971, 5 pp., EEA75-12805.
257.Marcotte, V. C., Sullivan, M. J. and Turene, F. E., "Metallurgical bonding of semiconductor pad terminals", IBM Tech. Disclosure Bull., 17, 1040 (1974), EEA78-5205.
258.Markstein, H. W., "Trends in automated wire bonding", Electron. Packag. & Prod., 15 (11), 39-45 (1975).
259.Marshall, C., "Low temperature methods of attachment of active devices to thick film circuits", Proceedings of the Conference on Hybrid Microelectronics, 1973, p. 75-81, EEA74-409.
260.Mika, M., "Microcircuit resistance pressure welding", Pr. Inst. Masz. Mat., 14 (1), 127-37 (1972), Polish, EEA76-8763.
261.Noll, D. A., "Method of joining (packaging)", Patent USA 3802987, Publ. April 1974.
262.Ortner, E. and Hartman, R. H., "O-bend lead for semiconductor packages", Patent USA 3449642, Publ. June 1969.
263.Pomerantz, D. L., "Method of bonding an insulator member to a passivating layer covering a surface of a semiconductor device", Patent USA 3595719, Publ. July 1971.
264.Rabut, E. W., "Arrangement for mounting semiconductive devices to heat sinks", Patent USA 3777238, Publ. December 1973.
265.Reichherzer, R., "Jointing methods for synthetic thermoplastic products", Technica, 22, 1907-11 (1973), German, EEA77-5153.
266.Riga, G., "TO-5 discoloration and bondability", Proceedings of the 11th Annual Reliability Physics Conference 1973, p. 26-32, EEA77-5781.
267.Ruggiero, E. M., "Method for bonding silicon semiconductor devices", Patent USA 3461462, Publ. August 1969.
268.Sayers, P., "Molecular bonding conductive films", Solid State Technol., 17 (9), 66-9 (1974), EEA78-2895.
269.Schoppe, C. G. and Pfeifer, H. J., "Termi-point-a new solderless connection technique", Feingerate Tech., 21 (5), 10-3 (1972), German, EEA75-27846.

270. Sherry, J. R., "Assembling plastic parts . . . without adhesives . . . without fasteners", Automation, 20 (11), 76-80 (1973), EEA77-13630.
271. Stevens, F., "Ribbon wire bonding", Electron. Prod. Methods & Equip., 4 (6), 13 (1975), EEA78-41729.
272. Wallis, G., "Field assisted glass sealing", Electrocompon. Sci. & Technol., 2, 45-53 (1975), EEA78-40480.
273. Wuich, W., "Adhesives for plastics", Elektron. Ind., 4 (12), EP137-40 (1973), German, EEA77-17834.
274. Yui, N., Hayashi, Y., Kano, T. and Teshima, H., "Wire bonding for high-frequency devices", Bull. Electrotech. Lab., 36, 41-50 (1972), Japanese, EEA75-36278.

VIII. Metal-Insulator Seals

1. Ceramic

275. Ameen, J. G., McBride, D. G. and Phillips, G. C., "Etching of high-alumina ceramics to promote copper adhesion", J. Electrochem. Soc., 120, 1518-22 (1973).
276. Anderson, N. C., "Method for making metal-to-ceramic seals", Patent USA 3736650, Publ. June 1973.
277. Andruszkiewicz, R., "Adhesive properties of AI-19 Ceramic", Pr. Osrodka Badaw.-Roxwojowego Elektron. Prozniowej, 2 (3), 75 (1974), Polish, EEA78-24425.
278. Bristow, R. H., "Ceramic-to-metal (Ag-Mn-Cu) seal", Patent USA 3700420, Publ. October 1972.
279. Bristow, R. H. and Jorgensen, P. J., "Method of making ceramic-to-metal seal", Patent USA 3736649 and Patent USA 3737977, Publ. June 1973.
280. Bronnes, R. L., Hughes, R. C. and Sweet, R. C., "Ceramic-to-metal bonding with sputtering as a metallization technique", Phillips Tech. Rev., 35, 209-11 (1975).
281. Burgess, J. F. and Neugebauer, C. A., "The gas-phase eutectic bonding of metals to ceramics for electronic applications", J. Electrochem. Soc., 121, 89C (1974), EEA77-26534.
282. Burgess, J. F., Neugebauer, C. A., Flanagan, G. and Moore, R. E., "Hybrid packages by the direct bonded copper process", Solid State Technol., 18 (5), 42-4 (1975), EEA78-33249.
283. Conwicke, J. A., "Sintering aids for producing alumina dielectric compositions cofireable with palladium", Patent USA 3801337, Publ. April 1974.
284. Crosby, G. E. and Shanefield, D. J., "Method for bonding a metal pattern to a substrate", Patent USA 3679472, Publ. July 1972.
285. Gutowski, R., Maslankiewicz, W. and Milosz, J., "Ceramic-metal joints with active metallic material", Elektronika, no. 12, 501-2 (1972), Polish, EEA76-11707.
286. Heidt, G. and Heimke, G., "Kinetics and mechanism of solderless metal-to-ceramic seals", Ber. Deut. Keram. Ges., 50, 303-7 (1973).

287. Hennicke, H. W. and Gruenling, H. W., "Process for the production of metal-ceramic (Al_2O_3) bond", Patent USA 3795041, Publ. March 1974.
288. Merz, J., Van Esdonk, J. and Janssen, J. F. M., "Metal-to-ceramic leadthrough", Patent USA 3854827, Publ. December 1974.
289. Moser, H. P. and Veth, G. M., "Compensator base for bonding lead-frames to ceramic substrates", Tech. Dig., no. 38, 21-2 (1975), EEA78-28834.
290. Klomp, J. T., "Adhesion bonding of ceramics to metals", Klei Keram., 21 (3), 58-64 (1971).
291. Klomp, J. T., "Recent developments for bonding ceramics to metals: I-II", Powder Met. Int., 3 (3), 142-6 (1971).
292. Klomp, J. T., "Bonding of metals to ceramics and glasses", Amer. Ceram. Soc. Bull., 51, 683-8 (1972).
293. Law, J. M., Stricker, A. A. and Von Kaenel, W., "Automatic pin insertion and bonding to a metalized pad on a substrate surface", Patent USA 3808681, Publ. May 1974.
294. Maslankiewicz, W., Milosz, J., Rybicki, K. and Gutowski, R., "On the mechanics of the creation of ceramic-metal sealed joints", Elektronika, 14, 225-9 (1973), Polish, EEA76-34662.
295. Popov, V. I., "Influence of doping elements introduced into copper on the specific electrical resistance, adhesion to substrate, thermal stability, and weldability of vacuum-condensed films", Izv. VUZ Fiz., no. 12, 25-31 (1973), Russian, EEA77-10274.
296. Schmidt-Bruecken, H. and Schlapp, W., "Bonding of metallic aluminum to alumina ceramic", Keram. Z., 25, 141-43 (1973).
297. Smith, B. R., "Composition and method of bonding gold to a ceramic substrate and a bonded gold article", Patent USA 3799890, Publ. March 1974.
298. Smith, B. R., "Gold composition and bonding gold to a ceramic substrate utilizing copper oxide and cadmium oxide", Patent USA 3799891, Publ. March 1974.
299. Takashio, H., "Alumina ceramic-to-metal seals by the Mo-Mn process", Yogyo Kyokai Shi, 79, 330-9 (1971).
300. Takashio, H., "Alumina ceramic-to-metal seals by the Mo process", Yogyo Kyokai Shi, 81 (3), 108-14 (1973).

2. Glass

301. Bernstein, M., "Method of bonding quartz to metal", Patent USA 3657076, Publ. April 1972.
302. Butler, M. F. and Corcia, J. T., "Petalite-spodumene-potassium silicate cement for bonding metal to glass", Patent USA 3832195, Publ. August 1974.
303. Correll, E. M., "Metal-to-glass-to-metal hermetic seal", Patent USA 3697823, Publ. October 1972.
304. English Electric Co., Ltd., "Glass-ceramic-metal seal", Patent UK 1312700, Publ. April 1973.

305.Eveyand, G. H. and Carson, J. B., "Method of sealing glass to metal", Patent USA 3632325, Publ. January 1972.
306.Francel, J., Minneman, L. C. and Nofziger, N. B., "Low-temperature sealant glass for sealing integrated circuit package parts", Patent USA 3778242, Publ. December 1973.
307.Galante, F. and Manfre, G., "Interface interaction in glass-metal composite production", Ceramurgia, 3 (1), 15-21 (1973).
308.Garceau, W. J., "Method for sealing glass to metal", Patent USA 3631589, Publ. January 1972.
309.Geyer, F., "Determination of the adhesive strength of glass-metal seals", Glastech. Ber., 46 (8), 153-55 (1973).
310.Hirayama, C., "Glass-to-metal seals comprising relatively high expansion metals", Patent USA 3804703, Publ. April 1974.
311.Horn, G. M., "Method of fabricating right spherical segment glass, shell-to-metal joint", Patent USA 3839109, Publ. October 1974.
312.Jacob, C. J. and Lawton, G. W., "Method of making electrical connections to a glass-encapsulated semiconductor device", Patent USA 3669734, Publ. June 1972.
313.Kim, C. and Tomozawa, M., "Glass-metal reaction in the electric field", Am. Ceram. Soc. Bull., 52, 380 (1973), EEA77-2498.
314.Koons, E. R., "Metal-to-glass-to-ceramic seal", Patent USA 3638076, Publ. January 1972.
315.Lucek, J. R., "Glass-ceramic to metal seals", Am. Ceram. Soc. Bull., 51, 349 (1972), EEA75-23775.
316.Matthews, H. N., "Ceramic-metal or glass-metal seals", Patent British 1294695, Publ. November 1972.
317.Osborne, J. F. and Pille, H. J., "Method of causing adherence of glass to gold", Patent USA 3728176, Publ. April 1973.
318.Pennebaker, W. B., "Improving adhesion between glass substrates and metal films", IBM Tech. Disclosure Bull., 16, 3426 (1974), EEA77-30354.
319.Root, F. A. and White, R. C., "Method of forming titanium metal-glass hermetic seals", Patent USA 3803875, Publ. April 1974.

IX. Interconnections

1. Single Layer

320.Anon., "Mounting and connection of integrated circuits", Electronic Engng., 41 (500), 17-20 (1969), EEA73-9634.
321.Arruda, J. and Peel, M., "Edgemount connectors - pluggability, reliability and standardization", 1973 IEEE International Convention and Exposition, Vol. VI, 50.2/9 pp., EEA77-2308.
322.Barlage, F. M., "Current induced breaking of thin film multiple circuit substrates into chips", Tech. Dig., no. 26, 7-8 (1972), EEA75-31850.
323.Barsuhn, H. E., Haug, W. and Krumm, H.. "Stacking and unstacking of substrates", IBM Tech. Disclosure Bull., 18, 442 (1975), EEA78-37580.

324. Bennett, B., "Fully repairable integrated circuit interconnections", Patent USA 3861023, Publ. January 1975, EEA78-28874.
325. Bingham, K. C. and Gillingham, A. G. A., "Hybrid techniques for computer circuits. III. Signal interconnection networks for integrated circuit chips", Proceedings of the Conference on Hybrid Microelectronics, 1973, p. 325-33, EEA77-142.
326. Boswell, D., "Thick films in relation to printed wiring", New Electron., 7 (10), 112, 115 (1974), EEA77-30369.
327. Burr, R. P. and Keog, R. J., "Simplified system of interconnections at high density wiring", Alta Frequenza, 40, 42-7 (1971), Italian, EEA74-17739.
328. Butt, S. H., "Developments in copper alloys for semiconductor connectors and packages", Proceedings of the 21st Electronic Components Conference, 1971, p. 103-14, EEA74-29967.
329. Calhoun, D. F. and McNamee, L. P., "A means of reducing custom LSI interconnection requirements", IEEE J. Solid-State Circuits, SC-7, 395-404 (1972), EEA75-35388.
330. Carbajal, B. C. and McMahon, W. R., "Corrosion resistant aluminum interconnect for plastic encapsulated integrated circuits", Proceedings of the 10th Electrical Insulation Conference, 1971, p. 61-3.
331. Chandragupta, K., "Electronic interconnection techniques", J. Inst. Eng. Electron. & Telecommun. Eng. Div., 53 (Pt. ET6), 252-7 (1973), EEA77-6112.
332. Clark, A., "From design to production in 'film wire'", Electronic Engng., 42 (507), 93-7 (1970), EEA73-20282.
333. Cunningham, J. A. and Fuller, C. R., "Metal contact (Mo) and interconnection system for nonhermetic enclosed semiconductor devices", Patent USA 3754168, Publ. August 1973.
334. Dhaka, V. A. and Kozik, A. F., "Interconnection metallurgy system for semiconductor devices", Patent USA 3823349, Publ. July 1974.
335. Donath, W. E., "Estimating expected connection length for LSI circuit layouts", IBM Tech. Disclosure Bull., 17, 1795-6 (1974), EEA78-15385.
336. Duncan, D. M., "Method of forming semiconductor device contacts", Patent USA 3740835, Publ. June 1973.
337. Edel, W. A. and Piriou, J., "Planarization of metal layers for interconnections on integrated circuits", IBM Tech. Disclosure Bull., 14 (12), 3837-8 (1972), EEA75-31815.
338. Farquharson, R. J., "Logic organization - its impact on interconnection density", IEEE International Convention, 1968, p. 204, EEA72-19391.
339. Fuller, C. R. and Ghate, P. B., "Method of forming contact and interconnect geometries for semiconductor devices and integrated circuits", Patent USA 3856648, Publ. December 1974.
340. Fuller, C. R. and Reinberg, A. R., "Contacting and connecting semiconductor devices in integrated circuits", Patent USA 3900944, Publ. August 1975, CA83-171683.

341. Ginsberg, G. L., "Connecting ICs into rigid printed wiring boards", Electron. Eng., 31 (5), 30, 33-4 (1972), EEA75-27955.
342. Guditz, E. A. and Burke, R. L., "Multichip integrated circuit memory with photoformed plated conductors", IEEE Trans. Parts, Hybrids & Packag., PHP-11, 89-96 (1975), EEA78-33191.
343. Ho, P. and Howard, J. K., "Electromigration resistant Al-Co alloy for narrow-line conductors", IBM Tech. Disclosure Bull., 18, 569-70 (1975), EEA78-37581.
344. Ho, P. and Howard, J. K., "Al-Co-Cu ternary metallurgy for narrow-line conductors", IBM Tech. Disclosure Bull., 18, 571 (1975), EEA78-37582.
345. Hu, K. C., "Application of polyimide film in chip packaging and interconnection", International Electronic Circuit Packaging Symposium, 1969, 14 pp., EEA73-27356.
346. Inman, M. N., "Interconnection panels: the alternative to p.c.b.s.", New Electron., 7 (6), 32, 34 (1974), EEA77-22622.
347. Isaacson, H., "Flexible circuitry interconnection matrix", Presented at Semiconductor Integrated Circuit Processing and Production Conference (abstracts), Anaheim, Calif., February 1971, EEA74-21791.
348. Kawanishi, H., Goto, S., Oyamada, T., Kato, H. and Kani, K., "A routing method of building block LSI", 7th Asilomar Conference on Circuits, Systems and Computers, 1973, p. 119-23, Publ. 1974, EEA77-30405.
349. Kernighan, B. W., Schweikert, D. G. and Persky, G., "An optimum channel-routing algorithm for polycell layouts of integrated circuits", Proceedings of the 10th Design Automation, 1973, p. 50-9, EEA77-2583.
350. Kerry, F. S., "The case for miniature bus bars", Electron. Packag. & Prod. (Int. Ed.), 7 (5), 27-9 (1971), EEA75-572.
351. Kirkpatrick, E. S. and Mayadas, A. F., "Theory of the resistivity of inhomogeneous conducting 'fine lines'", J. Appl. Phys., 44, 4370-7 (1973), EEA77-10314.
352. Komatinsky, R. R., "A high frequency microcircuit packaging and interconnection system", IEEE Trans. Parts, Hybrids & Packag., PHP-10, 110-15 (1974), EEA77-36317.
353. Larry, J. R. and Amin, R. B., "High adhesion, thick film palladium-silver conductors", 23rd Electronic Components Conference, 1973, p. 161-70, EEA76-25752.
354. Lawler, E. L., Levitt, K. N. and Lurner, I., "Module clustering to minimize delay in digital networks", IEEE Trans. Comput., C-18, 47-57 (1969), EEA72-19318.
355. Leinkram, C. Z. and Prom, R. L., "Microelectronic interconnection substrate", Patent USA 3607379, Publ. September 1971.
356. Lemon, T. H., "High tensile strength thick-film silver-palladium metallizations", Conference on Hybrid Microelectronics, 1975, p. 23-32, EEA78-41652.
357. Luttmer, W., "Methods of forming circuit interconnections", Patent USA 3835531, Publ. September 1974.

358.Markstein, H. W., "Interconnection update", Electron. Packag. & Prod., 15 (10), 29-34 (1975).
359.Marriott, C. L., Patzer, H. E., Hall, R. D., Caposell, C. D. and Klein, A. D., "High-strength aluminum interconnections for microelectronics packaging", Patent USA 3900353, Publ. August 1975, CA83-171680.
360.Naylor, R., "Digital IC applications - hybrid interconnection of digital circuits", Microelectron. & Reliab., 13, 411-12 (1974), EEA78-5379.
361.O'Brien, W. A. and Hansell, G., "Controlled impedance wiring techniques", Presented at Semiconductor Integrated Circuit Processing and Production Conference (abstracts), Anaheim, Calif., February 1971, EEA74-21573.
362.Owens, A. R. and Jaggers, K. A., "Pulse response of interconnections in silicon integrated circuits", Proc. Inst. Electr. Eng., 121, 541-7 (1974), EEA77-30398.
363.Pearce, G. R., "A new concept in printed wiring board/circuit package design and manufacture", Electron. Compon., 13, 282-3 (1972), EEA75-16093.
364.Pfahnl, A., "Thin film metallization and pattern generation techniques", Jap. J. Appl. Phys., Suppl. 2 (Pt. 1), A-857 (1974), EEA78-24444.
365.Quarmby, R. B., "Interconnection systems", New Electron., 7 (2), 70-1, 73 (1974), EEA77-18478.
366.Rausch, J. M., "Interconnecting with flexible circuitry", Electron. Packag. & Prod., 15 (5), 91-8 (1975).
367.Redwanz, J. O., "Semiconductor interconnecting system using conductive patterns bonded to thin flexible insulating films", Patent USA 3780352, Publ. December 1973.
368.Reed, B. S. and Wingo, D. T., "Packaged multilead semiconductor device with improved jumper connection", Patent USA 3436606, Publ. April 1969, EEA73-6809.
369.Richardson, D., "Approaches to multichip interconnections", IEEE International Convention Digest, 1971, p. 242-3, EEA74-21617.
370.Sawyer, N. J., Wells, D. C. and Frankland, H. G., "Interconnecting microwave integrated circuits in high temperature applications. I", Insul./Circuits, 21 (3), 47-9 (1975), EEA78-19937.
371.Sawyer, N. J., Wells, D. C. and Frankland, H. G., "Interconnecting microwave integrated circuits in high temperature applications. II", Insul./Circuits, 21 (4), 49-53 (1975), EEA78-24536.
372.Shannon, J. L., "Will interconnectable plug-ins meet your test and measurements needs better?", 1973 WESCON Technical Papers, Western Electronic Show and Convention, Vol. 17, 18/1-4 pp., EEA77-9273.
373.Shiba, H. and Tsunemitsu, H., "Method of forming electrical connections in a semiconductor integrated circuit", Patent USA 3741880, Publ. June 1973.
374.Socolovsky, A., "Diaphragm connection", Electron. Eng., 30 (2), 27-8 (1971), EEA74-21538.

375. Staller, J., "Backpanel wiring by automatic equipment", Presented at Semiconductor Integrated Circuit Processing and Production Conference (abstracts), Anaheim, Calif., February 1971, EEA74-21610.
376. Stein, S. J., Huang, C., Cang, L. and Schultz, G., "Recent advances in platinum-silver thick-film conductors", Solid State Technol., 18 (5), 25-33 (1975), EEA78-31185.
377. Terry, L. E. and Wilson, R. W., "Method of forming metalization backing for silicon wafer", Patent USA 3785892, Publ. January 1974.
378. Wurst, B., "A new wireless wiring technology for plug-in boards", Mess. U. Prufen, 6 (3), 195-6 (1970), German, EEA73-30520.

2. Multilayer

379. Abbas, S. A. and Weeks, W. T., "Analytical calculation metallized-hole resistance", IEEE Trans. Pts. Materials Packaging, PMP-6, 26-35 (1970), EEA73-20271.
380. Ardito, F. P. and Stockdale, R. W., "Making integral multilayer circuit boards with cable connection", IBM Tech. Disclosure Bull., 14, 701-2 (1971), EEA75-577.
381. Balyoz, J. and Gruodis, A. J., "Densified wiring channels design concept", IBM Tech. Disclosure Bull., 17, 1958-9 (1974), EEA78-19904.
382. Basseches, H. and Pfahnl, A., "Crossovers for interconnections on substrates", Proceedings of the Electronic Components Conference, 1969, p. 78-82, EEA72-23130.
383. Berglund, C. N., Nicollian, E. H., Tompsett, M. F. and Waggener, H. A., "Multiple-level metalization for integrated circuits", Patent USA 3837907, Publ. September 1974.
384. Bingham, K. C. and Naylor, R., "Multilayer interconnection techniques applied to high speed computer systems", IEEE Trans. Parts, Hybrids & Packag., PHP-10, 175-80 (1974), EEA78-2923.
385. Czechowska, Z., "Dielectric materials for thick-film circuits", Elektronika, 14, 515-18 (1973), Polish, EEA77-10291.
386. Detloff, T. N., "Multilayer interconnection boards for V.H.F. circuitry", 22nd Electronic Components Conference, 1972, p. 252-63.
387. Finch, W. F., Kaiser, H. L., Pakulski, F. J. and Reardon, W. E., "Integrated circuit interconnection and packaging", IBM Tech. Disclosure Bull., 13, 962-3 (1970), EEA74-4226.
388. Foerster, R. P., Heath, H. E. and Parks, H. L., "High speed logic packaging with planar coaxial 3-dimensional interconnections", Presented at Semiconductor Integrated Circuit Processing and Production Conference, Anaheim, Calif., February 1971, EEA74-21571.
389. Gargione, F., Keyzer, W. and Noel, G., "A multilayer, hybrid thin-film 1024 x 3 COS/MOS memory", New Electron., 6 (13), 25-29 (1973), EEA76-38320.

390. Gioia, J. C., "Multilayering for a low cost hermetic thick-film substrate package", Solid State Technol., 14 (5), 43-6 (1971), EEA74-25651.
391. Greig, W. J. and Soden, R. R., "Semiconductor device with multilevel metalization and method of making the same", Patent USA 3668484, Publ. June 1972.
392. Harada, S., Nonogaki, S., Oba, Y., Saiki, A. and others, "Multilayer interconnected structure for semiconductor integrated circuit and process for manufacturing", Patent USA 3801880, Publ. April 1974.
393. Hord, C. B., "'μ-fold'tm: a packaging technique for LSI and hybrid MSI assemblies", Proceedings of the Electronic Components Conference, 1969, p. 145-51, EEA72-23210.
394. Hou, H. S., "Application of uniform loading theory to circuit packaging and memory arrays in high-speed computers", IEEE Trans. Comput., C-21, 454-63 (1972), EEA75-16190.
395. Ilgenfritz, R. W., "Hybrid circuit technologies and multilayer packaging", Plating, 57, 135-7 (1970), EEA74-27295.
396. Ilgenfritz, R., Krohane, J. S. and Mogui, L., "The interconnection and packaging of thick film microcircuits", Proceedings of the Electronic Components Conference, 1970, p. 122-6, EEA74-452.
397. Isaak, H. R., Kanz, J. W. and Babiracki, F. G., "Development of large thick-film multilayer assemblies", International Microelectronics Symposium, 1971, 6-4, 18 pp., EEA75-12807.
398. Jongeans, A. and Pascual, C. M., "A simple method for construction of crossovers in hybrid circuits", Electron. & Fis. Apl., 16, 145-8 (1973), Spanish, EEA77-2617.
399. Keller, H. N. and Pfahni, A., "Technique for the fabrication of an air-isolated crossover", Patent USA 3783056, Publ. January 1974, CA80-75745.
400. Kikuchi, A. and Agatsuma, T., "Method of forming multilayer interconnections", Patent USA 3833434, Publ. September 1974.
401. Kobayashi, T. and Murase, K., "Characteristics of multilayered metallizations consisting of tantalum and aluminum on silicon and/or SiO_2", 24th Electronic Components Conference, 1974, p. 79-82, EEA77-34110.
402. Lindberg, F. A., Ponemone, S. J., Roach, J. A. and Klein, A. D., "Interconnection techniques for multilayer hybrid microcircuit packages", Proceedings of the Electronic Components Conference, 1969, p. 152-6, EEA72-23132.
403. McCormick, J. E., "Metals joining of electronic circuitry", In: Handbook of materials and processes for electronics, McGraw-Hill, New York, 116 pp. (1970), EEA74-30025.
404. Madas, I., "The assembly of integrated circuits. I", Finommechanika, 10, 108-16 (1971), Hungarian, EEA75-418.
405. Matcovich, T. J., "Interconnections in hybrid circuits", IEEE International Convention Digest, 1971, p. 240-1, EEA74-21616.

406. Maxwell, H. and Vinci, M., "The hard-wiring way to high densities", Electronic Packaging Prod., 8 (11), 38-43 (1968), EEA72-7453.
407. Miller, L. F., "Powder interconnections", International Electronic Circuit Packaging Symposium, 1969, 11 pp., EEA72-27445.
408. Mutoh, H., Mizokami, Y. and Matsui, H., "Multilayer metallization with the planar interconnect structure", Presented at Electrochemical Society Meeting, San Francisco, Calif., May 1974, EEA77-26498.
409. Mutoh, H., Mizokami, Y., Matsui, H., Hagiwara, S. and Ino, M., "Multilayer metallization with planar interconnect structure utilizing CVD Al_2O_3 films", J. Electrochem. Soc., 122, 987-92 (1975), EEA78-37643.
410. Noth, R. W., Pittman, G. D. and Ward, W. C., "Multilayer substrate repair method", IBM Tech. Disclosure Bull., 16 (9), 2789-90 (1974), EEA77-23079.
411. Okabe, T., Saiki, A., Nagata, M., Sato, K. and Makimoto, T., "A design approach for LSI using chip selection and circuit modification techniques", Electron. & Commun. Jap., 55 (11), 78-85 (1972), EEA76-38305.
412. Parks, H. R., "High density interconnections: packaging with volumetric efficiency", Electromech. Des., 16 (1), 36-40 (1972).
413. Pfahnl, A. and Berry, R. W., "Preparation and properties of beam crossovers for interconnections in thin film circuits", Thin Solid Films, 13, 51-5 (1972), EEA76-4706.
414. Platter, V. and Schwartz, G. C., "Monolithic studs for interlevel connectors", IBM Tech. Disclosure Bull., 17 (6), 1605-6 (1974), EEA78-11728.
415. Rioult, J. P. and Henri, D., "Polylithic approach of the large scale integration: preliminary studies of high density packaging realization", Onde Elect., 50, 186-91 (1970), French, EEA73-20149.
416. Riva, R., "Multilayer interconnections - technology", RC Riun. Ass. Elettrotec. Ital., 45 (2), 6.1.04/2 pp. (1970), Italian, EEA75-31819.
417. Sato, K., Harada, S., Saiki, A., Kimura, T., Okubo, T. and Mukai, K., "A novel planar multilevel interconnection technology utilizing polyimide", IEEE Trans. Parts, Hybrids & Packag., PHP-9, 176-80 (1973), EEA76-38310.
418. Schwartz, G. C., "An anodic process for forming the interconnection metallization for multilevel LSI", Presented at Electrochemical Society Meeting, San Francisco, Calif., May 1974, EEA77-23141.
419. Schwartz, G. C. and Platter, V., "An anodic process for forming planar interconnection metallization for multilevel LSI", J. Electrochem. Soc., 122, 1508-16 (1975).
420. Short, O. A., Hoffman, I. C. and Rosenberg, R. M., "Galvanic effects in thick film dielectric layers", Am. Ceram. Soc. Bull., 52, 713-14 (1973), EEA77-10287.
421. Smith, D. L., "Flexible multilayer interconnection board fabrication", IBM Tech. Disclosure Bull., 17, 990 (1974), EEA78-5168.

422.Takagi, M., Terada, C. and Nakayama, K., "A buried metallization and its application to a multilayer system", 1973 International Electron Devices Meeting Technical Digest, p. 269-71, EEA77-14619.
423.Vossen, J. L., Schnable, G. L. and Kern, W., "Processes for multilevel metallization", J. Vac. Sci. Technol., 11, 60-70 (1974), CA80-138002.

X. Metallized Ceramics

424.Balakrishma, P. and Mahapatra, S., "Electroless copper on alumina and other substrates", Plating, 60, 636-9 (1973), EEA77-385.
425.Buck, R. W., "Method for metalizing a ceramic", Patent USA 3801363, Publ. April 1974.
426.Campbell, A. E., Jr. and Jensen, A. O., "Metalized and brazed ceramics", Patent USA 3666429, Publ. May 1972.
427.Clark, D. E., Hench, L. L. and Bates, S. R., "Molybdenum metalizing on beryllia", Am. Ceram. Soc. Bull., 53, 473-7 (1974), EEA78-669.
428.Cole, S. S., Jr., "Metalizing material", Patent USA 3609105, Publ. September 1971.
429.Davis, R. H. and Deo, S. R., "An evaporation system for the metallization of bilevel substrates", J. Vac. Sci. & Technol., 11, 77-80 (1974), EEA77-38864.
430.Dillingham, R. P., "MECCA: a thick film variant", Am. Ceram. Soc. Bull., 52, 714 (1973), EEA77-10288.
431.Duffek, E. F., "Fine line patterning of metalized ceramics", Plating, 57, 37-42 (1970), EEA73-27268.
432.Gaither, J. F. and Wolf, E. A., "Method of metallizing ceramic parts", Tech. Dig., no. 28, 15-16 (1972), EEA76-84.
433.Groll, H. and Weidmann, W., "Deposition of metal-films on ceramic substrates for microwave-circuits by cathode-sputtering", Nachrichtentech. Z. (NTZ), 26, 448-53 (1973), German, EEA77-393.
434.Lavendel, H. W., "Metalized-ceramic/metal composites", Amer. Ceram. Soc. Bull., 53, 797-99, 808 (1974).
435.Schmeckenbecher, A. F., "Controlled plating of a corrosion barrier on metallized ceramic modules", IBM Tech. Disclosure Bull., 16, 3576 (1974), EEA77-38150.
436.Wilson, R. E., Jr. and Antonucci, R., "Soldered lines on metallized ceramic substrates", IBM Tech. Disclosure Bull., 16, 2886 (1974), EEA77-23080.

XI. Epoxies

437.Cadenhead, R. L., "Substrate attach epoxies", Solid State Technol., 18 (10), 53-5 (1975).
438.Davis, E. K. and Hansen, K. W., "Glasses for encapsulating semiconductor devices", Patent USA 3723835, Publ. March 1973.

439. Evans, D., Morgan, J. T. and Stapleton, G. B., "Epoxy resins for superconducting magnet encapsulation", Report RHEL/R251, Rutherford High Energy Lab., Chilton, Berks, England, 1972, 27 pp., EEA76-4405.
440. Foulke, K. W., "Thermal plastic radomes", 12th Symposium on Electromagnetic Windows, 1974, p. 151-5, EEA78-3350.
441. Harwood, M. G., Schofield, J. and Burrell, M. V., "Conducting polymers for die bonding of semiconductors to thick film circuits", Proceedings of the Conference on Hybrid Microelectronics, 1973, p. 63-70, EEA77-407.
442. Henderson, A. W., "Encapsulation materials: Plastics", Compon. Technol., 3 (3), 5-10 (1969), EEA72-12860.
443. Hirsch, H., "Resin systems for encapsulation of microelectronic packages", Solid State Technol., 13 (8), 48-55 (1970), EEA74-355.
444. Hnatek, E. R., "Epoxy B: a new IC encapsulant", Electron. Packag. & Prod. (Int. Ed.), 8 (4), 17-20 (1972), EEA75-39259.
445. Hnatek, E. R., "Epoxy package increases i.c. reliability at no extra cost", Electron. Engineering, 45 (540), 66-8 (1973), EEA76-16476.
446. Hooke, T. A., "Silicones in electronics - what they are and where they are used", Proc. Inst. Radio Electron. Eng. Aust., 32 (5), A11 (1971), EEA74-30020.
447. Kerlee, C., "Selecting encapsulants and potting compounds for electronic molecules", Proceedings of the 10th Electrical Insulation Conference, 1971, p. 166-70, EEA75-12835.
448. Kovel, W. and Stevens, T., "Silicone moulding compounds", Des. Electron., 8, 44-7 (1970), EEA74-495.
449. Kubo, E. and Urano, T., "Flame retardant epoxy molding compounds for encapsulation", Hitachi Hyoron, 56 (7), 83-8 (1974), Japanese, EEA78-67.
450. Lidgley, D., "Laminates for electronics", Electron. Components, 15 (1), 35-7 (1973), EEA76-38324.
451. Lovell, R. T., "FEP florocarbon film as an adhesive dielectric in electronic packaging", Insul./Circuits, 17 (11), 45-7 (1971), EEA75-9646.
452. McDonald, R. J., "High speed encapsulation using liquid injection molding with thermosetting silicone elastomers", Proceedings of the 10th Electrical Insulation Conference, 1971, p. 81-5.
453. McMillan, R. E., Misra, R. P. and Stroll, Z., "Diffusion barrier films for resin", Proceedings of the 10th Electrical Insulation Conference, 1971, p. 222-5.
454. Nolan, P. C. G., "Resin attachment of semiconductor dice on to film circuits", Microelectron. & Reliab., 12, 235-41 (1973), EEA76-35369.
455. Olberg, R. C., "The effects of epoxy encapsulant composition of semiconductor device stability", J. Electrochem. Soc., 118, 129-33 (1971), EEA74-15441.

456.Sack, R. and Frey, G., "Use of thermosetting plastics as material for small pressure resistant capsules", Electric, 24, 375-6 (1970), German, EEA74-497.
457.Salensky, G., "Epoxy resin systems for electrical/electronic encapsulation", Insul./Circuits, 18 (5), 19-25 (1972), EEA75-23428.
458.Segro, N. R. and Benchman, H. H., "New diallyl phthalate resins for low pressure molding encapsulation of electronic components", Proceedings of the 9th Electrical Insulation Conference, 1969, p. 98-100, EEA73-13585.
459.Simpson, W. E. and Pettican, K. A., "Plastics encapsulation of components", Electronic Engng., 42 (506), 50-6 (1970), EEA73-9636.
460.Stephenson, R. L., "Material selections for thermionic capsules", IEEE Conference Record of the Thermionic Conversion Specialist Conference, 1969, p. 404-10, EEA74-3808.
461.Tesla, N. P., "Semiconductor encapsulating composition", Patent UK 1200151, Publ. July 1970, EEA74-9868.
462.Thomas, G. L., "Semiconductor encapsulation glass", Patent USA 3493405, Publ. February 1970, EEA73-20218.
463.VanWert, B. and Waterman, W., "A new family of encapsulants: rigid solventless silicone casting resins", Proceedings of the 10th Electrical Insulation Conference, 1971, p. 42-5, EEA75-12836.
464.Zecher, R. F., "A further analysis of liquid resin molding", Proceedings of the 9th Electrical Insulation Conference, 1969, p. 54-6, EEA73-13583.

XII. Pastes/Screens

465.Austin, B. M., "Thick-film screen printing", Solid State Technol., 12 (6), 53-8 (1969), EEA73-514.
466.Aycock, T. W. and Wright, J. J., "Electrical resistor paste containing a small amount of charcoal", Patent USA 3711428, Publ. January 1973.
467.Cometta, M., "Technology of thick film circuits", Elettrificazione, no. 6, 270-6 (1975), Italian, EEA78-37590.
468.Cote, R., "Problems in screen printing and firing", Presented at Semiconductor Integrated Circuit Processing and Production Conference, Anaheim, Calif., February 1971, EEA74-21614.
469.Daiga, V. R., "Gold powder (paste)", Patent USA 3843379, Publ. October 1974.
470.Dubey, G. C., "Screens for screen printing of electronic circuits", Microelectron. & Reliab., 13, 203-7 (1974), EEA77-30367.
471.Epple, R., "Screen-printing paste", Patent German 2240168, Publ. February 1974, CA80-138485.
472.Franconville, F., Kurzweil, K. and Stalnecker, S. G., "Screen: essential tool for thick film printing", Solid State Technol., 17 (10), 61-8 (1974), EEA78-5351.

473. Horwood, R. J., "Towards a better understanding of screen print thickness control", Electrocompon. Sci. & Technol., 1 (2), 129-36 (1974), EEA78-19868.
474. Jolley, J. E., Weaver, R. V., Wells, H. J., Felten, J. J. and Roe, D. W., "Photoprintable materials and processing equipment for thick-film microcircuitry", Solid State Technol., 17 (5), 33-7 (1974), EEA77-30370.
475. Kurzweil, K. and Loughran, J., "Thick film pastes for multilayer use", IEEE Trans. Parts, Hybrids & Packag., PHP-9, 216-23 (1973), EEA77-14627.
476. Kusy, A., "An investigation of the relations between the technological parameters, phase constitution and electrical characteristics of thick Pd-Ag resistive films", Thin Solid Films, 17, 345-61 (1973), EEA76-38011.
477. Mase, S. and Watanake, T., "Multilayer metalized beryllia ceramics and a method for producing the same", Patent USA 3889041, Publ. June 1975, EEA78-41743.
478. Miller, L. F., "Paste transfer in the screening process", Solid State Technol., 12 (6), 46-52 (1969), EEA73-466.
479. Miller, L. F., "Screenability and rheology", Solid State Technol., 17 (10), 54-60 (1974), EEA78-5350.
480. Riemer, D. E., "Screen printing as a volumetric deposition process", Am. Ceram. Soc. Bull., 52 (9), 713 (1973), EEA77-10283.
481. Roback, S. M., "Thick film for consumer products", Electron. Prod. Methods & Equip., 2 (7), 38, 40, 42-3 (1973), EEA77-14579.
482. Schmeckenbecher, A. F., "Wiping arrangement (particle removal from metal paste)", IBM Tech. Disclosure Bull., 16, 3580 (1974), EEA77-38872.
483. Schneider, J., "Novel screen print stencils for thick film technique", Elektron. Ind., 5 (12), EP115 (1974), German, EEA78-15396.
484. Swayne, R. and Rao, U. V., "Thermal behavior of thick-film (pastes)", Solid State Technol., 16 (6), 45-47 (1973).
485. Trease, R. E. and Dietz, R. L., "Rheology of pastes in thick-film printing", Solid State Technol., 15 (1), 39-43 (1972), EEA75-27810.
486. Urfer, E. N., "Pastes for multilevel ceramics", Am. Ceram. Soc. Bull., 52, 713 (1974), EEA77-10284.
487. Urfer, E. N., "Simultaneous line and via screening", IBM Tech. Disclosure Bull., 16, 2890 (1974), EEA77-23082.
488. Walton, B., "Principles of thick film materials formulation", Proceedings of the Conference on Hybrid Microelectronics, 1973, p. 39-46, EEA-77389.
489. Webster, R., "Fine line screen printing yields as a function of physical design parameters", IEEE Trans. Manuf. Technol., MFT-4, 14-20 (1975), EEA78-37692.
490. Wolter, H. G., "Masks for screen-printed circuits", Patent German 2049146, Publ. April 1972, CA77-68576.

XIII. Resistors

491.Anon., "Zirconium boride resistors for hybrid integrated circuits", Microelectron. & Reliab., 14, 67 (1975), EEA78-15426.
492.Asama, K., Nishimura, Y. and Sasaki, H., "Study on the thick film resistance abrupt change by resin packaging", Fujitsu Sci. Tech. J., 6 (6), 133-46 (1970), EEA74-560.
493.Battista, M. A., Harbison, R. S., Freed, L. E., Nestork, W. J., Struck, J. R. and Tuman, D., "Resistor tracking on monolithic integrated circuits", IBM Tech. Disclosure Bull., 15, 423-4 (1972), EEA75-39202.
494.Bawa, S. C., Runthala, D. P. and Marathe, B. R., "Fabrication of thin film resistors and resistor networks by a selective etching process", J. Inst. Electron. & Telecommun. Eng., 20, 602-6 (1974), EEA78-33187.
495.Bennett, A. G., "Effect of rapidly changing voltages on thick-film resistors", Report BDX-613-1085-Rev., Bendix Corp., Kansas City, Mo., 1974, 29 pp., EEA78-28838.
496.Bianchi, D. A. and Cattaneo, A., "Evaluation of the noise index in laser trimmed thick film resistors", Conference on Hybrid Microelectronics, 1975, p. 205-12, EEA78-41656.
497.Bos, L. and VerHoef, J., "Preparation of stable, precise resistors by anodization adjust of tantalum nitride films", 1973 IEEE International Convention and Exposition, Vol. VI, 41.4/11 pp., EEA77-2596.
498.Bulger, G. A., "Stability analysis of laser trimmed thin film resistors", IEEE Trans. Parts, Hybrids & Packag., PHP-11, 172-7 (1975), EEA78-41664.
499.Camenzind, H. R. and Allison, D. F., "Bulk resistor and integrated circuit using it", Patent USA 3798753, Publ. March 1974, CA80-138416.
500.Degenkolb, R. S., Allington, T. R., Wang, Y. H. and Oakes, M., "Method of controlling resistance values of thick-film resistors", Patent USA 3793717, Publ. February 1974.
501.de Lacy, H., "The assessment of thick film resistor performance by high voltage pulsing", Conference on Hybrid Microelectronics, 1975, p. 219-28, EEA78-41658.
502.Faith, T. J., Jr., "High power density thin film resistors", IEEE Trans. Parts, Hybrids & Packag., PHP-11, 273-80 (1975).
503.Gangulee, A. and Kobliska, R. J., "Material for thin-film resistors", IBM Tech. Disclosure Bull., 17, 3093 (1975), EEA78-24443.
504.Hart, P. J., "Thick film resistor networks", Electron. Components, 16 (2), 41 (1974), EEA77-10290.
505.Heid, K., "Thin-film high sheet resistivity material", Solid State Technol., 16 (9), 56-58 (1973).
506.Hill, R. M., "Conduction theory applied to thick film resistors", Conference on Hybrid Microelectronics, 1975, p. 251-8, EEA78-41659.

507.Hof, G. J., "Effect of soldering flux on Ta_2N resistors", Report SLA-73-1045, Sandia Corp., Albuquerque, N. Mex., 1974, 18 pp., EEA78-28883.
508.Izumi, K., Masanobu, D. and Ariyoshi, H., "Properties of zirconium nitride film resistors deposited by reactive RF sputtering", IEEE Trans. Parts, Hybrids & Packag., PHP-11, 105-7 (1975), EEA78-33186.
509.Kamo, T., Yarmazaki, J. and Nakamura, M., "Tantalum aluminum nitride thin film resistors with high sheet resistance", Fujitsu Sci. & Tech. J., 10, 145-64 (1974), EEA77-36322.
510.Kapfer, V. C. and Bart, J. J., "Thin film nickel-chromium resistor failures in integrated circuits", Proceedings of the 10th Annual Conference on Reliability Physics, 1972, p. 175-82, EEA76-4708.
511.Kazmierowicz, C. W., "Where is the resistance in a thick film resistor network?", Am. Ceram. Soc. Bull., 52, 13 (1973), EEA77-10286.
512.Koltai, M. and Kollar, S., "Chip resistors", Hiradstech. Ipari Kut. Intez. Kozl., 13 (3), 31-44 (1973), Hungarian, EEA77-18865.
513.Law, H. T., "Trimming of thin film resistors by spark machining", Conference on Hybrid Microelectronics, 1975, p. 181-90, EEA78-41670.
514.Livermore, F. C. and Neathway, G. A., "Thin-film resistor anodization", Patent Canada 901178, Publ. May 1972, CA80-138414.
515.Luder, E., "Miniaturized circuits", Umsch. Wiss. & Tech., 73, 216-17 (1973), EEA76-26164.
516.Merker, K. and Priebsch, G. B., "Investigations into the effect of surface roughness of substrates on the behaviour of thin-film resistors for various thicknesses of CrNi films deposited by evaporation. II", Hermsdorfer Tech. Mitt., 13, 1125-30 (1973), German, EEA77-9819.
517.Merker, K. and Priebsch, G. B., "Investigations into the effect of surface roughness of substrates on the behaviour of thin-film resistors for various thicknesses of evaporated CrNi-films. III", Hermsdorfer Tech. Mitt., 13, 1131-5 (1973), German, EEA77-9820.
518.Mickelsen, R. A., "Tantalum thin film resistors by reactive evaporation", Patent USA 3874922, Publ. April 1975, EEA78-37603.
519.Mori, I., Narita, M. and Kasahara, Y., "Zirconium nitride film resistor", Patent Japan 73-39638, Publ. November 1973, CA80-138415.
520.Nicholas, K. H., Ford, R. A. and Beale, R. A., "Junction-isolated integrated-circuit resistor with crystal damage near isolation junction", Patent USA 3796929, Publ. March 1974.
521.Nowak, S., "The testing of the 3rd harmonic wave in thick-film resistors", Elektronika, 16, 151-4 (1975), Polish, EEA78-28837,
522.Nulchis, K., Ripka, G. and Szilagyi, M., "Some problems of the accuracy of thin film resistors", Finommech. & Mikrotech., 12, 300-4 (1973), Hungarian, EEA77-18845.

523.Olesen, S. T., "Switching resistors as microcomponents", EC-Nyt, no. 44, 11-12 (1975), Danish, EEA78-33184.
524.Oren, R., "High voltage IC's, transistors and resistors", 8th Convention of the Electrical and Electronics Engineers in Israel, 1973, p. 67, EEA77-6083.
525.Orlowska, I. and Wilamowski, B. M., "Design method of diffusion resistors", Oyo Buturi, 43 (Suppl.), 415-17 (1974), EEA78-2915.
526.Paulson, W. M., "Further studies on the reliability of thin-film nickel-chromium resistors", Proceedings of the 11th Annual Reliability Physics Conference 1973, p. 61-8, EEA77-5691.
527.Pitt, K., "Thick film circuits", Electron. Components, 16 (4), 19, 22-3 (1974), EEA77-14578.
528.Polinski, P. W., "Stability of thick-film resistors under high electromagnetic stress", Solid State Technol., 16 (5), 31-5 (1973), EEA76-34991.
529.Ripka, G., Papp, K. and Albrecht, M., "Design of thick film integrated circuits (resistors)", Hiradastechnika, 25, 353-2 (1974), Hungarian, EEA78-19869.
530.Salmela, A., "Resistor combination for digital systems", Elektron. Ind., 4, 258-9 (1973), German, EEA77-18840.
531.Samuel, A. and Vinikman, V., "Thin film Cr-SiO resistors", 8th Convention of the Electrical and Electronics Engineers in Israel, 1973, p. 47, Hebrew, EEA77-6062.
532.Sanders, T. J., "Fabrication of thin-film resistor crossovers for integrated circuits", Patent USA 3779841, Publ. December 1973.
533.Schiller, S., Heisig, U. and Panzer, S., "Electron-beam trimming of thin and thick film resistor networks", Solid State Technol., 18 (7), 38-44 (1975), EEA78-37600.
534.Shim, G. A., "Method of making a heat sinked resistor", Patent USA 3795046, Publ. March 1974.
535.Snopek, A., Felender, Z. and Pikes, J., "A method of selective etching in resistive multi-layer structures", Elektronika, 14, 474-6 (1973), Polish, EEA77-14581.
536.Taketa, Y. and Haradome, M., "High-frequency-discharge trimming of RuO_2-based thick film resistors: I. Affecting factors", IEEE Trans. Parts, Hybrids, Packag., PHP-9, 87-94 (1973).
537.Taketa, Y. and Haradome, M., "The basic reactions, compositions, and electrical properties of Pd-Ag thick-film resistors", IEEE Trans. Parts, Hybrids & Packag., PHP-9, 104-14 (1973), EEA76-29302.
538.Taketa, Y. and Haradome, M., "Stability and deterioration mechanism of thick film resistors", Microelectron. & Reliab., 13, 281-9 (1974), EEA77-38873.
539.Val'Dman, V. I. and Radionev, Y. A., "Some features of integrated-film microcircuit resistor designing", Report (unnumbered), Inst. Exact Mech. & Computing Tech. Acad. Sci., USSR, Moscow, 1971, 62 pp., Russian, EEA75-39212.
540.Van Loan, P. R., "Current trends in thick-film resistor technology", J. Can. Ceram. Soc., 40, 49-54 (1971).

541.Yamauchi, M. and Nagasaka, H., "Electrical properties of ceramics thick film resistors", J. Fac. Eng. Ibaraki Univ., no. 21, 99-102 (1973), Japanese, EEA78-672.

XIV. Capacitors

542.Buttoo, R. S. and Murthy, B. S., "Compatible thin film capacitors", Indian J. Technol., 12 (1), 13-15 (1974), EEA78-19933.
543.Eckert, L. J. and Payne, D. A., "Thick film dielectrics with glass bonded electrodes influence of interfacial reaction layers on composite impedance", Am. Ceram. Soc. Bull., 52, 377 (1973), EEA77-2682.
544.Evans, D., "Assembly of monolithic ceramic capacitors by remelting solder", Electron. & Microelectron. Ind., no. 200, 46-7 (1975), French, EEA78-24532.
545.Field, M. and Rapp, J. E., "Capacitor having thick-film glass-ceramic dielectric layer and method for manufacture", Patent USA 3864159, Publ. February 1975, EEA78-37595.
546.Huttemann, R. D., Morabito, J. M. and Gerstenberg, D., "Effect of light elements in (N, C, O) in tantalum on tantalum film capacitor properties", IEEE Trans. Parts, Hybrids & Packag., PHP-11, 67-72 (1975), EEA78-33245.
547.Isert, H., "Printed thick-film capacitors", Feinwerktech. & Messtech., 83, 173-8 (1975), German, EEA78-37592.
548.Kampa, J., Obuchowicz, E. and Petrus, K., "Thin film capacitors with Ta_2O_5 dielectric layer", Pr. Osrodka Badaw.-Rozwojowego Elektron. Prozniowej, 2 (2), 39-46 (1974), Polish, EEA78-2897.
549.Lobl, H. and Schmickl, H., "A new chip capacitor for film circuits", Component Rep., 8 (5), 124-6 (1973), EEA77-14623.
550.Mathiou, Y. G., Mitte, R. L., Olivier, E. B. and Robert, G. H., "Capacitor-type component fabrication by using thin film technology", Patent French 2249418, Publ. May 1975, CA83-211902.
551.Phillips, L. S., "Capacitors in hybrid circuits", Electron, no. 44, 30-5 (1974), EEA77-18863.
552.Ruehli, A. E. and Brennan, P. A., "Efficient capacitance calculations for three-dimensional multiconductor systems", IEEE Trans. Microwave Theory & Tech., MTT-21, 76-82 (1973), EEA76-8757.
553.Ruehli, A. E. and Brennan, P. A., "Accurate metalization capacitances for integrated circuits and packages", IEEE J. Solid-State Circuits, SC-8, 289-90 (1973), EEA76-29313.
554.Ruehli, A. E. and Brennan, P. A., "Capacitance models for integrated circuit metallization wires", IEEE J. Solid-State Circuits, SC-10, 530-6 (1975).
555.Teuschler, H. J., "Microelectronic components in thick film technique", Radio Fernsehen Elektron., 21, 262-4 (1972), German, EEA75-27806.

556.Yoshimura, S., Itoh, Y., Yasuda, M., Murakami, M., Takahashi, S. and Hasegawa, K., "An aluminum solid electrolytic capacitor with an organic semiconductor electrolyte", IEEE Trans. Parts, Hybrids & Packag., PHP-11, 315-20 (1975).
557.Zhikharev, Y. V., Kolchin, O. P. and Evsseva, I. A., "Surface quality of tantalum capacitor foil", Tsvet. Metal., no. 6, 64-7 (1973), Russian, CA79-117714.

XV. Laser Techniques
1. Resistor Trimming

558.Abenaim, D., "Functional trimming: discussion of a problem, description of a solution", Microelectronics, 5 (1), 33-9 (1973), EEA77-18866.
559.Anon., "Laser-beam trimming system for hybrid circuits", Elektron. Ind., 5, 14-15 (1974), German, EEA77-23162.
560.Anon., "Trimming of thick-film resistors by laser", Funk-Tech., no. 14, 501-4 (1974), German, EEA78-670.
561.Bader, R. F., "Automatic laser resistor trimming", Solid State Technol., 15 (9), 44-6 (1972), EEA75-846.
562.Britton, B., "Laser trimming of thick-film resistors", Electron. Equip. News, 15 (6), 52-4 (1973), EEA77-388.
563.Corwin, R. E., "The laser as a thick film tool", Am. Ceram. Soc. Bull., 52, 713 (1973), EEA77-10285.
564.Cozzens, W., "Galvanometer controlled laser trimming", Electron, no. 66, 37, 38, 40 (1975), EEA78-19873.
565.Gross, G. J., "The case for-laser trimming of hybrid resistors", Eval. Eng., 10 (5), 7 (1971), EEA75-27788.
566.Hurtig, G., III and Swenson, E. J., "An overview of laser functional trimming techniques", 24th Electronic Components Conference, 1974, p. 240-4, EEA77-34076.
567.Lejeune, B., "Adjustment of hybrid circuits by laser", Electron. & Microelectron. Ind., no. 172, 27-30 (1973), French, EEA76-29319.
568.Limann, O., "Laser trimming automation for hybrid circuits", Elektronik, 24 (9), 89 (1975), German, EEA78-41708.
569.Parisi, J. A., "Integrated resistor trimming", IBM Tech. Disclosure Bull., 17, 2878 (1975), EEA78-19917.
570.Parks, P. F., "Using a laser to produce prototype hybrid circuits", Am. Ceram. Soc. Bull., 52, 713 (1973), EEA77-10311.
571.Schomburg, R. A., "Measurement subsystem for an AC network laser trimmer", 24th Electronic Components Conference, 1974, p. 245-7, EEA77-34077.
572.Shah, A. S. and St. Pierre, E. J., "Two-pass laser cutting", IBM Tech. Disclosure Bull., 16, 3237 (1974), PA77-55536.
573.Steer, P., "Laser trimming of thick film circuits", Electron. Components, 16 (15), 14-15 (1974), EEA77-38871.
574.Stone, G. B., "Automatic laser resistor trimming: a systems approach to the design of an industrial laser system", Image Technol., 14 (6), 21-4 (1972), EEA76-19966.

575.Willis, J., "Techniques of laser trimming", Electron. Components, 16 (2), 35, 37, 39 (1974), EEA77-9817.

2. Insulator Processing

576.Anderson, L. C., Fraser, H. R., Jr. and Jemkins, J. B., "Laser micromachining of green ceramic dielectrics", Am. Ceram. Soc. Bull., 51, 652 (1972), EEA76-4682.
577.Avramchenko, P. F., Velichko, O. A., Garashchuk, V. P., Morarskii, V. E. and Molchan, I. V., "The laser welding of wires made of super-high purity aluminium", Autom. Weld., 27 (11), 66-7 (1974), EEA78-32161.
578.Balland, R., "Scribing ceramic substrates (laser application)", Electron. Prod. Methods & Equip., 4 (4), 6 (1975), EEA78-28826.
579.Briggs, T. H. and Frankson, R. W., "Routine analysis of alumina substrates by laser emission microprobe", Am. Ceram. Soc. Bull., 52 (4), 375 (1973), EEA77-2591.
580.Hensler, D., Soos, N. A., Hass, E. H., Frankson, R. W. and Rott, D. A., "Laser specular reflectometer for ceramic surface diagnostics", Amer. Ceram. Soc. Bull., 52, 191-4 (1973).
581.Illner, D., "Using lasers for processing problems in the electronics industry", Feingeraete Tech., 24 (1), 27-9 (1975), German, EEA78-15383.
582.Lambert, J. L., Raes, A., Coopmans, P. and Streydio, J. M., "Cutting sheet glass using a CO_2 laser", Silicates Ind., 38 (7-8), 145-7 (1973).
583.Longfellow, L., "Sawing alumina substrates with a CO_2 laser", Solid State Technol., 16 (8), 45-8 (1973), EEA76-38296.
584.Schroeder, D. H. and English, F. L., "A comparison of the strength of alumina substrates for different separation techniques", IEEE Trans. Parts, Hybrids & Packag., PHP-8, 4-6 (1972), EEA75-39306.
585.Steffen, J. and Hofer, A., "Method for the fabrication of holes in a workpiece by means of laser beams and apparatus for the performance of the method", Patent USA 3656988, Publ. April 1972.
586.Ulmer, W., "Forming of ceramic substrates by lasers", Sprechsaal Keram., Glas, Baustoffe, 106, 971-4, 976, 978 (1973).

3. Others

587.Anon., "Precision laser machining drilling by trepanning", Electro-Opt. Syst. Des., 5 (5), 31-3 (1973), EEA76-38886.
588.Anon., "Laser trimming on the chip", Electron. Prod. Methods & Equip., 4 (4), 8, 10 (1975), EEA78-33202.
589.Arruda, J. and Prifti, W., "Microelectronic package sealing with a laser", Electro-Opt. Syst. Des., 5 (5), 34-5 (1973), EEA76-38292.

590.Bette, H. P., "Computer-controlled laser systems and their application (to IC production)", Mess. & Pruef., no. 4, 239-42 (1973), German, EEA76-29321.
591.Friesem, A. A. and Vest, C. M., "Detection of micro fractures by holographic interferometry", Applied Optics, 8, 1253-4 (1969), PA73-5904.
592.Grant, R. M. and Brown, G. M., "Holographic nondestructive testing (HNDT)", Materials Evaluation, 27 (4), 79-84 (1969), PA72-42862.
593.Kuhn, L., Schuster, S. E., Zory, P. S., Jr., Lynch, G. W. and Parrish, J. T., "Experimental study of laser formed connections for LSI wafer personalization", IEEE J. Solid-State Circuits, SC-10, 219-28 (1975), EEA78-33212.
594.Lumley, R. M., "The laser as a manufacturing tool for the semiconductor industry", National Electronics Conference Proceedings, 24, 859-63 (1968).
595.National Research Development Corp., "Cutting processes employing a laser", Patent Brit. 1235653, Publ. June 1971.
596.Okamoto, E., "Laser use successfully simplifies manufacturing processes. I", JEE, no. 83, 56-8 (1973), EEA77-14628.
597.Oswald, D. R. and Munto, D. F., "A laser scan technique for electronic materials surface evaluation", J. Electron. Mater., 3, 225-42 (1974), EEA77-14102.
598.Paek, U. C. and Kestenbaum, A., "Thermal analysis of thin-film micromachining with lasers", J. Appl. Phys., 44, 2260-8 (1973), EEA76-23480.
599.Reese, W. E., "Use of pulse laser technique in failure analysis", Proceedings of the 11th Annual Reliability Physics Conference, 1973, p. 124-6, EEA77-6085.
600.Verbiest, R., "The industrial applications of the CO_2 laser", Energie, 200, 171-5 (1972), French, EEA76-79.
601.Wells, D. R., "NDT of sandwich structures by holographic interferometry", Materials Evaluation, 27, 225-31 (1969).
602.Whitehouse, D. R. and Ilgenfritz, R. W., "Laser beam technology for the microelectronics industry", Solid State Technol., 15 (7), 32-5 (1972), EEA75-35404.
603.Wiswall, C. E. and Maxwell, K. J., "Optimizing pulsed laser holographic NDT", Applied Optics, 9, 1724-5 (1970), PA73-63646.

XVI. Encapsulation

604.Abley, P., "Chemical removal of thermosets", Eng. Mater. & Des., 16 (3), 27-35 (1972), EEA76-11705.
605.Appen, A. A., "Preparation of heat-resistant inorganic coatings", Mitt. Ver. Deut. Emailfachleute, 20 (11), 116-20 (1972); 21 (3), 29-32 (1973), German.
606.Armantrout, G. A., Wichner, R. and Swierkowski, S. P., "Encapsulation of high-purity germanium detectors", IEEE Trans. Nucl. Sci., NS-21, 344-6 (1974), PA77-46817.

607. Associated Semiconductor Manufacturers Ltd., "Semiconductor housing", Patent UK 1144671, Publ. March 1969, EEA72-20307.
608. Berenbaum, L., "Coating narrow resistive lines to increase electromigration resistance", IBM Tech. Disclosure Bull., 17, 3280 (1975), EEA78-24474.
609. Bimson, A., "Modern techniques for controlling the quality of plastic-encapsulated integrated circuits for high-reliability applications", Mikroelektron., Manuskriptdruck Vort. Mikroelektron.-Kongr., 5th, 1972, p. 395-421, Publ. 1973, CA81-142546.
610. Boggs, R. N., "Encapsulated delicate components", Des. News, 26 (11), 42 (1971), EEA74-36545.
611. Bolger, J. C., "Encapsulating small, heat-sensitive components with quick-cure epoxy coating powders", Insul./Circuits, 20 (10), 39-42 (1974), EEA78-4792.
612. Brzezinski, J., "Pulverized plastic coatings. Methods of depositing coatings", Wiad. Elektrotech., 41, 448-51 (1973), Polish, EEA77-9281.
613. Carnes, R. A. and Taylor, D. W., "Encapsulated semiconductor device including a ceramic substrate", Patent USA 3439255, Publ. April 1969.
614. Colliver, D. J., "Microwave performance of glass-ceramic/copper encapsulations", Electron. Lett., 8, 569-70 (1972), EEA76-5411.
615. Coulson, K., "Electronics in resin", Electron, no. 50, 29-30, 32 (1974), EEA77-25629.
616. Delfs, H., "Hybrid currents - encapsulation of hybrid circuits", Int. Elektron. Rundsch., 25, 301-6 (1971), German, EEA75-6975.
617. Eisenmann, D. E., "Self-primed and thixotropically stabilized epoxy encapsulant", IBM Tech. Disclosure Bull., 17, 3630 (1975), EEA78-32157.
618. Flaskerud, P. and Mann, R., "Silver plated lead frames for large molded packages", Proceedings of the 12th Annual Reliability Physics Conference, 1974, p. 211-22.
619. Fleischmann, J. J., "Reciprocating screw encapsulation: the new method for plastic encapsulation", Insul./Circuits, 19 (3), 35-8 (1973), EEA76-22825.
620. Freemont, J. H. and Carrafa, J. L., "Take a look at thermoplastic polyesters for molded electrical/electronic components", Insul./Circuits, 20 (11), 47-9 (1974), EEA78-2213.
621. Friden Inc., "Encapsulated microcircuit", Patent UK 1165854, Publ. October 1969, EEA73-34387.
622. General Electric Co., "Encapsulated semiconductor component", Patent UK 1145434, Publ. March 1969, EEA72-19448.
623. General Electric Co., "Encapsulated electrical assemblies", Patent UK 1145450, Publ. March 1969, EEA72-19449.
624. Gregor, L. V. and Zuegel, M., "Glass-annealing process for encapsulating and stabilizing FET devices", Patent USA 3658678, Publ. April 1972.

625.Griffiths, G. W., "Practical aspects of fluid bed encapsulation for hybrid circuits", New Electron., 7 (13), 35, 37-8, 41 (1974), EEA77-36337.
626.Grubb, E. F., Hammer, F. W. and Monks, J. R., "Chemically strengthened glass-encapsulated glass articles and process therefor", Patent USA 3728095, Publ. April 1973.
627.Harnden, J. D., Jr., "Semiconductor device encapsulation", Patent UK 1362499, Publ. August 1974, CA81-178915.
628.Hasty, T. E., "Method for encapsulating discrete semiconductor chips", Patent USA 3739462, Publ. June 1973.
629.Heidenreich, P. J., "Glass encapsulated semiconductor devices", Patent USA 3518753, Publ. July 1970.
630.Herr, E. A., Fox, A. and Read, J. S., "Plastic encapsulated signal and power transistor reliability", IEEE Trans. Broadcast & Telev. Receivers, BTR-20, 117-48 (1974), EEA77-29336.
631.Hicks, J., "The A to Z of potted circuit modules. Taking the lid off of black box circuit functions", Control & Instrum., 7 (1), 46-7 (1975), EEA78-15379.
632.Hull, J. L., "Plastic encapsulation of hybrid microelectronic circuits", Electronic Commun., 3 (2), 18 (1968), EEA72-23165.
633.Intrator, A. M. and Adams, N., "Semiconductor device encapsulation", Patent UK 1377930, Publ. December 1974, CA83-20831.
634.Jakubowska, K., "The encapsulations for LSI integrated circuits", Elektronika, 14, 434-7 (1973), Polish, EEA77-2601.
635.Kern, W., "Glass-encapsulated semiconductor device", Patent USA 3755720, Publ. August 1973.
636.Khajezadeh, H., "Plastic-encapsulated semiconductor elements of high reliability", Patent German 2445250, Publ. April 1975, CA83-52095.
637.Kleber, W., "Electrostatic plastic coating - a highly productive anticorrosive procedure", Technik, 30, 251-4 (1975), German, EEA78-23273.
638.Larson, C. D. and Norring, B., "Processing components for encapsulation", Presented at the 21st Electronic Components Conference, Washington, D. C., May 1971, EEA74-29904.
639.Lee, H. R., "Semiconductor device with composite encapsulation", Patent USA 3496427, Publ. February 1970.
640.Licentia Patent Verwaltungs G.m.b.H., "Semiconductor components", Patent UK 1120577, Publ. July 1968, EEA72-4491.
641.Loeb, W. E., "Encapsulation by vacuum deposition of polymers", SPE J., 27 (9), 46 (1971), EEA75-7007.
642.Macdonald, A. R., "Engineer's guide to selection and operation of encapsulating systems", Insul./Circuits, 19 (5), 30-2 (1973), EEA76-31588.
643.Maguire, C. F., Koons, F. J. and Jarrett, Q. T., "Plastic encapsulated transistors", W. Elec. Eng., 14 (4), 44-51 (1970), EEA74-4923.
644.Martin, D. J., "Encapsulation with epoxy moulding compounds", Electron Engineering, 47 (569), 42-3 (1975), EEA78-32154.

645.Mayoux, C., Abd El Azim, A. and Guillaumon, J. C., "High-voltage insulating materials in spatial conditions, behaviour of embedded components", Conference on Electric Propulsion of Space Vehicles, 1973, p. 116-21, EEA77-10327.
646.Moxley, D. C., "Selective removal of encapsulant", Tech. Dig., no. 36, 27-8 (1974), EEA78-11775.
647.Newell, J. A., "Encapsulation in electronics", Electron, no. 50, 34, 36 (1974), EEA77-25630.
648.Nippon Electric KK, "Resin-moulded leadless high power semiconductor device", Patent UK 1395238, Publ. May 1975, EEA78-41637.
649.Nowacki, J. and Szyszej, K., "Possibilities of counteraction against formation of defects in plastics encapsulated semiconductor devices", Elektronika, no. 12, 503-5 (1972), Polish, EEA76-12238.
650.Nowalk, T. P. and Ferree, H. E., "Flat package encapsulation of electrical devices", Patent USA 3437887, Publ. April 1969, EEA73-4431.
651.Reich, B., "Plastic semiconductor devices and integrated circuits for military applications", Solid State Technol., 13 (1), 53-6 (1970), EEA73-21225.
652.Reich, B. and Hakim, E. B., "Can plastic semiconductor devices and microcircuits be used in military equipment?", Proceedings of the 1974 Annual Reliability and Maintainability Conference, p. 396-402, EEA77-22228.
653.Roberts, J. S., "Encapsulated light-activated semiconductor device", Patent USA 3796881, Publ. March 1974.
654.Rozdzial, P., "Encapsulating electronic subassemblies with various hardening resins", Polim. Tworzywa Wielkoczasteczkowe, 19, 627-31 (1974), Polish, EEA78-28830.
655.Rozdzial, P., "Encapsulation of electronic subassemblies with thermohardening resins. I", Polim. Tworzywa Wielkoczasteczkowe, 19, 551-7 (1974), Polish, EEA78-36599.
656.Ruetter, H. F. and Belardi, R. J., "Semiconductor device with protective glass sealing", Patent USA 3716765, Publ. February 1973.
657.Rybicky, J. and Marton, J. P., "A multicoat non-burnable protection system for electrical elements", IEEE Trans. Parts, Hybrids & Packag., PHP-11, 235-8 (1975), EEA78-40473.
658.Sakamoto, M., "Method for sealing (metal cap) a semiconductor element", Patent USA 3610870, Publ. October 1971.
659.Sandera, J., "Glass-coated semiconductor", Patent USA 3632432, Publ. January 1972.
660.Scheetz, R. J. and Sobota, S. M., "Conformal coating process to improve package reliability in adverse environment", Patent USA 3756852, Publ. September 1973.
661.Schwarz, H. and Plorin, S., "Semiconductor with encapsulating housing", Patent USA 3427510, Publ. February 1969, EEA72-20452.

662.Sealy, B. J. and D'Cruz, A. D. E., "Encapsulation of ion-implanted GaAs using native oxides", Electron. Lett., 11, 323-4 (1975), EEA78-37389.
663.Sealy, B. J. and Surridge, R. K., "A new thin film encapsulant for ion-implanted GaAs", Thin Solid Films, 26, L19-22 (1975), PA78-51133.
664.Semikron Gesellschaft fuer Gleichrichterbau and Elektronik mbH, "Manufacture of encapsulated semiconductor devices including semiconductor wafers", Patent British 1298766, Publ. December 1972.
665.Shimazaki, A., Kawata, N. and Koshishi, T., "Plastics encapsulation and low pressure transfer moulding for semiconductor", Fuji Elec. J., 43, 700-5 (1970), Japanese, EEA74-507.
666.Shulepova, T. G., "Use of powdered polymers as instrumentation coatings", Prib. & Sist. Upr., no. 5, 53-4 (1974), Russian, EEA78-70.
667.Siemens, A. G., "Encapsulated circuits", Patent UK 1122295, Publ. August 1968, EEA72-4490.
668.Sinnadurai, F. N., "The accelerated ageing of plastic encapsulated semiconductor devices in environments containing a high vapour pressure of water", Microelectron. & Reliab., 13, 23-7 (1974), EEA77-22191.
669.Slee, R. K. and Magnus, A. N., "Encapsulation of high reliability devices", Electron. Prod. Methods & Equip., 2 (8), 44, 47-8 (1973), EEA77-13628.
670.Smith, R. H., "Continuous powder coating encapsulation process", Proceedings of the 10th Electrical Insulation Conference, 1971, p. 233-5, EEA75-16091.
671.Starger, S. L., "Encapsulated semiconductor device having internal shielding", Patent USA 3469017, Publ. September 1969.
672.Swartz, G. A. and Chamberlain, R. E., "Encapsulated semiconductor device assembly", Patent USA 3783348, Publ. January 1974.
673.Szedon, J. R., Jackson, J. A. and Phillips, D. C., "Encapsulated solid state electronic devices having a sealed lead-encapsulant interface", Patent USA 3821099, Publ. June 1974, CA81-70135.
674.Szedon, J. R., Jackson, J. A. and Phillips, D. C., "Encapsulated solid state electronic devices having a sealed lead-encapsulant interface", Patent USA 3911475, Publ. October 1975, CA83-201127.
675.Tepper, H. F., "Packaging for a large memory system", IBM Tech. Disclosure Bull., 17, 987-8 (1974), EEA78-5382.
676.Thomsen, R. L., "Powder coating electronic components", Microelectronics, 5 (1), 45-8 (1973), EEA77-17829.
677.Utner, F., "Method of encapsulating end-contacted electrical components", Patent USA 3864798, Publ. February 1975, EEA78-27780.
678.Wakley, W. T. and Leeds, M. T., "Enclosure for (glass) encapsulating electronic components", Patent USA 3697666, Publ. October 1972.
679.Watson, A. C., "Microencapsulation", Science J., 6 (2), 62-8 (1970), EEA73-27353.

680.Whitby, F. J., Simpson, W. E. and Pettican, R. A., "Plastics encapsulation of components", Elect. Commun., 43 (3), 220-7 (1968), EEA72-301.
681.White, M. L., "Encapsulation of integrated circuits", Proc. IEEE, 57, 1610-16 (1969), EEA73-461.
682.White, M. L., "Encapsulating integrated circuits", Bell Lab. Rec., 52 (3), 78-83 (1974), EEA77-26493.
683.Wylie, J. R., "Selective flow coating of thin film integrated circuits", Tech. Dig., no. 36, 33-4 (1974), EEA78-11717.
684.Zido, J. E. and Whalen, T. A., "Resin encapsulated semiconductor device", Patent USA 3476987, Publ. November 1969, EEA73-31305.

XVII. Hermeticity

685.Adams, G. E., "Package hermeticity", Proceedings of the 11th Annual Reliability Physics Conference 1973, p. 95-7, EEA77-5783.
686.Araki, T., Hirayama, M., Fujine, N. and Kajimura, T., "A hermetically sealed 20 GHz integrated mixer", NEC Res. & Dev., no. 33, 81-5 (1974), EEA77-38382.
687.Blust, H. L. and Lindburg, N. L., "Plug sealing of hermetic enclosures", Patent USA 3826634, Publ. July 1974.
688.Bower, F. H., "Hermetic sealing of integrated circuit packages", Solid State Technol., 13 (8), 56-61 (1970), EEA74-356.
689.Brady, H. B., Kochis, J. L. and Tischler, O., "Hermetically sealed metal-film resistor", Patent USA 3688238, Publ. August 1972.
690.Brandi, H. W., "Hermetically sealed resistor", Patent USA 3643200, Publ. February 1972.
691.Delembre, A. E., Jarrige, G. R. and Nhiep, N. D., "Hermetic end seals for envelope of a metal-vapor lamp", Patent USA 3716744, Publ. February 1973.
692.Duensche, H. G., "Method of gastight sealing of semiconductor components", Patent USA 3717523, Publ. February 1973.
693.Fielder, W. A. and Kraut, A. J., "The hermetic sealing of large ceramic flat packages", International Hybrid Microelectronic Symposium, 1970, 7 pp., EEA74-29805.
694.Goldfarb, S., "Hermetic package sealing", IEEE International Convention Digest, 1971, p. 584-5, EEA74-25712.
695.Gordon, W. D. and Norman, R. S., "Process for forming hermetic seal for sheathed electrical conductors", Patent USA 3615324, Publ. October 1971.
696.Howard, L. F. and Sturgill, C. S., "Hermetically sealed electrolytic capacitor", Patent USA 3440495, Publ. April 1969, EEA73-9925.
697.Johnson, C. A., "Hermetically sealed electronic package", Patent USA 3753054, Publ. August 1973.
698.Laurent, P., "Coating for ceramic packages", IBM Tech. Disclosure Bull., 17, 2659 (1975), EEA78-23264.

699.Low, G. M., "Hermetically sealed semiconductor", Patent USA 3705255, Publ. December 1972.
700.Muller, G., "Glass for the hermetic encapsulation and passivation of Si-diodes", Elektron. Ind., 6 (4), EP21-4 (1975), German, EEA78-36591.
701.Noguchi, S. and Nanko, Y., "Hermetically sealed semiconductor device", Patent USA 3663868, Publ. May 1972.
702.Perrin, W. P. and Gentle, J., "Protective element for hermetically enclosed semiconductor devices", Patent USA 3487275, Publ. December 1969.
703.Pittwood, D. G., "Removable hermetic seal for electronic module packages", IBM Tech. Disclosure Bull., 17, 2634 (1975), EEA78-23263.
704.Postlehwaite, A. W., "Hermetic and non-hermetic packaging", Solid State Technol., 13 (8), 67-70, 75 (1970), EEA74-358.
705.Ramsey, T. H., Jr., "Critical parameters in glass sealed ceramic packages", Solid State Technol., 17 (9), 51-5, 59 (1974), EEA78-2216.
706.Rozdzial, P., "Hermetic encapsulation of electronic components with plastics", Elektronika, no. 7-8, 314-23 (1972), Polish, EEA75-39260.
707.Siemens AG, "Gastight sealing of semiconductor components", Patent British 1296629, Publ. November 1972.
708.Svec, J. J., "Sealing ceramic packages a furnace function", Ceram. Ind., 96 (2), 18-9 (1971).
709.Waldman, D. P., "Hermetic seals", IBM Tech. Disclosure Bull., 18, 98 (1975), EEA78-36602.
710.Wang, C. H. and Wakely, W. T., "Low temperature ceramic DIP sealing", Am. Ceram. Soc. Bull., 52, 716 (1973), EEA77-10321.
711.Zimmerman, D. D., "A new gold-tin alloy composition for hermetic package sealing and attachment of hybrid parts", Solid State Technol., 15 (1), 44-6 (1972), EEA75-27794.

XVIII. Fabrication Techniques

712.Anderson, C. E., "Method of fabrication of semiconductor device package", Patent USA 3716907, Publ. February 1973.
713.Anderson, L. C., "Method of fabricating multilayer circuits", Patent USA 3770529, Publ. November 1973.
714.Barsuhn, H. E., Dreckmann, U. T. and Haug, W. O., "Reducing the number of short circuits in double-layer wafer metallizations", IBM Tech. Disclosure Bull., 16, 3273 (1974), EEA77-30389.
715.Beall, R. J. and Zasio, J. J., "LSI chip package and method", Patent USA 3872583, Publ. March 1975.
716.Beyer, H. and Velsch, H. P., "The construction of the BBC-Electronic new module system", BBC Nachr. Germany, 50, 568-74 (1968), German, EEA72-7451.

717.Beyerlein, F. W., "A new assembly technology for integrated circuits and its applications in hybrid microelectronics", Presented at International Microelectronics Symposium, Chicago, Ill., October 1971.
718.Boswell, D. and Askwith, T. A., "Application of microengraving in the rapid manufacture of film circuit prototypes and close tolerance passive components", Conference on Integrated Circuits, 1967, p. 52-64, EEA73-480.
719.Budd, J. P. and Robson, W. K., "Product and process for cavity metalization of semiconductor packages", Patent USA 3793064, Publ. February 1974.
720.Bunker, T. D. and Thompson, D. F., "Fabrication of packages for integrated circuits", Patent USA 3689336, Publ. September 1972.
721.Calhoun, D. F. and Wolfe, G., "Status and future directions of pad relocation full wafer LSI", 23rd Electronic Components Conference, 1973, p. 7-14, EEA76-26153.
722.Chang, T. H., Wallman, B. A. and Hills, D. J., "Cutting down the size of microcircuits", Aust. Electron. Eng., 6 (7), 26-9 (1973), EEA77-18850.
723.Cichetti, M., "Tantalum thin film technology", Alta Freq., 42, 411-19 (1973), Italian, EEA77-10292.
724.Cozens, A. G., "Inter nepcon - assembly and repair of multi-chip modules", Electronic Engng., 41 (500), 22-4 (1969), EEA73-9555.
725.Cozens, A. G. and Teasdale, K., "Multi-chip modules - assembly and repair", Presented at Colloquium on Constructional Techniques Used in Microelectronics, London, England, November 1968, EEA73-3536.
726.Deces, G., "Use of bare semiconductor chips in hybrid microelectronics", Electron. & Microelectron. Ind., no. 195, 61-3 (1974), French, EEA78-15425.
727.DelMonte, L. A., "Fabrication of gold bumps for integrated circuit terminal contact", 23rd Electronic Components Conference, 1973, p. 21-5, EEA76-26155.
728.Docckal, J., "A new assembling technology that enables a three-dimensional connection of integrated circuits", Jemna Mech. & Opt., 19, 295-8 (1974), Czech, EEA78-5383.
729.Duckworth, R. G., "The manufacture of thin film circuits", Electron. Mfr., 17 (2), 9-16 (1972), EEA75-19281.
730.Early, R. C., "Thick film/flip chips - a systems approach", IEEE Trans. Manuf. Technol., MFT-4, 2-8 (1975), EEA78-33247.
731.Figuli, E. S., "Velocity gold plating of integrated circuits", Western Electric Eng., 18 (1), 18-24 (1974), EEA77-14633.
732.Fincham, F. D., "Method and apparatus for producing and packaging sealed tubular glass bodies", Patent USA 3734705, Publ. May 1973.
733.Foster, T. M. and Digiondomenico, O. J., "Thick film techniques for microwave integrated circuits", IEEE Trans. Parts, Hybrids & Packag., PHP-10, 88-94 (1974), EEA77-36332.
734.Fujimoto, K., "Method for manufacturing semiconductor devices", Patent USA 3461549, Publ. August 1969, EEA73-24729.

735.Furnival, T. J., "Method of making enclosure for a semiconductor device", Patent USA 3772764, Publ. November 1973.
736.Gaensslen, F. H., "Multichip packaging", IBM Tech. Disclosure Bull., 12, 1579-80 (1970), EEA73-27264.
737.Gallace, L. J., Gottesfeld, S. and Ingrassia, J., "IC fabrication improvements combat hostile environments", Electron. Engineering, 47 (568), 49-51 (1975), EEA78-33193.
738.Gejyo, T., Saito, T., Minagawa, S., Keikichi, M., et al., "Method for growing lead conductors on planar transistors", Patent USA 3796598, Publ. March 1974.
739.Gillis, T. B., "Low temperature processing in hybrid manufacturing increases survival rate", Insul./Circuits, 21 (6), 19-22 (1975), EEA78-33254.
740.Goldstein, D., "A prototype assembly facility for CerDip packaging", Solid State Technol., 15 (8), 33-40 (1972), EEA75-35406.
741.Greig, W. J. and Soden, R. R., "Semiconductor device with multilevel metalization and method of making", Patent USA 3668484, Publ. June 1972.
742.Guditz, E. A. and Burke, R. L., "Multichip integrated circuit memory with photoformed plated conductors", IEEE Trans. Parts, Hybrids & Packag., PHP-11, 89-96 (1975).
743.Hauck, J. and George, D., "Housing structure and method of manufacture for semiconductor device", Patent USA 3424852, Publ. January 1969, EEA72-15433.
744.Heinecke, R. A. H., "Integrated circuit", Patent German 2257201, Publ. June 1973, CA79-59197.
745.Hingorany, A. R., "Semiconductor device package having improved compatibility properties", Patent USA 3620668, Publ. December 1971.
746.Holmes, P. J. and Corkhill, J. R., "Thick-film integrated circuits. II", Electronic Compon., 10, 1331-3 (1969), EEA73-9590.
747.Ihochi, T. and Yamada, T., "Method for manufacturing a package of a semiconductor element", Patent USA 3729820, Publ. May 1973.
748.Ilgenfritz, R. W., Mogey, L. E. and Walter, D. W., "A high density thick film multilayer process for LSI circuits", IEEE Trans. Parts, Hybrids & Packag., PHP-10, 165-8 (1974), EEA78-2921.
749.Illyefalvi Vitez, Z., "Production of thin film networks by means of electroerosion micro-engraving", Hiradastechnika, 25, 363-5 (1974), Hungarian, EEA78-19874.
750.Johnston, S. A., Lyons, J. G. and Riedmayer, H. J., "Method of making integrated-circuit package", Patent USA 3837067, Publ. September 1974.
751.Jordan, J., "Hybrid assembly time reduced with purpose-designed chips", Electron. Eng., 44 (532), 57-61 (1972), EEA75-23387.
752.Kapnias, D. E., "Semiconductor package containing a dual epoxy and metal seal between a cove- and a substrate and method for forming the seal", Patent USA 3735211, Publ. May 1973.

753.Kennedy, B. W., "Integrated circuit package with lead structure and method of preparing", Patent USA 3778685, Publ. December 1973.
754.Kersey, R., "The fabrication of multilayer structures using various film technologies", 23rd Electronic Components Conference, 1973, p. 198-208, EEA76-26135.
755.Keys, L. K. and Russo, A. J., "The fabrication of large high conductor density multilayer screen-printed ceramic circuits", 22nd Electronic Components Conference, 1972, p. 231-42.
756.Kleineberg, H., "Production of multichip hybrid components for computers", Fermeldetechnik, 15 (1), 26 (1975), German, EEA78-24534.
757.Knau, H., Moll, V. and Sautter, D., "Process for manufacturing headerless encapsulated semiconductor devices", Patent USA 3491435, Publ. January 1970.
758.Kutch, G., Miller, L. F., Ofca, E. J. and Vogt, R. T., "Meshless moly-mask", IBM Tech. Disclosure Bull., 12, 2073-4 (1970), EEA73-35054.
759.McIntosh, C. M. and Schmeckenbecher, A. F., "Packaging of integrated circuits", IBM Tech. Disclosure Bull., 15, 1977-80 (1972), EEA76-12687.
760.Mandel, A. P., "Packaging microelectronic passive hybrid components", 8th Convention of the Electrical and Electronics Engineers in Israel, 1973, p. 89, EEA77-6113.
761.Manolescu, A. M., "Technologies and materials for the achievement of integrated thin-layer circuits", Autom. & Electron., 17 (1), 28-37 (1973), Romanian, EEA76-19952.
762.Manus, D. J., "Method for making semiconductor packaged devices and assemblies", Patent USA 3828425, Publ. August 1974.
763.Maxwell, H., "Mini wire-a practical high density electronic packaging technique", Insul./Circuits, 17 (2), 29-31 (1971), EEA74-17746.
764.Moyle, K. J. and Mann, R. S., "Semiconductor device packaging", Patent USA 3706840, Publ. December 1972, EEA76-35087.
765.Murphy, J., "Production packaging of very LSI chips", Presented at Semiconductor Integrated Circuit Processing and Production Conference, Anaheim, Calif., February 1971, EEA74-21600.
766.Mustonen, R., "Thin film technology and thin film materials in microelectronics", Saehkoe, 46, 564-9 (1973), Finnish, EEA77-10293.
767.Palermo, C. A. and Hartman, J. A., "MIC packaging using conductive epoxy and numerically controlled drill techniques", IEEE Trans. Parts, Hybrids & Packag., PHP-11, 157-8 (1975), EEA78-33253.
768.Parks, H. L., "Electrical circuit packaging structure and method of fabrication", Patent USA 3705332, Publ. December 1972.
769.Perott, P. G. and Ecclesia, F., "Modular electronic circuit assembly with improved subcomponent packaging assemblies", Patent USA 3478251, Publ. November 1969, EEA73-30541.

770. Persson, J., "Modular techniques - the economic answer", Focus, 157 (15), 8-10 (1970), EEA73-27309.
771. Pezzani, R., "Mounting techniques of power semiconductors", Electron. & Microelectron. Ind., no. 143, 57-63 (1971), French, EEA74-21663.
772. Photocircuits Corp., "Integrated circuit mounting method", Patent UK 1143531, Publ. February 1969, EEA72-19398.
773. Pittroff, L., "A guide to hybrid-circuit component compatibility", Electronics, 46 (1), 108-9 (1973), EEA76-8755.
774. Platzek, R. C., "MOS/LSI packaging problem or opportunity?", IEEE International Convention Digest, 1971, p. 246-7, EEA74-21618.
775. Prednis, L. J., "Computer controlled printed circuit assembly system", IEEE International Convention Digest, 1971, p. 452-3, EEA74-25829.
776. Reed, W. D. and Stoltz, R. A., "Methods for fabricating ceramic circuit boards with conductive through holes", Patent USA 3798060, Publ. March 1974.
777. Rhoades, W. T., "Packaging aspects of gigabit digital circuits", IEEE International Convention Digest, 1970, p. 258-9, EEA73-36995.
778. Rieser, R. G. and Chabal, J., "Method for laminating plastic to glass employing glass mold", Patent USA 3808077, Publ. April 1974.
779. Ripka, G., "The production technology of thick film hybrid integrated circuits on insulator substrate. I", Finommech. & Mikrotech., 12, 206-14 (1973), Hungarian, EEA77-6097.
780. Roberts, J. S., "Light activated semiconductor device", Patent USA 3590338, Publ. June 1971, EEA74-30860.
781. Rohloff, R. R., "Packaging of semiconductor devices", Patent USA 3691289, Publ. September 1972.
782. Salewski, G., "The production of hybrid circuits for the development of electronic equipment", Fermeldetechnik, 15 (1), 15-18 (1975), German, EEA78-24533.
783. Scapple, R. Y. and Keister, F. Z., "Fabrication aid for hybrid microcircuits", 24th Electronic Components Conference, 1974, p. 186-91, EEA77-34120.
784. Schmidt, J. G. and Lim, E., "Method for fabricating monolithic light-emitting semiconductor diodes and arrays thereof", Patent USA 3636617, Publ. January 1972.
785. Schuster, M. A. and Lytle, W. J., "Contacts for semiconductor devices, particularly integrated circuits, and methods of making", Patent USA 3840982, Publ. October 1974.
786. Scrupski, S. E., "ICs on film strip lend themselves to automatic handling by manufacturer and user, too", Electronics, 44 (3), 44-8 (1971), EEA74-9804.
787. Shallahamer, R. A., Naits, F. V. and Bernes, D. B., "Supporting and insulating means for junctured elements of small electrical components", Patent USA 3429981, Publ. February 1969, EEA73-3558.
788. Siemens, A. G., "Component mounting system", Patent UK 1144679, Publ. March 1969, EEA72-19447.

789.Spridco, J. P. and Mancewicz, T. A., "Method of making multilayer printed circuits", Patent USA 3605260, Publ. September 1971.
790.STC Ltd., "Integrated circuit package connector", Patent UK 1294639, Publ. November 1972.
791.Stotler, R. A. and White, H., "Packaging a billion operations per second computer", Presented at Semiconductor Integrated Circuit Processing and Production Conference, Anaheim, Calif., February 1971, EEA74-21670.
792.Tefft, E. G., "Glass-encapsulated semiconductor device fabrication process", Patent USA 3639975, Publ. February 1972.
793.Tiefert, K. H., "Semiconductor device manufacture (multilayer metal conductors)", Patent USA 3836446, Publ. September 1974.
794.Umbaugh, C. W., "Method in microcircuit package assembly providing non-abrasive electrically passive edges on integrated circuit chips", Patent USA 3838501, Publ. October 1974.
795.van der Drift, A., Gelling, W. G. and Rademakers, A., "Integrated circuits with leads on flexible tape", Philips Tech. Rev., 34, 85-95 (1974), EEA78-2916.
796.Wakely, W. T., "MSI/LSI packages: design problems and solutions", Presented at Semiconductor Integrated and Production Conference and Exhibition, Anaheim, Calif., February 1971, EEA74-21625.
797.Westinghouse Electric Corp., "Module assembly", Patent UK 1138278, Publ. December 1968, EEA72-9242.
798.Westinghouse Electric Corp., "Encapsulated semiconductor components", Patent UK 1201595, Publ. August 1970, EEA74-10451.
799.Wilson, G. P., "Microelectronic circuit technology in electronic systems: A radar designer's viewpoint", International Microelectronics Symposium, 1971, 2-3, 5 pp., EEA75-14453.

XIX. Types of Packages
1. Plastic

800.Bolster, W. N., "Epoxy conformal coatings solve physical and environmental packaging problems of electronic wristwatches economically", Insul./Circuits, 21 (10), 29-31 (1975), EEA78-40471.
801.Clark, R. D. and Smith, P. E., "The high-rel plastic package versus transfer molded encapsulation", Proceedings of the 10th Electrical Insulation Conference, 1971, p. 68-9, EEA75-12832.
802.Corradetti, M., "Plastic packages for m.o.s.-LSI: a reality of today", Presented at the 4th International Microelectronics Conference, Munich, Germany, November 1970, EEA74-12802.
803.Divoky, R. D., "Plastic headers in the manufacture of LED's, semiconductors", Insul./Circuits, 18 (8), 30-1 (1972), EEA75-35469.
804.Edfors, H. C., "Reliability considerations in plastic encapsulated microcircuits", Report RAC-TM-71-1, IIT Res. Inst., Chicago, Ill., 1971, 113 pp., EEA75-31824.

805.Flaskerud, P. D., "MOS devices in plastic packages - a reality", Presented at Semiconductor Integrated and Production Conference and Exhibition, Anaheim, Calif., February 1971, EEA74-21671.
806.Gallace, L. J. and Iara, J. S., "Evaluating reliability of plastic-packaged power transistors in consumer applications", IEEE Trans. Broadcast & Telev. Receivers, BTR-19, 194-204 (1973), EEA76-34661.
807.Khajezadeh, H., "High-reliability plastic package for integrated circuits", Proceedings of the 11th Annual Reliability Physics Conference, 1973, p. 236-44, EEA77-6089.
808.Nufer, R. W. and Pugleise, F. G., "Transfer circuitry module technology", Solid State Technol., 17 (10), 49-53, 72 (1974), EEA78-5341.
809.Roublot, R., "Five-watt integrated circuits", Electronique Industr., no. 135, 470-3 (1970), French, EEA73-34369.
810.Scrupski, S. E., "Plastic cavity packages for MOS LSI fill a very present need - but for how long?", Electronics, 45 (11), 102-6 (1972), EEA75-23366.
811.Smith, P. E., "Premolded microcircuit packages come on strong", Insul./Circuits, 20 (6), 17-19 (1974), EEA77-29337.

2. Ceramic

812.Anazawa, S., "Packaged semiconductor device including a housing in the form of a rectangular parallelepiped and ceramic rectangular base member", Patent USA 3801881, Publ. April 1974.
813.Aposporos, N. J., "Ceramic package for an integrated circuit", Patent USA 3665592, Publ. May 1972.
814.Asada, A., "Calculator's parts are integrated on a single glass substrate", Electronics, 48 (3), 109-11 (1975), EEA78-11768.
815.Bailey, J. T., "The case for using ceramics", Electron. Des., 22 (8), 74-7 (1974), EEA78-5340.
816.Carley, D. R., "Strip-line power transistor package", Patent USA 3478161, Publ. November 1969, EEA73-31410.
817.Class, W. H., "Thin-film circuit processing on dielectric substrates", Solid State Technol., 17 (8), 32-6 (1974), EEA77-38874.
818.Darrah, J. G., "New concepts in ceramic semiconductor packages", Presented at Semiconductor Integrated and Production Conference and Exhibition, Anaheim, Calif., February 1971, EEA74-21672.
819.Fisher, J. A., "Process for manufacturing integrated circuits on an alumina substrate", Patent USA 3783052, Publ. January 1974.
820.Furnival, T. J., "Cold-welded ceramic semiconductor package", Patent USA 3686540, Publ. August 1972.
821.Gardner, R. A. and Wilcox, D. L., "Mullite package for integrated circuit devices", Patent USA 3857923, Publ. December 1974.
822.Griffin, W. F. and Morena, A., Jr. "Ceramic package", Patent USA 3634600, Publ. January 1972.

823.Hill, G. J., "Ceramics for microcircuit applications", Proc. Brit. Ceram. Soc., no. 18, 247-62 (1970).
824.Ikeda, M., Takeoka, T., Nakata, T. and Ono, T., "Alumina ceramics for application in electronic field", Hitachi Hyoron, 56 (7), 66-70 (1974), Japanese, EEA78-664.
825.Jablonski, J. and Bielawski, W., "Assembly technology of the 'CERATAB' packaging of hybrid microcircuits", Elektronika, no. 5, 211-2 (1972), Polish, EEA75-31837.
826.Lynch, J. E., "Leadless ceramic package for integrated circuits", Patent USA 3641491, Publ. February 1972.
827.McAlexander, J. C., Jr., "Ceramic substrates for memory applications", Am. Ceram. Soc. Bull., 52 (4), 375 (1973), EEA77-2616.
828.Miyake, S. and Sakai, K., "Hermetic ceramic power package for high-frequency solid state device", Patent USA 3769560, Publ. October 1973.
829.Scrupski, S. E., "LSI packages make room at the top", Electronics, 45 (7), 119-20 (1972), EEA75-16743.
830.Shatz, S., "Semiconductor package of alumina and aluminum", Patent USA 3585272, Publ. June 1971.
831.Sundius, R. E., "A recommendation for ceramic over plastic IC packages for 1972 real-time commercial computer systems", Proceedings of the 10th Annual Conference on Reliability Physics, 1972, p. 102-4, EEA76-4699.

3. Multilayer Ceramic

832.Anon., "Development of multilayer ceramic packages for LSIs", In: Research and development awarded the Okochi memorial prize 1972, Tokyo, Japan: Okochi Memorial Found., p. 18-21 (1972), EEA76-26167.
833.Close, A. D., "Multilayer substrates utilizing thin-film and thick-film technologies", Solid State Technol., 14 (5), 38-42 (1971), EEA74-25650.
834.Cox, J. J., Jr., "Multilayer ceramic packaging", IEEE International Convention Digest, 1971, p. 562, EEA74-25584.
835.Fugardi, J. F. and Kaiser, H. D., "Multilayer ceramic by slurry casting over metallization", IBM Tech. Disclosure Bull., 16, 2887-8 (1974), EEA77-23081.
836.Gardner, R. A. and Nufer, R. W., "Properties of multilayer ceramic green sheets", Solid State Technol., 17 (5), 38-43 (1974), EEA77-30357.
837.Ghergia, V., "Multilayer ceramic monolithic integrated circuit technology", Electron. & Telecommun., 23, 163-71 (1974), Italian, EEA78-15422.
838.Griesemer, J. A., "Multilayer ceramic slide adapter for X-Y green sheet beam radiated 'via' punching", IBM Tech. Disclosure Bull., 16, 2894 (1974), EEA77-22234.
839.Haddad, M. C., "Metallizing multilayer ceramic top and bottom side", IBM Tech. Disclosure Bull., 16, 3612 (1974), EEA78-671.

840.Hargis, B. M., "Buried wire multilayer ceramics", Solid State Technol., 14 (5), 47-50 (1971), EEA74-25571.
841.Ihochi, T., "Screened multilayer ceramics for thick film hybrids", IEEE Trans. Parts, Hybrids & Packag., PHP-10, 115-19 (1974), EEA77-36334.
842.Kaiser, H. D., Pakulski, F. J. and Schmeckenbecher, A. F., "A fabrication technique for multilayer ceramic modules", Solid State Technol., 15 (5), 35-40 (1972), EEA75-27844.
843.Keller, W. R., Pirigyi, F. F., Cole, G. R. and Budd, J. P., "Multilayer alumina circuit boards", Proceedings of the Electronic Components Conference, 1969, p. 52-9, EEA72-23159.
844.Manner, H. and Metreaud, C. G., "Multilayer ceramic kerfless blanking and lamination", IBM Tech. Disclosure Bull., 27, 1377-7a (1974), EEA78-8319.
845.Noren, S., "Laminated ceramic high-frequency semiconductor package (BeO base)", Patent USA 3748544, Publ. July 1973.
846.Nufer, R. W., "Saturated vapor doctor blade assembly", IBM Tech. Disclosure Bull., 16, 2891 (1974), EEA77-22232.
847.Nufer, R. W., "Pressure separation of cast green ceramic", IBM Tech. Disclosure Bull., 16, 2892 (1974), EEA77-22233.
848.Phillips, G. C., Jr., "Fixture for fabricating complex substrate design from green sheet ceramics", IBM Tech. Disclosure Bull., 16, 3559 (1974), EEA78-68.
849.Surtani, K., "Screenable dielectric ML ceramic-metal seal", Presented at Semiconductor Integrated Circuit Processing and Production Conference, Anaheim, Calif., February 1971, EEA74-22785.
850.Wilcox, D. L., "Ceramics for packaging. I", Solid State Technol., 14 (1), 40-4, 48 (1971), EEA74-12800.
851.Wilcox, D. L., "Ceramics for packaging. II", Solid State Technol., 14 (2), 55-60 (1971), EEA74-17646.

4. Hybrid

852.Abbe, R. C., "The use of components in thick film hybrid system", Presented at Record of the Workshop on Industrial Electronics and Control Instrumentation for Thick-film Hybrid IC Technology, New York, March 1968, EEA72-10658.
853.Abbe, R. C., "Commercial application of thick film hybrid microelectronic packaging", Proceedings of the Electronic Components Conference, 1969, p. 231-41, EEA72-23211.
854.Altman, L., "Hybrid technology solves tough design problems", Electronics, 46 (12), 89 (1973), EEA76-26160.
855.Anon., "Hybrid IC - the perspective components of microelectronics", Sdelovaci Tech., 21, 385 (1973), Czech, EEA77-10319.
856.Anon., "Hybrid microcircuit design guide", Report SLA-74-0333, Sandia Labs., Albuquerque, N. Mex., 1974, 97 pp., EEA78-33251.
857.Ashman, W. G. and Chumbley, R., "Power control using hybrid technology", New Electron., 6 (6), 66 (1973), EEA76-16473.

858.Benoit, J. J., "Hybrid integration for hyperfrequencies: problems and solutions", Electron. & Microelectron. Ind., no. 205, 19-21 (1975), French, EEA78-33244.
859.Bielawski, W., "Power hybrid circuits", Elektronika, 15, 445-51 (1974), Polish, EEA78-11769.
860.Bigler, R. R., "Equipment designer's approach to packaging of hybrid microcircuits", Proceedings of the Electronic Components Conference, 1970, p. 12-15, EEA74-500.
861.Bingham, K. C. and Gurler, Y., "Hybrid techniques for computer circuits. II. The thick film aspects of a high thermal dissipation interconnection system for use with integrated circuits", Proceedings of the Conference on Hybrid Microelectronics, 1973, p. 313-24, EEA77-405.
862.Bingham, K. C. and Naylor, R., "Hybrid techniques for computer circuits. I. Interconnection hierarchy", Proceedings of the Conference of Hybrid Microelectronics, 1973, p. 305-12, EEA77-404.
863.Blanchut, R., "Hybrid circuits", Revue Polytech., no. 1304, 789 (1972), French, EEA75-35430.
864.Bower, F., "Hybrid IC assembly and packaging", Presented at Record of the Workshop on Industrial Electronics and Control Instrumentation for Thick-film Hybrid IC Technology, New York, March 1968, EEA72-10659.
865.Brady, T. E. and Hindermann, D. K., "A bilevel thin film hybrid circuit containing crossovers, resistors, capacitors, and integrated circuits", IEEE Trans. Parts, Hybrids & Packag., PHP-9, 181-5 (1973), EEA76-38317.
866.Brooke, G. and Baldwin, W. E. B., "Hybrid thick-film circuits: their design, application and manufacture", Wireless World, 79 (1449), 121-5 (1973), EEA76-16474.
867.Brown, C. E. and Stech, H. G., "Hybrid microelectronic packages", Solid State Technol., 14 (8), 44-5 (1971), EEA75-9608.
868.Burgess, J. F., Neugebauer, C. A. and Flanagan, G. T., "Hybrid packages by the direct bonded copper process", 24th Electronic Components Conference, 1974, p. 94-7, EEA77-34114.
869.Ciccio, J. A., Stuhlbarg, S. M. and Thun, R. E., "Hybrid LSI logic modules for aerospace", 22nd Electric Components Conference, 1972, p. 185-200.
870.Clay, F. A., Panousis, N. T. and Pierce, R. W., Jr., "Characterization of a chromium-gold deposition process for the production of thin film hybrid microcircuits", IEEE Trans. Parts, Hybrids & Packag., PHP-10, 258-62 (1974), EEA78-11771.
871.Cyzekely, G. S., "Reliability assurance criteria for film-hybrid microcircuits", 3rd Symposium on the Reliability of Electronics, 1973, p. 385-402, EEA77-6109.
872.Dalnoki, G. and Walton, G., "Hybrid integrated circuits on insulating substrates", Hiradastechnika, 23, 346-50 (1972), Hungarian, EEA76-8756.

873.Diel, B., "Multichip packages in hybrid film circuits", Siemens Electron. Components Bull., 7 (1), 13-6 (1972), EEA75-31841.
874.Dowling, T. J. and Colling, D. A., "A systems approach to hybrid packaging", Electro Technology, 83 (2), 38-42 (1969), EEA72-12861.
875.Forlani, F., "Thick film hybrid circuits", Alta Freq., 42, 420-9 (1973), Italian, EEA77-10310.
876.Fuller, G. L., Anderson, R. M., Jr. and Neudeck, G. W., "Undergraduate thick film hybrid circuit layout design and fabrication", IEEE Trans. Educ., E-16, 126-30 (1973), EEA76-29320.
877.Funk, W. and Schilz, W., "Thick film technique for hybrid integrated microwave circuits", Proceedings of the Conference on Hybrid Microelectronics, 1973, p. 129-38, EEA77-401.
878.Furhoven, T., van Marter, C. and Dunn, F., "HIC power amplifier for transducers", Presented at Semiconductor Integrated Circuit Processing and Production Conference, Anaheim, Calif., February 1971, EEA74-24361.
879.Galli, H., "Monolithic and hybrid integrated circuits", Elektroniker, 13 (10), EL6-10 (1974), German, EEA78-5377.
880.Gayet, A., "Thin film techniques in hybrid microcircuits. II", Elektron. Ind., 4 (4), 66-7 (1973), German, EEA76-29318.
881.Hampy, R. E., "A versatile Ta_2N-W-Au/SiO_2/Al/SiO_2 thin film hybrid microcircuit metallization system", IEEE Trans. Parts, Hybrids & Packag., PHP-11, 263-72 (1975).
882.Hass, B., McLeod, W. W. and Thun, R. E., "A method for the rapid and economical generation of hybrid LSI circuits", 24th Electronic Components Conference, 1974, p. 172-6, EEA77-34117.
883.Henderson, R., "The hybrid myth", Electron, no. 36, 69, 71-3 (1974), EEA77-10313.
884.Herren, H., "Thin-film hybrid ICs", Tech. Mitt. PTT, 51 (2), 74-8 (1973), German, EEA76-12682.
885.Hetherington, D. R., "Thick-film hybrids - what they can offer", New Electron., 8 (6), 64, 67-8 (1975), EEA78-24537.
886.Keister, F. Z. and Scapple, R. Y., "A thin-film multilayering technique for hybrid microcircuits", Solid State Technol., 17 (5), 44-7 (1974), EEA77-30420.
887.Knezich, A. G. and Goldfarb, S., "Hybrid voltage regulator module", International Microelectronics Symposium, 1971, 9-7, 8 pp., EEA75-12572.
888.Korn, G. A., "New packaging techniques for integrated-circuit hybrid computers: the APE II and LOCUST systems", Fifth International Analogue Computation Meetings, 1967, p. 41-7, Publ. 1968, EEA73-9637.
889.Lepoff, J., "Diodes for hybrid integrated circuits", Solid State Technol., 16 (8), 58-6 (1973), EEA76-38318.
890.Makos, W. S., "A new concept in system packaging utilizing hybrid technology", Proceedings of the IEEE 1974 National Aerospace and Electronics Conference, p. 237-43, EEA78-5381.
891.Mazenko, J. J., "A simple systems oriented hybrid packaging approach", 22nd Electronic Components Conference, 1972, p. 243-51.

892. Muller, J. H. W., "Hybrid integrated gyrator in dual-in-line package for universal application", IEEE Trans. Circuit Theory, CT-18, 726-8 (1971), EEA75-139.
893. Nesselroth, M. D. and Zeien, H., "Hybrid LSI array technology", Presented at Semiconductor Integrated Circuit Processing and Production Conference, Anaheim, Calif., February 1971, EEA74-21615.
894. Niwa, K., Nakamura, J., Murakawa, K. and Nakamura, M., "New alumina substrate for hybrid integrated circuits", IEEE Trans. Parts, Hybrids & Packag., PHP-10, 262-6 (1974), EEA78-11772.
895. Ohanian, M., "Hybrid microelectronic bonding and packaging metholodology", Presented at Microelectronics Comes of Age Symposium, San Francisco, Calif., August 1967, EEA72-12820.
896. Pilgrim, M. and Winterbottom, K., "Hybrid thin-film technology for microwave integrated circuits", Proceedings of the Conference on Hybrid Microelectronics, 1973, p. 153-9, EEA77-402.
897. Ripka, G., "The production technology of thick film hybrid integrated circuits on insulating substrate. II", Finommech. & Mikrotech., 12, 244-53 (1973), Hungarian, EEA77-14625.
898. Ripka, G. and Reczey, M., "The design of hybrid integrated thin film circuits on insulating substrate", Finommech. & Mikrotech., 12, 290-9 (1973), Hungarian, EEA77-18864.
899. Rivera, J., "Thick film technology", Rev. Esp. Electron., 21 (238), 42-5 (1974), Spanish, EEA78-5349.
900. Ruegg, F., "Why not use hybrids?", Electron. Des., 22 (14), 84-9 (1974), EEA78-15423.
901. Saddler, I. R., "Semiconductor device package with energy-absorbing layer", Patent USA 3735209, Publ. May 1973.
902. Smith, G. W. and Little, L. E., "Packaging of precision analog microcircuits", International Hybrid Microelectronic Symposium, 1970, 9 pp., EEA74-29734.
903. Staller, J. J., "Hybrid microelectronic packaging considerations", Electronic Commun., 3 (2), 17-18 (1968), EEA73-511.
904. Stroup, R. E., "Low-cost hybrid microelectronic packaging", International Hybrid Microelectronic Symposium, 1970, 6 pp., EEA74-29806.
905. Tapp, C. M. and Sharp, D. J., "Compilation of specifications describing hybrid microcircuit technology", Report SLA-74-0300, Sandia Labs., Albuquerque, N. Mex., 1974, 218 pp., EEA78-33250.
906. Williams, J., "Problems associated with hybrid thick film assembly", Presented at Semiconductor Integrated Circuit Processing and Production Conference, Anaheim, Calif., February 1971, EEA74-21612.
907. Winiger, F., "Thick-film hybrids", Tech. Mitt. PTT, 51 (2), 68-73 (1973), German, EEA76-12681.
908. Yamazaki, J., Kamo, T. and Nakamura, M., "High density thin film hybrid IC utilizing Ta-Al-N resistor and Ta_2-O_5-MnO_2 capacitor", 24th Electronic Components Conference, 1974, p. 111-17, EEA77-34088.

909.Yermolaev, Y. P., "Optimum topology of thick-film elements of hybrid integrated circuits", Izv. VUZ Radioelektron., 16 (4), 40-3 (1973), Russian, EEA76-26162.

5. Optoelectronic

910.Blank, J. M., "Weierstrass method applied to the optics of LED encapsulation", Solid State Technol., 14 (7), 43-8 (1971), EEA74-36969.
911.Callely, J. A., Jr. and MacDonald, T., "Encapsulated electroluminescent device", Patent USA 3509401, Publ. April 1970.
912.Clark, D. E. and Bailey, J. T., "LED's and light sensitive devices present challenges in packaging", Solid State Technol., 16 (11), 50-2 (1973), EEA77-10566.
913.Dowsett, J. and Smith, D. A., "The trend towards packaged crystal oscillators", Des. Electron., 8 (9), 20-3 (1971), EEA74-21378.
914.Giannotto, J. M., "Multiple crystal holder", Report ECOM-2972, Army Electronics Lab., Fort Monmouth, N. J., 1968, 19 pp., EEA73-3298.
915.Klein, R. M., "Encapsulation for light-emitting diodes", Patent USA 3834883, Publ. September 1974.
916.Norrie, J., "Packaging solid state displays", SID J., 10 (4), 5-8 (1973), EEA77-2908.
917.Sasaki, K. and Watanabe, S., "Method of sealing display cathodes in a glass envelope", Patent USA 3632324, Publ. January 1972.
918.Siegel, H. V., "Light-emitting semiconductor chips mounted in a slotted substrate forming a display apparatus", Patent USA 3573568, Publ. April 1971.
919.Wetterau, L. C., Jr., "Packaging optoelectronic devices", Presented at Semiconductor Integrated Circuit Processing and Production Conference, Anaheim, Calif., February 1971, EEA74-22262.

6. Others

920.Ahearn, J. J., Jr. and Taylor, W. L., "Thin skin packaging for electrical components", Patent USA 3474185, Publ. October 1969, EEA73-27364.
921.Bachner, F. J., "A new technique for interconnecting and packaging integrated circuits on flexible substrates", Insul./Circuits, 19 (10), 71 (1973), EEA77-410.
922.Balmer, T., "Sliding panels package electronic components", Des. News, 26 (7), 90 (1971), EEA74-29899.
923.Behrendt, D. L., "Electronic packaging for spacecraft", Insul./Circuits, 16 (13), 17-20 (1970), EEA74-16540.
924.Bixby, B. and Carver, L., "Passivated assembled circuits in a TO-3 package for power control", 1973 8th Annual Meeting of the IEEE Industry Applications Society, p. 93-8, EEA77-9592.
925.Choffart, P., "Tight cover for a semiconductor device", Patent USA 3433885, Publ. March 1969, EEA73-4253.

926.Cohn, C. E., "Considerations in the packaging of analog-digital converters", Rev. Sci. Instrum., 40, 512 (1969), EEA73-6807.
927.Crawford, A. B. and Turrin, R. H., "A packaged antenna for short-hop microwave radio systems", Bell Syst. Tech. J., 48, 1605-22 (1969), EEA73-4657.
928.Danielson, H., "Large thick-film multilayer substrates for DIL-packages", Proceedings of the Conference on Hybrid Microelectronics, 1973, p. 283-94, EEA77-387.
929.David, R. F. and Peluso, R. F., "A high-density packaging approach for integrated circuits", International Electronic Circuit Packaging Symposium, 1969, 7 pp., EEA73-27361.
930.Dibugnara, R., "High-power double-slug diode package", Patent USA 3844029, Publ. October 1974.
931.Donnerstein, R. L., Pao, H. C. and Thun, R. E., "Design criteria for optimized high density packaging of digital circuits", Proceedings of the Electronic Components Conference, 1970, p. 28-33, EEA74-501.
932.Dunaetz, R. A., "Spacecraft and high altitude insulation trends", Insulation, 14 (9), 49-52 (1969), EEA72-520.
933.Eddy, J. W., "PWC trunk circuits for the No. 1 EAX (Electronic Automatic Exchange)", GTE Autom. Elec. Tech. J., 13 (1), 10-18 (1972), EEA75-29182.
934.Ermolaev, Y. P., "A procedure for an optimum choice of electronic units", Izv. VUZ Radioelektronika, 12, 1300-5 (1969), Russian, EEA73-16707.
935.Forestieri, A. F. and Broder, J. D., "Improvements in silicon solar-cell cover glass assembly and packaging using FEP Teflon", Record of the 8th IEEE Photovoltaic Specialists Conference, 1970, p. 179-82, EEA74-9329.
936.Frigyes, I., "The effect of packaging on microwave integrated circuits", Presented at the International Conference on Advanced Microelectronics, Paris, France, April 1970, EEA74-1830.
937.Ginsberg, G. L., "Packaging enclosures and applications", Electron. Eng., 31 (7), 24, 26-8 (1972), EEA75-39258.
938.Goto, K. and Ito, Y., "Logic circuit package design", Rev. Elec. Commun. Lab., 19, 357-64 (1971), EEA74-27543.
939.Granberry, D., "Packaging techniques for microwave integrated circuits summary", Presented at Microelectronics Symposium, Clayton, Mo., September 1969, EEA73-17940.
940.Granberry, D. S. and Sterling, C. E., "Packaging an X-band microwave integrated circuit", Presented at the International Electron Devices Meeting, Washington, D. C., October 1967, EEA72-7456.
941.Greig, W. J. and Brown, R., "Packaging of multi-chip arrays", 22nd Electronic Components Conference, 1972, p. 169-80.
942.Greiling, P. T. and Laton, R. W., "Determination of semiconductor junction device package networks", IEEE Trans. Microwave Theory & Tech., MTT-22, 1140-5 (1974), EEA78-13211.

943.Grossman, S. E., "Leadless, pluggable IC packages reduce fabrication and repaid costs", Electronics, 46 (3), 83-9 (1973), EEA76-12683.
944.Guttmann, W., "Electronic component package and cover", Patent USA 3429980, Publ. February 1969, EEA72-3557.
945.Hall, E. E., "Electronic valve packaging for naval applications", Environmental Engng., no. 41, 11-15 (1969), EEA73-16721.
946.Hurst, S. L., "Specifications of threshold-logic gates for optimum S.S.I. logic packaging", Electron. Lett., 8, 514-5 (1972), EEA76-353.
947.IBM Corp., "Computer integrated circuit package system", Patent UK 1384220, Publ. February 1975, EEA78-33179.
948.Irwin, L. R. and Larsen, R. G., "Leadless packaging: A new concept for LSI systems", Presented at the 4th International Microelectronics Conference, Munich, Germany, November 1970, EEA74-12801.
949.Isaacson, H. M., "Flexible circuitry packaging at the module level", Electron. Packag. & Prod. (Int. Ed.), 7 (6), 27-30 (1971), EEA75-7058.
950.Isaak, H. R., Kanz, J. W. and Babiracki, E. G., "Development of large thick-film multilayer assemblies", IEEE Trans. Parts, Hybrids & Packag., PHP-8, 20-9 (1972), EEA75-27804.
951.Jones, R. J. and Armani, R. C., "Novel mounting technique for glass-package diodes", Patent USA 3757259, Publ. September 1973.
952.Joshi, K. C., "Wafer packaging with an interposer", IBM Tech. Disclosure Bull., 18, 352 (1975), EEA78-37632.
953.Kirby, E., "Backplane - the mechanical and electrical backbone", Electron. Eng., 31 (5), 39-40 (1972), EEA75-27843.
954.Krause, G., "Extremely high packaging densities by operating integrated circuits with r.f. power", Electron. Lett., 11, 327-8 (1975), EEA78-37608.
955.LaCombe, D. J. and Babcock, G. L., "Composite integrated circuits including semiconductor chips mounted on a common substrate with connections made through a dielectric encapsulator", Patent USA 3805375, Publ. April 1974.
956.Leppert, K. and Roth, W., "ES 902 - the fully integrated modular packaging system for industrial electronics equipment", Siemens Rev., 42, 333-8 (1975), EEA78-40474.
957.Liljenwall, E. T., "Packaging the pocket calculator", Hewlett-Packard J., 23 (10), 12-13 (1972), EEA75-31899.
958.McDermott, J., "In consumer packaging, it's modules, modules everywhere", Electron Des., 19 (15), 22-3 (1971), EEA74-33402.
959.Maegawa, H., Matsuda, J. and Takasu, M., "Flat packaging of magnetic bubble devices", 1974 Digests of the Intermag. Conference, 22-7/1 pp., EEA78-8749.
960.Markstein, H. W., "Packaging for detrimental environments", Electron. Packag. & Prod., 15 (8), 26-34 (1975).
961.Marx, D. W., "High-temperature semiconductor package", Patent USA 3566212, Publ. February 1971.

962.May, D. L., "Package for integrated-circuit chip", Patent USA 3774232, Publ. November 1973.
963.Mesler, L. L., "Semiconductor diode construction and mounting", Patent USA 3460002, Publ. August 1969, EEA73-21239.
964.Miller, D. J., "Techniques utilized in packaging spacecraft electronics", International Electronic Circuit Packaging Symposium, 1969, 8 pp., EEA73-29083.
965.Miller, D. J. and Weidner, G. G., "New maser-packaging approach", Proc. IEEE, 58, 181-2 (1970), EEA73-17899.
966.Murphy, H, E. and Steffe, W. C., "Monolithic integration for a camera control system", IEEE J. Solid-State Circuits, SC-8, 427-34 (1973), EEA77-6073.
967.Naybours, T. E., "Unique bi-fold microcircuit packaging design", Presented at the International Conference on Advanced Microelectronics, Paris, France, April 1970, EEA74-894.
968.Norrie, J. and Maltais, F., "IC packaging trends", 1973 IEEE International Convention and Exposition, Vol. VI, 50.3/6 pp., EEA77-2909.
969.Oddy, R. H., "A modern concept of circuit packaging", Sound & Vision Broadcast, 13 (1), 36-8 (1972).
970.Owens, R. P., "Mount-independent equivalent circuit of the S4 diode package", Electron. Lett., 17, 580-2 (1971), EEA74-37531.
971.Palermo, C. A. and Hartman, J. A., "MIC packaging using conductive epoxy and numerically controlled drill techniques", IEEE Trans. Parts, Hybrids & Packag., PHP-11, 157-8 (1975).
972.Parham, L., "Technology feasibility spacecraft technology report Volume I: Electronic packaging (Final report)", Report AD-673563, Martin Co., Denver, Colo., 1967, 323 pp., EEA72-16253.
973.Parker, B. J., "Microelectronics packaging", New Electron., 7 (2), 89, 91 (1974), EEA77-18871.
974.Parks, H. L., "Three dimensional planar coax packaging for microelectronic computer systems", Proceedings of the 21st Electronic Components Conference, 1971, p. 515-22, EEA74-29905.
975.Parks, H. L., "Three-dimensional IC packaging - with a twist", Electronics, 45 (3), 62-5 (1972), EEA75-9644.
976.Peel, M., "When to plug the ICs in", Electron. Eng., 31 (5), 36-7 (1972), EEA75-27842.
977.Pops, H. and Johnson, B. C., "Integrated circuit lead frame package", Patent USA 3854200, Publ. December 1974, EEA78-19952.
978.Read, W. S., "Electronic packaging and cabling of a Mars rough lander", International Electronic Circuit Packaging Symposium, 1969, 12 pp., EEA73-29082.
979.Riel, M. and Chalman, R., "Specifying special IC packages", Electron. Packag. & Prod., 15 (5), 65-8 (1975).
980.Rifkin, A. A., "A practical approach to packaging magnetic bubble devices", IEEE Trans. Magn., MAG-9, 429-33 (1973), EEA77-6244.
981.Ripka, G., "Packaging integrated circuits by moulding", Finommech. & Mikrotech., 13 (8), 250-5 (1974), Hungarian, EEA78-662.

982. Robertson, F. A., "Packaging techniques for modern microelectronics equipment", Electron. & Power, 17, 406-9 (1971), EEA75-414.

983. Rogers, B. C. and Wakely, W. T., "Diaphragm-connected leadless package for semiconductor devices", Patent USA 3621338, Publ. November 1971.

984. Roth, J. P., "Repairable large-scale integrated logic unit", IBM Tech. Disclosure Bull., 17, 1226-8 (1974), EEA78-5073.

985. Ryan, R. J., "Multilayer circuit board techniques", Patent USA 3756891, Publ. September 1973.

986. Saddler, I. R., "Semiconductor device package with energy-absorbing layer", Patent USA 3735209, Publ. May 1973.

987. Sato, H., Tanaka, M., Nakamura, M., Takahashi, S. and Kurono, H., "Package for mounting semiconductor device in microstrip line", Patent USA 3786375, Publ. January 1974.

988. Schebler, B. J., "Packaging of electronics for use at cryogenic temperatures", International Electronic Circuit Packaging Symposium, 1969, 9 pp., EEA73-27355.

989. Schmidt, J. L., "Packaging of micrologic elements", Patent USA 3562591, Publ. February 1971, EEA74-14847.

990. Schoberl, W., "H.F. power transistors: standard stripline package", Radio Mentor, 37 (2), 57-8 (1971), German, EEA74-15573.

991. Sear, B. E., "Packaging for high frequency", Electron. Eng., 31 (7), 20-1 (1972), EEA75-39256.

992. Socolovsky, A. and Ruth, S. B., "Packaging with integrated circuits", Electron. Eng., 31 (2), 51-2 (1972).

993. Spivak, R. R., "A low-inductance millimeter wave semiconductor package", IEEE Trans. Microwave Theory & Tech., MTT-19, 732-3 (1971), EEA74-27121.

994. Stoll, D. and Guthner, K., "A fully integrated construction concept for a subnanosecond computer", Presented at the 4th International Microelectronics Conference, Munich, Germany, November 1970, German, EEA74-12750.

995. Tamura, A., Tachibana, Y., Kasuga, K., Furuya, M., Kikukawa, H., Kurabe, H. and Tabata, K., "Packaging and assembly of D10 electronic switching system and its components", Hitachi Hyoron, 54 (11), 59-61 (1972), Japanese, EEA76-24343.

996. Tang, D. D., "High power varactor package for multichip application", Proc. IEEE, 57, 799-800 (1969), EEA72-17405.

997. Taynton, R., "Packaging of electrical equipment", Patent USA 3434014, Publ. March 1969, EEA73-6808.

998. Uberbacher, E. C., "Bubble memory package via eggerate", IBM Tech. Disclosure Bull., 14, 3378 (1972), EEA75-28072.

999. Urban, L., "Advanced concept for a microelectronic microwave receiver", International Electronic Circuit Packaging Symposium, 1969, 7 pp., EEA73-28360.

1000. Vogler, F., "Application of the AEG-Telefunken precision thick-film technique in commercial apparatus", Tech. Mitt. AEG-Telefunken., 62 (1), 29-31 (1972), German, EEA75-27811.

1001.Welsh, J. R., "Thyristors and diodes - the complete service from chip to heatsink", Electron, no. 45, 29, 31 (1974), EEA77-17828.
1002.Wendelburg, R. W., "What's missing in SCR's and power diodes", Sixth Annual Meeting of the IEEE Industry and General Applications Group, 1971, p. 897-9, EEA75-13732.
1003.Westrom, J. L., "The feasibility of packaging high-frequency microcircuits inside the magnetic core", IEEE Trans. Aerosp. & Electron. Syst., AES-7, 329-31 (1971), EEA74-21664.
1004.Whitman, M. E., "Package for electronic semiconductor devices", Patent USA 3848077, Publ. November 1974.
1005.Wood, J. R., "Packaging an electronic watch for minimum size and cost", Electron. Packag. & Prod., 15 (6), 81-6 (1975).
1006.Yanai, H., Kida, H., Yanagawa, T. and Trubaki, T., "Packaged semiconductor devices", Patent USA 3509434, Publ. April 1970.
1007.Yau, S. S., "Universal logic implementations", IEEE International Convention Digest, 1971, p. 374-5, EEA74-25514.

XX. Thermal Design

1. Cooling Techniques

1008.Anon., "Newcooling technique", Electron, no. 17, 23-4 (1972), EEA76-7775.
1009.Aung, W., "Heat transfer in electronic systems with emphasis on asymmetric heating", Bell Syst. Tech. J., 52, 907-25 (1973), EEA76-34660.
1010.Aung, W., Kessler, T. J. and Beitin, K, I., "Free convection cooling of electronic systems", IEEE Trans. Parts, Hybrids & Packag., PHP-9, 75 (1973), EEA76-28988.
1011.Boucher, S. G. and Paradis, L. R., "Experimental heat transfer investigations on modules mounting hybrid packages", 23rd Electronic Components Conference, 1973, p. 334-41, EEA76-29324.
1012.Dalley, M. J., "Cooling electronics. I", Eng. Mater. & Des., 18 (9), 29-30 (1974), EEA78-4789.
1013.Dilay, W. and Zulinski, E. J., "Cooling apparatus for semiconductor devices", Patent USA 3768548, Publ. October 1973.
1014.Dulnjev, G. N., "Heat dissipation in electronic apparatus of the second group", Feingeraete Tech., 22, 221 (1973), German, EEA76-31975.
1015.Finch, D. J., "Forced air cooling of dual-in-line packages", Marconi Rev., 35 (187), 260-71 (1972), EEA76-11706.
1016.Fitton, G. L., "Heat pipe practice", Electron. Prod. Methods & Equip., 2 (6), 13-5 (1973), EEA76-37737.
1017.Fitton, G. L., "Taking out the heat", Electron. Engineering, 45 (550), 65, 67-8 (1973), EEA77-5147.
1018.Gardner, J. A., Jr., "Liquid cooling safeguards high-power semiconductors", Electronics, 47 (4), 103-8 (1974), EEA77-17544.

1019. Harwick, N. E. and Walthausen, V., "Combined heat conductor and protective transistor package", Tech. Dig., no. 25, 39-40 (1972), EEA75-16089.
1020. Hawker Siddeley, Ltd., "Rapid cooling device for semiconductors", Patent UK 1394662, Publ. May 1975, EEA78-41401.
1021. Honnor, F. and Thomas, M. A., "Packaging and cooling problems associated with microelectronics equipment", Microelectronics and Reliability, 8 (4), 331-7 (1969), EEA73-13079.
1022. Hwang, U. P. and Simons, R. E., "Optimum thermal design of electronic packages cooled by free air convection", 1973 SWIEEECO Record of Technical Papers, p. 599-603, EEA76-25460.
1023. Laermer, L., "Air through hollow cards cools high-power LSI", Electronics, 47 (12), 113-18 (1974), EEA77-25628.
1024. Laermer, L. and Schenk, R. J., "Packaging structure having cooling means for a multiplicity of hermetic modules for integrated circuit chips", Patent USA 3706010, Publ. December 1972.
1025. Laing, N., "Cooling device for semiconductor components", Patent USA 3780356, Publ. December 1973.
1026. Nevala, D. J., "A guide to good air cooling (electronic equipment)", Electron. Des., 22 (4), 76-80 (1974), EEA77-25599.
1027. Noth, R. W., "Heat transfer from silicon chips and wafers", IBM Tech. Disclosure Bull., 17, 3544 (1975), PA78-62543.
1028. Parkes, J. F. A. and Preston, S. B., "Cooling integrated circuits in computers", Electron. & Power, 21, 369-72 (1975), EEA78-24378.
1029. Patrickson, P., "The heat pipe in electronics", Electron. Components, 16 (14), 22-3 (1974), EEA78-64.
1030. Rauhe, M. L., "Thermal output for high density packaging", 9th Annual Proceedings Reliability Physics, 1971, p. 204-10, EEA75-3984.
1031. Robbins, G. J. and Stankus, J. J., "Electrostatic cooling of electronic packages", IBM Tech. Disclosure Bull., 18, 101-2 (1975), EEA78-36595.
1032. Shardanand, "Compact photomultiplier housing with controlled cooling", Rev. Sci. Instrum., 43, 641-3 (1972), EEA75-19804.
1033. Shelpuk, B., "Thermal design. VIII. Heat exchangers cool hot plug-in pc boards", Electronics, 47 (13), 114-20 (1974), EEA77-29335.
1034. Snyder, C. L. and Hessinger, P. S., "Ceramic-based substrates for electronic circuits with improved heat-dissipating properties and circuits including the same", Patent USA 3694699, Publ. September 1972.
1035. Streb, A. J., "Heat pipes cool gear in restricted spaces", Electronics, 47 (25), 114-17 (1974), EEA78-4788.
1036. Tuck, G., "Cooling electronics. II", Eng. Mater. & Des., 18 (9), 31-2, 34 (1974), EEA78-4790.
1037. Veilleux, E. D., "Cold-plate design analysis for electronic equipment", J. Spacecr. & Rockets, 10 (10), 80-2 (1973), EEA77-13629.

1038.Veilleux, E. D., "A design guide to cooling cordwood modules for vacuum environments", IEEE Trans. Parts, Hybrids & Packag., PHP-10, 272-7 (1974), EEA78-10930.

1039.Watson, B. W., "Fusible phase-change temperature control for modern packaging design", Proceedings of the 21st Electronic Components Conference, 1971, p. 115-17, EEA74-29903.

2. Heat Sinks

1040.Anon., "Heatsink selection", Electron, no. 73, 19-20 (1975), EEA78-27775.

1041.Board, K., "Thermal properties of annular and array geometry semiconductor devices on composite heat sinks", Solid-State Electron., 16 (12), 1315-20 (1973), EEA77-5776.

1042.Brown, Boveri & Co., Ltd., "Heat sink for electrical or electronic components", Patent UK 1290506, Publ. September 1972.

1043.Chambers, A. E., "Semiconductor and heat sink fin assembly", Patent USA 3823351, Publ. July 1974.

1044.Chu, R. C. and Hwang, U. P., "Heat sink", IBM Tech. Disclosure Bull., 17, 3656-7 (1975), EEA78-27778.

1045.Connors, P. M., "Variable-area heat sink device", IBM Tech. Disclosure Bull., 17, 1016 (1974), EEA78-4791.

1046.Desmond, R. J. and Koenig, P. W., "Semiconductor device with a heat sink having a foot portion", Patent USA 3601667, Publ. August 1971.

1047.Dulnev, G. N. and Polshchikov, B. V., "Nonstationary thermal regime of an electronic unit with solid heat sinks", Izv. VUZ Priborostr., 16 (12), 89-93 (1973), Russian, EEA77-9271.

1048.Evan-hart, G. D., "Staggered-finger heat dissipators", Electron. Compon., 11, 1001-2 (1970), EEA74-498.

1049.Franklin, D. L., "Power-transistor design considerations for better forward-second-breakdown ($I_{s/b}$) characteristics", Sixth Annual Meeting of the IEEE Industry and General Applications Group, 1971, p. 851-4, EEA75-10256.

1050.Harwick, N. E. and Walthausen, V., "Combined heat conductor and protective transistor package", Tech. Dig., no. 25, 39-40 (1972).

1051.Joyce, W. B., "Thermal resistance of heat sinks with temperature-dependent conductivity", Solid-State Electron., 18, 321-2 (1975), EEA78-11368.

1052.Lootens, W. F., "Isolated heat-sink semiconductor device", Patent USA 3763403, Publ. October 1973.

1053.Markert, C., "Optimization of heat sinks for semiconductor elements", Feingeraete Tech., 23, 306-8 (1974), German, EEA77-38147.

1054.Marshall, S. B. and Dewey, R. F., "Plastic power ICs need skillful thermal design", Electronics, 46 (23), 102-4 (1973), EEA77-2597.

1055. Nufer, R. W., "High-density interconnectable package", IBM Tech. Disclosure Bull., 16, 2893 (1974), EEA77-23166.
1056. Oktay, S. and Schmeckenbecher, A. F., "Method for forming heat sinks on semiconductor device chips", Patent USA 3706127, Publ. December 1972.
1057. Oktay, S. and Schmeckenbecher, A. F., "Preparation and performance of dendritic heat sinks", J. Electrochem. Soc., 121, 912-18 (1974), EEA77-34097.
1058. Parsapour, H., "Modular heat sink", IBM Tech. Disclosure Bull., 17, 3313 (1975), EEA78-32156.
1059. Rozgonyi, G. A., DiLorenzo, J. V. and Heinlein, E., "Fabrication of GaAs IMPATT devices using electroplated heat sinks. II. Control of water curvature", J. Electrochem. Soc., 121, 426-30 (1974), EEA77-23001.
1060. Russell, E. M. and Thomson, I., "Embedded diamond heat sinks for avalanche diodes", Proc. IEEE, 60, 1014-5 (1972), EEA75-32635.
1061. Seal, M., "Method and means of making diamond heat sinks and heat sinks obtained by this method", Patent USA 3721289, Publ. March 1973.
1062. Streb, A. J., "The heat pipe: overcoming the thermal resistance barrier", Electron. Packag. & Prod. (Int. Ed.), 8 (1), 26-9 (1972).
1063. Tolar, N. J., "Fabrication of semiconductor devices having low thermal impedance bonds to heat sinks", Patent USA 3689993, Publ. September 1972.
1064. Van De Pol, D. and Tierney, J. K., "Free convection heat transfer from vertical fin-arrays", IEEE Trans. Parts, Hybrids & Packag., PHP-10, 267-71 (1974), EEA78-10929.
1065. Van de Water, P. W. M., "Semiconductor device having an improved heat sink arrangement", Patent USA 3629672, Publ. December 1971.
1066. Wheatley, C. F., Jr., "Heat sinking of semiconductor integrated circuit devices", Patent USA 3723833, Publ. March 1973.

3. Thermal Analysis

1067. Anwander, E. and Lawatsch, H., "Thermal measurements and thermal equivalent diagrams of semiconductor elements and cooling fins for the computer-based design and simulation of convertors", Elektrotech. Z. (ETZ) A, 96, 261-5 (1975), German, EEA78-28548.
1068. Baker, E., "Calculation of thermally induced mechanical stresses in encapsulated assemblies", IEEE Trans. Pts. Mater. & Packag., PMP-6, 121-8 (1970), EEA74-9867.
1069. Beck, J., "Temperature stabilization in hybrid circuits", Solid State Technol., 15 (5), 55-6 (1972), EEA75-27795.
1070. Bilotti, A. A., "Static temperature distribution in IC chips with isothermal heat sources", IEEE Trans. Electron Devices, ED-21, 217-26 (1974), EEA77-14591.

1071.Bokrinskaya, A. A., Skrynskii, V. S., Pankratova, N. B. and Pankratov, L. V., "Applicability of the thermal balance equation for the analysis of electrically heated components", Izv. VUZ Radioelektron., 17 (9), 55-9 (1974), Russian, EEA78-7856.
1072.Bolvin, R., "Thermal characteristics of ICs gain in importance", Electronics, 47 (22), 87-90 (1974), EEA78-31.
1073.Boucher, S., "Computer-aided thermal design of LSI packages", IEEE Trans. Parts, Hybrids & Packag., PHP-8, 44-50 (1972), EEA75-27392.
1074.Buchanan, R. C. and Reeber, M. D., "Thermal considerations in the design of hybrid microelectronic packages", Solid State Technol., 16 (2), 39-43 (1973).
1075.Chang, C. S., "Monolithic thermal protection circuit", IBM Tech. Disclosure Bull., 18, 402-3 (1975), EEA78-37636.
1076.Coats, T., "A computer-aided thermal analysis for microcircuits", Electron. Packag. & Prod., 15 (10), 63-7 (1975).
1077.Dean, D. J., "Microcircuit thermal design simplified by the superposition principle", Proceedings of the Conference on Hybrid Microelectronics, 1973, p. 83-99, EEA77-386.
1078.Ellison, G. N., "The effect of some composite structures on the thermal resistance of substrates and integrated circuit chips", IEEE Trans. Electron Devices, ED-20, 233-8 (1973), EEA76-16466.
1079.Gower, R. L., "Thermal characteristics of beam-lead chips", IEEE Trans. Parts, Hybrids & Packag., PHP-8, 39-43 (1972), EEA75-27751.
1080.Greenhouse, H. M. and McGill, R. L., "Design of temperature-controlled substrates for hybrid microcircuits", IEEE Trans. Parts, Hybrids & Packag., PHP-10, 137-45 (1974), EEA77-36336.
1081.Grossman, S. E., "The thermal demands of electronic design", Electronics, 46 (23), 98-101 (1973), EEA77-2621.
1082.Guckel, H., "Relation between internal thermal resistance of transistors and the method of alloying", Siemens Forsch.- & Entwicklungsber., 2, 293-5 (1973), German, EEA77-2553.
1083.Guntlow, V. T., Burks, D. P. and Marin, J. H., "Thermal design and packaging of hybrid integrated circuits", Electronic Compon., 10, 1045-8 (1969), EEA73-9587.
1084.Hambling, P. G. and Dean, D. J., "Microcircuit thermal design tables and computer program for two-dimensional layouts", Conference on Hybrid Microelectronics, 1975, p. 191-204, EEA78-41725.
1085.Hardwick, N. E., "Thermal impedance of ceramic packages using beam-lead IC chips", Proc. IEEE, 57, 1616-20 (1969), EEA73-462.
1086.Housman, R. J., "Thermal design problems with high speed ICs. II", Microelectronics, 1 (1), 45-9, 51-2 (1970), EEA74-349.
1087.Hughes, J. T. and Naylor, R., "Thermal considerations in hybrid forms of constructions", Proceedings of the Conference on Hybrid Microelectronics, 1973, p. 101-14, EEA77-400.
1088.Jensen, F., "Thermal analysis of electronic equipment", Elektronik, 12 (3), 27-30 (1974), Danish, EEA77-35569.

1089.Karaev, A. L., Akimov, V. N. and Uglov, A. A., "Temperature distribution during the welding of semiconductor devices", Fiz. & Khim. Obrab. Mater., no. 4, 22-7 (1974), Russian, EEA78-2666.
1090.Kerjilian, O. K. and Phillips, G. C., Jr., "Thermally enhanced multilayer ceramic substrate structure", IBM Tech. Disclosure Bull., 18, 353 (1975), EEA78-37575.
1091.Keyes, R. W., "Power dissipation in planar digital circuits", IEEE J. Solid-State Circuits, SC-10, 181 (1975), EEA78-19890.
1092.Kokkas, A. G., "Thermal analysis of multiple-layer structures", IEEE Trans. Electron Devices, ED-21, 674-81 (1974), EEA78-680.
1093.Kurochkin, V. V. and Matushkin, G. G., "A method of measuring the characteristics of self-overheating of integrated microcircuits", Avtometriya, no. 4, 101-8 (1972), Russian, EEA75-23362.
1094.Lage, J. A. and Yamez, R., "Approximate method for determining some thermal parameters of devices encapsulated in epoxy resins", Communicaciones, no. 14, 49-56 (1973), Spanish, EEA77-14622.
1095.McKee, H. B., "Heat pipe thermal control set point shift", J. Spacecr. & Rockets, 12, 191-2 (1975), EEA78-36598.
1096.McLaughlin, M. H. and Fitzrov, N. D., "Thermal chip evaluation of IC packaging", IEEE Trans. Parts, Hybrids & Packag., PHP-8, 39-44 (1972), EEA75-39179.
1097.Oktay, S., "Parametric study of temperature distributions in chips joined by controlled chip collapse techniques", Proceedings of the Electronic Components Conference, 1969, p. 429-52, EEA72-23140.
1098.Oktay, S., "Multi-fluid subdued boiling; theoretical analysis of multi-fluid interface bubbles", IBM J. Res. & Dev., 15, 342-54 (1971).
1099.Paivanas, J. A., "Stationary temperature profiles and heat flux distribution in a plastic-encapsulated circuit package", IBM J. Res. & Develop., 16, 292-302 (1972), EEA75-35393.
1100.Pikor, A. J., "Soldering flux for dissipation of thermal shock", Report TN-996, RCA, Princeton, N. J., 1975, 1 pp., EEA78-24543.
1101.Pogson, J. T. and Franklin, J. L., "Analysis and temperature control of hybrid microcircuits", Trans. ASME Ser. B, 95, 1048-52 (1973), EEA77-13822.
1102.Ratcliff, G., "The heat pipe in electronics", Electron, no. 73, 22-4 (1975), EEA78-27776.
1103.Roughan, P. E., "Thermal resistance of IC packages", IEEE Trans. Broadcast & Telev: Receivers, BTR-19, 113-6 (1973), EEA76-26117.
1104.Scapple, R. Y. and Keister, F. Z., "A simplified approach to hybrid thermal design", Solid State Technol., 16 (10), 51-4 (1973), EEA77-10317.

1105.Schlig, E. S., "Electrothermal considerations in display applications of light-emitting diodes", IEEE Trans. Electron Devices, ED-19, 847-51 (1972), EEA75-23892.
1106.Swayne, R. and Rao, U. V., "Thermal behavior of thick-films", Solid State Technol., 16 (6), 45-7 (1973), EEA76-35344.
1107.Sztankovics, L., "Temperature dependence of the electrical parameters of aluminium-oxide ceramics", Hiradastechnika, 26, 206-10 (1975), Hungarian, EEA78-41740.
1108.Tierney, J. K. and Koczkur, E., "Free convection heat transfer from a totally enclosed cabinet containing a simulated electronic equipment", IEEE Trans. Parts, Hybrids & Packag., PHP-7, 115-23 (1971), EEA75-488.
1109.Wilson, E. A., "Thermal analysis of integrated circuits packaging techniques", 23rd Electronic Components Conference, 1973, p. 328-33, EEA76-29297.
1110.Zaks, D. I. and Naumov, N. M., "Temperature field analysis of a semiconductor microcircuit", Izv. VUZ Radioelektron., 18 (1), 60-3 (1975), Russian, EEA78-19857.

XXI. Reliability and Failure Analysis

1. Metallization Failure

1111.Azarewicz, J. L. and Wrobel, T. F., "Radiation-stimulated failure mechanism in a dielectrically isolated integrated circuit", 24th Electronic Components Conference, 1974, p. 48-52, EEA77-34109.
1112.Black, J. R., "Device failure causes by high temperature. III", Electro Technology, 83 (1), 35-8 (1969), EEA72-9218.
1113.Black, J. R., "Electromigration failure modes in aluminum metallization for semiconductor devices", Proc. IEEE, 57, 1587-94 (1969).
1114.Brady, D. P. and Pfahnl, A., "Reliability of conductors and crossovers for film integrated circuits", Proceedings of the 11th Annual Conference on Reliability Physics, 1973, p. 89-90.
1115.Feldstein, N., "Reliability in printed circuitry metalization - a case for improved catalyzing systems", Plating, 60, 611-16 (1973), EEA77-174.
1116.Franz, H. E., "The effect of temperature shocks on the behaviour of microcracks in contact windows in integrated circuits", Presented at Joint Session on Electron Microscopy, Liege, Belgium, September 1973, EEA77-6094.
1117.Fresh, D. L. and Adolphsen, J. W., "SEM evaluation of metallization on semiconductors", Report TR-0074(4921)-2 SAMSO-TR-75-16, Aerospace Corp., El Segundo, Calif., 1974, 36 pp., EEA78-41404.
1118.Goldmann, L. S., "Getter to prevent corrosion", IBM Tech. Disclosure Bull., 17, 2879 (1975), EEA78-19936.

1119.Herr, E. A., "Reliability improvement with new passivation of solid encapsulated transistors", Proceedings of the 10th Annual Conference on Reliability Physics, 1972, p. 69-72, EEA76-4642.

1120.Hornung, A., "Reliability of molybdenum thin films in humid atmospheres", Proceedings of the 10th Annual Conference on Reliability Physics, 1972, p. 149-54, EEA76-4707.

1121.Kalmar, G., "Some characteristic kinds of failure of monolithic integrated circuits with special regard to metallization problems", Hiradastechnika, 24 (11), 345-7 (1973), Hungarian, EEA77-14590.

1122.Kapfer, V. C. and Bart, J. J., "Thin film nickel-chromium resistor failures in integrated circuits", Proceedings of the 10th Annual Conference on Reliability Physics, 1972, p. 175-82, EEA76-4708.

1123.Kirk, R. W. and Gurev, R. S., "A study of Al corrosion in hermetically sealed device package using mass spectrometry", Spring Meeting of the Electrochemical Society (extended abstracts), 1971, p. 157-8, EEA75-432.

1124.Paulson, W. M., "Further studies on the reliability of thin-film nickel-chromium resistors", Proceedings of the 11th Annual Conference on Reliability Physics, 1973, p. 61-8.

1125.Philofsky, E. M. and Ravi, K. V., "On measuring the mechanical properties of aluminum metallization and their relationship to reliability problems", Proceedings of the 11th Annual Reliability Physics Conference, 1973, p. 33-40, EEA77-5782.

1126.Rehrig, D. L., "Effect of deposition method on porosity in gold thin films", Plating, 61, 43-6 (1974), EEA77-18826.

1127.Reich, B., Hakin, E. B. and Malinowski, G., "Plastic encapsulated semiconductors: field performance versus laboratory testing", Proceedings of the 21st Electronic Components Conference, 1971, p. 11-16, EEA74-29902.

1128.Spriggs, R. S. and Cronshagen, A. H., "Metallization failures caused by organic adhesives used in hybrid microelectronic devices", Proceedings of the 10th Annual Conference on Reliability Physics, 1972, p. 201-3, EEA76-4720.

1129.Taketa, Y. and Haradome, M., "Mechanism of aging in Pd-Ag-thick-film resistors", IEEE Trans. Parts, Hybrids & Packag., PHP-9, 115-22 (1973), EEA76-29303.

1130.Tamburrino, A. I. and Kapfer, V. C., "Failure mechanism in plastic encapsulated microcircuits", Report RADC-TR-69-111, Rome Air Dev. Center, Griffiss AFB, N. Y., 1969, 34 pp., EEA73-20187.

1131.Verbokken, A. and Leenders, W., "Titanium-gold: high-reliability transistor metallization", Mullard Tech. Commun., 13 (124), 142-52 (1974), EEA78-2669.

1132.Wilson, R. W., "Factors affecting the reliability of metallization systems for semiconductor devices", J. Vac. Sci. & Technol., 12, 26 (1975), EEA78-24211.

1133.Wright, J. C., "Reliability improvements of plastic semiconductors using gold metallization", Proceedings of the 11th Annual Reliability Physics Conference, 1973, p. 224-9, EEA77-5793.

2. Bond Failure

1134.Adams, C. N., "A bonding-wire failure mode in plastic encapsulated integrated circuits", Proceedings of the 11th Annual Reliability Physics Conference, 1973, p. 41-4, EEA77-6084.
1135.Ailawadhi, M. L. and Kakar, A. K., "Torque and thermal cycling as methods of testing reliability of reflow-soldered chip to substrate joints", Proceedings of the 11th Annual Conference on Reliability Physics, 1973, p. 77-82.
1136.Andre, G. and Regnault, J., "Reliability problems related to plastic encapsulation of integrated circuits", Onde Elec., 52, 121-5 (1972), French, EEA75-19257.
1137.Anon., "Defects in wire bonds", Nat. Bur. Stand. Tech. News Bull., 55, 248-50 (1971), EEA75-10186.
1138.Anon., "Pursuing the reliable solder joint", Circuit Manufac., 15 (11), 50-4 (1975).
1139.Anstead, R. J. and Floyd, S. R., "Thermal effects on the integrity of aluminum to silicon contacts in silicon integrated circuits", IEEE Trans. on Electron Devices, ED-16, 381-6 (1968).
1140.Bascom, W. D. and Bitner, J. L., "Void reduction in large-area bonding of IC components", Solid State Technol., 18 (9), 37-9 (1975).
1141.Berg, H. and Hall, E. L., "Dissolution rates and reliability effects of Au, Ag, Ni and Cu in lead base solders (power transistor die bonding)", Proceedings of the 11th Annual Reliability Physics Conference, 1973, p. 10-20, EEA77-6004.
1142.Burks, D. P., "Reliability of solder-connected, thick-film hybrid integrated circuits", Proceedings of the 11th Annual Reliability Physics Conference, 1973, p. 93, EEA77-6102.
1143.Cohn, E., "Effects of wire quality and capillary maintenance on bonding reliability", Solid State Technol., 18 (9), 31-6 (1975).
1144.Condra, L. W., Svitak, J. J. and Pense, A. W., "The high temperature deformation properties of gold and thermocompression bonding", IEEE Trans. Parts, Hybrids & Packag., PHP-11, 290-5 (1975).
1145.Cox, R. H., Crosthwalt, D. L., Jr. and Dobrott, R. D., "The application of the scanning electron microscope to the development of high reliability semiconductor products", IEEE Trans. Electron Devices, ED-16, 376-80 (1969), EEA72-17152.
1146.Gregor, L. V., "Semiconductor chip left to improve interconnection reliability", IBM Tech. Disclosure Bull., 17, 2881-2 (1975), EEA78-19946.

1147.Goldman, L. S., "Optimizing cyclic fatigue life of controlled collapse chip joints", Proceedings of the Electronic Components Conference, 1969, p. 404-23, EEA69-23138.
1148.Guidici, D. C., "Ribbon wire versus round wire reliability for hybrid microcircuits (bonding)", IEEE Trans. Parts, Hybrids & Packag., PHP-11, 159-63 (1975), EEA78-33246.
1149.Harman, G. G., "Metallurgical failure modes of wire bonds", Proceedings of the 12th Annual Reliability Physics Conference, 1974, p. 131-41.
1150.Harman, G. G., "Metallurgical basis for the nondestructive wire-bond pull-test", Proceedings of the 12th Annual Reliability Physics Conference, 1974, p. 205-10.
1151.Horsting, C. W., "Purple plague and gold purity", Proceedings of the 10th Annual Conference on Reliability Physics, 1972, p. 155-8, EEA76-4719.
1152.Johnson, J. E., "Die bond failure modes", Proceedings of the 12th Annual Reliability Physics Conference, 1974, p. 150-4.
1153.Kahan, G. J., "Silver migration in glass dams between silver-palladium interconnections", IEEE Trans. Electr. Insul., EI-10, 86-94 (1975), EEA78-41643.
1154.Kanamori, S. and Takahashi, M., "Silicon die bond failure induced by nickel-silicon intermediate formation", Rev. Electr. Commun. Lab., 23, 552-8 (1975), EEA78-41587.
1155.Kinser, D. L., Graff, S. M. and Caruso, S. V., "The relationship between reliability and bonding techniques in hybrid systems", IEEE Trans. Parts, Hybrids & Packag., PHP-11, 195-201 (1975), EEA78-41712.
1156.Kirk, G. C., "Plastic package seal testing for semiconductors", Electron. Components, 16 (1), 19-20 (1974), EEA77-9967.
1157.Kulesza, F. W., "Electrically-conductive epoxies for die attach: a reliability test report", Microelectronics, 5 (1), 27-32 (1973), EEA77-18870.
1158.Linford, P. F. T. and Dobson, M., "Degradation under thermal soak of the strength of aluminium wire ultrasonically bonded to thick film gold (hybrid IC's)", Conference on Hybrid Microelectronics, 1975, p. 247-50, EEA78-41728.
1159.Munford, J. W., "The influence of several design and material variables on the propensity for solder joint cracking", IEEE Trans. Parts, Hybrids & Packag., PHP-11, 296-304 (1975).
1160.Nowakowski, M. F. and Villella, F., "Thermal excursion can cause bond problems", 9th Annual Proceedings Reliability Physics, 1971, p. 172-7, EEA75-3971.
1161.Oettinger, F. F. and Gladhill, R. L., "Thermal response measurements for semiconductor device die attachment evaluation", 1973 International Electron Devices Meeting Technical Digest, p. 47-50, EEA77-14244.
1162.Olson, H. C. and Knauss, G. L., "Environmental testing of soldered RF hybrid microcircuits", Report SLA-73-10849, Sandia Labs., Albuquerque, N. Mex., 1974, 35 pp., EEA78-41718.

1163.Panousis, N. T. and Bonham, H. B., "Bonding degradation in the tantalum nitride-chromium-gold metallization system", Proceedings of the 11th Annual Reliability Physics Conference, 1973, p. 21-5, EEA77-6063.
1164.Peck, D. S. and Zierdt, C. H., Jr., "Temperature-humidity acceleration of metal-electrolysis failure in semiconductor devices", Proceedings of the 11th Annual Reliability Physics Conference, 1973, p. 146-52, EEA77-5787.
1165.Pietrucha, B. M. and Reiss, E. M., "The reliability of epoxy as a die attach in digital and linear integrated circuits", Proceedings of the 12th Annual Reliability Physics Conference, 1974, p. 234-8.
1166.Ravi, K. V. and Philofsky, E. M., "Reliability improvement of wire bonds subjected to fatigue stresses", Proceedings of the 10th Annual Conference on Reliability Physics, 1972, p. 143-8, EEA76-4718.
1167.Reich, B. and Hakim, E. B., "Field reliability of plastic encapsulated transistors and integrated circuits", Microelectron. & Rel., 10, 461-5 (1971), EEA75-6964.
1168.Reich, B. and Hakim, E. B., "The effect of plastic packaging on interconnection reliability for military applications", 23rd Electronic Components Conference, 1973, p. 33-7, EEA76-26112.
1169.Riga, G., "TO-5 discoloration and bondability", Proceedings of the 11th Annual Conference of Reliability Physics, 1973, p. 26-32.
1170.Saito, M. and Hirota, S., "Effect of grain size on the life time of aluminum interconnections", Rev. Electr. Commun. Lab., 22, 678-94 (1974), EEA78-8603.
1171.Schafft, H. A., "Failure analysis of wire bonds", Proceedings of the 11th Annual Reliability Physics Conference, 1973, p. 98-104, EEA77-5784.
1172.Spriggs, R. S. and Cornshagen, A. H., "Wire-bond reliability in hybrid microcircuits", Proceedings of the 11th Annual Reliability Physics Conference, 1973, p. 83-8, EEA77-6099.
1173.Stanley, J. M., "Aging of coldweld-sealed quartz crystal units", Report ECOM-2946, Army Electronics Lab., Fort Monmouth, N. J., 1968, 34 pp., EEA72-10682.
1174.Wagner, R. G., "Chip components for hybrid microelectronics: a reliability status report", 3rd Symposium on the Reliability of Electronics, 1973, p. 417-23, EEA77-6110.
1175.Weirick, L. J., "A metallurgical analysis of stress-corrosion cracking of Kovar package lead", Solid State Technol., 18 (3), 25-30 (1975), EEA78-19866.
1176.Weiss, B. L., "A method of studying thermocompression bonding damage in GaAs using SEM", Microelectronics, 7 (1), 35-7 (1975).

3. Circuitry Failure

1177.Adams, C., "A bonding-wire failure mode in plastic-encapsulated integrated circuits", Proceedings of the 11th Annual Conf. on Reliability Physics, 1973, p. 41-4.

1178.Adams, J. D. and Ballew, D. D., "Integrated circuit quality and reliability considerations", 1973 WESCON Improving Producibility Workshop, 1/5 7 pp., EEA77-10309.

1179.Brauer, J. B., Kapfer, V. C. and Tamburrino, A. L., "Can plastic encapsulated microcircuits provide reliability with economy?", Microelectronics, 1 (1), 5-24 (1970), EEA74-348.

1180.Braun, L., "Electromigration testing - a current problem", Microelectron. & Reliab., 13, 215-28 (1974), EEA77-30396.

1181.Devaney, J. R., "Application of SEM to integrated circuit failure", Solid State Technol., 13 (3), 73-7 (1970).

1182.DiGiacomo, G., "Extraction of chloride in epoxy circuit package", IBM Tech. Disclosure Bull., 18, 179 (1975), EEA78-36596.

1183.English, A. T., Tai, K. L. and Turner, P. A., "Electromigration of Ti-Au thin-film conductors at 180°C", J. Appl. Phys., 45, 3757-67 (1974), EEA78-2890.

1184.Feldt, J. E. and Hnatek, E. R., "High reliability consumer IC's", Proceedings of the 10th Annual Conference on Reliability Physics, 1972, p. 78-81, EEA76-4695.

1185.Fetcher, B. C., "An eclectic integrated circuit reliability model", Report NASA-CR-125062, California Inst. Technol., Pasadena, 1972, 84 pp., EEA75-841.

1186.Flood, J. L., "Reliability aspects of plastic encapsulated integrated circuits", Proceedings of the 10th Annual Conference on Reliability Physics, 1972, p. 95-9, EEA76-4697.

1187.Fresh, D. L., "Reliability of interconnections on microcircuits", Report TR-0075(5901-02)-4 SAMSO-TR-75-92, Aerospace Corp., El Segundo, Calif., 1974, 9 pp., EEA78-41648.

1188.Gangulee, A., Tuxford, A. M., Romankiw, L. T. and Mayadas, A. F., "Structure and electrical properties of electrodeposited narrow gold lines", Electrochemical Society Fall Meeting (Extended abstracts), 1974, p. 282-3, EEA78-33181.

1189.Graczewski, M., "Methods of checking the reliability of a model series specialized circuits", Pr. Inst. Masz. Mat., 15 (1), 33-41 (1973), Polish, EEA77-10316.

1190.Keuren, E. V., Hendrickson, R. and Magyarics, R., "Circuit failure thresholds due to transient induced stresses", Conference on Electromagnetic Compatibility, 1975, p. 500-6, EEA78-28836.

1191.McQuarrie, P., "Corrosion in connection leads of semiconductor devices", SERT J., 8 (2), 43-4 (1974), EEA77-18572.

1192.McQuitty, D. W. and Lively, C. R., "IC failure and the packaging process", Electron. Packag. & Prod. (Int. Ed), 8 (1), 14-17 (1972), EEA75-15973.

1193. Mattana, G. U. and Piacentini, G. F., "Reliability demonstration of thin-film hybrid integrated circuits", Telettra, no. 2, 29-40 (1973), EEA76-35368.
1194. Petrikovits, I., "Reliability tests on thin-film hybrid microcircuits", Hitadstech. Ipari Kut. Intez. Kozl., 12 (1), 24-40 (1972), Hungarian, EEA75-39207.
1195. Saito, M., Anayama, H. and Shikama, S., "Reliability and failure analysis of semiconductor integrated circuits", Elec. Commun. Lab. Tech. J., 21, 1047-63 (1972), Japanese, EEA75-827.
1196. Sigsbee, R. A., "Failure model for electromigration", Proceedings of the 11th Annual Reliability Physics Conference, 1973, p. 301-5, EEA77-6056.
1197. Smith, J. S., "Analysis of electrical overstress failures", Proceedings of the 11th Annual Reliability Physics Conference, 1973, p. 105-7, EEA77-6048.
1198. Thornton, P. R., Hughes, K. A., Kyaw, H., Millward, C. and Sulway, D. V., "Failure analysis of microcircuitry by SEM", Microelectron. & Reliab., 6, 9-16 (1967).
1199. Vetter, R. A., "Analysis of integrated circuit failure modes and failure mechanisms derived from high temperature operating life tests", Proceedings of the 11th Annual Reliability Physics Conference, 1973, p. 133-7, EEA77-6104.
1200. Walczak, M. W., "IC failure rates and systems reliability prediction", Eval. Eng., 12 (1), 10-16 (1973), EEA76-19944.
1201. Yoshida, J., Honda, S. and Kojima, H., "Reliability (IC)", J. Inst. Electron. & Commun. Eng. Jap., 55, 597-601 (1972), Japanese, EEA75-39185.

4. Encapsulation Failure

1202. Beard, R., "Evolution of a high-reliability moulded dip", Proceedings of the 10th Annual Conference on Reliability Physics, 1972, p. 72-7, EEA76-4723.
1203. Doetsch, V., "Optimizing the reliability of semiconductor devices in plastic cases", Int. Elektron. Rundsch., 24 (7), 183-4 (1971), German, EEA74-29900.
1204. Edfors, H. C., "Failure mode control in plastic IC's", Proceedings of the Annual Reliability and Maintainability Symposium, 1972, p. 479-91, EEA75-15994.
1205. Fischer, F., "Moisture resistance of plastic packages for semiconductor devices", Presented at the Annual Reliability Physics Symposium, Las Vegas, Nev., April 1970, EEA70-23950.
1206. Gray, C. R., "Integrated circuits reliability: plastic versus ceramic dips", Proceedings of the 10th Annual Conference on Reliability Physics, 1972, p. 100-1, EEA76-4698.
1207. Grycan, J., "The reliability problems in semiconductor integrated circuits in plastic encapsulation", Elektronika, 14, 390-6 (1973), Polish, EEA77-2599.

1208.Hakim, E. B., "Plastic encapsulated semiconductor reliability-today", International Electronic Circuit Packaging Symposium, 1969, 7 pp., EEA73-28107.
1209.Hakim, E. B., Malinowski, G. and Holevinski, R., "Who wants reliable plastic semiconductors?", Proceedings of the 11th Annual Reliability Physics Conference 1973, p. 230-5, EEA77-5794.
1210.Halleck, M. C., "The I.C. plastic package: a simple method for predicting package performance", Proceedings of the 10th Annual Conference on Reliability Physics, 1972, p. 88-94, EEA76-4696.
1211.Hnatek, E. R., "New epoxy package increases IC reliability", Solid State Technol., 15 (11), 44-8 (1972), EEA76-8764.
1212.Kanou, K. and Shiomi, H., "Reliability of epoxy resin moisture-proofing of moulded silicon diode", Bull. Electrotech. Lab., 39, 10-13 (1975), Japanese, EEA78-19762.
1213.Knivila, A. R., "Effects of humidity and heat ageing on a new one-component silicone resin encapsulant", Proceedings of the 9th Electrical Insulation Conference, 1969, p. 111-4, EEA73-13082.
1214.Kunihiro, Y., "Plastic encapsulated integrated circuits - reliability", Electr. Commun. Lab. Tech. J., 23, 1225-42 (1974), EEA78-7855.
1215.Lampi, R. A., Roberts, N. D. and Costanza, F. A., "Automatic infrared radiometric scanning to detect flexible package seal defects", IEEE Trans. Ind. Electron. & Control Instrum., IECI-20, 205-12 (1973), EEA77-30.
1216.Lawson, R. W., "Accelerated testing of plastic encapsulated semiconductor components", Proceedings of the 12th Annual Conference on Reliability Physics, 1974, p. 243-9, CA82-67176.
1217.Martino, R. R., "Metallized hermetic plastic packages", Electron. Components, 14, 516 (1973), EEA76-25459.
1218.Reich, B. and Hakim, E. B., "Environmental factors governing field reliability of plastic transistors and integrated circuits", Proceedings of the 10th Annual Conference on Reliability Physics, 1972, p. 82-7, EEA76-4643.
1219.Reich, B., "Acceleration factors for plastic encapsulated semiconductor devices and their relationship to field performance", Microelectron. & Reliab., 14, 63-6 (1975), EEA78-15141.
1220.Stanley, A. G. and Rader, C. M., "Leak detection of integrated circuits and other semiconductor devices on multilayer circuit boards", Proceedings of the 12th Annual Reliability Physics Conference, 1974, p. 239-42.
1221.Stommel, A. M., "Evaluation of plastic-encapsulated semiconductors", Signal, 27 (6), 23-5 (1973), EEA78-19687.
1222.Watanabe, E. and Kanou, K., "Reliability to moisture-proof characteristics of epoxy resin coating films for electronic parts", Bull. Electrotech. Lab., 38, 541-50 (1974), EEA78-10928.

1223.Wittenberg, A. M. and Malee, I. F., "Mechanical stresses induced on encapsulated components", Proceedings of the 9th Electrical Insulation Conference, 1969, p. 101-4, EEA73-13080.

5. Other Failures

1224.Anayama, H., Terasaki, Y., Shikama, S. and Nakamura, M., "Reliability of semiconductor devices for DIPS-1", Rev. Electr. Commun. Lab., 21, 211-3 (1973), EEA76-37724.
1225.Anon., "Microcircuit device environmental data", Report RAC-MDED-0474, IIT Res. Inst., Chicago, Ill., 1974, 369 pp., EEA78-28835.
1226.Azarewicz, J. L. and Wrobel, T. F., "Radiation-stimulated failure mechanism in a dielectrically isolated integrated circuit", IEEE Trans. Parts, Hybrids & Packag., PHP-10, 159-64 (1974), EEA78-2907.
1227.Becker, P., "Failure analysis of linear ICs", Radio Mentor Electron., 40 (1), 220-3 (1974), German, EEA77-9243.
1228.Blanquart, P., "Quality and reliability control for LSI circuits", International Conference on Large Scale Integrated Circuits, 1974, p. 177-87, EEA78-11753.
1229.Blasberg, H. J., "The economics and reliability of modules in consumer electronics", Funkschau, 47 (6), 50-2 (1975), German, EEA78-27777.
1230.Chaudhari, P., Rosenberg, R. and Tu, K. N., "Preventing corrosion of contact materials in the presence of lead-tin solder", IBM Tech. Disclosure Bull., 17, 3753 (1975), EEA78-32164.
1231.DiCerto, J., "Poor packaging produces problems", Electron. Eng., 29 (9), 91-3 (1970), EEA74-341.
1232.Ebel, G. H. and Engelke, H. A., "Failure analysis of oxide defects", Proceedings of the 11th Annual Reliability Physics Conference 1973, p. 108-16, EEA77-5785.
1233.Eberhardt, W., Kalus, C. and Schleicher, L., "Investigation of semiconductor devices with the scanning electron microscope", Siemens Forsch. - & Entwicklungsber., 1, 315-21 (1972), German, EEA75-39194.
1234.Farrow, R. H. and Parker, G. W., "The failure physics approach to IC reliability", Microelectron. & Rel., 11, 151-7 (1972), EEA75-19254.
1235.Flood, J. L., "Quality assurance aspects of custom LSI", AGARD Lecture Series No. 75 Custom Design for Large Scale Integration (LSI), 1975, p. 6/1-26, EEA78-19931.
1236.Gabor, K., "Introduction of failure analysis into the development process of reliable integrated circuits", Hiradastechnika, 25, 241-8 (1974), EEA78-2904.
1237.Gerling, W., "Estimating failure-rates of semiconductor components", Qual. & Zuverlassigkeit, 19 (7), 152-7 (1974), German, EEA78-2671.

1238.Greenman, L. R., "Packaging requirements for optimized malfunction isolation by systematic substitution", IEEE Trans. Aerospace Electronic Syst., AES-5, 499-514 (1969), EEA72-14954.
1239.Hajdu, L. and Ligeti, S. R., "On the reliability problems and test of hybrid circuits", 3rd Symposium on the Reliability of Electronics, 1973, p. 93, EEA77-6106.
1240.Hakim, E. B. and Shappirio, J. R., "Failure mechanisms in gold metalized sealed junction devices", Solid State Technol., 18 (4), 66-8 (1975), EEA78-24499.
1241.Hall, T. C. and Webster, S. L., "Reliability aspects of module defoaming operation", 24th Electronic Components Conference, 1974, p. 25-8, EEA77-33499.
1242.Hart, A. R., McQuitty, D. W. and Wagner, N. K., "RF suspectibility of microelectronic components", Nav. Res. Rev., 27 (5-6), 12-16 (1974), EEA78-5345.
1243.Hurley, H. C., "Reliability evaluation of multilayer board, high-density packaging designs", Presented at Semiconductor Integrated Circuit Processing and Production Conference, Anaheim, Calif., February 1971, EEA74-21787.
1244.Kalinichenko, N. F., "Accelerated trials of integrated circuits", Dielektr. & Poluprovod, no. 6, 65-7 (1974), Russian, EEA78-24431.
1245.Katona, J., "Reliability investigations of integrated circuits", 3rd Symposium on the Reliability of Electronics, 1973, p. 181-91, EEA77-6049.
1246.Kemeny, A. P., "Life tests of SSI integrated circuits", Microelectron. & Reliab., 13, 119-42 (1974), EEA77-30395.
1247.Kosla, C. M. and Coenen, M. V., "Packaging industrial control electronics", Control Eng., 20 (11), 64-7 (1973), EEA77-13627.
1248.Lindwedel, J. H., "LSI reliability screen developments", 1972 IEEE Region Six (US Western region) Conference Record, p. 129-31, EEA76-4691.
1249.McMahon, R. E., "Some reliability considerations of large hybrids", Proceedings of the 11th Annual Reliability Physics Conference, 1973, p. 91, EEA77-6100.
1250.McQuitty, D. W., "Microelectronics reliability as related to materials and processing", 1972 IEEE Region Six (US Western region) Conference Record, p. 134-7, EEA76-4681.
1251.Miller, W. C. and Crain, D. W., "Reliability considerations in hybrid microcircuits", Microelectronics, 5 (1), 49-56 (1973), EEA77-18867.
1252.Mizusawa, T. and Satoh, H., "Critical operation margin test for LSI reliability assessment", Rev. Electr. Commun. Lab., 23, 546-51 (1975), EEA78-41646.
1253.Orlyuk, M. A. and Savchenko, Y. G., "Fault diagnosis in digital devices based on integrated circuits", Autom. & Remote Control, 34, 1179-82 (1973), EEA77-13973.

1254.Paddock, A. D. and Harper, J. G., "Tolerance of packaged MOS devices to sodium ion contamination", Spring Meeting of the Electrochemical Society (extended abstracts), 1971, p. 93-4, EEA74-37314.
1255.Pay, C., "Reliability in thick film and hybrid microcircuits", 3rd Symposium on the Reliability of Electronics, 1973, p. 321-32, EEA77-6107.
1256.Petrikovits, L., "Life tests of hybrid integrated circuits", 3rd Symposium on the Reliability of Electronics, 1973, p. 333-41, EEA77-6108.
1257.Platonov, B. D., Maiorov, S. A. and Bekker, Y. M., "Influence of several factors on the failure of semiconductor integrated circuit crystals", Izv. VUZ Priborostr., 18, 94-7 (1975), Russian, EEA78-24476.
1258.Roberts, B. C., "Influence of storage conditions on the behaviour of electronic components", Electron. Compon., 13, 539-48 (1972), EEA75-23408.
1259.Saeks, R., Singh, S. R. and Lin, R. W., "Fault isolation via components simulation", IEEE Trans. Circuit Theory, CT-19, 634-40 (1972), EEA72-35232.
1260.Sathan, H. P., "Thermal detection of hybrid integrated circuit faults", Tech. Dig., no. 34, 29-30 (1974), EEA77-38902.
1261.Savage, J., "Silver transfer through thick film insulants", Proceedings of the Conference on Hybrid Microelectronics, 1973, p. 335-40, EEA77-390.
1262.Shove, P. L., "The designer's need for information on the reliability of hybrids", Proceedings of the Conference on Hybrid Microelectronics, 1973, p. 247-9, EEA77-403.
1263.Smith, J. S., "Analysis of electrical-overstress failures", Proceedings of the 11th Annual Conference on Reliability Physics, 1973, p. 105-7.
1264.Stoll, H., "Reliability of electroceramics", Electro-Anzeiger, 25 (9), 167-8 (1972), German, EEA75-28108.
1265.Thomas, R. W. and Meyer, D. E., "Moisture in SC packages", Solid State Technol., 17 (9), 56-9 (1974), EEA78-2673.
1266.Totta, P. A., "IBM reliability experience with hybrid microcircuits", Proceedings of the 11th Annual Reliability Physics Conference, 1973, p. 92, EEA77-6101.
1267.Valko, I. P. and Ribenyi, A., "Factors influencing the reliability of integrated circuits", 3rd Symposium on the Reliability of Electronics, 1973, p. 403-14, EEA77-5600.
1268.Waite, G., "Engineering reliability", Proceedings of the 11th Annual Reliability Physics Conference, 1973, p. 94, EEA77-6103.
1269.Worotinski, W. and Konjajaw, W., "Protecting semiconductor components against electrostatic charges", Radio Fernsehen Elektron., 24, 440-1 (1975), German, EEA78-41399.
1270.Yanagisawa, T., Takahisa, K. and Shiomi, H., "Reliability of IC in thermal vacuum environment", Bull. Electrotech. Lab., 39, 80-6 (1975), Japanese, EEA78-19878.

1271.Zatz, S., "A new simplified method to measure moisture in micro enclosures", 24th Electronic Components Conference, 1974, p. 29-33, EEA77-33500.

XXII. Characterization

1. Substrates

1272.Aitken, J. E., Ladbrooke, P. H. and Potok, M. H. N., "Microwave measurement of the temperature coefficient of permittivity for sapphire and alumina", IEEE Trans. Microwave Theory & Tech., MTT-23, 526-9 (1975), EEA78-24535.

1273.Anjard, R. P., "Ceramic substrate strength testing", Solid State Technol., 17 (4), 96-8 (1974), EEA77-30356.

1274.Ball, J. H. and Olyphant, M., "Packaging and environmental studies of copper clad tfe-glass stripline", Presented at Proceedings of the 9th Electrical Insulation Conference, Boston, Mass., September 1969, EEA73-13083.

1275.Ball, J. H., Olyphant, M. and Van Dover, I. K., "Dependence of stripline performance on dielectric properties and packaging techniques", G-MTT International Microwave Symposium, 1969, p. 45-52, EEA72-23949.

1276.Belding, J. H. and Petacost, J. L., "Characterization of substrate surfaces from reactive alumina", Am. Ceram. Soc. Bull., 51, 361 (1972), EEA75-27731.

1277.Blodgett, W. E., "Ultrafine grain alumina substrates", Am. Ceram. Soc. Bull., 51, 361 (1972), EEA75-27730.

1278.Coleman, M. V. and Gurnett, G. E., "Surface area, structure and composition of debased alumina substrates", Conference on Hybrid Microelectronics, 1975, p. 1-16, EEA78-41650.

1279.Dentai, A. G., "Fluorescent high alumina ceramic substrates", Am. Ceram. Soc. Bull., 51, 681-2 (1972), EEA76-4704.

1280.Ehman, M. F., "Surface preparation of ceramic oxide crystals: Work damage and microhardness", J. Electrochem. Soc., 121, 1240-3 (1974), EEA78-665.

1281.Emery, J. K., "Surface characteristics of ceramic substrates for hybrid and microwave electronic circuits", Proceedings of a Symposium on the Science of Ceramic Machinery and Surface Finishing, 1970, p. 293-300, Publ. 1972, EEA75-39197.

1282.Ermert, H., "Accurate measurement of complex permittivity of low-loss MIC substrate slabs", Nachrichtentech. Z. (NTZ), 27, 43-6 (1974), EEA77-10315.

1283.Feltham, S. J. and Frankson, R. W., "Scanning electron microscope analysis of surface defects on alumina ceramic substrates", Solid State Technol., 16 (9), 59-61 (1973).

1284.Garault, Y., Guillon, F. and Jallageas, H., "An accurate method to measure the dielectric constant, and the loss tangent of m.i.c. substrates", International Conference on Electronics and Measurements, 1975, p. 172-82, EEA78-37566.

1285.Hellicar, N. J., "Characteristics of as-fired alumina substrates for thin film integrated circuitry", Am. Ceram. Soc. Bull., 52, 375 (1973), EEA77-2589.

1286.Hill, G. J., "The precise determination of the dielectric properties of alumina", IEEE Trans. Instrum. & Meas., IM-23, 443-6 (1974), EEA78-5343.

1287.Kamehara, N., Yamada, S. and Murakawa, K., "Surface roughness of alumina substrates", Jap. J. Appl. Phys., 14, 145-6 (1975), EEA78-11705.

1288.Lenzing, H. F., "Measurement of dielectric constant of ceramic substrates at microwave frequencies", Am. Ceram. Soc. Bull., 51, 361 (1972), EEA75-28103.

1289.McMahon, M. T., Jr., "Testing of metalization networks on insulative substrates supporting semiconductor chips", Patent USA 3746973, Publ. July 1973.

1290.Moran, P. L., "A technique for the testing of unassembled multilayer hybrid substrates", Conference on Hybrid Microelectronics, 1975, p. 229-40, EEA78-41726.

1291.Morzenti, P. T. and Shanefield, D. J., "Spinel segregation in high alumina thin film substrates", Am. Ceram. Soc. Bull., 52, 375 (1973), EEA77-2590.

1292.Niwa, K., Nakamura, J. and Murakawa, K., "Surface properties of new alumina substrates", Fujitsu Sci. & Tech. J., 10, 177-88 (1974), EEA77-30349.

1293.Paradis, E. L. and Shuskus, A. J., "Temperature determination of thin sapphire substrates", J. Vac. Sci. & Technol., 11, 824-5 (1974), PA78-13307.

1294.Rempes, P. E., Jr., Pratten, D. H. and Insley, R. H., "Process control dielectric strength testing of high alumina ceramics", Am. Ceram. Soc. Bull., 51, 414 (1972), EEA75-28105.

1295.Savage, J. and Davey, N., "Substrate bowing in thick film multilayered circuits", Conference on Hybrid Microelectronics, 1975, p. 17-22, EEA78-41651.

1296.Shanefield, D. J. and Mistler, R. E., "Fine grained alumina substrates: I, Manufacturing process", Amer. Ceram. Soc. Bull., 53, 416-20 (1974).

1297.Spathling, W., "Optical tests of the surface of alumina substrates", Radio Mentor Electron., 41 (2), 66-8 (1975), German, EEA78-19865.

1298.Sundahl, R. C. and Berrin, L., "Analysis and characterization of ceramic surfaces for electronic applications", Proceedings of a Symposium on the Science of Ceramic Machinery and Surface Finishing, 1970, p. 271-91, Publ. 1972, EEA75-39196.

1299.Swiss, W. R. and Gardner, R. A., "Controlling the sintering rate and shrinkage of ceramic materials", IBM Tech. Disclosure Bull., 16, 2889 (1974), EEA77-22231.

1300.Szedon, J. R. and Stelmak, J. P., "Effects of organic and inorganic dielectric films on semiconductor devices", IEEE Trans. Elect. Insulation, EI-5, 3-10 (1970), EEA73-17557.

1301. Sztankovics, L., "Purification of alumina to produce alumina-ceramics for the electronic industry", Hiradstech. Ipari Kut. Intez. Kozl., 13, 54-64 (1973), Hungarian, EEA77-6098.
1302. Sztankovics, L., "Flat-surface aluminium-oxide ceramics", Finommech. & Mikrotech., 13, 262-6 (1974), Hungarian, EEA78-8602.
1303. Tradeau, M. J., "Glass ceramic dielectrics", Onde Electr., 54, 393-8 (1974), French, EEA78-5389.
1304. Vahaviolos, S. J., "Real time detection of microcracks in brittle materials using stress wave emission (SWE) (ceramic substrates)", 24th Electronic Components Conference, 1974, p. 34-42, EEA77-34086.
1305. Watson, D. R., Davis, J. E., Briggs, T. H. and Ports, D. W., "High alumina substrates: two methods of determining compositional depth profiles", Am. Ceram. Soc. Bull., 52, 632, 635 (1973), EEA77-18830.
1306. Wilson, W. E. and Chaffin, R. J., "Radiation damage effects in microwave dielectric substrate materials", IEEE Trans. Nucl. Sci., NS-19, 82-5 (1972), EEA76-13582.

2. Bonds

1307. Bantz, W., "Ultrasonic wave inspection of bonded materials", Adhes. Age, 17 (2), 29 (1974), EEA77-29352.
1308. Bittner, T. L., "Testing multichip substrates", IBM Tech. Disclosure Bull., 17, 2584-5 (1975), EEA78-19934.
1309. Bohlen, H., "Contactless short/open testing of conductors using two electron beams", 6th International Congress on Microelectronics, 1974, 5 pp., EEA78-24433.
1310. Boore, E. J. and Sutter, D. M., "Testing bond strength of semiconductor device assemblies", Patent USA 3636755, Publ. January 1972.
1311. Clifford, J. R., Sega, R. M., Foos, G. D. and Throckmorton, A. A., "SEM examination of the Au-Al intermetallic on IC lead bonds", Scanning Electron Microscopy 1974, p. 979-86, EEA78-667.
1312. Etzel, J. G. and Munford, J. A., "Development of a weld quality monitor", Report NASA-TN-D-5304, NASA, Goddard Space Flight Center, Greenbelt, Md., 1969, 11 pp., EEA73-30424.
1313. Feinauer, D. and Heinzel, H., "The strength of soldered connections when soldering with tin-lead pastes suitable for screen printing processes", Funk-Tech., no. 5, 157-8 (1973), German, EEA76-19965.
1314. Gurev, H. and Khera, R. P., "Direct pull test for adherence of electroless nickel plated on silicon", Electrochemical Society Fall Meeting (Extended abstracts), 1974, p. 307, EEA78-32925.
1315. Hall, F. T. and Crecraft, D. I., "Bonded joints and non-destructive testing. NDT of resistance spots, roll-spot, stitch and seam welds", Non-Destruct. Test., 4, 181-91 (1971), EEA74-21685.
1316. Holyfield, S., "Interconnection network testing", Electron. Packag. & Prod., 15 (10), 36-44 (1975).

1317.Joshi, K. C., "Nondestructive bond quality monitor for resistance bonds", IBM Tech. Disclosure Bull., 14, 481 (1971), EEA74-38226.
1318.Kuhns, L. C., "Nondestructive testing of lead bonds", Tech. Dig., no. 33, 35 (1974), EEA77-23171.
1319.Leedy, K. O., "Scanning electron microscope examination of wire bands from high-reliability devices", Report NBS-TN-785, Nat. Bur. Stand., Washington, D. C., 1973, 28 pp., EEA77-18873.
1320.Pantanelli, G. P., "Measurement of the solderability of thick-film circuits; relationship between solder strength and a solderability test", Solid State Technol., 18 (10), 39-41 (1975).
1321.Paulk, J. I., Guyten, R. D., Simpson, D. I., Triplett, M. E., and Pace, J. V., III, "Hermetic seal evaluation for electronic components", Report NASA-CR-61225, Mississippi State Univ., State College, Ms., 1968, 109 pp., EEA72-7455.
1322.Ramsey, T. H., "Use of differential thermal analysis in controlling behavior of solder-glass seals in ceramic packages", Solid State Technol., 15 (1), 29-33 (1972), EEA75-28111.
1323.Rossoshinskii, A. A., Kislitsyn, V. M. and Alperovich, E. A., "The braze welding of semiconductor silicon devices", Autom. Weld., 24 (12), 33-6 (1971), EEA75-34951.
1324.Ruehli, A. E., "Electrical characterisation of interconnections for integrated circuits and packages", 8th Asilomar Conference on Circuits, Systems and Computers, 1974, p. 19-22, Publ. 1975, EEA78-24438.
1325.Schafft, H. A., "Methods for testing wire-bond electrical connections", Report NBS-TN-786, Nat. Bur. Stand., 1973, 17 pp., EEA77-18874.
1326.Sebeson, J. M., "Electron beam testing of circuit interconnections using collector stabilized conductor charging", J. Vac. Sci. & Technol., 10, 1060-3 (1973), EEA77-23101.
1327.Zakraysek, L., "Intermetallic characterization", Proceedings of the 11th Annual Reliability Physics Conference, 1973, p. 6-9, EEA77-5780.

3. Integrated Circuits

1328.Allen, D. P., "Logic tester uses single trial procedure to troubleshoot digital IC boards", Electronics, 46 (23), 89-93 (1973), EEA77-3750.
1329.Allen, D. P. and Lyons, N. P., "Selecting inputs to test digital circuits on complex IC boards", Electronics, 45 (15), 88-92 (1972), EEA75-27605.
1330.Anon., "The automatic circuit-board tester: a production tool like any other", Inter Electron., 27 (69), 36-7 (1972), French, EEA76-91.
1331.Ayers, R. L., "Short-circuit tester for module substrate", IBM Tech. Disclosure Bull., 16, 3312-13 (1974), EEA77-30346.

1332. Balasubramanian, P. S., Grazier, E., Henke, J. D. and Myers, M. J., "Testing LSI memory arrays using on-chip I/O shift register latches", IBM Tech. Disclosure Bull., 17, 2019-20 (1974), EEA78-15409.

1333. Bonar, R. G. and Millham, E. H., "Dynamic logic tester for large-scale integration", IBM Tech. Disclosure Bull., 27, 1426-7 (1974), EEA78-8616.

1334. Bowen, J., "Integrated circuit testing", Electron, no. 54, 59, 61 (1974), EEA77-38855.

1335. Campbell, J. F., Jr., "A new real-time function test generation system for complex LSI testing", IEEE Trans. Manuf. Technol., MFT-4, 52-5 (1975).

1336. Chew, B., "Scheduling an IC tester for maximum throughput", Electronics, 47 (17), 96-9 (1974), EEA77-38853.

1337. Citrin, D. A., "Electrical characterization of complex microcircuits", Report RADC-TR-72-145, General Electric Co., Pittsfield, Mass., 1972, 306 pp., EEA76-19941.

1338. Coker, B., "IC testing economics - in-house versus external testing devices", Eval. Eng., 14 (4), 26-7 (1975), EEA78-37563.

1339. Cozzi, L., "A universal MOS-LSI testing system", Electron. Prod. Methods & Equip., 4 (5), 9, 11 (1975), EEA78-37610.

1340. Dabrowski, G., "Testing digital integrated circuits", News Rohde & Schwarz, 15 (68), 19-21 (1975), EEA78-33162.

1341. Dollt, W., "Measurement problems the development of linear integrated circuits", Elektron. Anz., 6 (4), 79-82 (1974), German, EEA77-26514.

1342. Dykast, K., "Technical principles of testers for integrated circuits of modular numerically controlled instruments", Measuring Systems and Their Applications, EMISCON 73, p. 199-200, Publ. 1974, Czech, EEA78-19927.

1343. Fahrenkrog, D., Henninger, K. and Kaiser, E., "Packaging system and testing methods for electronic generator protection gear", Siemens Rev., 40, 159-62 (1973), EEA76-28324.

1344. Fasang, P. P., "A minicomputer-aided integrated circuit tester", 1973 IEEE Region 6 Conference on Minicomputers and Their Applications, p. 85-9, EEA77-2615.

1345. Gaspar, J., "Automatic testing of integrated circuits by the help of the MSI-tester", Meres. & Autom., 23, 132-5 (1975), Hungarian, EEA78-28823.

1346. Gerhard, A. R. and Perdick, L. P., "Computerized testing of thin-film circuit conductors", Solid State Technol., 14 (9), 41-7 (1971), EEA75-27793.

1347. Grimm, A. V., "LSI integrated circuit test development", Report BDX-613-1017, Bendix Corp., Kansas City, Mo., 1974, 62 pp., EEA78-24509.

1348. Hartmann, K., Kotyczka, W. and Strutt, M. J. O., "Equivalent networks for three different microwave bipolar transistor packages in the 2-10 GHz range", Electron. Lett., 7, 510-11 (1971), EEA74-37528.

1349.Hoffken, H. W. and Reiss, K., "A method to eliminate package influences from IC-fourpole measurements", Frequenz, 27, 230-7 (1973), German, EEA76-38309.
1350.Howe, L. D., Jr. and Zurla, F. A., Jr., "Troubleshooting large-scale integrated circuit units", IBM Tech. Disclosure Bull., 17, 1941-4 (1974), EEA78-15406.
1351.Kallmeyer, M., Kosanke, K., Schedewie, F., Solf, B. and Wagner, D., "Rapid, precise, computer-controlled measurement of X-Y coordinates", IBM J. Res. & Dev., 17, 490-9 (1973), EEA77-18833.
1352.Landy, K. and Toth, E., "IC testing equipment with laser", Hiradstech. Ipari Kut. Intez. Kozl., 15, 91-101 (1975), Hungarian, EEA78-28827.
1353.Lange, L. K. and Lindbloom, E., "Fault simulation method", IBM Tech. Disclosure Bull., 15, 662-3 (1972), EEA75-39205.
1354.Lee. Y. S. and Getsinger, W. J., "Characterization of packaged varactor diodes", IEEE GMTT International Microwave Symposium Digest of Technical Papers, 1971, p. 42-3, EEA74-36793.
1355.Lipman, J. A., "Rework of linear integrated circuits", Report AFWL-TR-74-53, Texas Instruments, Inc., Dallas, 1974, 72 pp., EEA78-24508.
1356.Longley, M. R., "Testing ICs. I. The dilemma of the commercial user", Microelectronics, 4 (4), 45-7 (1973), EEA76-29293.
1357.Loranger, J. A., "Testing ICs. II. Higher reliability can cost less", Microelectronics, 4 (4), 48-50 (1973), EEA76-29294.
1358.Luce, A. N., Jr., "Parameter characterization for integrated circuit evaluation", IBM Tech. Disclosure Bull., 17, 2289-90 (1975), EEA78-24464.
1359.McMahon, M. T., Jr., "AC performance test for large-scale integrated-circuit chips", Patent USA 3781670, Publ. December 1973.
1360.McMahon, M. T., Jr., "AC chip in-place test", IBM Tech. Disclosure Bull., 17, 1607-8 (1974), EEA78-11719.
1361.Miladinovic, B. and Lazovic, D., "One solution of automatic tester for final processing control in linear IC-s production", Automatika, 14, 251-4 (1973), Slovenian, EEA77-34090.
1362.Mow, W. C., "How to select an LSI tester for incoming inspection and evaluation testing", Eval. Eng., 11 (4), 4-5 (1972), EEA75-39172.
1363.Muller, K., "The strategy of the electrical functional test of integrated semiconductor stores", Z. Elektr. Inf. - & Energietech., 5, 137-43 (1975), German, EEA78-37565.
1364.Nakamo, H., "Testing (IC)", J. Inst. Electron. & Commun. Eng. Jap., 55, 593-6 (1972), Japanese, EEA75-39184.
1365.Narasimham, M. S., "Systems approach to diagnostics", IBM Tech. Disclosure Bull., 17, 3221-4 (1975), EEA78-24469.
1366.Nyblom, A., "Functional testing of IC's, a difficult art which pays dividends", Eltek. Aktuell Elektron., 17 (9), 66-9 (1974), Swedish, EEA78-2229.

1367.Parot, M., "Testing integrated circuits in situ", Toute Electron., no. 373, 19-22 (1973), French, EEA76-16464.
1368.Popovic, R. and Jevtic, M., "Automatic test system for a class of integrated circuits testing made by adaptation of an existing transistor and diode automatic test system", Automatika, 13, 330 (1972), Croatian, EEA76-29291.
1369.Predescu, J., "Tester built for less than $10 gives go/no go check of timer IC's", Electron. Des., 22 (11), 106 (1974), EEA78-15380.
1370.Routh, W. S., "Relationship of MOS circuit properties to LSI testing", Solid State Technol., 15 (3), 41-5 (1972), EEA75-27664.
1371.Runyon, S., "Focus on MSI/LSI testers", Electron. Des., 20 (197), 60-9 (1972), EEA75-40759.
1372.Saal, D., "Simple test arrangement for circuits", Elektron Int., no. 3, 80 (1975), German, EEA78-24379.
1373.Sands, S. A. and Wilcox, D. L., "Transmission line characteristics of conductors buried in ceramic", Proceedings of Electronic Components Conference, 1969, p. 83-6, EEA72-23254.
1374.Sheehan, L. J., III, "Watch out for traps in hybrid IC tests", Electron. Des., 20 (26), 58-60 (1972), EEA76-12680.
1375.Singh, S. and Singh, V. P., "A probabilistic approach for testing large scale integrated circuits", 7th Asilomar Conference on Circuits, Systems and Computers, 1973, p. 130-5, Publ. 1974, EEA77-30406.
1376.Siwek, K., "A minicomputer simulator for digital logic testing", Electron. Packag. & Prod., 15 (9), 91-5 (1975).
1377.Sjogren, A., "The logic testers hand instruments: Details, performance and price", Eltek. Aktuell Elektron., 17 (5-6), 38-47 (1974), Swedish, EEA77-36052.
1378.Spiralski, L., Skibinski, K. and Hasse, L., "Digital IC's manual testers", Pomiary Autom. Kontrola, 19, 269-72 (1973), Polish, EEA77-384.
1379.Strejcek, P., "The meters of integrated circuits", Sdelovaci Tech., 20, 203-4 (1972), Czech, EEA75-31849.
1380.Treetz, J., Hoppe, G. and Hansel, G., "Tester 1000 - Automatic Measuring Device for digital integrated circuits", Radio Fern-Sehen Elektron., 22, 280-6 (1973), German, EEA76-35337.
1381.van Iperen, B. B., "Impedance relations in a diode waveguide mount", Nachrichtentech. Fachber. (NTF) Germany, 35, 483-6 (1968), EEA72-15719.
1382.Williams, B., "L.S.I. automatic test equipment applied to dynamic microprocessor testing", New Electron., 7 (21), 50-2 (1974), EEA78-2912.
1383.Yereance, R. A., "New automatic inspection techniques", Eval. Eng., 10 (5), 26-7 (1971), EEA75-27850.

4. Coatings

1384.Bell, R. E., "Countdown to the 'right' plastic", Electron. Packag. & Prod. (Int. Ed.), 7 (6), 12-15 (1971), EEA75-7006.
1385.Benham, R., "The development of improved epoxy molding compounds for microcircuit encapsulation", Proceedings of the 10th Electrical Insulation Conference, 1971, p. 74-8, EEA75-12834.
1386.Cherry, B. W. and Wright, P. H., "The electrical resistance of epoxy resins", Aust. Telecommun. Res., 7 (1), 36-44 (1973), EEA77-2630.
1387.Christie, F. R., "Search for a humidity-resistant coating", Electronic Packaging Prod., 8 (11), 77-114 (1968), EEA72-7452.
1388.Davy, J. G., "Calculations for leak rates of hermetic packages", IEEE TRans. Parts, Hybrids & Packag., PHP-11, 177-89 (1975), EEA78-40472.
1389.Dewar, W. J. and Norring, W. D., "Flow characteristics of epoxy molding compounds and their relationship to product reliability", Presented at the Annual Reliability Physics Symposium (presentation abstracts), Las Vegas, Nev., April 1970, EEA73-27363.
1390.Harrison, J. C. and Rickard, E. F., "Plastics for long life microcircuit encapsulation. I. Material properties and possible failure mechanisms", Conference on Reliability in Electronics, 1969, p. 129-36, EEA73-9565.
1391.Kubota, Y., Uetake, T. and Saito, T., "Properties of epoxy resin as an encapsulating material", Rev. Elec. Commun. Lab., 19, 1026-37 (1971), EEA75-9648.
1392.Licari, J. J., Perkins, K. L. and Caruso, S. V., "Evaluation of electrically insulative adhesives for use in hybrid microcircuit fabrication", IEEE Trans. Parts, Hybrids & Packag., PHP-9, 199-207 (1973), EEA77-14626.
1393.Nagai, Y., Suzuki, T. and Wakabayashi, Y., "Development and evaluation of epoxy resin molding compounds for encapsulation use", Fujitsu Sci. & Tech. J., 11, 115-33 (1975), EEA78-36592.
1394.Nowacki, J. and Szyszej, K., "Reasons for the defects in plastics sealed semiconductor devices", Elektronika, no. 10, 428-31 (1972). Polish, EEA76-4407.
1395.Nufer, R. W. and Anderson, L. C., "Evaluation of epoxy encapsulants and their levels of contamination", Solid State Technol., 14 (8), 33-6 (1971), EEA75-9649.
1396.Pashley, R. D., Mayer, J. W., Eisen, F. H. and Welch, B., "Comparison of encapsulating dielectrics on ion-implanted GaAs", Presented at Electrochemical Society Meeting, San Francisco, Calif., May 1974, PA77-53616.
1397.Planting, P. J., "An approach for evaluating epoxy adhesives for use in hybrid microelectronics assembly", IEEE Trans. Parts, Hybrids & Packag., PHP-11, 305-11 (1975).

1398.Schafft, H. A., "ARPA/NBS Workshop II. Hermeticity testing for integrated circuits", Report NBS-SP-400-9, Nat. Bur. Stand., Washington, D. C., 1974, 36 pp., EEA78-15395.

1399.Suzuki, T., Nagai, Y. and Okutsu, S., "Development and evaluation of low pressure molding compounds for encapsulation use", Fujitsu Sci. Techn. J., 7, 165-93 (1971), EEA75-9645.

1400.Szedon, J. R., "An approach for evaluating polymer materials as protective coatings on hybrid microcircuits", IEEE Trans. Parts, Hybrids & Packag., PHP-10, 251-7 (1974), EEA78-11770.

1401.White, M. L., "The evaluation of the reliability of organic coatings for integrated circuits", Presented at the International Electron Devices Meeting, Washinton, D. C., October 1969, EEA73-20188.

5. Others

1402.Andrle, C., "Non-destructive testing of electronic component packages", TESLA Electron., 8 (2), 35-9 (1975), EEA78-32159.

1403.Angleton, J. L. and Webster, S. L., "Techniques for standardization of particle noise in electronic packages", Proceedings of the 12th Annual Reliability Physics Conference, 1974, p. 38-42.

1404.Anolick, E. S., "Automatic dielectric quality evaluating system", IBM Tech. Disclosure Bull., 17, 2918-19 (1975), EEA78-19919.

1405.Anon., "Reliable method for testing gross leaks in semiconductor component packages", Solid State Technol., 12 (8), 22 (1969), EEA73-3566.

1406.Baker, E., "Understanding cyclical thermal stress in electronic assemblies: the key to improving accelerated life testing", Insul./Circuits, 21 (11), 49-56 (1975).

1407.Blushinskii, A. B. and Presler, K. K., "A study of the (mechanical) loads acting on packaged electrical products while they are being transported", Elektrotekhnika, 40 (4), 50-2 (1969), Russian, EEA73-19445.

1408.Chen, W. T. and Kim, C. K., "The effect of package parasitics on a series connected GaAs IMPATT diode", IEEE Trans. Electron Devices, ED-21, 120-2 (1974), EEA77-14474.

1409.Clay, F. A., Panousis, N. T. and Pierce, R. W., "Characterization of a chromium-gold deposition process for the production of thin film hybrid microcircuits", 24th Electronic Components Conference, 1974, p. 98-104, EEA77-34087.

1410.Coleman, M. V., "Evaluation methods for the examination of thick film materials", Radio & Electron. Eng., 45 (3), 121-30 (1975), EEA78-19870.

1411.Dowling, T. J., "Aluminized expansion alloys for microelectronic packaging", Presented at Semiconductor Integrated Circuit Processing and Production Conference, Anaheim, Calif., February 1971, EEA74-21566.

1412. Ebel, G. H., "High temperature reverse bias testing at water level", Proceedings of the 10th Annual Conference on Reliability Physics, 1972, p. 204-7, EEA76-4721.
1413. Fahley, W. A., "Detection of random chip defects in monolithic microcircuits", 23rd Electronic Components Conference, 1973, p. 357-63, EEA76-29317.
1414. Franke, O., "Measuring technique of integrated circuits-present status and future prospects", Nachrichtentechnik, 20, 305-9 (1970), German, EEA73-37057.
1415. Hailes, L. and Crossland, W. A., "Thick films: The substrate-conductor relationship", Electron. Packag. & Prod. (Int. Ed.), 8 (4), 3-7 (1972), EEA75-39214.
1416. Hall, P. M., "Electrical resistance of metallized via holes", J. Vac. Sci. & Technol., 12, 88-91 (1975), EEA78-24445.
1417. Hampy, R. E. and Norwood, D. P., "Evaluation of conductive vias for hybrid microcircuits", Report SAND-74-0047, Sandia Corp., Albuquerque, N. Mex., 1974, 33 pp., EEA78-41717.
1418. Harrison, B., "Packaging materials and factors in packaging evaluation", Value Engng., 1, 113-16 (1968), EEA73-559.
1419. Haworth, R. F., "Packaging underwater electrical/electronic components on deep submergence vehicles", Insul./Circuits, 16, 25-31 (1970), EEA74-14844.
1420. Hays, D. A., "Control of stray effects in hybrid microcircuit design", IEEE Trans. Parts, Hybrids & Packag., PHP-10, 95-100 (1974), EEA77-36333.
1421. Herman, D. S., Schuster, M. A. and Gerber, R. M., "Hillock growth on vacuum deposited aluminum films (integrated circuit interconnections)", J. Vac. Sci. & Technol., 9, 515-9 (1972), EEA75-19409.
1422. Hitch, T. T., Whitaker, H. H., Botnick, E. M. and Goydish, B. L., "Chemical analyses of thick-film gold conductor inks", IEEE Trans. Parts, Hybrids & Packag., PHP-11, 248-52 (1975).
1423. Howard, R. T., "Methods for visual characterization of trimmed thick-film resistors", Solid State Technol., 14 (11), 39-45 (1971), EEA75-27809.
1424. Johnson, A. H., "Automatic inspection of via fill", IBM Tech. Disclosure Bull., 17, 3593-4 (1975), EEA78-32163.
1425. Monroe, J. W., "The effects of package parasitics on the stability of microwave negative resistance devices", IEEE Trans. Microwave Theory & Tech., MTT-21, 731-5 (1973), EEA77-94.
1426. Morabito, J. M., Thomas, J. H., III and Lesh, N. G., "Material characterization of Ti-Cu-Ni-Au(TCNA) - A new low-cost thin film conductor system", IEEE Trans. Parts, Hybrids & Packag., PHP-11, 253-62 (1975).
1427. Nelson, K. H., "Gas chromatographic detection of leaks in microelectronic packages", Microelectron. & Reliab., 8, 313-18 (1969), EEA73-13090.

1428.Onufer, R. J. and Stanley, E. C., "Using a rotational viscometer to characterize thick film materials - Part I", Insul./Circuits, 21 (13), 31-6 (1975).
1429.Peel, M. E. and Nickerson, E., "Environmental evaluation of contact finishes", Insul./Circuits, 21 (4), 55-63 (1975), EEA78-24541.
1430.Peppiette, G., "Temperature measurement on an IC-chip", Electron. Equip. News, 15 (11), 107 (1974), EEA77-27822.
1431.Piwczyk, B., "X-rays disclose an integrated circuit's defects", Telesis, 2 (7), 16-21 (1972), EEA76-12678.
1432.Ramsey, T. H., "Thermal and X-ray analyses of some electronic package sealing glasses", Am. Ceram. Soc. Bull., 50, 671-5 (1971), EEA74-36752.
1433.Roth, J. P., "Universal chip", IBM Tech. Disclosure Bull., 17, 1234-6 (1974), EEA78-5074.
1434.Stewart, R. and Neff, G., "Leak testing six million IC's", Presented at Semiconductor Integrated and Production Conference and Exhibition, Anaheim, Calif., February 1971, EEA74-21673.
1435.Stitch, M., Johnson, G. M., Kirk, B. P. and Brauer, J. B., "Microcircuit accelerated testing using high temperature operating tests", IEEE Trans. Reliab., R-24, 238-50 (1975), EEA78-41639.
1436.Weiske, K. and Linder, H., "Avoidance of damage to fine mechanical and electronic instruments in transport by means of adequately dimensioned packaging" ,Feing. Tech., 19 (3), 126-30 (1970), German, EEA73-20186.
1437.Winzer, G., Douklias, N. and Kraft, H. P., "Comments on filtering of defects in integrated circuits", Appl. Opt., 11, 1269 (1972), EEA75-23364.

XXIII. Computer-Aided Design

1438.Abos, I., Balint, L., Scsaurszki, P. and Revai, I., "Computer aided layout design of electronic and microwave circuits", In: Annual of the research institute for telecommunication 1973, TKI, Budapest, Hungary, p. 181-93 (1974), EEA78-15394.
1439.Bianco, B., Parodi, M. and Viviani, R., "Computerized analysis of packages for diodes operating in X-band", Alta Freq., 43 (12), 1038-9 (1974), EEA78-11087.
1440.Caughey, M., Luke, D., McCready, L. and Rombeck, H., "Circuit simulation by computer", Telesis, 2 (6), 17-24 (1972), EEA75-39195.
1441.Crocker, N. R., McGuffin, R. W., Naylor, R. and Vosper, H., "Computer aided placement and routing of high density chip interconnection systems", AGARD Conference Proceedings No. 130 Computer Aided Design for Electronic Circuits, 1973, p. 32/1-26, EEA77-5399.
1442.Dickerson, P. J., "Computer-aided design of thin-film anodization fixtures", Western Electric Eng., 19 (1), 24-9 (1975), EEA78-19877.

1443.Ehlbeck, H. W., "Automating material flow in semiconductor processing", Electron. Prod. Methods & Equip., 4 (1), 29-30, 32 (1975), EEA78-24154.

1444.Feller, A., "Design automation techniques for custom LSI arrays", AGARD Lecture Series No. 75 Custom Design for Large Scale Integration (LSI), 1975, p. 7/1-15, EEA78-19932.

1445.Geitz, G., Horbst, E. and Krause, H., "Graphic displays aid in the development of double-faced PC boards", Elektron. Rechenanlagen, 15 (4), 176-80 (1973), German, EEA77-172.

1446.Hyatt, G. P., "High speed digital computer implementation techniques", Presented at the International Electron Devices Meeting, Washington, D. C., October 1967, EEA72-10651.

1447.Karstadt, L. K., Burger, W. G. and Hsieh, C. M., "Impact analysis of large IC chips", 6th Asilomar Conference on Circuits and Systems, 1972, p. 462-6, Publ. 1973, EEA76-23504.

1448.Kelly, M. J., "CAD limits LSI", 8th Asilomar Conference on Circuits, Systems and Computers, 1974, p. 379-84, Publ. 1975, EEA78-24528.

1449.Klamet, H., "Computer-aided design of the layout of integrated circuits (CADLIC)", International Computing Symposium, 1973, p. 451-8, Publ. 1974, EEA77-14576.

1450.Kolias, J. T., "Computer connections and the fivefold increase", Electron. Packag. & Prod. (Int. Ed.), 8 (1), 10-13 (1972), EEA75-15995.

1451.Lusinchi, J. P., "Influence of CAD IC manufacturers users relationships", International Conference on Large Scale Integrated Circuits, 1974, p. 409-16, French, EEA78-11761.

1452.Mazurek, B. and Olszewski, T., "Program 'CYPR' for computer aided designing of the multilayer electronic packages including integrated and printed circuits", Pr. Przem. Inst. Telekomun., 25 (82), 40-51 (1975), Polish, EEA78-36600.

1453.Meadows, W. S., "CAD in circuit packaging", Electron. Equip. News, 15 (2), 37-9 (1973), EEA76-26114.

1454.Ozawa, Y., Murakami, M. and Suzuki, K., "Master slice LSI computer aided design system", 11th Design Automation Workshop, 1974, p. 19-25, EEA77-30412.

1455.Shah, R. R. and Yan, G., "Development of engineering application packages for interactive design of electronic systems", Report AECL-5228, Atomic Energy Canada Ltd., Chalk River, Ontario, 1975, 6 pp., PA78-40476.

1456.Shostack, K., "Design automation for digital logic packaging", 22nd Electronics Components Conference, 1972, p. 277-80.

1457.Sugiyama, N. and Hirano, M., "An automated integrated circuit layout design program of a systematic chip using building blocks", International Conference on Computer Aided Design, 1972, p. 293-8, EEA75-19267.

1458.Villers, P., "Interactive mechanical design automation", Presented at Semiconductor Integrated Circuit Processing and Production Conference, Anaheim, Calif., February 1971, EEA74-21666.

1459.Yoshida, K. and Ohta, T., "Topological layout of a monolithic integrated circuit", Electron. & Commun. Jap., 52 (12), 173-9 (1969), EEA73-9825.

XXIV. Tooling and Equipment

1460.Adams, A. L., Yearsley, G. A., Simmons, M. I. and Yager, B. P., "Automatic semiconductor bonding machine", Patent USA 3776447, Publ. December 1973.
1461.Altomen, R. W. and Smith, C. H., Jr., "Linear chip carrier", IBM Tech. Disclosure Bull., 14, 3676-7 (1972), EEA75-31814.
1462.Anon., "Picking the best case sealing equipment to fit your production", Mod. Mater. Handl., 29 (10), 75-80 (1974), EEA78-7859.
1463.Anon., "Modular equipments for micro-bonding", Finommech. & Mikrotech., 13, 326-33 (1974), Hungarian, EEA78-15137.
1464.Audi, R. D. and Hamel, M. A., "Lead frame forming and joining", IBM Tech. Disclosure Bull., 18, 65 (1975), EEA78-37332.
1465.Bauer, H. G. and Boettinger, V. E., "Vacuum holding and testing fixture", IBM Tech. Disclosure Bull., 16, 3289 (1974), EEA77-30352.
1466.Bodendorf, D. J., Olson, K. T., Trinko, J. P. and Winnard, J. R., "Active silicon chip carrier", IBM Tech. Disclosure Bull., 15, 656-7 (1972), EEA75-39208.
1467.Cordes, H., "A new piece of test equipment for digital circuits", Funkschau, 47 (4), 48-50 (1975), German, EEA78-28853.
1468.Cote, R. E., "Facilities, equipment, and manufacturing operations for circuit deposition and testing (thick film circuit manufacture)", Solid State Technol., 18 (1), 49-52 (1975), EEA78-15397.
1469.Devine, J. and McCormick, J. E., "The development of a torsional mode ultrasonic beam lead bonder", 22nd Electronic Components Conference, 1972, p. 321-4.
1470.Donaldson, P. E. K. and Sayer, E., "A vacuum centrifuge for void-free potting of implantable hybrid microcircuits in silicone", Med. & Biol. Eng., 13, 595-6 (1975), EEA78-41713.
1471.Duval, P., "Synthesis of leak detectors for semiconductor component sealing", Electron. & Microelectron. Ind., no. 173, 44-51 (1973), French, EEA76-28724.
1472.Fabenyi, E., "Screen printing and screen tensioning equipment for thick film circuits", Finommech. & Mikrotech., 14, 182-5 (1975), Hungarian, EEA78-37591.
1473.Fried, H., "Crystal soldering equipment", Finommechanika, 12 (1), 6-8 (1973), Hungarian, EEA76-16477.
1474.Furnival, T. J., "Apparatus for welding a cover to a tubular ceramic housing", Patent USA 3752382, Publ. August 1973.

1475.Gibbons, J., "Techniques for automatically controlled punching equipment", Presented at Semiconductor Integrated Circuit Processing and Production Conference, Anaheim, Calif., February 1971, EEA74-21690.
1476.Goldmann, L., Krall, B. and Wenskus, H., "Thermomechanical fatigue tester", IBM Tech. Disclosure Bull., 27, 1373-4 (1974), EEA78-8627.
1477.Hartleroad, R. J. and Garbowski, J. P., "Semiconductor loading apparatus for bonding", Patent USA 3887996, Publ. June 1975, EEA78-41732.
1478.Hom, A. B. H., "Improved bonding chuck for bonding integrated circuits", Report TN-112, RCA, Princeton, N. J., 1975, 2 pp., EEA78-41734.
1479.Hug, P., "Device prevents wire damage during circuit bonding", Insul./Circuits, 18 (9), 36 (1972), EEA75-39357.
1480.Jablonski, J., Bielawski, W. and Wedzricha, B., "Apparatus for applying zinc to conducting paths and substrate terminals of hybrid microcircuits", Elektronika, no. 4, 162-3 (1972), Polish, EEA75-27792.
1481.Jablonski, J. and Wedzicha, B., "Solder bath for conductive paths and joints on substrate plates of hybrid microcircuits", Elektronika, no. 5, 213 (1972), Polish, EEA75-31838.
1482.Jensen, N. F., "Apparatus for fastening contact pins to a substrate", Patent USA 3735466, Publ. May 1973.
1483.Johnson, R. F., Kotrch, G. S., Lau, J. A., Rivenburgh, D. I. and Romanosky, R. J., "Batch loading of solder buckets", IBM Tech. Disclosure Bull., 17, 2614 (1975), EEA78-24540.
1484.Jorschick, J. P. and McNaney, W. C., "Bidirectional storage helix", IBM Tech. Disclosure Bull., 16, 3155 (1974), EEA77-30010.
1485.Kato, H., Kawanishi, H., Goto, S., Oymada, T. and Kani, K., "On automated wire routing for building-block MOS LSI", Proceedings of the 1974 IEEE International Symposium on Circuits and Systems, p. 309-13, EEA77-38893.
1486.Khalilov, A. A., Khobotov, B. N. and Efremov, V. A., "Equipment for the sealing and inspection of hermeticity of semiconductor power devices", Weld. Prod., 20 (5), 83-4 (1973), EEA77-18573.
1487.Kimura, S. and Hojo, T., "Automatic bonding machine - a new assembly technique for reducing production cost (IC production)", JEE, no. 98, 44-6 (1975), EEA78-19950.
1488.Klocek, J. J., "Dispensing substrates into carriers", Tech. Dig., no. 37, 14 (1975), EEA78-24448.
1489.Kotrch, G. S., Lau, J. A. and Skarvinko, E. R., "Solder flux removal apparatus", IBM Tech. Disclosure Bull., 17, 3562-3 (1975), EEA78-32162.
1490.Kovacs, G., "Semi-automatic wiring apparatus", Finommechanika, 10 (15), 134-7 (1971), Hungarian, EEA75-575.

1491.Lambrecht, R. A., "System for molding electronic components", Patent USA 3739438, Publ. June 1973.
1492.Lau, J. A. and Skarvinko, E. R., "Cleaning device for high-density integrated circuit package", IBM Tech. Disclosure Bull., 17 (7), 1897-8 (1974), EEA78-15386.
1493.Lesch, G., "Soldering equipment for terminal frames", Finommech. & Mikrotech., 12, 341-3 (1973), Hungarian, EEA77-18842.
1494.McMahon, M. T., Jr., "Semiconductor structure for testing of metalization networks on insulative substrates supporting semiconductor chips", Patent USA 3803483, Publ. April 1974.
1495.Magee, R. A. and Rance, W. G., "Carrier and loader (for IC substrate conveyor)", IBM Tech. Disclosure Bull., 27, 1376 (1974), EEA78-8592.
1496.Mammano, R. A., "Dual-polarity IC regulators aid design and packaging", Electronics, 46 (4), 108-11 (1973), EEA76-11879.
1497.Marikar, Y. M. F. and Vasu, K. I., "Internal stress in electrodeposited thin films", J. Electrochem. Soc. India, 21 (1), 3-7 (1972), EEA75-19410.
1498.Mayhew, A. J. and Cheale, L. F., "Apparatus for the manufacture of printed circuits", Patent USA 3731615, Publ. May 1973.
1499.Miller, C. F. and Zenker, R. O., "Apparatus to sever (gold) wire bonded to semiconductor devices", Patent USA 3448910, Publ. June 1969.
1500.Miller, C. F., Zenker, R. O. and Durant, B. G., "Apparatus for effecting bonding of semiconductor parts", Patent USA 3451607, Publ. June 1969.
1501.Motorola Inc., "Encapsulating apparatus", Patent UK 1166603, Publ. October 1969, EEA73-27991.
1502.Odawara, G. and Nozaki, J., "Automated wire bonder", J. Fac. Eng. Univ. Tokyo B, 32, 637-43 (1974), EEA78-33220.
1503.Penick, L. C., "Device for testing the adherence of circuit elements to a substrate", Tech. Dig., no. 37, 21-2 (1975), EEA78-24374.
1504.Philips Electronic and Assoc. Ind. Ltd., "Semiconductor encapsulation mould", Patent UK 1125428, Publ. August 1968, EEA72-2315.
1505.Ramsey, H. J., "Apparatus for bonding semiconductor elements", Patent USA 3627190, Publ. December 1971.
1506.Renyei, K., "Development trends and standardization possibilities in electronic equipment mounting methods", Proceedings of Constronic '72 Conference on the Mechanical Aspects of Electronic Design. II., p. 509-17, German, EEA76-19123.
1507.Rokusek, L. F., "The use of eddy current drives on packaging equipment", Proceedings of the 1972 Seventh Annual Meeting of the IEEE Industry Applications Society, p. 671-8, EEA76-7659.
1508.Roman, W., "Methods for cutting integrated hybrid circuit substrates and semiconductor wafers", Elektronika, 15, 519-24 (1974), Polish, EEA78-15424.

1509. Russell, R. F. and Thompson, R., "Automatic manufacture of custom hybrid microelectronic circuits", Conference on Hybrid Microelectronics, 1975, p. 159-68, EEA78-41723.
1510. Smith, K. F. and Smith, K. D., "A computer controlled MTOS LSI tester", Comput. Design, 7 (9), 51-7 (1968), EEA72-4445.
1511. Sweitzer, S. E., "Coating apparatus including conveyor mask", Patent USA 3738315, Publ. June 1973.
1512. Tebbutt, M., "A review of equipment practice 1974", New Electron., 20 (7), 31-2, 35-6, 38, 40 (1974), EEA78-2214.
1513. Tompkins, R. F., "Module tray", IBM Tech. Disclosure Bull., 18, 410 (1975), EEA78-37579.
1514. Uetsuki, T., Okajima, M. and Yuasa, Y., "Monitoring system for coating a substrate", Patent USA 3892490, Publ. July 1975.
1515. Wagner, H. M., "Production equipment for the new packages", International Electronic Circuit Packaging Symposium, 1969, 13 pp., EEA73-27359.
1516. Wedzicha, B. and Bielawski, W., "Apparatus for the assembly of flat terminals in hybrid microcircuits microsoldering machine ML-180", Elektronika, no. 4, 160-1 (1972), Polish, EEA75-27791.
1517. Welff, B. B., "Apparatus for making hermetically sealed glazing units", Patent USA 3733237, Publ. May 1973.
1518. Westfall, W., "OSHA and the packaging machine", IEEE Trans. Ind. Appl. IA-10, 624-6 (1974), EEA78-69.
1519. Wiesler, M. and Macintyre, J. S., "Apparatus for (automatically) bonding semiconductive devices", Patent USA 3668044, Publ. June 1972.

XXV. Cost and Yield

1520. Anon., "Process yields 'as fired' precision ceramic parts", Insul./Circuits, 18 (10), 50-1 (1972), EEA76-80.
1521. Barnett, J., "Single-layer packaging slashes ceramic-DIP costs in half", Electronics, 46 (21), 119-21 (1973), EEA76-38319.
1522. Barooshian, C. T., "Economics/microelectronics", 1972 IEEE Region Six (US Western region) Conference Record, p. 183-6, EEA76-4722.
1523. Burkitt, A., "Easing problems with hybrids", Electron. Engineering, 47 (567), 60-1 (1975), EEA78-33243.
1524. Camenzind, H. R., "Custom integrated circuits with low tooling cost", 6th International Congress on Microelectronics, 1974, 7 pp., EEA78-24514.
1525. Capra, R. L. and McAtee, C. D., "Pre-packaging may be a solution for reducing costs", Elec. Light & Power, 50 (14), 58-60 (1972), EEA74-38082.
1526. Chesarek, R. H. and McNamee, L. P., "Interactive custom design of bipolar LSI wafers", 7th Asilomar Conference on Circuits Systems and Computers, 1973, p. 107-13, Publ. 1974, EEA77-30404.

1527. Conley, J. M., "How to prepare low cost artwork and masks for thick film hybrid integrated circuits", 1970 IEEE International Convention Digest, p. 416-17, EEA73-37100.
1528. Cook, J. W., "A new outlook on flatpack sealing yields", 22nd Electronic Components Conference, 1972, p. 220-5.
1529. Dryden, W. G., "Achieving consistent results with thick-film", Solid State Technol., 14 (9), 54-7 (1971), EEA75-27807.
1530. Fraioli, A. V., "Yield values in thick-film rheology", Solid State Technol., 17 (5), 48-50 (1974), EEA77-30371.
1531. Gaskell, D. C., "Low-cost integrated circuit mounting", Sch. Sci. Rev., 55, 344 (1973), PA77-26443.
1532. Gelder, E., "Technical and economic limits of monolithic ICs", Components Rep., 9 (2), 25-8 (1974), EEA77-36327.
1533. Hohmann, J., "The economics of custom MOS-LSI for low-volume users", Contr. Eng., 19, 58-61 (1972), EEA75-23365.
1534. Kaposi, A. A., "On the testability of digital hardware", Comput. Aided Des., 4 (4), 169-71 (1972), EEA75-31702.
1535. Lewis, N., "To test or not to test - what is it worth?", Microelectron. & Reliab., 11, 505-9 (1972), EEA76-8714.
1536. Loughran, J. and Kurzweil, K., "Economic considerations in multilayer thick film hybrids", IEEE Trans. Parts, Hybrids & Packag., PHP-10, 120-31 (1974), EEA77-36335.
1537. Marcello, J. W., "IC package standardization for greater profits", Presented at Semiconductor Integrated and Production Conference and Exhibition, Anaheim, Calif., February 1971, EEA74-21581.
1538. Medesha, A. L., "Effective costs in dig i/c packaging", Electron, no. 36, 65-9 (1973), EEA77-10272.
1539. Okabe, T., Nagata, M. and Shimada, S., "Analysis on yield of integrated circuits and a new expression for the yield", Electr. Eng. Jap., 92 (6), 135-41 (1972), EEA77-6050.
1540. Peel, M., "Selecting the optimum circuit-board package", Electronics, 46 (6), 115-7 (1973), EEA76-15706.
1541. Riemer, D. E., "Cost efficiency of thick-film conductors", Solid State Technol., 18 (10), 42-5 (1975).
1542. Turley, A. P. and Herman, D. S., "LSI yield projections based upon test pattern results: an application to multilevel metal structures", IEEE Trans. Parts, Hybrids & Packag., PHP-10, 230-4 (1974), EEA78-11702.
1543. Webster, R., "Fine line screen printing yields as a function of physical design parameters", IEEE Trans. Manuf. Technol., MFT-4, 14-20 (1975).

SUBJECT INDEX

In this book each chapter is designated by a capital letter and references are numbered starting with "1". Entries are given in this index by their number preceded by the letter of the chapter. When the entry is the title of a chapter its letter is given; and if it is the heading of a section the page is given.

MIX
Papier aus verantwortungsvollen Quellen
Paper from responsible sources
FSC® C105338

If you have any concerns about our products,
you can contact us on
ProductSafety@springernature.com

In case Publisher is established outside the EU,
the EU authorized representative is:
Springer Nature Customer Service Center GmbH
Europaplatz 3, 69115 Heidelberg, Germany

Printed by Libri Plureos GmbH
in Hamburg, Germany